Praise for *The Weather of the Future*

"A scorching vision of what life might be like in the warmer world that is already on its way. Although *Weather of the Future* sounds like an exercise in speculation, Ms. Cullen grounds her harrowing predictions—extrapolations, really—in 'the best available science' derived from an array of climate models, environmental data, and interviews with scientists. And her forecasts actually turn out to be an armature for discussing the fallout of climate change (from rising sea levels to more extreme weather) in an accessible, tactile fashion and for examining existing liabilities in various regions and cities, like overstretched infrastructure and dwindling water supplies." —*New York Times*

"[Cullen] accepts weather as a local matter, just as Tip O'Neill, longtime speaker of the House of Representatives, proclaimed all politics to be local. . . . *The Weather of the Future* uses a broad itinerary to illustrate the threats she perceives. It predicts more frequent and more violent storms, more hot spells, cold spells, droughts, famines, and huge waves of desperate refugees. . . . Cullen is likely to attract readers with an insistent style and quotes from people who claim to have been already damaged by global warming." —Associated Press

"Heidi Cullen is a groundbreaker. As the first scientist on national television whose full-time job was covering climate change, she ruffled many feathers reporting on the current impacts of global warming. Now in this important and timely book, Heidi breaks ground again, taking on the recent wave of climate-change skeptics by simplifying the misunderstood connection between weather and climate and bringing the true impact of the problem, literally, right to your front door."

—Laurie David, producer of the Academy Award–winning documentary *An Inconvenient Truth*

"If you're tired of being confused about climate change, buy this book. It's refreshingly readable, reliable, and Heidi Cullen is one of those gifted Americans who sees much 'more clearly on a cloudy day.'"

—*The Huffington Post*

"Heidi Cullen's beautifully crafted study provides the human detail that has been missing from most reports on climate science. . . . Cullen's sense of place is also exceptional; her climate scenarios are rooted in local landscapes and societies as much as in global models. This book sets a new benchmark for accessible writing on the likely weather of the future." —*New Scientist*

"Vivid and compelling, *The Weather of the Future* shows what life will be like in a warming world. Essential reading for anyone who's planning to inhabit the planet for the next few decades."
—Elizabeth Kolbert, author of *Field Notes from a Catastrophe*

"*The Weather of the Future* peers ahead at a world stricken by climate change. Using models to predict weather patterns, climatologist Heidi Cullen, a frequent contributor to The Weather Channel, explores seven regions and their grim futures: the Sahel in Africa, the Great Barrier Reef of Australia, California's Central Valley, two sites in Greenland, Bangladesh, and New York City. . . . The book is at its best and most insightful when it explores today's environment, such as regreening efforts in Niger."
—*Washington Post*

"In an accessible way, [Cullen] details the Earth's climate history and forecasts what might happen if we're not more careful."
—*New York Post*

"America's best-known climatologist, Heidi Cullen warns that carbon pollution is unequivocally causing irreversible planetary-scale climate damage that left unabated will trigger increasing heat waves, sustained droughts, extreme storms, and widespread coastal flooding in its wake. This is a forecast we must all heed."
—Larry Schweiger, president of the National Wildlife Federation

"Fact-filled and entertaining yet disturbing depiction of our world as temperatures rise. . . . A lively and troubling but not entirely doomsday scenario of our warmer future."
—*Kirkus Reviews*

"*The Weather of the Future* by Heidi Cullen presents a concise and readily accessible account of the science supporting the theory of global warming."
—*Chattanooga Times Free Press*

"Engrossing. . . . The book presents a surprisingly optimistic view of humanity's determination to come to terms with a daunting future."
—*Publishers Weekly*

"This is a woman to whom attention must be paid."
—*Booklist*

THE WEATHER OF THE FUTURE

THE WEATHER OF THE FUTURE

Heat Waves, Extreme Storms, and Other Scenes from a Climate-Changed Planet

Heidi Cullen

HARPER

NEW YORK • LONDON • TORONTO • SYDNEY

HARPER

A hardcover edition of this book was published in 2010 by HarperCollins Publishers.

The cover represents an artist's rendering of New York City with higher sea levels. Sea level rise of this magnitude will likely occur several centuries from now as the West Antarctic and Greenland ice sheets melt and higher temperatures warm and expand our oceans.

Special thanks to Lisa Ammerman for her skilled hand in making the maps throughout this book.

FIRST HARPER PAPERBACK PUBLISHED 2011.

Designed by Joseph Rutt

Library of Congress Cataloging-in-Publication Data is available upon request.

ISBN 978-0-06-172694-1 (pbk.)

11 12 13 14 15 ID/RRD 10 9 8 7 6 5 4 3 2 1

For my fellow scientists.
Never stop seeking the truth.

The sweetest thing in all my life has been the longing to find the place where all the beauty came from.

—C. S. Lewis

CONTENTS

LIST OF MAPS

INTRODUCTION

"What's the forecast?"

I heard this question a lot when I first started at The Weather Channel in 2003. People figured that if I worked at a 24-7 weather network, I must be a meteorologist. That was fine by me, I've always been a closet weather geek, and besides, who doesn't secretly want to be a meteorologist? There is something so appealing about going to work every day and predicting the future without the use of tarot cards or constellations.

I'm not sure how many people secretly want to be climatologists, but that's what I really am. And for anyone wondering what a climatologist is, here's a rough answer: Climatologists pick up where meteorologists leave off. We focus on weather timescales beyond the memory of the atmosphere, which is only about one week. And I guess you could say we also focus on timescales beyond human memory, which is shorter than you might think.

Climatologists look at patterns that range from months to hundreds, thousands, and even millions of years. The single most important and most obvious example of climate is the seasonal cycle, otherwise known as the four seasons. The seasons, a result of the 23.5° tilt of the Earth's axis, affect the weather dramatically. And the physics behind the seasons are well nailed down. Summer, the result of one hemisphere's being tilted closer to the sun, is warmer. And winter, the result of the other hemisphere's being tilted away from the sun, is colder. The forecast for the seasons follows the physics; this is why, if I issue a forecast in January that says it will be significantly warmer in six months, you probably won't think I'm a genius, but you'll believe me.

There are countless other patterns on our planet that influence the weather. Take El Niño, for example. Nicknamed by fishermen along the coast of Peru after the Christ child, El Niño is a warm ocean current that typically appears every few years around Christmastime and lasts for several months. From its home in the tropical Pacific Ocean, El Niño is powerful enough to influence weather across the entire world. Unlike the seasons, which are controlled by astronomical forces, El Niño results from processes that happen here on Earth and that don't come and go like celestial clockwork. Climatologists have come to understand that the physics of El Niño result from a series of complex interactions between the ocean and the atmosphere. Technically, El Niño (EN) describes the ocean component, whereas the atmospheric component is known as the Southern Oscillation (SO). That's why climatologists generally refer to it as ENSO.

During an ENSO event, the easterly trade winds weaken and the surface water off the coast of Peru and Ecuador warms up several degrees. This warmer water leads to increased evaporation, causing the air above it to rise and thereby affecting the winds. This conversation between ocean and atmosphere is nuanced and far-reaching. The atmosphere feels the influence of the warm ocean surface below it and conveys the message by shifts in tropical rainfall, which in turn affect wind patterns over much of the globe. For example, most El Niño winters are milder over western Canada and parts of the northern United States, and wetter over the southern United States from Texas to Florida. In other parts of the world, ENSO can bring drought to northern Australia, Indonesia, the Philippines, southeastern Africa, and northern Brazil. Heavier rainfall is often seen in coastal Ecuador, northwestern Peru, southern Brazil, central Argentina, and equatorial eastern Africa.

El Niño is just one of the ways climate can work itself into the weather. You might say meteorologists are obsessed with the atmosphere, whereas climatologists are obsessed with everything that *influences* the atmosphere. But in the end, we're all obsessed with this notion of predicting the future. The atmosphere may be where the weather lives, but it speaks to the ocean, the land, and sea ice on a regular basis. Consider them influential friends that are capable of forcing the atmosphere to behave in ways that are sometimes, as in the case of ENSO, predictable. The hope is that if scientists can untangle all the messy relationships at work within our climate system, we should be better able to keep people out of harm's way. The farther we can extend human memory, the longer out in time a society can see, and the better prepared we'll be for what's in the pipeline.

And that is where global warming enters the picture. If the four seasons are Mother Nature's most prominent signature within the climate system, then you might say that *global warming*, the term that refers to Earth's increasing temperature due to a buildup of greenhouse gases in the atmosphere, is humanity's most prominent signature. The big difference between global warming and other climate and weather phenomena is that in this case, we're the ones doing the talking. And greenhouse gases are the chatter we use to influence the behavior of the atmosphere.

Decades of study suggest that this conversation will slowly drown out all others, its influence cutting across all timescales and all regions of the planet. Global warming has already begun pushing around the timing of the four seasons, and ongoing research shows that it is also influencing the weather. Meanwhile, climate scientists have developed a robust understanding of the physics of this human interaction with the atmosphere. They have collected data and built predictive models of the climate system that are capable of looking into the past and—more important—the future. The forecast for Earth is in, and it's not good.

Part I of this book explains the cutting-edge science behind this long-term climate forecasting, demonstrating why the predictive models for next century should be trusted in the same way that you trust the forecast for tomorrow on your local news. Here we'll examine the relationship between our weather today and our forecasts down the road, looking at how climatologists assess the changing statistics of extreme weather events and how these changing statistics play into the long-term forecast.

We'll also look at the history of weather prediction and how it serves as the foundation of climate forecasts today. Weather forecasts and climate forecasts are based on the same principles of mathematics and physics. Yet they have inherent differences that allow weather forecasts to focus on short-term changes in the atmosphere, whereas climate forecasts focus on long-term changes to the entire system of ocean, land, and ice. Keep in mind: just as my initial forecast that July would be hotter than January didn't involve the weather on a specific day, so too a climate forecast for 2050 or 2100 looks at the big picture and not a specific day. A forecast for the year 2050 has the potential to be as meaningful and as useful as tomorrow's weather forecast; it's just used in a different way.

All this is important because if you don't trust the models, you won't believe the forecasts—and the forecasts are what Part II of this book is all about. In Part II, we'll look at the forty-year forecasts for a few important places around the world. To start, I asked climate scientists to list the places they thought were most vulnerable to the threat of global warming.

I then narrowed the list down to seven key examples. (I've included a fuller list of hot spots identified by climate scientists in Appendix 3.) I chose these seven places not necessarily because they're the most endangered places or because the stories they offer are the most dramatic, but instead because collectively they demonstrate a spectrum of risks that exist with climate change. By mid-century, not every part of the world will be affected by global warming in the same way. Each location I've chosen has its own Achilles' heel, a vulnerability that unabated climate change will expose and exploit until the place is forever altered. Taken together these vulnerabilities show the breadth of repercussions that climate change will bring. It is my hope that whether taken as individual stories or as a whole, the predictions found in this book will demonstrate that global warming will hit all of us in the places we love and the homes where we live.

If Hurricane Katrina taught us anything, it is that the worst-case scenario can happen. For the first time in human history, science has given us the ability to peer into a crystal ball of numbers and models and see what kind of a climate we'll be living in by mid-century if we continue to emit carbon at our current levels. I share this look at the future with people outside the scientific community not as a scare tactic or as hyperbole, but because only through such sharing will the world come to understand precisely what is at stake.

Let me show you what I mean. For several years, I've been giving lectures and seminars about climate change to a variety of groups all over the world. Sometimes I speak to scientists; sometimes the audience is primarily students and their parents; sometimes it's politicians and business executives. After one of my first seminars, several years ago, I was standing at the front of the lecture hall, putting my computer away, when I was approached by a man—probably in his late forties—who had a question. He had enjoyed my lecture and found that it opened his eyes to several new aspects of the science and impacts of climate change, but what he really wanted to know was this:

"Do you think I should sell my beach house?"

After he said this, I thought for a moment about how best to calculate the risk associated with owning beachfront property in the United States, factoring in our best estimates of impacts such as sea level rise, storm surge, and saltwater intrusion—just to name a few. The man waited patiently with an earnest, if slightly bemused, look on his face. I suspect he may have been joking with me, but I felt I owed him an answer.

It was then that something hit me: *This is the only way a lot of people can truly connect to the issue of climate change—via a long-term investment like real estate.*

The more I thought about his question, the more I realized that the scientific community had failed to communicate the threat of climate change in a way that made it real for people right now. We, as scientists, hadn't given people the proper tools to see that the impacts of climate change are visible right now and that they go far beyond melting ice caps.

I can honestly say that real estate is what comes up most often when I talk to people about global warming. While I've spent much of my research career looking at the *global* impacts of climate change, I fully understand that people want to see the *local* impacts. If people are going to understand what is really at stake, scientists have to find new ways to communicate the science, using data, images, and computer scenarios that convey more completely what climate change really looks like—both now and in the future. Beachfront property is only the tip of the iceberg.

This book is written with precisely that goal in mind. It's a book about climate science and climate scientists, but ultimately it lays bare the true stakes of climate change. It illustrates that doing nothing and remaining complacent are tantamount to accepting a future forty years down the road in which your town, your neighborhood, and even your backyard will not look the same. It is not an exaggeration when I say that no place on the planet will look the same forty years down the road if climate change continues. All weather is local, and as you'll see, in the future all climate change will be local, too.

PART I

YOUR WEATHER IS YOUR CLIMATE

1

CLIMATE AND WEATHER TOGETHER

It's better to build dams than to wait for the flood to come to its senses

—Mark Twain

The images were stark: a foreboding gray sky overhead, a turbulent river churning by in the background, and throngs of people—men and women—racing against time to save their town. With the temperature below freezing and snow from a late spring blizzard swirling around them, the volunteers worked hurriedly but efficiently to build makeshift levees, using millions of sandbags. Stuffed into snow boots and down coats, sons and daughters, mothers and fathers, grandmothers and grandfathers tossed bags weighted with sand to each other, each bag moving along to the next person, until at last the bag took its place standing guard alongside the swollen river.

This scene took place on the banks of the Red River during March and April 2009. As the late-season storm swept through, hydrologists at the North Central River Forecast Center warned that the Red River of the North, which runs through the towns of Fargo, North Dakota, and neighboring Moorhead, Minnesota, would crest at 43 feet: 24 feet above flood stage. The situation was tense for days, with the water rising at a seemingly unrelenting rate, but the communities along the river were equally unrelenting. They bagged and tossed around the clock, working in shifts in the frigid air to try to avoid a local catastrophe. People who could not pitch in with the actual bagging helped in other ways, making food, watching kids—banding together to do the work that everyone knew needed to be done. In the end, the job required 3.5 million sandbags and more than 350,000 cubic yards of dirt. Friends and neighbors as well as complete strangers had come together to build a makeshift levee that stretched more than 20 miles.

The Red River flooding of 2009 resulted in a community-wide effort to sandbag, thanks to flood forecasts that provided lifesaving information. Communities along the Red River prepared for more than a week as the U.S. National Weather Service continuously updated the predictions. The entire community around the river as well as state and local authorities came together to use the predictive information as effectively as possible. As the data changed and the severity of the problem rose, people did not sit around hoping that good intentions were enough; they came together to protect their future—even though there was uncertainty as to what exactly the river would do.

The Red River eventually crested at 40.82 feet. During the prolonged flooding, the river was above flood stage for sixty-one days. The U.S. Army Corps of Engineers alone estimates that it spent about $30 million to prevent more than $2 billion in damage. And as bad as the flooding was, the worst fears of the forecasters never came true.

Even so, hydrologists from the National Weather Service said this was the highest crest in more than 120 years of records on the Red River of the North. From the initial point of melt to the peak in the Fargo-Moorhead area, they said it was the fastest a flood like this had ever occurred. The speed, along with the 20 inches of snow that fell, overwhelmed the forecast models; this is why hydrological engineers—studying the flows and peaks of the flooding—are working to refine their forecasting methods. As they work to improve the flood forecast, the Army Corps of Engineers is developing a plan for permanent flood protection for the communities that faced record flooding—because if one thing is certain, there will be a next time.

Make no mistake: global warming increases the likelihood of floods such as the Red River flood. This brings me to a central question: if you know a flood is coming, are you going to wait until the water is at your door or are you going to run to the closest riverbank and start pouring sand into a bag?

Global warming has been called the "perfect problem"—perfect in the sense that it's hard to see and challenging to solve. It's hard to see because its signals elude most of our evolutionary panic buttons, save one—our analytical minds. Climate scientists may have built models and issued forecasts, which include mass extinction, submerged coastlines, and chronic food and water shortages; but look outside your window, and there is no sign of a storm fitting that description.

Psychologists say that humans are genetically wired to respond to palpable threats like a stampede of wild elephants or a gun at the back of the head. It's the abstract dangers, the ones we face in the distant future, like global warming, that are tough to wrap our arms around. I get that. I understand that looking at a forecast map for the year 2100, even with the chance of a global average temperature increase of 11°F and a 3-foot rise in global sea level, doesn't set off the requisite alarm bells. And I understand why global warming ranked at the bottom of a list of twenty national priorities in a recent poll by the Pew Research Center.[1] According to the Pew study, our collective list of concerns goes like this: the economy, jobs, terrorism, Social Security, education, energy, Medicare, health care, deficit reduction, health insurance, helping the poor, crime, moral decline, the military, tax cuts, environment, immigration lobbyists, trade policy, and global warming, in that order.

This isn't to say Americans aren't concerned about global warming. Several polls have made it clear that Americans get it; a majority of Americans now feel that global warming is real and that it's caused by human activities. But their concern has done little to alter how we prioritize the risks that global warming poses. Global warming seems less urgent than things staring us in the face. Ultimately, last is still last.

Psychologists chalk up the last-place finish to all the ways that global warming fails to connect with our emotions, our experiences, and our memories. For one, psychologists point to the fact that people have a "finite pool of worry." It's impossible to sustain concern about global warming when other worries, like an economic collapse or a home foreclosure, dive into the pool. Another issue, called the *single-action bias*,[2] is the human habit of taking just one action in response to a problem in situations where multiple solutions are required. For instance, buying your first compact fluorescent lightbulb or using a recycled bag seems to reduce or remove the feeling of worry or concern.

In essence, we aren't fully capable of processing global warming in the traditional human way. So we need to find a new way to look at it, a new way to understand it and break it down.

The traditional human way works something like this. According to cognitive psychologists, we have two different systems for processing risks.[3] One system is analytical. It involves evaluating data and statistics to come up with a careful internal cost-benefit analysis. It's all science. The other system is emotional and drawn from deep personal experience and human memory. This system processes the risk and converts it into a feeling. It makes a situation personal and immediate, and that is why it works quite well in the case of stampeding wild elephants or a gun at the head. These two systems are capable of describing the same event very differently. Research suggests that although the two processing systems operate in parallel, they are both more effective when they're able to interact. And in cases where the outputs from the two processing systems disagree, our emotions and memories usually win. The gun will always trump the numbers.

Or if the gun doesn't trump the numbers, it messes with them. Take, for example, the stock market. It's a classic example of the daily battle between reason and emotion. Data and statistics are fundamental to determining whether or not to buy or sell, but emotions clearly play a role, even when you least expect it. A paper published in the *Journal of Finance* in 2003 found a positive relationship between morning sunshine outside the stock exchange and market index stock returns that day at twenty-six stock exchanges internationally from 1928 to 1997.[4] So much for strictly rational price-setting, and a strong statement about the powerful influence of the weather.

Global warming has a lot of similarities with the stock market. The long-term temperature trend, like the long-term performance of the market, is up. But weather, like day-trading individual stocks, is highly volatile. And like the stock market, global warming is a textbook example of how a disconnect between the analytical and the emotional processing systems often results in a pretty lousy risk assessment. Your brain, after careful analytic consideration, is telling you that of course long-term drought, mass extinctions, and a rising sea level are serious concerns. But your gut just isn't feeling it. It's too far off, too impersonal.

Consequently, many of us are still struggling to see global warming. In fact, when asked to come up with a single, specific image of what global warming looks like, 74 percent of poll respondents see only one thing: melting ice. Although nearly six in ten think global warming is making weather events like droughts and storms more frequent, far fewer connect global warming with specific recent events. In their own personal experience, only 43 percent say weather patterns in the county where they live have become increasingly unstable over the past three years. Experience plays a large role in judging risk. But most of us, especially those in the younger generation, do not yet have experiences that we associate with the threat posed by climate change and cannot bring examples, good or bad, to the table. In fact, our brain is wired to assume that the future will be similar to what we have experienced so far.[5]

Yet having worked at The Weather Channel, I was continually awestruck by the extent to which people rallied around a weather forecast, whether it involved sandbagging in advance of the Red River flood or evacuating in advance of Hurricane Gustav. There's something inspiring about the way communities can pull together under extremely challenging circumstances. We're clearly quite good at processing the risks associated with extreme weather, and this is why it's so important for people to understand that their weather is their climate. Climate and global warming need to be built into our daily weather forecasts because by connecting climate and weather we can begin to work on our long-term memory and relate it to what's outside our window today. If climate is impersonal statistics, weather is personal experience. We need to reconnect them.

To understand how we can link climate and weather, it helps to explain why they aren't linked now. The short answer is time.

As climate forecasts and weather forecasts have evolved, they have been separated in the public mind because *weather* is concerned with the immediate whereas *climate* is more focused on the long term. We watch a weather report on Sunday night because we want to know what to expect during the week ahead. The climate forecast, which deals in timescales of months and years, often feels too remote and intangible (unless of course, real estate is involved). We might hear that scientists think this winter will be warmer or this summer will be hotter, but we wait to pass judgment until we can experience it for ourselves. Just as our brain is hardwired to perceive threats that are most immediate to us, we are hardwired to devote more energy to caring about the weather than to caring about the climate.

This separation between weather and climate has been reinforced by the national and local news media, which regularly devote a segment to forecasting tomorrow's weather, but rarely say anything about the climate forecast. It's not that the information isn't available; it's that the way the practice has evolved, we don't expect a climate forecast from our news outlets. As a result, we tend to separate the broad concepts of weather and climate—to see them as vastly different ideas when in reality the only big difference between them is time.

Your daily weather forecast is a function of what is happening in the atmosphere right now. We use the conditions of today (humidity, temperature, wind speed, atmospheric pressure, etc.) to help predict the weather of tomorrow. Meanwhile, climate forecasting gives a broader context to the weather we are currently experiencing. And that context is critical. It is also evolving as a result of greenhouse gas emissions. Think of your daily weather forecast and then average it over time and space—that's roughly what a climate forecast is communicating.

Because weather forecasting and climate forecasting focus on different timescales, their goals are not the same. Whereas weather forecasting is meant to tell you what to expect when you step outside in the morning, climate forecasting is focused on broad trends over time. Will there be a drought next summer? What is the risk of wildfire for the West? Will El Niño appear next year? Will the weather be hotter in 2050? In other words, although I can't tell you whether it will be raining on March 1, 2050, in Fargo, North Dakota, I can say that March, on average, will be warmer and that rainfall, on average, will be more intense.

But despite their different timescales, climate and weather forecasts are focused on achieving a similar result: the means to predict the future. Of course, the question then becomes what do we do with it. The weather forecast is so ingrained in our existence that we know very well how to act on it. If we hear on the radio in the morning that it's going to rain, we carry an umbrella. If we hear that the temperature is going to be unseasonably cool, then we pack a sweater. By definition, weather is a timescale we can't stop. With a weather forecast, we're working strictly on our defense.

However, with the climate forecast the necessary actions are not as straightforward, and this highlights some of the basic philosophical differences between weather and climate. I've come to view a long-range climate projection as an *anti-forecast* in the sense that it forecasts something you want to prevent. Think back to the Red River flood. Until now, we've been able to view extreme weather like flooding as an act of God. But science tells us that, owing to climate change, such floods will happen more often and we need to be prepared for them. I say that a climate forecast is an anti-forecast because it is in our power to prevent the forecast from happening. It represents only one possible future that could happen if we continue to burn fossil fuels as business-as-usual. The future is ultimately in our hands. And the situation is urgent because the longer we wait, the more climate change works its way into the weather, and once it's in the weather, it's there for good.

We are currently in a race against our own ability to intuitively trust what science is telling us, assess the risk of global warming, and predict the future. So when we look at a climate forecast out to 2100 and see temperatures upward of 11°F warmer and sea level 3 feet higher, we need to assess the risk as well as the different solutions necessary to prevent these outcomes. The challenge is to reduce greenhouse gas emissions, replace our energy infrastructure, and adapt to the warming already in the pipeline. And this is the complicated part.

By responding to and trusting the climate forecast, we will prevent it from coming true. Ninety-two percent of those surveyed in a Yale/George Mason poll said the nation should act to reduce global warming. In other words, the overwhelming majority of Americans think we should trust the long-term forecast. But 51 percent of Americans said that although we have the ability to stop global warming, they weren't sure if we actually would. They weren't convinced we'd be able to see and act on the forecast of global warming as the residents of Fargo, North Dakota, saw and acted on the flood forecast. For the people in Fargo, the risk was personal and the forecast was lifesaving.

Most Americans believe that we will not take steps to fix climate change until after it has begun to harm us personally. Unfortunately, by that point it will be too late. The climate system has time lags. And those time lags mean that the climate system doesn't respond immediately to all the extra greenhouse gases in the atmosphere. So, by the time you see it in the weather on a daily basis, it's too late to fix climate change. For most people, the fact that there is uncertainty surrounding the future threat of climate change means we should hold off on any expensive fixes—specifically, actions aimed at reducing greenhouse gas emissions from smokestacks and tailpipes—until we know more. Yet the people of Fargo, North Dakota, didn't wait to see if their town would be flooded; instead, they saw the forecast and started sandbagging. They knew instinctively that if you wait until the water is up to your waist, it's already too late.

2

SEEING CLIMATE CHANGE IN OUR PAST

It is not the strongest nor most intelligent of the species that survive; it is the one most adaptable to change.

—Charles Darwin, *On the Origin of Species*

Here's something that most climate scientists won't tell you about climate change: the Earth is going to be fine. As the history of climate change in this chapter shows, the Earth has gone through periods of warming and cooling in the past, and still it remains here. Unfortunately, you can't say the same for the species that occupy Earth—including us. The fact that the Earth will be fine doesn't necessarily mean that the human race will. The last 10,000 to 20,000 years have witnessed a period of dramatic growth in human civilization. Indeed, our growth during this time is unique among all species, but it has been highly dependent on the overall consistency of the climate.

In order to make predictions about the man-made climate change of the future and understand just how high the stakes are, we must first look to the natural climate change of the past. We humans see ourselves as highly adaptable creatures; indeed, whether or not we can endure the coming climate change hinges on our adaptability. However, this is not a given. As a species, we have never been forced to adapt to a global increase in temperature like the one we currently face. If climate change does indeed occur, the future for humans will become a lot less certain—we will be much like other animals before us who proved unable to adapt to changing climates.

Take the woolly mammoth, the unofficial mascot of the ice ages. To see a woolly mammoth is to see a climate that no longer exists, and it's been this way ever since people first started finding mammoth fossils.

Weighing 20,000 pounds, and with tusks 16 feet long, mammoths must have looked quite striking 20,000 years ago, as they strolled around what is now downtown Los Angeles. Obviously, this region was very different 20,000 years ago, and the climate of Earth was very different as well. It's a time scientists refer to as the Last Glacial Maximum (LGM). Major areas of the Earth were locked in relentless winter, covered in massive sheets of ice that grew in frigid strongholds to the north. Los Angeles was not covered by ice, but it was definitely influenced by the cold elsewhere. Forests, fields, and even mountains were no match for these vast sheets of ice, and they had the battle scars to prove it. The ice was voracious, drinking up the oceans and drawing down the sea level by almost 400 feet. But all this was not a problem for the woolly mammoth—quite the contrary. Woolly mammoths were well adapted to the cold climate of the LGM, with shaggy hair more than 3 feet long to protect them from frigid winds and a 3-inch layer of blubber to keep them warm.

This vast stretch of ice, known as the Laurentide ice sheet, buried what is today Canada, New England, the Midwest, and parts of Washington, Idaho, and Montana under a layer of ice more than 1 mile thick.[1] Just south of this vast ice sheet stretched a treeless tundra that was equally expansive. This was the summer home of the woolly mammoth. The mammoths nibbled on the coarse tundra grasses with their perfectly adapted, but probably not pearly white, teeth. Woolly mammoth teeth, in fact, were about 6 inches square, the biggest grinding teeth in the animal kingdom. Those teeth were perfectly suited to their mammoths' ice age vegetarian diet.

Evidence of mammoths has been found throughout the northern hemisphere. Early humans settling along the North Sea coast, sometime between 6,000 and 8,000 years ago, also encountered skeletal remains of the woolly mammoth. The low sea level at the end of the last ice age provided an exposed shelf between the Netherlands and England that the mammoths roamed freely across. When the sea level rose again, rough surf would have crushed the exposed skeletons of the woolly mammoths, but their teeth were tough as nails, and these would have survived and eventually washed up along the shore. All along the North Sea, storm waves would have tossed woolly mammoth fossils up onto the beach, like seashells, for Vikings to find.

I can't imagine what it must have been like finding a woolly mammoth tooth along the shore all those years ago. Such a discovery most certainly raised some tricky questions that would have been tough to answer at the time. An animal with teeth 6 inches wide? There was nothing walking around the North Sea coast at that time with teeth 6 inches wide. Had the North Sea settlers been able to ask their Stone Age ancestors who lived through the ice age 10,000 years earlier, these ancestors would have been able to explain everything. They had, in fact, hunted the woolly mammoth. But lacking a time machine, the early Viking settlers had to come up with their own story to explain the existence of these very large teeth.

On the basis of the size of the teeth, the Vikings calculated that the animal must have been more than 70 feet tall. In tribute they named their new home, in what is today Denmark and Germany, "Land of the Giants." They assumed the woolly mammoths were the children of an enormous ice giant to the north who had once ruled all of Scandinavia. The legend went on to say that when the ice giant was killed, his blood made the sea level rise and drowned all his furry children with the big teeth. This explanation, preserved today in Icelandic sagas, is the earliest recorded notion pointing to the existence of an ice age, and actually it's not that bad an explanation for what happened.

The early Vikings were some of the first people in recorded history to try to understand how and why climate change occurred. Perhaps one of the greatest misconceptions about climate change is the notion that studying it is something that began in the twentieth century. In fact, many of the first important discoveries about global warming were made during the 1800s.

A lot of people are surprised to learn that scientists have been working on the problem of global warming for well over 100 years. The key difference in the beginning, though, was that the scientists weren't studying humanity's role in the process: they were trying to understand something that for religious and cultural reasons was a dangerous idea at the time: perhaps the climate on Earth had not always been the same.

The base of climate science today comes from work that was done by these visionaries of the nineteenth century. What's so impressive about these pioneers is that they were able to see climate in ways no one had ever seen it before. They were trying to find answers to fundamental questions: Why is the sky blue? How old is our planet? Why were woolly mammoth bones popping up in the La Brea Tar Pits in Los Angeles? And so on. These scientists were starting from scratch in building a body of evidence about the Earth's climate. They had to frame the questions, devise the equipment, and then perform the experiments to come up with reproducible answers. Getting the planet to share its past is like pulling teeth. But, as it turns out, teeth had a lot to say.

In the 1800s when scientists once again began finding 6-inch teeth scattered across North America and Europe and into Siberia, they wanted to do something a little better than a Viking myth about an ice giant. They wanted to use the tools of science to build a rigorous explanation that could stand the test of time. Ironically, they ended up proving that the Vikings weren't too far off, at least with regard to the giant ice age.

In 1837 Louis Agassiz, a Swiss scientist, stood up before his colleagues at a conference in the Swiss town of Neuchâtel to present a theory suggesting that the Earth had indeed experienced an ice age. Like many others in his day, he had observed the glaciers of his native Switzerland and noticed the marks that these glaciers left behind: rocks with scratches and scars, mounds of debris called moraines that had been pushed up by glaciers, deep valleys, signs that large boulders had been carried long distances. Agassiz came to realize he was seeing classic signs of a process known as glaciation in places where there were no glaciers to be seen.

Agassiz was going up against an explanation that had come from the Bible. At the time, it was widely believed that *The Great Flood* was the only event with the power to do such heavy lifting.[2] The story of Noah's flood, with just a slight tweak, received almost unanimous support from the scientific community. The tweak had been provided by the great English geologist Charles Lyell. And it was required in order to overcome an inconsistency in the story. Lyell's revision to The Great Flood was his suggestion that the big boulders dropped off in strange places had, in fact, been transported by icebergs.

But that still left the issue of the strange scars on the rocks. Interestingly, plenty of local villagers at the time had already come to their own conclusions about these scars. Having grown up among glaciers, they didn't need a scientist to explain the origins of the strange scars. Throughout the towns and villages of Switzerland, it seems many people had already been convinced that the scratches and scars were the result of a flood of ice, not a flood of water as Lyell had suggested. They, in fact, had already come to accept the theory of a great ice age, just like the Vikings before them.

Despite the lukewarm reception of his presentation at the Swiss Society of Natural Sciences in Neuchâtel in 1837, Agassiz persisted.

In 1840, he even published a book called *Studies on Glaciers.*[3] With the help of numerous colleagues who had been convinced by his evidence and by the clarity of his argument, Agassiz fought hard to convince skeptics who clung to the theory of The Great Flood. Eventually, the overwhelming strength of the evidence won out. In the end, Agassiz had proved that there was a period of time when large areas of the Earth had been covered by ice sheets. Like the Swiss villagers, he had come up with the simplest and most consistent explanation. The ice age, he said, reached its maximum about 20,000 years ago, and then gave way to an eventual warming.

Let's go back to the woolly mammoth for a moment. The rise and fall of the woolly mammoth is linked to the rise and fall of the ice ages. The rise began around 300,000 years ago, as the Earth underwent a transition to a cooler climate. The peak of the last glacial period, the LGM, was about 20,000 years ago. After that, over a span of about 12,000 years, much of the ice melted, the sea level rose almost 400 feet, and the temperature rose about 11°F. Fossil evidence suggests that at the peak of the LGM, woolly mammoths could be found across Europe, Asia, and North America. They were so well adapted to the cold that during the last ice age, parts of Siberia may have had an average population density of about sixty woolly mammoths to every 40 square miles. But then, as the climate changed around them, they simply died out. As scientists processed the significance of this connection, they had to invent a word to explain the phenomenon. The word is *extinction*.

The woolly mammoth, that icon of the ice age, also became an icon of extinction. Before their extinction was recognized, no one had supposed that a robust species could simply disappear. So Louis Agassiz will always be credited not only with his theory of a "great ice age" but also with discovering extinction.

Over the years, Agassiz's theory of the ice age needed to be refined. The ice sheets were not as large as he had thought, and the ice age didn't arrive as suddenly as he had thought. Most important, there wasn't just one great ice age. In Scotland, plant fragments were found sandwiched between layers of glacial deposits. It became increasingly obvious that there had been not just one ice age but several large glaciations, one following another, separated by warm periods. Scientists came to understand that the Earth actually moved into and out of ice ages. And with that amazing discovery, a new crop of scientists began to work on a new theory they called *climate change*.

To grow a continental-scale ice sheet you need low temperatures. That much was clear. What wasn't so clear was how the temperatures had been lowered enough to permit the growth of ice on such a massive scale.

One important hypothesis of how the planet regulated its temperature was put forth by the French mathematician and physicist Joseph Fourier in 1824.[4] As a physicist, Fourier was interested in understanding some basic principles about the flow of heat around the planet. Specifically, he wanted to use the principles of physics to understand what sets the average surface temperature of Earth. It made perfect sense that the sun's rays warmed the surface of the Earth, but this left a nagging question: when light from the sun reaches the surface of the Earth and heats it up, why doesn't the Earth keep warming up until it's as hot as the sun? Why is the Earth's temperature set at roughly 59°F—the average temperature at its surface?

Fourier reasoned that there must be some balance between what the sun sends in and what the Earth sends back out, so he coined the term *planetary energy balance*, which is simply a way of saying that there is a balance between energy coming in from the sun and energy going back out to space. If the Earth continuously receives heat from the sun yet always has an average temperature hovering around 59°F, then it must be sending an equal amount of heat back to space. Fourier suggested that the Earth's surface must emit invisible infrared radiation that carries the extra heat back into space. Infrared radiation (IR), like sunlight, is a form of light. But it's a wavelength that our eyes can't see.

This was a good idea, but when he actually tried to calculate the planet's temperature using this effect, he got a temperature well below freezing. So he knew he must be missing something. To arrive at 59°F, the Earth's average temperature, Fourier realized that he needed the atmosphere to pick up the slack. And he discovered a phenomenon he called the *greenhouse effect*, a process whereby the gases in the Earth's atmosphere trap certain wavelengths of sunlight, not allowing them to escape back out to space. Like the glass in a greenhouse, these *greenhouse gases* let sunlight through on its way in from space, but intercept infrared light on its way back out.

In 1849, an Irish scientist, John Tyndall, was able to build on this idea. He had become obsessed with the glaciers he climbed while visiting the Alps on a vacation. Like many other scientists at the time, he wanted to understand how these massive sheets of ice formed and grew. He applied his personal observations of glaciers in the laboratory in 1859, when, at the age of thirty-nine, he began a series of innovative experiments.

Tyndall was intrigued by the concept of a *thermostat*. We know thermostats today as devices that regulate the temperature of a room by heating or cooling it. Tyndall devised an experiment to test whether the Earth's atmosphere might act like a thermostat, helping to control the planet's temperature. He reasoned that it might help explain how ice ages had blanketed parts of the Earth in the past.

For his experiment, Tyndall built a device, called a spectrophotometer, which he used to measure the amount of radiated heat (like the heat radiated from a stove) that gases such as water vapor, carbon dioxide, or ozone could absorb. His experiment showed that different gases in the atmosphere had different abilities to absorb and transmit heat. Some of the gases in the atmosphere—oxygen, nitrogen, and hydrogen—were essentially transparent to both sunlight and IR, but other gases were in fact opaque: they actually absorbed the IR, as if they were bricks in an oven. Those gases include carbon dioxide (CO_2) and also methane, nitrous oxide, and water vapor. These greenhouse gases are very good at absorbing infrared light. They spread heat back to the land and the oceans. They let sunlight through on its way in from space, but intercept IR on its way back out. Tyndall knew he was on to something. The fact that certain gases in the atmosphere could absorb IR implied a very clever natural thermostat, just as he had suspected. His top four candidates for a thermostat were water vapor, without which he said the Earth's surface would be "held fast in the iron grip of frost"; methane; ozone; and, of course, carbon dioxide.[5]

Tyndall's experiments proved that Fourier's greenhouse effect was real. They proved that nitrogen (78 percent) and oxygen (21 percent), the two main gases in the atmosphere, are not greenhouse gases, because a molecule of each of these elements has only two atoms, so it cannot absorb or radiate energy at IR wavelengths. However, water vapor, methane, and carbon dioxide, each of which is a molecule with three or more atoms, are excellent at trapping IR radiation. They absorb about 95 percent of the long-wave or IR radiation emitted from the surface. So, even though there are only trace amounts of these gases in the atmosphere, a little goes a long way toward making it really tough for all the heat to escape back into space. In other words, greenhouse gases in the atmosphere act as a secondary source of heat, in addition to the sun. And the greenhouse gases provide the additional warming that Fourier needed to explain that average temperature of 59°F.

Thanks to Tyndall, it is now accepted that visible light from the sun passes through the Earth's atmosphere without being blocked by CO_2. Only about 50 percent of incoming solar energy reaches the Earth's surface: about 30 percent is reflected by clouds and the Earth's surface (especially in icy regions), and about 15 percent is absorbed by water vapor. The sunlight that makes it to the Earth's surface is absorbed and reemitted at a longer wavelength, IR, that we cannot see, like heat from an oven. Carbon dioxide (like other heat-trapping gases, such as methane and water vapor) absorbs the IR and warms the air, which in turn warms the land and water below it. More carbon dioxide means more warming. This is where the concept of a natural thermostat becomes very powerful—mess with the amount of CO_2 in the atmosphere, and you're resetting the thermostat of the planet.

The idea was good, even profound, but the term *greenhouse effect* was not entirely accurate. Real greenhouses stay warm without a heater because the sun's rays shine in, warming the inside of the greenhouse, and the glass keeps the heat from escaping. But in reality the atmosphere is much more sophisticated than a greenhouse. Fourier had figured out something very important. He had figured out that the sun is not our only source of heat. The atmosphere, in fact, is a very powerful backup generator. This was yet another discovery on the road to understanding the relationship between temperature and carbon dioxide, a relationship that turns out to have profound implications for our climate.

Svante Arrhenius (1859–1927), a Swedish physicist and chemist, was another scientist who was smitten with ice ages. He took Tyndall's thermostat mechanism and ran with it, exploring whether the amount of CO_2 in the atmosphere could be fiddled with by an event such as a volcanic eruption. According to Tyndall's experiments, the additional carbon dioxide released by the volcano could conceivably raise the Earth's temperature, and Arrhenius wanted to see if that was actually true.

We refer to events or processes that result in changes to the climate as *forcings*. A volcanic eruption is an example of a natural forcing. A forcing can often result in an amplification (positive) or a reduction (negative) in the amount of change and often comes hand in hand with a *feedback*—a situation where some effect causes more of itself. In other words, if a forcing is the event that creates change, then the feedback amplifies that change. But keep in mind that a positive feedback is not positive in the sense of being good. *Positive* refers specifically to the direction of change, not to the desirability of the outcome. A negative feedback tends to reduce or stabilize a process, whereas a positive feedback tends to increase or magnify it.

Maybe, Arrhenius thought, this positive feedback mechanism was responsible for plunging the planet into an ice age. If the atmosphere were to dry out for some reason, the decreasing water vapor would hold less heat and the Earth would cool. Since cooler air holds less water vapor, the atmosphere would tend to dry more, amplifying the cooling. In addition, cooler temperatures would generally lead to increases in snow and ice, and so to yet another positive feedback. When snow and ice cover a region, such as the Arctic or Antarctica, their white, light-reflecting surface tends to bounce sunlight back out to space, helping to further reduce temperature. If regions covered by snow and ice expanded over more of North America and Europe, the climate would cool further while also increasing the ice sheets. Start with a drop in carbon dioxide, continue with a drop in temperature, add some snow and ice, and you've made an ice age.

Arrhenius thought his theory was quite solid, but he wanted to prove it mathematically. So he set about a series of grueling calculations that attempted to estimate the temperature response of changing levels of carbon dioxide in the atmosphere. These may have begun as "back of the envelope" calculations, but in 1896 he was confident enough to publish the work for his colleagues to read.[6] The end result of all of it was one simple number: 8°F.

That number represented roughly how much Arrhenius thought the Earth's average temperature would drop if the amount of CO_2 in the atmosphere fell by half. Once you factor in the positive feedbacks of water vapor, snow, and ice, an ice age seemed like a reasonable outcome. The only thing Arrhenius still needed was a mechanism for tinkering with atmospheric carbon dioxide, turning down the natural thermostat. And that is what led, in part, to the discovery of the carbon cycle.

Arrhenius asked a colleague, Arvid Högbom, to help him figure out how much carbon dioxide levels in the atmosphere might be able to change. Högbom had compiled estimates of how carbon dioxide flows through various parts of the planet, including emissions from volcanoes, absorption by the oceans, and so forth. This carbon cycle is a fundamental concept that is hugely important. If carbon dioxide really was the natural thermostat that scientists had been searching for, then the next crucial step would be to figure out how CO_2 cycles into and out of the ocean, the land, the atmosphere, and living matter such as plants and trees.

It turns out that carbon (the C in carbon dioxide) has the ability to cycle among a few different reservoirs. Relatively small amounts of carbon reside in the atmosphere, the ocean surface, and vegetation. A slightly larger amount is held in soils, and a much larger amount resides in the deep ocean. The biggest reservoir can be found in rocks and sediments. Carbon takes different chemical forms in different reservoirs. In the atmosphere, it is the gas carbon dioxide (CO_2).

The carbon cycle can be thought of, metaphorically, as a kind of reincarnation. This cycle is the great natural recycler of carbon atoms. The same carbon atoms in your body today have been used in countless other molecules for millions, even billions, of years. The wood burned in a fireplace last winter produced CO_2 that found its way into a tomato plant this spring. The borders are wide open and carbon cycles easily cross different zones. The atoms pair up, get into various substances for a while, come out of those, and go somewhere else—it is a continuous and ongoing cycle.

Here's a carbon cycle scenario. In phase one, volcanoes and hot springs transfer carbon from deep below the Earth's crust to the atmosphere. In phase two, the carbon dioxide is scrubbed from the atmosphere by a process called *chemical weathering.* Basically, when it rains, the rainwater combines with CO_2 in the atmosphere to form a weak acid, carbonic acid. That weak acid falls as rain and then chemically reacts with rocks, releasing carbon, which eventually makes its way into the ocean, where it is locked up in the shells of marine plankton.[7] After dying, the marine plankton eventually sink to the bottom and turn into rocks.

Here, the scenario gets really interesting. Experiments show that rates of chemical weathering are influenced by three environmental quantities: temperature, precipitation (rain and snow), and plant matter. Temperature, precipitation, and vegetation all act in a mutually reinforcing way to affect the rate of chemical weathering. The higher the temperature, the faster a rock is broken down by chemical weathering. Higher precipitation raises the level of groundwater held in soils and combines with CO_2 to form carbonic acid and more rapidly drive the weathering process. Remember that temperature and precipitation are linked; the amount of water vapor that air can hold rises with temperature. Likewise, the amount of vegetation is closely tied to temperature and precipitation. More rainfall means more vegetation, and more vegetation means more carbon stored in the soil.

So, carbon becomes the secret ingredient in adjusting the natural thermostat and changing the Earth's climate. The beauty of this mechanism is that it's a big loop. On the one hand, the speed of chemical weathering is tuned to the state of the Earth's climate. On the other hand, the climate is tuned to the rate at which CO_2 is pulled out of the atmosphere by chemical weathering. This is an example of a very sophisticated feedback loop.

Ultimately, chemical weathering is the most likely explanation for Earth's habitability over most of the 4.6 billion years of its existence. Any factor that heated Earth during any part of its history caused chemical weathering rates to increase. This increase, in turn, drew CO_2 out of the atmosphere at faster rates, and eventually resulted in a cooling to offset the warming. On the flip side, any factor that cooled Earth set off the opposite sequence of events. Chemical weathering constantly acts to moderate long-term climate changes by adjusting the CO_2 thermostat as needed. If positive feedbacks help push our climate into an ice age, chemical weathering helps to push us out of one.

As a result of chemical weathering, most of Earth's carbon is tied up below the surface in rocks and pools, including coal, oil, and natural gas. But now, of course, we humans are taking the coal, oil, and natural gas out of the ground and burning it, transferring long-stored carbon to the atmosphere. Nature's history tells us what to expect.

We tend to think of man-made global warming as a modern concept, something that has come into vogue in the last twenty years or so, but in reality this idea is more than 100 years old. As noted above, the notion that the global climate could be affected by human activities was first put forth by Svante Arrhenius in 1896. He based his proposal on his prediction that emissions of carbon dioxide from the burning of fossil fuels (i.e., coal, petroleum, and natural gas) and other combustion processes would alter atmospheric composition in ways that would lead to global warming. Arrhenius calculated how much the temperature of the Earth would drop if the amount of CO_2 in the atmosphere was halved; he also calculated the temperature increase to be expected from a doubling of CO_2 in the atmosphere—a rise of about 8°F.

More than a century later, the estimates from state-of-the-art climate models doing the same calculations to determine the increase in temperature due to a doubling of the CO_2 concentration show that the calculation by Arrhenius was in the right ballpark. The Fourth Assessment Report of the Intergovernmental Panel on Climate Change (IPCC) synthesized the results from eighteen climate models used by groups around the world to estimate climate sensitivity and its uncertainty. They estimated that a doubling of CO_2 would lead to an increase in global average temperature of about 5.4°F, with an uncertainty spanning the range from about 3.6°F to

8.1°F. It's amazing that Arrhenius, doing his calculations by hand and with very few data, came so close to the much more detailed calculations that can be done today.

Arrhenius's calculations, however, did have some shortcomings. For example, in estimating how long it would take for the CO_2 concentration in the atmosphere to double, he assumed that it would rise at a constant rate. With about 1.6 billion people on the planet in 1895 and with relatively small use of fossil fuels, Arrhenius predicted that it would take about 3,000 years for the atmospheric CO_2 concentration to double. Unfortunately, when scientists today factor in the quadrupling of world population since then and the increasing demand for energy, doubling is now projected before the end of this century unless substantial cutbacks in emissions are adopted by nations around the world. So, technically, Arrhenius was off by about 2,800 years. (Another of his doubtful predictions was that he firmly believed a warmer world would be a good thing.)

In Arrhenius's time, the impacts of global warming were mainly left to future investigation—the majority of scientists still needed to be convinced that the concentration of CO_2 in the atmosphere could vary, even over very long timescales, and that this variation could affect the climate. Scientists at the time were focused more on trying to understand the gradual shifts that took place over periods a thousand times longer than Arrhenius's estimate: those that accounted for alternating ice ages and warm periods and, in distant times (more than 65 million years ago), for the presence of dinosaurs. They couldn't even begin to wrap their minds around climate change on a human timescale of decades or centuries. Nobody thought there was any reason to worry about Arrhenius's hypothetical future warming, which he suggested would be caused by humans and their burning of fossil fuel. It was an idea that most experts at the time dismissed. Most scientists of the era believed that humanity was simply too small and too insignificant to influence the climate.

Fast-forward to the mid-1950s, and enter Charles David Keeling, a brilliant and passionate scientist who was then beginning his research career at Caltech. Keeling had become obsessed with carbon dioxide and wanted to understand what processes affected fluctuations in the amount of CO_2 in the atmosphere. Answering this question required an instrument that didn't exist, the equivalent of an ultra-accurate "atmospheric Breathalyzer." So Keeling built his own instrument and then spent months tinkering with it until it was as close to perfect as he could get at measuring the concentration of CO_2 in canisters with a range of values of known concentration.

Keeling tried his instrument out by measuring CO_2 concentrations in various locations around California and then comparing these samples in the lab against calibration gases. He began to notice that the samples he took in very pristine locations (i.e., spots where air came in off the Pacific Ocean) all yielded the same number. He suspected that he had identified the baseline concentration of CO_2 in the atmosphere; a clear signal that wasn't being contaminated by emissions from factories, farms, or uptake by forests and crops.

With this instrument, called a *gas chromatograph*, Keeling headed to the Scripps Institution of Oceanography to begin what is perhaps the single most important scientific contribution to the discovery of global warming. Keeling was on a mission to find out, once and for all, if CO_2 levels in the atmosphere were increasing. He would spend the next fifty years carefully tracking CO_2 and building, data point by data point, the finest instrumental record of the CO_2 concentration in the atmosphere, generating a time history that is now known by scientists as the Keeling curve.

The Keeling curve is a monthly record of atmospheric carbon dioxide levels that begins in 1958 and continues to today. The instrument Keeling built, the gas chromatograph, works by passing infrared (IR) light through a sample of air and measuring the amount of IR absorbed by the air. Because carbon dioxide is a greenhouse gas, Keeling knew that the more IR absorbed by the air, the higher the concentration of CO_2 in the air. Because CO_2 is found in very small concentrations, the gas chromatograph measures in terms of parts per million (ppm).

Keeling knew from his travels around California that he needed to make his measurements at a remote location that wouldn't be contaminated by local pollution. That's why he settled on Hawaii. Hawaii's big island is the site of the volcano Mauna Loa, and Keeling set up his CO_2 instrument near the top of Mauna Loa. Isolated in the middle of the Pacific Ocean and at more than 11,000 feet above sea level, the top of the Mauna Loa volcano is an ideal location to make measurements of atmospheric carbon dioxide that reflect global trends, but *not* local influences such as factories or forests that may boost or lower the carbon dioxide level within their vicinity. The sensors were positioned so that they sampled the incoming ocean breeze well above the thermal inversion layer; thus the air was not affected by nearby human activities, vegetation, or other factors on the island. Obviously, volcanoes are potentially a big source of CO_2, but Keeling took this into account when positioning his instrument, locating it upwind of Mauna Loa's vent and installing sensors to give alerts if the winds shift.

What he found was both disturbing and fascinating, creepy and profound. Keeling, using his Mauna Loa measurements, could see that with each passing year CO_2 levels were steadily moving upward. As the years passed and the Mauna Loa data accumulated, Keeling's CO_2 record became increasingly impressive, showing levels of carbon dioxide that were noticeably higher year after year after year. The first instrumental measurements indicated a CO_2 concentration of 315 ppm in 1958. The slow rise in its concentration over the first several years was enough to prompt a report from a panel of the President's Science Advisory Council to President Johnson in 1965, indicating that the early prediction that an increase in CO_2 could occur was correct and that global warming would indeed be expected to occur. This was the first instance when a document discussing global warming ended up in front of the president of the United States. It would not be the last.

In 2008, just over fifty years after Keeling started his observations, the concentration at Mauna Loa had reached 385 ppm. Keeling's measurements thus provided solid evidence that the atmospheric CO_2 concentration was increasing. If anything proved that Arrhenius had been on to something, it was these data.

One of the most striking aspects of the Keeling curve is a small CO_2 wiggle that takes place every year. For every little jump up, there is a little dip back down, so that the whole curve looks sawtoothed. This wiggle happens like clockwork and is timed with the seasons. In the northern hemisphere during fall and winter, plants and leaves die off and decay, releasing CO_2 back into the atmosphere and causing a small spike. And then during the spring and summer, when plants are taking CO_2 out of the atmosphere in order to grow, carbon dioxide levels drop. Hawaii, along with most of the planet's landmass, is situated in the northern hemisphere, so the seasonal trend in the Keeling curve is tracking the seasons in the northern hemisphere. The Keeling curve proved many important things at once. It proved that CO_2 levels in the atmosphere can indeed change and that they can change on very short timescales.

Keeling's record was the icing on the cake, and he rightly stands with Agassiz, Tyndall, and Arrhenius among the giants of climate science. He helped prove the reality of global warming by providing the data upon which the pioneering theories of Tyndall and Arrhenius could finally rest. As is the case in research science, Keeling's painstaking measurements have been verified and supplemented by many others. Measurements at about 100 other sites have confirmed the long-term trend shown by the Keeling curve, although no sites have a record as long as Mauna Loa. Other scientists have also extended the Keeling curve farther back in time, using measurements of CO_2 in air trapped in bubbles in polar ice and in mountain glaciers. Ice cores collected from Antarctica and Greenland can be used to reconstruct climate hundreds of thousands of years ago, showing that the preindustrial amount of CO_2—the level from A.D. 1000 to 1750—in the atmosphere was about 280 ppm, about 105 ppm below today's value. The record indicates that the concentation of CO_2 has increased about 36 percent in the last 150 years, with about half of that increase happening in the last three decades. In fact, the CO_2 concentration is now higher than any seen in at least the past 800,000 years—and probably many millions of years before the earliest ice core measurement.

Over the past century, the evidence has piled up in support of Arrhenius's explanation of global warming. As the evidence accumulates with each passing year, what was once a fringe hypothesis that sprang from the mind of a single scientist in Sweden is now part of the bedrock of scientific accomplishments. Unfortunately, scientific discoveries are not always good news. And there is a nagging fear among scientists that we'll prove ourselves to be not so different from the woolly mammoth, the symbol of a climate that no longer exists.

3

THE SCIENCE OF PREDICTION

If I have seen further, it is by standing on the shoulders of giants.

—Sir Isaac Newton

Prediction is an odd thing. Depending on your personality, predictions are a source of either comfort or anxiety. In one broad stroke, they have the power to reassure or destabilize. Predictions often give us an illusion of control in situations that are inherently out of our control. Nothing exemplifies this better than our relationship to weather forecasts. We can't stop the weather, but we can at least prepare for it. Ultimately, this preparation is what the science of prediction—be it climate or weather prediction—is all about.

A certain pleasure comes from knowing that meteorologists are generally right about the forecast, and a certain disappointment comes from finding out they got it wrong. Of course, we are not happy when predictions fail. Over the last fifty years, we have grown accustomed to the idea that the weather can be "predicted," so it feels like violation when a forecast turns out to be incorrect.

A big part of believing predictions like those in this book has to do with trusting and understanding the underlying data and models. Model simulations are the closest thing that scientists have to a crystal ball, and as a result data are the lifeblood of every prediction that weather and climate scientists make. At this point, weather prediction is so ingrained in our lives that we've stopped being skeptical about it. Even though the sun is shining, our experience tells us that we should trust the man or woman in front of the map who's gesturing at swirling shades of green behind it. Unfortunately, the same cannot be said for climate prediction; but as we will see here, the two are really not all that different.

Although weather prediction is now embedded in our psyche, the practice as we know it has been studied for only about 100 years. What we need to understand is that the mechanisms of weather predictions are very similar to those of climate predictions. So if we're comfortable trusting local forecasters' predictions about weather, we should probably think about trusting the predictions coming out of the country's climate laboratories.

The modern-day weather forecast originated on the battlefields of World War I. During that war, a young Quaker ambulance driver, Lewis Fry Richardson, fascinated by the possibility of seeing the weather before it happened, laid the groundwork for the daily weather forecasts that we all live by today.[1] Richardson, a true giant in weather forecasting, was also a pioneer in a branch of mathematics called *numerical analysis*. Numerical analysis looks for ways to find approximate solutions to problems that are too complicated to solve. It also serves as a bridge between people and computers. One of the key differences between people and computers is that computers can do arithmetic lightning fast. Humans, on the other hand, can come up with elegant mathematical equations to represent how the world works. Despite their elegance, those mathematical equations are hard to solve, and that's where numerical analysis comes in handy. Without it, computer models would not have been possible.

But before there were computers, there was Richardson. He was committed to the idea of generating the very first weather forecast using seven elegant mathematical equations developed by another giant in the field of meteorology, the Norwegian scientist Vilhelm Bjerknes. By the time of World War I, Bjerknes had come up with equations capable of describing the behavior of the atmosphere. The state of the atmosphere at any point could be described by seven values: (1) pressure, (2) temperature, (3) density, (4) water content, and wind—(5) east, (6) north, and (7) up. In essence, Bjerknes presented Richardson with seven complex calculus problems in need of transformation.

Richardson knew that the differential equations could be approximated and simplified using numerical analysis. And once the equations were simplified, he figured that he should be able to generate a weather forecast for central Europe. To do this, he divided the entire atmosphere into discrete columns measuring about 3° east–west and about 125 miles north–south; this division works out to about 12,000 columns on the surface and five rows in the atmosphere. If he calculated the value of each of the seven variables for each cell in the two columns over central Europe, he figured he'd have the first battlefield weather forecast.

Of course, at that time, Richardson did all his work by hand, in "offices" that can most charitably be described as airy—temporary rest camps with a view of the front lines of the fighting. Computers capable of doing the math were still a far-off dream, so with just pencil and paper, this driver with the Friends Ambulance Unit in France tackled the problem of weather prediction. Richardson himself was the computer. His forecast for central Europe was no small undertaking; he later wrote, "The scheme is complicated because the atmosphere is complicated." Even the simplified procedure required a maddening amount of arithmetic. There was so much arithmetic to be done that calculating a weather forecast just six hours out in time required about six months of work—rather late to be considered an actual forecast.

But Richardson was undaunted, expressing his dream that "someday in the dim future it will be possible to advance the computations faster than the weather advances." Of course, when he wrote these words, Richardson was imagining people doing the calculating. In the not too distant future, artificial computers would easily outrun time and see the future before it happened.

Unfortunately, the time it took to grind out the forecast calculations wasn't Richardson's only problem. The *initial conditions* he used to start the calculation were both incomplete and imprecise. He just didn't have all the observational data he needed to fully represent the physical state of the atmosphere. As a result, the first official weather forecast went down in history as a total bust, and with that bust came one of the cardinal rules of weather and climate prediction: your forecast is only as accurate as your data.

Still, though the forecast itself was off, much of what Richardson proposed was right. And luckily for all of us who count on reliable weather forecasts today, Richardson was brave enough to publish his ideas. However, he had to find his manuscript first—he had lost the sole copy during the Battle of Champagne in April 1917. He discovered it months later under a heap of coal. The book, eventually published in 1922, was called *Weather Prediction by Numerical Process*. And what at first appeared to be nothing more than a failed weather forecast is now widely considered one of the most profound books about meteorology ever written. Richardson had come up with a way to see into the future. But he couldn't do it alone. He needed computers.

Upon returning from the war, Richardson eventually quit meteorology when he realized that his work was being used for military purposes. A committed pacifist, this gentle giant actually destroyed some of his research to prevent it from being used by the military. He spent much of the rest of his life applying mathematics to the understanding of the causes of war. But as time went on, and the field of theoretical meteorology came of age, Richardson's early vision of a weather forecast was fully realized. Computers were being developed, and by the late 1940s the first successful numerical weather prediction was performed at the Institute for Advanced

Study in Princeton, New Jersey. By the 1950s routine weather forecasts were being produced; these used very simple models that did not take into account variables such as radiation and so led to some fairly large errors.

Yet in spite of these shortcomings, the computers proved very effective at predicting the weather, especially as more advanced forms of data collecting fed more accurate information into the models. Today, the North American Mesoscale (NAM) model developed by the National Weather Service—this is the model that The Weather Channel uses for its forecasts—takes about ninety minutes to ingest all the data (those very important initial conditions), and the actual computer calculations that provide the weather forecast out to eighty-four hours (3.5 days) take less than ninety minutes. So, give a model three hours, and it'll give you the weather for the entire country for the next three days.

As weather forecasts became more routine and forecasters' skill increased, scientists began to look for a new challenge; they began to look farther out in time. The goal was to build a model that represented the climate system. This was no small task. Weather models are concerned only about what's happening in the atmosphere. The atmosphere has a memory of roughly one week. That's why your local weather forecast goes out only about a week.

Climate models, however, needed to include much more. Scientists had to connect their mathematical version of the atmosphere to mathematical versions of the oceans, the land surface, and sea ice and biology. This was a vast expansion of weather prediction. And so, in the late 1940s, scientists, many of them meteorologists, set out to derive the mathematical equations that would describe the rest of the planet. They were building a computer model that would serve as a planetary stunt double. It would be an entirely new way of looking even farther into the future.

Under the direction of Joseph Smagorinsky at the U.S. Weather

Bureau in Washington, D.C., the work started with basic physics equations of fluids and energy, and then kept building from there. Syukuro Manabe, a Japanese meteorologist, arrived in the United States from Tokyo University in 1958 to help Smagorinsky. He began work on an atmospheric model that would include the basics: winds, rain, snow, and sun. He and Smagorinsky also included the greenhouse effect caused by both carbon dioxide and water vapor. This would allow them to eventually test what the increased carbon dioxide would do to the climate system.

In the meantime, building this "twin Earth" required understanding the nitty-gritty of how the world works. Manabe found himself in the library researching topics such as how different soils absorb water. By 1965 he and Smagorinsky had developed a three-dimensional model, which solved the basic equations for the atmosphere and was simple enough that the equations could be calculated efficiently. Still, it's important to keep in mind that this early model, and others, had no geography: no land and no oceans. Everything was averaged over bands of latitude, with continents and oceans mixed together to form a swamp that exchanged moisture with the atmosphere above it but was unable to absorb heat. All in all, the atmosphere generated by these models looked decent. The model output showed a realistic layered atmosphere, as well as a zone of rising air near the equator, and a subtropical band of deserts.

As the power of computers increased, climate modeling groups began popping up around the world. By the mid- to late 1960s, weather prediction models were already quite accurate at forecasting the weather three days in advance, and the field of meteorology was entering a more mature, operational phase. Also, climate models came to stand squarely on the shoulders of weather prediction models. A good weather forecast was of tremendous importance to the economy, and as a result, the field of weather prediction began to receive more funding. There was a concerted push to improve the data being used to initialize the models. The use of spy satellites for weather "reconnaissance" had been proposed as early as 1950. By 1960, the Department of Defense had used classified spy satellite technology to launch the first weather satellite. By 1969, the design of the Nimbus-3 satellite proved helpful in improving weather forecasts. The satellite's infrared (IR) detectors could measure the temperature of the atmosphere at various heights all over the world. Ironically, if we remember that Richardson was a pacifist, the science of weather prediction was benefiting from money and technology that originated in the military.

Even with the ongoing improvements in weather data and computer technology, the emerging field of climatology was struggling to avoid the old adage about computers, "Garbage in, garbage out." When trying to represent global climate, scientists encountered a mind-bogglingly complex system. This was an enormous intellectual challenge. In addition to an atmosphere, climate models include land surfaces, oceans, sea ice, and hydrology—variables that made climatology much more difficult for the primitive computers, and for the scientists crunching the numbers. Also, the climate models needed to run for a much longer time, since instead of dealing with the few days needed for a weather forecast, the scientists were trying to simulate over decades, centuries, and in some cases even thousands of years. These scientists were tackling an immense problem. They were, in a sense, building the Earth from scratch. But along the way they were coming to understand important differences between predicting the weather one day ahead and predicting the climate 100 years ahead.

As Richardson learned the hard way, good data are important. And in his case, not having the precise starting point or initial conditions of the atmosphere took an otherwise great weather forecast and put it on the road to ruin. This dependence on initial conditions showed just how valuable good data are: they enable useful forecasts to go out a week instead of only a few days. Through advances in technology, scientists were able to enhance their data and thus design more accurate weather forecasts, creating predictions that were more accurate and less vulnerable to variation than ever before.[2]

Interestingly, climate models and weather models are often one and the same. But while climate models simulate actual weather, their results are analyzed differently from weather models. Climate prediction is not nearly as dependent on initial conditions as weather prediction is. In other words, the climate at the end of this century won't care very much about the weather at the beginning of this century. Climate is not nearly as chaotic as weather (for example, we can easily predict that July will be hotter than January). Climate model output is often analyzed by studying the season-to-season, year-to-year, and even decade-to-decade evolution of the climate. Unlike weather forecast models, they never attempt to predict precisely what a single day will look like. Instead, they look at how the statistics of weather change.

This is a very important distinction between weather and climate models: for climate forecasts, the initial conditions in the atmosphere are not as important as the external forcings that have the ability to alter the character and types of weather (i.e., the statistics or what scientists would call the "distribution" of the weather) that make up the climate.

These forcings include, for instance, the Earth's distance from the sun; how many trees are growing on the surface of the Earth; and, of course, how much carbon dioxide is in the atmosphere. You can't use models to simulate changes in the climate unless you know what will happen to the forcings.

And then there are the actual equations that make up the model. Climate models are built from two types of equations. First, there is the physics, which comes in the form of elegant equations such as Newton's laws of motion and conservation of energy. Second, there are equations, known as parameterizations, that are derived from observations and attempt to represent our current understanding of certain aspects of climate and weather. The physics in these models is universal, whereas parameterizations can vary depending on the team building the model. Parameterizations are a way to estimate all the complicated interactions that have been observed in nature but whose physics can't be directly represented in models due to limitations in computer resources and speeds. Each model uses different parameterizations to approximate what it cannot represent directly. As a result, different models predict different degrees of warming.

Because parameterizations inevitably introduce uncertainty, climate assessments typically draw on the collective wisdom of about twenty climate model projections, making up an *ensemble* of model simulations. This ensemble approach gives a better estimate of reality than any one particular model (though some models are better than others). Choosing the ensemble average is a way of drawing on multiple models to reach a consensus, rather than relying on any single model. Weather forecasters do the same. The assumption is that the approximation errors among models tend to cancel each other when we average their projections. As a result, the common, most robust tendencies are captured.

As computational speed and observational data continued to increase and improve, climate model simulations began to look more and more like the real world. It was eventually clear that climate models were ready for prime time; they were good enough to work on the problem of global warming. Those grueling calculations that Arrhenius had labored over could now be done quickly and rather painlessly by computers.

In general, there are two types of climate model runs that test the impact of global warming on the climate system: *transient* runs and *equilibrium* runs. In a transient run, greenhouse gases are slowly added to the climate system and the model simulates the impact of the additional CO_2 at each time step. In an equilibrium run, the atmospheric CO_2 level is instantly doubled, and the model is run with the higher CO_2 level until the climate has fully adjusted to the forcings and has reached a new equilibrium. The global average change in surface temperature due to the doubling of CO_2 is a number referred to as *climate sensitivity.*

In 1967, Manabe's group carried out the first series of climate sensitivity experiments using a very simple equilibrium model that represented the atmosphere averaged over the entire globe. The goal was to estimate what the Earth's average temperature would be if the level of CO_2 in the atmosphere doubled. This was similar to what Arrhenius had done by hand in the 1890s when he estimated that the planet would warm about 8°F. Using his one-dimensional model, Manabe came up with a different number: about 3°F to 4°F. Later, in 1975, Manabe and his collaborator Richard Wetherald published an analysis using a more advanced model that they had designed. This time they came up with roughly 6°F. Though this number was also less than what Arrhenius had come up with, it was taken much more seriously, since it had been derived from methods and a model that were more rigorous than the earlier attempts.

By the end of the 1980s scientists were also working on transient runs of the climate system, testing their climate models with varying levels of CO_2 to see what the future might look like. And even when these climate models were still in their infancy, they pointed toward an interesting result. When oceans were included in this model world, they acted to delay the appearance of global warming in the atmosphere for a few decades. They did this by soaking up some of the extra heat. Some people see this time lag as a gift, in the sense that it allows us an opportunity to prepare for and adapt to the coming climate changes. But many see the time lag as a curse, because it gives us reasons to procrastinate.

And it's tempting to procrastinate if you don't trust the models. But in the 1980s, climate models were beginning to show a very interesting consistency. You could start twenty different models with twenty different initial conditions, but the runs would all converge when they estimated the change in average annual global temperatures. They would, of course, show random variations in weather patterns for a given region or season, but every single model got steadily warmer over time.

The problem with verification of such results is that it's not possible to jump to the end of the century to see if a climate model is any good. But scientists can get around this by using their models to simulate events that have already happened. This simulation is called hind-casting, and it's an efficient way to test whether a climate model is skillful. Successful hind-casting experiments boost our confidence that climate models can capture past events and therefore can serve as a decent guide to the future. By successfully hind-casting a number of past situations (the effects of volcanic eruptions, seasonal variations, etc.), we can increase confidence in model simulations of the future. Basically, we can't prove that the models are right until the future happens, but we can prove that the models function by using certain rigorous tests.

Scientists have performed hind-casting studies on several major events in climate history to test how well the models can reproduce the climate at those times. They've modeled the height of the last ice age about 20,000 years ago, known as the Last Glacial Maximum (LGM), as well as a regional cooling event in Europe and North America roughly 500 years ago, known as the Little Ice Age.

There are also a few, rare opportunities to run a climate model in *forecast mode*. In June 1991, the eruption of Mount Pinatubo in the Philippines provided a perfect natural climate experiment. Pinatubo had injected about 20 million tons of sulfate aerosols into the stratosphere and created the largest cloud of volcanic aerosol haze and the largest perturbation to the stratospheric aerosol layer since the eruption of Krakatau in 1883. The haze spread around the Earth in about three weeks and attained global coverage after about one year.

Jim Hansen, a leading climate scientist at NASA's Goddard Institute for Space Studies (GISS) in New York, recognized this as a great opportunity to perform a real-time experiment: to use a climate model to predict how the real world would respond before it actually responded. In other words, his team would use the model to make a climate forecast that could be proved correct or incorrect in a relatively short time. So Hansen and his team added the Mount Pinatubo eruption as a forcing to the GISS climate model and made a prediction about how much the planet would cool over the coming year: about 1°F globally. They also predicted that the cooling would be concentrated in the northern hemisphere and would last about a year. The test involved waiting to see how skillfully the model had captured the real-world cooling. In 1992 there was a pause in the long-term warming, much to the delight of those who were skeptical about global warming. The average global temperature dropped roughly 0.9°F. Roughly a year later, the cooling began to subside and the steady uptick in global temperature resumed. The results were in, and the climate models were proved correct.

Since Manabe's first experiment with doubled CO_2, equilibrium runs have been performed thousands of times using increasingly sophisticated models. Climate models have reached a level of maturity approaching, if not rivaling, that of weather models. Whereas Manabe's 1967 model was simply one big grid square meant to cover the entire planet, today's climate models have more than 1 million grid squares that cover the planet. Each grid square is about 70 miles by 70 miles, with twenty-six vertical layers in the atmosphere. The next generation of models will resolve down to 30 miles by 30 miles. And as computers get faster, the resolution will improve further. It's not impossible that models will one day be able to predict the climate for every square mile on the planet.

Computers have already become a lot faster. In the 1970s, a century of climate took more than a month to run. In the current version of the National Center for Atmospheric Research (NCAR) T85 model, a century's worth of climate takes as little as thirteen days. Keep in mind that these new models have not only smaller grid boxes but also much more realism, which requires doing more calculations in less time. Perhaps far more telling, despite the impressive advances in data collection, modeling, and computational strength, climate sensitivity hasn't changed very much. The climate sensitivity estimated by the top global climate models ranges from 3.6°F to 8.1°F for an atmosphere going from about 300 to 600 parts per million (ppm) of CO_2. This is not far different from Manabe's estimate of 6°F in 1975 or Arrhenius's calculation of 8°F in 1896. It raises the question: how many more times do we have to do this experiment before we believe the answer?

Beyond these specific temperature increases, climate models help us see that global warming isn't just what's going on at the poles; these models also reinforce trends that you yourself have probably noticed in your lifetime. In the United States, spring now arrives an average of ten days to two weeks earlier than it did twenty years ago. Many migratory bird species are arriving earlier. For example, a study of northeastern birds that migrate long distances found that birds wintering in the southern United States now arrive back in the Northeast an average of thirteen days earlier than they did during the first half of the last century. Snow cover is melting earlier. Plants are blooming almost two weeks earlier in spring. The ranges of many species in the United States have shifted northward and upward in elevation. For example, a study of Edith's checkerspot butterfly showed that 40 percent of the populations below 2,400 feet have disappeared, despite the availability of sufficient food and shelter. These are all further reflections of the warming taking place right now.

And then there are the results from climate models. Climate models help us to understand what is happening and why. The experiment itself is fairly straightforward. You take the observed temperature record of the past century and compare it with the temperature simulated by a climate model driven by *natural events* such as volcanic eruptions and *human activities* such as combustion of coal, oil, and natural gas. Accounting for just natural factors, the models simulate the behavior of what is called the *undisturbed climate system* for periods as long as thousands of years—if external conditions like solar radiation remain within their normal bounds for the whole period. In other words, the model simulates the climate of an Earth without us, an Earth undisturbed by burning fossil fuels and by deforestation.

When you take us out of the calculations, you take out all the greenhouse gas emissions human activities have caused since the industrial revolution provided fuel for our cars and factories and large expanses of forests were cleared for agriculture and development. The rationale is simple. If a climate model, run with only natural forcings, cannot re-create the strong warming since the 1970s, then the real world is currently doing something Mother Nature cannot do on her own. If you can establish this, then you've successfully established that the temperature trend is truly exceptional.

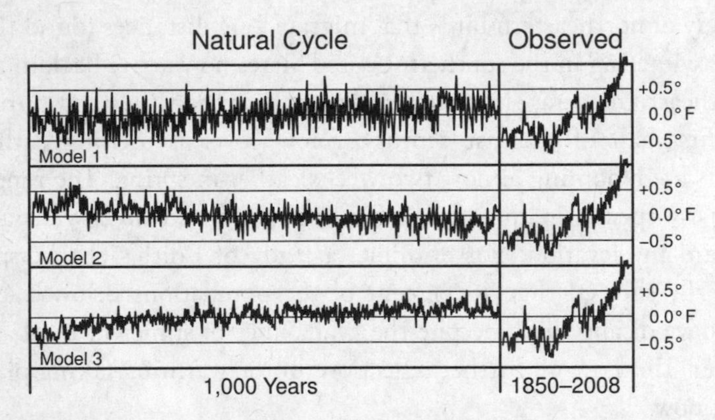

Natural variations in temperature of three different 1,000-year climate model simulations compared with observed data for 1850–2008. SOURCE: ADAPTED FROM STOUFFER ET AL., 1999, BY R. ZIEMLINSKI, CLIMATE CENTRAL.

The accompanying figure sums this up handily. It shows the natural variations in temperature of three different 1,000-year climate model simulations.[3] The variability in these different models is obvious: some periods are warmer, some cooler. But not one of these simulations captures any sign of an extended upward temperature trend. There isn't a single computer model simulation, called a *control run*, that exhibits a trend in global temperature as large or sustained as the observed temperature record. Hence, there is no way to explain the recent warming in terms of how the natural system has behaved over the last 1,000 years. If the recent warming trend were a result of natural forcings, then, assuming the models are correct, the model simulations would capture it and you would see a match between the observed record and the climate model. In fact, there isn't a single model that is able to produce a trend comparable to what we can see in the real world. Houston, we have a problem.

Climate models are not only important for showing us what could happen but are also a valuable tool for showing how much of it is our fault. With hind-casting, scientists can use climate models to isolate the physical fingerprint of human activity and figure out where the heightened levels of carbon in the atmosphere are coming from. Here's how it works. Different forcings—such as changes in solar radiation, volcano eruptions, or fluctuations in greenhouse gas concentrations—imprint different responses, or fingerprints, on the climate system. In the real world, these forcings are superimposed, one on top of another, making it difficult to assign blame to any single one. Therefore, climate models are used to make sense of the impact of each forcing, estimate the individual contribution of that forcing, and test whether it is responsible for the warming trend.

These forcings can be natural due to changes in solar radiation and volcanic eruptions, and they can be human-induced factors such as greenhouse gas concentrations. To repeat: climate models are used to calculate the fingerprint of each individual forcing, and thereby to distinguish how each forcing affects changes in temperature.

Take volcanoes. The idea that volcanoes affect climate has a long history. In 1784, Benjamin Franklin spoke of a constant dry fog all over Europe and North America that prevented the sun from doing its job and kept summer temperatures much chillier than usual. Franklin correctly attributed the dry fog to a large Icelandic volcano, called Laki, that erupted in 1783. In North America, the winter of 1784 was the longest and one of the coldest on record.

There was ice-skating in Charleston Harbor; a huge snowstorm hit the South; the Mississippi River froze at New Orleans; and there was ice in the Gulf of Mexico.

Scientists now know that volcanic eruptions, if large enough, can blast gas and dust into the lower stratosphere,[4] the layer of the atmosphere that begins about 6 miles above the Earth's surface. The strong winds at these altitudes, about 10 to 15 miles up, quickly disperse the volcanic material around the globe. The main gas emitted by the volcanoes, sulfur dioxide, eventually combines with oxygen and water to form sulfuric acid gas. This gas then condenses into fine droplets, or sulfate aerosols, that form a haze. The volcanic haze scatters some of the incoming sunlight back to space, and as a result temperature at the surface of the Earth, sometimes quite drastically, plummets for two to three years.

We've long known that solar radiation—like volcanoes—has the ability to affect global temperature, especially because the output of the sun is not constant. The sun has a well-established, roughly eleven-year cycle of *total solar irradiance*, during which its brightness changes over time. However, satellite measurements of total solar irradiance since 1979 show no increasing trend that could be responsible for global warming. The solar cycle is simply not strong enough to provide the temperature boost we have observed and measured. In addition, the eleven-year cycle is just that, a cycle—not a trend.[5] The sun is big and powerful, but its fingerprint simply does not match the observed warming. Its fingerprint is that of slight warming everywhere, including the stratosphere. If changes in solar output had been responsible for the recent climate warming, both the troposphere and the stratosphere would have warmed.[6]

No one is saying that solar variability and volcanic eruptions aren't important forms of climate forcing over the Earth's history. Climate model experiments show that the sun and volcanoes have indeed played an important role in changing temperature at timescales ranging from decades to centuries. In fact, climate model experiments show that prior to the industrial era, much of the variation in average temperatures in the northern hemisphere can be explained either as episodic cooling caused by large volcanic eruptions or as changes in the sun's output.

The problem is that changes in solar output and new volcanic eruptions simply are not powerful enough to generate the large temperature rise we're currently witnessing. All of the testing finds that these natural factors cannot explain the warming of recent decades. Climate models can accurately estimate how much warming these natural factors produce, and—to repeat—they do not have the strength to generate the temperature increase we're seeing.

The only climate models that are able to simulate the changes in temperature we saw in the twentieth century are those that include natural forcings as well as human forcings—such as greenhouse gases.

Hind-casting isn't the only way to prove that the carbon we're producing is raising temperatures. Interestingly, scientists have learned that not all carbon in the air is the same; in fact, the carbon that comes from us bears our distinct fingerprint, a chemical smoking gun that shows just how much of this problem really belongs to us.

It turns out that just as humans come into this world with unique sets of fingerprints, so too does carbon. Carbon enters the atmosphere from a lot of different places, and each place stamps the molecules of carbon dioxide with unique fingerprints before sending them off into the atmosphere. Volcanoes emit CO_2 into the atmosphere when they erupt; the soil and oceans release CO_2 into the atmosphere; and plants and trees give off carbon dioxide when they are cut or burned. Burning coal, oil, and natural gas releases carbon into the atmosphere to form carbon dioxide. When you have the right tools, distinguishing where an individual molecule of CO_2 comes from is not hard.

Tracing carbon is a bit like tracing a bullet back to the gun it was shot from, and as with a ballistic test that links bullets to a gun, it helps to understand that not all carbon is the same. Carbon atoms (like many atoms) have variations known as isotopes, and these different isotopes are found in varying amounts around the atmosphere. Some got there from the oceans, others from volcanoes, and others from us. Carbon 12, 13, and 14 are all examples of carbon isotopes that are found in the atmosphere, and each comes from a different combination of sources. Each source, to repeat, has a unique chemical fingerprint. Carbon from the oceans, the atmosphere, and the land contains a healthy mix of carbon 12 and carbon 14. But carbon from fossil fuels has almost no carbon 14 at all.

Scientists use an instrument called a mass spectrometer to measure the amounts of carbon isotopes in the atmosphere and track the origin of the carbon. The mass spectrometer is very precise; it knows exactly which isotope of carbon it is measuring because the different carbon isotopes have different masses. So, the mass spectrometer can distinguish a carbon 12 atom from a carbon 13 atom from a carbon 14 atom. With a spectrometer, scientists can trace where the CO_2 in the atmosphere originated by measuring the ratios of the different carbon isotopes. In other words, a spectrometer can say whether a sample of CO_2 came from the ocean or from a volcano or from burning a fossil fuel.

According to precise measurements from mass spectrometers at several locations around the globe, the carbon dioxide molecules currently in the atmosphere have very little carbon 13 and carbon 14. Using this chemical signature to trace the carbon back to its source tells us that the increase we're seeing in atmospheric CO_2 did not originate in the oceans. Carbon dioxide that outgases from the oceans is *not* depleted in carbon 13. This also means that the increase in atmospheric CO_2 did not originate from living plants and animals, because CO_2 from organisms is *not* depleted in carbon 14. The chemical fingerprints of the extra carbon dioxide in the atmosphere match only the fingerprints of coal, oil, natural gas, and deforestation because these are the only sources that produce carbon dioxide depleted in carbon 13 and carbon 14.

It's true that most of the carbon dioxide in the atmosphere today comes from natural sources. But most of the *additional* CO_2 that's been placed in the atmosphere over the last 250 years comes from us. And it's the additional CO_2 that's raising temperatures. In terms of molecules of carbon dioxide, roughly one out of every four CO_2 molecules in the atmosphere today was put there by us.

All this carbon fingerprinting and the various climate models add up to one inescapable reality: the predictions that scientists have been making for the last twenty years have been getting more accurate. Weather forecasts started out as shaky, debatable calculations but evolved into a system of forecasting that virtually everyone in the world now relies on; similarly, climate prediction has evolved to a point where its results are sounder than ever before. No matter how many different ways the scientists run it, the results come out the same—a warmer planet that's getting warmer as a result of our carbon emissions.

There's no realistic way to take comfort in what these numbers are telling us. The forecasts that the models lay out is dire, and even though we don't see them every night on our local news, we cannot ignore it. So if climate models can show us that the temperature is rising and that it's our fault, when will our weather start to reflect our predictions about the climate? The short answer is that it already has.

4

EXTREME WEATHER AUTOPSIES AND THE FORTY-YEAR FORECAST

I like watching basketball, but I'll admit that most of the time I can't keep up with it. Until the perfect moment when someone flies through the air and makes a basket, I can't see a damn thing. To me, the game is a lot of noise with very little signal. I'm much better off if I can watch a game with an aficionado. Watching the climate is no different. Learning how to see the climate system in play is like learning how to see the Lakers run a screen or watching a northeaster swing down from Canada. This is an art and a science, and it takes a trained ear to hear through the noise. Sometimes, hearing through the noise requires help in the form of a slow-motion instant replay. This is especially true when scientists attempt to understand the connection between global warming and weather that's happening to us right now.

The weather isn't what it used to be. In fact, all the data we've collected over the past fifty years point to the fact that the weather is getting more extreme.[1] But trying to isolate the fingerprint of global warming within the weather is much harder than isolating the fingerprint of global warming within the climate system. That doesn't mean it's not there; it just means that discerning climate change in the weather is a much noisier, more chaotic, and more complicated process. Ultimately, as in sports, the statistics can help us find the story buried beneath the noise. And climate scientists have come up with some very clever variations on using a slow-motion instant replay of the weather to help them understand how the statistics of extreme events are changing.

It turns out that you can use climate models as an instant replay to re-create a specific weather event. Think of this as like an autopsy, except that it's being performed on a specific extreme weather event. And although it can't determine the individual cause of the weather event, it can allow scientists to calculate the odds of such an event. Those odds can speak volumes if you know how to read them. The models can measure how much global warming shifted the odds in favor of allowing a specific type of weather event to happen. And, perhaps even more strikingly, the models allow us to see how those odds will play out and change in the future.

The European heat wave of 2003, an extreme weather event that killed more than 35,000 people, offers the best example of how climate models can help us see the global warming embedded within our weather. Public health officials were shocked at the scale of the human casualties caused by the heat. The largest number of casualties was in France, where almost 15,000 people perished in the first three weeks of August. Climate scientists were equally shocked at how far outside the range of historical temperature the heat wave registered. The summer of 2003 has been described as the biggest natural disaster in Europe on record.

The heat wave was dramatic. Temperatures in France soared to 104°F and remained unusually high for two weeks. There were extensive forest fires in Portugal, burning an estimated 1,500 square miles. Melting glaciers in the Alps caused avalanches and flash floods in Switzerland.

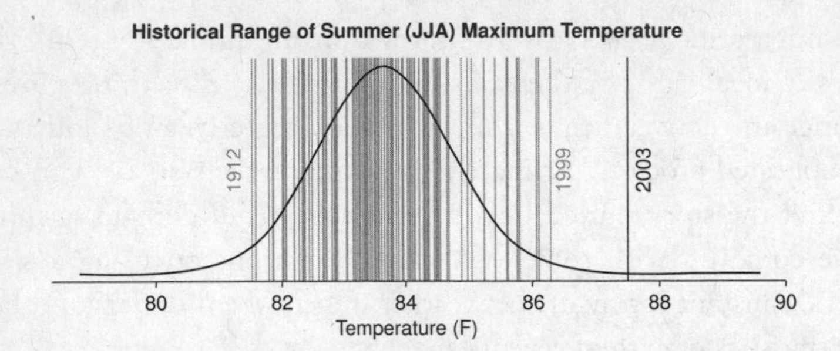

Each vertical line represents the mean summer temperature for a single year for the European region that spans [10W to 40E, 30N to 50N] over the period 1901 through 2003. Extreme values from the years 1912,1999, and 2003 are identified. SOURCE: ADAPTED FROM SCHÄR ET AL., 2004, BY C. TEBALDI, CLIMATE CENTRAL.

When we step back and compare the summer of 2003 with past summers, the picture becomes even more obvious. As you can see in the accompanying figure, there are a series of vertical lines that look rather like a bar code.[2] Each vertical line represents the mean summer temperature for a single year from the average of a region over Europe that spans [10W to 40E, 30N to 50N] over the period 1901 through 2003. Until the summer of 2003, the years 1912 and 1999 stood out at the edges as the most extreme temperatures in terms of hot and cold summers. Climate scientists estimate that the summer of 2003 was probably the hottest in Europe since at least A.D. 1500.

If climate is what you expect and weather is what you get, then the summer of 2003 was far outside what anyone would have expected. Statistically, in a natural climate system with no man-made CO_2 emissions, the chance of getting a summer as hot as 2003 would have been about once every thousand years, or one in 1,000.

The point of this weather autopsy isn't so much that the 2003 heat wave was, or was not, caused solely by global warming. Indeed, almost any weather event can occur on its own by chance in an unmodified climate. But using the climate models, it is possible to work out how much human activities may have *increased* the risk of the occurrence of such a heat wave. It's like smoking and lung cancer. People who don't smoke can still get the disease, but smoking one pack of cigarettes a day for twenty years increases your risk of developing lung cancer twentyfold. Thanks to some sophisticated climate models and well-honed statistical techniques, scientists can identify the *push* that global warming is giving the weather.

Step one was to re-create the 2003 heat wave in a high-resolution climate model using data observed during the heat wave. Scientists set up two sets of climate model experiments. One simulation included human-induced greenhouse gas emissions; the other simulation didn't include them. Like other climate model experiments, this one essentially created two worlds: a world with human influences and a world without them. By comparing the two, the scientists could look at what the risk of having a very hot summer is now and compare that with what the risk would have been if there hadn't been any human-influenced climate change. The difference between these two sets of odds would tell them just how much of a role humans played.

This weather autopsy, published in the science journal *Nature*,[3] showed that human influences had at least doubled the very rare chance of summers as hot as the one Europe experienced in 2003. The climate models showed that greenhouse gas emissions had contributed to an increase in such summers, from one in 1,000 years to at least one in 500 years and possibly one in 250 years.

What is perhaps most shocking is what happens when we run the models in *forecast* mode instead of autopsy mode. If the summer of 2003 had been a freak of nature, we could just chalk it up to chance. But the latest climate models paint a very bleak picture. According to their predictions, by the 2040s such summers will be happening every other year. And by the end of this century, people will look back wistfully to 2003 as a time when summers were much colder. Hindsight, as they say, is twenty-twenty.

In the United States, average annual temperature has risen more than 2°F during the past fifty years, and the temperature will continue to rise, depending on the amount of heat-trapping gases we emit globally.[4] Along with the general increase in average annual temperature, most of North America is experiencing more unusually hot days and nights, as well as heat waves.[5] Across the United States, ever since the record hot year of 1998, six of the last ten years (1998–2007) have had annual average temperatures that have made the record books, and these six years have ranked in the hottest 10 percent of all years since 1895 (when record keeping started). Also, the United States has seen fewer extremely cold days during the last few decades. In fact, the last ten years have brought fewer severe cold snaps than any other ten-year period since records began to be kept in 1895. There has been a decrease in frost days and a lengthening of the frost-free season over the past century as well.

By 2050, mid-range emissions scenarios predict that a day so hot that it is currently experienced only once every twenty years would occur every three years over much of the continental United States. By the end of the century, such a day would occur every other year, or more often. As for the cold, we'll be seeing even less of it. By 2100, the number of frost days averaged across North America is projected to decrease by one month; and decreases of more than two months are projected in some places.

The extreme weather of climate change is not limited to heat waves; the climate models suggest that other forms of extreme weather are also expected to increase. A warmer climate increases evaporation of water from land and oceans, and it allows more moisture to be held in the atmosphere. In other words, as the air gets warmer, it can hold more water vapor. Coupled with other warming-related changes, this additional moisture-holding capacity increases evaporation and will result in longer and more severe droughts in some areas and more flooding in others.

These trends are already beginning to be seen in the United States, and depending on where you are, you may have experienced them yourself. In the Northeast, the Midwest, and Alaska, the additional atmospheric moisture has contributed to more overall precipitation.

In the West and Southwest, the opposite is happening, as those areas have seen reductions in precipitation and increases in drought.

A warmer climate also means that it rains harder when it does rain. Diagnostic analyses have shown that with higher temperatures, a greater proportion of total precipitation comes from heavy precipitation events, such as blizzards and rainstorms. Heavy precipitation events averaged over North America have increased during the past fifty years, keeping pace with the increases in atmospheric water vapor that come from higher man-made carbon emissions.

Rain is just the start. The extreme weather produced by a warmer climate might include hurricanes. Because global warming results in warmer oceans, most experts on hurricanes agree that this warming of the ocean waters will make hurricanes more powerful. Hurricanes derive their energy from the evaporation of seawater, and water vapor evaporates more easily when it's warmer. The data suggest an increase in the number of more intense hurricanes in the North Atlantic during the next few decades. But more data are needed before scientists can be certain. Most climate models do predict that the strongest tropical cyclones will get stronger as global warming continues, but some models suggest that the total number of storms may actually decrease. There are several factors that influence the formation of tropical cyclones, including wind patterns, ocean currents, and local weather conditions. Any one of these factors might change in a warming world, in ways scientists are not yet able to predict.

But even if the jury is still out regarding the specific future of hurricanes, everyone is certain that damage will get worse. This will be due partly to population increases along the coast, and partly to the fact that global warming melts the ice caps at the poles, thereby raising the sea level. With a higher sea level come higher storm surges and more damage to our coastlines.

Ultimately, these extreme weather autopsies confirm something that many of us have long suspected: the weather is getting more extreme. The conditions have arisen for more major storms, longer droughts, and serious flooding, and they are getting worse.

These predictions and our seeming inability to heed their warming is a potential tragedy, reminiscent of Greek tragedy. Climate scientists seem to have become Cassandras, though the questions remain whether our forecasts will be heeded, and whether the harrowing scenarios of life on a warmer planet will come to pass.

The latter question is what Part II of this book is about. It explores the Achilles' heel of seven locations around the world, looking at how climate change will remind each one of its own frustrating vulnerabilities time after time if we humans continue to emit carbon at our current rates. Part II is about the unique perspective that each location offers on the risks of climate change to humankind. It's about using all the predictive tools that we've discussed—climate models, weather models, hind-casting, and extreme weather autopsies—to discuss the climate and weather scenarios for these various locations if global warming is allowed to continue unabated.

Just as the landscape of Earth is diverse and complex, so are the stories that specific landscapes will tell as climate change takes hold. These stories are the heart of Part II. Each chapter in Part II consists of two sections.

In the first section of each chapter, I've used climate models, environmental data, and—most important—the brightest scientific minds studying the climate in each location to help calculate and describe the specific risks each place will encounter over the coming decades. Despite all that math, models, and physics can tell us, these scientists are our most valuable tool for understanding the specific risks associated with each location. Because the climate of a place is so closely intertwined with the people who live there, it's nearly impossible to make an accurate prediction about the future without first understanding its most important initial condition: the human population. Along with all their measurements and readings, it is this understanding of the local population that these scientists strive for on a daily basis; the people's stories and experiences, what they've witnessed firsthand as they look climate change in the face every day, are as much a part of the predictions as the model-driven data. It is only by understanding the people who call each place home that we'll be able to predict how they'll react when climate change exposes their home to ever-increasing risks.

The second section of each chapter in Part II contains a series of predictions, based on a collection of climate models, which, taken together, offer a window into what the next forty years will look like for the place discussed in the chapter. The forecasts are works of fiction. I can't say whether the future will play out as I've described, but what I can say is that these predictions are based on the best available science derived from some of the cutting-edge climate models and real weather events from the historical record. In writing them, I have pored over reams of data with the aim of turning those numbers into stories, stories that give new visual guides to what climate change might look like in a place near you.

Each forecast begins with a glimpse of the regional climate during January and July based on two possible emissions scenarios. The scenarios are drawn from an average of at least fifteen different climate models, and each scenario makes different assumptions about future human activity—including greenhouse gas pollution, land-use alteration, technological development, and future economic development.

The first scenario is based on medium-high emissions. This scenario projects continuous population growth and uneven economic and technological growth. In it, the income gap between currently industrialized and developing parts of the world does not narrow. Heat-trapping emissions increase through the twenty-first century, and atmospheric CO_2 concentration approximately triples, relative to preindustrial levels, by 2100.

The second scenario is based on lower emissions. It characterizes a world with high economic growth and a global population that peaks by mid-century and then declines. There is a rapid shift toward less fossil fuel–intensive industries and the introduction of clean and resource-efficient technologies. Heat-trapping emissions peak at about mid-century and then decline. Atmospheric CO_2 concentration approximately doubles, relative to preindustrial levels, by 2100.

Some of these predictions have geopolitical implications; others have simply national ramifications. But one thing that's certain is that none of these scenarios will be happening in a vacuum. In a climate-changed Earth, every inch of land, ocean, and air will be affected. These stories represent some of the most dramatic and vulnerable locations, but they also represent places where the human species has been living for millennia, including some that will be rendered inhospitable by the changing climate. With regard to climate change, the worst-case scenario can be prevented through infrastructure investment (adaptation) and the adoption of clean energy technology and emissions reductions (mitigation). Some of these predictions examine what might happen if one location decides to adapt; others examine what might happen if a location refuses to change.

Ultimately, the basic assumptions driving these forecasts are that we will continue to burn fossil fuels, that the global population will continue to grow, and that because of both factors, greenhouse gases will continue to rise. The end result of this rise is that weather will not only become worse—it will become downright awful. Indeed, as the coming pages will show, the weather of the future and the way that weather affects life on Earth will be far worse than anything we've seen before.

Though these prophecies, as I've suggested, contain the seeds of a Greek tragedy, ultimately the forecasts also contain a kernel of hope, because unlike the prophecies in Greek tragedy, they are changeable. The forecasts paint a picture of just one *possible* future. While these forecasts, or indeed any forecasts, make certain assumptions about how trends will continue, one true variable they cannot approximate with much accuracy is our own behavior. We are the factor that could render all these predictions false, because we alone have the power to reduce our global carbon footprint. The question that we must now answer is how. In the end, these forecasts pose a question that is vital to our collective future: if we are really capable of forecasting the future and seeing the devastation of a changing climate in advance, will we act to prevent it? Can we rally around this forty-year forecast for the good of the world, or will we wait until the levees break before we decide to act?

PART II

THE WEATHER OF THE FUTURE

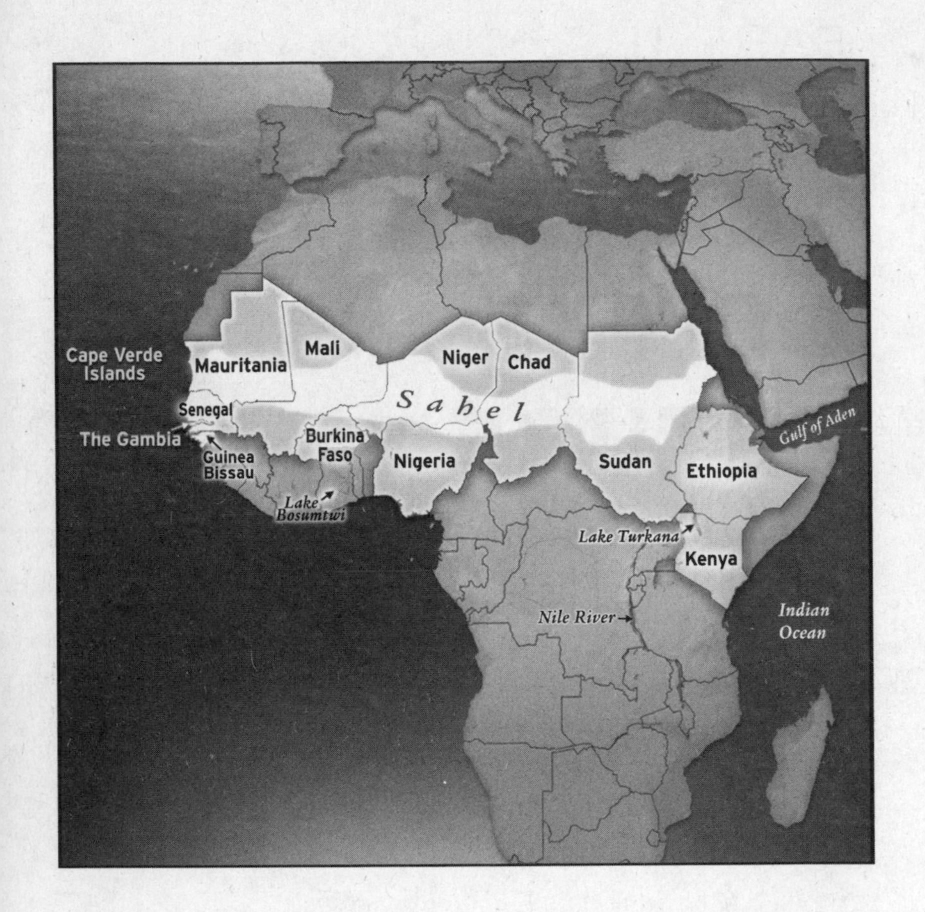

Cape Verde Islands
Mauritania
Mali
Niger
Chad
Sahel
Senegal
The Gambia
Guinea Bissau
Burkina Faso
Nigeria
Sudan
Ethiopia
Gulf of Aden
Lake Bosumtwi
Lake Turkana
Kenya
Nile River
Indian Ocean

5

THE SAHEL, AFRICA

The word *sahel* comes from the Arabic *sahil*, meaning *shore*, and like so many things about the Sahel region of Africa, it is ironic. This is because the Sahel is a parched shore that both unites and divides the Sahara desert to the north and central Africa's tropical rain forests to the south. This pairing of opposites is a recurring theme in the Sahel, home to nomads and farmers, Muslims and Christians, Arabs and Africans.

The Sahel is also a place where the past and the future are sharply defined by climate. A semiarid savanna stretching out over 2,400 miles from the Atlantic Ocean in the west to the Red Sea in the east, the Sahel is consistently identified as one of the most vulnerable places in the world to global warming. But the Sahel, perched just on the southern fringe of the Sahara desert, is no stranger to hostile and recurring extremes in climate. You might say the climate history of the Sahel is simply a battle between trees and sand, greens and tans, wet and dry. Climate scientists have tracked this battle over millions of years using evidence left behind on land and in the sea. If you could transport yourself back 10,000 years, you'd see that this was a time when trees, not sand, dominated the landscape of the Sahara. Evidence pulled from the bottom of the ocean indicates that the climate of the Sahara was not dry but overflowing with tropical grasslands and forests and dotted with large permanent lakes—some as large as the United Kingdom, such as Megalake, Chad.[1]

But eventually green gave way to tan, and trees were once again replaced by sand. Today the Sahara is the world's largest warm desert. The Sahel has also seen significant change, slipping frequently into periods of drought. A 3,000-year climate record pulled from the mud at the bottom of Lake Bosumtwi in Ghana (see the map) shows evidence of *mega-droughts* lasting for centuries.[2] These shifts in climate have done far more than transform forest into desert; they've also altered the course of human history[3]—which is where this story about the Sahel begins.

Human history starts sometime around 6 million or 7 million years ago, when at least one species of tree-dwelling ape left its forested habitat in Africa and became the first member of the human family, the Hominidae. The species that paleontologists call *Sahelanthropus tchadensis*, recently discovered in central Chad, is thought to be the "last common ancestor" linking humans to chimps. At some point, an adventurous hominid rose up on two legs, and eventually hominids evolved into an early genus that paleontologists call *Australopithecus*. More recently, they evolved into our genus, *Homo*. This matter is still an active debate, but many scientists believe that the need to adapt to open grasslands helped shape the evolution of *Australopithecus*. Recent evidence suggests that the further evolution of these early hominids can, at least in part, be linked to the steplike shift toward cooler, drier, and more open conditions in the Sahel region.

The evidence for these steplike shifts in climate comes from the ocean floor. Long cores containing mud drilled from the bottom of the Gulf of Aden, off the coast of east Africa, contain almost 10 million years of climate history.[4] These cores contain three thick layers of dust, which originated in the Sahel and was picked up and carried by the northeast trade winds and then eventually dumped into the ocean. These thick dust layers represent drier conditions over the Sahel, and they appear at roughly 2.8 million, 1.7 million, and 1 million years ago, which happen to be very important junctures in human history.

The first shift toward colder, drier conditions (roughly 2.8 million years ago) marked a definite transition of the climate of the Sahel from dense woodlands into more open grasslands. As a result, many species saw their forested home replaced by a vast expanse of savanna. Some plant and animal species probably shrank in numbers as a result of this habitat loss, but many other species simply became extinct. The increased competition for food brought about by a changing landscape would have intensified the pressure to adapt.

During this time, at least two new hominid branches emerged. These two new branches in the early human family tree include the genus *Homo* and the genus *Paranthropus*. Paleontologists suspect that *Homo* was more of a jack-of-all-trades, whereas *Paranthropus* was more of a specialist and retained many common features of *Australopithecus*. As it turned out, *Homo* was far more capable of adapting to the changing environment and managed to come up with new strategies to cope with the increased competition for food.

For example, the first evidence of stone tools—crude choppers and scrapers—appears in the fossil record about 2.6 million years ago. The discovery of tool cut marks on mammal bones provides evidence of meat processing and marrow extraction. It is possible that *Homo* exploited different habitats by exploiting different types of foods. For example, eating meat and marrow would have marked an improvement over a strictly vegetarian diet because, unlike nuts and seeds, meat was available year-round. It is worth noting that the appearance of a bigger, more powerful brain in *Homo* coincides with the time of drying between 2 and 3 million years ago. Fossil records from the Turkana basin straddling northern Kenya and southern Ethiopia suggest that *Homo* species were characterized by a more delicate frame and smaller cheek teeth than their *Australopithecus* ancestors, and indeed had much bigger brains. In time this branch evolved into modern humans.

The other branch of hominids, the genus *Paranthropus*, with its very large teeth and a specialized chewing apparatus: a saggital crest on top of the skull that supported strong jaw muscles, tried to exploit a fading environmental niche, ate a mostly vegetarian diet and used those strong teeth and jaw muscles to crush nuts and seeds and grind the coarse vegetable matter that could still be found in the denser patches of savanna along rivers. This niche strategy is thought to have left *Paranthropus* more vulnerable, struggling to adapt to the cooler, more arid landscape, which provided fewer vegetarian options.

Some scientists suggest that during the next shift toward colder, drier conditions, about 1.8 to 1.6 million years ago, the branches of both *Homo* and *Paranthropus* underwent further splitting and pruning. At this time, the species named *Homo habilis* became extinct and our direct ancestor, *Homo erectus*, first appears in the fossil record. Fossil evidence suggests that by the third dry spell, about 1 million years ago, the entire *Paranthropus* line had become extinct and *Homo erectus* emerged as the winner, going on to occupy sites in North Africa, Europe, and western Asia—and ultimately evolving into *Homo sapiens*, modern humans.[5] The rest, as they say, is history.

Today, the Sahel is home to more than 60 million people. It is a place that has become synonymous with the word *drought*. In the past 100 years alone, the region experienced three devastating droughts. The first stretched from 1910 to 1916; the second stretched from 1941 to 1945; and then came the worst of all droughts, a long period of sustained declining rainfall beginning in the late 1960s and known simply as *the desiccation*.[6]

Climatologists estimate that from the 1950s through the 1980s, the Sahel saw rainfall decrease by about 40 percent.[7] This drought is linked to the deaths of more than 100,000 people, mostly young children; and it set off a wave of migration from north to south, from rural areas to cities, and from inland to the coast.[8] As a result, squatter settlements and urban overcrowding, accompanied by rising unemployment, increased. Political instability and unrest intensified across many countries in the Sahel. The dessication is considered by many scientists to be one of the most striking examples of climate variability the world has seen. A looming question now facing climate scientists is: when will another drought of this magnitude occur? Another question is: if and when such a drought recurs, who will emerge as the winner—the people or the sand?

During the 1970s, striking images of this crisis made their way out of Africa and reached television screens and magazine covers all over the world. The pictures of barren landscapes and children with haunting eyes and distended bellies led to coordinated international humanitarian efforts to help reduce the suffering. The crisis also revived a long-standing debate within the scientific community over the fundamental causes of drought. The debate centered on the concept of *desertification*, a process whereby productive land is transformed into desert as a result of human mismanagement.[9]

The issue of desertification dates back to the 1930s, during colonial rule in west Africa. There was a growing concern that the Sahara desert might be slowly creeping into the Sahel. The colonial regimes blamed desertification on the African people, specifically on rapid population growth and poor agricultural practices. It was a new twist to an old story. Instead of studying the impact of climate on human history, scientists were studying the impact of human history on climate. Man-made landscape alterations caused by overgrazing, intensive agriculture, and the cutting of trees were offered as a possible cause of the Sahel's drought.[10] At the time, some studies went so far as to suggest that at least one-third of the planet's deserts were a result of human misuse of the land.

In some respects, framing the problem as man-made allowed it to seem more readily fixable. With upward of 750,000 people in Mali, Niger, and Mauritania totally dependent on food aid and more than 900,000 people in Chad severely affected by the lack of rainfall, the west African countries of Burkina Faso, Cape Verde,

Guinea-Bissau, The Gambia, Senegal, Mali, Niger, Mauritania, and Chad became a formal geopolitical entity defined by a shared goal of combating drought. Formed in 1973, the Permanent Interstate Committee for Drought Control has a mandate to invest in research to ensure food security and to reduce the impact of drought and desertification.

Just as they had done in the past, the people of the Sahel were looking for ways to adapt and survive in a changing landscape.

If you look at rainfall records for the 1950s and early 1960s—before the drought began—the weather was actually a little wetter than average. This short-lived boost in rainfall allowed many farmers to grow crops in the northern Sahel, a region that was usually not suitable for agriculture. This type of rainfall-related migration is an age-old adaptation strategy for the people of the Sahel. But when the rains stopped, these crops were the first to go.

During and after the drought of the late 1960s, the only way to survive was to expand cultivated land. To do this, farmers had to cut down trees. By 1975, much of the remaining natural woodland had been converted to farm fields to feed a rapidly growing population. But by clearing native trees and shrubs, farmers were exposing their fields to the fierce Sahara winds; this exposure resulted in plummeting crop yields, and windblown sand buried entire villages. In what would have been an ultimate irony, it was suggested that these attempts to sustain life were actually what led to so many deaths.

Before you can begin to unravel the causes of drought in the Sahel and see how climate change will make those droughts worse, you need to understand why it rains there in the first place.

"That short answer is the African monsoon," says Alessandra Giannini. Giannini is a climatologist at the International Research Institute for Climate and Society (IRI) at Columbia University and has spent the past several years studying rainfall, or the lack thereof, in the Sahel. "Rainfall in the Sahel results from the collision between two different air masses; the moisture-laden southwesterly winds originating over the Atlantic Ocean and the dry northeasterly trade winds coming off the African continent."

In other words, if you're lucky and the conditions are just right, the collision of these two air masses will take place squarely over the Sahel and usher in welcome thunderstorms and much-needed rainfall.[11] But as history has shown, the Sahel doesn't always get lucky. The reason is that the Sahel sits at the northern edge of the African monsoon—and in some years the monsoon simply isn't strong enough to muscle its way that far north. When this happens, the northern part of the Sahel might as well be the Sahara.

Overall, the northern tier of the Sahel is rarely lucky with respect to rainfall; it averages only about 4 to 8 inches a year. To the south, the situation is a little better, and rainfall averages between 24 and 28 inches a year.[12] But even so, there is tremendous variability from year to year. Some climate scientists argue that the concept of average rainfall doesn't even apply in the Sahel. Also, the rainy season is short and intense—typically centered on August and lasting no more than four months. That means the dry season is very long, in a place where more than 80 percent of the people make their living growing crops and grazing livestock. June through September is known simply as the *hunger season*—the period when the harvest from the previous year has been exhausted and the next season's harvest is not yet ripe.

This is exactly why Giannini and her colleagues at the IRI have been working to develop seasonal rainfall forecasts for the Sahel. But Giannini knows all too well that before you can offer a forecast, you need to understand the past. And in the Sahel, that means understanding the causes of the drought.

"Two competing drought mechanisms were being floated around, one *local* and one *remote*," explains Giannini. "The local mechanism involved land use change on the ground in the Sahel. The remote mechanism involved changing the temperature of the oceans. But no one could really demonstrate which was the stronger influence." The local mechanism pointed the finger at human activity—for example, deforestation and overgrazing. The remote mechanism suggested natural causes. In other words, there were two options: the drought was made either by man or by Mother Nature.

Evidence that local activities can lead to local droughts was first proposed in the 1970s by Jule Charney, an atmospheric scientist at the Massachusetts Institute of Technology. The *Charney hypothesis*, as it came to be known, suggested that deforestation and overgrazing literally cool the land surface and ultimately decrease clouds and rainfall. Climate models were used to test this idea, and that is when the local mechanism began to unravel. Models looking solely at deforestation were unable to produce the kind of large-scale drought that was actually taking place in the Sahel.[13] Furthermore, satellite pictures of the Sahel confirmed that the land surface hadn't been changed nearly enough to alter rainfall patterns. Strike one for this hypothesis of human causation.

Giannini was interested in testing the other possibility—the *remote* mechanism. In the 1980s, a group of researchers from the United Kingdom's Meteorological Office had confirmed that changes in ocean temperature played a big role in generating the Sahel drought.[14] It was just a question of how big.

"I wanted to understand a very simple question," explains Giannini. "I wanted to see how well a climate model could reproduce observed rainfall over the Sahel." So she used an atmospheric climate model that took only one real-world factor into account: the ocean surface temperature. The model didn't know anything about the Sahel and its history of deforestation, desertification, and land degradation. As far as Giannini's climate model was concerned, no one even lived in the Sahel. And as it turned out, that didn't matter. "I couldn't believe it. The model reproduced the observed rainfall beautifully," explains Giannini.[15] "We all know these models aren't perfect, but the connection between ocean surface temperature and drought in the Sahel was very compelling." So much for blaming the drought on the farmers.

Giannini discovered that ocean temperatures helped regulate the strength of the African monsoon. What was even more fascinating was how each ocean played a specific role. On a year-to-year basis the Pacific Ocean has an effect on the Sahel's rainfall, thanks to El Niño. During an El Niño event, the Sahel is typically expected to experience a drought, whereas during a La Niña event, when the tropical Pacific is cooler, the African monsoon is expected to strengthen and rainfall is expected to be more abundant. The Atlantic Ocean and the Indian Ocean affect the Sahel's rainfall over much longer periods, from decade to decade. A warming of the Indian Ocean means a drier Sahel. And in the Atlantic, where the relationship is somewhat trickier, overall, when the southern hemisphere warms more than the northern hemisphere the rain belt across the Sahel is attracted farther south, toward the warmed hemisphere. This effect dries the Sahel.

"Ultimately, we found that you could explain the Sahel drought, as well as its persistence, by looking to the ocean temperatures," says Giannini.

Understanding this connection to the oceans provides the basis for seasonal rainfall forecasts for the Sahel. In essence, these seasonal forecasts are not attempting to predict how much it will rain on a specific day in the Sahel; rather, they predict when the rainy season might fail altogether or when large-scale flooding is likely. By knowing the present ocean surface temperatures, you can use climate models to forecast how the oceans will probably evolve during the next several months. Climate forecasts are an average of many climate models; similarly, the IRI uses numerous climate models with different conditions in the atmosphere to average seasonal patterns of temperature and rainfall. These averages give the most accurate predictions for the coming season's climate; indeed, seasonal rainfall forecasts for the Sahel have been issued since 1997, providing significant help in drought planning and food security.

But although the climate models rely on ocean surface temperatures to forecast rainfall and temperature, Giannini is quick to add that human activity does still influence the severity of drought in the Sahel. "If you cut down enough trees in the Sahel, there's no doubt it's going to get warmer and drier," Giannini explains. "What people are doing down on the ground can amplify the drought signal." And it is now well accepted that the combined effects of population growth, deforestation, overgrazing, and lack of coherent environmental policies, along with a significant decrease in rainfall, resulted in the crisis of the 1960s and 1970s, which was unlike any the world had seen before.

Of course, there is another, broader human influence that goes beyond the behavior of the local population. Global warming has already warmed up ocean surface temperature by about 1°F during the past century, and given the already established relationship between ocean temperatures and droughts in the Sahel, this warming trend will almost certainly have a negative impact on the amount of future precipitation there.

Global warming is affecting the region in ways that are not yet fully understood. If anyone can figure it out, Isaac Held would be a good possibility. Held is a senior research scientist at a division of the U.S. National Oceanic and Atmospheric Administration (NOAA): its Geophysical Fluid Dynamics Laboratory (GFDL), a prominent climate modeling center in Princeton, New Jersey. Few people understand the complexity of rainfall in the Sahel better than Held. A member of the Intergovernmental Panel on Climate Change (IPCC), Held served as a lead author of the IPCC's Fourth Assessment Report chapter on regional climate projections. The IPCC's regional projections use fourteen state-of-the-art climate models to provide a glimpse into the future. The GFDL climate model is one of the best in the world. And if you believe this model's projections for the Sahel, you'll be very worried about the future.

Held knows why the GFDL model behaves as it does; he just doesn't know if the real world will behave the same way. Remember that today drought in the Sahel is very sensitive to the gradient between ocean temperature in the north and the south. "The models all agree that if you warm the oceans of the southern hemisphere with respect to the ocean of the northern hemisphere, you dry the Sahel," explains Held. "But if you warm the oceans uniformly, there's no consensus among the models."

With regard to total seasonal rainfall in the Sahel, the models actually diverge—some predict more rainfall, and some predict less. Most of the models produce only modest changes out to 2100, but there were two outliers—one projecting a very wet future Sahel and one projecting a very dry future Sahel. The GFDL model was the dry outlier. It projected that summer rainfall in the Sahel would decrease 30 percent or more by the year 2100. Needless to say, a rainfall reduction at that level would be catastrophic. "Our model dries the Sahel in response to uniform warming. And that's why we dry so strongly into the future. It's the global warming signal that dominates," explains Held. "We're still trying to understand it." But in the meantime, Held points to model simulations that show a far more robust response.

The models all agree that the Sahel is going to get warmer. The IPCC estimates a warming of roughly 6°F to 10°F by the end of the twenty-first century.[16] A recent study looking at future heat extremes shows that heat waves will become longer and more frequent in the Sahel. And by the end of this century, the heat index across all of northern Africa will spike from May through October.[17] It is expected that the people in the Sahel region will be the most vulnerable, experiencing up to 160 days per year in the twenty-first century with a significant chance of heatstroke. Today, there are up to 180 days of medium risk in the Sahel and no high-risk days. By the end of the century, the people living in the Sahel are projected to experience the most severe increases in sunstroke in the world.

Regarding rainfall, Held also points to a recent set of experiments by Michela Biasutti and Adam Sobel of Columbia University that show a robust response among all the different IPCC climate models.[18] This particular study looked at the *timing* of the rainfall season rather than at the average amount of rainfall. "They see a delayed onset of the rainy season in almost all the models," says Held. In other words, the rainy season of the Sahel is projected to start later and become shorter, with storms that will possibly be more intense. This is not good news. (Case in point: on September 1, 2009, Ouagadougou, the capital of Burkina Faso, was hit by an unprecedented storm that brought more than 10 inches of rain in just a few hours. Widespread flooding left nearly 130,000 people homeless; they sought shelter in churches, mosques, and schools.)

Despite these dramatic images, and despite the fears about how global warming may affect the lives of those in the Sahel, Alessandra Giannini tries to remain hopeful.

"Honestly, I don't like to play this doom and gloom card with respect to the future. As climate scientists, we often spend our lives looking at problems from afar. But in the case of the Sahel, when you look closely at what is happening on the ground, you will be able to see pockets of resilience, pockets of adaptation." Those pockets of resilience are proof that people working together have the potential to overcome the forces of nature. Just ask Chris Reij, a scientist at the Vrije Universiteit (VU) in Amsterdam who specializes in soil conservation.

Chris Reij spends his life actively looking for ways to promote adaptation that will help the Sahel weather a climate-changed world. "I must admit that I'm doing everything possible to abandon research now. I've changed from soil conservation research to development action," Reij says. "We clearly have enough information to act."

The reason to act is that the pockets of resilience Giannini mentioned are, in fact, pockets of trees, millions of them.[19] Satellite images and tree inventories have found that the Sahel has become greener over the past thirty years.[20] Needless to say, the exact cause of this greening is still not perfectly resolved. Some scientists think the Sahel is simply bouncing back from the gradual improvement in rainfall since 1988; others think global warming could actually be helping to boost rainfall totals and spur on vegetation growth.

But Reij thinks the cause is the farmers. "You could say we've come full circle with respect to our ideas about drought and desertification," explains Reij. "Farmers were *not* the cause of the drought, but they are a big part of the solution." Reij has worked in the Sahel since 1978 and has never been one to spread doom and gloom. "I think there are a lot more success stories in the Sahel than we tend to assume," he says.

One success story involves farmers in Niger. A desperately poor country twice the size of Texas, Niger has seen about 12.4 million acres of trees, shrubs, and crops replace what was once barren ground.[21] Barren ground is all too common in Niger; four-fifths of the country sits within the Sahara desert. As a result, the vast majority of Niger's rapidly growing population is concentrated in the southern part of the country, the portion that sits in the Sahel. Niger has one of the highest population growth rates in the world: 3.3 percent, amounting to about 450,000 new mouths to feed every year. Niger's population has doubled in the last twenty years and each woman bears, on average, about seven children. If this growth rate continues, there will be 56 million people living in Niger by 2050. Experts have already begun to question how a country with a very small band of cultivable land can continue to feed itself, given this population growth and the looming threat of drought.

And yet, because of these conditions, an adaptation strategy is already in the works. In the long battle between trees and sand, the trees have begun to gain some ground—thanks to the help of local farmers.

"The story about the farmers is not a technical story. It's about a social process. It's about farmers who broke with tradition and villages that organized themselves," explains Reij.

The tradition of land "cleaning" and tree removal became common in the 1930s, when the French colonial government pushed Nigerian farmers to grow crops for export. Another by-product of colonialism was the fact that all trees in Niger had been regarded as the property of the state; this gave farmers little incentive to protect them. Government foresters were tasked with managing the trees, but oversight was lax and as a result trees were chopped down for firewood or for construction, without regard for the environmental costs. The loss of tree cover also led to a fuelwood crisis. Poor households were forced to burn animal dung or crop residues instead of using these for compost; that practice reinforced the downward spiral of soil quality and crop yields.

Despite these long-standing habits, in the mid-1980s Reij and his coworkers noticed a new trend among the farmers of Niger; they had begun to cultivate the trees that were on their property.

"When we asked farmers in Niger why they had begun protecting and managing their on-farm trees, the first answer was, 'because we must fight the Sahara,' " explains Reij. "And to them 'fighting the Sahara' meant fighting dust storms." Early in the rainy season the winds from the north tend to pick up. During the drought of the 1960s and 1970s, the winds would bring tremendous amounts of sand with them. As Reij puts it, "The sand acts like a razor. And it was just cutting down their young crops and carrying off the topsoil."

Roughly 85 percent of the almost 14 million people who live in Niger subsist on rain-fed agriculture, with millet and sorghum making up more than 90 percent of the typical villager's diet. "The farmers would have to replant three or four times before a crop would eventually succeed," says Reij. And so, in the mid-1980s, farmers decided to do things differently. Instead of clearing the land of trees, they started to protect their trees, meticulously plowing around them when planting millet, sorghum, peanuts, and beans. "It's the local farmers who are the real heroes here," says Reij.

By 2007, somewhere between one-quarter and half of Niger's farmers were involved in regreening efforts, and it is estimated that at least 4.5 million people had seen the quality of their lives improve significantly.[22] Over time, farmers began to regard the trees in their fields as their property. And in recent years, the government has come to recognize the benefits of this strategy and has allowed individual farmers to own trees. Farmers now make money by selling parts of the tree: branches for fuel; leaves; seeds; fruit for food. Over time, those sales generate far more income than simply chopping down the tree for firewood. As a result, the farmers protect this source of income. Crop harvests have also risen. With the trees come better diets, improved nutrition, higher incomes, and an increased capacity to cope with drought. Many rural producers have doubled or tripled their incomes. In some villages, the annual hunger season no longer exists.

Three factors play a role in regreening, or what Reij refers to as farmer-managed natural regeneration (FMNR). "First, despite the deforestation that took place in the 1970s, there was still a rootstock in the subsoil. And that rootstock was still alive," explains Reij. "So as the rains gradually returned, and the soil and the trees were protected, the trees began regenerating." The second factor is livestock. "Livestock grazes and digests the seeds. That means when the seeds pass through the intestines of the livestock, they will germinate more easily." And as it turns out, not all supposedly barren soil is actually barren. "There's a beautiful word for this," explains Reij; "it's called the *seed memory* of the soil." Under the right conditions, seeds that have been dormant for more than a decade will suddenly start sprouting again. "And there you have the beginning of your trees," Reij says, smiling.

When Reij talks about a new green revolution in the Sahel, he means it literally. But it's not so much about planting trees—an expensive proposition that had been attempted unsuccessfully in the past—as it is about recycling them. In Niger, farmers have protected and managed about 200 million new trees during the past twenty-five years. The number of trees that have actually been planted in that same period is only about 65 million. "But of the 65 million trees that have been replanted, at best 20 percent have survived, leaving only about 12 million planted trees," explains Reij. "So there is a lesson to be drawn from this; tree planting can help, but protecting and managing natural regeneration is much cheaper and produces quicker and better results."

Reij says, "The point is, you can solve the problem of climate change and the problem of poverty in parallel. That's the nice thing." In essence, the trees set off a chain reaction that improves the local economy as well as the environment. "With more tree species in the system, you increase biodiversity. And at the same time, it means you produce more fodder, which means you can sustain more livestock," explains Reij. Perhaps the biggest benefit has come to many of the poorest members of Nigerian society—women and young men. "It's a lot easier for women, because they can prune those trees and they have the firewood immediately at hand," explains Reij. And as for the young men, the annual exodus to search for higher-paying jobs in urban areas has slowed, thanks to new opportunities to earn income in an expanded and diversified rural economy. "So you have all kinds of positive spin-offs," says Reij. He points to some recent economic research, which suggests that increasing agricultural production by 10 percent can reduce rural poverty by somewhere between 6 and 9 percent.

The trees also act like a buffer during worst-case scenarios, and such scenarios may happen more frequently in a climate-changed future. In 2005, a year when the western Sahel saw flooding, the rains failed yet again in Niger. Reij visited villages with trees and villages without trees. He found that famine was much less of a problem in villages with many trees than in villages with few trees.

Likewise, villages with many trees did not suffer any drought-related infant mortality. "People told us that because they had more trees, they could cut and prune trees and sell firewood on the market. They could then use that money to buy cereal to feed their children," says Reij. "No doubt it was a brutal situation, but at least the children made it through." Reij knows this is not what perfection looks like, but for right now, it's close enough.

Ultimately, evidence from Niger demonstrates how relatively small changes in human behavior can transform the regional ecology, restore biodiversity, and increase agricultural productivity. Reij thinks such behavioral changes may even help bring rain back to the Sahel. "If you put a thermometer into barren, sandy soil you immediately get 120°F. But just 1 meter away, where you have some surface cover, the temperature immediately drops to 109°F," says Reij. "And with a bit of luck, if you have vast areas of regreening, the question is: might that begin to have positive impacts on local rainfall as well?" For Reij, this might just be the perfect answer.

In June 2009, Reij helped launch the Sahel Regreening Initiative in Burkina Faso and Mali. He says, "We thought: why not start an initiative which tries to build upon and scale up the existing successes in Niger? When we talk about adaptation to climate change, I am convinced that reforestation is a fairly effective answer. You improve the environment, you improve agricultural production, and you reduce poverty. There is every reason to be hopeful. We have the wind at our back."

The question is whether that will be enough. This is a matter of which forcing the Sahel will be most sensitive to in the coming decades—the oceans, the land surface, or global warming. The oceans have probably dominated the climate of the Sahel for all of human history. But as Reij has shown, the land surface is also clearly very important, especially in the Sahel, where it has the ability to outfit the landscape and help offset some of the changes that climate change will bring.

Unfortunately for the Sahel, Niger's tree experiment is an isolated event. Some recent climate model simulations propose that until 2025, the impacts of land degradation and vegetation loss over sub-Saharan Africa may be even more important than global warming for understanding climate change.[23] Until 2025, these models show that a drier, warmer climate goes with decreased agricultural production, on the order of 5 to 20 percent. Peanuts, beans, maize, rice, and sorghum will be likely to have the biggest drops in yield.[24] A recent study focused on Mali found a 17 percent reduction in crop yields out to 2040. Scientists consider these studies red flags. They recommend immediate action to develop more heat-resistant crops for the Sahel. In Benin, for example, a shift to yams and manioc is suggested as one adaptation strategy. And then, of course, there are the trees.

Reij points again to those pockets of resilience. "These predictions for 2050 and beyond, a 5°F increase in temp, a 20 percent decrease in crop yields—in the context of a doubling population, that's quite dramatic. That's why I look for villages where farmers were able to depress temperatures by 5°F using trees, and villages where the use of simple conservation techniques helped increase crop yields by 40 or 50 percent or even more. You have to be able to find examples that show it's possible to counterbalance what lies ahead." The people of the Sahel have a lot riding on this battle between trees and sand.

As scientists continue to search for perfect answers, red flags are clear on the horizon. Held has been watching as more red flags appear with each passing year and each new research experiment. And as he settles in to prepare the next round of IPCC simulations for the Sahel, he has one big concern of his own. "My biggest concern is that the GFDL model turns out to be right," says Held. If that happens, then sand may have finally won the battle in the Sahel once and for all. Unless, of course, like Reij, we abandon the search for perfect answers and simply begin to fight back. That might be the biggest Sahelian irony of all.

Africa and the Sahel: The Forty-Year Forecast—Famine, Crop Loss, and Water Resources

NIAMEY, NIGER	TODAY		2050		2090	
Emissions Scenario	JAN.	JULY	JAN.	JULY	JAN.	JULY
Higher	75.8	85	78.5	88	82.5	92
Lower			78.5	87.4	80	88.9

Forecast
July 2015

There are few secrets in this parched, hostile landscape, especially about climate. As the models projected, the African climate was changing—becoming hotter and drier.

The trees were waging an all-out war against the brutality of the sun, and they seemed to be winning. Tens of millions of acres of Sahelian farmland had been planted with trees, and every leaf was viewed as a symbol of hope for the future. Agricultural productivity had increased, and there was actually a surplus of fuelwood. In the drier regions of Niger and Burkina Faso, people had begun reclaiming abandoned fields and getting new grain harvests by investing in simple water-harvesting technologies. Farmers were doing everything in their power to adapt to climate change, and the trees were their shield against the increasingly harsh climate. Through support from the Bill and Melinda Gates Foundation and OXFAM America, the regreening initiative had spread across the Sahel.

November 2022

But there was only so much the trees could do. Without significant infrastructure investment, farmers across the Sahel were alone in their battle against the climate. Across the Sahel, yields of peanuts, beans, maize, rice, and sorghum were beginning to fall. Those farmers able to obtain heat-resistant seeds from international aid groups fared better. And those who shifted to more drought-resistant crops, such as yams and manioc, actually managed to put some food on the table. But ultimately, without rainfall, it was no secret that the farmers, who had lovingly planted and tended these trees as if they were their children, would be forced to watch as they eventually fell victim to the inevitability that was climate change. Climate change was leaving more and more people with less and less.

This basic realization led to a growing resentment across the African continent, because it was also no secret that the rich had caused the problem while the poor of the world bore the brunt of the impact. The experts said it would only get worse, and their advice was simple—adopt aggressive emission reduction goals and take steps to help victims of climate change adapt. They said it was a matter of national security.

To return to 2010: As it turns out, half of all CO_2 emissions come from only about 10 percent of the world's population. And that 10 percent includes Saudi Arabian oil moguls and Chinese investment bankers, not just rich Americans—the operative word isn't *American*; it's *rich*. The atmosphere doesn't care whether you drive a Ferrari in Dubai or Shanghai or New York. All it sees is the CO_2. As a result, several experts have come up with ideas on how to even the CO_2 playing field—spread the CO_2 wealth, so to speak. One group of scientists recently suggested a Robin Hood idea that essentially takes emissions (by means of a cap and tax) from the rich and distributes them among the poor. These scientists see their plan as a way to help lift people out of poverty.[25]

After all, CO_2 is just another term for energy. The World Bank estimated that if people in the United States swapped their SUVs (about 40 million, total) for fuel-efficient compact cars, the change would free up space in the atmosphere for about 142 million tons of CO_2. If you could magically convert that CO_2 back into energy and give it to the poor, it would provide basic electricity to about 1.6 billion people. In essence, everyone in Africa could have lights and running water.[26] To repeat, it seemed like a good way to even the playing field—and to decrease resentment. But in the end, no one really wanted to give anything up for the sake of strangers.

In 2015, this basic inequity between rich and poor remained a serious problem. And climate change, described by national security experts as a *threat multiplier*,[27] was turning up the heat, literally. Conflict was spreading across the African continent like wildfire. You needed only look at the past few years for evidence. The instability in the Sahel—especially Darfur—showed how quickly disputes over access to water and food during times of drought became politicized. The climate problem magnified preexisting threats stemming from ethnic and religious conflicts. The former UN secretary-general, Ban Ki Moon, had made that very point in the *Washington Post* more than 15 years earlier:

> Almost invariably, we discuss Darfur in a convenient military and political shorthand—an ethnic conflict pitting Arab militias against black rebels and farmers. Look to its roots, though, and you discover a more complex dynamic. Amid the diverse social and political causes, the Darfur conflict began as an ecological crisis, arising at least in part from climate change.
>
> Two decades ago, the rains in southern Sudan began to fail. According to U.N. statistics, average precipitation has declined some 40 percent since the early 1980s. Scientists at first considered this to be an unfortunate quirk of nature. But subsequent investigation found that it coincided with a rise in temperatures of the Indian Ocean, disrupting seasonal monsoons. This suggests that the drying of sub-Saharan Africa derives, to some degree, from man-made global warming.[28]

Experts were warning that attempts by the United States to build a "hearts and minds" coalition against Islamist extremism were being undermined by climate. They listened carefully to the latest tape from Osama bin Laden, on which he railed once again about the inequities of global warming and CO_2 emissions.[29] Jobs for young men in North Africa, to take just one example, had been further reduced by warmer temperatures and declining rainfall, intensifying resentment and unrest.

Scientists had even modeled the connection between conflict and temperature, to prove the point.[30] The data were painfully clear: conflict increased in lockstep with temperature. And when you looked at conflict combined with climate model forecasts of future temperature trends, there was a roughly 50 percent increase in armed conflict—almost 400,000 additional battle deaths—by 2030. The need to reform the policies of African governments and foreign aid donors to deal with rising temperatures had become urgent.

March 2030

By 2030, piracy had become an epidemic on the order of HIV-AIDS. The piracy industry popped up after Somalia's central government collapsed in 1991. With no patrols along the shoreline, commercial fishing fleets from around the world came to plunder Somalia's tuna-rich waters. Initially, the pirates stepped in as a vigilante response to that illegal commercial fishing. Armed Somali fishermen confronted the crews of illegal fishing boats and demanded that they pay a tax. These were acts of desperation by local fishermen who had lost their livelihood. The entire country was, in fact, on the brink of starvation. It was not uncommon for pirates to smile when they were picked up by navy ships—they knew that they would at least get three square meals a day. But over time, these desperate bands of fishermen grew into something bigger, more organized, and much more sinister.

As rising ocean temperatures, pollution, and overfishing gradually erased their livelihood, more and more fishermen traded in their nets for machine guns and were hijacking any vessel they could catch: a sailboat, a yacht, an oil tanker, or a food ship chartered by the United Nations. Desperate times, they said, called for desperate measures. They said they had no other choice.

Because of all the hijackings, the waters off Somalia's coast were now considered the most dangerous shipping lanes in the world. The United Nations agreed to put a maritime peacekeeping force in place to patrol the waters, but there was only so much it could do. The main victims of the pirates were the Somali people. Nearly three out of every four Somalis had come to depend on food donations in order to survive. But the pirates were routinely overtaking the UN peacekeeping vessels, and the attacks made it very hard for the United Nations to keep sending provisions. Somali people were starving because the boats couldn't get through.

As of March 2030, no food ships had set sail from Mombasa to Mogadishu in months—the voyage was simply too dangerous. Despite the efforts of the international aid community and peacekeeping troops, conflict across Africa had increased along with the temperature. It looked more and more as though Africa was heading toward the status of a failed continent.

January 2050

After years of conflict, drought, and food shortages, Africa was finally able to capitalize on something it has in abundance—sunshine. The Desertec project had been on the table for years, but two key elements were always missing: European funding and the support of African countries. When increasing demand from China and India sent oil and gas prices into the stratosphere, the Germans were eventually able to pull the money together, on the promise that Desertec would offset Germany's dependence on Russian gas supplies. The Desertec consortium, brought together by Munich Re, the world's biggest reinsurer, consisted of some of Germany's biggest and most powerful companies, including Siemens and Deutsche Bank. The plant symbolized a way to solve the problems of climate change and energy security simultaneously—and if everything went according to plan, it might help Africans as well.

North African governments sold their desert in return for water. Let us use your desert to generate power, the Desertec consortium argued, and you can use our energy to desalinate seawater so as to irrigate crops that will help feed your growing populations. North Africa's demand for water had, in fact, increased by two-thirds—an amount that was far beyond the available supply. For Africa, energy security was far less of a concern than water security. The deal was straightforward—Desertec would generate electricity for export in return for desalinating seawater for Africa—and it was a deal that was very hard to pass up. So with little to lose, the North African countries signed on, making one request—the plant was to be renamed Desertec-Africa.

The science was there. Every year, each square yard of the Sahara desert receives more heat from the sun than would be obtained by burning two barrels of oil. The calculations showed that all of Europe's electricity could be made in an area just 150 miles across. The Desertec-Africa plant used a technology known as concentrated solar power (CSP). The sun's rays would be concentrated by the use of mirrors, to create heat. The heat would then be used to produce steam to drive steam turbines and electricity generators. The advantage a CSP plant like Desertec-Africa had over standard solar photovoltaic panels, which convert sunlight directly to electricity, was that it had heat storage tanks. The tanks were able to store heat during the day and then power steam turbines during overcast periods, bad weather, at night, or when there was a spike in demand.

The cost overruns were substantial, and sandstorms and warring factions in North Africa made construction of the Desertec-Africa project an ordeal. But thanks to an important innovation that allowed the plant to use less water, construction was kicked into high gear. Like a standard coal- or oil-fired power plant, a solar thermal station requires large amounts of cooling water—something that is nearly impossible to come by in the Sahara desert. The water condenses the steam after it goes through the generator's turbines. But the innovation allowed the Desertec-Africa plant to be fitted with an air-cooling system that cut water demand by up to 90 percent—a huge break for investors. While many still argued that Desertec-Africa would make Europe's energy supply a hostage to a politically unstable region and that Europe was unfairly exploiting Africa for its sunlight, the project went ahead.

By 2050, Desertec-Africa was producing about half of Europe's electricity, with a peak output of 400 gigawatts—roughly equivalent to the output of 400 coal-fired power stations. The electricity generated by Desertec-Africa reached Europe via high-voltage power lines and trans-Mediterranean links that went from Morocco to Spain across the Strait of Gibraltar; from Tunisia to Italy; from Libya to Greece; from Egypt to Turkey, via Cyprus; and from Algeria to France, via the Balearic Islands. Part of a wider European super-grid that conveyed power generated from wind turbines in the North Sea, hydroelectric dams in Scandinavia, geothermal activity in Iceland, and biofuels in eastern Europe, Desertec-Africa had helped reduce European emissions of CO_2 by about 80 percent. The consortium had hopes of expanding. After all, a patch less than 450 miles across the Sahara could meet the entire world's electricity needs.

South Pacific Ocean
GREAT BARRIER REEF
Magnetic Island
Townsville
Queensland

6

THE GREAT BARRIER REEF, AUSTRALIA

Joanie Kleypas and I first met in 2001, when I moved to Boulder, Colorado. I had just completed my postdoc at Columbia University and had decided to head west to begin work as a research scientist at the NCAR. Kleypas, a marine ecologist and geologist who uses climate models to study the health of coral reefs, had an office down the hall from mine.

At the time, I was studying a drought that had devastated a large stretch of central and southwest Asia. More than 60 million people across the region were in dire need of rain; and Afghanistan had been especially hard hit, as the drought came after two decades of political instability and economic isolation. I was looking at how large-scale climate patterns, like El Niño, might potentially help predict big droughts in the future and prevent such devastation. Whenever I needed a break, I'd stop in at Joanie's office to talk about her research on coral reefs. I remember thinking that Joanie was really lucky, because she studied beautiful places that were untouched by human activity. I loved listening to her talk about coral reefs and her travel to exotic places like Tahiti, the Caribbean, and, perhaps best of all, Australia's Great Barrier Reef.

I checked in with Kleypas recently to see how things were going with the Great Barrier Reef as well as to get a sense of what kind of impact global warming was expected to have on reefs in the coming years. She wasn't feeling so lucky. "I work on coral reefs, for God's sake. The entire coral science community is depressed," Kleypas admitted.

The Great Barrier Reef (GBR) is the largest tropical coral reef system in the world. It contains nearly 3,000 reefs built by more than 360 species of hard coral, and it attracts a wide variety of marine life. The GBR provides shelter to more than 400 types of sponges; 1,500 unique species of fish; 4,000 varieties of mollusks; 500 kinds of seaweed; and 800 types of echinoderms, which include starfish and sea urchins. It is bursting with activity, an oceanic metropolis analogous to Paris or New York. I grew up in New York, but given the choice, I would have much preferred growing up with a view of the GBR. It extends more than 1,200 miles along the northeast coast of Australia, and its area is about that of 70 million football fields, or half the size of Texas, which is where Kleypas is originally from.

As an undergraduate, Kleypas studied oceanography and marine biology at Lamar University in Beaumont, Texas. She had learned how to scuba-dive, thanks to her brother, and she dreamed of leaving the murky waters of the Gulf of Mexico for the majesty of the GBR. That dream came true when she received a Fulbright scholarship to study in Australia. She completed her PhD dissertation on the GBR and finally had a chance to see many of the things she only read about in textbooks. The GBR, home to six of the world's seven species of marine turtle, 30 percent of the world's soft coral varieties, and several hundred types of seabirds, became her backyard when she moved to Townsville, a small city of about 100,000 in the state of Queensland.

Townsville sits along the central part of the GBR and is a mecca for coral reef research. It's the site of James Cook University, where Kleypas studied, as well as the Australian Institute of Marine Science and the Great Barrier Reef Marine Park Authority. Townsville is about as close as you can get to the GBR and still breathe air. It is on the edge of the GBR lagoon, the reef just a thirty-minute ferry ride away. Kleypas did the reverse ferry commute for a while when she moved to Magnetic Island, population 2,107. More than half of this 20-square-mile island, known as "Maggie" to the locals, is designated a national park, so chances are you'll see more wallabies and koalas there than you will people. Kleypas rented a room from Lyndon DeVantier, one of the world's best field taxonomists for corals. "I would take my bike on the ferry and ride to the university in Townsville every day. It was an idyllic existence and I would have loved to have stayed," she says a little wistfully.

Kleypas began her career as a geologist interested in using corals to study changes in sea level. When she arrived at James Cook University, she and a team of researchers would go out for weeks at a time to remote sections of the GBR. They would cart out a small rig that allowed them to drill into the massive coral reef structures and extract cores. "Coral reefs are like big dipsticks," Kleypas explains. "They can tell us a lot about the geologic history of reef growth as sea level goes up and down." Despite being seasick from the strong currents and the 10- to 20-foot tides that swept the reef twice a day, Kleypas loved the GBR. And her hard work led to new insights about it. She and her team learned that since the end of the last ice age, the GBR had flourished or faltered depending on a delicate balance of conditions that included sea level, light, sediments, temperature, and circulation patterns. At the time, they weren't even thinking about carbon dioxide. They are now.

Corals have probably existed on the GBR for more than 25 million years.[1] The corals first formed during the geological era known as the Miocene. It was during the Miocene that India slammed into Asia and created the Himalayas. The Miocene also was a time when the Australian continent was on the move. As Australia tectonically made its way into the tropics, the shift to warmer ocean temperatures initiated the growth of some corals. Think of them as the very first version of the GBR, although back then the corals didn't form large structured reefs.

According to the Great Barrier Reef Marine Park Authority

(GBRMPA), the earliest record of complete reef structures dates back about 600,000 years. Research suggests that the current reef structure started growing above this older platform about 20,000 years ago, during the Last Glacial Maximum (LGM), the peak of the last ice age. At the time, much of the Earth's water was locked up in the form of ice, so the sea level was about 390 feet lower than it is today.

As the ice age came to an end, global temperature began to increase and the ice slowly retreated to the poles and mountaintops where it had originated. By around 13,000 years ago, corals began to move into the hills of what had been Australia's coastal plain but was now underwater. At the time, the sea level was still 200 feet lower than it is today, but the coastal plain had already been swallowed up by the sea: only a few islands protruded out of the water. As the little islands slowly became submerged beneath the ocean, the corals finally had a place to set up shop in earnest. Scientists estimate that the present-day, living reef structure is between 6,000 and 8,000 years old; in other words, it dates from the period during which the sea level is thought to have finally stabilized.

Think of the modern GBR as a living veneer draped over ancient limestone. "The reef is like a layer cake," Kleypas explains; "every time the sea level rises, the reef adds a new layer." Today, that big layer cake helps to feed hundreds of thousands of Australians, because it brings in about $6.9 billion annually from tourism and other sources. What Captain James Cook identified as the perfect spot for a prison colony in 1770, when he surveyed the place then called New Holland, is now recognized as a World Heritage Site. So much for first impressions.

It's lucky for the GBR that Australia drifted into the tropics during the Miocene. Of all regions of the ocean, corals like the tropics best. Kleypas calls the reefs in this range, the band that stretches from 30°N to 30°S, the *vacation* reefs. The GBR alone has become a vacation destination for more than 2 million people each year.

You can find corals outside the tropics, too. A sturdy, brave few have inched their way out of their traditional comfort zone into places like Japan and Bermuda. But the farther poleward you go, the less the corals can build and actually reach the size associated with a true reef community. And it's the reefs that contain all the staggeringly beautiful biodiversity. "They get puny as you move north," Kleypas explains. "You can find single corals growing along the North Atlantic coast. But you'll never find a reef." The corals simply don't like cooler temperatures; coolness slows their growth rate, and in some cases colder waters can actually kill them. It seems corals are a bit like Goldilocks; they don't want things to be too hot or too cold.

By the time she left Australia in 1991 to begin her job at NCAR, Kleypas had fallen in love with the reef ecosystem that she had been using to reconstruct ancient sea levels. And so she shifted her focus from using the reefs as a tool to just studying the reefs themselves. And in the meantime, she's become a reluctant expert on how global warming is affecting these magnificent ecosystems.

"I really do think the coral reef community is suffering from some form of depression," Kleypas says. "It's like this. Imagine you fall in love with the most beautiful, amazing person. And then that person comes down with cancer. It's an incredibly sad thing. I fell in love with the reefs, not their disease." But it's the disease, the rising level of carbon dioxide in the atmosphere, that has captured the attention of an entire generation of marine scientists who are intent on saving the reefs. Kleypas is someone who uses climate models to study the reefs,[2] but, ironically, she could never have predicted that her own research career would come to this. She never expected to be forecasting the eventual decline of coral reefs.

That's not to say everything was perfect with corals before the impact of global warming became glaringly evident. Coral reefs were already showing signs of stress due to local-scale impacts such as agricultural runoff and destructive overfishing practices that include bottom trawling and dynamiting. The overall decline in water quality, due to pollution from coastal development, didn't help matters either. "Basically, the reefs are in worse shape the closer they are to people. The farther out you go, the better they look," Kleypas says.

But the global-scale stress due to climate change is adding a new dimension and a new threat to the overall resilience of the coral reef ecosystem. Global warming affects corals in two ways. The first is temperature: the oceans are warming up. The second is ocean chemistry: the oceans are also becoming more acidic.

Corals begin their lives as soft-bodied larvae that float through the water and eventually settle on a hard surface. As they settle, they also partner up with marine microalgae called *zooxanthellae.* Corals, which are animals, and their microscopic plant roommates are one of the prime examples of what scientists call a *symbiotic relationship.* Once the coral has partnered with the microalgae, it then sets to work building its skeleton by pulling dissolved calcium carbonate compounds out of the ocean water. The limestone skeleton forms the physical structure we think of as the reef.

A reef is the result of colonies of corals building their skeletons, like a bricklayer laying bricks, steadily over thousands of years. Shacking up with the algae turns out to be a big asset, as corals gain a second helping of food on top of what they are able to pull directly out of the water column. That second helping turns out to be very big. The zooxanthellae can provide up to 90 percent of the corals' energy requirements. "It's like if we had algae growing on our skin," Kleypas explains. "Whenever we'd go out into the sun, we'd get a jolt of extra energy courtesy of that algae." The corals get their energy from the plant by means of photosynthesis. This symbiotic relationship provides the extra boost that allows the corals to grow so large and form such elaborate reef structures. By definition, the symbiotic relationship benefits both partners. The microalgae get nutrients in the form of waste released by the coral. And of course it is the relationship with the algae that makes corals so pretty. The tissues of corals themselves are clear. Most of the beautiful colors of the coral reef, which can range from the palest pink to darkest black, are a gift from the zooxanthellae.

It's true that coral reefs like warmth. Ideally, they are adapted to water temperatures ranging between 65°F and 90°F. But corals don't like a quick spike in temperature. Go a little above the range they're used to, and trouble starts. That trouble comes in the form of coral bleaching.[3]

Coral bleaching is the term scientists use to describe the loss of all or some algae and pigment by the coral. As the algae are ejected, the white calcium carbonate skeleton becomes visible through the translucent tissue layer. The coral is weakened because it has lost the food energy provided by the algae. Already, in many places of the world—such as the Maldives, the Seychelles, and Palau—coral bleaching has effectively destroyed more than 50 percent of reefs. In the Caribbean, the numbers are worse, with between 80 and 90 percent of the reefs destroyed by bleaching, disease, hurricanes, and a number of problems related to coastal development, fishing, and other human activities. And future climate model projections, the kind that Kleypas works on, indicate that coral bleaching events are expected to become more frequent and severe over the coming decades.

"Bleaching is what happens when the coral kicks the algae out," Kleypas notes. "The way I have come to understand it is that the coral can usually handle the hot water, at least until you turn on the lights. When you add sunlight on top of a spike in ocean temperatures, photosynthesis kicks into overdrive." One by-product of all this photosynthesis is the release of too many free oxygen radicals. Free radicals are the same agents involved in the process of aging in humans. Crank up the photosynthesis, and you crank up the free radical count. "All those free radicals hurt the coral," Kleypas says.

The coral can't handle the imbalance, so it responds by evicting its tenant: kicking out the algae.

"It's almost like diarrhea," she says. "We have symbiotic microflora in our gut, and diarrhea is a common biological response when there is an imbalance." The diarrhea then serves as a way to get rid of things that are normally good for us. This classic tale of a good relationship gone bad makes you wonder if it might be a metaphor for the larger relationship between the planet and us.

Almost every reef region in the world has suffered extensive stress and in some cases death from coral bleaching. In 1998, a severe bleaching episode took out upward of 16 percent of coral reefs worldwide. El Niño played an obvious role in that event. During El Niño conditions, ocean temperatures in the Indian Ocean and in the central to eastern Pacific Ocean increase. Along with the warmer ocean waters comes a stable mass of high pressure, exactly the kind of weather pattern that ushers in a prolonged period of hot, sunny days. What might look like perfect weather is actually a condition for extensive coral bleaching. El Niño turns the lights on in a very big way.

"As a general rule, corals on the Great Barrier Reef start bleaching once temperatures exceed the average yearly maximum by about 2°F," Kleypas explains. In some cases it takes a much higher temperature spike to cause bleaching. But scientists say the range is generally between 2°F and 4°F. The actual temperature at which bleaching starts depends on how healthy a reef is to begin with, and how many other stressors, such as pollution and overfishing, are present. But as a general rule, if you crank up the temperature, you will most certainly crank up the stress level.

Average annual ocean temperature around the GBR varies from about 68°F to 79°F. A bleaching event is mostly likely to happen in January, February, and March—the southern hemisphere's summer months. Conversely, June, July, and August—winter in the southern hemisphere—are cooler. The all-important maximum temperature, which sets the baseline for when bleaching can happen, ranges from about 79°F at the southern end to 86°F at the northern end.

The fact that there is a range within which bleaching can happen means that other factors, such as global warming and just plain bad luck, are important in determining the ultimate fate of a coral reef under stress. Add a heat wave or an El Niño event on top of a week where ocean temperatures are at a maximum, and you've drastically increased the risk of a coral bleaching event. Global warming is no different. It effectively starts you off at a point closer to the temperature at which corals bleach. Water temperature along the GBR has increased by about 0.7°F since 1850, and the central and southern portions of the reef have warmed up even more, about 1.2°F.

The mass bleaching event of 1998 represents a turning point, according to many experts. They say that mass coral bleaching events have increased in extent and severity worldwide over the last decade. Prior to 1998, many reef systems had never experienced a severe bleaching event. But since 1998, every region has seen severe bleaching, and many regions have experienced significant die-offs because of warmer ocean waters.

And although the GBR is recognized as one of the best-managed and best-maintained coral reef parks in the world, it too has felt the effects of severe bleaching. The 1998 event hit 50 percent of the GBR. Another bleaching event occurred in 2002; this time, 60 percent of the GBR was hit. Fortunately, the death toll was low, but about 5 percent of the GBR is gone. In 2006, another bleaching episode hit the GBR, but it was more localized.

The southern hemisphere summer of 2009 appears to have been another bad break for the GBR. On March 16, 2009, the Australian government reported that a "weather triple whammy" had led to yet another round of coral bleaching. Stifling heat in December, floods in January and February, and winds from the tropical cyclone Hamish arrived one after the other. Ocean temperatures across most of the reef rose 3.6°F to 5.4°F degrees above the December average.

Russell Reichelt, chairman of the Great Barrier Reef Marine Park Authority, said that the "triple whammy" raised serious concerns about global warming. "The forecasts are an increased frequency of extreme events," he said. All these factors individually cause stress to the GBR. But, Reichelt added, it was their combined impact that was most worrying. "Historically, the reef has been resilient to events like this, but it is rare, possibly unprecedented, to have three such events in such a short period of time."

Coral reefs are complex and stunningly beautiful ecosystems. And their beauty attracts significant tourist dollars: it is estimated that tourism associated with the GBR contributes more than $5 billion to the Australian economy. This figure doesn't include other sources of revenue, such as commercial fishing. There are five main commercial fisheries operating in the GBR that together catch about 26,000 tons of seafood each year, with a total gross value of more than $220 million. But the reefs also provide benefits that don't have a price tag. Coral reefs not only serve to protect Australia's fragile coastlines from storm damage but also have been used to make several anticancer drugs. So it's quite accurate to say that coral reefs save lives.

Ultimately, extreme weather on top of the long-term global warming trend spells trouble for the reefs as well as for Australia's economy. Climate and economic models predict losses to the GBR tourism industry of between $95.5 million and $293.5 million by 2020, as a result of bleaching-related damage. And when you factor in the costs of all the other risks that warmer temperatures pose for Australia—extended droughts, heat waves, wildfires, and so on—you begin to sense that Australia is in grave danger.

Janice Lough, a researcher at the Australian Institute of Marine Science, explains the global warming effect as follows: "This seemingly modest increase in baseline temperatures has been sufficient to take corals over the bleaching threshold in 1998, 2002, and again in 2006. Modeling of future impacts suggest that a 1.8°F to 3.6°F warming of the GBR would result in about 80 to 100 percent bleaching compared to about 50 percent in 1998 and 2002." In other words, global warming makes bad luck worse. Lough adds that maintaining the hard coral at the heart of the reefs requires corals to increase their upper thermal tolerance limits by 0.2°F to 1.8°F per decade. But how do you teach corals to become more heat tolerant?

Bleached corals aren't dead; they're just starving. The loss of their energy-providing zooxanthellae means they're not getting enough food. If the stressful conditions come to an end soon enough—that is, if the weather changes and temperatures become cooler again—the algae can come back, and the corals can survive the bleaching event. But corals that do survive a bleaching event come out of it in a weakened state. As a result, they'll be likely to experience reduced growth rates, decreased reproductive capacity, and increased susceptibility to diseases. Because it can take up to twenty years for reefs to fully recover, these recurring bleaching events are a kick in the teeth. Prolonged bleaching often leads to coral death.

Complicating matters is the fact that saving the reef from severe bleaching events requires patiently nursing it back to health. The recovery process is time-consuming and requires recolonization by coral larvae. Even under ideal conditions, coral recovery is slow and may take decades. You need sufficient connectivity of *source* reefs—reefs that export fertilized coral and fish eggs to other reefs downstream—as well as good water quality to make sure that the spawning and recruitment of larvae will succeed. Bleaching will actually kill the corals if the stresses are too severe or too persistent. This situation is really not so different from a prolonged drought. The condition of a coral when it enters a bleaching event is likely to determine its ability to survive the bleaching event.

Corals get a good bit of attention because a bleaching event is highly visible and because so much money is tied up in the reefs. But other parts of the reef ecosystem are also vulnerable to temperature. Seabird chicks have undergone severe die-offs during periods of unusually high sea temperatures. These die-offs, called *nesting failures*, result when parent birds can't get dinner for their chicks. The fish they prey on follow productivity zones that are temperature-dependent. And when these fish change location, the birds can't always find them. Sea turtles are also at risk. The sex ratio of turtle hatchlings is temperature-dependent, and continued warming could cause a significant bias toward females in future populations.

But again, temperature is just half of it. The other half of the global warming situation is ocean acidification (OA). Kleypas describes OA as the "silent problem" associated with increasing CO_2. It's also been described as "the other CO_2 problem." The other CO_2 problem has scientists very worried. And, ironically, it comes as the result of a favor the oceans are doing us. No good deed goes unpunished.

Atmospheric CO_2 is currently at 387 ppm, but it would actually be a lot higher if not for the oceans. Roughly 30 percent of the excess carbon dioxide released into the atmosphere by human activities since the industrial revolution has been absorbed by the oceans. That's the favor. If not for the ocean uptake, atmospheric CO_2 would be on the order of 450 ppm today, a level that would have led to even greater climate change than is already under way. But this favor provided by the oceans doesn't come cheap. It has led to a roughly 30 percent increase in the concentration of hydrogen ions through the process of OA.[4]

This process, OA, is simply what happens when you add carbon dioxide to seawater. The additional CO_2 causes a slight reduction in ocean pH, which is a measure of how acidic or basic a substance is. The pH scale ranges from 0 to 14. Pure water, for example, has a pH of 7 and is considered neutral. A pH less than 7, as in vinegar and lemon juice, is acidic. A pH greater than 7, as in ammonia or laundry detergent, is basic. The ocean is also slightly basic; the average pH of surface seawater today is about 8.1. But there is reason to believe that this is not the number it should be or the number it will remain.

The United States is the third-largest consumer of seafood in the world, with total consumer spending for fish and shellfish at about $60 billion per year. Coastal and marine commercial fishing generates as much as $30 billion per year, and nearly 70,000 jobs. Healthy coral reefs are the foundation of many of these viable fisheries. Needless to say, there are plenty of reasons to be worried about OA.

Although it is described as "silent," OA is a straightforward consequence of rising atmospheric CO_2. This condition doesn't have a lot of the uncertainties that plague some other climate change forecasts. It's freshman chemistry. And ocean pH is something we're good at measuring. Since the 1980s, pH measurements collected in the North Pacific Ocean (near Hawaii) and in the Atlantic Ocean (near Bermuda) are registering a decrease in pH of approximately 0.02 unit per decade. Since preindustrial times, the average pH of ocean surface water has fallen by approximately 0.1 unit, from approximately 8.2 to 8.1, and it is expected to decrease further, depending on how high CO_2 rises. If atmospheric CO_2 concentrations reach 800 ppm, pH is predicted to rise an additional 0.3 to 0.4 pH unit.

What worries scientists is that even a slight decrease in pH does something funky to ocean chemistry, specifically to the amount of *carbonate ions*, a very important form of carbon. Corals pull in carbonate ions and secrete calcium carbonate ($CaCO_3$). This is a process called *calcification*, and it uses the dissolved carbonate ions to form calcium carbonate minerals for shells and skeletal components. Once dissolved in seawater, CO_2 gas reacts with the water to form carbonic acid (H_2CO_3), which can then break apart by giving up hydrogen ions to form bicarbonate (HCO_3^-) and carbonate (CO_3^{2-}). Increasing the amount of carbon dioxide dissolved in the oceans has a nasty side effect: it decreases the amount of carbonate ions in the water. Fewer carbonate ions means less material for building such things as calcium carbonate reefs and clamshells.

A study published in the journal *Science* in 2009 seems to confirm this.[5] Experts at the Australian Institute of Marine Science in Townsville looked at coral samples from the GBR over the past twenty years to track changes in growth rates. Specifically, they measured the rate at which corals absorb calcium from seawater to build limestone skeletons. The study concluded that corals in the GBR are growing more slowly. The team of researchers investigated 328 colonies of massive *porites* corals from sixty-nine reefs covering coastal as well as oceanic locations spanning the entire length of the GBR. Because the *porites* coral lays down annual growth bands, it's possible to count back to a specific year and correlate the growth during that year with the sea surface temperature over the same time period. Ten of the cores dated back to 1572.

The researchers sliced up the cores and used X-rays to measure three growth values: skeletal density, annual growth rate, and calcification rate. The values for growth and density allowed them to calculate annual calcification. They found that between 1900 and 1970 calcification rates increased 5.4 percent. But that's when you could argue that the party ended. Calcification rates dropped 14.2 percent from 1990 to 2005, mainly owing to a slowdown in growth. Researchers measured calcification as decreasing from 0.56 inch per year to 0.49 inch per year. Scientists can't confirm yet that what they are seeing is indeed the impact of increased OA, as opposed to other stressors such as coastal pollution. But the fact that the effect is seen on inshore as well as offshore reefs suggests to them that the cause is more likely to be *global* (for example, temperature and ocean acidification) than local (for example, pollution). It appears that not just the economy but also the GBR is in a recession.

And the impact of OA isn't limited to corals. "There's some very cool new research out there about clown fish," Kleypas says. "Of course, it's also very depressing." The clown fish has become an iconic species ever since Disney's blockbuster *Finding Nemo* gave kids a look at the biodiversity of the GBR. Interestingly, Phil Munday and his colleagues at James Cook University found that OA affects Nemo's ability to find his way home. "Baby clown fish use their sense of smell to find a suitable habitat. And ocean acidification impacts their ability to differentiate between what is a suitable habitat and what isn't," Kleypas says. Recent research suggests that OA has impaired their sense of smell. "The baby fish aren't getting the signal that says, 'Bad habitat; don't go there!' and are less able to sense the proper habitat. And the researchers were not subjecting the fish to huge changes in pH. They were consistent with future projections," Kleypas adds.

Research by Ken Caldeira, an ecologist, and his team at Stanford's Carnegie Institution for Science suggests that ocean pH has not been more than 0.6 unit lower than today's levels during any time over the past 300 million years. Yet the results obtained with the Stanford climate model show that the continued release of fossil fuel CO_2 into the atmosphere could cause an eventual pH reduction of 0.7 unit over the next 300 years.[6] An unprecedented change in pH over 300 million years is a lot easier to handle than the same change over three centuries. When CO_2 changes over a time interval longer than 1 million years, ocean chemistry is buffered by interactions with carbonate minerals, and that buffering helps reduce the impact of acidification. Caldeira's research suggests we're talking about such a severe mismatch in timescales that adaptation is almost impossible.

When I asked Kleypas what she envisioned the GBR might look like by 2050, she said, "The distribution of reefs will be more patchy. Biodiversity will go down. There will be more algae and less hard coral. Erosion will become more noticeable. There will be fewer baby corals. It's not all going to be dead; . . . the deeper parts of reefs may fare better."

Climate models support this grim snapshot of the future. The models suggest that if CO_2 emissions stay as they are, average water temperature in the GBR could increase by another 3.6°F to 5.4°F by 2100.[7] And studies have suggested that these increasing baseline temperatures, combined with the likelihood of more extreme weather events, like heat waves and flooding, could result in mass bleaching events every two to three years. Recent modeling studies indicate that if atmospheric CO_2 levels hit 600 ppm, it will be very tough to save the corals. By 650 ppm, it will be impossible to save them.

Still, Kleypas thinks the overall outlook for impacts associated with increasing temperature is a little rosier than the outlook for impacts associated with OA. "We know corals can handle high temperature. We see them in the Red Sea and the Arabian Gulf. Corals can get used to warmer water, the same way people living in Arizona and Phoenix don't suffer as much from heat stress." In places like the Red Sea and the Arabian Gulf, corals don't bleach until they reach temperatures about 18°F higher than their summer maxima, a much higher threshold than for similar species located in cooler regions. But there is one problem: the projected rate and magnitude of temperature increase will quickly outpace the conditions under which coral reefs have adapted and flourished during the past 500,000 years. Experts are worried that corals won't be able to adapt fast enough to keep pace with even the most conservative projections for climate change. "The problem is," Kleypas adds, "it takes time for corals to get used to the increased temperature." And, unfortunately, time is not on their side.

With regard to finding an approach to help the coral adapt, coral bleaching reveals itself as the kind of problem where traditional management approaches that focus on minimizing or eliminating sources of stress don't help much. The ocean is not like the heated pool at a hotel or motel: coral reef managers are constrained by a frustrating inability to directly turn down the ocean temperature when their reefs start to overheat. Needless to say, when you can't control the single most significant factor affecting a bleaching event, you're dealing with a challenging environmental management problem. To address the problem, in 2006 the Great Barrier Reef Marine Park Authority published *A Reef Manager's Guide to Coral Bleaching.*[8]

Resilient reefs seem to share a few important qualities. One is location: they are located in a zone of cooler water. Some sites may have consistently cooler water because of upwelling or proximity to deep water. A second quality is shade: some reefs may be protected from bleaching because their exposure to the sun is limited by topographic or bathymetric features. Reefs shaded by cliffs or mountainous shorelines may also have a reduced risk of bleaching. Many reef areas are unlikely to have features that can provide shade, but fringing reef complexes around steep-sided limestone or volcanic islands, as in Palau and the Philippines, may have many shaded sites. A third quality is screening. Naturally turbid conditions may filter or screen sunlight, providing a measure of protection for corals exposed to anomalously warm water. Ongoing research suggests that organic matter in turbid areas may absorb ultraviolet (UV) wavelengths and screen sunlight. Corals at these sites may be less susceptible to bleaching.

The goal is to establish a network of marine protected areas (MPAs). If you can identify MPAs or reef areas that are likely to be more resistant to mass bleaching, then these are the places that have the best shot at survival in a warmer world. And if scientists can help set up a network of resilient coral reef refuges, then they can draw on these like a garden to reseed coral reefs that have been hurt by bleaching. In the context of mass coral bleaching, these refuges can serve as seed banks or source reefs for less resilient areas. But if the special reef refuges are to serve this role, they need to be effectively monitored and protected from local stressors such as anchor damage, overfishing, and pollution.

One such spot that scientists and conservationists are working hard to protect is an area called the Coral Triangle, which spans eastern Indonesia, parts of Malaysia, the Philippines, Papua New Guinea, Timor Leste, and the Solomon Islands and contains 53 percent of the world's corals. It's often compared to the Amazon rain forest of Brazil because it has such a high level of biodiversity. The Coral Triangle covers an area of 2.3 million square miles, about half the size of the United States. It has more than 600 reef-building coral species—75 percent of all species known to scientists—and more than 3,000 species of reef fish. It also has the greatest extent of mangrove forest of any region in the world. Both the mangroves and the coral reefs serve to protect fragile coastlines from damage by storms and tsunamis. Scientists feel that if they can protect this reef system, it will help them save other reef systems.

More than 120 million people live within the Coral Triangle, and about 2.25 million depend on its marine resources for their livelihood. Between the tuna fisheries and the tourism, the estimated total annual value of the coral reefs is $2.3 billion.

The MPAs are carefully selected areas where human development and exploitation of natural resources are regulated to protect species and habitats. By providing refuges for exploited fish stocks, MPAs provide benefits for commercial fisheries as well. Healthy fish stocks in MPAs replenish surrounding fishing grounds with eggs, larvae, and adult fish. Right now only 1 percent of the ocean is protected, compared with about 12 percent of the land. The question is whether the existing MPAs are in the right places and where we should put the next ones. Kleypas explains that finer-resolution climate models can help scientists select the optimal locations for MPAs. Implementing this principle in MPA design involves considering prevailing currents and adjacent non-reef areas. Linking MPAs along prevailing currents that carry larvae can replenish downstream reefs, increasing the probability of recovery at multiple coral reef sites. Adjacent non-reef areas are important to connectivity because they can become important staging areas for coral recruits as they move between reefs and into new areas.

"I call this high-CO_2 window the Noah's ark period. We have to save as many species as we can," Kleypas says. "We also need to help make the reefs more resilient. The reefs that are in the best shape today are the reefs with the best management practices," she adds. "I'd like to see some advances in coral reef restoration and coral farming." In essence, this involves managing the ocean more as we do the land. It's interesting to imagine coral reef farmers growing and tending to baby coral reefs. And this may be the best hope we have.

Another important part of management is monitoring the reefs using satellites. Coral Reef Watch, a program of the U.S. National Oceanic and Atmospheric Administration (NOAA), has developed tools to analyze satellite images and help reef managers assess the likelihood of mass coral bleaching events. It's a little like a weather forecast for your coral reef, and it includes maps and indexes that track how warm conditions are getting in the tropics.

The maps use satellite data to show the intensity and duration of spikes in sea surface temperature. If you can monitor the intensity and duration of heat stress, you can get a sense of where a mass bleaching event might occur and how bad it might be. Both the intensity and the duration of heat stress are important factors in predicting the onset and severity of a mass bleaching event. The monitoring tool tracks the anomalous sea surface temperature, the difference between the observed ocean temperature and the highest temperature expected for a specific location, based on long-term monthly averages. It provides a useful reference point that shows the extent to which current temperatures vary from those that the corals are accustomed to experiencing at that time of year.

Temperature anomalies of 2°F to 4°F extending over a period of several days to several weeks should alert managers that there is a medium to high risk of bleaching. The NOAA Coral Reef Watch program also developed Tropical Ocean Coral Bleaching Indices to provide additional nearly real-time information for twenty-four reef locations worldwide.[9] For each reef site, the closest 30-mile satellite data are extracted, including present sea surface temperature, degree heating weeks, climatology, and surface winds. Visual warnings are provided for each site when conditions reach levels known to trigger bleaching in vulnerable coral species. It's a bit like the way forest rangers track the potential for wildfires on land.

There are also monitoring programs that encourage laypeople to serve as scientists. The sheer size and remoteness of many reef areas can be a substantial challenge for reef managers wishing to detect the onset of bleaching and monitor bleaching-related impacts. Reef users can help managers keep an eye on the reef during periods of high risk. A program in the GRB called BleachWatch engages people who love the reefs by teaching them how to help monitor coral bleaching events. BleachWatch provides an early warning system for coral bleaching and forms part of the Coral Bleaching Response Plan of the Great Barrier Reef Marine Park Authority's (GBRMPA). The program is aimed at tour guides and allows them to go about their everyday work, be it guiding snorkel trails or diving, while taking a mental picture of their "home reef." Back on the vessel, staff members fill in the monitoring form and send it back to the GBRMPA at no postage cost. In return for the monitors' efforts, the GBRMPA analyzes the information and provides monthly site reports.

Experts like Kleypas also hope that high-resolution climate models might help in planning for the Noah's ark period. The models can be used to fast-forward in time and get a better sense of what the reefs might look like as temperatures go up and pH goes down. The models might also be able to serve as a tool that allows a better understanding of which species are going to make it and which are more vulnerable. The species that are most resilient and most likely to survive can potentially be used to reseed reefs that have suffered from bleaching. Kleypas says that right now, most models look only at the surface ocean temperature. But when you begin to study the ocean at depth, the models could help identify places where the reefs are likely to survive. The models could actually help managers target specific areas to protect—areas like the Coral Triangle that will probably be used to help rebuild the reefs that are having a harder time.

Despite all these efforts, some people remain worried that corals simply will not have the resilience or the adaptive strength required to get past this high-CO_2 window. Such people are calling for more dramatic measures. Some have recommended setting up an underwater repository of corals similar to the Svalbard Seed Bank, a cave on the Norwegian island of Spitsbergen that (as its name implies) preserves thousands of plant seeds from around the world. Svalbard is, in a sense, an underground "doomsday vault" built to serve as the ultimate safety net for the world's seed collections, protecting them from a wide range of threats. In a repository, the corals would be saved from rising temperatures and OA, but Kleypas sees this as a last resort.

With all these tools and programs coming online, Kleypas is trying very hard to stay positive. As a scientist she is pragmatic, but as someone who is passionate about coral reefs, she conveys an uncommon sense of hope. "Scientists are very introverted people by nature. We don't tend to be inspirational. We make future predictions based on the here and now. But I've been trying to give people hope. I hate to give a doomsday lecture and tell folks the reefs are all going to hell. People don't know what to do with that information."

On the other hand, she doesn't want to come across as a Pollyanna. "We've already entered into this window of high CO_2. So, we have to aim for mitigation. We can't just stabilize emissions. We need to then get CO_2 levels down. I like to tell people we don't know how high the CO_2 is going to be, because that level is fundamentally up to us."

Kleypas is not one to shy away from the possibility of genetic engineering. "Scientists are thinking about the symbiotic algae," she explains. "The question is: can we seed a reef with algae that are more resilient to temperature changes? One theory that has emerged is that the amount of bleaching that occurs at each reef may be influenced to some extent by the prevalence of stress-tolerant algae. And so scientists have begun surveying corals for the presence of stress-tolerant zooxanthellae within reefs. There's still so much to learn about these symbiotic algae, but identifying which algae are most stress-tolerant may help managers to assess the potential resilience of different sites. We know how to engineer resilience in terrestrial species, but we know much less how to do it in marine ecosystems. And we need to figure it out," she says with a sense of urgency. "This is one of the most remarkable times to be a scientist. Sure, we can just sit back and watch it happen and confirm that our predictions are coming true. But that would be embarrassing." I suppose it's fair to say that we've reached a point where we need to make our own luck.

The Great Barrier Reef, Australia: The Forty-Year Forecast—Coral Bleaching, Ocean Acidification, and Economic Struggle

<table>
<tr><th>TOWNSVILLE, AUSTRALIA</th><th colspan="2">TODAY</th><th colspan="2">2050</th><th colspan="2">2090</th></tr>
<tr><th>Emissions Scenario</th><th>JAN.</th><th>JULY</th><th>JAN.</th><th>JULY</th><th>JAN.</th><th>JULY</th></tr>
<tr><td>Higher</td><td rowspan="2">81.4</td><td rowspan="2">65.3</td><td>83.8</td><td>68.1</td><td>87</td><td>71.1</td></tr>
<tr><td>Lower</td><td>83.3</td><td>67.4</td><td>84.5</td><td>68.5</td></tr>
</table>

Forecast March 2017

For several months, squadrons of scuba divers from all over the world had been heading out into the unusually warm waters off the coast of Townsville, tanks of air strapped to their backs and monitoring checklists dangling from their wrists. The divers were roaming the seas in search of bleached corals—a terrible job for anyone who loves the reefs. El Niño set off a worldwide coral bleaching event affecting hundreds to thousands of miles of reefs simultaneously. This El Niño came when ocean temperatures were already warmer than average and caused severe to extreme bleaching even along the very carefully managed and monitored GBR—with the result that more than half of the colonies turned completely white. The divers went out each morning to identify sick corals and came back each evening hoping that park officials wouldn't need to come up with a catastrophic level for coral bleaching, too.

The Coral Reef Watch program set up by NOAA had been monitoring ocean surface temperature using satellites and was able to provide scientists and volunteers with almost up-to-the-minute information. Temperatures along the GBR ranged from 80°F to 84°F, and bleaching was widespread. Reports from volunteer groups, such as BleachWatch and Reef Check, warned that large sections of coral were bleaching at levels much higher than those seen in 1998, when waters heated by El Niño killed 15 percent of reefs worldwide. In the GBR alone, reef damage related to bleaching had caused losses to the tourism industry on the order of $250 million; and we had felt sick to think that when all was said and done, we might lose more than one-third of these beautiful reefs.

It was hoped that local solutions—for example, marine sanctuaries and volunteer monitoring efforts—might create lasting changes. But it's tough to manage global warming locally. And in the end, the root cause of bleaching, warmer ocean temperatures, could be addressed only by a worldwide effort to reduce CO_2 emissions. Scientists had hoped that the more remote reefs, safely isolated from human impact, might fare better. They were wrong. The bleaching event caused extensive mortality in nearly every coral reef region in the world. No man is an island—and no reef is safe from the long arm of climate change.

December 2019

As climate change continued to accelerate, the words *severe* and *extreme* were no longer adequate to describe the destructive potential of Australia's wildfires. That's why the residents of South Australia awoke on the first morning in December 2019 to face yet another sustained warning of catastrophic fire danger. The category *catastrophic* had been put in place by the Australian Bureau of Meteorology (BOM) in 2009, after the horrific wildfire called Black Saturday radically altered everyone's definition of how bad a wildfire could get.[10] The wildfires came on Saturday, February 7, 2009; and by the time the sun rose the next morning, they had killed 173 Australians and traumatized countless others. According to a report by the Victorian Bushfires Royal Commission,[11] the fires generated winds so strong "that trees appeared to have been screwed from the ground." At the time, no one dared imagine what Australia might look like if Black Saturday had gone on for a week. We found out in 2019. That catastrophic outbreak came to be known as Black December.

Fire marshals begged residents of the eastern Eyre Peninsula and the west coast districts in the state of South Australia to evacuate their homes immediately. If they had learned anything from Black Saturday—when many residents stayed on, hoping to defend their property but in the end losing their lives—it was that no one should attempt to be a hero. Fire authorities, however, had no official mandate, so they could not force people out of their homes; they could only beg the residents to leave.

It had been a brutal few months. In October, an ongoing drought had kicked up a thick wall of red dust that reduced visibility in Sydney to less than two city blocks. Snapshots taken of the Sydney Opera House—silent, ghostlike, and shrouded by a thick red veil—made their way around the globe, and the world looked on in fear and fascination. The ferocious wildfires, pervasive drought, and unbreathable air made Australia seem like hell on Earth. And for those who lived there it was.

In November, intense heat pushed the average temperature for the month into numbers never seen before. Towns throughout Victoria and southeast Australia were running 2°F to 4°F above the previous record, set in 2009. Melbourne was running 2°F above the record set during the previous November. The city's chief meteorologist summed it up: "Usually when you break records like these you break them by a tenth of a degree. But we're seeing we're two, three, or even four degrees above previous records. This is not natural." And by December, the heat and drought—together with low humidity and strong winds, created the perfect conditions for a catastrophic wildfire. Happy New Year, Australia.

January 2025

After the widespread bleaching event of 2017, the idea of setting up a coral bank began to gain traction. The Svalbard Global Seed Bank had proved to be successful. Why not try something similar with corals?

The Smithsonian Institution in Washington, D.C., finally received funding to set up the Smithsonian Global Coral Vault. Corals from tropical oceans were being placed in deep freeze at the Smithsonian to preserve them for posterity as they faced destruction from rising greenhouse gas levels. This *coral cryobank* would ultimately house hundreds of samples from each species. The funding came after new research suggested that most coral reefs would be largely dead by 2040, wiped out by a combination of rising temperatures and increasing acidity in the world's oceans. The affected areas included Australia's 1,600-mile GBR, Caribbean reefs, and reefs in the Coral Triangle—an area spanning Indonesia, the Philippines, Malaysia, Papua New Guinea, and East Timor. Carbon dioxide emissions had risen above the safe level for corals, and reefs around the world were showing the impact. The Smithsonian's vault was a matter of reverting to plan B. And its very desperation reflected the despair among scientists about rising CO_2 levels.

The coral vault applied a breakthrough deep-freeze technique developed by scientists to regenerate coral from frozen samples. The scientists took tiny biopsies from coral, froze them in liquid nitrogen at –330°F, and then thawed them to regenerate polyps. These scientists were proposing to do the same for every species of coral on the planet. There are about 1,800 known tropical corals and another 3,350 cold-water species. The Smithsonian would house about 1,000 samples of each coral in a large room in a subbasement of its museum in Washington, D.C. The facility was nicknamed the "Morgue of the Sea."

December 2050

The overall acidity of the ocean continued to increase. Corals reached the point where they were dissolving more quickly than they were growing. Consequently, many coral reefs were unsustainable. It was expected that pH levels would continue falling. By 2100, climate models forecast a further drop in pH of 0.3 to 0.5 unit—which would make the world's oceans more acidic than they had been in tens of millions of years. And while there might be thousands of coral polyps sitting in deep freeze at the Smithsonian, there was no ocean on the planet Earth that they could now call home.

7

CENTRAL VALLEY, CALIFORNIA

At the end of the movie *Pretty Woman*, after Richard Gere has climbed up a fire escape and rescued Julia Roberts from the drudgery of the real world—we hear a baritone voice declare, "This is Hollywood, the land of dreams. Some dreams come true, some don't. But keep on dreamin'." It's a classic Hollywood ending. Everyone is happy and full of hope for the future.

Hollywood may serve as the unofficial capital of dreams, but it's certainly not the only place in the Golden State that lies within the realm of unreality. The Central Valley, where the Sacramento–San Joaquin Delta is located, represents a very different kind of dream for the future, a kind that many scientists have come to see as sheer fantasy.

In the Sacramento–San Joaquin Delta, the Sacramento and San Joaquin rivers converge into canals, levees, streambeds, marshes, and peat islands. With an area that spans about 24 miles east to west and 48 miles north to south, the Delta is the hub of California's water supply system. The entire state—especially the rapidly growing and increasingly dry metropolitan areas, such as Los Angeles and San Diego to the south—depends on this very small area of land for its water. The dream is that the Delta will be able to supply enough clean, fresh water to help cities and crops increase forever, all without harm to the natural environment.

SACRAMENTO
SACRAMENTO
San Francisco Bay
SACRAMENTO RIVER
Suisun Marsh
SAN JOAQUIN RIVER

If this sounds too good to be true, that's because it is. Those cities now stretch from San Francisco and Silicon Valley in the north to Los Angeles and Orange County in the south. And the crops, including alfalfa and corn, grow on a version of *Fantasy Island.* Many islands in the Delta are 15 to 20 feet below sea level and are protected only by increasingly fragile levees. The Delta was a good dream, a very American dream. And it even came true, for a while.

But today, the Delta has become an example of how complicated and costly it can be to sustain a dream, especially when global warming is a factor. California's gold rush attracted a new generation of Americans who came west and saw endless possibilities. But endless possibilities and unbounded growth require infrastructure. And infrastructure comes at a cost, both for the economy and for the environment.

I recently spoke with an economist and an environmental engineer about the future of the Sacramento–San Joaquin Delta. Ellen Hanak, director of research at the Public Policy Institute of California (PPIC), and Jay Lund, an environmental engineering professor at the University of California-Davis, are part of a multidisciplinary team that includes biologists, economists, engineers, and a geologist. The team members are studying the Delta and attempting to salvage what remains of the dream. I spoke with Hanak and Lund about climate change and about two reports they had recently cowritten. The first report recognized that the Delta is in a crisis. The second report looked at ways to manage the Delta's water while protecting the Delta ecosystem. Management strategies that attempt to satisfy the competing interests of the Delta's economy and its environment have been discussed and debated for almost 100 years, and sometimes California's version of the battle between the North and the South has resulted.

Engineers are taught to find optimal solutions to problems. As applied to the Sacramento–San Joaquin Delta, *optimal* means finding a balance between environmental and economic interests, and between agricultural and urban users of water. *Optimal* doesn't mean perfect; it does mean that no one wins entirely. Fish will die and agricultural yields will decrease, but *optimal* means that the Delta will have a future. It also means writing a big check up front and making profound changes to the Delta landscape before Mother Nature or the climate makes its own changes for you.

"I'm not sure how it's going to play out. A lot of people are working to make sure nothing happens in the Delta, or that things only happen their way," says Lund.

The Central Valley stretches approximately 400 miles from north to south and is roughly the size of the state of Tennessee. Today, about 6.5 million people live in the Central Valley, which is considered the fastest-growing region in California. All the scientists I talked to when I was selecting the locations to write about in this book considered it a hot spot with regard to global warming.

The northern half is the Sacramento Valley, and the southern half is the San Joaquin Valley. The two halves meet at the Delta. California's capital, Sacramento, is located along the Sacramento River's banks. The rivers and their tributaries are harnessed by more than 100 large dams that produce the majority of California's hydropower. The Delta is part Frankenstein's monster, part natural wonder.

The Delta provides water for two out of three Californians, and for almost 4 million acres of farmland. At the Delta's western edge lies Suisun Marsh, and at its southern end are two prominent examples of California's water infrastructure: the Delta-Mendota Canal and the California Aqueduct. The canal and the aqueduct deliver water from upstream reservoirs and the melting mountain snowpack to cities and farms in every direction. The Delta is the hub of the state's water supply because it serves as the transit point for water. Whenever the Delta shows signs of its original personality and moves from being a predictable freshwater conveyance system toward being a vast tidal marsh, every effort is made to push it back into place, by repairing levees, releasing water from reservoirs, or reducing water exports. Needless to say, a lot of dreams rest on the earthen walls of the suboptimal canals and levees.

And that's the trouble with the Delta. Scientists say the canals and levees have become increasingly vulnerable to a catastrophic failure, whether it arrives abruptly in the form of an earthquake or slowly as the result of a rising sea level caused, in turn, by global warming. In any event, the scientists are nearly unanimous: the Delta is unsustainable.

At the start of the gold rush in the late 1840s, the levees provided a simple way to lock down the landscape and get more value out of the land. Lund, an expert on water resources in California, says, "If you look at the geological history of the Delta there was always a lot of variability." Native species, such as the delta smelt, came to depend on that variability. But there are now 1,300 miles of levees in the Delta and Suisun Marsh—they form a longer stretch than the entire California coastline—and the Delta's natural variability has been kept to an absolute minimum.

"The Delta is the direct result of rising sea level since the end of the last ice age," Lund explains. "During the last ice age, the Delta was to the west of the Golden Gate Bridge. But starting about 10,000 years ago, as sea level rose, the Delta moved inland. About 6,000 years ago, the Delta arrived at its current location, where, for the most part, it was able to keep up with sea level rise by building marshes. Sediment accumulation in the Delta kept up with the slow rise by forming thick deposits of peat." Peat is made of organic matter: decaying plants and animals that only partially decompose. This dead matter can't get enough oxygen to break down completely, because everything is waterlogged. "It took 6,000 years for that peat deposit to build, as one layer of new plant material grew on top of previous layers of peat," Lund says. Through this gradual process of flooding and rebuilding, a diverse, resilient ecosystem evolved. Then came the gold rush.

It was actually during the California gold rush that farmers stumbled on the Delta and struck their own kind of gold. The peat in the Delta was capable of producing excellent crops. But to farm the organic-rich soils, farmers first needed to drain the islands. After 6,000 years of continual flooding and rebuilding, the Delta was, for the first time, being pinned down. "This involved constructing levees around the islands, filling most tidal channels, and, most important, lowering local groundwater tables below crop root zones by constructing perimeter drains," Lund explains. "When you're located at the confluence of two major rivers, the dry period is when you want to grow your crops. And if you can keep the soil moist but not waterlogged all year long, then you've really struck gold."

The Delta was a perfect spot to settle and farm. Its proximity provided easy access to the miners and markets, and its soil was beyond comparison. "And so these natural levees were formalized and the islands were dried out," Lund explains. Today, the Delta grows more than 90 different crops, including wine grapes, pears, rice, corn, and tomatoes, producing more than $360 million annually in farm sales. But altering the landscape came at a price that has yet to be paid.

"The problem," Lund says, "is that peat soils are meant to be waterlogged. When they remain dry for long periods of time, they lose their integrity." So, every year you farm, you lose some soil. "Today, we're farming on borrowed time," Lund explains. "We're mining the soil until the islands fail." In just 150 years, about 6,000 years' worth of peat has been eroded. It's gone. By engineering the variability out of the system, we've attempted to pin something down that cannot be pinned down.

Pinning the Delta down also pushed it down, and so it is that much more vulnerable to earthquakes, flooding, and a rising sea level. When the levees were first constructed, no regulatory policies forced people to consider their impact on the Delta ecosystem. Farmers were doing their own engineering, and they were optimizing around just one quantity—agricultural yield. Hanak, the economist, explains, "Many of the water diversions upstream and within the Delta were made before the public demanded environmental protection."

Levees built 100 years ago confined water to channels and transformed the Delta from marshland into dry islands of land available for human use. The Delta islands started sinking when the marshlands, the source of all the fertile peat soil, were first drained. And the sinking—the scientific term is *subsidence*—continues to this day. There are now seventy-four islands in the Delta. Most of them are below sea level, many by more than 20 feet. Subsidence also increases seepage into the islands, raises the likelihood of levee failures, and increases the costs and consequences of catastrophic island flooding. Take your pick: earthquakes, floods, a rising sea level, subsidence, and urbanization all contribute to the increasing likelihood of multiple levee failures. Scientists, not known for hyperbole, describe this as a catastrophic failure.

The Delta has far more in common with New Orleans than with Hollywood. You could argue that when the Delta experienced a major levee break in June 2004, cracks in the dream were also beginning to appear. A year later, when the devastating effects of Katrina bore down on the old, inadequate levee system in New Orleans, Hanak, Lund, and their colleagues saw an indication of the future of the Delta. However, they saw an earthquake, not a hurricane.

As Lund, Hanak, and other scientists who study the Delta watched Hurricane Katrina bear down on New Orleans, they felt compelled to prevent something analogous from happening in their backyard. Scientists had long warned of the fate that awaited New Orleans if its infrastructure was not improved to prepare the city for a major hurricane. But political apathy, and perhaps human nature itself, prevented these scientists from making much headway.

Approximately 2 million people in the Central Valley count on levees for flood protection. And the capital city, Sacramento, which is among the fastest-growing cities in the United States, is the major metropolitan area at the highest risk of flooding. The problem is that Sacramento's infrastructure is inadequate; it doesn't meet even the minimal federal standards. Conservative estimates of potential flood damage to the Sacramento area alone exceed $25 billion. Sacramento is, of course, just one city among many in the Central Valley. Actually, Sacramento is well upstream of the Delta, and its land is above sea level, so failure in the Delta is unlikely to flood major population centers. Nevertheless, if the Delta goes, it can take a lot down with it. In addition to flooding large areas of lower-value agricultural land, it would cripple the delivery of water to the San Francisco Bay Area, Southern California, and the San Joaquin Valley.

So the Delta is now a big fishbowl—a fishbowl not very good for the native fish. The Delta is the habitat of more than fifty species of fish, including 75 percent of the state's commercial salmon catch. Today, the Delta supports what scientists kindly refer to as a *highly modified ecosystem*. Hanak explains, "We've engineered this system to the point where it's a lot more vulnerable. The invasive species like the artificial things we've created. But we're legally bound to protect the native species that aren't adapted to the new Delta we've created."

The Delta today is, in fact, a shadow of its former self. It resembles the Delta of the past only in that some of the original species, such as the delta smelt and chinook salmon, are still present. Invasive species, both plants and fish, now dominate the Delta's riprapped channels and islands; native species, including the delta smelt, the longfin smelt, and salmon, to put it mildly, are struggling.

The recent sharp decrease in the population of several prominent Delta fish species was a red flag for conservationists, many of whom believe that the entire Delta ecosystem is on the verge of collapse. Two species have already gone extinct in the Delta: the Sacramento perch, which needs to be reintroduced; and the thicktail chub, which has been globally extinct since 1957. Six Delta fish species are heading toward extinction: the southern green sturgeon, the longfin smelt, the delta smelt, the winter run chinook, the spring run chinook, and the Central Valley steelhead. Two species are in decline: the splittail and the late fall chinook.

With regard to global warming, every place has its own counterpart of the canary in the coal mine. In the Central Valley, the canary is most likely the tiny delta smelt, a translucent fish about the size of a human finger. In the fall of 2004, routine fish surveys registered sharp declines in the delta smelt, and it was listed as *threatened* under the Endangered Species Act. The number of delta smelt found in 2008 was the lowest in forty-two years of surveys. Federal scientists say the delta smelt is on the brink of extinction; some biologists conclude that it may be gone by 2010.

When scientists say the Delta's native ecosystem is collapsing, they mean it. A major factor contributing to the collapse is the pumps. These pumps are extremely powerful and can kill fish that get stuck inside them. In addition, the reduction of total outflow to the ocean and the disruption of natural flow patterns in what is now a complex network of channels in the Delta aren't helping matters. The situation has gotten so bad that in December 2007, U.S. District Judge Oliver Wanger imposed pumping restrictions to protect the delta smelt. The courts imposed 30 percent reductions on the amount of water that can be exported from the Delta by state and federal water projects. This means that even if water is available, it may not be delivered. On March 5, 2009, the state's Fish and Game Commission unanimously voted to list the longfin smelt, a relative of the delta smelt, as a threatened species under the California Endangered Species Act. The commission also voted to classify the delta smelt as an endangered species. (It had been listed as threatened since 1993.) The ruling to protect the environment has disrupted the water supply of much of the state.

The scientists concerned with the Delta hoped to do better than those who had been concerned with Katrina, by giving people a vision of what lay ahead. The title of their first report was *Envisioning Futures for the Sacramento–San Joaquin Delta*.[1]

"Back in the summer of 2005, I got in touch with these guys and we all agreed that someone needed to start looking at the Delta. It was a problem that required a multidisciplinary approach. There are so many moving parts," Hanak explains. "It was clear the Delta required more sophisticated long-term planning. And after years of neglect, the Delta ecosystem was showing signs that it had become unhinged. The crash of some Delta fish species between 2004 and 2005 helped focus attention."

Hanak says it was important to combine an economic perspective with an engineering perspective. "The way we try to look at it is that there are two primary objectives: water for humans and a healthy ecosystem for fish populations. If you do those two things . . . you might make the Delta more sustainable," Hanak says. "We wanted to give people a look at possible futures. We thought that it would help people to think long-term. We can't just tinker with the status quo."

Ask any scientists who study the Delta and they will tell you that it is likely to change significantly and abruptly within the next generation. A sudden catastrophic change would be a very hard landing indeed for those depending on the Delta. When a system pinned for more than a century swings loose, it's going to take a lot down with it. The scientists feel that their job is to help people see what a catastrophic failure would look like and then provide options to prevent the worst-case scenario.

The scientists might not be worried about a hurricane in the Delta, but they've got plenty of other scenarios that would result in a catastrophic failure. An earthquake and a flood are the two most likely scenarios, and both could take the levees out fairly quickly. According to recent calculations, the odds are roughly two in three that during the next fifty years either a large flood or a seismic event will affect the Delta. And the scientists say that these odds are a conservative estimate, for two obvious reasons. First, strain continues to accumulate on faults in the Bay Area, increasing the risk of seismic activity with each passing year. Second, the 100-year flood isn't what it used to be, thanks to global warming. The estimate of what a 100-year flood event in the Delta looks like is based on outdated hydrology data that don't adequately address the impact of climate change on the Delta. Scientists have already established that in recent years, climate change is causing much higher inflows from rivers feeding into the Delta.

When I asked Lund if it was fair to say that a catastrophic levee failure of some kind was guaranteed by the end of the century if nothing was done, he answered as any good engineer would: "Well, you can't give it a 100 percent probability. But I might put it at 99 percent."

Aside from the cause, the scientists see a lot of the same issues in the Delta that their counterparts had warned about with regard to Katrina. Lund and Hanak envision a levee failure that would directly threaten water supplies and affect thousands of roads, bridges, homes, and businesses at the same time.

An earthquake is the quickest way to take down the Delta, and even though earthquakes are geologic—not climate-based—events, the changing of the landscape through global warming will determine its resilience. It's what scientists call an unavoidable threat, and it's been on their radar for more than thirty years. There are at least five major faults close to the Delta, and these are capable of producing significant ground accelerations. The soils are poor enough and the levees weak enough that risk of failure due to liquefaction and settling is high. You can do a lot of seismic risk studies in thirty years, and all of them indicate a very high potential for major earthquakes in the region sometime in the near future. If there is a major earthquake—similar to the 1906 event in San

Francisco, which measured 8.25 on the Richter scale—many Delta island levees will fail simultaneously. And with each inch of rise in the sea level, the cost of such a failure increases significantly.

Even in a scenario involving a moderate-magnitude (6.0) earthquake, the seismic risk studies show a potential for multiple levee failures. The highest risk is in the western Delta, which is very close to several significant seismic sources and is already characterized by deep subsidence and poor foundations. There is a medium to high risk of catastrophic levee failures for almost all the central Delta as well. Scientists working for the Department of Water Resources (DWR) recently modeled the consequences of a catastrophic levee failure caused by a large earthquake. In one of the scenarios, the earthquake took out thirty levees, flooded sixteen islands, and cut off water exports for several months.[2]

But what the scientists fear most is something called the "Big Gulp." The name itself sums up the scenario. If the levees break, salt water from San Francisco Bay will come rushing in, proving that nature abhors a vacuum. Lund does a quick back-of-the-envelope calculation: "It would take as little as twelve hours for the salt water to begin intruding into the Delta."

In another earthquake scenario, scientists simulated a magnitude-6.5 earthquake that takes down twenty islands. This scenario shows a big tongue of salt water creeping in from the bay. Within thirty days, the Delta would be transformed into a saline estuary. The extra salt in the water would wreak havoc on the millions of people and the millions of acres of farmland that depend on Delta waters. Scientists estimate that a catastrophic failure of key levees would cost, in total, somewhere between $8 billion and $15 billion.

Even without catastrophic failure, rising sea levels will bring more salt into the Delta and significantly raise the cost of water treatment; raise the public health risks to the millions of Californians who rely on the Delta as a source of drinking water; and reduce the productivity of farms, which would be irrigated with increasingly salty water. An intrusion of salinity can be delayed for a time by releasing more freshwater into the Delta, as Lund explained, but it cannot be delayed indefinitely.

With regard to an earthquake, there is also the question of when. "If the earthquake happened during the summer when things are dry and water levels are low, it would get salty a whole lot more quickly," says Hanak. "All these islands are bowls and all of them would be filled with saltwater. If the earthquake happened during winter there would be more freshwater in the Delta to start with, so the Big Gulp would be a lot smaller." Timing, as they say, is everything.

Unfortunately, global warming is also a matter of timing. Global warming is insidious in that it discreetly fiddles with the timing of the Sierra Nevada snowpack. This snowpack is the true basis of California's water system. It's the state's largest surface reservoir; even though the water is in solid rather than liquid form for several months of the year. Snowmelt currently provides an annual average of 15 million acre-feet of water, slowly released between April and July each year. An acre-foot, as the term implies, is the amount of water needed to cover 1 acre 1 foot deep. More significantly, 1 acre-foot is 325,851 gallons, or about enough water to supply one to two California families for one year or to irrigate from one-eighth to one-third of California farmland, depending on the crop.

Mountain snowpack is like money in the bank. Much of California's water infrastructure was designed to capture the slow spring runoff and deliver it during the drier summer and fall months. However, warmer temperatures would cause the snow to melt earlier, thereby reducing summer supplies. Using a combination of historical data and climate models, the DWR projects that the Sierra snowpack will experience a 25 to 40 percent reduction from its historic average by 2050, thanks to global warming.

Over the past 150 years, monitored mountain glaciers have been shrinking. And with the earlier melting disrupting the timing, there are now more floods during the winter and worse droughts during the summer. A flood of salt water can of course do enormous damage, but a flood of freshwater isn't much better. Over the last fifty years, there has been a shift toward less snow and more rain in the Sierra Nevada. Climate change is also expected to result in warmer storms that bring less snowfall at lower elevations; this is more bad news for the snowpack and means, to use the metaphor above, less money in the bank for the state of California. These shifts have increased winter inflows to the Delta. Climate models indicate that this trend will continue, with even larger and more frequent floods in the future.

Whereas you might expect the melting to result in a short-term increase in the amount of water available in the "bank," the disrupted cycle is actually more likely to cause excess runoff, bringing flooding and an overflow of reservoirs not equipped to contain such large inflows of water. Steve Schneider, a climatologist at Stanford University, puts it this way: "Water managers have this horrible choice. You can leave the reservoirs full and hope you don't get a heat wave in February or March that melts snowpack early, floods productive farmland, and inundates surrounding communities."

This is exactly the kind of scenario that scientists expect. A week of unusually high temperatures leads to an exceptionally early snowmelt, which causes a big pulse of meltwater at a time when reservoirs are already full. If that week of heat is combined with a storm that brings heavy rainfall, significant flooding is likely.

Schneider continues with his explanation: "The other option is to play it safe and release water from the reservoirs, bringing the level down. Then you're praying for rain because you've got nothing to fight fires." Schneider adds, "Either way you're praying a lot." This, Schneider says, is why Californians have become deeply involved in climate legislation. "They don't want to deal with the increased risks of droughts and floods," he says, "Ideally, they'd be happier with a whole lot less climate change."

But California has already seen its share of climate change. During the last fifty years, winter and spring temperatures have been warmer, snowpack has been melting one to four weeks earlier, and flowers are blooming one to two weeks earlier. One of the fundamental problems with life on Earth is that natural resources, such as clean water for cities and for growing food, is not evenly distributed in space and time. As a result, we've tried to engineer the system to be more evenly distributed. Scientists even out the distribution by looking at past variability and building infrastructure to smooth away the bumps.

But global warming is working against us. So far, California is still betting that a system built using historical hydrological data can protect the future water supply and provide sufficient flood protection. But in the meantime, climate change has already redefined the hydrology of the Delta, making it quite clear that our past experience is not enough to serve as a guide. To think about it any differently would be dreaming.

A rising sea level makes the situation worse. If San Francisco Bay serves as the source of the Big Gulp, then a rise in sea level is the Little Gulp. During the past century, the sea level along California's coast has risen about 7 inches. It is projected to rise an additional 12 to 55 inches—or possibly even more, if large ice sheets melt—by the end of the century. Without immense investments to raise the Delta's levees, this rise in sea level will cause many levees to fail, pushing seawater into the Delta. A rising sea level can also increase the rate of saltwater intrusion into coastal aquifers; such intrusions would contaminate freshwater supplies. Even if the levees could be sustained, a higher sea level will increase the salinity of Delta waters. And the higher tide that sweeps in with a rising sea level would pose additional problems.

A higher sea level will make pumping water through the Delta increasingly unattractive and eventually infeasible. Even if the existing levee network could be maintained through unprecedented investments, the worsening water quality resulting from the rise in sea level would steadily reduce the economic value of water exports from within the Delta. The current costs of salinity in the Delta are already significant for agriculture and urban drinking water treatment in the southern Central Valley. More saline exports from the Delta will reduce the viability of agriculture in this region and increase costs of and health risks from drinking water from the Delta. With a continued rise in sea level, the volume of required outflows would continue to increase. Climate models suggest that by mid-century, the increased salinity of Delta waters will impose water treatment costs of about $300 million to $1 billion per year, every year.

Higher salinity will impose a direct water supply cost by requiring higher outflows to keep seawater away from the pumps. Scientists estimate that with a 1-foot rise in sea level, an annual average of at least 475,000 acre-feet of additional Delta outflow would have been required to maintain salinity conditions from 1981 to 2000 at the western edge of the Delta. That works out to about 10 percent of annual export volumes during the period. With an additional 3-foot rise in sea level late in this century, pumping through the Delta may no longer be practical. Steve Schneider says, "And that's why the Delta is the single most vulnerable place to sea level rise in the entire state. We've completely transformed the landscape! And we've transformed it in such a way that it has no resiliency. No one was thinking *sea level rise* when they designed this stuff. It used to be a giant marsh! Now we use it to grow alfalfa for animal feed."

But California has to play the cards it was dealt. Left to flow naturally through the state's rivers, most of the precipitation that falls in California would flow out to the Pacific Ocean either directly through the rivers of the north coast or through the San Francisco Bay via the Sacramento and San Joaquin rivers. This would leave the southern part of the state—which contains about two-thirds of the state's population—with little of California's available freshwater supplies.

Lund agrees that the Delta is at a tipping point. In 2008, the scientists I mentioned earlier teamed up again, and Lund was the lead author of a report titled *Comparing Futures for the Sacramento–San Joaquin Delta*.[3] The report looked at various options for preventing a catastrophic collapse of the levees. It was meant to offer realistic solutions. It was an opportunity to prevent what might otherwise be inevitable.

"It's a game of chicken," Lund says; "and I just hope they can stop playing chicken before the earthquake happens." Hanak adds, "The current situation in the Delta is bad for the economy and brutal for the fish. Yet folks in the Delta don't want to see any change. Our evidence suggests that is not possible. They need to be thinking proactively. And they're going to need help financially."

If the scientists' first report was an attempt to envision various future scenarios for the Delta, then their second report presented four possible ways to move the water around. Lund says, "If you want to have a significant amount of long-term water exports from the Delta and you want fish, you need a strategy. You can always screw that strategy up. You can engineer a system with the best intentions and make it worse. But if you do things right, the optimal strategy should at least be better than all the other strategies. Based on our analysis it looked like a peripheral canal was the optimal solution if you want to continue with significant exports." Hanak says, "It doesn't mean everyone agrees with us."

The proposed *peripheral canal* (a canal that goes around the periphery of the Delta) is fairly straightforward, and it isn't new. The first such proposal was defeated in a state referendum in 1982; Northern Californians turned out in force to vote against it. As a result, it remains controversial today. The scientists are now offering a proposal that improves on the original concept but conservation groups are still worried that if a peripheral canal is built, it will allow for even more unsustainable levels of water exports to the south. They might be right. As Lund says, you can screw up anything.

But if we don't screw it up, the canal would protect water exports in the event of a levee failure and would also help reduce the extent to which water flow is altered by the current configuration of the pumps. Pumping water directly out of the Delta clearly hurts the environment. Water diversion pumps in the southern Delta are so powerful that they actually make the Delta's maze-like channels flow backward. Hanak explains it this way: "You're not sucking water through the Delta to the pumps. The peripheral canal decouples the management of water for humans with the water for the Delta ecosystem. You bring the clean water directly to the pumps and you don't have to keep the Delta fresh. It'll get saltier in the fall. We pump all year round right now. You're also not sucking the fish down. We've been operating this system and pumping so much the rivers go backward. You're not as vulnerable to a catastrophic levee failure. You just have to repair the canal, not the entire levee system."

The scientists see the canal as an essential way to separate the state's water demand from a Delta environment under grave stress. Sixteen Delta fish species are being pushed toward extinction, in part, by this demand.

"If you're just worried about fish," Hanak explains, "you want absolutely no water exports. But the engineering question is: how much water can you take out of the system and not hurt the fish? If you're optimizing over both human needs and ecosystem needs, you'll see reductions in uses for things like low-margin agriculture. It will force us to cut back on lands that are least productive. It's hard to do. And that's why this is one of those things that has been off the table for a long time. The Bay Area actually depends more on the Delta than Southern California does. Most of the growth in the Bay Area has relied on water from the Delta. If Bay Area folks understood the implications, then I don't think they would object."

The peripheral canal would deliver water from the Sacramento River along the Delta's eastern edge and then down to the pumps in the south, in effect circumventing the Delta itself. That's why it is called peripheral. It requires building an earthen canal wider than two football fields and more than 40 miles long. The canal would give the Bay Area and Southern California a direct line to tap into the Sacramento River. The channel would divert some of the river's flow around the fragile Delta and on to existing pumps in the southern part of the Delta. From there, the river would continue to serve Los Angeles, San Diego, farms in the San Joaquin Valley, and much of the Bay Area.

Originally, the peripheral canal was seen only as a water conveyance. Now, many scientists also view it as a restoration tool and a hedge against disaster. But residents along the eastern edge of the Delta are concerned about the effects of a 40-mile canal with a footprint 1,000 feet wide. For this reason, scientists say that any consideration of a canal must first begin with a commitment to water conservation and efficiency efforts on a scale not yet attempted in California. Hanak says the price of the peripheral canal would be somewhere between $5 billion and $10 billion.

"That's not such a big problem," she says. "Water users can cover that. The current system is so unreliable and risky. The bigger question is who pays for improvements in fish habitat and for the mitigation for farmers who rely upon the Delta to make their living."

Lund adds, "It's about adaptive management. We don't really have any other choice when you begin to factor in the impacts of climate change, like sea level rise, and how these new forces will be pushing on the system. If we go ahead and build the peripheral canal, we've essentially constructed a whole new Delta. We're not going to know exactly how to operate it."

Adaptive management is an admission that pinning a situation down will never work, but it requires a willingness to make changes and to be flexible. "It's real water and it's real money to someone," says Lund. "It's very hard to do adaptive management because every experiment is real water and real money. But we need to do this."

That said, Lund still thinks the more likely possibility is that, as in the case of Hurricane Katrina, the decision regarding what to do will be made for us, something Lund calls "failing into a solution."

According to Lund, "The physics is not going to wait for the policy to be put in place. If you have the big earthquake, the event forces you into making the quick policy decision. And that quick policy decision could be a complete disaster. That's what I call failing into a solution." When I asked Lund if he thought it was likely that a good policy decision could be made in advance of the worst-case scenario, he wasn't necessarily optimistic. "It's incredibly difficult to reach a consensus. It asks too many people to give up too much."

But if an adaptive style of management were to go into effect, it could also deal with other big problems—the kind of problems that come with living in a place subject to extreme weather and extreme climate. For example, El Niño—the periodic warming of the Pacific Ocean—routinely inflicts millions of dollars' worth of damage on California. Losses caused by storms and floods during the 1997–1998 event alone were over $1 billion. Adaptive management would help reduce the state's vulnerability to El Niño events. Schneider says climate change is another problem that could be addressed. "The two big problems you need to keep in mind when you think about climate change in California," he says, "are wildfires and pollution."

California has the worst air quality in the United States. More than 90 percent of the population is in areas that violate the state's air quality standard for either ground-level ozone or airborne particulate matter. These two types of pollutants can cause or exacerbate a wide range of health problems, including asthma and heart attacks. They have also been shown to decrease lung function in children. Combined, ozone and particulate matter contribute to 8,800 deaths and $71 billion in health care costs every year. The connection with global warming is nothing more than simple chemistry. Higher temperatures increase the formation of ground-level ozone and particulate matter. Ambient ozone also reduces crop yields and harms the ecosystem. If global background ozone levels increase as projected in several future climate scenarios, it may become impossible to meet local air quality standards. Air pollution events will become more frequent, longer-lasting, and more intense as temperature increases. For example, scientists at the California Climate Change

Center forecast that the scenario based on moderate temperature increase almost doubles the number of days with weather conducive to ozone formation in the San Joaquin Valley, relative to today's conditions.

And then there are wildfires. Warmer temperatures, drier conditions, and increased winds could mean hotter wildfires that are harder to control. Aside from their destructive potential, wildfires also increase levels of fine particulate matter. Therefore, air quality could be further compromised by increases in wildfires, which emit fine particulate matter that can travel long distances, depending on wind conditions. The most recent analysis suggests that if greenhouse gas emissions are not significantly reduced, large wildfires could become up to 55 percent more frequent toward the end of the century. The California Regional Assessment notes an increase in the number and extent of areas burned by wildfires in recent years; and modeling the results under changing climate conditions suggests that fires may be hotter, move faster, and be more difficult to contain in the future.

When I ask Hanak what the Central Valley might look like by 2050 if we do nothing, she says, "The Delta water system will not be viable. You'll see a lot of open water, and the folks who depend on that water won't have access to it. We're likely to lose 1 million acres of San Joaquin farmland. It wouldn't be a devastation to the state economy, but it would certainly hurt that region, one of the country's most productive farming areas." In 2009 the California Climate Change Center released a report that builds on this picture. According to the report, water shortages across California are more likely in the future as snowmelt decreases, the climate warms, and the population grows. California's population is expected to grow from 35 million today to 55 million by 2050.

Climate models suggest that by the end of the century, late spring stream flow could decrease by one-third. Agricultural areas could be hard hit, with California's farmers losing as much as 25 percent of the water supply they need. Climate models indicate that by 2050 the average annual temperature in California will rise between 1°F and 2.3°F, depending on the level of global greenhouse gas emissions. And by the end of the century, if emissions proceed at a medium to high rate, temperatures in California are expected to rise somewhere between 4.7°F and 10.5°F. In a more optimistic scenario, with lower greenhouse gas emissions, temperatures still go up between 3°F and 5.6°F by 2100. If temperatures rise to the higher warming range, there could be as many as 100 more days per year with temperatures above 90°F in Los Angeles and above 95°F in Sacramento by 2100.[4]

In 2010, the peripheral canal will be under the spotlight as the proposal reaches the governor and the legislature. There will be lots to wrangle over, including whether and how to build it, as well as how the state will buy up land for 100,000 acres of environmental restoration in the Delta. A committee of government agencies is working on a related habitat conservation plan for the Delta, also due in 2010, which is expected to include a canal. The state Department of Water Resources is drafting an environmental impact report on canal options, also due for completion in 2010. Lund says, "You never know what will come out of the sausage machine that is politics." But what seems clear is that a classic Hollywood ending is unlikely.

Hanak and Lund will tell you that in order for the Delta to survive, it will need a substantial overhaul. They can offer what they see as an optimal solution, but they can't offer a perfect solution. The optimal solution means that some fish will still die and some farmers will lose their livelihood. It is not—to repeat—a Hollywood ending. The problem with engineering is that it provides a methodology for calculating your losses ahead of time. That means you have to take responsibility and accept those losses. Lund is not sure rational solutions are the inevitable choice, because people would rather simply hope for a happy ending. "The most likely outcome is we will not be able to decide in time. And our reports will be used as a hind-cast instead of a forecast. Our studies will serve as a nice example of what could have been done. And we will forever be known as the Katrina scientists of the Sacramento–San Joaquin Delta."

The Central Valley, California: The Forty-Year Forecast—Drought, Water Resources, and Agriculture Problems

SACRAMENTO, CALIFORNIA	TODAY		2050		2090	
Emissions Scenario	JAN.	JULY	JAN.	JULY	JAN.	JULY
Higher	45	76.1	47.5	80.2	50.5	84.0
Lower			47	79.3	48.3	80.8

Forecast
April 2017

By 2017, the future began to reveal itself to the residents of California. Temperatures continued creeping up, winters were becoming increasingly tepid, and spring came earlier and earlier. Summer seemed to last forever, whereas the Sierra snowpack was gone by June. None of this came as much of a surprise. It had always been fairly easy for the climate models to forecast temperature. Mostly, we were just relieved that it wasn't worse. At least the seasons were still pretty much the seasons, even if they were all tracking warmer than they should be. Those all-important winter storms that blew in from the North Pacific Ocean bringing rain and the occasional snowfall, part of the classic so-called Mediterranean seasonal precipitation pattern, were still around. We figured that as long as the timing of the rains remained about the same, California could focus on dealing with the heat.

It was in 2017 that a storm gathering in the Pacific would redefine how we looked at winter. It would also redefine how we looked at the ocean-atmosphere phenomenon called El Niño, a periodic warming of the equatorial Pacific Ocean. El Niño was now further juiced by all the extra heat in the atmosphere, and the winter storms it brought were expected to knock some of California's rainfall records out of the ballpark.

For the most part, El Niño events typically lasted about one year. But coral records from Palmyra Island in the tropical Pacific Ocean suggested that there were periods in the past when both the amplitude and the frequency of El Niño events changed abruptly.[5] Whatever the cause, the present El Niño seemed to be different. As the climate models struggled to describe how global warming might influence El Niño events in the future,[6] we watched as it changed right before our eyes.

The warming of sea surface temperatures in the equatorial Pacific was evident by early April 2017. And in early June, the NOAA Climate Prediction Center issued an El Niño Advisory stating that ocean temperatures in the central and eastern Pacific Ocean were the warmest since August 1997. Until now, the 1997–1998 El Niño had been one of the strongest in recorded history. It was lucky for California that we had learned from past mistakes. Following the large El Niño of 1982–1983, an event that went largely undetected until it kicked California in the teeth, federal money was used to make improvements to the observational network that fed into the climate models. With the installation of the TAO/TRITON array of moored buoys, scientists finally had more comprehensive data regarding the condition of the tropical Pacific Ocean. The data collected by means of this improved observing system supplied the climate models with very high-resolution snapshots of ongoing ocean conditions. The models, in turn, predicted that a strong El Niño was now growing in the Pacific and that it would continue to grow through early 2018 and perhaps even longer—with emphasis on the word *longer*.

By late spring 2017, scientists at NOAA issued a climate forecast for the approaching winter months, highlighting El Niño's probable impact on temperature and precipitation patterns in the United States. Mother Nature's forecasts, often more subtle, were not far behind. Luckily, there were plenty of old-timers, knowledgeable about wild creatures and the local environment, who were still capable of reading these more subtle messages with great skill. Phone calls came pouring in to the Marine Mammal Center in Sausalito about sightings of stranded sea lions. The sea lions, unable to find enough anchovies or herring, were malnourished. It was likely that the warming ocean temperatures associated with the growing El Niño were driving the fish away.

A lack of upwelling of cold water along California's coast was the key to many of the strange sightings. Wind-driven upwelling of cold, nutrient-rich deep water acts as a feeding trough. But El Niño turns down those winds, and that downturn reduces the upwelling. Reduced upwelling prevents nutrient input to surface waters, lowers the amount of plankton—tiny marine plants and animals—and messes with nearly everything that relies on the ocean for sustenance. Biologists studying salmon in the waters off San Francisco reported finding razor-thin sardines and anchovies, underweight for their size and probably undernourished. A brown pelican was spotted in a suburb of Phoenix, probably propelled by El Niño winds from the California coast all the way to Arizona. An official from the Maricopa Audubon Society explained that many starving birds had just given up and let the wind carry them.

According to NOAA's El Niño Watch, tropical El Niño conditions would continue to gain strength through January along the U.S. west coast, with sea surface temperature (SST) departures similar to those observed in December. The SSTs were 5°F to 6°F above normal off the northern California coast, more than 6°F above normal off the coast of San Francisco and Southern California, and 4°F to 5°F above normal off the coast of the Pacific Northwest.

By February the worst of the winter brought heavy rains, flooding, and mudslides in northern California, and there were flood warnings for the Russian and Napa rivers. Interstates were shut. Heavy surf, rising to 50 feet, eroded beaches. As much as 14 inches of rain fell in Santa Barbara. Hurricane-force winds of up to 95 miles per hour caused devastation. Residents of Sacramento stockpiled sandbags and cleaned out channels. And the 2 million people of the Central Valley who were counting on the levees for flood protection heaved a sigh of relief. The levees, for the most part, had survived the storm intact. But the overall frailty of the levee system was in plain sight, and the storm revived stalled discussions about the need for a peripheral canal. It was clear that something had to be done, and it was hoped that this El Niño might serve as the impetus.

After the first El Niño dissipated, climate models forecast one more round. In other words, the El Niño of 2017–2018 was technically the first of a pair. Surfers in Newport Beach had a field day. But while they were riding the best winter waves of the century, California itself was riding headlong into an economic meltdown. This time it was caused by a flood of water as opposed to a flood of bad mortgages. The state, still suffering from the bursting of the real estate bubble in 2009, was already strapped for cash and ill-equipped to make any significant infrastructure upgrades. There was talk of trying to float a $10 billion bond for the peripheral canal. But in the end, no one had the cash to support it.

There was one small gift amid all the wreckage. The El Niño rains generated a blooming of spring flowers across Death Valley. The winter storms that brought mudslides and death to Southern California dropped more than 10 inches of rain on this thirsty desert—five times more than usual—persuading the hearty wildflower seeds to blossom year after year. Death Valley hadn't seen such an array of flowers since the spring of 2005, almost fifteen years earlier. In March and April 2017, tourists came from around the world to see the landscape repainted in the rich blue of the desert lupine, the royal purple of the chia flower growing in clumps along the side of the road, the magnificent golden yellow of the California poppy and the desert dandelion, highlighted by the soft white petals of the tufted evening primrose scattered across the hillsides.

August 2027

By 2027, the dream that was California had begun to unravel. Temperatures across the Golden State, which had been rising since the 1950s, started to spike. August 2027 saw new record high temperatures being set almost every day. As one local meteorologist put it, "The good news is that the number of official heat waves in Sacramento this year is finally down, the bad news is that the heat wave that started in July is still going strong." Sacramento, along with much of the U.S. Southwest, was trapped in one long heat wave.

The intensity of this heat wave was crushing, and the duration of abnormally high maximum and minimum temperatures was reaching into every sector of California's economy. An all-time record for statewide energy consumption was set, and hundreds of heat-related deaths were reported. Meteorologists watched in awe as the semipermanent four-corners high-pressure system expanded to the west and north. Associated with this expanded high-pressure system was an unseasonably warm air mass settling over the region through the duration of the event, with an influx of monsoonal moisture from the Gulf of California and the Baja region. Typical mid-level winds from the southwest shifted to the southeast. This allowed for maximum transport of moisture over southern and central California, and served to lift minimum overnight temperatures higher. It also brought excessively humid conditions during the daytime hours—a situation not typically associated with California. In other words, California felt like Louisiana.

The heat wave proved devastating to the state's $3.2 billion wine-making industry. California, still the nation's largest wine producer and the fourth-largest wine producer worldwide, with its high-quality wines produced throughout the Napa and Sonoma valleys and along the northern and central coasts, had, in a sense, come under siege. High temperatures during the growing season caused grapes to ripen prematurely and reduced their quality. Specifically, an increase in the overall number of temperature extremes above 86°F during the growing season had shut down photosynthesis. Cooler spots, such as Mendocino and Monterey counties, were still hanging in. But the grape-growing regions of the Central Valley, with its already marginal conditions, were hit hard. California's losses due to heat were estimated at 30 percent and close to $1 billion.

It was a summer that seemed like science fiction and apparently wasn't set on the planet Earth. Mudslides became a big problem in areas deforested by wildfires, and these slides were almost impossible to escape. Airports were virtually forced to shut down as plane after plane was grounded, week after week, by decreased visibility from the wildfires. Air quality became unbearable. The entire state violated air quality standards for ground-level ozone (smog) and small particles. Air pollution in Los Angeles and the San Joaquin Valley was especially bad. The cost of managing an unruly climate had become crippling.

The models were right: over the past two decades, California's average temperature had risen somewhat more than 2°F. The temperature increase was further proof that a high-emissions scenario was happening. This meant that by mid-century, the serious changes would to start to kick in. And if nothing was done to ramp down emissions, by the end of the century, climate models projected that statewide average temperatures would rise more than 10°F. We couldn't even manage 2°F.

October 2040

By 2040, the seasons had become almost unrecognizable. The heat hung on as summer extended into a long, mind-numbingly brutal test of patience. Rainfall had become erratic, and a cool patch of tropical Pacific Ocean temperatures suggested that we were locked in something like a prolonged but sketchy La Niña event. For the U.S. Southwest, La Niña is synonymous with drought. Water was on everyone's mind.

Ironically, the original pact that governs water supply in the West was negotiated in 1922, during one of the wettest periods of the past 1,200 years. Paleoclimatologists, analyzing tree ring records across the Southwest, said it was only a matter of time before the western United States reverted to its old ways and the Colorado River became drier again. We knew that global warming would push the system back into drought harder and faster.

Climate change was indeed a game changer, especially for utilities and water managers who were trying to make sure everyone had a fair share. The pact of 1922 that allocated Colorado River water to California, Nevada, Arizona, Utah, Colorado, New Mexico, and Wyoming had come about in a very different world and did not fit readily into this new one. There was still plenty of argument about whether the new reality was in fact new or just an old reality come back to haunt us. But in any case, the reality was that it wasn't raining.

As the seasons became unfamiliar, so did the landscape. Its color and texture had begun to shift before our eyes. To start, high-elevation forests in California were receding. They just couldn't take the heat. Almost half were gone, replaced by grasslands. The desert, once lush with wildflowers, was being slowly overtaken by alien species such as red brome and buffle grasses—plant species native to Africa and the Mediterranean and able to do well in high temperatures. Not only did these noxious weeds outcompete some native species in the Sonoran Desert; they also fueled hot, cactus-killing fires. The magnificent saguaro cactus (*Carnegiea gigantea*)—the state flower of Arizona—and the Joshua tree (*Yucca brevifolia*) had become hard to find.[7] In fact, the majority of California's native species, numbering well over 3,000, were fading. It seemed as if everything was heading for the hills—but the hills were on fire.

One of the first lessons regarding climate change is that the conditions of the past can't predict the future. The Hoover Dam, completed in 1936 and once a proud symbol of American engineering power and vision, became a sorry reminder of our collective inability to change with the times, let alone stay one step ahead. Straddling the border between Nevada and Arizona, Hoover Dam was built to tame the flow of the magnificently wild Colorado River, which stretched over 1,450 miles from Colorado's Rocky Mountains to the Gulf of California. The Colorado River had supported the growth of cities such as Las Vegas and Phoenix, providing about 30 million people with water for drinking and irrigation. By the year 2040, Hoover Dam was empty, and a sort of bathtub ring—a thick white band of mineral deposits—marked the walls of Black Canyon. That ring showed where the waterline used to be—before the rain stopped but while the people kept pouring in.

Before the last drops were released a few months ago—to Las Vegas, Los Angeles, San Diego, Phoenix, Mexico, and some other places—our strategy had become hope and prayer. We hoped and prayed that life would just go back to the way it used to be, when there was still rain and we didn't have to think about water all the time. Casinos like the Bellagio and Mirage, with their beautiful fountains, lagoons, and waterfalls, reinforced our hope and made it easy for us to think that everything would be fine. My favorite was Planet Hollywood, where it rained every hour on the hour, day after day. The daily rain show became one of the most popular attractions—it almost seemed to prove that Sin City could not possibly be in the midst of a water crisis.

But Lake Mead and Lake Powell were always the biggest poker tables in Nevada. Back in 2009, water resource managers at the

Southern Nevada Water Authority were asking all the right questions about the cards they were holding. Was the drought one of the traditional droughts that the Colorado River had experienced in the past, or was it something very different? In just two weeks in April 2009, managers watched as Lake Powell lost the equivalent of 14 feet of snowpack. And by the summer of 2009, the reservoir's water level fell to its lowest point since 1965—back when Lake Powell was new and officials first diverted water from the Colorado River into it. By the end of August 2009, Lake Mead's elevation teetered at just 1,092 feet above sea level. Water resource experts knew they were betting on the future of the West. A mere 17 feet stood between hope and despair. By law, once the level dropped below 1,075 feet, the Southern Nevada Water Authority was required to find alternative sources of water. The future of water in the West, the future of great bottles of cabernet and zinfandel, the future of beautiful apricots and almonds and avocados, the future of life itself, depended on that 17 feet.

The models had been predicting that climate change could reduce the runoff that feeds the Colorado River between 5 and 25 percent by the middle of the century.[8] But there was a big difference between reduced runoff and empty reservoirs. That difference was a matter of managing the double whammy of climate change and population growth. One problem was that even under the most extensive drying scenario—a drop in runoff of 20 percent—water supplies wouldn't be affected immediately, because Lake Mead and Lake Powell, which could store up to 50 million acre-feet of water, provided a significant cushion. In fact, the total storage capacity of all the reservoirs on the Colorado exceeded 60 million acre-feet, almost four times the average annual flow on the river itself. As a result, scientists calculated the risk of complete reservoir depletion as low through 2026, making it easy to postpone action. But after that, all bets were off, and the risk soared. Good management could reduce the risk, but a policy of business as usual meant that under a worst-case climate scenario, by mid-century there was one chance in two of empty reservoirs in any given year. The scientists cautioned that if the water managers took aggressive steps to reduce downstream releases during periods of drought, it would still be possible to cut their losses by as much as one-third. But they warned that if the managers did nothing, as the West continued to warm, a drier Colorado River system could have a risk as high as one chance in two of completely depleting all of its reservoir storage by 2050. At the time, it was anyone's guess when or if Lake Mead would reach that point.

Looking back, the water resource managers did the best they could, considering that the country was in the midst of a recession. The national economic downturn reduced the utility's revenues just as they came to see the risk of climate change and wanted to take steps toward new infrastructure investments. Even so, they managed to allocate $800 million and embarked on a large-scale construction project to build a new intake pipe to pull water from Lake Mead. This project, known as the *third straw,* was a way to ensure that the utility could siphon water if the lake level dipped below 1,000 feet, the point at which the two existing intakes became useless. It was an excruciatingly difficult project, but they got it done. And it bought us extra time. They also sought permission to build a controversial $3.5 billion pipeline to transport groundwater from ranches in rural eastern Nevada. But the pipeline plan drew so much ire from ranchers and environmental groups that it never made its way out of court.

July 2050

Despite the fact that scientists had run countless simulations of an earthquake in California's Central Valley, the first twenty-four hours of the real thing were terrifying. It was a magnitude-6.5 earthquake, and it took down twenty islands in the San Joaquin Delta. As the levees collapsed, salt water from San Francisco Bay rushed in. The islands, now acting like bowls, filled up with salt water. It took a little more than thirteen hours for the salt water in San Francisco Bay to begin filling up the Delta. Unluckily, the earthquake happened during the summer, when the ground was dry and water levels were low. Just as predicted in an earlier scenario, within thirty days the Delta became a saline estuary; salt water ravaged millions of acres of farmland that had come to depend on freshwater from the Delta; and a catastrophic failure of important levees cost more than $16 billion. The president declared the Central Valley a federal disaster area. Federal and local lawmakers vowed to rebuild, and plans for a peripheral canal were finally put on the fast track.

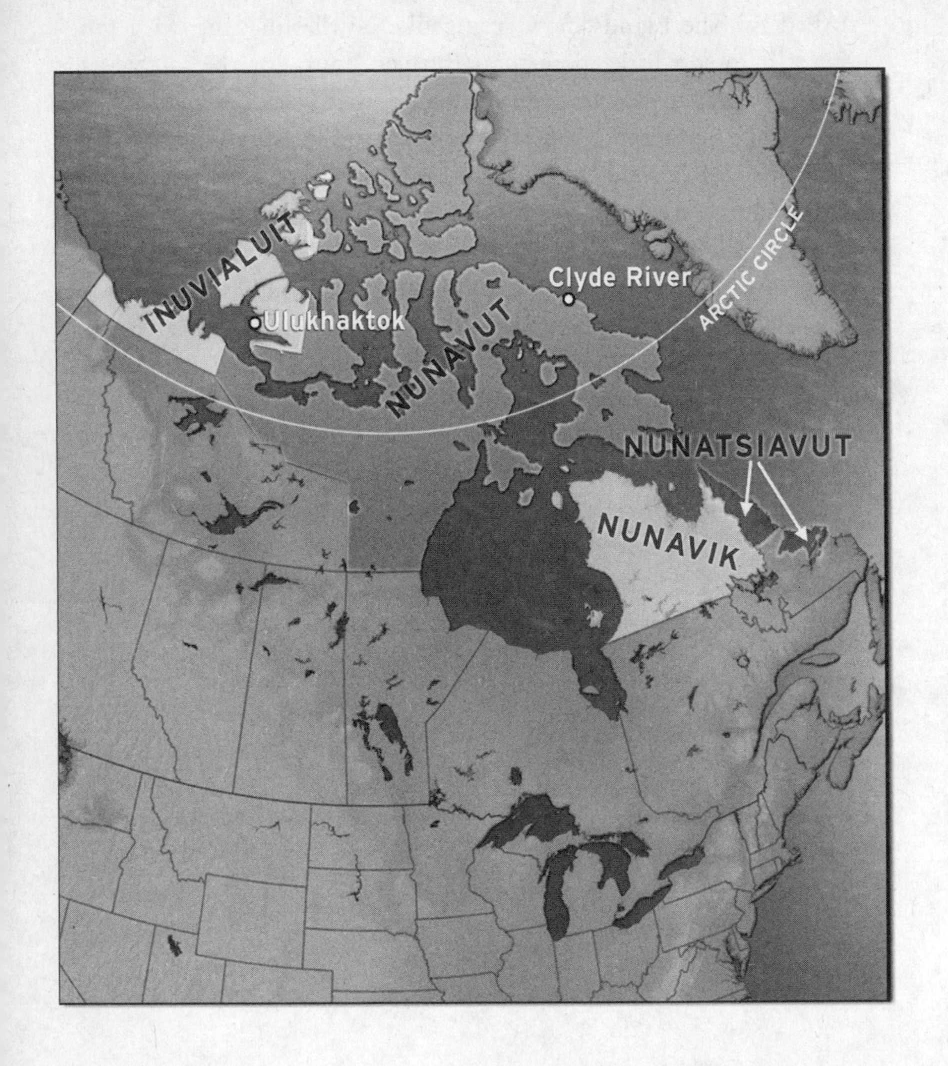
INUVIALUIT
Ulukhaktok
NUNAVUT
Clyde River
ARCTIC CIRCLE
NUNATSIAVUT
NUNAVIK

8

THE ARCTIC, PART ONE: INUIT NUNAAT, CANADA

Shari Gearheard had just returned from the 2009 Nunavut Quest dogsled race when I finally managed to catch up with her.

"I've now completed the Master Ninja course in sea ice travel and I'm 2 inches shorter for it," Gearheard said, laughing. Master Ninja is her description of an amazing six-week trip that took her 1,500 miles around Baffin Island. Her husband drove their dog team in the 400-mile race from Arctic Bay to Pond Inlet; and Gearheard, as part of the support crew, drove ahead on a snowmobile. This mix of tradition and technology—a dogsled and a snowmobile—defines life in the Arctic today. And it's safe to say that over those six weeks Gearheard saw, and felt, just about every kind of sea ice. "It gets pretty bumpy out there sometimes," she says. Like a lot of things in the Arctic, the sea ice is far more complex than it seems to be at first glance; this is why Gearheard finds it best to study the Arctic up close and in person.

If there were only two types of people in this world, summer and winter, Gearheard would definitely qualify as a winter person. "It started in childhood," she explains. "I loved winter and I loved the snow. When I was little, I used to get in trouble from my mom because one of my favorite things to do was to go by the side of our house where these big snowdrifts would form and I would just dig a big snow cave into the drift and go to sleep. She always worried that I would get trapped or something." Gearheard, now in her thirties, is a research scientist at the National Snow and Ice Data Center (NSIDC) at the University of Colorado in Boulder, where I suspect everyone else is a winter person, too. Gearheard telecommutes to NSIDC from Nunavut, in the Canadian Arctic, where she and her husband, Jake, live full time.

In Inuktitut, one of the local Inuit dialects spoken in Nunavut, Inuit means *the people* and Nunavut means *our land*. Nunavut is the largest and perhaps best-known of the four Inuit territories that make up the Canadian Arctic. Collectively, they are called the Inuit Nunaat. The other three territories are Nunavik in the northern portion of Quebec, Nunatsiavut along the coastal region of Labrador, and the Inuvialuit Settlement Region in the Northwest Territories. Gearheard's research is focused on collaborating with Inuit communities to document their knowledge of the environment and environmental change, and to link that knowledge with science. Part of her work includes understanding the impact of climate change on communities and how the Inuit are responding. You might say the dog team is part of her research, but it has also become part of her life.

"I've always loved sled dogs," she says. "When we moved here, good friends who have a dog team would take us out with them. And sometimes they would let us drive their dogs. We absolutely loved it." Gearheard and her husband thought about getting a team, but were intimidated by the amount of work involved. "Everything is made by hand," she explains: "the harnesses, the leads, the dog whip, along with learning all the commands. It's a major commitment." But one day, those same friends showed up and presented them with Siqaliq, a beautiful Inuit sled dog with an exceedingly large belly. "They said, 'Here you go; your team is going to be born in about two weeks. So you'd better get working,' " Gearheard says, laughing. And with that, she and her husband got to work.

They fixed up a dog pen and house for the puppies. Another

elder in the community gave them two more dogs; they purchased two from a local hunter; and soon their dog team was on its way. "Puppies grow so fast. So all of a sudden we had ten hungry dogs to feed," she adds. And needless to say, there's no Pet Smart in Nunavut. "Our dogs eat seal meat, so we've become novice seal hunters," Gearheard explains. "It's our life now. All of our free time is spent with our dogs." The dog team now has twenty stunning Inuit sled dogs, called *gimmiit* in Inuktitut; this is the same breed used by early Inuit to cross the Bering Strait 1,000 years ago. Actually, there are twenty sled dogs and one seal hunting, sea ice Master Ninja. Not bad for a kid from, as the Inuit say, *down south*.

For more than 5,000 years, the Inuit have occupied the vast territory stretching from the shores of the Chukchi Peninsula of Russia, east across Alaska and Canada, to the southeastern coast of Greenland. People from down south have come and gone over time, mostly to take things like whale, white fox, copper, or oil back home with them. In the meantime, the Inuit have taken this cold place, which sometimes seems barren and endlessly dark, and made a home out of all the ice and snow.

Inuit history is preserved and passed down through an oral tradition, the telling of stories. On the basis of this tradition, the Inuit believe that about 5,000 years ago, what they call the Sivullirmiut, or *first people*, began to move east from Alaska.[1] (Archaeologists use the terms *Pre-Dorset* and *Dorset* to describe these ancient Inuit people.) In less than 1,000 years, these early people traveled across the ice from the north coast of Alaska, across Canada, all the way to southern Greenland. Their camps were located in places where the hunting was good; and the bones they left behind suggest that the Sivullirmiut hunted seals, walrus, caribou, and ducks. Depending on the season, they collected clams, mussels, fish, seaweed, bird eggs, and berries. The Sivullirmiut also used delicate needles, made of bird bones, to make boots and clothes from the skins of seal, caribou, and polar bears. They made lamps from soapstone and used these for heat and light; perhaps they even cooked meat in soapstone pots. Unlike the Inuit of today, these early people didn't have dogsleds or large boats. Without dogsleds they were unable to cover long distances, and without large boats to hunt whales, their communities remained small.

By about A.D. 1000, the Thule people developed the large open skin boat known as the *umiaq*, as well as the harpoon that allowed them to begin hunting whales (or at least they are credited with having done so). The Thule people also used dogs and dogsleds for long-distance travel across the ice. Inuit tradition represents thousands of years of cultural developments through which the Inuit have come to master the very tough Arctic climate. Where many of us might see only vast sheets of ice, the Inuit saw endless possibilities. Lacking trees, they built igloos for winter housing out of snow, and they burned whale and seal blubber both for fuel and for lamps. They stretched sealskins over frameworks to build kayaks as well as umiaqs big enough to take out into unprotected water to hunt whales. The Inuit are the most flexible and sophisticated hunters in Arctic history and have adapted to shifts in climate with very little difficulty. This is why Gearheard and others are so interested to see how the Inuit, an ancient people who have built their lives around ice and snow, are responding to climate change.

Shifts in climate are not unfamiliar to the Inuit. It was during the Medieval Warm Period, from about A.D. 800 to 1300, that the Thule pushed east into northwestern Greenland from Canada. They were probably following the bowhead whale as the sea ice that had permanently closed off the channels between the northern Canadian islands during colder times finally began to melt in the summer. After the Medieval Warm Period came the Little Ice Age, and by the 1400s this brief return to colder conditions was well established. The Little Ice Age proved to be of little significance to the Inuit. The extra ice simply presented them with the opportunity to hunt ringed seals. In the meantime, at the Norse settlements in Greenland, settlers were still trying to grow hay and graze livestock.[2] Their inability to adapt to a changing climate ultimately proved fatal. Those early Viking colonies on Greenland have long since disappeared, suggesting that those of us from down south may have a thing or two to learn about resilience and adaptability from the Inuit.

Despite its harshness, the climate was never really a problem for the Inuit. It was the disease and alcohol brought by southern whalers and fur traders that almost took them out. Encounters between the Inuit and Europeans began in the late 1500s, when the first explorers sailed into the frigid waters of Davis Strait, Hudson Strait, and Hudson Bay. Although these first encounters were few and far between, they mark the eventual transition into what the Inuit call the *period of contact*—the period of *taking*. But the Europeans also *brought* a few things with them. Starting in the 1700s, the whalers and missionaries began to make their way north, and by 1850 they had become an almost permanent presence in the Arctic. The year-round settlement brought diseases such as smallpox and tuberculosis, which killed so many Inuit. The religion brought by the missionaries also left an impact on the Inuit. During his stay in the Frobisher Bay area in 1861–1862, the American explorer Charles Francis Hall wrote about the health of the Inuit and issued his own forecast:

> The days of the Inuit are numbered. There are very few of them left now. Fifty years may find them all passed away, without leaving one to tell that such a people ever lived.[3]

Needless to say, this forecast proved incorrect. But after a few centuries of contact, the Europeans had not only pushed the whale to the brink of extinction but also pushed the Inuit. As whales became harder to find, they were no longer profitable for commercial whalers; and for the Inuit, an important traditional source of food was endangered. Luckily, the market for whale oil and ambergris used in lamps and perfume was being replaced by a market for kerosene and synthetics. And as a result, some whaling captains and crews turned to trapping the arctic fox.

Today, through a process that began on April 1, 1999, with the establishment of Nunavut, the Canadian Arctic is officially recognized as the home of the Inuit in Canada. And since the establishment of Nunatsiavut in 2005, all four of the traditional Inuit territories have been covered by land claim agreements that establish regional autonomy within Canada. The Inuit may finally be in control of the Arctic; but as it turns out, the Arctic climate is being influenced from points farther south.

With only 152,000 people across the Arctic and limited industrial activity, there is little the Inuit can do to slow or stop global warming, because they contribute so little to total global greenhouse gas emissions.[4] But nonetheless, they are feeling its impact. The Inuit may have been able to withstand the smaller climate changes of the past, but what's happening in the present is a different story altogether.

Until recently, little was known about Inuit perspectives on climate change. In the mid-1990s, after speaking with elders and surveying the scientific literature, Gearheard found only a few references to Inuit knowledge of climate change. Also, she had begun to notice that although some researchers documented the impact of climate change on subsistence hunting practices, they didn't try to find out how community members *felt* about it. So Gearheard spent the next several years traveling back and forth between her home in the south and the Inuit communities of Igloolik, Baker Lake, and Clyde River in Nunavut. Through her conversations with hunters and elders—first using a translator and eventually starting to communicate on her own, speaking Inuktitut—she began to learn about Inuit knowledge and observations of weather, climate, and climate change. During one of her conversations, an elder in Igloolik, Zacharias Aqqiaruq, described the weather as *uggianaqtuq*. This Inuktitut word was later explained to Gearheard by a worker at the local research center:

> For example, I'm very close with my sister. Say I wasn't feeling myself one day and I went to go visit her. As soon as I walk in the room, or say something, she would know right away that something is wrong. She would ask me, "Is there something wrong with you?" She would say I was *uggianaqtuq*. I was not myself.
>
> —T. Iyerak, Igloolik, 2000[5]

Later, Gearheard would hear other opinions about the definition of *uggianaqtuq*. For example, it was said to be a reference to people fighting, tension, extreme heat, or something unseasonable or untimely. The root of the word may refer to a dog taking something in its mouth and shaking it. Another suggestion is that the word refers to something being eaten by lice. Though there are many suggested meanings for the word—*unexpected*, *unfamiliar*, *fighting*, *tension*, *unseasonable*, *untimely*, *being ripped apart*—interestingly, they all connect in some way to the changes Inuit have been experiencing in their environment in recent years. This was something Gearheard would hear again and again. The weather had become a stranger; it was no longer itself.

In 2004, after having worked with Inuit communities for a decade, Gearheard and her husband, Jake, made the decision to move north to Baffin Island and experience these changes and work with the Inuit firsthand. They live in Clyde River—Kangiqtugaapik in Inuktitut—a small Inuit community of about 850 people, and they are among the very few non-Inuit inhabitants who live there. In fact, they were the only non-Inuit team in the Nunavut Quest dogsled race. Clyde River is about 280 miles north of the Arctic Circle and is surrounded by some of the most dramatic fjords and cliffs in the world. The Arctic Circle is an imaginary line in latitude separating hard-core winter people from the rest of us. It is where the sun remains above the horizon at the summer solstice and remains below the horizon at the winter solstice. The Latin word *solstice* roughly translates into *sun stand still.* For Clyde River, that means the sun drops below the horizon around the middle of November and does not reappear again until the end of January. In December there is only about one hour of twilight each day. But come spring, Clyde River receives twenty-four hours of sunlight from the end of May until the end of August. It must be an amazing place in July, when the average high temperature is about 47°F and the sun never sets. But December, when darkness sets in and the average high is about -6°F, would be a different story.

Gearheard and her husband provide an interesting study in contrast. "There are some social issues in the community that are really tough," Gearheard explains. And it is these issues that her husband works on with other community members at Ilisaqsivik Society, a wellness and family resource center started by the community itself more than ten years ago. Issues such as substance abuse, domestic violence, poverty, and suicide are not uncommon in Inuit life. "Ilisaqsivik focuses on developing people's strengths and providing a lot of counseling and healing programs," Gearheard explains.

Compared with the longer timescales associated with climate change, these issues confront Inuit communities on a daily basis, and Gearheard is careful to keep her perspective. "Compared to something like, 'I can't feed my kids today,' climate change doesn't feel very immediate," says Gearheard. "Although I think people here care about climate change a lot, they definitely separate what is natural climate variability and what is not. I have often heard an elder say, 'We always had years when the sea ice was late or the sea ice broke up early, but it didn't happen eight years in a row.'"

> What [I] have noticed . . . in the last five to eight years, [is that] when it should be freezing up . . . it becomes overcast,

> snow starts falling for a long period of time . . . that affects freeze-up . . . whenever it's overcast the temperature rises a bit, freeze-up doesn't occur as quickly.
>
> —N. Arnatsiaq, Igloolik, 2004[6]

The climate isn't the only thing changing in the Arctic. Within a single lifetime, the Inuit have gone from living off the land to a wage economy. They are attempting to balance two very different lifestyles: one of sled dogs and subsistence hunting, the other of ski-doos and soda pop. Almost all of us can relate to this clash of tradition and modern convenience.

"The food here is insanely expensive," says Gearheard. "A loaf of bread can cost $8; a small box of Tide costs $35." Much of the healthier food is more reasonably priced because it's subsidized. But soda pop, for example, is not subsidized. "A can of pop can be $5," Gearheard says, "especially when you get into summer and supplies are dwindling. The supply ship only comes once a year." As a result, Gearheard and her husband, along with many families in Clyde River and throughout the Arctic, have come to rely on a commercial air freight service called Food Mail that is based in Quebec.[7] "They basically do your shopping for you. You send a list by e-mail or fax and they shop it all up and get it together and package it," Gearheard says. Food Mail has a contract with the Canadian government, which subsidizes the service. "So you send in your list on Monday and it comes on Thursday," explains Gearheard. "You can get all kinds of fresh vegetables, essentially everything you can get down south."

"In the past, prior to settlement, people didn't have a choice. It wasn't a question of whether you wanted to hunt," says Tristan Pearce, a graduate student finishing his PhD at the University of Guelph in Ontario, Canada. "It was a question of whether you wanted to eat." But now the Inuit are faced with choosing between their traditional lifestyle and modern conveniences and the result is often a blend of the two. "Now it's a question of do you want to eat *country foods* like seal, musk-ox and whale or do you want to go to the store and buy processed foods."

Pearce works in the town of Ulukhaktok, formerly Holman, in the Inuvialuit Settlement Region. Ulukhaktok is a coastal community of approximately 430 people located on the west coast of Victoria Island in the Northwest Territories. It evolved as a permanent settlement starting in 1939, with the establishment of a Hudson's Bay Company (HBC) trading post and a Roman Catholic mission near the location of the current settlement. Throughout the 1940s and 1950s, the regional population continued to live in isolated hunting and trapping camps and came to Ulukhaktok several times a year to trade furs and socialize. The federal government shipped three housing units to Ulukhaktok in 1960 and another four in 1961. In the years to follow, some families moved to Ulukhaktok permanently, but others live there seasonally. Snowmobiles, satellite television, Christian churches, and a wage economy all brought profound social change to this group, traditionally known as the Copper Inuit. The Copper Inuit speak Inuinnaqtun, and Western Inuit from the Mackenzie Delta region who also live in Ulukhaktok speak Inuvialuktun. But now English is the dominant language for younger people.

Despite this modernization, the Inuit are the ultimate survivors, technologically sophisticated, extremely adaptable, and yet traditional at their core. Thanks to research by Gearheard, Pearce, and others, traditional Inuit knowledge of climate and weather is gaining more attention and much-deserved respect from within the broader scientific community.

For a long time, traditional knowledge of weather and climate was classified as anecdotal by western science, *anecdotal* being a code word for *unscientific*. Yet as we learn more about traditional knowledge of climate, and collaborate with indigenous peoples, scientists have gained a deeper understanding of the interconnections within our climate system. It turns out that old-fashioned firsthand observations have a very important place in the high-tech world of modern science.

A perfect example of how traditional knowledge works its way into Western science involves potato farmers in the Andes and El Niño. Climate scientists and anthropologists had long heard stories about traditional forecasts developed by Indian farmers in the Andes Mountains of Peru and Bolivia, dating back to the late sixteenth century. The anecdote went like this. Potato farmers would meet in small groups at each winter solstice in late June (the southern hemisphere winter) to discuss the planting date for the potato crop. Then, in total darkness, during the longest and coldest nights of the year, they would climb to a mountaintop in order to see the Pleiades, a star cluster in the constellation Taurus. The Pleiades are visible low in the sky to the northeast just before dawn. In years when the Pleiades looked big and bright, the farmers would plant potatoes at the usual time. But in years when the Pleiades looked small and dim, they would expect the rains to arrive late and be sparse, and so they would postpone planting by several weeks. The farmers were using the appearance of the Pleiades to forecast the timing and quantity of precipitation during the rainy season, several months later, beginning in October and extending through March.

Mark Cane, a professor at Columbia University, and John Chiang who at the time was his graduate student and is now a professor at Berkeley, were able to find the physics underlying this traditional forecast by using modern satellite data. Cane and Chiang knew there was a strong link between El Niño and precipitation in South America. Rainfall over the Andes is lower during El Niño years. This relationship is obvious for the three months of the year with the highest rainfall: December, January, and February. More important, rainfall in October is also diminished by El Niño, and this suggests that the rainy season starts later during El Niño years. But how could El Niño alter not just the weather but also the apparent brightness of the Pleiades in June, four months before the rainy season begins?

This was where modern technology became very useful. Using satellites capable of measuring cloud cover from space, Chiang showed that there was an increase in high clouds during El Niño years. Specifically, high clouds just to the northeast of the Andean highlands increased during late June, interfering with observers' view of the Pleiades. In other words, the brightness of the Pleiades in late June indeed correlates with rainfall during the growing season for potatoes the following October through March.[8] This climate forecast is one of many that have come to the attention of scientists, and it reinforces the importance and significance of traditional knowledge.

In the Arctic, traditional forecasts always come back to snow and ice. And unlike the Andean farmers, who were focused on potato yields, the Inuit are interested in elements that affect hunting. For example, they look at the time it takes for sea ice to reach a certain thickness, at wind strength, and at the relative timing of sea ice breakup and animal migrations. "Ice is extremely important because it is essentially the highway over which the Inuit travel to hunt," says Pearce. Freeze-up generally occurs between the end of October and mid-November, and breakup usually occurs in late June or early July.[9]

Inuit travel by snow machine towing large sleds called *alliaks* in Inuinnaqtun or by dog team over the sea ice, river ice, and lake ice to reach a number of hunting areas, both on the land and on the ice. The coldest months of the year, December through March, are considered by community members to be the safest for travel, owing to the thickness and stability of the ice. During the winter, hunters in Ulukhaktok use the sea ice to hunt seals and polar bears, trap foxes, get to musk ox harvesting areas, and travel to neighboring communities. Traveling on the sea ice is inherently dangerous because of rough ice, cracks, open water leads, and storms, but hunters manage these risks by taking precautions and applying their knowledge of local ice conditions.

In the last several years, however, changes in the climate have altered and in some cases increased the magnitude and frequency of hazards that hunters have to deal with. In particular, some areas of sea ice, over which harvesters are accustomed to travel, are no longer stable, and in some instances the ice has not formed, because of strong winds and milder temperatures. Even experienced harvesters have encountered hazards in what are thought to be safe travel areas. Hunters are now often taking risks to travel on the sea ice even when it is melting or thin to reach hunting areas.[10] Recently, several hunters have been stranded, injured, or forced to take alternative travel routes, or have had to deal with lost or damaged equipment (such as snow machines breaking through the ice) as a result of unexpected changes in weather and sea ice conditions.[11] These are many of the same the types of issues Gearheard and her husband had to confront during their 400-mile dogsled race. I wonder if her mom still worries.

"Typically, before going out on the land or sea ice, hunters consult with other hunters and elders about sea ice and weather conditions. They observe the height and form of clouds, the brightness and movement of stars, and the direction and strength of wind, to attempt to forecast the weather in order to decide if it is safe to travel," explains Pearce. Hunters also often consult satellite imagery of the sea ice and weather forecasts available on the Internet, merging traditional and new weather forecasting techniques. As for navigation, compasses are of little use in the Arctic because of its proximity to the magnetic north pole, and hunters often rely on traditional techniques such as observing snowdrifts, memorizing landforms, and using celestial navigation (stars) to guide them to their destination.

"For example, when hunters travel on the sea ice, they look at the direction and form of snowdrifts that are created by the wind. Snowdrifts indicate the prevailing wind, which hunters use to identify the direction they are traveling. When it is dark out or there is a blizzard, hunters can use snowdrifts to guide their dog team or snow machine in the direction they want to go, either crossing the snowdrifts or traveling alongside them, a traditional cruise control," Pearce says.

> When I was younger I remember that the ice freezes at the end of September, or the first week of October . . . now it freezes [in] late October, even [the] first week of November.
>
> —H. Ittusardjuat, Igloolik, 2004[12]

> Long ago the cold gradually set in and the ice gets thicker. Now [there are] long spells of strong winds and the ocean can't freeze up.
>
> —H. Ittusardjuat, Igloolik, 2004[13]

"I think that the clouds and the winds are two of the most common things that people use. And people talk a lot about how those have changed," Gearheard explains. "What's really interesting about the traditional forecasting is there's no general set of rules that people use," she adds. "What I've learned is that it's very individual. People might use the same indicators, but the way they use them or read them is a little bit different for each person. The cloud formations, what kind of cloud they are, what direction they're moving. And they also observe the different levels like lower clouds and upper clouds and how the clouds are formed in relation to what wind is blowing at the time." In the Arctic, where the weather is a matter of life and death, each forecast is very personal.

This traditional knowledge is passed down through the generations and combined with repeated personal experience and observations. Each generation learns how to evaluate risks, what preparations to make before going out on the ice to hunt, and what to do in an emergency. Yet although the Inuit use traditional techniques to forecast the weather, a lot of people will also turn on the television or radio to check the weather forecast before they head out on the ice. "Most hunters use a blend of both traditional and new weather forecasting and navigational techniques, with the traditional techniques still holding precedence when push comes to shove," says Pearce.

As this blend of the old and the new takes over, the younger generation loses its connection to traditional knowledge. Pearce attributes the decline partially to the southern educational system that has made its way into the Arctic. Hunting and traditional knowledge are not things teachers from the south bring with them. The less time the members of the younger generation spend hunting, the greater their dependence on wage employment, and the more they become separated from their elders. But the disconnect is also a matter of lack of "necessity." After all, who needs to learn how to predict the weather if television does it for you?

Consequently, certain skills necessary for safe and successful harvesting are not being learned; these include traditional forms of navigation, knowledge of wildlife migration patterns, and the ability to make snow shelters. Also, weaker social networks compromise the ability to cope with changing climatic conditions. Although new technology and institutions can help fill the gap, these sometimes serve to further erode traditional knowledge.

> It is more dangerous [for the younger generation] because they don't know the conditions, what to avoid.
>
> —Kautaq Joseph, Arctic Bay[14]

> I think we have lost the skills so much. I mean, what would have not been dangerous for a man 50 years ago is now dangerous . . . because we have lost so many skills.
>
> —James Ungalak, Igloolik[15]

> We go to areas where we wouldn't normally go because we are assured [by the GPS] we will know where we are. . . . We [also] take more chances.
>
> —Nick Arnatsiaq, Igloolik[16]

> The dog teams know the thin ice and the thicker ice so [people] know that they can walk through thin ice. Snowmobile doesn't say, "Alert! This is thin ice." So it's more dangerous [by snowmobile] than by dog team.
>
> —Herve Paniaq, Igloolik, 2004[17]

> If you don't know the traditional knowledge, you won't last very long: you will freeze to death if you don't know how to survive.
>
> —David Kalluk, Arctic Bay[18]

"Something that I find interesting," Pearce says, "is that when I ask young hunters if they have traveled on the sea ice on their own or as the leader of a group, it's seldom that someone under thirty-four says yes. It's more common for someone to say that they have been out with their grandpa or other relative but not on their own.

"The sea ice has always been dangerous, but climate change has exacerbated some risks, and hazards associated with travel on the sea ice are becoming more common, such as thin and unstable ice in areas where it is expected to be thick. We're seeing very experienced hunters go through the ice."

But the problem is not just that the younger and the older generations are not communicating; it's that some traditional ways of forecasting are beginning to fail.

> The foundation of the wind has changed, it's gone. The wind will now come from any direction, any time of day. Before you could predict [the wind] but not anymore: [the wind] will be from the south and then the same day the wind shifts direction.
>
> —David Aqiaruq, Igloolik, 2006[19]

> The weather nowadays is unpredictable. You can check the five-day forecast but that doesn't mean that's the weather you're going to get.
>
> —G. Lundie, Churchill, 2006[20]

"Inuit do recognize that there is knowledge erosion, but for weather forecasting the problem is not that they don't know how to do it anymore; many people do," Gearheard explains. "It's that their forecasting techniques no longer fit the weather that's happening now. And because forecasting the weather is literally a matter of life and death, they don't want to teach it because it might put someone in harm's way."

Perhaps because of what Gearheard, Pearce, and others have seen personally, many scientists believe that cultural preservation, along with housing and infrastructure improvements, is an important way to help the Inuit simultaneously tackle the issues of climate change and cultural erosion.

The Inuit still hunt narwhals, ringed seals, walrus, beluga whales, arctic char, caribou, polar bears, and a variety of migratory birds.[21] But they see changes in wildlife, too. Ringed seals, for instance, are believed to be particularly susceptible to climate change. They depend on the sea ice for pupping and snow cover in spring to build their birth lairs.[22] All these conditions will be affected by climate change. Climate change is also likely to increase harvesting pressure on seals; easily accessible year-round near most Nunavut communities, seals become "fallback" prey when hunters cannot reach hunting grounds for other species. Caribou are also believed to be susceptible to climate change.

For the Inuit, climate change means greater risk and greater uncertainty. Their own observations have indicated changes in temperature and precipitation, permafrost, coastal erosion, and ice instability.[23] For coastal communities such as Shishmaref in Alaska, perhaps the prime example of climate change in the Arctic, sea ice serves as a buffer against battering waves during severe Arctic storms. Inuit knowledge is already evolving in response to climate change. The increasing unpredictability of the weather and sea ice is becoming part of the collective social memory.

> My aunt, Mable Tooli, said [to me]: "The Earth is faster now." She was not meaning that the time is moving fast these days or that events are going faster. But she was talking about how all this weather is changing. Back in the old days they could predict the weather by observing the stars, the sky, and other events. The old people think that back then they could predict the weather pattern for a few days in advance. Not anymore! And my aunt was saying that because the weather patterns are [changing] so fast now, those predictions cannot be made anymore. The weather patterns are changing so quickly she could think the Earth is moving faster now.
>
> —Caleb Pungowiyi, 2000[24]

Scientists will tell you that climate change is happening faster in the Arctic than anywhere else on the planet. They have been tracking the big picture across the Arctic using every high-tech tool available to them, including satellites and radar. Average temperature has risen almost twice as fast across the Arctic as in the rest of the world during the past few decades. And it's not just the temperature that is moving fast. There is also a widespread melting of glaciers—and a thawing of permafrost, ground that was until now permanently frozen. Permafrost has warmed almost 3.5°F in recent decades. The warming is affecting villages' water supply, sewage systems, and infrastructure, as pipes sit aboveground and are vulnerable to any shifts associated with permafrost. The tree line, another boundary distinguishing the Arctic from points south, is creeping north. And snow cover has decreased about 10 percent during the past thirty years. Winter temperature in Alaska and western Canada has increased about 5°F to 7°F during the past fifty years. Winter is becoming shorter and warmer, and that is not good news if you're a winter person.

There also is the sea ice. The physics of much of this fast change goes back to the ice. First, as Arctic snow and ice melt, darker land and ocean surfaces open up. They absorb a lot more of the sun's energy, and therefore the temperature spikes up faster. This is a process called *Arctic amplification*. The temperature is amplified by the loss of ice. And so it follows that one of the coldest places on the planet is warming up fastest.

Scientists have been carefully tracking the extent of sea ice using satellites since 1978. The extent of Arctic sea ice has a natural cycle: it grows and shrinks with the seasons. Twenty-five years ago, the seasonal range usually peaked in March at about 6 million square miles and shrank to about 3 million square miles at the end of the summer melt season in September.

As the climate grows warmer, the summer melt season lengthens and intensifies, resulting in less sea ice at summer's end. With nearly twenty-four hours of sunlight hitting areas of open water, the ocean heats up more and the open area grows larger. Summer eats into autumn and winter, and sea ice formation, critical for insulating the warmer ocean from the cooler atmosphere, is delayed. Less sea ice at the end of summer pushes more heat into the atmosphere in autumn. The seasons are out of sync. It is only once the sea ice forms that heat exchange is finally capped. With continued summer ice loss and more ocean heat gain, Arctic amplification will eventually eat its way into winter.

Aside from Arctic amplification, there are other reasons the Arctic is seeing such rapid change. One reason has to do with the fact that the air in the Arctic holds less moisture (because it's so cold). Because of this lack of moisture, a greater fraction of the energy that comes from increasing concentrations of greenhouse gases can go directly into warming the atmosphere. In places that are more humid, such as the tropics, the energy is split between warming the air and evaporating the moisture. In other words, in the Arctic, CO_2 has one less job to do. It can focus strictly on warming things up.

There's been a lot of talk within the scientific community about when the Arctic might become ice-free in summer. Since the start of the modern satellite era in late 1978, the extent of Arctic ice has shown a downward trend across all months, with the largest decrease occurring in September. The decline in September is about 12 percent per decade. And starting in about 2002, the pace of melt seems to have picked up. The extreme seasonal minima of 2007 and 2008 reinforce this tendency. Since 2002, scientists have watched in what can best be described as shock as the September decline has continued. According to analyses by the U.S. National Snow and Ice Data Center, where Gearheard works, the September minimum was set in 2007, but 2008 entered the record books as the second-lowest of the satellite era, and probably the second-lowest of at least a century.

Scientists also track the overall thickness of the ice, because this has significant implications for what the Arctic will look like in the future. Ice thickness also has its own rhythm. Back in the 1980s, when the system was still stable and running like clockwork, about 40 percent of the ice in April would consist of young, fairly thin ice that had formed the previous autumn and winter. The other 60 percent consisted of thicker ice that had survived one or more melt seasons. Generally, the older the ice, the more melt seasons it has survived, and the thicker it is. This distribution would change with the seasons. Most of the young, thin ice melted to form open water areas while some of the older ice thinned in summer, and only a bit of it actually melted. The ice that survived the summer thickened again the next autumn and winter. Over the course of each year, ice growth exceeded melt. That doesn't happen anymore.

In recent years, old ice has been harder to come by. The thickness distribution has shifted in favor of the young, thin ice. Because the ice is thinner at the start of the melt season, open water areas develop earlier than before and become more extensive through the summer. As a result, the amplification that boosts the melting is even stronger. The past few years have seen the distribution shift to even thinner spring ice, resulting in even larger open water areas absorbing solar radiation, and an even stronger feedback. The Inuit are seeing it, too.

"People here in Clyde River will say that they've lost three to four weeks of sea ice already," says Gearheard. "The sea ice is so important in terms of hunting and travel. But people so far have been able to cope with it. But it's things like the weather changing and losing their ability to forecast that is really hitting home. That is a significant impact," Gearheard adds.

"You can already see some of the ways the Inuit are adapting," says Pearce. "For example, we're already seeing people invest in larger boats with larger motors because the open water season has lengthened. Some people see a real opportunity to having a boat because they can travel to fall and spring hunting grounds despite the absence of sea ice."

Inuit have no choice but to adapt. And Gearheard and Pearce feel strongly that involving Inuit communities in the research process will help them develop the necessary adaptation strategies. With that in mind, Gearheard and some other community members in Clyde River developed the Igliniit project. The project is part of the Inuit Sea Ice Use and Occupancy Project. An International Polar Year project, Igliniit brings together Inuit knowledge with engineering and cutting-edge technology. In Inuktitut, *igliniit* refers to the trails that are routinely traveled by hunters and other members of the community.

The location, use, condition, and changes in these *igliniit* over time and space can help both community members and other researchers learn a great deal about the Arctic. Engineers and Inuit hunters have come together to design a new, integrated GPS system that can be mounted on snowmobiles. The GPS automatically logs the location of the snow machine every thirty seconds, providing geo-referenced waypoints that can later be mapped to produce the travelers' routes on a map. In addition to tracking routes, the Igliniit system logs weather conditions (temperature, humidity, pressure, etc.) and the observations of hunters (animals, sea ice features, hazards, place-names, etc.) through a customized computer screen that is in Inuktitut and has a user-friendly icon interface.

The hunters carry digital cameras that allow them to take pictures of conditions as well as make videos. The data and images can then be downloaded and turned into maps. These maps show the routes of individual snowmobiles, along with the geo-referenced observations of the hunters and weather conditions. When you overlay maps of different hunters over time and space, you end up with a valuable picture of meteorological data and Inuit land, ice, and resource use. At the moment, the Igliniit system is being tested on snowmobiles and dog teams, but Gearheard hopes it will eventually be made available to boats and all-terrain vehicles.

Gearheard and her team hope the Igliniit system will provide useful information for communities like Clyde River. Hunters can print out their own maps to keep a record of their travel routes and hunting spots. And collectively, the community can use the maps to see where its members have had the most hunting success, changes in animal populations, changes in snow conditions, connections between weather conditions and travel conditions, and locations of hazards.

"In traditional Inuit society people share food," explains Pearce. "It can be community to community as well as household to household. For example, when Ulukhaktok has limited access to caribou, a nearby community in Nunavut will help them out and send caribou. Or if Sachs Harbor has geese or Tuktoyaktuk has beluga whale and Ulukhaktok doesn't, they will send geese and *muktuk* (whale meat). Whereas Ulukhaktok might have a really good char run that year, so they will send char."

These maps might also help preserve traditional values. In addition, community leaders can use the maps for matters related to planning or negotiating land use. The weather information is also extremely valuable, as weather data are spotty in the Arctic, restricted to a limited number of weather stations. Last, Igliniit has the potential to serve a role in search and rescue operations. Igliniit maps can be used to keep up-to-date records of sea ice, land, and water hazards logged by hunters, and these maps can be easily shared with other hunters. Gearheard's team is currently looking into the possibility of incorporating live tracking or personal locator beacons (PLBs) into the Igliniit system, so that anyone carrying an Igliniit unit can send out a warning signal in order to be tracked quickly.

"You can look at climate change in isolation, but I think it would be a real disappointment," says Pearce. That is exactly why he, Gearheard, and others are working to connect climate change to other issues facing Arctic communities. By connecting climate change to issues like education, sustainable development, and alleviating poverty, you can reduce vulnerability across the board. When you find ways to strengthen Inuit communities today, you can also strengthen Inuit communities in the future.

"I'm always hearing the words 'in the future' or 'future projections,' " says Pearce. "But when I am in the Arctic, I am constantly struck by the fact that the future has arrived, and Inuit are living it. It's happening now." Under the scenario of the Intergovernmental Panel on Climate Change (IPCC) for "business as usual" greenhouse gas emissions, simulations from the current generation of coupled global climate models indicate that the Arctic will warm between 7°F and 13°F over the next 100 years. And the Arctic Ocean could become seasonally ice-free, or nearly so, sometime after 2040.

Whereas most climate model simulations show the September loss of sea ice increasing in the coming decades, this acceleration seems to have already begun. In other words, when you compare the present with future projections from climate models, we are already on the fast track of climate change. The Earth, as the Inuit say, is faster now.

I asked Gearheard what she imagined the Inuit might be doing fifty years from now if all the summer sea ice disappears. "I guess when I think about the future, I just think about the people that I know right now and try to imagine what they would be doing. And I can't imagine that the people I know would just stop hunting or stop going out on the land. Every person that I know who is active in that way would still be doing it. They might be hunting something different. They might be hunting somewhere else. They might be selling their ski-doo and buying a boat. But no matter what happens, they would still be Inuit and they would still be out there hunting."

This raises a question: what happens to the Inuit if the snow and ice go away? "That's a really big question," says Gearheard. But she sees the connection to snow and ice and tradition even in very young children in Clyde River. "If you ask them to draw a picture of their family, you'll very often get a picture of their family on a dogsled traveling over the ice. Even if they don't have one," Gearheard says. "The snow, the ice, it's all wrapped up in there."

"The truth is," says Pearce, "in the Arctic, people have a strong connection with their environment. This is their home; this is where their life is. They are going to adapt to new conditions regardless if any action is taken against climate change. They are not going anywhere. They make that very clear. If you ask an Inuit in Ulukhaktok what their future aspirations are, they will tell you that they plan on staying in the Arctic. End of story." Sea ice or no sea ice. I guess that also applies to the rest of us.

9

THE ARCTIC, PART TWO: GREENLAND

Erik the Red may have deserved to go to hell, but instead he went to Greenland. In A.D. 982, after being banished from Norway and then Iceland for murder, the infamous Viking explorer headed west. Only 500 miles from the shores of Iceland, he discovered a beautiful island with rich pastureland tucked into deep fjords overflowing with crystal blue water. When his term of banishment expired three years later, he returned to Iceland, in the hope of finding a few brave souls willing to join him and settle this little slice of heaven he had discovered. Erik the Red christened his new home *Greenland*.

It has been suggested that the naming of Greenland is a very early example of bait-and-switch advertising—a deceitful way to get warm bodies to a cold place. The right name may be all it takes to sell people on a dream of a better life and a brighter future. On the other hand, although Erik the Red seems to have been a murderer, he wasn't necessarily a liar. You might say that in naming Greenland, he told a white lie. At the time, Greenland was actually much greener than it is today. From about A.D. 800 to 1300, the Medieval Warm Period, Greenland's climate was much milder, and the southern part of Greenland, where Erik the Red settled, did indeed have meadows lush with grass, willows, and wild berries.

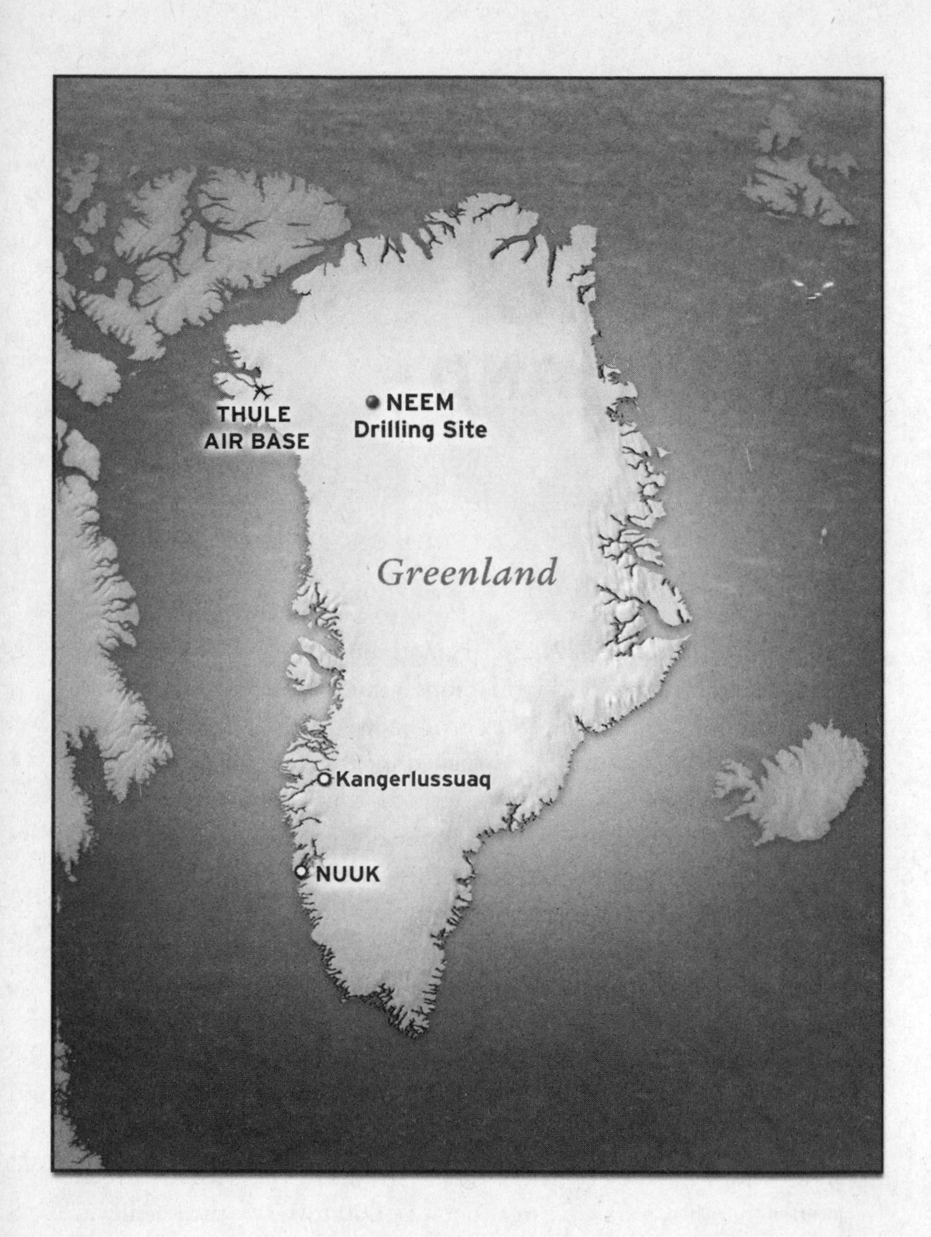
THULE
AIR BASE
NEEM
Drilling Site
Greenland
Kangerlussuaq
NUUK

In any case, Erik the Red must have been a good pitchman, because in 985 he led a fleet of twenty-five Viking longships to settle two new colonies on Greenland. During the next ten years, as the news of free pastureland traveled back to Iceland, three more ships carrying hopeful settlers set sail for Greenland. And by the year 1000, virtually all the land suitable for farms in the Western and Eastern settlements of Greenland had been claimed. About 1,000 people lived at the Western Settlement and 4,000 people at the Eastern Settlement, which, despite its name, was located about 300 miles to the south.[1] Erik the Red had successfully converted 5,000 Icelanders into Greenlanders, but he certainly hadn't led them to the promised land.

Over time, Erik the Red's original white lie grew whiter. Summers were becoming shorter and cooler, and winters were downright frigid—even by Viking standards, which were quite harsh. This limited the amount of time the cattle, sheep, goats, and horses could be kept outside to pasture and increased the need for winter fodder. As the temperature dropped, the amount of sea ice increased. The sea ice became a frozen barrier, making it more and more difficult for ships to pass. As a result, trade and communication with Europe and Scandinavia were choked off and Greenland became increasingly isolated.

The cooler climate also brought the Inuit down from the north and into more regular contact with the Norse colony. Their relationship might have served as an impetus for change, pushing the Viking settlers to find new ways to deal with the cold; but in fact it only brought more problems. The little archaeological evidence that exists suggests that there was violence between the two groups. The changing climate had ushered in a period now known as the Little Ice Age. And after 500 years of settlement, the Viking colony, unable to adapt to the cooler conditions and unwilling to supplement Scandinavian tradition with Inuit coping strategies, eventually collapsed. The last written record of the Norse Greenlanders comes from a marriage in the church of Hvalsey in 1408. The church still stands today.

In May 1721, Hans Egede, a thirty-five-year-old Lutheran missionary, received permission from Frederick IV of Denmark to search for Erik the Red's lost colony. No word had come out of Greenland for more than 300 years, and Egede feared that the Viking colony was lost or, perhaps worse, that the colonists had lost their faith. And so Egede and his wife set sail from Bergen, Norway, and headed for Greenland, where they intended to set up a mission. Upon their arrival, Egede found no Norse survivors. He did, however, find the Inuit. And so he started his mission among them.

Egede, called the apostle of the Eskimos, spent fifteen years in Greenland. He studied the Inuit language and tried his hand at translating Christian texts, a task that required the ability to adapt the text in such a way that his teachings would resonate with the Inuit experience. For instance, the Inuit did not eat bread—a fact that made the Lord's Prayer rather cryptic. Egede made one small but critical adjustment and wrote, "Give us this day our daily harbor seal." It seems that Egede decided to leave the concept of hell unaltered, despite the cold climate. The Inuit learned about a very hot place where sinners were sent for eternity.

Today, new settlers are traveling to Greenland in search of the promised land. But they come in corporate jets rather than Viking longships, and this time the Inuit are happy to see them. Greenland figuratively and (as I will explain later) literally is on the rise.

Gold and diamond prospectors are heading to the southern part of Greenland. Alcoa, the U.S. aluminum giant, is preparing to build a smelter powered by hydroelectric energy in Maniitsoq, on Greenland's west coast. In addition to precious metals and diamonds, Greenland also has oil and natural gas. The United States Geological Survey (USGS) recently assessed the area north of the Arctic Circle and concluded that about 30 percent of the world's undiscovered gas and 13 percent of the undiscovered oil may be found there, mostly offshore and under less than 550 yards of water.[2] The USGS estimates that the Arctic could hold 90 billion barrels of oil and 1,670 trillion cubic feet of gas, much of it off Greenland. East Greenland alone is estimated to contain 8.9 billion barrels of oil. Exxon Mobil and Chevron are already increasing their exploration. But there's a catch.

Greenland is hidden under a 1.6-mile-thick layer of ice known as the Greenland Ice Sheet (GIS). And with more than 80 percent of the island essentially on ice, much of Greenland's potential wealth is more or less wishful thinking. Greenland may not be the promised land quite yet, but it is a land full of promises.

A Danish protectorate since 1721, Greenland has long sought to sever ties with its benevolent colonizer, a colonizer that supports the island with an annual grant of $600 million. On June 21, 2009, Greenland came one step closer to eventual independence from Denmark by voting in a new era of self-governance. Under the new self-government agreement, Greenland will keep half of all proceeds from oil and mineral finds. At first it will also continue to receive the $600 million annual grant from Denmark, but as petroleum and mineral revenues increase, the grant will be reduced—and it will continue to be reduced until it hits zero. Greenland can choose to secede from Denmark anytime along the way. As the ice melts, the money will rush in—or so the thinking goes.

If a changing climate helps speed that process up, then many Greenlanders say bring it on. For the people of Greenland, 90 percent of whom are indigenous Inuit, the question of how quickly Greenland's ice will melt is not merely an abstraction; it represents freedom. However, there's another catch. If the GIS melted, exposing all the riches beneath it, it would also raise sea level worldwide by 23 feet.[3] This is why the loss of the GIS represents one of the worst-case scenarios with regard to global warming.

There is a rather straightforward accounting system as it applies to global warming in Greenland. There are pros and cons, pluses and minuses, gains and losses. The larger question is: how will these add up? Warmer winters are already making life tough for traditional communities that hunt and rely on predictable sea ice (see Chapter 8). Hunters who use the sea ice for hunting and travel have found themselves idle when the ice fails to form and the whales, seals, and birds they hunt shift their migratory routes. Melting permafrost is buckling roads and airport runways, raising costs for the mining companies that are seeking aluminum, diamonds, gold, zinc, and more. But the warmer weather also encourages tourism, and the loss of ice means that the ship transport season in the Arctic is easier and longer. Fishermen report a rise in some fish stocks, including cod and halibut, as a result of the warm-water currents that now flood into Disko Bay. Shops in the island's capital, Nuuk, have even begun to offer homegrown potatoes and broccoli—crops you don't necessarily associate with Greenland. Whether you think in terms of dollars, temperature, glaciers, or even broccoli, there are plenty of people who like what they see.[4] Ultimately, though, the issue comes back to the ice. And you can argue that whatever factors control the amount of ice control Greenland's destiny. That is where the scientists come in.

Scientists have been trying to understand Greenland's ice for a long time. There is no doubt that Greenland fluctuates between warmer and cooler, wetter and drier, greener and whiter. Hans Egede Saabye—who was the grandson of Hans Egede and was also a missionary—first noticed this.[5] Saabye evidently had a keen eye for changes in weather and climate. He was an *observationalist* in the classic sense of the word, collecting data and monitoring change. Saabye wrote down these observations in his diary, and one of his comments was particularly perceptive: "In Greenland, all winters are severe, yet they are not alike. . . . When the winter in Denmark was severe, as we perceive it, the winter in Greenland in its manner was mild, and conversely." Scientists now understand that Saabye was describing an important atmospheric pattern called the North Atlantic Oscillation (NAO).

The NAO is a phenomenon that affects weather and climate from North America to Siberia and from the Arctic to the equator.[6] It is a dominant mode of natural climate variability; and as with El Niño, scientists are working to develop a long-range seasonal forecast for it. The main feature of the NAO is a seesaw of atmospheric pressure between a persistent high-pressure cell over the Azores and an equally persistent low-pressure cell over Iceland. The NAO index, which swings between positive and negative, represents a measure of the relative strength of these two pressure systems. Depending on the phase, the NAO can bring large changes in surface air temperature, winds, storminess, and precipitation.[7] The NAO is most pronounced during the winter, and that is why the connection to the ice exists. In the case of Greenland, it appears that snow falling in the center of the island is linked to a negative NAO index.[8] But the NAO is just one of many factors affecting Greenland's ice; that is why ice sheets are a very complicated matter.

When scientists talk about the GIS, they are talking about *mass balance*. This concept implies that the amount of snow falling in the middle of Greenland is balanced by the amount melting along the sides. During the early 1990s, the GIS was in the Zen-like state of mass balance.[9] Today, scientists are talking about *mass loss* because all signs point to the fact that the GIS is melting much faster than it's growing.[10] Mass loss is caused by a combination of two factors. There is *melt*, which can result from an increase in temperature and is caused by a variety of factors. There is also a process called *calving*. Calving, a much slower process than melting, takes place when glaciers flow into the sea and eventually break away from the coast. A recent study conducted with the Ice, Cloud, and Land Elevation Satellite (IceSat)—a NASA satellite that uses lasers to calculate change in elevation—found that the glaciers are indeed speeding up where they flow into the sea.[11] Scientists think that the most likely cause of faster glacier flow is warm ocean currents reaching the coast and melting the glacier front.[12] But calving is so poorly understood that it remains one of the most unpredictable components of future rises in the sea level.

Since 2002, Greenland has come under the watchful eye of another NASA satellite mission: the Gravity Recovery and Climate Experiment, or GRACE, which some scientists call *amazing GRACE.* This mission doesn't see continents or oceans so much as it sees gravity. Since its launch in 2002, GRACE has been acquiring ultraprecise measurements of Greenland's mass loss. Scott Luthcke, a geophysicist at NASA's Goddard Space Flight Center in Greenbelt, Maryland, describes GRACE like this. "Imagine you caught a big fish and you wanted to weigh it. One way you could do it is you could go out and buy a scale. But you could also just use a spring," Luthcke begins. "The distance that the spring stretches when you hang the fish from it is a representation of how much the fish weighs. That's pretty much what GRACE is doing." Of course, GRACE, which is able to measure changes as small as the width of a human hair, is doing it (as noted above) with high precision, using microwaves that essentially act as the spring to measure the mass of the fish—the GIS. Technically, GRACE, two identical satellites separated by a distance of 137 miles, is doing it from a polar orbit 310 miles above the Earth's surface.

The beauty, or perhaps the amazing part, of GRACE is that every month since 2003 it has flown over the GIS, keeping track of its comings and goings, its growing and melting. When you fly over the same place again and again, you can measure how much it's pulled apart and compressed together and how much the mass on the ground has changed. Unfortunately, in the case of the GIS, the fish is slowly disappearing.

Luthcke is an observationalist in the purest sense. He just happens to be working with some of the most sophisticated measuring devices ever built. Data from GRACE shows that Greenland lost about 200 billion tons of water per year from July 2003 to

July 2008. That works out to the equivalent of Lake Erie draining into the ocean every two years. Needless to say Greenland is losing weight.

This enormous weight loss is a result of melting and thinning at the margins along Greenland's coast[13] as well as a surging (or acceleration) of outlet glaciers into the sea.[14] One example of an outlet glacier is the Jakobshavn Isbrae glacier.[15] Jakobshavn Isbrae is Greenland's largest outlet glacier, draining about 6.5 percent of the GIS area. It has been surveyed repeatedly since 1991 and has been accelerating since the mid-1990s. That said, this tremendous crumbling and melting along Greenland's coastal margins is partly compensated for by some mass gain in the interior of the island, a gain that is controlled in part by the North Atlantic Oscillation.[16] Even so, the GIS has shrunk so much in recent years that the underlying bedrock, like a ship emptying of its cargo, is lurching up at a rate of about 1.6 inches each year. Greenland is indeed rising.

J. P. Steffensen, a scientist at the Center for Ice and Climate at the University of Copenhagen's Niels Bohr Institute, agrees that an amount of meltwater equivalent to Lake Erie every two years is pretty impressive, but it's virtually nothing compared with what he's seen happen in his ice cores. "We've seen climate changes that would have wiped off life, changes that are just mind-boggling fast," says Steffensen. "We literally see the ice age ending from one year to the next," Steffensen says. "It's pretty scary."

Steffensen reads ice the way most of us read a book. And there is one chapter that worries him. "The Earth has always had climate changes. The problem is that right now we have a climate change which is permanent," he explains. "Even though nature has done it by itself before, we should not put ourselves into the position of kicking the climate system. That's my little worry."

Steffensen is the field operation manager for the North Greenland Eemian Ice Drilling Project (NEEM). The NEEM camp, located in the far north of Greenland (see map), brings together scientists from fourteen nations including the United States and is aimed at retrieving a core of solid ice 1.6 miles long. The goal is to unlock the climate history trapped inside tiny air bubbles from the ancient atmosphere. You could say Steffensen is old school. He prefers data to models, hardware to software, and ice cores to satellites. "There's a much-admired saying in our community," Steffensen says with a smile: "There's no substitute for data." That is where the ice cores come in. "I mean, when you do climate models, you have to realize that the models do not include the things you don't know. With ice core data, we can see things happening in the climate system that the models so far have never been able to capture. We reveal the climate history," Steffensen says.

He gives an example of two rapid warming periods he sees in the ice that constitute a profound climate shift: one at 14,700 and one at 11,700 years ago.[17] It's thought that the shift must have come from the atmosphere, which is far more nimble than the slowly churning oceans. If the atmosphere is capable of suddenly flipping into a new pattern, then this new pattern could have contributed to the rapid warming of the entire northern hemisphere.[18] Taken together, these two pulses of rapid warmth pushed the Earth's climate out of the last ice age and into the Holocene, our current warm period. The part that scares Steffensen is that during the ice age, the Earth's climate was far more unstable than it has been of late. The Holocene, in addition to being warm, is known for its rather remarkable stability. In this regard, observations of the real-world climate system and simulations from climate models have yet to overlap.

Although climate models help scientists better understand the complex mechanisms that create rapid climate shifts, the models can't seem to actually make such shifts happen. Typically, when a climate model tries to simulate the abrupt natural climatic shifts of the distant past, those shifts end up taking more than 100 years to occur—which is hardly abrupt.[19] This suggests that some aspects of the physics still need to be worked out. And Steffensen is hoping that data from the NEEM ice core can be of some use. This specific ice core will allow scientists to see the climate of Greenland over the past 130,000 years, and isolate a fascinating but poorly documented interglacial period known as the Eemian. During the Eemian, Greenland's temperature was about 5°F to 9°F warmer than it is today, so this period is a meaningful analogue for future climate.[20] Global sea level rise during the Eemian is also a matter of great interest, as it's likely that the sea level was between 13 and 20 feet higher. Steffensen and his colleagues at NEEM are looking at the Eemian to learn how quickly the ice covering Greenland might melt when the climate is that much warmer.

The NEEM project officially began in 2007, when Steffensen and some colleagues dragged equipment from the previous drill site—a camp called NGRIP—over to the NEEM drill site. Steffensen has carefully selected this new drill site on the basis of radar profiling of the internal ice layers and the bedrock topography. If an ice core really is like a book, then in this case the book is *War and Peace*, and of the 1.6 miles of ice each chapter or year of climate history during the Eemian is about one-third inch long. Ice cores, like tree rings, allow you to reconstruct climate on an annual basis. With each passing year, snow falling on central Greenland lays down a distinct layer, trapping bubbles of atmospheric gas, dust, and other impurities and gradually compacting into ice that captures an ancient climate record stretching back hundreds of thousands of years.

"Ice cores serve as a remarkable archive of past climate and atmosphere because of the bubbles of air that are trapped in the ice," explains Jeff Severinghaus, a climate scientist at the Scripps Institution of Oceanography in La Jolla, California. "And the beautiful thing about an ice core is that it has all of these different indicators: atmosphere composition, temperature, mean ocean temperature, dust." The oldest ice cores now go back about 800,000 years, and scientists are optimistic about pushing this method back even farther. Many of the scientists are involved in International Partnerships in Ice Core Sciences, which is aiming to retrieve a 1.5-million-year-old ice core. "And what's so remarkable is that you can still answer questions down to the year," says Severinghaus. "It's really like pulling back the veil."

What worries many scientists is that behind the veil they might find a threshold or a tipping point past which the GIS becomes perpetually out of balance, unstable, locked in a persistent state of wasting away.[21] Right now, the temperature range assigned to that tipping point is large because there is a high degree of uncertainty. The IPCC puts the temperature increase at somewhere between 3°F and 8°F above the long-term average. Given the 1.3°F of warming we've already put into the system, it's a temperature range that could easily be in the cards during this century.

A fundamental problem is that existing models of ice sheets are unable to explain the speed of the recent changes in the GIS that GRACE and IceSat are observing. In other words, the models cannot reproduce the data. Scientists such as Scott Luthcke are seeing things happen in Greenland right now that, technically, the models don't show as happening for another thirty years.[22] Even if some temperature threshold is passed, the IPCC gives a 1,000-year timescale for a total collapse of the GIS. But, given the inability of current models to simulate the rapid disappearance of continental ice right now, let alone at the end of the last ice age, a lower limit of 300 years is conceivable.[23]

I met up with Steffensen, Severinghaus, and other scientists from the NEEM project in Kangerlussuaq, a former Cold War outpost for the U.S. Army and now the site of Greenland's major international airport. Steffensen, who is Danish, has spent much of his life studying Greenland's ice. "I totaled it up," he says, taking a puff on his pipe. "I'm fifty-two years old now and I've been to Greenland twenty-three times. That works out to spending almost six years on the ice. I guess you could say the ice went straight into my belly and it stayed there."

Kangerlussuaq also serves as the staging ground for the NEEM scientists who are flying north to the ice camp. This time Steffensen will stay behind to manage the logistics and make sure the ice cores get safely onto the bright red Greenland Airlines jets bound for Copenhagen. But he is no stranger to life in a remote field camp. His first season on the ice was in 1980. "It was a marriage for life," he says solemnly. That statement turns out be more true than I realized. Steffensen's wife is a fellow scientist, Dorthe Dahl-Jensen, also a professor at the University of Copenhagen and the project leader at the NEEM drill site. She shares with Steffensen the difficult task of coordinating the drill teams and the scientists, as well as making sure the ice cores get safely from the drill camp to laboratories around the world. "But my main interest is really the ice," she says. "So, I normally find the time every day to go down into the science trench and work with the core samples, because that's really my heart," she says with a shy smile.

If the NEEM project is successful, it will be the first complete record of the Eemian. None of the former deep ice cores from Greenland contain complete and undisturbed layers from this warm period, because the layers had either melted or been disturbed by ice flow close to the bedrock. "The last several ice cores in Greenland tried to get this interglacial period but didn't quite succeed," explains Jeff Severinghaus. "NEEM is really trying to get a record of the last time that the Earth was warmer than today," he explains. "So the Eemian is really an analogue of what our future looks like under global warming. It's a very, very realistic scenario for what we may experience in the next 100 to 200 years."

"We know that even though it was warmer in Greenland, it wasn't warm enough for the whole Greenland Ice Sheet to disintegrate," explains Dahl-Jensen. "And that's something that is pretty hotly debated." Specifically, this is the question of the tipping point, or how much warming we would need in the future before the GIS would totally disappear. Scientists already know that shrinkage of the GIS during the Eemian contributed an estimated 6 to 10 feet to the global rise in sea level, although a widespread ice cap still remained over portions of Greenland.[24] "Our earlier results tell us that during the Eemian, the Greenland Ice Sheet was about 30 percent smaller. And that tells us that roughly 3 to 7 feet of global sea level rise came from the Greenland Ice Sheet alone," Dahl-Jensen says. "We can also see that when the climate is warmer, it is also very stable. And that's another big debate. If we aggressively warm the climate, will it shift back into an unstable regime?"

If the weight of 130,000 years of climate history isn't enough pressure on these delicate layers of ice, the hopes of the climate science community are also bearing down. "I feel a sense of awe when I am able to peer into the deep, deep past time," explains Severinghaus. "It's very hard to put into words, but it's really quite a sense of excitement and wonder and mystery." And perhaps just a touch of dread.

The NEEM field camp accommodates about thirty researchers and technicians from May to August. "It's kind of like a frontier outpost up here," says Vasilii Petrenko, a scientist at the University of Colorado. "It's a very simple life. We're working for about fifteen hours a day most days. But we take breaks. We come back for lunch; we come back to warm up sometimes, get some tea and cookies. And everybody gathers in camp for dinner." The food at NEEM gets universal raves. "I can honestly say that I eat better here than I do at home. It's a very calorie-rich diet, but you need it here. Once you've spent a couple of days in the science trench, your body starts adjusting to the cold, and producing more heat, so our calorie requirements just skyrocket. I probably eat twice as much here as I do normally at home."

The field camp may look like a frontier outpost on the surface, but the scientists are engaged in some very sophisticated climate research in the underground science trench dug from the snow 30 feet below the surface where the ice core drilling and initial processing is done.

Developing a climate history involves reading many types of measurements from the wide variety of particles that get trapped.

Oxygen isotopes are a proxy for local temperature; excess deuterium is a proxy for ocean surface temperature; dust and calcium originate from low-latitude Asian deserts; sodium indicates marine sea salt. Impurities in the ice reflect the impurity load of the atmosphere of the past, and gas bubbles trapped between the snow crystals contain samples of the actual atmosphere, reflecting the amount of greenhouse gases such as carbon dioxide. The crystal structure of ice and the content of biological material also provide information about past climatic conditions. Volcanic eruptions can be used to date the ice. A peak of volcanic dust in the ice core allows you to match it to a volcanic eruption and have an independent estimate of age.

"All of these kinds of different indicators are on exactly the same timescale, so you can really make detailed comparisons between one indicator and another," explains Severinghaus. "Carbon dioxide is a strong greenhouse gas. So, one of the things that we see in the ice cores is a strong correlation between carbon dioxide levels and temperatures. So at times of warm temperatures, carbon dioxide is high; at times of cold temperatures, carbon dioxide is low—which reinforces what science has been showing recently: that carbon dioxide does cause warming. During the Eemian, carbon dioxide was definitely lower than today, a lot lower than today," says Petrenko.

"It suggests that today's carbon dioxide levels are entering the danger zone, basically. The Earth hasn't seen these levels of CO_2 for millions of years, which means that we are headed for a climate that is beyond anything that the ice cores can show," Petrenko continues. With regard to the future and the still unwritten chapters of climate history, past data can take you only so far. "The Earth is not a system that you should do experiments on," says Severinghaus. "Better to look at the experiments that Mother Nature did on her own, in the past, and study the results of those experiments. That's why looking at ice cores is incredibly valuable," he adds.

Back in Kangerlussuaq, I meet up with Steffensen before the long flight home. He reflects on the work of the NEEM researchers and on the looming issue of climate change. "We have to get used to this word *change*," he says. "That's why we have a past, why we have a future—time is flowing forward. We should never strive to re-create the past. That's impossible, because nothing will ever be as it has been. So my point is we should look forward with hope. But," he adds after a moment, "we should never forget that nature can also turn dreadful." Just ask Erik the Red.

The Arctic: The Forty-Year Forecast—Ice Melt, Mineral Resources, and a Hospitable Arctic Circle

CLYDE RIVER, NUNAVUT	TODAY		2050		2090	
Emissions Scenario	JAN.	JULY	JAN.	JULY	JAN.	JULY
Higher Emissions	–18.6	38.9	–12.6	40.6	–3.7	42.7
Lower Emissions			–13.9	40.2	–10.4	41.1

KANGERLUSSUAQ, GREENLAND	TODAY		2050		2090	
Emissions Scenario	JAN.	JULY	JAN.	JULY	JAN.	JULY
A2	–0.6	49	3.2	51.4	8.5	54.2
B1			3.2	51	5.3	51.8

Forecast June 2011

The northern coastline of Alaska was changing. Midway between Point Barrow and Prudhoe Bay, a stretch of coastal cliffs as long as a football field was being devoured by the ocean every three years. As bigger and bigger waves pounded away at the shoreline and warm seawater chipped away at their base, the 12-foot-high bluffs lining Prudhoe Bay toppled into the Beaufort Sea—more than 30 feet disappeared every year.[25] Up here, climate change was a triple threat, involving warmer oceans, stronger waves, and shrinking sea ice.

Arctic sea ice was now declining at a rate of almost 12 percent per decade. And to make matters worse, less than 20 percent of the ice cover was more than two years old—the lowest amount ever recorded since satellite measurements began. The young ice, so thin and fragile, didn't stand a chance.

Despite the dramatic changes taking place across the Arctic, it wasn't the satellite data or the retreat of sea ice that transformed many skeptics into believers. It was the Russians. When the Russians planted a titanium flag at the bottom of the Arctic Ocean, in order to lay claim to the possibly vast oil and mineral deposits, many skeptics took notice. Still, it was unclear how best to think about what was happening in the Arctic. Was this like the race to the moon—a matter of national pride requiring us to beat the Russians, or anyone else? Or was the Arctic a modern-day version of the Wild West?

In any event, there were enough old-fashioned border disputes to keep everyone busy. There was a border dispute between Canada and the United States over the legal status of the Northwest Passage; there was a dispute between Greenland and Denmark over economic and political independence; and there was a dispute among all Arctic nations over access to mineral rights. These are not things you bother to fight about unless you are fundamentally convinced that the ice will be gone someday. And yet, ironically, countries that had originally denied climate change were now scrambling over resources that would be valuable only if the climate models proved to be correct. This was proof that even sophisticated countries were still better at grasping short-term opportunities than responding rationally to long-term threats. There is nothing like a good old-fashioned gold rush to turn people into believers.

In 1946, the U.S. government was so impressed with the strategic potential of Greenland—the world's largest island—that it secretly attempted to buy this island from Denmark for $100 million. By 2011, the United States was wishing it had offered a lot more. The money to be made from aluminum production alone was enough to make your head spin. Alcoa had dammed up two rivers in West Greenland and built one of the world's largest aluminum smelters—mining about 340,000 tons a year. But Alcoa had to share its profits. Part of the deal was that the Home Rule Government of Greenland and Alcoa jointly owned the hydro power stations, the transmission lines, and the smelter plant. The people of Greenland knew a good opportunity when they saw it. And for them, climate change was a path to freedom. There was money to be made in the new Greenland. It was just a question of who was going to make it.

May 2022

As the sea ice continued to melt, the disputes between Canada and the United States over control of the Northwest Passage increased. Canada maintained that the passage lay inside its territorial waters allowing it to exercise control over all ship traffic; the United States wanted the passage to be classified as an international sea-lane, outside any one nation's jurisdiction. Despite the controversy, the Canadian government staked its claim to the Arctic waters. It constructed new naval bases in the area and ordered a new fleet of Arctic patrol boats. It also established underwater listening posts for submarines and ships. In the end, Asia, Europe, Russia, and the United States refused to budge and the Northwest and Northeast passages were deemed international waters. Canada was, however, able to collect a small fee—about one-fifth of the $4 billion generated annually by the Suez Canal—for maintenance.

The retreating sea ice also opened a potential for deepwater drilling for oil and gas deposits. The Russians were holding tight to their claim that a large portion of the Arctic was their geological territory. But the gold rush philosophy of the "Wild North" could more accurately be described as "First come, first served."

As claims were staked and deals were made, fishermen a little farther south went in search of missing cod. The North Sea—the turbulent pocket of ocean bordered by the United Kingdom and Scandinavia (among others)—had always been a very fertile fishing ground. A little more than a century ago it yielded almost 20 percent of the world's fish harvest.

But by 2022, the North Sea was, on average, 3°F warmer than it had been fifty years before. The warmer temperature had chased away all the plankton that young cod eat in the spring. Just as the fishermen were searching for the cod, the cod were searching for the plankton. And the plankton were looking for cooler water.[26] In this new world, everybody was looking for something.

Because of climate change, nothing was where it was supposed to be, so international fishery management was a bit of a mess. In the meantime, the fish-and-chip shops in England were buying their cod from trawlers sailing the coasts of Iceland. The fishermen were trying to scrape by on shrimp and whelks (large marine snails). And the jellyfish had decided that the North Sea felt just right.

The telltale signs of a shifting climate were quite obvious in other ways, too. Across the lands of the Inuit, the formerly pale brown, treeless landscape had begun to turn dark green; spruce, larch, and fir trees were popping up from Sachs Harbor to Clyde River to Iqaluit. But the warmer temperatures meant serious trouble for any species that relied on the ice. Hunting had become increasingly difficult as the delicate food web continued to unravel. Seals and walrus depended on cod, the cod depended on crustaceans, the crustaceans depended on algae, and the algae depended on the ice to provide a home. In essence, the hunt for seals and walrus had become a hunt for ice. The amount of time hunters were able to spend out on the ice had shrunk from months to weeks to days as vast areas of open water made traditional hunting grounds inaccessible. Hunters were often forced to shoot some of their sled dogs because they had no walrus or seal meat to feed them. The "ice highway" the hunters had relied upon for centuries was closing down permanently.

Overall, the ground itself had become more and more unstable as permafrost thawed and gave way. Roads buckled, sewer and waterlines burst, home foundations sagged, and trees tipped over in random directions as if they were drunk. The permafrost was turning into a soggy sponge. There were countless attempts to find clever ways to keep the permafrost frozen, including refrigerated slabs and insulated carpets. But in the end chemistry always won. The heat was unstoppable.

September 2032

The year 2040 was when climate scientists had collectively predicted the Arctic would be fully ice-free in summer. We all knew this was a conservative estimate. Most sea ice models, despite improvements in the physics and better satellite data, had repeatedly underestimated the speed of Arctic ice melt over the past decade. It was just a question of how much the models were underestimating the melt.

In the end, as natural climate variations such as the North Atlantic Oscillation and the human-induced long-term trend played tug-of-war, the models were off by only eight years. By 2032, summer sea ice in the Arctic had all but disappeared. As they patiently waited for the ice to recede, shipping companies had been carefully planning for the opening of the Northwest and Northeast passages. They reengineered the next generation of Arctic-ready cargo ships and trained crews in how to deal with the cold. When the time finally came, they were ready to take a test drive in the Northwest and Northeast passages.

The trip from Yokohama to New York via the Northwest Passage was 2,200 miles shorter than by the Suez Canal. And the Northeast Passage to Europe via the North Sea would save about 4,200 miles between Yokohama and Rotterdam. From Singapore to Rotterdam, it was about 1,300 miles shorter than going through the very expensive Suez Canal. But despite the focus on miles, the experts were quick to point out that it was *time* they really wanted to save. They estimated that the Northeast Passage could shave off about seven days of travel time.

But still, a lot of risks came with Arctic shipping routes. The worst was the need for absolutely open water so the ships would be able to maintain speed. If a ship hit even a small chunk of ice at 22 knots, the crew would have to deal with a hefty hole. Needless to say, there were still plenty of small chunks of ice floating around that were difficult to spot, even with the new remote sensing equipment. The first major oil spill occurred just a few weeks after the new routes opened, when an empty container ship heading back to Yokohama from New York collided with a chunk of ice moving in for winter. This was bad news all around; but it was especially bad for the fish, as they had fewer and fewer places to escape to.

The Norwegians had always billed their fish as the "purest fish in the world," thanks to the clean, cold Arctic waters. Uncontaminated fish were almost impossible to come by, and the Norwegians had been charging a premium. Needless to say, this wasn't a claim they could make anymore.

The warming had brought cod, herring, halibut, and haddock north in search of food, but it also brought oil and gas tankers and container ships from all over the world. In addition to rice from China and cars from Japan, the ships brought various contaminants and diseases that fouled the Arctic waters. Pollution and disease devastated the fishing stock. The Norwegian defense minister said in a speech at an international meeting in Moscow to discuss how to handle the collapse of the fisheries, "It used to be we had a world with plenty of food but bad distribution. Now we have a world where we have plenty of distribution options, but not enough food." That pretty much summed up the situation.

Meanwhile the Canadian government was moving full speed ahead with its plans to become the final resting ground for carbon dioxide. Carbon capture and storage facilities were built in British Columbia, Alberta, and Saskatchewan—which had been identified as offering the right combination of high-volume CO_2 emission facilities located close to abundant geologic storage sites.[27] The CO_2 would be liquefied and then injected into depleted oil and gas reservoirs, or into saline aquifers located more than 1 mile below the surface. The Canadians figured that they could sell space in their reservoirs and aquifers to the Americans as an additional source of income. Oddly, despite all the innovation during the past forty years, no one could figure out how to make CO_2 anything other than an expensive nuisance—the climatic equivalent of nuclear waste.

October 2050

It's tempting to say that places like Greenland and Nunavut had been living their version of the American dream. The decision to move forward with the aluminum and zinc mines, the shipping, the oil tankers, and the natural gas pipelines turned Nuuk and Iqaluit—the capitals of Greenland and Nunavut, respectively—into fashionable international cities complete with four-star hotels and fine restaurants. But the boom had begun to show signs of busting—or perhaps more accurately, the boom was melting. Infrastructure costs to combat crumbling roads and sagging buildings were out of control. The permafrost was neither permanent nor frozen. And methane had begun to burst out at the Arctic's seams.

Methane hydrates, essentially natural gas trapped in ice crystals,[28] had become the next big thing in the Arctic, drawing investors from Dubai, Russia, the United States, and elsewhere. Imagine a snowball on fire and you've got a decent picture of what a methane hydrate looks like. Exploration for methane hydrates during the 2010s and 2020s had given way to large-scale production of natural gas during the 2030s. That was when the money really started to pour in. It was estimated that more energy was locked up in methane hydrates than in all other known fossil fuels combined. That was all most investors needed to hear.

The "Wild North" was now also dubbed "Saudi North." Hydrate deposits more than a half mile thick were found scattered under permafrost all over the Arctic, including the North Slope of Alaska, the Mackenzie River delta of Canada's Northwest Territories, and the Messoyakha gas field of western Siberia. Exploration also yielded several productive sites beneath the ocean floor at water depths greater than about 1,600 feet.

But as with everything else, the situation wasn't that simple. Methane brought tremendous wealth to the Arctic, but it also brought trouble. As temperatures continued to warm, methane hydrates, both in Arctic permafrost and beneath the oceans at continental margins, destabilized.[29] In other words, methane—a greenhouse gas with twenty-three times the heat-trapping capacity of CO_2, began pouring out of the Arctic. As the permafrost melted, the methane destabilization acted as a runaway feedback and further increased global warming. There was evidence that something like this had happened in the past, 635 million years ago, when unzippering the methane reservoir had warmed the Earth dramatically.[30]

Researchers had long warned that methane might be the final trigger setting off a climate change time bomb. New calculations showed that the levels of methane emissions from northern wetlands were going up year after year. And the potential for further warming was upward of several degrees—a scenario that frightened even the most stubborn skeptics.[31] In the end, many of us came to think that this new Arctic might after all have more in common with the barren lunar landscape.

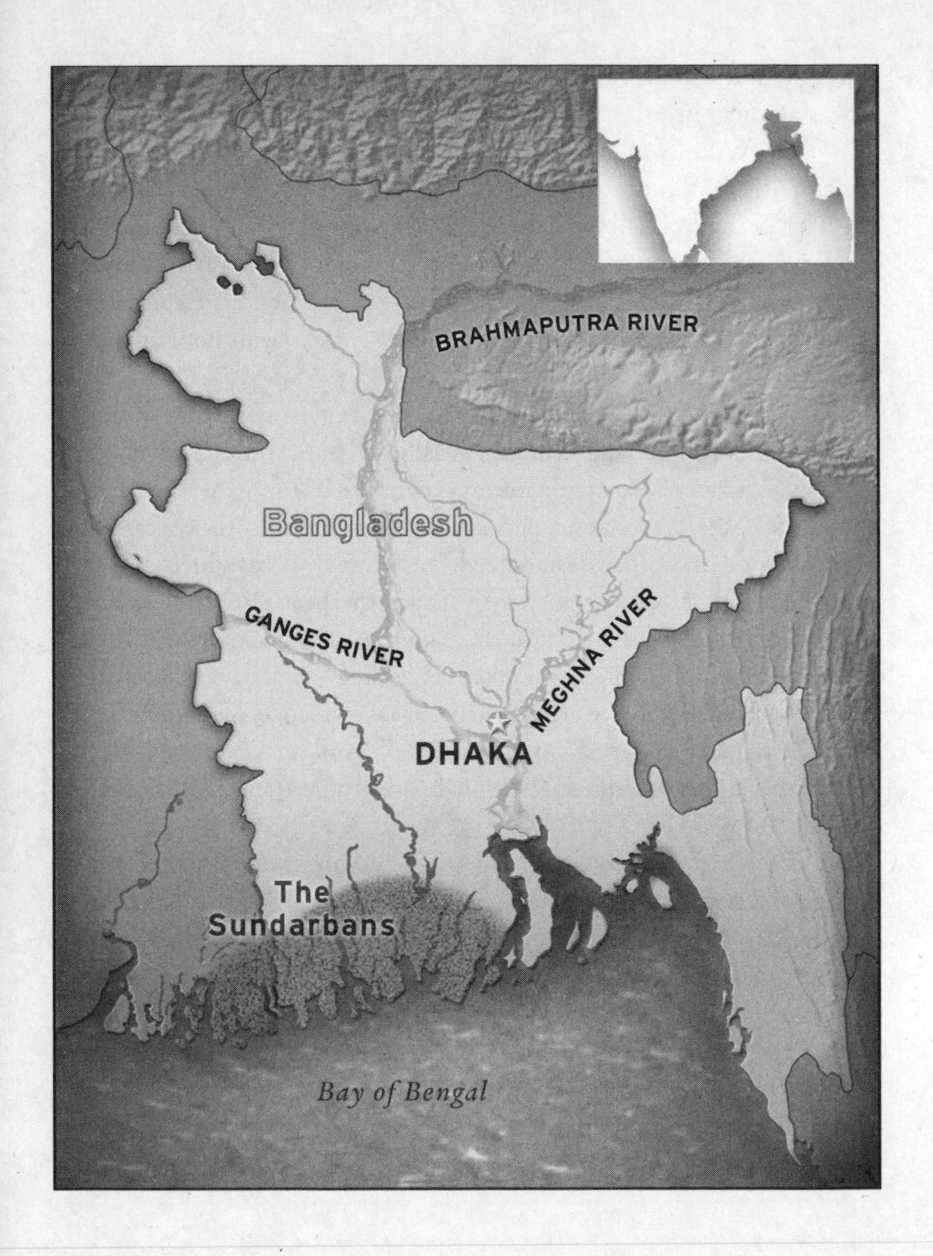

BRAHMAPUTRA RIVER
Bangladesh
GANGES RIVER
MEGHNA RIVER
DHAKA
The
Sundarbans
Bay of Bengal

10

DHAKA, BANGLADESH

It's hard to say why you fall in love with a person, or with a place. It's the same with science: sometimes your research is a lifelong passion, but sometimes a problem suddenly assaults you out of the blue and forces you to work on it—for the rest of your life. In the case of Peter Webster, an atmospheric scientist at the Georgia Institute of Technology, you might say his research on floods in Bangladesh began with a dare that led to a blind date that ended in a committed relationship. But it's probably better to just have him tell the story.

"It started off in a very strange way," he explains. "I was at a meeting in Bangkok in late 1998. We had just published a paper on the Indian Ocean Dipole." Webster's specialty is ocean-atmosphere interactions. The ocean and the atmosphere can interact in important and predictable ways that play a role in climate and weather: El Niño is a prime example of ocean-atmosphere interactions. Webster has a PhD from the Massachusetts Institute of Technology (MIT) and was among a crop of very successful graduate students who came out of MIT in the 1970s. Many of them, including Webster, worked with Jule Charney, a legend in numerical weather prediction. Webster is beginning to develop into a legend as well. He's received numerous scientific honors, including two of the most prestigious awards in the geosciences: the Carl Gustav Rossby Research Award from the American Meteorological Society (AMS) and the Adrian Gill Medal from the Royal Society. The wall of his office at the Georgia Institute of Technology in downtown Atlanta, where I met with him to talk about his research in Bangladesh, is a striking reminder of what the walls of accomplished people generally look like.

If you study climate science, you'll soon learn that there are a lot of *dipoles*: places where some quantity, such as atmospheric pressure or temperature, flips and flops between high and low or hot and cold. A dipole is a classic example of an ocean-atmosphere interaction. In the case of the Indian Ocean Dipole (IOD), ocean temperatures in the eastern and the western Indian Ocean flip between warm and cool, with the rains following the warm ocean temperatures.[1] The IOD also sets up what Webster calls "a seesaw of sea level" in the Bay of Bengal. Dipoles are important drivers of climate variability, and if you understand how they work, they can often help predict when a drought or a flood might be coming your way.

At the conference in Bangkok, Webster presented a paper suggesting a large-scale climate connection between the IOD, sea level in the Bay of Bengal, and the monsoon rains. This is exactly the kind of association climatologists search for, to better understand the physics of a system. Webster's research showed that when sea level in the Bay of Bengal was high, so was the risk of flooding: that is, the two events were highly correlated. "And in 1998," Webster explains, "the sea level in the Bay of Bengal was about a foot higher than normal."

At the time of his talk, Bangladesh was at the tail end of what came to be known as "the flood of the century." Sixty percent of the country had been flooded for more than three months, from July through September 1998. Dhaka, the capital of Bangladesh, was under 6 feet of water.[2] When all was said and done, the flood of 1998 caused 1,100 deaths; inundated nearly 39,000 square miles; made 30 million people homeless; damaged 500,000 homes; caused heavy losses to infrastructure;[3] and resulted in $2.8 billion in damages. Mother Nature has never cut Bangladesh a lot of slack.

Bangladesh has a problematic geography, and the floods are just the beginning. The geographic setting of Bangladesh makes the country highly vulnerable to many kinds of natural disasters. In addition to floods, it has experienced tropical cyclones, droughts, tornadoes, earthquakes, water contaminated by arsenic, and landslides. Bangladesh has seen more natural disasters than one might expect in such a small country—it is only about the size of the state of Iowa. But whereas Iowa has a population of 3 million, Bangladesh has a population of more than 162 million. That works out to some 2,900 people per square mile, making Bangladesh one of the most densely populated countries in the world.

Another problem is that two-thirds of Bangladesh is less than 17 feet above sea level; only in the extreme northwest will you find an elevation of more than 100 feet. And in this small, densely packed, low-lying country there are 230 rivers. Three of them—the Ganges, the Brahmaputra, and the Meghna rivers—come together to form a large floodplain. Eighty percent of Bangladesh sits within that floodplain, and everyone who lives there knows that in any given year, roughly one-quarter of the country will be flooded. And everyone also knows that every few years Bangladesh will experience a severe flood that inundates more than 70 percent of the country. That's easy to predict. The hard part is predicting the details of where and when the floods will come. But one detail scientists are certain of is that climate change will make the floods worse.

And the floods already have the power to devastate. Farmers and fishermen can easily lose a year's worth of income during a single flood. These were the people Webster was trying to reach with his weather forecasts. "My research had suggested that the floods could be related to sea level in the Bay of Bengal. It was as if the floods came whenever the drain in the Bay got clogged up," Webster explains. So, at the end of his talk in Bangkok Webster made a rather provocative announcement: "We can now understand why Bangladesh has floods. And in understanding these large-scale controls, we can forecast them." Those were mighty big words—and Webster knew it.

The rains in Bangladesh begin in May, when the southwest trade winds, known as the *monsoons*, are drawn to the Indian subcontinent by the intense heat and consequent low pressure over Pakistan. The trade winds blow across the North Indian Ocean, picking up moisture along the way, before they head into Bangladesh and go through to the Himalayas. When the winds hit the side of the Himalayas, it begins to rain as a result of a process called *orographic uplift*. As the air travels up the side of the mountain, it cools, forcing the moisture to condense and fall out as rain. Basically, the rain continues until early October. During these months, the total rainfall varies from 4 feet in the northwest of Bangladesh to 11 feet in coastal areas, and to more than 16 feet in the northeast. Needless to say, this is one of the rainiest places in the world.

"And so I kept getting these e-mails saying 'Do you really believe you can predict the floods?' I said, 'Sure.' And I was very cocky about that," Webster says smiling. "Then they asked if I'd be interested in a grant to work on forecasting the floods. I said, 'Yes, of course.' So we wrote a proposal to develop a flood forecasting system for Bangladesh."

Professors, as a rule, don't turn down grant opportunities. And Webster saw this grant as a dare that had the potential to save hundreds or even thousands of lives each year. "I thought it would be a very easy problem, because all you really need for a flood forecast are four quantities: the forecasted rainfall, the sea level in the Bay of Bengal, and the levels of the two major rivers, the Ganges and the Brahmaputra, all the way to India," Webster explains. "In fact, I could never understand why the Bangladeshis hadn't already done it." So Webster and his team went off to collect the ingredients for their flood forecast model. This was his blind date with Bangladesh.

"I remember the first time I went into the villages," Webster says, thinking back. "I asked a farmer to tell me about the flooding. We were standing in his rice paddy, and he said, 'These fields here always flood because we are in the lowlands.' " Farmers living in these lowland areas have adapted to the floods by building their houses on raised mounds and adjusting the way they farm.

"Naturally, I asked him if he would take me to the highlands. At which point we walked a little ways over and up a bank of no more than 2 or 3 feet. He said, 'Here, these are the highlands.' " Webster says in amazement, "We were literally standing in a slightly raised paddy field. I would have never noticed the difference." But the farmer knew the difference very well. During a flood, it was the highland crop that might have a chance of surviving, saving him and his family from starvation.

When Webster and his team got to Dhaka to assemble the forecast model, they soon discovered that India provided no stream gauge data to Bangladesh for the rivers that originated up north in India. "That explained why no one had ever issued a forecast. No one had any idea what the conditions were upstream," Webster explains. Bangladesh may have a lot of rivers, but more than 90 percent of the water in those rivers comes from outside its borders. "The best we could do was a forecast out a day or two. And even that wasn't any good, because the upstream conditions are so critical," Webster says.

Data are the lifeblood of good research and the principal ingredient in a reliable forecast of weather or climate. That's why scientists work so hard at setting up international observing networks. The climate system knows nothing of national borders; and just as Lewis Fry Richardson discovered when he started the science of weather prediction on a battlefield during World War I, a model run with lousy data will give you a lousy forecast. Garbage in, garbage out.

"So we decided to build a hydrological model for both the Brahmaputra and the Ganges River basin that could *estimate* the streamflow upstream," Webster explains. They obtained data and weather forecasts from the European Centre for Medium-Range Weather Forecasts (ECMWF) and fed them into the hydrological models. Their model also incorporates estimates of precipitation from two satellite-based systems, along with streamflow measurements of rivers inside Bangladesh. "We basically used the rainfall forecasts from ECMWF to feed the rivers with rain. And then we used satellite data to calibrate the rainfall. That's how we worked around the problem. And we came up with a solid scheme for a ten-day forecast. Our first forecast was in 2003, and it was pretty darn good," Webster says.

In the summer of 2004, Webster and his team generated ten-day forecasts showing that the Brahmaputra River would be likely to flood on two occasions in July.[4] They were right. The 2004 floods inundated almost 40 percent of the country. But unfortunately their forecast had not been much help to the local people. At the time, the flood forecast still hadn't been fully integrated into Bangladeshi warning systems, and more than 500 people in Bangladesh and India died in the rising waters. So Webster and his team went about setting up a communication network. They worked with the Flood Forecasting Warning Center (FFWC) in Bangladesh and the Asian Disaster Preparedness Center to develop a network to distribute the forecasts directly to people in five districts along the Brahmaputra and Ganges rivers, including impoverished families living on islands that were known as river chars.

"It's all done using webs of cell phones, and it reaches over 100,000 people," Webster says. Cell phones are everywhere, even in remote villages. "People are told that a flood of a certain level is coming at certain time. And that allows farmers to harvest early and protect their seeds," he explains. Once the communication channels were improved, his forecasts began saving lives and livelihoods. "An economist we were working with calculated that people were saving, on average, 25 percent in infrastructure and household damage by knowing the flood was coming," Webster says. "And saving 25 percent of everything is a lot."

By 2008, after ten years of research and outreach, Webster and his team had proved they could make a skillful forecast. The next step was to operationalize the process. Webster wanted to assemble a team at the FFWC to produce and issue the forecasts. But the situation wasn't that simple. "I had a rather disturbing interaction with the director of the FFWC," Webster says. In May 2008, just as Bangladesh was entering the flood season, Webster set up a workshop in Dhaka where he taught meteorologists at the FFWC how to generate the ten-day flood forecasts.

"But the director told me the FFWC was hesitant to give out a ten-day forecast, even though we had demonstrated predictability," Webster continues. "We argued and he said, 'But what if the forecast is wrong?' " The director told Webster that the FFWC didn't want to be held responsible for any losses. "So I said to him, 'If you saw someone crossing a road just as a truck appeared over the horizon, would you wait to warn the poor soul until there was no chance for him to get away?' It was an unpleasant conversation."

Webster was shocked that someone would turn down an opportunity to give people enough lead time to prepare for and, it was hoped, prevent a disaster. "I think it's morally wrong. You can't do that." Webster remains committed to getting the forecast out to people in rural areas, people who need as much time as the science of forecasting can allow. "Right now, the FFWC has agreed to monitor the ten-day forecasts but will not issue them," he says.

It strikes me that maybe Webster's story isn't so different from the larger fight to communicate the risks of global warming. A forecast for 2050 is not so different from a ten-day forecast for flooding. Neither is perfect, but both are critical to making informed decisions. And many would argue that both carry the same moral obligation.

Omar Rahman is someone who has come to sense this moral obligation. Rahman is a demographer and physician who studies social networks and urbanization. "I'm not a climate change person," Rahman explains over the phone from his home in Dhaka. "I got interested in climate change from a different perspective. I am interested in migration." As it happens, Rahman is himself a living study in migration.

"I basically left Bangladesh to go to college in the 1970s and stayed in the United States for twenty-eight years," Rahman says. After studying biochemistry at Harvard, he received a medical degree from Northwestern University. In 1996, he returned to Harvard as an assistant professor of demography and epidemiology at the School of Public Health. His work focused on the concept of resilience—specifically, the resilience of people living in rural Bangladesh.

"I found it odd that I was sitting in Cambridge writing about rural Bangladesh," he says. Rahman had written extensively about issues of development and had begun to feel a pull back home. "I was an academic writing papers that very few people would read," he explains. He began to question the impact of his research. The pull of a place can be quite powerful, and Rahman began to contemplate returning to Bangladesh.

"A friend of mine gave me a very good piece of advice," Rahman says. "He told me, 'If you go back home, you need to go back for good. Don't just test the waters, because you'll never stay.' " Rahman explains, "Developing countries are difficult places to work, and constant comparison is detrimental to staying." And then came what might have been the last straw. "We were told we had to move out of our apartment in Cambridge," Rahman says, laughing. "And so, in some ways, it came down to either moving to the suburbs or moving back to Bangladesh." Rahman, his wife, and their two children moved back to Dhaka in 2003, the same year Webster issued his first flood forecast.

"Ultimately, I am interested in the people part of this whole thing," Rahman says. And it is the human part of the problem of climate change in Bangladesh that makes everyone very nervous.

The most widely used estimate of how many people around the world could become *climate refugees*, a term heavy with political and moral overtones, is 200 million by 2050. To put that in perspective, about 1 million Irish immigrants came to the United States because of the potato famine during the late 1840s. These projections about climate refugees are based on a very crude formula for estimating migration, so it's safe the say the numbers are still fuzzy. Modeling people's behavior is a lot harder than modeling a flood. But whatever the number ends up being, it's likely to be of a magnitude not seen before in human history. And in Bangladesh alone, the exodus is estimated to be in the millions. Projections range from 6 million to 15 million by 2050.

Most of the migration in Bangladesh right now is internal. People are moving from coastal and rural areas to cities such as Dhaka. The reasons for leaving the place where you were born are varied. However, "I will honestly tell you that right now, most of the migration is economic," Rahman says. "None of the migration is driven by concern about climate change. That will come in twenty to thirty years."

But some observers argue that climate migration is happening already, and that it's being blamed on the weather instead of the climate. More severe floods and droughts are hitting the landless and poor farmers in the same villages that Webster visited. And the floods and droughts, which have always occurred in Bangladesh, will grow worse with time. Increased soil erosion and saltwater intrusion in coastal areas will make it more difficult to farm and work the nets for shrimp fry, leaving people with few options but to migrate. Many will end up in the slums of Dhaka. In the end, of course, it's always the economy. In rural Bangladesh, the weather is the economy. And if you believe the climate models, the weather will get worse.

By 2050, the population of Bangladesh will have grown from about 162 million people today to more than 220 million.[5] Today, more than 13 million people live in Dhaka. It's the fastest-growing megacity in the world. And every year, slightly more than 400,000 people in Bangladesh move to the capital, hoping to find a better life. Nearly 15,000 new cars were sold in Dhaka in 2008, a record high. There may be plenty of people and cars, but there are acute shortages of just about everything else. There are no sidewalks.

There is no mass transit system. And right now, there is enough power for only about 35 percent of the population. When I spoke with Omar Rahman on the telephone, he had been without power for eight hours that day. As he said, comparison with the developed world is detrimental. But nonetheless, when people along the coast who are unable to grow rice or work the nets to catch shrimp fry make the comparison between Dhaka and their own situation, they will still decide that Dhaka holds the keys to a better life. And by 2050, this megacity with very little energy, transportation, and water infrastructure is expected to be the home of more than 40 million people.[6]

Experts like Rahman worry about how Dhaka will cope with the rapid and unplanned urbanization in Bangladesh. Dhaka is not immune from the problems of geography that plague the rest of Bangladesh. The city is located in the coastal zone and is just as vulnerable as the rest of the zone to floods, storms, and tropical cyclones. Drainage is already a serious problem, and sewers routinely overflow during the monsoon season. And the slums, situated in the lowest-lying parts of the city, are even more vulnerable. The millions of poor who have settled there are crowded into metal shacks with no running water. They have merely traded one form of vulnerability for another.

"The impact of rapid urbanization is huge," Rahman says. "We are trying to model it because we know there is no way to stop it. By 2050, Bangladesh will have transformed from a population where only 5 percent of people lived in urban areas in the 1970s to a population where 50 percent of people live in urban areas." Rahman explains that the growth in Dhaka is a microcosm of the urbanization taking place across Africa and Asia. The UN Development Program estimates that by 2050, 70 percent of the world's population will be living in urban areas.

The next step in the migration pattern is across national borders. "When most people think of these issues," says Webster. "They mostly think about how terrible it's going to be for the people there. But this is an enormous problem from a national security standpoint, too. Because all of a sudden you have 200 million people who are displaced, people who have become climate refugees. Where do they go? They go to India. They go to Myanmar. But they won't be very happy people," Webster concludes.

For national security experts, migration is one aspect of climate change that evokes a real sense of dread. In the global landscape the argument is simple: what happens in Bangladesh doesn't stay in Bangladesh. The United Nations Department of Economic and Social Affairs estimates that 1.1 million migrants will enter the United States each year between now and 2050. Many are expected to come from Bangladesh.

It's also estimated that more than 10 million Bangladeshis have already made the move to India during the past twenty years. This issue is a constant source of tension between the two nations, and climate change isn't helping. India, for its part, sees climate change as bringing multiple threats, and apparently one threat is people. And so India is in the process of building a fence to keep them out. "I think they were going to build a fence anyway," says Rahman. "The data on this isn't clear, but I think the fence was ultimately built for political reasons. And the climate refugee argument is being used as an excuse," Rahman adds.

India maintains that the purpose of the fence is to protect the country against smuggling and terrorism as well as illegal immigration, claiming that about 5 million Bangladeshis are in India illegally. This is a number the government of Bangladesh is quick to contest. The fence runs along India's porous 2,500-mile border with Bangladesh. It is high, and it's made of heavily reinforced barbed wire. Climate change may not have created the fence but provides a plausible reason to continue building it. Still, as the Indian government works to complete the fence in the hope of keeping people out, the problem is not so much the people as the climate. The fence won't stop the floods, the cyclones, the droughts, or the rising sea level. The forecast for 2050 is going to require a lot more than a barbed wire fence.

Water, now and always, is at the heart of Bangladesh's problems. The country suffers from both too much and too little water.

The scenarios for 2050 and beyond predict that this water problem will worsen, owing to a number of factors. Rising temperatures and decreasing winter precipitation will bring more drought. Rising sea level will bring salt water into the rice paddies and rob Bangladesh of its agricultural land. Floods that result from increased snowmelt and a stronger monsoon will happen more frequently and last longer. And more intense cyclones will batter the coast, an economic hub and cultural treasure, with higher storm surge. This small country has always been vulnerable, and climate change will make it more vulnerable. It doesn't take a climate scientist to figure that out. Just ask Omar Rahman.

"I came into this field as somewhat of a climate skeptic," Rahman admits. "As a scientist, I always want to see the data. But I have to say I was convinced. These were serious people, people who were not prone to exaggeration. And the data spoke for itself," Rahman says. The Intergovernmental Panel on Climate Change (IPCC), the scientific group responsible for building the data and models that convinced Rahman, has issued a very strong statement about the changes that are taking place in Bangladesh. Temperatures in Bangladesh have already increased. The Fourth Assessment report indicates an increasing trend of about 1.8°F in May and 0.9°F in November during the fourteen-year period from 1985 to 1998. Annual average temperature in South Asia (5°N to 30°N, 65°E to 100°E) is projected to increase 3.2°F by 2050 and 5.6°F by 2100, according to the IPCC Fourth Assessment Report. The seasonal values for South Asia are shown in the accompanying table. Temperatures in Bangladesh are projected to increase 1.8°F and

2.5°F by 2030 and 2050, respectively, according to a recent assessment by the Bangladesh Centre for Advanced Studies (BCAS).

	2020s	2050s	2080s
Dec.–Feb.	2.2ºF	5.8ºF	9.7ºF
Mar.–May	2.2ºF	5.4ºF	9.4ºF
Jun.–Aug.	0.9ºF	3.0ºF	5.6ºF
Sept.–Nov.	1.3ºF	4.3ºF	7.6ºF

Projected changes in surface air temperature for South Asia under the highest future emission trajectory (A1F1) for three time slices: 2020s, 2050s, and 2080s. (SOURCE: IPCC FOURTH ASSESSMENT REPORT)

These warmer temperatures, as well as black soot from local factories and cook stoves, spell trouble for the Himalayan glaciers, the largest body of ice outside the polar caps and a critical source of freshwater throughout Asia. About 15,000 Himalayan glaciers support perennial rivers including the Ganges and the Brahmaputra that, in turn, provide a lifeline to millions of people in Bangladesh and across Asia. But the story is complicated by the shielding effect of pebbles and rocks. Satellite images show that while the majority of glaciers in the Himalayas were shrinking about half of the glaciers in the Karakoram region of the northwestern Himalayas were actually stable or even advancing.[7]

The roughly 20-mile-long Gangotri glacier has been receding alarmingly in recent years. Between 1842 and 1935, it was receding at an average rate of 24 feet every year; the average rate of recession between 1985 and 2001 was about 75 feet per year. The current trends of glacial melt suggest that the Ganges and Brahmaputra, which crisscross the northern Indian plain, could run dry during the summer months as a consequence of climate change. A fence won't be of much help when that happens.

Whereas Rahman may have initially been a skeptic, the people who live in rural Bangladesh have watched the climate change with their own eyes. According to a study carried out by the International Union for Conservation of Nature (IUCN) in Bangladesh, people in rural communities are reporting excessive and erratic rainfall, an increase in the number of flash floods, temperature variation, changes in seasonal cycles, and an increased occurrence of droughts and dry spells. These effects are likely to worsen, and adaptation strategies are urgently required. The question is: how bad does a situation need to get before it makes people leave?

One complication is the fact that, like floods, droughts in Bangladesh are seasonal. Depending on their timing, they can devastate crops, especially in the northwestern region, which generally has lower rainfall than the rest of the country. Drought brings significant hardship to poor agricultural laborers and others who cannot find work. In these areas, unemployment leading to seasonal hunger is often a problem, especially in the months leading up to the November–December rice harvest. If the entire crop fails because of drought, the situation for poor people can become critical. The IPCC predicts lower and more erratic rainfall, resulting in increasing droughts, especially in the drier northern and western regions of the country.

Too much water is no better. The problem is, again, the timing. More than 80 percent of the roughly 7 feet of annual precipitation in Bangladesh comes during the monsoon period. Most of the climate models estimate that precipitation will increase during the summer monsoon. The reason is fairly straightforward. As the temperature increases, air over land will warm more than air over oceans in the summer. This will deepen the low-pressure system over land, which happens anyway in the summer, and will enhance the monsoon. Moreover, there is a link between Eurasian snow cover and the strength of the monsoon; when snow cover retreats—as it is expected to do with the higher temperatures—the monsoon strengthens.

Climate models indicate a general increase in the intensity of heavy rainfall events in the future, with large increases over the Arabian Sea, the tropical Indian Ocean, northern Pakistan, northwest India, northeast India, Bangladesh, and Myanmar.[8] Scientists expect a roughly 10 percent increase in rainfall during monsoon the season by 2030, while dry seasons could see harsher droughts.

The problem for Bangladesh is that two things are happening at once. The warmer temperatures are intensifying the monsoon and they are melting the glaciers; and unfortunately the melting season happens to coincide with the monsoon season. Rapid glacier melt will mean more water flowing down the Ganges and Brahmaputra rivers during the monsoon months, causing even more devastating floods. Webster has been studying the changing frequency of flood events. "We've been looking at the flow of the Brahmaputra and the Ganges out to the year 2100. Two things happen. You see more flood events and the flood levels are much higher," Webster explains. Scientists say once-in-twenty-year floods are already occurring about every four years. However, in the long term, as the water in the rivers disappears, the result will be more severe droughts.

Although floods and droughts are a serious concern, the most serious issue of all is a rising sea level. The 146 million people living within about 3 feet of mean sea level worldwide are at risk from the projected rise in sea level over the coming century. An even greater number—268 million living within about 16 feet—are at risk when the added impact of storm surge is considered. Moreover, these numbers are rising, owing to the combination of a growing population and its coastward migration. A 3.3-foot rise in sea level would inundate about 20 percent of Bangladesh's total land, directly threatening 11 percent of the population with inundation (this figure is based on current population distribution). In addition, the backwater and increased river flow from sea level rise could affect 60 percent of the population.[9]

Oddly, the rise in sea level is the one estimate for which the cautious ways of the IPCC that Rahman so appreciated are wrong. In an effort to address the uncertainty, the IPCC chose to go with the most conservative estimate, which some would say is wildly conservative. The IPCC Fourth Assessment Report estimated global sea level rise over the coming century in a range from 11 inches to 31 inches. But it left room by stating, "The upper values of the ranges given are not to be considered upper bounds . . . for global sea level rise because existing models are unable to account for uncertainties such as changes in ice sheet flow."

Even with these conservative estimates, the IPCC Fourth Assessment Report still projects that rising sea levels could wipe out more cultivated land in Bangladesh than anywhere else in the world. In Bangladesh, you can drive 60 miles inland from the coast, and you'll go up only a few feet in elevation. When you get about 150 miles away from the shore, the land finally starts to rise up another 50 to 65 feet. I recently talked with experts who are preparing the next IPCC report; they said the projected value rise in sea level is actually closer to 5 feet, instead of the previously estimated 31 inches. It's tough to wrap your mind around that. The factor driving migration is the land loss in coastal areas that will result from the rising sea level. "Somewhere between 20 to 25 percent of Bangladesh will be inundated in the next fifty years," Rahman says. In Bangladesh, even the land is leaving.

The main impact of a rising sea level would be salinity ingress, causing the rivers in the coastal belt to become brackish or saline. This would have a serious impact on food production. In Bangladesh, production of rice and wheat might drop by 8 percent and 32 percent, respectively, by the year 2050.

Rising salinity levels as brackish water inundates cropland could hurt rice and wheat production. Overexploitation of groundwater in many countries of Asia has resulted in a drop in its level, leading to an ingress of seawater in coastal areas and making the subsurface water saline. India, China, and Bangladesh are especially susceptible to increasing salinity of their groundwater as well as surface water resources, especially along the coast, due to increases in sea level as a direct impact of global warming. The Meteorological Research Center at the South Asian Association for Regional Cooperation carried out a study on the recent rise in coastal sea level in Bangladesh. The study used twenty-two years of tidal data from three coastal stations. It revealed that the rate of rise in sea level during the last twenty-two years is many times higher than the mean rate of global rise over 100 years; this suggests that regional subsidence could be making the situation worse.

There are also the tropical cyclones. According to the UN Development Program, Bangladesh is the most vulnerable country in the world to cyclones. Scientists report that 2007 was the worst year on record for intense hurricanes in Bangladesh. "The worst-case scenario for Bangladesh is rising sea levels and increased floods," says Webster. "But you add to that, of course, increased intensity of hurricanes in the spring and fall. So that would be the triple whammy.

"I'm not quite sure if Bangladesh is an adaptable country," Webster continues. "Imagine a country the size of Iowa becoming half the size of Iowa with double the population." That is the long-term forecast for Bangladesh. "Ultimately," Webster explains, "they are running out of land."

And some of the land is so beautiful. The Sundarbans—the wetlands region straddling the coasts of western Bangladesh and neighboring India—was formed by the deposition of materials from the Ganges, Brahmaputra, and Meghna rivers. If the Sundarbans is lost, the habitat for several valuable species will also be lost. An 18-inch rise in sea level would inundate 75 percent of the Sundarbans; a 26-inch rise could inundate all of the system. That would threaten what is now the single largest mangrove area in the world and is designated a World Heritage Site. The name *Sundarbans* means *the beautiful forest* in Bengali. The mangroves in this forest, within the delta of the Ganges, Brahmaputra, and Meghna rivers on the Bay of Bengal, act as natural buffers against tropical cyclones and as a filtration system for estuarine water and freshwater. They also serve as nurseries for many marine invertebrate species and fish. The Sundarbans mangrove forests are well known for their biodiversity, including 260 bird species, Indian otters, spotted deer, wild boars, fiddler crabs, mud crabs, three marine lizard species, five marine turtle species, and several threatened species, such as the estuarine crocodile, the Indian python and the famous Bengal tiger. It was for these reasons that the Sundarbans National Park, India, and the Bangladesh part of the Sundarbans were added to the World Heritage List in 1987 and 1997, respectively.[10]

The rise in sea level and the decreased availability of freshwater—particularly during winter, when rainfall will be less—will cause an inland intrusion of saline water. As a result, many mangrove species, intolerant of increased salinity, may be threatened. In addition, the highly dense human settlements just outside the mangrove area will restrict the migration of the mangrove areas to less saline land. The mangroves are caught in the middle. The shrinking of the mangrove areas will have an effect on the country's economy. Many industries that depend on raw materials from the Sundarbans will be threatened with closure, and large-scale unemployment could result. A project of the United Nations Development Program (UNDP) has evaluated the cost of building about 1,375 miles of protective storm and flood embankments that would supposedly provide the same level of protection as the Sundarbans mangroves. The capital investment was estimated at about $294 million and the yearly maintenance budget at $6 million—much more than the amount currently spent on the conservation of the mangrove forests in the area.

Webster may be a skeptic who thinks Bangladesh may not be adaptable, but Omar Rahman has no choice but to find adaptation strategies that will work for this country. It is his home. When I ask him what the year 2050 might look like if we stick with the status quo he is quick to respond. "I think if climate change is not taken seriously, if the predictions are right and sea level rises more than one meter, we would see an almost unimaginable catastrophe. It's the worst-case scenario," says Rahman.

He thinks the rapid urbanization exacerbates the problem. When 50 percent of a country is urban, it is no longer a rural country. And Rahman agrees that food production is problematic. "Right now the country is almost self-sufficient in rice production—which is pretty amazing, given that a country the size of Iowa can feed over 160 million people. But we will lose a lot of acreage, and that means we'll have to import a lot of rice, like Japan. From an economist's point of view, that's going to be a big shock, but not something that is unapproachable, if we start thinking about it now. Better than if we sit and wait.

"It's unfortunate that we seem to be beset by natural disasters. But when I think about the story of Bangladesh, I think it is a story of hope," Rahman says.

Bangladesh arose in 1971 from a civil war. At the time, the U.S. secretary of state, Henry Kissinger, called Bangladesh an "international basket case."

"Now I like to say Bangladesh has gone from being known as an international basket case to a Bengal tiger," Rahman says. Until the worldwide economic slump that began in 2008, Bangladesh's economy was growing at a pace not far behind India's; Rahman attributed this to a developing culture of entrepreneurship and a thriving garment industry. Garments are the main export for Bangladesh, making up more than half of total exports.

"I don't think anyone would have predicted that we would have become an entrepreneurial culture. The garment industry has bred a new class of entrepreneurs," Rahman explains. In 2007, the World Bank predicted that Bangladesh could join the ranks of middle-income countries within two decades. "The best thing that's happened to us is that we don't have oil or mineral wealth," Rahman says. "We've had to develop our people. And we've done spectacularly well, despite our natural disasters—famine, you name it, we've had it."

The economic growth has happened because of the resilience and the tenacity of the Bangladeshi people. But Rahman would still like to see Bangladesh become more sustainable; that's one of the reasons he came back.

"As a demographer I'm interested in structure. It has had the most successful family program. In the mid-1970s women had seven children; now it's about 2.6." Estimates project that the population should stabilize at around 250 million. That still is a lot people to feed on a plot of land that is only the size of Iowa, and which is expected to shrink by one-fourth, owing to a rising sea level, by the middle of the century.

Rahman has an idea that he thinks might help kick-start adaptation programs: an international center for climate change adaptation that's actually located in the developing world. As he says, "You know you are underdeveloped when most of the literature about your country is written by people outside your country. We decided to set up the center *inside* the developing world."

Rahman believes students will learn more in the living laboratory of Bangladesh than in a sterile classroom in Cambridge about what vulnerable countries need to tackle climate change. Rather than getting hung up on the fence, the country needs to build embankments. It needs cyclone shelters and research on rice. And it needs to address the already explosive internal migration to Dhaka.

"I said: Look. We as a university would like to sponsor a coordinated set of activities. Train the next generation of scientists. Not enough people have the right kind of skills. It will be international. Draw people from all over the world as well as from inside Bangladesh. We will train people to think on timescales climate change requires," Rahman says. The program at the Independent University, in Dhaka, where Rahman is provost, will be a twelve- to eighteen-month master's program. Students will spend time out in the field, as Webster did. And their research will cover all aspects of the problem. Climate change is a problem that will require engineers and computer scientists as well as chemists and anthropologists to solve. Climate change is not a problem that can be constrained in a straitjacket.

"As an educator, the best part is that you get a chance to mold young people. For decades now, people have gone to America because it offered so many opportunities. It's changing now. There is an insularity. America has always struggled with being open versus retreating into itself. I benefited so much so I don't want to see that change. The thing that I miss the most is the ability to reinvent yourself. That's the best part of American culture. America is the most egalitarian place in the world. It's the sense that you can be anyone. You don't feel that you're worse off. I want Bangladesh to move to that."

Rahman seems to be suggesting that if he can help his students reinvent Bangladesh, they may in turn reinvent themselves. This is what happens when a trained psychiatrist works on global warming.

As Rahman looks to the future, he sees three adaptation options for Bangladesh: retreat, accommodation, and protection. "In view of the high population density and shortage of land, retreat is not possible. We should pursue the two other options. Some of the adaptation options are: raising of forest all along the coast, protection of mangrove forests, changing cropping pattern and variety in the coastal area, construction of embankments where feasible, construction of 'safe shelters' for emergency situations like extreme events, etc. In fact, many of these options are already in operation—on a limited scale, though.

"Adaptation has crosscuttings of different disciplines, and hence a multidisciplinary and integrated approach need to be taken up to reduce vulnerability. Adaptation will require thinking big and small—for instance, changing cropping patterns and developing new seeds able to survive in the changed climatic conditions. Part of our attempt is to train leaders to prepare before it happens. . . . You need to mobilize people and you need to empower them," Rahman says.

In the end, it's the people who have the potential to do this. They will improve the adaptive capacity of the country. Bangladesh may have a problematic geography, but it has great people—people who refuse to let the West define them as hopeless.

Leaders here estimate that it will cost $500 million just to raise embankments in some areas about 8 inches, a level that by the time construction is complete might not be high enough to keep growing storm surges at bay. Adaptation is not sexy or cheap. Scientists at the Bangladesh Rice Research Institute are working to develop a strain of rice that can withstand higher salinity levels.[11] Adaptation will require infrastructure investments across the board. Bangladesh needs to build embankments and cyclone shelters. The government says the country's power-generating capacity is at a maximum, 4,000 megawatts, which covers only 35 percent of the total population. The newly elected government has vowed to increase power generation in order to boost economic development. Rahman estimates that by 2015, the demand for electricity in Dhaka will rise to 10,000 megawatts. That, Rahman says, creates an enormous opportunity for clean energy projects to promote energy efficiency and renewables at a household level.

But he noted that resistance to spending precious dollars on more expensive low-carbon technologies in Bangladesh remains strong. Here, economic growth and fighting poverty remain the top priorities. "We are one of the most negligible emitters of greenhouse gas," Rahman says. Bangladesh, currently one of the poorest countries on Earth, has virtually no hand in causing climate change. The average Bangladeshi emits about one-third ton of carbon dioxide each year—a lot less than the roughly 20 tons emitted annually by the average American. At the global level, Bangladesh emits less than 0.2 percent of world total. To put that in perspective, the city of New York alone emits about 0.25 percent of the world's total greenhouse gases.

As Rahman says, "Cooking stoves account for almost 20 percent of emissions in Bangladesh. Cooking stoves. This is the level of industrialization we're talking about." Rahman still has hopes for megacities; he says that leaders need to start viewing land use and other aspects of city planning as critical components of preparing for climate change. "Properly managed, urbanization can be a good thing," he said. "Improving urban management is itself an adaptation strategy."

For people in Bangladesh, climate change is not a theoretical, academic, or distant concern. It is a question of survival. It is a question of infrastructure. It is a question of water and energy. It is a question of believing the forecast for 2050 and beyond.

At the end of our long phone conversation, Rahman couldn't help making a comparison between his old life in the United States and his new life in Bangladesh, "I lived in the United States for twenty-eight years, and there are things I miss. There is no question about that. I would have had more things if I stayed in the United States: a bigger house, a bigger car. But even that . . . I don't regret for an instant. I think I've come at a very exciting time. But there is no question in my mind that I made the right decision. This is not just about me; it's about something larger than myself. I don't think I could say that if I had stayed in the United States."

I guess sometimes you do get to choose whom you fall in love with.

Bangladesh: The Forty-Year Forecast—Sea-Level Rise, Floods, and Climate Refugees

DHAKA, BANGLADESH	TODAY		2050		2090	
Emissions Scenario	**JAN.**	**JULY**	**JAN.**	**JULY**	**JAN.**	**JULY**
Higher	65.6	83.7	68.4	86.2	72.9	88.8
Lower			68.3	85.6	69.7	86.6

Forecast
January 2016

Nowhere was the issue of water more problematic than in South Asia. From the 29,029-foot Mount Everest in the Himalayas down to the lush, swampy mangrove forests of the Sundarbans, water was being held hostage by climate change. And that meant more than 1.3 billion people, dependent on the good graces of the climate system to deliver life-sustaining water in a timely and dependable manner, had a very big problem on their hands.

The problem came in several forms and started at the very top, in the Himalayas—mountains that stretch from Pakistan to India, China, Nepal, and Bhutan. The more than 15,000 glaciers that have covered the Himalayas for millennia bear a significant responsibility—they feed Asia's nine largest rivers, including the Ganges, Indus, Brahmaputra, Mekong, Yangtze, and Yellow, and bring a steady supply of pure, cool water to the people of South Asia. The problem was that the glaciers draped over these majestic mountains were retreating at an alarming rate. Scientists estimated that most were pulling back between tens and hundreds of feet each year; this rate made the Himalayan glaciers the fastest-melting glaciers in the world. In this very remote and beautiful place, the sound of global warming had become deafening.

The sound itself came from what is known as a glacial lake outburst flood (GLOF). Glacial lakes, which form as a result of a melting glacier, had become overwhelmed by meltwater. Every few minutes the chemistry of global warming showed off its handiwork: somewhere along the 1,500-mile mountain chain, rising temperatures ripped heavy chunks of ice loose from glacier after glacier. As the ice came loose, it crashed down, adding more and more water to already overflowing glacial lakes. Eventually, the lakes had no choice but to burst—releasing huge quantities of water. By 2016, every country in the Himalayan region had suffered from glacial lake outburst flooding.[12]

The melt rate had begun to increase in the early 1990s. It was then that the now infamous Luggye glacier in Bhutan—retreating more than 520 feet a year—finally broke off, on October 7, 1994. The lake burst open, releasing more than 4 billion gallons of water down the Pho River, killing 21 people, and wiping out entire villages and farms.[13] Floods like that were almost routine by 2016.

After countless floods, local villagers in Bhutan and across the Himalayas took matters into their own hands. They organized a small army of workers to combat the effects of climate change. Local officials estimated that by reducing glacial lake levels by 15 to 20 feet, they might be able prevent catastrophic flooding. The cost of widening just one lake ran upward of several million dollars. For the most part, the money was provided by the Least Developed Nations Fund—a special fund set up by the United Nations Framework Convention on Climate Change to help the world's poorest nations adapt to climate change. In Bhutan the work was done without the help of heavy machinery—the workers used just picks and shovels. In their first year, people from across Bhutan working at the Thorthormi glacier managed to lower the lake's level by 35 inches. But it would take years to get the lake to a safe level. And while there was a great sense of accomplishment with each inch the lake dropped, people realized that the risk of flooding would eventually be dwarfed by the problem standing in line behind it. The bigger catastrophe would come with the eventual disappearance of the glaciers. After the floods came the drought.

April 2022

Problems with water came from the mountains, but they also came from under the Earth. Groundwater—formed by the natural percolation of rain and snow into soil and stored in pockets of porous rock—was being depleted at an alarming rate. Satellites operated by NASA first revealed that groundwater levels in northern India had been declining by as much as 1 foot per year over the past two decades—a completely unsustainable rate. The Indians denied this, but the satellites did not understand politics. They clearly showed that 26 cubic miles of groundwater had disappeared from aquifers in areas of Haryana, Punjab, Rajasthan, and the nation's capital territory, Delhi, over the past four years alone—enough water to fill Lake Mead, the largest man-made reservoir in the United States, three times over.[14] The Indian government said that U.S. satellites should mind their own business.

The reductions of streamflow and groundwater did little to improve relations between India and Pakistan. A meeting was called between high-level Indian and Pakistani officials to renegotiate the Indus Waters Treaty, first signed by the two countries in 1960. Upon signing the treaty, the countries had agreed that the six primary rivers of the Indus basin would be split evenly between India and Pakistan. For more than sixty years the treaty withstood the strain of wars between India and Pakistan. But now that groundwater had also become scarcer and scarcer, Pakistan believed it was being taken advantage of and requested a larger share of Indus water. The fact that both countries had nuclear capabilities lent a chilling new dimension to the water negotiations.

August 2026

Woes involving water in the south came in a different form. In early August, millions of Bangladeshis had been marooned or displaced by floodwaters. The death toll currently stood at more than 5,000, but it was expected to rise. The floods had been coming more often, just as the models had predicted. Everyone was grateful for the seasonal and twenty-day flood forecasts routinely issued to Bangladeshis by the Asian Disaster Preparedness Center. At least these forecasts gave people a chance to prepare. With advance notice, they could postpone planting or hurry to harvest some or all of their crops, move livestock to safety, encircle fishponds with nets to prevent fish from escaping, and stock food and other supplies. It was something. And it allowed them to continue living in the place they loved.

In addition to relying on the flood forecasts, villagers also engaged in simple solutions they hoped would help decrease their vulnerability. Home foundations and frames were constructed using lightweight composite materials that could bend but would not break during a storm. Women wove these fibers from jute, one of Bangladesh's common plants, with recycled plastics to form strong building material. The people of Bangladesh did everything in their power to stay. They even used materials that would float on the rising tide of a coastal surge—hoping these might serve as life rafts when the next flood came. But in the end, it was not their choice to make. With each successive flood more and more people began to pack what was left of their possessions and leave. First the numbers were in the thousands; then they gradually increased to the hundreds of thousands. A steady rain of refugees poured down on Dhaka—a city already overcrowded. Men and women sometimes arrived with dreams of finding better jobs and better lives. But many had left their villages with no dream other than to save the lives of their children—too many had been lost to the floodwaters already.

September 2039

In India, the groundwater situation worsened. Officials were forced to begin illegally withdrawing more water from the Indus River than they had been allotted. In response, Pakistan threatened to call in troops along the border.

As Himalayan glaciers retreated and groundwater was further drawn down, sea level continued its steady rise.[15] As the sea level crept higher and higher, the saltwater front traveled hundreds of miles upstream, and the salinity in surface water increased almost sixtyfold. The increase in salinity altered soil quality and nutrient loads. Simply put, the salt water was killing the trees. Down in the Sundarbans, the mangroves were dying. And the ripple effect of climate change was not hard to predict. The mangrove forests of the Sundarbans contained one of the last remaining populations of wild Bengal tigers left in the world.

July 2050

Despite plenty of competition, South Asia remained the most food-insecure place on planet. Rice and wheat yields continued to spiral downward because of high temperatures and low water supply.[16] Governments around the region had tried to pool resources and engage in some of the less expensive adaptation measures. The hope was to moderate the predicted crop shortfalls and keep as many people fed as possible. As a result, planting dates were shifted and farmers switched to existing drought- and heat-tolerant crop varieties. Money was also spent on the development of new crop varieties and the expansion of irrigated areas. These measures showed some the biggest benefits. But in the end, there is only so much you can do without water. With an additional 130 million people pushed from food insecurity to famine, a mass exodus was under way. As part of an international agreement, the United States and the European Union agreed to take in millions of the hungry and displaced. And after years of fighting over what to call them, they were now officially known as climate refugees.

In Bangladesh, the water problem was surreal. With almost 25 percent of the country underwater—as a result of rising seas, recurrent flooding from increasingly malicious tropical cyclones, and the slow and deadly seepage of saline water into wells and fields—Bangladesh had become a wasteland. One by one, millions of men gathered their families and left their mud-caked villages, never to return. Many of them crossed illegally into India, hoping to find construction work in Assam and West Bengal.

Others, like Hassan and his family, fled to Dhaka. In his village Hassan had been a proud man, able to support his family with his handiwork. Now he was shining shoes on the streets of this megacity, whose population had swelled to more than 40 million. Hassan and his family were among the last to leave their village. It took him several months to persuade his wife to move away from the only place she had ever known. But finally, she gave in. He promised to take her to the Dhaka Zoo. She had always dreamed of seeing the Bengal tigers up close. The zoo was now the only place they could be found.

NEW YORK STATE
Delaware Watershed
Catskill Watershed
Croton Watershed
The Bronx
Manhattan
Brooklyn
Queens
Staten Island

11

NEW YORK, NEW YORK

New Year's 2000 is one of the few moments in my life that I remember with great precision. I was standing on the corner of Fifty-Ninth Street and Columbus Avenue, freezing cold but happy to have company. My roommate had decided, uncharacteristically, that she was going to Times Square with some friends, and so I tagged along not wanting to be home alone in the apartment. It was almost midnight, so Fifty-Ninth Street was as close as we could get to Times Square. Despite frigid temperatures and months of media hype with dire predictions of power outages, bank runs, and other random catastrophes, there we all were. Hundreds of thousands of people from all over the world, ready to hug a stranger and ring in the New Year. As always, there was a sense that the New Year—in this case, it was also celebrated as the start of a new millennium—held the possibility of something better. And as far as I could tell, the competing sense that we might also witness human civilization crushed by a little bug was not lost on anyone.

That bug, of course, was the Y2K bug, the millenium bug. It was nothing more and nothing less than a computer bug resulting from the practice in early computer program design of representing the year with two digits. In fact, the term *Y2K* itself was born out of a good programmer's relentless pursuit of efficiency. David Eddy, one of an army of programmers who worked on fixing the problem is credited with coining this term. He says he coined it on June 12, 1995,[1] in an absentminded e-mail. "Being a good programmer," he explains, "I'm a minimalist typist. And Y2K was simply 60 percent less effort/cheaper to type than year 2000." Funny the way good intentions can come back to haunt us.

Actually, Y2K was wrapped up in something much bigger. "Y2K coincided with the end of the millennium, so it became somewhat of a Rorschach blot for our collective anxiety about the future. The greater the number of 0s in a year, the more we freak out," explains Paul Saffo, a technology forecaster based in Silicon Valley who was among the first to push businesses to take Y2K seriously. He adds, "Y2K tapped into some pretty apocalyptic stuff. And in that sense I think it has some similarities with climate change." Saffo is a consulting professor in the School of Engineering at Stanford University, where he teaches forecasting and the impact of technological change on the future. "Like Y2K, climate change is a technology problem that resonates with millennial anxieties." I guess that would make climate change *Y2K 2.0*.

Still, there are some important differences. Government and business spent on the order of $100 billion dollars fixing Y2K, but the problem of climate change is a lot bigger and a lot harder to solve. "And ultimately, the Y2K story ends happily with a bunch of geeks saving the world from a stupid problem the geeks themselves created," Saffo says. He thinks it will take a lot more than an army of geeks to fix the climate bug. And, of course, it's not clear that the climate story will have a happy ending.

As a technology forecaster, Saffo helped persuade the business community to get to work quickly on Y2K. "Actually, it was pretty simple. I told them, 'This is not hype. You can either fix the bug now, or you can wait until the last minute. But the longer you wait, the more expensive it will be to fix and the tougher it will be to hire people to fix it,'" Saffo explains. By then, businesses had already been running their own tests. And the outcome of the tests, which consisted of nothing more complicated than advancing their computer clocks out in time to the year 2000, suggested that Y2K was indeed a problem. When the clocks got to the year 2000, their computers stopped working. That's what you might call a straightforward modeling experiment.

Even so, some businesses underreacted to Y2K at first, and then, just as Saffo had warned, they spent more money than they should have scrambling to fix the bug in their software. "I liked to use the sailboat analogy," Saffo explains. "I'd say, 'Imagine you've got a sailboat and you need to sail around an island. You can start to circle when you're still a mile from shore and it will be easy. But if you wait until you're only 100 meters away, there will be rocks and reefs. There will be a lot more drama.'"

But ultimately, the business sector wasn't worried about drama so much as it was lured by potential opportunity.

"What really persuaded them in the end," Saffo says, "was that we presented Y2K as an opportunity. We said, 'Don't just solve the Y2K problem; use this as an opportunity to improve your business.'" I guess in the end, we all live in the hope of a better future. The problem with climate change is that it presents a set of different futures and forces us to choose one: continue "business as usual" and live on a hot planet with rising seas or change course and rebuild our energy and transportation infrastructure. Either way, we will have to pay. And either way, Saffo's sailboat analogy still applies.

When you're talking about the future impact of climate change on New York City, Saffo's sailboat analogy requires some modification: just pretend you're on the island instead of in the sailboat. But you're not in one of the buildings that make up the city's famous skyline or in one of the yellow cabs that snarl traffic in the streets and avenues. Instead, you're in a rather unassuming building above a place called Tom's Restaurant.

I am convinced that being a scientist in New York City is a very special experience, because New York is an *anti*–ivory tower.

Alternate side of the street parking forces the scientists who work at the Goddard Institute for Space Studies (GISS) to dash out of seminars and move their cars: parking tickets pile up if these scientists don't pay attention to the real world. And the ground floor of GISS, one of the most important climate modeling centers in the world, is home to Tom's Restaurant, a greasy spoon and Columbia University hangout made famous by the television show *Seinfeld*. I took classes at GISS as a grad student, and I found that the smell of french fries permeates the entire building. I don't know how GISS scientists can get anything done.

It's in this setting that Cynthia Rosenzweig works. Rosenzweig, a senior research scientist who heads up the Climate Impacts Group at GISS, hopes Americans can be convinced that, as with Y2K, fixing the climate bug is an opportunity to be seized sooner rather than later. And she's spent her career proving that climate change is not hype.

Rosenzweig came to GISS as a young graduate student to work on agriculture in the early 1980s. "I arrived at a time when GISS was developing some of the first global climate model projections. Jim was the director," she says. Jim is James Hansen, a well-known climate scientist and an outspoken advocate of reducing emissions. Hansen is still the director of GISS, which still has its headquarters in a nondescript building on the corner of 112th Street and Broadway.

Rosenzweig's graduate work was in agronomy, and perhaps because she entered the climate community as somewhat of an outsider, she has a talent for connecting the mathematicians, atmospheric scientists, and physicists who know the climate change projections with the economists, policy makers, and engineers who have to figure out what to do about this issue. Rosenzweig has come up with a way to study the messy business of climate impacts and offer a range of solutions.

"As scientists, we had all done so much work on the land-based resources, the ecosystems, the agriculture. We were racing to figure out what climate change would do to our ice caps, our forests, our food supply. And I began to think we were missing something really important: we were missing what climate change would do to us," Rosenzweig says. "Over 50 percent of the world's population lives in cities. I'd lived in Manhattan almost my entire life and I began to realize that we had better find out how climate change is going to affect cities, because that's where the people are." And like a sailboat navigating a turn around an island, Rosenzweig's research began to shift from studying the impact of climate change on nature to the impact of climate change on *human nature*.

In the year 2000, Rosenzweig was tapped to lead the Metropolitan East Coast Assessment,[2] one of eighteen research projects that came to be known as the National Assessment. The goal of each regional assessment was to understand the impact of climate change on infrastructure and people. Rosenzweig, also a coordinating lead author of the IPCC Fourth Assessment Report, was able to apply that experience to the study of climate impact being done for New York. And in August 2008, much to Rosenzweig's delight, the Metro East Coast Assessment led to the creation of the New York City Panel on Climate Change (NPCC), modeled on the IPCC. "New York City has its very own IPCC," she explains. "There is no other city in the world that can say that."

Cities cover less than 1 percent of the Earth's surface, but they hold half the population and produce about 70 percent of the total greenhouse gas emissions. That's why focusing on reducing emissions in cities is so important. Actually, despite their big collective carbon footprint, the cities' reliance on mass transit and smaller, stacked living spaces makes them very energy-efficient. For example, the carbon footprint of the average New Yorker is less than one-third the size of the average American's carbon footprint.[3]

Another compelling reason to tackle global warming in cities is that they are very vulnerable to a changing climate. And when you live in a city that is also an island, you've got even bigger worries. The NPCC, using data and models to project future climate change for New York City, has identified some of the most serious potential risks to New York's infrastructure. After tallying up all the risks, the panel members hand off their assessment to the mayor and the New York City Climate Change Adaptation Task Force. It's up to them to decide what the city should do about climate change. The task force consists of thirty-eight city, state, and federal agencies; regional public authorities; and private companies that operate, maintain, or regulate critical infrastructure. And as Rosenzweig sees it, New York City has decided to fix the climate bug now. The city has decided to see climate change as an opportunity, just like Y2K.

In February 2009, the NPCC released its latest report, *Climate Risk Information*. This presents a picture of what New York could look like in the future under different scenarios for greenhouse gas emissions. The conclusion is simple: the more greenhouse gases we emit globally, the more problems New York will have locally. The report shows New York as climate scientists like Rosenzweig see it. It digs, in detail, into all the gritty infrastructure issues the city is facing: everything from sewer pipes backing up to power lines sagging to airport runways flooding. As with most things in life, the details, the fine print, will get you in the end.

The report is fascinating to read if you're an infrastructure junkie like me, and Rosenzweig admits that she has become almost obsessed. "I have been all around the world working on climate change, but it became so much more real to me when I came to grips with it here in New York," she explains. There's an old saying: what you work on works on you. And as far as New York is concerned, rising temperature, increased risk of flooding, and a rise in sea level are working to make it more vulnerable.

"For New York, climate change means blackouts. That's just pure and simple," says Steve Hammer. Hammer is the director of the Urban Energy Program at Columbia University's Center for Energy, Marine Transportation, and Public Policy. He's been working on a statewide project, looking at the impact of climate change on New York state's energy supply. You can't get around it: energy is intimately connected with temperature. And the temperature in New York is going up.

The average annual temperature in New York City from 1971 to 2000 was approximately 55°F. And if you look over the long term, you'll see an upward trend. Since 1900, the annual average temperature in New York City has risen 2.5°F. There is plenty of natural variability in the record, but nonetheless, this long-term trend is interesting. And when you look at summer heat, it gets even more interesting. From 1971 to 2000, New York City averaged about fourteen days a year with temperatures over 90°F, and there were about two heat waves a summer. A bona fide heat wave is defined as three or more consecutive days with maximum temperatures above 90°F. A 100°F day in New York is actually a rarity: less than one day per year hits that level.

Of course, the number of extreme events in any given year varies a lot. For example, in 2002 New York City experienced temperatures of 90°F or higher on 33 days. But two years later, in 2004, there were only two days of 90°F or higher. What's interesting is that seven of the ten years with the most days over 90°F have occurred since 1980. Climate models suggest that the frequency and duration of heat waves will continue to increase unless greenhouse gas emissions are sharply reduced. You can also use climate models to estimate how the statistics of hot days will change in the future, on the basis of different emissions scenarios. Even with the scenario involving modest greenhouse gas emissions, known as A1B (see the accompanying table), high temperatures will steadily go up, forcing people to turn up the air-conditioning. By the end of the century, 100°F days will no longer be such a rarity.[4]

That's why Hammer, as someone who makes policy regarding energy, is worried about climate change.

"Energy systems are generally rated for a certain temperature and power load. If you keep running your power plant full blast for ten days during a heat wave, that's when things begin to break down. We know AC demand is going up and we know heat extremes are

Temperature	1971–2000	2020s	2050s	2080s
90°F	14	23–29	29–45	37–64
100°F	0.4	0.6–1	1–4	2–9

Number of hot days in New York over four different time periods, based on observed data and an ensemble of sixteen climate models and three emissions scenarios. These values represent the central range (67 percent) of the model output. (SOURCE: NEW YORK CITY PANEL ON CLIMATE CHANGE)

going up. That's why the city is extraordinarily vulnerable," says Hammer.

And as the number of hot days begins to increase, materials begin to break down: concrete, bridges, rail lines. More and more strain is placed on the materials that make up New York's extensive infrastructure.

"For example, the capacity to transmit electricity over power lines drops with higher temperature due to increased resistance," explains Hammer.

The NPCC looked at temperature projections over the coming century. The analysts used sixteen climate models and three emissions scenarios—each scenario assuming a different human reality in the future. The first emissions scenario describes a world with rapid population growth and limited sharing of technology. The second emissions scenario describes a world where the effects of economic growth are partially offset by the introduction of new technologies and decreases in global population after 2050. The third emissions scenario describes a world where global population grows to about 9 billion by 2050 but then declines to about 7 billion by the end of the century. This is also a scenario in which society places emphasis on clean, efficient technologies that reduce the growth of greenhouse gas emissions. As a result, it has the lowest greenhouse gas emissions of the three, with emissions beginning to decrease by 2040.

Keep in mind that the range of temperatures predicted for New

York is mostly a reflection of these different emissions scenarios. If emissions are kept down, New York will be likely to stay along the lower end of the temperature range. But if nothing is done to reduce emissions, New York will be likely to see temperatures that are even higher. According to the NPCC, New York's average temperature is expected to increase by about 1.5°F to 3°F by the 2020s, 3°F to 5°F by the 2050s, and 4°F to 7.5°F by the 2080s. In other words, by 2080, the overall climate of New York City will be more like that of Raleigh, North Carolina, or Norfolk, Virginia, if greenhouse gas emissions aren't sharply reduced in the coming years.

Although this may sound good to people who love the heat, the problem is that New York's energy infrastructure wasn't built with Raleigh's climate in mind.

"The energy risk becomes very apparent when you look at what percentage of the overall power load in the state is currently going toward air-conditioning," says Hammer. "Not too long ago, that number represented approximately 2 percent of total electricity load. But the thing is, turning on the AC creates peak power demand problems. As that demand increases, you'll need to dramatically increase the amount of total power generation capacity available around the state. You can't expect to satisfy peak demand increases by drawing in power from other places, as demand increases will be a regional phenomenon," says Hammer.

If they're not made more energy efficient, cities have the potential to become trapped in a vicious circle with regard to climate. More heat extremes lead to increased energy demand, which leads to more heat extremes. That's why fixing the climate bug is so crucial.

There are two ways to do it. One way is called *mitigation*: you can fix the climate bug by reducing greenhouse gas emissions. Another way is known as *adaptation*: you can cope with the climate problem by fixing the infrastructure. Rosenzweig and the rest of the panel on climate change are trying to show New York how to do both. "We're trying to create this road map whereby climate information can bear on other areas of society," Rosenzweig says. The thinking is that if people can see what New York might look like in the future, they will opt to avoid the unnecessary drama, of which rising temperature is just one example.

"I know it's old-fashioned. But I am very much into win-win solutions. And anything we do today will help us today," Rosenzweig adds.

That ended up being true of Y2K as well. Saffo says you can credit the millennium bug for the swift rebound of New York City's computing systems after the attacks of 9/11. "Y2K forced Wall Street to make upgrades. Wall Street had a Y2K drill. They practiced that drill and it paid off," Saffo says. The system redundancies developed in anticipation of Y2K allowed the city's transportation and telecommunications sectors to provide service despite the enormous damage on 9/11. Those redundant networks and contingency plans put in place by an army of geeks led to an opportunity that may well have saved lives.

"And guess what? We are *not* perfectly adapted to the climate extremes of today!" says Rosenzweig. "That's why everything we do to adapt now is going to help us right now."

The power grid isn't the only area that remains unsuited to the climate extremes of today. Flood protection is another glaring example that Rosenzweig cites to show how ill-suited we are for the present, let alone the future.

"The Saw Mill River Parkway"—a major traffic artery that connects New York City with the suburbs to its north—"floods now every time there's a heavy rain," says Rosenzweig. "So why don't we, as a region get organized and realize, Hey, these types of events are going to happen more frequently. There is no need for everyone to pile onto the parkway and sit in these big puddles. We're smarter than that. It could be as simple as having everyone telecommute that day." It turns out that not all adaptations are expensive, and not all involve infrastructure. A lot of the issue is just how we choose to manage and operate the infrastructure.

Rosenzweig is really worried about flooding. "Without a doubt,

New York's biggest vulnerability is enhanced coastal flooding due to sea level rise. The sea level rise is guaranteed. It's unidirectional. Just like the warming," says Rosenzweig. When you live on an island, a little sea level rise goes a long way. But this isn't a problem just for the people who live along Manhattan's coastline. A lot of New York City's critical infrastructure sits less than 10 feet above mean sea level; and experts like Rae Zimmerman, a member of the NPCC and a professor of planning and public administration at New York University, say that anything below 10 feet is vulnerable to flooding during major storm events.[5]

If you look at Zimmerman's list of vulnerable transportation infrastructure, you'll see dozens of well-known places, including the Canal Street Subway Station in Chinatown, which is 8.7 feet above sea level; the Christopher Street Subway Station in Greenwich Village, which is almost 15 feet below sea level; the New York entrance to the Holland Tunnel, at 9.5 feet above sea level; and LaGuardia Airport, at only 6.8 feet above sea level. (See Appendix 2 for more details.)

Until about 150 years ago, sea level had been rising along the east coast of the United States at a rate of about 0.34 to 0.43 inch per decade. It had been rising at this rate since the end of the last ice age, mostly because of regional subsidence or sinking, as the Earth's crust still slowly readjusted to the melting of the ice sheets. But within the past 150 years, as global temperatures have increased, regional sea level has been rising more rapidly. At present, rates of rise in the sea level in New York City range between 0.86 inch and 1.5 inches per decade. The long-term average rate since 1900 is 1.2 inches per decade. In Lower Manhattan, the water at the Battery has risen more than 1 foot during the last century.

As a result, the 1-in-100-years flood, or the flood that has a 1 percent chance of occurring in any given year, will happen about every eighty years instead. The current 1-in-100-years flood can produce a sea surge of approximately 8.6 feet for much of New York City, and that surge height is shifting upward just as the chance of such a flood is shifting upward. Flood statistics in general are changing. By the end of the twenty-first century, the kind of coastal floods that at present occur about once per decade may occur every other year. According to the World Bank Climate Resilient Cities report, a 100-year flood may increase from once in eighty years, where it is today, to once in forty-three years by the 2020s and to once in nineteen years by the 2050s.

"This goes beyond the climate forecasts," says Rosenzweig. "This is fundamentally about our values. This is about who we are as a city." Rosenzweig has already been a host for visiting Dutch flood control experts. "We have a lot to learn from the Dutch about adaptation planning," she says. The Netherlands has waged a long battle against the sea, and this struggle has clearly shaped Dutch values. The Dutch have lost enough battles to be very concerned about climate change.

One of the Netherlands' biggest battles came on the evening of January 31, 1953, when a high-tide storm breached its famous dikes in more than 450 places. You could argue that this flood reshaped the political, environmental, and psychological landscape of the nation as much as it reshaped the land. More than 1,800 people died, many as they slept. More than 47,000 homes and buildings were swept away.

Twenty days after that devastating flood, the Delta Plan was launched. Dutch politicians set in motion a $3 billion, thirty-year program to end the threat from the sea once and for all. The Dutch decided that their strongest sea defenses would be designed to stand up against a storm so strong it would occur only once in 10,000 years. The river levee and dike systems were built to withstand a 1,250-year storm. The country built an elaborate network of dikes, man-made islands, and a 1.5-mile stretch of sixty-two gates to control the entry and exit of North Sea waters into and out of the low-lying southwestern provinces. The Delta Plan is one of the largest construction efforts in human history and is considered by the

American Society of Civil Engineers (ASCE) as one of the seven wonders of the modern world.

New York—like the rest of the United States—doesn't get nearly that kind of praise from the ASCE. In fact, in its 2009 Infrastructure Report Card, the ASCE gives America's total infrastructure a D. In New York State, ASCE's most serious concern is bridges, roads, and mass transit. The engineers found that 46 percent of New York's major roads are in poor or mediocre condition, 42 percent of New York's bridges are structurally deficient or functionally obsolete, and 45 percent of New York's major urban highways are congested. In addition, there are 391 high-hazard dams in New York. A *high-hazard dam* is defined as one whose failure would cause loss of life and significant property damage. Forty-eight of New York's 5,089 dams are also in need of rehabilitation to meet the state's applicable safety standards. One explanation for all this is that by 2030, just about all of New York's major infrastructure networks will be more than a century old.

As problematic as the dams and bridges are, they're only part of New York City's infrastructure problem. The city's first subway line opened in 1904, the same year as the first New Year's Eve bash in Times Square, and the subway signaling technology that's still in use today was built before World War II. The energy grid was started in the 1880s, and two of the city's water tunnels were completed before 1936. (A third is being built now and is expected to be finished in 2020.)

In short, New York is an old city facing new problems. In April 2007, the city launched a comprehensive sustainability plan: PlaNYC 2030.[6] The 2007 plan now covers both mitigation and adaptation. New York is fighting its own battle with the sea—on two fronts.

In New York, the big storms come in two types: hurricanes and northeasters. Hurricanes are more likely to cause the 1-in-100-years and the 1-in-500-years floods. Northeasters, the ferocious winter storms named for the continuous, strong northeasterly winds blowing cold air down from the Arctic, are the main source of the 1-in-10-year coastal floods.

"Tell me when the next hurricane's going to hit and I'll tell you how soon we have a big problem in New York," says Hammer. As with the Dutch in 1953, predicting when the next big hurricane will hit Manhattan is impossible. All you can do is build, knowing it's going to happen again at some point.

Historically, hurricanes have hit New York (see Appendix 2) between July and October. On the basis of this short record, the National Hurricane Center estimates the *return period* for a category 1 hurricane in New York at about once every twenty years. A category 1 hurricane has sustained winds of between 74 and 95 miles per hour on the Saffir-Simpson hurricane wind scale. A return period of twenty years for a category 1 hurricane at a given location means that on average during the previous 100 years, a category 1 or greater hurricane has passed within 86 miles of this location about five times. The return period for a category 3 hurricane, which has sustained winds of 111 to 130 miles per hour, is roughly once in seventy years for New York. To put this period in perspective, it's about one in ten for Miami.

According to a 1995 study by the U.S. Army Corps of Engineers, a category 3 hurricane in New York could create a surge of up to 16 feet at LaGuardia Airport, 21 feet at the Lincoln Tunnel entrance, 24 feet at the Battery Tunnel, and 25 feet at John F. Kennedy International Airport. And that's with sea level measurements as of 1995. The impact could be even greater if the storm hit at high tide, as was the case in the Netherlands in 1953. The Army Corps of Engineers estimates that as many as 3 million people would need to be evacuated from New York City.

"The Dutch have decided a 1-in-10,000 year standard is right for them," says Rosenzweig. "And we have to focus on doing what's right for New York. But we need to decide now." A category 3 hurricane is capable of producing sustained winds of more than 111 miles per hour, but the current building code requires windows to withstand only gusts of 110 miles an hour. Add to that the fact that wetlands in and around New York City, the natural sponges that help absorb some of the damage done by hurricanes, have shrunk by almost 90 percent over the past century, owing to a combination of development and storm damage. Numerous decisions need to be made. For example, some scientists have suggested that New Yorkers follow the lead of the Dutch and construct storm surge barriers. These experts have simulated storm surge to evaluate the effectiveness of barriers at several points in New York Harbor. Two historical storms were evaluated—Hurricane Floyd and the northeaster of December 1992—and in both simulations the barriers were shown to be operationally effective. But such barriers are very expensive and won't provide protection everywhere. The decision is up to New York.

Hurricanes illustrate some of the most dramatic risks to New York's infrastructure, but there are plenty of everyday weather events that will become increasingly problematic for the city's support systems—most notably the guts of New York, its sewer system. New York City's drainage and wastewater system is extensive. It consists of about 6,600 miles of sewers, 130,000 catch basins, almost 100 pumping stations, and fourteen water pollution control plants.[7] The sewer system was designed to minimize standing water on roadways and streets, is mostly gravity based, and has been built over hundreds of years. The sunk-cost investment in the city's sewer systems is enormous, and the problem is that there's almost no flexibility to modify existing piping to install bigger pipes.

Rae Zimmerman is not only a member of NPCC but also the director of the Institute for Civil Infrastructure Systems at New York University's Robert F. Wagner Graduate School of Public Service. "Sewage water and storm water use the same pipes. They do it for economics, but it makes them more vulnerable in the event that something happens," she explains. In addition to the cost and disruption, the time to effect changes would be extremely long. More space would need to be found within the maze of subsurface utilities below the streets, and pumping might be needed in some instances to convey storm and wastewater flows. Despite these obstacles, some changes must be made to prevent flooding of streets and basements.

Like a lot of older cities in the country, New York City has a single, combined sewer system that handles sanitary waste as well as storm water. When it's not raining, sewage treatment plants can handle all the sewage and clean it up. But when it rains, the vast amount of rainwater that goes into the sewers exceeds their capacity, so some of it has to be released into the rivers untreated. If rainfall becomes more intense—as observed data and climate models suggest will happen—the sewer system could be overwhelmed. That would result in more flooding of streets and basements, and more untreated waste would enter rivers.

Rising sea level will also be a factor. "New York City has regulators. So when the tides go out, the pipes—the storm sewers—are exposed and the water flows out of them and into the river. But when the tide comes in and there's flooding, the regulators shut, or else you would have all the river water flooding New York City. Now, if there is sea level rise to the point where those regulators are always shut, there is no place for any storm water to go, and it will all spill onto the street," Zimmerman explains.

There are some simple and cheap solutions that can help alleviate the strain. Staten Island is already preserving its natural drainage corridors, called *bluebelts*. These bluebelts are nothing more than streams, ponds, and other wetland areas that act to convey, store, and filter storm water, with the added benefit of providing open spaces and diverse wildlife habitats. Bluebelts have been shown to save tens of millions of dollars in infrastructure costs, compared with conventional sewers. And in Hendrix Creek, a tributary to Jamaica Bay, ribbed mussel beds have been reintroduced to test this mollusk's ability to improve the water quality of tributaries around combined sewer overflow outfalls.

As always with climate change, there is the problem of too much water and too little water. Approximately 90 percent of the city's water supply is from the Catskill and Delaware systems. New York City's drinking water originates from a watershed about 2,000 square miles in area, situated 125 miles north of the city. It provides 1.1 billion gallons per day to 8.2 million city residents plus an additional 1 million people upstate. The system has a network of nineteen reservoirs and three controlled lakes throughout the Croton watershed east of the Hudson River and the Catskill and Delaware watersheds west of the Hudson. Some of the new problems associated with climate change could compromise the existing water supply and treatment systems.

As the temperature increases, more precipitation will fall as rain than as snow; consequently, there will be less storage and therefore reduced inflows to reservoirs during the spring season. Peak snowmelt in the Catskills is already shifting to earlier in the year. A recent study shows that during the period from 1952 to 2005 it shifted from early April to late March. In this scenario, the juggling act between floods and water supply gets harder. Lower reservoir levels can protect against a sudden flood, but low levels may reduce the statistical probability that water resources will be available to the city. A drought watch is declared when there is no better than a 50 percent probability that either the Catskill or the Delaware reservoir system will be filled by the following June 1. This definition is based on records going back to 1927—but such records are not an accurate predictor of the future.

New York uses about 1,060 million gallons of water per day (mgd). (In more familiar terms, this is 1.06 billion.) But demand can easily rise to more than 2,000 mgd (2 billion) during heat waves. For example, August 2, 2006, was the third successive day with temperatures in the 90s and humidity over 70 percent. The daily flow was 1,560 mgd (1.56 billion), but the peak flow reached

2,020 mgd (2.02 billion). During a heat wave in the city, illegal hydrant usage jumps up, just at a time when water demand is at its highest. And the ability of the existing aqueducts to refill the city's main distribution reservoir is pushed to the limit. You can usually expect water levels to go down and water pressures throughout the entire system to become quite weak. That spells trouble for firefighters, increases the number of complaints about low water pressure on the upper floors of buildings, and increases sediment resuspension within water mains. You can almost hear the city groaning under the strain.

For all the drama it is capable of creating, climate change is ultimately about a million boring little fixes. But as Rosenzweig notes, these boring little fixes can have a profound impact. One actual example of an adaptation strategy is installing more fire hydrant locks. And in May 2008, the Climate Action and Assessment Plan published by the Department of Environmental Protection called for the installation of better hydrant-locking mechanisms.

More frequent droughts will begin to work in combination with rising sea level in ways that will affect the water supply. "All these systems are intertwined," Rosenzweig explains. "Climate change impacts in urban areas like New York are completely integrated. That's why our science teams have to be integrated. If they're not, we get the wrong answer." In New York, it is essential to take a multidisciplinary approach that involves scientists with many different backgrounds, because they can all come at the problem from a different point of view.

As Rosenzweig explains, "We meet every month with our full team of scientists. The hydrologists tend to be very confident. They often say, 'Don't worry, we can handle it, climate variability is our middle name.' And so we were discussing the fact that the Palmer Drought Index, a commonly used index used to measure drought severity, shows more frequent droughts in the New York region.

The hydrologists said, 'Don't worry, we have a pipe that goes into the Hudson River at Chelsea, which is a little town up the river, so we're just going to supplement the supply by taking in water from the Hudson River at Chelsea.' And a colleague from the NPCC waves her hand in the back of the room and says, 'But we just calculated that when you include the impacts of sea level rise, the salt front in the tidal estuary would swing all the way up to Chelsea.'

"This is a classic example of why we need all these experts in the same room. Sure, we can take in additional water from the Hudson River at Chelsea. But if it's salt water, that's not going to help us very much."

PlaNYC has set a goal to reduce New York City's greenhouse gas emissions 30 percent by 2030. There are four principal strategies in the mitigation plan: avoid sprawl, generate clean power, make buildings more energy efficient, and create sustainable transportation. New York City's population is expected to grow from 8.36 million today to about 9.1 million by 2030. Scientists agree that far deeper emissions reductions, on the order of 80 percent, will be necessary by 2050 if we are to stabilize global temperatures. In New York, the reduction plan will focus mainly on buildings, which contribute 80 percent of greenhouse gas emissions, compared with about 35 percent nationally. And 85 percent of the 950,000 buildings in New York City in 2030 already exist today. The city has promised to do a carbon inventory every year to track its progress.

Zimmerman likes to point out that New York shouldn't focus solely on climate change. "I am a firm believer in the fact that we can be addressing national security, climate change, and sustainability simultaneously. In fact, I don't think we should be doing any of this stuff separately," she says. Zimmerman would like to see the city diversify and decentralize a lot of its infrastructure. "If we generated more electricity from solar and wind, we'd be reducing greenhouse gas emissions and we'd also be reducing the consequences of terrorist attacks because the production of energy would be more decentralized. The goal is to have a building-by-building supply," she says. "I believe in decentralized infrastructure, but that doesn't mean decentralized cities. I believe in cities!"

In keeping with this building-by-building approach, the rooftops of New York City hold a lot of hidden potential. One study cited by PlaNYC calculates that if all the rooftops in the city were covered with solar panels, they could produce nearly 18 percent of the city's energy needs during daytime hours. In New York, roughly 40 percent of the carbon footprint comes from electricity consumed, and another 40 percent comes from the heating fuels burned directly in buildings. So making buildings more efficient is a major part of the strategy to reduce carbon emissions. It can also reduce air pollution. Nearly one-third of locally produced particulate matter in our air comes from heating fuel. Public health can improve quickly as a result of efforts that improve air quality and building efficiency.

New York may be looking to learn from others, but it's also hoping to serve as an example of best practices. Steve Hammer also serves as an adviser to the Energy Smart Cities mayoral training program being developed by the Joint U.S.–China Cooperation on Clean Energy, a nonprofit with offices in Shanghai and Beijing. The training program is a three-year initiative that will introduce Chinese mayors to the best practices of the West, with the goal of helping cities reduce their energy intensity by 20 percent by 2010. Energy intensity, the ratio of energy use to output, is a way to measure the overall energy efficiency of an economy. Hammer brought Dr. Rit Aggarwala, the lead architect of the PlaNYC report, to China to speak at the first training session, along with a dozen other international experts.

In general, Hammer says, climate change is getting much more attention in China. "About four years ago, there was one wind turbine blade manufacturer that had set up shop in China. Today there are over seventy making blades and other turbine parts. The market is shifting very rapidly here," says Hammer. But for all this market growth, Hammer acknowledges there is much to be done at the local level. "The mayors I've met generally agree they can do a better job at local sustainability matters. Many would like to take this on, but they need help in identifying strategies that are appropriate for their city. China's central government could help prime this pump by investing in local energy planning, much like the Obama administration has started to do in the United States."

Swift action is essential, as the rate of urban growth in China is dramatic. "The country will build 50,000 new skyscrapers in the next twenty years, and what could be more important than making them sustainable? In 1980, Shanghai had 112 buildings more than eight stories high. Now it has 13,000 of them," says Hammer. "It's like an almost infinite Manhattan."

"The question I get asked the most," Rosenzweig says, "is: am I optimistic or pessimistic about the future? I really feel climate change is the issue that is challenging human beings to become sustainable. It's an issue that is finally big enough and destructive enough that it forces us to pay attention. And if sustainability becomes a full-fledged universal value, we'll be OK." Rosenzweig, even knowing everything she knows, is optimistic.

Paul Saffo is not. He says we're still stuck in a debate between two camps over what exactly to do about climate change and sustainability. "On the one side you have the engineers and on the other side you have the Druids. I love them both," he says. "The problem is that the engineers say let's fix our way out of this; let's use technology and go faster in the future. The Druids say let's go slower. Let's go back to a time when things were smaller and simpler. The climatologists can't seem to decide which they are."

But maybe that's the point. Maybe the climatologists know we have to do both. We have to look forward and we have to look back.

In 1609, 400 years ago, the English explorer Henry Hudson sailed into New York Bay. Hudson had been hired by the Dutch East India Company to find a northeast all-water trade route to Asia. On his ship, the *Half Moon*, Hudson and his crew left Amsterdam and sailed along the coast of Norway until they hit ferocious weather and sea ice. Rather than return empty-handed, and disobeying orders to search only for a northeast route, Hudson made a 3,000-mile detour in search of warmer weather and the dream of finding a southwest route to Asia through North America. He never found the southwest route, but he did make another discovery, the island of Manhattan. The Dutch colonial settlement and fur trading outpost of New Amsterdam served as the capital of New Netherland until it later fell under British rule and was renamed New York. New York wasn't the dream Hudson had been chasing, but it has come to represent the hopes and dreams of millions.

"I guess," says Rosenzweig, "everyone has the New York of their dreams. For me, it's a climate-resilient city." The year 2009 marked the second anniversary of PlaNYC.[8] On Earth Day, April 22, Mayor Bloomberg announced that eighty-five of the plan's 127 initiatives were on time or ahead of schedule. As part of PlaNYC, the city had converted 15 percent of the yellow taxi fleet to hybrid vehicles; planted 174,189 trees across the five boroughs; acquired 13,500 acres of land to protect the upstate water supply; saved 327 tons of nitrogen oxide (NO_x) per year by means of retrofits to the Staten Island ferry fleet; and started twenty storm water retention pilot projects. Maybe Rosenzweig is right to be optimistic.

New York, New York: The Forty-Year Forecast—Hurricanes, Infrastructure, and Sea-Level Rise

NEW YORK, NEW YORK	TODAY		2050		2090	
Emissions Scenario	JAN.	JULY	JAN.	JULY	JAN.	JULY
Higher	31.6	74.7	34.9	78.1	38.4	81.9
Lower			34	77.4	35.4	78.6

Forecast
September 2013

You could say New York deserved a lucky break. It had already been slapped around enough by tropical storms. Back in September 2004, the remnants of Hurricane Frances had flooded its subways and stranded passengers. And in September 1999, Hurricane Floyd, by then weakened to a tropical storm, had dumped more than 10 inches of rain on the city, causing mudslides on the bluffs overlooking the Hudson River near the Tappan Zee Bridge. There were still plenty of people who remembered Hurricane Donna, the category 3 storm that pounded New York City on September 12, 1960, with sustained winds of more than 90 miles per hour. Donna had flooded lower Manhattan almost to waist level on West and Cortlandt Streets—the southwest corner of what later became the site of the World Trade Center. The last thing New York needed was another hurricane.

An average hurricane season has eleven named storms and six hurricanes, including two major hurricanes. The United States Landfalling Hurricane Probability Project put the risk that New York would be hit by a major hurricane (category 3 or more) by 2050 at 90 percent.[9] We all knew that eventually, with or without global warming, a major hurricane was going to hit New York. The question was simply when the Atlantic Ocean would start to play hardball again.

The period from 2009 to 2012 was a stretch with the fewest named storms and hurricanes since 1997—thanks, in part, to an El Niño in 2009–2010. El Niño produced strong wind shear across the tropical Atlantic, which meant fewer and shorter-lived storms. It almost seemed as though, after producing Katrina in 2005, the Atlantic Ocean had gone into semiretirement. If hurricane seasons were anything like baseball, then the Atlantic seemed to be in a very welcome slump. For four years in a row, no major hurricanes had hit the United States. And as for baseball: the Yankees, in their new stadium, went on a winning spree of four World Series in a row.

But all good things must come to an end. And in September 2013, with the Yankees not even looking to play in the postseason, the Atlantic Ocean reawakened and one specific hurricane seemed to be in a New York state of mind. After beginning as a garden-variety low-pressure system moving off the coast of west Africa, the storm that eventually came to be known as Hurricane Homer gathered strength as it crossed over unusually warm tropical Atlantic waters. The warm ocean water acted like a heat pump, fueling the hurricane and causing it to increase in intensity. Many models still struggled to predict exactly how climate change would affect hurricanes, but there was general agreement that warmer water meant more intense storms.

The NOAA GOES-12 satellite recorded the storm's every move, and initially the National Hurricane Center issued a watch for Miami, expecting the storm to hit there. Miami's residents stockpiled supplies, boarded windows, and secured boats—as usual. But then Homer took a turn north and surprised everyone when it started speeding up. The Bermuda High, a large area of high pressure in the Atlantic, pushed the storm up the coast as warm water provided fuel to the system. In time, the hurricane would achieve a record-breaking forward speed of 75 miles per hour. As the system raced up the coastline of the Carolinas, the revised track forecast issued by the National Hurricane Center warned of a category 3 storm making its way directly toward New York City. Here was a system that bore a striking resemblance to an event that had almost happened back in 1938, when the Long Island Express, which ultimately missed New York City by just 75 miles, did tremendous damage up and down the northeast coast. The only difference was that this one didn't look as though it was going to miss.

The models suggested that New York was about to get swallowed by storm surge. Surge levels for Hurricane Homer had been calculated by the U.S. Army Corps of Engineers using NOAA's

SLOSH model, and in a worst-case scenario Homer was likely to create a surge of up to 25 feet at John F. Kennedy Airport, 21 feet at the Lincoln Tunnel entrance, 24 feet at the Battery, and 16 feet at LaGuardia Airport. The U.S. Army Corps of Engineers estimated that nearly 30 percent of the south side of Manhattan would be flooded. The storm surge flooding would threaten billions of dollars of property. Rising sea level was already a factor, as each seemingly small increase in sea level gave the hurricane a longer, more destructive reach into the city. Since 1900, sea level in New York City had been rising at rate of about 1.2 inches per decade. Hurricane Homer plus this rise meant more storm-related coastal flooding, more inundation of wetlands, more structural damage, and more money lost.

Because of the projected track and the fact that the highest, most destructive winds lay to the right of the storm's eye, it was anticipated that Homer could pass directly over New York with gusts over 150 miles per hour—shattering the glass in skyscrapers and sending razor-sharp shards raining to the ground. In addition, the counterclockwise, westerly flowing wind would funnel the surge waters into New York City harbor.

As the storm swept up the coast, we all became experts on the history of hurricanes in New York. We knew that there had been only two honest-to-goodness direct hits in New York City in recent history—the Great September Gale of 1815, and a storm that came on September 3, 1821, and made landfall at Jamaica Bay. Both were category 3 storms and both did extensive damage. With widespread flooding in lower Manhattan as far north as Canal Street, the 1821 hurricane set the record for the highest storm surge in Manhattan—nearly 13 feet. One question was whether Hurricane Homer would go down in the history books as the third direct hit. Another question was what, if anything, could New Yorkers do to prepare themselves.

For the most part, people simply jumped ship and left the city. It ended up as the largest peacetime evacuation since Hurricane Floyd. As New Yorkers headed through bridges and tunnels, transit workers stayed behind, trying to ready pumps and prep storms drains. It felt rather like rearranging deck chairs on the *Titanic*, but there was little else to do. There was simply no infrastructure to deal with a storm of this magnitude. People did what they could and then simply sat back, prayed, and vowed that they wouldn't allow the city to be this vulnerable again.

Fortunately, the Bermuda High shifted, and the hurricane began to shift course: the center of the storm headed out to the open sea. And New York, which could have been swallowed up whole and spat out by this storm, got off relatively easy.

That said, there was still significant damage to sift through. There was severe coastal erosion and heavy street flooding. The Rockaway beaches nearly vanished, owing to the high winds and storm surge. The sewer system was completely unprepared for the volume of rain that fell in such a short time. The water began pooling at the street corners, and then gradually rose to inflict damage on parked cars and storefronts. Many owners of the city's famous facades had boarded up their windows to protect the glass from the wind, but the water was one thing that no one could have prepared for.

Though no buildings were submerged, and even though the storm was a near miss, the water inflicted excessive damage across the southern third of the island. Many people who lived in basement apartments returned to the city to find their possessions soaking wet and now had to contend with an unfortunate and unhealthy problem: mold.

On top of all this, September set a new record in New York for warmth. In general, temperatures usually associated with August were now stretching deeper and deeper into September; and now New York was in the grip of a late summer heat wave. With electricity out in many parts of the city and nighttime lows still in the 80s, people felt overwhelmed by weather.

Trees were yet another casualty. Tree-lined blocks from the Upper West Side to Park Slope, Brooklyn, were stripped of their greenery. Picturesque streets suddenly looked as if they had encountered a wood chipper, with the pavement and cars covered in brown and green shrapnel. Trees were down all over Central Park; the cleanup and replanting would eventually cost several million dollars. The parts of the park that were hit hardest were closed to the public for several months, making for an unusually quiet Central Park in the fall.

And then there was the subway system. Storm tides overtopped some of the region's seawalls for only a few hours, but they still managed to flood the subway as well as the PATH train systems at the station in Hoboken, New Jersey, shutting down these transportation systems for almost a week. This shutdown made it very difficult for those who had left the city to get back home and start the cleanup. The stations that everyone knew were vulnerable to flooding in an extreme weather event proved to be just that. Because of its proximity to the Hudson River and its depth below sea level, the Christopher Street subway station was shut down longest, with water covering the tracks for several weeks after the hurricane.

Despite the tens of millions of dollars in damage, New Yorkers realized how lucky they were. But when they slowly began to file back into the city, their questions followed them home. How could a major metropolitan area like New York be so vulnerable? Why wasn't more being done to replace century-old infrastructure? If the Yankees could build a new stadium for $1.3 billion, why couldn't the mayor and the Metropolitan Transit Authority make basic improvements to the subway? After all, the Dutch and the British had already spent billions fortifying their cities against the growing threat of storms and rising sea level.

January 2014

The high-resolution model projections of what *could* have happened to New York got under everyone's skin. Deep down, we all knew it was only a matter of time before something really happened. The city, with its elderly infrastructure and vulnerable coast, needed help. Other cities were already adapting to climate change. Boston had elevated a sewage treatment plant to keep it from being flooded; this project was based on projections from scientists at Harvard regarding the rise in sea level. And Chicago, through its Green Roof and Cool Grants Program, encouraged rooftop gardens and reflective roofs to help keep the temperature down and ease heat waves. New York, of course, had its own plans in place, and now the hurricane that almost was had given everyone a new sense of what would eventually come to be.

Organizations and volunteers across the city began to implement strategies that had been laid out by the New York City Climate Change Adaptation Task Force. Given the climate change projections performed by Columbia University's Center for Climate Systems Research and NASA's Goddard Institute for Space Studies, we understood that the city was probably headed for a 3°F to 5°F increase in temperature, a 2.5 to 7.5 percent increase in precipitation, and a 6- to 12-inch rise in sea level by 2050. We worked to rebuild the city with that climate in mind.

And so began a coordinated plan to adapt the city's roads, bridges, and tunnels; its mass-transit network; its water and sewer systems; its gas and steam production and distribution systems; its telecommunication networks; and other critical infrastructure to be able to deal with the likelihood of more extreme weather. There were plans to modify dam infrastructure and allow for water releases in anticipation of a storm, and to inventory existing tide gates and identify priority locations most vulnerable to a rise in sea level and to storm surges. There were programs that focused on the long-term viability of New York's water and sewer systems. There were land acquisition programs aimed at protecting the watershed, as well as a plan for new water quality infrastructure. A complete facelift for the city would cost an unthinkable sum. So the work was done piecemeal and at the local level by concerned groups of public and private organizations. I'm not really sure how long a New York minute is, but we were now thinking about a New York century. It was as if the entire city had become obsessed with adapting to climate change. Go figure.

April 2017

It was a good thing that we did begin to chip away at the city's infrastructure shortcomings. In April 2017, four freak storms each dumped more than 7.5 inches of rain in upper Manhattan in one day, setting new records for daily accumulation. Another three days saw between 4 inches and 6 inches of rain at locations across the city within a four-hour period. In some areas, 4 inches of rain fell in one hour alone. The city's existing storm water conveyance system had been designed for only 1.75 inches of rainfall per hour. We added "new storm drains" to the city's wish list. Needless to say, the April storms flooded the subway, too. It was obvious that we also needed to focus our attention underground.

There were at least three situations that could take out the subways: (1) flooding of the tracks over the third rail; (2) water pouring through street-level vents, leading to a smoke condition; and (3) flood impacts on the signaling system. As rainwater seeped through tunnel walls and headed down subway grates and stairwells, sump pumps in 280 pump rooms next to the subway tracks would pull the water back up to street level.[10] That water then naturally flowed toward the storm drains. But more often than not, the storm drains were unable to handle the flow of water. They were designed to take away only so much water. Eventually, the water would make its way onto the subway tracks, hitting the all-important third rail. The 600 volts of electricity running through the third rail would cause the water to boil and set all the floating debris on fire. In addition, the water would short-circuit the electrical signals and switches, making it impossible for train operators to know when it was safe to stop or go. The subway system had two critical sources of power: the direct current propelling the trains, and the alternating current powering the signals. Even if the direct current was spared, the trains couldn't run without signals.

It took time, but the New York City Transit Authority installed additional pumps and built new storage tanks for processing rainwater runoff. It also developed a computer system to better monitor storage during times of overflow. Also, the Department of Environmental Protection moved electrical equipment, such as pump motors and circuit breakers at the Rockaway Wastewater Treatment Plant in Queens, from 25 feet below sea level to 14 feet above sea level. The beach nourishment projects that had been going on since 1924 to prevent coastal erosion were also stepped up in the Rockaways as a way of dealing with the effects of a rising sea level. Depositing more sand on the beach strengthened the defense that the beach provided during coastal flooding.

But in the long term, it became increasingly clear that a new network of storm surge barriers would need to be constructed in order to protect the city. As the sea level rose, New York became even more vulnerable to storm surge flooding. It would take high-water levels of only 4.9 to 5.7 feet above mean sea level to cause flooding over some of the southern Manhattan seawalls. Global warming was expected to increase the rate at which sea level rose: from about 1 foot per century to between 1.6 and 2.5 feet per century.

It was suggested that four barriers 30 feet high and able to withstand a 1,000-year flood event should be constructed: two off the coast of Staten Island, one to the northeast and one to the southwest; one off the coast of the Bronx, protecting LaGuardia Airport; and one off Breezy Point in the Far Rockaways to help protect John F. Kennedy Airport. The barriers would need to operate with as little as one hour's warning, closing each gate in fifteen minutes and the complete barrier in thirty minutes. The construction time was estimated to be eight years.

July 2027

As the models projected, temperatures across the city had been steadily creeping up. The daily highs were getting higher, and so were the nightly lows. By the 2020s everyone realized that summers were longer and hotter than ever before. That's why it came as no big surprise when New York broke the record for the longest stretch of 90-degree days ever recorded. For twenty-one days straight, New York's aging power grid was brought to its knees. The increase in peak electricity load when people cranked up their air conditioners resulted in routine brownouts as well as an increase in costs associated with cooling water for power plant operations. Even with the broad mandate for infrastructural repair that had been handed down after Hurricane Homer, the grid proved too complex to overhaul. Unfortunately, this delay had deadly consequences, which began with a lightning strike.

The lightning strike that set off the New York blackout of 2027 caused the loss of two transmission lines and a subsequent loss of power from the nuclear plant at Indian Point. New York Power Pool Operators called for Con Edison operators to "shed load."

Because of the power failure, LaGuardia and Kennedy airports were closed for about eight hours, automobile tunnels were closed owing to lack of ventilation, and 10,000 people had to be evacuated from the subway. Con Ed called the shutdown an "act of God," enraging the politicians in City Hall, most of whom said that the utility was guilty of gross negligence for not working to better prepare the power grid for a heat wave that had been years in the making. In many neighborhoods, veterans of the Northeast blackout of 2003 headed to the streets at the first sign of darkness. But many of them did not find the same spirit. In poor neighborhoods across the city, looting and arson erupted.

January 2039

The storm surge barriers were over budget and not even close to being on schedule, but they were finally done. This was the eight-year project that took twenty-two years to complete. Regardless of the complaints and infighting, we were all just glad to see the job finally done. And there was more good news. The runways at Kennedy, LaGuardia, and Newark airports were raised in anticipation of higher flood levels. There was also a plan in place to move the West Side Highway inland. Coastal areas were rezoned for parks and recreational uses, not high-density residential development.

None of it was cheap, and plenty of people began to worry that it was overkill. Elevating single-family homes in Long Island by 2 feet could cost anywhere from $22 to $62 per square foot. Additional seawalls cost about $5 million per mile.[11] For the country as a whole, it was estimated that building seawalls to protect the United States from coastal flooding would cost from $46 billion to $146 billion.[12] Scientists had also come up with ways to incorporate mangroves and sea grasses into the design of seawalls to improve their environmental impact and make them look better, too.

It was strange to see the climate projections playing out before our eyes. What used to be the 1-in-10-years coastal flood, fifty years ago, now came every other year. And the 1-in-100-years coastal flood happened four times more often. It was a good thing we had raised those wastewater treatment plants.

Over the years, new research had begun to suggest that the rise in sea level along New York's coast would be much higher than we had originally projected. The extra bump in sea level came because the Gulf Stream had thermally expanded and was slowing down as a result of warmer ocean surface temperatures.[13] The new estimates for 2050, once you included all the sources of the rise in sea level—from Greenland, from Antarctica, from glaciers and ice caps, and from thermal expansion—as well as the dynamic effects, could be as high as 3 feet.[14] And adding as little as 1.5 feet of sea level rise to the storm surge of a category 3 hurricane on a worst-case storm track would devastate many parts of the city—the Rockaways, Coney Island, much of southern Brooklyn and Queens, portions of Long Island City, Astoria, Flushing Meadows–Corona Park, Queens, Lower Manhattan, and eastern Staten Island from Great Kills Harbor north to the Verrazano Bridge would be underwater. Thank God we had built those storm surge barriers and seawalls.

August 2050

Not everyone loves New York. But those who do love it love it intensely. And through some combination of luck and high-tech ingenuity, those who loved the city ultimately saved it. In 2050, when Hurricane Xavier—a category 4 monster, which sprang up from the bathtub that the Atlantic had become finally arrived—people sat back and watched it like the World Series. We knew we had a home team advantage, just like the Yankees.

EPILOGUE

THE TRILLIONTH TON

The large fields and acres produced no grain
The flooded fields produced no fish
The watered gardens produced no honey and wine
The heavy clouds did not rain. . . .
On its plains where grew fine plants
"lamentation reeds" now grew.

—"The Curse of Akkad," c. 4110 BP

"The Curse of Akkad," an epic poem known as a *city lament,* is believed to have been written by a Sumerian priest after the collapse of the Akkadian empire, about 4,200 years ago. The lament is considered a work of literature, although some archaeologists believe it describes actual events that detail the fall of the world's first empire. Whatever it is, the lament is probably one reason why I became a climatologist. I wanted to know if some of those details included an abrupt change in climate.

In its heyday 4,300 years ago, the Akkadian empire, under the rule of Sargon of Akkad, stretched across Mesopotamia from the Persian Gulf to the headwaters of the Tigris and Euphrates rivers. In addition to deserving the title "world's first emperor," Sargon probably also deserves credit for the world's first merger and acquisition. He successfully merged the remote agricultural hinterlands of northern Mesopotamia with the urban Babylonian city-states such as Kish, and Ur and Uruk in the south. He acquired those the old-fashioned way, with a large army.

It was a good setup, while it lasted. Northern Mesopotamia was prized for its rain-fed agriculture and served as a critical supplier of grain to the economic hubs to the south. Tell Leilan, a provincial northern capital, was part of the breadbasket Sargon had come to depend on. It served as a central processing point, distributing grain throughout his growing empire, as well as his growing army. Today, Tell Leilan, in northeastern Syria, is the site of a small Kurdish village. Tucked away in a corner of the modern Tell Leilan rests what remains of this great ancient capital. Harvey Weiss, a professor at Yale University, returns here every few years to excavate. I had been working with Weiss on the collapse of Akkadia for more than four years before I ever saw what remains of this ancient Akkadian capital.

You could say that my experiences, until I joined Weiss and his team of archaeologists on a dig in the summer of 1999, had been only vicarious. I had studied ancient Near Eastern history, modern Middle East policy, Arabic, and of course, climate. I was what you might call book smart. I could tell you the average high temperature and wind speed for northeastern Syria in July, but that didn't mean much until I stood on a dig site for eight hours a day in 110°F heat taking in mouthfuls of windblown dust. I may have been there in my head, but the real experience was a whole different ball game. As with most things in life, including global warming, there is something important to be said for personal experience.

Weiss's previous excavations had shown that between 4,600 and 4,400 years ago, Tell Leilan grew about sixfold in size, from 37 acres to more than 200 acres.[1] The city's residential quarters showed signs of urban planning, including straight streets lined with potsherds and with drainage lanes; and its acropolis contained several storerooms for grain distribution. But sometime around 4,200 years ago, this society began to fall apart and archaeological evidence indicates a mass exodus from Tell Leilan to points south. Tell Leilan was abandoned and sat empty for some 300 years. It was probably during this abandonment period that the city lament was composed. After less than 100 years, the world's first empire was gone.

At least, this was the story when I signed on to research the Akkadian collapse. Weiss had assembled enough archaeological evidence to suggest, at least to him, that the Akkadian empire had collapsed abruptly because of a rapidly changing climate. However, he had no data on climate to support this theory. And without such data, the story of Tell Leilan was just that, a sad story told in the form of an epic poem. With the help and guidance of Peter deMenocal, one of my PhD advisers at Columbia University, I obtained the top 6 feet of a long ocean sediment core that had been pulled up from the bottom of the Gulf of Oman. We intended to use the core to reconstruct the climate of Mesopotamia over the geologic period known as the Holocene, which spans the last 10,000 years. By taking samples at half-inch intervals down the entire length of the core, I was able to build a history of Mesopotamian climate in 100-year steps, using only dried mud from the bottom of the sea. The ocean floor remembers everything.

Mesopotamia is a notoriously dusty place, and the dust there is predominantly composed of a mineral, dolomite. Mesopotamia is also a notoriously windy place. When its steady southwest wind, called the *shamal*, kicks up, that dust is transported and dumped into the Persian Gulf and the Gulf of Oman. Technically, my mud from the Gulf of Oman was actually Mesopotamian dust. Everything has to come from somewhere.

I spent my first semester of grad school in the lab at Lamont-Doherty Earth Observatory, crushing dried core samples and measuring the concentration of dolomite in each sample. The goal was to reconstruct a history of Mesopotamian drought. On the basis of soil characteristics and prevailing winds, the more dolomite in my mud sample, the more drought in Mesopotamia, and the more trouble for Sargon.

Grinding, measuring, and running about 200 samples consumed

the better part of my life for more than a year. When the work was finally complete, I had an elegant spike about one-third of the way down the core to show for my effort.[2] According to the X-ray diffractometer (XRD) analysis, this spike meant the amount of dolomite in that one particular sample was six times higher than it had been throughout the rest of the Holocene. That spike suggested that a severe drought had gripped Mesopotamia for more than a century. And according to the AMS dates, the spike in dolomite occurred very close to 4,200 years ago, at about the time a weary Sumerian priest was sitting down to write an epic poem of collapse. Weiss had taken the city lament at its word, and our ocean sediment core suggests he was right. The world's first empire stood for just 100 years. In the end, it amounted to less than 1 inch of mud at the bottom of the ocean.

The United Nations Framework Convention on Climate Change (UNFCCC), an international environmental treaty crafted at the Earth Summit, held in Rio de Janeiro in 1992, was ratified by 192 countries, including the United States.[3] The stated objective of the Convention (Article 2) is to stabilize atmospheric concentrations of greenhouse gas at a low enough level to "prevent dangerous human interference with the climate system." The words *dangerous human interference*, carefully selected by diplomats and policy makers, are not the language of scientists.

"One of the things that has always been difficult about the concept of dangerous human interference is that it involves value judgments," explains Susan Solomon, a senior scientist with NOAA. But in 1996, using the best available science, the members of the European Union decided to make a value judgment. They agreed that a temperature increase of 3.6°F above the preindustrial global average temperature constituted dangerous human interference with the climate system. And in order to avoid severe, widespread impacts they argued that there was a need to keep global warming below that level.

Solomon knows that workings of the policy world quite well. She cochaired Working Group 1 for the IPCC Fourth Assessment Report, and research done earlier in her career on the ozone hole provided the scientific foundation for the UN Montreal Protocol, the international agreement to protect the ozone layer. Solomon had helped prove the existence and cause of the ozone hole after leading an expedition to McMurdo Sound, Antarctica, in 1986. The ozone hole appears in the early spring, so Solomon and her team spent months in Antarctica, enduring brutally cold temperatures and nearly twenty-four-hour darkness, in order to observe the hole as it formed. She and her team were able to gather enough data to provide the first evidence that enhanced levels of chlorine oxide from the chlorofluorocarbons (CFCs) were the primary cause of the ozone hole.[4] The CFCs, stable compounds used in refrigerators, in air conditioners, and as a propellant in aerosol cans, were reacting with the clouds in the stratosphere to destroy the protective ozone layer. Solomon and her colleagues had the data to prove it.

Global warming is somewhat less straightforward. "How do you decide what a dangerous level of human interference means?" Solomon asks. "It's been a real challenge as far as what science can do to inform that process without becoming, frankly, unscientific," she explains. "It's not as clear-cut as observing an Antarctic ozone hole form." That's why Solomon chose to focus on another principle of the UNFCCC. Article 3 of the Convention emphasizes "threats of serious or *irreversible* damage."

"And I can tell you from the amount of time I spent around the diplomatic folks during the IPCC process that when they say 'serious *or* irreversible damage,' that's what they mean. If they meant to say serious *and* irreversible, they would have said so. They probably spent five days negotiating whether it was going to be serious 'or' or serious 'and,' " she adds. Solomon was struck by the word *irreversible*. "Irreversible is something that doesn't involve any value judgments. Something is either irreversible or it's not," she says. "That is a piece of the problem that science can inform." And in a recent research paper, Solomon looked at the extent to which human influence on the climate is irreversible. What she discovered surprised even her.

Until recently, most scientists were working under the assumption that if we went cold turkey and brought CO_2 emissions to zero, CO_2 concentrations measured in parts per million (ppm) in the atmosphere would peak and then fall most of the way down toward preindustrial levels in about 100 to 200 years, with the warming decreasing along with them. Solomon and her colleagues, using a climate model known as an Earth-system model of intermediate complexity (EMIC), wanted to see for themselves how long it would take for the concentration and climate to head down. An EMIC is not as fancy as a general circulation model, but it has the advantage of being fast. So Solomon was able to run very long simulations of the Earth's climate and see what the atmosphere remembered of us 1,000 years from now, in the year 3000. The experiment tested what would happen if CO_2 emissions suddenly stopped after peaking at different concentrations, ranging from 450 to 1,200 ppm.[5] In their model, CO_2 levels dropped so slowly that by the year 3000 the atmospheric concentration was still substantially above preindustrial levels. Global temperatures also stayed high. The atmosphere turns out to have a better memory than we thought.

"Somebody said to me recently that I've introduced two words into this debate that just weren't there before. And they are *unequivocal* and *irreversible*," Solomon says. "And that's not a bad set of words." It was Solomon, armed with her thesaurus, who introduced the now famous word *unequivocal* into the IPCC Working Group I Summary for Policymakers.[6] For scientists, the statement reads like Hemingway, concise and powerful:

> Warming of the climate system is *unequivocal*, as is now evident from observations of increases in global average air and ocean temperatures, widespread melting of snow and ice and rising global average sea level.

The statement reflects the kind of pure, scientific analysis and multiple lines of evidence that Solomon is passionate about. She is someone who draws a sharp line in the sand between science and politics.[7] Her personal opinions, she has said repeatedly, have no place in the policy arena.

"When we came up with *unequivocal*, some people said it wouldn't stick. It's too complicated. We thought a lot about what word to use. And we played around with words like *incontrovertible* and *undeniable*. But we didn't want to use those, because they were too political in their tone. They are not the kind of words you would see in a scientific paper," Solomon explains. "We went with it because it's not a statement about how others are reacting to the science, it's a very internal statement about the nature of the evidence," she says. This was a word that represented science, not politics.

"The fact that you have independent measurements not just of temperature, but also of sea level rise, and retreat of Arctic sea ice, and retreat of glaciers worldwide, increases in water vapor, all of which fit what we know should be happening on a warming planet," Solomon says—"that is the reason we were able to use this word." The science, in other words, speaks for itself.

Solomon's second word is no less powerful. "Article 3 of the framework convention is a recognition that things that are irreversible deserve special attention. Because it means you can't back out of them," she explains. But aside from death and taxes, *irreversibility* is not a concept we Americans tend to embrace. "The thing that makes this tricky is that just about any other pollution problem—acid rain, smog, DDT—works in a straightforward way. When you stop emitting, the problem goes away. The thing that's really hard here is that we've recognized it's not going to work that way with global warming. We're turning the dial on the global temperature as if it were a thermostat. But it only cranks one way. We don't get to turn it back down," she says. History has not been kind to those who have failed to understand the physics of the irreversible. Just look at a symbol of irreversibility, Easter Island.

Settled by Polynesians sometime around A.D. 900, Easter Island is an isolated 66-square-mile chunk of volcanic land situated in the middle of the Pacific Ocean between Peru and Australia. The volcanic tuff was carved to create the enormous stone statues, called *maoi*, that made Easter Island famous. The Polynesians used tree trunks to transport and erect the *maoi*. In fact, trees made much of life on the remote island possible. And archaeological evidence suggests that at one time Easter Island had a diverse forest. The bark of certain trees was used to make rope or beaten into cloth; other trees were used to build canoes, or to make harpoons.

The canoes and harpoons were essential, as the common dolphin, a porpoise weighing up to 165 pounds, was the main source of meat on the island. The Polynesians hunted with harpoons from large wooden canoes, far from the shore. Palms were probably the most important trees. The trunks provided sap that could be fermented to make wine, honey, and sugar. The fronds were ideal for thatching houses, and for making baskets, mats, and boat sails. It would seem as if the only things not made of wood on Easter Island were the *maoi*, which are thought to represent high-ranking ancestors. Wood was used for so many aspects of life on the island that by 1722, when the Dutch explorer Jacob Rogeveen arrived on the island, he saw no trees more than 10 feet tall. The gradual deforestation was nearly complete. Today, Easter Island serves as one of the most extreme examples of forest destruction in all of history.

Deforestation made it impossible for the Polynesians to build the canoes that allowed them to catch porpoises. Land birds, with no trees to nest in, disappeared. Palm nuts, Malay apples, and all other wild fruits were gone. The deforestation also led to soil erosion that resulted in reduced agricultural yields. It is believed that deforestation set off a chain of events that eventually led to collapse. The population dropped sharply and the construction of the *maoi* ceased. The only things left to eat on the island were rats and people. In his book *Collapse*, Jared Diamond says:

> I have often asked myself, "What did the Easter Islander who cut down the last palm tree say while he was doing it?" Like modern loggers, did he shout "Jobs, not trees!"? Or: "Technology will solve our problems, never fear, we'll find a substitute for wood"? Or: "We don't have proof that there aren't palms somewhere else on Easter, we need more research, your proposed ban on logging is premature and driven by fear mongering"? Similar questions arise for every society that has inadvertently damaged its environment, including ours.[8]

This begs the question: with regard to global warming, what is the equivalent of cutting down the last palm tree? The answer may not be simple, but if you accept the meaning of the words *unequivocal* and *irreversible*, and you accept the implied value judgment of 3.6°F warming, you can come up with a fairly good proxy. "If you accept that, you can now calculate how many gigatons of carbon you can emit and keep warming below 3.6°F," Solomon says. Scientists can boil it down to one number. And that number is 1 trillion.

Myles Allen—a professor at the University of Oxford—and his colleagues found that if we could limit all CO_2 emissions from fossil fuels and changes in land use to 1 trillion tons of carbon in total, there would be a good chance that the climate would not warm more than 3.6°F above its preindustrial range.[9] A companion study, by Malte Meinshausen at the Potsdam Institute, found that the world would have to limit emissions of all greenhouse gases, not just CO_2, to the equivalent of 400 gigatons of carbon between 2000 and 2050 in order to stand a 75 percent chance of avoiding more than 3.6°F of warming.[10] The other greenhouse gases, such as methane and nitrous oxide, are expected to produce as much warming as 125 gigatons of carbon in the form of CO_2; that means emissions of CO_2 itself over the next forty years have to add up to less than 275 gigatons of carbon. No one said this was going to be easy.

That is especially true when you consider how much CO_2 we've put up there already. Since the start of the industrial revolution about 250 years ago, we've burned about half of the 1 trillion tons. Global emissions currently average about 9 billion tons a year, and they're rising. In May 2009, the Energy Information Administration (EIA) released *International Energy Outlook 2009*, an annual report that projects energy trends through 2030. The report concluded that with no further policies to reduce CO_2 emissions, total global emissions will reach about 11 billion tons of carbon by 2030. The jump in emissions is attributed to a projected 44 percent increase in energy demand by 2030, much of it produced from fossil fuels. Renewable energy is expected to grow fastest, but fossil fuels will continue to serve as the dominant source of energy to meet the growing demand coming from developing nations. The report says 94 percent of the increase in industrial energy use between now and 2030 is expected to take place in developing countries; Brazil, Russia, India, and China are expected to account for two-thirds of that growth.[11]

The implications are simple: the more CO_2 we dump into the atmosphere, the warmer it gets, and the more serious *and* irreversible the damage. If carbon emissions were trees, then the more we've cut down by 2020, the fewer will be left to cut by 2050. In essence, the 1 trillion-ton limit allows the world to follow its current trend for about forty more years before having to quit carbon cold turkey, all at once. Somewhere out there, in a coal seam in Wyoming or an oil field in Saudi Arabia, sits the trillionth ton. Unless scientists come up with some way to suck CO_2 out of the atmosphere or a cheap form of renewable energy, the fate of that trillionth ton rests in the hands of policy makers, not scientists.

In December 2009, seventeen years after the Earth Summit in

Rio de Janeiro, policy makers and scientists gathered once again at the Conference of the Parties (COP 15) in Copenhagen, Denmark. Since the ratification of the UNFCCC, they have been using these conferences to assess progress and establish legally binding obligations for countries to reduce their greenhouse gas emissions. It was at COP 3 in 1997 that the Kyoto Protocol first set binding targets for reducing greenhouse gas emissions. The overall goal for COP 15 was to establish a new global climate agreement to replace the Kyoto Protocol when it expires in 2012. Dozens of countries, including China and the United States—the top two carbon polluters—came to Copenhagen with proposals to cut their emissions. But in the end, world leaders left the meeting with a deal that was seen as weak and lacking in detail. Although it did set up the first major program of aid to help poorer nations adapt to climate change, it offered few specifics in terms of pollution reductions. Upon the announcement of the deal, a team of experts led by a professor at MIT made a quick calculation that converted the language of policy makers into the language of scientists. As it stood, the deal was likely to result in a 5.7°F rise in average global temperature. In the end, the words of science—*unequivocal* and *irreversible*—were still not powerful enough to shift the forecast for the future.

APPENDIX 1

UNITED STATES CLIMATE CHANGE ALMANAC

Clearly, climate change will leave no part of the world untouched; but just as all the places I wrote about have their own indicators showing that climate change is taking place, so does your town or city. What follows here is an almanac that includes current and future temperature trends for a variety of cities around the United States. I'll start with a look at the big picture in the United States.

1. Total Monthly Record High versus Low Temperatures for the United States

One way to get an idea of how climate change is making itself felt in terms of day-to-day weather in the United States is to look at the total number of daily record high and low temperatures that have been set around the country. A *record daily high* means that the high temperature recorded at a specific weather station was higher on a specific day than on that same day in previous years. A *record daily low* means that the lowest temperature on a specific day at a specific weather station was lower than on that same day in previous years.

In a recent study, scientists gathered and tallied records from across the United States to get a better picture of the long-term changes in record highs and record lows.[1] They analyzed millions of daily high and low temperature readings taken over a period of six decades at about 1,800 weather stations across the country. The temperature measurements were collected by the National Climatic Data Center of the National Oceanic and Atmospheric Administration (NOAA) and underwent a strict quality control process that can pick out potential problems, such as missing data and inconsistent readings caused by changes in thermometers and station locations. What makes the tables below quite fascinating is that there is such a large discrepancy between the total number of record high temperatures and the total number of record low temperatures. Technically, if the Earth's temperature was not increasing, you would expect that the number of record daily highs and lows being set each year should be about even. But that is far from the case. For the period from January 1, 2000, to December 20, 2009, the continental United States set 294,276 record highs and only 145,498 record lows. And if you look back over the past sixty years, that picture is reinforced. The ratio of record daily high to record daily low temperatures was almost one to one in the 1950s but has been rising steadily since the 1980s.

As you'll notice in the tables below, there is still a lot of variability from year to year. In fact, October 2009 serves as an excellent example of a month that was quite cold: the number of record lows was 1,908 while the number of record highs was only 1,017. On average, October 2009 was the third-coolest October since record keeping began in 1895. The average October temperature of 50.8°F was 4.0°F below the twentieth-century average. But when you step back and look at the big picture, global land and ocean surface temperature for October 2009 was the sixth-warmest on record, with an anomaly of 1.03°F above the twentieth-century average of 57.1°F. The lesson here is that there will always be natural variability in space and time!

What's also interesting is that the current two-to-one ratio of highs versus lows across the country comes from the fact that there is a relatively small number of record lows. In other words, much of the nation's warming is taking place at night, and, as a result, temperatures dip down less often to set new record lows. This finding is completely consistent with climate models showing that higher overnight lows are to be expected as the planet warms.

In addition to looking at historical temperatures in recent decades, scientists also used a computer model to simulate how record high and low temperatures are likely to change in the future. The model indicates that if greenhouse gas emissions continue to increase in a "business-as-usual" scenario, the U.S. ratio of daily record high to record low temperatures would increase to about twenty to one by 2050 and fifty to one by 2100.

Number of Record High and Low Temperatures for the United States for 2000

Month	2000 High	2000 Low
January	4,480	110
February	3,758	402
March	4,259	189
April	2,082	934
May	6,287	887
June	2,222	1,681
July	2,953	1,949
August	4,295	979
September	6,306	2,813
October	2,123	4,251
November	689	2,810
December	601	1,866
Total	40,055	18,871

Number of Record High and Low Temperatures for the United States for 2001

Month	2001 High	2001 Low
January	1,299	1,152
February	1,339	466
March	732	877
April	3,202	1,942
May	4,830	1,713
June	1,633	1,356
July	2,038	1,797
August	3,699	503
September	2,246	1,586
October	2,109	2,262
November	4,001	339
December	1,724	241
Total	**28,852**	**14,234**

Number of Record High and Low Temperatures for the United States for 2002

Month	2002 High	2002 Low
January	4,814	495
February	2,033	1,261
March	1,208	3,956
April	5,518	1,530
May	2,242	6,487
June	2,687	935
July	4,079	623
August	3,833	2,299
September	2,176	358
October	1,142	2,314
November	2,144	1,514
December	1,817	728
Total	**33,693**	**22,500**

Number of Record High and Low Temperatures for the United States for 2003

Month	2003 High	2003 Low
January	3,969	917
February	1,176	1,227
March	2,113	1,276
April	2,530	948
May	2,896	1,299
June	1,157	1,409
July	4,852	923
August	3,630	326
September	1,677	657
October	7,699	966
November	2,925	1,777
December	1,080	260
Total	**35,704**	**11,985**

Number of Record High and Low Temperatures for the United States for 2004

Month	2004 High	2004 Low
January	1,475	1,124
February	547	834
March	6,242	365
April	1,958	623
May	2,524	1,275
June	1,253	1,752
July	742	2,509
August	724	5,186
September	1,122	1,025
October	1,514	517
November	796	589
December	1,801	1,281
Total	**20,698**	**17,080**

Number of Record High and Low Temperatures for the United States for 2005

Month	2005 High	2005 Low
January	3,805	743
February	2,539	158
March	2,266	599
April	1,341	1,727
May	2,627	2,290
June	1,556	836
July	3,714	1,354
August	2,286	649
September	4,341	642
October	2,426	898
November	4,595	826
December	2,553	1,986
Total	**34,049**	**12,708**

Number of Record High and Low Temperatures for the United States for 2006

Month	2006 High	2006 Low
January	3,805	743
February	2,539	158
March	2,266	599
April	1,341	1,727
May	2,627	2,290
June	1,556	836
July	3,714	1,354
August	2,286	649
September	4,341	642
October	2,426	898
November	4,595	826
December	2,553	1,986
Total	**34,049**	**12,708**

Number of Record High and Low Temperatures for the United States for 2007

Month	2007 High	2007 Low
January	2,279	1,531
February	872	1,779
March	6,956	933
April	1,556	3,145
May	2,494	889
June	1,852	988
July	3,165	1,429
August	6,641	519
September	3,014	1,237
October	4,145	549
November	1,570	267
December	1,832	390
Total	**36,376**	**13,656**

Number of Record High and Low Temperatures for the United States for 2008

Month	2008 High	2008 Low
January	1,813	528
February	1,399	533
March	490	540
April	389	1,734
May	1,281	1,036
June	1,585	534
July	494	658
August	949	519
September	507	555
October	733	1,099
November	1,564	883
December	1,654	556
Total	**12,858**	**9,175**

Number of Record High and Low Temperatures for the United States for 2009

Month	2009 High	2009 Low
January	1,292	642
February	1,556	392
March	1,696	855
April	1,602	746
May	1,074	847
June	1,167	538
July	1,274	1,904
August	1,144	1,205
September	1,028	360
October	1,017	1,908
November	1,096	143
December	264	1,086
Total	**14,210**	**10,626**

Total Number of Record High and Low Temperatures for the United States from 2000 to 2009

Year	High	Low
2000	40,055	18,871
2001	28,852	14,234
2002	33,693	22,500
2003	35,704	11,985
2004	20,698	17,080
2005	34,049	12,708
2006	37,465	14,762
2007	36,376	13,656
2008	12,858	9,175
2009	14,210	10,626
Total	**293,960**	**145,597**

2. Increasing Number of Hot Days in U.S. Cities

The bar plots that follow illustrate how extremely hot days are likely to become more common and more intense as the overall climate warms during the next century. They also help demonstrate that a general warming of the climate has significant implications for the number and severity of extreme weather and climate events: in this case, heat waves.

For selected large cities or regions in the United States, the plots compare the average number of extremely hot days observed during the summer months in the twentieth century with projections for extremely hot days in the middle and end of the twenty-first century. Climate research indicates that heat waves are likely to be more stifling, and potentially more deadly, in coming decades as climate change progresses. For many locations, heat waves are projected to be more frequent, more intense, and longer lasting, and extreme heat events that are currently considered rare will become more common in coming years.[2]

Extreme weather and climate events can cause significant damage, and heat waves are considered public health emergencies because of their effects on human health. Hot temperatures contribute to increased emergency room visits and hospital admissions for cardiovascular disease and can cause heatstroke and other life-threatening conditions. The elderly are particularly vulnerable to extreme heat.

Heat waves such as the Chicago heat wave of 1995 and the European heat wave of 2003, which killed an estimated 50,000 people, have proved especially deadly to vulnerable populations, including the elderly and persons with respiratory illnesses.[3]

Scientists at Climate Central conducted their own special analysis to generate the values in these graphics, using techniques and general climate projections that are well established in the peer-reviewed scientific literature.[4]

The projections for extreme heat in the years 2050 and 2090

are based on an average of twelve computer models that simulate climate. For these projections, Climate Central used a scenario of moderate-high greenhouse gas emissions, which currently appears optimistic, since global emissions have exceeded this scenario in recent years. Climate Central used a common technique to translate large-scale climate information from the computer models to provide useful information about local and regional conditions. This method involves calculating differences between time series data from current and future global climate model simulations and then adding these changes to time series of observed climate data.

Scientists at Climate Central first identified weather observation stations closest to each city, as well as the closest point in the output of computer models, which is known as a *grid point*. For the station data, Climate Central examined temperature information for the summer months (June, July, and August) during two twenty-year periods to determine how extreme heat events have evolved during the twentieth century. Those periods were 1951–1970 and 1981–2000.

For the computer model data, Climate Central looked at two future twenty-year periods of projected maximum temperature data, 2046–2064 and 2081–2100, as well as the simulated current climate period 1981–2000. These data shed light on how extreme heat events may evolve as the climate changes. The analyses are based on recent data from weather stations, regional-scale outputs from climate projection models, and a common technique for deducing best-guess local climate projections from regional projections. This method involves calculating differences between current and future global climate model simulations, and applying them to observed climate data from the same vicinity.

For a given month, Climate Central calculated changes in the twenty-year average monthly maximum temperature between the two periods in the station data and between future time periods in the computer simulations and the current climate simulation. This provided a comparison of the twenty-year climatology at the end of the twentieth century and the earlier period 1951–1970. It also permitted a comparison of the model-simulated average monthly maximum temperature in 2046–2064 with that of 1981–2000, and the end-of-the-century period 2081–2100 with 1981–2000.

Because data from twelve different computer models were used, Climate Central took a model average of the differences in simulated temperature changes between the time periods. Next, Climate Central turned to the station observations and used the daily temperature data for the period 1981–2000 to determine a climatology of daily temperature values. From this climatology came the set of numbers to go into the second bar of each group of four bars. Each group pertains to a given temperature threshold of 90°F, 95°F, and 100°F. Climate Central counted how many days exceeded the temperature threshold during each of the twenty months in the climatology and then averaged those numbers.

Climate Central generated the other bars similarly, simply shifting the climatology of daily maximum temperatures for the period 1981–2000 by the temperature changes computed from the model simulations (or the 1951–1970 observed data) and repeating the count on the new sets of daily maximum temperatures that were generated.

This last step created new, simulated data for each city for twenty Augusts in the middle of the twenty-first century. The same method that was used with actual 1981–2000 temperatures to estimate the average number of days over each temperature threshold in this future scenario was then applied.

The resulting projections give long-term averages, not predictions for any individual year; actual outcomes will vary significantly from year to year, owing to the natural variability of climate. Furthermore, because the modeling and methods used involve uncertainty, the projections should be taken as best guesses within a range of uncertainty. True long-term averages will be likely to prove somewhat higher or lower than the projections here. However, all twelve models are unanimous in projecting increased hot days (relative to the present) by the middle of the twenty-first century.

All model outputs used were based on a scenario of medium-high greenhouse gas emissions, called A1B by the Intergovernmental Panel on Climate Change. Carbon dioxide and other greenhouse gas emissions during the decade spanning 2000–2010 have already exceeded the A1B scenario, so the projections here are conservative and represent a future in which greenhouse gas emissions are reduced compared with the current trend.

The number of hot July and August days (above 90°F, 95°F, and 100°F) in twenty U.S. cities over four time periods (1951–1970, 1981–2000, 2046–2064, 2081–2100) are presented below. Cities are listed in order of population size. For a given month (in this case, July or August), changes in the twenty-year average monthly number of high temperature are represented by four columns. The first two columns represent two distinct twenty-year periods (1951–1970, labeled 1960; and 1981–2000, labeled 1990) in the station data. The second two columns represent best guesses about the future, with column three representing the twenty-year climatology in 2046–2064 (labeled 2050) compared with the twenty-year climatology in 1981–2000, both model derived; and column four representing the twenty-year climatology in 2081–2100 (labeled 2090) compared with the twenty-year climatology in 1981–2000, both model derived.

New York

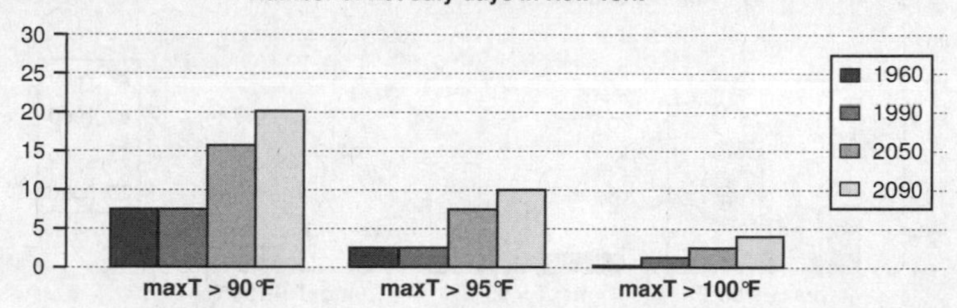

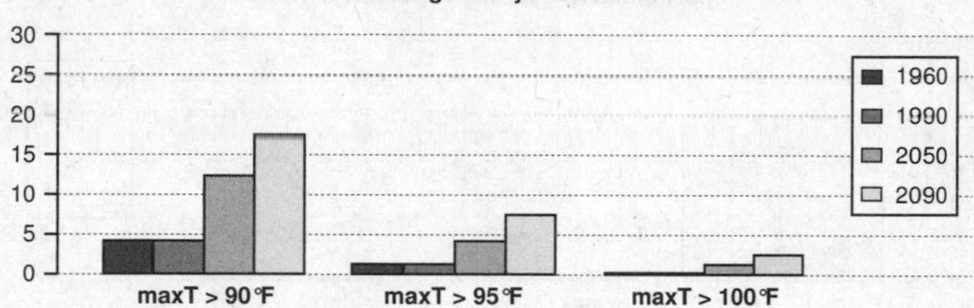

Los Angeles

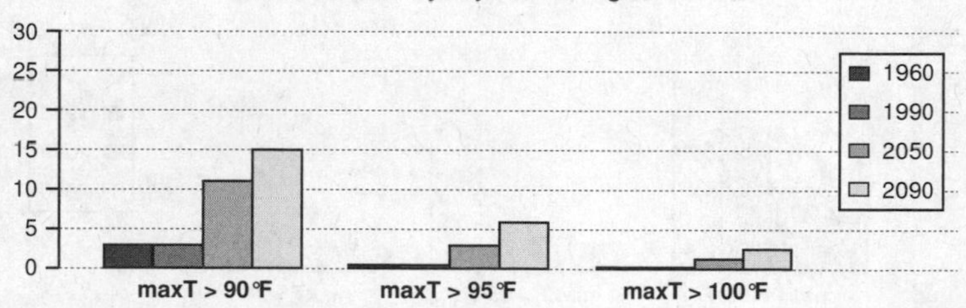

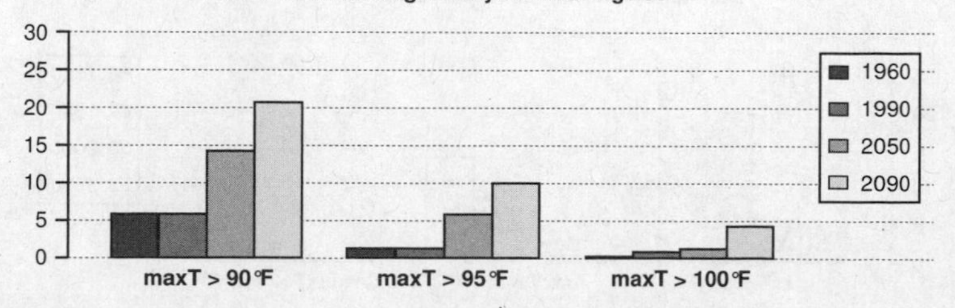

Chicago

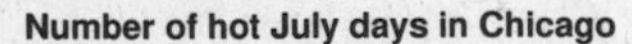

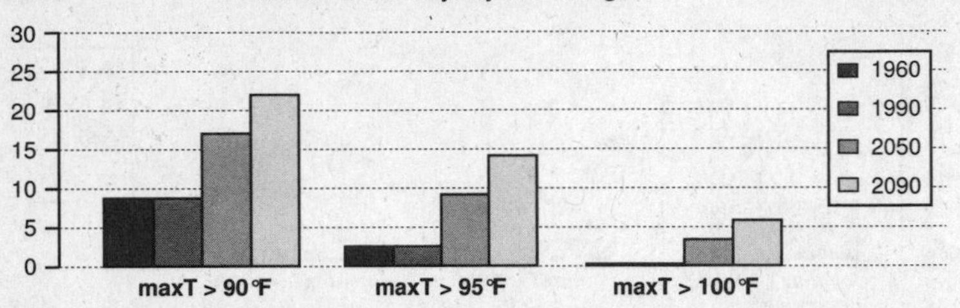

Number of hot August days in Chicago

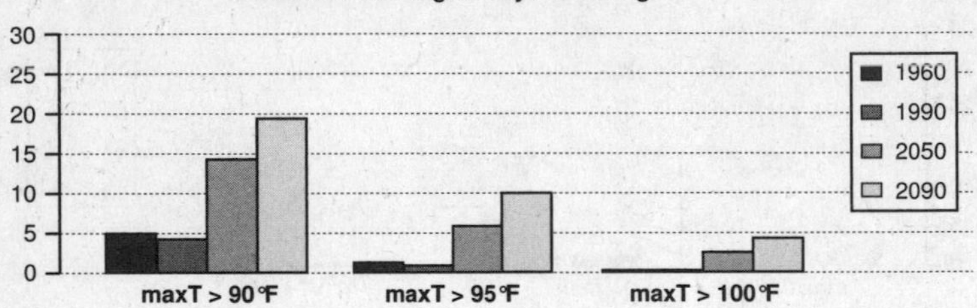

Philadelphia

Number of hot July days in Philadelphia

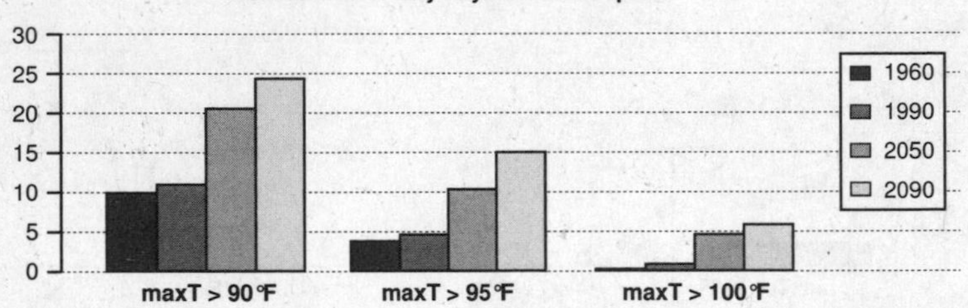

Number of hot August days in Philadelphia

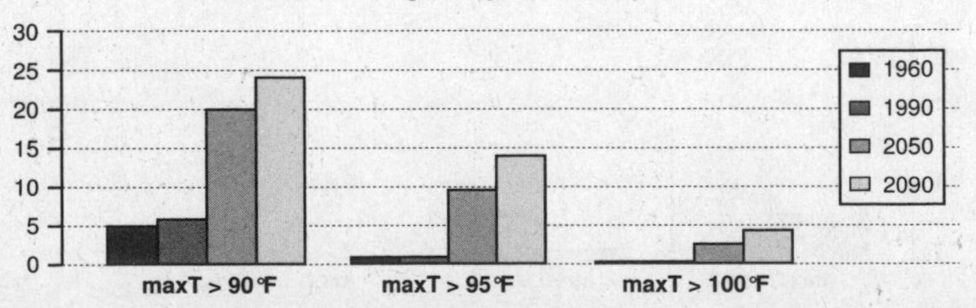

Dallas

Number of hot July days in Dallas

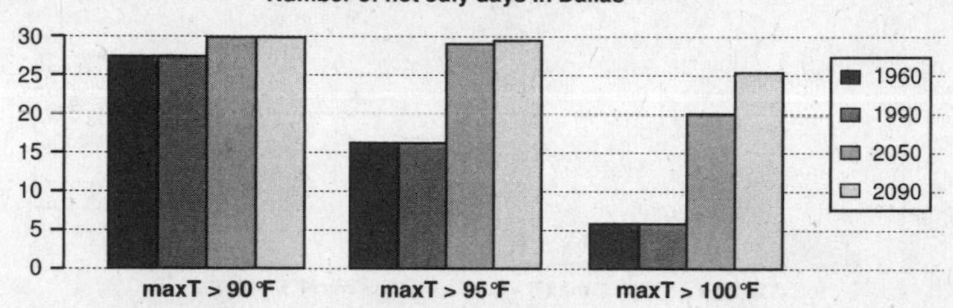

Number of hot August days in Dallas

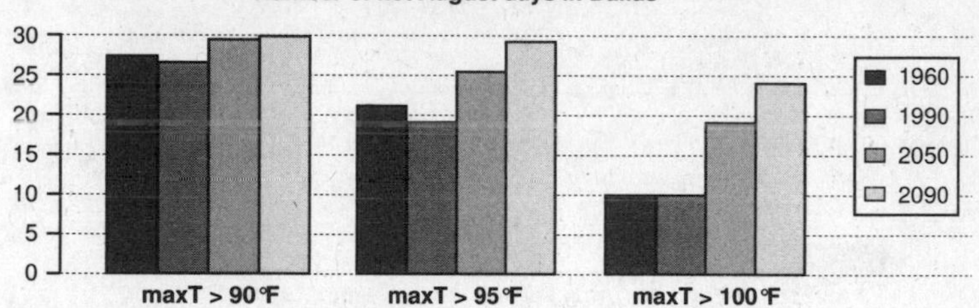

San Francisco

Number of hot July days in San Francisco

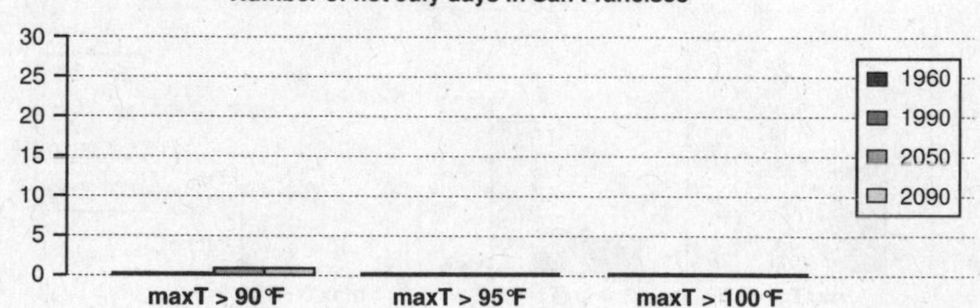

Number of hot August days in San Francisco

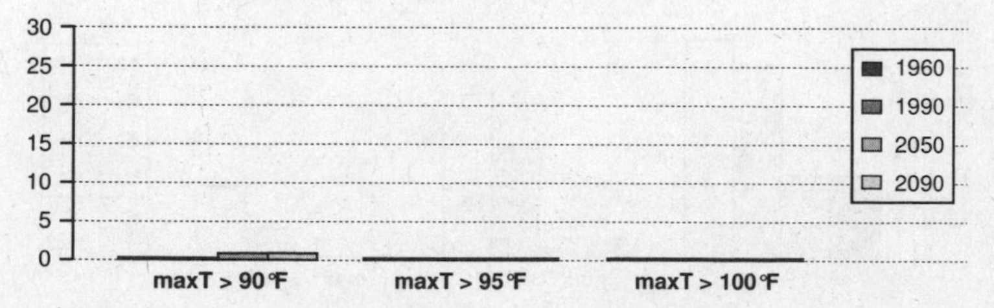

Boston

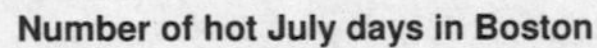

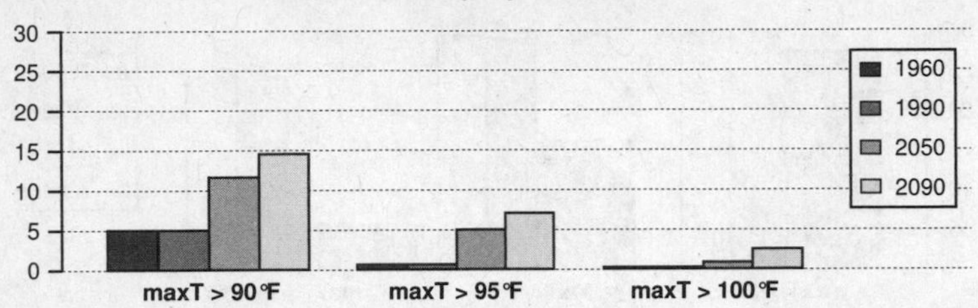

Number of hot August days in Boston

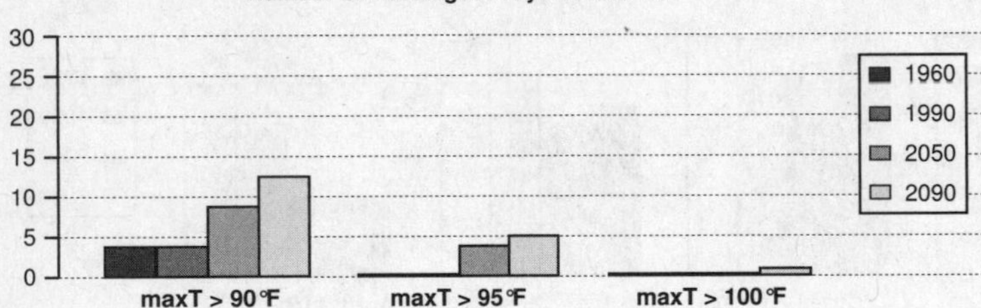

Atlanta

Number of hot July days in Atlanta

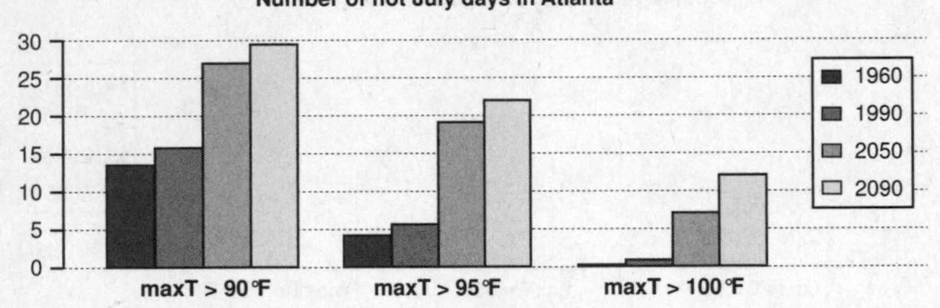

Number of hot August days in Atlanta

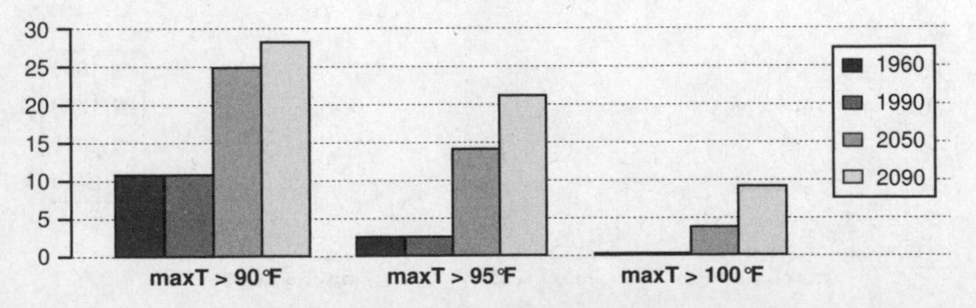

Washington, D.C.

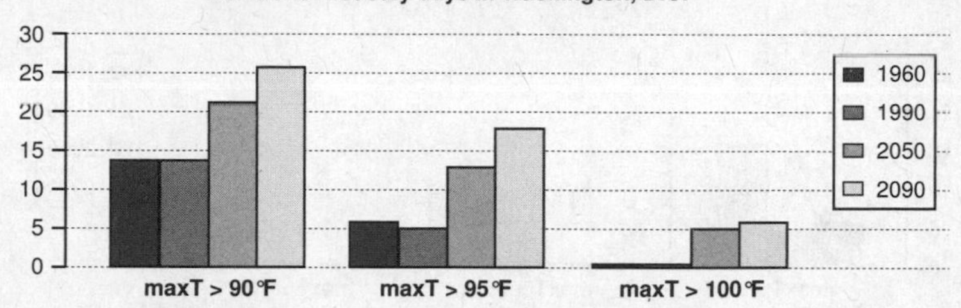

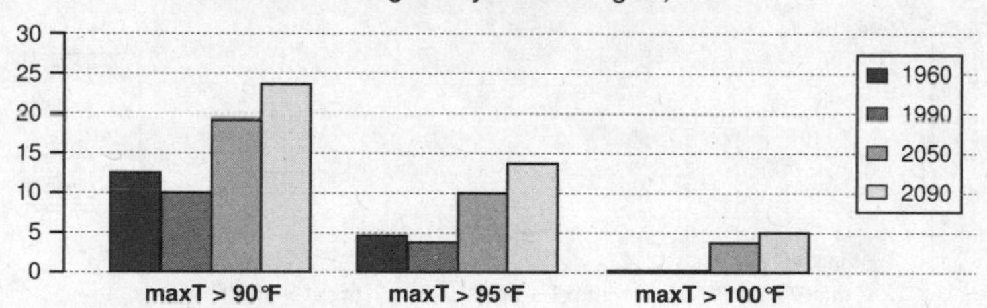

Houston

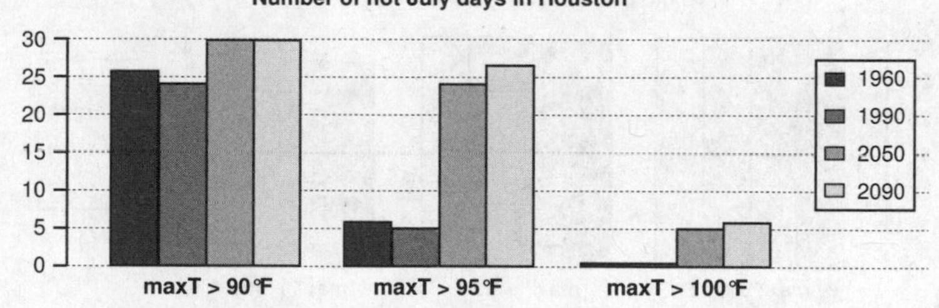

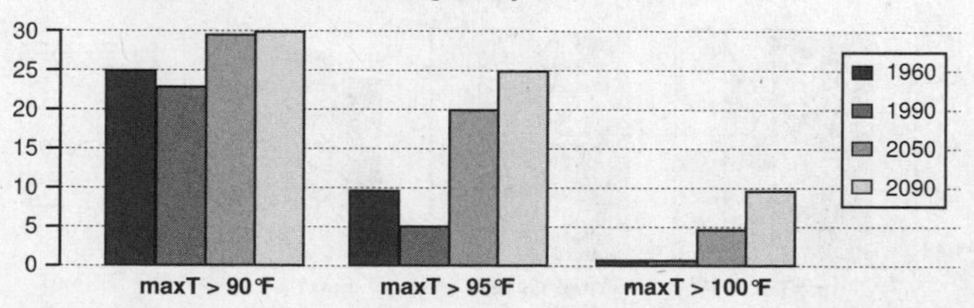

Detroit

Number of hot July days in Detroit

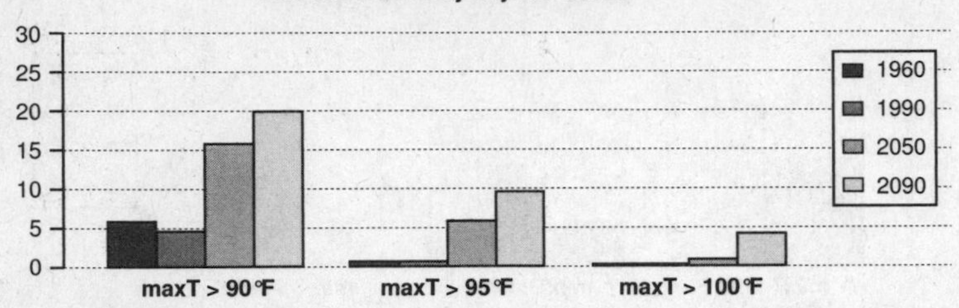

Number of hot August days in Detroit

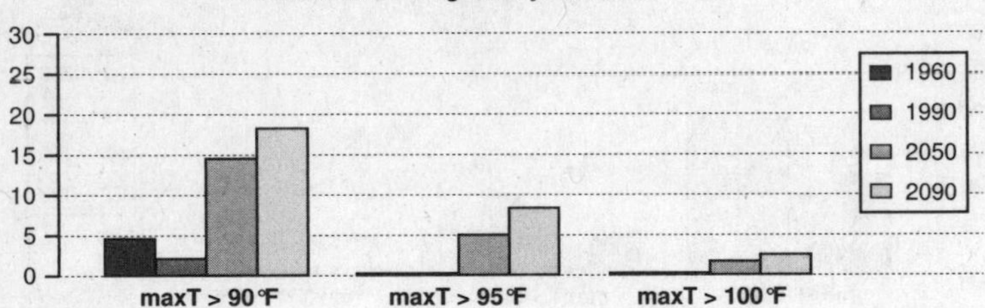

Phoenix

Number of hot July days in Phoenix

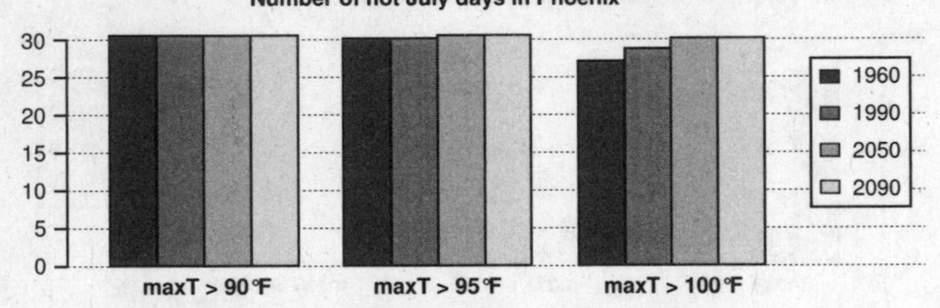

Number of hot August days in Phoenix

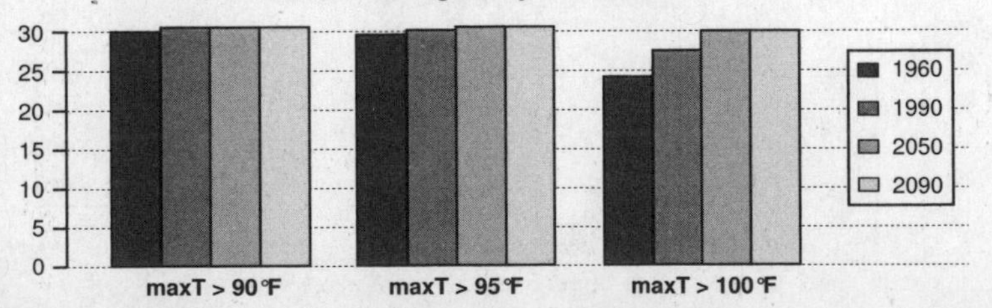

Tampa

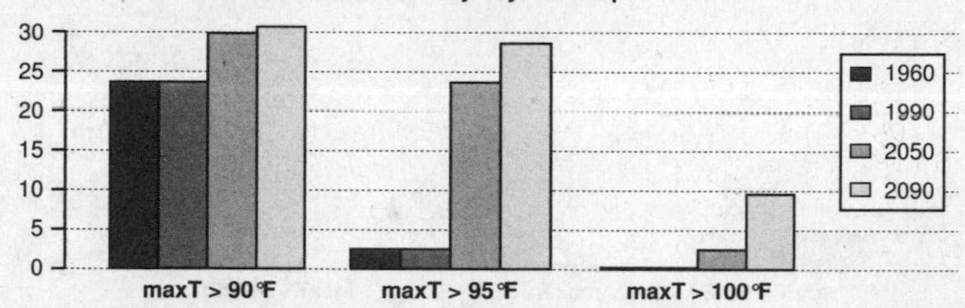

Number of hot August days in Tampa

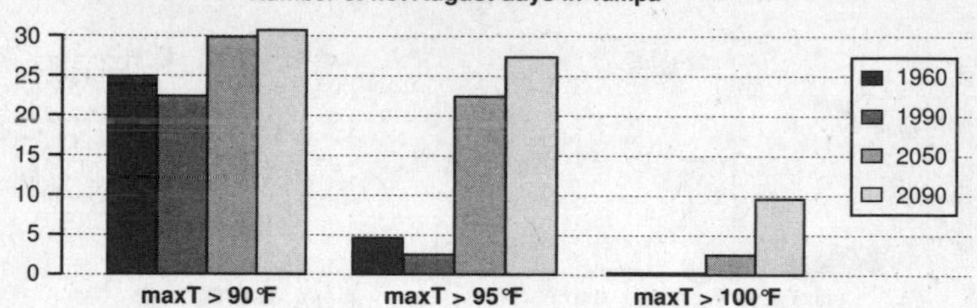

Seattle

Number of hot July days in Seattle

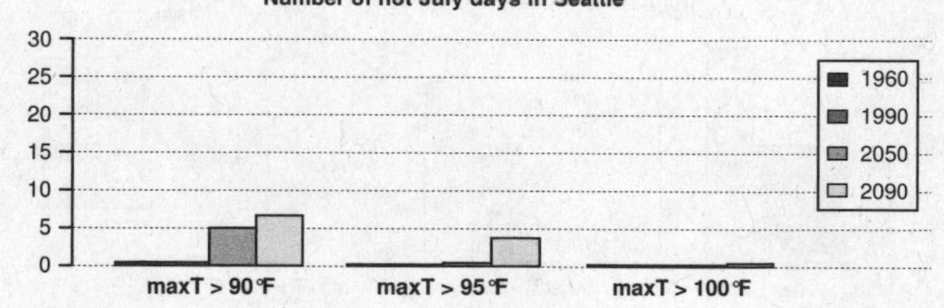

Number of hot August days in Seattle

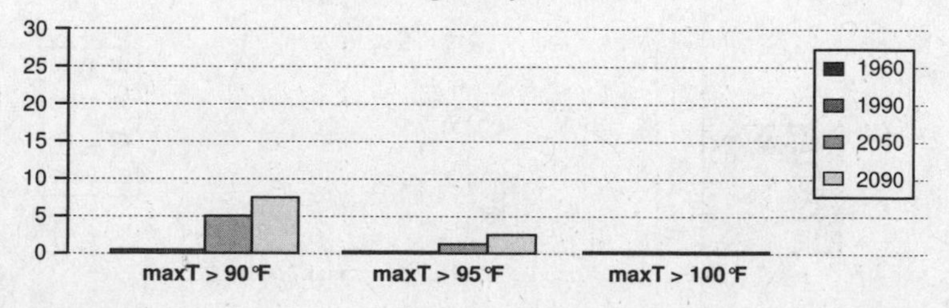

Minneapolis

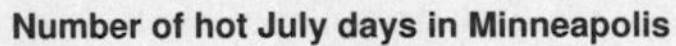

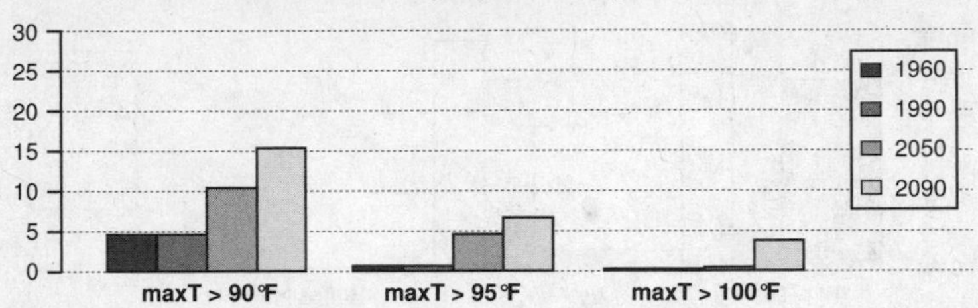

Number of hot August days in Minneapolis

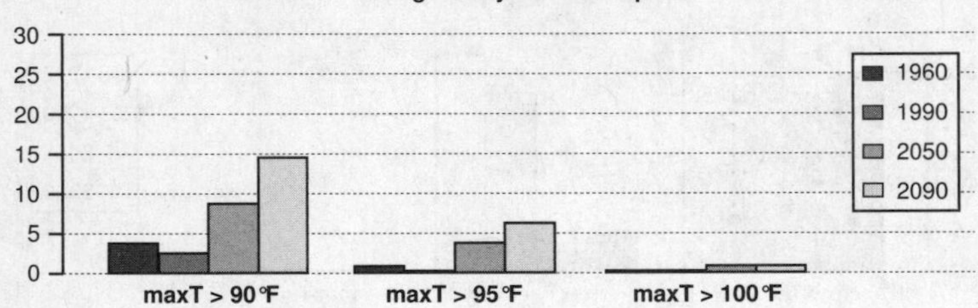

Miami

Number of hot July days in Miami

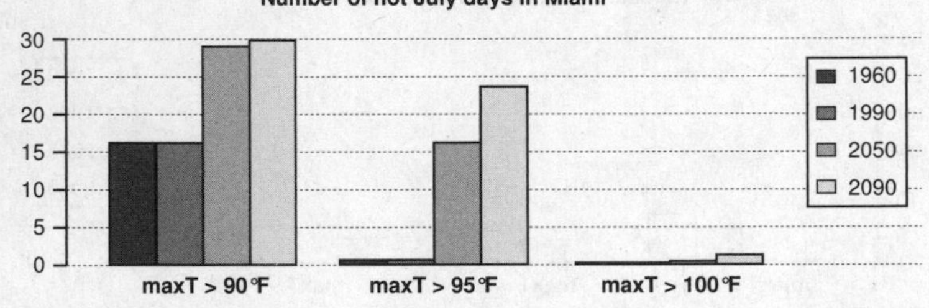

Number of hot August days in Miami

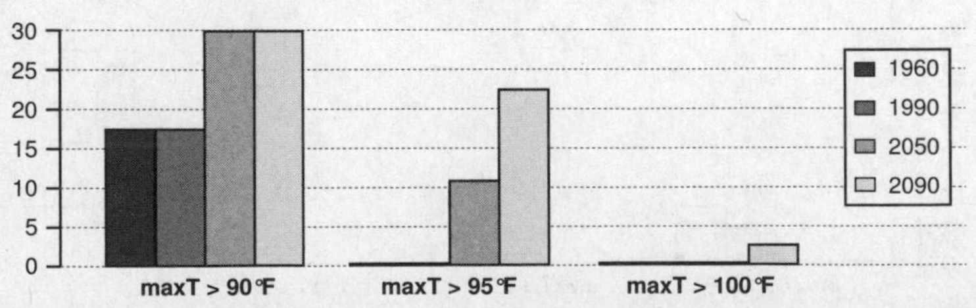

Cleveland

Number of hot July days in Cleveland

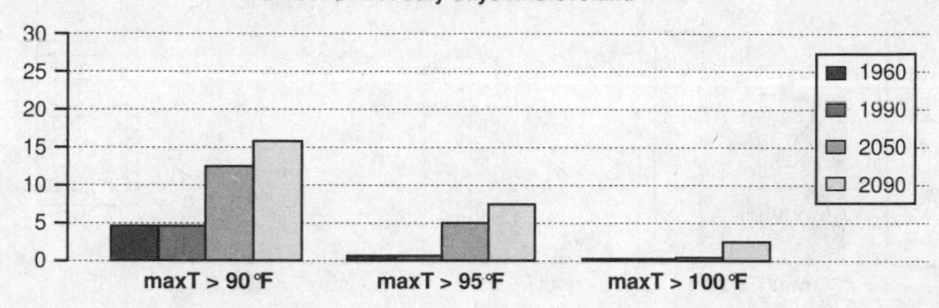

Number of hot August days in Cleveland

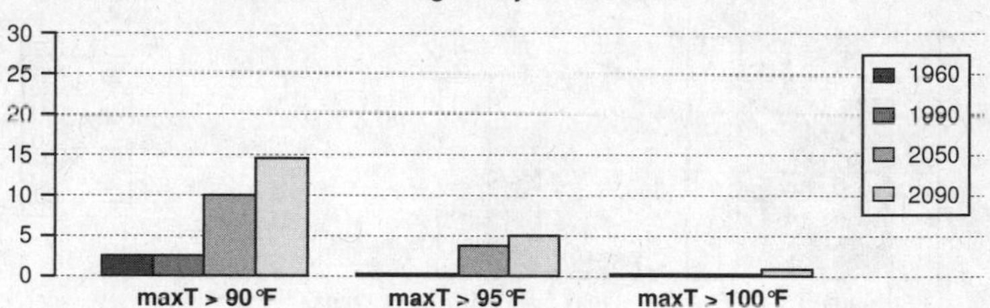

Denver

Number of hot July days in Denver

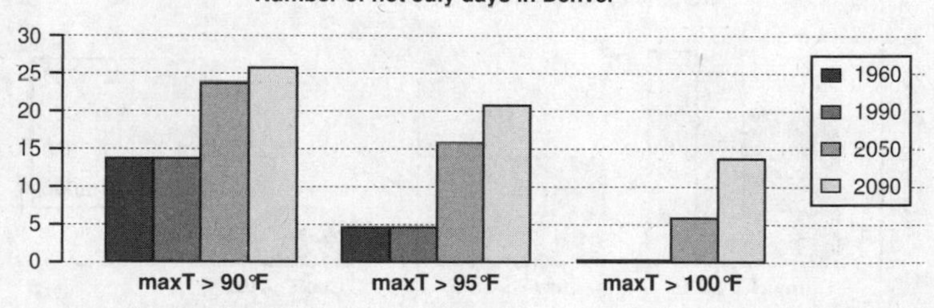

Number of hot August days in Denver

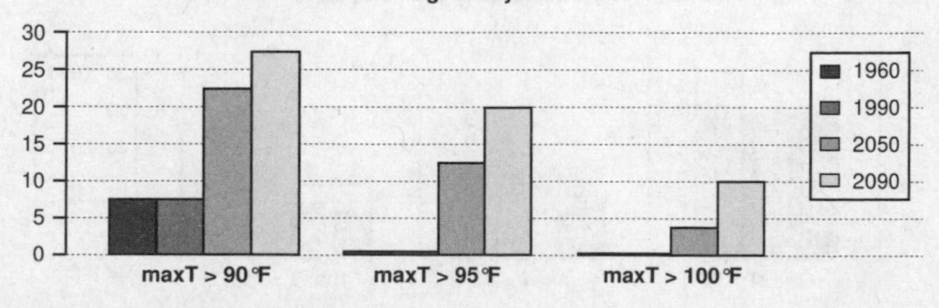

Orlando

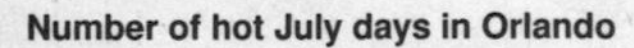

Number of hot July days in Orlando

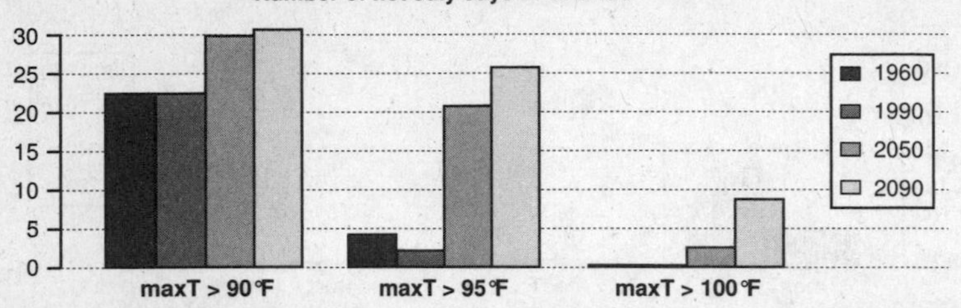

Number of hot August days in Orlando

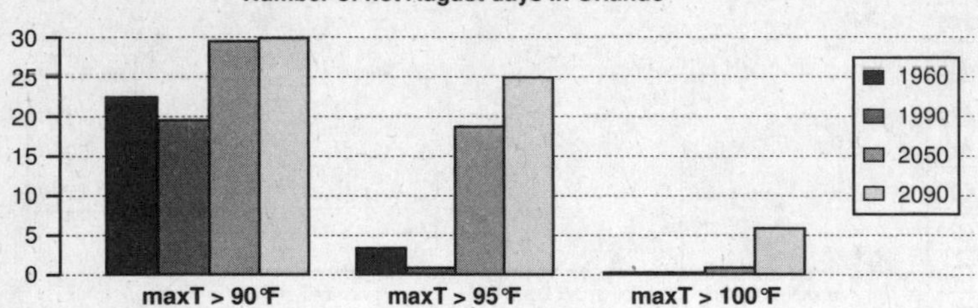

Sacramento

Number of hot July days in Sacramento

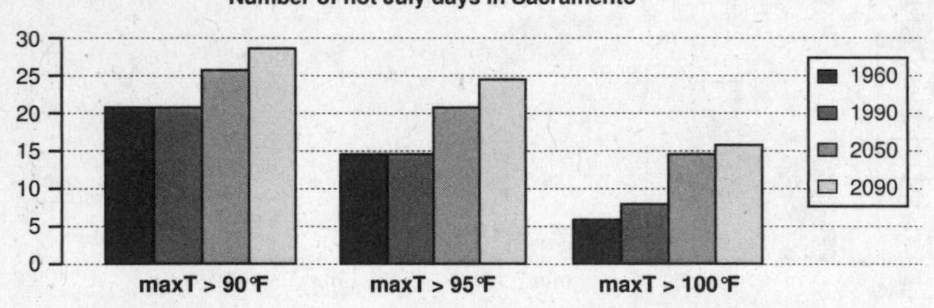

Number of hot August days in Sacramento

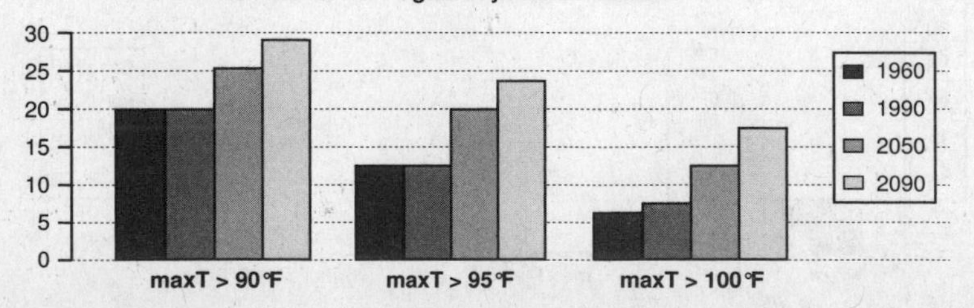

3. Temperature Trends by State

It's always useful to look at the long-term temperature trend at a more regional level. For example, the table below shows how much each state in the continental United States has warmed, on average, over the period 1976–2005. The first column shows the long-term trend averaged over the entire year (January to December). The second column shows the long-term trend for *winter* (December to February). You'll notice that winters have warmed significantly.

1976–2005	Annual	Winter	1976–2005	Annual	Winter
Alabama	1.41 °F*	3.81 °F*	Nevada	1.58 °F*	0.4 °F
Arizona	1.97 °F*	0.62 °F	New Hampshire	1.76 °F*	3.22 °F*
Arkansas	1.06 °F*	3.75 °F*	New Jersey	1.85 °F*	4.13 °F*
California	1.56 °F*	0.27 °F	New Mexico	1.73 °F*	1.59 °F
Colorado	1.64 °F*	2.31 °F	New York	2.18 °F*	3.94 °F*
Connecticut	1.57 °F*	3.58 °F*	North Carolina	1.32 °F*	3.28 °F*
Delaware	1.77 °F*	4.04 °F*	North Dakota	0.89 °F	4.58 °F
Florida	1.21 °F*	2.79 °F*	Ohio	1.89 °F*	5.2 °F*
Georgia	0.76 °F	2.82 °F*	Oklahoma	0.58 °F	3.4 °F*
Idaho	2.12 °F*	2.74 °F	Oregon	1.42 °F*	1.45 °F
Illinois	2.01 °F*	6.26 °F*	Pennsylvania	2.14 °F*	4.48 °F*
Indiana	1.96 °F*	5.86 °F*	Rhode Island	1.96 °F*	4.24 °F*
Iowa	0.93 °F	4.8 °F*	South Carolina	1.07 °F*	2.86 °F*
Kansas	0.61 °F	3.47 °F	South Dakota	1.11 °F	5.25 °F
Kentucky	2.06 °F*	5.47 °F*	Tennessee	1.81 °F*	4.8 °F*
Louisiana	1.65 °F*	3.79 °F*	Texas	1.32 °F*	3.09 °F*
Maine	0.64 °F	1.75 °F	Utah	2.14 °F*	2.31 °F
Maryland	1.61 °F*	3.92 °F*	Vermont	2.37 °F*	3.74 °F*
Massachusetts	1.58 °F*	3.56 °F*	Virginia	1.04 °F*	3.28 °F*
Michigan	1.91 °F *	4.75 °F *	Washington	2.23 °F*	3.06 °F*
Minnesota	1.48 °F	5.7 °F*	Washington, D.C.	1.11 °F*	3.31 °F*
Mississippi	1.3 °F*	3.59 °F*	West Virginia	1.65 °F*	4.15 °F*
Missouri	1.2 °F	5.03 °F*	Wisconsin	1.96 °F*	6.01 °F*
Montana	1.54 °F*	3.97 °F	Wyoming	1.52 °F*	2.79 °F
Nebraska	1.45 °F	4.97 °F*			

*Asterisk denotes statistical significance at the 90 percent confidence level. (SOURCE: C. TEBALDI, CLIMATE CENTRAL)

APPENDIX 2

NEW YORK STATISTICS

New York City Area: Hurricanes

Date	Name	Category*
Sept. 3–5, 1815	Great September Gale of 1815	3
Sept. 3, 1821	unnamed	1–2
Sept. 1869	New England Storm	1
Aug. 23, 1893	Eastern New England Storm	1
Aug. 23, 1893	Midnight Storm	1–2
Sept. 21, 1938	Long Island Express/New England Storm	3
Sept. 15, 1944	unnamed	1
Aug. 1954	Carol	3
Sept. 12, 1960	Donna	3
Sept. 21, 1961	Esther	1–2
June 1972	Agnes	1
Aug. 10, 1976	Belle	1
Sept. 27, 1985	Gloria	2–3
Aug. 1991	Bob	2
Sept. 1999	Floyd	2

*Category column represents the estimated storm maximum; not necessarily experienced in the region of New York City. (SOURCE: NPCC, CLIMATE RISK INFORMATION)

New York City Transportation Facilities Potentially Most Vulnerable to Inundation from Climate Change

Facility	Elevation (feet)
Metropolitan Transit Authority (MTA)	
Christopher Street Station (1 line)	–14.6
Canal Street Station (A, C, E lines)	8.7
South Ferry Station (1 line)	9.1
Long Island Railroad Far Rockaway Station	9.2
Verrazano Narrows Bridge	8
MetroNorth Hudson Line tracks, Croton River	7.0–7.5
Port Authority of New York and New Jersey	
Holland Tunnel NY Entrance	9.5
LaGuardia Airport	6.8
Port Newark and Elizabeth	9.6
NYC Department of Transportation	
Battery Park Tunnel	9
FDR Drive above 59th Street	6

(Source: R. Zimmerman[5])

APPENDIX 3

THE WORLD'S MOST VULNERABLE PLACES

On the basis of my discussions with dozens of climate experts, I selected just a few vulnerable places from around the world to showcase the specific regional risks associated with climate change. There are many vulnerable places that I was, of course, unable to discuss. But Mike MacCracken, the chief scientist for Climate Change Programs at the Climate Institute, has assembled an excellent list of the top ten prevailing threats associated with climate change as well as examples of the places that are most vulnerable to these threats.[6] The threats are listed, in no specific order, below.

1. Sea, Salt, and Storms

River deltas, bays, and estuaries: Dhaka, Cairo, New Orleans, Sacramento–San Joaquin Delta, Hong Kong, Chesapeake Bay, New York, London.

Low-lying coastal plains: Miami, Charleston, Boston, Long Island, Amsterdam, Venice, Rotterdam, Venice.

Barrier islands: Alaskan villages, North Carolina coast.

Island nations: Fiji, Tahiti, Tuvalu.

2. Acidification

Coral reefs and atolls: Maldives; Great Barrier Reef; American trust territories; Key West, Florida.

3. Water

Mediterranean environments: Sacramento, San Diego; Los Angeles; Atlanta; Las Vegas; Phoenix; Albuquerque; Dakar; Lima, Peru; Quito, Ecuador; La Paz, Bolivia; Sana'a, Yemen.

4. Snowmelt and Runoff

Snow-fed rivers amid forested mountains: Many large cities of India, Pakistan, and China; Portland and the Pacific Northwest; Sacramento–San Joaquin river basin; downstream from the Alps.

5. Fire and Beetles

Arid regions: Western United States, Canada (Alberta, British Columbia); Spain; Portugal.

6. Food and Mass Migration

Agricultural plains: United States Great Plains; Australian coastal regions; Tijuana, Mexico; Lagos, Africa; Nairobi, Kenya.

7. Permafrost Thaw

Seasonal freezing: Fairbanks, Alaska; northern Canada in general; Siberia.

8. Heat

Hot and humid (summer) weather regimes: Texas generally, New York, Chicago, Paris, southern China.

9. Hurricanes and Typhoons

Tropical cyclone paths: New Orleans, New York, Miami, Charleston, Chesapeake Bay, Hong Kong, Tokyo and other cities in Japan, Shanghai, Manila.

10. Dead Zones, Pollution, and Disease

Rio de la Plata/Buenos Aires: hypoxia due to upstream changes in land use and increased severe precipitation events. Mexico City: air pollution.

NOTES

1. CLIMATE AND WEATHER TOGETHER

1. Kohut, A., Keeter, S., Doherty, C., Dimock, M., and Remez, M. Economy, Jobs Trump All Other Policy Priorities (Pew Research Center for the People and the Press, Washington, DC, 2009).
2. Weber, E. U., The Utility of Measuring and Modeling Perceived Risk, in *Choice, Decision, and Measurement: Essays in Honor of R. Duncan Luce*, edited by A. A. J. Marely (Lawrence Erlbaum Associates, Mahwah, NJ, 1997), pp. 45–57.
3. Weber, E., Experience-Based and Description-Based Perceptions of Long-Term Risk: Why Global Warming Does Not Scare Us (Yet). *Climatic Change* 77 (1–2), 103–120 (2006). See also Marx, S. M. et al., Communication and Mental Processes: Experiential and Analytic Processing of Uncertain Climate Information. *Global Environmental Change* 17 (1), 47–58 (2007).
4. Hirshleifer, D., and Shumway, T., Good Day Sunshine: Stock Returns and the Weather. *Journal of Finance* 58 (3), 1009–1032 (2003).
5. Sunstein, C. R., The Availability Heuristic, Intuitive Cost-Benefit Analysis, and Climate Change. *Climatic Change* 77 (1–2), 195–210 (2006).

2. SEEING CLIMATE CHANGE IN OUR PAST

1. Imbrie, J., and Imbrie, K. P., *Ice Ages: Solving the Mystery* (Enslow Publishers, Short Hills, NJ, 1979).
2. Weart, S. R. ed., *The Discovery of Global Warming* (Harvard University Press, Cambridge, MA, 2003).
3. Agassiz, L., *Etudes sur les glaciers* (privately published, Neuchâtel, 1840).
4. Fourier, J., Remarques générales sur les températures du Globe Terrestre et des espaces planétaires. *Annales de Chemie et de Physique* 27, 136–167 (1824).
5. Burchfield, J. A., John Tyndall—A Biographical Sketch, in *John Tyndall, Essays on a Natural Philosopher* (Royal Dublin Society, Dublin, 1981).
6. Arrhenius, S., On the Influence of Carbonic Acid in the Air upon the Temperature of the Ground. *Philosophical Magazine and Journal of Science* 41, 237–275 (1896).
7. Ruddiman, W. F., *Earth's Climate: Past and Future* (W. H. Freeman and Company, New York, 2001).

3. THE SCIENCE OF PREDICTION

1. Lynch, P., *The Emergence of Numerical Weather Prediction: Richardson's Dream* (Cambridge University Press, New York, 2006).

2. Somerville, R. C. J., *The Forgiving Air: Understanding Environmental Change,* 2nd ed. (American Meteorological Society, Boston, 2008).
3. Hegerl, G. C., and Zwiers, F. W., Understanding and Attributing Climate Change, in *Climate Change 2007: The Physical Science Basis. Contribution of Working Group I to the Fourth Assessment Report of the Intergovernmental Panel on Climate Change,* edited by S. Solomon et al. (Cambridge University Press, Cambridge, U.K., 2007). See also Stouffer, R. J., Hegerl, G. C., and Tett, S. F. B., A Comparison of Surface Air Temperature Variability in Three 1000-Year Coupled Ocean-Atmosphere Model Integrations. *Journal of Climate* 13, 513–537 (1999).
4. Hansen, J., et al., A Pinatubo Climate Modeling Investigation, in *The Mount Pinatubo Eruption: Effects on the Atmosphere and Climate,* edited by G. Fiocco, D. Fua, and G. Visconti (Springer-Verlag, Heidelberg, Germany, 1996), pp. 233–272.
5. Duffy, P. B., Santer, B. D., and Wigley, T. M. L., Solar Variability Does Not Explain Late-20th-Century Warming. *Physics Today* (January 2009), 48–49 (2009). See also North, G. R., Wu, Q., and Stevens, M., in *Solar Variability and Its Effect on Climate,* edited by J. M. Pap and P. Fox (American Geophysical Union, Washington, DC, 2004), Vol. Geophysical Monograph 141.
6. Thompson, D. W. J., and Solomon, S., Recent Stratospheric Climate Trends as Evidenced in Radiosonde Data: Global Structure and Tropospheric Linkages. *Journal of Climate* 18, 4785–4795 (2005). See also Ramaswamy, V. et al., Anthropogenic and Natural Influences in the Evolution of Lower Stratospheric Cooling. *Science* 311 (5764), 1138–1141 (2006).

4. EXTREME WEATHER AUTOPSIES AND THE FORTY-YEAR FORECAST

1. Karl, T. R., et al., *Weather and Climate Extremes in a Changing Climate: Regions of Focus* (U.S. Climate Change Science Program, 2008).
2. Schär, C., et al., The Role of Increasing Temperature Variability for European Summer Heat Waves. *Nature* 427 (6972), 332–336 (2004).
3. Stott, P. A., Stone, D. A., and Allen, M. R., Human Contribution to the European Heat Wave of 2003. *Nature* 432 (7017), 610–614 (2004).
4. Karl, T. R., Melillo, J., and Peterson, T. C., *Global Climate Change Impacts in the United States* (U.S. Climate Change Science Program, 2009).
5. Meehl, G. A., and Tebaldi, C., More Intense, More Frequent, and Longer Lasting Heat Waves in the 21st Century. *Science* 305 (5686), 994–997 (2004).

5. THE SAHEL, AFRICA

1. deMenocal, P. B., et al., Abrupt Onset and Termination of the African Humid Period: Rapid Climate Responses to Gradual Insolation Forcing. *Quaternary Science Reviews* 19, 347–361 (2000). See also deMenocal, P. B., Ortiz, J., Guilerson, T., and Sarnthein, M., Coherent High—and Low-Latitude Climate Variability During the Holocene Warm Period. *Science* 288, 2198–2202 (2000).
2. Shanahan, T. M., et al., Atlantic Forcing of Persistent Drought in West Africa. *Science* 324, 377–380 (2009).
3. Bobe, R., and Behrensmeyer, A. K., The Expansion of Grassland Ecosystems in Africa in Relation to Mammalian Evolution and the Origin of the Genus Homo.

Palaeogeography, Palaeoclimatology, Palaeoecology 207, 399–420 (2004). See also deMenocal, P. B., Plio-Pleistocene African Climate. *Science* 270 (5233), 53–59 (1995).

4. Feakins, S. J., deMenocal, P. B., and Eglinton, T. I., Biomarker Records of Late Neogene Changes in Northeast African Vegetation. *Geology* 33 (12), 977–980 (2005).
5. deMenocal, P. B., African Climate Change and Faunal Evolution During the Pliocene-Pleistocene. *Earth and Planetary Science Letters* 220, 3–24 (2004).
6. Kandji, S. T., Verchot, L., and Mackensen, J., 2006.
7. Held, I., Delworth, T. L., Lu, J., Findell, K. L., and Knutson, T. R., Simulation of Sahel Drought in the 20th and 21st Centuries. *PNAS* 102 (50), 17891–17896 (2005).
8. Hulme, M., Climatic Perspectives on Sahelian Dessication: 1973–1998. *Global Environmental Change* 11 (19–29) (2001).
9. Herrmann, S. M., and Hutchinson, C. F., The Changing Contexts of the Desertification Debate. *Journal of Arid Environments* 63, 538–555 (2005).
10. Olsson, L., and Hall-Beyer, M., Greening of the Sahel in *Encyclopedia of Earth*, edited by C. J. Cleveland (Environmental Information Coalition, National Council for Science and the Environment, Washington, DC, 2008).
11. Hayward, D. F., and Oguntoyinbo, J. S., *The Climatology of West Africa* (Barnes & Noble, 1987).
12. Nicholson, S. E., Climatic Variations in the Sahel and Other African Regions During the Past Five Centuries. *Journal of Arid Environments* 1 (3–24) (1978).
13. Xue, Y., and Shukla, K., The Influence of Land Surface Properties on Sahel Climate. Part I: Desertification. *Journal of Climate* 6, 2232–2245 (1993).
14. Folland, C. K., Palmer, T. N., and Parker, D. E., Sahel Rainfall and Worldwide Sea Temperatures, 1901–85. *Nature* 302, 602–607 (1986).
15. Giannini, A., Saravanan, R., and Chang, P., Oceanic Forcing of Sahel Rainfall on Interannual to Interdecadal Time Scales. *Science* 302 (1027–1030) (2003).
16. IPCC, 2007. See also Patricola, C. M., and Cook, K. H., Northern African Climate at the End of the Twenty-First Century: An Integrated Application of Regional and Global Climate Models. *Climate Dynamics* (2009).
17. Biasutti, M., and Sobel, A. H., Delayed Seasonal Cycle and African Monsoon in a Warmer Climate. *Geophys. Res. Lett* submitted (2009).
18. Paeth, H., and Thamm, H. P., Regional Modelling of Future African Climate North of 15 Degrees S Including Greenhouse Warming and Land Degradation. *Climatic Change* 83 (3), 401–427 (2007).
19. Polgreen, L. In Niger, Trees and Crops Turn Back the Desert, in *The New York Times* (February 2007).
20. Reij, C., Tappan, G., and Belemire, A., Changing Land Management Practices and Vegetation on the Central Plateau of Burkina Faso (1968–2002). *Journal of Arid Environments* 63, 642–659 (2005).
21. Bierbaum, R., Fay, M., "World Development Report 2010: Development and Change" (The World Bank, 2010). See also WRI, 2008.
22. Reij, C., and Smaling, E. M. A., Analyzing Successes in Agriculture and Land Management in Sub-Saharan Africa: Is Macro-Level Gloom Obscuring Positive Micro-Level Change? *Land Use Policy* 25, 410–420 (2008).
23. Paeth, H., Born, K., Girmes, R., Podzun, R., and Jacob, D., Regional Climate

Change in Tropical and Northern Africa due to Greenhouse Forcing and Land Use Changes. *Journal of Climate* 22, 114–132 (2009).
24. Butt, T. A., McCarl, B. A., Angerer, J., Dyke, P. T., and Stuth, J. W., The Economic Food Security Implications of Climate Change in Mali. *Climatic Change* 68, 355–378 (2005).
25. Sullivan, G. R. et al., National Security and the Threat of Climate Change, edited by S. Goodman (CNA, Alexandria, VA, 2007), pp. 63.
26. Chakravarty, S. et al., Sharing Global CO2 Emission Reductions Among One Billion High Emitters. *PNAS* 106 (43), 11884–11888 (2009).
27. Bierbaum, R., Fay, M., "World Development Report 2010: Development and Change" (The World Bank, 2010).
28. Ki Moon, B., A Climate Culprit in Darfur, in *The Washington Post* (June 16, 2007).
29. Mabey, N. Delivering Climate Security: International World Responses to a Climate Changed World, in *Whitehall Paper*, edited by RUSI (Routledge Journals, Oxford, U.K., 2008), Vol. 69.
30. Burke, M. B. B., Miguel, E., Satynath, S., Dykema, J. A., and Lobell, D., Warming Increases the Risk of Civil War in Africa. *PNAS* 106 (49), 20670–20674 (2009).

6. THE GREAT BARRIER REEF, AUSTRALIA

1. Veron, J. E. N., *A Reef in Time: The Great Barrier Reef from Beginning to End* (Belknap Press of Harvard University Press, Cambridge, MA, 2008).
2. Hoegh-Guldberg, O. et al., Coral Reefs under Rapid Climate Change and Ocean Acidification. *Science* 318 (5857), 1737–1742 (2007). See also Kleypas, J. A., Buddemeier, R. W., and Gattuso, J. P., The Future of Coral Reefs in an Age of Global Change. *International Journal of Earth Sciences* 90, 426 (2001). See also Kleypas, J. A., and Eakin, C. M., Scientists' Perceptions of Threats to Coral Reefs: Results of a Survey of Coral Reef Researchers. *Bulletin of Marine Science* 80 (No. 2), 419–436 (2007).
3. Carpenter, K. E., et al., One-Third of Reef-Building Corals Face Elevated Extinction Risk from Climate Change and Local Impacts. *Science* 321 (5888), 560–563 (2008).
4. Scott C. Doney, V. J. F., Richard A. Feely, Joan A. Kleypas, Ocean Acidification: The Other CO_2 Problem. *Annual Review of Marine Science* Vol. 1, 169–192 (2009).
5. De'ath, G., Lough, J. M., and Fabricius, K. E., Declining Coral Calcification on the Great Barrier Reef. *Science* 323 (5910), 116–119 (2009).
6. Caldeira, K., and Wickett, M. E., Ocean Model Predictions of Chemistry Changes from Carbon Dioxide Emissions to the Atmosphere and Ocean. *Journal of Geophysical Research Oceans* 110, C09S04 (2005). See also Caldeira, K., and Wickett, M. E., Anthropogenic Carbon and Ocean pH. *Nature* 425, 365 (2003).
7. Lough, J. M., 10th Anniversary Review: A Changing Climate for Coral Reefs. *Journal of Environmental Monitoring* 10 (1), 21–29 (2008).
8. Marshall, P., and Schuttenberg, H., *A Reef Manager's Guide to Coral Bleaching* (Australian Government Great Barrier Reef Marine Park Authority, Townsville, Australia, 2006), p. 176.
9. NOAA, Available at http://coralreefwatch.noaa.gov/ (2009).
10. Henson, B., Heat, Fire, and Fear in Australia, in *UCAR Magazine* (UCAR, 2009).
11. Interim Report, edited by State Government of Victoria (Victorian Bushfires Royal Commission, Victoria, Australia 2009).

7. CENTRAL VALLEY, CALIFORNIA

1. Lund, J., et al., *Envisioning Futures for the Sacramento-San Jaoquin Delta* (Public Policy Institute of California, San Francisco, CA, 2007).
2. Vicuna, S., Hanemann, M., Dale, L., "Economic Impacts of Delta Levee Failure Due to Climate Change: A Scenario Analysis" (University of California, Berkeley for the California Energy Commission, PIER Energy-Related Environmental Research, 2006).
3. Lund, J., et al., *Comparing Futures for the Sacramento-San Joaquin Delta* (Public Policy Institute of California, San Francisco, CA, 2008), p. 147.
4. Mastrandrea, M. D., Tebaldi, C., Snyder, C. P., Schneider, S. H., "Current and Future Impacts of Extreme Events in California" (PIER Technical Report CEC-500-2009-026-D, 2009).
5. Cobb, K., Charles, C. D., Cheng, H., and Edwards, R. L., El Nino/Southern Oscillation and Tropical Pacific Climate during the Last Millennium. *Nature* 424, 271–276 (2003).
6. Guilyardi, E., et al., Understanding El Niño in Ocean-Atmosphere General Circulation Models: Progress and Challenges. *BAMS* 3, 325–340 (2009).
7. Luers, A. L., Cayan, D. R., Franco, G., Hanemann, M., Croes, B., "Our Changing Climate: Assessing the Risks to California" (The California Climate Change Center, 2006).
8. Rajagopalan, B., et al., Water Supply Risk on the Colorado River: Can Management Mitigate? *Water Resources Research* 45 (8) (2009).

8. THE ARCTIC, PART ONE: INUIT NUNAAT

1. *ITK 5000 Years of Inuit History and Heritage*, edited by Inuit Tapiriit Kanatami (Inuit Tapiriit Kanatami, Ottawa, Canada, 2009).
2. Diamond, J., *Collapse: How Societies Choose to Fail or Succeed* (Penguin Group, New York, 2005).
3. Hall, C. F., *Life with the Esquimaux* (Sampson Low, Son and Marston, London, 1864).
4. Ford, J., et al., Reducing Vulnerability to Climate Change in the Arctic: The Case of Nunavut, Canada. *Arctic* 60 (2), 150–166 (2007).
5. Fox, S., "These Are Things That Are Really Happening": Inuit Perspectives on the Evidence and Impacts of Climate Change in Nunavut, in *The Earth Is Faster Now: Indigenous Observations of Arctic Environmental Change*, edited by I. Krupnik and D. Jolly (Arctic Research Consortium of the United States, Fairbanks, Alaska, 2002), pp. 12–53.
6. Laidler, G., Ford, J., Gough, W. A., and Ikummaq, T., Assessing Inuit Vulnerability to Sea Ice Change in Igloolik, Nunavut. *Climatic Change* (in press, 2009).
7. INAC, The Food Mail Program, Available at http://www.ainc-inac.gc.ca/nth/fon/fm/index-eng.asp (2009).
8. Orlove, B., Chiang, J. C. H., and Cane, M. A., Ethnoclimatology in the Andes: A Cross-Disciplinary Study Uncovers a Scientific Basis for the Scheme Andean Potato Farmers Traditionally Use to Predict the Coming Rains. *American Scientist* 90 (5), 428–435 (2002). See also Orlove, B. S., Chiang, J. C. H., and Cane, M. A., Forecasting Andean Rainfall and Crop Yield from the Influence of El Niño on Pleiades Visibility. *Nature* 403, 68–71 (2000). See also Ford, J. D., Pearce, T., Gilligan, J.,

Smit, B., and Oakes, J., Climate Change and Hazards Associated with Ice Use in Northern Canada. *Arctic, Antarctic, and Alpine Research* 40 (4), 647–659 (2008).

9. Pearce, T., et al., Inuit Vulnerability and Adaptive Capacity to Climate Change in Ulukhaktok, Northwest Territories, Canada. *Polar Record* 45 (1–21) (2009).
10. Condon, R., Ogina, J., and Elders, H., *The Northern Copper Inuit: A History* (University of Toronto Press, Toronto, 1996).
11. NWMB, "The Nunavut Wildlife Harvest Study Final Report" (Iqualuit: National Wildlife Management Board, 2004).
12. Ainley, D. G., Tynan, C. T., and Stirling, I., Sea Ice: A Critical Habitat for Polar Marine Animals and Birds, in *Sea Ice: An Introduction to Its Physics, Chemistry, Biology, and Geology.*, edited by D. N. Thomas and G. S. Dieckmann (Blackwell Science, Oxford, U.K. 2003), pp. 240–266.
13. Ferguson, S. H., Climate Change and Ringed Seal (Phoca hispida) Recruitment in Western Hudson Bay. *Marine Mammal Science* 21 (1), 121–135 (2005).
14. Krupnik, I., and Jolly, D., eds., *The Earth Is Faster Now: Indigenous Observations of Arctic Environmental Change* (Arctic Research Consortium of the United States, Fairbanks, Alaska, 2002).
15. Manabe, S., and Stouffer, R. J., Sensitivity of a Global Climate Model to an Increase of CO2 in the Atmosphere. *Journal of Geophysical Research* 85 (C10), 5529–5554 (1980).
16. Holland, M. M., and Bitz, C. M., Polar Amplification of Climate Change in Coupled Models. *Climate Dynamics* 21 (221–232) (2003).
17. Serreze, M. C., Barrett, A. P., Stroeve, J. C., Kindig, D. N., and Holland, M. M., The Emergence of Surface-Based Arctic Amplification. *The Cryosphere* 3, 11–19 (2009).
18. NSIDC, Arctic Sea Ice News and Analysis (2008).
19. Serreze, M., and Stroeve, J. C., Standing on the Brink. *Nature Reports Climate Change* (2008).
20. Gearheard, S., "Igliniit Project" (International Polar Year, 2009).
21. Stroeve, J. C., Holland, M. M., Meier, W., Scambos, T., and Serreze, M., Arctic Sea Ice Decline: Faster than Forecast. *Geophysical Research Letters* 34 (L09501) (2007).
22. Maslanik, J. A., et al., A Younger, Thinner Arctic Ice Cover: Increased Potential for Rapid, Extensive Sea-Ice Loss. *Geophysical Research Letters* 34 (L24501) (2007).
23. Zhang, X., and Walsh, J., *Journal of Climate* 19 (2006).
24. Holland, M. M., Bitz, C. M., and Tremblay, B., Future Abrupt Reductions in the Summer Arctic Sea Ice. *Geophysical Research Letters* 33 (L23503) (2006).

9. THE ARCTIC, PART TWO: GREENLAND

1. Diamond, J., *Collapse: How Societies Choose to Fail or Succeed* (Penguin Group, New York, 2005).
2. Gautier, D. L., et al., Assessment of Undiscovered Oil and Gas in the Arctic. *Science* 324 (5931), 1175–1179 (2009).
3. Lenton, T. M., et al., Tipping Elements in the Earth's Climate System. *105* (6), 1786–1793 (2008).
4. Funk, M., Greenland Rising, in *Outside* (2009).
5. Saabye, H. E., Brudstykker sd en dagbog holden i Groenland i aarene 1770–1778, in *Medd. Gronland*, edited by H. Oesterman (1942).

6. van Loon, H., and Rogers, J. C., The See-Saw in Winter Temperatures Between Greenland and Northern Europe. Part I. *Monthly Weather Review* 104 (1978).
7. Hurrell, J. W., and Deser, C., North Atlantic Climate Variability: The Role of the North Atlantic Oscillation. *Journal of Marine Systems* 78 (1), 28–41 (2009).
8. Johannessen, O. M., Khvorostovsky, K., Miles, M. W., and Bobylev, L. P., Recent Ice-Sheet Growth in the Interior of Greenland. *Science Express* (2005).
9. Zwally, H. J., et al., Mass Changes of the Greenland and Antarctic Ice Sheets and Shelves and Contributions to Sea-Level Rise: 1992–2002. *Journal of Glaciology* 51 (175), 509–527 (2005).
10. Hanna, E., et al., Runoff and Mass Balance of the Greenland Ice Sheet: 1958–2003. *Journal of Geophysical Research* 110 (D13108) (2005).
11. Pritchard, H. D., Arthern, R. J., Vaughan, D. J., and Edwards, L. A., Extensive Dynamic Thinning on the Margins of the Greenland and Antarctic Ice Sheets. *Nature* 461 (7263) (2009).
12. Bindschadler, R. *Science* 311, 1720–1721 (2006).
13. Krabill, W., et al., Greenland Ice Sheet: High-Elevation Balance and Peripheral Thinning. *Science* 289, 428–430 (2000).
14. Rignot, E., and Kanagaratnam, P., Changes in the Velocity Structure of the Greenland Ice Sheet. *Science* 311 (5763), 986–990 (2006).
15. Ekstrom, G., Nettles, M., and Tasi, V. C., Seasonality and Increasing Frequency of Greenland Glacial Earthquakes. *Science* 311, 1756–1758 (2006).
16. Joughin, I., Abdalati, W., and Fahnestock, M., Large Fluctuations in Speed on Greenland's Jakobshavn Isbræ Glacier. *Nature* 432, 608–610 (2004).
17. Steffensen, J. P., et al., High-Resolution Greenland Ice Core Data Show Abrupt Climate Change Happens in Few Years. *Science* 321 (5889), 680–684 (2008).
18. Fluckiger, J., Climate Change: Did You Say "Fast"? *Science* 321 (5889), 650–651 (2008).
19. Stouffer, R. J., et al., GFDL's CM2 Global Coupled Climate Models. Part IV: Idealized Climate Response. *Journal of Climate* 19 (5), 723–740 (2006). See also Fluckiger, J., Knutti, R., White, J. W. C., and Renssen, H., Modeled Seasonality of Glacial Abrupt Climate Events. *Climate Dynamics* 31, 633–645 (2008).
20. Kukla, G. J., et al., Last Interglacial Climates. *Quaternary Research* 58, 2–13 (2002). See also Oppo, D. W., McManus, J. F., and Cullen, J. L., Evolution and Demise of the Last Interglacial Warmth in the Subpolar North Atlantic. *Quaternary Science Reviews* 25, 3268–3277 (2006).
21. Huybrechts, P., and De Wolde, J., The Dynamic Response of the Greenland and Antarctic Ice Sheets to Multiple-Century Climatic Warming. *Journal of Climate* 12, 2169–2188 (1999).
22. Stroeve, J., Holland, M. M., Meier, W., Scambos, T., and Serreze, M., Arctic Sea Ice Decline: Faster than Forecast. *Geophysical Research Letters* 34, L09501 (2007).
23. Hansen, J. E., A Slippery Slope: How Much Global Warming Constitutes "Dangerous Anthropogenic Interference"? An Editorial Essay. *Climatic Change* 68, 269–279 (2005).
24. Otto-Bliesner, B. L., et al., Simulating Arctic Climate Warmth and Icefield Retreat in the Last Interglaciation. *Science* 311, 1751–1753 (2006).

25. Portions of Arctic Coastline Eroding, No End in Sight, Says New Study (*Science-Daily*, December 17, 2009).
26. Kirby, R. R., and Beaugrand, G., Trophic Amplification of Climate Warming. *Proceedings of the Royal Society B* 276 (1676) (2009).
27. *ICON Carbon Dioxide Capture and Storage Initiative: A Canadian Clean Energy Opportunity*, edited by ICON (2009), p. 16.
28. *DOE Methane Hydrate: Future Energy Within Our Grasp*, edited by Office of Fossil Energy (U.S. Department of Energy, Washington, DC, 2007).
29. Walter, K. M., Zimov, S. A., Chanton, J. P., Verbyla, D., and Chapin III, F. S., Methane Bubbling from Siberian Thaw Lakes as a Positive Feedback to Climate Warming. *Nature* 443, 71–75 (2006).
30. Kennedy, M., Mrofka, D., and von der Borch, C., Snowball Earth Termination by Destabilization of Equatorial Permafrost Methane Clathrate. *Nature* 453, 642–645 (2008).
31. Walter Anthony, K., Methane: A Menace Surfaces, in *Scientific American* (2009), Vol. 301, pp. 68–75.

10. DHAKA, BANGLADESH

1. Webster, P. J., Moore, A., Loschnigg, J., and Leban, M., Coupled Ocean-Atmosphere Dynamics in the Indian Ocean during 1997–98. *Nature* 401, 356–360 (1999).
2. Del Ninno, C., Dorosh, P. A., Smith, L. C., and Roy, D., 2001.
3. *NAPA National Adaptation Programme of Action* (NAPA) (Ministry of Environment and Forest Government of the People's Republic of Bangladesh, 2005), p. 46.
4. Webster, P. J., et al., A Three-Tier Overlapping Prediction Scheme: Tools for Strategic and Tactical Decisions in the Developing World, in *Predictability of Weather and Climate*, edited by T. N. Palmer (Cambridge University Press, 2006), pp. 645–673.
5. *UNDP World Population Prospects: The 2008 Revision* (United Nations Population Division, 2008).
6. *MEF Bangladesh Climate Change Strategy and Action Plan 2008* (Ministry of Environment and Forests Government of the People's Republic of Bangladesh, Dhaka, Bangladesh, 2008), p. 68.
7. Scherler, D., Bookhagen, B., and Strecker, M. R., "Spatially variable response of Himalayan glaciers to climate change affected by debris cover." *Nature Geoscience* (2011).
8. Christensen, J. H., et al., *2007: Regional Climate Projections in Climate Change 2007: The Physical Science Basis. Contribution of Working Group I to the Fourth Assessment Report of the Intergovernmental Panel on Climate Change*, edited by S. Solomon et al. (Cambridge, U.K., 2007).
9. Agrawala, S., Ota, T., Ahmed, A. U., Smith, J., van Aalst, M., "Development and Climate Change in Bangladesh: Focus on Coastal Flooding and The Sundarbans" (Organisation for Economic Co-operation and Development, 2003).
10. Colette, A. *Case Studies on Climate Change and World Heritage* (UNESCO, Paris, 2007), p. 79.
11. Friedman, L., Bangladesh: Where the Climate Exodus Begins, in *Greenwire* (2009).
12. Mool, P. K., et al., *Inventory of Glaciers, Glacial Lakes, and Glacial Lake Outburst Floods: Monitoring and Early Warning Systems in the Hindu Kush-Himalayan Region*, edited by ICIMOD (ICIMOD, Bhutan, 2001), p. 254.

13. Nayar, A., When the Ice Melts. *Nature* 461, 1042–1046 (2009).
14. Rodell, M., Velicogna, I., and Famiglietti, J.S., Satellite-Based Estimates of Groundwater Depletion in India. *Nature* 460, 999–1002 (2009).
15. Cruz, R. V., et al., Asia. *Climate Change 2007: Impacts, Adaptation and Vulnerability. Contribution of Working Group II to the Fourth Assessment Report of the Intergovernmental Panel on Climate Change*, edited by M.L. Parry et al. (Cambridge, U.K., 2007), pp. 469–506.
16. Lobell, D. B., et al., Prioritizing Climate Change Adaptation Needs for Food Security in 2030. *Science* 319, 607–610 (2008).

11. NEW YORK, NEW YORK

1. Eddy, D. Y2K Discussion List, in Y2K *Discussion List*, edited by Peter de Jager's (1995).
2. Rosenzweig, C., and Solecki, W., eds., *Metro East Coast, Report for the U.S. Global Change Research Program* (Columbia Earth Institute, New York, 2001).
3. DEP, "Report 1: Assessment and Action Plan – A Report Based on the Ongoing Work of the DEP Climate Change Task Force" (The New York City Department of Environment Protection Climate Change Program, 2008).
4. Rosenzweig, C., et al., Climate Risk Information, in *New York City Panel on Climate Change* (New York, 2009).
5. Rosenzweig, C., and Solecki, W., Climate Change and a Global City: The Potential Consequences of Climate Variability and Change, in *Metro East Coast Assessment* (Columbia Earth Institute, New York, 2001), p. 207.
6. *PlaNYC PlaNYC: A Greener, Greater, New York* (Mayor's Office of Long-Term Planning and Sustainability, New York, NY, 2007).
7. Rosenzweig, C., et al., Managing Climate Change Risks in New York City's Water System: Assessment and Adaptation Planning. *Mitigation and Adaptation Strategies for Global Change* 12 (8), 1391–1409 (2007).
8. *PlaNYC PlaNYC Progress Report 2009* (Mayor's Office of Long-Term Planning and Sustainability, New York, NY, 2009), p. 43.
9. Gray, W., *Landfall Probability Table in United States Landfalling Hurricane Probability Project*, edited by Tropical Meteorology Research Project (Fort Collins, CO, 2009).
10. Chan, S., Why the Subways Flood, in *The New York Times* (August 8, 2007).
11. USGS National Assessment of Coastal Vulnerability to Future Sea-level Rise, in *USGS Fact Sheet* FS-076–00 (Washington, DC, 2000).
12. Ackerman, F., and Stanton, E. A., *Climate Change and the U.S. Economy: The Costs of Inaction* (Tufts University and Stockholm Environment Institute, Medford, MA, 2008).
13. Yin, J., Schlesinger, M. E., and Stouffer, R. J., Model Projections of Rapid Sea-level Rise on the Northeast Coast of the United States. *Nature Geoscience* 2, 262–266 (2009).
14. Kopp, R. E., Personal Communication (Princeton, NJ, Dec. 2, 2009).

EPILOGUE

1. Weiss, H., et al., The Genesis and Collapse of 3rd Millenium North Mesopotamian Civilization. *Science* 261, 994–1004 (1993).

2. Cullen, H. M., and deMenocal, P. B., The Possible Role of Climate in the Collapse of the Akkadian Empire: Evidence from the Deep Sea. *Geology* 20 (8), 379–382 (2000).
3. *UNFCCC United Nations Framework Convention on Climate Change* (United Nations Environment Programme/World Meteorological Organization Information Unit on Climate Change [IUCC] on behalf of the Interim Secretariat of the Convention, Switzerland, 2002).
4. Solomon, S., Garcia, R. R., Rowland, S. F., and Wuebbles, D. J., On the Depletion of Antarctic Ozone. *Nature* 321, 755–758 (1986).
5. Solomon, S., Plattner, G., and Friedlingstein, P., Irreversible Climate Change Due to Carbon Dioxide Emissions. *PNAS* 106 (6), 1704–1709 (2009).
6. IPCC Physical Science Basis. Contribution of Working Group I to the Fourth Assessment Report of the Intergovernmental Panel on Climate Change in *Climate Change 2007*, edited by S. Solomon et al. (Cambridge University Press, Cambridge, 2007), p. 996.
7. Revkin, A. C., Scientist at Work: Susan Solomon; Melding Science and Diplomacy to Run a Global Climate Review, in *The New York Times* (February 2007).
8. Diamond, J., *Collapse: How Societies Choose to Fail or Succeed* (Penguin Group, New York, 2005).
9. Frame, D. J., et al., Warming Caused by Cumulative Carbon Emissions Towards the Trillionth Tonne. *Nature* 458, 1163–1166 (2009).
10. Meinshausen, M., et al., Greenhouse-Gas Emission Targets for Limiting Global Warming to 2°C. *Nature* 458, 1158–1162 (2009).
11. *EIA International Energy Outlook 2009* (EIA, 2009), p. 274.

APPENDIX

1. Meehl, G. A., Tebaldi, C., Walton, G., Easterling, D., and McDaniel, L., The Relative Increase of Record High Maximum Temperatures Compared to Record Low Minimum Temperatures, in the U.S. *Geophysical Research Letters* (in press).
2. *CCSP Weather and Climate Extremes in a Changing Climate, Regions of Focus: North America, Hawaii, Caribbean and U.S. Pacific Islands*, edited by T. R. Karl et al. (Department of Commerce, NOAA's National Climatic Data Center, Washington, DC, 2008), p. 164.
3. Meehl, G. A., and Tebaldi, C., More Intense, More Frequent, and Longer Lasting Heat Waves in the 21st Century. *Science* 305 (5686), 994–997 (2004).
4. Arnell, N. W., *Global Warming, River Flows and Water Resources* (Wiley, Chichester, U.K., 1996). See also Gleick, P. H., Methods for Evaluating the Regional Hydrologic Effects of Global Climate Changes. *Journal of Hydrology* 88, 97–116 (1986).
5. Zimmernan, R., and Cusker, M., Institutional Decision-making, in *Climate Change and a Global City: The Potential Consequences of Climate Variability and Change. Metro East Coast*, edited by C. Rosenzweig and W. D. Solecki (Columbia Earth Institute New York, 2001).
6. MacCracken, M. C., Moore, F., and Topping, J. C., eds., *Sudden and Disruptive Climate Change* (Earthscan, London, 2008). See also MacCracken, M. C., Prospects for Future Climate Change Reasons for Early Action. *Journal of the Air and Waste Management Association* 58, 735–786 (2008).

ACKNOWLEDGMENTS

Rob Socolow, a physics professor at Princeton University who has thought long and hard about how to solve the problem of global warming, adapted the words of Winston Churchill to describe the challenge the world is currently facing. "Never has the work of so few scientists led to so much being asked of so many." I'm struck by these words, because it isn't very often that a scientific finding so thoroughly challenges the foundation on which modern society rests. The discovery of global warming has certainly resulted in an enormous request on behalf of the scientific community to change the way we do business. And although it is not a problem that can be solved overnight, I firmly believe we will get there.

I would like to sincerely thank my colleagues at Climate Central: Berrien Moore, Joanne Graziano, Ben Strauss, Michael Lemonick, Eric Larson, Claudia Tebaldi, Remik Ziemlinski, Nicole Heller, Phil Duffy, Jessica Harrop, Paul Ferlita, Iveta Weinberg, and Andrew Freedman. Science journalism has been hard hit in the past year, and I applaud the hard work of my fellow scientists and journalists engaged in the task of communicating climate science and technology. Climate Central was established to provide unbiased information about climate science in order to help people better understand what the future may hold if we continue to have an infrastructure based only on fossil fuels. Without the guidance and support of Climate Central's founding board members, this organization could not have moved from concept to reality. For engaging in this vision of a new model of science journalism, I would like to thank Wendy Schmidt; Joe Sciortino and the Schmidt Family Foundation; Steve Pacala at Princeton University; Gus Speth at Yale University; and

Jane Lubchenco, who has long since retired from the board and become chief administrator at NOAA. I would also like to thank Climate Central's original board members: Richard Somerville at the Scripps Institution of Oceanography; Michael Oppenheimer at Princeton University; Sally Benson, Pamela Matson, and Jon Krosnick at Stanford University; Skip Lupia at the University of Michigan; Mary Evelyn Tucker at Yale University; and John Holdren, who has since retired from the board to serve as director of the Office of Science and Technology Policy.

My most heartfelt thanks to the scientists who shared their lives and their time with me: Stephen Schneider at Stanford University; Susan Solomon at NOAA's Earth System Research Laboratory; Joanie Kleypas at the National Center for Atmospheric Research; Cynthia Rosenzweig at NASA's Goddard Institute for Space Studies; Steve Hammer at Columbia University; Ellen Hanak at the Public Policy Institute of California; Jay Lund at the University of California, Davis; Peter Webster at the Georgia Institute of Technology; Omar Rahman at Independent University in Dhaka, Bangladesh; Tristan Pearce at the University of Guelph; Shari Gearheard of the National Snow and Ice Data Center at the University of Colorado at Boulder; Alessandra Giannini at the International Research Institute for Climate and Society; Chris Reij at Vrije Universiteit; Isaac Held at NOAA's Geophysical Fluid Dynamics Laboratory (GFDL); J. P. Steffensen and Dorthe Dahl-Jensen at the University of Copenhagen; Scott Luthcke at the NASA Goddard Space Flight Center; and John Chiang at the University of California, Berkeley.

I also need to thank Mike MacCracken of the Climate Institute for carefully reading through this manuscript; his knowledge of climate science is truly encyclopedic. In addition, I'm grateful to Jerry Meehl and Peter Gent at NCAR, Ron Stouffer at NOAA's GFDL, Peter deMenocal at the Lamont-Doherty Earth Observatory of Columbia University, Vasilii Petrenko at the University of Colorado, Michael Bender at Princeton University, and Phil Arkin at the University of Maryland—all very busy scientists who were willing to lend their expertise and feedback.

I'm so very grateful to Marla Hoppenfeld and everyone involved in this project at HarperCollins Publishers. And I am especially grateful to Matt Harper, my editor—without his continued support and guidance, this book would have never become a reality. My thanks also go out to Lisa Sharkey for helping to shepherd this project through a tough economic landscape.

I must also thank my wonderful and supportive family: William and Heidi Cullen; Rosemary Wenke and Jim Honrine; and Stephen, Donna, Matt, and Keith Cullen. If it weren't for my mom and dad, who came to my rescue on the weekends to help around the house (and bake apple pie!), this project would have been a whole lot harder if not altogether impossible.

And finally, the project would have been truly impossible without the love, support, and friendship of my husband, Drew. He is my rock and my hero. And to our pooches, Homer and Emma, here's to finally taking some long walks again!

Heidi Cullen
Princeton, New Jersey, January 2010

INDEX

About the Author

Dr. Heidi Cullen is a senior research scientist with Climate Central, a nonprofit research organization through which she reports on climate for outlets including *PBS NewsHour*, Time .com, and The Weather Channel. Before joining Climate Central, Dr. Cullen served as The Weather Channel's first on-air climate expert and helped create *Forecast Earth*, the first weekly television series to focus on issues related to climate change and the environment. She is a visiting lecturer at Princeton University, a member of the American Geophysical Union and the American Meteorological Society, and an associate editor of the journal *Weather, Climate, and Society*. She has appeared on *Good Morning America*, *The CBS Evening News with Katie Couric*, *The View*, and *Larry King Live*. She holds a B.S. in engineering and a Ph.D. in climatology from Columbia University and lives with her husband and two dogs in Princeton, New Jersey.

About the Author

The Editor

Leland S. Person is Professor of English at the University of Cincinnati. He is the author of *The Cambridge Introduction to Nathaniel Hawthorne*, *Henry James and the Suspense of Masculinity*, and *Aesthetic Headaches: Women and a Masculine Poetics in Poe, Melville, and Hawthorne*, and many articles on nineteenth-century American writers. His most recent book is *The American Novel to 1870* (coedited with Gerald Kennedy), vol. 5 in the 12-volume *Oxford History of the Novel in English*.

NORTON CRITICAL EDITIONS
AMERICAN REALISM & REFORM

ALCOTT, LITTLE WOMEN
ALGER, RAGGED DICK
ANDERSON, WINESBURG, OHIO
CHESNUTT, THE CONJURE STORIES
CHESNUTT, THE MARROW OF TRADITION
CHOPIN, THE AWAKENING
CRANE, MAGGIE: A GIRL OF THE STREETS
CRANE, THE RED BADGE OF COURAGE
DOUGLASS, NARRATIVE OF THE LIFE OF FREDERICK DOUGLASS
DREISER, SISTER CARRIE
DU BOIS, THE SOULS OF BLACK FOLK
EMERSON, EMERSON'S PROSE AND POETRY
FULLER, WOMAN IN THE NINETEENTH CENTURY
HAWTHORNE, THE BLITHEDALE ROMANCE
HAWTHORNE, THE HOUSE OF THE SEVEN GABLES
HAWTHORNE, NATHANIEL HAWTHORNE'S TALES
HAWTHORNE, THE SCARLET LETTER AND OTHER WRITINGS
HOWELLS, THE RISE OF SILAS LAPHAM
JACOBS, INCIDENTS IN THE LIFE OF A SLAVE GIRL
JAMES, THE AMBASSADORS
JAMES, THE AMERICAN
JAMES, THE PORTRAIT OF A LADY
JAMES, TALES OF HENRY JAMES
JAMES, THE TURN OF THE SCREW
JAMES, THE WINGS OF THE DOVE
LINCOLN, LINCOLN'S SELECTED WRITINGS
MELVILLE, THE CONFIDENCE-MAN
MELVILLE, MELVILLE'S SHORT NOVELS
MELVILLE, MOBY-DICK
MELVILLE, PIERRE
MONTGOMERY, ANNE OF GREEN GABLES
MOODIE, ROUGHING IT IN THE BUSH
NORRIS, MCTEAGUE
NORTHUP, TWELVE YEARS A SLAVE
POE, THE SELECTED WRITINGS OF EDGAR ALLAN POE
RIIS, HOW THE OTHER HALF LIVES
SINCLAIR, THE JUNGLE
STOWE, UNCLE TOM'S CABIN
THOREAU, WALDEN, CIVIL DISOBEDIENCE, AND OTHER WRITINGS
TWAIN, ADVENTURES OF HUCKLEBERRY FINN
TWAIN, THE ADVENTURES OF TOM SAWYER
TWAIN, A CONNECTICUT YANKEE IN KING ARTHUR'S COURT
TWAIN, PUDD'NHEAD WILSON AND THOSE EXTRAORDINARY TWINS
WASHINGTON, UP FROM SLAVERY
WHARTON, THE AGE OF INNOCENCE
WHARTON, ETHAN FROME
WHARTON, THE HOUSE OF MIRTH
WHITMAN, LEAVES OF GRASS AND OTHER WRITINGS

A NORTON CRITICAL EDITION

Nathaniel Hawthorne
THE SCARLET LETTER
AND OTHER WRITINGS

AUTHORITATIVE TEXTS

CONTEXTS

CRITICISM

SECOND EDITION

Edited by

LELAND S. PERSON
UNIVERSITY OF CINCINNATI

W • W • NORTON & COMPANY • *New York* • *London*

For Nina Baym
in token of my admiration for her career

W. W. Norton & Company has been independent since its founding in 1923, when William Warder Norton and Mary D. Herter Norton first published lectures delivered at the People's Institute, the adult education division of New York City's Cooper Union. The firm soon expanded its program beyond the Institute, publishing books by celebrated academics from America and abroad. By midcentury, the two major pillars of Norton's publishing program—trade books and college texts—were firmly established. In the 1950s, the Norton family transferred control of the company to its employees, and today—with a staff of four hundred and a comparable number of trade, college, and professional titles published each year—W. W. Norton & Company stands as the largest and oldest publishing house owned wholly by its employees.

Manufacturing by LSC Communications
Book design by Antonina Krass
Production supervisor: Elizabeth Marotta

Library of Congress Cataloging-in-Publication Data

Names: Hawthorne, Nathaniel, 1804–1864 author. | Person, Leland S. editor.
Title: The scarlet letter and other writings : authoritative texts, contexts, criticism / Nathaniel Hawthorne; edited by Leland S. Person, University of Cincinnati.
Description: Second edition. | New York : W W Norton & Company, 2016. | Series: A norton critical edition | Includes bibliographical references.
Identifiers: LCCN 2016036763 | **ISBN 9780393264890 (pbk.)**
Subjects: LCSH: Boston (Mass.)—History—Colonial period, ca. 1600–1775—Fiction. | Triangles (Interpersonal relations)—Fiction. | Illegitimate children—Fiction. | Women immigrants—Fiction. | Puritans—Fiction. | Hawthorne, Nathaniel, 1804–1864. Scarlet letter.
Classification: LCC PS1868.A2 P47 2016 | DDC 813/.3—dc23
LC record available at https://lccn.loc.gov/2016036763

W. W. Norton & Company, Inc., 500 Fifth Avenue, New York, N.Y. 10110
www.wwnorton.com
W. W. Norton & Company Ltd., 15 Carlisle Street, London W1D 3BS

6 7 8 9 0

Contents

Preface

This second edition of *The Scarlet Letter and Other Writings* appears 167 years after the original publication of the novel (on March 16, 1850). *The Scarlet Letter* sold close to 6,000 copies in its first year of publication. More than 200 years later, the first Norton edition of *The Scarlet Letter and Other Writings* has been one of the best-selling Norton Critical Editions, with more credit due the author than the editor. For copy text, I have again chosen to use the third edition, published in September 1850 by Ticknor, Reed, and Fields, because that edition, the first set in stereotype plates, was the basis of subsequent printings in Hawthorne's lifetime and so represents the text that most nineteenth-century readers actually read.

I have retained all five of Hawthorne's short prose works that seem most pertinent as harbingers of *The Scarlet Letter*. I hope my inclusion of Hawthorne's "Mrs. Hutchinson" has helped that sketch receive a wider readership, especially in view of Hawthorne's famous observation in chapter 1 of *The Scarlet Letter* that the wild rose bush outside the prison door had "sprung up under the feet of the sainted Ann Hutchinson." "Endicott and the Red Cross" includes Hawthorne's earliest mention of a woman wearing a scarlet letter and helps augment the Puritan context of the later novel. "The Minister's Black Veil" can be considered an early trying out of a character and situation Hawthorne would develop in his portrait of Arthur Dimmesdale. "Young Goodman Brown," arguably Hawthorne's most thoughtful treatment of the Salem witchcraft hysteria of 1692, also helps to highlight his use of setting and symbolic spaces in *The Scarlet Letter*, especially the lure of the forest. In its treatment of Georgiana's relationship to Aylmer and its fixation on marked women's bodies, "The Birthmark" anticipates Hester Prynne's situation and can provide important additional material for gender-oriented analyses of *The Scarlet Letter*.

In addition to those short works, I have also provided pertinent samples from Hawthorne's letters and notebooks—letters he wrote at the time he was composing and publishing the novel and notebook entries that formed the basis, often in only slightly revised form, for passages in the narrative. I have expanded the passage from Hawthorne's campaign biography of Franklin Pierce, which is often cited by scholars interested in Hawthorne's attitude toward slavery and abolition, because I think reading Hawthorne's assessment of Pierce's views on abolition in its full context has the potential to color our views of both men's politics. I have included selected early reviews of *The Scarlet Letter*, including the little-known, arguably feminist review by Jane Swisshelm that Robert S. Levine, who uses that review as the linchpin for a fascinating examination of nineteenth-century women's responses to the novel, was kind enough to edit for the first edition.

Deciding which critical essays to include in this edition has been a wonderfully enjoyable task, as well as an impossible one. My temptation in editing the first edition and even more so in editing the second was to keep adding scholarly works. For obvious reasons, I fought that temptation and forced myself to substitute some new essays for old ones. This second edition includes ten new essays. I have tried as much as possible to provide examples of most recent approaches to the novel and the shorter works, especially those emphasizing the relationship between the novel and its nineteenth-century context. I have also included an essay on film versions of the novel, especially the 1995 version directed by Roland Joffé and starring Demi Moore, Gary Oldham, and Robert Duvall. Jamie Barlowe's feminist and largely appreciative assessment of that film also includes brief accounts of all previous film versions.

Preparing the footnotes for this edition has been a fascinating experience, an opportunity to read Hawthorne in a new way—a hypertextual way. I discovered many live links, or at least found many opportunities to create live links, and many chances to enter portals that took me into new places. I have relied on many different sources, online and print. For definitions I have consulted the *Oxford English Dictionary* and Merriam-Webster's online dictionary. For the many Biblical passages to which Hawthorne refers, the King James Bible available at the University of Virginia's Electronic Text Center (http://etext.lib.virginia.edu/kjv.browse.html). For biographical information on Hawthorne and his family, I have consulted James R. Mellow, *Nathaniel Hawthorne in His Times;* Arlin Turner, *Nathaniel Hawthorne: A Biography*; and Brenda Wineapple, *Hawthorne: A Life;* as well as Stephen Nissenbaum, "The Firing of Nathaniel Hawthorne"; Margaret Moore, *The Salem World of Nathaniel Hawthorne;* and Joseph B. Felt's *Annals of Salem from Its First Settlement* (1827). For information on the settlement and early governance of the Massachusetts Bay Colony, Caleb Snow, *History of Boston* (Boston, 1825); John Winthrop, *The Journal of John Winthrop, 1630–1649,* ed. Richard S. Dunn, James Savage, and Laetitia Yeandle (Cambridge: Harvard UP, 1996); William Bradford, *Bradford's History of Plymouth Plantation, 1606–1646*, ed. William T. Davis (New York: Scribner's, 1923); Robert Emmet Wall, Jr., *Massachusetts Bay: The Crucial Decade, 1640–1650* (New Haven: Yale UP, 1972); Alfred A. Cave, *The Pequot War* (Amherst: U of Massachusetts P, 1996). On Anne Hutchinson, David D. Hall, *The Antinomian Controversy, 1636–1638* (Middletown: Wesleyan UP, 1968); Amy Schrager Lang, *Prophetic Woman: Anne Hutchinson and the Problem of Dissent in the Literature of New England.* For information on English history, especially the first half of the seventeenth century, Samuel R. Gardiner, *History of England from the Accession of James I to the Outbreak of the Civil War, 1603–1642*, 10 vols. (New York: Longmans, Green, 1909). For additional information about William Prynne, I have read William Lamont, *Puritanism and Historical Controversy* (London: University College of London Press, 1996), 15–25, and Mukhtar Ali Isani, "Hawthorne and the Branding of William Prynne," *New England Quarterly* 45 (1972): 182–95. For additional information on the Thomas Overbury murder case, Alfred S. Reid, *The Yellow Ruff and The Scarlet Letter: A Source for Hawthorne's Novel* (Gainesville: U of Florida P, 1955); Anne Somerset, *Unnatural Murder: Poison at the Court of James I*

(London: Weidenfield & Nicolson, 1997); and David Lindley, *The Trials of Frances Howard: Fact and Fiction at the Court of King James* (New York: Routledge, 1993).

My career-long interest in Hawthorne has been aided by many scholars too numerous to list here. Members of the Nathaniel Hawthorne Society, a wonderful community of scholars, have been a constant source of inspiration. I want to give special thanks to Terence Martin and Nina Baym for their mentorship, friendship, and especially for the examples they provided about how to write about Hawthorne. Thanks also to Larry J. Reynolds, who urged me to take on this project. I want to thank my colleagues at the University of Cincinnati, Jon Kamholtz and Jay Twomey, for important pieces of information about Hawthorne's references. Thanks to Geri Hinkle-Wesseling, Kevin Newman, and Devore Nixon for their help in preparing the manuscript for this second Norton Critical Edition. Special thanks to Julia Reidhead and Carol Bemis at W. W. Norton, who gave me this opportunity.

Note on the Texts

For copy text of *The Scarlet Letter,* I have chosen to use the third edition of the novel, published in September 1850 by Ticknor, Reed, and Fields, because that edition, the first set in stereotype plates (by Hobart & Robins and the New England Type and Stereotype Foundry), was the basis of subsequent printings in Hawthorne's lifetime and so represents the text that most nineteenth-century readers actually read. I have collated that edition with the first edition and the Centenary Edition published by Ohio State University Press. There is no evidence that Hawthorne examined proofs for any edition of the novel, and since no manuscript pages in Hawthorne's hand survive, Hawthorne's final intentions for the text are impossible to determine.

Differences among these three editions are minor but worth noting. The Centenary text has "roundabout" in several places, where the first and third editions have "round about" (e.g., pages 19–20, 88, and 150 of this volume); I have kept the two-word form, "round about." The first edition has "heaped up," while the third and Centenary editions have "heaped-up" (p. 23); I have used "heaped-up." The third edition is missing a word in line 36 on p. 30; like the editors of the Centenary Edition, I have inserted the word "form" where it seems called for. The first and third editions have "midday" in line 38 on p. 44; the Centenary Edition has "mid-day." I have kept the word as "midday." The Centenary Edition has "stedfastly" in line 7 on p. 47, in line 18 on p. 142, and in line 49 on p. 145; I have preserved "steadfastly," as in the first and third editions. Hawthorne wrote both "elfish" and "elvish" in *The Scarlet Letter.* The Centenary Edition editors regularized the spelling to "elfish" in every case, but I see no reason to do so and therefore have preserved Hawthorne's different renderings ("elvish" in line 4 on p. 63; "elfish" in line 4 on p. 64; "elvish" in line 3 on p. 96; "elvish" in line 17 on p. 97). In line 20 on p. 66, the first edition has no comma between "occurrences" and "remembered." The third and Centenary editions insert the comma, and I have preserved it. In line 15 on p. 70, the Centenary Edition has "hall-window," while the first and third editions have "hall window," which I have preserved. In line 1 on p. 111, the first edition has "die," while the third and Centenary editions have "dye," which I have preserved. In line 5 on p. 114, the Centenary Edition has an exclamation point after "time," but the first and third editions use a comma; I have kept the comma. The first edition has no comma after "painfully" in line 31 on p. 146; I have preserved the comma, as in the third and Centenary editions. In line 31 on p. 154, the first edition has "sobre-hued," the Centenary Edition has "sombre-hued," while the third edition has "sober-hued," which I have preserved.

For "Mrs. Hutchinson," I have used the original version of the sketch published in the Salem *Gazette*, 7 December 1830. For "Endicott and the Red Cross" and "The Minister's Black Veil," *Twice-Told Tales*, vol. 1 of the Riverside Press edition (Boston: Houghton, Mifflin, 1882). And for "Young Goodman Brown" and "The Birth-mark," *Mosses from an Old Manse*, vol. 2 of the Riverside Press edition (Boston: Houghton, Mifflin, 1882).

The Texts of
THE SCARLET LETTER AND OTHER WRITINGS

THE

SCARLET LETTER,

A ROMANCE.

BY

NATHANIEL HAWTHORNE.

BOSTON:
TICKNOR, REED, AND FIELDS.
M DCCC.L.

Preface to the Second Edition.[1]

Much to the author's surprise, and (if he may say so without additional offence) considerably to his amusement, he finds that his sketch of official life, introductory to THE SCARLET LETTER, has created an unprecedented excitement in the respectable community immediately around him. It could hardly have been more violent, indeed, had he burned down the Custom-House, and quenched its last smoking ember in the blood of a certain venerable personage, against whom he is supposed to cherish a peculiar malevolence. As the public disapprobation would weigh very heavily on him, were he conscious of deserving it, the author begs leave to say, that he has carefully read over the introductory pages, with a purpose to alter or expunge whatever might be found amiss, and to make the best reparation in his power for the atrocities of which he has been adjudged guilty. But it appears to him, that the only remarkable features of the sketch are its frank and genuine good-humor, and the general accuracy with which he has conveyed his sincere impressions of the characters therein described. As to enmity, or ill-feeling of any kind, personal or political, he utterly disclaims such motives. The sketch might, perhaps, have been wholly omitted, without loss to the public or detriment to the book; but, having undertaken to write it, he conceives that it could not have been done in a better or a kindlier spirit, nor, so far as his abilities availed, with a livelier effect of truth.

The author is constrained, therefore, to republish his introductory sketch without the change of a word.

SALEM, *March* 30, 1850.

1. For the second edition of *The Scarlet Letter*, Hawthorne added this brief preface—an unrepentant response to the anger that his original "Custom-House" preface had provoked from many Salem residents.

Contents.

The Custom-House

Introductory to "The Scarlet Letter"

It is a little remarkable, that—though disinclined to talk overmuch of myself and my affairs at the fireside, and to my personal friends—an autobiographical impulse should twice in my life have taken possession of me, in addressing the public. The first time was three or four years since, when I favored the reader—inexcusably, and for no earthly reason, that either the indulgent reader or the intrusive author could imagine—with a description of my way of life in the deep quietude of an Old Manse.[1] And now—because, beyond my deserts, I was happy enough to find a listener or two on the former occasion—I again seize the public by the button, and talk of my three years' experience in a Custom-House. The example of the famous "P. P., Clerk of this Parish," was never more faithfully followed.[2] The truth seems to be, however, that, when he casts his leaves forth upon the wind, the author addresses, not the many who will fling aside his volume, or never take it up, but the few who will understand him, better than most of his schoolmates or lifemates. Some authors, indeed, do far more than this, and indulge themselves in such confidential depths of revelation as could fittingly be addressed, only and exclusively, to the one heart and mind, of perfect sympathy; as if the printed book, thrown at large on the wide world, were certain to find out the divided segment of the writer's own nature, and complete his circle of existence by bringing him into communion with it. It is scarcely decorous, however, to speak all, even where we speak impersonally. But, as thoughts are frozen and utterance benumbed, unless the speaker stand in some true relation with his audience, it may be pardonable to imagine that a friend, a kind and apprehensive, though not the closest friend, is listening to our talk; and then, a native reserve being thawed by this genial consciousness, we may prate of the circumstances that lie around us, and even of ourself, but still keep the inmost Me behind its veil. To this extent, and within these limits, an author, methinks, may be autobiographical, without violating either the reader's rights or his own.

It will be seen, likewise, that this Custom-House sketch has a certain propriety, of a kind always recognized in literature, as explaining how a

1. On their wedding day (July 9, 1842), Nathaniel and Sophia Hawthorne moved into the Old Manse in Concord, Massachusetts. The house had been built by Ralph Waldo Emerson's grandfather, William Emerson. The Hawthornes lived there until the fall of 1845, when they moved in with Hawthorne's mother and sisters in Salem. Hawthorne's second short-story collection, *Mosses from an Old Manse* (1846), included a prefatory essay, "The Old Manse."
2. In 1728, Alexander Pope published "Memoirs of P. P., Clerk of This Parish," a satire on contemporary memoirs. Literary London assumed that Pope meant to burlesque especially *The History of His Own Times* (1724), a posthumous publication by Gilbert Burnett, Lord Bishop of Salisbury.

large portion of the following pages came into my possession, and as offering proofs of the authenticity of a narrative therein contained. This, in fact,—a desire to put myself in my true position as editor, or very little more, of the most prolix among the tales that make up my volume,—this, and no other, is my true reason for assuming a personal relation with the public.[3] In accomplishing the main purpose, it has appeared allowable, by a few extra touches, to give a faint representation of a mode of life not heretofore described, together with some of the characters that move in it, among whom the author happened to make one.

In my native town of Salem, at the head of what, half a century ago, in the days of old King Derby,[4] was a bustling wharf,—but which is now burdened with decayed wooden warehouses, and exhibits few or no symptoms of commercial life; except, perhaps, a bark or brig, halfway down its melancholy length, discharging hides; or, nearer at hand, a Nova Scotia schooner, pitching out her cargo of fire-wood,—at the head, I say, of this dilapidated wharf, which the tide often overflows, and along which, at the base and in the rear of the row of buildings, the track of many languid years is seen in a border of unthrifty grass,—here, with a view from its front windows adown this not very enlivening prospect, and thence across the harbor, stands a spacious edifice of brick. From the loftiest point of its roof, during precisely three and a half hours of each forenoon, floats or droops, in breeze or calm, the banner of the republic; but with the thirteen stripes turned vertically, instead of horizontally, and thus indicating that a civil, and not a military post of Uncle Sam's government, is here established. Its front is ornamented with a portico of half a dozen wooden pillars, supporting a balcony, beneath which a flight of wide granite steps descends towards the street. Over the entrance hovers an enormous specimen of the American eagle, with outspread wings, a shield before her breast, and, if I recollect aright, a bunch of intermingled thunderbolts and barbed arrows in each claw. With the customary infirmity of temper that characterizes this unhappy fowl, she appears, by the fierceness of her beak and eye, and the general truculency of her attitude, to threaten mischief to the inoffensive community; and especially to warn all citizens, careful of their safety, against intruding on the premises which she overshadows with her wings. Nevertheless, vixenly as she looks, many people are seeking, at this very moment, to shelter themselves under the wing of the federal eagle; imagining, I presume, that her bosom has all the softness and snugness of an eiderdown pillow. But she has no great tenderness, even in her best of moods, and, sooner or later,—oftener soon than late,—is apt to fling off her nestlings, with a scratch of her claw, a dab of her beak, or a rankling wound from her barbed arrows.

The pavement round about the above-described edifice—which we may as well name at once as the Custom-House of the port—has grass enough growing in its chinks to show that it has not, of late days, been

3. Hawthorne initially intended *The Scarlet Letter* to be a short story and part of another collection, tentatively entitled *Old Time Legends*. The story grew long enough that his editor, James T. Fields, encouraged him to publish it separately as a novel, but he did not remove several references in "The Custom-House" to the original project.
4. Elias Haskett Derby (1739–1799), Salem merchant and ship owner.

worn by any multitudinous resort of business. In some months of the year, however, there often chances a forenoon when affairs move onward with a livelier tread. Such occasions might remind the elderly citizen of that period, before the last war with England,[5] when Salem was a port by itself; not scorned, as she is now, by her own merchants and ship-owners, who permit her wharves to crumble to ruin, while their ventures go to swell, needlessly and imperceptibly, the mighty flood of commerce at New York or Boston. On some such morning, when three or four vessels happen to have arrived at once,—usually from Africa or South America,—or to be on the verge of their departure thitherward, there is a sound of frequent feet, passing briskly up and down the granite steps. Here, before his own wife has greeted him, you may greet the sea-flushed shipmaster, just in port, with his vessel's papers under his arm, in a tarnished tin box. Here, too, comes his owner, cheerful or somber, gracious or in the sulks, accordingly as his scheme of the now accomplished voyage has been realized in merchandise that will readily be turned to gold, or has buried him under a bulk of incommodities, such as nobody will care to rid him of. Here, likewise,—the germ of the wrinkle-browed, grizzly-bearded, care-worn merchant,—we have the smart young clerk, who gets the taste of traffic as a wolf-cub does of blood, and already sends adventures in his master's ships, when he had better be sailing mimic-boats upon a mill-pond. Another figure in the scene is the outward-bound sailor in quest of a protection; or the recently arrived one, pale and feeble, seeking a passport to the hospital. Nor must we forget the captains of the rusty little schooners that bring fire-wood from the British provinces; a rough-looking set of tarpaulins, without the alertness of the Yankee aspect, but contributing an item of no slight importance to our decaying trade.

Cluster all these individuals together, as they sometimes were, with other miscellaneous ones to diversify the group, and, for the time being, it made the Custom-House a stirring scene. More frequently, however, on ascending the steps, you would discern—in the entry, if it were summer time, or in their appropriate rooms, if wintry or inclement weather—a row of venerable figures, sitting in old-fashioned chairs, which were tipped on their hind legs back against the wall. Oftentimes they were asleep, but occasionally might be heard talking together, in voices between speech and a snore, and with that lack of energy that distinguishes the occupants of alms-houses, and all other human beings who depend for subsistence on charity, on monopolized labor, or anything else but their own independent exertions. These old gentlemen—seated, like Matthew, at the receipt of customs, but not very liable to be summoned thence, like him, for apostolic errands—were Custom-House officers.[6]

Furthermore, on the left hand as you enter the front door, is a certain room or office, about fifteen feet square, and of a lofty height; with two of its arched windows commanding a view of the aforesaid dilapidated wharf, and the third looking across a narrow lane, and along a portion of

5. The War of 1812 (1812–14).
6. See Matthew 9.9: "And as Jesus passed forth from thence, he saw a man, named Matthew, sitting at the receipt of custom: and he saith unto him, Follow me. And he arose, and followed him."

Derby-street. All three give glimpses of the shops of grocers, block-makers, slop-sellers, and ship-chandlers; around the doors of which are generally to be seen, laughing and gossiping, clusters of old salts, and such other wharf-rats as haunt the Wapping[7] of a seaport. The room itself is cobwebbed, and dingy with old paint; its floor is strewn with gray sand, in a fashion that has elsewhere fallen into long disuse; and it is easy to conclude, from the general slovenliness of the place, that this is a sanctuary into which womankind, with her tools of magic, the broom and mop, has very infrequent access. In the way of furniture, there is a stove with a voluminous funnel; an old pine desk, with a three-legged stool beside it; two or three wooden-bottom chairs, exceedingly decrepit and infirm; and—not to forget the library—on some shelves, a score or two of volumes of the Acts of Congress, and a bulky Digest of the Revenue Laws. A tin pipe ascends through the ceiling, and forms a medium of vocal communication with other parts of the edifice. And here, some six months ago,—pacing from corner to corner, or lounging on the long-legged stool, with his elbow on the desk, and his eyes wandering up and down the columns of the morning newspaper,—you might have recognized, honored reader, the same individual who welcomed you into his cheery little study, where the sunshine glimmered so pleasantly through the willow branches, on the western side of the Old Manse. But now, should you go thither to seek him, you would inquire in vain for the Locofoco Surveyor.[8] The besom of reform has swept him out of office; and a worthier successor wears his dignity, and pockets his emoluments.

This old town of Salem—my native place, though I have dwelt much away from it, both in boyhood and maturer years—possesses, or did possess, a hold on my affections, the force of which I have never realized during my seasons of actual residence here. Indeed, so far as its physical aspect is concerned, with its flat, unvaried surface, covered chiefly with wooden houses, few or none of which pretend to architectural beauty,—its irregularity, which is neither picturesque nor quaint, but only tame,—its long and lazy street, lounging wearisomely through the whole extent of the peninsula, with Gallows Hill and New Guinea at one end, and a view of the alms-house at the other,[9]—such being the features of my native town, it would be quite as reasonable to form a sentimental attachment to a disarranged checker-board. And yet, though invariably happiest elsewhere, there is within me a feeling for old Salem, which, in lack of a better phrase, I must be content to call affection. The sentiment is probably assignable to the deep and aged roots which my family has struck into the soil. It is now nearly two centuries and a quarter since the original Briton, the earliest emigrant of my name, made his appearance in the wild and forest-bordered settlement, which has since become a city.[1]

7. A suburb of East London on the river Thames, Wapping provided the main entrance to the London docks. *Slop-sellers*: "Slop" could refer to several things, including inexpensive clothes.
8. A reform-minded faction of the Democratic party in New York, called the Equal Rights party. *Locofoco*: a type of friction match that party members used for illumination when conservative Democrats turned out the lights in Tammany Hall before one of their meetings.
9. Gallows Hill, at the southern edge of Salem, is the area where nineteen "witches" were hanged in the summer of 1692.
1. Hawthorne's paternal great-great-great grandfather, William Hathorne (c. 1606–1681), became a notable public figure in Salem after he settled there in 1636. He served on the Board of Selectmen for many years and fought in King Philip's War. Hawthorne added the "w" to the family name when he was in his twenties.

And here his descendants have been born and died, and have mingled their earthy substance with the soil; until no small portion of it must necessarily be akin to the mortal frame wherewith, for a little while, I walk the streets. In part, therefore, the attachment which I speak of is the mere sensuous sympathy of dust for dust. Few of my countrymen can know what it is; nor, as frequent transplantation is perhaps better for the stock, need they consider it desirable to know.

But the sentiment has likewise its moral quality. The figure of that first ancestor, invested by family tradition with a dim and dusky grandeur, was present to my boyish imagination, as far back as I can remember. It still haunts me, and induces a sort of home-feeling with the past, which I scarcely claim in reference to the present phase of the town. I seem to have a stronger claim to a residence here on account of this grave, bearded, sable-cloaked and steeple-crowned progenitor,—who came so early, with his Bible and his sword, and trode the unworn street with such a stately port, and made so large a figure, as a man of war and peace,—a stronger claim than for myself, whose name is seldom heard and my face hardly known. He was a soldier, legislator, judge; he was a ruler in the Church; he had all the Puritanic traits, both good and evil. He was likewise a bitter persecutor; as witness the Quakers, who have remembered him in their histories, and relate an incident of his hard severity towards a woman of their sect, which will last longer, it is to be feared, than any record of his better deeds, although these were many.[2] His son, too, inherited the persecuting spirit, and made himself so conspicuous in the martyrdom of the witches, that their blood may fairly be said to have left a stain upon him.[3] So deep a stain, indeed, that his old dry bones, in the Charter-street burial-ground, must still retain it, if they have not crumbled utterly to dust! I know not whether these ancestors of mine bethought themselves to repent, and ask pardon of heaven for their cruelties; or whether they are now groaning under the heavy consequences of them, in another state of being. At all events, I, the present writer, as their representative, hereby take shame upon myself for their sakes, and pray that any curse incurred by them—as I have heard, and as the dreary and unprosperous condition of the race, for many a long year back, would argue to exist—may be now and henceforth removed.

Doubtless, however, either of these stern and black-browed Puritans would have thought it quite a sufficient retribution for his sins, that, after so long a lapse of years, the old trunk of the family tree, with so much venerable moss upon it, should have borne, as its top-most bough, an idler like myself. No aim, that I have ever cherished, would they recognize as laudable; no success of mine—if my life, beyond its domestic scope, had ever been brightened by success—would they deem otherwise than worthless, if not positively disgraceful. "What is he?" murmurs one gray shadow of my forefathers to the other. "A writer of story-books! What kind of a business in life,—what mode of glorifying God, or being serviceable to mankind in his day and generation,—may that be? Why, the

2. William Hathorne ordered a Quaker woman, Ann Coleman, to be whipped through the streets of Salem.
3. John Hathorne (1641–1717), William's son and Hawthorne's great-great grandfather, presided at the Salem witch trials.

degenerate fellow might as well have been a fiddler!" Such are the compliments bandied between my great-grandsires and myself, across the gulf of time! And yet, let them scorn me as they will, strong traits of their nature have intertwined themselves with mine.

Planted deep, in the town's earliest infancy and childhood, by these two earnest and energetic men, the race has ever since subsisted here; always, too, in respectability; never, so far as I have known, disgraced by a single unworthy member; but seldom or never, on the other hand, after the first two generations, performing any memorable deed, or so much as putting forward a claim to public notice. Gradually, they have sunk almost out of sight; as old houses, here and there about the streets, get covered half-way to the eaves by the accumulation of new soil. From father to son, for above a hundred years, they followed the sea; a gray-headed shipmaster, in each generation, retiring from the quarter-deck to the homestead, while a boy of fourteen took the hereditary place before the mast, confronting the salt spray and the gale, which had blustered against his sire and grandsire.[4] The boy, also, in due time, passed from the forecastle to the cabin, spent a tempestuous manhood, and returned from his world-wanderings, to grow old, and die, and mingle his dust with the natal earth. This long connection of a family with one spot, as its place of birth and burial, creates a kindred between the human being and the locality, quite independent of any charm in the scenery or moral circumstances that surround him. It is not love, but instinct. The new inhabitant—who came himself from a foreign land, or whose father or grandfather came—has little claim to be called a Salemite; he has no conception of the oyster-like tenacity with which an old settler, over whom his third century is creeping, clings to the spot where his successive generations have been imbedded. It is no matter that the place is joyless for him; that he is weary of the old wooden houses, the mud and dust, the dead level of site and sentiment, the chill east wind, and the chillest of social atmospheres;—all these, and whatever faults besides he may see or imagine, are nothing to the purpose. The spell survives, and just as powerfully as if the natal spot were an earthly paradise. So has it been in my case. I felt it almost as a destiny to make Salem my home; so that the mould of features and cast of character which had all along been familiar here—ever, as one representative of the race lay down in his grave, another assuming, as it were, his sentry-march along the main street—might still in my little day be seen and recognized in the old town. Nevertheless, this very sentiment is an evidence that the connection, which has become an unhealthy one, should at last be severed. Human nature will not flourish, any more than a potato, if it be planted and replanted, for too long a series of generations, in the same worn-out soil. My children have had other birthplaces, and, so far as their fortunes may be within my control, shall strike their roots into unaccustomed earth.

On emerging from the Old Manse, it was chiefly this strange, indolent, unjoyous attachment for my native town, that brought me to fill a place in Uncle Sam's brick edifice, when I might as well, or better, have

4. Hawthorne's father, Nathaniel Hathorne, was a sea captain who died of yellow fever in Suriname in 1808, when Hawthorne was not quite four.

gone somewhere else.[5] My doom was on me. It was not the first time, nor the second, that I had gone away,—as it seemed, permanently,—but yet returned, like the bad half-penny; or as if Salem were for me the inevitable centre of the universe. So, one fine morning, I ascended the flight of granite steps, with the President's commission in my pocket, and was introduced to the corps of gentlemen who were to aid me in my weighty responsibility, as chief executive officer of the Custom-House.

I doubt greatly—or, rather, I do not doubt at all—whether any public functionary of the United States, either in the civil or military line, has ever had such a patriarchal body of veterans under his orders as myself. The whereabouts of the Oldest Inhabitant was at once settled, when I looked at them. For upwards of twenty years before this epoch, the independent position of the Collector had kept the Salem Custom-House out of the whirlpool of political vicissitude, which makes the tenure of office generally so fragile. A soldier,—New England's most distinguished soldier,—he stood firmly on the pedestal of his gallant services; and, himself secure in the wise liberality of the successive administrations through which he had held office, he had been the safety of his subordinates in many an hour of danger and heart-quake. General Miller was radically conservative; a man over whose kindly nature habit had no slight influence; attaching himself strongly to familiar faces, and with difficulty moved to change, even when change might have brought unquestionable improvement.[6] Thus, on taking charge of my department, I found few but aged men. They were ancient sea-captains, for the most part, who, after being tost on every sea, and standing up sturdily against life's tempestuous blast, had finally drifted into this quiet nook; where, with little to disturb them, except the periodical terrors of a Presidential election, they one and all acquired a new lease of existence. Though by no means less liable than their fellow-men to age and infirmity, they had evidently some talisman or other that kept death at bay. Two or three of their number, as I was assured, being gouty and rheumatic, or perhaps bed-ridden, never dreamed of making their appearance at the Custom-House, during a large part of the year; but, after a torpid winter, would creep out into the warm sunshine of May or June, go lazily about what they termed duty, and, at their own leisure and convenience, betake themselves to bed again. I must plead guilty to the charge of abbreviating the official breath of more than one of these venerable servants of the republic. They were allowed, on my representation, to rest from their arduous labors, and soon afterwards—as if their sole principle of life had been zeal for their country's service; as I verily believe it was—withdrew to a better world. It is a pious consolation to me, that, through my interference, a sufficient space was allowed them for repentance of the evil and corrupt practices, into which, as a matter of course, every Custom-House officer must be supposed to fall. Neither the front nor the back entrance of the Custom-House opens on the road to Paradise.

5. Hawthorne was appointed surveyor of the Salem Custom House by President James K. Polk and took office in the spring of 1846. His annual salary was $1,200 (about $36,000 in 2017 dollars).
6. General James F. Miller (1776–1851) fought in the War of 1812, served as the first territorial governor of Arkansas (1819–25), and held the position of collector of the Port of Salem from 1825 until 1849.

The greater part of my officers were Whigs.[7] It was well for their venerable brotherhood that the new Surveyor was not a politician, and though a faithful Democrat in principle, neither received nor held his office with any reference to political services.[8] Had it been otherwise,—had an active politician been put into this influential post, to assume the easy task of making head against a Whig Collector, whose infirmities withheld him from the personal administration of his office,—hardly a man of the old corps would have drawn the breath of official life, within a month after the exterminating angel had come up the Custom-House steps. According to the received code in such matters, it would have been nothing short of duty, in a politician, to bring every one of those white heads under the axe of the guillotine. It was plain enough to discern, that the old fellows dreaded some such discourtesy at my hands. It pained, and at the same time amused me, to behold the terrors that attended my advent; to see a furrowed cheek, weather-beaten by half a century of storm, turn ashy pale at the glance of so harmless an individual as myself; to detect, as one or another addressed me, the tremor of a voice, which, in long-past days, had been wont to bellow through a speaking-trumpet, hoarsely enough to frighten Boreas[9] himself to silence. They knew, these excellent old persons, that, by all established rule,—and, as regarded some of them, weighed by their own lack of efficiency for business,—they ought to have given place to younger men, more orthodox in politics, and altogether fitter than themselves to serve our common Uncle. I knew it too, but could never quite find in my heart to act upon the knowledge. Much and deservedly to my own discredit, therefore, and considerably to the detriment of my official conscience, they continued, during my incumbency, to creep about the wharves, and loiter up and down the Custom-House steps. They spent a good deal of time, also, asleep in their accustomed corners, with their chairs tilted back against the wall; awaking, however, once or twice in a forenoon, to bore one another with the several thousandth repetition of old sea-stories, and mouldy jokes, that had grown to be pass-words and countersigns among them.

The discovery was soon made, I imagine, that the new Surveyor had no great harm in him. So, with lightsome hearts, and the happy consciousness of being usefully employed,—in their own behalf, at least, if not for our beloved country,—these good old gentlemen went through the various formalities of office. Sagaciously, under their spectacles, did they peep into the holds of vessels! Mighty was their fuss about little matters, and marvellous, sometimes, the obtuseness that allowed greater ones to slip between their fingers! Whenever such a mischance occurred,—when a wagon-load of valuable merchandise had been smuggled ashore, at noon-day, perhaps, and directly beneath their unsuspicious noses,—nothing

7. The Democrats and Whigs were the two major political parties in the middle of the nineteenth century until a coalition of Whig, Free-Soil, and abolitionist northern Democrats formed the Republican party in the 1850s. Abraham Lincoln was elected the first Republican president in 1860.
8. Hawthorne had a vested interest in claiming little interest in politics. He was fired from his position as surveyor after Zachary Taylor became president in 1848. He lobbied to retain his position on the grounds that he was apolitical, but local Whigs, led by the Reverend Charles W. Upham, claimed with some truth that he had been more politically active than he acknowledged.
9. God of the north wind in Greek mythology.

could exceed the vigilance and alacrity with which they proceeded to lock, and double-lock, and secure with tape and sealing-wax, all the avenues of the delinquent vessel. Instead of a reprimand for their previous negligence, the case seemed rather to require an eulogium on their praiseworthy caution, after the mischief had happened; a grateful recognition of the promptitude of their zeal, the moment that there was no longer any remedy.

Unless people are more than commonly disagreeable, it is my foolish habit to contract a kindness for them. The better part of my companion's character, if it have a better part, is that which usually comes uppermost in my regard, and forms the type whereby I recognize the man. As most of these old Custom-House officers had good traits, and as my position in reference to them, being paternal and protective, was favorable to the growth of friendly sentiments, I soon grew to like them all. It was pleasant, in the summer forenoons,—when the fervent heat, that almost liquefied the rest of the human family, merely communicated a genial warmth to their half-torpid systems,—it was pleasant to hear them chatting in the back entry, a row of them all tipped against the wall, as usual; while the frozen witticisms of past generations were thawed out, and came bubbling with laughter from their lips. Externally, the jollity of aged men has much in common with the mirth of children; the intellect, any more than a deep sense of humor, has little to do with the matter; it is, with both, a gleam that plays upon the surface, and imparts a sunny and cheery aspect alike to the green branch, and gray, mouldering trunk. In one case, however, it is real sunshine; in the other, it more resembles the phosphorescent glow of decaying wood.

It would be sad injustice, the reader must understand, to represent all my excellent old friends as in their dotage. In the first place, my coadjutors were not invariably old; there were men among them in their strength and prime, of marked ability and energy, and altogether superior to the sluggish and dependent mode of life on which their evil stars had cast them. Then, moreover, the white locks of age were sometimes found to be the thatch of an intellectual tenement in good repair. But, as respects the majority of my corps of veterans, there will be no wrong done, if I characterize them generally as a set of wearisome old souls, who had gathered nothing worth preservation from their varied experience of life. They seemed to have flung away all the golden grain of practical wisdom, which they had enjoyed so many opportunities of harvesting, and most carefully to have stored their memories with the husks. They spoke with far more interest and unction of their morning's breakfast, or yesterday's, to-day's, or to-morrow's dinner, than of the shipwreck of forty or fifty years ago, and all the world's wonders which they had witnessed with their youthful eyes.

The father of the Custom-House—the patriarch, not only of this little squad of officials, but, I am bold to say, of the respectable body of tide-waiters all over the United States—was a certain permanent Inspector.[1] He might truly be termed a legitimate son of the revenue system, dyed in

1. William Lee served as inspector. According to Margaret Moore, Lee's daughters never forgave Hawthorne for his "trivializing" sketch, and Hawthorne later regretted having written disparagingly about him. See Moore, 182.

the wool, or, rather, born in the purple; since his sire, a Revolutionary colonel, and formerly collector of the port, had created an office for him, and appointed him to fill it, at a period of the early ages which few living men can now remember. This Inspector, when I first knew him, was a man of four-score years, or thereabouts, and certainly one of the most wonderful specimens of winter-green that you would be likely to discover in a lifetime's search. With his florid cheek, his compact figure, smartly arrayed in a bright-buttoned blue coat, his brisk and vigorous step, and his hale and hearty aspect, altogether he seemed—not young, indeed—but a kind of new contrivance of Mother Nature in the shape of man, whom age and infirmity had no business to touch. His voice and laugh, which perpetually reëchoed through the Custom-House, had nothing of the tremulous quaver and cackle of an old man's utterance; they came strutting out of his lungs, like the crow of a cock, or the blast of a clarion. Looking at him merely as an animal,—and there was very little else to look at,—he was a most satisfactory object, from the thorough healthfulness and wholesomeness of his system, and his capacity, at that extreme age, to enjoy all, or nearly all, the delights which he had ever aimed at, or conceived of. The careless security of his life in the Custom-House, on a regular income, and with but slight and infrequent apprehensions of removal, had no doubt contributed to make time pass lightly over him. The original and more potent causes, however, lay in the rare perfection of his animal nature, the moderate proportion of intellect, and the very trifling admixture of moral and spiritual ingredients; these latter qualities, indeed, being in barely enough measure to keep the old gentleman from walking on all-fours. He possessed no power of thought, no depth of feeling, no troublesome sensibilities; nothing, in short, but a few commonplace instincts, which, aided by the cheerful temper that grew inevitably out of his physical well-being, did duty very respectably, and to general acceptance, in lieu of a heart. He had been the husband of three wives, all long since dead; the father of twenty children, most of whom, at every age of childhood or maturity, had likewise returned to dust. Here, one would suppose, might have been sorrow enough to imbue the sunniest disposition, through and through, with a sable tinge. Not so with our old Inspector! One brief sigh sufficed to carry off the entire burden of these dismal reminiscences. The next moment, he was as ready for sport as any unbreeched infant; far readier than the Collector's junior clerk, who, at nineteen years, was much the elder and graver man of the two.

I used to watch and study this patriarchal personage with, I think, livelier curiosity, than any other form of humanity there presented to my notice. He was, in truth, a rare phenomenon; so perfect, in one point of view; so shallow, so delusive, so impalpable, such an absolute nonentity, in every other. My conclusion was that he had no soul, no heart, no mind; nothing, as I have already said, but instincts; and yet, withal, so cunningly had the few materials of his character been put together, that there was no painful perception of deficiency, but, on my part, an entire contentment with what I found in him. It might be difficult—and it was so—to conceive how he should exist hereafter, so earthly and sensuous did he seem; but surely his existence here, admitting that it was to terminate with his last breath, had been not unkindly given; with no higher moral responsibilities than the beasts of the field, but with a larger scope of enjoyment than theirs, and with all their blessed immunity from the dreariness and duskiness of age.

One point, in which he had vastly the advantage over his four-footed brethren, was his ability to recollect the good dinners which it had made no small portion of the happiness of his life to eat. His gourmandism was a highly agreeable trait; and to hear him talk of roast-meat was as appetizing as a pickle or an oyster. As he possessed no higher attribute, and neither sacrificed nor vitiated any spiritual endowment by devoting all his energies and ingenuities to subserve the delight and profit of his maw, it always pleased and satisfied me to hear him expatiate on fish, poultry, and butcher's meat, and the most eligible methods of preparing them for the table. His reminiscences of good cheer, however ancient the date of the actual banquet, seemed to bring the savor of pig or turkey under one's very nostrils. There were flavors on his palate, that had lingered there not less than sixty or seventy years, and were still apparently as fresh as that of the mutton-chop which he had just devoured for his breakfast. I have heard him smack his lips over dinners, every guest at which, except himself, had long been food for worms. It was marvellous to observe how the ghosts of bygone meals were continually rising up before him; not in anger or retribution, but as if grateful for his former appreciation, and seeking to repudiate an endless series of enjoyment, at once shadowy and sensual. A tender-loin of beef, a hind-quarter of veal, a spare-rib of pork, a particular chicken, or a remarkably praiseworthy turkey, which had perhaps adorned his board in the days of the elder Adams, would be remembered;[2] while all the subsequent experience of our race, and all the events that brightened or darkened his individual career, had gone over him with as little permanent effect as the passing breeze. The chief tragic event of the old man's life, so far as I could judge, was his mishap with a certain goose, which lived and died some twenty or forty years ago; a goose of most promising figure, but which, at table, proved so inveterately tough that the carving-knife would make no impression on its carcass, and it could only be divided with an axe and handsaw.

But it is time to quit this sketch; on which, however, I should be glad to dwell at considerably more length, because, of all men whom I have ever known, this individual was fittest to be a Custom-House officer. Most persons, owing to causes which I may not have space to hint at, suffer moral detriment from this peculiar mode of life. The old Inspector was incapable of it; and, were he to continue in office to the end of time, would be just as good as he was then, and sit down to dinner with just as good an appetite.

There is one likeness, without which my gallery of Custom-House portraits would be strangely incomplete; but which my comparatively few opportunities for observation enable me to sketch only in the merest outline. It is that of the Collector, our gallant old General, who, after his brilliant military service, subsequently to which he had ruled over a wild Western territory, had come hither, twenty years before, to spend the decline of his varied and honorable life. The brave soldier had already numbered, nearly or quite, his threescore years and ten, and was pursuing the remainder of his earthly march, burdened with infirmities which even the

2. John Adams (1735–1826), second president (1797–1801) of the United States.

martial music of his own spirit-stirring recollections could do little towards lightening. The step was palsied now, that had been foremost in the charge. It was only with the assistance of a servant, and by leaning his hand heavily on the iron balustrade, that he could slowly and painfully ascend the Custom-House steps, and, with a toilsome progress across the floor, attain his customary chair beside the fireplace. There he used to sit, gazing with a somewhat dim serenity of aspect at the figures that came and went; amid the rustle of papers, the administering of oaths, the discussion of business, and the casual talk of the office; all which sounds and circumstances seemed but indistinctly to impress his senses, and hardly to make their way into his inner sphere of contemplation. His countenance, in this repose, was mild and kindly. If his notice was sought, an expression of courtesy and interest gleamed out upon his features; proving that there was light within him, and that it was only the outward medium of the intellectual lamp that obstructed the rays in their passage. The closer you penetrated to the substance of his mind, the sounder it appeared. When no longer called upon to speak, or listen, either of which operations cost him an evident effort, his face would briefly subside into its former not uncheerful quietude. It was not painful to behold this look; for, though dim, it had not the imbecility of decaying age. The framework of his nature, originally strong and massive, was not yet crumbled into ruin.

To observe and define his character, however, under such disadvantages, was as difficult a task as to trace out and build up anew, in imagination, an old fortress, like Ticonderoga, from a view of its gray and broken ruins.[3] Here and there, perchance, the walls may remain almost complete; but elsewhere may be only a shapeless mound, cumbrous with its very strength, and overgrown, through long years of peace and neglect, with grass and alien weeds.

Nevertheless, looking at the old warrior with affection,—for, slight as was the communication between us, my feeling towards him, like that of all bipeds and quadrupeds who knew him, might not improperly be termed so,—I could discern the main points of his portrait. It was marked with the noble and heroic qualities which showed it to be not by a mere accident, but of good right, that he had won a distinguished name. His spirit could never, I conceive, have been characterized by an uneasy activity; it must, at any period of his life, have required an impulse to set him in motion; but, once stirred up, with obstacles to overcome, and an adequate object to be attained, it was not in the man to give out or fail. The heat that had formerly pervaded his nature, and which was not yet extinct, was never of the kind that flashes and flickers in a blaze; but, rather, a deep, red glow, as of iron in a furnace. Weight, solidity, firmness; this was the expression of his repose, even in such decay as had crept untimely over him, at the period of which I speak. But I could imagine, even then, that, under some excitement which should go deeply into his consciousness,—roused by a trumpet-peal, loud enough to awaken all of his energies that were not dead, but only slumbering,—he was yet capable of flinging off his infirmities like a sick man's gown, dropping

3. One of the first major battles of the Revolutionary War occurred on May 10, 1775, when Ethan Allen led his Green Mountain Boys in a successful attack on British troops at Fort Ticonderoga, in what is now upstate New York.

the staff of age to seize a battle-sword, and starting up once more a warrior. And, in so intense a moment, his demeanor would have still been calm. Such an exhibition, however, was but to be pictured in fancy; not to be anticipated, nor desired. What I saw in him—as evidently as the indestructible ramparts of Old Ticonderoga, already cited as the most appropriate simile—were the features of stubborn and ponderous endurance, which might well have amounted to obstinacy in his earlier days; of integrity, that, like most of his other endowments, lay in a somewhat heavy mass, and was just as unmalleable and unmanageable as a ton of iron ore; and of benevolence, which, fiercely as he led the bayonets on at Chippewa or Fort Erie, I take to be of quite as genuine a stamp as what actuates any or all the polemical philanthropists of the age.[4] He had slain men with his own hand, for aught I know;—certainly, they had fallen, like blades of grass at the sweep of the scythe, before the charge to which his spirit imparted its triumphant energy;—but, be that as it might, there was never in his heart so much cruelty as would have brushed the down off a butterfly's wing. I have not known the man, to whose innate kindliness I would more confidently make an appeal.

Many characteristics—and those, too, which contribute not the least forcibly to impart resemblance in a sketch—must have vanished, or been obscured, before I met the General. All merely graceful attributes are usually the most evanescent; nor does Nature adorn the human ruin with blossoms of new beauty, that have their roots and proper nutriment only in the chinks and crevices of decay, as she sows wall-flowers over the ruined fortress of Ticonderoga. Still, even in respect of grace and beauty, there were points well worth noting. A ray of humor, now and then, would make its way through the veil of dim obstruction, and glimmer pleasantly upon our faces. A trait of native elegance, seldom seen in the masculine character after childhood or early youth, was shown in the General's fondness for the sight and fragrance of flowers. An old soldier might be supposed to prize only the bloody laurel on his brow; but here was one, who seemed to have a young girl's appreciation of the floral tribe.

There, beside the fireplace, the brave old General used to sit; while the Surveyor—though seldom, when it could be avoided, taking upon himself the difficult task of engaging him in conversation—was fond of standing at a distance, and watching his quiet and almost slumberous countenance. He seemed away from us, although we saw him but a few yards off; remote, though we passed close beside his chair; unattainable, though we might have stretched forth our hands and touched his own. It might be that he lived a more real life within his thoughts, than amid the unappropriate environment of the Collector's office. The evolutions of the parade; the tumult of the battle; the flourish of old, heroic music, heard thirty years before;—such scenes and sounds, perhaps, were all alive before his intellectual sense. Meanwhile, the merchants and ship-masters, the spruce clerks and uncouth sailors, entered and departed; the bustle of this commercial and Custom-House life kept up its little murmur round

4. In the War of 1812, American troops won important battles at Fort Erie and at Chippewa (near Niagara Falls) in the summer of 1814. General Miller fought at the Battle of Chippewa (July 5, 1814), the day after Hawthorne's tenth birthday.

about him; and neither with the men nor their affairs did the General appear to sustain the most distant relation. He was as much out of place as an old sword—now rusty, but which had flashed once in the battle's front, and showed still a bright gleam along its blade—would have been, among the inkstands, paper-folders, and mahogany rulers, on the Deputy Collector's desk.

There was one thing that much aided me in renewing and recreating the stalwart soldier of the Niagara frontier,—the man of true and simple energy. It was the recollection of those memorable words of his,—"I'll try, Sir!"—spoken on the very verge of a desperate and heroic enterprise, and breathing the soul and spirit of New England hardihood, comprehending all perils, and encountering all.[5] If, in our country, valor were rewarded by heraldic honor, this phrase—which it seems so easy to speak, but which only he, with such a task of danger and glory before him, has ever spoken—would be the best and fit test of all mottoes for the General's shield of arms.

It contributes greatly towards a man's moral and intellectual health, to be brought into habits of companionship with individuals unlike himself, who care little for his pursuits, and whose sphere and abilities he must go out of himself to appreciate. The accidents of my life have often afforded me this advantage, but never with more fullness and variety than during my continuance in office. There was one man, especially, the observation of whose character gave me a new idea of talent.[6] His gifts were emphatically those of a man of business; prompt, acute, clear-minded; with an eye that saw through all perplexities, and a faculty of arrangement that made them vanish, as by the waving of an enchanter's wand. Bred up from boyhood in the Custom-House, it was his proper field of activity; and the many intricacies of business, so harassing to the interloper, presented themselves before him with the regularity of a perfectly comprehended system. In my contemplation, he stood as the ideal of his class. He was, indeed, the Custom-House in himself; or, at all events, the main spring that kept its variously revolving wheels in motion; for, in an institution like this, where its officers are appointed to subserve their own profit and convenience, and seldom with a leading reference to their fitness for the duty to be performed, they must perforce seek elsewhere the dexterity which is not in them. Thus, by an inevitable necessity, as a magnet attracts steel-filings, so did our man of business draw to himself the difficulties which everybody met with. With an easy condescension, and kind forbearance towards our stupidity,—which, to his order of mind, must have seemed little short of crime,—would he forthwith, by the merest touch of his finger, make the incomprehensible as clear as daylight. The merchants valued him not less than we, his esoteric friends. His integrity was perfect; it was a law of nature with him, rather than a choice or a principle; nor can it be otherwise than the main condition of an intellect so remarkably clear and accurate as his, to be honest and regular in the administration of affairs. A

5. When General Winfield Scott commanded General Miller to take a British battery at Lundy's Lane on July 25, 1814, he reportedly replied, "I'll try, Sir!"

6. Zachariah Burchmore (1809–1884), secretary of the Democratic party in Salem, served as Custom House clerk and was one of Hawthorne's allies in his effort to retain his job. Burchmore was fired from his position shortly after Hawthorne's own dismissal. See Nissenbaum, 73–75.

stain on his conscience, as to anything that came within the range of his vocation, would trouble such a man very much in the same way, though to a far greater degree, than an error in the balance of an account, or an ink-blot on the fair page of a book of record. Here, in a word,—and it is a rare instance in my life,—I had met with a person thoroughly adapted to the situation which he held.

Such were some of the people with whom I now found myself connected. I took it in good part, at the hands of Providence, that I was thrown into a position so little akin to my past habits; and set myself seriously to gather from it whatever profit was to be had. After my fellowship of toil and impracticable schemes with the dreamy brethren of Brook Farm;[7] after living for three years within the subtile influence of an intellect like Emerson's; after those wild, free days on the Assabeth, indulging fantastic speculations, beside our fire of fallen boughs, with Ellery Channing; after talking with Thoreau about pine-trees and Indian relics, in his hermitage at Walden; after growing fastidious by sympathy with the classic refinement of Hillard's culture; after becoming imbued with poetic sentiment at Longfellow's hearth-stone;—it was time, at length, that I should exercise other faculties of my nature, and nourish myself with food for which I had hitherto had little appetite. Even the old Inspector was desirable, as a change of diet, to a man who had known Alcott.[8] I looked upon it as an evidence, in some measure, of a system naturally well balanced, and lacking no essential part of a thorough organization, that, with such associates to remember, I could mingle at once with men of altogether different qualities, and never murmur at the change.

Literature, its exertions and objects, were now of little moment in my regard. I cared not, at this period, for books; they were apart from me. Nature,—except it were human nature,—the nature that is developed in earth and sky, was, in one sense, hidden from me; and all the imaginative delight, wherewith it had been spiritualized, passed away out of my mind. A gift, a faculty, if it had not departed, was suspended and inanimate within me. There would have been something sad, unutterably dreary, in all this, had I not been conscious that it lay at my own option to recall whatever was valuable in the past. It might be true, indeed, that this was a life which could not, with impunity, be lived too long; else, it might make me permanently other than I had been, without transforming me into any shape which it would be worth my while to take. But I never considered it as other than a transitory life. There was always a prophetic instinct, a low whisper in my ear, that, within no long period, and whenever a new change of custom should be essential to my good, a change would come.

7. George Ripley (1802–1880) founded the utopian community Brook Farm in 1841 at a site near West Roxbury, Massachusetts, just west of Boston. Hawthorne invested $1,000 (the price of two shares) in the experiment and lived at the farm for several months (April–October) in 1841, but he quickly grew disenchanted. His experiences at Brook Farm form the pretext for *The Blithedale Romance* (1852).
8. Hawthorne recalls his three years at the Old Manse in Concord. He lived about a mile from Ralph Waldo Emerson (1803–1882), enjoyed the company of William Ellery Channing (1818–1901) and Henry David Thoreau (1817–1862), whose Walden cabin he visited shortly before leaving Concord in 1845. Boston attorney George Hillard (1808–1879) became a good friend. Hawthorne knew poet Henry Wadsworth Longfellow (1807–1882) from their days as students at Bowdoin College. Bronson Alcott (1799–1888), father of Louisa May Alcott, was a prominent, if eccentric, Concord intellectual.

Meanwhile, there I was, a Surveyor of the Revenue, and, so far as I have been able to understand, as good a Surveyor as need be. A man of thought, fancy, and sensibility, (had he ten times the Surveyor's proportion of those qualities,) may, at any time, be a man of affairs, if he will only choose to give himself the trouble. My fellow-officers, and the merchants and sea-captains with whom my official duties brought me into any manner of connection, viewed me in no other light, and probably knew me in no other character. None of them, I presume, had ever read a page of my inditing, or would have cared a fig the more for me, if they had read them all; nor would it have mended the matter, in the least, had those same unprofitable pages been written with a pen like that of Burns or of Chaucer, each of whom was a Custom-House officer in his day, as well as I.[9] It is a good lesson—though it may often be a hard one—for a man who has dreamed of literary fame, and of making for himself a rank among the world's dignitaries by such means, to step aside out of the narrow circle in which his claims are recognized, and to find how utterly devoid of significance, beyond that circle, is all that he achieves, and all he aims at. I know not that I especially needed the lesson, either in the way of warning or rebuke; but, at any rate, I learned it thoroughly; nor, it gives me pleasure to reflect, did the truth, as it came home to my perception, ever cost me a pang, or require to be thrown off in a sigh. In the way of literary talk, it is true, the Naval Officer—an excellent fellow, who came into office with me and went out only a little later—would often engage me in a discussion about one or the other of his favorite topics, Napoleon or Shakspeare. The Collector's junior clerk, too,—a young gentleman who, it was whispered, occasionally covered a sheet of Uncle Sam's letter-paper with what (at the distance of a few yards) looked very much like poetry,—used now and then to speak to me of books, as matters with which I might possibly be conversant. This was my all of lettered intercourse; and it was quite sufficient for my necessities.

No longer seeking nor caring that my name should be blazoned abroad on title-pages, I smiled to think that it had now another kind of vogue. The Custom-House marker imprinted it, with a stencil and black paint, on pepper-bags, and baskets of anatto, and cigar-boxes, and bales of all kinds of dutiable merchandise, in testimony that these commodities had paid the impost, and gone regularly through the office.[1] Borne on such queer vehicle of fame, a knowledge of my existence, so far as a name conveys it, was carried where it had never been before, and, I hope, will never go again.

But the past was not dead. Once in a great while, the thoughts, that had seemed so vital and so active, yet had been put to rest so quietly, revived again. One of the most remarkable occasions, when the habit of bygone days awoke in me, was that which brings it within the law of literary propriety to offer the public the sketch which I am now writing.

In the second story of the Custom-House, there is a large room, in which the brick-work and naked rafters have never been covered with paneling and plaster. The edifice—originally projected on a scale adapted

9. Poets Robert Burns and Geoffrey Chaucer had held jobs similar to Hawthorne's at the Custom House.

1. Hawthorne refers to the seal that labeled each box, "Salem / N Hawthorne / Sur[iname] / 1847." *Anatto*: a small evergreen tree whose seeds are used to produce an orange-red dye.

to the old commercial enterprise of the port, and with an idea of subsequent prosperity destined never to be realized—contains far more space than its occupants know what to do with. This airy hall, therefore, over the Collector's apartments, remains unfinished to this day, and, in spite of the aged cobwebs that festoon its dusky beams, appears still to await the labor of the carpenter and mason. At one end of the room, in a recess, were a number of barrels, piled one upon another, containing bundles of official documents. Large quantities of similar rubbish lay lumbering the floor. It was sorrowful to think how many days, and weeks, and months, and years of toil, had been wasted on these musty papers, which were now only an encumbrance on earth, and were hidden away in this forgotten corner, never more to be glanced at by human eyes. But, then, what reams of other manuscripts—filled not with the dulness of official formalities, but with the thought of inventive brains and the rich effusion of deep hearts—had gone equally to oblivion; and that, moreover, without serving a purpose in their day, as these heaped-up papers had, and—saddest of all—without purchasing for their writers the comfortable livelihood which the clerks of the Custom-House had gained by these worthless scratchings of the pen! Yet not altogether worthless, perhaps, as materials of local history. Here, no doubt, statistics of the former commerce of Salem might be discovered, and memorials of her princely merchants,—old King Derby,—old Billy Gray,—old Simon Forrester,—and many another magnate in his day; whose powdered head, however, was scarcely in the tomb, before his mountain-pile of wealth began to dwindle.[2] The founders of the greater part of the families which now compose the aristocracy of Salem might here be traced, from the petty and obscure beginnings of their traffic, at periods generally much posterior to the Revolution, upward to what their children look upon as long-established rank.

Prior to the Revolution, there is a dearth of records; the earlier documents and archives of the Custom-House having, probably, been carried off to Halifax, when all the King's officials accompanied the British army in its flight from Boston. It has often been a matter of regret with me; for, going back, perhaps, to the days of the Protectorate,[3] those papers must have contained many references to forgotten or remembered men, and to antique customs, which would have affected me with the same pleasure as when I used to pick up Indian arrowheads in the field near the Old Manse.

But, one idle and rainy day, it was my fortune to make a discovery of some little interest. Poking and burrowing into the heaped-up rubbish in the corner; unfolding one and another document, and reading the names of vessels that had long ago foundered at sea or rotted at the wharves, and those of merchants, never heard of now on 'Change,[4] nor very readily decipherable on their mossy tomb-stones; glancing at such matters with

2. Simon Forrester (1748–1817), a wealthy Salem merchant and ship owner, and brother-in-law to Hawthorne's father. William Gray (1750–1825), a very wealthy Massachusetts merchant who began his career in Salem before settling in Boston and being elected lieutenant governor of Massachusetts.
3. Name given to the English government under Oliver Cromwell that formed in the aftermath of the English Civil War (1642–49) and the execution of King Charles I. The Protectorate ended in 1660 with the restoration of the monarchy and the ascension of Charles II to the throne.
4. Merchant's Exchange in Boston.

the saddened, weary, half-reluctant interest which we bestow on the corpse of dead activity,—and exerting my fancy, sluggish with little use, to raise up from these dry bones an image of the old town's brighter aspect, when India was a new region, and only Salem knew the way thither,—I chanced to lay my hand on a small package, carefully done up in a piece of ancient yellow parchment. This envelope had the air of an official record of some period long past, when clerks engrossed their stiff and formal chirography on more substantial materials than at present. There was something about it that quickened an instinctive curiosity, and made me undo the faded red tape, that tied up the package, with the sense that a treasure would here be brought to light. Unbending the rigid folds of the parchment cover, I found it to be a commission, under the hand and seal of Governor Shirley, in favor of one Jonathan Pue, as Surveyor of his Majesty's Customs for the port of Salem, in the Province of Massachusetts Bay.[5] I remembered to have read (probably in Felt's Annals) a notice of the decease of Mr. Surveyor Pue, about fourscore years ago; and likewise, in a newspaper of recent times, an account of the digging up of his remains in the little grave-yard of St. Peter's Church, during the renewal of that edifice. Nothing, if I rightly call to mind, was left of my respected predecessor, save an imperfect skeleton, and some fragments of apparel, and a wig of majestic frizzle; which, unlike the head that it once adorned, was in very satisfactory preservation. But, on examining the papers which the parchment commission served to envelop, I found more traces of Mr. Pue's mental part, and the internal operations of his head, than the frizzled wig had contained of the venerable skull itself.

They were documents, in short, not official, but of a private nature, or, at least, written in his private capacity, and apparently with his own hand. I could account for their being included in the heap of Custom-House lumber only by the fact, that Mr. Pue's death had happened suddenly; and that these papers, which he probably kept in his official desk, had never come to the knowledge of his heirs, or were supposed to relate to the business of the revenue. On the transfer of the archives to Halifax, this package, proving to be of no public concern, was left behind, and had remained ever since unopened.

The ancient Surveyor—being little molested, I suppose, at that early day, with business pertaining to his office—seems to have devoted some of his many leisure hours to researches as a local antiquarian, and other inquisitions of a similar nature. These supplied material for petty activity to a mind that would otherwise have been eaten up with rust. A portion of his facts, by the by, did me good service in the preparation of the article entitled "MAIN STREET," included in the present volume.[6] The remainder may perhaps be applied to purposes equally valuable, hereafter; or not impossibly may be worked up, so far as they go, into a regular history of Salem, should my veneration for the natal soil ever impel me to so pious a task. Meanwhile, they shall be at the command of any gentleman,

5. William Shirley (1694–1771) was royal governor of Massachusetts from 1741 to 1756. According to Joseph B. Felt's *Annals of Salem,* Jonathan Pue was appointed surveyor of Salem in 1752.
6. Hawthorne originally intended "Main Street," a lengthy sketch about Salem, to form part of *Old Time Legends.* He ended up publishing the sketch in *Aesthetic Papers* (1849), a collection edited by Sophia's sister, Elizabeth Palmer Peabody.

inclined, and competent, to take the unprofitable labor off my hands. As a final disposition, I contemplate depositing them with the Essex Historical Society.

But the object that most drew my attention, in the mysterious package, was a certain affair of fine red cloth, much worn and faded. There were traces about it of gold embroidery, which, however, was greatly frayed and defaced; so that none, or very little, of the glitter was left.[7] It had been wrought, as was easy to perceive, with wonderful skill of needlework; and the stitch (as I am assured by ladies conversant with such mysteries) gives evidence of a now forgotten art, not to be recovered even by the process of picking out the threads. This rag of scarlet cloth,—for time, and wear, and a sacrilegious moth, had reduced it to little other than a rag,—on careful examination, assumed the shape of a letter. It was the capital letter A. By an accurate measurement, each limb proved to be precisely three inches and a quarter in length. It had been intended, there could be no doubt, as an ornamental article of dress; but how it was to be worn, or what rank, honor, and dignity, in by-past times, were signified by it, was a riddle which (so evanescent are the fashions of the world in these particulars) I saw little hope of solving. And yet it strangely interested me. My eyes fastened themselves upon the old scarlet letter, and would not be turned aside. Certainly, there was some deep meaning in it, most worthy of interpretation, and which, as it were, streamed forth from the mystic symbol, subtly communicating itself to my sensibilities, but evading the analysis of my mind.

While thus perplexed,—and cogitating, among other hypotheses, whether the letter might not have been one of those decorations which the white men used to contrive, in order to take the eyes of Indians,—I happened to place it on my breast. It seemed to me,—the reader may smile, but must not doubt my word,—it seemed to me, then, that I experienced a sensation not altogether physical, yet almost so, as of burning heat; and as if the letter were not of red cloth, but red-hot iron. I shuddered, and involuntarily let it fall upon the floor.

In the absorbing contemplation of the scarlet letter, I had hitherto neglected to examine a small roll of dingy paper, around which it had been twisted. This I now opened, and had the satisfaction to find, recorded by the old Surveyor's pen, a reasonably complete explanation of the whole affair. There were several foolscap sheets, containing many particulars respecting the life and conversation of one Hester Prynne, who appeared to have been rather a noteworthy personage in the view of our ancestors. She had flourished during the period between the early days of Massachusetts and the close of the seventeenth century.[8] Aged persons, alive in the time of Mr. Surveyor Pue, and from whose oral testimony he had made up his narrative, remembered her, in their youth, as a very old, but not decrepit woman, of a stately and solemn aspect. It

7. Hawthorne may be thinking of his own early story, "Endicott and the Red Cross" (1837), which includes a woman wearing a scarlet letter that she has embroidered with gold thread. See p. 165 of this Norton Critical Edition.

8. Approximately from 1640 to 1690. Hester Prynne must have arrived in Boston in 1640 since she has been in residence for two years when the novel opens in 1642. Roughly speaking, her life encompassed the period between 1620, the year the Pilgrims emigrated to Plymouth, and the Salem witch trials in 1692.

had been her habit, from an almost immemorial date, to go about the country as a kind of voluntary nurse, and doing whatever miscellaneous good she might; taking upon herself, likewise, to give advice in all matters, especially those of the heart; by which means, as a person of such propensities inevitably must, she gained from many people the reverence due to an angel, but, I should imagine, was looked upon by others as an intruder and a nuisance. Prying further into the manuscript, I found the record of other doings and sufferings of this singular woman, for most of which the reader is referred to the story entitled "THE SCARLET LETTER"; and it should be borne carefully in mind, that the main facts of that story are authorized and authenticated by the document of Mr. Surveyor Pue. The original papers, together with the scarlet letter itself,—a most curious relic,—are still in my possession, and shall be freely exhibited to whomsoever, induced by the great interest of the narrative, may desire a sight of them.[9] I must not be understood as affirming, that, in the dressing up of the tale, and imagining the motives and modes of passion that influenced the characters who figure in it, I have invariably confined myself within the limits of the old Surveyor's half a dozen sheets of foolscap. On the contrary, I have allowed myself, as to such points, nearly or altogether as much license as if the facts had been entirely of my own invention. What I contend for is the authenticity of the outline.

This incident recalled my mind, in some degree, to its old track. There seemed to be here the ground-work of a tale. It impressed me as if the ancient Surveyor, in his garb of a hundred years gone by, and wearing his immortal wig,—which was buried with him, but did not perish in the grave,—had met me in the deserted chamber of the Custom-House. In his port was the dignity of one who had borne his Majesty's commission, and who was therefore illuminated by a ray of the splendor that shone so dazzlingly about the throne. How unlike, alas! the hang-dog look of a republican official, who, as the servant of the people, feels himself less than the least, and below the lowest, of his masters. With his own ghostly hand, the obscurely seen but majestic figure had imparted to me the scarlet symbol, and the little roll of explanatory manuscript. With his own ghostly voice, he had exhorted me, on the sacred consideration of my filial duty and reverence towards him,—who might reasonably regard himself as my official ancestor,—to bring his mouldy and moth-eaten lucubrations before the public. "Do this," said the ghost of Mr. Surveyor Pue, emphatically nodding the head that looked so imposing within its memorable wig, "do this, and the profit shall be all your own! You will shortly need it; for it is not in your days as it was in mine, when a man's office was a life-lease, and oftentimes an heirloom. But, I charge you, in this matter of old Mistress Prynne, give to your predecessor's memory the credit which will be rightfully due!" And I said to the ghost of Mr. Surveyor Pue,—"I will!"

On Hester Prynne's story, therefore, I bestowed much thought. It was the subject of my meditations for many an hour, while pacing to and fro across my room, or traversing, with a hundred-fold repetition, the long extent from the front-door of the Custom-House to the side-entrance, and back again. Great were the weariness and annoyance of the old

9. There is no evidence that the scarlet letter or these papers ever existed outside Hawthorne's imagination.

Inspector and the Weighers and Gaugers, whose slumbers were disturbed by the unmercifully lengthened tramp of my passing and returning footsteps. Remembering their own former habits, they used to say that the Surveyor was walking the quarter-deck. They probably fancied that my sole object—and, indeed, the sole object for which a sane man could ever put himself into voluntary motion—was, to get an appetite for dinner. And to say the truth, an appetite, sharpened by the east wind that generally blew along the passage, was the only valuable result of so much indefatigable exercise. So little adapted is the atmosphere of a Custom-House to the delicate harvest of fancy and sensibility, that, had I remained there through ten Presidencies yet to come, I doubt whether the tale of "The Scarlet Letter" would ever have been brought before the public eye. My imagination was a tarnished mirror. It would not reflect, or only with miserable dimness, the figures with which I did my best to people it. The characters of the narrative would not be warmed and rendered malleable by any heat that I could kindle at my intellectual forge. They would take neither the glow of passion nor the tenderness of sentiment, but retained all the rigidity of dead corpses, and stared me in the face with a fixed and ghastly grin of contemptuous defiance. "What have you to do with us?" that expression seemed to say. "The little power you might once have possessed over the tribe of unrealities is gone! You have bartered it for a pittance of the public gold. Go, then, and earn your wages!" In short, the almost torpid creatures of my own fancy twitted me with imbecility, and not without fair occasion.

It was not merely during the three hours and a half which Uncle Sam claimed as his share of my daily life, that this wretched numbness held possession of me. It went with me on my sea-shore walks, and rambles into the country, whenever—which was seldom and reluctantly—I bestirred myself to seek that invigorating charm of Nature, which used to give me such freshness and activity of thought, the moment that I stepped across the threshold of the Old Manse. The same torpor, as regarded the capacity for intellectual effort, accompanied me home, and weighed upon me in the chamber which I most absurdly termed my study. Nor did it quit me, when, late at night, I sat in the deserted parlor, lighted only by the glimmering coal-fire and the moon, striving to picture forth imaginary scenes, which, the next day, might flow out on the brightening page in many-hued description.

If the imaginative faculty refused to act at such an hour, it might well be deemed a hopeless case. Moonlight, in a familiar room, falling so white upon the carpet, and showing all its figures so distinctly,—making every object so minutely visible, yet so unlike a morning or noontide visibility,—is a medium the most suitable for a romance-writer to get acquainted with his illusive guests.[1] There is the little domestic scenery of the well-known apartment; the chairs, with each its separate individuality; the centre-table, sustaining a work-basket, a volume or two, and an extinguished lamp; the sofa; the book-case; the picture on the wall;—all these details, so completely seen, are so spiritualized by the unusual light, that they seem to lose their actual substance, and become things of intellect.

1. Compare the passage from Hawthorne's notebook on pp. 204–05 of this Norton Critical Edition.

Nothing is too small or too trifling to undergo this change, and acquire dignity thereby. A child's shoe; the doll, seated in her little wicker carriage; the hobby-horse;—whatever, in a word, has been used or played with, during the day, is now invested with a quality of strangeness and remoteness, though still almost as vividly present as by daylight. Thus, therefore, the floor of our familiar room has become a neutral territory, somewhere between the real world and fairy-land, where the Actual and the Imaginary may meet, and each imbue itself with the nature of the other. Ghosts might enter here, without affrighting us. It would be too much in keeping with the scene to excite surprise, were we to look about us and discover a form, beloved, but gone hence, now sitting quietly in a streak of this magic moonshine, with an aspect that would make us doubt whether it had returned from afar, or had never once stirred from our fireside.

The somewhat dim coal-fire has an essential influence in producing the effect which I would describe. It throws its unobtrusive tinge throughout the room, with a faint ruddiness upon the walls and ceiling, and a reflected gleam from the polish of the furniture. This warmer light mingles itself with the cold spirituality of the moon-beams, and communicates, as it were, a heart and sensibilities of human tenderness to the forms which fancy summons up. It converts them from snow-images into men and women. Glancing at the looking-glass, we behold—deep within its haunted verge—the smouldering glow of the half-extinguished anthracite, the white moonbeams on the floor, and a repetition of all the gleam and shadow of the picture, with one remove further from the actual, and nearer to the imaginative. Then, at such an hour, and with this scene before him, if a man, sitting all alone, cannot dream strange things, and make them look like truth, he need never try to write romances.

But, for myself, during the whole of my Custom-House experience, moonlight and sunshine, and the glow of fire-light, were just alike in my regard; and neither of them was of one whit more avail than the twinkle of a tallow-candle. An entire class of susceptibilities, and a gift connected with them,—of no great richness or value, but the best I had,—was gone from me.

It is my belief, however, that, had I attempted a different order of composition, my faculties would not have been found so pointless and inefficacious. I might, for instance, have contented myself with writing out the narratives of a veteran shipmaster, one of the Inspectors, whom I should be most ungrateful not to mention, since scarcely a day passed that he did not stir me to laughter and admiration by his marvelous gifts as a story-teller. Could I have preserved the picturesque force of his style, and the humorous coloring which nature taught him how to throw over his descriptions, the result, I honestly believe, would have been something new in literature. Or I might readily have found a more serious task. It was a folly, with the materiality of this daily life pressing so intrusively upon me, to attempt to fling myself back into another age; or to insist on creating the semblance of a world out of airy matter, when, at every moment, the impalpable beauty of my soap-bubble was broken by the rude contact of some actual circumstance. The wiser effort would have been, to diffuse thought and imagination through the opaque substance of to-day, and thus to make it a bright transparency; to spiritualize the burden that began to weigh so heavily; to seek, resolutely, the true and indestructible value that lay hidden in the petty and wearisome incidents, and ordinary characters, with which I was now conversant. The fault was mine. The page of life that was spread out before me seemed dull and commonplace, only because I had not fathomed its deeper import. A better book than I shall ever write was there; leaf after leaf presenting itself to me, just as it was written out by the reality of the flitting hour, and vanishing as fast as written, only because my brain wanted the insight and my hand the cunning to transcribe it. At some future day, it may be, I shall remember a few scattered fragments and broken paragraphs, and write them down, and find the letters turn to gold upon the page.

These perceptions have come too late. At the instant, I was only conscious that what would have been a pleasure once was now a hopeless toil. There was no occasion to make much moan about this state of affairs. I had ceased to be a writer of tolerably poor tales and essays, and had become a tolerably good Surveyor of the Customs. That was all. But, nevertheless; it is anything but agreeable to be haunted by a suspicion that one's intellect is dwindling away; or exhaling, without your consciousness, like ether out of a phial; so that, at every glance, you find a smaller and less volatile residuum. Of the fact, there could be no doubt; and, examining myself and others, I was led to conclusions, in reference to the effect of public office on the character, not very favorable to the mode of life in question. In some other form, perhaps, I may hereafter develop these effects. Suffice it here to say, that a Custom-House officer, of long continuance, can hardly be a very praiseworthy or respectable personage, for many reasons; one of them, the tenure by which he holds his situation, and another, the very nature of his business, which—though, I trust, an honest one—is of such a sort that he does not share in the united effort of mankind.

An effect—which I believe to be observable, more or less, in every individual who has occupied the position—is, that, while he leans on the mighty arm of the Republic, his own proper strength departs from him. He loses, in an extent proportioned to the weakness or force of his original nature, the capability of self-support. If he possess an unusual share of native energy, or the enervating magic of place do not operate too long upon him, his forfeited powers may be redeemable. The ejected officer—fortunate in the unkindly shove that sends him forth betimes, to struggle amid a struggling world—may return to himself, and become all that he has ever been. But this seldom happens. He usually keeps his ground just long enough for his own ruin, and is then thrust out, with sinews all unstrung, to totter along the difficult footpath of life as he best may. Conscious of his own infirmity,—that his tempered steel and elasticity are lost,—he forever afterwards looks wistfully about him in quest of support external to himself. His pervading and continual hope—a hallucination, which, in the face of all discouragement, and making light of impossibilities, haunts him while he lives, and, I fancy, like the convulsive throes of the cholera, torments him for a brief space after death—is, that finally, and in no long time, by some happy coincidence of circumstances, he shall be restored to office. This faith, more than anything else, steals the pith and availability out of whatever enterprise he may dream of undertaking. Why should he toil and moil, and be at so much trouble to pick himself up out of the mud, when, in a little while hence, the strong arm of his Uncle will raise and support him? Why should he work for his living here, or go to dig gold in California, when he is so soon to be made happy, at monthly intervals, with a little pile of glittering coin out of his Uncle's pocket?[2] It is sadly curious to observe how slight a taste of office suffices to infect a poor fellow with this singular disease. Uncle Sam's gold—meaning no disrespect to the worthy old gentleman—has, in this respect, a quality of enchantment like that of the Devil's wages. Whoever touches it should look well to himself, or he may find the bargain to go hard against him, involving, if not his soul, yet many of its better attributes; its sturdy force, its courage and constancy, its truth, its self-reliance, and all that gives the emphasis to manly character.

Here was a fine prospect in the distance! Not that the Surveyor brought the lesson home to himself, or admitted that he could be so utterly undone, either by continuance in office, or ejectment. Yet my reflections were not the most comfortable. I began to grow melancholy and restless; continually prying into my mind, to discover which of its poor properties were gone, and what degree of detriment had already accrued to the remainder. I endeavored to calculate how much longer I could stay in the Custom-House, and yet go forth a man. To confess the truth, it was my greatest apprehension,—as it would never be a measure of policy to turn out so quiet an individual as myself, and it being hardly in the nature of a public officer to resign,—it was my chief trouble, therefore, that I was likely to grow gray and decrepit in the Surveyorship, and become much such another animal as the old Inspector. Might it not, in the tedious lapse of official life that lay before me, finally be with me as it was with this venerable friend,—to make the dinner-hour the nucleus of the day, and to spend the rest of it, as an old dog spends it, asleep in the sunshine or in the shade? A dreary look-forward this, for a man who felt it to be the best definition of happiness to live throughout the whole range of his faculties and sensibilities! But, all this while, I was giving myself very unnecessary alarm. Providence had meditated better things for me than I could possibly imagine for myself.

A remarkable event of the third year of my Surveyorship—to adopt the tone of "P. P."—was the election of General Taylor to the Presidency.[3] It is essential, in order to form a complete estimate of the advantages of official life, to view the incumbent at the in-coming of a hostile administration. His position is then one of the most singularly irksome, and, in every contingency, disagreeable, that a wretched mortal can possibly occupy; with seldom an alternative of good, on either hand, although what presents itself to him as the worst event may very probably be the best. But it is a strange experience, to a man of pride and sensibility, to know that his interests are within the control of individuals who neither love nor understand him, and by whom, since one or the other must needs happen, he would rather be injured than obliged. Strange, too, for one who has kept his calmness throughout the contest, to observe the bloodthirstiness that is developed in the hour of triumph, and to be conscious that he is himself

2. Gold was discovered at Sutter's Mill in California in January 1848, and the first wave of gold seekers arrived in 1849.
3. General Zachary Taylor, a Whig, was elected president in 1848 and fired Hawthorne from his surveyor job in the early summer of 1849.

among its objects! There are few uglier traits of human nature than this tendency—which I now witnessed in men no worse than their neighbors—to grow cruel, merely because they possessed the power of inflicting harm. If the guillotine, as applied to office-holders, were a literal fact, instead of one of the most apt of metaphors, it is my sincere belief, that the active members of the victorious party were sufficiently excited to have chopped off all our heads, and have thanked Heaven for the opportunity![4] It appears to me—who have been a calm and curious observer, as well in victory as defeat—that this fierce and bitter spirit of malice and revenge has never distinguished the many triumphs of my own party as it now did that of the Whigs. The Democrats take the offices, as a general rule, because they need them, and because the practice of many years has made it the law of political warfare, which, unless a different system be proclaimed, it were weakness and cowardice to murmur at. But the long habit of victory has made them generous. They know how to spare, when they see occasion; and when they strike, the axe may be sharp, indeed, but its edge is seldom poisoned with ill-will; nor is it their custom ignominiously to kick the head which they have just struck off.

In short, unpleasant as was my predicament, at best, I saw much reason to congratulate myself that I was on the losing side, rather than the triumphant one. If, heretofore, I had been none of the warmest of partisans, I began now, at this season of peril and adversity, to be pretty acutely sensible with which party my predilections lay; nor was it without something like regret and shame, that, according to a reasonable calculation of chances, I saw my own prospect of retaining office to be better than those of my Democratic brethren. But who can see an inch into futurity, beyond his nose? My own head was the first that fell!

The moment when a man's head drops off is seldom or never, I am inclined to think, precisely the most agreeable of his life. Nevertheless, like the greater part of our misfortunes, even so serious a contingency brings its remedy and consolation with it, if the sufferer will but make the best, rather than the worst, of the accident which has befallen him. In my particular case, the consolatory topics were close at hand, and, indeed, had suggested themselves to my meditations a considerable time before it was requisite to use them. In view of my previous weariness of office, and vague thoughts of resignation, my fortune somewhat resembled that of a person who should entertain an idea of committing suicide, and, although beyond his hopes, meet with the good hap to be murdered. In the Custom-House, as before in the Old Manse, I had spent three years; a term long enough to rest a weary brain; long enough to break off old intellectual habits, and make room for new ones; long enough, and too long, to have lived in an unnatural state, doing what was really of no advantage nor delight to any human being, and withholding myself from toil that would, at least, have stilled an unquiet impulse in me. Then, moreover, as regarded his unceremonious ejectment, the late Surveyor was not altogether ill-pleased to be recognized by the Whigs as an enemy; since his

4. As Larry J. Reynolds notes in an essay reprinted on pp. 484–99 of this Norton Critical Edition, Zachary Taylor's political appointments in 1849 were reported in Democratic papers as beheadings of Democratic party members. The guillotine, infamous for its use during the French Revolution, decapitated its victims.

inactivity in political affairs,—his tendency to roam, at will, in that broad and quiet field where all mankind may meet, rather than confine himself to those narrow paths where brethren of the same household must diverge from one another,—had sometimes made it questionable with his brother Democrats whether he was a friend. Now, after he had won the crown of martyrdom, (though with no longer a head to wear it on,) the point might be looked upon as settled. Finally, little heroic as he was, it seemed more decorous to be overthrown in the downfall of the party with which he had been content to stand, than to remain a forlorn survivor, when so many worthier men were falling; and, at last, after subsisting for four years on the mercy of a hostile administration, to be compelled then to define his position anew, and claim the yet more humiliating mercy of a friendly one.

Meanwhile the press had taken up my affair, and kept me, for a week or two, careering through the public prints, in my decapitated state, like Irving's Headless Horseman; ghastly and grim, and longing to be buried, as a politically dead man ought.[5] So much for my figurative self. The real human being, all this time, with his head safely on his shoulders, had brought himself to the comfortable conclusion that everything was for the best; and, making an investment in ink, paper, and steel-pens, had opened his long-disused writing-desk, and was again a literary man.

Now it was, that the lucubrations of my ancient predecessor, Mr. Surveyor Pue, came into play. Rusty through long idleness, some little space was requisite before my intellectual machinery could be brought to work upon the tale, with an effect in any degree satisfactory. Even yet, though my thoughts were ultimately much absorbed in the task, it wears, to my eye, a stern and sombre aspect; too much ungladdened by genial sunshine; too little relieved by the tender and familiar influences which soften almost every scene of nature and real life, and, undoubtedly, should soften every picture of them. This uncaptivating effect is perhaps due to the period of hardly accomplished revolution, and still seething turmoil, in which the story shaped itself. It is no indication, however, of a lack of cheerfulness in the writer's mind; for he was happier, while straying through the gloom of these sunless fantasies, than at any time since he had quitted the Old Manse. Some of the briefer articles, which contribute to make up the volume, have likewise been written since my involuntary withdrawal from the toils and honors of public life, and the remainder are gleaned from annuals and magazines, of such antique date that they have gone round the circle, and come back to novelty again.* Keeping up the metaphor of the political guillotine, the whole may be considered as the POSTHUMOUS PAPERS OF A DECAPITATED SURVEYOR; and the sketch which I am now bringing to a close, if too autobiographical for a modest person to publish in his lifetime, will readily be excused in a gentleman who writes from beyond the grave. Peace be with all the world! My blessing on my friends! My forgiveness to my enemies! For I am in the realm of quiet!

The life of the Custom-House lies like a dream behind me. The old Inspector,—who, by the by, I regret to say, was overthrown and killed by

5. A reference to Washington Irving's "The Legend of Sleepy Hollow" (1820), which includes the Headless Horseman (in keeping with Hawthorne's decapitation theme).

* At the time of writing this article, the author intended to publish, along with "The Scarlet Letter," several shorter tales and sketches. These it has been thought advisable to defer [*Hawthorne's note*].

a horse, some time ago; else he would certainly have lived forever,—he, and all those other venerable personages who sat with him at the receipt of custom, are but shadows in my view; white-headed and wrinkled images, which my fancy used to sport with, and has now flung aside forever. The merchants,—Pingree, Phillips, Shepard, Upton, Kimball, Bertram, Hunt,—these, and many other names, which had such a classic familiarity for my ear six months ago,—these men of traffic, who seemed to occupy so important a position in the world,—how little time has it required to disconnect me from them all, not merely in act, but recollection! It is with an effort that I recall the figures and appellations of these few. Soon, likewise, my old native town will loom upon me through the haze of memory, a mist brooding over and around it; as if it were no portion of the real earth, but an overgrown village in cloud-land, with only imaginary inhabitants to people its wooden houses, and walk its homely lanes, and the unpicturesque prolixity of its main street. Henceforth, it ceases to be a reality of my life. I am a citizen of somewhere else. My good townspeople will not much regret me; for—though it has been as dear an object as any, in my literary efforts, to be of some importance in their eyes, and to win myself a pleasant memory in this abode and burial-place of so many of my forefathers—there has never been, for me, the genial atmosphere which a literary man requires, in order to ripen the best harvest of his mind. I shall do better amongst other faces; and these familiar ones, it need hardly be said, will do just as well without me.

It may be, however,—O, transporting and triumphant thought!—that that the great-grandchildren of the present race may sometimes think kindly of the scribbler of bygone days, when the antiquary of days to come, among the sites memorable in the town's history, shall point out the locality of THE TOWN PUMP![6]

6. Hawthorne had published another Salem sketch, "A Rill from the Town Pump," in 1835.

1) 1st person narrator

- Self insert - Hauthrawe
- Historical

- Self promoting and depretating

2) intro theme of supernaturals

outsider

3) Outsider

4) corrupt Government

5) HOTHORN HATES PURITAINS?!!

↳ pro outsider

The Scarlet Letter

I. The Prison-Door

A throng of bearded men, in sad-colored garments, and gray, steeple-crowned hats, intermixed with women, some wearing hoods, and others bareheaded, was assembled in front of a wooden edifice, the door of which was heavily timbered with oak, and studded with iron spikes.

The founders of a new colony, whatever Utopia of human virtue and happiness they might originally project, have invariably recognized it among their earliest practical necessities to allot a portion of the virgin soil as a cemetery, and another portion as the site of a prison. In accordance with this rule, it may safely be assumed that the forefathers of Boston had built the first prison-house somewhere in the vicinity of Cornhill, almost as seasonably as they marked out the first burial-ground, on Isaac Johnson's lot, and round about his grave, which subsequently became the nucleus of all the congregated sepulchres in the old church-yard of King's Chapel.[1] Certain it is, that, some fifteen or twenty years after the settlement of the town, the wooden jail was already marked with weather-stains and other indications of age, which gave a yet darker aspect to its beetle-browed and gloomy front.[2] The rust on the ponderous iron-work of its oaken door looked more antique than anything else in the New World. Like all that pertains to crime, it seemed never to have known a youthful era. Before this ugly edifice, and between it and the wheel-track of the street, was a grass-plot, much overgrown with burdock, pig-weed, apple-peru, and such unsightly vegetation, which evidently found something congenial in the soil that had so early borne the black flower of civilized society, a prison. But, on one side of the portal, and rooted almost at the threshold, was a wild rose-bush, covered, in this month of June, with its delicate gems, which might be imagined to offer their fragrance and fragile beauty to the prisoner as he went in, and to the condemned criminal as he came forth to his doom, in token that the deep heart of Nature could pity and be kind to him.

This rose-bush, by a strange chance, has been kept alive in history; but whether it had merely survived out of the stern old wilderness, so long

1. Caleb Snow calls Isaac Johnson the "father of Boston" because he was instrumental in encouraging John Winthrop's party to settle on the south side of the Charles River. According to Snow, Johnson died in 1630 and asked to be buried at the southwest corner of his property, the site of the old courthouse. This is the origin of the first burying ground to which Hawthorne refers.
2. See Charles Ryskamp's essay in this Norton Critical Edition (p. 282). Hawthorne sets the novel between 1642 and 1649. Governor John Winthrop, whose death occurs in chapter 12, died on March 26, 1649. Richard Bellingham was governor of the Massachusetts Bay Colony in 1642, but he had in fact just been replaced by Winthrop at the time Hester stands on the scaffold.

after the fall of the gigantic pines and oaks that originally over-shadowed it,—or whether, as there is fair authority for believing, it had sprung up under the footsteps of the sainted Ann Hutchinson, as she entered the prison-door,—we shall not take upon us to determine.[3] Finding it so directly on the threshold of our narrative, which is now about to issue from that inauspicious portal, we could hardly do otherwise than pluck one of its flowers, and present it to the reader. It may serve, let us hope, to symbolize some sweet moral blossom, that may be found along the track, or relieve the darkening close of a tale of human frailty and sorrow.

II. The Market-Place

The grass-plot before the jail, in Prison-lane, on a certain summer morning, not less than two centuries ago, was occupied by a pretty large number of the inhabitants of Boston; all with their eyes intently fastened on the iron-clamped oaken door. Amongst any other population, or at a later period in the history of New England, the grim rigidity that petrified the bearded physiognomies of these good people would have augured some awful business in hand. It could have betokened nothing short of the anticipated execution of some noted culprit, on whom the sentence of a legal tribunal had but confirmed the verdict of public sentiment. But, in that early severity of the Puritan character, an inference of this kind could not so indubitably be drawn. It might be that a sluggish bond-servant, or an undutiful child, whom his parents had given over to the civil authority, was to be corrected at the whipping-post. It might be, that an Antinomian,[1] a Quaker, or other heterodox religionist, was to be scourged out of the town, or an idle and vagrant Indian, whom the white man's fire-water had made riotous about the streets, was to be driven with stripes into the shadow of the forest. It might be, too, that a witch, like old Mistress Hibbins, the bitter-tempered widow of the magistrate, was to die upon the gallows.[2] In either case, there was very much the same solemnity of demeanor on the part of the spectators; as befitted a people amongst whom religion and law were almost identical, and in whose character both were so thoroughly interfused, that the mildest and the severest acts of public discipline were alike made venerable and awful. Meagre, indeed, and cold, was the sympathy that a transgressor might look for, from such bystanders, at the scaffold. On the other hand, a penalty which, in our days, would infer a degree of mocking infamy and ridicule, might then be invested with almost as stern a dignity as the punishment of death itself.

It was a circumstance to be noted, on the summer morning when our story begins its course, that the women, of whom there were several in the crowd, appeared to take a peculiar interest in whatever penal infliction might be expected to ensue. The age had not so much refinement, that

3. Anne Hutchinson (1591–1643) was banished from Massachusetts (to Rhode Island) in 1638 for unlawful preaching and, in Governor John Winthrop's words, for "being a woman not fit for our society." She had been hosting prayer meetings for women and had reproached most of the ministers, including Reverend John Wilson, for preaching a covenant of works rather than a covenant of grace—behavior, in Winthrop's terms, not "fitting for your sex."
1. One who believes that faith alone can merit salvation and, conversely, that neither good nor evil works affect salvation.
2. Ann Hibbins, sister of Governor Richard Bellingham, was executed as a witch in 1656.

any sense of impropriety restrained the wearers of petticoat and farthingale from stepping forth into the public ways, and wedging their not unsubstantial persons, if occasion were, into the throng nearest to the scaffold at an execution. Morally, as well as materially, there was a coarser fibre in those wives and maidens of old English birth and breeding, than in their fair descendants, separated from them by a series of six or seven generations; for, throughout that chain of ancestry, every successive mother has transmitted to her child a fainter bloom, a more delicate and briefer beauty, and a slighter physical frame, if not a character of less force and solidity, than her own. The women who were now standing about the prison-door stood within less than half a century of the period when the man-like Elizabeth had been the not altogether unsuitable representative of the sex.[3] They were her countrywomen; and the beef and ale of their native land, with a moral diet not a whit more refined, entered largely into their composition. The bright morning sun, therefore, shone on broad shoulders and well-developed busts, and on round and ruddy cheeks, that had ripened in the far-off island, and had hardly yet grown paler or thinner in the atmosphere of New England. There was, moreover, a boldness and rotundity of speech among these matrons, as most of them seemed to be, that would startle us at the present day, whether in respect to its purport or its volume of tone.

"Goodwives," said a hard-featured dame of fifty, "I'll tell ye a piece of my mind. It would be greatly for the public behoof, if we women, being of mature age and church-members in good repute, should have the handling of such malefactresses as this Hester Prynne.[4] What think ye, gossips? If the hussy stood up for judgment before us five, that are now here in a knot together, would she come off with such a sentence as the worshipful magistrates have awarded? Marry, I trow not!"[5]

"People say," said another, "that the Reverend Master Dimmesdale, her godly pastor, takes it very grievously to heart that such a scandal should have come upon his congregation."

"The magistrates are God-fearing gentlemen, but merciful over-much,—that is a truth," added a third-autumnal matron. "At the very least, they should have put the brand of a hot iron on Hester Prynne's forehead. Madam Hester would have winced at that, I warrant me. But she,—the naughty baggage,—little will she care what they put upon the bodice of her gown!

dichotomy

3. Queen Elizabeth I ruled England from 1558 to 1603. She was succeeded by James I, who ruled from 1603 to 1625. Charles I was king of England during most of the period in which Hawthorne sets the novel, although he was beheaded on January 30, 1649, six months before the main action of the novel comes to a close.
4. Hawthorne probably took the name "Prynne" from William Prynne (1600–1669), a vehement anti-Catholic Puritan, who devoted a lengthy book, *Histriomastix; A Scourge of Stage Players* (1632), to castigating Englishmen (and implicity King Charles) for attending plays. When he published diatribes against Archbishop of Canterbury William Laud, whom he considered a Catholic in disguise, he was punished by having his ears cut off and the letters "SL" (for "Seditious Libeller") burnt into his cheeks. Hawthorne undoubtedly appreciated the coincidence of writing a story about a similar punishment that bore the same initials (see Isani and Alfred S. Reid in Selected Bibliography). Uncannily anticipating Hester's alteration of the scarlet letter and its meaning, William Prynne responded to his branding by composing a Latin distich (a pair of verse lines, or couplet), "in which he interpreted the S L which he now bore indelibly on his cheeks as *Stigmata Laudis*, the Scars of Laud" (Samuel R. Gardiner, *History of England from the Accession of James I to the Outbreak of the Civil War, 1603–1642*, 10 vols. [New York: Longmans, Green, 1909], 8:232). Isani notes the feud between Prynne and Archbishop Laud, one of whose protégés was William Chillingworth.
5. Think or believe not.

Why, look you, she may cover it with a brooch, or such like heathenish adornment, and so walk the streets as brave as ever!"

"Ah, but," interposed, more softly, a young wife, holding a child by the hand, "let her cover the mark as she will, the pang of it will be always in her heart."

"What do we talk of marks and brands, whether on the bodice of her gown, or the flesh of her forehead?" cried another female, the ugliest as well as the most pitiless of these self-constituted judges. "This woman has brought shame upon us all, and ought to die.[6] Is there not law for it? Truly there is, both in the Scripture and the statute-book. Then let the magistrates, who have made it of no effect, thank themselves if their own wives and daughters go astray!"[7]

"Mercy on us, goodwife," exclaimed a man in the crowd, "is there no virtue in woman, save what springs from a wholesome fear of the gallows? That is the hardest word yet! Hush, now, gossips! for the lock is turning in the prison door, and here comes Mistress Prynne herself."

The door of the jail being flung open from within, there appeared, in the first place, like a black shadow emerging into sunshine, the grim and grisly presence of the town-beadle, with a sword by his side, and his staff of office in his hand. This personage prefigured and represented in his aspect the whole dismal severity of the Puritanic code of law, which it was his business to administer in its final and closest application to the offender. Stretching forth the official staff in his left hand, he laid his right upon the shoulder of a young woman, whom he thus drew forward; until, on the threshold of the prison-door, she repelled him, by an action marked with natural dignity and force of character, and stepped into the open air, as if by her own free will. She bore in her arms a child, a baby of some three months old, who winked and turned aside its little face from the too vivid light of day; because its existence, heretofore, had brought it acquainted only with the gray twilight of a dungeon, or other darksome apartment of the prison.

When the young woman—the mother of this child—stood fully revealed before the crowd, it seemed to be her first impulse to clasp the infant closely to her bosom; not so much by an impulse of motherly affection, as that she might thereby conceal a certain token, which was wrought or fastened into her dress. In a moment, however, wisely judging that one token of her shame would but poorly serve to hide another, she took the baby on

6. John Winthrop notes that Mary Latham of Plymouth Colony and James Britton were condemned to die for adultery in March 1644. Winthrop explains that Mary Latham had been rejected by a young man she loved, vowed to marry the "next that came to her," and ended up "matched with an ancient man" for whom she had no affection (*Journal*, 500–501). Charles Boewe and Murray G. Murphy (see Bibliography) also note the case of Salem's Hester Craford, who in 1688 was ordered to be "severely whipped" for fornicating with John Wedg. The judgment, which was carried out by William Hathorne, was suspended for a month or so because of the birth of their child.

7. Magistrates were elected officials of the Massachusetts Bay Colony, whose government was organized, by the Royal Charter, under a governor, a deputy governor, and a group of assistants (elected from among the freemen, who were church members and stockholders of the company). Although they were elected, the magistrates formed an exclusive group of the wealthiest and highest-born settlers. Shortly before the novel opens, as Robert Emmet Wall explains, the colonists had approved a "Body of Liberties" that gave slightly more power to the whole body of freemen and even nonfreemen. During the years of the novel's action (1642–49) the magistrates faced more democratic challenges to their authority. Hawthorne's repeated references to the magistrates help to generalize the colonists' challenges to government authority beyond Hester's individual case (*Massachusetts Bay: The Crucial Decade*, 1640–1650 [New Haven: Yale UP, 1972], 18).

her arm, and, with a burning blush, and yet a haughty smile, and a glance that would not be abashed, looked around at her townspeople and neighbors. On the breast of her gown, in fine red cloth, surrounded with an elaborate embroidery and fantastic flourishes of gold thread, appeared the letter A. It was so artistically done, and with so much fertility and gorgeous luxuriance of fancy, that it had all the effect of a last and fitting decoration to the apparel which she wore; and which was of a splendor in accordance with the taste of the age, but greatly beyond what was allowed by the sumptuary regulations of the colony.[8]

The young woman was tall, with a figure of perfect elegance on a large scale. She had dark and abundant hair, so glossy that it threw off the sunshine with a gleam, and a face which, besides being beautiful from regularity of feature and richness of complexion, had the impressiveness belonging to a marked brow and deep black eyes. She was lady-like, too, after the manner of the feminine gentility of those days; characterized by a certain state and dignity, rather than by the delicate, evanescent, and indescribable grace, which is now recognized as its indication. And never had Hester Prynne appeared more lady-like, in the antique interpretation of the term, than as she issued from the prison. Those who had before known her, and had expected to behold her dimmed and obscured by a disastrous cloud, were astonished, and even startled, to perceive how her beauty shone out, and made a halo of the misfortune and ignominy in which she was enveloped. It may be true, that, to a sensitive observer, there was something exquisitely painful in it. Her attire, which, indeed, she had wrought for the occasion, in prison, and had modeled much after her own fancy, seemed to express the attitude of her spirit, the desperate recklessness of her mood, by its wild and picturesque peculiarity. But the point which drew all eyes, and, as it were, transfigured the wearer,—so that both men and women, who had been familiarly acquainted with Hester Prynne, were now impressed as if they beheld her for the first time,—was that SCARLET LETTER, so fantastically embroidered and illuminated upon her bosom. It had the effect of a spell, taking her out of the ordinary relations with humanity, and enclosing her in a sphere by herself.

"She hath good skill at her needle, that's certain," remarked one of her female spectators; "but did ever a woman, before this brazen hussy, contrive such a way of showing it! Why, gossips, what is it but to laugh in the faces of our godly magistrates, and make a pride out of what they, worthy gentlemen, meant for a punishment?"

"It were well," muttered the most iron-visaged of the old dames, "if we stripped Madam Hester's rich gown off her dainty shoulders; and as for the red letter, which she hath stitched so curiously, I'll bestow a rag of mine own rheumatic flannel, to make a fitter one!"

"O, peace, neighbors, peace!" whispered their youngest companion; "do not let her hear you! Not a stitch in that embroidered letter, but she has felt it in her heart."

The grim beadle now made a gesture with his staff.

8. In his *History of Boston,* Caleb Snow mentions the Puritans' regulation of fashion in the observation that the Reverend John Cotton "found it necessary to exert his influence to suppress superfluous and unnecessarily expensive fashions. . . . Gold or silver laces, girdles, or hatbands, embroidered caps, immoderate great veils and immoderate great sleeves incurred special disapprobation" (55).

"Make way, good people, make way, in the King's name!" cried he. "Open a passage; and, I promise ye, Mistress Prynne shall be set where man, woman and child, may have a fair sight of her brave apparel, from this time till an hour past meridian. A blessing on the righteous Colony of the Massachusetts, where iniquity is dragged out into the sunshine! Come along, Madam Hester, and show your scarlet letter in the market-place!"

A lane was forthwith opened through the crowd of spectators. Preceded by the beadle, and attended by an irregular procession of stern-browed men and unkindly visaged women, Hester Prynne set forth towards the place appointed for her punishment. A crowd of eager and curious schoolboys, understanding little of the matter in hand, except that it gave them a half-holiday, ran before her progress, turning their heads continually to stare into her face, and at the winking baby in her arms, and at the ignominious letter on her breast. It was no great distance, in those days, from the prison-door to the market-place. Measured by the prisoner's experience, however, it might be reckoned a journey of some length; for, haughty as her demeanor was, she perchance underwent an agony from every footstep of those that thronged to see her, as if her heart had been flung into the street for them all to spurn and trample upon. In our nature, however, there is a provision, alike marvellous and merciful, that the sufferer should never know the intensity of what he endures by its present torture, but chiefly by the pang that rankles after it. With almost a serene deportment, therefore, Hester Prynne passed through this portion of her ordeal, and came to a sort of scaffold, at the western extremity of the market-place. It stood nearly beneath the eaves of Boston's earliest church, and appeared to be a fixture there.

In fact, this scaffold constituted a portion of a penal machine, which now, for two or three generations past, has been merely historical and traditionary among us, but was held, in the old time, to be as effectual an agent, in the promotion of good citizenship, as ever was the guillotine among the terrorists of France. It was, in short, the platform of the pillory; and above it rose the framework of that instrument of discipline, so fashioned as to confine the human head in its tight grasp, and thus hold it up to the public gaze. The very ideal of ignominy was embodied and made manifest in this contrivance of wood and iron. There can be no outrage, methinks, against our common nature,—whatever be the delinquencies of the individual,—no outrage more flagrant than to forbid the culprit to hide his face for shame; as it was the essence of this punishment to do. In Hester Prynne's instance, however, as not unfrequently in other cases, her sentence bore, that she should stand a certain time upon the platform, but without undergoing that gripe about the neck and confinement of the head, the proneness to which was the most devilish characteristic of this ugly engine. Knowing well her part, she ascended a flight of wooden steps, and was thus displayed to the surrounding multitude, at about the height of a man's shoulders above the street.

Had there been a Papist among the crowd of Puritans, he might have seen in this beautiful woman, so picturesque in her attire and mien, and with the infant at her bosom, an object to remind him of the image of Divine Maternity, which so many illustrious painters have vied with one another to represent; something which should remind him, indeed, but only by contrast, of that sacred image of sinless motherhood, whose infant was to redeem the

the time

Mother Mary Divine God

world. Here, there was the taint of deepest sin in the most sacred quality of human life, working such effect, that the world was only the darker for this woman's beauty, and the more lost for the infant that she had borne.

The scene was not without a mixture of awe, such as must always invest the spectacle of guilt and shame in a fellow-creature, before society shall have grown corrupt enough to smile, instead of shuddering, at it. The witnesses of Hester Prynne's disgrace had not yet passed beyond their simplicity. They were stern enough to look upon her death, had that been the sentence, without a murmur at its severity, but had none of the heartlessness of another social state, which would find only a theme for jest in an exhibition like the present. Even had there been a disposition to turn the matter into ridicule, it must have been repressed and overpowered by the solemn presence of men no less dignified than the Governor, and several of his counsellors, a judge, a general, and the ministers of the town; all of whom sat or stood in a balcony of the meeting-house, looking down upon the platform. When such personages could constitute a part of the spectacle, without risking the majesty or reverence of rank and office, it was safely to be inferred that the infliction of a legal sentence would have an earnest and effectual meaning. Accordingly, the crowd was sombre and grave. The unhappy culprit sustained herself as best a woman might, under the heavy weight of a thousand unrelenting eyes, all fastened upon her, and concentred at her bosom. It was almost intolerable to be borne. Of an impulsive and passionate nature, she had fortified herself to encounter the stings and venomous stabs of public contumely, wreaking itself in every variety of insult; but there was a quality so much more terrible in the solemn mood of the popular mind, that she longed rather to behold all those rigid countenances contorted with scornful merriment, and herself the object. Had a roar of laughter burst from the multitude,—each man, each woman, each little shrill-voiced child, contributing their individual parts,—Hester Prynne might have repaid them all with a bitter and disdainful smile. But, under the leaden infliction which it was her doom to endure, she felt, at moments, as if she must needs shriek out with the full power of her lungs, and cast herself from the scaffold down upon the ground, or else go mad at once.

Yet there were intervals when the whole scene, in which she was the most conspicuous object, seemed to vanish from her eyes, or, at least, glimmered indistinctly before them, like a mass of imperfectly shaped and spectral images. Her mind, and especially her memory, was preternaturally active, and kept bringing up other scenes than this roughly hewn street of a little town, on the edge of the Western wilderness; other faces than were lowering upon her from beneath the brims of those steeple-crowned hats. Reminiscences, the most trifling and immaterial, passages of infancy and school-days, sports, childish quarrels, and the little domestic traits of her maiden years, came swarming back upon her, intermingled with recollections of whatever was gravest in her subsequent life; one picture precisely as vivid as another; as if all were of similar importance, or all alike a play. Possibly, it was an instinctive device of her spirit, to relieve itself, by the exhibition of these phantasmagoric forms, from the cruel weight and hardness of the reality.

Be that as it might, the scaffold of the pillory was a point of view that revealed to Hester Prynne the entire track along which she had been treading, since her happy infancy. Standing on that miserable eminence, she saw again her native village, in Old England, and her paternal home; a decayed house of gray stone, with a poverty-stricken aspect, but retaining a half-obliterated shield of arms over the portal, in token of antique gentility. She saw her father's face, with its bald brow, and reverend white beard, that flowed over the old-fashioned Elizabethan ruff; her mother's, too, with the look of heedful and anxious love which it always wore in her remembrance, and which, even since her death, had so often laid the impediment of a gentle remonstrance in her daughter's pathway. She saw her own face, glowing with girlish beauty, and illuminating all the interior of the dusky mirror in which she had been wont to gaze at it. There she beheld another countenance, of a man well stricken in years, a pale, thin, scholar-like visage, with eyes dim and bleared by the lamplight that had served them to pore over many ponderous books. Yet those same bleared optics had a strange, penetrating power, when it was their owner's purpose to read the human soul. This figure of the study and the cloister, as Hester Prynne's womanly fancy failed not to recall, was slightly deformed, with the left shoulder a trifle higher than the right. Next rose before her, in memory's picture-gallery, the intricate and narrow thoroughfares, the tall, gray houses, the huge cathedrals, and the public edifices, ancient in date and quaint in architecture, of a Continental city;[9] where a new life had awaited her, still in connection with the misshapen scholar; a new life, but feeding itself on time-worn materials, like a tuft of green moss on a crumbling wall. Lastly, in lieu of these shifting scenes, came back the rude market-place of the Puritan settlement, with all the townspeople assembled and leveling their stern regards at Hester Prynne,—yes, at herself,—who stood on the scaffold of the pillory, an infant on her arm, and the letter A, in scarlet, fantastically embroidered with gold thread, upon her bosom!

Could it be true? She clutched the child so fiercely to her breast, that it sent forth a cry; she turned her eyes downward at the scarlet letter, and even touched it with her finger, to assure herself that the infant and the shame were real. Yes!—these were her realities,—all else had vanished!

III. The Recognition

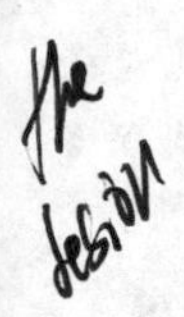

From this intense consciousness of being the object of severe and universal observation, the wearer of the scarlet letter was at length relieved, by discerning, on the outskirts of the crowd, a figure which irresistibly took possession of her thoughts. An Indian, in his native garb, was standing there; but the red men were not so infrequent visitors of the English settlements, that one of them would have attracted any notice from Hester Prynne, at such a time; much less would he have excluded all other objects and ideas from her mind. By the Indian's side, and evidently sustaining a companionship with him, stood a white man, clad in a strange disarray of civilized and savage costume.

He was small in stature, with a furrowed visage, which, as yet, could hardly be termed aged. There was a remarkable intelligence in his

9. Amsterdam, where the Puritans had lived after leaving England.

features, as of a person who had so cultivated his mental part that it could not fail to mould the physical to itself, and become manifest by unmistakable tokens. Although, by a seemingly careless arrangement of his heterogeneous garb, he had endeavored to conceal or abate the peculiarity, it was sufficiently evident to Hester Prynne, that one of this man's shoulders rose higher than the other. Again, at the first instant of perceiving that thin visage, and the slight deformity of the figure, she pressed her infant to her bosom, with so convulsive a force that the poor babe uttered another cry of pain. But the mother did not seem to hear it.

At his arrival in the market-place, and some time before she saw him, the stranger had bent his eyes on Hester Prynne. It was carelessly, at first, like a man chiefly accustomed to look inward, and to whom external matters are of little value and import, unless they bear relation to something within his mind. Very soon, however, his look became keen and penetrative. A writhing horror twisted itself across his features, like a snake gliding swiftly over them, and making one little pause, with all its wreathed intervolutions in open sight. His face darkened with some powerful emotion, which, nevertheless, he so instantaneously controlled by an effort of his will, that, save at a single moment, its expression might have passed for calmness. After a brief space, the convulsion grew almost imperceptible, and finally subsided into the depths of his nature. When he found the eyes of Hester Prynne fastened on his own, and saw that she appeared to recognize him, he slowly and calmly raised his finger, made a gesture with it in the air, and laid it on his lips.

Then, touching the shoulder of a townsman who stood next to him, he addressed him, in a formal and courteous manner.

"I pray you, good Sir," said he, "who is this woman?—and wherefore is she here set up to public shame?"

"You must needs be a stranger in this region, friend," answered the townsman, looking curiously at the questioner and his savage companion, "else you would surely have heard of Mistress Hester Prynne, and her evil doings. She hath raised a great scandal, I promise you, in godly Master Dimmesdale's church."

"You say truly," replied the other. "I am a stranger, and have been a wanderer, sorely against my will. I have met with grievous mishaps by sea and land, and have been long held in bonds among the heathen-folk, to the southward; and am now brought hither by this Indian, to be redeemed out of my captivity. Will it please you, therefore, to tell me of Hester Prynne's,—have I her name rightly?—of this woman's offences, and what has brought her to yonder scaffold?"

"Truly, friend; and methinks it must gladden your heart, after your troubles and sojourn in the wilderness," said the townsman, "to find yourself, at length, in a land where iniquity is searched out, and punished in the sight of rulers and people; as here in our godly New England. Yonder woman, Sir, you must know, was the wife of a certain learned man, English by birth, but who had long dwelt in Amsterdam, whence, some good time agone, he was minded to cross over and cast in his lot with us of the Massachusetts. To this purpose, he sent his wife before him, remaining himself to look after some necessary affairs. Marry, good Sir, in some two years, or less, that the woman has been a dweller here in Boston, no tidings have come of this learned gentleman, Master Prynne; and his young wife, look you, being left to her own misguidance——"

"Ah!—aha!—I conceive you," said the stranger, with a bitter smile. "So learned a man as you speak of should have learned this too in his books. And who, by your favor, Sir, may be the father of yonder babe—it is some three or four months old, I should judge—which Mistress Prynne is holding in her arms?"

"Of a truth, friend, that matter remaineth a riddle; and the Daniel who shall expound it is yet a-wanting,[1] answered the townsman. "Madam Hester absolutely refuseth to speak, and the magistrates have laid their heads together in vain. Peradventure the guilty one stands looking on at this sad spectacle, unknown of man, and forgetting that God sees him."

"The learned man," observed the stranger, with another smile, "should come himself, to look into the mystery."

"It behooves him well, if he be still in life," responded the townsman. "Now, good Sir, our Massachusetts magistracy, bethinking themselves that this woman is youthful and fair, and doubtless was strongly tempted to her fall;—and that, moreover, as is most likely, her husband may be at the bottom of the sea;—they have not been bold to put in force the extremity of our righteous law against her. The penalty thereof is death. But in their great mercy and tenderness of heart, they have doomed Mistress Prynne to stand only a space of three hours on the platform of the pillory, and then and thereafter, for the remainder of her natural life, to wear a mark of shame upon her bosom."

"A wise sentence!" remarked the stranger, gravely bowing his head. "Thus she will be a living sermon against sin, until the ignominious letter be engraved upon her tomb-stone. It irks me, nevertheless, that the partner of her iniquity should not, at least, stand on the scaffold by her side. But he will be known!—he will be known!—he will be known!"

He bowed courteously to the communicative townsman, and, whispering a few words to his Indian attendant, they both made their way through the crowd.

While this passed, Hester Prynne had been standing on her pedestal, still with a fixed gaze towards the stranger; so fixed a gaze, that, at moments of intense absorption, all other objects in the visible world seemed to vanish, leaving only him and her. Such an interview, perhaps, would have been more terrible than even to meet him as she now did, with the hot, midday sun burning down upon her face, and lighting up its shame; with the scarlet token of infamy on her breast; with the sin-born infant in her arms; with a whole people, drawn forth as to a festival, staring at the features that should have been seen only in the quiet gleam of the fireside, in the happy shadow of a home, or beneath a matronly veil, at church. Dreadful as it was, she was conscious of a shelter in the presence of these thousand witnesses. It was better to stand thus, with so many betwixt him and her, than to greet him, face to face, they two alone. She fled for refuge, as it were, to the public exposure, and dreaded the moment when its protection should be withdrawn from her. Involved in these thoughts, she

1. See Daniel 5.12: "Forasmuch as an excellent spirit, and knowledge, and understanding, interpreting of dreams, and shewing of hard sentences, and dissolving of doubts, were found in the same Daniel, whom the king named Belteshazzar: now let Daniel be called, and he will shew the interpretation."

scarcely heard a voice behind her, until it had repeated her name more than once, in a loud and solemn tone, audible to the whole multitude.

"Hearken unto me, Hester Prynne!" said the voice.

It has already been noticed, that directly over the platform on which Hester Prynne stood was a kind of balcony, or open gallery, appended to the meeting-house. It was the place whence proclamations were wont to be made, amidst an assemblage of the magistracy, with all the ceremonial that attended such public observances in those days. Here, to witness the scene which we are describing, sat Governor Bellingham himself, with four sergeants about his chair, bearing halberds, as a guard of honor.[2] He wore a dark feather in his hat, a border of embroidery on his cloak, and a black velvet tunic beneath; a gentleman advanced in years, with a hard experience written in his wrinkles. He was not ill fitted to be the head and representative of a community, which owed its origin and progress, and its present state of development, not to the impulses of youth, but to the stern and tempered energies of manhood, and the sombre sagacity of age; accomplishing so much, precisely because it imagined and hoped so little. The other eminent characters, by whom the chief ruler was surrounded, were distinguished by a dignity of mien, belonging to a period when the forms of authority were felt to possess the sacredness of Divine institutions. They were, doubtless, good men, just, and sage. But, out of the whole human family, it would not have been easy to select the same number of wise and virtuous persons, who should be less capable of sitting in judgment on an erring woman's heart, and disentangling its mesh of good and evil, than the sages of rigid aspect towards whom Hester Prynne now turned her face. She seemed conscious, indeed, that whatever sympathy she might expect lay in the larger and warmer heart of the multitude; for, as she lifted her eyes towards the balcony, the unhappy woman grew pale and trembled.

The voice which had called her attention was that of the reverend and famous John Wilson, the eldest clergyman of Boston, a great scholar, like most of his contemporaries in the profession, and withal a man of kind and genial spirit.[3] This last attribute, however, had been less carefully developed than his intellectual gifts, and was, in truth, rather a matter of shame than self-congratulation with him. There he stood, with a border of grizzled locks beneath his skull-cap; while his gray eyes, accustomed to the shaded light of his study, were winking, like those of Hester's infant, in the unadulterated sunshine. He looked like the darkly engraved portraits which we see prefixed to old volumes of sermons; and had no more right

2. As Charles Ryskamp points out in his essay in this Norton Critical Edition (p. 281), Hawthorne takes a small liberty with dates. Bellingham's term ended in May 1642, a month or so before this scene occurs. John Winthrop would have been governor in June 1642. Hawthorne's decision to identify Bellingham as governor is ironic, because Bellingham was prosecuted in June 1642 by the General Court for improper behavior. According to Winthrop's journal, Bellingham had "obtained" for himself a wife (Penelope Pelham) who was already "contracted to a friend of his, who lodged in his house," and he had "married himself"—that is, conducted their marriage ceremony. Given Hester's later justification of her behavior, Bellingham's rationale for his behavior—"the strength of his affection, and that she was not absolutely promised to the other gentleman"—seems especially ironic (Winthrop, *Journal*, 367). Bellingham would seem to have little moral authority to judge Hester Prynne.

3. John Wilson (1588–1667), one of the first settlers of the Massachusetts Bay Colony, pronounced the sentence of excommunication upon Anne Hutchinson just before her banishment in 1638, labeling her a heathen and a leper (D. D. Hall, 388).

than one of those portraits would have, to step forth, as he now did, and meddle with a question of human guilt, passion and anguish.

"Hester Prynne," said the clergyman, "I have striven with my young brother here, under whose preaching of the word you have been privileged to sit,"—here Mr. Wilson laid his hand on the shoulder of a pale young man beside him,—"I have sought, I say, to persuade this godly youth, that he should deal with you, here in the face of Heaven, and before these wise and upright rulers, and in hearing of all the people, as touching the vileness and blackness of your sin. Knowing your natural temper better than I, he could the better judge what arguments to use, whether of tenderness or terror, such as might prevail over your hardness and obstinacy; insomuch that you should no longer hide the name of him who tempted you to this grievous fall. But he opposes to me, (with a young man's over-softness, albeit wise beyond his years,) that it were wronging the very nature of woman to force her to lay open her heart's secrets in such broad daylight, and in presence of so great a multitude. Truly, as I sought to convince him, the shame lay in the commission of the sin, and not in the showing of it forth. What say you to it, once again, brother Dimmesdale? Must it be thou, or I, that shall deal with this poor sinner's soul?"

There was a murmur among the dignified and reverend occupants of the balcony; and Governor Bellingham gave expression to its purport, speaking in an authoritative voice, although tempered with respect towards the youthful clergyman whom he addressed.

"Good Master Dimmesdale," said he, "the responsibility of this woman's soul lies greatly with you. It behooves you, therefore, to exhort her to repentance, and to confession, as a proof and consequence thereof."

The directness of this appeal drew the eyes of the whole crowd upon the Reverend Mr. Dimmesdale; a young clergyman, who had come from one of the great English universities,[4] bringing all the learning of the age into our wild forest-land. His eloquence and religious fervor had already given the earnest of high eminence in his profession. He was a person of very striking aspect, with a white, lofty, and impending brow, large, brown, melancholy eyes, and a mouth which, unless when he forcibly compressed it, was apt to be tremulous, expressing both nervous sensibility and a vast power of self-restraint. Notwithstanding his high native gifts and scholar-like attainments, there was an air about this young minister,—an apprehensive, a startled, a half-frightened look,—as of a being who felt himself quite astray and at a loss in the pathway of human existence, and could only be at ease in some seclusion of his own. Therefore, so far as his duties would permit, he trod in the shadowy by-paths, and thus kept himself simple and childlike; coming forth, when occasion was, with a freshness, and fragrance, and dewy purity of thought, which, as many people said, affected them like the speech of an angel.

Such was the young man whom the Reverend Mr. Wilson and the Governor had introduced so openly to the public notice, bidding him speak, in the hearing of all men, to that mystery of a woman's soul, so sacred even in its pollution. The trying nature of his position drove the blood from his cheek, and made his lips tremulous.

4. Oxford. See p. 76.

"Speak to the woman, my brother," said Mr. Wilson. "It is of moment to her soul, and therefore, as the worshipful Governor says, momentous to thine own, in whose charge hers is. Exhort her to confess the truth!"

The Reverend Mr. Dimmesdale bent his head, in silent prayer, as it seemed, and then came forward.

"Hester Prynne," said he, leaning over the balcony, and looking down steadfastly into her eyes, "thou hearest what this good man says, and seest the accountability under which I labor. If thou feelest it to be for thy soul's peace, and that thy earthly punishment will thereby be made more effectual to salvation, I charge thee to speak out the name of thy fellow-sinner and fellow-sufferer! Be not silent from any mistaken pity and tenderness for him; for, believe me, Hester, though he were to step down from a high place, and stand there beside thee, on thy pedestal of shame, yet better were it so, than to hide a guilty heart through life. What can thy silence do for him, except it tempt him—yea, compel him, as it were—to add hypocrisy to sin? Heaven hath granted thee an open ignominy, that thereby thou mayest work out an open triumph over the evil within thee, and the sorrow without. Take heed how thou deniest to him—who, perchance, hath not the courage to grasp it for himself—the bitter, but wholesome, cup that is now presented to thy lips!"

The young pastor's voice was tremulously sweet, rich, deep, and broken. The feeling that it so evidently manifested, rather than the direct purport of the words, caused it to vibrate within all hearts, and brought the listeners into one accord of sympathy. Even the poor baby, at Hester's bosom, was affected by the same influence; for it directed its hitherto vacant gaze towards Mr. Dimmesdale, and held up its little arms, with a half pleased, half plaintive murmur. So powerful seemed the minister's appeal, that the people could not believe but that Hester Prynne would speak out the guilty name; or else that the guilty one himself, in whatever high or lowly place he stood, would be drawn forth by an inward and inevitable necessity, and compelled to ascend the scaffold.

Hester shook her head.

"Woman, transgress not beyond the limits of Heaven's mercy!" cried the Reverend Mr. Wilson, more harshly than before. "That little babe hath been gifted with a voice, to second and confirm the counsel which thou hast heard. Speak out the name! That, and thy repentance, may avail to take the scarlet letter off thy breast."

"Never!" replied Hester Prynne, looking, not at Mr. Wilson, but into the deep and troubled eyes of the younger clergyman. "It is too deeply branded. Ye cannot take it off. And would that I might endure his agony, as well as mine!"

"Speak, woman!" said another voice, coldly and sternly, proceeding from the crowd about the scaffold. "Speak; and give your child a father!"

"I will not speak!" answered Hester, turning pale as death, but responding to this voice, which she too surely recognized. "And my child must seek a heavenly Father; she shall never know an earthly one!"

"She will not speak!" murmured Mr. Dimmesdale, who, leaning over the balcony, with his hand upon his heart, had awaited the result of his appeal. He now drew back, with a long respiration. "Wondrous strength and generosity of a woman's heart! She will not speak!"

Discerning the impracticable state of the poor culprit's mind, the elder clergyman, who had carefully prepared himself for the occasion, addressed to the multitude a discourse on sin, in all its branches, but with continual reference to the ignominious letter. So forcibly did he dwell upon this symbol, for the hour or more during which his periods were rolling over the people's heads, that it assumed new terrors in their imagination, and seemed to derive its scarlet hue from the flames of the infernal pit. Hester Prynne, meanwhile, kept her place upon the pedestal of shame, with glazed eyes, and an air of weary indifference. She had borne, that morning, all that nature could endure; and as her temperament was not of the order that escapes from too intense suffering by a swoon, her spirit could only shelter itself beneath a stony crust of insensibility, while the faculties of animal life remained entire. In this state, the voice of the preacher thundered remorselessly, but unavailingly, upon her ears. The infant, during the latter portion of her ordeal, pierced the air with its wailings and screams; she strove to hush it, mechanically, but seemed scarcely to sympathize with its trouble. With the same hard demeanor, she was led back to prison, and vanished from the public gaze within its iron-clamped portal. It was whispered, by those who peered after her, that the scarlet letter threw a lurid gleam along the dark passage-way of the interior.

IV. The Interview

After her return to the prison, Hester Prynne was found to be in a state of nervous excitement that demanded constant watchfulness, lest she should perpetrate violence on herself, or do some half-frenzied mischief to the poor babe. As night approached, it proving impossible to quell her insubordination by rebuke or threats of punishment, Master Brackett, the jailer,[1] thought fit to introduce a physician. He described him as a man of skill in all Christian modes of physical science, and likewise familiar with whatever the savage people could teach, in respect to medicinal herbs and roots that grew in the forest. To say the truth, there was much need of professional assistance, not merely for Hester herself, but still more urgently for the child; who, drawing its sustenance from the maternal bosom, seemed to have drank in with it all the turmoil, the anguish and despair, which pervaded the mother's system. It now writhed in convulsions of pain, and was a forcible type, in its little frame, of the moral agony which Hester Prynne had borne throughout the day.

Closely following the jailer into the dismal apartment, appeared that individual, of singular aspect, whose presence in the crowd had been of such deep interest to the wearer of the scarlet letter. He was lodged in the prison, not as suspected of any offence, but as the most convenient and suitable mode of disposing of him, until the magistrates should have conferred with the Indian sagamores respecting his ransom. His name was announced as Roger Chillingworth. The jailer, after ushering him into the room, remained a moment, marvelling at the comparative quiet that followed his entrance; for Hester Prynne had immediately become as still as death, although the child continued to moan.

1. Caleb Snow mentions Brackett as "prison keeper," beginning in 1633 (116).

"Prithee, friend, leave me alone with my patient," said the practitioner. "Trust me, good jailer, you shall briefly have peace in your house; and, I promise you, Mistress Prynne shall hereafter be more amenable to just authority than you may have found her heretofore."

"Nay, if your worship can accomplish that," answered Master Brackett, "I shall own you for a man of skill indeed! Verily, the woman hath been like a possessed one; and there lacks little, that I should take in hand to drive Satan out of her with stripes."

The stranger had entered the room with the characteristic quietude of the profession to which he announced himself as belonging. Nor did his demeanor change, when the withdrawal of the prison-keeper left him face to face with the woman, whose absorbed notice of him, in the crowd, had intimated so close a relation between himself and her. His first care was given to the child; whose cries, indeed, as she lay writhing on the trundle-bed, made it of peremptory necessity to postpone all other business to the task of soothing her. He examined the infant carefully, and then proceeded to unclasp a leathern case, which he took from beneath his dress. It appeared to contain medical preparations, one of which he mingled with a cup of water.

"My old studies in alchemy,"[2] observed he, "and my sojourn, for above a year past, among a people well versed in the kindly properties of simples, have made a better physician of me than many that claim the medical degree. Here, woman! The child is yours,—she is none of mine,—neither will she recognize my voice or aspect as a father's. Administer this draught, therefore, with thine own hand."

Hester repelled the offered medicine, at the same time gazing with strongly marked apprehension into his face.

"Wouldst thou avenge thyself on the innocent babe?" whispered she.

"Foolish woman!" responded the physician, half coldly, half soothingly. "What should ail me, to harm this misbegotten and miserable babe? The medicine is potent for good; and were it my child,—yea, mine own, as well as thine!—I could do no better for it."

As she still hesitated, being, in fact, in no reasonable state of mind, he took the infant in his arms, and himself administered the draught. It soon proved its efficacy, and redeemed the leech's[3] pledge. The moans of the little patient subsided; its convulsive tossings gradually ceased; and, in a few moments, as is the custom of young children after relief from pain, it sank into a profound and dewy slumber. The physician, as he had a fair right to be termed, next bestowed his attention on the mother. With calm and intent scrutiny, he felt her pulse, looked into her eyes,—a gaze that made her heart shrink and shudder, because so familiar, and yet so strange and cold,—and, finally, satisfied with his investigation, proceeded to mingle another draught.

"I know not Lethe nor Nepenthe," remarked he; "but I have learned many new secrets in the wilderness, and here is one of them,—a recipe that an Indian taught me, in requital of some lessons of my own, that

2. Experiments for turning common metals, such as lead, into gold. Aylmer in Hawthorne's "The Birthmark" is also associated with alchemy.
3. An archaic term for physician; in the seventeenth century, it was applied in ordinary prose only to veterinary practitioners (*OED*).

were as old as Paracelsus. Drink it! It may be less soothing than a sinless conscience. That I cannot give thee. But it will calm the swell and heaving of thy passion, like oil thrown on the waves of a tempestuous sea."

He presented the cup to Hester, who received it with a slow, earnest look into his face; not precisely a look of fear, yet full of doubt and questioning, as to what his purposes might be. She looked also at her slumbering child.

"I have thought of death," said she,—"have wished for it,—would even have prayed for it, were it fit that such as I should pray for anything. Yet, if death be in this cup, I bid thee think again, ere thou beholdest me quaff it. See! It is even now at my lips."

"Drink, then," replied he, still with the same cold composure. "Dost thou know me so little, Hester Prynne? Are my purposes wont to be so shallow? Even if I imagine a scheme of vengeance, what could I do better for my object than to let thee live,—than to give thee medicines against all harm and peril of life,—so that this burning shame may still blaze upon thy bosom?" As he spoke, he laid his long forefinger on the scarlet letter, which forthwith seemed to scorch into Hester's breast, as if it had been red-hot. He noticed her involuntary gesture, and smiled. "Live, therefore, and bear about thy doom with thee, in the eyes of men and women,—in the eyes of him whom thou didst call thy husband,—in the eyes of yonder child! And, that thou mayest live, take off this draught."

Without further expostulation or delay, Hester Prynne drained the cup, and, at the motion of the man of skill, seated herself on the bed where the child was sleeping; while he drew the only chair which the room afforded, and took his own seat beside her. She could not but tremble at these preparations; for she felt that—having now done all that humanity, or principle, or, if so it were, a refined cruelty, impelled him to do, for the relief of physical suffering—he was next to treat with her as the man whom she had most deeply and irreparably injured.

"Hester," said he, "I ask not wherefore, nor how, thou hast fallen into the pit, or say, rather, thou hast ascended to the pedestal of infamy, on which I found thee. The reason is not far to seek. It was my folly, and thy weakness. I,—a man of thought,—the book-worm of great libraries,—a man already in decay, having given my best years to feed the hungry dream of knowledge,—what had I to do with youth and beauty like thine own! Misshapen from my birth-hour, how could I delude myself with the idea that intellectual gifts might veil physical deformity in a young girl's fantasy! Men call me wise. If sages were ever wise in their own behoof, I might have foreseen all this. I might have known that, as I came out of the vast and dismal forest, and entered this settlement of Christian men, the very first object to meet my eyes would be thyself, Hester Prynne, standing up, a statue of ignominy, before the people. Nay, from the moment when we came down the old church-steps together, a married pair, I might have beheld the bale-fire of that scarlet letter blazing at the end of our path!"

"Thou knowest," said Hester,—for, depressed as she was, she could not endure this last quiet stab at the token of her shame,—"thou knowest that I was frank with thee. I felt no love, nor feigned any."

4. A sixteenth-century Swiss alchemist (1493–1541). *Lethe*: in Greek mythology, one of the rivers of Hades, associated with forgetfulness. *Nepenthe*: a drug that purportedly assuaged grief.

"True," replied he. "It was my folly! I have said it. But, up to that epoch of my life, I had lived in vain. The world had been so cheerless! My heart was a habitation large enough for many guests, but lonely and chill, and without a household fire. I longed to kindle one! It seemed not so wild a dream,—old as I was, and sombre as I was, and misshapen as I was,—that the simple bliss, which is scattered far and wide, for all mankind to gather up, might yet be mine. And so, Hester, I drew thee into my heart, into its innermost chamber, and sought to warm thee by the warmth which thy presence made there!"

"I have greatly wronged thee," murmured Hester.

"We have wronged each other," answered he. "Mine was the first wrong, when I betrayed thy budding youth into a false and unnatural relation with my decay. Therefore, as a man who has not thought and philosophized in vain, I seek no vengeance, plot no evil against thee. Between thee and me, the scale hangs fairly balanced. But, Hester, the man lives who has wronged us both! Who is he?"

"Ask me not!" replied Hester Prynne, looking firmly into his face. "That thou shalt never know!"

"Never, sayest thou?" rejoined he, with a smile of dark and self-relying intelligence. "Never know him! Believe me, Hester, there are few things,—whether in the outward world, or to a certain depth, in the invisible sphere of thought,—few things hidden from the man who devotes himself earnestly and unreservedly to the solution of a mystery. Thou mayest cover up thy secret from the prying multitude. Thou mayest conceal it, too, from the ministers and magistrates, even as thou didst this day, when they sought to wrench the name out of thy heart, and give thee a partner on thy pedestal. But, as for me, I come to the inquest with other senses than they possess. I shall seek this man, as I have sought truth in books; as I have sought gold in alchemy. There is a sympathy that will make me conscious of him. I shall see him tremble. I shall feel myself shudder, suddenly and unawares. Sooner or later, he must needs be mine!"

The eyes of the wrinkled scholar glowed so intensely upon her, that Hester Prynne clasped her hands over her heart, dreading lest he should read the secret there at once.

"Thou wilt not reveal his name? Not the less he is mine," resumed he, with a look of confidence, as if destiny were at one with him. "He bears no letter of infamy wrought into his garment, as thou dost; but I shall read it on his heart. Yet fear not for him! Think not that I shall interfere with Heaven's own method of retribution, or, to my own loss, betray him to the gripe of human law. Neither do thou imagine that I shall contrive aught against his life; no, nor against his fame, if, as I judge, he be a man of fair repute. Let him live! Let him hide himself in outward honor, if he may! Not the less he shall be mine!"

"Thy acts are like mercy," said Hester, bewildered and appalled. "But thy words interpret thee as a terror!"

"One thing, thou that wast my wife, I would enjoin upon thee," continued the scholar. "Thou hast kept the secret of thy paramour. Keep, likewise, mine! There are none in this land that know me. Breathe not, to any human soul, that thou didst ever call me husband! Here, on this wild outskirt of the earth, I shall pitch my tent; for, elsewhere a wanderer, and isolated from human interests, I find here a woman, a man, a child, amongst whom and myself there exist the closest ligaments. No matter whether of love or hate; no matter whether of right or wrong! Thou and thine, Hester Prynne, belong to me. My home is where thou art, and where he is. But betray me not!"

"Wherefore dost thou desire it?" inquired Hester, shrinking, she hardly knew why, from this secret bond. "Why not announce thyself openly, and cast me off at once?"

"It may be," he replied, "because I will not encounter the dishonor that besmirches the husband of a faithless woman. It may be for other reasons. Enough, it is my purpose to live and die unknown. Let, therefore, thy husband be to the world as one already dead, and of whom no tidings shall ever come. Recognize me not, by word, by sign, by look! Breathe not the secret, above all, to the man thou wottest of. Shouldst thou fail me in this, beware! His fame, his position, his life, will be in my hands. Beware!"

"I will keep thy secret, as I have his," said Hester.

"Swear it!" rejoined he.

And she took the oath.

"And now, Mistress Prynne," said old Roger Chillingworth, as he was hereafter to be named, "I leave thee alone; alone with thy infant, and the scarlet letter! How is it, Hester? Doth thy sentence bind thee to wear the token in thy sleep? Art thou not afraid of nightmares and hideous dreams?"

"Why dost thou smile so at me?" inquired Hester, troubled at the expression of his eyes. "Art thou like the Black Man that haunts the forest round about us? Hast thou enticed me into a bond that will prove the ruin of my soul?"

"Not thy soul," he answered, with another smile. "No, not thine!"

V. Hester at Her Needle

Hester Prynne's term of confinement was now at an end. Her prison-door was thrown open, and she came forth into the sunshine, which, falling on all alike, seemed, to her sick and morbid heart, as if meant for no other purpose than to reveal the scarlet letter on her breast. Perhaps there was a more real torture in her first unattended footsteps from the threshold of the prison, than even in the procession and spectacle that have been described, where she was made the common infamy, at which all mankind was summoned to point its finger. Then, she was supported by an unnatural tension of the nerves, and by all the combative energy of her character, which enabled her to convert the scene into a kind of lurid triumph. It was, moreover, a separate and insulated event, to occur but once in her lifetime, and to meet which, therefore, reckless of economy, she might call up the vital strength that would have sufficed for many quiet years. The very law that condemned her—a giant of stern features, but with vigor to support, as well as to annihilate, in his iron arm—had held her up, through the terrible ordeal of her ignominy. But now, with this unattended walk from her prison-door, began the daily custom; and she must either sustain and carry it forward by the ordinary resources of her nature, or sink beneath it. She could no longer borrow from the future to help her through the present grief. To-morrow would bring its own trial with it; so would the next day, and so would the next; each its own trial, and yet the very same that was now so unutterably grievous to be borne. The days of the far-off future would toil onward, still with the same burden for her to take up, and bear along with her, but never to fling down; for the accumulating days, and added years, would pile up their misery upon the heap of shame. Throughout them all, giving up her individuality, she would become the general symbol at which the preacher and moralist might point, and in which they might vivify and embody their images of woman's frailty and sinful passion. Thus the young and pure would be taught to look at her, with the scarlet letter flaming on her breast,—at her, the child of honorable parents,—at her, the mother of a babe, that would hereafter be a woman,—at her, who had once been innocent,—as the figure, the body, the reality of sin. And over her grave, the infamy that she must carry thither would be her only monument.

It may seem marvellous, that, with the world before her,—kept by no restrictive clause of her condemnation within the limits of the Puritan settlement, so remote and so obscure,—free to return to her birthplace, or to any other European land, and there hide her character and identity under a new exterior, as completely as if emerging into another state of being,—and having also the passes of the dark, inscrutable forest open to her, where the wildness of her nature might assimilate itself with a people whose customs and life were alien from the law that had condemned her,—it may seem marvellous, that this woman should still call that place her home, where, and where only, she must needs be the type of shame. But there is a fatality, a feeling so irresistible and inevitable that it has the force of doom, which almost invariably compels human beings to linger around and haunt, ghost-like, the spot where some great and marked event has given the color to their lifetime; and still the more irresistibly, the darker the tinge that saddens it. Her sin, her ignominy, were the roots which she had struck into the soil. It was as if a new birth, with stronger assimilations than the first, had converted the forest-land, still so uncongenial to every other pilgrim and wanderer, into Hester Prynne's wild and dreary, but life-long home. All other scenes of earth—even that village of rural England, where happy infancy and stainless maidenhood seemed yet to be in her mother's keeping, like garments put off long ago—were foreign to her, in comparison. The chain that bound her here was of iron links, and galling to her inmost soul, but could never be broken.

It might be, too,—doubtless it was so, although she hid the secret from herself, and grew pale whenever it struggled out of her heart, like a serpent from its hole,—it might be that another feeling kept her within the scene and pathway that had been so fatal. There dwelt, there trode the feet of one with whom she deemed herself connected in a union, that, unrecognized on earth, would bring them together before the bar of final judgment, and make that their marriage-altar, for a joint futurity of endless retribution. Over and over again, the tempter of souls had thrust this idea upon Hester's contemplation, and laughed at the passionate and desperate joy with which she seized, and then strove to cast it from her. She barely looked the idea in the face, and hastened to bar it in its dungeon. What she compelled herself to believe,—what, finally, she reasoned upon, as her motive for continuing a resident of New England,—was half a truth, and half a self-delusion. Here, she said to herself, had been the scene of her guilt, and here should be the scene of her earthly punishment; and so, perchance, the torture of her daily shame would at length purge her soul, and work out another purity than that which she had lost; more saint-like, because the result of martyrdom.

Hester Prynne, therefore, did not flee. On the outskirts of the town, within the verge of the peninsula, but not in close vicinity to any other habitation, there was a small thatched cottage. It had been built by an earlier settler, and abandoned, because the soil about it was too sterile for cultivation, while its comparative remoteness put it out of the sphere of that social activity which already marked the habits of the emigrants. It stood on the shore, looking across a basin of the sea at the forest-covered hills, towards the west. A clump of scrubby trees, such as alone grew on the peninsula, did not so much conceal the cottage from view, as seem to denote that here was some object which would fain have been, or at least ought to be, concealed. In this little, lonesome dwelling, with some slender means that she possessed, and by the license of the magistrates, who still kept an inquisitorial watch over her, Hester established herself, with her infant child. A mystic shadow of suspicion immediately attached itself to the spot. Children, too young to comprehend wherefore this woman should be shut out from the sphere of human charities, would creep nigh enough to behold her plying her needle at the cottage-window, or standing in the doorway, or laboring in her little garden, or coming forth along the pathway that led townward; and, discerning the scarlet letter on her breast, would scamper off with a strange, contagious fear.

Lonely as was Hester's situation, and without a friend on earth who dared to show himself, she, however, incurred no risk of want. She possessed an art that sufficed, even in a land that afforded comparatively little scope for its exercise, to supply food for her thriving infant and herself. It was the art—then, as now, almost the only one within a woman's grasp—of needle-work. She bore on her breast, in the curiously embroidered letter, a specimen of her delicate and imaginative skill, of which the dames of a court might gladly have availed themselves, to add the richer and more spiritual adornment of human ingenuity to their fabrics of silk and gold. Here, indeed, in the sable simplicity that generally characterized the Puritanic modes of dress, there might be an infrequent call for the finer productions of her handiwork. Yet the taste of the age, demanding whatever was elaborate in compositions of this kind, did not fail to extend its influence over our stern progenitors, who had cast behind them so many fashions which it might seem harder to dispense with. Public ceremonies, such as ordinations, the installation of magistrates, and all that could give majesty to the forms in which a new government manifested itself to the people, were, as a matter of policy, marked by a stately and well-conducted ceremonial, and a sombre, but yet a studied magnificence. Deep ruffs, painfully wrought bands, and gorgeously embroidered gloves, were all deemed necessary to the official state of men assuming the reins of power; and were readily allowed to individuals dignified by rank or wealth, even while sumptuary laws forbade these and similar extravagances to the plebeian order. In the array of funerals, too,—whether for the apparel of the dead body, or to typify, by manifold emblematic devices of sable cloth and snowy lawn, the sorrow of the survivors,—there was a frequent and characteristic demand for such labor as Hester Prynne could supply. Baby-linen—for babies then wore robes of state—afforded still another possibility of toil and emolument.

By degrees, nor very slowly, her handiwork became what would now be termed the fashion. Whether from commiseration for a woman of so miserable a destiny; or from the morbid curiosity that gives a fictitious value even to common or worthless things; or by whatever other intangible circumstance was then, as now, sufficient to bestow, on some persons, what others might seek in vain; or because Hester really filled a gap which must otherwise have remained vacant; it is certain that she had ready and fairly requited employment for as many hours as she saw fit to occupy with her needle. Vanity, it may be, chose to mortify itself, by putting on, for ceremonials of pomp and state, the garments that had been wrought by her sinful hands. Her needle-work was seen on the ruff of the Governor; military men wore it on their scarfs, and the minister on his band; it decked the baby's little cap; it was shut up, to be mildewed and moulder away, in the coffins of the dead. But it is not recorded that, in a single instance, her skill was called in aid to embroider the white veil which was to cover the pure blushes of a bride. The exception indicated the ever relentless vigor with which society frowned upon her sin.

Hester sought not to acquire anything beyond a subsistence, of the plainest and most ascetic description, for herself, and a simple abundance for her child. Her own dress was of the coarsest materials and the most sombre hue; with only that one ornament,—the scarlet letter,—which it was her doom to wear. The child's attire, on the other hand, was distinguished by a fanciful, or, we might rather say, a fantastic ingenuity, which served, indeed, to heighten the airy charm that early began to develop itself in the little girl, but which appeared to have also a deeper meaning. We may speak further of it hereafter. Except for that small expenditure in the decoration of her infant, Hester bestowed all her superfluous means in charity, on wretches less miserable than herself, and who not unfrequently insulted the hand that fed them. Much of the time, which she might readily have applied to the better efforts of her art, she employed in making coarse garments for the poor. It is probable that there was an idea of penance in this mode of occupation, and that she offered up a real sacrifice of enjoyment, in devoting so many hours to such rude handiwork. She had in her nature a rich, voluptuous, Oriental characteristic,—a taste for the gorgeously beautiful, which, save in the exquisite productions of her needle, found nothing else, in all the possibilities of her life, to exercise itself upon. Women derive a pleasure, incomprehensible to the other sex, from the delicate toil of the needle. To Hester Prynne it might have been a mode of expressing, and therefore soothing, the passion of her life. Like all other joys, she rejected it as sin. This morbid meddling of conscience with an immaterial matter betokened, it is to be feared, no genuine and steadfast penitence, but something doubtful, something that might be deeply wrong, beneath.

In this manner, Hester Prynne came to have a part to perform in the world. With her native energy of character, and rare capacity, it could not entirely cast her off, although it had set a mark upon her, more intolerable to a woman's heart than that which branded the brow of Cain.[1] In all her intercourse with society, however, there was nothing that made her feel as if she belonged to it. Every gesture, every word, and even the silence of those with whom she came in contact, implied, and often expressed, that she was banished, and as much alone as if she inhabited another sphere, or communicated with the common nature by other organs and senses than the rest of human kind. She stood apart from moral interests, yet close beside them, like a ghost that revisits the familiar fireside, and can no longer make itself seen or felt; no more smile with the household joy, nor mourn with the kindred sorrow; or, should it succeed in manifesting its forbidden sympathy, awakening only terror and horrible repugnance. These emotions, in fact, and its bitterest scorn besides, seemed to be the sole portion that she retained in the universal heart. It was not an age of delicacy; and her position, although she understood it well, and was in little danger of forgetting it, was often brought before her vivid self-perception, like a new anguish, by the rudest touch upon the tenderest spot. The poor, as we have already said, whom she sought out to be the objects of her bounty, often reviled the hand that was stretched forth to succor them. Dames of elevated rank, likewise, whose doors she entered in the way of her occupation, were accustomed to distil drops of bitterness into her heart; sometimes through that alchemy of quiet malice, by which women can concoct a subtile poison from ordinary trifles; and sometimes, also, by a coarser expression, that fell upon the sufferer's defenceless breast like a rough blow upon an ulcerated wound. Hester had schooled herself long and well; she never responded to these attacks, save by a flush of crimson that rose irrepressibly over her pale cheek, and again subsided into the depths of her bosom. She was patient,—a martyr, indeed,—but she forebore to pray for her enemies; lest, in spite of her forgiving aspirations, the words of the blessing should stubbornly twist themselves into a curse.

Continually, and in a thousand other ways, did she feel the innumerable throbs of anguish that had been so cunningly contrived for her by the undying, the ever-active sentence of the Puritan tribunal. Clergymen paused in the street to address words of exhortation, that brought a crowd, with its mingled grin and frown, around the poor, sinful woman. If she entered a church, trusting to share the Sabbath smile of the Universal Father, it was often her mishap to find herself the text of the discourse. She grew to have a dread of children; for they had imbibed from their parents a vague idea of something horrible in this dreary woman, gliding silently through the town, with never any companion but one only child. Therefore, first allowing her to pass, they pursued her at a distance with shrill cries, and the utterance of a word that had no distinct purport to their own minds, but was none the less terrible to her, as proceeding from lips that babbled it unconsciously. It seemed to argue so wide a diffusion of her shame, that all nature knew of it; it could have caused her no deeper pang, had the leaves of the trees whispered the dark story among themselves,—had the summer breeze murmured about it—had

1. See Genesis 4.1–16. For killing his brother, Abel, Cain was sentenced to be "a fugitive and a vagabond." When Cain protested that he would be killed, "the LORD set a mark upon Cain, lest any finding him should kill him" (4.15).

the wintry blast shrieked it aloud! Another peculiar torture was felt in the gaze of a new eye. When strangers looked curiously at the scarlet letter,—and none ever failed to do so,—they branded it afresh into Hester's soul; so that, oftentimes, she could scarcely refrain, yet always did refrain, from covering the symbol with her hand. But then, again, an accustomed eye had likewise its own anguish to inflict. Its cool stare of familiarity was intolerable. From first to last, in short, Hester Prynne had always this dreadful agony in feeling a human eye upon the token; the spot never grew callous; it seemed, on the contrary, to grow more sensitive with daily torture.

But sometimes, once in many days, or perchance in many months, she felt an eye—a human eye—upon the ignominious brand, that seemed to give a momentary relief, as if half of her agony were shared. The next instant, back it all rushed again, with still a deeper throb of pain; for, in that brief interval, she had sinned anew. Had Hester sinned alone?

Her imagination was somewhat affected, and, had she been of a softer moral and intellectual fibre, would have been still more so, by the strange and solitary anguish of her life. Walking to and fro, with those lonely footsteps, in the little world with which she was outwardly connected, it now and then appeared to Hester,—if altogether fancy, it was nevertheless too potent to be resisted,—she felt or fancied, then, that the scarlet letter had endowed her with a new sense. She shuddered to believe, yet could not help believing, that it gave her a sympathetic knowledge of the hidden sin in other hearts. She was terror-stricken by the revelations that were thus made. What were they? Could they be other than the insidious whispers of the bad angel, who would fain have persuaded the struggling woman, as yet only half his victim, that the outward guise of purity was but a lie, and that, if truth were everywhere to be shown, a scarlet letter would blaze forth on many a bosom besides Hester Prynne's? Or, must she receive those intimations—so obscure, yet so distinct—as truth? In all her miserable experience, there was nothing else so awful and so loathsome as this sense. It perplexed, as well as shocked her, by the irreverent inopportuneness of the occasions that brought it into vivid action. Sometimes the red infamy upon her breast would give a sympathetic throb, as she passed near a venerable minister or magistrate, the model of piety and justice, to whom that age of antique reverence looked up, as to a mortal man in fellowship with angels. "What evil thing is at hand?" would Hester say to herself. Lifting her reluctant eyes, there would be nothing human within the scope of view, save the form of this earthly saint! Again, a mystic sisterhood would contumaciously assert itself, as she met the sanctified frown of some matron, who, according to the rumor of all tongues, had kept cold snow within her bosom throughout life. That unsunned snow in the matron's bosom, and the burning shame on Hester Prynne's,—what had the two in common? Or, once more, the electric thrill would give her warning,—"Behold, Hester, here is a companion!"—and, looking up, she would detect the eyes of a young maiden glancing at the scarlet letter, shyly and aside, and quickly averted, with a faint, chill crimson in her cheeks; as if her purity were somewhat sullied by that momentary glance. O Fiend, whose talisman was that fatal symbol, wouldst thou leave nothing, whether in youth or age, for this poor sinner to revere?—such loss of faith is ever one of the saddest results of sin. Be it accepted as a proof that all was not corrupt in this poor victim of her own frailty, and man's hard law, that Hester Prynne yet struggled to believe that no fellow-mortal was guilty like herself.

The vulgar, who, in those dreary old times, were always contributing a grotesque horror to what interested their imaginations, had a story about the scarlet letter which we might readily work up into a terrific legend. They averred, that the symbol was not mere scarlet cloth, tinged in an earthly dye-pot, but was red-hot with infernal fire, and could be seen glowing all alight, whenever Hester Prynne walked abroad in the night-time. And we must needs say, it seared Hester's bosom so deeply, that perhaps there was more truth in the rumor than our modern incredulity may be inclined to admit.

VI. Pearl

We have as yet hardly spoken of the infant; that little creature, whose innocent life had sprung, by the inscrutable decree of Providence, a lovely and immortal flower, out of the rank luxuriance of a guilty passion. How strange it seemed to the sad woman, as she watched the growth, and the beauty that became every day more brilliant, and the intelligence that threw its quivering sunshine over the tiny features of this child! Her Pearl!—For so had Hester called her; not as a name expressive of her aspect, which had nothing of the calm, white, unimpassioned luster that would be indicated by the comparison. But she named the infant "Pearl," as being of great price,—purchased with all she had,—her mother's only treasure![1] How strange, indeed! Man had marked this woman's sin by a scarlet letter, which had such potent and disastrous efficacy that no human sympathy could reach her, save it were sinful like herself. God, as a direct consequence of the sin which man thus punished, had given her a lovely child, whose place was on that same dishonored bosom, to connect her parent forever with the race and descent of mortals, and to be finally a blessed soul in heaven! Yet these thoughts affected Hester Prynne less with hope than apprehension. She knew that her deed had been evil; she could have no faith, therefore, that its result would be good. Day after day, she looked fearfully into the child's expanding nature; ever dreading to detect some dark and wild peculiarity, that should correspond with the guiltiness to which she owed her being.

Certainly, there was no physical defect. By its perfect shape, its vigor, and its natural dexterity in the use of all its untried limbs, the infant was worthy to have been brought forth in Eden; worthy to have been left there, to be the plaything of the angels, after the world's first parents were driven out. The child had a native grace which does not invariably coëxist with faultless beauty; its attire, however simple, always impressed the beholder as if it were the very garb that precisely became it best. But little Pearl was not clad in rustic weeds. Her mother, with a morbid purpose that may be better understood hereafter, had bought the richest

1. See Matthew 13.45–46: "The kingdom of heaven is like unto a merchant man, seeking goodly pearls: Who, when he had found one pearl of great price, went and sold all that he had, and bought it."

tissues that could be procured, and allowed her imaginative faculty its full play in the arrangement and decoration of the dresses which the child wore, before the public eye. So magnificent was the small figure, when thus arrayed, and such was the splendor of Pearl's own proper beauty, shining through the gorgeous robes which might have extinguished a paler loveliness, that there was an absolute circle of radiance around her, on the darksome cottage floor. And yet a russet gown, torn and soiled with the child's rude play, made a picture of her just as perfect. Pearl's aspect was imbued with a spell of infinite variety; in this one child there were many children, comprehending the full scope between the wild-flower prettiness of a peasant-baby, and the pomp, in little, of an infant princess. Throughout all, however, there was a trait of passion, a certain depth of hue, which she never lost; and if, in any of her changes, she had grown fainter or paler, she would have ceased to be herself;—it would have been no longer Pearl!

This outward mutability indicated, and did not more than fairly express, the various properties of her inner life. Her nature appeared to possess depth, too, as well as variety; but—or else Hester's fears deceived her—it lacked reference and adaptation to the world into which she was born. The child could not be made amenable to rules. In giving her existence, a great law had been broken; and the result was a being whose elements were perhaps beautiful and brilliant, but all in disorder; or with an order peculiar to themselves, amidst which the point of variety and arrangement was difficult or impossible to be discovered. Hester could only account for the child's character—and even then most vaguely and imperfectly—by recalling what she herself had been, during that momentous period while Pearl was imbibing her soul from the spiritual world, and her bodily frame from its material of earth. The mother's impassioned state had been the medium through which were transmitted to the unborn infant the rays of its moral life; and, however white and clear originally, they had taken the deep stains of crimson and gold, the fiery lustre, the black shadow, and the untempered light, of the intervening substance. Above all, the warfare of Hester's spirit, at that epoch, was perpetuated in Pearl. She could recognize her wild, desperate, defiant mood, the flightiness of her temper, and even some of the very cloud-shapes of gloom and despondency that had brooded in her heart. They were now illuminated by the morning radiance of a young child's disposition, but, later in the day of earthly existence, might be prolific of the storm and whirlwind.

The discipline of the family, in those days, was of a far more rigid kind than now. The frown, the harsh rebuke, the frequent application of the rod, enjoined by Scriptural authority, were used, not merely in the way of punishment for actual offences, but as a wholesome regimen for the growth and promotion of all childish virtues.[2] Hester Prynne, nevertheless, the lonely mother of this one child, ran little risk of erring on the side of undue severity. Mindful, however, of her own errors and misfortunes, she early sought to impose a tender, but strict control over the infant immortality that was committed to her charge. But the task was beyond her skill. After testing both smiles and frowns, and proving that

2. See Proverbs 13.24: "He that spareth his rod hateth his son: but he that loveth him chasteneth him betimes."

neither mode of treatment possessed any calculable influence, Hester was ultimately compelled to stand aside, and permit the child to be swayed by her own impulses. Physical compulsion or restraint was effectual, of course, while it lasted. As to any other kind of discipline, whether addressed to her mind or heart, little Pearl might or might not be within its reach, in accordance with the caprice that ruled the moment. Her mother, while Pearl was yet an infant, grew acquainted with a certain peculiar look, that warned her when it would be labor thrown away to insist, persuade, or plead. It was a look so intelligent, yet inexplicable, so perverse, sometimes so malicious, but generally accompanied by a wild flow of spirits, that Hester could not help questioning, at such moments, whether Pearl was a human child.[3] She seemed rather an airy sprite, which, after playing its fantastic sports for a little while upon the cottage-floor, would flit away with a mocking smile. Whenever that look appeared in her wild, bright, deeply black eyes, it invested her with a strange remoteness and intangibility; it was as if she were hovering in the air and might vanish, like a glimmering light, that comes we know not whence, and goes we know not whither. Beholding it, Hester was constrained to rush towards the child,—to pursue the little elf in the flight which she invariably began,—to snatch her to her bosom, with a close pressure and earnest kisses,—not so much from overflowing love, as to assure herself that Pearl was flesh and blood, and not utterly delusive. But Pearl's laugh, when she was caught, though full of merriment and music, made her mother more doubtful than before.

Heart-smitten at this bewildering and baffling spell, that so often came between herself and her sole treasure, whom she had bought so dear, and who was all her world, Hester sometimes burst into passionate tears. Then, perhaps,—for there was no foreseeing how it might affect her,—Pearl would frown, and clench her little fist, and harden her small features into a stern, unsympathizing look of discontent. Not seldom, she would laugh anew, and louder than before, like a thing incapable and unintelligent of human sorrow. Or—but this more rarely happened—she would be convulsed with a rage of grief, and sob out her love for her mother, in broken words, and seem intent on proving that she had a heart, by breaking it. Yet Hester was hardly safe in confiding herself to that gusty tenderness; it passed, as suddenly as it came. Brooding over all these matters, the mother felt like one who has evoked a spirit, but, by some irregularity in the process of conjuration, has failed to win the master-word that should control this new and incomprehensible intelligence. Her only real comfort was when the child lay in the placidity of sleep. Then she was sure of her, and tasted hours of quiet, sad, delicious happiness; until—perhaps with that perverse expression glimmering from beneath her opening lids—little Pearl awoke!

How soon—with what strange rapidity, indeed!—did Pearl arrive at an age that was capable of social intercourse, beyond the mother's ever-ready smile and nonsense-words![4] And then what a happiness would it have been, could Hester Prynne have heard her clear, bird-like voice mingling

3. Compare the passage from Hawthorne's notebook on p. 206 of this Norton Critical Edition.
4. Hawthorne uses Pearl's aging in this chapter to move the novel forward approximately three years.

with the uproar of other childish voices, and have distinguished and unraveled her own darling's tones, amid all the entangled outcry of a group of sportive children! But this could never be. Pearl was a born outcast of the infantile world. An imp of evil, emblem and product of sin, she had no right among christened infants. Nothing was more remarkable than the instinct, as it seemed, with which the child comprehended her loneliness; the destiny that had drawn an inviolable circle round about her; the whole peculiarity, in short, of her position in respect to other children. Never, since her release from prison, had Hester met the public gaze without her. In all her walks about the town, Pearl, too, was there; first as the babe in arms, and afterwards as the little girl, small companion of her mother, holding a forefinger with her whole grasp, and tripping along at the rate of three or four footsteps to one of Hester's. She saw the children of the settlement, on the grassy margin of the street, or at the domestic thresholds, disporting themselves in such grim fashion as the Puritanic nurture would permit; playing at going to church, perchance; or at scourging Quakers; or taking scalps in a sham-fight with the Indians; or scaring one another with freaks of imitative witchcraft. Pearl saw, and gazed intently, but never sought to make acquaintance. If spoken to, she would not speak again. If the children gathered about her, as they sometimes did, Pearl would grow positively terrible in her puny wrath, snatching up stones to fling at them, with shrill, incoherent exclamations, that made her mother tremble, because they had so much the sound of a witch's anathemas in some unknown tongue.[5]

The truth was, that the little Puritans, being of the most intolerant brood that ever lived, had got a vague idea of something outlandish, unearthly, or at variance with ordinary fashions, in the mother and child; and therefore scorned them in their hearts, and not unfrequently reviled them with their tongues. Pearl felt the sentiment, and requited it with the bitterest hatred that can be supposed to rankle in a childish bosom. These outbreaks of a fierce temper had a kind of value, and even comfort, for her mother; because there was at least an intelligible earnestness in the mood, instead of the fitful caprice that so often thwarted her in the child's manifestations. It appalled her, nevertheless, to discern here, again, a shadowy reflection of the evil that had existed in herself. All this enmity and passion had Pearl inherited, by inalienable right, out of Hester's heart. Mother and daughter stood together in the same circle of seclusion from human society; and in the nature of the child seemed to be perpetuated those unquiet elements that had distracted Hester Prynne before Pearl's birth, but had since begun to be soothed away by the softening influences of maternity.

At home, within and around her mother's cottage, Pearl wanted not a wide and various circle of acquaintance. The spell of life went forth from her ever creative spirit, and communicated itself to a thousand objects, as a torch kindles a flame wherever it may be applied. The unlikeliest materials,—a stick, a bunch of rags, a flower,—were the puppets of Pearl's witchcraft, and, without undergoing any outward change, became spiritually adapted to whatever drama occupied the stage of her inner world. Her

5. In Hawthorne's "The Gentle Boy" (1832), the Quaker child Ilbrahim is persecuted and ultimately killed by Puritan children.

one baby-voice served a multitude of imaginary personages, old and young, to talk withal. The pine-trees, aged, black and solemn, and flinging groans and other melancholy utterances on the breeze, needed little transformation to figure as Puritan elders; the ugliest weeds of the garden were their children, whom Pearl smote down and uprooted, most unmercifully. It was wonderful, the vast variety of forms into which she threw her intellect, with no continuity, indeed, but darting up and dancing, always in a state of preternatural activity,—soon sinking down, as if exhausted by so rapid and feverish a tide of life,—and succeeded by other shapes of a similar wild energy. It was like nothing so much as the phantasmagoric play of the northern lights. In the mere exercise of the fancy, however, and the sportiveness of a growing mind, there might be little more than was observable in other children of bright faculties; except as Pearl, in the dearth of human playmates, was thrown more upon the visionary throng which she created. The singularity lay in the hostile feelings with which the child regarded all these offspring of her own heart and mind. She never created a friend, but seemed always to be sowing broadcast the dragon's teeth, whence sprung a harvest of armed enemies, against whom she rushed to battle.[6] It was inexpressibly sad—then what depth of sorrow to a mother, who felt in her own heart the cause!—to observe, in one so young, this constant recognition of an adverse world, and so fierce a training of the energies that were to make good her cause, in the contest that must ensue.

Gazing at Pearl, Hester Prynne often dropped her work upon her knees, and cried out with an agony which she would fain have hidden, but which made utterance for itself, betwixt speech and a groan,—"O Father in Heaven;—if Thou art still my Father,—what is this being which I have brought into the world!" And Pearl, overhearing the ejaculation, or aware, through some more subtile channel, of those throbs of anguish, would turn her vivid and beautiful little face upon her mother, smile with sprite-like intelligence, and resume her play.

One peculiarity of the child's deportment remains yet to be told. The very first thing which she had noticed, in her life, was—what?—not the mother's smile, responding to it, as other babies do, by that faint, embryo smile of the little mouth, remembered so doubtfully afterwards, and with such fond discussion whether it were indeed a smile. By no means! But that first object of which Pearl seemed to become aware was—shall we say it?—the scarlet letter on Hester's bosom! One day, as her mother stooped over the cradle, the infant's eyes had been caught by the glimmering of the gold embroidery about the letter; and, putting up her little hand, she grasped at it, smiling, not doubtfully, but with a decided gleam, that gave her face the look of a much older child. Then, gasping for breath, did Hester Prynne clutch the fatal token, instinctively endeavoring to tear it away; so infinite was the torture inflicted by the intelligent touch of Pearl's baby-hand. Again, as if her mother's agonized gesture were meant only to make sport for her, did little Pearl look into her eyes, and smile! From that epoch, except when the child was asleep, Hester had never felt a moment's safety; not a moment's calm enjoyment of her. Weeks, it is true, would sometimes elapse, during which

6. In Greek mythology, Cadmus planted dragon's, or serpent's, teeth in the ground, and soldiers sprang up from the seeds.

Pearl's gaze might never once be fixed upon the scarlet letter; but then, again, it would come at unawares, like the stroke of sudden death, and always with that peculiar smile, and odd expression of the eyes.

Once, this freakish, elvish cast came into the child's eyes, while Hester was looking at her own image in them, as mothers are fond of doing; and, suddenly,—for women in solitude, and with troubled hearts, are pestered with unaccountable delusions,—she fancied that she beheld, not her own miniature portrait, but another face, in the small black mirror of Pearl's eye. It was a face, fiend-like, full of smiling malice, yet bearing the semblance of features that she had known full well, though seldom with a smile, and never with malice in them. It was as if an evil spirit possessed the child, and had just then peeped forth in mockery. Many a time afterwards had Hester been tortured, though less vividly, by the same illusion.[7]

In the afternoon of a certain summer's day, after Pearl grew big enough to run about, she amused herself with gathering handfuls of wild-flowers, and flinging them, one by one, at her mother's bosom; dancing up and down, like a little elf, whenever she hit the scarlet letter. Hester's first motion had been to cover her bosom with her clasped hands. But, whether from pride or resignation, or a feeling that her penance might best be wrought out by this unutterable pain, she resisted the impulse, and sat erect, pale as death, looking sadly into little Pearl's wild eyes. Still came the battery of flowers, almost invariably hitting the mark, and covering the mother's breast with hurts for which she could find no balm in this world, nor knew how to seek it in another. At last, her shot being all expended, the child stood still and gazed at Hester, with that little, laughing image of a fiend peeping out—or, whether it peeped or no, her mother so imagined it—from the unsearchable abyss of her black eyes.

"Child, what art thou?" cried the mother.

"O, I am your little Pearl!" answered the child.

But, while she said it, Pearl laughed, and began to dance up and down, with the humorsome gesticulation of a little imp, whose next freak might be to fly up the chimney.

"Art thou my child, in very truth?" asked Hester.

Nor did she put the question altogether idly, but, for the moment, with a portion of genuine earnestness; for, such was Pearl's wonderful intelligence, that her mother half doubted whether she were not acquainted with the secret spell of her existence, and might not now reveal herself.

"Yes; I am little Pearl!" repeated the child, continuing her antics.

"Thou art not my child! Thou art no Pearl of mine!" said the mother, half playfully; for it was often the case that a sportive impulse came over her, in the midst of her deepest suffering. "Tell me, then, what thou art, and who sent thee hither?"

"Tell me, mother!" said the child, seriously, coming up to Hester, and pressing herself close to her knees. "Do thou tell me!"

"Thy Heavenly Father sent thee!" answered Hester Prynne.

But she said it with a hesitation that did not escape the acuteness of the child. Whether moved only by her ordinary freakishness, or because an evil spirit prompted her, she put up her small forefinger, and touched the scarlet letter.

7. Compare the passage in Hawthorne's notebook on pp. 204–05 of this Norton Critical Edition.

"He did not send me!" cried she, positively. "I have no Heavenly Father!"

"Hush, Pearl, hush! Thou must not talk so!" answered the mother, suppressing a groan. "He sent us all into this world. He sent even me, thy mother. Then, much more, thee! Or, if not, thou strange and elfish child, whence didst thou come?"

"Tell me! Tell me!" repeated Pearl, no longer seriously, but laughing, and capering about the floor. "It is thou that must tell me!"

But Hester could not resolve the query, being herself in a dismal labyrinth of doubt. She remembered—betwixt a smile and a shudder—the talk of the neighboring townspeople; who, seeking vainly elsewhere for the child's paternity, and observing some of her odd attributes, had given out that poor little Pearl was a demon offspring; such as, ever since old Catholic times, had occasionally been seen on earth, through the agency of their mother's sin, and to promote some foul and wicked purpose. Luther, according to the scandal of his monkish enemies, was a brat of that hellish breed; nor was Pearl the only child to whom this inauspicious origin was assigned, among the New England Puritans.[8]

VII. The Governor's Hall

Hester Prynne went, one day, to the mansion of Governor Bellingham, with a pair of gloves, which she had fringed and embroidered to his order, and which were to be worn on some great occasion of state; for, though the chances of a popular election had caused this former ruler to descend a step or two from the highest rank, he still held an honorable and influential place among the colonial magistracy.[1]

Another and far more important reason than the delivery of a pair of embroidered gloves impelled Hester, at this time, to seek an interview with a personage of so much power and activity in the affairs of the settlement. It had reached her ears, that there was a design on the part of some of the leading inhabitants, cherishing the more rigid order of principles in religion and government, to deprive her of her child. On the supposition that Pearl, as already hinted, was of demon origin, these good people not unreasonably argued that a Christian interest in the mother's soul required them to remove such a stumbling-block from her path. If the child, on the other hand, were really capable of moral and religious growth, and possessed the elements of ultimate salvation, then, surely, it would enjoy all the fairer prospect of these advantages, by being transferred to wiser and better guardianship than Hester Prynne's. Among those who promoted the design, Governor Bellingham was said to be one of the most busy. It may appear singular, and, indeed, not a little ludicrous, that an affair of this kind, which, in later days, would have been referred to no higher jurisdiction than that of the selectmen of the town, should

8. Martin Luther (1483–1546), leader of the Protestant Reformation. Hawthorne refers to a legend that Luther's mother had intercourse with the devil. Heiko A. Oberman says that the rumor persisted into the nineteenth century; see *Luther: Man Between God and the Devil* (New Haven: Yale UP, 1989), 88.

1. Richard Bellingham's term as governor ended in May 1642. He would be reelected in 1654 and again in 1665.

then have been a question publicly discussed, and on which statesmen of eminence took sides. At that epoch of pristine simplicity, however, matters of even slighter public interest, and of far less intrinsic weight, than the welfare of Hester and her child, were strangely mixed up with the deliberations of legislators and acts of state. The period was hardly, if at all, earlier than that of our story, when a dispute concerning the right of property in a pig, not only caused a fierce and bitter contest in the legislative body of the colony, but resulted in an important modification of the framework itself of the legislature.[2]

Full of concern, therefore,—but so conscious of her own right that it seemed scarcely an unequal match between the public, on the one side, and a lonely woman, backed by the sympathies of nature, on the other,—Hester Prynne set forth from her solitary cottage. Little Pearl, of course, was her companion. She was now of an age to run lightly along by her mother's side, and, constantly in motion, from morn till sunset, could have accomplished a much longer journey than that before her. Often, nevertheless, more from caprice than necessity, she demanded to be taken up in arms; but was soon as imperious to be set down again, and frisked onward before Hester on the grassy pathway, with many a harmless trip and tumble. We have spoken of Pearl's rich and luxuriant beauty; a beauty that shone with deep and vivid tints; a bright complexion, eyes possessing intensity both of depth and glow, and hair already of a deep, glossy brown, and which, in after years, would be nearly akin to black. There was fire in her and throughout her; she seemed the unpremeditated offshoot of a passionate moment. Her mother, in contriving the child's garb, had allowed the gorgeous tendencies of her imagination their full play; arraying her in a crimson velvet tunic, of a peculiar cut, abundantly embroidered with fantasies and flourishes of gold thread. So much strength of coloring, which must have given a wan and pallid aspect to cheeks of a fainter bloom, was admirably adapted to Pearl's beauty, and made her the very brightest little jet of flame that ever danced upon the earth.

But it was a remarkable attribute of this garb, and, indeed, of the child's whole appearance, that it irresistibly and inevitably reminded the beholder of the token which Hester Prynne was doomed to wear upon her bosom. It was the scarlet letter in another form; the scarlet letter endowed with life! The mother herself—as if the red ignominy were so deeply scorched into her brain that all her conceptions assumed its form—had carefully wrought out the similitude; lavishing many hours of morbid ingenuity, to create an analogy between the object of her affection and the emblem of her guilt and torture. But, in truth, Pearl was the one, as well as the other; and only in consequence of that identity had Hester contrived so perfectly to represent the scarlet letter in her appearance.

2. Caleb Snow discusses this incident at length. A "thoughtless pig strayed from its owner, one good Mrs. Sherman," and "wandered through the town, breaking into every body's corn as its hunger dictated." The pig was apparently kept by a Captain Keayne for nearly a year before he decided to kill it. At that point, the owner of the stray pig demanded restitution, which Keayne refused. The matter went to church court, which found in Keayne's favor. The case was appealed several times, and eventually caused a rift between the magistrates and their deputies such that two houses of government were created. As Snow concludes, "this was the origin of our present Senate" (*History of Boston*, 95–97).

As the two wayfarers came within the precincts of the town, the children of the Puritans looked up from their play,—or what passed for play with those sombre little urchins,—and spake gravely one to another:—

"Behold, verily, there is the woman of the scarlet letter; and, of a truth, moreover, there is the likeness of the scarlet letter running along by her side! Come, therefore, and let us fling mud at them!"

But Pearl, who was a dauntless child, after frowning, stamping her foot, and shaking her little hand with a variety of threatening gestures, suddenly made a rush at the knot of her enemies, and put them all to flight. She resembled, in her fierce pursuit of them, an infant pestilence,—the scarlet fever, or some such half-fledged angel of judgment,—whose mission was to punish the sins of the rising generation. She screamed and shouted, too, with a terrific volume of sound, which, doubtless, caused the hearts of the fugitives to quake within them. The victory accomplished, Pearl returned quietly to her mother, and looked up, smiling, into her face.

Without further adventure, they reached the dwelling of Governor Bellingham. This was a large wooden house, built in a fashion of which there are specimens still extant in the streets of our elder towns; now moss-grown, crumbling to decay, and melancholy at heart with the many sorrowful or joyful occurrences, remembered or forgotten, that have happened, and passed away, within their dusky chambers. Then, however, there was the freshness of the passing year on its exterior, and the cheerfulness, gleaming forth from the sunny windows, of a human habitation, into which death had never entered. It had, indeed, a very cheery aspect; the walls being overspread with a kind of stucco, in which fragments of broken glass were plentifully intermixed; so that, when the sunshine fell aslant-wise over the front of the edifice, it glittered and sparkled as if diamonds had been flung against it by the double handful. The brilliancy might have befitted Aladdin's palace, rather than the mansion of a grave old Puritan ruler. It was further decorated with strange and seemingly cabalistic figures and diagrams, suitable to the quaint taste of the age, which had been drawn in the stucco when newly laid on, and had now grown hard and durable, for the admiration of after times.

Pearl, looking at this bright wonder of a house, began to caper and dance, and imperatively required that the whole breadth of sunshine should be stripped off its front, and given her to play with.

"No, my little Pearl!" said her mother. "Thou must gather thine own sunshine. I have none to give thee!"

They approached the door; which was of an arched form, and flanked on each side by a narrow tower or projection of the edifice, in both of which were lattice-windows, with wooden shutters to close over them at need. Lifting the iron hammer that hung at the portal, Hester Prynne gave a summons, which was answered by one of the Governor's bond-servants; a free-born Englishman, but now a seven years' slave. During that term he was to be the property of his master, and as much a commodity of bargain and sale as an ox, or a joint-stool. The serf wore the blue coat, which was the customary garb of serving-men at that period, and long before, in the old hereditary halls of England.

"Is the worshipful Governor Bellingham within?" inquired Hester.

"Yea, forsooth," replied the bond-servant, staring with wide-open eyes at the scarlet letter, which, being a new-comer in the country, he had never before seen. "Yea, his honorable worship is within. But he hath a godly minister or two with him, and likewise a leech. Ye may not see his worship now."

"Nevertheless, I will enter," answered Hester Prynne; and the bond-servant, perhaps judging from the decision of her air, and the glittering symbol in her bosom, that she was a great lady in the land, offered no opposition.

So the mother and little Pearl were admitted into the hall of entrance. With many variations, suggested by the nature of his building-materials, diversity of climate, and a different mode of social life, Governor Bellingham had planned his new habitation after the residences of gentlemen of fair estate in his native land. Here, then, was a wide and reasonably lofty hall, extending through the whole depth of the house, and forming a medium of general communication, more or less directly, with all the other apartments. At one extremity, this spacious room was lighted by the windows of the two towers, which formed a small recess on either side of the portal. At the other end, though partly muffled by a curtain, it was more powerfully illuminated by one of those embowed hall-windows which we read of in old books, and which was provided with a keep and cushioned seat. Here, on the cushion, lay a folio tome, probably of the Chronicles of England, or other such substantial literature; even as, in our own days, we scatter gilded volumes on the centre-table, to be turned over by the casual guest. The furniture of the hall consisted of some ponderous chairs, the backs of which were elaborately carved with wreaths of oaken flowers; and likewise a table in the same taste; the whole being of the Elizabethan age, or perhaps earlier, and heirlooms, transferred hither from the Governor's paternal home. On the table—in token that the sentiment of old English hospitality had not been left behind—stood a large pewter tankard, at the bottom of which, had Hester or Pearl peeped into it, they might have seen the frothy remnant of a recent draught of ale.

On the wall hung a row of portraits, representing the forefathers of the Bellingham lineage, some with armor on their breasts, and others with stately ruffs and robes of peace. All were characterized by the sternness and severity which old portraits so invariably put on; as if they were the ghosts, rather than the pictures, of departed worthies, and were gazing with harsh and intolerant criticism at the pursuits and enjoyments of living men.

At about the centre of the oaken panels, that lined the hall, was suspended a suit of mail, not, like the pictures, an ancestral relic, but of the most modern date; for it had been manufactured by a skilful armorer in London, the same year in which Governor Bellingham came over to New England. There was a steel head-piece, a cuirass, a gorget, and greaves, with a pair of gauntlets and a sword hanging beneath; all, and especially the helmet and breastplate, so highly burnished as to glow with white radiance, and scatter an illumination everywhere about upon the floor.[3] This bright panoply was not meant for mere idle show, but had been worn by the Governor on many a solemn muster and training field, and

3. Hawthorne lists several pieces of armor: *cuirass* (breastplate), *gorget* (throat protector), *greaves* (shin guards), and *gauntlets* (gloves).

had glittered, moreover, at the head of a regiment in the Pequod war.[4] For, though bred a lawyer, and accustomed to speak of Bacon, Coke, Noye, and Finch, as his professional associates,[5] the exigencies of this new country had transformed Governor Bellingham into a soldier, as well as a statesman and ruler.

Little Pearl—who was as greatly pleased with the gleaming armor as she had been with the glittering frontispiece of the house—spent some time looking into the polished mirror of the breastplate.

"Mother," cried she, "I see you here. Look! Look!"

Hester looked, by way of humoring the child; and she saw that, owing to the peculiar effect of this convex mirror, the scarlet letter was represented in exaggerated and gigantic proportions, so as to be greatly the most prominent feature of her appearance. In truth, she seemed absolutely hidden behind it. Pearl pointed upward, also, at a similar picture in the head-piece; smiling at her mother, with the elfish intelligence that was so familiar an expression on her small physiognomy. That look of naughty merriment was likewise reflected in the mirror, with so much breadth and intensity of effect, that it made Hester Prynne feel as if it could not be the image of her own child, but of an imp who was seeking to mould itself into Pearl's shape.

"Come along, Pearl," said she, drawing her away. "Come and look into this fair garden. It may be, we shall see flowers there; more beautiful ones than we find in the woods."

Pearl, accordingly, ran to the bow-window, at the further end of the hall, and looked along the vista of a garden-walk, carpeted with closely shaven grass, and bordered with some rude and immature attempt at shrubbery. But the proprietor appeared already to have relinquished, as hopeless, the effort to perpetuate on this side of the Atlantic, in a hard soil and amid the close struggle for subsistence, the native English taste for ornamental gardening. Cabbages grew in plain sight; and a pumpkin-vine, rooted at some distance, had run across the intervening space, and deposited one of its gigantic products directly beneath the half-window; as if to warn the Governor that this great lump of vegetable gold was as

4. Histories of the Pequot War do not mention Bellingham's participation in the war, which occurred during the late spring and summer of 1637. Hawthorne undoubtedly knew of the internal political wrangling among the Massachusetts Bay contingent, the result, as Alfred Cave explains, of the Puritans' giving high military rank to magistrates such as Winthrop and Dudley while the actual field commanders received lesser ranks. The man who actually led the Massachusetts Bay forces, Captain John Underhill, complained about this practice. Underhill also strongly supported Anne Hutchinson, an allegiance that cost him his rank and position after the war. Hawthorne may, then, be making a visual joke at Bellingham's expense by placing an empty suit of armor on the wall of his house. The Pequots were nearly wiped out by a coalition of Puritan forces from Massachusetts and Connecticut. These forces set fire to the Pequot fort at Mystic (May 26, 1637) and killed more than 500 members of the tribe, including women and children.

5. Francis Bacon, Sir Edward Coke, William Noy, and Sir John Finch played prominent political roles during the reigns of King James I and King Charles I, and each played a part in one of the sensational legal cases (the Thomas Overbury murder case and William Prynne's prosecution) that lurk in the background of *The Scarlet Letter.* Finch (1584–1634), for example, served as chief justice of the Common Pleas (1634–1640) and in that role prosecuted Prynne for libel. As Samuel R. Gardiner notes, it was Finch who "savagely added a wish that Prynne should be branded on the cheeks with the letters S.L., as a Seditious Libeller" (8:229). Noy (1577–1634), King Charles I's attorney general from 1632 until 1634, was also involved in Prynne's prosecution (Gardiner 7:334). Bacon (1561–1626), member of the House of Commons, is considered one of the founders of modern science. He and Coke (1552–1634), also a prominent member of the House of Commons and political rival of Bacon, prosecuted the murderers of Thomas Overbury, including Lord (Robert Carr) and Lady (Frances Howard) Somerset. For more on that case see n. 5, p. 80, of this Norton Critical Edition.

rich an ornament as New England earth would offer him. There were a few rose-bushes, however, and a number of apple-trees, probably the descendants of those planted by the Reverend Mr. Blackstone, the first settler of the peninsula; that half mythological personage, who rides through our early annals, seated on the back of a bull.[6]

Pearl, seeing the rose-bushes, began to cry for a red rose, and would not be pacified.

"Hush, child, hush!" said her mother, earnestly. "Do not cry, dear little Pearl! I hear voices in the garden. The Governor is coming, and gentlemen along with him!"

In fact, adown the vista of the garden avenue, a number of persons were seen approaching towards the house. Pearl, in utter scorn of her mother's attempt to quiet her, gave an eldritch scream, and then became silent; not from any notion of obedience, but because the quick and mobile curiosity of her disposition was excited by the appearance of these new personages.

VIII. The Elf-Child and the Minister

Governor Bellingham, in a loose gown and easy cap,—such as elderly gentlemen loved to endue themselves with, in their domestic privacy,—walked foremost, and appeared to be showing off his estate, and expatiating on his projected improvements. The wide circumference of an elaborate ruff, beneath his gray beard, in the antiquated fashion of King James' reign, caused his head to look not a little like that of John the Baptist in a charger.[1] The impression made by his aspect, so rigid and severe, and frost-bitten with more than autumnal age, was hardly in keeping with the appliances of worldly enjoyment wherewith he had evidently done his utmost to surround himself. But it is an error to suppose that our grave forefathers—though accustomed to speak and think of human existence as a state merely of trial and warfare, and though unfeignedly prepared to sacrifice goods and life at the behest of duty—made it a matter of conscience to reject such means of comfort, or even luxury, as lay fairly within their grasp. This creed was never taught, for instance, by the venerable pastor, John Wilson, whose beard, white as a snow-drift, was seen over Governor Bellingham's shoulder; while its wearer suggested that pears and peaches might yet be naturalized in the New England climate, and that purple grapes might possibly be compelled to flourish, against the sunny garden-wall. The old clergyman, nurtured at the rich bosom of the English Church, had a long-established and legitimate taste for all good and comfortable things; and however stern he might show

6. Blackstone was an early settler who built his house on a peninsula at the mouth of the Charles River. According to Caleb Snow, in April 1633 Blackstone was granted fifty acres in return for the peninsula, called Blackstone's Neck (50). Snow calls Blackstone a "very eccentrick character" (51) and reports that when he grew old and unable to travel on foot, "he used to ride on a bull, which he had tamed and tutored to that use" (53).

1. See Mark 6.21–29. The daughter of Herodias asked King Herod for the head of John the Baptist on a charger (platter): "And immediately the king sent an executioner, and commanded his head to be brought: and he went and beheaded him in the prison / And brought his head in a charger, and gave it to the damsel: and the damsel gave it to her mother / And when his disciples heard of it, they came and took up his corpse, and laid it in a tomb."

himself in the pulpit, or in his public reproof of such transgressions as that of Hester Prynne, still, the genial benevolence of his private life had won him warmer affection than was accorded to any of his professional contemporaries.

Behind the Governor and Mr. Wilson came two other guests; one, the Reverend Arthur Dimmesdale, whom the reader may remember, as having taken a brief and reluctant part in the scene of Hester Prynne's disgrace; and, in close companionship with him, old Roger Chillingworth, a person of great skill in physic, who, for two or three years past, had been settled in the town. It was understood that this learned man was the physician as well as friend of the young minister, whose health had severely suffered, of late, by his too unreserved self-sacrifice to the labors and duties of the pastoral relation.

The Governor, in advance of his visitors, ascended one or two steps, and, throwing open the leaves of the great hall window, found himself close to little Pearl. The shadow of the curtain fell on Hester Prynne, and partially concealed her.

"What have we here?" said Governor Bellingham, looking with surprise at the scarlet little figure before him. "I profess, I have never seen the like, since my days of vanity, in old King James' time, when I was wont to esteem it a high favor to be admitted to a court mask! There used to be a swarm of these small apparitions, in holiday time; and we called them children of the Lord of Misrule.[2] But how gat such a guest into my hall?"

"Ay, indeed!" cried good old Mr. Wilson. "What little bird of scarlet plumage may this be? Methinks I have seen just such figures, when the sun has been shining through a richly painted window, and tracing out the golden and crimson images across the floor. But that was in the old land. Prithee, young one, who art thou, and what has ailed thy mother to bedizen thee in this strange fashion? Art thou a Christian child,—ha? Dost know thy catechism? Or art thou one of those naughty elfs or fairies, whom we thought to have left behind us, with other relics of Papistry, in merry old England?"

"I am mother's child," answered the scarlet vision, "and my name is Pearl!"

"Pearl?—Ruby, rather!—or Coral!—or Red Rose, at the very least, judging from thy hue!" responded the old minister, putting forth his hand in a vain attempt to pat little Pearl on the cheek. "But where is this mother of thine? Ah! I see," he added; and, turning to Governor Bellingham, whispered, "This is the selfsame child of whom we have held speech together; and behold here the unhappy woman, Hester Prynne, her mother!"

"Sayest thou so?" cried the Governor. "Nay, we might have judged that such a child's mother must needs be a scarlet woman, and a worthy type of her of Babylon![3] But she comes at a good time; and we will look into this matter forthwith."

2. The master of Christmas revels in medieval England, associated with pleasure and rule-breaking.
3. Bellingham compares Hester to the Whore of Babylon. See Revelation 17.4–5: "And the woman was arrayed in purple and scarlet colour, and decked with gold and precious stones and pearls, having a golden cup in her hand full of abominations and filthiness of her fornication: / And upon her forehead was a name written, MYSTERY, BABYLON THE GREAT, THE MOTHER OF HARLOTS AND ABOMINATIONS OF THE EARTH."

Governor Bellingham stepped through the window into the hall, followed by his three guests.

"Hester Prynne," said he, fixing his naturally stern regard on the wearer of the scarlet letter, "there hath been much question concerning thee, of late. The point hath been weightily discussed, whether we, that are of authority and influence, do well discharge our consciences by trusting an immortal soul, such as there is in yonder child, to the guidance of one who hath stumbled and fallen, amid the pitfalls of this world. Speak thou, the child's own mother! Were it not, thinkest thou, for thy little one's temporal and eternal welfare, that she be taken out of thy charge, and clad soberly, and disciplined strictly, and instructed in the truths of heaven and earth? What canst thou do for the child, in this kind?"

"I can teach my little Pearl what I have learned from this!" answered Hester Prynne, laying her finger on the red token.

"Woman, it is thy badge of shame!" replied the stern magistrate. "It is because of the stain which that letter indicates, that we would transfer thy child to other hands."

"Nevertheless," said the mother, calmly, though growing more pale, "this badge hath taught me,—it daily teaches me,—it is teaching me at this moment,—lessons whereof my child may be the wiser and better, albeit they can profit nothing to myself."

"We will judge warily," said Bellingham, "and look well what we are about to do. Good Master Wilson, I pray you, examine this Pearl,—since that is her name,—and see whether she hath had such Christian nurture as befits a child of her age."

The old minister seated himself in an arm-chair, and made an effort to draw Pearl betwixt his knees. But the child, unaccustomed to the touch or familiarity of any but her mother, escaped through the open window, and stood on the upper step, looking like a wild tropical bird, of rich plumage, ready to take flight into the upper air. Mr. Wilson, not a little astonished at this outbreak,—for he was a grandfatherly sort of personage, and usually a vast favorite with children,—essayed, however, to proceed with the examination.

"Pearl," said he, with great solemnity, "thou must take heed to instruction, that so, in due season, thou mayest wear in thy bosom the pearl of great price. Canst thou tell me, my child, who made thee?"

Now Pearl knew well enough who made her; for Hester Prynne, the daughter of a pious home, very soon after her talk with the child about her Heavenly Father, had begun to inform her of those truths which the human spirit, at whatever stage of immaturity, imbibes with such eager interest. Pearl, therefore, so large were the attainments of her three years' lifetime, could have borne a fair examination in the New England Primer, or the first column of the Westminster Catechisms, although unacquainted with the outward form of either of those celebrated works.[4] But

4. *The New England Primer*, first published probably in 1690, was designed to teach children the alphabet. The best-known verse is undoubtedly the one for the letter A: "In ADAM'S Fall / We sinned all." The verse for the letter Q might also be pertinent: "Queen Esther sues / And saves the Jews." *The Westminster Catechism*, which was published in 1647, included a series of more than 100 questions and answers based on Biblical passages. The closest question to the one Reverend Wilson asks Pearl is "How did God create man?" The answer, from Genesis, is that God created man in His own image. Given the time of the novel's action, Pearl could not have "borne a fair examination" in either of these works.

that perversity, which all children have more or less of, and of which little Pearl had a ten-fold portion, now, at the most inopportune moment, took thorough possession of her, and closed her lips, or impelled her to speak words amiss. After putting her finger in her mouth, with many ungracious refusals to answer good Mr. Wilson's question, the child finally announced that she had not been made at all, but had been plucked by her mother off the bush of wild roses that grew by the prison-door.

This fantasy was probably suggested by the near proximity of the Governor's red roses, as Pearl stood outside of the window; together with her recollection of the prison rose-bush, which she had passed in coming hither.

Old Roger Chillingworth, with a smile on his face, whispered something in the young clergyman's ear. Hester Prynne looked at the man of skill, and even then, with her fate hanging in the balance, was startled to perceive what a change had come over his features,—how much uglier they were,—how his dark complexion seemed to have grown duskier, and his figure more misshapen,—since the days when she had familiarly known him. She met his eyes for an instant, but was immediately constrained to give all her attention to the scene now going forward.

"This is awful!" cried the Governor, slowly recovering from the astonishment into which Pearl's response had thrown him. "Here is a child of three years old, and she cannot tell who made her! Without question, she is equally in the dark as to her soul, its present depravity, and future destiny! Methinks, gentlemen, we need inquire no further."

Hester caught hold of Pearl, and drew her forcibly into her arms, confronting the old Puritan magistrate with almost a fierce expression. Alone in the world, cast off by it, and with this sole treasure to keep her heart alive, she felt that she possessed indefeasible rights against the world, and was ready to defend them to the death.

"God gave me the child!" cried she. "He gave her in requital of all things else, which ye had taken from me. She is my happiness!—she is my torture, none the less! Pearl keeps me here in life! Pearl punishes me too! See ye not, she is the scarlet letter, only capable of being loved, and so endowed with a million-fold the power of retribution for my sin? Ye shall not take her! I will die first!"

"My poor woman," said the not unkind old minister, "the child shall be well cared for!—far better than thou canst do it."

"God gave her into my keeping," repeated Hester Prynne, raising her voice almost to a shriek. "I will not give her up!"—And here, by a sudden impulse, she turned to the young clergyman, Mr. Dimmesdale, at whom, up to this moment, she had seemed hardly so much as once to direct her eyes.—"Speak thou for me!" cried she. "Thou wast my pastor, and hadst charge of my soul, and knowest me better than these men can. I will not lose the child! Speak for me! Thou knowest,—for thou hast sympathies which these men lack!—thou knowest what is in my heart, and what are a mother's rights, and how much the stronger they are, when that mother has but her child and the scarlet letter! Look thou to it! I will not lose the child! Look to it!"

At this wild and singular appeal, which indicated that Hester Prynne's situation had provoked her to little less than madness, the young minister at once came forward, pale, and holding his hand over his heart, as was his custom whenever his peculiarly nervous temperament was thrown into agitation. He looked now more careworn and emaciated than as we described him at the scene of Hester's public ignominy; and whether it were his failing health, or whatever the cause might be, his large dark eyes had a world of pain in their troubled and melancholy depth.

"There is truth in what she says," began the minister, with a voice sweet, tremulous, but powerful, insomuch that the hall reëchoed, and the hollow armor rang with it,—"truth in what Hester says, and in the feeling which inspires her! God gave her the child, and gave her, too, an instinctive knowledge of its nature and requirements,—both seemingly so peculiar,—which no other mortal being can possess. And, moreover, is there not a quality of awful sacredness in the relation between this mother and this child?"

"Ay!—how is that, good Master Dimmesdale?" interrupted the Governor. "Make that plain, I pray you!"

"It must be even so," resumed the minister. "For, if we deem it otherwise, do we not thereby say that the Heavenly Father, the Creator of all flesh, hath lightly recognized a deed of sin, and made of no account the distinction between unhallowed lust and holy love? This child of its father's guilt and its mother's shame hath come from the hand of God, to work in many ways upon her heart, who pleads so earnestly, and with such bitterness of spirit, the right to keep her. It was meant for a blessing; for the one blessing of her life! It was meant, doubtless, as the mother herself hath told us, for a retribution too; a torture to be felt at many an unthought of moment; a pang, a sting, an ever-recurring agony, in the midst of a troubled joy! Hath she not expressed this thought in the garb of the poor child, so forcibly reminding us of that red symbol which sears her bosom?"

"Well said, again!" cried good Mr. Wilson. "I feared the woman had no better thought than to make a mountebank[5] of her child!"

"O, not so!—not so!" continued Mr. Dimmesdale. "She recognizes, believe me, the solemn miracle which God hath wrought, in the existence of that child. And may she feel, too,—what, methinks, is the very truth—that this boon was meant, above all things else, to keep the mother's soul alive, and to preserve her from blacker depths of sin into which Satan might else have sought to plunge her! Therefore it is good for this poor, sinful woman that she hath an infant immortality, a being capable of eternal joy or sorrow, confided to her care,—to be trained up by her to righteousness,—to remind her, at every moment, of her fall,—but yet to teach her, as it were by the Creator's sacred pledge, that, if she bring the child to heaven, the child also will bring its parent thither! Herein is the sinful mother happier than the sinful father. For Hester Prynne's sake, then, and no less for the poor child's sake, let us leave them as Providence hath seen fit to place them!"

"You speak, my friend, with a strange earnestness," said old Roger Chillingworth, smiling at him.

"And there is a weighty import in what my young brother hath spoken," added the Reverend Mr. Wilson. "What say you, worshipful Master Bellingham? Hath he not pleaded well for the poor woman?"

"Indeed hath he," answered the magistrate, "and hath adduced such arguments, that we will even leave the matter as it now stands; so long, at

5. Charlatan or pretender.

least, as there shall be no further scandal in the woman. Care must be had, nevertheless, to put the child to due and stated examination in the catechism, at thy hands or Master Dimmesdale's. Moreover, at a proper season, the tithing-men must take heed that she go both to school and to meeting."[6]

The young minister, on ceasing to speak, had withdrawn a few steps from the group, and stood with his face partially concealed in the heavy folds of the window-curtain; while the shadow of his figure, which the sunlight cast upon the floor, was tremulous with the vehemence of his appeal. Pearl, that wild and flighty little elf, stole softly towards him, and taking his hand in the grasp of both her own, laid her cheek against it; a caress so tender, and withal so unobtrusive, that her mother, who was looking on, asked herself,—"Is that my Pearl?" Yet she knew that there was love in the child's heart, although it mostly revealed itself in passion, and hardly twice in her lifetime had been softened by such gentleness as now. The minister,—for, save the long-sought regards of woman, nothing is sweeter than these marks of childish preference, accorded spontaneously by a spiritual instinct, and therefore seeming to imply in us something truly worthy to be loved,—the minister looked round, laid his hand on the child's head, hesitated an instant, and then kissed her brow. Little Pearl's unwonted mood of sentiment lasted no longer; she laughed, and went capering down the hall, so airily, that old Mr. Wilson raised a question whether even her tiptoes touched the floor.

"The little baggage hath witchcraft in her, I profess," said he to Mr. Dimmesdale. "She needs no old woman's broomstick to fly withal!"

"A strange child!" remarked old Roger Chillingworth. "It is easy to see the mother's part in her. Would it be beyond a philosopher's research, think ye, gentlemen, to analyze that child's nature, and, from its make and mould, to give a shrewd guess at the father?"

"Nay; it would be sinful, in such a question, to follow the clew of profane philosophy," said Mr. Wilson. "Better to fast and pray upon it; and still better, it may be, to leave the mystery as we find it, unless Providence reveal it of its own accord. Thereby, every good Christian man hath a title to show a father's kindness towards the poor, deserted babe."

The affair being so satisfactorily concluded, Hester Prynne, with Pearl, departed from the house. As they descended the steps, it is averred that the lattice of a chamber-window was thrown open, and forth into the sunny day was thrust the face of Mistress Hibbins, Governor Bellingham's bitter-tempered sister, and the same who, a few years later, was executed as a witch.[7]

"Hist, hist!" said she, while her ill-omened physiognomy seemed to cast a shadow over the cheerful newness of the house. "Wilt thou go with us to-night? There will be a merry company in the forest; and I well-nigh promised the Black Man that comely Hester Prynne should make one."[8]

"Make my excuse to him, so please you!" answered Hester, with a triumphant smile. "I must tarry at home, and keep watch over my little

6. Tithing men collect church taxes—typically, one tenth of one's income.
7. As noted, Ann Hibbins was executed as a witch in 1656.
8. Compare the forest scene in "Young Goodman Brown," in which both Brown and his wife, Faith, apparently meet the devil, or the Black Man, in the forest (see pp. 175–77 of this Norton Critical Edition).

Pearl. Had they taken her from me, I would willingly have gone with thee into the forest, and signed my name in the Black Man's book too, and that with mine own blood!"

"We shall have thee there anon!" said the witch-lady, frowning, as she drew back her head.

But here—if we suppose this interview betwixt Mistress Hibbins and Hester Prynne to be authentic, and not a parable—was already an illustration of the young minister's argument against sundering the relation of a fallen mother to the offspring of her frailty. Even thus early had the child saved her from Satan's snare.

IX. The Leech

Under the appellation of Roger Chillingworth, the reader will remember, was hidden another name, which its former wearer had resolved should never more be spoken. It has been related, how, in the crowd that witnessed Hester Prynne's ignominious exposure, stood a man, elderly, travel-worn, who, just emerging from the perilous wilderness, beheld the woman, in whom he hoped to find embodied the warmth and cheerfulness of home, set up as a type of sin before the people. Her matronly fame was trodden under all men's feet. Infamy was babbling around her in the public market-place. For her kindred, should the tidings ever reach them, and for the companions of her unspotted life, there remained nothing but the contagion of her dishonor; which would not fail to be distributed in strict accordance and proportion with the intimacy and sacredness of their previous relationship. Then why—since the choice was with himself—should the individual, whose connection with the fallen woman had been the most intimate and sacred of them all, come forward to vindicate his claim to an inheritance so little desirable? He resolved not to be pilloried beside her on her pedestal of shame. Unknown to all but Hester Prynne, and possessing the lock and key of her silence, he chose to withdraw his name from the roll of mankind, and, as regarded his former ties and interests, to vanish out of life as completely as if he indeed lay at the bottom of the ocean, whither rumor had long ago consigned him. This purpose once effected, new interests would immediately spring up, and likewise a new purpose; dark, it is true, if not guilty, but of force enough to engage the full strength of his faculties.

In pursuance of this resolve, he took up his residence in the Puritan town, as Roger Chillingworth, without other introduction than the learning and intelligence of which he possessed more than a common measure. As his studies, at a previous period of his life, had made him extensively acquainted with the medical science of the day, it was as a physician that he presented himself, and as such was cordially received. Skilful men, of the medical and chirurgical[1] profession, were of rare occurrence in the colony. They seldom, it would appear, partook of the religious zeal that brought other emigrants across the Atlantic. In their researches into the human frame, it may be that the higher and more subtile faculties of such men were materialized, and that they lost the spiritual view of existence

1. Surgical (archaic form).

amid the intricacies of that wondrous mechanism, which seemed to involve art enough to comprise all of life within itself. At all events, the health of the good town of Boston, so far as medicine had aught to do with it, had hitherto lain in the guardianship of an aged deacon and apothecary, whose piety and godly deportment were stronger testimonials in his favor than any that he could have produced in the shape of a diploma. The only surgeon was one who combined the occasional exercise of that noble art with the daily and habitual flourish of a razor. To such a professional body Roger Chillingworth was a brilliant acquisition. He soon manifested his familiarity with the ponderous and imposing machinery of antique physic; in which every remedy contained a multitude of far-fetched and heterogeneous ingredients, as elaborately compounded as if the proposed result had been the Elixir of Life.[2] In his Indian captivity, moreover, he had gained much knowledge of the properties of native herbs and roots; nor did he conceal from his patients, that these simple medicines, Nature's boon to the untutored savage, had quite as large a share of his own confidence as the European pharmacopœia, which so many learned doctors had spent centuries in elaborating.

This learned stranger was exemplary, as regarded, at least, the outward forms of a religious life, and, early after his arrival, had chosen for his spiritual guide the Reverend Mr. Dimmesdale. The young divine, whose scholar-like renown still lived in Oxford, was considered by his more fervent admirers as little less than a heavenly-ordained apostle, destined, should he live and labor for the ordinary term of life, to do as great deeds for the now feeble New England Church, as the early Fathers had achieved for the infancy of the Christian faith. About this period, however, the health of Mr. Dimmesdale had evidently begun to fail. By those best acquainted with his habits, the paleness of the young minister's cheek was accounted for by his too earnest devotion to study, his scrupulous fulfilment of parochial duty, and, more than all, by the fasts and vigils of which he made a frequent practice, in order to keep the grossness of this earthly state from clogging and obscuring his spiritual lamp. Some declared, that, if Mr. Dimmesdale were really going to die, it was cause enough, that the world was not worthy to be any longer trodden by his feet. He himself, on the other hand, with characteristic humility, avowed his belief, that, if Providence should see fit to remove him, it would be because of his own unworthiness to perform its humblest mission here on earth. With all this difference of opinion as to the cause of his decline, there could be no question of the fact. His form grew emaciated; his voice, though still rich and sweet, had a certain melancholy prophecy of decay in it; he was often observed, on any slight alarm or other sudden accident, to put his hand over his heart, with first a flush and then a paleness, indicative of pain.

Such was the young clergyman's condition, and so imminent the prospect that his dawning light would be extinguished, all untimely, when Roger Chillingworth made his advent to the town. His first entry on the scene, few people could tell whence, dropping down, as it were, out of the sky, or starting from the nether earth, had an aspect of mystery, which was easily heightened to the miraculous. He was now known to be a man

2. From alchemy; a potion for making one immortal.

of skill; it was observed that he gathered herbs, and the blossoms of wild-flowers, and dug up roots, and plucked off twigs from the forest-trees, like one acquainted with hidden virtues in what was valueless to common eyes. He was heard to speak of Sir Kenelm Digby,[3] and other famous men,—whose scientific attainments were esteemed hardly less than supernatural,—as having been his correspondents or associates. Why, with such rank in the learned world, had he come hither? What could he, whose sphere was in great cities, be seeking in the wilderness? In answer to this query, a rumor gained ground,—and, however absurd, was entertained by some very sensible people,—that Heaven had wrought an absolute miracle, by transporting an eminent Doctor of Physic, from a German university, bodily through the air, and setting him down at the door of Mr. Dimmesdale's study! Individuals of wiser faith, indeed, who knew that Heaven promotes its purposes without aiming at the stage-effect of what is called miraculous interposition, were inclined to see a providential hand in Roger Chillingworth's so opportune arrival.

This idea was countenanced by the strong interest which the physician ever manifested in the young clergyman; he attached himself to him as a parishioner, and sought to win a friendly regard and confidence from his naturally reserved sensibility. He expressed great alarm at his pastor's state of health, but was anxious to attempt the cure, and, if early undertaken, seemed not despondent of a favorable result. The elders, the deacons, the motherly dames, and the young and fair maidens, of Mr. Dimmesdale's flock, were alike importunate that he should make trial of the physician's frankly offered skill. Mr. Dimmesdale gently repelled their entreaties.

"I need no medicine," said he.

But how could the young minister say so, when, with every successive Sabbath, his cheek was paler and thinner, and his voice more tremulous than before,—when it had now become a constant habit, rather than a casual gesture, to press his hand over his heart? Was he weary of his labors? Did he wish to die? These questions were solemnly propounded to Mr. Dimmesdale by the elder ministers of Boston and the deacons of his church, who, to use their own phrase, "dealt with him" on the sin of rejecting the aid which Providence so manifestly held out. He listened in silence, and finally promised to confer with the physician.

"Were it God's will," said the Reverend Mr. Dimmesdale, when, in fulfilment of this pledge, he requested old Roger Chillingworth's professional advice, "I could be well content, that my labors, and my sorrows, and my sins, and my pains, should shortly end with me, and what is earthly of them be buried in my grave, and the spiritual go with me to my eternal state, rather than that you should put your skill to the proof in my behalf."

"Ah," replied Roger Chillingworth, with that quietness which, whether imposed or natural, marked all his deportment, "it is thus that a young clergyman is apt to speak. Youthful men, not having taken a deep root, give up their hold of life so easily! And saintly men, who walk with God on earth, would fain be away, to walk with him on the golden pavements of the New Jerusalem."

3. Sir Kenelm Digby (1603–1665), physical scientist and naval commander, authored *Nature of Bodies* and *The Immortality of Reasonable Souls* in 1644. An alchemist and astrologer, he would have shared interests with Chillingworth.

"Nay," rejoined the young minister, putting his hand to his heart, with a flush of pain flitting over his brow, "were I worthier to walk there, I could be better content to toil here."

"Good men ever interpret themselves too meanly," said the physician.

In this manner, the mysterious old Roger Chillingworth became the medical adviser of the Reverend Mr. Dimmesdale. As not only the disease interested the physician, but he was strongly moved to look into the character and qualities of the patient, these two men, so different in age, came gradually to spend much time together. For the sake of the minister's health, and to enable the leech to gather plants with healing balm in them, they took long walks on the sea-shore, or in the forest; mingling various talk with the plash and murmur of the waves, and the solemn wind-anthem among the tree-tops. Often, likewise, one was the guest of the other, in his place of study and retirement. There was a fascination for the minister in the company of the man of science, in whom he recognized an intellectual cultivation of no moderate depth or scope; together with a range and freedom of ideas, that he would have vainly looked for among the members of his own profession. In truth, he was startled, if not shocked, to find this attribute in the physician. Mr. Dimmesdale was a true priest, a true religionist, with the reverential sentiment largely developed, and an order of mind that impelled itself powerfully along the track of a creed, and wore its passage continually deeper with the lapse of time. In no state of society would he have been what is called a man of liberal views; it would always be essential to his peace to feel the pressure of a faith about him, supporting, while it confined him within its iron framework. Not the less, however, though with a tremulous enjoyment, did he feel the occasional relief of looking at the universe through the medium of another kind of intellect than those with which he habitually held converse. It was as if a window were thrown open, admitting a freer atmosphere into the close and stifled study, where his life was wasting itself away, amid lamp-light, or obstructed day-beams, and the musty fragrance, be it sensual or moral, that exhales from books. But the air was too fresh and chill to be long breathed with comfort. So the minister, and the physician with him, withdrew again within the limits of what their church defined as orthodox.

Thus Roger Chillingworth scrutinized his patient carefully, both as he saw him in his ordinary life, keeping an accustomed pathway in the range of thoughts familiar to him, and as he appeared when thrown amidst other moral scenery, the novelty of which might call out something new to the surface of his character. He deemed it essential, it would seem, to know the man, before attempting to do him good. Wherever there is a heart and an intellect, the diseases of the physical frame are tinged with the peculiarities of these. In Arthur Dimmesdale, thought and imagination were so active, and sensibility so intense, that the bodily infirmity would be likely to have its ground-work there. So Roger Chillingworth—the man of skill, the kind and friendly physician—strove to go deep into his patient's bosom, delving among his principles, prying into his recollections, and probing everything with a cautious touch, like a treasure-seeker in a dark cavern. Few secrets can escape an investigator, who has opportunity and license to undertake such a quest, and skill to follow it up. A man burdened with a secret should especially avoid the intimacy of his physician. If the latter possess native sagacity, and a nameless something more,—let us call it intuition; if he show no intrusive egotism, nor disagreeably prominent characteristics of his own; if he have the power, which must be born with him, to bring his mind into such affinity with his patient's, that this last shall unawares have spoken what he imagines himself only to have thought; if such revelations be received without tumult, and acknowledged not so often by an uttered sympathy as by silence, an inarticulate breath, and here and there a word, to indicate that all is understood; if to these qualifications of a confidant be joined the advantages afforded by his recognized character as a physician;—then, at some inevitable moment, will the soul of the sufferer be dissolved, and flow forth in a dark, but transparent stream, bringing all its mysteries into the daylight.

Roger Chillingworth possessed all, or most, of the attributes above enumerated. Nevertheless, time went on; a kind of intimacy, as we have said, grew up between these two cultivated minds, which had as wide a field as the whole sphere of human thought and study, to meet upon; they discussed every topic of ethics and religion, of public affairs, and private character; they talked much, on both sides, of matters that seemed personal to themselves; and yet no secret, such as the physician fancied must exist there, ever stole out of the minister's consciousness into his companion's ear. The latter had his suspicions, indeed, that even the nature of Mr. Dimmesdale's bodily disease had never fairly been revealed to him. It was a strange reserve!

After a time, at a hint from Roger Chillingworth, the friends of Mr. Dimmesdale effected an arrangement by which the two were lodged in the same house; so that every ebb and flow of the minister's life-tide might pass under the eye of his anxious and attached physician. There was much joy throughout the town, when this greatly desirable object was attained. It was held to be the best possible measure for the young clergyman's welfare; unless, indeed, as often urged by such as felt authorized to do so, he had selected some one of the many blooming damsels, spiritually devoted to him, to become his devoted wife. This latter step, however, there was no present prospect that Arthur Dimmesdale would be prevailed upon to take; he rejected all suggestions of the kind, as if priestly celibacy were one of his articles of church-discipline. Doomed by his own choice, therefore, as Mr. Dimmesdale so evidently was, to eat his unsavory morsel always at another's board, and endure the life-long chill which must be his lot who seeks to warm himself only at another's fireside, it truly seemed that this sagacious, experienced, benevolent old physician, with his concord of paternal and reverential love for the young pastor, was the very man, of all mankind, to be constantly within reach of his voice.

The new abode of the two friends was with a pious widow, of good social rank, who dwelt in a house covering pretty nearly the site on which the venerable structure of King's Chapel has since been built. It had the grave-yard, originally Isaac Johnson's home-field, on one side, and so was well adapted to call up serious reflections, suited to their respective employments, in both minister and man of physic. The motherly care of the good widow assigned to Mr. Dimmesdale a front apartment, with a sunny exposure, and heavy window-curtains, to create a noontide shadow, when desirable. The walls were hung round with tapestry, said to be from the Gobelin looms, and, at all events, representing the Scriptural story of David and Bathsheba, and Nathan the Prophet, in colors still unfaded, but which made the fair woman of the scene almost as grimly picturesque as the woe-denouncing seer.[4] Here, the pale clergyman piled up his library, rich with parchment-bound folios of the Fathers, and the lore of Rabbis, and monkish erudition, of which the Protestant divines, even while they vilified and decried that class of writers, were yet constrained often to avail themselves. On the other side of the house, old Roger Chillingworth arranged his study and laboratory; not such as a modern man of science would reckon even tolerably complete, but provided with a distilling apparatus, and the means of compounding drugs and chemicals, which the practised alchemist knew well how to turn to purpose. With such commodiousness of situation, these two learned persons sat themselves down, each in his own domain, yet familiarly passing from one apartment to the other, and bestowing a mutual and not incurious inspection into one another's business.

And the Reverend Arthur Dimmesdale's best discerning friends, as we have intimated, very reasonably imagined that the hand of Providence had done all this, for the purpose—besought in so many public, and domestic, and secret prayers—of restoring the young minister to health. But—it must now be said—another portion of the community had latterly begun to take its own view of the relation betwixt Mr. Dimmesdale and the mysterious old physician. When an uninstructed multitude attempts to see with its eyes, it is exceedingly apt to be deceived. When, however, it forms its judgment, as it usually does, on the intuitions of its great and warm heart, the conclusions thus attained are often so profound and so unerring, as to possess the character of truths supernaturally revealed. The people, in the case of which we speak, could justify its prejudice against Roger Chillingworth by no fact or argument worthy of serious refutation. There was an aged handicraftsman, it is true, who had been a citizen of London at the period of Sir Thomas Overbury's murder, now some thirty years agone; he testified to having seen the physician, under some other name, which the narrator of the story had now forgotten, in company with Doctor Forman, the famous old conjurer, who was implicated in the affair of Overbury.[5]

4. The widow's choice of tapestries could hardly be more appropriate or better designed to keep Dimmesdale's guilt before his eyes. See 2 Samuel 11–12. David committed adultery with Bathsheba, wife of Uriah the Hittite. He then sent Uriah to die in battle and married Bathsheba, but the Lord sent Nathan the Prophet to chastise and punish David. In his comment upon a case involving William Hathorne (who in 1641 argued for punishment according to the letter of the law), John Winthrop used the example of David and Bathsheba to argue that punishments for crimes should be discretionary because God himself had not chosen to put David and Bathsheba to death for their adultery (*Journal,* 381–82).

5. The murder of Sir Thomas Overbury (1581–1613) was a sensational sex scandal involving adultery and murder during the reign of James I. Overbury protested the plan of his friend, Robert Carr, to marry Frances Howard, wife of the Earl of Essex, who sought an annulment of her marriage. Lady Essex claimed that her husband was impotent, and she and Carr (Viscount of Rochester) entered into an adulterous affair, even though a gynecological exam (probably of a substitute her family supplied) vouched for her virginity. Overbury was imprisoned by King James in 1613 and apparently poisoned to death. Frances Howard received an annulment, and married Carr. Two years later, six of the plotters were tried and convicted of Overbury's murder, and four were executed. Robert Carr and Frances Howard, now Lord and Lady Somerset, were convicted and imprisoned but then pardoned by King James. The prosecution emphasized adultery at Howard's trial in order to establish motive for the murder. Sir Edward Coke (see n. 5, p. 68) drew a connection to David and Bathsheba. Frances Howard consulted Dr. Simon Forman, a wizard, astrologer, and fortune-teller, to whom she confided her fear of having to "lie" with her husband and her sexual desire for Carr. Forman was also a prolific adulterer who kept meticulous records of his conquests. Alfred S. Reid, *The Yellow Ruff and The Scarlet Letter: A Source of Hawthorne's Novel* (1955), offers an extended treatment of connections between these events and Hawthorne's novel.

Two or three individuals hinted, that the man of skill, during his Indian captivity, had enlarged his medical attainments by joining in the incantations of the savage priests; who were universally acknowledged to be powerful enchanters, often performing seemingly miraculous cures by their skill in the black art. A large number—and many of these were persons of such sober sense and practical observation that their opinions would have been valuable, in other matters—affirmed that Roger Chillingworth's aspect had undergone a remarkable change while he had dwelt in town, and especially since his abode with Mr. Dimmesdale. At first, his expression had been calm, meditative, scholar-like. Now, there was something ugly and evil in his face, which they had not previously noticed, and which grew still the more obvious to sight, the oftener they looked upon him. According to the vulgar idea, the fire in his laboratory had been brought from the lower regions, and was fed with infernal fuel; and so, as might be expected, his visage was getting sooty with the smoke.

To sum up the matter, it grew to be a widely diffused opinion, that the Reverend Arthur Dimmesdale, like many other personages of especial sanctity, in all ages of the Christian world, was haunted either by Satan himself, or Satan's emissary, in the guise of old Roger Chillingworth. This diabolical agent had the Divine permission, for a season, to burrow into the clergyman's intimacy, and plot against his soul. No sensible man, it was confessed, could doubt on which side the victory would turn. The people looked, with an unshaken hope, to see the minister come forth out of the conflict, transfigured with the glory which he would unquestionably win. Meanwhile, nevertheless, it was sad to think of the perchance mortal agony through which he must struggle towards his triumph.

Alas! to judge from the gloom and terror in the depths of the poor minister's eyes, the battle was a sore one, and the victory anything but secure.

X. The Leech and His Patient

Old Roger Chillingworth, throughout life, had been calm in temperament, kindly, though not of warm affections, but ever, and in all his relations with the world, a pure and upright man. He had begun an investigation, as he imagined, with the severe and equal integrity of a judge, desirous only of truth, even as if the question involved no more than the air-drawn lines and figures of a geometrical problem, instead of human passions, and wrongs inflicted on himself. But, as he proceeded, a terrible fascination, a kind of fierce, though still calm, necessity seized the old man within its gripe, and never set him free again, until he had done all its bidding. He now dug into the poor clergyman's heart, like a miner searching for gold; or, rather, like a sexton delving into a grave, possibly in quest of a jewel that had been buried on the dead man's bosom, but likely to find nothing save mortality and corruption. Alas for his own soul, if these were what he sought!

Sometimes, a light glimmered out of the physician's eyes, burning blue and ominous, like the reflection of a furnace, or, let us say, like one of those gleams of ghastly fire that darted from Bunyan's awful door-way in the hill-side, and quivered on the pilgrim's face.[1] The soil where this dark miner was working had perchance shown indications that encouraged him.

"This man," said he, at one such moment, to himself, "pure as they deem him,—all spiritual as he seems,—hath inherited a strong animal nature from his father or his mother. Let us dig a little further in the direction of this vein!"

Then, after long search into the minister's dim interior, and turning over many precious materials, in the shape of high aspirations for the welfare of his race, warm love of souls, pure sentiments, natural piety, strengthened by thought and study, and illuminated by revelation,—all of which invaluable gold was perhaps no better than rubbish to the seeker,—he would turn back, discouraged, and begin his quest towards another point. He groped along as stealthily, with as cautious a tread, and as wary an outlook, as a thief entering a chamber where a man lies only half asleep,—or, it may be, broad awake,—with purpose to steal the very treasure which this man guards as the apple of his eye. In spite of his premeditated carefulness, the floor would now and then creak; his garments would rustle; the shadow of his presence, in a forbidden proximity, would be thrown across his victim. In other words, Mr. Dimmesdale, whose sensibility of nerve often produced the effect of spiritual intuition, would become vaguely aware that something inimical to his peace had thrust itself into relation with him. But old Roger Chillingworth, too, had perceptions that were almost intuitive; and when the minister threw his startled eyes towards him, there the physician sat; his kind, watchful, sympathizing, but never intrusive friend.

Yet Mr. Dimmesdale would perhaps have seen this individual's character more perfectly, if a certain morbidness, to which sick hearts are liable, had not rendered him suspicious of all mankind. Trusting no man as his friend, he could not recognize his enemy when the latter actually appeared. He therefore still kept up a familiar intercourse with him, daily receiving the old physician in his study; or visiting the laboratory, and, for recreation's sake, watching the processes by which weeds were converted into drugs of potency.

One day, leaning his forehead on his hand, and his elbow on the sill of the open window, that looked towards the grave-yard, he talked with Roger Chillingworth, while the old man was examining a bundle of unsightly plants.

"Where," asked he, with a look askance at them,—for it was the clergyman's peculiarity that he seldom, now-a-days, looked straightforth at any object, whether human or inanimate—"where, my kind doctor, did you gather those herbs, with such a dark, flabby leaf?"

"Even in the grave-yard here at hand," answered the physician, continuing his employment. "They are new to me. I found them growing on a grave, which bore no tomb-stone, nor other memorial of the dead man, save these ugly weeds, that have taken upon themselves to keep him in remembrance. They grew out of his heart, and typify, it may be, some hideous secret that was buried with him, and which he had done better to confess during his lifetime."

1. In John Bunyan's allegorical *Pilgrim's Progress* (1678), Christian encounters fire flashing out of a hill on his journey from the City of Destruction to the Celestial City.

"Perchance," said Mr. Dimmesdale, "he earnestly desired it, but could not."

"And wherefore?" rejoined the physician. "Wherefore not; since all the powers of nature call so earnestly for the confession of sin, that these black weeds have sprung up out of a buried heart, to make manifest an unspoken crime?"

"That, good Sir, is but a fantasy of yours," replied the minister. "There can be, if I forebode aright, no power, short of the Divine mercy, to disclose, whether by uttered words, or by type or emblem, the secrets that may be buried with a human heart. The heart, making itself guilty of such secrets, must perforce hold them, until the day when all hidden things shall be revealed. Nor have I so read or interpreted Holy Writ, as to understand that the disclosure of human thoughts and deeds, then to be made, is intended as a part of the retribution. That, surely, were a shallow view of it. No; these revelations, unless I greatly err, are meant merely to promote the intellectual satisfaction of all intelligent beings, who will stand waiting, on that day, to see the dark problem of this life made plain. A knowledge of men's hearts will be needful to the completest solution of that problem. And I conceive, moreover, that the hearts holding such miserable secrets as you speak of will yield them up, at that last day, not with reluctance, but with a joy unutterable."

"Then why not reveal them here?" asked Roger Chillingworth, glancing quietly aside at the minister. "Why should not the guilty ones sooner avail themselves of this unutterable solace?"

"They mostly do," said the clergyman, griping hard at his breast, as if afflicted with an importunate throb of pain. "Many, many a poor soul hath given its confidence to me, not only on the death-bed, but while strong in life, and fair in reputation. And ever, after such an outpouring, O, what a relief have I witnessed in those sinful brethren! even as in one who at last draws free air, after long stifling with his own polluted breath. How can it be otherwise? Why should a wretched man, guilty, we will say, of murder, prefer to keep the dead corpse buried in his own heart, rather than fling it forth at once, and let the universe take care of it!"

"Yet some men bury their secrets thus," observed the calm physician.

"True; there are such men," answered Mr. Dimmesdale. "But, not to suggest more obvious reasons, it may be that they are kept silent by the very constitution of their nature. Or,—can we not suppose it?—guilty as they may be, retaining, nevertheless, a zeal for God's glory and man's welfare, they shrink from displaying themselves black and filthy in the view of men; because, thenceforward, no good can be achieved by them; no evil of the past be redeemed by better service. So, to their own unutterable torment, they go about among their fellow-creatures, looking pure as new-fallen snow; while their hearts are all speckled and spotted with iniquity of which they cannot rid themselves."

"These men deceive themselves," said Roger Chillingworth, with somewhat more emphasis than usual, and making a slight gesture with his forefinger. "They fear to take up the shame that rightfully belongs to them. Their love for man, their zeal for God's service,—these holy impulses may or may not coëxist in their hearts with the evil inmates to which their guilt has unbarred the door, and which must needs propagate a hellish breed within them. But, if they seek to glorify God, let them not lift heavenward their unclean hands! If they would serve their fellow-men, let them do it by making manifest the power and reality of conscience, in constraining them to penitential self-abasement! Wouldst thou have me to believe, O wise and pious friend, that a false show can be better—can be more for God's glory, or man's welfare—than God's own truth? Trust me, such men deceive themselves!"

"It may be so," said the young clergyman, indifferently, as waiving a discussion that he considered irrelevant or unseasonable. He had a ready faculty, indeed, of escaping from any topic that agitated his too sensitive and nervous temperament.—"But, now, I would ask of my well-skilled physician, whether, in good sooth, he deems me to have profited by his kindly care of this weak frame of mine?"

Before Roger Chillingworth could answer, they heard the clear, wild laughter of a young child's voice, proceeding from the adjacent burial-ground. Looking instinctively from the open window,—for it was summer-time,—the minister beheld Hester Prynne and little Pearl passing along the foot-path that traversed the enclosure. Pearl looked as beautiful as the day, but was in one of those moods of perverse merriment which, whenever they occurred, seemed to remove her entirely out of the sphere of sympathy or human contact. She now skipped irreverently from one grave to another; until, coming to the broad, flat, armorial tomb-stone of a departed worthy,—perhaps of Isaac Johnson himself,—she began to dance upon it. In reply to her mother's command and entreaty that she would behave more decorously, little Pearl paused to gather the prickly burrs from a tall burdock which grew beside the tomb. Taking a handful of these, she arranged them along the lines of the scarlet letter that decorated the maternal bosom, to which the burrs, as their nature was, tenaciously adhered. Hester did not pluck them off.

Roger Chillingworth had by this time approached the window, and smiled grimly down.

"There is no law, nor reverence for authority, no regard for human ordinances or opinions, right or wrong, mixed up with that child's composition," remarked he, as much to himself as to his companion. "I saw her, the other day, bespatter the Governor himself with water, at the cattle-trough in Spring-lane. What, in Heaven's name, is she? Is the imp altogether evil? Hath she affections? Hath she any discoverable principle of being?"

"None,—save the freedom of a broken law," answered Mr. Dimmesdale, in a quiet way, as if he had been discussing the point within himself. "Whether capable of good, I know not."

The child probably overheard their voices; for, looking up to the window, with a bright, but naughty smile of mirth and intelligence, she threw one of the prickly burrs at the Reverend Mr. Dimmesdale. The sensitive clergyman shrunk, with nervous dread, from the light missile. Detecting his emotion, Pearl clapped her little hands, in the most extravagant ecstasy. Hester Prynne, likewise, had involuntarily looked up; and all these four persons, old and young, regarded one another in silence, till the child laughed aloud, and shouted,—"Come away, mother! Come away, or yonder old Black Man will catch you! He hath got hold of the minister already. Come away, mother, or he will catch you! But he cannot catch little Pearl!"

So she drew her mother away, skipping, dancing, and frisking fantastically, among the hillocks of the dead people, like a creature that had nothing in common with a bygone and buried generation, nor owned herself akin to it. It was as if she had been made afresh, out of new elements, and must perforce be permitted to live her own life, and be a law unto herself, without her eccentricities being reckoned to her for a crime.

"There goes a woman," resumed Roger Chillingworth, after a pause, "who, be her demerits what they may, hath none of that mystery of hidden sinfulness which you deem so grievous to be borne. Is Hester Prynne the less miserable, think you, for that scarlet letter on her breast?"

"I do verily believe it," answered the clergyman. "Nevertheless, I cannot answer for her. There was a look of pain in her face, which I would gladly have been spared the sight of. But still, methinks, it must needs be better for the sufferer to be free to show his pain, as this poor woman Hester is, than to cover it all up in his heart."

There was another pause; and the physician began anew to examine and arrange the plants which he had gathered.

"You inquired of me, a little time agone," said he, at length, "my judgment as touching your health."

"I did," answered the clergyman, "and would gladly learn it. Speak frankly, I pray you, be it for life or death."

"Freely, then, and plainly," said the physician, still busy with his plants, but keeping a wary eye on Mr. Dimmesdale, "the disorder is a strange one; not so much in itself, nor as outwardly manifested,—in so far, at least, as the symptoms have been laid open to my observation. Looking daily at you, my good Sir, and watching the tokens of your aspect, now for months gone by, I should deem you a man sore sick, it may be, yet not so sick but that an instructed and watchful physician might well hope to cure you. But—I know not what to say—the disease is what I seem to know, yet know it not."

"You speak in riddles, learned Sir," said the pale minister, glancing aside out of the window.

"Then, to speak more plainly," continued the physician, "and I crave pardon, Sir,—should it seem to require pardon,—for this needful plainness of my speech. Let me ask,—as your friend,—as one having charge, under Providence, of your life and physical well-being,—hath all the operation of this disorder been fairly laid open and recounted to me?"

"How can you question it?" asked the minister. "Surely, it were child's play, to call in a physician, and then hide the sore!"

"You would tell me, then, that I know all?" said Roger Chillingworth, deliberately, and fixing an eye, bright with intense and concentrated intelligence, on the minister's face. "Be it so! But, again! He to whom only the outward and physical evil is laid open, knoweth, oftentimes, but half the evil which he is called upon to cure. A bodily disease, which we look upon as whole and entire within itself, may, after all, be but a symptom of some ailment in the spiritual part. Your pardon, once again, good Sir, if my speech give the shadow of offence. You, Sir, of all men whom I have known, are he whose body is the closest conjoined, and imbued, and identified, so to speak, with the spirit whereof it is the instrument."

"Then I need ask no further," said the clergyman, somewhat hastily rising from his chair. "You deal not, I take it, in medicine for the soul!"

"Thus, a sickness," continued Roger Chillingworth, going on, in an unaltered tone, without heeding the interruption,—but standing up, and confronting the emaciated and white-cheeked minister, with his low, dark, and misshapen figure,—"a sickness, a sore place, if we may so call it, in your spirit, hath immediately its appropriate manifestation in your bodily frame. Would you, therefore, that your physician heal the bodily evil? How may this be, unless you first lay open to him the wound or trouble in your soul?"

"No!—not to thee!—not to an earthly physician!" cried Mr. Dimmesdale, passionately, and turning his eyes, full and bright, and with a kind of fierceness, on old Roger Chillingworth. "Not to thee! But, if it be the soul's disease, then do I commit myself to the one Physician of the soul! He, if it stand with his good pleasure, can cure; or he can kill! Let him do with me as, in his justice and wisdom; he shall see good. But who art thou, that meddlest in this matter?—that dares thrust himself between the sufferer and his God?"

With a frantic gesture, he rushed out of the room.

"It is as well to have made this step," said Roger Chillingworth to himself, looking after the minister, with a grave smile. "There is nothing lost. We shall be friends again anon. But see, now, how passion takes hold upon this man, and hurrieth him out of himself! As with one passion, so with another! He hath done a wild thing ere now, this pious Master Dimmesdale, in the hot passion of his heart!"

It proved not difficult to reëstablish the intimacy of the two companions, on the same footing and in the same degree as heretofore. The young clergyman, after a few hours of privacy, was sensible that the disorder of his nerves had hurried him into an unseemly outbreak of temper, which there had been nothing in the physician's words to excuse or palliate. He marvelled, indeed, at the violence with which he had thrust back the kind old man, when merely proffering the advice which it was his duty to bestow, and which the minister himself had expressly sought. With these remorseful feelings, he lost no time in making the amplest apologies, and besought his friend still to continue the care, which, if not successful in restoring him to health, had, in all probability, been the means of prolonging his feeble existence to that hour. Roger Chillingworth readily assented, and went on with his medical supervision of the minister; doing his best for him, in all good faith, but always quitting the patient's apartment, at the close of a professional interview, with a mysterious and puzzled smile upon his lips. This expression was invisible in Mr. Dimmesdale's presence, but grew strongly evident as the physician crossed the threshold.

"A rare case!" he muttered. "I must needs look deeper into it. A strange sympathy betwixt soul and body! Were it only for the art's sake, I must search this matter to the bottom!"

It came to pass, not long after the scene above recorded, that the Reverend Mr. Dimmesdale, at noon-day, and entirely unawares, fell into a deep, deep slumber, sitting in his chair, with a large black-letter volume open before him on the table. It must have been a work of vast ability in the somniferous school of literature. The profound depth of the minister's repose was the more remarkable, inasmuch as he was one of those persons whose sleep, ordinarily, is as light, as fitful, and as easily scared away, as a small bird hopping on a twig. To such an unwonted remoteness, however, had his spirit now withdrawn into itself, that he stirred not in his chair, when old Roger Chillingworth, without any extraordinary precaution, came into the room. The physician advanced directly in front of his patient, laid his hand upon his bosom, and thrust aside the vestment, that, hitherto, had always covered it even from the professional eye.

Then, indeed, Mr. Dimmesdale shuddered, and slightly stirred.

After a brief pause, the physician turned away.

But, with what a wild look of wonder, joy, and horror! With what a ghastly rapture, as it were, too mighty to be expressed only by the eye and features, and therefore bursting forth through the whole ugliness of his figure, and making itself even riotously manifest by the extravagant gestures with which he threw up his arms towards the ceiling, and stamped his foot upon the floor! Had a man seen old Roger Chillingworth, at that moment of his ecstacy, he would have had no need to ask how Satan comports himself, when a precious human soul is lost to heaven, and won into his kingdom.

But what distinguished the physician's ecstacy from Satan's was the trait of wonder in it!

XI. The Interior of a Heart

After the incident last described, the intercourse between the clergyman and the physician, though externally the same, was really of another character than it had previously been. The intellect of Roger Chillingworth had now a sufficiently plain path before it. It was not, indeed, precisely that which he had laid out for himself to tread. Calm, gentle, passionless, as he appeared, there was yet, we fear, a quiet depth of malice, hitherto latent, but active now, in this unfortunate old man, which led him to imagine a more intimate revenge than any mortal had ever wreaked upon an enemy. To make himself the one trusted friend, to whom should be confided all the fear, the remorse, the agony, the ineffectual repentance, the backward rush of sinful thoughts, expelled in vain! All that guilty sorrow, hidden from the world, whose great heart would have pitied and forgiven, to be revealed to him, the Pitiless, to him, the Unforgiving! All that dark treasure to be lavished on the very man, to whom nothing else could so adequately pay the debt of vengeance!

The clergyman's shy and sensitive reserve had balked this scheme. Roger Chillingworth, however, was inclined to be hardly, if at all, less satisfied with the aspect of affairs, which Providence—using the avenger and his victim for its own purposes, and, perchance, pardoning, where it seemed most to punish—had substituted for his black devices. A revelation, he could almost say, had been granted to him. It mattered little, for his object, whether celestial, or from what other region. By its aid, in all the subsequent relations betwixt him and Mr. Dimmesdale, not merely the external presence, but the very inmost soul, of the latter, seemed to be brought out before his eyes, so that he could see and comprehend its every movement. He became, thenceforth, not a spectator only, but a chief actor, in the poor minister's interior world. He could play upon him as he chose. Would he arouse him with a throb of agony? The victim was forever on the rack; it needed only to know the spring that controlled the engine;—and the physician knew it well! Would he startle him with sudden fear? As at the waving of a magician's wand, uprose a grisly phantom,—uprose a thousand phantoms,—in many shapes, of death, or more awful shame, all flocking round about the clergyman, and pointing with their fingers at his breast!

All this was accomplished with a subtlety so perfect, that the minister, though he had constantly a dim perception of some evil influence watching over him, could never gain a knowledge of its actual nature. True, he looked doubtfully, fearfully,—even, at times, with horror and the bitterness of hatred,—at the deformed figure of the old physician. His gestures, his gait, his grizzled beard, his slightest and most indifferent acts, the very fashion of his garments, were odious in the clergyman's sight; a token implicitly to be relied on, of a deeper antipathy in the breast of the latter than he was willing to acknowledge to himself. For, as it was impossible to assign a reason for such distrust and abhorrence, so Mr. Dimmesdale, conscious that the poison of one morbid spot was infecting his heart's entire substance, attributed all his presentiments to no other cause. He took himself to task for his bad sympathies in reference to Roger Chillingworth, disregarded the lesson that he should have drawn from them, and did his best to root them out. Unable to accomplish this, he nevertheless, as a matter of principle, continued his habits of social familiarity with the old man, and thus gave him constant opportunities for perfecting the purpose to which—poor, forlorn creature that he was, and more wretched than his victim—the avenger had devoted himself.

While thus suffering under bodily disease, and gnawed and tortured by some black trouble of the soul, and given over to the machinations of his deadliest enemy, the Reverend Mr. Dimmesdale had achieved a brilliant popularity in his sacred office. He won it, indeed, in great part, by his sorrows. His intellectual gifts, his moral perceptions, his power of experiencing and communicating emotion, were kept in a state of preternatural activity by the prick and anguish of his daily life. His fame, though still on its upward slope, already overshadowed the soberer reputations of his fellow-clergymen, eminent as several of them were. There were scholars among them, who had spent more years in acquiring abstruse lore, connected with the divine profession, than Mr. Dimmesdale had lived; and who might well, therefore, be more profoundly versed in such solid and valuable attainments than their youthful brother. There were men, too, of a sturdier texture of mind then his, and endowed with a far greater share of shrewd, hard, iron, or granite understanding; which, duly mingled with a fair proportion of doctrinal ingredient, constitutes a highly respectable, efficacious, and unamiable variety of the clerical species. There were others, again, true saintly fathers, whose faculties had been elaborated by weary toil among their books, and by patient thought, and etherealized, moreover, by spiritual communications with the better world, into which their purity of life had almost introduced these holy personages, with their garments of mortality still clinging to them. All that they lacked was the gift that descended upon the chosen disciples at Pentecost, in tongues of flame; symbolizing, it would seem, not the power of speech in foreign and unknown languages, but that of addressing the whole human brotherhood in the heart's native language. These fathers, otherwise so apostolic, lacked Heaven's last and rarest attestation of their office, the Tongue of Flame.[1] They would have vainly sought—had they ever dreamed of seeking—to express the highest truths through the humblest medium of familiar words and images. Their voices came down, afar and indistinctly, from the upper heights where they habitually dwelt.

Not improbably, it was to this latter class of men that Mr. Dimmesdale, by many of his traits of character, naturally belonged. To the high mountain-peaks of faith and sanctity he would have climbed, had not the tendency been thwarted by the burden, whatever it might be, of crime or anguish, beneath which it was his doom to totter. It kept him down, on a level with the lowest; him, the man of ethereal attributes, whose voice the angels might else have listened to and answered! But this very burden it was, that gave him sympathies so intimate with the sinful brotherhood of mankind; so that his heart vibrated in unison with theirs, and received their pain into itself, and sent its own throb of pain through a thousand other hearts, in gushes of sad, persuasive eloquence. Oftenest persuasive, but sometimes terrible! The people knew not the power that moved them thus. They deemed the young clergyman a miracle of holiness. They fancied him the mouth-piece of Heaven's messages of wisdom, and rebuke, and love. In their eyes, the very ground on which he trod was sanctified. The virgins of his church grew pale around him, victims of a passion so imbued with religious sentiment that they imagined it to be all religion, and brought it openly, in their white bosoms, as their most acceptable sacrifice before the altar. The aged members of his flock, beholding Mr. Dimmesdale's frame so feeble, while they were themselves so rugged in their infirmity, believed that he would go heavenward before them, and enjoined it upon their children, that their old bones should be buried close to their young pastor's holy grave. And, all this time, perchance, when poor Mr. Dimmesdale was thinking of his grave, he questioned with himself whether the grass would ever grow on it, because an accursed thing must there be buried!

It is inconceivable, the agony with which this public veneration tortured him! It was his genuine impulse to adore the truth, and to reckon all things shadow-like, and utterly devoid of weight or value, that had not its divine essence as the life within their life. Then, what was he?—a substance?—or the dimmest of all shadows? He longed to speak out, from his own pulpit, at the full height of his voice, and tell the people what he was. "I, whom you behold in these black garments of the priesthood,—I, who ascend the sacred desk, and turn my pale face heavenward, taking upon myself to hold communion, in your behalf, with the Most High Omniscience,—I, in whose daily life you discern the sanctity of Enoch,[2]—I, whose footsteps, as you suppose, leave a gleam along my earthly track, whereby the pilgrims that shall come after me may be guided to the regions of the blest,—I, who have laid the hand of baptism

1. See Acts 2.1–11: "And when the day of Pentecost was fully come . . . suddenly there came a sound from Heaven as of a rushing mighty wind, and it filled all the house where they were sitting. / And there appeared unto them cloven tongues like as of fire, and it sat upon each of them. / And they were all filled with the Holy Ghost, and began to speak with other tongues, as the Spirit gave them utterance" (1–4).
2. Enoch offers one of several examples of the power of faith. See Hebrews 11.5: "By faith Enoch was translated that he should not see death." Others mentioned in Hebrews 11 are Abel, Noah, Abraham, and Sarah.

baptism is important

very important

upon your children,—I, who have breathed the parting prayer over your dying friends, to whom the Amen sounded faintly from a world which they had quitted,—I, your pastor, whom you so reverence and trust, am utterly a pollution and a lie!"

More than once, Mr. Dimmesdale had gone into the pulpit, with a purpose never to come down its steps, until he should have spoken words like the above. More than once, he had cleared his throat, and drawn in the long, deep, and tremulous breath, which, when sent forth again, would come burdened with the black secret of his soul. More than once—nay, more than a hundred times—he had actually spoken! Spoken! But how? He had told his hearers that he was altogether vile, a viler companion of the vilest, the worst of sinners, an abomination, a thing of unimaginable iniquity; and that the only wonder was, that they did not see his wretched body shrivelled up before their eyes, by the burning wrath of the Almighty! Could there be plainer speech than this? Would not the people start up in their seats, by a simultaneous impulse, and tear him down out of the pulpit which he defiled? Not so, indeed! They heard it all, and did but reverence him the more. They little guessed what deadly purport lurked in those self-condemning words. "The godly youth!" said they among themselves. "The saint on earth! Alas, if he discern such sinfulness in his own white soul, what horrid spectacle would he behold in thine or mine!" The minister well knew—subtle, but remorseful hypocrite that he was!—the light in which his vague confession would be viewed. He had striven to put a cheat upon himself by making the avowal of a guilty conscience, but had gained only one other sin, and a self-acknowledged shame, without the momentary relief of being self-deceived. He had spoken the very truth, and transformed it into the veriest falsehood. And yet, by the constitution of his nature, he loved the truth, and loathed the lie, as few men ever did. Therefore, above all things else, he loathed his miserable self!

His inward trouble drove him to practices more in accordance with the old, corrupted faith of Rome, than with the better light of the church in which he had been born and bred. In Mr. Dimmesdale's secret closet, under lock and key, there was a bloody scourge.[3] Oftentimes, this Protestant and Puritan divine had plied it on his own shoulders; laughing bitterly at himself the while, and smiting so much the more pitilessly because of that bitter laugh. It was his custom, too, as it has been that of many other pious Puritans, to fast,—not, however, like them, in order to purify the body and render it the fitter medium of celestial illumination, but rigorously, and until his knees trembled beneath him, as an act of penance. He kept vigils, likewise, night after night, sometimes in utter darkness; sometimes with a glimmering lamp; and sometimes, viewing his own face in a looking-glass, by the most powerful light which he could throw upon it. He thus typified the constant introspection wherewith he tortured, but could not purify, himself. In these lengthened vigils, his brain often reeled, and visions seemed to flit before him; perhaps seen doubtfully, and by a faint light of their own, in the remote dimness of the chamber, or more vividly, and close beside him, within the looking-glass.

3. A bloody whip. See 2 Maccabees 9.11: "Here therefore, being plagued, he began to leave off his great pride, and to come to the knowledge of himself by the scourge of God, his pain increasing every moment."

Now it was a herd of diabolic shapes, that grinned and mocked at the pale minister, and beckoned him away with them; now a group of shining angels, who flew upward heavily, as sorrow-laden, but grew more ethereal as they rose. Now came the dead friends of his youth, and his white-bearded father, with a saint-like frown, and his mother, turning her face away as she passed by. Ghost of a mother,—thinnest fantasy of a mother,—methinks she might yet have thrown a pitying glance towards her son! And now, through the chamber which these spectral thoughts had made so ghastly, glided Hester Prynne, leading along little Pearl, in her scarlet garb, and pointing her forefinger, first at the scarlet letter on her bosom, and then at the clergyman's own breast.

None of these visions ever quite deluded him. At any moment, by an effort of his will, he could discern substances through their misty lack of substance, and convince himself that they were not solid in their nature, like yonder table of carved oak, or that big, square, leathern-bound and brazen-clasped volume of divinity. But, for all that, they were, in one sense, the truest and most substantial things which the poor minister now dealt with. It is the unspeakable misery of a life so false as his, that it steals the pith and substance out of whatever realities there are around us, and which were meant by Heaven to be the spirit's joy and nutriment. To the untrue man, the whole universe is false,—it is impalpable,—it shrinks to nothing within his grasp. And he himself, in so far as he shows himself in a false light, becomes a shadow, or, indeed, ceases to exist. The only truth that continued to give Mr. Dimmesdale a real existence on this earth, was the anguish in his inmost soul, and the undissembled expression of it in his aspect. Had he once found power to smile, and wear a face of gayety, there would have been no such man!

On one of those ugly nights, which we have faintly hinted at, but forborne to picture forth, the minister started from his chair. A new thought had struck him. There might be a moment's peace in it. Attiring himself with as much care as if it had been for public worship, and precisely in the same manner, he stole softly down the staircase, undid the door, and issued forth.

XII. The Minister's Vigil

Walking in the shadow of a dream, as it were, and perhaps actually under the influence of a species of somnambulism, Mr. Dimmesdale reached the spot, where, now so long since, Hester Prynne had lived through her first hours of public ignominy. The same platform or scaffold, black and weather-stained with the storm or sunshine of seven long years, and foot-worn, too, with the tread of many culprits who had since ascended it, remained standing beneath the balcony of the meeting-house. The minister went up the steps.

It was an obscure night of early May.[1] An unvaried pall of cloud muffled the whole expanse of sky from zenith to horizon. If the same multitude

1. John Winthrop died March 26, 1649, so Hawthorne is taking a small liberty with history by moving this scene to May. Much as he had done in chapter 6, Hawthorne uses chapter 11 to advance the action of the novel—in this case, four more years.

which had stood as eye-witnesses while Hester Prynne sustained her punishment could now have been summoned forth, they would have discerned no face above the platform, nor hardly the outline of a human shape, in the dark gray of the midnight. But the town was all asleep. There was no peril of discovery. The minister might stand there, if it so pleased him, until morning should redden in the east, without other risk than that the dank and chill night-air would creep into his frame, and stiffen his joints with rheumatism, and clog his throat with catarrh and cough; thereby defrauding the expectant audience of to-morrow's prayer and sermon. No eye could see him, save that ever-wakeful one which had seen him in his closet, wielding the bloody scourge. Why, then, had he come hither? Was it but the mockery of penitence? A mockery, indeed, but in which his soul trifled with itself! A mockery at which angels blushed and wept, while fiends rejoiced, with jeering laughter! He had been driven hither by the impulse of that Remorse which dogged him everywhere, and whose own sister and closely linked companion was that Cowardice which invariably drew him back, with her tremulous gripe, just when the other impulse had hurried him to the verge of a disclosure. Poor, miserable man! what right had infirmity like his to burden itself with crime? Crime is for the iron-nerved, who have their choice either to endure it, or, if it press too hard, to exert their fierce and savage strength for a good purpose, and fling it off at once! This feeble and most sensitive of spirits could do neither, yet continually did one thing or another, which intertwined, in the same inextricable knot, the agony of heaven-defying guilt and vain repentance.

And thus, while standing on the scaffold, in this vain show of expiation, Mr. Dimmesdale was overcome with a great horror of mind, as if the universe were gazing at a scarlet token on his naked breast, right over his heart. On that spot, in very truth, there was, and there had long been, the gnawing and poisonous tooth of bodily pain. Without any effort of his will, or power to restrain himself, he shrieked aloud; an outcry that went pealing through the night, and was beaten back from one house to another, and reverberated from the hills in the background; as if a company of devils, detecting so much misery and terror in it, had made a plaything of the sound, and were bandying it to and fro.

"It is done!" muttered the minister, covering his face with his hands. "The whole town will awake, and hurry forth, and find me here!"

But it was not so. The shriek had perhaps sounded with a far greater power, to his own startled ears, than it actually possessed. The town did not awake; or, if it did, the drowsy slumberers mistook the cry either for something frightful in a dream, or for the noise of witches; whose voices, at that period, were often heard to pass over the settlements or lonely cottages, as they rode with Satan through the air. The clergyman, therefore, hearing no symptoms of disturbance, uncovered his eyes and looked about him. At one of the chamber-windows of Governor Bellingham's mansion, which stood at some distance, on the line of another street, he beheld the appearance of the old magistrate himself, with a lamp in his hand, a white night-cap on his head, and a long white gown enveloping his figure. He looked like a ghost, evoked unseasonably from the grave. The cry had evidently startled him. At another window of the same house, moreover, appeared old Mistress Hibbins, the Governor's sister, also with a lamp, which, even thus far off, revealed the expression of her sour and discontented face. She thrust forth her head from the lattice, and looked anxiously upward. Beyond the shadow of a doubt, this venerable witch-lady had heard Mr. Dimmesdale's outcry, and interpreted it, with its multitudinous echoes and reverberations, as the clamor of the fiends and night-hags, with whom she was well known to make excursions into the forest.

Detecting the gleam of Governor Bellingham's lamp, the old lady quickly extinguished her own, and vanished. Possibly, she went up among the clouds. The minister saw nothing further of her motions. The magistrate, after a wary observation of the darkness—into which, nevertheless, he could see but little further than he might into a millstone—retired from the window.

The minister grew comparatively calm. His eyes, however, were soon greeted by a little, glimmering light, which, at first a long way off, was approaching up the street. It threw a gleam of recognition on here a post, and there a garden-fence, and here a latticed window-pane, and there a pump, with its full trough of water, and here, again, an arched door of oak, with an iron knocker, and a rough log for the doorstep. The Reverend Mr. Dimmesdale noted all these minute particulars, even while firmly convinced that the doom of his existence was stealing onward, in the footsteps which he now heard; and that the gleam of the lantern would fall upon him, in a few moments more, and reveal his long-hidden secret. As the light drew nearer, he beheld, within its illuminated circle, his brother clergyman,—or, to speak more accurately, his professional father, as well as highly valued friend,—the Reverend Mr. Wilson; who, as Mr. Dimmesdale now conjectured, had been praying at the bedside of some dying man. And so he had. The good old minister came freshly from the death-chamber of Governor Winthrop, who had passed from earth to heaven within that very hour. And now, surrounded, like the saint-like personages of olden times, with a radiant halo, that glorified him amid this gloomy night of sin,—as if the departed Governor had left him an inheritance of his glory, or as if he had caught upon himself the distant shine of the celestial city, while looking thitherward to see the triumphant pilgrim pass within its gates,—now, in short, good Father Wilson was moving homeward, aiding his footsteps with a lighted lantern! The glimmer of this luminary suggested the above conceits to Mr. Dimmesdale, who smiled,—nay, almost laughed at them,—and then wondered if he were going mad.

As the Reverend Mr. Wilson passed beside the scaffold, closely muffling his Geneva cloak about him with one arm, and holding the lantern before his breast with the other, the minister could hardly restrain himself from speaking.

"A good evening to you, venerable Father Wilson! Come up hither, I pray you, and pass a pleasant hour with me!"

Good heavens! Had Mr. Dimmesdale actually spoken? For one instant, he believed that these words had passed his lips. But they were uttered only within his imagination. The venerable Father Wilson continued to step slowly onward, looking carefully at the muddy pathway before his feet, and never once turning his head towards the guilty platform. When the light of the glimmering lantern had faded quite away, the minister discovered, by the faintness which came over him, that the last few moments had been a crisis of terrible anxiety; although his mind had made an involuntary effort to relieve itself by a kind of lurid playfulness.

Shortly afterwards, the like grisly sense of the humorous again stole in among the solemn phantoms of his thought. He felt his limbs growing stiff with the unaccustomed chilliness of the night, and doubted whether he should be able to descend the steps of the scaffold. Morning would break, and find him there. The neighborhood would begin to rouse itself. The earliest riser, coming forth in the dim twilight, would perceive a vaguely defined figure aloft on the place of shame; and, half crazed betwixt alarm and curiosity, would go, knocking from door to door, summoning all the people to behold the ghost—as he needs must think it—of some defunct transgressor. A dusky tumult would flap its wings from one house to another. Then—the morning light still waxing stronger—old patriarchs would rise up in great haste, each in his flannel gown, and matronly dames, without pausing to put off their night-gear. The whole tribe of decorous personages, who had never heretofore been seen with a single hair of their heads awry, would start into public view, with the disorder of a nightmare in their aspects. Old Governor Bellingham would come grimly forth, with his King James' ruff fastened askew; and Mistress Hibbins, with some twigs of the forest clinging to her skirts, and looking sourer than ever, as having hardly got a wink of sleep after her night ride; and good Father Wilson, too, after spending half the night at a death-bed, and liking ill to be disturbed, thus early, out of his dreams about the glorified saints. Hither, likewise, would come the elders and deacons of Mr. Dimmesdale's church, and the young virgins who so idolized their minister, and had made a shrine for him in their white bosoms; which now, by the by, in their hurry and confusion, they would scantly have given themselves time to cover with their kerchiefs. All people, in a word, would come stumbling over their thresholds, and turning up their amazed and horror-stricken visages around the scaffold. Whom would they discern there, with the red eastern light upon his brow? Whom, but the Reverend Arthur Dimmesdale, half frozen to death, overwhelmed with shame, and standing where Hester Prynne had stood!

Carried away by the grotesque horror of this picture, the minister, unawares, and to his own infinite alarm, burst into a great peal of laughter. It was immediately responded to by a light, airy, childish laugh, in which with a thrill of the heart,—but he knew not whether of exquisite pain, or pleasure as acute,—he recognized the tones of little Pearl.

"Pearl! Little Pearl!" cried he, after a moment's pause; then, suppressing his voice,—Hester! Hester Prynne! Are you there?"

"Yes; it is Hester Prynne!" she replied, in a tone of surprise; and the minister heard her footsteps approaching from the sidewalk, along which she had been passing. "It is I, and my little Pearl."

"Whence come you, Hester?" asked the minister. "What sent you hither?"

"I have been watching at a death-bed," answered Hester Prynne;—"at Governor Winthrop's death-bed, and have taken his measure for a robe, and am now going homeward to my dwelling."

"Come up hither, Hester, thou and little Pearl," said the Reverend Mr. Dimmesdale. "Ye have both been here before, but I was not with you. Come up hither once again, and we will stand all three together!"

She silently ascended the steps, and stood on the platform, holding little Pearl by the hand. The minister felt for the child's other hand, and took it. The moment that he did so, there came what seemed a tumultuous rush of new life, other life than his own, pouring like a torrent into his heart, and hurrying through all his veins, as if the mother and the child were communicating their vital warmth to his half-torpid system. The three formed an electric chain.

"Minister!" whispered little Pearl.

"What wouldst thou say, child?" asked Mr. Dimmesdale.

"Wilt thou stand here with mother and me, to-morrow noontide?" inquired Pearl.

"Nay; not so, my little Pearl," answered the minister; for, with the new energy of the moment, all the dread of public exposure, that had so long been the anguish of his life, had returned upon him; and he was already trembling at the conjunction in which—with a strange joy, nevertheless—he now found himself. "Not so, my child. I shall, indeed, stand with thy mother and thee one other day, but not to-morrow."

Pearl laughed, and attempted to pull away her hand. But the minister held it fast.

"A moment longer, my child!" said he.

"But wilt thou promise," asked Pearl, "to take my hand, and mother's hand, to-morrow noontide?"

"Not then, Pearl," said the minister, "but another time."

"And what other time?" persisted the child.

"At the great judgment day," whispered the minister,—and, strangely enough, the sense that he was a professional teacher of the truth impelled him to answer the child so. "Then, and there, before the judgment-seat, thy mother, and thou, and I, must stand together. But the daylight of this world shall not see our meeting!"

Pearl laughed again.

But, before Mr. Dimmesdale had done speaking, a light gleamed far and wide over all the muffled sky. It was doubtless caused by one of those meteors, which the night-watcher may so often observe burning out to waste, in the vacant regions of the atmosphere. So powerful was its radiance, that it thoroughly illuminated the dense medium of cloud betwixt the sky and earth. The great vault brightened, like the dome of an immense lamp. It showed the familiar scene of the street, with the distinctness of mid-day, but also with the awfulness that is always imparted to familiar objects by an unaccustomed light. The wooden houses, with their jutting stories and quaint gable-peaks; the doorsteps and thresholds, with the early grass springing up about them; the garden-plots, black with freshly turned earth; the wheel-track, little worn, and, even in the market-place, margined with green on either side;—all were visible, but with a singularity of aspect that seemed to give another moral interpretation to the things of this world than they had ever borne before. And there stood the minister, with his hand over his heart; and Hester Prynne, with the embroidered letter glimmering on her bosom; and little Pearl, herself a symbol, and the connecting link between those two. They stood in the noon of that strange and solemn splendor, as if it were the light that is to reveal all secrets, and the daybreak that shall unite all who belong to one another.

There was witchcraft in little Pearl's eyes; and her face, as she glanced upward at the minister, wore that naughty smile which made its expression frequently so elvish. She withdrew her hand from Mr. Dimmesdale's and pointed across the street. But he clasped both his hands over his breast, and cast his eyes towards the zenith.

Nothing was more common, in those days, than to interpret all meteoric appearances, and other natural phenomena, that occurred with less regularity than the rise and set of sun and moon, as so many revelations from a supernatural source. Thus, a blazing spear, a sword of flame, a bow, or a sheaf of arrows, seen in the midnight sky, prefigured Indian warfare. Pestilence was known to have been foreboded by a shower of crimson light. We doubt whether any marked event, for good or evil, ever befell New England, from its settlement down to Revolutionary times, of which the inhabitants had not been previously warned by some spectacle of this nature. Not seldom, it had been seen by multitudes. Oftener, however, its credibility rested on the faith of some lonely eye-witness, who beheld the wonder through the colored, magnifying, and distorting medium of his imagination, and shaped it more distinctly in his after-thought. It was, indeed, a majestic idea, that the destiny of nations should be revealed, in these awful hieroglyphics, on the cope of heaven. A scroll so wide might not be deemed too expansive for Providence to write a people's doom upon. The belief was a favorite one with our forefathers, as betokening that their infant commonwealth was under a celestial guardianship of peculiar intimacy and strictness. But what shall we say, when an individual discovers a revelation, addressed to himself alone, on the same vast sheet of record! In such a case, it could only be the symptom of a highly disordered mental state, when a man, rendered morbidly self-contemplative by long, intense, and secret pain, had extended his egotism over the whole expanse of nature, until the firmament itself should appear no more than a fitting page for his soul's history and fate!

We impute it, therefore, solely to the disease in his own eye and heart, that the minister, looking upward to the zenith, beheld there the appearance of an immense letter,—the letter A,—marked out in lines of dull red light. Not but the meteor may have shown itself at that point, burning duskily through a veil of cloud; but with no such shape as his guilty imagination gave it; or, at least, with so little definiteness, that another's guilt might have seen another symbol in it.

There was a singular circumstance that characterized Mr. Dimmesdale's psychological state, at this moment. All the time that he gazed upward to the zenith, he was, nevertheless, perfectly aware that little Pearl was pointing her finger towards old Roger Chillingworth, who stood at no great distance from the scaffold. The minister appeared to see him, with the same glance that discerned the miraculous letter. To his features, as to all other objects, the meteoric light imparted a new expression; or it might well be that the physician was not careful then, as at all other times, to hide the malevolence with which he looked upon his victim. Certainly, if the meteor kindled up the sky, and disclosed the earth, with an awfulness that admonished Hester Prynne and the clergyman of the day of judgment, then might Roger Chillingworth have passed with them for the arch-fiend, standing there with a smile and scowl, to claim his own. So vivid was the expression, or so intense the minister's perception of it, that it seemed still to remain painted on the darkness, after the meteor had vanished, with an effect as if the street and all things else were at once annihilated.

"Who is that man, Hester?" gasped Mr. Dimmesdale, overcome with terror. "I shiver at him! Dost thou know the man? I hate him, Hester!"

She remembered her oath, and was silent.

"I tell thee, my soul shivers at him!" muttered the minister again. "Who is he? Who is he? Canst thou do nothing for me? I have a nameless horror of the man!"

"Minister," said little Pearl, "I can tell thee who he is!"

"Quickly, then, child!" said the minister, bending his ear close to her lips. "Quickly!—and as low as thou canst whisper."

Pearl mumbled something into his ear, that sounded, indeed, like human language, but was only such gibberish as children may be heard amusing themselves with, by the hour together. At all events, if it involved any secret information in regard to old Roger Chillingworth, it was in a tongue unknown to the erudite clergyman, and did but increase the bewilderment of his mind. The elvish child then laughed aloud.

"Dost thou mock me now?" said the minister.

"Thou was not bold!—thou wast not true!"—answered the child. "Thou wouldst not promise to take my hand, and mother's hand, to-morrow noontide!"

"Worthy Sir," answered the physician, who had now advanced to the foot of the platform. "Pious Master Dimmesdale! can this be you? Well, well, indeed! We men of study, whose heads are in our books, have need to be straitly looked after! We dream in our waking moments, and walk in our sleep. Come, good Sir, and my dear friend, I pray you, let me lead you home!"

"How knewest thou that I was here?" asked the minister, fearfully.

"Verily, and in good faith," answered Roger Chillingworth, "I knew nothing of the matter. I had spent the better part of the night at the bedside of the worshipful Governor Winthrop, doing what my poor skill might to give him ease. He going home to a better world, I, likewise, was on my way homeward, when this strange light shone out. Come with me, I beseech you, Reverend Sir; else you will be poorly able to do Sabbath duty to-morrow. Aha! see now, how they trouble the brain,—these books!—these books! You should study less, good Sir, and take a little pastime; or these night-whimseys will grow upon you."

"I will go home with you," said Mr. Dimmesdale.

With a chill despondency, like one awaking, all nerveless, from an ugly dream, he yielded himself to the physician, and was led away.

The next day, however, being the Sabbath, he preached a discourse which was held to be the richest and most powerful, and the most replete with heavenly influences, that had ever proceeded from his lips. Souls, it is said, more souls than one, were brought to the truth by the efficacy of that sermon, and vowed within themselves to cherish a holy gratitude towards Mr. Dimmesdale throughout the long hereafter. But, as he came down the pulpit steps, the gray-bearded sexton met him, holding up a black glove, which the minister recognized as his own.

"It was found," said the sexton, "this morning, on the scaffold where evil-doers are set up to public shame. Satan dropped it there, I take it, intending a scurrilous jest against your reverence. But, indeed, he was blind and foolish, as he ever and always is. A pure hand needs no glove to cover it!"

"Thank you, my good friend," said the minister, gravely, but startled at heart; for, so confused was his remembrance, that he had almost brought himself to look at the events of the past night as visionary. "Yes, it seems to be my glove, indeed!"

"And, since Satan saw fit to steal it, your reverence must needs handle him without gloves, henceforward," remarked the old sexton, grimly smiling. "But did your reverence hear of the portent that was seen last night?—a great red letter in the sky,—the letter A, which we interpret to stand for Angel. For, as our good Governor Winthrop was made an angel this past night, it was doubtless held fit that there should be some notice thereof!"

"No," answered the minister, "I had not heard of it."

XIII. Another View of Hester

In her late singular interview with Mr. Dimmesdale, Hester Prynne was shocked at the condition to which she found the clergyman reduced. His nerve seemed absolutely destroyed. His moral force was abased into more than childish weakness. It grovelled helpless on the ground, even while his intellectual faculties retained their pristine strength, or had perhaps acquired a morbid energy, which disease only could have given them. With her knowledge of a train of circumstances hidden from all others, she could readily infer that, besides the legitimate action of his own conscience, a terrible machinery had been brought to bear, and was still operating, on Mr. Dimmesdale's well-being and repose. Knowing what this poor, fallen man had once been, her whole soul was moved by the shuddering terror with which he had appealed to her,—the outcast woman,—for support against his instinctively discovered enemy. She decided, moreover, that he had a right to her utmost aid. Little accustomed, in her long seclusion from society, to measure her ideas of right and wrong by any standard external to herself, Hester saw—or seemed to see—that there lay a responsibility upon her, in reference to the clergyman, which she owed to no other, nor to the whole world besides. The links that united her to the rest of human kind—links of flowers, or silk, or gold, or whatever the material—had all been broken. Here was the iron link of mutual crime, which neither he nor she could break. Like all other ties, it brought along with it its obligations.

Hester Prynne did not now occupy precisely the same position in which we beheld her during the earlier periods of her ignominy. Years had come and gone. Pearl was now seven years old. Her mother, with the scarlet letter on her breast, glittering in its fantastic embroidery, had long been a familiar object to the townspeople. As is apt to be the case when a person stands out in any prominence before the community, and, at the same time, interferes neither with public nor individual interests and convenience, a species of general regard had ultimately grown up in reference to Hester Prynne. It is to the credit of human nature, that, except where its selfishness is brought into play, it loves more readily than it hates. Hatred, by a gradual and quiet process, will even be transformed to love, unless the change be impeded by a continually new irritation of the original feeling of hostility. In this matter of Hester Prynne, there was neither irritation nor irksomeness. She never battled with the public, but submitted, uncomplainingly, to its worst usage; she made no claim upon it, in requital for what she suffered; she did not weigh upon its sympathies. Then, also, the blameless purity of her life during all these years in which she had been set apart to infamy, was reckoned largely in her favor. With nothing now to lose, in the sight of mankind, and with no hope, and seemingly no wish, of gaining anything, it could only be a genuine regard for virtue that had brought back the poor wanderer to its paths.

It was perceived, too, that while Hester never put forward even the humblest title to share in the world's privileges,—further than to breathe the common air, and earn daily bread for little Pearl and herself by the faithful labor of her hands,—she was quick to acknowledge her sisterhood with the race of man, whenever benefits were to be conferred. None so ready as she to give of her little substance to every demand of poverty; even though the bitter-hearted pauper threw back a gibe in requital of the food brought regularly to his door, or the garments wrought for him by the fingers that could have embroidered a monarch's robe. None so self-devoted as Hester, when pestilence stalked through the town. In all seasons of calamity, indeed, whether general or of individuals, the outcast of society at once found her place. She came, not as a guest, but as a rightful inmate, into the household that was darkened by trouble; as if its gloomy twilight were a medium in which she was entitled to hold intercourse with her fellow-creatures. There glimmered the embroidered letter, with comfort in its unearthly ray. Elsewhere the token of sin, it was the taper of the sick-chamber. It had even thrown its gleam, in the sufferer's hard extremity, across the verge of time. It had shown him where to set his foot, while the light of earth was fast becoming dim, and ere the light of futurity could reach him. In such emergencies, Hester's nature showed itself warm and rich; a well-spring of human tenderness, unfailing to every real demand, and inexhaustible by the largest. Her breast, with its badge of shame, was but the softer pillow for the head that needed one. She was self-ordained a Sister of Mercy; or, we may rather say, the world's heavy hand had so ordained her, when neither the world nor she looked forward to this result. The letter was the symbol of her calling. Such helpfulness was found in her,—so much power to do, and power to sympathize,—that many people refused to interpret the scarlet A by its original signification. They said that it meant Able; so strong was Hester Prynne, with a woman's strength.

It was only the darkened house that could contain her. When sunshine came again, she was not there. Her shadow had faded across the threshold. The helpful inmate had departed, without one backward glance to gather up the meed of gratitude, if any were in the hearts of those whom she had served so zealously. Meeting them in the street, she never raised her head to receive their greeting. If they were resolute to accost her, she laid her finger on the scarlet letter, and passed on. This might be pride, but was so like humility, that it produced all the softening influence of the latter quality on the public mind. The public is despotic in its temper; it is capable of denying common justice, when too strenuously demanded as a right; but quite as frequently it awards more than justice, when the appeal is made, as despots love to have it made, entirely to its generosity. Interpreting Hester Prynne's deportment as an appeal of this nature, society was inclined to show its former victim a more benign countenance than she cared to be favored with, or, perchance, than she deserved.

The rulers, and the wise and learned men of the community, were longer in acknowledging the influence of Hester's good qualities than the people. The prejudices which they shared in common with the latter were fortified in themselves by an iron framework of reasoning, that made it a far tougher labor to expel them. Day by day, nevertheless, their sour and rigid wrinkles were relaxing into something which, in the due course of years, might grow to be an expression of almost benevolence. Thus it was with the men of rank, on whom their eminent position imposed the guardianship of the public morals. Individuals in private life, meanwhile, had quite forgiven Hester Prynne for her frailty; nay, more, they had begun to look upon the scarlet letter as the token, not of that one sin, for which she had borne so long and dreary a penance, but of her many good deeds since. "Do you see that woman with the embroidered badge?" they would say to strangers. "It is our Hester,—the town's own Hester,—who is so kind to the poor, so helpful to the sick, so comfortable to the afflicted!" Then, it is true, the propensity of human nature to tell the very worst of itself, when embodied in the person of another, would constrain them to whisper the black scandal of bygone years. It was none the less a fact, however, that, in the eyes of the very men who spoke thus, the scarlet letter had the effect of the cross on a nun's bosom. It imparted to the wearer a kind of sacredness, which enabled her to walk securely amid all peril. Had she fallen among thieves, it would have kept her safe. It was reported, and believed by many, that an Indian had drawn his arrow against the badge, and that the missile struck it, but fell harmless to the ground.

The effect of the symbol—or, rather, of the position in respect to society that was indicated by it—on the mind of Hester Prynne herself, was powerful and peculiar. All the light and graceful foliage of her character had been withered up by this red-hot brand, and had long ago fallen away, leaving a bare and harsh outline, which might have been repulsive, had she possessed friends or companions to be repelled by it. Even the attractiveness of her person had undergone a similar change. It might be partly owing to the studied austerity of her dress, and partly to the lack of demonstration in her manners. It was a sad transformation, too, that her rich and luxuriant hair had either been cut off, or was so completely hidden by a cap, that not a shining lock of it ever once gushed into the sunshine. It was due in part to all these causes, but still more to something else, that there seemed to be no longer anything in Hester's face for Love to dwell upon; nothing in Hester's form, though majestic and statue-like, that Passion would ever dream of clasping in its embrace; nothing in Hester's bosom, to make it ever again the pillow of Affection. Some attribute had departed from her, the permanence of which had been essential to keep her a woman. Such is frequently the fate, and such the stern development, of the feminine character and person, when the woman has encountered, and lived through, an experience of peculiar severity. If she be all tenderness, she will die. If she survive, the tenderness will either be crushed out of her, or—and the outward semblance is the same—crushed so deeply into her heart that it can never show itself more. The latter is perhaps the truest theory. She who has once been woman, and ceased to be so, might at any moment become a woman again, if there were only the magic touch to effect the transfiguration. We shall see whether Hester Prynne were ever afterwards so touched, and so transfigured.

Much of the marble coldness of Hester's impression was to be attributed to the circumstance, that her life had turned, in a great measure, from passion and feeling, to thought. Standing alone in the world,—alone, as to any dependence on society, and with little Pearl to be guided and protected,—alone, and hopeless of retrieving her position, even had she not scorned to consider it desirable,—she cast away the fragments of a broken chain. The world's law was no law for her mind. It was an age in which the human intellect, newly emancipated, had taken a more active and a wider range than for many centuries before. Men of the sword had overthrown nobles and kings. Men bolder than these had overthrown and rearranged—not actually, but within the sphere of theory, which was their most real abode—the whole system of ancient prejudice, wherewith was linked much of ancient principle. Hester Prynne imbibed this spirit. She assumed a freedom of speculation, then common enough on the other side of the Atlantic, but which our forefathers, had they known it, would have held to be a deadlier crime than that stigmatized by the scarlet letter.[1] In her lonesome cottage, by the sea-shore, thoughts visited her, such as dared to enter no other dwelling in New England; shadowy guests, that would have been as perilous as demons to their entertainer, could they have been seen so much as knocking at her door.

It is remarkable, that persons who speculate the most boldly often conform with the most perfect quietude to the external regulations of society. The thought suffices them, without investing itself in the flesh and blood of action. So it seemed to be with Hester. Yet, had little Pearl never come to her from the spiritual world, it might have been far otherwise. Then, she might have come down to us in history, hand in hand with Ann Hutchinson, as the foundress of a religious sect. She might, in one of her phases, have been a prophetess. She might, and not improbably would, have suffered death from the stern tribunals of the period, for attempting to undermine the foundations of the Puritan establishment. But, in the education of her child, the mother's enthusiasm of thought had something to wreak itself upon. Providence, in the person of this little girl, had assigned to Hester's charge the germ and blossom of womanhood, to be cherished and developed amid a host of difficulties. Everything was against her. The world was hostile. The child's own nature had something wrong in it, which continually betokened that she had been born amiss,—the effluence of her mother's lawless passion,—and often impelled Hester to ask, in bitterness of heart, whether it were for ill or good that the poor little creature had been born at all.

Indeed, the same dark question often rose into her mind, with reference to the whole race of womanhood. Was existence worth accepting, even to the happiest among them? As concerned her own individual existence, she had long ago decided in the negative, and dismissed the point as settled. A

1. Hawthorne is probably referring to the English Civil War, whose dates parallel those of the novel. By May 1649, King Charles I had been overthrown and beheaded (in January 1649). Hawthorne invites us to see Hester as an American version of such British revolutionaries as Oliver Cromwell.

tendency to speculation, though it may keep woman quiet, as it does man, yet makes her sad. She discerns, it may be, such a hopeless task before her. As a first step, the whole system of society is to be torn down, and built up anew. Then, the very nature of the opposite sex, or its long hereditary habit, which has become like nature, is to be essentially modified, before woman can be allowed to assume what seems a fair and suitable position. Finally, all other difficulties being obviated, woman cannot take advantage of these preliminary reforms, until she herself shall have undergone a still mightier change; in which, perhaps, the ethereal essence, wherein she has her truest life, will be found to have evaporated. A woman never overcomes these problems by any exercise of thought. They are not to be solved, or only in one way. If her heart chance to come upper-most, they vanish. Thus, Hester Prynne, whose heart had lost its regular and healthy throb, wandered without a clew in the dark labyrinth of mind; now turned aside by an insurmountable precipice; now starting back from a deep chasm. There was wild and ghastly scenery all around her, and a home and comfort nowhere. At times, a fearful doubt strove to possess her soul, whether it were not better to send Pearl at once to heaven, and go herself to such futurity as Eternal Justice should provide.

The scarlet letter had not done its office.

Now, however, her interview with the Reverend Mr. Dimmesdale, on the night of his vigil, had given her a new theme of reflection, and held up to her an object that appeared worthy of any exertion and sacrifice for its attainment. She had witnessed the intense misery beneath which the minister struggled, or, to speak more accurately, had ceased to struggle. She saw that he stood on the verge of lunacy, if he had not already stepped across it. It was impossible to doubt, that, whatever painful efficacy there might be in the secret sting of remorse, a deadlier venom had been infused into it by the hand that proffered relief. A secret enemy had been continually by his side, under the semblance of a friend and helper, and had availed himself of the opportunities thus afforded for tampering with the delicate springs of Mr. Dimmesdale's nature. Hester could not but ask herself, whether there had not originally been a defect of truth, courage and loyalty, on her own part, in allowing the minister to be thrown into a position where so much evil was to be foreboded, and nothing auspicious to be hoped. Her only justification lay in the fact, that she had been able to discern no method of rescuing him from a blacker ruin than had overwhelmed herself, except by acquiescing in Roger Chillingworth's scheme of disguise. Under that impulse, she had made her choice, and had chosen, as it now appeared, the more wretched alternative of the two. She determined to redeem her error, so far as it might yet be possible. Strengthened by years of hard and solemn trial, she felt herself no longer so inadequate to cope with Roger Chillingworth as on that night, abased by sin, and half maddened by the ignominy that was still new, when they had talked together in the prison-chamber. She had climbed her way, since then, to a higher point. The old man, on the other hand, had brought himself nearer to her level, or perhaps below it, by the revenge which he had stooped for.

In fine, Hester Prynne resolved to meet her former husband, and do what might be in her power for the rescue of the victim on whom he had so evidently set his gripe. The occasion was not long to seek. One afternoon, walking with Pearl in a retired part of the peninsula, she beheld the old physician, with a basket on one arm, and a staff in the other hand, stooping along the ground, in quest of roots and herbs to concoct his medicines withal.

XIV. Hester and the Physician

Hester bade little Pearl run down to the margin of the water, and play with the shells and tangled seaweed, until she should have talked awhile with yonder gatherer of herbs. So the child flew away like a bird, and, making bare her small white feet, went pattering along the moist margin of the sea. Here and there she came to a full stop, and peeped curiously into a pool, left by the retiring tide as a mirror for Pearl to see her face in. Forth peeped at her, out of the pool, with dark, glistening curls around her head, and an elf-smile in her eyes, the image of a little maid, whom Pearl, having no other playmate, invited to take her hand, and run a race with her. But the visionary little maid, on her part, beckoned likewise, as if to say,—"This is a better place! Come thou into the pool!" And Pearl, stepping in, mid-leg deep, beheld her own white feet at the bottom; while, out of a still lower depth, came the gleam of a kind of fragmentary smile, floating to and fro in the agitated water.

Meanwhile, her mother had accosted the physician.

"I would speak a word with you," said she,—"a word that concerns us much."

"Aha! And is it Mistress Hester that has a word for old Roger Chillingworth?" answered he, raising himself from his stooping posture. "With all my heart! Why, Mistress, I hear good tidings of you, on all hands! No longer ago than yester-eve, a magistrate, a wise and godly man, was discoursing of your affairs, Mistress Hester, and whispered me that there had been question concerning you in the council. It was debated whether or no, with safety to the common weal, yonder scarlet letter might be taken off your bosom. On my life, Hester, I made my entreaty to the worshipful magistrate that it might be done forthwith!"

"It lies not in the pleasure of the magistrates to take off this badge," calmly replied Hester. "Were I worthy to be quit of it, it would fall away of its own nature, or be transformed into something that should speak a different purport."

"Nay, then, wear it, if it suit you better," rejoined he. "A woman must needs follow her own fancy, touching the adornment of her person. The letter is gayly embroidered, and shows right bravely on your bosom!"

All this while, Hester had been looking steadily at the old man, and was shocked, as well as wonder-smitten, to discern what a change had been wrought upon him within the past seven years. It was not so much that he had grown older; for though the traces of advancing life were visible, he bore his age well, and seemed to retain a wiry vigor and alertness. But the former aspect of an intellectual and studious man, calm and quiet, which was what she best remembered in him, had altogether vanished, and been succeeded by an eager, searching, almost fierce, yet carefully guarded look. It seemed to be his wish and purpose to mask this expression with a smile; but the latter played him false, and flickered over his visage so derisively, that the spectator could see his blackness all the better for it. Ever and anon, too, there came a glare of red light out of his eyes; as if the old man's soul were on fire, and kept on smouldering duskily within his breast, until, by some casual puff of passion, it was blown into a momentary flame. This he repressed, as speedily as possible, and strove to look as if nothing of the kind had happened.

In a word, old Roger Chillingworth was a striking evidence of man's faculty of transforming himself into a devil, if he will only, for a reasonable space of time, undertake a devil's office. This unhappy person had effected such a transformation, by devoting himself, for seven years, to the constant analysis of a heart full of torture, and deriving his enjoyment thence, and adding fuel to those fiery tortures which he analyzed and gloated over.

The scarlet letter burned on Hester Prynne's bosom. Here was another ruin, the responsibility of which came partly home to her.

"What see you in my face," asked the physician, "that you look at it so earnestly?"

"Something that would make me weep, if there were any tears bitter enough for it," answered she. "But let it pass! It is of yonder miserable man that I would speak."

"And what of him?" cried Roger Chillingworth, eagerly, as if he loved the topic, and were glad of an opportunity to discuss it with the only person of whom he could make a confidant. "Not to hide the truth, Mistress Hester, my thoughts happen just now to be busy with the gentleman. So speak freely; and I will make answer."

"When we last spake together," said Hester, "now seven years ago, it was your pleasure to extort a promise of secrecy, as touching the former relation betwixt yourself and me. As the life and good fame of yonder man were in your hands, there seemed no choice to me, save to be silent, in accordance with your behest. Yet it was not without heavy misgivings that I thus bound myself; for, having cast off all duty towards other human beings, there remained a duty towards him; and something whispered me that I was betraying it, in pledging myself to keep your counsel. Since that day, no man is so near to him as you. You tread behind his every footstep. You are beside him, sleeping and waking. You search his thoughts. You burrow and rankle in his heart! Your clutch is on his life, and you cause him to die daily a living death; and still he knows you not. In permitting this, I have surely acted a false part by the only man to whom the power was left me to be true!"

"What choice had you?" asked Roger Chillingworth. "My finger, pointed at this man, would have hurled him from his pulpit into a dungeon,—thence, peradventure, to the gallows!"

"It had been better so!" said Hester Prynne.

"What evil have I done the man?" asked Roger Chillingworth again. "I tell thee, Hester Prynne, the richest fee that ever physician earned from monarch could not have bought such care as I have wasted on this miserable priest! But for my aid, his life would have burned away in torments, within the first two years after the perpetration of his crime and thine. For, Hester, his spirit lacked the strength that could have borne up, as thine has, beneath a burden like thy scarlet letter. O, I could reveal a goodly secret! But enough! What art can do, I have exhausted on him. That he now breathes, and creeps about on earth, is owing all to me!"

"Better he had died at once!" said Hester Prynne.

"Yea, woman, thou sayest truly!" cried old Roger Chillingworth, letting the lurid fire of his heart blaze out before her eyes. "Better had he died at once! Never did mortal suffer what this man has suffered. And all, all, in the sight of his worst enemy! He has been conscious of me. He has felt an influence dwelling always upon him like a curse. He knew, by some spiritual sense,—for the Creator never made another being so sensitive as this,—he knew that no friendly hand was pulling at his heart-strings, and that an eye was looking curiously into him, which sought only evil, and found it. But he knew not that the eye and hand were mine! With the superstition common to his brotherhood, he fancied himself given over to a fiend, to be tortured with frightful dreams, and desperate thoughts, the sting of remorse, and despair of pardon; as a foretaste of what awaits him beyond the grave. But it was the constant shadow of my presence!—the closest propinquity of the man whom he had most vilely wronged!—and who had grown to exist only by this perpetual poison of the direst revenge! Yea, indeed!—he did not err!—there was a fiend at his elbow! A mortal man, with once a human heart, has become a fiend for his especial torment!"

The unfortunate physician, while uttering these words, lifted his hands with a look of horror, as if he had beheld some frightful shape, which he could not recognize, usurping the place of his own image in a glass. It was one of those moments—which sometimes occur only at the interval of years—when a man's moral aspect is faithfully revealed to his mind's eye. Not improbably, he had never before viewed himself as he did now.

"Hast thou not tortured him enough?" said Hester, noticing the old man's look. "Has he not paid thee all?"

"No!—no!—He has but increased the debt!" answered the physician; and as he proceeded, his manner lost its fiercer characteristics, and subsided into gloom. "Dost thou remember me, Hester, as I was nine years agone? Even then, I was in the autumn of my days, nor was it the early autumn. But all my life had been made up of earnest, studious, thoughtful, quiet years, bestowed faithfully for the increase of mine own knowledge, and faithfully, too, though this latter object was but casual to the other,—faithfully for the advancement of human welfare. No life had been more peaceful and innocent than mine; few lives so rich with benefits conferred. Dost thou remember me? Was I not, though you might deem me cold, nevertheless a man thoughtful for others, craving little for himself,—kind, true, just, and of constant, if not warm affections? Was I not all this?"

"All this, and more," said Hester.

"And what am I now?" demanded he, looking into her face, and permitting the whole evil within him to be written on his features. "I have already told thee what I am! A fiend! Who made me so?"

"It was myself!" cried Hester, shuddering. "It was I, not less than he. Why hast thou not avenged thyself on me?"

"I have left thee to the scarlet letter," replied Roger Chillingworth. "If that have not avenged me, I can do no more!"

He laid his finger on it, with a smile.

"It has avenged thee!" answered Hester Prynne.

"I judged no less," said the physician. "And now, what wouldst thou with me touching this man?"

"I must reveal the secret," answered Hester, firmly. "He must discern thee in thy true character. What may be the result, I know not. But this long debt of confidence, due from me to him, whose bane and ruin I have been, shall at length be paid. So far as concerns the overthrow or preservation of his fair fame and his earthly state, and perchance his life, he is in thy hands. Nor do I,—whom the scarlet letter had disciplined to truth, though it be the truth of red-hot iron, entering into the soul,—nor do I perceive such advantage in his living any longer a life of ghastly emptiness, that I shall stoop to implore thy mercy. Do with him as thou wilt! There is no good for him,—no good for me,—no good for thee! There is no good for little Pearl! There is no path to guide us out of this dismal maze!"

"Woman, I could well-nigh pity thee!" said Roger Chillingworth, unable to restrain a thrill of admiration too; for there was a quality almost majestic in the despair which she expressed. "Thou hadst great elements. Peradventure, hadst thou met earlier with a better love than mine, this evil had not been. I pity thee, for the good that has been wasted in thy nature!"

"And I thee," answered Hester Prynne, "for the hatred that has transformed a wise and just man to a fiend! Wilt thou yet purge it out of thee, and be once more human? If not for his sake, then doubly for thine own! Forgive, and leave his further retribution to the Power that claims it! I said, but now, that there could be no good event for him, or thee, or me, who are here wandering together in this gloomy maze of evil, and stumbling, at every step, over the guilt wherewith we have strewn our path. It is not so! There might be good for thee, and thee alone, since thou hast been deeply wronged, and hast it at thy will to pardon. Wilt thou give up that only privilege? Wilt thou reject that priceless benefit?"

"Peace, Hester, peace!" replied the old man, with gloomy sternness. "It is not granted me to pardon. I have no such power as thou tellest me of. My old faith, long forgotten, comes back to me, and explains all that we do, and all we suffer. By thy first step awry, thou didst plant the germ of evil; but since that moment, it has all been a dark necessity. Ye that have wronged me are not sinful, save in a kind of typical illusion; neither am I fiend-like, who have snatched a fiend's office from his hands. It is our fate. Let the black flower blossom as it may! Now go thy ways, and deal as thou wilt with yonder man."

He waived his hand, and betook himself again to his employment of gathering herbs.

XV. Hester and Pearl

So Roger Chillingworth—a deformed old figure, with a face that haunted men's memories longer than they liked—took leave of Hester Prynne, and went stooping away along the earth. He gathered here and there an herb, or grubbed up a root, and put it into the basket on his arm. His gray beard almost touched the ground, as he crept onward. Hester gazed after him a little while, looking with a half fantastic curiosity to see whether the tender grass of early spring would not be blighted beneath him, and show the wavering track of his footsteps, sere and brown, across its cheerful verdure. She wondered what sort of herbs they were, which the old man was so sedulous to gather. Would not the earth, quickened to an evil purpose by the sympathy of his eye, greet him with poisonous shrubs, of species hitherto unknown, that would start up under his fingers? Or might it suffice him, that every wholesome growth should be converted into something deleterious and malignant at his touch? Did the sun, which shone so brightly everywhere else, really fall upon him? Or was there, as it rather seemed, a circle of ominous shadow moving along with his deformity, whichever way he turned himself? And whither was he now going? Would he not suddenly sink into the earth, leaving a barren and blasted spot, where, in due course of time, would be seen deadly nightshade, dogwood, henbane,[1] and whatever else of vegetable wickedness the climate could produce, all flourishing with hideous luxuriance? Or would he spread bat's wings and flee away, looking so much the uglier, the higher he rose towards heaven?

"Be it sin or no," said Hester Prynne, bitterly, as she still gazed after him, "I hate the man!"

She upbraided herself for the sentiment, but could not overcome or lessen it. Attempting to do so, she thought of those long-past days, in a distant land, when he used to emerge at eventide from the seclusion of his study, and sit down in the fire-light of their home, and in the light of her nuptial smile. He needed to bask himself in that smile, he said, in order that the chill of so many lonely hours among his books might be taken off the scholar's heart. Such scenes had once appeared not otherwise than happy, but now, as viewed through the dismal medium of her subsequent life, they classed themselves among her ugliest remembrances. She marveled how such scenes could have been! She marveled how she could ever have been wrought upon to marry him! She deemed it her crime most to be repented of, that she had ever endured, and reciprocated, the lukewarm grasp of his hand, and had suffered the smile of her lips and eyes to mingle and melt into his own. And it seemed a fouler offence committed by Roger Chillingworth, than any which had since been done him, that, in the time when her heart knew no better, he had persuaded her to fancy herself happy by his side.

"Yes, I hate him!" repeated Hester, more bitterly than before. "He betrayed me! He has done me worse wrong than I did him!"

Let men tremble to win the hand of woman, unless they win along with it the utmost passion of her heart! Else it may be their miserable fortune, as it was Roger Chillingworth's, when some mightier touch than their own may have awakened all her sensibilities, to be reproached even for the calm content, the marble image of happiness, which they will have imposed upon her as the warm reality. But Hester ought long ago to have done with this injustice. What did it betoken? Had seven long years, under the torture of the scarlet letter, inflicted so much of misery, and wrought out no repentance?

The emotions of that brief space, while she stood gazing after the crooked figure of old Roger Chillingworth, threw a dark light on Hester's

1. The poison (*Hyoscyamus niger*) Claudius poured into King Hamlet's ear in Shakespeare's play. *Nightshade* (*Atropa belladonna*): contains atropine.

state of mind, revealing much that she might not otherwise have acknowledged to herself.

He being gone, she summoned back her child.

"Pearl! Little Pearl! Where are you?"

Pearl, whose activity of spirit never flagged, had been at no loss for amusement while her mother talked with the old gatherer of herbs. At first, as already told, she had flirted fancifully with her own image in a pool of water, beckoning the phantom forth, and—as it declined to venture—seeking a passage for herself into its sphere of impalpable earth and unattainable sky. Soon finding, however, that either she or the image was unreal, she turned elsewhere for better pastime. She made little boats out of birch-bark, and freighted them with snail-shells, and sent out more ventures on the mighty deep than any merchant in New England; but the larger part of them foundered near the shore. She seized a live horse-shoe by the tail, and made prize of several five-fingers,[2] and laid out a jelly-fish to melt in the warm sun. Then she took up the white foam, that streaked the line of the advancing tide, and threw it upon the breeze, scampering after it, with winged footsteps, to catch the great snow-flakes ere they fell. Perceiving a flock of beach-birds, that fed and fluttered along the shore, the naughty child picked up her apron full of pebbles, and, creeping from rock to rock after these small sea-fowl, displayed remarkable dexterity in pelting them. One little gray bird, with a white breast, Pearl was almost sure, had been hit by a pebble, and fluttered away with a broken wing. But then the elf-child sighed, and gave up her sport; because it grieved her to have done harm to a little being that was as wild as the sea-breeze, or as wild as Pearl herself.

Her final employment was to gather sea-weed, of various kinds, and make herself a scarf, or mantle, and a head-dress, and thus assume the aspect of a little mermaid. She inherited her mother's gift for devising drapery and costume. As the last touch to her mermaid's garb, Pearl took some eel-grass, and imitated, as best she could, on her own bosom, the decoration with which she was so familiar on her mother's. A letter,—the letter A,—but freshly green, instead of scarlet! The child bent her chin upon her breast, and contemplated this device with strange interest; even as if the one only thing for which she had been sent into the world was to make out its hidden import.

"I wonder if mother will ask me what it means?" thought Pearl.

Just then, she heard her mother's voice, and flitting along as lightly as one of the little sea-birds, appeared before Hester Prynne, dancing, laughing, and pointing her finger to the ornament upon her bosom.

"My little Pearl," said Hester, after a moment's silence, "the green letter, and on thy childish bosom, has no purport. But dost thou know, my child, what this letter means which thy mother is doomed to wear?"

"Yes, mother," said the child. "It is the great letter A. Thou hast taught me in the horn-book."[3]

Hester looked steadily into her little face; but, though there was that singular expression which she had so often remarked in her black eyes,

2. Starfish. *Horse-shoe*: horseshoe crab.
3. An early reader made by gluing a piece of parchment to a board and covering it with a piece of transparent horn.

she could not satisfy herself whether Pearl really attached any meaning to the symbol. She felt a morbid desire to ascertain the point.

"Dost thou know, child, wherefore thy mother wears this letter?"

"Truly do I!" answered Pearl, looking brightly into her mother's face. "It is for the same reason that the minister keeps his hand over his heart!"

"And what reason is that?" asked Hester, half smiling at the absurd incongruity of the child's observation; but, on second thoughts, turning pale. "What has the letter to do with any heart, save mine?"

"Nay, mother, I have told all I know," said Pearl, more seriously than she was wont to speak. "Ask yonder old man whom thou hast been talking with! It may be he can tell. But in good earnest now, mother dear, what does this scarlet letter mean?—and why dost thou wear it on thy bosom?—and why does the minister keep his hand over his heart?"

She took her mother's hand in both her own, and gazed into her eyes with an earnestness that was seldom seen in her wild and capricious character. The thought occurred to Hester, that the child might really be seeking to approach her with child-like confidence, and doing what she could, and as intelligently as she knew how, to establish a meeting-point of sympathy. It showed Pearl in an unwonted aspect. Heretofore, the mother, while loving her child with the intensity of a sole affection, had schooled herself to hope for little other return than the waywardness of an April breeze; which spends its time in airy sport, and has its gusts of inexplicable passion, and is petulant in its best of moods, and chills oftener than caresses you, when you take it to your bosom; in requital of which misdemeanors, it will sometimes, of its own vague purpose, kiss your cheek with a kind of doubtful tenderness, and play gently with your hair, and then begone about its other idle business, leaving a dreamy pleasure at your heart. And this, moreover, was a mother's estimate of the child's disposition. Any other observer might have seen few but unamiable traits, and have given them a far darker coloring. But now the idea came strongly into Hester's mind, that Pearl, with her remarkable precocity and acuteness, might already have approached the age when she could be made a friend, and intrusted with as much of her mother's sorrows as could be imparted, without irreverence either to the parent or the child. In the little chaos of Pearl's character, there might be seen emerging—and could have been, from the very first—the steadfast principles of an unflinching courage,—an uncontrollable will,—a sturdy pride, which might be disciplined into self-respect,—and a bitter scorn of many things, which, when examined, might be found to have the taint of falsehood in them. She possessed affections, too, though hitherto acrid and disagreeable, as are the richest flavors of unripe fruit. With all these sterling attributes, thought Hester, the evil which she inherited from her mother must be great indeed, if a noble woman do not grow out of this elfish child.

Pearl's inevitable tendency to hover about the enigma of the scarlet letter seemed an innate quality of her being. From the earliest epoch of her conscious life, she had entered upon this as her appointed mission. Hester had often fancied that Providence had a design of justice and retribution, in endowing the child with this marked propensity; but never, until now, had she bethought herself to ask, whether, linked with that design, there might not likewise be a purpose of mercy and beneficence. If little Pearl were entertained with faith and trust, as a spirit messenger no less than an earthly child, might it not be her errand to soothe away the sorrow that lay cold in her mother's heart, and converted it into a tomb?—and to help her to overcome the passion, once so wild, and even yet neither dead nor asleep, but only imprisoned within the same tomb-like heart?

Such were some of the thoughts that now stirred in Hester's mind, with as much vivacity of impression as if they had actually been whispered into her ear. And there was little Pearl, all this while, holding her mother's hand in both her own, and turning her face upward, while she put these searching questions, once, and again, and still a third time.

"What does the letter mean, mother?—and why dost thou wear it?—and why does the minister keep his hand over his heart?"

"What shall I say?" thought Hester to herself. "No! If this be the price of the child's sympathy, I cannot pay it."

Then she spoke aloud.

"Silly Pearl," said she, "what questions are these? There are many things in this world that a child must not ask about. What know I of the minister's heart? And as for the scarlet letter, I wear it for the sake of its gold thread."

In all the seven bygone years, Hester Prynne had never before been false to the symbol on her bosom. It may be that it was the talisman of a stern and severe, but yet a guardian spirit, who now forsook her; as recognizing that, in spite of his strict watch over her heart, some new evil had crept into it, or some old one had never been expelled. As for little Pearl, the earnestness soon passed out of her face.

But the child did not see fit to let the matter drop. Two or three times, as her mother and she went homeward, and as often at suppertime, and while Hester was putting her to bed, and once after she seemed to be fairly asleep, Pearl looked up, with mischief gleaming in her black eyes.

"Mother," said she, "what does the scarlet letter mean?"

And the next morning, the first indication the child gave of being awake was by popping up her head from the pillow, and making that other inquiry, which she had so unaccountably connected with her investigations about the scarlet letter:—

"Mother!—Mother!—Why does the minister keep his hand over his heart?"

"Hold thy tongue, naughty child!" answered her mother, with an asperity that she had never permitted to herself before. "Do not tease me; else I shall shut thee into the dark closet!"

XVI. A Forest Walk

Hester Prynne remained constant in her resolve to make known to Mr. Dimmesdale, at whatever risk of present pain or ulterior consequences, the true character of the man who had crept into his intimacy. For several days, however, she vainly sought an opportunity of addressing him in some of the meditative walks which she knew him to be in the habit of taking, along the shores of the peninsula, or on the wooded hills of the neighboring country. There would have been no scandal, indeed, nor peril to the holy whiteness of the clergyman's good fame, had she visited him in his own study; where many a penitent, ere now, had confessed sins of perhaps as deep a dye as the one betokened by the scarlet letter. But, partly that she dreaded the secret or undisguised interference of old Roger Chillingworth, and partly that her conscious heart imputed suspicion where none could have been felt, and partly that both the minister and she would need the whole wide world to breathe in, while they talked together,—for all these reasons, Hester never thought of meeting him in any narrower privacy than beneath the open sky.

At last, while attending in a sick-chamber, whither the Reverend Mr. Dimmesdale had been summoned to make a prayer, she learnt that he had gone, the day before, to visit the Apostle Eliot, among his Indian converts.[1] He would probably return, by a certain hour, in the afternoon of the morrow. Betimes, therefore, the next day, Hester took little Pearl,—who was necessarily the companion of all her mother's expeditions, however inconvenient her presence,—and set forth.

The road, after the two wayfarers had crossed from the peninsula to the mainland, was no other than a foot-path. It straggled onward into the mystery of the primeval forest. This hemmed it in so narrowly, and stood so black and dense on either side, and disclosed such imperfect glimpses of the sky above, that, to Hester's mind, it imaged not amiss the moral wilderness in which she had so long been wandering. The day was chill and somber. Overhead was a gay expanse of cloud, slightly stirred, however, by a breeze; so that a gleam of flickering sunshine might now and then be seen at its solitary play along the path. This flitting cheerfulness was always at the further extremity of some long vista through the forest. The sportive sunlight—feebly sportive, at best, in the predominant pensiveness of the day and scene—withdrew itself as they came nigh, and left the spots where it had danced the drearier, because they had hoped to find them bright.

"Mother," said little Pearl, "the sunshine does not love you. It runs away and hides itself, because it is afraid of something on your bosom. Now, see! There it is, playing, a good way off. Stand you here, and let me run and catch it. I am but a child. It will not flee from me; for I wear nothing on my bosom yet!"

"Nor ever will, my child, I hope," said Hester.

"And why not, mother?" asked Pearl, stopping short, just at the beginning of her race. "Will not it come of its own accord, when I am a woman grown?"

"Run away, child," answered her mother, "and catch the sunshine! It will soon be gone."

Pearl set forth, at a great pace, and, as Hester smiled to perceive, did actually catch the sunshine, and stood laughing in the midst of it, all brightened by its splendor, and scintillating with the vivacity excited by rapid motion. The light lingered about the lonely child, as if glad of such a playmate, until her mother had drawn almost nigh enough to step into the magic circle too.

"It will go now," said Pearl, shaking her head.

1. John Eliot (1604–1690), commonly called "Apostle Eliot," sought to convert Massachusetts Indians to Christianity. He translated the Bible into Algonquin beginning in the 1640s. He also testified against Anne Hutchinson at her trial. Hawthorne's *Grandfather's Chair* (1840) includes a chapter on Eliot.

"See!" answered Hester, smiling. "Now I can stretch out my hand, and grasp some of it."

As she attempted to do so, the sunshine vanished; or, to judge from the bright expression that was dancing on Pearl's features, her mother could have fancied that the child had absorbed it into herself, and would give it forth again, with a gleam about her path, as they should plunge into some gloomier shade. There was no other attribute that so much impressed her with a sense of new and untransmitted vigor in Pearl's nature, as this never-failing vivacity of spirits; she had not the disease of sadness, which almost all children, in these latter days, inherit, with the scrofula,[2] from the troubles of their ancestors. Perhaps this too was a disease, and but the reflex of the wild energy with which Hester had fought against her sorrows, before Pearl's birth. It was certainly a doubtful charm, imparting a hard, metallic lustre to the child's character. She wanted—what some people want throughout life—a grief that should deeply touch her, and thus humanize and make her capable of sympathy. But there was time enough yet for little Pearl.

"Come, my child!" said Hester, looking about her from the spot where Pearl had stood still in the sunshine. "We will sit down a little way within the wood, and rest ourselves."

"I am not aweary, mother," replied the little girl. "But you may sit down, if you will tell me a story meanwhile."

"A story, child!" said Hester. "And about what?"

"O, a story about the Black Man," answered Pearl, taking hold of her mother's gown, and looking up, half earnestly, half mischievously, into her face. "How he haunts this forest, and carries a book with him,—a big, heavy book, with iron clasps; and how this ugly Black Man offers his book and an iron pen to everybody that meets him here among the trees; and they are to write their names with their own blood. And then he sets his mark on their bosoms! Didst thou ever meet the Black Man, mother?"

"And who told you this story, Pearl?" asked her mother, recognizing a common superstition of the period.

"It was the old dame in the chimney-corner, at the house where you watched last night," said the child. "But she fancied me asleep while she was talking of it. She said that a thousand and a thousand people had met him here, and had written in his book, and have his mark on them. And that ugly-tempered lady, old Mistress Hibbins, was one. And, mother, the old dame said that this scarlet letter was the Black Man's mark on thee, and that it glows like a red flame when thou meetest him at midnight, here in the dark wood. Is it true, mother? And dost thou go to meet him in the night-time?"

"Didst thou ever awake, and find thy mother gone?" asked Hester.

"Not that I remember," said the child. "If thou fearest to leave me in our cottage, thou mightest take me along with thee. I would very gladly go! But, mother, tell me now! Is there such a Black Man? And didst thou ever meet him? And is this his mark?"

"Wilt thou let me be at peace, if I once tell thee?" asked her mother.

"Yes, if thou tellest me all," answered Pearl.

2. A tubercular neck infection.

"Once in my life I met the Black Man!" said her mother. "This scarlet letter is his mark!"

Thus conversing, they entered sufficiently deep into the wood to secure themselves from the observation of any casual passenger along the forest track. Here they sat down on a luxuriant heap of moss; which, at some epoch of the preceding century, had been a gigantic pine, with its roots and trunk in the darksome shade, and its head aloft in the upper atmosphere. It was a little dell where they had seated themselves, with a leaf-strewn bank rising gently on either side, and a brook flowing through the midst, over a bed of fallen and drowned leaves. The trees impending over it had flung down great branches, from time to time, which choked up the current, and compelled it to form eddies and black depths at some points; while, in its swifter and livelier passages, there appeared a channel-way of pebbles, and brown, sparkling sand. Letting the eyes follow along the course of the stream, they could catch the reflected light from its water, at some short distance within the forest, but soon lost all traces of it amid the bewilderment of tree-trunks and underbrush, and here and there a huge rock covered over with gray lichens. All these giant trees and boulders of granite seemed intent on making a mystery of the course of this small brook; fearing, perhaps, that, with its never-ceasing loquacity, it should whisper tales out of the heart of the old forest whence it flowed, or mirror its revelations on the smooth surface of a pool. Continually, indeed, as it stole onward, the streamlet kept up a babble, kind, quiet, soothing, but melancholy, like the voice of a young child that was spending its infancy without playfulness, and knew not how to be merry among sad acquaintance and events of sombre hue.

"O brook! O foolish and tiresome little brook!" cried Pearl, after listening awhile to its talk. "Why art thou so sad? Pluck up a spirit, and do not be all the time sighing and murmuring!"

But the brook, in the course of its little lifetime among the forest-trees, had gone through so solemn an experience that it could not help talking about it, and seemed to have nothing else to say. Pearl resembled the brook, inasmuch as the current of her life gushed from a well-spring as mysterious, and had flowed through scenes shadowed as heavily with gloom. But, unlike the little stream, she danced and sparkled, and prattled airily along her course.

"What does this sad little brook say, mother?" inquired she.

"If thou hadst a sorrow of thine own, the brook might tell thee of it," answered her mother, "even as it is telling me of mine! But now, Pearl, I hear a foot-step along the path, and the noise of one putting aside the branches. I would have thee betake thyself to play, and leave me to speak with him that comes yonder."

"Is it the Black Man?" asked Pearl.

"Wilt thou go and play, child?" repeated her mother. "But do not stray far into the wood. And take heed that thou come at my first call."

"Yes, mother," answered Pearl. "But if it be the Black Man, wilt thou not let me stay a moment, and look at him, with his big book under his arm?"

"Go, silly child!" said her mother, impatiently. "It is no Black Man! Thou canst see him now, through the trees. It is the minister!"

"And so it is!" said the child. "And, mother, he has his hand over his heart! Is it because, when the minister wrote his name in the book, the Black Man set his mark in that place? But why does he not wear it outside his bosom, as thou dost, mother?"

"Go now, child, and thou shalt tease me as thou wilt another time," cried Hester Prynne. "But do not stray far. Keep where thou canst hear the babble of the brook."

The child went singing away, following up the current of the brook, and striving to mingle a more lightsome cadence with its melancholy voice. But the little stream would not be comforted, and still kept telling its unintelligible secret of some very mournful mystery that had happened—or making a prophetic lamentation about something that was yet to happen—within the verge of the dismal forest. So Pearl, who had enough of shadow in her own little life, chose to break off all acquaintance with this repining brook. She set herself, therefore, to gathering violets and wood-anemones, and some scarlet columbines that she found growing in the crevices of a high rock.

When her elf-child had departed, Hester Prynne made a step or two towards the track that led through the forest, but still remained under the deep shadow of the trees. She beheld the minister advancing along the path, entirely alone, and leaning on a staff which he had cut by the wayside. He looked haggard and feeble, and betrayed a nerveless despondency in his air, which had never so remarkably characterized him in his walks about the settlement, nor in any other situation where he deemed himself liable to notice. Here it was wofully visible, in this intense seclusion of the forest, which of itself would have been a heavy trial to the spirits. There was a listlessness in his gait; as if he saw no reason for taking one step further, nor felt any desire to do so, but would have been glad, could he be glad of anything, to fling himself down at the root of the nearest tree, and lie there passive, forevermore. The leaves might bestrew him, and the soil gradually accumulate and form a little hillock over his frame, no matter whether there were life in it or no. Death was too definite an object to be wished for, or avoided.

To Hester's eye, the Reverend Mr. Dimmesdale exhibited no symptom of positive and vivacious suffering, except that, as little Pearl had remarked, he kept his hand over his heart.

XVII. The Pastor and His Parishioner

Slowly as the minister walked, he had almost gone by, before Hester Prynne could gather voice enough to attract his observation. At length, she succeeded.

"Arthur Dimmesdale!" she said, faintly at first; then louder, but hoarsely. "Arthur Dimmesdale!"

"Who speaks?" answered the minister.

Gathering himself quickly up, he stood more erect, like a man taken by surprise in a mood to which he was reluctant to have witnesses. Throwing his eyes anxiously in the direction of the voice, he indistinctly beheld a form under the trees, clad in garments so sombre, and so little relieved from the gray twilight into which the clouded sky and the heavy foliage had darkened the noontide, that he knew not whether it were a woman or a shadow. It may be, that his pathway through life was haunted thus, by a spectre that had stolen out from among his thoughts.

He made a step nigher, and discovered the scarlet letter.

"Hester! Hester Prynne!" said he. "Is it thou? Art thou in life?"

"Even so!" she answered. "In such life as has been mine these seven years past! And thou, Arthur Dimmesdale, dost thou yet live?"

It was no wonder that they thus questioned one another's actual and bodily existence, and even doubted of their own. So strangely did they meet, in the dim wood, that it was like the first encounter, in the world beyond the grave, of two spirits who had been intimately connected in their former life, but now stood coldly shuddering, in mutual dread; as not yet familiar with their state, nor wonted to the companionship of disembodied beings. Each a ghost, and awe-stricken at the other ghost! They were awe-stricken likewise at themselves; because the crisis flung back to them their consciousness, and revealed to each heart its history and experience, as life never does, except at such breathless epochs. The soul beheld its features in the mirror of the passing moment. It was with fear, and tremulously, and, as it were, by a slow, reluctant necessity, that Arthur Dimmesdale put forth his hand, chill as death, and touched the chill hand of Hester Prynne. The grasp, cold as it was, took away what was dreariest in the interview. They now felt themselves, at least, inhabitants of the same sphere.

Without a word more spoken,—neither he nor she assuming the guidance, but with an unexpressed consent,—they glided back into the shadow of the woods, whence Hester had emerged, and sat down on the heap of moss where she and Pearl had before been sitting. When they found voice to speak, it was, at first, only to utter remarks and inquiries such as any two acquaintance might have made, about the gloomy sky, the threatening storm, and, next, the health of each. Thus they went onward, not boldly, but step by step, into the themes that were brooding deepest in their hearts. So long estranged by fate and circumstances, they needed something slight and casual to run before, and throw open the doors of intercourse, so that their real thoughts might be led across the threshold.

After a while, the minister fixed his eyes on Hester Prynne's.

"Hester," said he, "hast thou found peace?"

She smiled drearily, looking down upon her bosom.

"Hast thou?" she asked.

"None!—nothing but despair!" he answered. "What else could I look for, being what I am, and leading such a life as mine? Were I an atheist,—a man devoid of conscience,—a wretch with coarse and brutal instincts,—I might have found peace, long ere now. Nay, I never should have lost it! But, as matters stand with my soul, whatever of good capacity there originally was in me, all of God's gifts that were the choicest have become the ministers of spiritual torment. Hester, I am most miserable!"

"The people reverence thee," said Hester. "And surely thou workest good among them! Doth this bring thee no comfort?"

"More misery, Hester!—only the more misery!" answered the clergyman, with a bitter smile. "As concerns the good which I may appear to do, I have no faith in it. It must needs be a delusion. What can a ruined soul, like mine, effect towards the redemption of other souls?—or a polluted soul, towards their purification? And as for the people's reverence, would that it were turned to scorn and hatred! Canst thou deem it, Hester, a consolation, that I must stand up in my pulpit, and meet so many eyes turned upward to my face, as if the light of heaven were beaming from it!—must see my flock hungry for the truth, and listening to my words as if a tongue of Pentecost were speaking!—and then look inward, and discern the black reality of what they idolize? I have laughed, in bitterness and agony of heart, at the contrast between what I seem and what I am! And Satan laughs at it!"

"You wrong yourself in this," said Hester, gently. "You have deeply and sorely repented. Your sin is left behind you, in the days long past. Your present life is not less holy, in very truth, than it seems in people's eyes. Is there no reality in the penitence thus sealed and witnessed by good works?[1] And wherefore should it not bring you peace?"

"No, Hester, no!" replied the clergyman. "There is no substance in it! It is cold and dead, and can do nothing for me! Of penance, I have had enough! Of penitence, there has been none! Else, I should long ago have thrown off these garments of mock holiness, and have shown myself to mankind as they will see me at the judgment-seat. Happy are you, Hester, that wear the scarlet letter openly upon your bosom! Mine burns in secret! Thou little knowest what a relief it is, after the torment of a seven years' cheat, to look into an eye that recognizes me for what I am! Had I one friend,—or were it my worst enemy!—to whom, when sickened with the praises of all other men, I could daily betake myself, and be known as the vilest of all sinners, methinks my soul might keep itself alive thereby. Even thus much of truth would save me! But, now, it is all falsehood!—all emptiness!—all death!"

Hester Prynne looked into his face, but hesitated to speak. Yet, uttering his long-restrained emotions so vehemently as he did, his words here offered her the very point of circumstances in which to interpose what she came to say. She conquered her fears, and spoke.

"Such a friend as thou hast even now wished for," said she, "with whom to weep over thy sin, thou hast in me, the partner of it!"—Again she hesitated, but brought out the words with an effort.—"Thou hast long had such an enemy, and dwellest with him, under the same roof!"

The minister started to his feet, gasping for breath, and clutching at his heart, as if he would have torn it out of his bosom.

"Ha! What sayest thou!" cried he. "An enemy! And under mine own roof! What mean you?"

Hester Prynne was now fully sensible of the deep injury for which she was responsible to this unhappy man, in permitting him to lie for so many years, or, indeed, for a single moment, at the mercy of one whose purposes could not be other than malevolent. The very contiguity of his enemy, beneath whatever mask the latter might conceal himself, was enough to disturb the magnetic sphere of a being so sensitive as Arthur Dimmesdale. There had been a period when Hester was less alive to this consideration; or, perhaps, in the misanthropy of her own trouble, she left the minister

1. Anne Hutchinson had been charged with claiming that several Puritan ministers, including John Wilson, had preached a covenant of works, rather than a covenant of grace, as a way to salvation.

to bear what she might picture to herself as a more tolerable doom. But of late, since the night of his vigil, all her sympathies towards him had been both softened and invigorated. She now read his heart more accurately. She doubted not, that the continual presence of Roger Chillingworth,—the secret poison of his malignity, infecting all the air about him,—and his authorized interference, as a physician, with the minister's physical and spiritual infirmities,—that these bad opportunities had been turned to a cruel purpose. By means of them, the sufferer's conscience had been kept in an irritated state, the tendency of which was, not to cure by wholesome pain, but to disorganize and corrupt his spiritual being. Its result, on earth, could hardly fail to be insanity, and hereafter, that eternal alienation from the Good and True, of which madness is perhaps the earthly type.

Such was the ruin to which she had brought the man, once,—nay, why should we not speak it?—still so passionately loved! Hester felt that the sacrifice of the clergyman's good name, and death itself, as she had already told Roger Chillingworth, would have been infinitely preferable to the alternative which she had taken upon herself to choose. And now, rather than have had this grievous wrong to confess, she would gladly have lain down on the forest-leaves, and died there, at Arthur Dimmesdale's feet.

"O Arthur," cried she, "forgive me! In all things else, I have striven to be true! Truth was the one virtue which I might have held fast, and did hold fast, through all extremity; save when thy good,—thy life,—thy fame,—were put in question! Then I consented to a deception. But a lie is never good, even though death threaten on the other side! Dost thou not see what I would say? That old man!—the physician!—he whom they call Roger Chillingworth!—he was my husband!"

The minister looked at her, for an instant, with all that violence of passion, which—intermixed, in more shapes than one, with his higher, purer, softer qualities,—was, in fact, the portion of him which the Devil claimed, and through which he sought to win the rest. Never was there a blacker or a fiercer frown than Hester now encountered. For the brief space that it lasted, it was a dark transfiguration. But his character had been so much enfeebled by suffering, that even its lower energies were incapable of more than a temporary struggle. He sank down on the ground, and buried his face in his hands.

"I might have known it," murmured he. "I did know it! Was not the secret told me, in the natural recoil of my heart, at the first sight of him, and as often as I have seen him since? Why did I not understand? O Hester Prynne, thou little, little knowest all the horror of this thing! And the shame!—the indelicacy!—the horrible ugliness of this exposure of a sick and guilty heart to the very eye that would gloat over it! Woman, woman, thou art accountable for this! I cannot forgive thee!"

"Thou shalt forgive me!" cried Hester, flinging herself on the fallen leaves beside him. "Let God punish! Thou shalt forgive!"

With sudden and desperate tenderness, she threw her arms around him, and pressed his head against her bosom; little caring though his cheek rested on the scarlet letter. He would have released himself, but strove in vain to do so. Hester would not set him free, lest he should look her sternly in the face. All the world had frowned on her,—for seven long years had it frowned upon this lonely woman,—and still she bore it all, nor ever once turned away her firm, sad eyes. Heaven, likewise, had frowned upon her, and she had not died. But the frown of this pale, weak, sinful, and sorrow-stricken man was what Hester could not bear, and live!

"Wilt thou yet forgive me!" she repeated, over and over again. "Wilt thou not frown? Wilt thou forgive?"

"I do forgive you, Hester," replied the minister, at length, with a deep utterance, out of an abyss of sadness, but no anger. "I freely forgive you now. May God forgive us both! We are not, Hester, the worst sinners in the world. There is one worse than even the polluted priest! That old man's revenge has been blacker than my sin. He has violated, in cold blood, the sanctity of the human heart. Thou and I, Hester, never did so!"

"Never, never!" whispered she. "What we did had a consecration of its own. We felt it so! We said so to each other! Hast thou forgotten it?"

"Hush, Hester!" said Arthur Dimmesdale, rising from the ground. "No; I have not forgotten!"

They sat down again, side by side, and hand clasped in hand, on the mossy trunk of the fallen tree. Life had never brought them a gloomier hour; it was the point whither their pathway had so long been tending, and darkening ever, as it stole along;—and yet it enclosed a charm that made them linger upon it, and claim another, and another, and, after all, another moment. The forest was obscure around them, and creaked with a blast that was passing through it. The boughs were tossing heavily above their heads; while one solemn old tree groaned dolefully to another, as if telling the sad story of the pair that sat beneath, or constrained to forebode evil to come.

And yet they lingered. How dreary looked the forest-track that led backward to the settlement, where Hester Prynne must take up again the burden of her ignominy, and the minister the hollow mockery of his good name! So they lingered an instant longer. No golden light had ever been so precious as the gloom of this dark forest. Here, seen only by his eyes, the scarlet letter need not burn into the bosom of the fallen woman! Here, seen only by her eyes, Arthur Dimmesdale, false to God and man, might be, for one moment, true!

He started at a thought that suddenly occurred to him.

"Hester," cried he, "here is a new horror! Roger Chillingworth knows your purpose to reveal his true character. Will he continue, then, to keep our secret? What will now be the course of his revenge?"

"There is a strange secrecy in his nature," replied Hester, thoughtfully; "and it has grown upon him by the hidden practices of his revenge. I deem it not likely that he will betray the secret. He will doubtless seek other means of satiating his dark passion."

"And I!—how am I to live longer, breathing the same air with this deadly enemy?" exclaimed Arthur Dimmesdale, shrinking within himself, and pressing his hand nervously against his heart,—a gesture that had grown involuntary with him. "Think for me, Hester! Thou art strong. Resolve for me!"

"Thou must dwell no longer with this man," said Hester, slowly and firmly. "Thy heart must be no longer under his evil eye!"

"It were far worse than death!" replied the minister. "But how to avoid it? What choice remains to me? Shall I lie down again on these withered leaves, where I cast myself when thou didst tell me what he was? Must I sink down there, and die at once?"

"Alas, what a ruin has befallen thee!" said Hester, with the tears gushing into her eyes. "Wilt thou die for very weakness? There is no other cause!"

"The judgment of God is on me," answered the conscience-stricken priest. "It is too mighty for me to struggle with!"

"Heaven would show mercy," rejoined Hester, "hadst thou but the strength to take advantage of it."

"Be thou strong for me!" answered he. "Advise me what to do."

"Is the world, then, so narrow?" exclaimed Hester Prynne, fixing her deep eyes on the minister's, and instinctively exercising a magnetic power over a spirit so shattered and subdued that it could hardly hold itself erect. "Doth the universe lie within the compass of yonder town, which only a little time ago was but a leaf-strewn desert, as lonely as this around us? Whither leads yonder forest track? Backward to the settlement, thou sayest! Yes; but onward, too! Deeper it goes, and deeper, into the wilderness, less plainly to be seen at every step; until, some few miles hence, the yellow leaves will show no vestige of the white man's tread. There thou art free! So brief a journey would bring thee from a world where thou hast been most wretched, to one where thou mayest still be happy! Is there not shade enough in all this boundless forest to hide thy heart from the gaze of Roger Chillingworth?"

"Yes, Hester; but only under the fallen leaves!" replied the minister, with a sad smile.

"Then there is the broad pathway of the sea!" continued Hester. "It brought thee hither. If thou so choose, it will bear thee back again. In our native land, whether in some remote rural village or in vast London,—or, surely, in Germany, in France, in pleasant Italy,—thou wouldst be beyond his power and knowledge! And what hast thou to do with all these iron men, and their opinions? They have kept thy better part in bondage too long already!"

"It cannot be!" answered the minister, listening as if he were called upon to realize a dream. "I am powerless to go! Wretched and sinful as I am, I have had no other thought than to drag on my earthly existence in the sphere where Providence hath placed me. Lost as my own soul is, I would still do what I may for other human souls! I dare not quit my post, though an unfaithful sentinel, whose sure reward is death and dishonor, when his dreary watch shall come to an end!"

"Thou art crushed under this seven years' weight of misery," replied Hester, fervently resolved to buoy him up with her own energy. "But thou shalt leave it all behind thee! It shall not cumber thy steps, as thou treadest along the forest-path; neither shalt thou freight the ship with it, if thou prefer to cross the sea. Leave this wreck and ruin here where it hath happened. Meddle no more with it! Begin all anew! Hast thou exhausted possibility in the failure of this one trial? Not so! The future is yet full of trial and success. There is happiness to be enjoyed! There is good to be done! Exchange this false life of thine for a true one. Be, if thy spirit summon thee to such a mission, the teacher and apostle of the red men. Or,—as is more thy nature,—be a scholar and a sage among the wisest and the most renowned of the cultivated world. Preach! Write!

Act! Do anything, save to lie down and die! Give up this name of Arthur Dimmesdale, and make thyself another, and a high one, such as thou canst wear without fear or shame. Why shouldst thou tarry so much as one other day in the torments that have so gnawed into thy life!—that have made thee feeble to will and to do!—that will leave thee powerless even to repent! Up, and away!"

"O Hester!" cried Arthur Dimmesdale, in whose eyes a fitful light, kindled by her enthusiasm, flashed up and died away, "thou tellest of running a race to a man whose kness are tottering beneath him! I must die here! There is not the strength or courage left me to venture into the wide, strange, difficult world, alone!"

It was the last expression of the despondency of a broken spirit. He lacked energy to grasp the better fortune that seemed within his reach.

He repeated the word.

"Alone, Hester!"

"Thou shalt not go alone!" answered she, in a deep whisper.

Then, all was spoken!

XVIII. A Flood of Sunshine

Arthur Dimmesdale gazed into Hester's face with a look in which hope and joy shone out, indeed, but with fear betwixt them, and a kind of horror at her boldness, who had spoken what he vaguely hinted at, but dared not speak.

But Hester Prynne, with a mind of native courage and activity, and for so long a period not merely estranged, but outlawed, from society, had habituated herself to such latitude of speculation as was altogether foreign to the clergyman. She had wandered, without rule or guidance, in a moral wilderness; as vast, as intricate and shadowy, as the untamed forest, amid the gloom of which they were now holding a colloquy that was to decide their fate. Her intellect and heart had their home, as it were, in desert places, where she roamed as freely as the wild Indian in his woods. For years past she had looked from this estranged point of view at human institutions, and whatever priests or legislators had established; criticising all with hardly more reverence than the Indian would feel for the clerical band, the judicial robe, the pillory, the gallows, the fireside, or the church. The tendency of her fate and fortunes had been to set her free. The scarlet letter was her passport into regions where other women dared not tread. Shame, Despair, Solitude! These had been her teachers,—stern and wild ones,—and they had made her strong, but taught her much amiss.

The minister, on the other hand, had never gone through an experience calculated to lead him beyond the scope of generally received laws; although, in a single instance, he had so fearfully transgressed one of the most sacred of them. But this had been a sin of passion, not of principle, nor even purpose. Since that wretched epoch, he had watched, with morbid zeal and minuteness, not his acts,—for those it was easy to arrange,—but each breath of emotion, and his every thought. At the head of the social system, as the clergymen of that day stood, he was only the more trammelled by its regulations, its principles, and even its prejudices. As a priest, the framework of his order inevitably hemmed him in. As a man who had once sinned, but who kept his conscience all alive and painfully sensitive by the fretting of an unhealed wound, he might have been supposed safer within the line of virtue than if he had never sinned at all.

Thus, we seem to see that, as regarded Hester Prynne, the whole seven years of outlaw and ignominy had been little other than a preparation for this very hour. But Arthur Dimmesdale! Were such a man once more to fall, what plea could be urged in extenuation of his crime? None; unless it avail him somewhat, that he was broken down by long and exquisite suffering; that his mind was darkened and confused by the very remorse which harrowed it; that, between fleeing as an avowed criminal, and remaining as a hypocrite, conscience might find it hard to strike the balance; that it was human to avoid the peril of death and infamy, and the inscrutable machinations of an enemy; that, finally, to this poor pilgrim, on his dreary and desert path, faint, sick, miserable, there appeared a glimpse of human affection and sympathy, a new life, and a true one, in exchange for the heavy doom which he was now expiating. And be the stern and sad truth spoken, that the breach which guilt has once made into the human soul is never, in this mortal state, repaired. It may be watched and guarded; so that the enemy shall not force his way again into the citadel, and might even, in his subsequent assaults, select some other avenue, in preference to that where he had formerly succeeded. But there is still the ruined wall, and, near it, the stealthy tread of the foe that would win over again his unforgotten triumph.

The struggle, if there were one, need not be described. Let it suffice, that the clergyman resolved to flee, and not alone.

"If, in all these past seven years," thought he, "I could recall one instant of peace or hope, I would yet endure, for the sake of that earnest of Heaven's mercy. But now,—since I am irrevocably doomed,—wherefore should I not snatch the solace allowed to the condemned culprit before his execution? Or, if this be the path to a better life, as Hester would persuade me, I surely give up no fairer prospect by pursuing it! Neither can I any longer live without her companionship; so powerful is she to sustain,—so tender to soothe! O Thou to whom I dare not lift mine eyes, wilt Thou yet pardon me!"

"Thou wilt go!" said Hester, calmly, as he met her glance.

The decision once made, a glow of strange enjoyment threw its flickering brightness over the trouble of his breast. It was the exhilarating effect—upon a prisoner just escaped from the dungeon of his own heart—of breathing the wild, free atmosphere of an unredeemed, unchristianized, lawless region. His spirit rose, as it were, with a bound, and attained a nearer prospect of the sky, than throughout all the misery which had kept him grovelling on the earth. Of a deeply religious temperament, there was inevitably a tinge of the devotional in his mood.

"Do I feel joy again?" cried he, wondering at himself. "Methought the germ of it was dead in me! O Hester, thou art my better angel! I seem to have flung myself—sick, sin-stained, and sorrow-blackened—down upon these forest-leaves, and to have risen up all made anew, and with new powers to glorify Him that hath been merciful! This is already the better life! Why did we not find it sooner?"

"Let us not look back," answered Hester Prynne. "The past is gone! Wherefore should we linger upon it now? See! With this symbol, I undo it all, and make it as it had never been!"

So speaking, she undid the clasp that fastened the scarlet letter, and, taking it from her bosom, threw it to a distance among the withered leaves. The mystic token alighted on the hither verge of the stream. With a hand's breadth further flight it would have fallen into the water, and have given the little brook another woe to carry onward, besides the unintelligible tale which it still kept murmuring about. But there lay the embroidered letter, glittering like a lost jewel, which some ill-fated wanderer might pick up, and thenceforth be haunted by strange phantoms of guilt, sinkings of the heart, and unaccountable misfortune.

The stigma gone, Hester heaved a long, deep sigh, in which the burden of shame and anguish departed from her spirit. O exquisite relief! She had not known the weight, until she felt the freedom! By another impulse, she took off the formal cap that confined her hair; and down it fell upon her shoulders, dark and rich, with at once a shadow and a light in its abundance, and imparting the charm of softness to her features. There played around her mouth, and beamed out of her eyes, a radiant and tender smile, that seemed gushing from the very heart of womanhood. A crimson flush was glowing on her cheek, that had been long so pale. Her sex, her youth, and the whole richness of her beauty, came back from what men call the irrevocable past, and clustered themselves, with her maiden hope, and a happiness before unknown, within the magic circle of this hour. And, as if the gloom of the earth and sky had been but the effluence of these two mortal hearts, it vanished with their sorrow. All at once, as with a sudden smile of heaven, forth burst the sunshine, pouring a very flood into the obscure forest, gladdening each green leaf, transmuting the yellow fallen ones to gold, and gleaming adown the gray trunks of the solemn trees. The objects that had made a shadow hitherto, embodied the brightness now. The course of the little brook might be traced by its merry gleam afar into the wood's heart of mystery, which had become a mystery of joy.

Such was the sympathy of Nature—that wild, heathen Nature of the forest, never subjugated by human law, nor illumined by higher truth—with the bliss of these two spirits! Love, whether newly born, or aroused from a death-like slumber, must always create a sunshine, filling the heart so full of radiance, that it overflows upon the outward world. Had the forest still kept its gloom, it would have been bright in Hester's eyes, and bright in Arthur Dimmesdale's!

Hester looked at him with the thrill of another joy.

"Thou must know Pearl!" said she. "Our little Pearl! Thou hast seen her,—yes, I know it!—but thou wilt see her now with other eyes. She is a strange child! I hardly comprehend her! But thou wilt love her dearly, as I do, and wilt advise me how to deal with her."

"Dost thou think the child will be glad to know me?" asked the minister, somewhat uneasily. "I have long shrunk from children, because they often show a distrust,—a backwardness to be familiar with me. I have even been afraid of little Pearl!"

"Ah, that was sad!" answered the mother. "But she will love thee dearly, and thou her. She is not far off. I will call her! Pearl! Pearl!"

"I see the child," observed the minister. "Yonder she is, standing in a streak of sunshine, a good way off, on the other side of the brook. So thou thinkest the child will love me?"

Hester smiled, and again called to Pearl, who was visible, at some distance, as the minister had described her, like a bright-apparelled vision, in a sunbeam, which fell down upon her through an arch of boughs. The ray quivered to and fro, making her figure dim or distinct,—now like a real child, now like a child's spirit,—as the splendor went and came again. She heard her mother's voice, and approached slowly through the forest.

Pearl had not found the hour pass wearisomely, while her mother sat talking with the clergyman. The great black forest—stern as it showed itself to those who brought the guilt and troubles of the world into its bosom—became the playmate of the lonely infant, as well as it knew how. Sombre as it was, it put on the kindest of its moods to welcome her. It offered her the partridge-berries, the growth of the preceding autumn, but ripening only in the spring, and now red as drops of blood upon the withered leaves. These Pearl gathered, and was pleased with their wild flavor. The small denizens of the wilderness hardly took pains to move out of her path. A partridge, indeed, with a brood of ten behind her, ran forward threateningly, but soon repented of her fierceness, and clucked to her young ones not to be afraid. A pigeon, alone on a low branch, allowed Pearl to come beneath, and uttered a sound as much of greeting as alarm. A squirrel, from the lofty depths of his domestic tree, chattered either in anger or merriment,—for a squirrel is such a choleric and humorous little personage, that it is hard to distinguish between his moods,—so he chattered at the child, and flung down a nut upon her head. It was a last year's nut, and already gnawed by his sharp tooth. A fox, startled from his sleep by her light footstep on the leaves, looked inquisitively at Pearl, as doubting whether it were better to steal off, or renew his nap on the same spot. A wolf, it is said,—but here the tale has surely lapsed into the improbable,—came up, and smelt of Pearl's robe, and offered his savage head to be patted by her hand. The truth seems to be, however, that the mother-forest, and these wild things which it nourished, all recognized a kindred wildness in the human child.

And she was gentler here than in the grassy-margined streets of the settlement, or in her mother's cottage. The flowers appeared to know it; and one and another whispered as she passed, "Adorn thyself with me, thou beautiful child, adorn thyself with me!"—and, to please them, Pearl gathered the violets, and anemones, and columbines, and some twigs of the freshest green, which the old trees held down before her eyes. With these she decorated her hair, and her young waist, and became a nymph-child, or an infant dryad,[1] or whatever else was in closest sympathy with the antique wood. In such guise had Pearl adorned herself, when she heard her mother's voice, and came slowly back.

Slowly; for she saw the clergyman!

1. Wood nymph.

XIX. The Child at the Brook-Side

"Thou wilt love her dearly," repeated Hester Prynne, as she and the minister sat watching little Pearl. "Dost thou not think her beautiful? And see with what natural skill she has made those simple flowers adorn her! Had she gathered pearls, and diamonds, and rubies, in the wood, they could not have become her better. She is a splendid child! But I know whose brow she has!"

"Dost thou know, Hester," said Arthur Dimmesdale, with an unquiet smile, "that this dear child, tripping about always at thy side, hath caused me many an alarm? Methought—O Hester, what a thought is that, and how terrible to dread it!—that my own features were partly repeated in her face, and so strikingly that the world might see them! But she is mostly thine!"

"No, no! Not mostly!" answered the mother, with a tender smile. "A little longer, and thou needest not to be afraid to trace whose child she is. But how strangely beautiful she looks, with those wild flowers in her hair! It is as if one of the fairies, whom we left in our dear old England, had decked her out to meet us."

It was with a feeling which neither of them had ever before experienced, that they sat and watched Pearl's slow advance. In her was visible the tie that united them. She had been offered to the world, these seven years past, as the living hieroglyphic, in which was revealed the secret they so darkly sought to hide,—all written in this symbol,—all plainly manifest,—had there been a prophet or magician skilled to read the character of flame! And Pearl was the oneness of their being. Be the foregone evil what it might, how could they doubt that their earthly lives and future destinies were conjoined, when they beheld at once the material union, and the spiritual idea, in whom they met, and were to dwell immortally together? Thoughts like these—and perhaps other thoughts, which they did not acknowledge or define—threw an awe about the child, as she came onward.

"Let her see nothing strange—no passion nor eagerness—in thy way of accosting her," whispered Hester. "Our Pearl is a fitful and fantastic little elf, sometimes. Especially, she is seldom tolerant of emotion, when she does not fully comprehend the why and wherefore. But the child hath strong affections! She loves me, and will love thee!"

"Thou canst not think," said the minister, glancing aside at Hester Prynne, "how my heart dreads this interview, and yearns for it! But, in truth, as I already told thee, children are not readily won to be familiar with me. They will not climb my knee, nor prattle in my ear, nor answer to my smile; but stand apart, and eye me strangely. Even little babes, when I take them in my arms, weep bitterly. Yet Pearl, twice in her little lifetime, hath been kind to me! The first time,—thou knowest it well! The last was when thou ledst her with thee to the house of yonder stern old Governor."

"And thou didst plead so bravely in her behalf and mine!" answered the mother. "I remember it; and so shall little Pearl. Fear nothing! She may be strange and shy at first, but will soon learn to love thee!"

By this time Pearl had reached the margin of the brook, and stood on the further side, gazing silently at Hester and the clergyman, who still sat together on the mossy tree-trunk, waiting to receive her. Just where she had paused, the brook chanced to form a pool, so smooth and quiet that it reflected a perfect image of her little figure, with all the brilliant picturesqueness of her beauty, in its adornment of flowers and wreathed foliage, but more refined and spiritualized than the reality. This image, so nearly identical with the living Pearl, seemed to communicate somewhat of its own shadowy and intangible quality to the child herself. It was strange, the way in which Pearl stood, looking so steadfastly at them through the dim medium of the forest-gloom; herself, meanwhile, all glorified with a ray of sunshine, that was attracted thitherward as by a certain sympathy. In the brook beneath stood another child,—another and the same,—with likewise its ray of golden light. Hester felt herself, in some indistinct and tantalizing manner, estranged from Pearl; as if the child, in her lonely ramble through the forest, had strayed out of the sphere in which she and her mother dwelt together, and was now vainly seeking to return to it.

There was both truth and error in the impression; the child and mother were estranged, but through Hester's fault, not Pearl's. Since the latter rambled from her side, another inmate had been admitted within the circle of the mother's feelings, and so modified the aspect of them all, that Pearl, the returning wanderer, could not find her wonted place, and hardly knew where she was.

"I have a strange fancy," observed the sensitive minister, "that this brook is the boundary between two worlds, and that thou canst never meet thy Pearl again. Or is she an elfish spirit, who, as the legends of our childhood taught us, is forbidden to cross a running stream? Pray hasten her; for this delay has already imparted a tremor to my nerves."

"Come, dearest child!" said Hester, encouragingly, and stretching out both her arms. "How slow thou art! When hast thou been so sluggish before now? Here is a friend of mine, who must be thy friend also. Thou wilt have twice as much love, henceforward, as thy mother alone could give thee! Leap across the brook, and come to us. Thou canst leap like a young deer!"

Pearl, without responding in any manner to these honey-sweet expressions, remained on the other side of the brook. Now she fixed her bright, wild eyes on her mother, now on the minister, and now included them both in the same glance; as if to detect and explain to herself the relation which they bore to one another. For some unaccountable reason, as Arthur Dimmesdale felt the child's eyes upon himself, his hand—with that gesture so habitual as to have become involuntary—stole over his heart. At length, assuming a singular air of authority, Pearl stretched out her hand, with the small forefinger extended, and pointing evidently towards her mother's breast. And beneath, in the mirror of the brook, there was the flower-girdled and sunny image of little Pearl, pointing her small forefinger too.

"Thou strange child, why dost thou not come to me?" exclaimed Hester.

Pearl still pointed with her forefinger; and a frown gathered on her brow; the more impressive from the childish, the almost baby-like aspect of the features that conveyed it. As her mother still kept beckoning to her, and arraying her face in a holiday suit of unaccustomed smiles, the child stamped her foot with a yet more imperious look and gesture. In the brook, again, was the fantastic beauty of the image, with its reflected frown, its pointed finger, and imperious gesture, giving emphasis to the aspect of little Pearl.

"Hasten, Pearl; or I shall be angry with thee!" cried Hester Prynne, who, however inured to such behavior on the elf-child's part at other seasons, was naturally anxious for a more seemly deportment now. "Leap across the brook, naughty child, and run hither! Else I must come to thee!"

But Pearl, not a whit startled at her mother's threats, any more than mollified by her entreaties, now suddenly burst into a fit of passion, gesticulating violently, and throwing her small figure into the most extravagant contortions. She accompanied this wild outbreak with piercing shrieks, which the woods reverberated on all sides; so that, alone as she was in her childish and unreasonable wrath, it seemed as if a hidden multitude were lending her their sympathy and encouragement. Seen in the brook, once more, was the shadowy wrath of Pearl's image, crowned and girdled with flowers, but stamping its foot, wildly gesticulating, and, in the midst of all, still pointing its small forefinger at Hester's bosom!

"I see what ails the child," whispered Hester to the clergyman, and turning pale in spite of a strong effort to conceal her trouble and annoyance. "Children will not abide any, the slightest, change in the accustomed aspect of things that are daily before their eyes. Pearl misses something which she has always seen me wear!"

"I pray you," answered the minister, "if thou hast any means of pacifying the child, do it forthwith! Save it were the cankered wrath of an old witch, like Mistress Hibbins," added he, attempting to smile, "I know nothing that I would not sooner encounter than this passion in a child. In Pearl's young beauty, as in the wrinkled witch, it has a preternatural effect. Pacify her, if thou lovest me!"

Hester turned again towards Pearl, with a crimson blush upon her cheek, a conscious glance aside at the clergyman, and then a heavy sigh; while, even before she had time to speak, the blush yielded to a deadly pallor.

"Pearl," said she, sadly, "look down at thy feet! There!—before thee!—on the hither side of the brook!"

The child turned her eyes to the point indicated; and there lay the scarlet letter, so close upon the margin of the stream, that the gold embroidery was reflected in it.

"Bring it hither!" said Hester.

"Come thou and take it up!" answered Pearl.

"Was ever such a child!" observed Hester, aside to the minister. "O, I have much to tell thee about her! But, in very truth, she is right as regards this hateful token. I must bear its torture yet a little longer,—only a few days longer,—until we shall have left this region, and look back hither as to a land which we have dreamed of. The forest cannot hide it! The mid-ocean shall take it from my hand, and swallow it up forever!"

With these words, she advanced to the margin of the brook, took up the scarlet letter, and fastened it again into her bosom. Hopefully, but a moment ago, as Hester had spoken of drowning it in the deep sea, there was a sense of inevitable doom upon her, as she thus received back this deadly symbol from the hand of fate. She had flung it into infinite space!—she had drawn an hour's free breath!—and here again was the scarlet misery, glittering on the old spot! So it ever is, whether thus typified or no, that an evil deed invests itself with the character of doom. Hester next gathered up the heavy tresses of her hair, and confined them beneath her cap. As if there were a withering spell in the sad letter, her beauty, the warmth and richness of her womanhood, departed, like fading sunshine; and a gray shadow seemed to fall across her.

When the dreary change was wrought, she extended her hand to Pearl. "Dost thou know thy mother now, child?" asked she, reproachfully, but with a subdued tone. "Wilt thou come across the brook, and own thy mother, now that she has her shame upon her,—now that she is sad?"

"Yes; now I will!" answered the child, bounding across the brook, and clasping Hester in her arms. "Now thou art my mother indeed! And I am thy little Pearl!"

In a mood of tenderness that was not usual with her, she drew down her mother's head, and kissed her brow and both her cheeks. But then—by a kind of necessity that always impelled this child to alloy whatever comfort she might chance to give with a throb of anguish—Pearl put up her mouth, and kissed the scarlet letter too!

"That was not kind!" said Hester. "When thou hast shown me a little love, thou mockest me!"

"Why doth the minister sit yonder?" asked Pearl.

"He waits to welcome thee," replied her mother. "Come thou, and entreat his blessing! He loves thee, my little Pearl, and loves thy mother too. Wilt thou not love him? Come! he longs to greet thee!"

"Doth he love us?" said Pearl, looking up, with acute intelligence, into her mother's face. "Will he go back with us, hand in hand, we three together, into the town?"

"Not now, dear child," answered Hester. "But in days to come he will walk hand in hand with us. We will have a home and fireside of our own; and thou shalt sit upon his knee; and he will teach thee many things, and love thee dearly. Thou wilt love him; wilt thou not?"

"And will he always keep his hand over his heart?" inquired Pearl.

"Foolish child, what a question is that!" exclaimed her mother. "Come and ask his blessing!"

But, whether influenced by the jealousy that seems instinctive with every petted child towards a dangerous rival, or from whatever caprice of her freakish nature, Pearl would show no favor to the clergyman. It was only by an exertion of force that her mother brought her up to him, hanging back, and manifesting her reluctance by odd grimaces; of which, ever since her babyhood, she had possessed a singular variety, and could transform her mobile physiognomy into a series of different aspects, with a new mischief in them, each and all. The minister—painfully embarrassed, but hoping that a kiss might prove a talisman to admit him into the child's kindlier regards—bent forward, and impressed one on her brow. Hereupon, Pearl broke away from her mother, and running to the brook, stooped over it, and bathed her forehead, until the unwelcome kiss was quite washed off, and diffused through a long lapse of the gliding water. She then remained apart, silently watching Hester and the clergyman; while they talked together, and made such arrangements as were suggested by their new position, and the purposes soon to be fulfilled.

And now this fateful interview had come to a close. The dell was to be left a solitude among its dark, old trees, which, with their multitudinous tongues, would whisper long of what had passed there, and no mortal be the wiser. And the melancholy brook would add this other tale to the mystery with which its little heart was already overburdened, and whereof it still kept up a murmuring babble, with not a whit more cheerfulness of tone than for ages heretofore.

XX. The Minister in a Maze

As the minister departed, in advance of Hester Prynne and little Pearl, he threw a backward glance; half expecting that he should discover only some faintly traced features or outline of the mother and the child, slowly fading into the twilight of the woods. So great a vicissitude in his life could not at once be received as real. But there was Hester, clad in her gray robe, still standing beside the tree-trunk, which some blast had overthrown a long antiquity ago, and which time had ever since been covering with moss, so that these two fated ones, with earth's heaviest burden on them, might there sit down together, and find a single hour's rest and solace. And there was Pearl, too, lightly dancing from the margin of the brook,—now that the intrusive third person was gone,—and taking her old place by her mother's side. So the minister had not fallen asleep, and dreamed!

In order to free his mind from this indistinctness and duplicity of impression, which vexed it with a strange disquietude, he recalled and more thoroughly defined the plans which Hester and himself had sketched for their departure. It had been determined between them, that the Old World, with its crowds and cities, offered them a more eligible shelter and concealment than the wilds of New England, or all America, with its alternatives of an Indian wigwam, or the few settlements of Europeans, scattered thinly along the seaboard. Not to speak of the clergyman's health, so inadequate to sustain the hardships of a forest life, his native gifts, his culture, and his entire development, would secure him a home only in the midst of civilization and refinement; the higher the state, the more delicately adapted to it the man. In furtherance of this choice, it so happened that a ship lay in the harbor; one of those questionable cruisers, frequent at that day, which, without being absolutely outlaws of the deep, yet roamed over its surface with a remarkable irresponsibility of character. This vessel had recently arrived from the Spanish Main, and, within three days' time, would sail for Bristol. Hester Prynne—whose vocation, as a self-enlisted Sister of Charity, had brought her acquainted with the captain and crew—could take upon herself to secure the passage of two individuals and a child, with all the secrecy which circumstances rendered more than desirable.

The minister had inquired of Hester, with no little interest, the precise time at which the vessel might be expected to depart. It would probably be on the fourth day from the present. "That is most fortunate!" he had then said to himself. Now, why the Reverend Mr. Dimmesdale considered it so very fortunate, we hesitate to reveal. Nevertheless,—to hold nothing back from the reader,—it was because, on the third day from the present, he was to preach the Election Sermon; and, as such an occasion formed an honorable epoch in the life of a New England clergyman, he could not have chanced upon a more suitable mode and time of terminating his professional career.[1] "At least, they shall say of me," thought this exemplary man, "that I leave no public duty unperformed, nor ill performed!" Sad, indeed, that an introspection so profound and acute as this poor minister's should be so miserably deceived! We have had, and may still have, worse things to tell of him; but none, we apprehend, so pitiably weak; no evidence, at once so slight and irrefragable, of a subtle disease, that had long since begun to eat into the real substance of his character. No man, for any considerable period, can wear one face to himself, and another to the multitude, without finally getting bewildered as to which may be the true.

The excitement of Mr. Dimmesdale's feelings, as he returned from his interview with Hester, lent him unaccustomed physical energy, and hurried him townward at a rapid pace. The pathway among the woods seemed wilder, more uncouth with its rude natural obstacles, and less trodden by the foot of man, than he remembered it on his outward journey. But he leaped across the plashy places, thrust himself through the clinging underbrush, climbed the ascent, plunged into the hollow, and overcame, in short, all the difficulties of the track, with an unweariable activity that astonished him. He could not but recall how feebly, and with what frequent pauses for breath, he had toiled over the same ground, only two days before. As he drew near the town, he took an impression of change from the series of familiar objects that presented themselves. It seemed not yesterday, not one, nor two, but many days, or even years ago, since he had quitted them. There, indeed, was each former trace of the street, as he remembered it, and all the peculiarities of the houses, with the due multitude of gable-peaks, and a weather-cock at every point where his memory suggested one. Not the less, however, came this importunately obtrusive sense of change. The same was true as regarded the acquaintances whom he met, and all the well-known shapes of human life, about the little town. They looked neither older nor younger now; the beards of the aged were no whiter, nor could the creeping babe of yesterday walk on his feet to-day; it was impossible to describe in what respect they differed from the individuals on whom he had so recently bestowed a parting glance; and yet the minister's deepest sense seemed to inform him of their mutability. A similar impression struck him most remarkably, as he passed under the walls of his own church. The edifice had so very strange, and yet so familiar, an aspect, that Mr. Dimmesdale's mind vibrated between two ideas; either that he had seen it only in a dream hitherto, or that he was merely dreaming about it now.[2]

This phenomenon, in the various shapes which it assumed, indicated no external change, but so sudden and important a change in the spectator of the familiar scene, that the intervening space of a single day had operated on his consciousness like the lapse of years. The minister's own will, and Hester's will, and the fate that grew between them, had wrought this transformation. It was the same town as heretofore; but the same minister returned not from the forest. He might have said to the friends

1. The Puritans elected a new governor every year. In this church state, Dimmesdale has been accorded the honor of preaching a sermon as part of the inaugural celebration. Although Hawthorne does not identify the new governor, John Endicott was elected in 1649.
2. Compare the experiences of young Goodman Brown after he returns from the forest (see p. 178 of this Norton Critical Edition).

who greeted him,—"I am not the man for whom you take me! I left him yonder in the forest, withdrawn into a secret dell, by a mossy tree-trunk, and near a melancholy brook! Go, seek your minister, and see if his emaciated figure, his thin cheek, his white, heavy, pain-wrinkled brow, be not flung down there, like a cast-off garment!" His friends, no doubt, would still have insisted with him,—"Thou art thyself the man!"—but the error would have been their own, not his.

Before Mr. Dimmesdale reached home, his inner man gave him other evidences of a revolution in the sphere of thought and feeling. In truth, nothing short of a total change of dynasty and moral code, in that interior kingdom, was adequate to account for the impulses now communicated to the unfortunate and startled minister. At every step he was incited to do some strange, wild, wicked thing or other, with a sense that it would be at once involuntary and intentional; in spite of himself, yet growing out of a profounder self than that which opposed the impulse. For instance, he met one of his own deacons. The good old man addressed him with the paternal affection and patriarchal privilege, which his venerable age, his upright and holy character, and his station in the Church, entitled him to use; and, conjoined with this, the deep, almost worshipping respect, which the minister's professional and private claims alike demanded. Never was there a more beautiful example of how the majesty of age and wisdom may comport with the obeisance and respect enjoined upon it, as from a lower social rank, and inferior order of endowment, towards a higher. Now, during a conversation of some two or three moments between the Reverend Mr. Dimmesdale and this excellent and hoary-bearded deacon, it was only by the most careful self-control that the former could refrain from uttering certain blasphemous suggestions that rose into his mind, respecting the communion-supper. He absolutely trembled and turned pale as ashes, lest his tongue should wag itself, in utterance of these horrible matters, and plead his own consent for so doing, without his having fairly given it. And, even with this terror in his heart, he could hardly avoid laughing, to imagine how the sanctified old patriarchal deacon would have been petrified by his minister's impiety!

Again, another incident of the same nature. Hurrying along the street, the Reverend Mr. Dimmesdale encountered the eldest female member of his church; a most pious and exemplary old dame; poor, widowed, lonely, and with a heart as full of reminiscences about her dead husband and children, and her dead friends of long ago, as a burial-ground is full of storied grave-stones. Yet all this, which would else have been such heavy sorrow, was made almost a solemn joy to her devout old soul, by religious consolations and the truths of Scripture, wherewith she had fed herself continually for more than thirty years. And, since Mr. Dimmesdale had taken her in charge, the good grandam's chief earthly comfort—which, unless it had been likewise a heavenly comfort, could have been none at all—was to meet her pastor, whether casually, or of set purpose, and be refreshed with a word of warm, fragrant, heaven-breathing Gospel truth, from his beloved lips, into her dulled, but rapturously attentive ear. But, on this occasion, up to the moment of putting his lips to the old woman's ear, Mr. Dimmesdale, as the great enemy of souls would have it, could recall no text of Scripture, nor aught else, except a brief, pithy, and, as it then appeared to him, unanswerable argument against the immortality of the human soul. The instilment thereof into her mind would probably have caused this aged sister to drop down dead, at once, as by the effect of an intensely poisonous infusion. What he really did whisper, the minister could never afterwards recollect. There was, perhaps, a fortunate disorder in his utterance, which failed to impart any distinct idea to the good widow's comprehension, or which Providence interpreted after a method of its own. Assuredly, as the minister looked back, he beheld an expression of divine gratitude and ecstasy that seemed like the shine of the celestial city on her face, so wrinkled and ashy pale.

Again, a third instance. After parting from the old church-member he met the youngest sister of them all. It was a maiden newly won—and won by the Reverend Mr. Dimmesdale's own sermon, on the Sabbath after his vigil—to barter the transitory pleasures of the world for the heavenly hope, that was to assume brighter substance as life grew dark around her, and which would gild the utter gloom with final glory. She was fair and pure as a lily that had bloomed in Paradise. The minister knew well that he was himself enshrined within the stainless sanctity of her heart, which hung its snowy curtains about his image, imparting to religion the warmth of love, and to love a religious purity. Satan, that afternoon, had surely led the poor young girl away from her mother's side, and thrown her into the pathway of this sorely tempted, or—shall we not rather say?—this lost and desperate man. As she drew nigh, the arch-fiend whispered him to condense into small compass and drop into her tender bosom a germ of evil that would be sure to blossom darkly soon, and bear black fruit betimes. Such was his sense of power over this virgin soul, trusting him as she did, that the minister felt potent to blight all the field of innocence with but one wicked look, and develop all its opposite with but a word. So—with a mightier struggle than he had yet sustained—he held his Geneva cloak before his face, and hurried onward, making no sign of recognition, and leaving the young sister to digest his rudeness as she might. She ransacked her conscience,—which was full of harmless little matters, like her pocket or her work-bag,—and took herself to task, poor thing! for a thousand imaginary faults; and went about her household duties with swollen eyelids the next morning.

Before the minister had time to celebrate his victory over this last temptation, he was conscious of another impulse, more ludicrous, and almost as horrible. It was,—we blush to tell it,—it was to stop short in the road, and teach some very wicked words to a knot of little Puritan children who were playing there, and had but just begun to talk. Denying himself this freak, as unworthy of his cloth, he met a drunken seaman, one of the ship's crew from the Spanish Main. And, here, since he had so valiantly forborne all other wickedness, poor Mr. Dimmesdale longed, at least, to shake hands with the tarry black-guard, and recreate himself with a few improper jests, such as dissolute sailors so abound with, and a volley of good, round, solid, satisfactory, and heaven-defying oaths! It was not so much a better principle, as partly his natural good taste, and still more his buckramed[3] habit of clerical decorum, that carried him safely through the latter crisis.

3. Stiffened.

"What is it that haunts and tempts me thus?" cried the minister to himself, at length, pausing in the street, and striking his hand against his forehead. "Am I mad? or am I given over utterly to the fiend? Did I make a contract with him in the forest, and sign it with my blood? And does he now summon me to its fulfilment, by suggesting the performance of every wickedness which his most foul imagination can conceive?"

At the moment when the Reverend Mr. Dimmesdale thus communed with himself, and struck his forehead with his hand, old Mistress Hibbins, the reputed witch-lady, is said to have been passing by. She made a very grand appearance; having on a high head-dress, a rich gown of velvet, and a ruff done up with the famous yellow starch, of which Ann Turner, her especial friend, had taught her the secret, before this last good lady had been hanged for Sir Thomas Overbury's murder.[4] Whether the witch had read the minister's thoughts, or no, she came to a full stop, looked shrewdly into his face, smiled craftily, and—though little given to converse with clergymen—began a conversation.

"So, reverend Sir, you have made a visit into the forest," observed the witch-lady, nodding her high head-dress at him. "The next time, I pray you to allow me only a fair warning, and I shall be proud to bear you company. Without taking overmuch upon myself, my good word will go far towards gaining any strange gentleman a fair reception from yonder potentate you wot of!"

"I profess, madam," answered the clergyman, with a grave obeisance, such as the lady's rank demanded, and his own good-breeding made imperative,—"I profess, on my conscience and character, that I am utterly bewildered as touching the purport of your words! I went not into the forest to seek a potentate; neither do I, at any future time, design a visit thither, with a view to gaining the favor of such personage. My one sufficient object was to greet that pious friend of mine, the Apostle Eliot, and rejoice with him over the many precious souls he hath won from heathendom!"

"Ha, ha, ha!" cackled the old witch-lady, still nodding her high head-dress at the minister. "Well, well, we must needs talk thus in the daytime! You carry it off like an old hand! But at midnight, and in the forest, we shall have other talk together!"

She passed on with her aged stateliness, but often turning back her head and smiling at him, like one willing to recognize a secret intimacy of connection.

"Have I then sold myself," thought the minister, "to the fiend whom, if men say true, this yellow-starched and velveted old hag has chosen for her prince and master!"

The wretched minister! He had made a bargain very like it! Tempted by a dream of happiness, he had yielded himself, with deliberate choice, as he had never done before, to what he knew was deadly sin. And the infectious poison of that sin had been thus rapidly diffused throughout his moral system. It had stupefied all blessed impulses, and awakened into vivid life the whole brotherhood of bad ones. Scorn, bitterness, unprovoked malignity,

4. Anne Turner was one of the conspirators convicted of the murder of Thomas Overbury. See n. 5, p. 80, for more information. Mrs. Turner helped Frances Howard arrange assignations with Carr and confessed to providing the poison to Overbury's jailer, who gave it to Overbury.

gratuitous desire of ill, ridicule of whatever was good and holy, all awoke, to tempt, even while they frightened him. And his encounter with old Mistress Hibbins, if it were a real incident, did but show his sympathy and fellowship with wicked mortals, and the world of perverted spirits.

He had, by this time, reached his dwelling, on the edge of the burial-ground, and, hastening up the stairs, took refuge in his study. The minister was glad to have reached this shelter, without first betraying himself to the world by any of those strange and wicked eccentricities to which he had been continually impelled while passing through the streets. He entered the accustomed room, and looked around him on its books, its windows, its fireplace, and the tapestried comfort of the walls,[5] with the same perception of strangeness that had haunted him throughout his walk from the forest-dell into the town, and thitherward. Here he had studied and written; here, gone through fast and vigil, and come forth half alive; here, striven to pray; here, borne a hundred thousand agonies! There was the Bible, in its rich old Hebrew, with Moses and the Prophets speaking to him, and God's voice through all! There, on the table, with the inky pen beside it, was an unfinished sermon, with a sentence broken in the midst, where his thoughts had ceased to gush out upon the page, two days before. He knew that it was himself, the thin and white-cheeked minister, who had done and suffered these things, and written thus far into the Election Sermon! But he seemed to stand apart, and eye this former self with scornful, pitying, but half-envious curiosity. That self was gone. Another man had returned out of the forest; a wiser one; with a knowledge of hidden mysteries which the simplicity of the former never could have reached. A bitter kind of knowledge that!

While occupied with these reflections, a knock came at the door of the study, and the minister said, "Come in!"—not wholly devoid of an idea that he might behold an evil spirit. And so he did! It was old Roger Chillingworth that entered. The minister stood, white and speechless, with one hand on the Hebrew Scriptures, and the other spread upon his breast.

"Welcome home, reverend Sir," said the physician. "And how found you that godly man, the Apostle Eliot? But methinks, dear Sir, you look pale; as if the travel through the wilderness had been too sore for you. Will not my aid be requisite to put you in heart and strength to preach your Election Sermon?"

"Nay, I think not so," rejoined the Reverend Mr. Dimmesdale. "My journey, and the sight of the holy Apostle yonder, and the free air which I have breathed, have done me good, after so long confinement in my study. I think to need no more of your drugs, my kind physician, good though they be, and administered by a friendly hand."

All this time, Roger Chillingworth was looking at the minister with the grave and intent regard of a physician towards his patient. But, in spite of this outward show, the latter was almost convinced of the old man's knowledge, or, at least, his confident suspicion, with respect to his own interview with Hester Prynne. The physician knew then, that, in the minister's regard, he was no longer a trusted friend, but his bitterest enemy. So much being known, it would appear natural that a part of it

5. A joke, assuming the David and Bathsheba tapestry still adorns Dimmesdale's wall. See n. 4, p. 80.

should be expressed. It is singular, however, how long a time often passes before words embody things; and with what security two persons, who choose to avoid a certain subject, may approach its very verge, and retire without disturbing it. Thus, the minister felt no apprehension that Roger Chillingworth would touch, in express words, upon the real position which they sustained towards one another. Yet did the physician, in his dark way, creep frightfully near the secret.

"Were it not better," said he, "that you use my poor skill to-night? Verily, dear Sir, we must take pains to make you strong and vigorous for this occasion of the Election discourse. The people look for great things from you; apprehending that another year may come about, and find their pastor gone."

"Yea, to another world," replied the minister, with pious resignation. "Heaven grant it be a better one; for, in good sooth, I hardly think to tarry with my flock through the flitting seasons of another year! But, touching your medicine, kind Sir, in my present frame of body, I need it not."

"I joy to hear it," answered the physician. "It may be that my remedies, so long administered in vain, begin now to take due effect. Happy man were I, and well deserving of New England's gratitude, could I achieve this cure!"

"I thank you from my heart, most watchful friend," said the Reverend Mr. Dimmesdale, with a solemn smile. "I thank you, and can but requite your good deeds with my prayers."

"A good man's prayers are golden recompense!" rejoined old Roger Chillingworth, as he took his leave. "Yea, they are the current gold coin of the New Jerusalem, with the King's own mint-mark on them!"

Left alone, the minister summoned a servant of the house, and requested food, which, being set before him, he ate with ravenous appetite. Then, flinging the already written pages of the Election Sermon into the fire, he forthwith began another, which he wrote with such an impulsive flow of thought and emotion, that he fancied himself inspired; and only wondered that Heaven should see fit to transmit the grand and solemn music of its oracles through so foul an organ-pipe as he. However, leaving that mystery to solve itself, or go unsolved forever, he drove his task onward, with earnest haste and ecstasy. Thus the night fled away, as if it were a winged steed, and he careering on it; morning came, and peeped, blushing, through the curtains; and at last sunrise threw a golden beam into the study, and laid it right across the minister's bedazzled eyes. There he was, with the pen still between his fingers, and a vast, immeasurable tract of written space behind him!

XXI. The New England Holiday

Betimes in the morning of the day on which the new Governor was to receive his office at the hands of the people, Hester Prynne and little Pearl came into the market-place. It was already thronged with the craftsmen and other plebeian inhabitants of the town, in considerable numbers; among whom, likewise, were many rough figures, whose attire of deer-skins marked them as belonging to some of the forest settlements, which surrounded the little metropolis of the colony.

On this public holiday, as on all other occasions, for seven years past, Hester was clad in a garment of coarse gray cloth. Not more by its hue than by some indescribable peculiarity in its fashion, it had the effect of making her fade personally out of sight and outline; while, again, the scarlet letter brought her back from this twilight indistinctness, and revealed her under the moral aspect of its own illumination. Her face, so long familiar to the townspeople, showed the marble quietude which they were accustomed to behold there. It was like a mask; or, rather, like the frozen calmness of a dead woman's features; owing this dreary resemblance to the fact that Hester was actually dead, in respect to any claim of sympathy, and had departed out of the world with which she still seemed to mingle.

It might be, on this one day, that there was an expression unseen before, nor, indeed, vivid enough to be detected now; unless some preternaturally gifted observer should have first read the heart, and have afterwards sought a corresponding development in the countenance and mien. Such a spiritual seer might have conceived, that, after sustaining the gaze of the multitude through seven miserable years as a necessity, a penance, and something which it was a stern religion to endure, she now, for one last time more, encountered it freely and voluntarily, in order to convert what had so long been agony into a kind of triumph. "Look your last on the scarlet letter and its wearer!"—the people's victim and life-long bond-slave, as they fancied her, might say to them. "Yet a little while, and she will be beyond your reach! A few hours longer, and the deep, mysterious ocean will quench and hide forever the symbol which ye have caused to burn upon her bosom!" Nor were it an inconsistency too improbable to be assigned to human nature, should we suppose a feeling of regret in Hester's mind, at the moment when she was about to win her freedom from the pain which had been thus deeply incorporated with her being. Might there not be an irresistible desire to quaff a last, long, breathless draught of the cup of wormwood and aloes, with which nearly all her years of womanhood had been perpetually flavored? The wine of life, henceforth to be presented to her lips, must be indeed rich, delicious, and exhilarating, in its chased[1] and golden beaker; or else leave an inevitable and weary languor, after the lees of bitterness wherewith she had been drugged, as with a cordial of intensest potency.

Pearl was decked out with airy gayety. It would have been impossible to guess that this bright and sunny apparition owed its existence to the shape of gloomy gray; or that a fancy, at once so gorgeous and so delicate as must have been requisite to contrive the child's apparel, was the same that had achieved a task perhaps more difficult, in imparting so distinct a peculiarity to Hester's simple robe. The dress, so proper was it to little Pearl, seemed an effluence, or inevitable development and outward manifestation of her character, no more to be separated from her than the many-hued brilliancy from a butterfly's wing, or the painted glory from the leaf of a bright flower. As with these, so with the child; her garb was all of one idea with her nature. On this eventful day, moreover, there was a certain singular inquietude and excitement in her mood, resembling nothing so much as the shimmer of a diamond, that sparkles and flashes with the varied throbbings of the breast on which it is displayed. Children

1. Engraved or embossed.

have always a sympathy in the agitations of those connected with them; always, especially, a sense of any trouble or impending revolution, of whatever kind, in domestic circumstances; and therefore Pearl, who was the gem on her mother's unquiet bosom, betrayed, by the very dance of her spirits, the emotions which none could detect in the marble passiveness of Hester's brow.

This effervescence made her flit with a birdlike movement, rather than walk by her mother's side. She broke continually into shouts of a wild, inarticulate, and sometimes piercing music. When they reached the market-place, she became still more restless, on perceiving the stir and bustle that enlivened the spot; for it was usually more like the broad and lonesome green before a village meeting-house, than the centre of a town's business.

"Why, what is this, mother?" cried she. "Wherefore have all the people left their work to-day? Is it a play-day for the whole world? See, there is the black-smith! He has washed his sooty face, and put on his Sabbath-day clothes, and looks as if he would gladly be merry, if any kind body would only teach him how! And there is Master Brackett, the old jailer, nodding and smiling at me. Why does he do so, mother?"

"He remembers thee a little babe, my child," answered Hester.

"He should not nod and smile at me, for all that,—the black, grim, ugly-eyed old man!" said Pearl. "He may nod at thee, if he will; for thou art clad in gray, and wearest the scarlet letter. But see, mother, how many faces of strange people, and Indians among them, and sailors! What have they all come to do, here in the market-place?"

"They wait to see the procession pass," said Hester. "For the Governor and the magistrates are to go by, and the ministers, and all the great people and good people, with the music and the soldiers marching before them."

"And will the minister be there?" asked Pearl. "And will he hold out both his hands to me, as when thou ledst me to him from the brookside?"

"He will be there, child," answered her mother. "But he will not greet thee to-day; nor must thou greet him."

"What a strange, sad man is he!" said the child, as if speaking partly to herself. "In the dark night-time he calls us to him, and holds thy hand and mine, as when we stood with him on the scaffold yonder! And in the deep forest, where only the old trees can hear, and the strip of sky see it, he talks with thee, sitting on a heap of moss! And he kisses my forehead, too, so that the little brook would hardly wash it off! But here, in the sunny day, and among all the people, he knows us not; nor must we know him! A strange, sad man is he, with his hand always over his heart!"

"Be quiet, Pearl! Thou understandest not these things," said her mother. "Think not now of the minister, but look about thee, and see how cheery is every-body's face to-day. The children have come from their schools, and the grown people from their workshops and their fields, on purpose to be happy. For, to-day, a new man is beginning to rule over them; and so—as has been the custom of mankind ever since a nation was first gathered—they make merry and rejoice; as if a good and golden year were at length to pass over the poor old world!"

It was as Hester said, in regard to the unwonted jollity that brightened the faces of the people. Into this festal season of the year—as it already was, and continued to be during the greater part of two centuries—the Puritans compressed whatever mirth and public joy they deemed allowable to human infirmity; thereby so far dispelling the customary cloud, that, for the space of a single holiday, they appeared scarcely more grave than most other communities at a period of general affliction.

But we perhaps exaggerate the gray or sable tinge, which undoubtedly characterized the mood and manners of the age. The persons now in the market-place of Boston had not been born to an inheritance of Puritanic gloom. They were native Englishmen, whose fathers had lived in the sunny richness of the Elizabethan epoch; a time when the life of England, viewed as one great mass, would appear to have been as stately magnificent, and joyous, as the world has ever witnessed. Had they followed their hereditary taste, the New England settlers would have illustrated all events of public importance by bonfires, banquets, pageantries, and processions. Nor would it have been impracticable, in the observance of majestic ceremonies, to combine mirthful recreation with solemnity, and give, as it were, a grotesque and brilliant embroidery to the great robe of state, which a nation, at such festivals, puts on. There was some shadow of an attempt of this kind in the mode of celebrating the day on which the political year of the colony commenced. The dim reflection of a remembered splendor, a colorless and manifold diluted repetition of what they had beheld in proud old London,—we will not say at a royal coronation, but at a Lord Mayor's show[2]—might be traced in the customs which our forefathers instituted, with reference to the annual installation of magistrates. The fathers and founders of the commonwealth—the statesman, the priest, and the soldier—deemed it a duty then to assume the outward state and majesty, which, in accordance with antique style, was looked upon as the proper garb of public or social eminence. All came forth, to move in procession before the people's eye, and thus impart a needed dignity to the simple framework of a government so newly constructed.

Then, too, the people were countenanced, if not encouraged, in relaxing the severe and close application to their various modes of rugged industry, which, at all other times, seemed of the same piece and material with their religion. Here, it is true, were none of the appliances which popular merriment would so readily have found in the England of Elizabeth's time, or that of James;—no rude shows of a theatrical kind; no minstrel, with his harp and legendary ballad, nor gleeman, with an ape dancing to his music; no juggler, with his tricks of mimic witchcraft; no Merry Andrew,[3] to stir up the multitude with jests, perhaps hundreds of years old, but still effective, by their appeals to the very broadest sources of mirthful sympathy. All such professors of the several branches of jocularity would have been sternly repressed, not only by the rigid discipline of law, but by the general sentiment which gives law its vitality. Not the less, however, the great, honest face of the people smiled, grimly, perhaps, but widely too. Nor were sports wanting, such as the colonists had witnessed, and shared in, long ago, at the country fairs and on the village-greens of England; and which it was thought well to keep alive on this new soil, for the sake of the courage and manliness that were essential in them. Wrestling-matches, in the different fashions of Cornwall and Devonshire, were seen here and there about the market-place;

2. In London, Lord Mayor's Day (November 9) featured a similar procession and celebration, as the lord mayor processed to and from Westminster, where he received from the lord chancellor the assent of the Crown to his election (*OED*).
3. Clown or entertainer. *Gleeman*: minstrel.

in one corner, there was a friendly bout at quarterstaff; and—what attracted most interest of all—on the platform of the pillory, already so noted in our pages, two masters of defence were commencing an exhibition with the buckler and broadsword.[4] But, much to the disappointment of the crowd, this latter business was broken off by the interposition of the town beadle, who had no idea of permitting the majesty of the law to be violated by such an abuse of one of its consecrated places.

It may not be too much to affirm, on the whole, (the people being then in the first stages of joyless deportment, and the offspring of sires who had known how to be merry, in their day,) that they would compare favorably, in point of holiday keeping, with their descendants, even at so long an interval as ourselves. Their immediate posterity, the generation next to the early emigrants, wore the blackest shade of Puritanism, and so darkened the national visage with it, that all the subsequent years have not sufficed to clear it up. We have yet to learn again the forgotten art of gayety.

The picture of human life in the market-place, though its general tint was the sad gray, brown, or black of the English emigrants, was yet enlivened by some diversity of hue. A party of Indians—in their savage finery of curiously embroidered deer-skin robes, wampum-belts, red and yellow ochre, and feathers, and armed with the bow and arrow and stone-headed spear—stood apart, with countenances of inflexible gravity, beyond what even the Puritan aspect could attain. Nor, wild as were these painted barbarians, were they the wildest feature of the scene. This distinction could more justly be claimed by some mariners,—a part of the crew of the vessel from the Spanish Main,—who had come ashore to see the humors of Election Day. They were rough-looking desperadoes, with sun-blackened faces, and an immensity of beard; their wide, short trousers were confined about the waist by belts, often clasped with a rough plate of gold, and sustaining always a long knife, and, in some instances, a sword. From beneath their broad-brimmed hats of palm-leaf, gleamed eyes which, even in good nature and merriment, had a kind of animal ferocity. They transgressed, without fear or scruple, the rules of behavior that were binding on all others; smoking tobacco under the beadle's very nose, although each whiff would have cost a townsman a shilling; and quaffing, at their pleasure, draughts of wine or aqua-vitæ from pocket-flasks, which they freely tendered to the gaping crowd around them. It remarkably characterized the incomplete morality of the age, rigid as we call it, that a license was allowed the seafaring class, not merely for their freaks on shore, but for far more desperate deeds on their proper element. The sailor of that day would go near to be arraigned as a pirate in our own. There could be little doubt, for instance, that this very ship's crew, though no unfavorable specimens of the nautical brotherhood, had been guilty, as we should phrase it, of depredations on the Spanish commerce, such as would have perilled all their necks in a modern court of justice.

But the sea, in those old times, heaved, swelled and foamed, very much at its own will, or subject only to the tempestuous wind, with hardly any attempts at regulation by human law. The buccaneer on the wave might relinquish his calling, and become at once, if he chose, a man of probity

4. Large, wide sword designed for cutting. *Quarterstaff*: long pole used as a weapon. *Buckler*: small shield.

and piety on land; nor, even in the full career of his reckless life, was he regarded as a personage with whom it was disreputable to traffic, or casually associate. Thus, the Puritan elders, in their black cloaks, starched bands, and steeple-crowned hats, smiled not unbenignantly at the clamor and rude deportment of these jolly seafaring men; and it excited neither surprise nor animadversion, when so reputable a citizen as old Roger Chillingworth, the physician, was seen to enter the market-place, in close and familiar talk with the commander of the questionable vessel.

The latter was by far the most showy and gallant figure, so far as apparel went, anywhere to be seen among the multitude. He wore a profusion of ribbons on his garment, and gold lace on his hat, which was also encircled by a gold chain, and surmounted with a feather. There was a sword at his side, and a sword-cut on his forehead, which, by the arrangement of his hair, he seemed anxious rather to display than hide. A landsman could hardly have worn this garb and shown this face, and worn and shown them both with such a galliard[5] air, without undergoing stern question before a magistrate, and probably incurring fine or imprisonment, or perhaps an exhibition in the stocks. As regarded the shipmaster, however, all was looked upon as pertaining to the character, as to a fish his glistening scales.

After parting from the physician, the commander of the Bristol ship strolled idly through the market-place; until, happening to approach the spot where Hester Prynne was standing, he appeared to recognize, and did not hesitate to address her. As was usually the case wherever Hester stood, a small vacant area—a sort of magic circle—had formed itself about her, into which, though the people were elbowing one another at a little distance, none ventured, or felt disposed to intrude. It was a forcible type of the moral solitude in which the scarlet letter enveloped its fated wearer; partly by her own reserve, and partly by the instinctive, though no longer so unkindly, withdrawal of her fellow-creatures. Now, if never before, it answered a good purpose, by enabling Hester and the seaman to speak together without risk of being overheard; and so changed was Hester Prynne's repute before the public, that the matron in town most eminent for rigid morality could not have held such intercourse with less result of scandal than herself.

"So, mistress," said the mariner, "I must bid the steward make ready one more berth than you bargained for! No fear of scurvy or ship-fever, this voyage! What with the ship's surgeon and this other doctor, our only danger will be from drug or pill; more by token, as there is a lot of apothecary's stuff aboard, which I traded for with a Spanish vessel."

"What mean you?" inquired Hester, startled more than she permitted to appear. "Have you another passenger?"

"Why, know you not," cried the shipmaster, "that this physician here—Chillingworth, he calls himself—is minded to try my cabin-fare with you? Ay, ay, you must have known it; for he tells me he is of your party, and a close friend to the gentleman you spoke of,—he that is in peril from these sour old Puritan rulers!"

"They know each other well, indeed," replied Hester, with a mien of calmness, though in the utmost consternation. "They have long dwelt together."

5. Spirited, gay.

Nothing further passed between the mariner and Hester Prynne. But, at that instant, she beheld old Roger Chillingworth himself, standing in the remotest corner of the market-place, and smiling on her; a smile which—across the wide and bustling square, and through all the talk and laughter, and various thoughts, moods, and interests of the crowd—conveyed secret and fearful meaning.

XXII. The Procession

Before Hester Prynne could call together her thoughts, and consider what was practicable to be done in this new and startling aspect of affairs, the sound of military music was heard approaching along a contiguous street. It denoted the advance of the procession of magistrates and citizens, on its way towards the meeting-house; where, in compliance with a custom thus early established, and ever since observed, the Reverend Mr. Dimmesdale was to deliver an Election Sermon.

Soon the head of the procession showed itself, with a slow and stately march, turning a corner, and making its way across the market-place. First came the music. It comprised a variety of instruments, perhaps imperfectly adapted to one another, and played with no great skill; but yet attaining the great object for which the harmony of drum and clarion addresses itself to the multitude,—that of imparting a higher and more heroic air to the scene of life that passes before the eye. Little Pearl at first clapped her hands, but then lost, for an instant, the restless agitation that had kept her in a continual effervescence throughout the morning; she gazed silently, and seemed to be borne upward, like a floating sea-bird, on the long heaves and swells of sound. But she was brought back to her former mood by the shimmer of the sunshine on the weapons and bright armor of the military company, which followed after the music, and formed the honorary escort of the procession. This body of soldiery—which still sustains a corporate existence, and marches down from past ages with an ancient and honorable fame—was composed of no mercenary materials. Its ranks were filled with gentlemen, who felt the stirrings of martial impulse, and sought to establish a kind of College of Arms, where, as in an association of Knights Templars,[1] they might learn the science, and, so far as peaceful exercise would teach them, the practices of war. The high estimation then placed upon the military character might be seen in the lofty port of each individual member of the company. Some of them, indeed, by their services in the Low Countries[2] and on other fields of European warfare, had fairly won their title to assume the name and pomp of soldiership. The entire array, moreover, clad in burnished steel, and with plumage nodding over their bright morions,[3] had a brilliancy of effect which no modern display can aspire to equal.

1. A military order founded in the Catholic church early in the twelfth century. Members wore white habits with a distinctive red cross. They grew in power and wealth, becoming a transnational order answerable only to papal authority. In 1307, King Philip IV of France had all Templars arrested and tried for heresy upon accusations of denying Christ, spitting on the Cross, and sodomy. The order was formally dissolved in 1312. By the nineteenth century, the Knights Templar were popularly associated with occult magic (astrology and alchemy) and freemasonry.
2. The area now comprising the Netherlands, Luxembourg, and Belgium.
3. Brimmed helmets.

And yet the men of civil eminence, who came immediately behind the military escort, were better worth a thoughtful observer's eye. Even in outward demeanor, they showed a stamp of majesty that made the warrior's haughty stride look vulgar, if not absurd. It was an age when what we call talent had far less consideration than now, but the massive materials which produce stability and dignity of character a great deal more. The people possessed, by hereditary right, the quality of reverence; which, in their descendants, if it survive at all, exists in smaller proportion, and with a vastly diminished force, in the selection and estimate of public men. The change may be for good or ill, and is partly, perhaps, for both. In that old day, the English settler on these rude shores—having left king, nobles, and all degrees of awful rank behind, while still the faculty and necessity of reverence were strong in him—bestowed it on the white hair and venerable brow of age; on long-tried integrity; on solid wisdom and sad-colored experience; on endowments of that grave and weighty order which gives the idea of permanence, and comes under the general definition of respectability. These primitive statesmen, therefore,—Bradstreet, Endicott, Dudley, Bellingham, and their compeers,—who were elevated to power by the early choice of the people, seem to have been not often brilliant, but distinguished by a ponderous sobriety, rather than activity of intellect.[4] They had fortitude and self-reliance, and, in time of difficulty or peril, stood up for the welfare of the state like a line of cliffs against a tempestuous tide. The traits of character here indicated were well represented in the square cast of countenance and large physical development of the new colonial magistrates. So far as a demeanor of natural authority was concerned, the mother country need not have been ashamed to see these foremost men of an actual democracy adopted into the House of Peers, or made the Privy Council[5] of the sovereign.

Next in order to the magistrates came the young and eminently distinguished divine, from whose lips the religious discourse of the anniversary was expected. His was the profession, at that era, in which intellectual ability displayed itself far more than in political life; for—leaving a higher motive out of the question—it offered inducements powerful enough, in the almost worshipping respect of the community, to win the most aspiring ambition into its service. Even political power—as in the case of Increase Mather[6]—was within the grasp of a successful priest.

It was the observation of those who beheld him now, that never, since Mr. Dimmesdale first set his foot on the New England shore, had he exhibited such energy as was seen in the gait and air with which he kept his pace in the procession. There was no feebleness of step, as at other times; his frame was not bent; nor did his hand rest ominously upon his

4. Simon Bradstreet (1604–1697), John Endicott (1588–1665), Thomas Dudley (1576–1653), and Richard Bellingham (1592–1672) each served as governor of Massachusetts Bay Colony. During the seven years encompassed by the novel, Bellingham, Endicott, Dudley, and John Winthrop served as governor. Bradstreet was not elected governor until 1679. His wife, poet Anne Bradstreet, was Thomas Dudley's daughter. Bradstreet, Dudley, Endicott, and Winthrop interrogated Anne Hutchinson at her General Court appearance in November 1637.
5. A group of advisors selected by the king. *House of Peers*: House of Lords, the second chamber of British Parliament; the other is the House of Commons.
6. A prominent minister, Increase Mather (1639–1723) was the first president of Harvard College. In referring to Mather's political power, Hawthorne probably refers to Mather's responsibility for negotiating a new charter for the colony. Mather and his son, Cotton, were both involved in the Salem witch trials of 1692.

heart. Yet, if the clergyman were rightly viewed, his strength seemed not of the body. It might be spiritual, and imparted to him by angelic ministrations. It might be the exhilaration of that potent cordial, which is distilled only in the furnace-glow of earnest and long-continued thought. Or, perchance, his sensitive temperament was invigorated by the loud and piercing music, that swelled heavenward, and uplifted him on its ascending wave. Nevertheless, so abstracted was his look, it might be questioned whether Mr. Dimmesdale even heard the music. There was his body, moving onward, and with an unaccustomed force. But where was his mind? Far and deep in its own region, busying itself, with preternatural activity, to marshal a procession of stately thoughts that were soon to issue thence; and so he saw nothing, heard nothing, knew nothing, of what was around him; but the spiritual element took up the feeble frame, and carried it along, unconscious of the burden, and converting it to spirit like itself. Men of uncommon intellect, who have grown morbid, possess this occasional power of mighty effort, into which they throw the life of many days, and then are lifeless for as many more.

Hester Prynne, gazing steadfastly at the clergyman, felt a dreary influence come over her, but wherefore or whence she knew not; unless that he seemed so remote from her own sphere, and utterly beyond her reach. One glance of recognition, she had imagined, must needs pass between them. She thought of the dim forest, with its little dell of solitude, and love, and anguish, and the mossy tree-trunk, where, sitting hand in hand, they had mingled their sad and passionate talk with the melancholy murmur of the brook. How deeply had they known each other then! And was this the man? She hardly knew him now! He, moving proudly past, enveloped, as it were, in the rich music, with the procession of majestic and venerable fathers; he, so unattainable in his worldly position, and still more so in that far vista of his unsympathizing thoughts, through which she now beheld him! Her spirit sank with the idea that all must have been a delusion, and that, vividly as she had dreamed it, there could be no real bond betwixt the clergyman and herself. And thus much of woman was there in Hester, that she could scarcely forgive him,—least of all now, when the heavy footstep of their approaching Fate might be heard, nearer, nearer, nearer!—for being able so completely to withdraw himself from their mutual world; while she groped darkly, and stretched forth her cold hands, and found him not.

Pearl either saw and responded to her mother's feelings, or herself felt the remoteness and intangibility that had fallen around the minister. While the procession passed, the child was uneasy, fluttering up and down, like a bird on the point of taking flight. When the whole had gone by, she looked up into Hester's face.

"Mother," said she, "was that the same minister that kissed me by the brook?"

"Hold thy peace, dear little Pearl!" whispered her mother. "We must not always talk in the market-place of what happens to us in the forest."

"I could not be sure that it was he; so strange he looked," continued the child. "Else I would have run to him, and bid him kiss me now, before all the people; even as he did yonder among the dark old trees. What would the minister have said, mother? Would he have clapped his hand over his heart, and scowled on me, and bid me begone?"

"What should he say, Pearl," answered Hester, "save that it was no time to kiss, and that kisses are not to be given in the market-place? Well for thee, foolish child, that thou didst not speak to him!"

Another shade of the same sentiment, in reference to Mr. Dimmesdale, was expressed by a person whose eccentricities—or insanity, as we should term it—led her to do what few of the townspeople would have ventured on; to begin a conversation with the wearer of the scarlet letter, in public. It was Mistress Hibbins, who, arrayed in great magnificence, with a triple ruff, a broidered stomacher,[7] a gown of rich velvet, and a gold-headed cane, had come forth to see the procession. As this ancient lady had the renown (which subsequently cost her no less a price than her life) of being a principal actor in all the works of necromancy that were continually going forward, the crowd gave way before her, and seemed to fear the touch of her garment, as if it carried the plague among its gorgeous folds. Seen in conjunction with Hester Prynne,—kindly as so many now felt towards the latter,—the dread inspired by Mistress Hibbins was doubled, and caused a general movement from that part of the market-place in which the two women stood.

"Now, what mortal imagination could conceive it!" whispered the old lady, confidentially, to Hester. "Yonder divine man! That saint on earth, as the people uphold him to be, and as—I must needs say—he really looks! Who, now, that saw him pass in the procession, would think how little while it is since he went forth out of his study,—chewing a Hebrew text of Scripture in his mouth, I warrant,—to take an airing in the forest! Aha! we know what that means, Hester Prynne! But, truly, forsooth, I find it hard to believe him the same man. Many a church-member saw I, walking behind the music, that has danced in the same measure with me, when Somebody was fiddler, and, it might be, an Indian powwow or a Lapland[8] wizard changing hands with us! That is but a trifle, when a woman knows the world. But this minister! Couldst thou surely tell, Hester, whether he was the same man that encountered thee on the forest-path?"

"Madam, I know not of what you speak," answered Hester Prynne, feeling Mistress Hibbins to be of infirm mind; yet strangely startled and awe-stricken by the confidence with which she affirmed a personal connection between so many persons (herself among them) and the Evil One. "It is not for me to talk lightly of a learned and pious minister of the Word, like the Reverend Mr. Dimmesdale!"

"Fie, woman, fie!" cried the old lady, shaking her finger at Hester. "Dost thou think I have been to the forest so many times, and have yet no skill to judge who else has been there? Yea; though no leaf of the wild garlands, which they wore while they danced, be left in their hair! I know thee, Hester; for I behold the token. We may all see it in the sunshine; and it glows like a red flame in the dark. Thou wearest it openly; so there need be no question about that. But this minister! Let me tell thee, in thine ear! When the Black Man sees one of his own servants, signed and sealed, so shy of owning to the bond as is the Reverend Mr. Dimmesdale, he hath a way of ordering matters so that the mark shall be disclosed in open daylight to the eyes of all the world! What is it

7. Waistcoat.
8. Northernmost Sweden, Norway, and Finland.

that the minister seeks to hide, with his hand always over his heart? Ha, Hester Prynne!"

"What is it, good Mistress Hibbins?" eagerly asked little Pearl. "Hast thou seen it?"

"No matter, darling!" responded Mistress Hibbins, making Pearl a profound reverence. "Thou thyself wilt see it, one time or another. They say, child, thou art of the lineage of the Prince of the Air! Wilt thou ride with me, some fine night, to see thy father? Then thou shalt know wherefore the minister keeps his hand over his heart!"

Laughing so shrilly that all the market-place could hear her, the weird old gentlewoman took her departure.

By this time the preliminary prayer had been offered in the meeting-house, and the accents of the Reverend Mr. Dimmesdale were heard commencing his discourse. An irresistible feeling kept Hester near the spot. As the sacred edifice was too much thronged to admit another auditor, she took up her position close beside the scaffold of the pillory. It was in sufficient proximity to bring the whole sermon to her ears, in the shape of an indistinct, but varied, murmur and flow of the minister's very peculiar voice.

This vocal organ was in itself a rich endowment; insomuch that a listener, comprehending nothing of the language in which the preacher spoke, might still have been swayed to and fro by the mere tone and cadence. Like all other music, it breathed passion and pathos, and emotions high or tender, in a tongue native to the human heart, wherever educated. Muffled as the sound was by its passage through the church-walls, Hester Prynne listened with such intentness, and sympathized so intimately, that the sermon had throughout a meaning for her, entirely apart from its indistinguishable words. These, perhaps, if more distinctly heard, might have been only a grosser medium, and have clogged the spiritual sense. Now she caught the low undertone, as of the wind sinking down to repose itself; then ascended with it, as it rose through progressive gradations of sweetness and power, until its volume seemed to envelop her with an atmosphere of awe and solemn grandeur. And yet, majestic as the voice sometimes became, there was forever in it an essential character of plaintiveness. A loud or low expression of anguish,—the whisper, or the shriek, as it might be conceived, of suffering humanity, that touched a sensibility in every bosom! At times this deep strain of pathos was all that could be heard, and scarcely heard, sighing amid a desolate silence. But even when the minister's voice grew high and commanding,—when it gushed irrepressibly upward,—when it assumed its utmost breadth and power, so overfilling the church as to burst its way through the solid walls, and diffuse itself in the open air,—still, if the auditor listened intently, and for the purpose, he could detect the same cry of pain. What was it? The complaint of a human heart, sorrow-laden, perchance guilty, telling its secret, whether of guilt or sorrow, to the great heart of mankind; beseeching its sympathy or forgiveness,—at every moment,—in each accent,—and never in vain! It was this profound and continual undertone that gave the clergyman his most appropriate power.

During all this time, Hester stood, statue-like, at the foot of the scaffold. If the minister's voice had not kept her there, there would nevertheless have been an inevitable magnetism in that spot, whence she dated the first hour of her life of ignominy. There was a sense within her,—too ill-defined to be made a thought, but weighing heavily on her mind,—that her whole orb of life, both before and after, was connected with this spot, as with the one point that gave it unity.

Little Pearl, meanwhile, had quitted her mother's side, and was playing at her own will about the market-place. She made the sombre crowd cheerful by her erratic and glistening ray; even as a bird of bright plumage illuminates a whole tree of dusky foliage, by darting to and fro, half seen and half concealed amid the twilight of the clustering leaves. She had an undulating, but, oftentimes, a sharp and irregular movement. It indicated the restless vivacity of her spirit, which to-day was doubly indefatigable in its tiptoe dance, because it was played upon and vibrated with her mother's disquietude. Whenever Pearl saw anything to excite her ever active and wandering curiosity, she flew thitherward, and, as we might say, seized upon that man or thing as her own property, so far as she desired it; but without yielding the minutest degree of control over her motions in requital. The Puritans looked on, and, if they smiled, were none the less inclined to pronounce the child a demon offspring, from the indescribable charm of beauty and eccentricity that shone through her little figure, and sparkled with its activity. She ran and looked the wild Indian in the face; and he grew conscious of a nature wilder than his own. Thence, with native audacity, but still with a reserve as characteristic, she flew into the midst of a group of mariners, the swarthy-cheeked wild men of the ocean, as the Indians were of the land; and they gazed wonderingly and admiringly at Pearl, as if a flake of the sea-foam had taken the shape of a little maid, and were gifted with a soul of the sea-fire, that flashes beneath the prow in the night-time.

One of these seafaring men—the shipmaster, indeed, who had spoken to Hester Prynne—was so smitten with Pearl's aspect, that he attempted to lay hands upon her, with purpose to snatch a kiss. Finding it as impossible to touch her as to catch a humming-bird in the air, he took from his hat the gold chain that was twisted about it, and threw it to the child. Pearl immediately twined it around her neck and waist, with such happy skill, that, once seen there, it became a part of her, and it was difficult to imagine her without it.

"Thy mother is yonder woman with the scarlet letter," said the seaman. "Wilt thou carry her a message from me?"

"If the message pleases me, I will," answered Pearl.

"Then tell her," rejoined he, "that I spake again with the black-a-visaged, hump-shouldered old doctor, and he engages to bring his friend, the gentleman she wots of, aboard with him. So let thy mother take no thought, save for herself and thee. Wilt thou tell her this, thou witch-baby?"

"Mistress Hibbins says my father is the Prince of the Air!" cried Pearl, with a naughty smile. "If thou callest me that ill name, I shall tell him of thee; and he will chase thy ship with a tempest!"

Pursuing a zigzag course across the market-place, the child returned to her mother, and communicated what the mariner had said. Hester's strong, calm, steadfastly enduring spirit almost sank, at last, on beholding this dark and grim countenance of an inevitable doom, which—at the moment when a passage seemed to open for the minister and herself out of their labyrinth of misery—showed itself, with an unrelenting smile, right in the midst of their path.

With her mind harassed by the terrible perplexity in which the shipmaster's intelligence involved her, she was also subjected to another trial. There were many people present, from the country round about, who had often heard of the scarlet letter, and to whom it had been made terrific by a hundred false or exaggerated rumors, but who had never beheld it with their own bodily eyes. These, after exhausting other modes of amusement, now thronged about Hester Prynne with rude and boorish intrusiveness. Unscrupulous as it was, however, it could not bring them nearer than a circuit of several yards. At that distance they accordingly stood, fixed there by the centrifugal force of the repugnance which the mystic symbol inspired. The whole gang of sailors, likewise, observing the press of spectators, and learning the purport of the scarlet letter, came and thrust their sunburnt and desperado-looking faces into the ring. Even the Indians were affected by a sort of cold shadow of the white man's curiosity, and, gliding through the crowd, fastened their snake-like black eyes on Hester's bosom; conceiving, perhaps, that the wearer of this brilliantly embroidered badge must needs be a personage of high dignity among her people. Lastly the inhabitants of the town (their own interest in this worn-out subject languidly reviving itself, by sympathy with what they saw others feel) lounged idly to the same quarter, and tormented Hester Prynne, perhaps more than all the rest, with their cool, well-acquainted gaze at her familiar shame. Hester saw and recognized the self-same faces of that group of matrons, who had awaited her forthcoming from the prison-door, seven years ago; all save one, the youngest and only compassionate among them, whose burial-robe she had since made. At the final hour, when she was so soon to fling aside the burning letter, it had strangely become the centre of more remark and excitement, and was thus made to sear her breast more painfully, than at any time since the first day she put it on.

While Hester stood in that magic circle of ignominy, where the cunning cruelty of her sentence seemed to have fixed her forever, the admirable preacher was looking down from the sacred pulpit upon an audience, whose very inmost spirits had yielded to his control. The sainted minister in the church! The woman of the scarlet letter in the market-place! What imagination would have been irreverent enough to surmise that the same scorching stigma was on them both!

XXIII. The Revelation of the Scarlet Letter

The eloquent voice, on which the souls of the listening audience had been borne aloft as on the swelling waves of the sea, at length came to a pause. There was a momentary silence, profound as what should follow the utterance of oracles. Then ensued a murmur and half-hushed tumult; as if the auditors, released from the high spell that had transported them into the region of another's mind, were returning into themselves, with all their awe and wonder still heavy on them. In a moment more, the crowd began to gush forth from the doors of the church. Now that there was an end, they needed other breath, more fit to support the gross and earthly life into which they relapsed, than that atmosphere which the preacher had converted into words of flame, and had burdened with the rich fragrance of his thought.

In the open air their rapture broke into speech. The street and the market-place absolutely babbled, from side to side, with applauses of the minister. His hearers could not rest until they had told one another of what each knew better than he could tell or hear. According to their united testimony, never had man spoken in so wise, so high, and so holy a spirit, as he that spake this day; nor had inspiration ever breathed through mortal lips more evidently than it did through his. Its influence could be seen, as it were, descending upon him, and possessing him, and continually lifting him out of the written discourse that lay before him, and filling him with ideas that must have been as marvellous to himself as to his audience. His subject, it appeared, had been the relation between the Deity and the communities of mankind, with a special reference to the New England which they were here planting in the wilderness. And, as he drew towards the close, a spirit as of prophecy had come upon him, constraining him to its purpose as mightily as the old prophets of Israel were constrained; only with this difference, that, whereas the Jewish seers had denounced judgments and ruin on their country, it was his mission to foretell a high and glorious destiny for the newly gathered people of the Lord. But, throughout it all, and through the whole discourse, there had been a certain deep, sad undertone of pathos, which could not be interpreted otherwise than as the natural regret of one soon to pass away. Yes; their minister whom they so loved—and who so loved them all, that he could not depart heavenward without a sigh—had the foreboding of untimely death upon him, and would soon leave them in their tears! This idea of his transitory stay on earth gave the last emphasis to the effect which the preacher had produced; it was as if an angel, in his passage to the skies, had shaken his bright wings over the people for an instant,—at once a shadow and a splendor,—and had shed down a shower of golden truths upon them.

Thus, there had come to the Reverend Mr. Dimmesdale—as to most men, in their various spheres, though seldom recognized until they see it far behind them—an epoch of life more brilliant and full of triumph than any previous one, or than any which could hereafter be. He stood, at this moment, on the very proudest eminence of superiority, to which the gifts of intellect, rich lore, prevailing eloquence, and a reputation of whitest sanctity, could exalt a clergyman in New England's earliest days, when the professional character was of itself a lofty pedestal. Such was the position which the minister occupied, as he bowed his head forward on the cushions of the pulpit, at the close of his Election Sermon. Meanwhile Hester Prynne was standing beside the scaffold of the pillory, with the scarlet letter still burning on her breast!

Now was heard again the clangor of the music, and the measured tramp of the military escort, issuing from the church-door. The procession was to be marshalled thence to the town-hall, where a solemn banquet would complete the ceremonies of the day.

Once more, therefore, the train of venerable and majestic fathers was seen moving through a broad pathway of the people, who drew back reverently, on either side, as the Governor and magistrates, the old and wise men, the holy ministers, and all that were eminent and renowned, advanced into the midst of them. When they were fairly in the market-place, their presence was greeted by a shout. This—though doubtless it might acquire additional force and volume from the childlike loyalty which the age awarded to its rulers—was felt to be an irrepressible outburst of enthusiasm kindled in the auditors by that high strain of eloquence which was yet reverberating in their ears. Each felt the impulse in himself, and, in the same breath, caught it from his neighbor. Within the church, it had hardly been kept down; beneath the sky, it pealed upward to the zenith. There were human beings enough, and enough of highly wrought and symphonious feeling, to produce that more impressive sound than the organ tones of the blast, or the thunder, or the roar of the sea; even that mighty swell of many voices, blended into one great voice by the universal impulse which makes likewise one vast heart out of the many. Never, from the soil of New England, had gone up such a shout! Never, on New England soil, had stood the man so honored by his mortal brethren as the preacher!

How fared it with him then? Were there not the brilliant particles of a halo in the air about his head? So etherealized by spirit as he was, and so apotheosized by worshipping admirers, did his footsteps, in the procession, really tread upon the dust of earth?

As the ranks of military men and civil fathers moved onward, all eyes were turned towards the point where the minister was seen to approach among them. The shout died into a murmur, as one portion of the crowd after another obtained a glimpse of him. How feeble and pale he looked, amid all his triumph! The energy—or say, rather, the inspiration which had held him up, until he should have delivered the sacred message that brought its own strength along with it from heaven—was withdrawn, now that it had so faithfully performed its office. The glow, which they had just before beheld burning on his cheek, was extinguished, like a flame that sinks down hopelessly among the late-decaying embers. It seemed hardly the face of a man alive, with such a deathlike hue; it was hardly a man with life in him, that tottered on his path so nervelessly, yet tottered, and did not fall!

One of his clerical brethren,—it was the venerable John Wilson,—observing the state in which Mr. Dimmesdale was left by the retiring wave of intellect and sensibility, stepped forward hastily to offer his support. The minister tremulously, but decidedly, repelled the old man's arm. He still walked onward, if that movement could be so described, which rather resembled the wavering effort of an infant, with its mother's arms in view, outstretched to tempt him forward. And now, almost imperceptible as were the latter steps of his progress, he had come opposite the well-remembered and weather-darkened scaffold, where, long since, with all that dreary lapse of time between, Hester Prynne had encountered the world's ignominious stare. There stood Hester, holding little Pearl by the hand! And there was the scarlet letter on her breast! The minister here made a pause; although the music still played the stately and rejoicing march to which the procession moved. It summoned him onward,—onward to the festival!—but here he made a pause.

Bellingham, for the last few moments, had kept an anxious eye upon him. He now left his own place in the procession, and advanced to give assistance; judging, from Mr. Dimmesdale's aspect, that he must otherwise inevitably fall. But there was something in the latter's expression that warned back the magistrate, although a man not readily obeying the vague intimations that pass from one spirit to another. The crowd, meanwhile, looked on with awe and wonder. This earthly faintness was, in their view, only another phase of the minister's celestial strength; nor would it have seemed a miracle too high to be wrought for one so holy, had he ascended before their eyes, waxing dimmer and brighter, and fading at last into the light of heaven!

He turned towards the scaffold, and stretched forth his arms.

"Hester," said he, "come hither! Come, my little Pearl!"

It was a ghastly look with which he regarded them; but there was something at once tender and strangely triumphant in it. The child, with the bird-like motion which was one of her characteristics, flew to him, and clasped her arms about his knees. Hester Prynne—slowly, as if impelled by inevitable fate, and against her strongest will—likewise drew near, but paused before she reached him. At this instant, old Roger Chillingworth thrust himself through the crowd,—or, perhaps, so dark, disturbed and evil, was his look, he rose up out of some nether region,—to snatch back his victim from what he sought to do! Be that as it might, the old man rushed forward, and caught the minister by the arm.

"Madman, hold! what is your purpose?" whispered he. "Wave back that woman! Cast off this child! All shall be well! Do not blacken your fame, and perish in dishonor! I can yet save you! Would you bring infamy on your sacred profession?"

"Ha, tempter! Methinks thou art too late!" answered the minister, encountering his eye, fearfully, but firmly. "Thy power is not what it was! With God's help, I shall escape thee now!"

He again extended his hand to the woman of the scarlet letter.

"Hester Prynne," cried he, with a piercing earnestness, "in the name of Him, so terrible and so merciful, who gives me grace, at this last moment, to do what—for my own heavy sin and miserable agony—I withheld myself from doing seven years ago, come hither now, and twine thy strength about me! Thy strength, Hester; but let it be guided by the will which God hath granted me! This wretched and wronged old man is opposing it with all his might!—with all his own might, and the fiend's! Come, Hester, come! Support me up yonder scaffold!"

The crowd was in a tumult. The men of rank and dignity, who stood more immediately around the clergyman, were so taken by surprise, and so perplexed as to the purport of what they saw,—unable to receive the explanation which most readily presented itself, or to imagine any other,—that they remained silent and inactive spectators of the judgment which Providence seemed about to work. They beheld the minister, leaning on Hester's shoulder, and supported by her arm around him, approach the scaffold, and ascend its steps; while still the little hand of the sin-born child was clasped in his. Old Roger Chillingworth followed, as one intimately connected with the drama of guilt and sorrow in which they had all been actors, and well entitled, therefore, to be present, at its closing scene.

"Hadst thou sought the whole earth over," said he, looking darkly at the clergyman, "there was no one place so secret,—no high place nor lowly place, where thou couldst have escaped me,—save on this very scaffold!"

"Thanks be to Him who hath led me hither!" answered the minister.

Yet he trembled, and turned to Hester with an expression of doubt and anxiety in his eyes, not the less evidently betrayed, that there was a feeble smile upon his lips.

"Is not this better," murmured he, "than what we dreamed of in the forest?"

"I know not! I know not!" she hurriedly replied. "Better? Yea; so we may both die, and little Pearl die with us!"

"For thee and Pearl, be it as God shall order," said the minister; "and God is merciful! Let me now do the will which he hath made plain before my sight. For, Hester, I am a dying man. So let me make haste to take my shame upon me!"

Partly supported by Hester Prynne, and holding one hand of little Pearl's, the Reverend Mr. Dimmesdale turned to the dignified and venerable rulers; to the holy ministers, who were his brethren; to the people, whose great heart was thoroughly appalled, yet overflowing with tearful sympathy, as knowing that some deep life-matter—which, if full of sin, was full of anguish and repentance likewise—was now to be laid open to them. The sun, but little past its meridian, shone down upon the clergyman, and gave a distinctness to his figure, as he stood out from all the earth, to put in his plea of guilty at the bar of Eternal Justice.

"People of New England!" cried he, with a voice that rose over them, high, solemn, and majestic,—yet had always a tremor through it, and sometimes a shriek, struggling up out of a fathomless depth of remorse and woe,—"ye, that have loved me!—ye, that have deemed me holy!—behold me here, the one sinner of the world! At last!—at last!—I stand upon the spot where, seven years since, I should have stood; here, with this woman, whose arm, more than the little strength wherewith I have crept hitherward, sustains me, at this dreadful moment, from grovelling down upon my face! Lo, the scarlet letter which Hester wears! Ye have all shuddered at it! Wherever her walk hath been,—wherever, so miserably burdened, she may have hoped to find repose,—it hath cast a lurid gleam of awe and horrible repugnance round about her. But there stood one in the midst of you, at whose brand of sin and infamy ye have not shuddered!"

It seemed, at this point, as if the minister must leave the remainder of his secret undisclosed. But he fought back the bodily weakness,—and, still more, the faintness of heart,—that was striving for the mastery with him. He threw off all assistance, and stepped passionately forward a pace before the woman and the child.

"It was on him!" he continued, with a kind of fierceness; so determined was he to speak out the whole. "God's eye beheld it! The angels were forever pointing at it! The Devil knew it well, and fretted it continually with the touch of his burning finger! But he hid it cunningly from men, and walked among you with the mien of a spirit, mournful, because so pure in a sinful world!—and sad, because he missed his heavenly kindred! Now, at the death-hour, he stands up before you! He bids you look again at Hester's scarlet letter! He tells you, that, with all its mysterious horror, it is but the shadow of what he bears on his own breast, and that even this, his own red stigma, is no more than the type of what has seared his inmost heart! Stand any here that question God's judgment on a sinner? Behold! Behold a dreadful witness of it!"

With a convulsive motion, he tore away the ministerial band from before his breast. It was revealed! But it were irreverent to describe that revelation. For an instant, the gaze of the horror-stricken multitude was concentred on the ghastly miracle; while the minister stood, with a flush of triumph in his face, as one who, in the crisis of acutest pain, had won a victory. Then, down he sank upon the scaffold! Hester partly raised him, and supported his head against her bosom. Old Roger Chillingworth knelt down beside him, with a blank, dull countenance, out of which the life seemed to have departed.

"Thou hast escaped me!" he repeated more than once. "Thou hast escaped me!"

"May God forgive thee!" said the minister. "Thou, too, hast deeply sinned!"

He withdrew his dying eyes from the old man, and fixed them on the woman and the child.

"My little Pearl," said he, feebly,—and there was a sweet and gentle smile over his face, as of a spirit sinking into deep repose; nay, now that the burden was removed, it seemed almost as if he would be sportive with the child,—"dear little Pearl, wilt thou kiss me now? Thou wouldst not, yonder, in the forest! But now thou wilt?"

Pearl kissed his lips. A spell was broken. The great scene of grief, in which the wild infant bore a part, had developed all her sympathies; and as her tears fell upon her father's cheek, they were the pledge that she would grow up amid human joy and sorrow, nor forever do battle with the world, but be a woman in it. Towards her mother, too, Pearl's errand as a messenger of anguish was all fulfilled.

"Hester," said the clergyman, "farewell!"

"Shall we not meet again?" whispered she, bending her face down close to his. "Shall we not spend our immortal life together? Surely, surely, we have ransomed one another, with all this woe! Thou lookest far into eternity, with those bright dying eyes! Then tell me what thou seest?"

"Hush, Hester, hush!" said he, with tremulous solemnity. "The law we broke!—the sin here so awfully revealed!—let these alone be in thy thoughts! I fear! I fear! It may be, that, when we forgot our God,—when we violated our reverence each for the other's soul,—it was thenceforth vain to hope that we could meet hereafter, in an everlasting and pure reunion. God knows; and He is merciful! He hath proved his mercy, most of all, in my afflictions. By giving me this burning torture to bear upon my breast! By sending yonder dark and terrible old man, to keep the torture always at red-heat! By bringing me hither, to die this death of triumphant ignominy before the people! Had either of these agonies been wanting, I had been lost forever! Praised be his name! His will be done! Farewell!"

That final word came forth with the minister's expiring breath. The multitude, silent till then, broke out in a strange, deep voice of awe and wonder, which could not as yet find utterance, save in this murmur that rolled so heavily after the departed spirit.

XXIV. Conclusion

After many days, when time sufficed for the people to arrange their thoughts in reference to the foregoing scene, there was more than one account of what had been witnessed on the scaffold.

Most of the spectators testified to having seen, on the breast of the unhappy minister, a SCARLET LETTER—the very semblance of that worn by Hester Prynne—imprinted in the flesh. As regarded its origin, there were various explanations, all of which must necessarily have been conjectural. Some affirmed that the Reverend Mr. Dimmesdale, on the very day when Hester Prynne first wore her ignominious badge, had begun a course of penance,—which he afterwards, in so many futile methods, followed out,—by inflicting a hideous torture on himself. Others contended that the stigma had not been produced until a long time subsequent, when old Roger Chillingworth, being a potent necromancer, had caused it to appear, through the agency of magic and poisonous drugs. Others, again,—and those best able to appreciate the minister's peculiar sensibility, and the wonderful operation of his spirit upon the body,—whispered their belief, that the awful symbol was the effect of the ever active tooth of remorse, gnawing from the inmost heart outwardly, and at last manifesting Heaven's dreadful judgment by the visible presence of the letter. The reader may choose among these theories. We have thrown all the light we could acquire upon the portent, and would gladly, now that it has done its office, erase its deep print out of our own brain; where long meditation has fixed it in very undesirable distinctness.

It is singular, nevertheless, that certain persons, who were spectators of the whole scene, and professed never once to have removed their eyes from the Reverend Mr. Dimmesdale, denied that there was any mark whatever on his breast, more than on a new-born infant's. Neither, by their report, had his dying words acknowledged, nor even remotely implied, any, the slightest connection, on his part, with the guilt for which Hester Prynne had so long worn the scarlet letter. According to these highly respectable witnesses, the minister, conscious that he was dying,—conscious, also, that the reverence of the multitude placed him already among saints and angels,—had desired, by yielding up his breath in the arms of that fallen woman, to express to the world how utterly nugatory[1] is the choicest of man's own righteousness. After exhausting life in his efforts for mankind's spiritual good, he had made the manner of his death a parable, in order to impress on his admirers the mighty and mournful lesson, that, in the view of Infinite Purity, we are sinners all alike. It was to teach them, that the holiest among us has but attained so far above his fellows as to discern more clearly the Mercy which looks down, and repudiate more utterly the phantom of human merit, which would look aspiringly upward. Without disputing a truth so momentous, we must be allowed to consider this version of Mr. Dimmesdale's story as only an instance of that stubborn fidelity with which a man's friends—and especially a clergyman's—will sometimes uphold his character, when proofs, clear as the mid-day sunshine on the scarlet letter, establish him a false and sin-stained creature of the dust.

1. Worthless.

The authority which we have chiefly followed,—a manuscript of old date, drawn up from the verbal testimony of individuals, some of whom had known Hester Prynne, while others had heard the tale from contemporary witnesses,—fully confirms the view taken in the foregoing pages. Among many morals which press upon us from the poor minister's miserable experience, we put only this into a sentence:—"Be true! Be true! Be true! Show freely to the world, if not your worst, yet some trait whereby the worst may be inferred!"

Nothing was more remarkable than the change which took place, almost immediately after Mr. Dimmesdale's death, in the appearance and demeanor of the old man known as Roger Chillingworth. All his strength and energy—all his vital and intellectual force—seemed at once to desert him; insomuch that he positively withered up, shrivelled away, and almost vanished from mortal sight, like an uprooted weed that lies wilting in the sun. This unhappy man had made the very principle of his life to consist in the pursuit and systematic exercise of revenge; and when, by its completest triumph and consummation, that evil principle was left with no further material to support it, when, in short, there was no more Devil's work on earth for him to do, it only remained for the unhumanized mortal to betake himself whither his Master would find him tasks enough, and pay him his wages duly. But, to all these shadowy beings, so long our near acquaintances,—as well Roger Chillingworth as his companions,—we would fain be merciful. It is a curious subject of observation and inquiry, whether hatred and love be not the same thing at bottom. Each, in its utmost development, supposes a high degree of intimacy and heart-knowledge; each renders one individual dependent for the food of his affections and spiritual life upon another; each leaves the passionate lover, or the no less passionate hater, forlorn and desolate by the withdrawal of his subject. Philosophically considered, therefore, the two passions seem essentially the same, except that one happens to be seen in a celestial radiance, and the other in a dusky and lurid glow. In the spiritual world, the old physician and the minister—mutual victims as they have been—may, unawares, have found their earthly stock of hatred and antipathy transmuted into golden love.

Leaving this discussion apart, we have a matter of business to communicate to the reader. At old Roger Chillingworth's decease, (which took place within the year,) and by his last will and testament, of which Governor Bellingham and the Reverend Mr. Wilson were executors, he bequeathed a very considerable amount of property, both here and in England, to little Pearl, the daughter of Hester Prynne.

So Pearl—the elf-child,—the demon offspring, as some people, up to that epoch, persisted in considering her,—became the richest heiress of her day, in the New World. Not improbably, this circumstance wrought a very material change in the public estimation; and, had the mother and child remained here, little Pearl, at a marriageable period of life, might have mingled her wild blood with the lineage of the devoutest Puritan among them all. But, in no long time after the physician's death, the wearer of the scarlet letter disappeared, and Pearl along with her. For many years, though a vague report would now and then find its way across the sea,—like a shapeless piece of driftwood tost ashore, with the initials of a name upon it,—yet no tidings of them unquestionably authentic were received.

The story of the scarlet letter grew into a legend. Its spell, however, was still potent, and kept the scaffold awful where the poor minister had died, and likewise the cottage by the sea-shore, where Hester Prynne had dwelt. Near this latter spot, one afternoon, some children were at play, when they beheld a tall woman, in a gray robe, approach the cottage-door. In all those years it had never once been opened; but either she unlocked it, or the decaying wood and iron yielded to her hand, or she glided shadow-like through these impediments,—and, at all events, went in.

On the threshold she paused,—turned partly round,—for, perchance, the idea of entering all alone, and all so changed, the home of so intense a former life, was more dreary and desolate than even she could bear. But her hesitation was only for an instant, though long enough to display a scarlet letter on her breast.

And Hester Prynne had returned, and taken up her long-forsaken shame! But where was little Pearl? If still alive, she must now have been in the flush and bloom of early womanhood. None knew—nor ever learned, with the fulness of perfect certainty—whether the elfchild had gone thus untimely to a maiden grave; or whether her wild, rich nature had been softened and subdued, and made capable of a woman's gentle happiness. But, through the remainder of Hester's life, there were indications that the recluse of the scarlet letter was the object of love and interest with some inhabitant of another land. Letters came, with armorial seals upon them, though of bearings unknown to English heraldry. In the cottage there were articles of comfort and luxury, such as Hester never cared to use, but which only wealth could have purchased, and affection have imagined for her. There were trifles, too, little ornaments, beautiful tokens of a continual remembrance, that must have been wrought by delicate fingers at the impulse of a fond heart. And, once, Hester was seen embroidering a baby-garment, with such a lavish richness of golden fancy as would have raised a public tumult, had any infant, thus apparelled, been shown to our sober-hued community.

In fine, the gossips of that day believed,—and Mr. Surveyor Pue, who made investigations a century later, believed,—and one of his recent successors in office, moreover, faithfully believes,—that Pearl was not only alive, but married, and happy, and mindful of her mother; and that she would most joyfully have entertained that sad and lonely mother at her fireside.

But there was a more real life for Hester Prynne, here, in New England, than in that unknown region where Pearl had found a home. Here had been her sin; here, her sorrow; and here was yet to be her penitence. She had returned, therefore, and resumed,—of her own free will, for not the sternest magistrate of that iron period would have imposed it,—resumed the symbol of which we have related so dark a tale. Never afterwards did it quit her bosom. But, in the lapse of the toilsome, thoughtful, and self-devoted years that made up Hester's life, the scarlet letter ceased to be a stigma which attracted the world's scorn and bitterness, and became a type of something to be sorrowed over, and looked upon with awe, yet with reverence too. And, as Hester Prynne had no selfish ends, nor lived in any measure for her own profit and enjoyment, people brought all their sorrows and perplexities, and besought her counsel, as one who had herself gone through a mighty trouble. Women, more especially,—in the continually recurring trials of wounded, wasted, wronged, misplaced, or erring and sinful passion,—or with the dreary burden of a heart unyielded, because unvalued and unsought,—came to Hester's cottage, demanding why they were so wretched, and what the remedy! Hester comforted and counselled them, as best she might.[2] She assured them, too, of her firm belief, that, at some brighter period, when the world should have grown ripe for it, in Heaven's own time, a new truth would be revealed, in order to establish the whole relation between man and woman on a surer ground of mutual happiness.[3] Earlier in life, Hester had vainly imagined that she herself might be the destined prophetess, but had long since recognized the impossibility that any mission of divine and mysterious truth should be confided to a woman stained with sin, bowed down with shame, or even burdened with a life-long sorrow. The angel and apostle of the coming revelation must be a woman, indeed, but lofty, pure, and beautiful; and wise, moreover, not through dusky grief, but the ethereal medium of joy; and showing how sacred love should make us happy, by the truest test of a life successful to such an end!

So said Hester Prynne, and glanced her sad eyes downward at the scarlet letter. And, after many, many years, a new grave was delved, near an old and sunken one, in that burial-ground beside which King's Chapel has since been built. It was near that old and sunken grave, yet with a space between, as if the dust of the two sleepers had no right to mingle. Yet one tombstone served for both. All around, there were monuments carved with armorial bearings; and on this simple slab of slate—as the curious investigator may still discern, and perplex himself with the purport—there appeared the semblance of an engraved escutcheon. It bore a device, a herald's wording of which might serve for a motto and brief description of our now concluded legend; so somber is it, and relieved only by one ever-glowing point of light gloomier than the shadow:—

"ON A FIELD, SABLE, THE LETTER A, GULES."[4]

2. It is worth recalling that Anne Hutchinson had been charged with unlawfully hosting weekly meetings for women. Under interrogation by John Winthrop, Hutchinson cited the New Testament passage in Titus that the "older women should instruct the younger." See Titus 2.3–5: "The aged women likewise, that they be in behaviour as becometh holiness, not false accusers, not given to much wine, teachers of good things; / That they may teach the young women to be sober, to love their husbands, to love their children, / To be discreet, chaste, keepers at home, good, obedient to their own husbands, that the word of God be not blasphemed."
3. Compare the passage in Hawthorne's campaign biography of Franklin Pierce in this Norton Critical Edition (pp. 220–21). As Thomas Mitchell argues in *Hawthorne's Fuller Mystery*, Hawthorne may also be echoing passages in Margaret Fuller's *Woman in the Nineteenth Century* (1845). Early in the book Fuller comments, "Yet, then and only then, will mankind be ripe for this, when inward and outward freedom for woman as much as for man shall be acknowledged as a right, not yielded as a concession" (20). Near the conclusion she observes, "And will not she soon appear? The woman who shall vindicate their birthright for all women; who shall teach them what to claim, and how to use what they obtain? Shall not her name be for her era Victoria, for her country and life Virginia. Yet predictions are rash; she herself must teach us to give her the fitting name" (104). See Margaret Fuller, *Woman in the Nineteenth Century*, ed. Larry J. Reynolds (New York: Norton, 1998).
4. "On a black background, the letter A in red." *Sable*: one of the colors of English heraldry, indicated by horizontal and vertical lines crossing each other. *Gules*: the color red, represented by vertical lines (*OED*).

OTHER WRITINGS

Mrs. Hutchinson†

The character of this female suggests a train of thought which will form as natural an introduction to her story as most of the prefaces to Gay's Fables or the tales of Prior,[1] besides that the general soundness of the moral may excuse any want of present applicability. We will not look for a living resemblance of Mrs. Hutchinson, though the search might not be altogether fruitless.—But there are portentous indications, changes gradually taking place in the habits and feelings of the gentle sex, which seem to threaten our posterity with many of those public women, whereof one was a burthen too grievous for our fathers. The press, however, is now the medium through which feminine ambition chiefly manifests itself, and we will not anticipate the period, (trusting to be gone hence ere it arrive,) when fair orators shall be as numerous as the fair authors of our own day. The hastiest glance may show, how much of the texture and body of cis-atlantic literature is the work of those slender fingers, from which only a light and fanciful embroidery has heretofore been required, that might sparkle upon the garment without enfeebling the web. Woman's intellect should never give the tone to that of man, and even her morality is not exactly the material for masculine virtue. A false liberality which mistakes the strong division lines of Nature for arbitrary distinctions, and a courtesy, which might polish criticism but should never soften it, have done their best to add a girlish feebleness to the tottering infancy of our literature. The evil is likely to be a growing one. As yet, the great body of American women are a domestic race; but when a continuance of ill-judged incitements shall have turned their hearts away from the fire-side, there are obvious circumstances which will render female pens more numerous and more prolific than those of men, though but equally encouraged; and (limited of course by the scanty support of the public, but increasing indefinitely within those limits) the ink-stained Amazons will expel their rivals by actual pressure, and petticoats wave triumphant over all the field.[2] But, allowing that such forebodings are slightly exaggerated, is it good for woman's self that the path of feverish hope, of tremulous success, of bitter and ignominious disappointment, should be left wide open to her? Is the prize worth her having if she win it? Fame does not increase the peculiar respect which men pay to female excellence, and there is a delicacy, (even in rude bosoms, where few would think to find it) that perceives, or fancies, a sort of impropriety in the display of woman's naked mind to the gaze of the world, with indications by which its inmost secrets may be searched out. In fine, criticism should examine with a stricter, instead of a more indulgent eye, the merits of females at its bar, because they are to justify themselves for an irregularity which men do not commit in appearing there; and woman, when she feels the impulse of genius like a command of Heaven within her, should be aware that she is relinquishing a part of the

† From the original version of the sketch published in the Salem *Gazette*, 7 Dec. 1830: 4.

1. John Gay (1685–1732) published his *Fables* in 1727. An expanded version appeared in 1738. Matthew Prior (1664–1721) is perhaps best-known for *The Hind and the Panther Transvers'd to the Story of the Country and the City Mouse* (1687).
2. This characterization of women writers anticipates by twenty-five years Hawthorne's notorious remark in a January 19, 1855, letter to George Ticknor that "America is now wholly given over to a d——d mob of scribbling women" with whose "trash" the "public taste is occupied."

loveliness of her sex, and obey the inward voice with sorrowing reluctance, like the Arabian maid who bewailed the gift of Prophecy. Hinting thus imperfectly at sentiments which may be developed on a future occasion, we proceed to consider the celebrated subject of this sketch.

Mrs. Hutchinson was a woman of extraordinary talent and strong imagination, whom the latter quality, following the general direction taken by the enthusiasm of the times, prompted to stand forth as a reformer in religion. In her native country, she had shown symptoms of irregular and daring thought, but, chiefly by the influence of a favorite pastor, was restrained from open indiscretion. On the removal of this clergyman, becoming dissatisfied with the ministry under which she lived, she was drawn in by the great tide of Puritan emigration, and visited Massachusetts within a few years after its first settlement. But she bore trouble in her own bosom, and could find no peace in this chosen land.—She soon began to promulgate strange and dangerous opinions, tending, in the peculiar situation of the colony, and from the principles which were its basis and indispensable for its temporary support, to eat into its very existence. We shall endeavor to give a more practical idea of this part of her course.

It is a summer evening. The dusk has settled heavily upon the woods, the waves, and the Trimontane peninsula,[3] increasing that dismal aspect of the embryo town which was said to have drawn tears of despondency from Mrs. Hutchinson, though she believed that her mission thither was divine. The houses, straw-thatched and lowly roofed, stand irregularly along streets that are yet roughened by the roots of the trees, as if the forest, departing at the approach of man, had left its reluctant foot prints behind. Most of the dwellings are lonely and silent; from a few we may hear the reading of some sacred text, or the quiet voice of prayer; but nearly all the sombre life of the scene is collected near the extremity of the village. A crowd of hooded women, and of men in steeple-hats and close cropt hair, are assembled at the door and open windows of a house newly built. An earnest expression glows in every face, and some press inward as if the bread of life were to be dealt forth, and they feared to lose their share, while others would fain hold them back, but enter with them since they may not be restrained. We also will go in, edging through the thronged door-way to an apartment which occupies the whole breadth of the house. At the upper end, behind a table on which are placed the Scriptures and two glimmering lamps, we see a woman, plainly attired, as befits her ripened years; her hair, complexion, and eyes are dark, the latter somewhat dull and heavy, but kindling up with a gradual brightness. Let us look round upon the hearers. At her right hand, his countenance suiting well with the gloomy light which discovers it, stands Vane the youthful governor, preferred by a hasty judgment of the people over all the wise and hoary heads that had preceded him to New-England.[4] In his mysterious eyes we may read a dark enthusiasm, akin to that of the woman whose cause he has espoused, combined with a shrewd worldly foresight, which tells him that her doctrines will be productive of change and tumult, the elements of his power and delight. On her left, yet slightly drawn back so as to evince a less decided support, is Cotton, no young and hot enthusiast,

3. The site of Boston on the south side of the Charles River.
4. Henry Vane (1613–1662), who supported Hutchinson, served as governor in 1636–37 and thus preceded John Winthrop, who presided at Hutchinson's "trial."

but a mild, grave man in the decline of life, deep in all the learning of the age, and sanctified in heart and made venerable in feature by the long exercise of his holy profession.[5] He also is deceived by the strange fire now laid upon the altar, and he alone among his brethren is excepted in the denunciation of the new Apostle, as sealed and set apart by Heaven to the work of the ministry. Others of the priesthood stand full in front of the woman, striving to beat her down with brows of wrinkled iron, and whispering sternly and significantly among themselves, as she unfolds her seditious doctrines and grows warm in their support. Foremost is Hugh Peters, full of holy wrath, and scarce containing himself from rushing forward to convict her of damnable heresies; there also is Ward, meditating a reply of empty puns, and quaint antitheses, and tinkling jests that puzzle us with nothing but a sound.[6] The audience are variously affected, but none indifferent. On the foreheads of the aged, the mature, and strong-minded, you may generally read steadfast disapprobation, though here and there is one, whose faith seems shaken in those whom he had trusted for years; the females, on the other hand, are shuddering and weeping, and at times they cast a desolate look of fear around them; while the young men lean forward, fiery and impatient, fit instruments for whatever rash deed may be suggested. And what is the eloquence that gives rise to all these passions? The woman tells them, (and cites texts from the Holy Book to prove her words,) that they have put their trust in unregenerated and uncommissioned men, and have followed them into the wilderness for naught. Therefore their hearts are turning from those whom they had chosen to lead them to Heaven, and they feel like children who have been enticed far from home, and see the features of their guides change all at once, assuming a fiendish shape in some frightful solitude.

These proceedings of Mrs. Hutchinson could not long be endured by the provincial government. The present was a most remarkable case, in which religious freedom was wholly inconsistent with public safety, and where the principles of an illiberal age indicated the very course which must have been pursued by worldly policy and enlightened wisdom. Unity of faith was the star that had guided these people over the deep, and a diversity of sects would either have scattered them from the land to which they had as yet so few attachments, or perhaps have excited a diminutive civil war among those who had come so far to worship together. The opposition to what may be termed the established church had now lost its chief support, by the removal of Vane from office and his departure for England, and Mr. Cotton began to have that light in regard to his errors, which will sometimes break in upon the wisest and most pious men, when their opinions are unhappily discordant with those of the Powers that be. A Synod, the first in New England, was speedily assembled, and pronounced its condemnation of the obnoxious doctrines.[7] Mrs. Hutchinson was next summoned before the supreme civil tribunal, at which, however, the most eminent of the clergy

5. John Cotton (1584–1652) was centrally involved in the antinomian (see n. 1, p. 36) controversy of 1636–38, which resulted in Hutchinson's banishment from the colony. Cotton was called as a witness at Hutchinson's trial and testified that he had not heard her claim that his "brother" ministers were preaching a covenant of works.
6. Hugh Peters (1598–1660) replaced Roger Williams as pastor in Salem. Nathaniel Ward (1578–1652), author of *The Simple Cobler of Aggawam* (1647), wrote the *Body of Liberties* (1641), which constrained the power of Massachusetts Bay Colony magistrates.
7. The Synod convened on August 30, 1637. John Winthrop was reelected governor in 1637, 1638, and 1639.

were present, and appear to have taken a very active part as witnesses and advisers. We shall here resume the more picturesque style of narration.

It is a place of humble aspect where the Elders of the people are met, sitting in judgment upon the disturber of Israel. The floor of the low and narrow hall is laid with planks hewn by the axe,—the beams of the roof still wear the rugged bark with which they grew up in the forest, and the hearth is formed of one broad unhammered stone, heaped with logs that roll their blaze and smoke up a chimney of wood and clay. A sleety shower beats fitfully against the windows, driven by the November blast, which comes howling onward from the northern desert, the boisterous and unwelcome herald of a New England winter.[8] Rude benches are arranged across the apartment and along its sides, occupied by men whose piety and learning might have entitled them to seats in those high Councils of the ancient Church, whence opinions were sent forth to confirm or supersede the Gospel in the belief of the whole world and of posterity.—Here are collected all those blessed Fathers of the land, who rank in our veneration next to the Evangelists of Holy Writ, and here also are many, unpurified from the fiercest errors of the age and ready to propagate the religion of peace by violence. In the highest place sits Winthrop, a man by whom the innocent and the guilty might alike desire to be judged, the first confiding in his integrity and wisdom, the latter hoping in his mildness. Next is Endicott, who would stand with his drawn sword at the gate of Heaven, and resist to the death all pilgrims thither, except they travelled his own path.[9] The infant eyes of one in this assembly beheld the faggots blazing round the martyrs, in bloody Mary's time;[1] in later life he dwelt long at Leyden, with the first who went from England for conscience sake; and now, in his weary age, it matters little where he lies down to die. There are others whose hearts were smitten in the high meridian of ambitious hope, and whose dreams still tempt them with the pomp of the old world and the din of its crowded cities, gleaming and echoing over the deep. In the midst, and in the centre of all eyes, we see the Woman. She stands loftily before her judges, with a determined brow, and, unknown to herself, there is a flash of carnal pride half hidden in her eye, as she surveys the many learned and famous men whom her doctrines have put in fear. They question her, and her answers are ready and acute; she reasons with them shrewdly, and brings scripture in support of every argument; the deepest controversialists of that scholastic day find here a woman, whom all their trained and sharpened intellects are inadequate to foil. But by the excitement of the contest, her heart is made to rise and swell within her, and she bursts forth into eloquence. She tells them of the long unquietness which she had endured in England, perceiving the corruption of the church, and yearning for a purer and more perfect light, and how, in a day of solitary prayer, that light was given; she claims for herself the peculiar power of distinguishing between the chosen of man and the Seated of Heaven, and affirms that her gifted eye can see the glory round the foreheads of the Saints, sojourning in their mortal state.—She declares herself commissioned to separate the true shepherds from the false, and denounces

8. The date was November 2, 1637.
9. John Endicott was deputy governor of Massachusetts Bay Colony in 1637.
1. Queen Mary—or "Bloody Mary," a fiercely anti-Protestant Catholic—ruled England 1553–58. She was succeeded by Queen Elizabeth I, a Protestant.

present and future judgments on the land, if she be disturbed in her celestial errand. Thus the accusations are proved from her own mouth. Her judges hesitate, and some speak faintly in her defence; but, with a few dissenting voices, sentence is pronounced, bidding her go out from among them, and trouble the land no more.

Mrs. Hutchinson's adherents throughout the colony were now disarmed, and she proceeded to Rhode Island, an accustomed refuge for the exiles of Massachusetts, in all seasons of persecution.[2] Her enemies believed that the anger of Heaven was following her, of which Governor Winthrop does not disdain to record a notable instance, very interesting in a scientific point of view, but fitter for his old and homely narrative than for modern repetition. In a little time, also, she lost her husband, who is mentioned in history only as attending her footsteps, and whom we may conclude to have been (like most husbands of celebrated women) a mere insignificant appendage of his mightier wife.[3] She now grew uneasy among the Rhode-Island colonists, whose liberality towards her, at an era when liberality was not esteemed a christian virtue, probably arose from a comparative insolicitude on religious matters, more distasteful to Mrs. Hutchinson than even the uncompromising narrowness of the Puritans. Her final movement was to lead her family within the limits of the Dutch Jurisdiction, where, having felled the trees of a virgin soil, she became herself the virtual head, civil and ecclesiastical, of a little colony.

Perhaps here she found the repose, hitherto so vainly sought. Secluded from all whose faith she could not govern, surrounded by the dependents over whom she held an unlimited influence, agitated by none of the tumultuous billows which were left swelling behind her, we may suppose, that, in the stillness of Nature, her heart was stilled. But her impressive story was to have an awful close. Her last scene is as difficult to be portrayed as a shipwreck, where the shrieks of the victims die unheard along a desolate sea, and a shapeless mass of agony is all that can be brought home to the imagination. The savage foe was on the watch for blood. Sixteen persons assembled at the evening prayer; in the deep midnight, their cry rang through the forest; and daylight dawned upon the lifeless clay of all but one. It was a circumstance not to be unnoticed by our stern ancestors, in considering the fate of her who had so troubled their religion, that an infant daughter, the sole survivor amid the terrible destruction of her mother's household, was bred in a barbarous faith, and never learned the way to the Christian's Heaven. Yet we will hope, that there the mother and the child have met.[4]

Endicott and the Red Cross[†]

At noon of an autumnal day, more than two centuries ago, the English colors were displayed by the standard-bearer of the Salem trainband, which

2. After Roger Williams was banished in 1635, he settled in Rhode Island. Hutchinson was sentenced in winter but allowed to postpone her departure until spring.
3. William Hutchinson died in 1642, whereupon Anne Hutchinson moved her family to New Amsterdam (New York).
4. Hutchinson and all but one of the six children who had accompanied her died in an Indian attack in 1643.
† From *Twice-Told Tales*, vol. 1 of the Riverside Press edition (Boston: Houghton, Mifflin, 1882), pp. 485–94.

had mustered for martial exercise under the orders of John Endicott.[1] It was a period when the religious exiles were accustomed often to buckle on their armor, and practise the handling of their weapons of war. Since the first settlement of New England, its prospects had never been so dismal. The dissensions between Charles the First and his subjects were then, and for several years afterwards, confined to the floor of Parliament.[2] The measures of the King and ministry were rendered more tyrannically violent by an opposition, which had not yet acquired sufficient confidence in its own strength to resist royal injustice with the sword. The bigoted and haughty primate, Laud, Archbishop of Canterbury, controlled the religious affairs of the realm, and was consequently invested with powers which might have wrought the utter ruin of the two Puritan colonies, Plymouth and Massachusetts.[3] There is evidence on record that our forefathers perceived their danger, but were resolved that their infant country should not fall without a struggle, even beneath the giant strength of the King's right arm.

Such was the aspect of the times when the folds of the English banner, with the Red Cross in its field, were flung out over a company of Puritans.[4] Their leader, the famous Endicott, was a man of stern and resolute countenance, the effect of which was heightened by a grizzled beard that swept the upper portion of his breastplate. This piece of armor was so highly polished that the whole surrounding scene had its image in the glittering steel. The central object in the mirrored picture was an edifice of humble architecture with neither steeple nor bell to proclaim it—what nevertheless it was—the house of prayer. A token of the perils of the wilderness was seen in the grim head of a wolf, which had just been slain within the precincts of the town, and according to the regular mode of claiming the bounty, was nailed on the porch of the meeting-house. The blood was still plashing on the doorstep. There happened to be visible, at the same noontide hour, so many other characteristics of the times and manners of the Puritans, that we must endeavor to represent them in a sketch, though far less vividly than they were reflected in the polished breastplate of John Endicott.

In close vicinity to the sacred edifice appeared that important engine of Puritanic authority, the whipping-post—with the soil around it well trodden by the feet of evil doers, who had there been disciplined. At one corner of the meeting-house was the pillory, and at the other the stocks; and, by a singular good fortune for our sketch, the head of an Episcopalian and suspected Catholic was grotesquely incased in the former machine; while a fellow-criminal, who had boisterously quaffed a health to the king, was confined by the legs in the latter. Side by side, on the meeting-house steps, stood a male and a female figure. The man was a tall, lean, haggard personification of fanaticism, bearing on his breast this label,—A WANTON GOSPELLER—which betokened that he had dared to give interpretations of Holy Writ unsanctioned by the infallible judgment of the civil and religious

1. The incident on which Hawthorne based this tale occurred in November 1634.
2. King Charles I of England had ascended to the throne in 1625 upon the death of his father, James I. As a result of the English Civil War (1642–49), he would be overthrown and beheaded. Puritanism arose in opposition to the Church of England, which Puritans accused of leaning toward Catholicism. Charles I tried to suppress Puritan and Presbyterian dissent within the Church.
3. Archbishop of Canterbury William Laud (1573–1645) headed the Church of England and, along with King Charles I, became the target of Puritan protest.
4. The English flag, with its red cross, symbolized the power of the Church of England.

rulers. His aspect showed no lack of zeal to maintain his heterodoxies, even at the stake. The woman wore a cleft stick on her tongue, in appropriate retribution for having wagged that unruly member against the elders of the church; and her countenance and gestures gave much cause to apprehend that, the moment the stick should be removed, a repetition of the offence would demand new ingenuity in chastising it.[5]

The above-mentioned individuals had been sentenced to undergo their various modes of ignominy, for the space of one hour at noonday. But among the crowd were several whose punishment would be life-long; some, whose ears had been cropped, like those of puppy dogs; others, whose cheeks had been branded with the initials of their misdemeanors; one, with his nostrils slit and seared; and another, with a halter about his neck, which he was forbidden ever to take off, or to conceal beneath his garments. Methinks he must have been grievously tempted to affix the other end of the rope to some convenient beam or bough. There was likewise a young woman, with no mean share of beauty, whose doom it was to wear the letter A on the breast of her gown, in the eyes of all the world and her own children. And even her own children knew what that initial signified. Sporting with her infamy, the lost and desperate creature had embroidered the fatal token in scarlet cloth, with golden thread and the nicest art of needlework; so that the capital A might have been thought to mean Admirable, or anything rather than Adulteress.[6]

Let not the reader argue, from any of these evidences of iniquity, that the times of the Puritans were more vicious than our own, when, as we pass along the very street of this sketch, we discern no badge of infamy on man or woman. It was the policy of our ancestors to search out even the most secret sins, and expose them to shame, without fear or favor, in the broadest light of the noonday sun. Were such the custom now, perchance we might find materials for a no less piquant sketch than the above.

Except the malefactors whom we have described, and the diseased or infirm persons, the whole male population of the town, between sixteen years and sixty, were seen in the ranks of the trainband. A few stately savages, in all the pomp and dignity of the primeval Indian, stood gazing at the spectacle. Their flint-headed arrows were but childish weapons compared with the matchlocks[7] of the Puritans, and would have rattled harmlessly against the steel caps and hammered iron breastplates which inclosed each soldier in an individual fortress. The valiant John Endicott glanced with an eye of pride at his sturdy followers, and prepared to renew the martial toils of the day.

"Come, my stout hearts!" quoth he, drawing his sword. "Let us show these poor heathen that we can handle our weapons like men of might. Well for them, if they put us not to prove it in earnest!"

5. The gruesome punishments that Hawthorne details in these two paragraphs have been gathered from various times and places in the seventeenth-century history of Massachusetts Bay Colony. In his *Annals of Salem*, for example, Joseph Felt mentions branding with letters, cropping ears, putting a cleft stick on the tongue, and placing individuals in the stocks and the pillory.
6. Hawthorne's "camera" lingers on this anonymous woman, whose situation provides many seeds that germinate and sprout in *The Scarlet Letter*—the woman's beauty, the letter itself, with its embroidery in gold thread and its subtle change in meaning at the woman's hand, as well as the negotiation of meaning between the woman and the people observing her.
7. Muskets fired by holding a match over a hole in the breech.

The iron-breasted company straightened their line, and each man drew the heavy butt of his matchlock close to his left foot, thus awaiting the orders of the captain. But, as Endicott glanced right and left along the front, he discovered a personage at some little distance with whom it behooved him to hold a parley. It was an elderly gentleman, wearing a black cloak and band, and a high-crowned hat, beneath which was a velvet skull-cap, the whole being the garb of a Puritan minister.[8] This reverend person bore a staff which seemed to have been recently cut in the forest, and his shoes were bemired as if he had been travelling on foot through the swamps of the wilderness. His aspect was perfectly that of a pilgrim, heightened also by an apostolic dignity. Just as Endicott perceived him he laid aside his staff, and stooped to drink at a bubbling fountain which gushed into the sunshine about a score of yards from the corner of the meeting-house. But, ere the good man drank, he turned his face heavenward in thankfulness, and then, holding back his gray beard with one hand, he scooped up his simple draught in the hollow of the other.

"What, ho! good Mr. Williams," shouted Endicott. "You are welcome back again to our town of peace. How does our worthy Governor Winthrop?[9] And what news from Boston?"

"The Governor hath his health, worshipful Sir," answered Roger Williams, now resuming his staff, and drawing near. "And for the news, here is a letter, which, knowing I was to travel hitherward to-day, his Excellency committed to my charge. Belike it contains tidings of much import; for a ship arrived yesterday from England."

Mr. Williams, the minister of Salem and of course known to all the spectators, had now reached the spot where Endicott was standing under the banner of his company, and put the Governor's epistle into his hand. The broad seal was impressed with Winthrop's coat of arms. Endicott hastily unclosed the letter and began to read, while, as his eye passed down the page, a wrathful change came over his manly countenance. The blood glowed through it, till it seemed to be kindling with an internal heat; nor was it unnatural to suppose that his breastplate would likewise become red-hot with the angry fire of the bosom which it covered. Arriving at the conclusion, he shook the letter fiercely in his hand, so that it rustled as loud as the flag above his head.

"Black tidings these, Mr. Williams," said he; "blacker never came to New England. Doubtless you know their purport?"

"Yea, truly," replied Roger Williams; "for the Governor consulted, respecting this matter, with my brethren in the ministry at Boston; and my opinion was likewise asked. And his Excellency entreats you by me, that the news be not suddenly noised abroad, lest the people be stirred up unto some outbreak, and thereby give the King and the Archbishop a handle against us."[1]

8. Roger Williams (1603–1683) became pastor at Salem in 1634, but he was hardly an "elderly gentleman," being only thirty at the time. Less than a year later, Williams was banished from the colony and, rather than be deported to England, fled to what is now Rhode Island, where he lived in close proximity to the Narragansett Indians.
9. Thomas Dudley, not John Winthrop, was governor of Massachusetts Bay Colony in November 1634. Winthrop's term had ended in May, and he was not reelected until 1637.
1. Actually, Williams was more radical than Endicott. He denied, for example, that King Charles had any right to grant title to Indian lands.

"The Governor is a wise man—a wise man, and a meek and moderate," said Endicott, setting his teeth grimly. "Nevertheless, I must do according to my own best judgment. There is neither man, woman, nor child in New England, but has a concern as dear as life in these tidings; and if John Endicott's voice be loud enough, man, woman, and child shall hear them. Soldiers, wheel into a hollow square! Ho, good people! Here are news for one and all of you."

The soldiers closed in around their captain; and he and Roger Williams stood together under the banner of the Red Cross; while the women and the aged men pressed forward, and the mothers held up their children to look Endicott in the face. A few taps of the drum gave signal for silence and attention.

"Fellow-soldiers,—fellow-exiles," began Endicott, speaking under strong excitement, yet powerfully restraining it, "wherefore did ye leave your native country? Wherefore, I say, have we left the green and fertile fields, the cottages, or, perchance, the old gray halls, where we were born and bred, the churchyards where our forefathers lie buried? Wherefore have we come hither to set up our own tombstones in a wilderness? A howling wilderness it is! The wolf and the bear meet us within halloo of our dwellings. The savage lieth in wait for us in the dismal shadow of the woods. The stubborn roots of the trees break our ploughshares, when we would till the earth. Our children cry for bread, and we must dig in the sands of the sea-shore to satisfy them. Wherefore, I say again, have we sought this country of a rugged soil and wintry sky? Was it not for the enjoyment of our civil rights? Was it not for liberty to worship God according to our conscience?"

"Call you this liberty of conscience?" interrupted a voice on the steps of the meeting-house.

It was the Wanton Gospeller. A sad and quiet smile flitted across the mild visage of Roger Williams. But Endicott, in the excitement of the moment, shook his sword wrathfully at the culprit—an ominous gesture from a man like him.

"What hast thou to do with conscience, thou knave?" cried he. "I said liberty to worship God, not license to profane and ridicule him. Break not in upon my speech, or I will lay thee neck and heels[2] till this time to-morrow! Hearken to me, friends, nor heed that accursed rhapsodist. As I was saying, we have sacrificed all things, and have come to a land whereof the old world hath scarcely heard, that we might make a new world unto ourselves, and painfully seek a path from hence to heaven. But what think ye now? This son of a Scotch tyrant—this grandson of a Papistical and adulterous Scotch woman, whose death proved that a golden crown doth not always save an anointed head from the block"—[3]

"Nay, brother, nay," interposed Mr. Williams; "thy words are not meet for a secret chamber, far less for a public street."

"Hold thy peace, Roger Williams!" answered Endicott, imperiously. "My spirit is wiser than thine for the business now in hand. I tell ye, fellow-exiles, that Charles of England, and Laud, our bitterest persecutor, arch-priest of

2. Place the man in the pillory, where his neck and feet would be secured.
3. King James I (of Scotland) was the father of King Charles I, who succeeded him in 1625. Mary (Stuart), a Catholic born in 1542, became queen of Scotland when only six days old. She was executed by order of Queen Elizabeth in 1587.

Canterbury, are resolute to pursue us even hither. They are taking counsel, saith this letter, to send over a governor-general, in whose breast shall be deposited all the law and equity of the land. They are minded, also, to establish the idolatrous forms of English Episcopacy; so that, when Laud shall kiss the Pope's toe, as cardinal of Rome, he may deliver New England, bound hand and foot, into the power of his master!"

A deep groan from the auditors,—a sound of wrath, as well as fear and sorrow,—responded to this intelligence.

"Look ye to it, brethren," resumed Endicott, with increasing energy. "If this king and this arch-prelate have their will, we shall briefly behold a cross on the spire of this tabernacle which we have builded, and a high altar within its walls, with wax tapers burning round it at noonday. We shall hear the sacring bell, and the voices of the Romish priests saying the mass. But think ye, Christian men, that these abominations may be suffered without a sword drawn? without a shot fired? without blood spilt, yea, on the very stairs of the pulpit? No,—be ye strong of hand and stout of heart! Here we stand on our own soil, which we have bought with our goods, which we have won with our swords, which we have cleared with our axes, which we have tilled with the sweat of our brows, which we have sanctified with our prayers to the God that brought us hither! Who shall enslave us here? What have we to do with this mitred prelate,—with this crowned king? What have we to do with England?"

Endicott gazed round at the excited countenances of the people, now full of his own spirit, and then turned suddenly to the standard-bearer, who stood close behind him.

"Officer, lower your banner!" said he.

The officer obeyed; and, brandishing his sword, Endicott thrust it through the cloth, and, with his left hand, rent the Red Cross completely out of the banner. He then waved the tattered ensign above his head.[4]

"Sacrilegious wretch!" cried the high-churchman in the pillory, unable longer to restrain himself, "thou hast rejected the symbol of our holy religion!"

"Treason, treason!" roared the royalist in the stocks. "He hath defaced the King's banner!"

"Before God and man, I will avouch the deed," answered Endicott. "Beat a flourish, drummer!—shout, soldiers and people!—in honor of the ensign of New England. Neither Pope nor Tyrant hath part in it now!"

With a cry of triumph, the people gave their sanction to one of the boldest exploits which our history records. And forever honored be the name of Endicott! We look back through the mist of ages, and recognize in the rending of the Red Cross from New England's banner the first omen of that deliverance which our fathers consummated after the bones of the stern Puritan had lain more than a century in the dust.[5]

4. Endicott defaced the English flag in November 1634. As John Winthrop notes, he was "admonished" for giving the English an excuse for thinking "ill" of the colonists and was "disabled" for one year from holding any public office. The court declined any heavier sentence on the grounds that Endicott acted from "tenderness of conscience" rather than any "evil intent," and he was restored to his position as magistrate at the next election, in 1636.

5. Hawthorne suggests that Endicott's actions look forward to the Revolutionary War.

Young Goodman Brown†

Young Goodman Brown[1] came forth at sunset into the street at Salem village; but put his head back, after crossing the threshold, to exchange a parting kiss with his young wife. And Faith, as the wife was aptly named, thrust her own pretty head into the street, letting the wind play with the pink ribbons of her cap while she called to Goodman Brown.

"Dearest heart," whispered she, softly and rather sadly, when her lips were close to his ear, "prithee[2] put off your journey until sunrise and sleep in your own bed to-night. A lone woman is troubled with such dreams and such thoughts that she's afeard of herself sometimes. Pray tarry with me this night, dear husband, of all nights in the year."

"My love and my Faith," replied young Goodman Brown, "of all nights in the year, this one night must I tarry away from thee. My journey, as thou callest it, forth and back again, must needs be done 'twixt now and sunrise. What, my sweet, pretty wife, dost thou doubt me already, and we but three months married?"

"Then God bless you!" said Faith, with the pink ribbons; "and may you find all well when you come back."

"Amen!" cried Goodman Brown. "Say thy prayers, dear Faith, and go to bed at dusk, and no harm will come to thee."

So they parted; and the young man pursued his way until, being about to turn the corner by the meeting-house, he looked back and saw the head of Faith still peeping after him with a melancholy air, in spite of her pink ribbons.

"Poor little Faith!" thought he, for his heart smote him. "What a wretch am I to leave her on such an errand! She talks of dreams, too. Methought as she spoke there was trouble in her face, as if a dream had warned her what work is to be done to-night. But no, no; 't would kill her to think it. Well, she's a blessed angel on earth; and after this one night I'll cling to her skirts and follow her to heaven."

With this excellent resolve for the future, Goodman Brown felt himself justified in making more haste on his present evil purpose. He had taken a dreary road, darkened by all the gloomiest trees of the forest, which barely stood aside to let the narrow path creep through, and closed immediately behind. It was all as lonely as could be; and there is this peculiarity in such a solitude, that the traveller knows not who may be concealed by the innumerable trunks and the thick boughs overhead; so that with lonely footsteps he may yet be passing through an unseen multitude.

"There may be a devilish Indian behind every tree," said Goodman Brown to himself; and he glanced fearfully behind him as he added, "What if the devil himself should be at my very elbow!"

His head being turned back, he passed a crook of the road, and, looking forward again, beheld the figure of a man, in grave and decent attire,

† From *Mosses from an Old Manse*, vol. 2 of the Riverside Press edition (Boston: Houghton, Mifflin, 1882), pp. 89–106.

1. Goodman, like Goody, is a common term for an individual, along the lines of "Mister."
2. The phrase "pray thee" often becomes "prithee."

seated at the foot of an old tree. He arose at Goodman Brown's approach and walked onward side by side with him.

"You are late, Goodman Brown," said he. "The clock of the Old South was striking as I came through Boston, and that is full fifteen minutes agone."[3]

"Faith kept me back a while," replied the young man, with a tremor in his voice, caused by the sudden appearance of his companion, though not wholly unexpected.

It was now deep dusk in the forest, and deepest in that part of it where these two were journeying. As nearly as could be discerned, the second traveller was about fifty years old, apparently in the same rank of life as Goodman Brown, and bearing a considerable resemblance to him, though perhaps more in expression than features. Still they might have been taken for father and son. And yet, though the elder person was as simply clad as the younger, and as simple in manner too, he had an indescribable air of one who knew the world, and who would not have felt abashed at the governor's dinner table or in King William's court, were it possible that his affairs should call him thither.[4] But the only thing about him that could be fixed upon as remarkable was his staff, which bore the likeness of a great black snake, so curiously wrought that it might almost be seen to twist and wriggle itself like a living serpent. This, of course, must have been an ocular deception, assisted by the uncertain light.

"Come, Goodman Brown," cried his fellow-traveller, "this is a dull pace for the beginning of a journey. Take my staff, if you are so soon weary."

"Friend," said the other, exchanging his slow pace for a full stop, "having kept covenant by meeting thee here, it is my purpose now to return whence I came. I have scruples touching the matter thou wot'st of."[5]

"Sayest thou so?" replied he of the serpent, smiling apart. "Let us walk on, nevertheless, reasoning as we go; and if I convince thee not thou shalt turn back. We are but a little way in the forest yet."

"Too far! too far!" exclaimed the goodman, unconsciously resuming his walk. "My father never went into the woods on such an errand, nor his father before him. We have been a race of honest men and good Christians since the days of the martyrs; and shall I be the first of the name of Brown that ever took this path and kept"—

"Such company, thou wouldst say," observed the elder person, interpreting his pause. "Well said, Goodman Brown! I have been as well acquainted with your family as with ever a one among the Puritans; and that's no trifle to say. I helped your grandfather, the constable, when he lashed the Quaker woman so smartly through the streets of Salem;[6] and it was I that brought your father a pitch-pine knot, kindled at my own hearth, to set fire to an Indian village, in King Philip's war.[7] They were

3. Old South Meeting House, built in 1729, was a gathering place for American revolutionaries—the place, for example, where the idea for the Boston Tea Party was publicly discussed.
4. King William III ruled England 1689–1702 and so was king during the Salem witch trials, in 1692.
5. An abbreviation of "wotest," "to know."
6. As Hawthorne also notes in the "Custom-House" introduction to *The Scarlet Letter*, his paternal great-great-grandfather, William Hathorne (1606–1681), had ordered a Quaker woman, Ann Coleman, to be whipped through the streets of Salem.
7. In his *Annals of Salem*, Joseph Felt notes that Hawthorne's great-great uncle, Captain William Hathorne (son of the William Hathorne mentioned in n. 6 above), and his company surprised four hundred Indians at Cocheco during King Philip's War. He captured two hundred Indians and sent them to Boston; seven or eight were sentenced to "immediate death, and the rest were transported and sold as slaves" (2: 507).

my good friends, both; and many a pleasant walk have we had along this path, and returned merrily after midnight. I would fain be friends with you for their sake."

"If it be as thou sayest," replied Goodman Brown, "I marvel they never spoke of these matters; or, verily, I marvel not, seeing that the least rumor of the sort would have driven them from New England. We are a people of prayer, and good works to boot, and abide no such wickedness."

"Wickedness or not," said the traveller with the twisted staff, "I have a very general acquaintance here in New England. The deacons of many a church have drunk the communion wine with me; the select-men of divers towns make me their chairman; and a majority of the Great and General Court are firm supporters of my interest. The governor and I, too—But these are state secrets."[8]

"Can this be so?" cried Goodman Brown, with a stare of amazement at his undisturbed companion. "Howbeit, I have nothing to do with the governor and council; they have their own ways, and are no rule for a simple husbandman like me. But, were I to go on with thee, how should I meet the eye of that good old man, our minister, at Salem village? Oh, his voice would make me tremble both Sabbath day and lecture day."

Thus far the elder traveller had listened with due gravity; but now burst into a fit of irrepressible mirth, shaking himself so violently that his snake-like staff actually seemed to wriggle in sympathy.

"Ha! ha! ha!" shouted he again and again; then composing himself, "Well, go on, Goodman Brown, go on; but, prithee, don't kill me with laughing."

"Well, then, to end the matter at once," said Goodman Brown, considerably nettled, "there is my wife, Faith. It would break her dear little heart; and I'd rather break my own."

"Nay, if that be the case," answered the other, "e'en go thy ways, Goodman Brown. I would not for twenty old women like the one hobbling before us that Faith should come to any harm."

As he spoke he pointed his staff at a female figure on the path, in whom Goodman Brown recognized a very pious and exemplary dame, who had taught him his catechism in youth, and was still his moral and spiritual adviser, jointly with the minister and Deacon Gookin.[9]

"A marvel, truly, that Goody Cloyse[1] should be so far in the wilderness at nightfall," said he. "But with your leave, friend, I shall take a cut through the woods until we have left this Christian woman behind. Being a stranger to you, she might ask whom I was consorting with and whither I was going."

"Be it so," said his fellow-traveller. "Betake you to the woods, and let me keep the path."

Accordingly the young man turned aside, but took care to watch his companion, who advanced softly along the road until he had come within a staff's length of the old dame. She, meanwhile, was making the best of her way, with singular speed for so aged a woman, and mumbling

8. The General Court was the ruling body of the Massachusetts Bay Colony. William Phips was appointed governor during the Salem witch trials, in 1692, and eventually put an end to the proceedings.
9. Daniel Gookin (1612–1687) served as superintendent of all Massachusetts Indians from 1656 until his death. He supported Apostle Eliot in his missionary work.
1. Sarah Cloyce was accused of witchcraft and imprisoned during the Salem witch trials. She was not executed.

some indistinct words—a prayer, doubtless—as she went. The traveller put forth his staff and touched her withered neck with what seemed the serpent's tail.

"The devil!" screamed the pious old lady.

"Then Goody Cloyse knows her old friend?" observed the traveller, confronting her and leaning on his writhing stick.

"Ah, forsooth, and is it your worship indeed?" cried the good dame. "Yea, truly is it, and in the very image of my old gossip, Goodman Brown, the grandfather of the silly fellow that now is. But—would your worship believe it?—my broomstick hath strangely disappeared, stolen, as I suspect, by that unhanged witch, Goody Cory,[2] and that, too, when I was all anointed with the juice of smallage, and cinquefoil, and wolf's bane"—[3]

"Mingled with fine wheat and the fat of a new-born babe," said the shape of old Goodman Brown.

"Ah, your worship knows the recipe," cried the old lady, cackling aloud. "So, as I was saying, being all ready for the meeting, and no horse to ride on, I made up my mind to foot it; for they tell me there is a nice young man to be taken into communion to-night. But now your good worship will lend me your arm, and we shall be there in a twinkling."

"That can hardly be," answered her friend. "I may not spare you my arm, Goody Cloyse; but here is my staff, if you will."

So saying, he threw it down at her feet, where, perhaps, it assumed life, being one of the rods which its owner had formerly lent to the Egyptian magi.[4] Of this fact, however, Goodman Brown could not take cognizance. He had cast up his eyes in astonishment, and, looking down again, beheld neither Goody Cloyse nor the serpentine staff, but his fellow-traveller alone, who waited for him as calmly as if nothing had happened.

"That old woman taught me my catechism," said the young man; and there was a world of meaning in this simple comment.

They continued to walk onward, while the elder traveller exhorted his companion to make good speed and persevere in the path, discoursing so aptly that his arguments seemed rather to spring up in the bosom of his auditor than to be suggested by himself. As they went, he plucked a branch of maple to serve for a walking stick, and began to strip it of the twigs and little boughs, which were wet with evening dew. The moment his fingers touched them they became strangely withered and dried up as with a week's sunshine. Thus the pair proceeded, at a good free pace, until suddenly, in a gloomy hollow of the road, Goodman Brown sat himself down on the stump of a tree and refused to go any farther.

"Friend," said he, stubbornly, "my mind is made up. Not another step will I budge on this errand. What if a wretched old woman do choose to

2. Deliberately or not, Hawthorne deviates from the historical record and switches the identities of two women. Sarah Cloyce is undoubtedly the "unhanged witch" to whom he refers. As noted above, she was arrested and imprisoned for witchcraft, but Governor Phips dissolved the special court in October 1692 before she could be executed. Martha Cory was hanged on September 22, 1692, part of the last group of accused witches to be executed. Her husband, Giles, was pressed to death on September 19 for refusing to stand trial.
3. Hawthorne lists ingredients for a witch's flying potion. *Smallage*, a form of wild celery. *Cinquefoil*, or potentilla: in the rose family. *Wolf's-bane*, or monkshood: a poisonous herb.
4. See Exodus 7:9–10: "When Pharaoh shall speak unto you, saying, Shew a miracle for you: then thou shalt say unto Aaron, Take thy rod, and cast it before Pharaoh, and it shall become a serpent. / And Moses and Aaron went in unto Pharaoh, and they did so as the LORD had commanded: and Aaron cast down his rod before Pharaoh, and before his servants, and it became a serpent."

go to the devil when I thought she was going to heaven: is that any reason why I should quit my dear Faith and go after her?"

"You will think better of this by and by," said his acquaintance, composedly. "Sit here and rest yourself a while; and when you feel like moving again, there is my staff to help you along."

Without more words, he threw his companion the maple stick, and was as speedily out of sight as if he had vanished into the deepening gloom. The young man sat a few moments by the roadside, applauding himself greatly, and thinking with how clear a conscience he should meet the minister in his morning walk, nor shrink from the eye of good old Deacon Gookin. And what calm sleep would be his that very night, which was to have been spent so wickedly, but so purely and sweetly now, in the arms of Faith! Amidst these pleasant and praiseworthy meditations, Goodman Brown heard the tramp of horses along the road, and deemed it advisable to conceal himself within the verge of the forest, conscious of the guilty purpose that had brought him thither, though now so happily turned from it.

On came the hoof tramps and the voices of the riders, two grave old voices, conversing soberly as they drew near. These mingled sounds appeared to pass along the road, within a few yards of the young man's hiding-place; but, owing doubtless to the depth of the gloom at that particular spot, neither the travellers nor their steeds were visible. Though their figures brushed the small boughs by the wayside, it could not be seen that they intercepted, even for a moment, the faint gleam from the strip of bright sky athwart which they must have passed. Goodman Brown alternately crouched and stood on tiptoe, pulling aside the branches and thrusting forth his head as far as he durst without discerning so much as a shadow. It vexed him the more, because he could have sworn, were such a thing possible, that he recognized the voices of the minister and Deacon Gookin, jogging along quietly, as they were wont to do, when bound to some ordination or ecclesiastical council. While yet within hearing, one of the riders stopped to pluck a switch.

"Of the two, reverend sir," said the voice like the deacon's, "I had rather miss an ordination dinner than to-night's meeting. They tell me that some of our community are to be here from Falmouth and beyond, and others from Connecticut and Rhode Island, besides several of the Indian powwows,[5] who, after their fashion, know almost as much deviltry as the best of us. Moreover, there is a goodly young woman to be taken into communion."

"Mighty well, Deacon Gookin!" replied the solemn old tones of the minister. "Spur up, or we shall be late. Nothing can be done, you know, until I get on the ground."

The hoofs clattered again; and the voices, talking so strangely in the empty air, passed on through the forest, where no church had ever been gathered or solitary Christian prayed. Whither, then, could these holy men be journeying so deep into the heathen wilderness? Young Goodman Brown caught hold of a tree for support, being ready to sink down on the ground, faint and overburdened with the heavy sickness of his heart. He looked up to the sky, doubting whether there really was a heaven above him. Yet there was the blue arch, and the stars brightening in it.

5. Medicine men.

"With heaven above and Faith below, I will yet stand firm against the devil!" cried Goodman Brown.

While he still gazed upward into the deep arch of the firmament and had lifted his hands to pray, a cloud, though no wind was stirring, hurried across the zenith and hid the brightening stars. The blue sky was still visible, except directly overhead, where this black mass of cloud was sweeping swiftly northward. Aloft in the air, as if from the depths of the cloud, came a confused and doubtful sound of voices. Once the listener fancied that he could distinguish the accents of towns-people of his own, men and women, both pious and ungodly, many of whom he had met at the communion table, and had seen others rioting at the tavern. The next moment, so indistinct were the sounds, he doubted whether he had heard aught but the murmur of the old forest, whispering without a wind. Then came a stronger swell of those familiar tones, heard daily in the sunshine at Salem village, but never until now from a cloud of night. There was one voice, of a young woman, uttering lamentations, yet with an uncertain sorrow, and entreating for some favor, which, perhaps, it would grieve her to obtain; and all the unseen multitude, both saints and sinners, seemed to encourage her onward.

"Faith!" shouted Goodman Brown, in a voice of agony and desperation; and the echoes of the forest mocked him, crying, "Faith! Faith!" as if bewildered wretches were seeking her all through the wilderness.

The cry of grief, rage, and terror was yet piercing the night, when the unhappy husband held his breath for a response. There was a scream, drowned immediately in a louder murmur of voices, fading into far-off laughter, as the dark cloud swept away, leaving the clear and silent sky above Goodman Brown. But something fluttered lightly down through the air and caught on the branch of a tree. The young man seized it, and beheld a pink ribbon.

"My Faith is gone!" cried he, after one stupefied moment. "There is no good on earth; and sin is but a name. Come, devil; for to thee is this world given."

And, maddened with despair, so that he laughed loud and long, did Goodman Brown grasp his staff and set forth again, at such a rate that he seemed to fly along the forest path rather than to walk or run. The road grew wilder and drearier and more faintly traced, and vanished at length, leaving him in the heart of the dark wilderness, still rushing onward with the instinct that guides mortal man to evil. The whole forest was peopled with frightful sounds—the creaking of the trees, the howling of wild beasts, and the yell of Indians; while sometimes the wind tolled like a distant church bell, and sometimes gave a broad roar around the traveller, as if all Nature were laughing him to scorn. But he was himself the chief horror of the scene, and shrank not from its other horrors.

"Ha! ha! ha!" roared Goodman Brown when the wind laughed at him. "Let us hear which will laugh loudest. Think not to frighten me with your deviltry. Come witch, come wizard, come Indian powwow, come devil himself, and here comes Goodman Brown. You may as well fear him as he fear you."

In truth, all through the haunted forest there could be nothing more frightful than the figure of Goodman Brown. On he flew among the black pines, brandishing his staff with frenzied gestures, now giving vent to an inspiration of horrid blasphemy, and now shouting forth such laughter as set all the echoes of the forest laughing like demons around him. The fiend in his own shape is less hideous than when he rages in the breast of man. Thus sped the demoniac on his course, until, quivering among the trees, he saw a red light before him, as when the felled trunks and branches of a clearing have been set on fire, and throw up their lurid blaze against the sky, at the hour of midnight. He paused, in a lull of the tempest that had driven him onward, and heard the swell of what seemed a hymn, rolling solemnly from a distance with the weight of many voices. He knew the tune; it was a familiar one in the choir of the village meeting-house. The verse died heavily away, and was lengthened by a chorus, not of human voices, but of all the sounds of the benighted wilderness pealing in awful harmony together. Goodman Brown cried out, and his cry was lost to his own ear by its unison with the cry of the desert.

In the interval of silence he stole forward until the light glared full upon his eyes. At one extremity of an open space, hemmed in by the dark wall of the forest, arose a rock, bearing some rude, natural resemblance either to an altar or a pulpit, and surrounded by four blazing pines, their tops aflame, their stems untouched, like candles at an evening meeting. The mass of foliage that had overgrown the summit of the rock was all on fire, blazing high into the night and fitfully illuminating the whole field. Each pendent twig and leafy festoon was in a blaze. As the red light arose and fell, a numerous congregation alternately shone forth, then disappeared in shadow, and again grew, as it were, out of the darkness, peopling the heart of the solitary woods at once.

"A grave and dark-clad company," quoth Goodman Brown.

In truth they were such. Among them, quivering to and fro between gloom and splendor, appeared faces that would be seen next day at the council board of the province, and others which, Sabbath after Sabbath, looked devoutly heavenward, and benignantly over the crowded pews, from the holiest pulpits in the land. Some affirm that the lady of the governor was there.[6] At least there were high dames well known to her, and wives of honored husbands, and widows, a great multitude, and ancient maidens, all of excellent repute, and fair young girls, who trembled lest their mothers should espy them. Either the sudden gleams of light flashing over the obscure field bedazzled Goodman Brown, or he recognized a score of the church members of Salem village famous for their especial sanctity. Good old Deacon Gookin had arrived, and waited at the skirts of that venerable saint, his revered pastor. But, irreverently consorting with these grave, reputable, and pious people, these elders of the church, these chaste dames and dewy virgins, there were men of dissolute lives and women of spotted fame, wretches given over to all mean and filthy vice, and suspected even of horrid crimes. It was strange to see that the good shrank not from the wicked, nor were the sinners abashed by the saints. Scattered also among their pale-faced enemies were the Indian priests, or powwows, who had often scared their native forest with more hideous incantations than any known to English witchcraft.

6. Mary Phips, wife of Governor William Phips, was repeatedly accused of witchcraft. See Robert Calef's *More Wonders of the Invisible World* and Hawthorne's *Grandfather's Chair*.

"But where is Faith?" thought Goodman Brown; and, as hope came into his heart, he trembled.

Another verse of the hymn arose, a slow and mournful strain, such as the pious love, but joined to words which expressed all that our nature can conceive of sin, and darkly hinted at far more. Unfathomable to mere mortals is the lore of fiends. Verse after verse was sung; and still the chorus of the desert swelled between like the deepest tone of a mighty organ; and with the final peal of that dreadful anthem there came a sound, as if the roaring wind, the rushing streams, the howling beasts, and every other voice of the unconcerted wilderness were mingling and according with the voice of guilty man in homage to the prince of all. The four blazing pines threw up a loftier flame, and obscurely discovered shapes and visages of horror on the smoke wreaths above the impious assembly. At the same moment the fire on the rock shot redly forth and formed a glowing arch above its base, where now appeared a figure. With reverence be it spoken, the figure bore no slight similitude, both in garb and manner, to some grave divine of the New England churches.

"Bring forth the converts!" cried a voice that echoed through the field and rolled into the forest.

At the word, Goodman Brown stepped forth from the shadow of the trees and approached the congregation, with whom he felt a loathful brotherhood by the sympathy of all that was wicked in his heart. He could have well-nigh sworn that the shape of his own dead father beckoned him to advance, looking downward from a smoke wreath, while a woman, with dim features of despair, threw out her hand to warn him back.[7] Was it his mother? But he had no power to retreat one step, nor to resist, even in thought, when the minister and good old Deacon Gookin seized his arms and led him to the blazing rock. Thither came also the slender form of a veiled female, led between Goody Cloyse, that pious teacher of the catechism, and Martha Carrier, who had received the devil's promise to be queen of hell. A rampant hag was she.[8] And there stood the proselytes beneath the canopy of fire.

"Welcome, my children," said the dark figure, "to the communion of your race. Ye have found thus young your nature and your destiny. My children, look behind you!"

They turned; and flashing forth, as it were, in a sheet of flame, the fiend worshippers were seen; the smile of welcome gleamed darkly on every visage.

"There," resumed the sable form, "are all whom ye have reverenced from youth. Ye deemed them holier than yourselves, and shrank from your own

7. The cases against the witches depended on "specter evidence"—claims that the shapes or specters of the witches had tormented the accusers by pinching, biting, and tempting them. A key question was whether the devil could take over the "shape" of and impersonate an innocent person without that person's consent. Eventually, as doubts about the trials spread, a consensus developed that the devil could in fact impersonate an innocent person, casting into doubt nearly all the convictions. As Governor Phips explained in a February 21, 1693, letter to the Earl of Nottingham, "Mr. Increase Mathew [Mather] and several other Divines did give it as their Judgment that the Devil might afflict in the shape of an innocent person and that the look and touch of the suspected persons was not sufficient proof against them" (Boyer and Nissenbaum, 121).

8. Martha Carrier was hanged for witchcraft on August 19, 1692. In his *Wonders of the Invisible World* (1693), Cotton Mather refers to her as a "Rampant Hag" whom the Devil had promised should be "Queen of Hell."

sin, contrasting it with their lives of righteousness and prayerful aspirations heavenward. Yet here are they all in my worshipping assembly. This night it shall be granted you to know their secret deeds: how hoary-bearded elders of the church have whispered wanton words to the young maids of their households; how many a woman, eager for widows' weeds, has given her husband a drink at bedtime and let him sleep his last sleep in her bosom; how beardless youths have made haste to inherit their fathers' wealth; and how fair damsels—blush not, sweet ones—have dug little graves in the garden, and bidden me, the sole guest, to an infant's funeral. By the sympathy of your human hearts for sin ye shall scent out all the places—whether in church, bed-chamber, street, field, or forest—where crime has been committed, and shall exult to behold the whole earth one stain of guilt, one mighty blood spot. Far more than this. It shall be yours to penetrate, in every bosom, the deep mystery of sin, the fountain of all wicked arts, and which inexhaustibly supplies more evil impulses than human power—than my power at its utmost—can make manifest in deeds. And now, my children, look upon each other."

They did so; and, by the blaze of the hell-kindled torches, the wretched man beheld his Faith, and the wife her husband, trembling before that unhallowed altar.

"Lo, there ye stand, my children," said the figure, in a deep and solemn tone, almost sad with its despairing awfulness, as if his once angelic nature could yet mourn for our miserable race. "Depending upon one another's hearts, ye had still hoped that virtue were not all a dream.[9] Now are ye undeceived. Evil is the nature of mankind. Evil must be your only happiness. Welcome again, my children, to the communion of your race."

"Welcome," repeated the fiend worshippers, in one cry of despair and triumph.

And there they stood, the only pair, as it seemed, who were yet hesitating on the verge of wickedness in this dark world. A basin was hollowed, naturally, in the rock. Did it contain water, reddened by the lurid light? or was it blood? or, perchance, a liquid flame? Herein did the shape of evil dip his hand and prepare to lay the mark of baptism upon their foreheads, that they might be partakers of the mystery of sin, more conscious of the secret guilt of others, both in deed and thought, than they could now be of their own. The husband cast one look at his pale wife, and Faith at him. What polluted wretches would the next glance show them to each other, shuddering alike at what they disclosed and what they saw!

"Faith! Faith!" cried the husband, "look up to heaven, and resist the wicked one."

Whether Faith obeyed he knew not. Hardly had he spoken when he found himself amid calm night and solitude, listening to a roar of the wind which died heavily away through the forest. He staggered against the rock, and felt it chill and damp; while a hanging twig, that had been all on fire, besprinkled his cheek with the coldest dew.

9. When John Hathorne questioned Rebecca Nurse, who was being tried for witchcraft, he reproached her for her lack of emotion: "It is very awful for all to see these agonies, and you, an old professor [of witchcraft], thus charged with contracting with the devil by the effects of it, and yet to see you stand with dry eyes when there are so many wet." Her reply: "You do not know my heart" (see Boyer and Nissenbaum, 24).

The next morning young Goodman Brown came slowly into the street of Salem village, staring around him like a bewildered man. The good old minister was taking a walk along the graveyard to get an appetite for breakfast and meditate his sermon, and bestowed a blessing, as he passed, on Goodman Brown. He shrank from the venerable saint as if to avoid an anathema. Old Deacon Gookin was at domestic worship, and the holy words of his prayer were heard through the open window. "What God doth the wizard pray to?" quoth Goodman Brown. Goody Cloyse, that excellent old Christian, stood in the early sunshine at her own lattice, catechizing a little girl who had brought her a pint of morning's milk. Goodman Brown snatched away the child as from the grasp of the fiend himself. Turning the corner by the meeting-house, he spied the head of Faith, with the pink ribbons, gazing anxiously forth, and bursting into such joy at sight of him that she skipped along the street and almost kissed her husband before the whole village. But Goodman Brown looked sternly and sadly into her face, and passed on without a greeting.

Had Goodman Brown fallen asleep in the forest and only dreamed a wild dream of a witch-meeting?

Be it so if you will; but, alas! it was a dream of evil omen for young Goodman Brown. A stern, a sad, a darkly meditative, a distrustful, if not a desperate man did he become from the night of that fearful dream. On the Sabbath day, when the congregation were singing a holy psalm, he could not listen because an anthem of sin rushed loudly upon his ear and drowned all the blessed strain. When the minister spoke from the pulpit with power and fervid eloquence, and, with his hand on the open Bible, of the sacred truths of our religion, and of saint-like lives and triumphant deaths, and of future bliss or misery unutterable, then did Goodman Brown turn pale, dreading lest the roof should thunder down upon the gray blasphemer and his hearers. Often, awaking suddenly at midnight, he shrank from the bosom of Faith; and at morning or eventide, when the family knelt down at prayer, he scowled and muttered to himself, and gazed sternly at his wife, and turned away. And when he had lived long, and was borne to his grave a hoary corpse, followed by Faith, an aged woman, and children and grandchildren, a goodly procession, besides neighbors not a few, they carved no hopeful verse upon his tombstone, for his dying hour was gloom.

The Minister's Black Veil†

*A Parable.**

The sexton stood in the porch of Milford[1] meeting-house, pulling busily at the bell-rope. The old people of the village came stooping along the street. Children, with bright faces, tripped merrily beside their parents,

† From *Twice-Told Tales*, vol. 1 of the Riverside Press edition (Boston: Houghton, Mifflin, 1882), pp. 52–69.

* Another clergyman in New England, Mr. Joseph Moody, of York, Maine, who died about eighty years since, made himself remarkable by the same eccentricity that is here related of the Reverend Mr. Hooper. In his case, however, the symbol had a different import. In early life he had accidentally killed a beloved friend; and from that day till the hour of his own death, he hid his face from men. [*Hawthorne's note*]

1. Milford, Massachusetts, about thirty miles west of Boston.

or mimicked a graver gait, in the conscious dignity of their Sunday clothes. Spruce bachelors looked sidelong at the pretty maidens, and fancied that the Sabbath sunshine made them prettier than on week days. When the throng had mostly streamed into the porch, the sexton began to toll the bell, keeping his eye on the Reverend Mr. Hooper's door. The first glimpse of the clergyman's figure was the signal for the bell to cease its summons.

"But what has good Parson Hooper got upon his face?" cried the sexton in astonishment.

All within hearing immediately turned about, and beheld the semblance of Mr. Hooper, pacing slowly his meditative way towards the meeting-house. With one accord they started, expressing more wonder than if some strange minister were coming to dust the cushions of Mr. Hooper's pulpit.

"Are you sure it is our parson?" inquired Goodman Gray of the sexton.

"Of a certainty it is good Mr. Hooper," replied the sexton. "He was to have exchanged pulpits with Parson Shute, of Westbury; but Parson Shute sent to excuse himself yesterday, being to preach a funeral sermon."

The cause of so much amazement may appear sufficiently slight. Mr. Hooper, a gentlemanly person, of about thirty, though still a bachelor, was dressed with due clerical neatness, as if a careful wife had starched his band, and brushed the weekly dust from his Sunday's garb. There was but one thing remarkable in his appearance. Swathed about his forehead, and hanging down over his face, so low as to be shaken by his breath, Mr. Hooper had on a black veil. On a nearer view it seemed to consist of two folds of crape, which entirely concealed his features, except the mouth and chin, but probably did not intercept his sight, further than to give a darkened aspect to all living and inanimate things. With this gloomy shade before him, good Mr. Hooper walked onward, at a slow and quiet pace, stooping somewhat, and looking on the ground, as is customary with abstracted men, yet nodding kindly to those of his parishioners who still waited on the meeting-house steps. But so wonder-struck were they that his greeting hardly met with a return.

"I can't really feel as if good Mr. Hooper's face was behind that piece of crape," said the sexton.

"I don't like it," muttered an old woman, as she hobbled into the meeting-house. "He has changed himself into something awful, only by hiding his face."

"Our parson has gone mad!" cried Goodman Gray, following him across the threshold.

A rumor of some unaccountable phenomenon had preceded Mr. Hooper into the meeting-house, and set all the congregation astir. Few could refrain from twisting their heads towards the door; many stood upright, and turned directly about; while several little boys clambered upon the seats, and came down again with a terrible racket. There was a general bustle; a rustling of the women's gowns and shuffling of the men's feet, greatly at variance with that hushed repose which should attend the entrance of the minister. But Mr. Hooper appeared not to notice the perturbation of his people. He entered with an almost noiseless step, bent his head mildly to the pews on each side, and bowed as he passed his oldest parishioner, a white-haired great-grandsire, who occupied an arm-chair in the centre of the aisle. It was strange to observe how slowly this venerable man became conscious of something singular in the appearance of his pastor. He seemed not fully to partake of the prevailing wonder, till Mr. Hooper had ascended the stairs, and showed himself in the pulpit, face to face with his congregation, except for the black veil. That mysterious emblem was never once withdrawn. It shook with his measured breath, as he gave out the psalm; it threw its obscurity between him and the holy page, as he read the Scriptures; and while he prayed, the veil lay heavily on his uplifted countenance. Did he seek to hide it from the dread Being whom he was addressing?

Such was the effect of this simple piece of crape, that more than one woman of delicate nerves was forced to leave the meeting-house. Yet perhaps the pale-faced congregation was almost as fearful a sight to the minister, as his black veil to them.

Mr. Hooper had the reputation of a good preacher, but not an energetic one: he strove to win his people heavenward by mild, persuasive influences, rather than to drive them thither by the thunders of the Word. The sermon which he now delivered was marked by the same characteristics of style and manner as the general series of his pulpit oratory. But there was something, either in the sentiment of the discourse itself, or in the imagination of the auditors, which made it greatly the most powerful effort that they had ever heard from their pastor's lips. It was tinged, rather more darkly than usual, with the gentle gloom of Mr. Hooper's temperament. The subject had reference to secret sin, and those sad mysteries which we hide from our nearest and dearest, and would fain conceal from our own consciousness, even forgetting that the Omniscient can detect them. A subtle power was breathed into his words. Each member of the congregation, the most innocent girl, and the man of hardened breast, felt as if the preacher had crept upon them, behind his awful veil, and discovered their hoarded iniquity of deed or thought. Many spread their clasped hands on their bosoms. There was nothing terrible in what Mr. Hooper said, at least, no violence; and yet, with every tremor of his melancholy voice, the hearers quaked. An unsought pathos came hand in hand with awe. So sensible were the audience of some unwonted attribute in their minister, that they longed for a breath of wind to blow aside the veil, almost believing that a stranger's visage would be discovered, though the form, gesture, and voice were those of Mr. Hooper.

At the close of the services, the people hurried out with indecorous confusion, eager to communicate their pent-up amazement, and conscious of lighter spirits the moment they lost sight of the black veil. Some gathered in little circles, huddled closely together, with their mouths all whispering in the centre; some went homeward alone, wrapt in silent meditation; some talked loudly, and profaned the Sabbath day with ostentatious laughter. A few shook their sagacious heads, intimating that they could penetrate the mystery; while one or two affirmed that there was no mystery at all, but only that Mr. Hooper's eyes were so weakened by the midnight lamp, as to require a shade. After a brief interval, forth came good Mr. Hooper also, in the rear of his flock. Turning his veiled face from one group to another, he paid due reverence to the hoary heads, saluted the middle aged with kind dignity as their friend and spiritual guide, greeted the young with mingled authority and love, and laid his hands on the little children's heads to bless them. Such was always his custom on the Sabbath day. Strange and bewildered looks repaid him for his courtesy. None, as on former occasions, aspired to the honor of walking by their pastor's side. Old Squire Saunders,[2] doubtless by an accidental lapse of memory, neglected to invite Mr. Hooper to his table, where the good clergyman had been wont to bless the food, almost every Sunday since his settlement. He returned, therefore, to the parsonage, and, at the moment of closing the door, was observed to look back upon the people, all of whom had their eyes fixed upon the minister. A sad smile gleamed faintly from beneath the black veil, and flickered about his mouth, glimmering as he disappeared.

"How strange," said a lady, "that a simple black veil, such as any woman might wear on her bonnet, should become such a terrible thing on Mr. Hooper's face!"

"Something must surely be amiss with Mr. Hooper's intellects," observed her husband, the physician of the village. "But the strangest part of the affair is the effect of this vagary, even on a sober-minded man like myself. The black veil, though it covers only our pastor's face, throws its influence over his whole person, and makes him ghostlike from head to foot. Do you not feel it so?"

"Truly do I," replied the lady; "and I would not be alone with him for the world. I wonder he is not afraid to be alone with himself!"

"Men sometimes are so," said her husband.

The afternoon service was attended with similar circumstances. At its conclusion, the bell tolled for the funeral of a young lady. The relatives and friends were assembled in the house, and the more distant acquaintances stood about the door, speaking of the good qualities of the deceased, when their talk was interrupted by the appearance of Mr. Hooper, still covered with his black veil. It was now an appropriate emblem. The clergyman stepped into the room where the corpse was laid, and bent over the coffin, to take a last farewell of his deceased parishioner. As he stooped, the veil hung straight down from his forehead, so that, if her eyelids had not been closed forever, the dead maiden might have seen his face. Could Mr. Hooper be fearful of her glance, that he so hastily caught back the black veil? A person who watched the interview between the dead and living, scrupled not to affirm, that, at the instant when the clergyman's features were disclosed, the corpse had slightly shuddered, rustling the shroud and muslin cap, though the countenance retained the composure of death.[3] A superstitious old woman was the only witness of this prodigy. From the coffin Mr. Hooper passed into the chamber of the mourners, and thence to the head of the staircase, to make the funeral prayer. It was a tender and heart-dissolving prayer, full of sorrow, yet so imbued with celestial hopes, that the music of a heavenly harp, swept by the fingers of the dead, seemed

2. Michael J. Colacurcio links Old Squire Saunders to "Poor Richard" Saunders in Benjamin Franklin's *Almanac* and *Way to Wealth* (*The Province of Piety*, 351–53).
3. Hawthorne refers to the folk superstition that a corpse will move or sometimes bleed from the nose in the presence of the murderer. Edgar Allan Poe apparently caught the hint because he claimed that the "*true* import of the narrative" involves a "crime of dark dye, (having reference to the 'young lady')." See review of *Twice-Told Tales*, *Graham's Magazine* 20 (May 1842): 298–300. Although most critics interpret the reference to the "young lady" as a hint that the minister may have murdered the woman whose funeral he presided over, several critics note that the "young Lady" could also be Elizabeth. Thanks to Poe scholar John Gruesser for pointing out the latter possibility.

faintly to be heard among the saddest accents of the minister. The people trembled, though they but darkly understood him when he prayed that they, and himself, and all of mortal race, might be ready, as he trusted this young maiden had been, for the dreadful hour that should snatch the veil from their faces. The bearers went heavily forth, and the mourners followed, saddening all the street, with the dead before them, and Mr. Hooper in his black veil behind.

"Why do you look back?" said one in the procession to his partner.

"I had a fancy," replied she, "that the minister and the maiden's spirit were walking hand in hand."

"And so had I, at the same moment," said the other.

That night, the handsomest couple in Milford village were to be joined in wedlock. Though reckoned a melancholy man, Mr. Hooper had a placid cheerfulness for such occasions, which often excited a sympathetic smile where livelier merriment would have been thrown away. There was no quality of his disposition which made him more beloved than this. The company at the wedding awaited his arrival with impatience, trusting that the strange awe, which had gathered over him throughout the day, would now be dispelled. But such was not the result. When Mr. Hooper came, the first thing that their eyes rested on was the same horrible black veil, which had added deeper gloom to the funeral, and could portend nothing but evil to the wedding. Such was its immediate effect on the guests that a cloud seemed to have rolled duskily from beneath the black crape, and dimmed the light of the candles. The bridal pair stood up before the minister. But the bride's cold fingers quivered in the tremulous hand of the bridegroom, and her deathlike paleness caused a whisper that the maiden who had been buried a few hours before was come from her grave to be married. If ever another wedding were so dismal, it was that famous one where they tolled the wedding knell.[4] After performing the ceremony, Mr. Hooper raised a glass of wine to his lips, wishing happiness to the new-married couple in a strain of mild pleasantry that ought to have brightened the features of the guests, like a cheerful gleam from the hearth. At that instant, catching a glimpse of his figure in the looking-glass, the black veil involved his own spirit in the horror with which it overwhelmed all others. His frame shuddered, his lips grew white, he spilt the untasted wine upon the carpet, and rushed forth into the darkness. For the Earth, too, had on her Black Veil.

The next day, the whole village of Milford talked of little else than Parson Hooper's black veil. That, and the mystery concealed behind it, supplied a topic for discussion between acquaintances meeting in the street, and good women gossiping at their open windows. It was the first item of news that the tavern-keeper told to his guests. The children babbled of it on their way to school. One imitative little imp covered his face with an old black handkerchief, thereby so affrighting his playmates that the panic seized himself, and he well-nigh lost his wits by his own waggery.

It was remarkable that of all the busybodies and impertinent people in the parish, not one ventured to put the plain question to Mr. Hooper, wherefore he did this thing. Hitherto, whenever there appeared the

4. Hawthorne refers to his own tale "The Wedding Knell" (1836), in which a funeral bell begins to toll as the bride enters the church.

slightest call for such interference, he had never lacked advisers, nor shown himself averse to be guided by their judgment. If he erred at all, it was by so painful a degree of self-distrust, that even the mildest censure would lead him to consider an indifferent action as a crime. Yet, though so well acquainted with this amiable weakness, no individual among his parishioners chose to make the black veil a subject of friendly remonstrance. There was a feeling of dread, neither plainly confessed nor carefully concealed, which caused each to shift the responsibility upon another, till at length it was found expedient to send a deputation of the church, in order to deal with Mr. Hooper about the mystery, before it should grow into a scandal. Never did an embassy so ill discharge its duties. The minister received them with friendly courtesy, but became silent, after they were seated, leaving to his visitors the whole burden of introducing their important business. The topic, it might be supposed, was obvious enough. There was the black veil swathed round Mr. Hooper's forehead, and concealing every feature above his placid mouth, on which, at times, they could perceive the glimmering of a melancholy smile. But that piece of crape, to their imagination, seemed to hang down before his heart, the symbol of a fearful secret between him and them. Were the veil but cast aside, they might speak freely of it, but not till then. Thus they sat a considerable time, speechless, confused, and shrinking uneasily from Mr. Hooper's eye, which they felt to be fixed upon them with an invisible glance. Finally, the deputies returned abashed to their constituents, pronouncing the matter too weighty to be handled, except by a council of the churches, if, indeed, it might not require a general synod.[5]

But there was one person in the village unappalled by the awe with which the black veil had impressed all beside herself. When the deputies returned without an explanation, or even venturing to demand one, she, with the calm energy of her character, determined to chase away the strange cloud that appeared to be settling round Mr. Hooper, every moment more darkly than before. As his plighted wife, it should be her privilege to know what the black veil concealed. At the minister's first visit, therefore, she entered upon the subject with a direct simplicity, which made the task easier both for him and her. After he had seated himself, she fixed her eyes steadfastly upon the veil, but could discern nothing of the dreadful gloom that had so overawed the multitude: it was but a double fold of crape, hanging down from his forehead to his mouth, and slightly stirring with his breath.

"No," said she aloud, and smiling, "there is nothing terrible in this piece of crape, except that it hides a face which I am always glad to look upon. Come, good sir, let the sun shine from behind the cloud. First lay aside your black veil: then tell me why you put it on."

Mr. Hooper's smile glimmered faintly.

"There is an hour to come," said he, "when all of us shall cast aside our veils. Take it not amiss, beloved friend, if I wear this piece of crape till then."[6]

5. A church court, as in the case of Anne Hutchinson, when a synod was convened to examine and judge her.
6. Judgment Day.

"Your words are a mystery, too," returned the young lady. "Take away the veil from them, at least."

"Elizabeth, I will," said he, "so far as my vow may suffer me. Know, then, this veil is a type and a symbol, and I am bound to wear it ever, both in light and darkness, in solitude and before the gaze of multitudes, and as with my familiar friends. No mortal eye will see it withdrawn. This dismal shade must separate me from the world: even you, Elizabeth, can never come behind it!"

"What grievous affliction hath befallen you," she earnestly inquired, "that you should thus darken your eyes forever?"

"If it be a sign of mourning," replied Mr. Hooper, "I, perhaps, like most other mortals, have sorrows dark enough to be typified by a black veil."

"But what if the world will not believe that it is the type of an innocent sorrow?" urged Elizabeth. "Beloved and respected as you are, there may be whispers that you hide your face under the consciousness of secret sin. For the sake of your holy office, do away this scandal!"

The color rose into her cheeks as she intimated the nature of the rumors that were already abroad in the village. But Mr. Hooper's mildness did not forsake him. He even smiled again—that same sad smile, which always appeared like a faint glimmering of light, proceeding from the obscurity beneath the veil.

"If I hide my face for sorrow, there is cause enough," he merely replied; "and if I cover it for secret sin, what mortal might not do the same?"

And with this gentle, but unconquerable obstinacy did he resist all her entreaties. At length Elizabeth sat silent. For a few moments she appeared lost in thought, considering, probably, what new methods might be tried to withdraw her lover from so dark a fantasy, which, if it had no other meaning, was perhaps a symptom of mental disease. Though of a firmer character than his own, the tears rolled down her cheeks. But, in an instant, as it were, a new feeling took the place of sorrow: her eyes were fixed insensibly on the black veil, when, like a sudden twilight in the air, its terrors fell around her. She arose, and stood trembling before him.

"And do you feel it then, at last?" said he mournfully.

She made no reply, but covered her eyes with her hand, and turned to leave the room. He rushed forward and caught her arm.

"Have patience with me, Elizabeth!" cried he, passionately. "Do not desert me, though this veil must be between us here on earth. Be mine, and hereafter there shall be no veil over my face, no darkness between our souls! It is but a mortal veil—it is not for eternity! O! you know not how lonely I am, and how frightened, to be alone behind my black veil. Do not leave me in this miserable obscurity forever!"

"Lift the veil but once, and look me in the face," said she.

"Never! It cannot be!" replied Mr. Hooper.

"Then farewell!" said Elizabeth.

She withdrew her arm from his grasp, and slowly departed, pausing at the door, to give one long shuddering gaze, that seemed almost to penetrate the mystery of the black veil. But, even amid his grief, Mr. Hooper smiled to think that only a material emblem had separated him from happiness, though the horrors, which it shadowed forth, must be drawn darkly between the fondest of lovers.

From that time no attempts were made to remove Mr. Hooper's black veil, or, by a direct appeal, to discover the secret which it was supposed to hide. By persons who claimed a superiority to popular prejudice, it was reckoned merely an eccentric whim, such as often mingles with the sober actions of men otherwise rational, and tinges them all with its own semblance of insanity. But with the multitude, good Mr. Hooper was irreparably a bugbear.[7] He could not walk the street with any peace of mind, so conscious was he that the gentle and timid would turn aside to avoid him, and that others would make it a point of hardihood to throw themselves in his way. The impertinence of the latter class compelled him to give up his customary walk at sunset to the burial ground; for when he leaned pensively over the gate, there would always be faces behind the gravestones, peeping at his black veil. A fable went the rounds that the stare of the dead people drove him thence. It grieved him, to the very depth of his kind heart, to observe how the children fled from his approach, breaking up their merriest sports, while his melancholy figure was yet afar off. Their instinctive dread caused him to feel more strongly than aught else, that a preternatural horror was interwoven with the threads of the black crape. In truth, his own antipathy to the veil was known to be so great, that he never willingly passed before a mirror, nor stooped to drink at a still fountain, lest, in its peaceful bosom, he should be affrighted by himself. This was what gave plausibility to the whispers, that Mr. Hooper's conscience tortured him for some great crime too horrible to be entirely concealed, or otherwise than so obscurely intimated. Thus, from beneath the black veil, there rolled a cloud into the sunshine, an ambiguity of sin or sorrow, which enveloped the poor minister, so that love or sympathy could never reach him. It was said that ghost and fiend consorted with him there. With self-shudderings and outward terrors, he walked continually in its shadow, groping darkly within his own soul, or gazing through a medium that saddened the whole world. Even the lawless wind, it was believed, respected his dreadful secret, and never blew aside the veil. But still good Mr. Hooper sadly smiled at the pale visages of the worldly throng as he passed by.

Among all its bad influences, the black veil had the one desirable effect, of making its wearer a very efficient clergyman. By the aid of his mysterious emblem—for there was no other apparent cause—he became a man of awful power over souls that were in agony for sin. His converts always regarded him with a dread peculiar to themselves, affirming, though but figuratively, that, before he brought them to celestial light, they had been with him behind the black veil. Its gloom, indeed, enabled him to sympathize with all dark affections. Dying sinners cried aloud for Mr. Hooper, and would not yield their breath till he appeared; though ever, as he stooped to whisper consolation, they shuddered at the veiled face so near their own. Such were the terrors of the black veil, even when Death had bared his visage! Strangers came long distances to attend service at his church, with the mere idle purpose of gazing at his figure, because it was forbidden them to behold his face. But many were made to quake ere they departed! Once, during Governor Belcher's administration, Mr. Hooper

7. A goblin or specter.

was appointed to preach the election sermon.[8] Covered with his black veil, he stood before the chief magistrate, the council, and the representatives, and wrought so deep an impression, that the legislative measures of that year were characterized by all the gloom and piety of our earliest ancestral sway.

In this manner Mr. Hooper spent a long life, irreproachable in outward act, yet shrouded in dismal suspicions; kind and loving, though unloved, and dimly feared; a man apart from men, shunned in their health and joy, but ever summoned to their aid in mortal anguish. As years wore on, shedding their snows above his sable veil, he acquired a name throughout the New England churches, and they called him Father Hooper. Nearly all his parishioners, who were of mature age when he was settled, had been borne away by many a funeral: he had one congregation in the church, and a more crowded one in the churchyard; and having wrought so late into the evening, and done his work so well, it was now good Father Hooper's turn to rest.

Several persons were visible by the shaded candlelight, in the death chamber of the old clergyman. Natural connections he had none. But there was the decorously grave, though unmoved physician, seeking only to mitigate the last pangs of the patient whom he could not save. There were the deacons, and other eminently pious members of his church. There, also, was the Reverend Mr. Clark, of Westbury, a young and zealous divine, who had ridden in haste to pray by the bedside of the expiring minister.[9] There was the nurse, no hired handmaiden of death, but one whose calm affection had endured thus long in secrecy, in solitude, amid the chill of age, and would not perish, even at the dying hour. Who, but Elizabeth! And there lay the hoary head of good Father Hooper upon the death pillow, with the black veil still swathed about his brow, and reaching down over his face, so that each more difficult gasp of his faint breath caused it to stir. All through life that piece of crape had hung between him and the world: it had separated him from cheerful brotherhood and woman's love, and kept him in that saddest of all prisons, his own heart; and still it lay upon his face, as if to deepen the gloom of his darksome chamber, and shade him from the sunshine of eternity.

For some time previous, his mind had been confused, wavering doubtfully between the past and the present, and hovering forward, as it were, at intervals, into the indistinctness of the world to come. There had been feverish turns, which tossed him from side to side, and wore away what little strength he had. But in his most convulsive struggles, and in the wildest vagaries of his intellect, when no other thought retained its sober influence, he still showed an awful solicitude lest the black veil should slip aside. Even if his bewildered soul could have forgotten, there was a faithful woman at his pillow, who, with averted eyes, would have covered that aged face, which she had last beheld in the comeliness of manhood. At length the death-stricken old man lay quietly in the torpor of mental

8. Jonathan Belcher (1681–1757), royal governor of Massachusetts and New Hampshire (1730–41) and later royal governor of New Jersey (1747–57) and founder of Princeton University. As in *The Scarlet Letter*, Hawthorne refers to the tradition of having a religious service to celebrate the election of a new governor. Being asked to preach the election sermon was a great honor.
9. Michael J. Colacurcio considers Peter Clark the most likely source for Hawthorne's Reverend Mr. Clark (*Province of Piety*, 355–57). In 1739, Peter Clark preached the official election sermon before Governor Belcher.

and bodily exhaustion, with an imperceptible pulse, and breath that grew fainter and fainter, except when a long, deep, and irregular inspiration seemed to prelude the flight of his spirit.

The minister of Westbury approached the bedside.

"Venerable Father Hooper," said he, "the moment of your release is at hand. Are you ready for the lifting of the veil that shuts in time from eternity?"

Father Hooper at first replied merely by a feeble motion of his head; then, apprehensive, perhaps, that his meaning might be doubtful, he exerted himself to speak.

"Yea," said he, in faint accents, "my soul hath a patient weariness until that veil be lifted."

"And is it fitting," resumed the Reverend Mr. Clark, "that a man so given to prayer, of such a blameless example, holy in deed and thought, so far as mortal judgment may pronounce; is it fitting that a father in the church should leave a shadow on his memory, that may seem to blacken a life so pure? I pray you, my venerable brother, let not this thing be! Suffer us to be gladdened by your triumphant aspect as you go to your reward. Before the veil of eternity be lifted, let me cast aside this black veil from your face!"

And thus speaking, the Reverend Mr. Clark bent forward to reveal the mystery of so many years. But, exerting a sudden energy, that made all the beholders stand aghast, Father Hooper snatched both his hands from beneath the bedclothes, and pressed them strongly on the black veil, resolute to struggle, if the minister of Westbury would contend with a dying man.

"Never!" cried the veiled clergyman. "On earth, never!"

"Dark old man!" exclaimed the affrighted minister, "with what horrible crime upon your soul are you now passing to the judgment?"

Father Hooper's breath heaved; it rattled in his throat; but, with a mighty effort, grasping forward with his hands, he caught hold of life, and held it back till he should speak. He even raised himself in bed; and there he sat, shivering with the arms of death around him, while the black veil hung down, awful, at that last moment, in the gathered terrors of a lifetime. And yet the faint, sad smile, so often there, now seemed to glimmer from its obscurity, and linger on Father Hooper's lips.

"Why do you tremble at me alone?" cried he, turning his veiled face round the circle of pale spectators. "Tremble also at each other! Have men avoided me, and women shown no pity, and children screamed and fled, only for my black veil? What, but the mystery which it obscurely typifies, has made this piece of crape so awful? When the friend shows his inmost heart to his friend; the lover to his best beloved; when man does not vainly shrink from the eye of his Creator, loathsomely treasuring up the secret of his sin; then deem me a monster, for the symbol beneath which I have lived, and die! I look around me, and, lo! on every visage a Black Veil!"

While his auditors shrank from one another, in mutual affright, Father Hooper fell back upon his pillow, a veiled corpse, with a faint smile lingering on the lips. Still veiled, they laid him in his coffin, and a veiled corpse they bore him to the grave. The grass of many years has sprung up and withered on that grave, the burial stone is moss-grown, and good Mr. Hooper's face is dust; but awful is still the thought that it mouldered beneath the Black Veil!

The Birthmark†

In the latter part of the last century there lived a man of science, an eminent proficient in every branch of natural philosophy, who not long before our story opens had made experience of a spiritual affinity more attractive than any chemical one. He had left his laboratory to the care of an assistant, cleared his fine countenance from the furnace smoke, washed the stain of acids from his fingers, and persuaded a beautiful woman to become his wife. In those days when the comparatively recent discovery of electricity and other kindred mysteries of Nature seemed to open paths into the region of miracle, it was not unusual for the love of science to rival the love of woman in its depth and absorbing energy. The higher intellect, the imagination, the spirit, and even the heart might all find their congenial aliment[1] in pursuits which, as some of their ardent votaries believed, would ascend from one step of powerful intelligence to another, until the philosopher should lay his hand on the secret of creative force and perhaps make new worlds for himself. We know not whether Aylmer possessed this degree of faith in man's ultimate control over Nature. He had devoted himself, however, too unreservedly to scientific studies ever to be weaned from them by any second passion. His love for his young wife might prove the stronger of the two; but it could only be by intertwining itself with his love of science, and uniting the strength of the latter to his own.

Such a union accordingly took place, and was attended with truly remarkable consequences and a deeply impressive moral. One day, very soon after their marriage, Aylmer sat gazing at his wife with a trouble in his countenance that grew stronger until he spoke.

"Georgiana," said he, "has it never occurred to you that the mark upon your cheek might be removed?"

"No, indeed," said she, smiling; but perceiving the seriousness of his manner, she blushed deeply. "To tell you the truth it has been so often called a charm that I was simple enough to imagine it might be so."

"Ah, upon another face perhaps it might," replied her husband; "but never on yours. No, dearest Georgiana, you came so nearly perfect from the hand of Nature that this slightest possible defect, which we hesitate whether to term a defect or a beauty, shocks me, as being the visible mark of earthly imperfection."

"Shocks you, my husband!" cried Georgiana, deeply hurt; at first reddening with momentary anger, but then bursting into tears. "Then why did you take me from my mother's side? You cannot love what shocks you!"

To explain this conversation it must be mentioned that in the centre of Georgiana's left cheek there was a singular mark, deeply interwoven, as it were, with the texture and substance of her face. In the usual state of her complexion—a healthy though delicate bloom—the mark wore a tint of deeper crimson, which imperfectly defined its shape amid the surrounding rosiness. When she blushed it gradually became more indistinct, and finally vanished amid the triumphant rush of blood that bathed the whole

† From *Mosses from an Old Manse*, vol. 2 of the Riverside Press edition (Boston: Houghton, Mifflin, 1882), pp. 47–69.

1. Food, sustenance.

cheek with its brilliant glow. But if any shifting motion caused her to turn pale there was the mark again, a crimson stain upon the snow, in what Aylmer sometimes deemed an almost fearful distinctness. Its shape bore not a little similarity to the human hand, though of the smallest pygmy size. Georgiana's lovers were wont to say that some fairy at her birth hour had laid her tiny hand upon the infant's cheek, and left this impress there in token of the magic endowments that were to give her such sway over all hearts. Many a desperate swain would have risked life for the privilege of pressing his lips to the mysterious hand. It must not be concealed, however, that the impression wrought by this fairy sign manual varied exceedingly, according to the difference of temperament in the beholders. Some fastidious persons—but they were exclusively of her own sex—affirmed that the bloody hand, as they chose to call it, quite destroyed the effect of Georgiana's beauty, and rendered her countenance even hideous. But it would be as reasonable to say that one of those small blue stains which sometimes occur in the purest statuary marble would convert the Eve of Powers to a monster.[2] Masculine observers, if the birthmark did not heighten their admiration, contented themselves with wishing it away, that the world might possess one living specimen of ideal loveliness without the semblance of a flaw. After his marriage,—for he thought little or nothing of the matter before,—Aylmer discovered that this was the case with himself.

Had she been less beautiful,—if Envy's self could have found aught else to sneer at,—he might have felt his affection heightened by the prettiness of this mimic hand, now vaguely portrayed, now lost, now stealing forth again and glimmering to and fro with every pulse of emotion that throbbed within her heart; but seeing her otherwise so perfect, he found this one defect grow more and more intolerable with every moment of their united lives. It was the fatal flaw of humanity which Nature, in one shape or another, stamps ineffaceably on all her productions, either to imply that they are temporary and finite, or that their perfection must be wrought by toil and pain. The crimson hand expressed the ineludible gripe in which mortality clutches the highest and purest of earthly mould, degrading them into kindred with the lowest, and even with the very brutes, like whom their visible frames return to dust. In this manner, selecting it as the symbol of his wife's liability to sin, sorrow, decay, and death, Aylmer's somber imagination was not long in rendering the birthmark a frightful object, causing him more trouble and horror than ever Georgiana's beauty, whether of soul or sense, had given him delight.

At all the seasons which should have been their happiest, he invariably and without intending it, nay, in spite of a purpose to the contrary, reverted to this one disastrous topic. Trifling as it at first appeared, it so connected itself with innumerable trains of thought and modes of feeling that it became the central point of all. With the morning twilight Aylmer opened his eyes upon his wife's face and recognized the symbol of imperfection; and when they sat together at the evening hearth his eyes wandered stealthily to her cheek, and beheld, flickering with the blaze of the wood

2. Vermont-born sculptor Hiram Powers (1805–1873), whom Hawthorne would get to know when he lived in Florence, was best-known for his *Greek Slave*, a controversial sculpture that toured the United States in 1847–48. Hawthorne probably refers to the sculpture *Eve Tempted* (1842), Powers's first full-length nude (National Museum of American Art in Washington, D.C.).

fire, the spectral hand that wrote mortality where he would fain have worshipped. Georgiana soon learned to shudder at his gaze. It needed but a glance with the peculiar expression that his face often wore to change the roses of her cheek into a deathlike paleness, amid which the crimson hand was brought strongly out, like a bas-relief of ruby on the whitest marble.

Late one night when the lights were growing dim, so as hardly to betray the stain on the poor wife's cheek, she herself, for the first time, voluntarily took up the subject.

"Do you remember, my dear Aylmer," said she, with a feeble attempt at a smile, "have you any recollection of a dream last night about this odious hand?"

"None! none whatever!" replied Aylmer, starting; but then he added, in a dry, cold tone, affected for the sake of concealing the real depth of his emotion, "I might well dream of it; for before I fell asleep it had taken a pretty firm hold of my fancy."

"And you did dream of it?" continued Georgiana, hastily; for she dreaded lest a gush of tears should interrupt what she had to say. "A terrible dream! I wonder that you can forget it. Is it possible to forget this one expression?—'It is in her heart now; we must have it out!' Reflect, my husband; for by all means I would have you recall that dream."

The mind is in a sad state when Sleep, the all-involving, cannot confine her specters within the dim region of her sway, but suffers them to break forth, affrighting this actual life with secrets that perchance belong to a deeper one. Aylmer now remembered his dream. He had fancied himself with his servant Aminadab,[3] attempting an operation for the removal of the birthmark; but the deeper went the knife, the deeper sank the hand, until at length its tiny grasp appeared to have caught hold of Georgiana's heart; whence, however, her husband was inexorably resolved to cut or wrench it away.

When the dream had shaped itself perfectly in his memory, Aylmer sat in his wife's presence with a guilty feeling. Truth often finds its way to the mind close muffled in robes of sleep, and then speaks with uncompromising directness of matters in regard to which we practise an unconscious self-deception during our waking moments. Until now he had not been aware of the tyrannizing influence acquired by one idea over his mind, and of the lengths which he might find in his heart to go for the sake of giving himself peace.

"Aylmer," resumed Georgiana, solemnly, "I know not what may be the cost to both of us to rid me of this fatal birthmark. Perhaps its removal may cause cureless deformity; or it may be the stain goes as deep as life itself. Again: do we know that there is a possibility, on any terms, of unclasping the firm gripe of this little hand which was laid upon me before I came into the world?"

"Dearest Georgiana, I have spent much thought upon the subject," hastily interrupted Aylmer. "I am convinced of the perfect practicability of its removal."

"If there be the remotest possibility of it," continued Georgiana, "let the attempt be made at whatever risk. Danger is nothing to me; for life,

3. See Matthew 1.4. As critics have pointed out, spelled backwards, "Aminadab" becomes "bad anima" (evil female spirit).

while this hateful mark makes me the object of your horror and disgust,—life is a burden which I would fling down with joy. Either remove this dreadful hand, or take my wretched life! You have deep science. All the world bears witness of it. You have achieved great wonders. Cannot you remove this little, little mark, which I cover with the tips of two small fingers? Is this beyond your power, for the sake of your own peace, and to save your poor wife from madness?"

"Noblest, dearest, tenderest wife," cried Aylmer, rapturously, "doubt not my power. I have already given this matter the deepest thought—thought which might almost have enlightened me to create a being less perfect than yourself. Georgiana, you have led me deeper than ever into the heart of science. I feel myself fully competent to render this dear cheek as faultless as its fellow; and then, most beloved, what will be my triumph when I shall have corrected what Nature left imperfect in her fairest work! Even Pygmalion, when his sculptured woman assumed life, felt not greater ecstasy than mine will be."[4]

"It is resolved, then," said Georgiana, faintly smiling. "And, Aylmer, spare me not, though you should find the birthmark take refuge in my heart at last."

Her husband tenderly kissed her cheek—her right cheek—not that which bore the impress of the crimson hand.

The next day Aylmer apprised his wife of a plan that he had formed whereby he might have opportunity for the intense thought and constant watchfulness which the proposed operation would require; while Georgiana, likewise, would enjoy the perfect repose essential to its success. They were to seclude themselves in the extensive apartments occupied by Aylmer as a laboratory, and where, during his toilsome youth, he had made discoveries in the elemental powers of Nature that had roused the admiration of all the learned societies in Europe. Seated calmly in this laboratory, the pale philosopher had investigated the secrets of the highest cloud region and of the profoundest mines; he had satisfied himself of the causes that kindled and kept alive the fires of the volcano; and had explained the mystery of fountains, and how it is that they gush forth, some so bright and pure, and others with such rich medicinal virtues, from the dark bosom of the earth. Here, too, at an earlier period, he had studied the wonders of the human frame, and attempted to fathom the very process by which Nature assimilates all her precious influences from earth and air, and from the spiritual world, to create and foster man, her masterpiece. The latter pursuit, however, Aylmer had long laid aside in unwilling recognition of the truth—against which all seekers sooner or later stumble—that our great creative Mother, while she amuses us with apparently working in the broadest sunshine, is yet severely careful to keep her own secrets, and, in spite of her pretended openness, shows us nothing but results. She permits us, indeed, to mar, but seldom to mend, and, like a jealous patentee, on no account to make. Now, however, Aylmer resumed these half-forgotten investigations; not, of course, with such hopes or wishes as first suggested them; but because they involved

4. In Greek mythology, Pygmalion sculpted a female figure, Galatea, with whom he fell in love. The goddess Aphrodite brought Galatea to life, Pygmalion married her, and they conceived a son, Paphos. See Book 10 of Ovid's *Metamorphoses*. Also see the well-known painting *Pygmalion and Galatea*, by Jean-Léon Gérôme (1824–1904).

much physiological truth and lay in the path of his proposed scheme for the treatment of Georgiana.

As he led her over the threshold of the laboratory, Georgiana was cold and tremulous. Aylmer looked cheerfully into her face, with intent to reassure her, but was so startled with the intense glow of the birthmark upon the whiteness of her cheek that he could not restrain a strong convulsive shudder. His wife fainted.

"Aminadab! Aminadab!" shouted Aylmer, stamping violently on the floor.

Forthwith there issued from an inner apartment a man of low stature, but bulky frame, with shaggy hair hanging about his visage, which was grimed with the vapors of the furnace. This personage had been Aylmer's underworker during his whole scientific career, and was admirably fitted for that office by his great mechanical readiness, and the skill with which, while incapable of comprehending a single principle, he executed all the details of his master's experiments. With his vast strength, his shaggy hair, his smoky aspect, and the indescribable earthiness that incrusted him, he seemed to represent man's physical nature; while Aylmer's slender figure, and pale, intellectual face, were no less apt a type of the spiritual element.

"Throw open the door of the boudoir, Aminadab," said Aylmer, "and burn a pastil."[5]

"Yes, master," answered Aminadab, looking intently at the lifeless form of Georgiana; and then he muttered to himself, "If she were my wife, I'd never part with that birthmark."

When Georgiana recovered consciousness she found herself breathing an atmosphere of penetrating fragrance, the gentle potency of which had recalled her from her deathlike faintness. The scene around her looked like enchantment. Aylmer had converted those smoky, dingy, somber rooms, where he had spent his brightest years in recondite pursuits, into a series of beautiful apartments not unfit to be the secluded abode of a lovely woman. The walls were hung with gorgeous curtains, which imparted the combination of grandeur and grace that no other species of adornment can achieve; and as they fell from the ceiling to the floor, their rich and ponderous folds, concealing all angles and straight lines, appeared to shut in the scene from infinite space. For aught Georgiana knew, it might be a pavilion among the clouds. And Aylmer, excluding the sunshine, which would have interfered with his chemical processes, had supplied its place with perfumed lamps, emitting flames of various hue, but all uniting in a soft, impurpled radiance. He now knelt by his wife's side, watching her earnestly, but without alarm; for he was confident in his science, and felt that he could draw a magic circle round her within which no evil might intrude.

"Where am I? Ah, I remember," said Georgiana, faintly; and she placed her hand over her cheek to hide the terrible mark from her husband's eyes.

"Fear not, dearest!" exclaimed he. "Do not shrink from me! Believe me, Georgiana, I even rejoice in this single imperfection, since it will be such a rapture to remove it."

"Oh, spare me!" sadly replied his wife. "Pray do not look at it again. I never can forget that convulsive shudder."

5. An aromatic paste.

In order to soothe Georgiana, and, as it were, to release her mind from the burden of actual things, Aylmer now put in practice some of the light and playful secrets which science had taught him among its profounder lore. Airy figures, absolutely bodiless ideas, and forms of unsubstantial beauty came and danced before her, imprinting their momentary footsteps on beams of light. Though she had some indistinct idea of the method of these optical phenomena, still the illusion was almost perfect enough to warrant the belief that her husband possessed sway over the spiritual world. Then again, when she felt a wish to look forth from her seclusion, immediately, as if her thoughts were answered, the procession of external existence flitted across a screen. The scenery and the figures of actual life were perfectly represented, but with that bewitching, yet indescribable difference which always makes a picture, an image, or a shadow so much more attractive than the original. When wearied of this, Aylmer bade her cast her eyes upon a vessel containing a quantity of earth. She did so, with little interest at first; but was soon startled to perceive the germ of a plant shooting upward from the soil. Then came the slender stalk; the leaves gradually unfolded themselves; and amid them was a perfect and lovely flower.

"It is magical!" cried Georgiana. "I dare not touch it."

"Nay, pluck it," answered Aylmer,—"pluck it, and inhale its brief perfume while you may. The flower will wither in a few moments and leave nothing save its brown seed vessels; but thence may be perpetuated a race as ephemeral as itself."

But Georgiana had no sooner touched the flower than the whole plant suffered a blight, its leaves turning coal-black as if by the agency of fire.

"There was too powerful a stimulus," said Aylmer, thoughtfully.

To make up for this abortive experiment, he proposed to take her portrait by a scientific process of his own invention. It was to be effected by rays of light striking upon a polished plate of metal. Georgiana assented; but, on looking at the result, was affrighted to find the features of the portrait blurred and indefinable; while the minute figure of a hand appeared where the cheek should have been. Aylmer snatched the metallic plate and threw it into a jar of corrosive acid.[6]

Soon, however, he forgot these mortifying failures. In the intervals of study and chemical experiment he came to her flushed and exhausted, but seemed invigorated by her presence, and spoke in glowing language of the resources of his art. He gave a history of the long dynasty of the alchemists, who spent so many ages in quest of the universal solvent by which the golden principle might be elicited from all things vile and base. Aylmer appeared to believe that, by the plainest scientific logic, it was altogether within the limits of possibility to discover this long-sought medium; "but," he added, "a philosopher who should go deep enough to acquire the power would attain too lofty a wisdom to stoop to the exercise of it." Not less singular were his opinions in regard to the elixir vitæ. He more than intimated that it was at his option to concoct a liquid that should prolong life for years, perhaps interminably; but that it would

6. Hawthorne is describing daguerreotypy, an early form of photography developed in France and very popular in America in the 1840s. Daguerreotypes were made by focusing an image on a silver-plated copper sheet and then treating the plate with mercury vapors. Hawthorne used daguerreotypy more extensively in *The House of the Seven Gables* (1851).

produce a discord in Nature which all the world, and chiefly the quaffer of the immortal nostrum, would find cause to curse.

"Aylmer, are you in earnest?" asked Georgiana, looking at him with amazement and fear. "It is terrible to possess such power, or even to dream of possessing it."

"Oh, do not tremble, my love," said her husband. "I would not wrong either you or myself by working such inharmonious effects upon our lives; but I would have you consider how trifling, in comparison, is the skill requisite to remove this little hand."

At the mention of the birthmark, Georgiana, as usual, shrank as if a redhot iron had touched her cheek.

Again Aylmer applied himself to his labors. She could hear his voice in the distant furnace room giving directions to Aminadab, whose harsh, uncouth, misshapen tones were audible in response, more like the grunt or growl of a brute than human speech. After hours of absence, Aylmer reappeared and proposed that she should now examine his cabinet of chemical products and natural treasures of the earth. Among the former he showed her a small vial, in which, he remarked, was contained a gentle yet most powerful fragrance, capable of impregnating all the breezes that blow across a kingdom. They were of inestimable value, the contents of that little vial; and, as he said so, he threw some of the perfume into the air and filled the room with piercing and invigorating delight.

"And what is this?" asked Georgiana, pointing to a small crystal globe containing a gold-colored liquid. "It is so beautiful to the eye that I could imagine it the elixir of life."

"In one sense it is," replied Aylmer; "or, rather, the elixir of immortality. It is the most precious poison that ever was concocted in this world. By its aid I could apportion the lifetime of any mortal at whom you might point your finger. The strength of the dose would determine whether he were to linger out years, or drop dead in the midst of a breath. No king on his guarded throne could keep his life if I, in my private station, should deem that the welfare of millions justified me in depriving him of it."

"Why do you keep such a terrific drug?" inquired Georgiana in horror.

"Do not mistrust me, dearest," said her husband, smiling; "its virtuous potency is yet greater than its harmful one. But see! here is a powerful cosmetic. With a few drops of this in a vase of water, freckles may be washed away as easily as the hands are cleansed. A stronger infusion would take the blood out of the cheek, and leave the rosiest beauty a pale ghost."

"Is it with this lotion that you intend to bathe my cheek?" asked Georgiana, anxiously.

"Oh, no," hastily replied her husband; "this is merely superficial. Your case demands a remedy that shall go deeper."

In his interviews with Georgiana, Aylmer generally made minute inquiries as to her sensations and whether the confinement of the rooms and the temperature of the atmosphere agreed with her. These questions had such a particular drift that Georgiana began to conjecture that she was already subjected to certain physical influences, either breathed in with the fragrant air or taken with her food. She fancied likewise, but it might be altogether fancy, that there was a stirring up of her system—a strange, indefinite sensation creeping through her veins, and tingling, half painfully, half pleasurably, at her heart. Still, whenever she dared to look into the mirror, there she beheld herself pale as a white rose and with the crimson birthmark stamped upon her cheek. Not even Aylmer now hated it so much as she.

To dispel the tedium of the hours which her husband found it necessary to devote to the processes of combination and analysis, Georgiana turned over the volumes of his scientific library. In many dark old tomes she met with chapters full of romance and poetry. They were the works of the philosophers of the middle ages, such as Albertus Magnus, Cornelius Agrippa, Paracelsus, and the famous friar who created the prophetic Brazen Head.[7] All these antique naturalists stood in advance of their centuries, yet were imbued with some of their credulity, and therefore were believed, and perhaps imagined themselves to have acquired from the investigation of Nature a power above Nature, and from physics a sway over the spiritual world. Hardly less curious and imaginative were the early volumes of the Transactions of the Royal Society,[8] in which the members, knowing little of the limits of natural possibility, were continually recording wonders or proposing methods whereby wonders might be wrought.

But to Georgiana the most engrossing volume was a large folio from her husband's own hand, in which he had recorded every experiment of his scientific career, its original aim, the methods adopted for its development, and its final success or failure, with the circumstances to which either event was attributable. The book, in truth, was both the history and emblem of his ardent, ambitious, imaginative, yet practical and laborious life. He handled physical details as if there were nothing beyond them; yet spiritualized them all, and redeemed himself from materialism by his strong and eager aspiration towards the infinite. In his grasp the veriest clod of earth assumed a soul. Georgiana, as she read, reverenced Aylmer and loved him more profoundly than ever, but with a less entire dependence on his judgment than heretofore. Much as he had accomplished, she could not but observe that his most splendid successes were almost invariably failures, if compared with the ideal at which he aimed. His brightest diamonds were the merest pebbles, and felt to be so by himself, in comparison with the inestimable gems which lay hidden beyond his reach. The volume, rich with achievements that had won renown for its author, was yet as melancholy a record as ever mortal hand had penned. It was the sad confession and continual exemplification of the shortcomings of the composite man, the spirit burdened with clay and working in matter, and of the despair that assails the higher nature at finding itself so miserably thwarted by the earthly part. Perhaps every man of genius in whatever sphere might recognize the image of his own experience in Aylmer's journal.

So deeply did these reflections affect Georgiana that she laid her face upon the open volume and burst into tears. In this situation she was found by her husband.

7. Roger Bacon (c. 1214–1294), friar reputed to have created a talking head out of brass. Albertus Magnus (c. 1220–1280), Dominican bishop and philosopher best known as a teacher of St. Thomas Aquinas and as a proponent of Aristotelianism at the University of Paris. Cornelius Agrippa (1486–1535), German alchemist notable for his works on occult philosophy. Paracelsus (1493–1541), Swiss alchemist.

8. The oldest scientific journal in continuous publication, *Philosophical Transactions* of the Royal Society of London was first published in 1665.

"It is dangerous to read in a sorcerer's books," said he with a smile, though his countenance was uneasy and displeased. "Georgiana, there are pages in that volume which I can scarcely glance over and keep my senses. Take heed lest it prove as detrimental to you."

"It has made me worship you more than ever," said she.

"Ah, wait for this one success," rejoined he, "then worship me if you will. I shall deem myself hardly unworthy of it. But come, I have sought you for the luxury of your voice. Sing to me, dearest."

So she poured out the liquid music of her voice to quench the thirst of his spirit. He then took his leave with a boyish exuberance of gayety, assuring her that her seclusion would endure but a little longer, and that the result was already certain. Scarcely had he departed when Georgiana felt irresistibly impelled to follow him. She had forgotten to inform Aylmer of a symptom which for two or three hours past had begun to excite her attention. It was a sensation in the fatal birthmark, not painful, but which induced a restlessness throughout her system. Hastening after her husband, she intruded for the first time into the laboratory.

The first thing that struck her eye was the furnace, that hot and feverish worker, with the intense glow of its fire, which by the quantities of soot clustered above it seemed to have been burning for ages. There was a distilling apparatus in full operation. Around the room were retorts, tubes, cylinders, crucibles, and other apparatus of chemical research. An electrical machine stood ready for immediate use. The atmosphere felt oppressively close, and was tainted with gaseous odors which had been tormented forth by the processes of science. The severe and homely simplicity of the apartment, with its naked walls and brick pavement, looked strange, accustomed as Georgiana had become to the fantastic elegance of her boudoir. But what chiefly, indeed almost solely, drew her attention, was the aspect of Aylmer himself.

He was pale as death, anxious and absorbed, and hung over the furnace as if it depended upon his utmost watchfulness whether the liquid which it was distilling should be the draught of immortal happiness or misery. How different from the sanguine and joyous mien that he had assumed for Georgiana's encouragement!

"Carefully now, Aminadab; carefully, thou human machine; carefully, thou man of clay!" muttered Aylmer, more to himself than his assistant. "Now, if there be a thought too much or too little, it is all over."

"Ho! ho!" mumbled Aminadab. "Look, master! look!"

Aylmer raised his eyes hastily, and at first reddened, then grew paler than ever, on beholding Georgiana. He rushed towards her and seized her arm with a gripe that left the print of his fingers upon it.

"Why do you come hither? Have you no trust in your husband?" cried he, impetuously. "Would you throw the blight of that fatal birthmark over my labors? It is not well done. Go, prying woman, go!"

"Nay, Aylmer," said Georgiana with the firmness of which she possessed no stinted endowment, "it is not you that have a right to complain. You mistrust your wife; you have concealed the anxiety with which you watch the development of this experiment. Think not so unworthily of me, my husband. Tell me all the risk we run, and fear not that I shall shrink; for my share in it is far less than your own."

"No, no, Georgiana!" said Aylmer, impatiently; "it must not be."

"I submit," replied she calmly. "And, Aylmer, I shall quaff whatever draught you bring me; but it will be on the same principle that would induce me to take a dose of poison if offered by your hand."

"My noble wife," said Aylmer, deeply moved, "I knew not the height and depth of your nature until now. Nothing shall be concealed. Know, then, that this crimson hand, superficial as it seems, has clutched its grasp into your being with a strength of which I had no previous conception. I have already administered agents powerful enough to do aught except to change your entire physical system. Only one thing remains to be tried. If that fail us we are ruined."

"Why did you hesitate to tell me this?" asked she.

"Because, Georgiana," said Aylmer, in a low voice, "there is danger."

"Danger? There is but one danger—that this horrible stigma shall be left upon my cheek!" cried Georgiana. "Remove it, remove it, whatever be the cost, or we shall both go mad!"

"Heaven knows your words are too true," said Aylmer, sadly. "And now, dearest, return to your boudoir. In a little while all will be tested."

He conducted her back and took leave of her with a solemn tenderness which spoke far more than his words how much was now at stake. After his departure Georgiana became rapt in musings. She considered the character of Aylmer, and did it completer justice than at any previous moment. Her heart exulted, while it trembled, at his honorable love—so pure and lofty that it would accept nothing less than perfection nor miserably make itself contented with an earthlier nature than he had dreamed of. She felt how much more precious was such a sentiment than that meaner kind which would have borne with the imperfection for her sake, and have been guilty of treason to holy love by degrading its perfect idea to the level of the actual; and with her whole spirit she prayed that, for a single moment, she might satisfy his highest and deepest conception. Longer than one moment she well knew it could not be; for his spirit was ever on the march, ever ascending, and each instant required something that was beyond the scope of the instant before.

The sound of her husband's footsteps aroused her. He bore a crystal goblet containing a liquor colorless as water, but bright enough to be the draught of immortality. Aylmer was pale; but it seemed rather the consequence of a highly-wrought state of mind and tension of spirit than of fear or doubt.

"The concoction of the draught has been perfect," said he, in answer to Georgiana's look. "Unless all my science have deceived me, it cannot fail."

"Save on your account, my dearest Aylmer," observed his wife, "I might wish to put off this birthmark of mortality by relinquishing mortality itself in preference to any other mode. Life is but a sad possession to those who have attained precisely the degree of moral advancement at which I stand. Were I weaker and blinder it might be happiness. Were I stronger, it might be endured hopefully. But, being what I find myself, methinks I am of all mortals the most fit to die."

"You are fit for heaven without tasting death!" replied her husband. "But why do we speak of dying? The draught cannot fail. Behold its effect upon this plant."

On the window seat there stood a geranium diseased with yellow blotches, which had overspread all its leaves. Aylmer poured a small quantity of the liquid upon the soil in which it grew. In a little time, when the roots of the plant had taken up the moisture, the unsightly blotches began to be extinguished in a living verdure.

"There needed no proof," said Georgiana, quietly. "Give me the goblet. I joyfully stake all upon your word."

"Drink, then, thou lofty creature!" exclaimed Aylmer, with fervid admiration. "There is no taint of imperfection on thy spirit. Thy sensible frame, too, shall soon be all perfect."

She quaffed the liquid and returned the goblet to his hand.

"It is grateful," said she with a placid smile. "Methinks it is like water from a heavenly fountain; for it contains I know not what of unobtrusive fragrance and deliciousness. It allays a feverish thirst that had parched me for many days. Now, dearest, let me sleep. My earthly senses are closing over my spirit like the leaves around the heart of a rose at sunset."

She spoke the last words with a gentle reluctance, as if it required almost more energy than she could command to pronounce the faint and lingering syllables. Scarcely had they loitered through her lips ere she was lost in slumber. Aylmer sat by her side, watching her aspect with the emotions proper to a man the whole value of whose existence was involved in the process now to be tested. Mingled with this mood, however, was the philosophic investigation characteristic of the man of science. Not the minutest symptom escaped him. A heightened flush of the cheek, a slight irregularity of breath, a quiver of the eyelid, a hardly perceptible tremor through the frame,—such were the details which, as the moments passed, he wrote down in his folio volume. Intense thought had set its stamp upon every previous page of that volume, but the thoughts of years were all concentrated upon the last.

While thus employed, he failed not to gaze often at the fatal hand, and not without a shudder. Yet once, by a strange and unaccountable impulse, he pressed it with his lips. His spirit recoiled, however, in the very act; and Georgiana, out of the midst of her deep sleep, moved uneasily and murmured as if in remonstrance. Again Aylmer resumed his watch. Nor was it without avail. The crimson hand, which at first had been strongly visible upon the marble paleness of Georgiana's cheek, now grew more faintly outlined. She remained not less pale than ever; but the birthmark, with every breath that came and went, lost somewhat of its former distinctness. Its presence had been awful; its departure was more awful still. Watch the stain of the rainbow fading out of the sky, and you will know how that mysterious symbol passed away.

"By Heaven! It is well-nigh gone!" said Aylmer to himself, in almost irrepressible ecstasy. "I can scarcely trace it now. Success! Success! And now it is like the faintest rose color. The lightest flush of blood across her cheek would overcome it. But she is so pale!"

He drew aside the window curtain and suffered the light of natural day to fall into the room and rest upon her cheek. At the same time he heard a gross, hoarse chuckle, which he had long known as his servant Aminadab's expression of delight.

"Ah, clod! ah, earthly mass!" cried Aylmer, laughing in a sort of frenzy, "you have served me well! Matter and spirit—earth and heaven—have both done their part in this! Laugh, thing of the senses! You have earned the right to laugh."

These exclamations broke Georgiana's sleep. She slowly unclosed her eyes and gazed into the mirror which her husband had arranged for that purpose. A faint smile flitted over her lips when she recognized how barely perceptible was now that crimson hand which had once blazed forth with such disastrous brilliancy as to scare away all their happiness. But then her eyes sought Aylmer's face with a trouble and anxiety that he could by no means account for.

"My poor Aylmer!" murmured she.

"Poor? Nay, richest, happiest, most favored!" exclaimed he. "My peerless bride, it is successful! You are perfect!"

"My poor Aylmer," she repeated, with a more than human tenderness, "you have aimed loftily; you have done nobly. Do not repent that with so high and pure a feeling, you have rejected the best the earth could offer. Aylmer, dearest Aylmer, I am dying!"

Alas! It was too true! The fatal hand had grappled with the mystery of life, and was the bond by which an angelic spirit kept itself in union with a mortal frame. As the last crimson tint of the birthmark—that sole token of human imperfection—faded from her cheek, the parting breath of the now perfect woman passed into the atmosphere, and her soul, lingering a moment near her husband, took its heavenward flight. Then a hoarse, chuckling laugh was heard again! Thus ever does the gross fatality of earth exult in its invariable triumph over the immortal essence which, in this dim sphere of half development, demands the completeness of a higher state. Yet, had Aylmer reached a profounder wisdom, he need not thus have flung away the happiness which would have woven his mortal life of the selfsame texture with the celestial. The momentary circumstance was too strong for him; he failed to look beyond the shadowy scope of time, and, living once for all in eternity, to find the perfect future in the present.

CONTEXTS

Passages from Hawthorne's Notebooks and Letters

NATHANIEL HAWTHORNE

From American Notebooks†

January 4, 1839

Letters in the shape of figures of men, &c. At a distance, the words composed by the letters are alone distinguishable. Close at hand, the figures alone are seen, and not distinguished as letters. Thus things may have a positive, a relative, and a composite meaning, according to the point of view.

October 27, 1841

To symbolize moral or spiritual disease by disease of the body;—thus, when a person committed any sin, it might cause a sore to appear on the body;—this to be wrought out.

1844

The Unpardonable Sin might consist in a want of love and reverence for the Human Soul; in consequence of which, the investigator pried into its dark depths, not with a hope or purpose of making it better, but from a cold philosophical curosity,—content that it should be wicked in whatever kind or degree, and only desiring to study it out. Would not this, in other words, be the separation of the intellect from the heart?

Late 1844 or early 1845

The life of a woman, who, by the old colony law, was condemned always to wear the letter A, sewed on her garment, in token of her having committed adultery.

August 9, 1845

In the eyes of a young child, or other innocent person, the image of a cherub or an angel to be seen peeping out;—in those of a vicious person, a devil.

† From *The American Notebooks*, vol. 8 of *The Centenary Edition of the Works of Nathaniel Hawthorne*, ed. Claude M. Simpson (Columbus: Ohio State UP, 1972), pp. 183, 222, 251, 254, 270, 340, 342, 343, 383–84, 412–13, 420–21, 430. Reprinted by permission of Ohio State University Press.

[Monday, August 22nd, 1842]

I took a walk through the woods, yesterday afternoon, to Mr. Emerson's, with a book which Margaret Fuller had left behind her, after a call on Saturday eve. I missed the nearest way, and wandered into a very secluded portion of the forest—for forest it might justly be called, so dense and sombre was the shade of oaks and pines. Once I wandered into a tract so overgrown with bushes and underbrush that I could scarcely force a passage through. * * *

* * *

After leaving the book at Mr. Emerson's, I returned through the woods, and entering Sleepy Hollow, I perceived a lady reclining near the path which bends along its verge. It was Margaret herself. She had been there the whole afternoon, meditating or reading; for she had a book in her hand, with some strange title, which I did not understand and have forgotten. She said that nobody had broken her solitude, and was just giving utterance to a theory that no inhabitant of Concord ever visited Sleepy Hollow, when we saw a whole group of people entering the sacred precincts. Most of them followed a path that led them remote from us; but an old man passed near us, and smiled to see Margaret lying on the ground, and me sitting by her side. He made some remark about the beauty of the afternoon, and withdrew himself into the shadow of the wood. Then we talked about Autumn—and about the pleasures of getting lost in the woods—and about the crows, whose voices Margaret had heard—and about the experiences of early childhood, whose influence remains upon the character after the collection of them has passed away—and about the sight of mountains from a distance, and the view from their summits—and about other matters of high and low philosophy. In the midst of our talk, we heard footsteps above us, on the high bank; and while the intruder was still hidden among the trees, he called to Margaret, of whom he had gotten a glimpse. Then he emerged from the green shade; and, behold, it was Mr. Emerson, who, in spite of his clerical consecration, had found no better way of spending the Sabbath than to ramble among the woods. He appeared to have had a pleasant time; for he said that there were Muses in the woods to-day, and whispers to be heard in the breezes. It being now nearly six o'clock, we separated, Mr. Emerson and Margaret towards his house, and I towards mine, where my little wife was very busy getting tea. By the bye, Mr. Emerson gave me an invitation to dinner to-day, to be complied with or not, as might suit my convenience at the time; and it happens not to suit.

[Friday,] October 13th, 1848

During this moon, I have two or three evenings, sat sometime in our sitting-room, without light, except from the coal-fire and the moon. Moonlight produces a very beautiful effect in the room; falling so white upon the carpet, and showing its figures so distinctly; and making all the room so visible, and yet so different from a morning or noontide visibility. There are all the familiar things;—every chair, the tables, the couch, the bookcase, all the things that we are accustomed to in the daytime; but now it seems as if we were remembering them through a lapse of years rather than seeing them with the immediate eye. A child's shoe—the doll, sitting in her little wicker-carriage—all objects, that have been used or played with during the day, though still as familiar as ever, are invested with something like strangeness and remoteness. I cannot in any measure express it. Then the somewhat dim coal-fire throws its unobtrusive tinge through the room—a faint ruddiness upon the wall—which has a not unpleasant effect in taking from the colder spirituality of the moonbeams. Between both these lights, such a medium is created that the room seems just fit for the ghosts of persons very dear, who have lived in the room with us, to glide noiselessly in, and sit quietly down, without affrighting us. It would be like a matter of course, to look round, and find some familiar form in one of the chairs. If one of the white curtains happen to be down before the windows, the moonlight makes a delicate tracery with the branches of the trees, the leaves somewhat thinned by the progress of autumn, but still pretty abundant. It is strange how utterly I have failed to give anything of the effect of moonlight in a room.

The fire-light diffuses a mild, heart-warm influence through the room; but is scarcely visible, unless you particularly look for it—and then you become conscious of a faint tinge upon the cieling [*sic*], of a reflected gleam from the mahogany furniture; and if your eyes fall on the glass, deep within it you perceive the glow of the burning anthracite.

I hate to leave such a scene; and when retiring to bed, after closing the sitting-room door, I re-open it, again and again, to peep back at the warm, cheerful, solemn repose, the white light, the faint ruddiness, the dimness,—all like a dream, and which makes me feel as if I were in a conscious dream.

PASSAGES ABOUT UNA

January 28, 1849

* * * Una is now teazing her mother to go out to walk with her, and will not accept of the compromise of going out to play in the yard by herself, but continually recurs to the subject—"Mother, I wish you would take me one little walk:—Mother, if I could but take one little walk—Mother, if you would but let me go with Dora." She is certainly a most pertinacious teaser. Mamma has now proposed that Una shall walk up and down the street by herself—to which she consents with alacrity, never, I believe, having been so far trusted before. "How I do want to go out in the street alone!" says she. It is one step forward, in her existence. So she is dressed up in her purple pelisse, white satin bonnet, with a green veil, and white muff, with woolen gaiters on her well-calved pedestals; in which garb she looks anything but beautiful. Her beauty is the most flitting, transitory, most uncertain and unaccountable affair, that ever had a real existence; it beams out when nobody expects it; it has mysteriously passed away, when you think yourself sure of it;—if you glance sideways at her, you perhaps think it is illuminating her face, but, turning full round to enjoy it, it is gone again. Her mother sees it much oftener than I do; yet, neither is the revelation always withheld from me. When really visible, it is rare and precious as the vision of an angel; it is a transfiguration—a grace, delicacy, an ethereal fineness, which, at once, in my secret soul, makes me give up all severe

opinions that I may have begun to form respecting her. It is but fair to conclude, that, on these occasions, we see her real soul; when she seems less lovely, we merely see something external. But, in truth, one manifestation belongs to her as much as another; for, before the establishment of principles, what is character but the series and succession of moods?

February 1, 1849

* * * It is a very good discipline for Una to carry a book on her head; not merely physical discipline, but moral as well; for it implies a restraint upon her usual giddy, impetuous demeanor. She soon, however, begins to move with great strides, and sudden jerks, and to tumble about in extravagant postures;—a very unfortunate tendency that she has; for she is never graceful or beautiful, except when perfectly quiet. Violence—exhibitions of passions—strong expression of any kind—destroy her beauty. Her voice, face, gestures—every manifestation, in short—becomes disagreeable.

The children have been playing ball together; and Una, heated by the violence with which she plays, sits down on the floor, and complains grievously of warmth—opens her breast. This is the physical manifestation of the evil spirit that struggles for the mastery of her; he is not a spirit at all, but an earthy monster, who lays his grasp on her spinal marrow, her brain, and other parts of her body that lie in closest contiguity to her soul; so that the soul has the discredit of these evil deeds. She is recovered now, and is bounding across the room with a light and graceful motion; but soon sinks down on the floor, complaining of being tired. Her mood, to-day, is less tempestuous than usual—yet it has no settled level.

[Monday,] July 30th [1849], 1/2 past 10 o'clock

Another bright forenoon, warmer than yesterday, with flies buzzing through the sunny air. Mother still lives, but is gradually growing weaker, and appears to be scarcely sensible. Julian is playing quietly about, and is now out of doors, probably hanging on the gate. Una takes a strong and strange interest in poor mother's condition, and can hardly be kept out of the chamber—endeavoring to thrust herself into the door, whenever it is opened, and continually teazing me to be permitted to go up. This is partly the intense curiosity of her active mind—partly, I suppose, natural affection. I know not what she supposes is to be the final result to which grandmamma is approaching. She talks of her being soon to go to God, and probably thinks that she will be taken away bodily. Would to God it were to be so! Faith and trust would be far easier than they are now. But, to return to Una, there is something that almost frightens me about the child—I know not whether elfish or angelic, but, at all events, supernatural. She steps so boldly into the midst of everything, shrinks from nothing, has such a comprehension of everything, seems at times to have but little delicacy, and anon shows that she possesses the finest essence of it; now so hard, now so tender; now so perfectly unreasonable, soon again so wise. In short, I now and then catch an aspect of her, in which I cannot believe her to be my own human child, but a spirit strangely mingled with good and evil, haunting the house where I dwell. The little boy is always the same child, and never varies in his relation to me.

From Letters[†]

To H. W. Longfellow, Cambridge

Custom-House,
Salem. Novr 11, '47.

Dear Longfellow,

I have read Evangeline with more pleasure than it would be decorous to express. It cannot fail, I think, to prove the most triumphant of all your successes. Everybody likes it. I wrote a notice of it for our democratic paper, which Conolly edits; but he has not inserted it[1]—why I know not, unless he considers it unworthy of the subject; as it undoubtedly was. But let him write a better if he can. I have heard the poem—and other of your poems, the Wreck of the Hesperus[2] among them—discussed here in the Custom-House. It was very queer, and would have amused you much.

How seldom we meet! It would do me good to see you occasionally; but my duties, official, marital, and paternal, keep me pretty constantly at home; and when I do happen to have a day of leisure, it might chance to be a day of occupation with you—so I do not come.[3] I live at No. 14 Mall-street now. May I not hope to see you there?

I am trying to resume my pen;[4] but the influences of my situation and customary associates are so anti-literary, that I know not whether I shall succeed. Whenever I sit alone, or walk alone, I find myself dreaming about stories, as of old; but these forenoons in the Custom House undo all that the afternoons and evenings have done. I should be happier if I could write—also, I should like to add something to my income, which, though tolerable, is a tight fit. If you can suggest any work of pure literary drudgery, I am the very man for it.

I have heard nothing of Hillard, since his departure.[5] Cannot you tell me something about him?

Your friend,
Nathl Hawthorne.

† From *The Letters, 1843–1853*, vol. 16 of *The Centenary Edition of the Works of Nathaniel Hawthorne*, ed. Thomas Woodson, L. Neal Smith, and Norman Holmes Pearson (Columbus: Ohio State UP, 1985), pp. 215–16, 263–65, 269–71, 279–82, 305–06, 307–08, 311–13, 329–30, 386–87. Reprinted by permission of Ohio State University Press.

1. NH's notice appeared in the Salem *Advertiser*, November 13: it is rpt. in Randall Stewart, "Hawthorne's Contributions to *The Salem Advertiser*," *American Literature*, v (1934), 333–35. Longfellow's reply of November 29 to NH is in *The Letters of Henry Wadsworth Longfellow: 1844–1856*, Volume III, ed. Andrew R. Hilen (Cambridge: Belknap Press, 1967), pp. 145–46.
2. In Longfellow's *Ballads and Other Poems* (1841).
3. Longfellow had invited NH to dine at Nahant in August, but had received no answer from him; see *Letters*, III, 146.
4. The next day, SH [Sophia Hawthorne] wrote to her mother that NH had begun to write, and on November 23 informed her: "My husband began retiring to his study on the first November and writes every afternoon" (MSS, Berg [Collection, New York Public Library]).
5. [George] Hillard [classmate at Bowdoin] had sailed for England on July 1, and reached Italy on September 2. During his trip, he sent long descriptive letters to his wife that were shared by his friends George Ticknor, Sumner, and Longfellow, and were eventually transformed into his popular *Six Months in Italy* (1853). See Longfellow, *The Letters of Henry Wadsworth Longfellow,* ed. Andrew Hilen (Cambridge: Harvard UP, 1966–82), III, 131–32, 139, 155–56, 167–69.

To G. S. Hillard, Boston

Salem, March 5th. 1849.

Dear Hillard,

It is a very long time since I have held converse with you by tongue or pen; but I have thought of you none the less, and have enjoyed Europe with you,[1] and rejoiced at your safe return. My present object in writing is one with which you would hardly imagine yourself to have anything to do.

I am informed that there is to be a strong effort among the politicians here to remove me from office, and that my successor is already marked out. I do not think that this ought to be done; for I was not appointed to office as a reward for political services, nor have I acted as a politician since. A large portion of the local Democratic party look coldly on me, for not having used the influence of my position to obtain the removal of whigs—which I might have done, but which I in no case did. Neither was my own appointment made at the expense of a Whig; for my predecessor[2] was appointed by Tyler, in his latter days, and called himself a Democrat. Nor can any charge of inattention to duty, or other official misconduct, be brought against me; or, if so, I could easily refute it. There is therefore no ground for disturbing me, except on the most truculent party system. All this, however, will be of little avail with the slang-whangers[3]—the vote-distributors—the Jack Cades[4]—who assume to decide upon these matters, after a political triumph; and as to any literary claims of mine, they would not weigh a feather, nor be thought worth weighing at all.

But it seems to me that an inoffensive man of letters—having obtained a pitiful little office on no other plea than his pitiful little literature—ought not to be left to the mercy of these thick-skulled and no-hearted ruffians. It is for this that I now write to you. There are men in Boston—Mr Rufus Choate,[5] for instance—whose favorable influence with the administration would make it impossible to remove me, and whose support and sympathy might fairly be claimed in my behalf—not on the ground that I am a very good writer, but because I gained my position, such as it is, by my literary character, and have done nothing to forfeit

1. Presumably NH had read Hillard's letters from Europe. Hillard had returned to Boston on October 19, 1848.
2. Nehemiah Brown was surveyor, 1843–46. In a letter to President Polk, November 1, 1845, Zachariah Burchmore, secretary of the Essex County Democratic Committee, and A. D. Wait, chairman, stressed that "N. Brown, the person who now holds the office of surveyor, is at present moment Whig Sheriff in Essex County," and that Brown had "always been" a Whig (MS, National Archives). See Paul Cortissoz, "The Political Life of Nathaniel Hawthorne," Diss. New York U, 1955, 58. NH elaborates his contrary claim in his letter of June 18, 1849 [see pp. 211–13 of this Norton Critical Edition—*editor's note*].
3. Noisy or abusive talkers or writers.
4. Cade was the leader of a rebellion in southern England in 1450. As dramatized in Shakespeare's *II Henry VI,* Act IV, he is a ranting, anti-intellectual demagogue who orders the beheading of a learned and loyal nobleman.
5. A famous lawyer (1799–1859) who had been an organizer of the Whig party in Massachusetts, and had served as a congressman, 1830–34, and U.S. senator, 1841–45. See *Dictionary of American Biography*, ed. Allen Johnson and Dumas Malone (New York: Scribner, 1943); Jean V. Matthews, *Rufus Choate: The Law and Civic Virtue* (Philadelphia: Temple UP, 1980). In a letter to Louisa, October 5, 1848, SH described a recent evening rally for Choate in Salem that she and NH had observed, and said that she was "very glad . . . to hear the heavy hurrah" (MS, Berg). Hillard, a Whig, persuaded Choate to write on NH's behalf to Secretary Meredith immediately after his dismissal, on June 9. Choate characterized NH as "a writer of rare beauty, & merit & fame, a person of the purest character, & in politics perfectly quiet & silent" (MS, National Archives; see Nevins, p. 128).

that tenure. I do not think that you can have any objection to bringing this matter under the consideration of such men; but if you do so object, I am sure it will be for some good reason, and therefore beg you not to stir in it. I do not want any great fuss to be made; the whole thing is not worth it; but I should like to have the Administration enlightened by a few such testimonials as would take my name out of the list of ordinary office-holders, and at least prevent any hasty action. I think, too, that the letters (if you obtain any) had better contain no allusion to the proposed attack on me, as it may possibly fall through of itself. Certainly, the general feeling here in Salem would be in my favor; but I have seen too much of the modes of political action to lay any great stress on that.

Be pleased on no account to mention this matter to any Salem man, however friendly to me he may profess himself. If any movement on my part were heard of, it would precipitate their assault.

So much for business. I do not let myself be disturbed by these things, but employ my leisure hours in writing, and go on as quietly as ever. I see that Longfellow has written a prose-tale.[6] How indefatigable he is!—and how adventurous! Well he may be, for he never fails.

Remember me to Mrs. Hillard. Sophia is well, and our children continue to flourish famously. Why do you not come to see us?

Your friend,
Nathl Hawthorne.

To H. W. Longfellow, Cambridge

[Salem] Custom House, June 5th. 1849

Dear Longfellow,

I meant to have written you before now about Kavanaugh,[1] but have had no quiet time, during my letter-writing hours; and now the freshness of my thoughts has exhaled away. It is a most precious and rare book—as fragrant as a bunch of flowers, and as simple as one flower. A true picture of life, moreover—as true as those reflections of the trees and banks that I used to see in the Concord, but refined to a higher degree than they; as if the reflection were itself reflected. Nobody but yourself would dare to write so quiet a book; nor could any other succeed in it. It is entirely original; a book by itself; a true work of genius, if ever there were one; and yet I should not wonder if many people (God confound them!) were to see no such matter in it. In fact, I doubt whether hardly anybody else has enjoyed it so much as I; although I have heard or seen none but favorable opinions.

I should like to have written a long notice of it, and would have done so for the Salem Advertiser; but, on the strength of my notice of Evangeline and some half-dozen other books, I have been accused of a connection with the editorship of that paper, and of writing political articles—which I never did one single time in my whole life.[2] I must confess, it stirs up a

6. *Kavanagh: A Tale*, to be published May 12.
1. Longfellow had sent NH an inscribed copy, dated May 19 (now in Berg).
2. Eben N. Walton (1825–1907), editor of the *Advertiser* since 1847, was to write to NH on June 30 to deny this charge, in a letter intended to be forwarded to Washington as evidence. Walton stated that "only two articles from your pen have appeared in its columns, one a notice of a dramatic company, the other a notice of Longfellow's 'Evangeline' " (Nevins, pp. 129–30).

little of the devil within me, to find myself hunted by these political bloodhounds. If they succeed in getting me out of office, I will surely immolate one or two of them. Not that poor monster of a Conolly,[3] whom I desire only to bury in oblivion, far out of my own remembrance. Nor any of the common political brawlers, who work on their own level, and can conceive of no higher ground than what they occupy. But if there be among them (as there must be, if they succeed) some men who claim a higher position, and ought to know better, I may perhaps select a victim,[4] and let fall one little drop of venom on his heart, that shall make him writhe before the grin of the multitude for a considerable time to come. This I will do, not as an act of individual vengeance, but in your behalf as well as mine, because he will have violated the sanctity of the priesthood to which we both, in our different degrees, belong. I do not claim to be a poet; and yet I cannot but feel that some of the sacredness of that character adheres to me, and ought to be respected in me, unless I step out of its immunities, or make it a plea for violating any of the rules of ordinary life. When other people concede me this privilege, I never think that I possess it; but when they disregard it, the consciousness makes itself felt. If they will pay no reverence to the imaginative power when it causes herbs of grace and sweet-scented flowers to spring up along their pathway, then they should be taught what it can do in the way of producing nettles, skunk-cabbage, deadly night-shade, wolf's bane, dog-wood.[5] If they will not be grateful for its works of beauty and beneficence, then let them dread it as a pervasive and penetrating mischief, that can reach them at their firesides and in their bedchambers, follow them to far countries, and make their very graves refuse to hide them. I have often thought that there must be a good deal of enjoyment in writing personal satire; but, never having felt the slightest ill-will towards any human being, I have hitherto been debarred from this peculiar source of pleasure. I almost hope I shall be turned out, so as to have an opportunity of trying it. I cannot help smiling in anticipation of the astonishment of some of these local magnates here, who suppose themselves quite out of the reach of any retribution on my part.

I have spent a good deal of time in Boston, within a few weeks; my two children having been ill of the scarlet-fever there; and the little boy was in quite an alarming way. I could not have submitted in the least, had it gone ill with him; but God spared me that trial—and there are no real misfortunes, save such as that. Other troubles may irritate me superficially; nothing else can go near the heart.

3. Horace Conolly had been chairman of the Second Congressional District Democratic Committee, and had written to President Polk in 1845 supporting NH's candidacy for surveyor. Now he had joined the Whigs, and was to participate in the unanimous resolution on July 3 by the Whig Ward Committees and the Government of the Taylor Club to approve NH's removal. See Nevins, pp. 99–100, 105–6.
4. SH's letters to her father, June 10, and to her mother, June 21, suggest that the most likely victims were Charles W. Upham; Richard Saltonstall Rogers (1792–1873), a merchant; Nathaniel Silsbee, Jr. (1804–1881), a merchant, brother of NH's former friend Mary Silsbee Sparks, and current mayor of Salem; and George Humphrey Devereux (1809–1878), a lawyer (Lathrop, *Memories*, p. 96; MS, Berg).
5. Compare the narrator's (or Hester's) speculation about Chillingworth's grave, "where, in due course of time, would be seen deadly nightshade, dogwood, henbane, and whatever else of vegetable wickedness the climate could produce, all flourishing with hideous luxuriance" (*SL*, pp. 175–76) [p. 107 of this Norton Critical Edition—*editor's note*].

I mean to come and dine with you, the next time you invite me; and Hillard said he would come too.[6] Do not let it be within a week, however; for Bridge and his wife expect to be here in the course of that time.

Please to present my regards to Mrs. Longfellow, and believe me

ever your friend,
Nath Hawthorne.

To G. S. Hillard, Boston

Salem, June 18th, 1849.

My dear Hillard,

There is an article respecting me in the Boston Atlas of Saturday, which seems to require some notice from my pen;[1] and I choose to give my answer in the form of a letter to yourself, because I would be understood as speaking with a more than common carefulness in regard to the accuracy of what I say.[2] For, what a man should I be, my dear Hillard, if I could dream of connecting your stainless integrity, and honorable name, with any statement which I did not believe to be strictly true!

The article first charges me with never having received the approbation of the Democrats of Salem for the Surveyorship; an accusation which I do not think it necessary, just at this time, to repel. As respects the imputation of having been an office-seeker, I would say, that while residing at Concord, I was earnestly and repeatedly urged to become a candidate for the post office in Salem, by a person who claimed to be the representative of the great majority of the local Democratic party. My consent being reluctantly given, the attempt was made and failed; not from any defect in me, as a candidate, but because the incumbent—my present esteemed friend, Dr. Brown—contrary to what had been told me, was an excellent officer, and had the great bulk of the party with him. Subsequently, without solicitition on my part, two offices were successively tendered to me by Mr Bancroft, each of larger emolument than the one which it afterwards suited me to take.

The article further says, that my predecessor in the Surveyorship was a Whig. Mr. Nehemiah Brown, the gentleman in question, obtained the office through the following succession of changes:—Mr. Daniels,[3] a Whig, appointed in 1840, had been succeeded, after the Tyler revolution, by the late Mr. Edward Palfrey,[4] a Democrat, who held the office for a considerable

6. Longfellow suggested June 21. He noted in his journal, June 10: "Sumner dined with us; and we discussed Hawthorne's dismissal from the Custom House in terms not very complimentary to General Taylor and his cabinet" (MS, Harvard; *Life of Henry Wadsworth Longfellow: With Extracts from His Journals and Correspondence*, Volume II, ed. Samuel Longfellow [Boston: Ticknor and Company, 1886], p. 142).
1. The militantly Whig *Atlas* published in an editorial on June 16 an anonymous letter, perhaps by Charles Upham, attacking NH's alleged political innocence. Stephen Nissenbaum, "The Firing of Nathaniel Hawthorne," *Essex Institute Historical Collection*, CXIV (1978): 67.
2. This letter was published in the Boston *Advertiser* "at the request of a friend," as the prefatory note stated. The Whig *Advertiser* "would not be understood as expressing any opinion on the question of Mr. Hawthorne's removal from office" and published the letter "because we consider it due to the writer, that he should have the opportunity to lay before the public his own statement of his position." Elizabeth Peabody wrote to her mother from Boston, June 15: "Hawthorne was here today & wrote a letter. . . . He seems to be all in a rouse" (MS, Antioch).
3. Stephen Daniels (1798–1872), officially surveyor, 1841–43, now operated a grocery store.
4. Edward Palfrey, surveyor, 1838–41.

time during the recess of the Senate. The nomination of Mr. Palfrey not being confirmed, Mr. George W. Mullet,[5] another Democrat, was nominated by President Tyler, and likewise rejected by the Senate. The President, in this emergency, having no opportunity to take the wishes of the local party, and the session drawing to a close, nominated Mr. Brown, who, then and subsequently, was one of that peculiar class of politicians styled Tyler Democrats.[6] I refer, in proof of his democracy, to the records and members of the Hickory Club. I refer to a crowd of witnesses, as well Whigs as Democrats. I refer, among others—and am most happy so to do—to a gentleman now very prominent and active in our local politics, the Rev. Charles Wentworth Upham, who told me, in presence of David Roberts, Esq., that I need never fear removal under a Whig administration, inasmuch as my appointment had not displaced a Whig. Lastly, I refer, frankly and fearlessly, and with entire confidence in his response, to Mr. Nehemiah Brown himself.

In the second year of President Polk's administration, Mr. Brown was removed, and succeeded by myself—not on any charge derogatory to his character—but simply because, as was the predicament of many other Tyler Democrats, his appointment had not been based on any mode of selection by the local party.

I am further accused of having been an active politician, while in office; in proof of which, it is averred that I have been a member, during two years, of the Democratic Town Committee, and a delegate, last year, to the Democratic State Convention. As respects the latter, I do not remember ever being chosen a delegate to that, or any convention, and certainly never was present at one, in my whole life. I do remember having seen my name, in the Salem Advertiser, as a member of the Democratic Town Committee; but I never was otherwise notified of the fact, never attended a meeting, never acted officially, and have no other knowledge of my membership than having seen my name, as aforesaid[7] I never in my life walked in a torch light procession, and—I am almost tempted to say—would hardly have done any thing so little in accordance with my tastes and character, had the result of the Presidential election depended on it. My contributions to the Salem Advertiser have been a few notices of books, and other miscellaneous paragraphs, perhaps a dozen in all; never a single line of politics. I have ceased, for upwards of three years, to write for the Democratic Review, and never did write a political article for that, or any other journal or newspaper; nor an article that had the remotest reference to politics, with the single exception of a biographical sketch of Cilley, written at the request of the editor, as a tribute to the memory of an early and very dear friend.

The article further insinuates, as I apprehend it, the charge of fraud or dishonesty against me, and refers for proof to the Blue Book,[8] where, as it

5. Mullett (1809–1893) had been chairman of the Democratic Committee of Salem.
6. John Tyler, a Virginia aristocrat elected vice-president by the Whigs in 1840, had become president upon W. H. Harrison's death in April 1841. He soon quarreled with Whig leaders, and they expelled him from their party. Tyler thereupon led a return of disaffected Whigs to the Democratic party, forming an alliance with conservative, pro-slavery Democrats.
7. NH was listed as a member of the Democratic Town Committee in the Salem *Advertiser,* November 9, 1846, and November 6, 1847. He was listed as a delegate to the Democratic State Convention in Worchester in the *Advertiser* on August 30, 1848 (Cortissoz, p. 85).
8. Of the U.S. Treasury Department.

affirms, the Democratic officers of the Custom House, appear to have received larger amounts than the Whigs.[9] In reply, I have merely to state that the emoluments of the officers are strictly and necessarily commensurate with the amount of service rendered; and that, in all matters relating to this point, I have been under the constant supervision, as well as general direction, of Colonel Miller, a Whig, the Deputy Collector, and now the Collector of the port.

I have thus, I believe, responded to all the charges, point by point. I am happy that my accuser has given me the opportunity, and should have been still more so, had he come forward under his own name, and met me, face to face, before the public. But, now, if he be a gentleman,—as not improbably he may be—he will be willing, I trust, to acknowledge, that the slanders of private animosity and the distorting medium of party prejudice may have deceived him, as to my position, my conduct, and my character. This frank acknowledgement is all I ask.

Affectionately yours,
Nathl Hawthorne.

George S. Hillard, Esq, Boston.

To J. T. Fields, Boston

Salem, Jan. 15th 1850.

My dear Fields,

I send you, at last, the manuscript portion of my volume; not quite all of it, however, for there are three chapters still to be written of "The Scarlet Letter." I have been much delayed by illness in my family and other interruptions. Perhaps you will not like the book nor think well of its prospects with the public. If so (I need not say) I shall not consider you under any obligation to publish it. 'The Scarlet Letter' is rather a delicate subject to write upon, but in the way in which I have treated it, it appears to me there can be no objections on that score. The article entitled "Custom House" is introductory to the volume, so please read it first. In the process of writing, all political and official turmoil has subsided within me, so that I have not felt inclined to execute justice on any of my enemies. I have not yet struck out a title, but may possibly hit on one before I close the package. If not, there need be no running title of the book over each page, but only of the individual articles. Calculating the page of the new volume at the size of that of the 'Mosses,' I can supply 400 and probably more. "The Scarlet Letter," I suppose, will make half of that number; otherwise, the calculation may fall a little short, though I think not.

Very truly yours,
Nathl Hawthorne.

P.S. The proof-sheets will need to be revised by the author. I write such an infernal hand that this is absolutely indispensable.

9. The charge is detailed in Upham's "Memorial" to Secretary Meredith of July 6 (Nevins, p. 116; Cortissoz, p. 270).

If my wife approves—whom I have made the umpire in the matter—I shall call the book Old-Time Legends; together with *sketches, experimental and ideal.* I believe we must consider the book christened as above. Of course, it will be called simply "Old-Time Legends," and the rest of the title will be printed in small capitals. I wish I could have brought a definition of the whole book within the compass of a single phrase, but it is impossible. If you think it essentially a bad title, I will make further trials.

To J. T. Fields, Boston

Salem, January 20th. 1850.

My dear Fields,

I am truly glad that you like the introduction; for I was rather afraid that it might appear absurd and impertinent to be talking about myself, when nobody, that I know of, has requested any information on that subject.

As regards the size of the book, I have been thinking a good deal about it. Considered merely as a matter of taste and beauty, the form of publication which you recommend seems to me much preferable to that of the 'Mosses.[1] In the present case, however, I have some doubts of the expediency; because, if the book is made up entirely of 'The Scarlet Letter,' it will be too sombre. I found it impossible to relieve the shadows of the story with so much light as I would gladly have thrown in. Keeping so close to its point as the tale does, and diversified no otherwise than by turning different sides of the same dark idea to the reader's eye, it will weary very many people, and disgust some. Is it safe, then, to stake the fate of the book entirely on this one chance? A hunter loads his gun with a bullet and several buck-shot; and, following his sagacious example, it was my purpose to conjoin the one long story with half a dozen shorter ones; so that, failing to kill the public outright with my biggest and heaviest lump of lead, I might have other chances with the smaller bits, individually and in the aggregate.

However, I am willing to leave these considerations to your judgment, and should not be sorry to have you decide for the separate publication.

In this latter event, it appears to me that the only proper title for the book would be 'The Scarlet Letter'; for 'The Custom House' is merely introductory—an entrance-hall to the magnificent edifice which I throw open to my guests. It would be funny, if, seeing the further passages so dark and dismal, they should all choose to stop there!

If 'The Scarlet Letter' is to be the title, would it not be well to print it on the title-page in red ink? I am not quite sure about the good taste of so doing; but it would certainly be piquant and appropriate—and, I think, attractive to the great gull whom we are endeavoring to circumvent.

Very truly Yours,
Nathl Hawthorne

J. T. Fields, Esq.

1. In *Yesterdays with Authors*, p. 51, James T. Fields claimed that "after reading the first chapters of the story," he persuaded NH "to elaborate it, and publish it as a separate work."

[*An undated draft of the letter:*]

As regards the book, I have been thinking and considering—I was rather afraid that it appears sagacious absurd and impertinent to have some doubts, of the introduction to the book, which you recommend. I have found it impossible to relieve the shadows of the story with so much light as I would gladly stake the fate of the book entirely on the public. However, I am willing to leave these considerations to your judgment, and should not be sorry to have you decide for the separate publication.

If the Judgment Letter is to be the title—print it on the title page in red ink. I think that the only proper title for the book would be the Scarlet Letter. I am quite sure about the taste of so doing. I think it is attractive and appropriate—

To Horatio Bridge, Portsmouth

Salem, Feby 4th. 1850.

Dear Bridge,

I finished my book only yesterday; one end being in the press in Boston, while the other was in my head here in Salem—so that, as you see, the story is at least fourteen miles long.

I should make you a thousand apologies for being so negligent a correspondent; if you did not know me of old, and as you have tolerated me so many years, I do not fear that you will give me up now. The fact is, I have a natural abhorrence of pen and ink, and nothing short of absolute necessity ever drives me to them.[1]

My book, the publisher tells me, will not be out before April.[2] He speaks of it in tremendous terms of approbation; so does Mrs Hawthorne, to whom I read the conclusion, last night. It broke her heart and sent her to bed with a grievous headache—which I look upon as triumphant success![3] Judging from its effect on her and the publisher, I may calculate on what bowlers call a 'ten-strike.'[4] Yet I do not make any such calculation. Some portions of the book are powerfully written; but my writings do not, nor ever will, appeal to the broadest class of sympathies, and therefore will not attain a very wide popularity. Some like them very much; others care nothing for them, and see nothing in them. There is an introduction to this book—giving a sketch of my Custom-House life, with an imaginative touch here and there—which perhaps may be more

1. In a letter to John Jay of June 12, 1849, Bridge had mentioned NH's "peculiarly retiring disposition" that "prevents him from applying to anyone for literary employment, or authorizing his friends to do so." See Lease, "Hawthorne and *Blackwood's* in 1849," *Jahrbuch für Amerikastudien*, XIV (1969): 153.
2. *SL* was published March 16.
3. SH wrote to Mary Mann on February 12: "I do not know what you will think of the Romance. It is most powerful, & contains a moral as terrific & stunning as a thunder bolt. It shows that the Law cannot be broken" (MS, Berg). See also *English Notebooks*, vols. 21 and 22 *of the Centenary Edition of the Works of Nathaniel Hawthorne*, ed. Thomas Woodson and Bill Ellis (Columbus: Ohio State UP, 1997), p. 225.
4. Five years later Hawthorne would enter another account of this moment in his *English Notebooks.* He recalls his "emotions when I read the last scene of the *Scarlet Letter* to my wife, just after writing it—tried to read it, rather, for my voice swelled and heaved, as if I were tossed up and down on an ocean, as it subsided after a storm. But I was in a very nervous state, then, having gone through a great diversity and severity of emotion, for many months past. I think I have never overcome my own adamant in any other instance" [*editor of this Norton Critical Edition*].

widely attractive than the main narrative. The latter lacks sunshine. To tell you the truth it is—(I hope Mrs. Bridge is not present)—it is positively a h–ll-fired story, into which I found it almost impossible to throw any cheering light.

This house on Goose Creek,[5] which you tell me of, looks really attractive; but I am afraid there must be a flaw somewhere. I like the rent amazingly. I wish you would look at it, and form your own judgement, and report accordingly; and should you decide favorably I will come myself and see it. But if it appears ineligible to you, I shall let the matter rest there; it being inconvenient for me to leave home—partly because funds are to be husbanded at this juncture of my affairs; and partly because I can ill spare the time, as the winter is the season when my brain-work is chiefly accomplished.

As regards the African Journals, the next time I see my publisher, I will make inquiries and ask advice as to how it is best to use them up. It appeared to me that the additional matter would not be copious enough to make a volume by itself; so that the basis would have to be the African Cruiser. If I settle down by you, you could easily talk a volume, which I would write down.

I should like to give up the house which I now occupy, at the beginning of April; and must soon make a decision as to where I shall go.[6] I long to get into the country; for my health, latterly, is not quite what it has been, for many years past. I should not long stand such a life of bodily inactivity and mental exertion as I have led for the last few months. An hour or two of daily labor in a garden, and a daily ramble in country air or on the seashore, would keep all right. Here, I hardly go out once a week. Do not allude to this matter in your letters to me; as my wife already sermonizes me quite sufficiently on my habits—and I never own up to not feeling perfectly well. Neither do I feel anywise ill, but only a lack of physical vigor and energy, which re-acts upon the mind. I detest this town so much that I hate to go into the streets, or to have the people see me. Anywhere else, I shall at once be entirely another man.

With our best regards to Mrs. Bridge, I remain,

truly Your friend,
Nath Hawthorne

To Horatio Bridge, Portsmouth

Salem, April 13th. 1850.

Dear Bridge,

I am glad you like the Scarlet Letter; it would have been a sad matter indeed, if I had missed the favorable award of my oldest and friendliest

5. NH had first written to Bridge around August 3, 1849, asking him to look for a house in or near Portsmouth. Bridge replied August 6, describing a house "two stories in height; has twelve acres of good land attached; is a mile from the corner in Kittery" (MS, Berg; see John D. Gordan, "Nathaniel Hawthorne, The Years of Fulfillment, 1804–1853," *Bulletin of the New York Public Library*, 59 [1955], p. 216).

6. SH wrote to Mary Mann on February 12 that since NH wanted seclusion they were considering a house at "the Upper Falls" (Essex Falls), and one at Hamilton, villages in Essex County; on the fourteenth she wrote to her mother that they were considering a house in West Cambridge, near the Longfellows. She also inquired of Longfellow about Caroline Tappan's "Red House" in Lenox, presently occupied by S. G. Ward's farmer, Luther Butler (MSS, Berg).

critic. The other day, I met with your notice of 'Twice-told Tales,' for the Augusta Age;[1] and I really think that nothing better has been said about them since. This book has been highly successful; the first edition having been exhausted in ten days, and the second (5000 copies in all) promising to go off rapidly.

As to the Salem people, I really thought that I had been exceedingly good-natured in my treatment of them. They certainly do not deserve good usage at my hands, after permitting me—(their most distinguished citizen; for they have no other that was ever heard of beyond the limits of the Congressional district)—after permitting me to be deliberately lied down, not merely once, but at two separate attacks, on two false indictments, without hardly a voice being raised in my behalf; and then sending one of the false witnesses to Congress, others to the State legislature, and choosing another as their Mayor.[2] I feel an infinite contempt for them, and probably have expressed more of it than I intended; for my preliminary chapter has caused the greatest uproar that ever happened here since witch-times. If I escape from town without being tarred-and-feathered, I shall consider it good luck. I wish they *would* tar-and-feather me—it would be such an entirely novel kind of distinction for a literary man! And from such judges as my fellow-citizens, I should look upon it as a higher honor than a laurel-crown.

I have taken a cottage in Lenox, and mean to take up my residence there, about the first of May. In the interim, my wife and children are going to stay in Boston, and nothing could be more agreeable to myself than to spend a week or so with you; so that your invitation comes extremely apropos. In fact, I was on the point of writing to propose a visit. We shall remove our household gods from this infernal locality, tomorrow or next day. I will leave my family at Dr. Peabody's, and come to Portsmouth on Friday of this week—or on Saturday at furthest, unless prevented from coming at all. I shall take the train that leaves Boston at 11 o' clock; so, if you happen to be in Portsmouth, that afternoon, please to look after me. I am very glad of this opportunity of seeing you; for I am afraid you will never find your way to Lenox.

I thank Mrs. Bridge for her good wishes as respects my future removals from office; but I should be sorry to anticipate such bad fortune as being ever again appointed to one.[3]

Truly Your friend,
Nathl Hawthorne

1. Bridge's review had appeared in his hometown newspaper in 1837, but has not yet been found. Franklin Pierce, to whom Bridge had sent the review at the time of its appearance, may have given it to NH in Boston in mid-February. See SH to her mother, February 16, 1850 (MS, Berg).
2. Daniel P. King was reelected to a fourth term in the U.S. House of Representatives in 1848, and Charles W. Upham had served in the Massachusetts House of Representatives from 1840 to 1849. Nathaniel Silsbee, Jr., was mayor of Salem.
3. Bridge had written to Franklin Pierce, November 4, 1849: "Perhaps [NH] may remain [at Lenox] unless at the end of this Administration, he should have a good office tendered him. He deserves it richly, for his removal has made a great flutter in the Whig flock. I trust that this change will be beneficial to H. in two ways—first by making him work with his pen; and then by giving him an office which will enable him to lay up something beyond a bare support. The Democrats will be sure to remember him for his removal showed how popular and how deserving he is" (MS, Library of Congress).

To J. T. Fields, Boston

Lenox, Jany 27. 1851

Dear Fields,

I intend to put the House of the Seven Gables into the express man's hands to-day; so that, if you do not soon receive it, you may conclude that it has miscarried—in which case, I shall not consent to the Universe existing a moment longer. I have no copy of it, except the wildest scribble of a first draught; so that it could never be restored.[1]

It has met with extraordinary success from that portion of the public to whose judgement it has been submitted;—viz, from my wife.[2] I likewise prefer it to the Scarlet Letter; but an author's opinion of his book, just after completing it, is worth little or nothing; he being then in the hot or cold fit of a fever, and certain to rate it too high or too low.

It has undoubtedly one disadvantage, in being brought so close to the present time; whereby its romantic improbabilities become more glaring.

I deem it indispensable that the proof-sheets should be sent me for correction. It will cause some delay, no doubt, but probably not much more than if I lived at Salem. At all events, I don't see how it can be helped. My autography is sometimes villainously blind; and it is odd enough that wherever the printers do mistake a word, it is just the very jewel of a word, worth all the rest of the Dictionary. When the Twice-told Tales are ready, I wish you would give a copy to Longfellow and another to Hillard on my part. I should be glad to have two or three sent hither.

I observe, in one of your catalogues, that you advertise a handsome edition of the New Testament.[3] Will you be kind enough to send it to me, when next you are making up a packet? Did not I suggest to you, last summer, the publication of the Bible, in ten or twelve 12mo. volumes? I think it would have great success; and, at least (but, as a bookseller, I suppose this is the very smallest of your cares) it would result in the salvation of a great many souls, who will never find their way to Heaven, if left to learn it from the inconvenient editions of the Scriptures, now in use. It is very singular that this form of publishing the Bible, in a single bulky or closely-printed volume, should be so long continued. It was first adopted I suppose as being the universal mode of publication at the time when the Bible was translated. Shakespeare, and the other old dramatists and poets, were first published in the same form; but all of them have long since been broken into dozens and scores of portable and readable volumes—and why not the Bible?[4]

1. Fields had written on January 22, "it will be a great thing to get out the Vol. before March as at that time I shall go to the South & intend to sell a great many copies among the Book-sellers. Take this with yr plans if you please & decide with me it will be well on receipt of this, if you are all ready, to send me at once the Mss." (MS, Berg).
2. SH wrote to her mother on this date of *The House of the Seven Gables* (*HSG*): "its depth of wisdom, its high tone, the flowers of Paradise scattered over all the dark places . . ." (MS, Berg).
3. Probably *The New Testament of Our Lord and Saviour Jesus Christ,* published by Ticknor in 1842 and 1844. See *The Cost Books of Ticknor and Fields and Their Predecessors, 1832–1858,* ed. Warren S. Tryon and William Charvat (New York: Bibliographical Society of America, 1949), p. 73.
4. Fields replied on January 30 that such a plan "would be very well for a rich concern able and willing to lose a vast sum of money in the Cause of human salvation. We have the will but not the lucre" (MS, Berg).

I congratulate you on the laurels (additional to those already encircling your brow) which you will gain as poet of the P. B. K.[5] I am very proud of my publisher.

Those Osgood monumental scamps have paid me nothing.

Somebody has written to condole with me on an attack in the Church Review, in reference to the Introduction to the Scarlet Letter, and the work itself.[6] If really good, I should be glad to see it; but unless particularly so, I do not care about it. I think it essential to my success as an author, to have some bitter enemies.

The certificate of deposit of one hundred dollars was duly received. I had something else to say, but have forgotten what.

Truly Yours,
Nathl Hawthorne.

NATHANIEL HAWTHORNE

From Life of Franklin Pierce†

When the series of measures known under the collective term of The Compromise were passed by Congress, in 1850, and put to so searching a test, here at the north, the reverence of the people for the constitution, and their attachment to the Union, General Pierce was true to the principles which he had long ago avowed. At an early period of his congressional service, he had made known, with the perfect frankness of his character, those opinions upon the slavery question which he has never since seen occasion to change in the slightest degree. There is an unbroken consistency in his action with regard to this matter. It is entirely of a piece, from his first entrance upon public life until the moment when he came forward, while many were faltering, to throw the great weight of his character and influence into the scale in favor of those measures through which it was intended to redeem the pledges of the constitution, and to preserve and renew the old love and harmony among the sisterhood of states. His approval embraced the whole series of these acts, as well those which bore hard upon northern views and sentiments as those in which the south deemed itself to have made more than reciprocal concessions.

No friend nor enemy, that knew Franklin Pierce, would have expected him to act otherwise. With his view of the whole subject, whether looking at it through the medium of his conscience, his feelings, or his intellect, it was impossible for him not to take his stand as the unshaken advocate of Union, and of the mutual steps of compromise which that great object unquestionably demanded. The fiercest, the least scrupulous, and the most consistent of those who battle against slavery recognize the same fact that he does. They see that merely human wisdom and human efforts cannot subvert it except by tearing to pieces the constitution, breaking

5. Fields's poem, delivered before the Harvard chapter of Phi Beta Kappa July 26, 1853, has never been published.
6. L. W. Mansfield had written on January 22. See *Memories*, pp. 140–41. Arthur Cleveland Coxe, "The Writings of Hawthorne," *Church Review*, III (January, 1851), 489–511. See Faust, pp. 79–85: *The Critical Heritage*, pp. 179–84.
† From *Life of Franklin Pierce* (Boston: Ticknor, Reed, and Fields, 1852), pp. 110–19.

the pledges which it sanctions, and severing into distracted fragments that common country which Providence brought into one nation, through a continued miracle of almost two hundred years, from the first settlement of the American wilderness until the revolution. In the days when, a young member of Congress, he first raised his voice against agitation, Pierce saw these perils and their consequences. He considered, too, that the evil would be certain, while the good was, at best, a contingency, and (to the clear, practical foresight with which he looked into the future) scarcely so much as that, attended, as the movement was and must be, during its progress, with the aggravated injury of those whose condition it aimed to ameliorate, and terminating, in its possible triumph,—if such possibility there were,—with the ruin of two races which now dwelt together in greater peace and affection, it is not too much to say, than had ever elsewhere existed between the taskmaster and the serf.

Of course, there is another view of all these matters. The theorist may take that view in his closet; the philanthropist by profession may strive to act upon it uncompromisingly, amid the tumult and warfare of his life. But the statesman of practical sagacity—who loves his country as it is, and evolves good from things as they exist, and who demands to feel his firm grasp upon a better reality before he quits the one already gained—will be likely here, with all the greatest statesmen of America, to stand in the attitude of a conservative. Such, at all events, will be the attitude of Franklin Pierce. We have sketched some of the influences amid which he grew up, inheriting his father's love of country, mindful of the old patriot's valor in so many conflicts of the revolution, and having close before his eyes the example of brothers and relatives, more than one of whom have bled for America, both at the extremest north and farthest south; himself, too, in early manhood, serving the Union in its legislative halls, and, at a maturer age, leading his fellow-citizens, his brethren, from the widest-sundered states, to redden the same battle fields with their kindred blood, to unite their breath into one shout of victory, and perhaps to sleep, side by side, with the same sod over them. Such a man, with such hereditary recollections, and such a personal experience, must not narrow himself to adopt the cause of one section of his native country against another. He will stand up, as he has always stood, among the patriots of the whole land. And if the work of anti-slavery agitation, which, it is undeniable, leaves most men who earnestly engage in it with only half a country in their affections—if this work must be done, let others do it.

Those northern men, therefore, who deem the great cause of human welfare all represented and involved in this present hostility against southern institutions, and who conceive that the world stands still except so far as that goes forward—these, it may be allowed, can scarcely give their sympathy or their confidence to the subject of this memoir. But there is still another view, and probably as wise a one. It looks upon slavery as one of those evils which divine Providence does not leave to be remedied by human contrivances, but which, in its own good time, by some means impossible to be anticipated, but of the simplest and easiest operation, when all its uses shall have been fulfilled, it causes to vanish like a dream. There is no instance, in all history, of the human will and intellect having perfected any great moral reform by methods which it adapted to that end; but the progress of the world, at every step, leaves some evil or wrong on the path behind it, which the wisest of mankind, of their own set purpose, could never have found the way to rectify. Whatever contributes to the great cause of good, contributes to all its subdivisions and varieties; and, on this score, the lover of his race, the enthusiast, the philanthropist of whatever theory, might lend his aid to put a man, like the one before us, into the leadership of the world's affairs.

How firm and conscientious was General Pierce's support of the Compromise, may be estimated from his conduct in reference to the reverend John Atwood. In the foregoing pages it has come oftener in our way to illustrate the bland and prepossessing features of General Pierce's character, than those sterner ones which must necessarily form the bones, so to speak, the massive skeleton, of any man who retains an upright attitude amidst the sinister influences of public life. The transaction now alluded to affords a favorable opportunity for indicating some of these latter traits.

In October, 1850, a democratic convention, held at Concord, nominated Mr. Atwood as the party's regular candidate for governor. The Compromise, then recent, was inevitably a prominent element in the discussions of the convention; and a series of resolutions were adopted, bearing reference to this great subject, fully and unreservedly indorsing the measures comprehended under it, and declaring the principles on which the Democracy of the state was about to engage in the gubernatorial contest. Mr. Atwood accepted the nomination, acceding to the platform thus tendered him, taking exceptions to none of the individual resolutions, and, of course, pledging himself to the whole by the very act of assuming the candidacy, which was predicated upon them.

The reverend candidate, we should conceive, is a well-meaning, and probably an amiable man. In ordinary circumstances, he would, doubtless, have gone through the canvass triumphantly, and have administered the high office to which he aspired with no discredit to the party that had placed him at its head. But the disturbed state of the public mind on the Compromise Question rendered the season a very critical one; and Mr. Atwood, unfortunately, had that fatal weakness of character, which, however respectably it may pass in quiet times, is always bound to make itself pitiably manifest under the pressure of a crisis. A letter was addressed to him by a committee, representing the party opposed to the Compromise, and with whom, it may be supposed, were included those who held the more thorough-going degrees of anti-slavery sentiment. The purpose of the letter was to draw out an expression of Mr. Atwood's opinion on the abolition movement generally, and with an especial reference to the Fugitive Slave Law, and whether, as chief magistrate of the state, he would favor any attempt for its repeal. In an answer of considerable length, the candidate expressed sentiments that brought him unquestionably within the Free Soil pale, and favored his correspondents, moreover, with a pretty decided judgment as to the unconstitutional, unjust, and oppressive character of the Fugitive Slave act.

During a space of about two months, this very important document was kept from the public eye. Rumors of its existence, however, became gradually noised abroad, and necessarily attracted the attention of Mr. Atwood's democratic friends. Inquiries being made, he acknowledged the existence of the letter, but averred that it had never been delivered, that it was merely a rough draught, and that he had hitherto kept it within his own control, with a view to more careful consideration. In accordance with the advice of friends, he expressed a determination, and apparently in good faith, to suppress the letter, and thus to sever all connection with the anti-slavery party. This, however, was now beyond his power. A copy of the letter had been taken; it was published, with high commendations, in the anti-slavery newspapers; and Mr. Atwood was exhibited in the awkward predicament of directly avowing sentiments on the one hand which he had implicitly disavowed, on the other, of accepting a nomination based on principles diametrically opposite.

The candidate appears to have apprehended this disclosure, and he hurried to Concord, and sought counsel of General Pierce, with whom he was on terms of personal kindness, and between whom and himself, heretofore, there had never been a shade of political difference. An interview with the general and one or two other gentlemen ensued. Mr. Atwood was cautioned against saying or writing a word that might be repugnant to his feelings or his principles; but, voluntarily, and at his own suggestion, he now wrote, for publication, a second letter, in which he retracted every objectionable feature of his former one, and took decided ground in favor of the Compromise, including all its individual measures. Had he adhered to this latter position, he might have come out of the affair, if not with the credit of consistency, yet, at least, as a successful candidate in the impending election. But his evil fate, or, rather, the natural infirmity of his character, was not so to be thrown off. The very next day, unhappily, he fell into the hands of some of his anti-slavery friends, to whom he avowed a constant adherence to the principles of his first letter, describing the second as having been drawn from him by importunity, in an excited state of his mind, and without a full realization of its purport.

It would be needlessly cruel to Mr. Atwood to trace, with minuteness, the further details of this affair. It is impossible to withhold from him a certain sympathy, or to avoid feeling that a very worthy man, as the world goes, had entangled himself in an inextricable knot of duplicity and tergiversation, by an ill-advised effort to be two opposite things at once. For the sake of true manhood, we gladly turn to consider the course adopted by General Pierce.

The election for governor was now at a distance of only a few weeks; and it could not be otherwise than a most hazardous movement for the democratic party, at so late a period, to discard a candidate with whom the people had become familiar. It involved nothing less than the imminent peril of that political supremacy which the party had so long enjoyed. With Mr. Atwood as candidate, success might still be considered certain. To a short-sighted and a weak man, it would have appeared the obvious policy to patch up the difficulty, and, at all events, to conquer, under whatever leadership, and with whatever allies. But it was one of those junctures which test the difference between the man of principle and the mere politician—the man of moral courage and him who yields to temporary expediency. General Pierce could not consent that his party should gain a nominal triumph, at the expense of what he looked upon as its real integrity and life. With this view of the matter, he had no hesitation in his course; nor could the motives which otherwise would have been strongest with him—pity for the situation of an unfortunate individual, a personal friend, a democrat, as Mr. Atwood describes himself, of nearly fifty years' standing—incline him to mercy, where it would have been fatal to his sense of right. He took decided ground against Mr. Atwood. The convention met again, and nominated another candidate. Mr. Atwood went into the field as the candidate of the anti-slavery party, drew off a sufficient body of democrats to defeat the election by the people, but was himself defeated in the legislature.

Thus, after exhibiting to the eyes of mankind (or such portion of mankind as chanced to be looking in that direction) the absurd spectacle of a gentleman of extremely moderate stride attempting a feat that would have baffled a Colossus,—to support himself, namely, on both margins of the impassable chasm that has always divided the anti-slavery faction from the New Hampshire Democracy,—this ill-fated man attempted first to throw himself upon one side of the gulf, then on the other, and finally tumbled headlong into the bottomless depth between. His case presents a painful, but very curious and instructive instance of the troubles that beset weakness, in those emergencies which demand steadfast moral strength and energy—of which latter type of manly character there can be no truer example than Franklin Pierce.

CRITICISM

Nineteenth-Century Reviews of *The Scarlet Letter*

EVERT A. DUYCKINCK

The Scarlet Letter†

Mr. Hawthorne introduces his new story to the public, the longest of all that he has yet published, and most worthy in this way to be called a romance, with one of those pleasant personal descriptions which are the most charming of his compositions, and of which we had so happy an example in the preface to his last collection, the Mosses from an Old Manse. In these narratives everything seems to fall happily into its place. The style is simple and flowing, the observation accurate and acute; persons and things are represented in their minutest shades, and difficult traits of character presented with an instinct which art might be proud to imitate. They are, in fine, little cabinet pictures exquisitely painted. The readers of the Twice Told Tales will know the pictures to which we allude. They have not, we are sure, forgotten Little Annie's Ramble, or the Sights from a Steeple. This is the Hawthorne of the present day in the sunshine. There is another Hawthorne less companionable, of sterner Puritan aspect, with the shadow of the past over him, a reviver of witchcrafts and of those dark agencies of evil which lurk in the human soul, and which even now represent the old gloomy historic era in the microcosm and eternity of the individual; and this Hawthorne is called to mind by such tales as the Minister's Black Veil or the Old Maid in the Winding Sheet, and reappears in the Scarlet Letter, a romance. Romantic in sooth! Such romance as you may read in the intensest sermons of old Puritan divines, or in the mouldy pages of that Marrow of Divinity, the ascetic Jeremy Taylor.[1]

The Scarlet Letter is a psychological romance. The hardiest Mrs. Malaprop[2] would never venture to call it a novel. It is a tale of remorse, a study of character in which the human heart is anatomized, carefully, elaborately, and with striking poetic and dramatic power. Its incidents are simply these. A woman in the early days of Boston becomes the subject of the discipline of the court of those times, and is condemned to

† From *Literary World* (March 30, 1850). A few months later, Duyckinck would publish Herman Melville's well-known review essay, "Hawthorne and His Mosses," in the *Literary World*. Melville would suggest the existence of "two Hawthornes" (one light, one dark) in very similar terms to those Duyckinck uses early in this review.

1. Jeremy Taylor (1613–1667), English bishop and theologian.
2. Character in Richard Sheridan's comedy *The Rivals* (1775), whose humor resulted from substituting an incorrect but similar-sounding word.

stand in the pillory and wear henceforth, in token of her shame, the scarlet letter A attached to her bosom. She carries her child with her to the pillory. Its other parent is unknown. At this opening scene her husband from whom she had been separated in Europe, preceding him by ship across the Atlantic, reappears from the forest, whither he had been thrown by shipwreck on his arrival. He was a man of a cold intellectual temperament, and devotes his life thereafter to search for his wife's guilty partner and a fiendish revenge. The young clergyman of the town, a man of a devont sensibility and warmth of heart, is the victim, as this Mephistophilean old physician fixes himself by his side to watch over him and protect his health, an object of great solicitude to his parishioners, and, in reality, to detect his suspected secret and gloat over his tortures. This slow, cool, devilish purpose, like the concoction of some sublimated hell broth, is perfected gradually and inevitably. The wayward, elfish child, a concentration of guilt and passion, binds the interests of the parties together, but throws little sunshine over the scene. These are all the characters, with some casual introductions of the grim personages and manners of the period, unless we add the scarlet letter, which, in Hawthorne's hands, skilled to these allegorical, typical semblances, becomes vitalized as the rest. It is the hero of the volume. The denouement is the death of the clergyman on a day of public festivity, after a public confession in the arms of the pilloried, branded woman. But few as are these main incidents thus briefly told, the action of the story, or its passion, is "long, obscure, and infinite." It is a drama in which thoughts are acts. The material has been thoroughly fused in the writer's mind, and springs forth an entire, perfect creation. We know of no American tales except some of the early ones of Mr. Dana,[3] which approach it in conscientious completeness. Nothing is slurred over, superfluous, or defective. The story is grouped in scenes simply arranged, but with artistic power, yet without any of those painful impressions which the use of the words, as it is the fashion to use them, "grouping" and "artistic" excite, suggesting artifice and effort at the expense of nature and ease.

Mr. Hawthorne has, in fine, shown extraordinary power in this volume, great feeling and discrimination, a subtle knowledge of character in its secret springs and outer manifestations. He blends, too, a delicate fancy with this metaphysical insight. We would instance the chapter towards the close, entitled "The Minister in a Maze," where the effects of a diabolic temptation are curiously depicted, or "The Minister's Vigil," the night scene in the pillory. The atmosphere of the piece also is perfect. It has the mystic element, the weird forest influences of the old Puritan discipline and era. Yet there is no affrightment which belongs purely to history, which has not its echo even in the unlike and perversely commonplace customhouse of Salem. Then for the moral. Though severe, it is wholesome; and is a sounder bit of Puritan divinity than we have been of late accustomed to hear from the degenerate successors of Cotton Mather. We hardly know another writer who has lived so much among the new school who would have handled this delicate subject without an infusion of George

3. Richard Henry Dana (1815–1882), American writer best-known for *Two Years Before the Mast* (1840).

Sand.[4] The spirit of his old Puritan ancestors, to whom he refers in the preface, lives in Nathaniel Hawthorne.

We will not mar the integrity of the Scarlet Letter by quoting detached passages. Its simple and perfect unity forbids this. Hardly will the introductory sketch bear this treatment without exposing the writer to some false impressions; but as evidence of the possession of a style faithfully and humorously reflective of the scenes of the passing hour, which we earnestly wish he may pursue in future volumes, we may give one or two separable sketches.

There is a fine, natural portrait of General Miller, the collector; equal in its way to the Old Inspector, the self-sufficing gourmand lately presented in our journal; and there are other officials as well done. A page, however, of as general application, and of as sound profit as any in this office-seeking age, is that which details, in its mental bearing, "The Paralysis of Office."

* * *

The personal situation of Nathaniel Hawthorne—in whom the city by his removal lost an indifferent official, and the world regained a good author—is amusingly presented in this memoir of "A Decapitated Surveyor."

* * *

And a literary man long may he remain, an honor and a support to the craft, of genuine worth and fidelity, to whom no word is idle, given to the world no truer product of the American soil, though of a peculiar culture, than Nathaniel Hawthorne.

EDWIN PERCY WHIPPLE

Review of New Books†

In this beautiful and touching romance Hawthorne has produced something really worthy of the fine and deep genius which lies within him. The "Twice Told Tales," and "Mosses from an Old Manse," are composed simply of sketches and stories, and although such sketches and stories as few living men could write, they are rather indications of the possibilities of his mind than realizations of its native power, penetration, and creativeness. In "The Scarlet Letter" we have a complete work, evincing a true artist's certainty of touch and expression in the exhibition of characters and events, and a keen-sighted and far-sighted vision into the essence and purpose of spiritual laws. There is a profound philosophy underlying the story which will escape many of the readers whose attention is engrossed by the narrative.

4. George Sand (1804–1876), pseudonym of Amandine-Aurore-Lucile Dupin, French Romantic novelist, noted for her numerous love affairs, and considered risqué by nineteenth-century standards.

† From *Graham's Magazine* 36.5 (May 1850): 345–46.

The book is prefaced by some fifty pages of autobiographical matter, relating to the author, his native city of Salem, and the Custom House, from which he was ousted by the Whigs. These pages, instinct with the vital spirit of humor, show how rich and exhaustless a fountain of mirth Hawthorne has at his command. The whole representation has the dreamy yet distinct remoteness of the purely comic ideal. The view of Salem streets; the picture of the old Custom House at the head of Derby's wharf, with its torpid officers on a summer's afternoon, their chairs all tipped against the wall, chatting about old stories, "while the frozen witticisms of past generations were thawed out, and came bubbling with laughter from their lips"—the delineation of the old Inspector, whose "reminiscences of good cheer, however ancient the date of the actual banquet, seemed to bring the savor of pig or turkey under one's very nostrils," and on whose palate there were flavors "which had lingered there not less than sixty or seventy years, and were still apparently as fresh as that of the mutton-chop which he had just devoured for his breakfast," and the grand view of the stout Collector, in his aged heroism, with the honors of Chippewa and Fort Erie on his brow, are all encircled with that visionary atmosphere which proves the humorist to be a poet, and indicates that his pictures are drawn from the images which observation has left on his imagination. The whole introduction, indeed, is worthy of a place among the essays of Addison and Charles Lamb.[1]

With regard to "The Scarlet Letter," the readers of Hawthorne might have expected an exquisitely written story; expansive in sentiment, and suggestive in characterization, but they will hardly be prepared for a novel of so much tragic interest and tragic power, so deep in thought and so condensed in style, as is here presented to them. It evinces equal genius in the region of great passions and elusive emotions, and bears on every page the evidence of a mind thoroughly alive, watching patiently the movements of morbid hearts when stirred by strange experiences, and piercing, by its imaginative power, directly through all the externals to the core of things. The fault of the book, if fault it have, is the almost morbid intensity with which the characters are realized, and the consequent lack of sufficient geniality in the delineation. A portion of the pain of the author's own heart is communicated to the reader, and although there is great pleasure received while reading the volume, the general impression left by it is not satisfying to the artistic sense. Beauty bends to power throughout the work, and therefore the power displayed is not always beautiful. There is a strange fascination to a man of contemplative genius in the psychological details of a strange crime like that which forms the plot of the Scarlet Letter, and he is therefore apt to become, like Hawthorne, too painfully anatomical in his exhibition of them.

If there be, however, a comparative lack of relief to the painful emotions which the novel excites, owing to the intensity with which the author concentrates attention on the working of dark passions, it must be confessed that the moral purpose of the book is made more definite by this very deficiency. The most abandoned libertine could not read the volume

1. Charles Lamb (1775–1834), English essayist and poet, most famous for his collection *Essays of Elia* (1823, 1833). Joseph Addison (1672–1719), English writer and poet, coauthor and coeditor, with Richard Steele, of the *Tatler* and the *Spectator*.

without being thrilled into something like virtuous resolution, and the roué would find that the deep-seeing eye of the novelist had mastered the whole philosophy of that guilt of which practical roués are but childish disciples. To another class of readers, those who have theories of seduction and adultery modeled after the French school of novelists, and whom libertinism is of the brain, the volume may afford matter for very instructive and edifying contemplation; for, in truth, Hawthorne, in The Scarlet Letter, has utterly undermined the whole philosophy on which the French novels rest, by seeing farther and deeper into the essence both of conventional and moral laws; and he has given the results of his insight, not in disquisitions and criticisms, but in representations more powerful even than those of Sue, Dumas,[2] or George Sand. He has made his guilty parties end, not as his own fancy or his own benevolent sympathies might dictate, but as the spiritual laws, lying back of all persons, dictated to him. In this respect there is hardly a novel in English literature more purely objective.

As everybody will read "The Scarlet Letter," it would be impertinent to give a synopsis of the plot. The principal characters, Dimmesdale, Chillingworth, Hester, and little Pearl, all indicate a firm grasp of individualities, although from the peculiar method of the story, they are developed more in the way of logical analysis than by events. The descriptive portions of the novel are in a high degree picturesque and vivid, bringing the scenes directly home to the heart and imagination, and indicating a clear vision of the life as well as forms of nature. Little Pearl is perhaps Hawthorne's finest poetical creation, and is the very perfection of ideal impishness.

In common, we trust, with the rest of mankind, we regretted Hawthorne's dismissal from the Custom House, but if that event compels him to exert his genius in the production of such books as the present, we shall be inclined to class the Honorable Secretary of the Treasury among the great philanthropists. In his next work we hope to have a romance equal to The Scarlet Letter in pathos and power, but more relieved by touches of that beautiful and peculiar humor, so serene and so searching, in which he excels almost all living writers.

ANNE W. ABBOTT

The Scarlet Letter†

That there is something not unpleasing to us in the misfortunes of our best friends, is a maxim we have always spurned, as a libel on human nature. But we must be allowed, in behalf of Mr. Hawthorne's friend and gossip, the literary public, to rejoice in the event—a "removal" from the office of Surveyor of the Customs for the port of Salem,—which has brought him back to our admiring, and, we modestly hope, congenial society, from associations and environments which have confessedly been detrimental to his genius, and to those qualities of heart, which, by an unconscious

2. Alexandre Dumas (1802–1870), French historical novelist best known for *The Three Musketeers* (1844) and *The Count of Monte Cristo* (1845). Eugène Sue (1804–1857), French novelist of the Parisian underworld and slum life.

† From *North American Review* 71.148 (July 1850): 135–48.

revelation through his style, like the involuntary betrayal of character in a man's face and manners, have won the affection of other than personal friends. We are truly grieved at the savage "scratches" our phœnix has received from the claws of the national eagle, scratches gratuitous and unprovoked, whereby his plumage remains not a little ruffled, if his breast be not very deeply lacerated. We hope we do not see tendencies to *self immolation* in the introductory chapter to this volume. It seems suicidal to a most enviable fame, to show the fine countenance of the sometime denizen of Concord Parsonage, once so serene and full of thought, and at the same time so attractively arch, now cloudy and peevish, or dressed in sardonic smiles, which would scare away the enthusiasm of less hearty admirers than those he "holds by the button." The pinnacle on which the "conscience of the beautiful" has placed our author's graceful image is high enough, however, to make slight changes from the wear and tear of out-door elements, highway dust, and political vandalism, little noticed by those accustomed to look lovingly up to it. Yet they cannot be expected to regret a "removal," which has saved those finer and more delicate traits, in which genius peculiarly manifests itself, from being worn away by rough contact, or obliterated by imperceptible degrees through the influence of the atmosphere.

Mr. Hawthorne's serious apprehensions on this subject are thus candidly expressed:—

> "I began to grow melancholy and restless; continually prying into my mind, to discover which of its poor properties were gone, and what degree of detriment had already accrued to the remainder. I endeavored to calculate how much longer I could stay in the Custom House, and yet go forth a man. To confess the truth, it was my greatest apprehension,—as it would never be a measure of policy to turn out so quiet an individual as myself, and it being hardly in the nature of a public officer to resign,—it was my chief trouble, therefore, that I was likely to grow gray and decrepit in the Surveyorship, and become much such another animal as the old Inspector. Might it not, in the tedious lapse of official life that lay before me, finally be with me as it was with this venerable friend,—to make the dinner hour the nucleus of the day, and to spend the rest of it, as an old dog spends it, asleep in the sunshine or the shade? A dreary look-forward this, for a man who felt it to be the best definition of happiness to live throughout the whole range of his faculties and sensibilities! But, all this while, I was giving myself very unnecessary alarm. Providence had meditated better things for me than I could possibly imagine for myself."

A man who has so rare an individuality to lose may well shudder at the idea of becoming a soulless machine, a sort of official scarecrow, having only so much of manly semblance left as will suffice to warn plunderers from the property of "Uncle Sam." Haunted by the horror of mental annihilation, it is not wonderful that he should look askance at the drowsy row of officials, as they reclined uneasily in tilted chairs, and should measure their mental torpidity by the length of time they had been subjected to the soul-exhaling process in which he had not yet got beyond the conscious stage. It was in pure apprehension, let us charitably hope, and not in a satirical, and far less a malicious, mood, that he describes one of them as retaining barely enough of the moral and spiritual nature to keep him from going upon all fours, and possessing neither soul, heart, nor mind more worthy of immortality than the spirit of the beast, which "goeth downward." Judging his aged colleagues thus, well might the young publican, as yet spiritually alive, stand aghast! A man may be excusable for starving his *intellect*, if Providence has thrown him into a situation where its dainty palate cannot be gratified. But for the well being of his *moral nature*, he is more strictly responsible, and has no right, under any circumstances, to remain in a position where, from causes beyond his control, his conscience is deprived of its supremacy over the will, and policy or expediency, whether public or selfish, placed upon its throne. "Most men," says our honest author, "suffer moral detriment from this mode of life," from causes which, (having just devoted four pages to a full-length caricature,) he had not space to hint at, except in the following pithy admonition to the aspirants after a place in the Blue Book.

> "Uncle Sam's gold—meaning no disrespect to the worthy old gentleman—has, in this respect, a quality of enchantment, like that of the Devil's wages. Whoever touches it should look well to himself, or he may find the bargain to go hard against him, involving, if not his soul, yet many of his better attributes; its sturdy force, its courage and constancy, its truth, its self-reliance, and all that gives the emphasis to manly character."

It was great gain for a man like Mr. Hawthorne to depart this truly unprofitable life; but we wish that his demise had been quiet and Christian, and not by violence. We regret that any of the bitterness of heart engendered by the political battle, and by his subsequent decapitulation without being judged by his peers, should have come with him to a purer and higher state of existence. That a head should fall, and even receive "an ignominious kick," is but a common accident in a party struggle, and would be of no more consequence to the world in Mr. Hawthorne's case than any other, (the metaphorical head not including brains,) provided the spirit had suffered no material injury in the encounter. Of that, however, we have no means of judging, except by comparing this book of recent production with his former writings. Of the "stern and sombre" pictures of the world and human life, external and internal, found in the Scarlet Letter, we shall speak anon. The preface claims some farther notice.

One would conclude, that the mother on whose bosom the writer was cherished in his urchinhood had behaved herself like a very stepmother towards him, showing a vulgar preference of those sons who have gathered, and thrown into her lap, gifts more substantial than garlands and laurel wreaths. This appears from his reluctant and half ashamed confession of attachment to her, and his disrespectful remarks upon her homely and commonplace features, her chilly and unsocial disposition, and those marks of decay and premature age which needed not to be pointed out. The portrait is like, no doubt; but we cannot help imagining the ire of the ancient dame at the unfilial satire. Indeed, a faint echo of the voice of her indignation has arrived at our ears. She complains, that, in anatomizing the characters of his former associates for the entertainment of the public, he has used the scalpel on some subjects, who, though they could not defend themselves, might possibly wince; and that all who came under his hand, living or dead, had probably relatives among his readers, whose affections might be wounded.

Setting this consideration apart, we confess that, to our individual taste, this naughty chapter is more piquant than any thing in the book; the style is racy and pungent, not elaborately witty, but stimulating the reader's attention agreeably by original turns of expression, and unhackneyed combinations of words, falling naturally into their places, as if of their own accord, and not obtained by far seeking and impressment into the service. The sketch of General Miller is airily and lightly done; no other artist could have given so much character to each fine drawn line as to render the impression almost as distinct to the reader's fancy as a portrait drawn by rays of light is to the bodily vision. Another specimen of his word painting, the lonely parlor seen by the moonlight melting into the warmer glow of the fire, while it reminds us of Cowper's much quoted and admired verse, has truly a great deal more of genuine poetry in it. The delineations of wharf scenery, and of the Custom House, with their appropriate figures and personages, are worthy of the pen of Dickens; and really, so far as mere style is concerned, Mr. Hawthorne has no reason to thank us for the compliment; he has the finer touch, if not more genial feeling, of the two. Indeed, if we except a few expressions which savor somewhat strongly of his late unpoetical associations, and the favorite metaphor of the guillotine, which, however apt, is not particularly agreeable to the imagination in such detail, we like the preface better than the tale.

No one who has taken up the Scarlet Letter will willingly lay it down till he has finished it; and he will do well not to pause, for he cannot resume the story where he left it. He should give himself up to the magic power of the style, without stopping to open wide the eyes of his good sense and judgment, and shake off the spell; or half the weird beauty will disappear like a "dissolving view." To be sure, when he closes the book, he will feel very much like the giddy and bewildered patient who is just awaking from his first experiment of the effects of sulphuric ether. The soul has been floating or flying between earth and heaven, with dim ideas of pain and pleasure strangely mingled, and all things earthly swimming dizzily and dreamily, yet most beautiful, before the half shut eye. That the author himself felt this sort of intoxication as well as the willing subjects of his enchantment, we think, is evident in many pages of the last half of the volume. His imagination has sometimes taken him fairly off his feet, insomuch that he seems almost to doubt if there be any firm ground at all,—if we may so judge from such mist-born ideas as the following.

> "But, to all these shadowy beings, so long our near acquaintances,—as well Roger Chillingworth as his companions,—we would fain be merciful. It is a curious subject of observation and inquiry, whether hatred and love be not the same thing at bottom. Each, in its utmost development, supposes a high degree of intimacy and heart-knowledge; each renders one individual dependent for the food of his affections and spiritual life upon another; each leaves the passionate lover, or the no less passionate hater, forlorn and desolate by the withdrawal of his object. Philosophically considered, therefore, the two passions seem essentially the same, except the one happens to be seen in a celestial radiance, and the other in a dusky and lurid glow. In the spiritual world, the old physician and the

minister—mutual victims as they have been—may, unawares, have found their earthly stock of hatred and antipathy transmuted into golden love."

Thus devils and angels are alike beautiful, when seen through the magic glass; and they stand side by side in heaven, however the former may be supposed to have come there. As for Roger Chillingworth, he seems to have so little in common with man, he is such a gnome-like phantasm, such an unnatural personification of an abstract idea, that we should be puzzled to assign him a place among angels, men, or devils. He is no more a man than Mr. Dombey,[1] who sinks down a mere *caput mortuum*, as soon as pride, the only animating principle, is withdrawn. These same "shadowy beings" are much like "the changeling the fairies made o' a benweed." Hester at first strongly excites our pity, for she suffers like an immortal being; and our interest in her continues only while we have hope for her soul, that its baptism of tears will reclaim it from the foul stain which has been cast upon it. We see her humble, meek, self-denying, charitable, and heart-wrung with anxiety for the moral welfare of her wayward child. But anon her humility catches a new tint, and we find it pride; and so a vague unreality steals by degrees over all her most humanizing traits—we lose our confidence in all—and finally, like Undine,[2] she disappoints us, and shows the dream-land origin and nature, when we were looking to behold a Christian.

There is rather more power, and better keeping, in the character of Dimmesdale. But here again we are cheated into a false regard and interest, partly perhaps by the associations thrown around him without the intention of the author, and possibly contrary to it, by our habitual respect for the sacred order, and by our faith in religion, where it has once been rooted in the heart. We are told repeatedly, that the Christian element yet pervades his character and guides his efforts; but it seems strangely wanting. "High aspirations for the welfare of his race, warm love of souls, pure sentiments, natural piety, strengthened by thought and study, and illuminated by revelation—all of which invaluable gold was little better than rubbish" to Roger Chillingworth, are little better than rubbish at all, for any use to be made of them in the story. Mere suffering, aimless and without effect for purification or blessing to the soul, we do not find in God's moral world. The sting that follows crime is most severe in the purest conscience and the tenderest heart, in mercy, not in vengeance, surely; and we can conceive of any cause constantly exerting itself without its appropriate effects, as soon as of a seven years' agony without penitence. But here every pang is wasted. A most obstinate and unhuman passion, or a most unwearying conscience it must be, neither being worn out, or made worse or better, by such a prolonged application of the scourge. Penitence may indeed be life-long; but as for this, we are to understand that there is no penitence about it. We finally get to be quite of the author's mind, that "the only truth that continued to give Mr. Dimmesdale a real existence on this earth, was the anguish in his inmost soul, and the undissembled expression of it in his aspect. Had he once found

1. The rich, arrogant father in Charles Dickens's *Dombey and Son* (1848), who is finally humbled by the loss of his business, wife, and son.
2. A mythical female water nymph who acquires a soul through marriage to a mortal.

power to smile, and wear an aspect of gayety, there had been no such man." He duly exhales at the first gleam of hope, an uncertain and delusive beam, but fatal to his misty existence. From that time he is a fantasy, an opium dream, his faith a vapor, his reverence blasphemy, his charity mockery, his sanctity impurity, his love of souls a ludicrous impulse to teach little boys bad words; and nothing is left to bar the utterance of "a volley of good, round, solid, satisfactory, heaven-defying oaths," (a phrase which seems to smack its lips with a strange *goût!*) but good taste and the mere outward shell, "the buckramed habit of clerical decorum." The only conclusion is, that the shell never possessed any thing real,—never was the Rev. Arthur Dimmesdale, as we have foolishly endeavored to suppose; that he was but a changeling, or an imp in grave apparel, not an erring, and consequently suffering human being, with a heart still upright enough to find the burden of conscious unworthiness and undeserved praise more intolerable than open ignominy and shame, and refraining from relieving his withering conscience from its load of unwilling hypocrisy, if partly from fear, more from the wish to be yet an instrument of good to others, not an example of evil which should weaken their faith in religion. The closing scene, where the satanic phase of the character is again exchanged for the saintly, and the pillory platform is made the stage for a triumphant *coup de théatre*, seems to us more than a failure.

But Little Pearl—gem of the purest water—what shall we say of her? That if perfect truth to childish and human nature can make her a mortal, she is so; and immortal, if the highest creations of genius have any claim to immortality. Let the author throw what light he will upon her, from his magical prism, she retains her perfect and vivid human individuality. When he would have us call her elvish and implike, we persist in seeing only a capricious, roguish, untamed child, such as many a mother has looked upon with awe, and a feeling of helpless incapacity to rule. Every motion, every feature, every word and tiny shout, every naughty scream and wild laugh, come to us as if our very senses were conscious of them. The child is a true child, the only genuine and consistent mortal in the book; and wherever she crosses the dark and gloomy track of the story, she refreshes our spirit with pure truth and radiant beauty, and brings to grateful remembrance the like ministry of gladsome childhood, in some of the saddest scenes of actual life. We feel at once that the author must have a "Little Pearl" of his own, whose portrait, consciously or unconsciously, his pen sketches out. Not that we would deny to Mr. Hawthorne the power to call up any shape, angel or goblin, and present it before his readers in a striking and vivid light. But there is something more than imagination in the picture of "Little Pearl." The heart takes a part in it, and puts in certain inimitable touches of nature here and there, such as fancy never dreamed of, and only a long and loving observation of the ways of childhood could suggest. The most characteristic traits are so interwoven with the story, (on which we do not care to dwell,) that it is not easy to extract a paragraph which will convey much of the charming image to our readers. The most convenient passage for our purpose is the description of Little Pearl playing upon the sea-shore. We take in the figure of the old man as a dark back-ground, or contrast, to heighten the effect.

* * *

[Quotes passage in which Pearl plays on the shore (see pp. 102–03).]

Here follows a dialogue in the spirit of the idea that runs through the book,—that revenge may exist without any overt act of vengeance that could be called such, and that a man who refrains from avenging himself, may be more diabolical in his very forbearance than he who in his passionate rage inflicts what evil he may upon his enemy; the former having that spirit of cold hate which could gloat for years, or forever, over the agonies of remorse and despair, over the anguish bodily and mental, and consequent death or madness, of a fellow man, and never relent—never for a moment be moved to pity. This master passion of hatred, swallowing up all that is undevilish and human in Roger Chillingworth, makes him a pure abstraction at last, a sort of mythical fury, a match for Alecto the Unceasing.

* * *

[Quotes a long passage describing Hester's meeting with Chillingworth in the forest, followed by a long passage describing Pearl at play (see pp. 103–04, 106–07, 108).]

We know of no writer who better understands and combines the elements of the picturesque in writing than Mr. Hawthorne. His style may be compared to a sheet of transparent water, reflecting from its surface blue skies, nodding woods, and the smallest spray or flower that peeps over grassy margin; while in its clear yet mysterious depths we espy rarer and stranger things, which we must dive for, if we would examine. Whether they might prove gems or pebbles, when taken out of the fluctuating medium through which the sun-gleams reach them, is of no consequence to the effect. Every thing charms the eye and ear, and nothing looks like art and pains-taking. There is a naturalness and a continuous flow of expression in Mr. Hawthorne's books, that makes them delightful to read, especially in this our day, when the fear of triteness drives some writers, (even those who might otherwise avoid that reproach,) to adopt an abrupt and dislocated style, administering to our jaded attention frequent thumps and twitches, by means of outlandish idioms and forced inversions, and now and then flinging at our heads an incomprehensible, break-jaw word, which uncivilized missile stuns us to a full stop, and an appeal to authority. No authority can be found, however, which affords any remedy or redress against determined outlaws. After bumping over "rocks and ridges, and grid-iron bridges," in one of these prosaic latter-day omnibuses, how pleasant it is to move over flowery turf upon a spirited, but properly trained Pegasus, who occasionally uses his wings, and skims along a little above *terra firma*, but not with an alarming preference for cloudland or rarefied air. One cannot but wonder, by the way, that the master of such a wizard power over language as Mr. Hawthorne manifests should not choose a less revolting subject than this of the Scarlet Letter, to which fine writing seems as inappropriate as fine embroidery. The ugliness of pollution and vice is no more relieved by it than the gloom of the prison is by the rose tree at its door. There are some palliative expressions used, which cannot, even as a matter of taste, be approved.

Regarding the book simply as a picture of the olden time, we have no fault to find with costume or circumstance. All the particulars given us, (and he is not wearisomely anxious to multiply them to show his research,) are in good keeping and perspective, all in softened outlines and neutral tint, except the ever fresh and unworn image of childhood, which stands out from the canvas in the gorgeously attired "Little Pearl." He forbears to mention the ghastly gallows-tree, which stood hard by the pillory and whipping-post, at the city gates, and which one would think might have been banished with them from the precincts of Boston, and from the predilections of the community of whose opinions it is the focus. When a people have opened their eyes to the fact, that it is not the best way of discountenancing vice to harden it to exposure and shame, and make it brazen-faced, reckless, and impudent, they might also be convinced, it would seem, that respect for human life would not be promoted by publicly violating it, and making a spectacle, or a newspaper theme, of the mental agony and dying struggles of a human being, and of him least fit, in the common belief, to be thus hurried to his account. "Blood for blood!" We are shocked at the revengeful custom among uncivilized tribes, when it bears the aspect of private revenge, because the executioners must be of the kindred of the slain. How much does the legal retribution in kind, which civilized man exacts, differ in reality from the custom of the savage? The law undertakes to avenge its own dignity, to use a popular phrase; that is, it regards the community as one great family, and constitutes itself the avenger of blood in its behalf. It is not punishment, but retaliation, which does not contemplate the reform of the offender as well as the prevention of crime; and where it wholly loses the remedial element, and cuts off the opportunity for repentance which God's mercy allows, it is worthy of a barbarous, not a Christian, social alliance. What sort of combination for mutual safety is it, too, when no man feels safe, because fortuitous circumstances, ingeniously bound into a chain, may so entangle Truth that she cannot bestir herself to rescue us from the doom which the judgment of twelve fallible men pronounces, and our protector, the law, executes upon us?

But we are losing sight of Mr. Hawthorne's book, and of the old Puritan settlers, as he portrays them with few, but clearly cut and expressive, lines. In these sketchy groupings, Governor Bellingham is the only prominent figure, with the Rev. John Wilson behind him, "his beard, white as a snow-drift, seen over the Governor's shoulder."

> "Here, to witness the scene which we are describing, sat Governor Bellingham himself, with four sergeants about his chair, bearing halberds as a guard of honor. He wore a dark feather in his hat, a border of embroidery on his cloak, and a black velvet tunic beneath; a gentleman advanced in years, and with a hard experience written in his wrinkles. He was not ill-fitted to be the head and representative of a community, which owed its origin and progress, and its present state of development, not to the impulses of youth, but to the stern and tempered energies of manhood, and the sombre sagacity of age; accomplishing so much, precisely because it imagined and hoped so little."

With this portrait, we close our remarks on the book, which we should not have criticized at so great length, had we admired it less. We hope to be forgiven, if in any instance our strictures have approached the limits of what may be considered personal. We would not willingly trench upon the right which an individual may claim, in common courtesy, not to have his private qualities or personal features discussed to his face, with everybody looking on. But Mr. Hawthorne's example in the preface, and the condescending familiarity of the attitude he assumes therein, are at once our occasion and our apology.

JANE SWISSHELM

The Scarlet Letter†

This appears to be *the* romance of the day, and is decidedly a curiosity. The author opens with a most humorous description of the Custom House at Salem, and its numerous officers, including himself, during his time of service, or rather leisure, in the establishment. While there employed in serving our mutual uncle Samuel, he spent a portion of his time burrowing amongst a pile of old manuscript which had accumulated in one of the unfinished rooms, and one day found a roll in which he discovered a bit of old worm-eaten scarlet embroidery, which had once been in the form of the letter A. The manuscript was found to contain an outline history of this curious shred; and our author tells of his many futile endeavors to weave of this material a romance. When he was about giving up in despair, Gen. Taylor's election sealed his political death; he was released from the task of doing nothing, and then comes the story. We incline to think much of the fame of the book was acquired by the introduction, which contains most amusing portraits of persons still living, or recently dead, and many decided political hits. But the tale itself is unique, and told in a most masterly manner.

It opens with a description of the jail at Boston in the early times, its dingy walls and iron-barred portal, with the wild rose-bush which grew close by. On a bright morning a solemn and expectant crowd had assembled here, from whose looks and demeanor one might have expected an execution. Presently from the dingy doorway the Beadle comes, leading forward a woman of commanding mien and surpassing beauty, clad in sombre gray, and on her bosom the letter A blazing in embroidery of scarlet and gold. In her arms she carried an infant of two months, which, upon seeing the crowd, she clasps so as to cover the badge upon her breast. Then, as if remembering the folly of covering one emblem of her disgrace with another, she set the child on one arm, and with a firm step proceeded to the pillory, which she ascended, and where she was secured in the usual position which prevented the hiding of the face. Here for many hours she sat bearing the reproving looks of the assembled throng, and listening to lectures from the clergy and elders of the church, and

† From the *Saturday Visiter* (Sept. 28, 1850): 146. Edited by Robert S. Levine for the first edition of this Norton Critical Edition.

exhortations to reveal the name of her partner in sin. This she refused, and the Rev. Arthur Dimmesdale, her pastor, a young man of great beauty and talent, and of commanding reputation for sanctity, is called upon to address her, which he does in a most impressive speech, begging her to name him who had caused her fall, and thus aid him in coming to repentance—reminding her that as he was too weak to acknowledge his sin she would do him a great service by preventing his living in hypocrisy, &c. &c.—to all which she replies by a short and peremptory refusal to implicate any one. He turns away with an apostrophe to the constancy of woman, which leads the reader to suspect he himself is the sinner thus screened; but of this the spectators never dream; none except one old man who has just arrived with an Indian, that comes to receive a ransom for him who has long been his prisoner. This old man, Hester Prynne, the heroine on the scaffold, sees and recognizes as her husband, a preacher, whom she had preceded to the colony two years, and who had been thought dead. When she is released from the pillory and taken back to prison, the old man visits her in the character of physician, and endeavors to wring from her the name of her seducer. She is firm on this point, but finally promises under oath not to reveal their previous connection, but to permit him to live under an assumed name. He already suspected Rev. Dimmesdale, and in the character of physician and friend devoted his life to a most subtle and fiend-like vengeance. When Hester is released from prison she takes up her abode in a lonely cabin on the seaside, where she maintains herself and child by needlework. Her poetic imagination and inventive genius find outlet in her employment, and her embroidery becomes the fashion. Her child is a perfect incarnation of the spirit of beauty—a wild, fitful, impulsive little sprite, who was even in babyhood attracted by the blazing insignia on her mother's sombre dress. This becomes a bitter portion of Hester's punishment, and every fit of passion in little Pearl the author attributes to the circumstances of her birth—paints the child's fitful spirit as a mark of the Divine displeasure, on account of the law broken by her parents. Whenever mother and child appear they are greeted by the Puritans, old and young, with cold and silent contempt, or hootings and epithets of infamy. If they appeared in church, all shrank from them, and the language of every one was, "Come not near me! I am holier than thou!" In no crowd did Hester stand in fear of being jostled. She was the moral leper whom none might dare to touch—the blazing emblem of the virtuous indignation of an entire community. Yet Hester went quietly on her way. Was any sick, or suffering in great distress, Hester was there to minister to every want. Even scorn and insults from those she aided, did not drive her from their side while aid was wanted, but that time past she never recognized them more. Did gratitude prompt them to notice her kindly in public, she laid her finger on the scarlet letter—the emblem of her shame, and passed on in silence. So the years sped, and after a while she acquired the title of "our Hester," and many said the A upon her breast meant Able, she was so strong to assist and comfort. In the mean time poor Dimmesdale underwent most terrible penance from the serpent-cunning of his old tormentor—the lashings of conscience, and the enthusiastic admiration of his parishioners. He becomes a monomaniac, and one night at midnight and during a storm, he goes and mounts the pillory, there alone and unseen to undergo the ordeal Hester had passed. Here Hester and Pearl find him as they return from a death-bed. They go up and sit with him there, and the old doctor comes to witness the scene. This showed Hester the state of abject misery to which her weak lover was reduced, and she resolves to free him from the fangs of the old serpent, the doctor. So she meets him in the forest and reveals the identity of the doctor and her former husband—advises, urges him to fly to Europe, and offers to accompany him. A plan is fixed upon, and he falls into a state of fiendish excitement, which to us appears somewhat preposterous. He is strangely impelled to blaspheme and swear, indulge in brutal jests, and mock at every thing he believes to be sacred! In this frame of mind he composes a sermon to be preached on the occasion of the installation of a new governor. This sermon is a miracle, and electrifies collected thousands. In the crowd without stands Hester, at the foot of the pillory, within sound of his voice, and surrounded by the circle of infamy which kept all from approaching within some yards. Here she learns their plan of flight has been discovered and frustrated by the old doctor, and stood in her despair when her lover came out of the church, tottering and pale, surrounded by admiring, almost worshipping thousands. When he sees Hester he approaches and asks her to aid him in ascending the scaffold. They and Pearl go up, and there to the electrified crowd he proclaims his guilt, and dies. The old doctor had now nothing to live for, and soon died, leaving Pearl heiress to a large fortune. Pearl's nature appears changed from the time of her father's death, and she becomes gentle, affectionate—comprehensible. She and her mother disappear for some years, and then Hester returns to the cabin alone. It is supposed from signs that Pearl is the wife of some nobleman in a foreign land, but Hester voluntarily returns, takes up her badge of shame, lives and dies in the cabin by the sea side, and finally sleeps beside her lover.

When one has read the book the query is, "what did the author mean! What moral lesson did he want to inculcate? What philosophy did he want to teach?"

If he meant to teach the sinfulness of Hester's sin—the great and divine obligation and sanctity of a legal marriage contract, and the monstrous depravity of a union sanctioned only by affection, his book is the most sublime failure of the age. Hester Prynne stands morally, as Saul did physically amongst his contemporaries, the head and shoulders taller than the tallest. She is the most glorious creation of fiction that has ever crossed our path. We never dreamed of any thing so sublime as the moral force and grandeur of her character. Scott's Jeanie Deans sinks into insignificance beside her. Jane Eyer [*sic*] is a chip floating with the current of popular opinion, while Hester rows her boat up from the brink of Niagara, and lands at Buffalo as calm and self-possessed as ordinary people from a ride on the "raging canal." The Divines and Elders and Governors and Magistrates and honorably married dames of her day, look like pasteboard puppets beside breathing men and women, when they come in contact with "their Hester." What one instinctively blames her for is, that she did not save her poor imbecile lover from the insane persecutions of the old sinner who was putting him to death by slow tortures. She should have protected Dimmesdale as well as kept his secret. It was not like herself to desert him, and leave him in the embrace of such a wily old serpent.

As for the author's lame attempts to make Pearl a punishment sent to her mother, we never saw a mother who would not be happy to be so punished—never knew a child who did not give fifty times the evidence of being sent in wrath. If any argument, pro or con, could be drawn from Pearl, she was surely a special evidence of the Divine approbation of the law which governed her birth. If such a little "jet of flame" as Pearl can be considered a sign of a broken law, we do wonder what Hawthorne thinks of the royal idiots whose existence testifies to the validity and legality of the pompous marriage rites of Queens and Empresses?

If Hawthorne really wants to teach the lesson ostensibly written on the pages of his book, he had better try again. For our part if we knew there was such another woman as Hester Prynne in Boston now, we should travel all the way there to pay our respects, while the honorable characters of the book are such poor affairs it would scarce be worth while throwing a mud-ball at the best of them.

ORESTES BROWNSON

Literary Notices and Criticisms†

Mr. Hawthorne is a writer endowed with a large share of genius, and in the species of literature he cultivates has no rival in this country, unless it be Washington Irving. His *Twice-told Tales*, his *Mosses from an Old Manse*, and other contributions to the periodical press, have made him familiarly known, and endeared him to a large circle of readers. The work before us is the largest and most elaborate of the romances he has as yet published, and no one can read half a dozen pages of it without feeling that none but a man of true genius and a highly cultivated mind could have written it. It is a work of rare, we may say of fearful power, and to the great body of our countrymen who have no well defined religious belief, and no fixed principles of virtue, it will be deeply interesting and highly pleasing.

We have neither the space nor the inclination to attempt an analysis of Mr. Hawthorne's genius, after the manner of the fashionable criticism of the day. Mere literature for its own sake we do not prize, and we are more disposed to analyze an author's work than the author himself. Men are not for us mere psychological phenomena, to be studied, classed, and labelled. They are moral and accountable beings, and we look only to the moral and religious effect of their works. Genius perverted, or employed in perverting others, has no charms for us, and we turn away from it with sorrow and disgust. We are not among those who join in the worship of passion, or even of intellect. God gave us our faculties to be employed in his service, and in that of our fellow-creatures for his sake, and our only legitimate office as critics is to inquire, when a book is sent us for review, if its author in producing it has so employed them.

Mr. Hawthorne, according to the popular standard of morals in this age and this community, can hardly be said to pervert God's gifts, or to exert an immoral influence. Yet his work is far from being unobjectionable. The

† From *Brownson's Quarterly* 4 (Oct. 1850): 528–32.

story is told with great naturalness, ease, grace, and delicacy, but it is a story that should not have been told. It is a story of crime, of an adulteress and her accomplice, a meek and gifted and highly popular Puritan minister in our early colonial days,—a purely imaginary story, though not altogether improbable. Crimes like the one imagined were not unknown even in the golden days of Puritanism, and are perhaps more common among the descendants of the Puritans than it is at all pleasant to believe; but they are not fit subjects for popular literature, and moral health is not promoted by leading the imagination to dwell on them. There is an unsound state of public morals when the novelist is permitted, without a scorching rebuke, to select such crimes, and invest them with all the fascinations of genius, and all the charms of a highly polished style. In a moral community such crimes are spoken of as rarely as possible, and when spoken of at all, it is always in terms which render them loathsome, and repel the imagination.

Nor is the conduct of the story better than the story itself. The author makes the guilty parties suffer, and suffer intensely, but he nowhere manages so as to make their sufferings excite the horror of his readers for their crime. The adulteress suffers not from remorse, but from regret, and from the disgrace to which her crime has exposed her, in her being condemned to wear emblazoned on her dress the Scarlet Letter which proclaims to all the deed she has committed. The minister, her accomplice, suffers also, horribly, and feels all his life after the same terrible letter branded on his heart, but not from the fact of the crime itself, but from the consciousness of not being what he seems to the world, from his having permitted the partner in his guilt to be disgraced, to be punished, without his having the manliness to avow his share in the guilt, and to bear his share of the punishment. Neither ever really repents of the criminal deed; nay, neither ever regards it as really criminal, and both seem to hold it to have been laudable, because they *loved* one another,—as if the love itself were not illicit, and highly criminal. No man has the right to love another man's wife, and no married woman has the right to love any man but her husband. Mr. Hawthorne in the present case seeks to excuse Hester Prynne, a married woman, for loving the Puritan minister, on the ground that she had no love for her husband, and it is hard that a woman should not have some one to love; but this only aggravated her guilt, because she was not only forbidden to love the minister, but commanded to love her husband, whom she had vowed to love, honor, cherish, and obey. The modern doctrine that represents the affections as fatal, and wholly withdrawn from voluntary control, and then allows us to plead them in justification of neglect of duty and breach of the most positive precepts of both the natural and the revealed law, cannot be too severely reprobated.

Human nature is frail, and it is necessary for every one who standeth to take heed lest he fall. Compassion for the fallen is a duty which we all owe, in consideration of our own failings, and especially in consideration of the infinite mercy our God has manifested to his erring and sinful children. But however binding may be this duty, we are never to forget that sin is sin, and that it is pardonable only through the great mercy of God, on condition of the sincere repentance of the sinner. But in the present case neither of the guilty parties repents of the sin, neither exclaims with the royal prophet, who had himself fallen into the sin of adultery and murder, *Misere mei Deus, secundum magnam misericordiam; et secundum multitudinem miserationum tuarum, dele iniquitatem meam. Amplius lava me ab iniquitate mea; et a peccato munda me. Quoniam iniquitatem meam cognosco, et peccatum meum contra me est semper.*[1] They hug their illicit love; they cherish their sin; and after the lapse of seven years are ready, and actually agree, to depart into a foreign country, where they may indulge it without disguise and without restraint. Even to the last, even when the minister, driven by his agony, goes so far as to throw off the mask of hypocrisy, and openly confess his crime, he shows no sign of repentance, or that he regarded his deed as criminal.

The Christian who reads *The Scarlet Letter* cannot fail to perceive that the author is wholly ignorant of Christian asceticism, and that the highest principle of action he recognizes is pride. In both the criminals, the long and intense agony they are represented as suffering springs not from remorse, from the consciousness of having offended God, but mainly from the feeling, especially on the part of the minister, that they have failed to maintain the integrity of their character. They have lowered themselves in their own estimation, and cannot longer hold up their heads in society as honest people. It is not their conscience that is wounded, but their pride. *He* cannot bear to think that he wears a disguise, that he cannot be the open, frank, stainless character he had from his youth aspired to be, and *she*, that she is driven from society, lives a solitary outcast, and has nothing to console her but her fidelity to her paramour. There is nothing Christian, nothing really moral, here. The very pride itself is a sin; and pride often a greater sin than that which it restrains us from committing. There are thousands of men and women too proud to commit carnal sins, and to the indomitable pride of our Puritan ancestors we may attribute no small share of their external morality and decorum. It may almost be said, that, if they had less of that external morality and decorum, their case would be less desperate; and often the violation of them, or failure to maintain them, by which their pride receives a shock, and their self-complacency is shaken, becomes the occasion, under the grace of God, of their conversion to truth and holiness. As long as they maintain their self-complacency, are satisfied with themselves, and feel that they have outraged none of the decencies of life, no argument can reach them, no admonition can startle them, no exhortation can move them. Proud of their supposed virtue, free from all self-reproach, they are as placid as a summer morning, pass through life without a cloud to mar their serenity, and die as gently and as sweetly as the infant falling asleep in its mother's arms. We have met with these people, and after laboring in vain to waken them to a sense of their actual condition, till completely discouraged, we have been tempted to say, Would that you might commit some overt act, that should startle you from your sleep, and make you feel how far pride is from being either a virtue, or the safeguard of virtue,—or convince you of your own insufficiency for yourselves, and your absolute need of Divine grace. Mr. Hawthorne seems never to have learned that pride is not only sin, but the root of all sin, and that humility is not only a virtue, but the root of all virtue. No genuine contrition or repentance ever springs from pride, and the sorrow for sin

1. Psalm 51.1–3: "Have mercy upon me, O God, according to thy loving kindness: according unto the multitude of thy tender mercies blot out my transgressions. / Wash me thoroughly from mine iniquity, and cleanse me from my sin. / For I acknowledge my transgressions: and my sin is ever before me."

because it mortifies our pride, or lessens us in our own eyes, is nothing but the effect of pride. All true remorse, all genuine repentance, springs from humility, and is sorrow for having offended God, not sorrow for having offended ourselves.

Mr. Hawthorne also mistakes entirely the effect of Christian pardon upon the interior state of the sinner. He seems entirely ignorant of the religion that can restore peace to the sinner,—true, inward peace, we mean. He would persuade us, that Hester had found pardon, and yet he shows us that she had found no inward peace. Something like this is common among popular Protestant writers, who, in speaking of great sinners among Catholics that have made themselves monks or hermits to expiate their sins by devoting themselves to prayer, and mortification, and the duties of religion, represent them as always devoured by remorse, and suffering in their interior agony almost the pains of the damned. An instance of this is the Hermit of Engeddi in Sir Walter Scott's *Talisman*. These men know nothing either of true remorse, or of the effect of Divine pardon. They draw from their imagination, enlightened, or rather darkened, by their own experience. Their speculations are based on the supposition that the sinner's remorse is the effect of wounded pride, and that during life the wound can never be healed. All this is false. The remorse does not spring from wounded pride, and the greatest sinner who really repents, who really does penance, never fails to find interior peace. The mortifications he practises are not prompted by his interior agony, nor designed to bring peace to his soul; they are a discipline to guard against his relapse, and an expiation that his interior peace already found, and his overflowing love to God for his superabounding mercy, lead him to offer to God, in union with that made by his blessed Lord and Master on the cross.

Again, Mr. Hawthorne mistakes the character of confession. He does well to recognize and insist on its necessity; but he is wrong in supposing that its office is simply to disburden the mind by communicating its secrets to another, to restore the sinner to his self-complacency, and to relieve him from the charge of cowardice and hypocrisy. Confession is a duty we owe to God, and a means, not of restoring us to our self-complacency, but of restoring us to the favor of God, and reëstablishing us in his friendship. The work before us is full of mistakes of this sort, in those portions where the author really means to speak like a Christian, and therefore we are obliged to condemn it, where we acquit him of all unchristian intention.

As a picture of the old Puritans, taken from the position of a moderate transcendentalist and liberal of the modern school, the work has its merits; but as little as we sympathize with those stern old Popery-haters, we do not regard the picture as at all just. We should commend where the author condemns, and condemn where he commends. Their treatment of the adulteress was far more Christian than his ridicule of it. But enough of fault-finding, and as we have no praise, except what we have given, to offer, we here close this brief notice.

ARTHUR CLEVELAND COXE

The Writings of Hawthorne†

Current Literature, in America, has generally been forced to depend, for criticism, upon personal partiality or personal spleen. We have had very little reviewing on principle; almost none with the pure motive of building up a sound and healthful literature for our country, by cultivating merit, correcting erratic genius, abasing assumption and imposture, and insisting on the fundamental importance of certain great elements, without which no literature can be either beneficial or enduring. Our reviews have, accordingly, exercised very little influence over public taste. They have been rather tolerated than approved; and, for the most part, have led a very precarious existence, rather as attempts than as achievements; creditable make-believes; tolerable domestic imitations of the imported article; well enough in their way, but untrustworthy for opinion, and worthless for taste. Their reviewals of contemporary authors have too commonly been a mere daubing of untempered mortar, or else a deliberate assault, with intent to kill. In either case the reviewer has betrayed himself, as writing, not for the public, but for the satisfaction or the irritation of the author; and the game of mock reviewing has become as notorious as that of mock auctions. The intelligent public hears the hammering and the outcry, but has got used to it, and passes by. Nobody's opinion of a book is the more or less favorable for anything that can be said in this or that periodical. * * *

So it must be, however, till our periodicals become something more than repositories of sophomorical eulogy, or ribaldry, upon literary toys and trifles. Reviews are superfluous, except as they represent a want, which they undertake to supply, from competent resources, and in an earnest spirit of accomplishing an honorable purpose. We make no apology, therefore, for becoming reviewers, when we acknowledge our earnest hope, not only that we may do something to assist the literary and theological studies of Anglo-American Churchmen, but that we may make the voice of the Church more audible to the American public in general, and thus may exercise, for the benefit of popular authors, some salutary influence upon public taste. Our mission—to borrow a little cant from the times—is, indeed, rather religious than literary; yet, in an age when literature makes very free with religion, we must be pardoned for supposing that religion owes some attention to literature. * * * We know not the literary world, except from a distant view, and have nothing in common with its aims or its occupations; but we think it high time that the literary world should learn that Churchmen are, in a very large proportion, their readers and book-buyers, and that the tastes and principles of Churchmen have as good a right to be respected as those of Puritans and Socialists. It is in this relation to our subject that we have taken up the clever and popular writings of Hawthorne; and we propose to consider them, without any attempt to give them a formal review, just in the free

† From *The Church Review* 3.4 (Jan. 1851): 489–511. Coxe's well-known condemnation of *The Scarlet Letter* also includes a lengthy review of *Mosses from an Old Manse*.

and conversational manner which is permitted to table-talk or social intercourse; and if we can thus afford our author a candid exhibition of the impressions he is producing on a large, but quiet portion of the community, and prompt him to a future career more worthy of their entire regard, we shall feel that we have done the State, as well as the Church, some service; and no anxiety for our reputation as critics shall spoil our appetite for a smoking plum-pudding at Christmas.

In taking up Mr. Hawthorne's volumes, we are happy to particularize our general professions of impartiality, and to describe ourselves as heartily his well-wishers, knowing nothing either of him or his works, beyond what is patent to all men, in his own published confessions, or in other publications of the popular character. True we must own to a little prejudice against him, as a conspicuous member of the Bay School, but, in counterpoise, we must put in a profession of a specific feeling in his favor, as at all events one of the best of them, the very Irving of Down-East. He is one of the few Bays whose freest egotism seldom moves our disgust, and whom we are, in truth, disposed to thank for gossiping at random about himself and friends, as if every one knew both him and them, and were anxiously watching them with telescopes and lorgnettes. In fact, we were particularly interested in his graphic description of that ancient seat of witchcraft in which he tells us he was born, for having had forefathers of our own among the broad-banded and steeple-crowned worthies of old Salem, we were glad to learn, more than geographies and gazetteers are wont to tell us, of its appearance and present condition. Nay, we began to feel a degree of cousinry with our author, in spite of ourselves, when, in an old family record of a marriage not very remotely connected with our own existence, we found the name of his ancestor, *Colonel Hathorn*, familiarly mentioned, with those of other Salemites who hasted to the wedding, in the year of Grace 1713, and were there gravely lectured, over their sack-posset, by godly Master Noyes, the Puritan parson. With such, and many other feelings in our author's favor, we take up his works. In fact, who can resist a pleasant influence in his behalf, exhaling from his very name, redolent as it is of guilelessness and springtide, and rich with associations derived from old ballads and madrigals that celebrate the garden-like agriculture of England? In faint suggestiveness too of "Hawthornden," it has a flavor of Scots poesy and the English drama; of Drummond and of Jonson; and if some patriot Pope or Gifford wants a name whose easy lubricity of pronunciation just suits a flowing line, who would not wish that Hawthorne's might be paired with Irving's, as indissolubly as Beaumont's with Fletcher's, and that the twain might be freely allowed to rank as the *lucida sidera* of our literary horizon? It is not for want of a predisposition to admire and praise our author and his performances, that we shall be obliged to say many things in a different humor.

* * *

* * * It is chiefly, in hopes, to save our author from embarking largely into this business of Fescennine romance, that we enter upon a brief examination of his latest and most ambitious production, "The Scarlet Letter."

The success which seems to have attended this bold advance of Hawthorne, and the encouragement which has been dealt out by some professed critics, to its worst symptoms of malice prepense, may very naturally lead, if unbalanced by a moderate dissent, to his further compromise of his literary character. We are glad, therefore, that "The Scarlet Letter" is, after all, little more than an experiment, and need not be regarded as a step necessarily fatal. It is an attempt to rise from the composition of petty tales, to the historical novel; and we use the expression *an attempt*, with no disparaging significance, for it is confessedly a trial of strength only just beyond some former efforts, and was designed as part of a series. It may properly be called a novel, because it has all the ground-work, and might have been very easily elaborated into the details, usually included in the term; and we call it *historical*, because its scene-painting is in a great degree true to a period of our Colonial history, which ought to be more fully delineated. We wish Mr. Hawthorne would devote the powers which he only partly discloses in this book, to a large and truthful portraiture of that period, with the patriotic purpose of making us better acquainted with the stern old worthies, and all the *dramatis personæ* of those times, with their yet surviving habits, recollections, and yearnings, derived from maternal England. Here is, in fact, a rich and even yet an unexplored field for historic imagination; and touches are given in "The Scarlet Letter," to secret springs of romantic thought, which opened unexpected and delightful episodes to our fancy, as we were borne along by the tale. Here a maiden reminiscence, and here a grave ecclesiastical retrospection, clouding the brow of the Puritan colonists, as they still remembered home, in their wilderness of lasting exile! Now a lingering relic of Elizabethan fashion in dress, and now a turn of expression, betraying the deep traces of education under influences renounced and foresworn, but still instinctively prevalent!

Time has just enough mellowed the facts, and genealogical research has made them just enough familiar, for their employment as material for descriptive fiction; and the New England colonies might now be made as picturesquely real to our perception, as the Knickerbocker tales have made the Dutch settlements of the Hudson. This, however, can never be done by the polemical pen of a blind partisan of the Puritans; it demands Irving's humorously insinuating gravity, and all his benevolent satire, with a large share of honest sympathy for at least the earnestness of wrong-headed enthusiasm. We are stimulated to this suggestion by the very life-like and striking manner in which the days of Governor Winthrop are sketched in the book before us, by the beautiful picture the author has given us of the venerable old pastor Wilson, and by the outline portraits he has thrown in, of several of their contemporaries. We like him all the better for his tenderness of the less exceptionable features of the Puritan character; but we are hardly sure that we like his flings at their failings. If it should provoke a smile to find us sensitive in this matter, our consistency may be very briefly demonstrated. True, we have our own fun with the follies of the Puritans; it is our inseparable privilege as Churchmen, thus to compensate ourselves for many a scar which their frolics have left on our comeliness. But when a degenerate Puritan, whose Socinian conscience is but the skimmed-milk of their creamy fanaticism, allows such a conscience to curdle within him, in dyspeptic acidulation, and then belches forth derision at the sour piety of his forefathers—we snuff at him, with an honest scorn, knowing very well that he likes the Puritans for their worst enormities, and hates them only for their redeeming merits.

The Puritan rebelling against the wholesome discipline of that Ecclesiastical Law, which Hooker has demonstrated, with Newtonian evidence, to be but a moral system of central light with its dependent order and illumination; the Puritan with his rough heel and tough heart, mounted upon altars, and hacking down crosses, and sepulchres, and memorials of the dead; the Puritan with his axe on an Archbishop's neck, or holding up in his hand the bleeding head of a martyred king; the Puritan in all this guilt, has his warmest praise, and his prompt witness that he allows the deeds of his fathers, and is ready to fill up the measure of their iniquity; but the Puritans, with a blessed inconsistency, repeating liturgic doxologies to the triune GOD, or, by the domestic hearth, bowing down with momentary conformity, to invoke the name of Jesus, whom the Church had taught him to adore as an atoning Saviour—these are the Puritans at whom the driveler wags his head, and shoots out his tongue! We would not laugh in that man's company. No—no! we heartily dislike the Puritans, so far as they were Puritan; but even in them we recognize many good old English virtues, which Puritanism could not kill. They were in part our ancestors, and though we would not accept the bequest of their enthusiasm, we are not ashamed of many things to which they clung, with principle quite as characteristic. We see no harm in a reverent joke now and then, at an abstract Puritan, in spite of our duty to our progenitors, and Hudibras shall still be our companion, when, at times, the mental bow requires fresh elasticity, and bids us relax its string. There is, after all, something of human kindness, in taking out an old grudge in the comfort of a hearty, side-shaking laugh, and we think we are never freer from bitterness of spirit, than when we contemplate the Banbury zealot hanging his cat on Monday, and reflect that Strafford and Montrose fell victims to the same mania that destroyed poor puss. But there is another view of the same Puritan, which even a Churchman may charitably allow himself to respect, and when precisely that view is chosen by his degenerate offspring for unfilial derision, we own to a sympathy for the grim old Genevan features, at which their seventh reproduction turns up a repugnant nose; for sure we are that the young Ham is gloating over his father's nakedness, with far less of sorrow for the ebriety of a parent, than of satisfaction in the degradation of an orthodox patriarch. Now without asserting that it is so, we are not quite so sure, as we would like to be, that our author is not venting something of this spirit against the Puritans, in his rich delineation of "godly Master Dimmesdale," and the sorely abused confidence of his flock. There is a provoking concealment of the author's motive, from the beginning to the end of the story; we wonder what he would be at; whether he is making fun of all religion, or only giving a fair hint of the essential sensualism of enthusiasm. But, in short, we are astonished at the kind of incident which he has selected for romance. It may be such incidents were too common, to be wholly out of the question, in a history of the times, but it seems to us that good taste might be pardoned for not giving them prominence in fiction. In deference to the assertions of a very acute analyst, who has written ably on the subject of colonization, we are inclined to think, as we have said before, that barbarism was indeed "the first danger" of the pilgrim settlers. Of a period nearly contemporary with that of Mr. Hawthorne's narrative, an habitual eulogist has recorded that "on going to its Church and court records, we discover mournful evidences of incontinence, even in the respectable families; as if, being cut off from the more refined pleasures of society, their baser passions had burnt away the restraints of delicacy, and their growing coarseness of manners had allowed them finally to seek, in these baser passions, the spring of their enjoyments." We are sorry to be told so, by so unexceptionable a witness.* We had supposed, with the Roman satirist, that purity might at least be credited to those primitive days, when a Saturnian simplicity was necessarily revived in primeval forests, by the New England colonists:

> Quippe aliter tunc orbe novo, cœloque recenti
> Vivebant homines.[1]

but a Puritan doctor in divinity publishes the contrary, and a Salemite novelist selects the intrigue of an adulterous minister, as the groundwork of his ideal of those times! We may acknowledge, with reluctance, the historical fidelity of the picture, which retailers of fact and fiction thus concur in framing, but we cannot but wonder that a novelist should select, of all features of the period, that which reflects most discredit upon the cradle of his country, and which is in itself so revolting, and so incapable of receiving decoration from narrative genius.

And this brings inquiry to its point. Why has our author selected such a theme? Why, amid all the suggestive incidents of life in a wilderness; of a retreat from civilization to which, in every individual case, a thousand circumstances must have concurred to reconcile human nature with estrangement from home and country; or amid the historical connections of our history with Jesuit adventure, savage invasion, regicide outlawry, and French aggression, should the taste of Mr. Hawthorne have preferred as the proper material for romance, the nauseous amour of a Puritan pastor, with a frail creature of his charge, whose mind is represented as far more debauched than her body? Is it, in short, because a running undertide of filth has become as requisite to a romance, as death in the fifth act to a tragedy? Is the French era actually begun in our literature? And is the flesh, as well as the world and the devil, to be henceforth dished up in fashionable novels, and discussed at parties, by spinsters and their beaux, with as unconcealed a relish as they give to the vanilla in their ice cream? We would be slow to believe it, and we hope our author would not willingly have it so, yet we honestly believe that "the Scarlet Letter" has already done not a little to degrade our literature, and to encourage social licentiousness: it has started other pens on like enterprises, and has loosed the restraint of many tongues, that have made it an apology for "the evil communications which corrupt good manners." We are painfully tempted to believe that it is a book made for the market, and that the market has made it merchantable, as they do game, by letting everybody understand that the commodity is in high condition, and smells strongly of incipient putrefaction.

We shall entirely mislead our reader if we give him to suppose that "the Scarlet Letter" is coarse in its details, or indecent in its phraseology. This

* Barbarism the first Danger, by H. Bushnell, D. D [*Coxe's note*].

1. "Indeed, then, when earth was new and heaven young, men lived differently" (Juvenal, Satire 6).

very article of our own, is far less suited to ears polite, than any page of the romance before us; and the reason is, we call things by their right names, while the romance never hints the shocking words that belong to its things, but, like Mephistophiles, insinuates that the arch-fiend himself is a very tolerable sort of person, if nobody would call him Mr. Devil. We have heard of persons who could not bear the reading of some Old Testament Lessons in the service of the Church: such persons would be delighted with our author's story; and damsels who shrink at the reading of the Decalogue, would probably luxuriate in bathing their imagination in the crystal of its delicate sensuality. The language of our author, like patent blacking, "would not soil the whitest linen," and yet the composition itself, would suffice, if well laid on, to Ethiopize the snowiest conscience that ever sat like a swan upon that mirror of heaven, a Christian maiden's imagination. We are not sure we speak quite strong enough, when we say, that we would much rather listen to the coarsest scene of Goldsmith's "Vicar," read aloud by a sister or daughter, than to hear from such lips, the perfectly chaste language of a scene in "the Scarlet Letter," in which a married wife and her reverend paramour, with their unfortunate offspring, are introduced as the actors, and in which the whole tendency of the conversation is to suggest a sympathy for their sin, and an anxiety that they may be able to accomplish a successful escape beyond the seas, to some country where their shameful commerce may be perpetuated. Now, in Goldsmith's story there are very coarse words, but we do not remember anything that saps the foundations of the moral sense, or that goes to create unavoidable sympathy with unrepenting sorrow, and deliberate, premeditated sin. The "Vicar of Wakefield" is sometimes coarsely virtuous, but "the Scarlet Letter" is delicately immoral.

There is no better proof of the bad tendency of a work, than some unintentional betrayal on the part of a young female reader, of an instinctive consciousness against it, to which she has done violence, by reading it through. In a beautiful region of New England, where stage-coaches are not yet among things that were, we found ourselves, last summer, one of a traveling party, to which we were entirely a stranger, consisting of young ladies fresh from boarding-school, with the proverbial bread-and-butter look of innocence in their faces, and a nursery thickness about their tongues. Their benevolent uncle sat outside upon the driver's box, and ours was a seat next to a worshipful old dowager, who seemed to bear some matronly relation to the whole coach-load, with the single exception of ourselves. In such a situation it was ours to keep silence, and we soon relapsed into nothingness and a semi-slumberous doze. Meanwhile our young friends were animated and talkative, and as we were approaching the seat of a College, their literature soon began to expose itself. They were evidently familiar with the Milliners' Magazines in general, and even with Graham's and Harper's. They had read James, and they had read Dickens; and at last their criticisms rose to Irving and Walter Scott, whose various merits they discussed with an artless anxiety to settle forever the question whether the one was not "a charming composer," and the other "a truly beautiful writer." Poor girls! had they imagined how much harmless amusement they were furnishing to their drowsy, dusty, and very unentertaining fellow traveler, they might, quite possibly, have escaped both his praise and his censure! They came at last to Longfellow and Bryant, and rhythmically regaled us with the "muffled drum" of the one, and the somewhat familiar opinion of the other, that

"Truth crushed to earth will rise again."

And so they came to Hawthorne, of whose "Scarlet Letter" we then knew very little, and that little was favorable, as we had seen several high encomiums of its style. We expected a quotation from the "Celestial Railroad," for we were traveling at a rate which naturally raised the era of railroads in one's estimation, by rule of contrary; but no—the girls went straight to "the Scarlet Letter." We soon discovered that one Hester Prynne was the heroine, and that she had been made to stand in the pillory, as, indeed, her surname might have led one to anticipate. We discovered that there was a mysterious little child in the question, that she was a sweet little darling, and that her "sweet, pretty little name," was "Pearl." We discovered that mother and child had a meeting, in a wood, with a very fascinating young preacher, and that there was a hateful creature named Chillingworth, who persecuted the said preacher, very perseveringly. Finally, it appeared that Hester Prynne was, in fact, Mrs. Hester Chillingworth, and that the hateful old creature aforesaid had a very natural dislike to the degradation of his spouse, and quite as natural a hatred of the wolf in sheep's clothing who had wrought her ruin. All this leaked out in conversation, little by little, on the hypothesis of our protracted somnolency. There was a very gradual approximation to the point, till one inquired—"didn't you think, from the first, that he was the one?" A modest looking creature, who evidently had not read the story, artlessly inquired—"what one?"—and then there was a titter at the child's simplicity, in the midst of which we ventured to be quite awake, and to discover by the scarlet blush that began to circulate, that the young ladies were not unconscious to themselves that reading "the Scarlet Letter" was a thing to be ashamed of. These schoolgirls had, in fact, done injury to their young sense of delicacy, by devouring such a dirty story; and after talking about it before folk, inadvertently, they had enough of mother Eve in them, to know that they were ridiculous, and that shame was their best retreat.

Now it would not have been so if they had merely exhibited a familiarity with "the Heart of Mid-Lothian,"[2] and yet there is more mention of the foul sin in its pages, than there is in "the Scarlet Letter." Where then is the difference? It consists in this—that the holy innocence of Jeanie Deans, and not the shame of Effie, is the burthen of that story, and that neither Effie's fall is made to look like virtue, nor the truly honorable agony of her stern old father, in bewailing his daughter's ruin, made a joke, by the insinuation that it was quite gratuitous. But in Hawthorne's tale, the lady's frailty is philosophized into a natural and necessary result of the Scriptural law of marriage, which, by holding her irrevocably to her vows, as plighted to a dried up old bookworm, in her silly girlhood, is viewed as making her heart an easy victim to the adulterer. The sin of her seducer too, seems to be considered as lying not so much in the deed itself, as in his long concealment of it, and, in fact, the whole moral of the tale is given in the words—"Be true—be true," as if sincerity in sin were

2. *The Heart of Midlothian* (1818), novel by Sir Walter Scott. When Effie Deans is sentenced to death for murdering her child, her sister Jeanie obtains a pardon from Queen Caroline.

virtue, and as if "Be clean—be clean," were not the more fitting conclusion. "The untrue man" is, in short, the hang-dog of the narrative, and the unclean one is made a very interesting sort of a person, and as the two qualities are united in the hero, their composition creates the interest of his character. Shelley himself never imagined a more dissolute conversation than that in which the polluted minister comforts himself with the thought, that the revenge of the injured husband is worse than his own sin in instigating it. "Thou and I never did so, Hester"—he suggests: and she responds—"never, never! What we did had *a consecration of its own*, we felt it so—we said so to each other!" This is a little too much—it carries the Bay-theory a little too far for our stomach! "Hush, Hester!" is the sickish rejoinder; and fie, Mr. Hawthorne! is the weakest token of our disgust that we can utter. The poor bemired hero and heroine of the story should not have been seen wallowing in their filth, at such a rate as this.

We suppose this sort of sentiment must be charged to the doctrines enforced at "Brook-farm," although "Brook-farm" itself could never have been Mr. Hawthorne's home, had not other influences prepared him for such a Bedlam. At all events, this is no mere slip of the pen; it is the essential morality of the work. If types, and letters, and words can convey an author's idea, he has given us the key to the whole, in a very plain intimation that the Gospel has not set the relations of man and woman where they should be, and that a new Gospel is needed to supersede the seventh commandment, and the bond of Matrimony. Here it is, in full: our readers shall see what the world may expect from Hawthorne, if he is not stopped short, in such brothelry. Look at this conclusion:—

"*Women*—in the continually recurring trials of wounded, wasted, wronged, misplaced, or erring and sinful passion, or with the dreary burden of a heart unyielded, because unvalued and unsought—came to Hester's cottage, demanding why they were so wretched, and what the remedy! Hester comforted and counseled them as best she might. She assured them too *of her firm belief*, that, at some brighter period, when the world should have grown ripe for it, in Heaven's own time, *a new truth would be revealed, in order to establish the whole relation between man and woman on a surer ground of mutual happiness*."

This is intelligible English; but are Americans content that such should be the English of their literature? This is the question on which we have endeavored to deliver our own earnest convictions, and on which we hope to unite the suffrages of all virtuous persons, in sympathy with the abhorrence we so unhesitatingly express. To think of making such speculations the amusement of the daughters of America! The late Convention of females at Boston, to assert the "rights of woman," may show us that there are already some, who think the world is even now *ripe for it*; and safe as we may suppose our own fair relatives to be above such a low contagion, we must remember that to a woman, the very suggestion of a mode of life for her, as preferable to that which the Gospel has made the glorious sphere of her duties and her joys, is an insult and a degradation, to which no one that loves her would allow her to be exposed.

We assure Mr. Hawthorne, in conclusion, that nothing less than an earnest wish that his future career may redeem this misstep, and prove a blessing to his country, has tempted us to enter upon a criticism so little suited to our tastes, as that of his late production. We commend to his attention the remarks of Mr. Alison, on cotemporary popularity, to be found in the review of Bossuet. We would see him, too, rising to a place among those immortal authors who have "clothed the lessons of religion in the burning words of genius;" and let him be assured, that, however great his momentary success, there is no lasting reputation for such an one as he is, except as it is founded on real worth, and fidelity to the morals of the Gospel. The time is past, when mere authorship provokes posthumous attention; there are too many who write with ease, and too many who publish books, in our times, for an author to be considered anything extraordinary. Poems perish in newspapers, now-a-days, which, at one time, would have made, at least, a name for biographical dictionaries; and stories lie dead in the pages of magazines, which would once have secured their author a mention with posterity. Hereafter those only will be thought of, who have enbalmed their writings in the hearts and lives of a few, at least, who learned from them to love truth and follow virtue. The age of "mute inglorious Miltons," is as dead as the age of chivalry. Everybody can write, and everybody can publish. But still, the wise are few; and it is only the wise, who can attain, in any worthy sense, to shine as the stars forever.

AMORY DWIGHT MAYO

The Works of Nathaniel Hawthorne†

* * *In the remarks we propose to make on this author, there are no pretensions to an exhaustive criticism. It is more than a spring-day's journey to walk around the boundaries, and explore all the paths of his remarkable mind; and he who would attempt to do it, may be compelled to reverse his decision by a deeper insight into his books, or a new manifestation of his power. We only say that we have faithfully read the half-dozen volumes by which he is known, and will try to convey the impression they leave upon our mind. The task of describing their contents and quoting fine passages, we leave to the newspapers. We write for those who have read, and, like ourselves, wish to talk an hour of "things seen and heard" in this new world of genius.

The first thing which attracts our notice upon these pages, is the acuteness and extent of the writer's power of observation. His eye adjusts itself to objects beyond and within the ordinary circle of vision. Wherever he looks, he sees distinctly, and the sweep of his gaze comprehends a wide area. His perception of beauty in nature is singularly keen and comprehensive. He paints an object in the light as with sunbeams, while the shadow or the transition to it are transferred with equal fidelity. We think the works of few poets will present so accurate and extensive portraiture of nature as his; of living nature, for he sees the characteristic points in a landscape, and in a high degree possesses the Spenserian power of transforming, by one magic word, a lifeless pictured catalogue of natural objects, to an actual breathing and moving scene. Perhaps he dwells with more fondness upon the minute and evanescent, than the grand and substantial in nature. Yet he is not incompetent to interpret its noble appearances.

† From *Universalist Quarterly* 8 (July 1851): 272–93.

The same clearness of vision he carries into life. He has a vivid perception of historical events and periods, no less than of the actual existence of to-day. He pictures a street with such fidelity that we walk upon its pavements, elbow our way through its shadowy throngs, and raise our voices above the clatter of omnibus wheels, to shout into ghostly ears. Whether it be little Ned Higgins offering his cent to Hepzibah[1] for the gingerbread Jim Crow—"the one that has not a broken foot," or Peter Goldthwaite[2] swinging his axe through the cloud of dust in his own attic, or Hester Prynne walking to the scene of public exposure, wearing the scarlet letter upon her bosom, or Judge Pyncheon sitting dead in the low-studded room, with the mouse at his foot, and grimalkin looking in at the window, and the fly walking across his naked eyeball, or the crowd of dancers in the hall of the old Province-House, and the shadowy procession of governors down its steps,—everywhere is the same wonderful daguerreo-typing of the facts, and the same reproduction of the essential principle of life, which is the chief fact of the spectacle.

His eye does not fail, but increases in power, when directed to the world of thought and feeling. Mr. Hawthorne's books are tables of spiritual statistics, embracing the natural history of the human mind in its ordinary, but oftener in its extraordinary conditions. One would think he had ransacked the experience of all the men, women and children in his neighborhood, been the chaplain of the state prison and madhouse, beside having telegraphic intelligence of the mental state of every out-of-the-way, queer creature in the land. He is a man to whom we would not care to talk an hour if we had any secrets, for as sure as we have a tongue, without invitation we should tell them all to him, even on the top of a mail coach, or in the Merchants' Exchange. If he ever seems to overlook common traits of character, or ordinary states of mind, it is only from the absorbing interest attached to half developed germs of individuality, and flitting or profound spiritual appearances. Like all soul-gazers, he loves to walk along the dim labyrinthine passages of the mind, and poke his head into its cobwebbed closets, and clamber up stairways which are peculiarly unsafe. At all the critical moments of life,—when a man is trying to choke down his confession of love, or holds his first child in his arms, or topples over into the gulf of some terrible sin, he is sure to be a spectator. Yet in moments of less intensity he loves to hear a child prattle, or the old gravestone-cutter of Martha's Vineyard gossip over his epitaphs, or the six vagabonds in the moving show-cart retail their miscellaneous experience.

But he looks further than this. Nature, life, and the soul are only the foreground in his perspective, for beyond them he sees those spiritual laws, which sweep down ages of time, athwart the world, cross each other without confusion and almost annihilate human freedom by their fatal execution. All things are guarded by these relentless keepers, and if they ever escape for a moment, a million of eyes track the fugitives, and an unseen power compels them to walk of their own accord back through their open prison doors. We believe the faculty which perceives the invisible highways of God running up and down the universe, is the author's

1. Ned Higgins and Hepzibah Pyncheon are characters in Hawthorne's *The House of the Seven Gables* (1851).
2. The title character in Hawthorne's tale "Peter Goldthwaite's Treasure" (1838).

rarest gift,—more than any other, modifies the ordinary operations of his mind, and is the source of the chief characteristics in his method of delineation.

* * *

* * * Now and then a man is born who can look straight down into the spirit without searing his eyeballs, witness this awful conflict of law and will, trace its results, know the impotence as well as the strength of man, and not lose his balance of mind, or health of affections. One poet alone looked through the length and breadth and depth of life, and unscared by the awful spectacle, with the ease and joy of a little child, wove a few of its groupings into dramas, by which all the poets in the world now swear. But this could not be until a greater than man had affirmed the paternity of God, and by the revelation of Christianity at once secured the welfare of the race, and widened to infinity the possibilities of art. Christianity alone made a Shakespeare possible, for no human creature without peril of insanity, could have looked so deeply into the very heart of existence, unless guarded by its love and faith, and arched over by its firmament of immortal hopes.

Among the few American writers who have the peculiarity of genius which consists of insight into this fact of the soul, Mr. Hawthorne occupies a prominent position. We are convinced that the rarest quality of his mind is the power of tracing the relations of spiritual laws to character. He looks at the soul, life, and nature, from the stand-point of Providence. He follows the track of one of God's mental or moral laws. Every thing which appears along its borders is minutely investigated, though sometimes appreciated rather for its nearness to his path than its own value. If we mistake not, this is the clue to all his works. Even his lightest tale gains a peculiarity of treatment and depth of tone from it, though the tendency of his mind is better perceived in his more elaborate works. Wherever he goes, whoever he meets, or whatever may be the scenes amid which he mingles, this thought is uppermost:—How are these things related to each other, and to those great spiritual agencies which underlie and encompass them? Whatever else Mr. Hawthorne may be, and we do not deny to him great versatility of powers, he is, more than any thing else, a seer.

His view of human nature determines his treatment of individuals. He can hardly be called a truthful delineator of character. His men and women have the elements of life, though not arranged in harmonious proportions. Our interest is concentrated upon the point, in the nature of each, where the battle is raging between human will and spiritual laws. How far has the man obeyed or disobeyed these rules of life, by what process is he receiving reward or retribution, does he accept or resist it, and how are other men implicated in his fate,—are the chief objects of inquiry. A few remarkable exceptions to this mode of treatment will not disturb this assertion. His people either enlist our admiration by a single intense devotion to some high purpose, or compel our sympathies by their struggles to escape the ruin which their own sins or errors have invited. They are analyzed, rather than created, and we obtain from them the impression received from a crowd elated beyond measure by some absorbing enthusiasm, or writhing under the infliction of some terrible chastisement. An artist who should fill his gallery with portraits of inmates of lunatic asylums, disciples of Miller in their ascension robes, and orators at the top of their happiest climax, would hardly be regarded as a correct delineation of the human face, neither will consummate skill of analysis and execution redeem the ghostly family of our author's spiritual off-spring from the charge of untruth to a healthy nature.

And his pictures of life are generally from the same point of view. He shows us a street, a domestic circle, a public assembly, or a whole village, describes them with wonderful fidelity, yet just as we think we have them securely located upon solid ground, by one magic sentence the whole is transmuted to a symbolic picture, and a witch element in the atmosphere makes us doubt whether we are not in dream-land. Even the beautiful introductions to the "Mosses from an old Manse," and the "Scarlet Letter," are tinged with this peculiarity. The Concord there, is hardly the one that appears from the rail-track; and the Salem custom-house and its inmates, are judged from that point whence we all should put on a somewhat sorry aspect, whence the mercantile life of New England especially will not bear severe criticism. An observing reader will be struck by this tendency to symbolism on every page of these books.

So nature is regarded oftener by him in its relations to the human mind, and mental and moral laws, than as existing for any independent purpose. His exquisite pencilling of her beautiful scenes is generally illustrative of the person who is the central figure of the landscape. His winds howl a warning through open doors on the advent of some critical moment. "Alice's Posies," and "The Pyncheon Elm," the garden of "The House of the Seven Gables," the wood where Hester and Dimmesdale talked, and the midnight sky, seen by the minister from the pillory, all prefigure in outline and detail the spiritual states of those who lived among them. No strange cat walks across a path, no mouse ventures out of the wainscot, no robin sings upon the house-top, or bee dives into a squash blossom, which does not know its business and fulfil its destiny, in his drama.

This prominent tendency in Mr. Hawthorne's mind, at times assumes the form of disease. Doubtless the most profound, and from an angel's point of view, the truest estimate of man, life and nature, is that in which they are woven into a spectacle illustrative of spiritual laws. God is indeed "above all, through all, and in all," yet this doctrine must be held in connection with all we have before said of man's consciousness of freedom, or it becomes a false statement of our relations to Providence. No man can reconcile this apparent contradiction, but any man knows that he must stand by himself, or go adrift to foolishness and ruin. Therefore this Providential view of life cannot legitimately occupy the foreground in a correct delineation of existence, but should rather be the mountain range, and the horizon line, and the forces beneath the surface; and he who ignores the more obvious relations of the spirit, to live always near its central blaze, must obtain and impart false impressions, and become an unfit medium for its complete interpretation. So is it to a degree with Mr. Hawthorne. A tendency to disease in his nature, appears in the fearful intensity of his narratives. There is also a sort of unnaturalness in his world. It is seen not in the noon-day sun, so often as by moon-beams, and by auroral or volcanic lights. All that he describes may and does actually happen, but something else happens, by the omission of which we fail sometimes to acknowledge the reality of his delineation. This tendency appears in many of the tales in the "Mosses from an Old Manse," and reached its climax in the "Scarlet Letter." In "The House of the Seven Gables," we see the author struggling out of its grasp, with a vigor which we believe ensures a final recovery.

The constitution of Mr. Hawthorne's mind, in other respects, is admirably calculated to fit him for his primary office of seer. For all danger of that godless or misanthropic spirit, which so often destroys men who know much of human nature, is averted by his great affections. He follows the track of a spiritual law into the darkest or wildest scene, without losing his faith in God, or his love for humanity. With an impressibility that makes him alive even to the buttons of his overcoat, with the quickest insight through motives, and the sophistry of sin, and an overpowering sense of the ludicrous, he never loses his human sympathies. He looks upon the spectacle of existence, with the same pensive smile always upon his face, changing only to a more touching gleam of joy and sadness. He is one whose eye we feel upon our souls, yet to whom we cheerfully confide their treasures. His humor, too, seems only a part of his great love, so innocently does it play over the surface of every thing he touches; it is beyond our power to analyze it, and were it not, we are sure we should hardly risk the loss of a tithe of the pleasure we receive from it, by the attempt. We know of no modern writer who holds us so completely at his will in this respect; not Emerson, or Lamb. Neither he, nor his readers, can ever be thrown into utter desperation, for thought of the very absurdity of the position.

* * *

Perhaps four years were never spent to better purpose, than those in Mr. Hawthorne's life, between the publication of "Mosses from an Old Manse," and "The Scarlet Letter." The only account we have of them, is in the sketch of the "Custom House," which introduces the latter work. Like most men of genius, our author is not disposed to do full justice to those influences which have powerfully contributed to the growth of his mind. Often when such men are receiving and appropriating most rapidly, they are tormented with a nervous suspicion of the decay of their power. But never was such want of faith more signally rebuked, than in the writer we are reviewing, for we suspect that "spacious edifice of brick" has seldom been turned to so good use, as by this man who looked through its machinery and its occupants to the facts which they unconsciously represent. The portrait of this place is wonderfully vivid, and from the author's point of observation, doubtless true.

The story so gracefully introduced, is the most remarkable of Mr. Hawthorne's works, whether we consider felicity of plot, sustained interest of development, analysis of character, or the witchery of a style which invests the whole with a strange, ethereal beauty. These qualities of the book are so evident, that we now desire to go beneath them to those which make it, in many respects, the most powerful imaginative work of the present era of English literature. No reader possessing the slightest portion of spiritual insight, can fail to perceive that the chief value of this romance is religious. It is an attempt to delineate the involved action of spiritual laws, and their effects upon individual character, with an occasional glimpse into the organization of society. Of course it has been a puzzle to the critics, and a pebble between the teeth of the divines, transcending the artificial rules of the former, and making sad work with the creeds and buckram moralities of the latter.

Standing as "The Scarlet Letter" does, at the junction of several moral highways, it is not easy to grasp the central idea around which it instinctively arranged itself in the author's mind. The most obvious fact upon its pages is, that the only safety for a human soul consists in appearing to be exactly what it is. If holy, it must not wrench itself out of its sphere to become a part in any satanic spectacle; if corrupt, it must heroically stand upon the low ground of its own sinfulness, and rise through penitence and righteousness.

This law of life is exhibited in the contrasted characters of Dimmesdale and Hester. Whatever errors of head or heart, or infelicity of circumstances, prevent Hester from fully realizing the Christian ideal of repentance, she sternly respects her moral relations to society. She embroiders the badge of her own infamy, and without complaint submits to isolation, the pity, scorn and indifference of the world, and the withering of her own nature under the blaze of a noonday exposure to the hot sun of social displeasure; she turns her face toward humanity, and begins the life-long task of beating up to virtue against the pitiless storm which overthrows so many an offender. If the impending fate of the minister forces her to catch at the sole hope of escaping from her penance, and the closing scenes of the drama are necessary to make her an angel of mercy to the very community she had outraged by the sin of her youth, we may in mercy impute her falterings to that infirmity of our nature, which its greatest interpreter has represented by the concession of Isabella to the artifice of Mariana, and the untruth of Desdemona. As far as human fidelity to a spiritual law can go, did Hester live out the fact of the correspondence of seeming and being. Not so with the less heroic partner of her guilt. We cannot deny that all the arguments which may be used to palliate insincerity apply to Dimmesdale. The voluntary step he must take by confession, was from a more than mortal elevation to a more than human abasement. His constitutional weakness, too, is an excusing circumstance, and especially the genuineness of his repentance up to a certain point. Yet the radical vice of his soul was not submission to his passions, but cowardice; and the reflex action of this cowardice disarranged his whole life, placing him in false relations to the community and the woman he had wronged, and laying open his naked heart to the eye of the demon that was the appointed agent of his final ruin. Of the value of these two persons, considered as accurate delineations of character, nothing very flattering can be said. We see them in the midst of conflict, and in the strife of soul and law many wonderful revelations of human nature appear. Yet a strict fidelity to the engrossing object of the book, renders the author unfaithful to individual humanity. Dimmesdale and Hester are the incarnate action and reaction of the law of sincerity.

Another fact which appears in this book, is the downward tendency of sin; once let a soul be untrue, even though half in ignorance of its duty, and its world is disorganized, so that every step in its new path involves it in greater difficulties. The cardinal error, in this maze of guilt and wretchedness, is Hester's marriage with Chillingworth. She committed that sin which women are every day repeating, though never without retribution, as certain, if not as visible as hers, of giving her hand to a man she did not entirely love. There are souls great and good enough to stand firmly against the recoil of such an act, but Hester was not one of these. Her true husband at last came, and she could only give him a guilty love. By her fatal error she had cut herself off from the power to bless him by her affection as long as God should keep her in the bonds of a false marriage. The proclivity of her former error drove her on to sin again with more obvious consequences, if not with deeper guilt. And then came, in rapid succession, the ruin of Dimmesdale, the transformation of Chillingworth, the transmission of a diseased nature to her child, and the wide spread scandal of a whole community.

And growing out of this act, and its retribution, is the whole question of the relation of the sexes, and the organization of society. The author does not grapple with these intricate problems, though he knows as much of the falsity of what is called marriage, and the unnatural position of woman, as those who are more ready to undertake the cure of the world. And the hypocrisy of Dimmesdale, and the searing of heart in Hester, point to a social state in which purity will exist in connection with a mercy which shall throw no artificial obstacles in the way of a sinner's repentance.

Another fact more perplexing to a Christian moralist is here illustrated,—that a certain experience in sin enlarges the spiritual energies and the power to move the souls of men to noble results. The effects of Dimmesdale's preaching are perfectly credible, and moral, although he stood in false relations to those he addressed. True, the limitation at last came in his public exposure, yet we had almost said he could not have left his mark so deep upon the conscience of that community, had he lived and died otherwise. And Hester's error was the downward step in the winding stair leading to a higher elevation. This feature of the work, so far from being a blemish, is only a proof of the writer's insight, and healthy moral philosophy. He has portrayed sin with all its terrible consequences, yet given the other side of a problem which must excite our wonder, rebuke our shallow theories, and direct us to an all-embracing, infinite love for its solution.

In the character of Chillingworth appears another law,—the danger of cherishing a merely intellectual interest in the human soul. The Leech, is a man of diseased mental acuteness, changed to a demon by yielding to an unholy curiosity. Seduced by the opportunity to know the nature of Dimmesdale, he is drawn to the discovery of the fatal secret,—a discovery which he is not strong enough to bear. His character and fate are an awful rebuke to that insatiable desire for soul-gazing, which is the besetting devil of many men. Our human nature is too sacred to be applied to such uses, and he who enters its guarded enclosure from the mere impulse for intellectual analysis, risks his own soul as surely as he outrages that of another.

Passing from these points of the book to its general moral tone, we find the author's delineation of spiritual laws equalled by his healthy and profound religious sentiment. In justice to human nature, he shows all the palliative circumstances to guilt, while he is sternly true to eternal facts of morality. It is not improper for a novelist to do the former, if he leave the latter uppermost in the mind of the reader. Throughout the work we have not once detected the writer in a concession to that sophistical philanthropy, which, from the vantage-ground of mercy, would pry up the foundation of all religious obligation. His book is a fine contrast to the volumes of a class of modern novelists, who with a large developement of the humane sentiment, and an alarming briskness at catching the palliations of transgression, seem to have lost the sense of immutable moral distinctions. One side of Mr. Hawthorne's mind would furnish the heads of several first class French romancers. It may be that some of his statements on the side of destiny are too strong, and that human will appears to have a play too limited in his world, yet we look upon such passages rather as exaggerations of his idea of the omnipotence of God's law, than as indications of an irreligious fatalism.

We have already noticed the tendency to a symbolical view of nature and life, in this author's genius. In "The Scarlet Letter," it supplies the complete frame-work of the story—the age and social state in which the drama is cast being merely subsidiary to it. The gleam of the symbolical letter invests every object with a typical aspect. The lonely shores along which the minister walked, the wood in which he met Hester, the pillory and the street lit up by Mr. Wilson's lantern, are seen in this mysterious relation to the characters and plot of the story. But all the symbolism of the tale concentrates in the witch-child, Pearl. She seems to absorb and render back, by each development of her versatile being, the secret nature of every thing with which she comes in contact. She is the microcosm of the whole history with its surroundings. As a poetical creation, we know not where to look for her equal in modern literature. She is the companion of Mignon and Little Nell,[3] more original in conception than either, if not as strong in her hold upon our affections.

As a work of art, this book has great merits, shaded by a few conspicuous faults. We cannot too much admire the skill with which the tangled skein of counteracting law and character is unravelled, the compact arrangement and suggestive disposition of the parts. The analysis of character is also inimitable, and the style, is a fit dress for the strange and terrible history it rehearses. Yet we shall be disappointed if we look for any remarkable delineation of character, or portraiture of historical manners. There is a certain ghastliness about the people and life of the book, which comes from its exclusively subjective character and absence of humor. The world it describes is untrue to actual existence; for, although such a tragedy may be acting itself in many a spot upon earth, yet it is hidden more deeply beneath the surface of existence than this, modified by a thousand trivialities, and joys, and humorous interludes of humanity. No puritan city ever held such a throng as stalks through the "Scarlet Letter;" even in a well conducted mad-house, life is not so lurid and intense. The author's love for symbolism occasionally amounts to a ridiculous melodramatic perversity, as when it fathers such things as the minister's hand over his heart, and the hideous disfigurement of his bosom, Dame Hibbins from Gov. Bellingham's window screeching after Hester to go into the forest and sign the black man's book, and the meteoric "A" seen upon the sky during the mid-night vigil.

* * *

3. Child character in Charles Dickens's *The Old Curiosity Shop* (1841). *Mignon*: a fairylike child in Goethe's *Wilhelm Meister's Apprentice* (1796).

Review Essay

ROBERT S. LEVINE

Antebellum Feminists on Hawthorne: Reconsidering the Reception of *The Scarlet Letter*†

A number of antebellum feminists greatly admired *The Scarlet Letter,* but readers of the standard reception histories of Hawthorne would be unable to find a record of their responses. J. Donald Crowley's *Nathaniel Hawthorne: The Critical Heritage* includes only one selection from a nineteenth-century American woman writer, a review of *The Scarlet Letter* by Anne W. Abbott, the pious daughter of a Massachusetts clergyman. Gary Scharnhorst's *The Critical Response to Nathaniel Hawthorne's* The Scarlet Letter fails to include an example from any nineteenth-century American woman writer; and in the section on *The Scarlet Letter* in John L. Idol, Jr., and Buford Jones's *Nathaniel Hawthorne: The Contemporary Reviews,* the only response by an American woman writer (either in the reprinted reviews or the checklist of other reviews) is Anne W. Abbott's. The review by Abbott, which is regularly cited by Hawthorne critics, is worth our brief attention. Appearing in the *North American Review* shortly after the publication of *The Scarlet Letter,* Abbott's review offers an appreciation of Hawthorne's portrayal of "mental torpidity" in the prefatory "The Custom-House," but for the most part focuses on Hawthorne's failure to invest the novel with a proper Christian spirit. The initially appealing Hester Prynne, Abbott says, finally "disappoints us, and shows the dreamland origin and nature, when we were looking to behold a Christian." Abbott had hoped to find a truer Christian in Dimmesdale, but here, too, she is disappointed, proclaiming that "the Christian element . . . seems strangely wanting" in the minister. Though she is more tolerant and admiring of the book than most other writers of the religious press, there are no great differences between Abbott's response and, say, that of the clergyman Arthur Cleveland Coxe, who stated with respect to *The Scarlet Letter* in the *Church Review and Ecclesiastical Register*: "We protest against any toleration to a popular and gifted writer, when he perpetrates bad morals."[1]

† Written for the first edition of this Norton Critical Edition. Reprinted with the author's permission.

1. Anne W. Abbott, "The Scarlet Letter," *North American Review* 71 (July 1850), rpt. in *Nathaniel Hawthorne: The Contemporary Reviews,* ed. John L. Idol, Jr., and Buford Jones (New York: Cambridge UP, 1994), 129–30, and on pp. 231–39 of this Norton Critical Edition; Arthur Cleveland Coxe, "The Writings of Hawthorne," *Church Review and Ecclesiastical Register* 3 (January 1851), rpt. in Idol and Jones, *Nathaniel Hawthorne,* 146, and on pp. 246–54 of this Norton Critical Edition. See also *Nathaniel Hawthorne: The Critical Heritage,* ed J. Donald Crowley (1970; rpt. London and New York: Routledge, 1997); and *The Critical Response to Nathaniel Hawthorne's* The Scarlet Letter, ed. Gary Scharnhorst (New York: Greenwood P, 1992).

In a useful recent volume, *Hawthorne and Women: Engendering and Expanding the Hawthorne Tradition*, the editors John L. Idol, Jr., and Melinda M. Ponder present essays that illuminate Hawthorne's interactions with women writers and intellectuals, such as Margaret Fuller and Elizabeth Palmer Peabody, and his influence on a range of women fiction writers, including Harriet Beecher Stowe, Elizabeth Stoddard, and Rebecca Harding Davis. But with only one important exception, which I will discuss below, this revisionary volume ultimately works to perpetuate the notion that there were virtually no antebellum feminists who responded to *The Scarlet Letter* in the wake of its publication, and that those women of the 1850s who did respond to Hawthorne's novels responded in the manner of Abbott. Thus Idol and Ponder assert in their introduction that the contemporary reviews of Hawthorne written by women "differ little from those written by men. In subjects treated, style, and perceptions of Hawthorne's genius and character, women reviewers generally responded to Hawthorne's fiction as if they were integral members of the New England (or national) clerisy."[2]

Ironically, the apparent existence of only conventional responses to Hawthorne's romances from antebellum women writers has contributed to the notion that Hawthorne, seemingly ignored by more radical women, was as conventional and moralistic as the critics like Abbott who attacked him. In a recent polemic, *The Scarlet Mob of Scribblers: Rereading Hester Prynne*, Jamie Barlowe suggests as much, taking Hawthorne's conventionality as a given and thus excoriating male critics who "admire the duplicitous radical subversion of men like Nathaniel Hawthorne and hold up as a model his male fantasy of a radical, subversive woman, Hester Prynne."[3] But how "duplicitous" was Hawthorne in his conception of Hester as a subversive? Michael T. Gilmore has argued that Hester's "dissident side . . . associates her with antebellum feminism,"[4] and it is the contention of this reception study that Gilmore's view of such an association is no male fantasy. The noted, antebellum feminists Jane Swisshelm, Amelia Bloomer, Grace Greenwood, and Charlotte Forten all admired *The Scarlet Letter*, and yet their responses to the novel have not made a mark on the historical record. Some of the fault may lie with the nature of the "contemporary responses" book, for such books tend to focus on full-length reviews at the expense of shorter reviews and different forms of contemporary responses, such as letters and journals or essays not principally devoted to reviewing particular works. In her polemic, Barlowe attacks what she regards as a male Hawthorne industry for overlooking the responses of twentieth-century women writers to *The Scarlet Letter*. One could argue similarly that, with respect to the responses of

2. John L. Idol, Jr., and Melinda M. Ponder, eds., *Hawthorne and Women: Engendering and Expanding the Hawthorne Tradition* (Amherst: U of Massachusetts P, 1999), 13.
3. Jamie Barlowe, *The Scarlet Mob of Scribblers: Rereading Hester Prynne* (Carbondale: Southern Illinois UP, 2000), 18. On the important role of male elites in the making of Hawthorne's canonical reputation, see also Jane Tompkins, "Masterpiece Theater: The Politics of Hawthorne's Literary Reputation," in *Sensational Designs: The Cultural Work of American Fiction, 1790–1860* (New York: Oxford UP, 1985) and Richard Brodhead, *The School of Hawthorne* (New York: Oxford UP, 1986).
4. Michael T. Gilmore, "Hawthorne and the Making of the Middle Class," in *Rethinking Class: Literary Studies and Social Formations*, ed. Wai Chee Dimock and Gilmore (New York: Columbia UP, 1994), 227.

antebellum feminists, the predominately male critics of twentieth-century Hawthorne studies (and women critics as well, including Barlowe) did not care to take note of, or did not know how to look for, such responses. How else to explain the neglect until 1999 (and beyond) of the popular writer Jane Swisshelm's extensive review of *The Scarlet Letter*, which is among the most incisive short essays on the novel ever published?

Jane Grey Swisshelm (1815–1884) was born Jane Grey Cannon in Pittsburgh and emerged as a major voice in women's rights, antislavery, and temperance reform. In 1836 she married the Methodist farmer James Swisshelm, a man she came to regard as a tyrant. She left him on several occasions and in the early 1840s began writing feminist essays for local newspapers. In 1847 she founded and edited the Pittsburgh *Saturday Visiter* [*sic*] as an antislavery and women's rights paper, building a circulation of around 6,000 subscribers by 1850. Swisshelm remarked in her autobiography that men were outraged that "a woman had started a political paper," but according to her probably overstated testimony, the "*Visiter* had thousands of readers scattered over every State and Territory in the nation, in England and the Canadas. It was quoted more perhaps than any other paper in the country."[5] In 1852 she sold her interest in the paper while remaining an important contributor. Continuing to find life with her husband intolerable, she left him for good in 1857, moving with her six-year-old daughter to St. Cloud, Minnesota, where she founded the *St. Cloud Visiter*, one of the most vigorously antislavery newspapers in the western territories. Unfazed when a proslavery group destroyed her printing press in 1858, she started up the *St. Cloud Democrat* in its place. In 1863 she volunteered to serve as a nurse in the Union army, and after the war she retired to Swissvale, Pennsylvania, where she eventually wrote her memoir, *Half a Century* (1880).

As a writer and social critic, Swisshelm is best on display in her most popular book, *Letters to Country Girls* (1853), which collects a number of her weekly opinion pieces from the *Saturday Visiter*. Deploying a comic; aggressive, "impudent" voice similar to that of the popular columnist Fanny Fern, Swisshelm sought to remedy what she termed the "masculine-superiority fever," "a deeply-rooted and unsightly cancer, which disfigures the entire face of the body-politic." Complaining of men's efforts to keep women at home in "'a woman's place,'" Swisshelm asserted about women who attempt to do something more or different: "let her aspire to turn editor, public speaker, doctor, lawyer—take up any profession or avocation which is deemed honorable and requires talent, and O! bring the Cologne, get a cambric kerchief and a feather fan, unloose his corsets and take off his cravat!"[6] The same sort of spirited antipatriarchal energy informed her 1850 review of *The Scarlet Letter* in the September 28 issue of the Pittsburgh *Saturday Visiter*. That review had not been reprinted or even mentioned in Hawthorne criticism until 1999, when it was reprinted as an Appendix to Idol and Ponder's *Hawthorne and Women*. Oddly, Idol

5. Jane Grey Swisshelm, *Half a Century* (Chicago: Jansen, McClurg & Company, 1880), 113, 123. For additional biographical background on Swisshelm, see *Crusader and Feminist: Letters of Jane Grey Swisshelm*, 1858–1865, ed. Arthur J. Larsen (1934; rpt. Westport, Connecticut: Hyperion P, 1976).

6. Jane G. Swisshelm, *Letters to Country Girls* (New York: John R. Riker, 1853), 79, 77, 78.

and Ponder supply no introduction to the review, no biographical or historical contextualization. Moreover, not one of the essays in the volume mentions the review, which seems to have had no impact on Idol and Ponder's conception of antebellum women's responses to Hawthorne's fiction as elaborated in their introduction. The review is thus both there and not there in *Hawthorne and Women;* I have transcribed the review from its original source in the *Saturday Visiter* and it is printed in its entirety in this Norton Critical Edition.[7]

Swisshelm's review of *The Scarlet Letter* is notable for what could be termed its Hester-centric reading of the novel. Rather than morally condemn Hester, Swisshelm hails her as "the most glorious creation of fiction that has ever crossed our path,"[8] a woman who, in a world of timorous and judgmental men, bravely acts on her romantic and sexual desires. Thus Swisshelm begins her review by making it clear that the genius of Hawthorne's romance lies not in the gossipy and political "The Custom-House," but in the "unique" tale central to the novel itself, "told in a most masterly manner." Having made great claims for the novel, Swisshelm, following the conventions of the time, then launches into an extended, approximately 1,200-word reading of the plot, which of course is not simply a plot summary but a reorchestration of the story lines and motifs that had most engaged her, particularly Hester's refusal to succumb to a bad marriage. In her columns in the Pittsburgh *Saturday Visiter,* Swisshelm regularly insisted that marriage should be "a spirit union . . . designed to make one out of two" and thus railed against marriages in which women were subordinated to a brutal, unloving master.[9] Swisshelm's rehearsal of plot has little to say about Dimmesdale and Chillingworth. Instead, the emphasis is on Hester's sexuality, her "commanding mien and surpassing beauty," her relation to Pearl, and her efforts to negotiate her place in the hostile Puritan community. There is considerable psychological acuity in Swisshelm's reading of Hester's efforts to find new outlets for her "poetic imagination and inventive genius" and sympathy for the plight of the reviled adulterous mother and for the child herself. As Swisshelm nicely remarks about the transformed Pearl at the end of the novel: "she becomes gentle, affectionate—comprehensible." Swisshelm remains puzzled and even annoyed by Hawthorne's depiction of Hester's voluntary return to Boston to once again take up the scarlet letter. Why would Hester capitulate in such a way? The implicit answer supplied by Swisshelm helps to keep the novel true to her sense of its romantic vision: the return will enable Hester at her death to "sleep[ ] beside her lover."

In her review, then, Swisshelm celebrates a novel that, perhaps despite itself, depicts the bankruptcy of a loveless marriage, raises questions about a woman's legal obligation to remain confined within such a marriage, and supports a woman's desire for sexual fulfillment (even if the result is

7. See pp. 239–42. It needs to be noted that there are over twenty transcription errors in the 1999 reprinting of Swisshelm's review in Idol and Ponder's *Hawthorne and Women,* pp. 288–91. Among the errors are "is even" for "was even," "administer" for "minister," "the state of abject misery" for "the abject state," "persecutions" for "insane persecutions," and "no one" for "none."
8. All citations from Jane Swisshelm's "The Scarlet Letter" are from the original publication in the *Saturday Visiter* (28 Sept. 1850): 2 (p. 146 of the 1850 volume).
9. See the Pittsburgh *Saturday Visiter* (24 Nov. 1849): 2; and for Swisshelm's related comments on marriage, see her editorials in the *Saturday Visiter,* especially 1 June 1850, 27 July 1850, and 3 Aug. 1850.

a child out of wedlock). My "perhaps" is deliberate here, for it gets at Swisshelm's own uncertainty about Hawthorne's didactic aims, as suggested by her repeated use of "if" when considering whether his politics might more resemble someone like Anne Abbott's than her own. "If he meant to teach the sinfulness of Hester's sin—the great and divine obligation and sanctity of a legal marriage contract, and the monstrous depravity of a union sanctioned only by affection," she writes, "his book is the most sublime failure of the age." But she goes on to suggest that it would be absurd to condemn Hester as a sinner, drawing on Hawthorne's romance to support her argument. So perhaps Hawthorne did not mean to condemn her; perhaps he meant to argue for the "moral force and grandeur of [Hester's] character." As for the men of the novel, Swisshelm perceives little more than intolerant Puritan leaders, a "poor imbecile lover" (Dimmesdale), and an "insane . . . old sinner" (Chillingworth). "If," she comically writes, "such a little 'jet of flame' as Pearl can be considered a sign of a broken law, we do wonder what Hawthorne thinks of the royal idiots whose existence testifies to the validity and legality of the pompous marriage rites of Queens and Empresses." Instead of regarding Pearl as punishment for her mother's adulterous sexuality, Swisshelm, quite unlike Abbott, sees "Divine approbation of the law which governed her birth." When Swisshelm writes in her concluding paragraph that "[i]f Hawthorne really wants to teach the lesson ostensibly written on the pages of his book, he had better try again," the suggestion is that if Hawthorne did *not* want to teach that lesson, then he got things just right. Ultimately what Swisshelm emphasizes in *The Scarlet Letter* through her "if"'s is Hawthorne's conflict between his attraction to the subversive Hester and his desire to contain her. In her own way, Swisshelm anticipates by 125 years or so the critical insights of Nina Baym and other feminist readers of the novel who, in the words of Robert K. Martin, regard Hawthorne as speaking "his Romantic desire through Hester as he also speaks his sense of guilt and shame."[1] The shape of Hawthorne criticism over the course of the twentieth century could well have had a significantly different arc had more readers known of Swisshelm's pioneering review before 1999.

Swisshelm is not the only antebellum feminist neglected by Hawthorne scholars. Two months before Swisshelm published her review in the *Saturday Visiter*, there appeared an admiring literary notice of *The Scarlet Letter* in the July 1850 issue of Amelia Bloomer's *The Lily: A Ladies Journal, Devoted to Temperance and Literature.* I have been unable to find any mention of this evocative notice in Hawthorne bibliography or scholarship. Best known for her championing of dress reform for women (she promoted Turkish-style pantaloons, which came to be known as "Bloomers"), Amelia Jenks Bloomer (1818–1894) during most of her life as a reformer was primarily committed to temperance. From her feminist perspective, temperance reform promised to put a check on patriarchal power, which she

1. Robert K. Martin, "Hester Prynne, *C'est Moi:* Nathaniel Hawthorne and the Anxieties of Gender," in *Engendering Men: The Question of Male Feminist Criticism*, ed. Joseph A. Boone and Michael Cadden (NewYork: Routledge, 1990),129. The only discussion of Swisshelm's review of *The Scarlet Letter* that I have been able to locate is in the historian Peter F. Walker's *Moral Choices: Memory, Desire, and Imagination in Nineteenth-Century American Abolition* (Baton Rouge: Louisiana State UP, 1978), 133–35. Walker reads the review, and indeed Swisshelm's entire career, mainly in terms of her troubled marriage; in his rendering, her, political engagement is a kind of "symptom" of the marriage.

conceived of both literally and metaphorically as a form of intoxication that threatened the stability of the home. Born Amelia Jenks in Homer, New York, she married the Quaker antislavery lawyer Dexter Bloomer in 1840, and shortly thereafter began contributing temperance writings to local newspapers and journals. She attended the Seneca Falls Woman's Rights Convention in 1848 and around the same time helped to form a women's temperance organization in Seneca Falls. One year later she founded *The Lily*, the second newspaper, after Swisshelm's, to be established and edited entirely by a woman. The paper had a subscription base of around 4,000 and remained a vigorous forum for women's reform causes until Bloomer sold the paper in 1855. Subsequently she moved to Iowa with her husband, where she continued her temperance work until her death in 1894.

Though *The Lily* announced itself as a journal devoted to temperance and literature, Bloomer printed virtually no literary reviews or notices. One of the few was of Hawthorne's *The Scarlet Letter*, which appeared on the front page of the issue of July 1850, followed immediately, on the front and second page, by a reprinting of Hawthorne's "A Rill from the Town-Pump." The notice, offered here in its entirety, came in the form of a letter from "T" to Amelia Bloomer:

> MRS. BLOOMER—The perusal of "The Scarlet Letter," a new work by NATHANIEL HAWTHORNE, has naturally recalled to mind the earlier productions by the same author. Some time ago some of his most beautiful sketches were collected and published, with the title of "Twice Told Tales," which have taken a high place among the literary works of our day, and would, alone, mark the author as one possessed of rare mental endowments. Although a writer of prose, Hawthorne always sees Nature with the eye of a poet, and yet, his descriptions are so accurate and finished, he may fairly be ranked among her true interpreters. His pictures have all the truthfulness of the Daguerreotype, and, at the same time, that warm and golden light upon them which the hand of genius alone can impart. Judging of the merits of the productions, from the impressions they have left upon me to this day, I may safely say one of the best is that entitled "A Rill from the Town Pump"; and I ask for it a place in your monthly, because, while it shows the peculiar power Hawthorne possesses of throwing a charm around even the commonest subject, it carries with it a lesson in morals which few can forget, and is, in fact, one of the best temperance lectures extant. "A Vision of the Fountain," another of his productions, which describes water gushing forth, pure and bright, from its home in the woods, untrammelled by the hand of art, yet adorned with the highest order of natural beauty, would not be inappropriate to your columns; but your readers shall first admire the speech supposed to have been delivered by a resident of the old town of Salem, Mass, the birth-place of at least one great orator. T.[2]

Of course the large irony of this notice is that it identifies the worth of *The Scarlet Letter* in the very terms that Hawthorne in "The Custom-House" claims to have rejected: in relation to his popular "A Rill from the

2. *The Lily: A Ladies Journal, Devoted to Temperance and Literature,* 2 (1 July 1850): 49; "A Rill from the Town-Pump" appeared on pp. 49–50 (which were actually pp. 1–2 of the July issue).

Town-Pump" (1835). For Bloomer, "T," and presumably other feminist-temperance women associated with *The Lily*, "Rill," in which a talking town pump makes a millennialist case for the value of pure water, was not simply a quaint, sentimental, evangelical, or "puritanical" piece of temperance writing, but a work that engaged issues of health, body, and gender. In short, "Rill" for these readers remained a charged aesthetic and political document well worth reprinting fifteen years after its initial publication. Hawthorne wrote "Rill" at a time when temperance reform was predominately in the hands of elites who were concerned about controlling male workers' drinking at the newly developing factories. Bloomer republished the sketch shortly after the Seneca Falls Convention had helped to redefine temperance as a major concern of women reformers.

To a certain extent, then, it can be argued that Bloomer, through "T"'s notice and the reprinting of "A Rill from the Town-Pump," appropriated *The Scarlet Letter* for her own feminist-temperance purposes, but such an argument would obscure the fact that Hawthorne's antipatriarchal politics of the body may not have been considerably different from hers. In *The Lily*, Bloomer printed numerous accounts of the connections between male drinking and abusive husbands, beginning with the inaugural issue of January 1, 1849, when she remarked in a squib on the "fearful sight" of a young woman marrying a man "who loves to linger around the wine cup": "wealth, talents, fame can never gild the drunkard's home, nor sooth the sorrows of a drunkard's wife." Jane Swisshelm herself wrote in the October 1850 issue of *The Lily* that the "drunkard," by giving "himself up to a base, sordid selfish appetite . . . [is] a monster!"[3] In *The Scarlet Letter* Hawthorne works metaphorically to present Chillingworth as a kind of monster driven by an intemperate appetite for revenge. In his next novel, *The House of the Seven Gables* (1851), Hawthorne conceptualizes issues of temperance more explicitly in the manner of writers for *The Lily*, portraying Judge Pyncheon as a husband literally intoxicated by such stimulants as wine and coffee, who reveals the full extent of his intemperate patriarchal authority immediately after his marriage: "the lady got her death-blow in the honey-moon, and never smiled again, because her husband compelled her to serve him with coffee, every morning, at his bedside, in token of fealty to her liege-lord and master."[4] When Hawthorne mentions in the same paragraph that the wife died three or four years later, the suggestion is that her death can be taken as a form of rebellious suicide.

The notice on *The Scarlet Letter* in *The Lily*, along with the reprinting of "Rill," would thus have worked to focus attention on the politics of gender in the novel, creating sympathy for Hester as a "victim" of intemperate patriarchal authority. Though "T" does not develop an explicit

3. *The Lily* 1 (1 Jan. 1849): 3; "Mrs. Swisshelm's Opinion of Drunkards," *The Lily*, 2 (1 Oct. 1850): 78. On women and temperance, see Carol Mattingly, ed., *Water Drops from Women Writers: A Temperance Reader* (Carbondale: Southern Illinois UP, 2001). Though Swisshelm would contribute a number of pieces to *The Lily*, she remained distrustful of two key aspects of Bloomer's feminism. She believed that Bloomer's dress reforms mistakenly attempted to cover up what she regarded as the very real differences between the sexes; and she believed that Bloomer put too much emphasis on women's rights at the expense of antislavery; see Swisshelm, *Half a Century*, chap. 29.
4. Nathaniel Hawthorne, *The House of the Seven Gables*, ed, William Charvat et al. (Columbus: Ohio State UP, 1965), 123.

argument along those lines, we can extrapolate such a reading from her praise of "natural beauty," which in the novel is regularly associated with Hester. For "T," that which is most valued is "water gushing forth, untrammelled by the hand of art, yet adorned with the highest order of natural beauty." In *The Scarlet Letter*, there is a clear divide between the artifice of the town and the naturalness of the woods, with the town regularly associated with the patriarchal authority of the Puritan masters and the woods with the antipatriarchal revolutionism of Mistress Hibbins and Hester. True, the fact that Hester is also an artist makes the gender dichotomy that I am suggesting somewhat tenuous. But in her celebration of the natural, "T" does suggest an alliance with the Hester whose "beauty shone out" on the scaffold at the opening of the novel and who seven years later manages to restore that natural beauty during her forest walk with Dimmesdale. In this great revolutionary moment, Hester tosses aside the scarlet letter and declares to Dimmesdale, "What we did had a consecration of its own." At which point, Hawthorne wryly points to the dynamic of gender and interpretation by remarking that the "whole richness of her beauty came back from what men call the irrevocable past."[5] In her ability in the woods to state her views and desires and annunciate a millennial vision of freedom, Hester has, to push the metaphor just a bit, achieved a natural gushing forth of feeling that links her to the purity celebrated by "T" in both "Rill" and "A Vision of the Fountain."

As the favorable notice in *The Lily* suggests, Hawthorne was regarded by some activist women as a critic of patriarchal authority. But did this writer who spoke to women who were concerned about the slavery of the bottle speak to women concerned about slavery on the southern plantation? Jane Swisshelm, who advocated both temperance and antislavery, was certainly responsive to *The Scarlet Letter*, even as she suspected that Hawthorne may have fled from some of the radical implications of his novel. Was such a flight the result of his contradictory relation to slavery? In an influential article of 1986, Jonathan Arac linked the compromise of the novel's ending (Hester's willingness to return to Boston and keep herself in a sort of irresolute relation to patriarchal civil authority) to the Compromise of 1850, which was generally supported by Hawthorne's friends in the Democratic party. In the compromise of *The Scarlet Letter* and in the Compromise of 1850, Arac argues, there was an effort to balance conservative and radical tendencies into a somewhat mystified functional order.[6] Far more critical of Hawthorne's Democratic politics is Jean Fagan Yellin, who criticizes him for his failure to "exhort his readers to act to end human bondage, as William Lloyd Garrison had been doing in his newspaper the *Liberator* every Friday since 1831."[7] In her essay "Hawthorne and the Slavery Question," Yellin adduces evidence of Hawthorne's racism and unwillingness to challenge slavery head-on, but her comparison to Garrison is tendentious, given that virtually no antebellum white American author was engaged in antislavery at the level of Garrison (who, it is worth recalling, was regarded by Frederick Douglass

5. Nathaniel Hawthorne, *The Scarlet Letter*, pp. 39, 118, and 122 in this Norton Critical Edition.
6. Jonathan Arac, "The Politics of *The Scarlet Letter*," in *Ideology and Classic American Literature*, ed. Sacvan Bercovitch and Myra Jehlen (New York: Cambridge UP, 1986), 247–66.
7. Jean Fagan Yellin, "Hawthorne and the Slavery Question," in *A Historical Guide to Nathaniel Hawthorne*, ed. Larry J. Reynolds (New York: Oxford UP, 2001), 137.

and others as a paternalistic racist).[8] Viewed from our present moment, Hawthorne indeed failed the test of slavery, but viewed from within the context of his times, there is compelling evidence that feminist-abolitionists held him in high regard, perhaps because they didn't see him as such a failure after all. I will be examining two antislavery women writers who praised *The Scarlet Letter*, but first it would be useful to consider a notable attack on Hawthorne's politics of slavery, which appeared in the leading African American newspaper of the time.

Yellin's criticisms of Hawthorne were anticipated in an anonymous squib in an 1855 issue of *Frederick Douglass' Paper*. At the conclusion of the essay, which praised Emerson for his antislavery politics, the writer (perhaps Douglass) states: "The fact is, that with the exception of Mr. Hawthorne, every New England author who is likely to be heard of a hundred years hence, is in favor of freedom."[9] During the 1850s such criticism of Hawthorne's politics was rather unusual, in part, I would argue, because his fictions suggested a propensity in favor of freedom, with the mesmerical motifs of *The House of the Seven Gables* and *The Blithedale Romance*, for instance, conveying a genuine revulsion towards patriarchal enslavers. The propensity for freedom discernible in his romances was not merely rhetorical, for there is evidence that Hawthorne, despite his hostility toward abolitionists, was opposed to slavery. In a letter to Longfellow of 8 May 1851, Hawthorne voiced his objection to the Fugitive Slave Law, and in his 1852 campaign biography of Franklin Pierce, he called slavery one of the "evils" of the world. Despite these beliefs, Hawthorne, as he stated in an infamous passage in the Pierce biography, argued that the actual job of ending slavery should be left to the mysterious workings of "divine Providence."[1] Hawthorne's resistance to political antislavery had much to do with the fact that his closest friends dating back to his Bowdoin College days were active members of the Democratic party—a party that, it is worth remembering, despite its support for the Compromise of 1850, was radical at its inception in its hostility toward class privilege. As Yellin points out, Hawthorne's friendship with Pierce in particular got him into trouble during the Civil War, when he chose to dedicate *Our Old Home* (1863) to his friend, who was rightly regarded in antislavery circles as having been sympathetic to southern slave interests during his presidency. Hawthorne's Unionist acquaintances were appalled by this apparent sign of disloyalty, as they had been appalled a year earlier by his skeptical remarks on Abraham Lincoln in his 1862 *Atlantic* essay, "Chiefly about War-Matters."

The relatively mild contempt expressed toward Hawthorne in *Frederick Douglass' Paper* is actually not representative of the views of the temperance women and feminist-abolitionists who commented on *The Scarlet Letter* during the 1850s, and it is notable that at least two antislavery women writers of the period, Grace Greenwood and Charlotte Forten,

8. See Frederick Douglass, *My Bondage and My Freedom*, ed. William L. Andrews (1855; Urbana: U of Illinois P, 1987), 220.
9. *Frederick Douglass' Paper* (23 Feb. 1855): 2.
1. See Hawthorne, *The Letters*, 1843–1853, ed. Thomas Woodson, L. Neal Smith, and Norman Holmes Pearson (Columbus: Ohio State UP, 1985), 431; Hawthorne, *Life of Franklin Pierce* (1852), in *Miscellanies: Biographical and Other Sketches and Letters by Nathaniel Hawthorne* (Boston and New York: Houghton Mifflin Company, 1900), 166 (p. 220 of this Norton Critical Edition).

regarded Hawthorne favorably within the context of their own antislavery politics. Greenwood's remarks on *The Scarlet Letter* appeared in the *National Era*, the antislavery newspaper based in Washington, D.C., which would become famous when it serialized *Uncle Tom's Cabin* in 1851–52. Hawthorne was treated with great respect in this paper throughout the 1850s, beginning with an enthusiastic review of *The Scarlet Letter* in an April 1850 issue. Because the anonymous review (perhaps written by Greenwood) has hitherto not been noted or reprinted in the various collections of contemporary responses to Hawthorne's fiction, I offer it here in its entirety:

> The introductory chapter to this work, containing a description of the Custom House, its life and inmates, reminds one of the happiest and quaintest portraitures of Charles Lamb. The Romance itself is complete in design and execution. Simple in its action, with few incidents, and few characters, it is yet rich in thought, in feeling, and in philosophy. The story is one of crime and its punishment, dating back a century ago, the scene being laid in the colony of Massachusetts. The characters, in the exhibition of which the author displays extraordinary psychological skill, are Hester Prynne, a woman, young, beautiful, proud, daring, full of passion, of inexhaustible energy of character, branded with a crime that banishes her from the sympathies of society, amidst which she lives and moves, all alone, with the symbol of her shame blazoned upon her:—a minister, of brilliant endowments, fearing God, loving man; but suffering for years the gnawing pangs of deadly remorse for a sin, he would not repeat, but dares not confess—and a student, advanced in years, once amiable, devoted to the pursuit of knowledge, but transformed into a fiend by cherishing and feeding a terrible appetite for revenge, which in the end works the death of the minister through the influence of spiritual torture, utterly ruins himself, and blasts the earthly hopes of the unfortunate woman. In the subtle, thrilling analysis of these characters, and of the workings of the passions and sentiments by which they were controlled, consists the extraordinary power of this singular romance.[2]

As would be the case in Swisshelm's review published later in the year, the anonymous writer celebrates Hawthorne's psychological acuity, focusing on the splendor of his creation of Hester Prynne. The clear expression of sympathy for Hester works to condemn a society that fails to offer a similar sympathy, and thus, by extension, raises questions about reviewers like Anne Abbott, who ultimately want to bring the novel into the realm of a conventional Christian moral ethic. The review is extraordinary, then, for its utter lack of moral judgment on the behavior of a woman who has an adulterous affair resulting in a child. Moreover, the antislavery reviewer's description of Hester as "branded" implicitly links her to the slaves, many of whom were branded, thus anticipating by around 150 years Leland S. Person's argument that Hester "resembles the slave mothers like Harriet Jacobs," and that "in representing her maternally, Hawthorne shows more sympathy and ironic understanding of the politics of her

2. "THE SCARLET LETTER," *National Era* 4.171 (11 April 1850): 59.

motherhood than his nineteenth-and twentieth-century detractors have allowed."[3]

Later in 1850, around the time of Swisshelm's review, the *National Era* printed two travel letters that commented favorably on *The Scarlet Letter*. These were the work of the writer Grace Greenwood, the pen name of Sarah Jane Lippincott (1823–1904). Born Sara Jane Clarke in Pompey, New York, she began writing under the pen name of Grace Greenwood in the early 1840s, publishing essays on domesticity and reform in *Godey's Lady's Book* and numerous other newspapers and journals. She opposed the Mexican War, capital punishment, African colonization, and slavery. Her poem "The Leap from the Long Bridge" (1851) helped to inspire William Wells Brown's *Clotel* (1853), the first published novel by an African American; and during the Civil War she earned President Lincoln's praise for her visits to Union hospitals and camps. Her antislavery politics were on display in many of her letters to the *National Era*, but she also published letters that simply focused on her travels. In a letter of 18 September 1850 from Lynn, Massachusetts, for instance, Greenwood described her pleasant visit to the Gloucester seashore and her subsequent visit to Boston, where she attended the theater and visited the studio of the painter C. G. Thompson. At the studio, she admired his great portrait of Hawthorne, and her reflections on Hawthorne's physiognomy allowed her to comment on his widely discussed recent novel, *The Scarlet Letter*. She writes of Thompson's oil painting:

> In the deep, dark eye of Hawthorne lies the secret of that wonderful mastery—that half-beautiful, half-fearful power—that strange, weird-like fascination, which so enchain one in "THE SCARLET LETTER"; while, in the warm fulness and quiet scorn of the lips, we re-read that memorable "Preface," wherein the play of delicate fancy and a delicious humor alternated with cold, sharp strokes of merciless satire.
>
> Mr. Hawthorne is, according to this portrait, a singularly handsome man, but his face wears an expression of unconsciousness, or rather disdain, of his beauty.[4]

Granted, Greenwood doesn't say all that much about Hawthorne in this short description. But it is worth noticing that Greenwood and, as we will see, Charlotte Forten place great emphasis on what can be learned about an author by looking at his portrait. As in Hawthorne's *House*, there is a sense that a portrait, whether a painting or a daguerreotype, can supply access to something more "natural," "real," and essential than the living, breathing personage viewed in the midst of daily activities. Looking into Hawthorne's dark eye, Greenwood discovers the "secret" of the "wonderful mastery" that "so enchain one" in *The Scarlet Letter*. In the context of the aesthetic vocabulary of the time, that enchaining power would have been the power of sympathy. I would posit, then, on the basis of this short reflection, that Greenwood's viewing of the portrait helped her better to

3. Leland S. Person, "The Dark Labyrinth of Mind: Hawthorne, Hester, and the Ironies of Racial Mothering." *Studies in American Fiction* 29 (2001): 44.

4. "Letters from Grace Greenwood. From the Shore. Lynn, September 18, 1850," *National Era* 4.195 (26 Sept. 1850): 155. For a reproduction of Cephas Thompson's magnificent 1850 oil painting of Hawthorne, see Rita K. Gollin, *Portraits of Nathaniel Hawthorne: An Iconography* (DeKalb: Northern Illinois UP, 1983), 30.

understand Hawthorne's ability to sympathize with a figure like Hester Prynne who, like Hawthorne at the Custom-House, was a "beauty" who became a victim of patriarchal authority. Two months later Greenwood presented patriarchal authority in terms more specific to slavery in a highly political letter from Boston of 9 November 1850. In the midst of a long letter on the injustices of the Fugitive Slave Bill, Greenwood took the occasion not to lambaste Hawthorne for his Democratic politics, but to note his forthcoming publications: "I have a bit of literary intelligence for you. The title of the forthcoming romance, by the author of THE SCARLET LETTER, is 'THE HOUSE OF SEVEN GABLES [*sic*].' Is it not quaint and Hawthornish? Another work by this most delightful author is a volume for children, entitled 'TRUE STORIES FROM HISTORY AND BIOGRAPHY.' This Ticknor is bringing out in fine style."[5]

The fact that Hawthorne and Greenwood had the same publisher in George Ticknor may help to explain Greenwood's graciousness toward Hawthorne and, in the early 1850s, Hawthorne's toward Greenwood. (Hawthorne would eventually have unkind things to say about Greenwood in letters of 1854 and 1856.) Hawthorne, for instance, wrote Greenwood on 17 April 1852 to praise her 1852 *Greenwood Leaves: A Collection of Sketches and Letters, Second Series*, commenting that her letters "are the best that any woman writes—of course, better than any man's." In her excellent assessment of the relationship between Greenwood and Hawthorne, Nina Baym speculates that Hawthorne may have not even read *Greenwood Leaves*, and had simply offered his praise as a form of male noblesse oblige. But she goes on to argue that if this man, who regularly instructed his wife that women should remain out of the public eye, had read *Leaves*, "he could not possibly have escaped awareness of their openly political and fundamentally public nature." Baym thus concludes that there is much more in common between Hawthorne and Greenwood than Hawthorne would want to concede. I would add to Baym's analysis another reason for Hawthorne's warm letter of 1852 to Greenwood: that even if he had not read *Greenwood Leaves*, he would have read or heard about Greenwood's flattering mentions of him in the *National Era*. (*Greenwood Leaves* reprints a number of Greenwood's letters in the *National Era*, but *not* the two she wrote about Hawthorne.) And I would underscore the significance of the fact that those flattering mentions appeared in a journal devoted to antislavery.[6]

The African American diarist and antislavery reformer Charlotte L. Forten [Grimké] was also quite taken with *The Scarlet Letter*, commenting on it both in her private journals and in a published newspaper report on Salem. The granddaughter of the prosperous sailmaker and antislavery activist James Forten, Charlotte Forten (1837–1914) was born in

5. "Letter of Grace Greenwood. Boston, November 9, 1850," *National Era* 4.203 (21 Nov. 1850): 18.
6. Hawthorne, *The Letters, 1843–1853*, 532; Nina Baym, "Again and Again, the Scribbling Women," in *Hawthorne and Women*, 31–32. See also Hawthorne, *The Letters, 1853–1856*, ed. Thomas Woodson, L. Neal Smith, and Norman Holmes Pearson (Columbus: Ohio State UP, 1987), 166, 456–57. Throughout the 1850s, the *National Era* would continue to write only good things about Hawthorne, even noting in 1857, after he left his Democratic patronage position as consul to Liverpool: "[A]s the 'Scarlet Letter' followed his retirement from one office, we may expect, as an early result of this resignation, a book in no wise inferior to that remarkable romance in power and popularity" (11.558 [10 Sept. 1857]: 147).

Philadelphia and raised by her father, Robert Forten, an abolitionist whose wife died three years after Charlotte's birth. Objecting to Philadelphia's segregated schools, Robert sent Charlotte to Salem in 1853, and there she lived with the antislavery activists Charles Lenox Remond and his wife, Amy Matilda Remond, while attending Higginson Grammar School. She began keeping her diary in 1854 and published several poems and essays in the late 1850s. Moving back and forth between Salem and Philadelphia, she taught for a while at Epes Grammar School in Salem, and eventually got a federal teaching position in 1862 at Port Royal, St. Helena Island, where she taught recently freed South Carolina slaves. Until her diaries were published in the twentieth century, her most famous piece of writing was her two-part essay about that teaching experience, "Life on the Sea Islands," which appeared in the May and June 1864 issues of the *Atlantic Monthly*.[7] She married the Presbyterian minister Francis J. Grimké in 1878 and eventually settled with him in Washington, D.C., where she remained active in African American political and intellectual affairs.

Shortly after moving to Salem in 1853, Forten became a close friend of Hawthorne's sister Elizabeth. Perhaps because of that friendship, Forten's journals register a regular, enthusiastic reading of Hawthorne, taking special note of *The Scarlet Letter, The House of the Seven Gables*, and *Tanglewood Tales*, even as she is responding to the political controversies of the day. Outraged by the Massachusetts Supreme Court's ruling against the fugitive slave Anthony Burns, Forten wrote on 2 June 1854: "To-day Massachusetts has again been disgraced; again she has showed her submissions to the Slave Power; and Oh! with what deep sorrow do we think of what will doubtless be the fate of that poor man, when he is again consigned to the horrors of slavery." One month later, in an entry of 17 July 1854, she remarked with anger and dismay on the pervasive antiblack racism of "my native land—where I am hated and oppressed because God has given me a *dark skin*. How did this cruel this absurd prejudice ever exist?"[8] Between these 1854 entries on Anthony Burns and antiblack racism, she wrote her most substantial entry on Hawthorne, which has nothing but good things to say about this prominent supporter of the Democratic party.

Like Grace Greenwood, Forten responds to Hawthorne through a reading of a portrait, in this case a daguerreotype. She is shown the daguerreotype and another Hawthorne portrait (perhaps an 1840 oil painting by Charles Osgood) by his sister Elizabeth, whom she describes as having an "eerie, spectral look which instantly brought to my mind the poem of 'The Ancient Mariner.'" Forten responds quite differently to the supposedly gloomy Nathaniel Hawthorne:

> I have seen to-day a portrait of Hawthorne, one of the finest that has ever been taken of him. He has a splendid head. That noble, expansive brow bears the unmistakeable impress of genius and superior intellect. And in the depths of those dark, expressive eyes there is a

7. Charlotte L. Forten's "Life on the Sea Islands" may be found in William L. Andrews, *Classic African American Women's Narratives* (New York: Oxford UP, 2003), 364–91.
8. *The Journals of Charlotte Forten Grimké*, ed. Brenda Stevenson (New York: Oxford UP, 1988), 65, 87.

> strange, mysterious influence which one feels in reading his works, and which I felt most forcible when reading that thrilling story "The Scarlet Letter." Yet there is in his countenance no trace of that gloom which pervades some of his writings; particularly that strange tale "The Unpardonable Sin" and many of the "Twice Told Tales." After reading them, I had pictured the author to myself as very dark and gloomy-looking. But I was agreeably disappointed. Grave, earnest, thoughtful, he appears but not gloomy. His sister, who, with much kindness showed me his portrait, is very singular-looking. . . . She showed me another portrait of Hawthorne taken when he was very young. His countenance, though glowing with genius, has more of the careless, sanguine expression of youth than profound, elevated thought which distinguishes his maturer years, and gives to his fine face and to his deeply interesting writings that mysterious charm which is felt and acknowledged by all.[9]

In many respects, what is most interesting about this reflection is what is not there: criticism of Hawthorne's politics. Instead, as in Greenwood's comments on Thompson's painting, one discerns an admiration for a writer whose fiction, one has to assume, was not viewed as radically at odds with her commitment to antislavery. That said, what mainly comes across in Forten's remarks is a lack of concern about connections between politics and art. Forten's Hawthorne, specifically the Hawthorne of *The Scarlet Letter*, is a playful, mysterious, and even elusive presence whose novel cannot be reduced to a single meaning, political or otherwise. Taking an interpretive position quite different from that of Yellin over a century later, she seems simply moved by a text that she cannot quite pin down, but that certainly speaks in inspiring ways to a reader who was also concerned about racism and slavery in her contemporary culture. Compared to Swisshelm, Bloomer, and even Greenwood, then, Forten seems the least willing to appropriate the novel for a specific end, ultimately admiring the novel for its hard-to-describe otherness and aesthetic power. One suspects that if she had thought the novel's main objective was to discipline the heroine and reinforce white patriarchy, she would have been a far less appreciative reader.

The following year, Forten commented again on *The Scarlet Letter* in her diary. In an entry of 26 November 1855, she remarked: "Saw for the first time the Custom House of which I read Hawthorne's descriptions in the introduction to that thrilling story—the 'Scarlet Letter.' I should have known it at once by the description. I wonder that I have not visited it before." What is striking here, in addition to Forten's terming of the novel as "thrilling," is her understanding of Hawthorne's investment in the real, despite his protestations to the contrary in the prefaces of his three romances of the 1850s. She links *The Scarlet Letter* to the real again in her first published essay, "Glimpses of New England," which appeared anonymously in the 1858 *National Anti-Slavery Standard*. In a somewhat veiled reference to *The House of the Seven Gables*, she refers to "the Old Witch House," which, she says, "has recently been defaced and

9. *Journals of Charlotte Forten Grimké*, 84–85.

desecrated by the erection of an apothecary shop in front of one of its wings." But she is explicit with her reference to *The Scarlet Letter*, which draws from her diary: "On Derby street, a street of wharves, stands the old Custom-House, which Hawthorne has so minutely described in his introduction to 'The Scarlet Letter.'" Again, her admiration for Hawthorne's fidelity to the real dominates both short mentions of his romances in an essay that, unsurprisingly, given its publication in an antislavery newspaper, has as one of its main concerns the situation of antislavery in Salem. The essay concludes by lamenting that it is only "the faithful few, too, few alas!" who fight the antislavery struggle.[1] Given her interest in the real, and her politics of antislavery, it is somewhat surprising that Forten, unlike the anonymous critic in *Frederick Douglass' Paper*, can continue to read Hawthorne, the known Democrat, the author of the biography of Franklin Pierce, so enthusiastically. Some might regret Forten's naiveté in ignoring connections between Hawthorne's politics and art; I find inspiring her ability to read across the color line and against the grain of a political reductionism. At the very least, Forten's appreciations of Hawthorne should help to prompt critics toward more capacious and unpredictable explorations of Hawthorne's relationship to the slave culture of his time.

Writing about neglected African American perspectives in antebellum slave culture, Frederick Douglass stressed the importance of bringing African Americans more to the center of historical narratives by attending to "marks, traces, possibles, and probabilities."[2] In my survey of responses to *The Scarlet Letter* by the antebellum feminists Swisshelm, Bloomer, Greenwood, and Forten, I have similarly considered marks, traces, possibilities, and probabilities, and I would suggest that there are in all likelihood many more marks and traces still to be detected. We do not yet have a full grasp of the reception of *The Scarlet Letter* in the many strata of antebellum culture. Swisshelm in particular represents and anticipates an important school of criticism that is both exhilarated by Hawthorne's bold conception of the rebellious Hester and skeptical about the extent of his investment in the containment strategies of his novel. In a "Postscript" appended to one of the best studies that we have of *The Scarlet Letter*'s containment strategies, Sacvan Bercovitch suggests the possibility of what he terms an "aversive" reading of the novel, one that, going against the grain of his overall argument about the novel as "an agent of socialization," emphasizes dissent over containment and takes as a measure of Hawthorne's own identification with the radicalism of Hester "unerased traces of contradiction in the novel—undesired silences that . . . do not quite succeed in silencing conflict."[3] Such a reading, I have been suggesting, is both hinted at and sanctioned by the enthusiastic response of the antebellum feminists under consideration in this essay. These commentators appreciate an imaginative writer who,

1. *Journals of Charlotte Forten Grimké*,145; [Charlotte Forten], "Glimpses of New England, *National Anti-Slavery Standard* (19 June 1858): 4.
2. Frederick Douglass, "The Heroic Slave" (1853), in *The Oxford Frederick Douglass Reader*, ed. William L. Andrews (New York: Oxford UP, 1996), 132.
3. Sacvan Bercovitch, *The Office of the Scarlet Letter* (Baltimore: Johns Hopkins UP, 1991), xii, 155, 157.

despite his possible containment strategies, was responsive to feminist issues of the time, and thus could be read in the context of their own political projects, or, as would appear to be the case with Forten, in an awed dis-relation from any particular political project. Their Hawthorne needs to become a more central constituent of our Hawthorne.

Puritan Background and Sources

CHARLES RYSKAMP

The New England Sources of *The Scarlet Letter*†

After all the careful studies of the origins of Hawthorne's tales and the extensive inquiry into the English sources of *The Scarlet Letter*,[1] it is surprising that the American sources for the factual background of his most famous novel have been largely unnoticed. As would seem only natural, Hawthorne used the most creditable history of Boston available to him at that time, and one which is still an important source for the identification of houses of the early settlers and for landmarks in the city. The book is Dr. Caleb H. Snow's *History of Boston*. Study and comparison of the many histories read by Hawthorne reveal his repeated use of it for authentication of the setting of *The Scarlet Letter*. Consequently, for the most part this article will be concerned with Snow's book.

If we are to see the accurate background Hawthorne created, some works other than Snow's must also be mentioned, and the structure of time as well as place must be established. Then it will become apparent that although Hawthorne usually demanded authentic details of colonial history, some small changes were necessary in his portrayal of New England in the 1640's. These were not made because of lack of knowledge of the facts, nor merely by whim, but according to definite purposes—so that the plot would develop smoothly to produce the grand and simple balance of the book as we know it.

During the "solitary years," 1825–37, Hawthorne was "deeply engaged in reading everything he could lay his hands on. It was said in those days that he had read every book in the Athenaeum . . ."[2] Yet no scholar has studied his notebooks without expressing surprise at the exceptionally few remarks there on his reading. Infrequently one will find a bit of "curious information, sometimes with, more often without, a notation of the source; and some of these passages find their way into his creative

† From *American Literature* 31.1 (1959): 257–72. Page numbers in square brackets refer to this Norton Critical Edition.

1. I shall make no reference to the English sources of *The Scarlet Letter* which have been investigated by Alfred S. Reid in *The Yellow Ruff and The Scarlet Letter* (Gainesville, 1955) and in his edition of *Sir Thomas Overbury's Vision . . . and Other English Sources of Nathaniel Hawthorne's "The Scarlet Letter"* (Gainesville, 1957). Most of this article was written before the publication of Reid's books. It may serve, however, as a complement or corrective to the central thesis put forth by Reid: "that accounts of the murder of Sir Thomas Overbury were Hawthorne's principal sources in composing *The Scarlet Letter*" (*The Yellow Ruff*, 112).
2. James T. Fields, *Yesterdays with Authors* (Boston, 1900), 47. For a list of books which Hawthorne borrowed from the Salem Athenaeum, see Marion L. Kesselring, *Hawthorne's Reading 1828–1850* (New York, 1949). All of my sources are included in this list, except the second edition (1845) of Felt's *Annals of Salem*.

work."[3] But for the most part Hawthorne did not reveal clues concerning the books he read and used in his own stories. About half of his writings deal in some way with colonial American history, and Professor Turner believes that "Hawthorne's indebtedness to the history of New England was a good deal larger than has ordinarily been supposed."[4] Certainly in *The Scarlet Letter* the indebtedness was much more direct than has hitherto been known.

Any work on the exact sources would have been almost impossible if it had not been for Hawthorne's particular use of the New England annals. Most of these are similar in content. The later historian builds on those preceding, who, in turn, must inevitably base all history on the chronicles, diaries, and records of the first settlers. Occasionally an annalist turns up a hitherto unpublished fact, a new relationship, a fresh description. It is these that Hawthorne seizes upon for his stories, for they would, of course, strike the mind of one who had read almost all the histories, and who was intimate with the fundamentals of colonial New England government.

As a young bachelor in Salem Hawthorne, according to his future sister-in-law, Elizabeth Peabody, "made himself thoroughly acquainted with the ancient history of Salem, and especially with the witchcraft era."[5] This meant that he studied Increase Mather's *Illustrious Providences* and Cotton Mather's *Magnalia Christi Americana*. He read the local histories of all the important New England towns. He read—and mentioned in his works—Bancroft's *History of the United States*, Hutchinson's *History of Massachusetts*, Snow's *History of Boston*, Felt's *Annals of Salem*, and Winthrop's *Journal*.[6] His son reported that Hawthorne pored over the daily records of the past: newspapers, magazines, chronicles, English state trials, "all manner of lists of things. . . . The forgotten volumes of the New England Annalists were favorites of his, and he drew not a little material from them."[7] He used these works to establish verisimilitude and greater materiality for his own books. His reading was perhaps most often chosen to help him—as he wrote to Longfellow—"give a life-like semblance to such shadowy stuff"[8] as formed his romances. Basically it was an old method of achieving reality, most successfully accomplished in his own day by Scott; but for Hawthorne the ultimate effects were quite different. Here and there Hawthorne reported actual places, incidents, and people—historical facts—and these were united with the creations of his mind. His explicitly stated aim in *The Scarlet Letter* was that "the Actual and the Imaginary may meet, and each imbue itself with the nature of the other" (55) [28]. His audience should recognize "the authenticity of the outline" (52) [26] of the novel, and this would help them to accept the actuality of the passion and guilt which it contained. For the author himself, the strongest reality of outline or scene was in the past, especially the history of New England.

3. *The American Notebooks*, ed. Randall Stewart (New Haven, 1932), xxxii.
4. H. Arlin Turner, "Hawthorne's Literary Borrowings," *PMLA*, LI, 545 (June, 1936).
5. Moncure D. Conway, *Life of Nathaniel Hawthorne* (New York, 1890), 31.
6. Edward Dawson, *Hawthorne's Knowledge and Use of New England History: A Study of Sources* (Nashville, Tenn., 1939), 5–6; Turner, 551.
7. Julian Hawthorne, *Hawthorne Reading* (Cleveland, 1902), 107–108, 111, 132. Hawthorne's sister Elizabeth wrote to James T. Fields: "There was [at the Athenaeum] also much that related to the early History of New England. . . . I think if you looked over a file of old Colonial Newspapers you would not be surprised at the fascination my brother found in them. There were a few volumes in the Salem Athenaeum; he always complained because there were no more" (Randall Stewart, "Recollections of Hawthorne by His Sister Elizabeth," *American Literature*, XVI, 324, 330, Jan., 1945).
8. *The American Notebooks*, xlii.

The time scheme of the plot of *The Scarlet Letter* may be dated definitely. In chapter 12, "The Minister's Vigil," the event which brings the various characters together is the death of Governor Winthrop. From the records we know that the old magistrate died on March 26, 1649.[9] However, Hawthorne gives the occasion as Saturday, "an obscure night of early May" (179) [91]. Some suggestions may be made as reasons for changing the date. It would be difficult to have a night-long vigil in the cold, blustery month of March without serious plot complications. The rigidly conceived last chapters of the book require a short period of time to be dramatically and psychologically effective. The mounting tension in the mind and heart of the Reverend Mr. Dimmesdale cries for release, for revelation of his secret sin. Hawthorne realized that for a powerful climax, not more than a week, or two weeks at the most, should elapse between the night of Winthrop's death, when Dimmesdale stood on the scaffold, and the public announcement of his sin to the crowd on Election Day. The Election Day (275) [134] and the Election Sermons (257) [128] were well known and traditionally established in the early colony in the months of May or June.[1] (The election of 1649, at which John Endicott became governor, was held on May 2.) Consequently Hawthorne was forced to choose between two historical events, more than a month apart. He wisely selected May, rather than March, 1649, for the time of the action of the last half of the book (chapters 12–23).

The minister's expiatory watch on the scaffold is just seven years after Hester Prynne first faced the hostile Puritans on the same platform (179, 194, 205) [91, 98, 103]. Therefore, the first four chapters of *The Scarlet Letter* may be placed in June, 1642 (see 68) [36]. Hawthorne says that at this time Bellingham was governor (85–86) [45]. Again one does not find perfect historical accuracy; if it were so, then Winthrop would have been governor, for Bellingham had finished his term of office just one month before.[2] A possible reason for Hawthorne's choice of Bellingham will be discussed later.

The next major scene—that in which Hester Prynne goes to the mansion of Bellingham—takes place three years later (1645).[3] Hawthorne

9. William Allen, *An American Biographical and Historical Dictionary* (Cambridge, Mass., 1809), 616; Caleb H. Snow, *A History of Boston* (Boston, 1825), 104; Thomas Hutchinson, *The History of Massachusetts* (Salem, 1795), I, 142.
1. John Winthrop, *The History of New England from 1630 to 1649* (Boston, 1825–1826), II, 31, 218 (a note on p. 31 states that the charter of 1629 provided for a general election on "the last Wednesday in Easter term yearly"; after 1691, on the last Wednesday of May); also Daniel Neal, *The History of New-England . . . to . . . 1700* (London, 1747), II, 252. Speaking of New England festivals, Neal writes: "their Grand Festivals are the Day of the annual Election of Magistrates at *Boston*, which is the latter End of *May*; and the Commencement at *Cambridge*, which is the last *Wednesday* in *July*, when Business is pretty much laid aside, and the People are as chearful among their Friends and Neighbours, as the *English* are at *Christmas*." Note Hawthorne's description of Election Day (*The Scarlet Letter*, [137]): "Had they followed their hereditary taste, the New England settlers would have illustrated all events of public importance by bonfires, banquets, pageantries and processions. . . . There was some shadow of an attempt of this kind in the mode of celebrating the day on which the political year of the colony commenced. The dim reflection of a remembered splendor, a colorless and manifold diluted repetition of what they had beheld in proud old London . . . might be traced in the customs which our forefathers instituted, with reference to the annual installation of magistrates."
2. Winthrop, II, 31: June 2, 1641, Richard Bellingham elected governor. Winthrop, II, 63: May 18, 1642, John Winthrop elected governor.
3. *The Scarlet Letter*, 138 [71]: "Pearl, therefore, so large were the attainments of her three years' lifetime, could have borne a fair examination in the New England Primer, or the first column of the Westminster Catechisms, although unacquainted with the outward form of either of those celebrated works." The Westminster Catechisms were not formulated until 1647; the New England Primer was first brought out ca. 1690.

correctly observes: "though the chances of a popular election had caused this former ruler to descend a step or two from the highest rank, he still held an honorable and influential place among the colonial magistracy" (125) [64].[4] From the description of the garden of Bellingham's house we know that the time of the year was late summer (132–133) [68–69].

With these references to time, as Edward Dawson has suggested,[5] we can divide the major action of the novel as follows:

Act One

i. Chapters 1–3. The Market-Place, Boston. A June morning, 1642.
ii. Chapter 4. The Prison, Boston. Afternoon of the same day.

Act Two

Chapters 7–8. The home of Richard Bellingham, Boston. Late summer, 1645.

Act Three

i. Chapter 12. The Market-Place. Saturday night, early May, 1649.
ii. Chapters 14–15. The sea coast, "a retired part of the peninsula" (202) [103]. Several days later.
iii. Chapters 16–19. The forest. Several days later.

Act Four

Chapters 21–23. The Market-Place. Three days later (see 257) [128].

The place of each action is just as carefully described as is the time. Hawthorne's picture of Boston is done with precise authenticity. A detailed street-by-street and house-by-house description of the city in 1650 is given by Snow in his *History of Boston*. It is certainly the most complete history of the early days in any work available to Hawthorne. Whether he had an early map of Boston cannot be known, but it is doubtful that any existed from the year 1650. However, the City of Boston Records, 1634–1660, and the "Book of Possessions" with the reconstructed maps (made in 1903–1905 by George Lamb, based on the original records)[6] prove conclusively the exactness of the descriptions written by Snow and Hawthorne.

Hawthorne locates the first scene of *The Scarlet Letter* in this way:

> . . . it may safely be assumed that the forefathers of Boston had built the first prison-house somewhere in the vicinity of Cornhill, almost as seasonably as they marked out the first burial-ground, on Isaac Johnson's lot, and round about his grave, which subsequently became the nucleus of all the congregated sepulchres in the old churchyard of King's Chapel. (67) [35].[7]

> It was no great distance, in those days, from the prison-door to the market-place. . . . Hester Prynne . . . came to a sort of scaffold, at the

4. Winthrop, II, 220: on May 14, 1645, Thomas Dudley had been elected governor.
5. I am largely indebted to Dawson, 17, for this time scheme.
6. For the drawing of the map reproduced with this essay, I am grateful to Professor W. F. Shellman, Jr., of the School of Architecture, Princeton University.
7. Concerning Isaac Johnson, Snow writes: "According to his particular desire expressed on his death bed, he was buried at the Southwest corner of the lot, and the people exhibited their attachment to him, by ordering their bodies to be buried near him. This was the origin of the first burying place, at present the Chapel burial ground" (37).

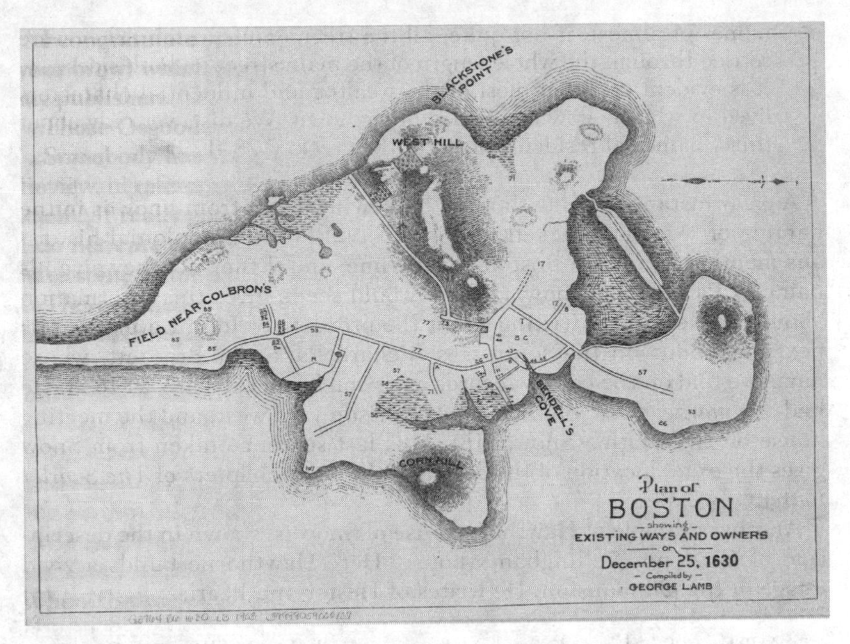

Courtesy of the Norman B. Leventhal Map Center at the Boston Public Library.

> western extremity of the market-place. It stood nearly beneath the eaves of Boston's earliest church, and appeared to be a fixture there. (75–76) [40][8]

Snow says that in 1650 Governor Bellingham and the Rev. John Wilson lived on one side of the Market-place and Church Square (Snow, 117). Near Spring Lane on the other side of the Square (mentioned by Hawthorne when little Pearl says, "I saw her, the other day, bespatter the Governor himself with water, at the cattle-trough in Spring Lane," 164) [84] was the home of Governor Winthrop (Snow, 108). All the action of *The Scarlet Letter* set in Boston is thus centered in the heart of the city. This, as Snow takes great pains to point out, was where all the leading townsmen lived. He writes:

> It has been so often repeated that it is now generally believed the north part of the town was at that period the most populous. We are convinced that the idea is erroneous. . . . The book of possessions records the estates of about 250, the number of their houses, barns, gardens, and sometimes the measurement of their lands. It seems to embrace the period from 1640 to 1650, and we conclude, gives us the

8. Justin Winsor, in *The Memorial History of Boston* (Boston, 1881), I, 506, 539, writes: "The whipping-post appears as a land-mark in the Boston records in 1639, and the frequent sentences to be whipped must have made the post entirely familiar to the town. It stood in front of the First Church, and was probably thought to be as necessary to good discipline as a police-station now is . . . The stocks stood sometimes near the whipping-post. . . . And here, at last, before the very door of the sanctuary, perhaps to show that the Church and State went hand-in-hand in precept and penalty, stood the first whipping-post,—no unimportant adjunct of Puritan life."

> names of almost, if not quite, all the freemen of Boston. They were settled through the whole length of the main street on both sides. . . . It is evident too, that most of the wealthy and influential characters lived in what is now the centre of the town. We discover only about thirty names of residents north of the creek. (128–129)

A clear instance of Hawthorne's borrowing a fact from Snow is in the naming of "Master Brackett, the jailer" (92) [48]. Few colonial historians mention a jailer in Boston at this time, and if they do, they give his name as Parker. But Snow, alone it would seem, gives this information about Brackett, after writing about the property of John Leverett: "His next neighbour on the south was Richard Parker or Brackett, whose name we find on the colony records as prison keeper so early as 1638. He had *'the market stead'* on the east, the prison yard west, and the meeting house on the south" (Snow, 116). This last sentence taken from Snow gives the exact location of the action of the early chapters of *The Scarlet Letter.*

Another example of Hawthorne's use of Snow is, shown in the description of Governor Bellingham's house. Here Hawthorne builds a vivid image of the old mansion. He writes of Hester and Pearl:

> Without further adventure, they reached the dwelling of Governor Bellingham. This was a large wooden house, built in a fashion of which there are specimens still extant in the streets of our older towns. . . . It had, indeed, a very cheery aspect; the walls being overspread with a kind of stucco, in which fragments of broken glass were plentifully intermixed; so that, when the sunshine fell aslantwise over the front of the edifice, it glittered and sparkled as if diamonds had been flung against it by the double handful. . . . It was further decorated with strange and seemingly cabalistic figures and diagrams, suitable to the quaint taste of the age, which had been drawn in the stucco when newly laid on, and had now grown hard and durable, for the admiration of after times. (128–129) [66][9]

There are almost no representations of the first settlers' houses in the New England annals. But Snow on one occasion does print an old plate showing an "Ancient building at the corner of Ann-Street and Market-Square" (166). And he describes the house in a way which bears a remarkable resemblance to the sketch written by Hawthorne twenty-five years later:

> This, says a description furnished by a friend, is perhaps the only wooden building now standing in the city to show what was considered elegance of architecture here, a century and a half ago. . . . The outside is covered with plastering, or what is commonly called roughcast. But instead of pebbles, which are generally used at the present day to make a hard surface on the mortar, broken glass was used. This glass appears like that of common junk bottles, broken into pieces of about half an inch diameter, . . . This surface was also variegated with ornamental squares, diamonds and flowers-de-luce. (167)[1]

9. Hawthorne also accurately noted that Governor Bellingham was "bred a lawyer." Snow writes of Bellingham: "He was by education a lawyer" (159).

1. For a possible source for details concerning the interior of Bellingham's house, the front door, knocker, etc., see Joseph B. Felt, *Annals of Salem* (2nd ed.; Salem, 1845), I, 403–406.

Snow is also the only historian who tells the story of Mrs. Sherman's pig in order to bring out its effect upon the early Massachusetts government.[2] Hawthorne, with his characteristic interest in the unusual fact from the past, refers to this strange incident:

> At that epoch of pristine simplicity, however, matters of even slighter public interest, and of far less intrinsic weight, than the welfare of Hester and her child, were strangely mixed up with the deliberations of legislators and acts of state. The period was hardly, if at all, earlier than that of our story, when a dispute concerning the right of property in a pig not only caused a fierce and bitter contest in the legislative body of the colony, but resulted in an important modification of the framework itself of the legislature. (126) [65]

In his version of the story Snow said that the incident "gave rise to a change also in regard to the Assistants" (95) and that because of the confusion and dissatisfaction over the decision of the court, "provision was made for some cases in which, if the two houses differed, it was agreed that the major vote of the whole should be decisive. This was the origin of our present Senate" (96).

The characters named in *The Scarlet Letter*—other than Hester, Pearl, Chillingworth, and Dimmesdale, for whom we can find no real historical bases—were actual figures in history. The fictional protagonists of the action move and gain their being in part through their realistic meetings with well-known people of colonial Boston. Even the fantastic Pearl grows somewhat more substantial in the light of the legend and story of her primitive world. She is seen, for example, against the silhouette of the earlier Mr. Blackstone. When describing Bellingham's garden Hawthorne relates: "There were a few rose-bushes, however, and a number of apple-trees, probably the descendants of those planted by the Reverend Mr. Blackstone, the first settler of the peninsula; that half-mythological personage, who rides through our early annals, seated on the back of a bull" (133) [69]. Snow had said:

> By right of previous possession, Mr. Blackstone had a title to proprietorship in the whole peninsula. It was in fact for a time called Blackstone's neck. . . . Mr. Blackstone was a very eccentrick character. He was a man of learning, and had received episcopal ordination in England. . . . It was not very long before Mr. Blackstone found that there might be more than one kind of nonconformity, and was virtually obliged to leave the remainder of his estate here. . . . Let the cause of his removal have been what it may, certain it is that he went and settled by the Pawtucket river. . . . At this his new plantation he lived uninterrupted for many years, and there raised an orchard, the first that ever bore apples in Rhode Island. He had the first of the sort called yellow sweetings, that were ever in the world, and is said to have planted the first orchard in Massachusetts also. . . . Though he was far from agreeing in opinion with Roger Williams, he used frequently to go to Providence to preach the gospel; and to encourage his younger hearers, while he gratified his own benevolent disposition, he would give them of his apples, which were the first they ever

2. Snow, 95–96. Hutchinson, I, 135–136 also refers to the incident, but not in this particular way.

> saw. It was said that when he grew old and unable to travel on foot, not having any horse, he used to ride on a bull, which he had tamed and tutored to that use. (50–53)

This account is taken virtually word for word from a series of articles called "The Historical Account of the Planting and Growth of Providence" published in the Providence *Gazette* (January 12 to March 30, 1765).[3] However, Snow adds to this narrative the application to Boston, which would be of special interest to Hawthorne (the phrase, "and is said to have planted the first orchard in Massachusetts also").

The only minor characters that are developed to such an extent that they become in any way memorable figures are Mrs. Hibbins and the Rev. John Wilson. Hawthorne's use of Mrs. Hibbins shows again a precise interest in the byways of Boston history. He describes the costume of the "reputed witch-lady" carefully (264, 286) [132, 143]. He refers to her as "Governor Bellingham's bitter-tempered sister, . . . the same who, a few years later, was executed as a witch" (44) [74]. And again, during the minister's vigil, Hawthorne writes that Dimmesdale beheld "at one of the chamber-windows of Governor Bellingham's mansion . . . the appearance of the old magistrate himself. . . . At another window of the same house, moreover, appeared old Mistress Hibbins, the Governor's sister . . ." (181) [92]. In Snow's book there is this account of Mrs. Ann Hibbins:

> The most remarkable occurrence in the colony in the year 1655 was the trial and condemnation of Mrs. Ann Hibbins of Boston for witchcraft. Her husband, who died July 23, 1654, was an agent for the colony in England, several years one of the assistants, and a merchant of note in the town; but losses in the latter part of his life had reduced his estate, and increased the natural crabbedness of his wife's temper, which made her turbulent and quarrelsome, and brought her under church censures, and at length rendered her so odious to her neighbours as to cause some of them to accuse her of witchcraft. The jury brought her in guilty, but the magistrates refused to accept the verdict; so the cause came to the general court, where the popular clamour prevailed against her, and the miserable old lady was condemned and executed in June 1656. (140)[4]

There seems to be only one source for Hawthorne's reference to Mrs. Hibbins as Bellingham's sister. That is in a footnote by James Savage in the 1825 edition of John Winthrop's *History of New England,* and it was this edition that Hawthorne borrowed from the Salem Athenaeum.[5] Savage writes that Mrs. Hibbins "suffered the punishment of death, for the ridiculous crime, the year after her husband's decease; her brother, Bellingham, not exerting, perhaps, his highest influence for her preservation."[6] Hawthorne leads the reader to assume that Mrs. Hibbins, nine years before

3. These were reprinted in the Massachusetts Historical Society's *Collections*, 2nd Ser., IX, 166–203 (1820).
4. This is almost a literal copy from Hutchinson, I, 173. See also William Hubbard, "A General History of New England," Massachusetts Historical Society *Collections*, 2nd Ser., V, 574 (1815); Winthrop, I, 321.
5. Kesselring, 64.
6. Winthrop, I, 321 n. This contradicts Julian Hawthorne's observation: "As for Mistress Hibbins, history describes her as Bellingham's relative, but does not say that she was his sister, as is stated in the 'Romance'" ("Scenes of Hawthorne's Romances," *Century Magazine,* XXVIII, 391, July, 1884).

the death of her husband, is living at the home of her brother. Hawthorne uses this relationship between Bellingham and Mrs. Hibbins in order to have fewer stage directions and explanations. It helps him to establish a more realistic unity in the tale. It partially explains the presence of the various people at the Market-Place the night of the minister's vigil, since Bellingham's house was just north of the scaffold. It also suggests why Bellingham is the governor chosen for the opening scenes of the novel, to prevent the plot from becoming encumbered with too many minor figures.

The Reverend John Wilson's description is sympathetically done, and it is for the most part historically accurate. Hawthorne presents him as "the reverend and famous John Wilson, the eldest clergyman of Boston, a great scholar, like most of his contemporaries in the profession, and withal a man of kind and genial spirit" (86) [45]. Cotton Mather,[7] William Hubbard,[8] and Caleb Snow testify to his remarkable "compassion for the distressed and . . . affection for all" (Snow, 156). William Allen, in his *American Biographical and Historical Dictionary,* writes that "Mr. Wilson was one of the most humble, pious, and benevolent men of the age, in which he lived. Kind affections and zeal were the prominent traits in his character. . . . Every one loved him. . . ."[9] Hawthorne, to gain dramatic opposition to Dimmesdale, makes the preacher seem older than he really was. He pictures the man of fifty-seven as "the venerable pastor, John Wilson . . . [with a] beard, white as a snow-drift"; and later, as the "good old minister" (182) [93].

Hawthorne's description of Puritan costuming has been substantiated by twentieth-century research. Although the elders of the colonial church dressed in "sad-colored garments, and gray, steeple-crowned hats" (67) [35][1] and preached simplicity of dress, Hawthorne recognized that "the church attendants never followed that preaching."[2] "Lists of Apparell" left by the old colonists in their wills, inventories of estates, ships' bills of lading, laws telling what must *not* be worn, ministers' sermons denouncing excessive ornamentation in dress, and portraits of the leaders prove that "little of the extreme Puritan is found in the dress of the first Boston colonists."[3] Alice Morse Earle, after going over the lists of clothing brought by the Puritans, concludes:

> From all this cheerful and ample dress, this might well be a Cavalier emigration; in truth, the apparel supplied as an outfit to the Virginia planters (who are generally supposed to be far more given over to rich dress) is not as full nor as costly as this apparel of Massachusetts Bay. In this as in every comparison I make, I find little to indicate any difference between Puritan and Cavalier in quantity of

7. *Magnalia Christi Americana* (London, 1702), bk. III, 46.
8. Hubbard, 604.
9. Allen, 613. The Reverend John Wilson was born in 1588; he died in 1667.
1. The phrase, "steeple-crowned hats," is used by Hawthorne each time he describes the dress of the Puritan elders (*The Scarlet Letter,* 24, 67, 79, 278) [11, 35, 41, 139]. The only source that I have been able to find for this particular phrase is in an essay on hats in a series of articles on clothing worn in former times: Joseph Moser, "Vestiges, Collected and Recollected, Number XXIV," *European Magazine,* XLV, 409–415 (1804). The Charge-Books of the Salem Athenaeum show that Hawthorne read the magazine in which this article appeared. Moser wrote about the "elevated and solemn beavers of the Puritans" (414) and the "high and steeple-crowned hats, probably from an idea, that the conjunction of Church and State was necessary to exalt their archetype in the manner that it was exalted" (411).
2. Alice Morse Earle, *Two Centuries of Costume in America* (New York, 1903), I, 8.
3. Earle, I, 13.

> garments, in quality, or cost—or, indeed, in form. The differences in England were much exaggerated in print; in America they often existed wholly in men's notions of what a Puritan must be. (I, 34)

Hawthorne's descriptions agree with the early annals. The embroideries and bright colors worn by Pearl, the silks and velvets of Mrs. Hibbins, Hester's needlework—the laces, "deep ruffs . . . and gorgeously embroidered gloves"—were, as he said, "readily allowed to individuals dignified by rank or wealth, even while sumptuary laws forbade these and similar extravagances to the plebeian order" (105–106) [54]. The Court in 1651 had recorded "its utter detestation and dislike that men or women of mean condition should take upon them the garb of Gentlemen, by wearing gold or silver lace . . . which, though allowable to persons of greater Estates or more liberal Education, yet we cannot but judge it intolerable in persons of such like condition."[4] Hawthorne's attempt to create an authentic picture of the seventeenth century is shown in *The American Notebooks* where he describes the "Dress of an old woman, 1656."[5] But all of Hawthorne's description is significant beyond the demands of verisimilitude. In *The Scarlet Letter* he is repeating the impressions which are characteristic of his tales: the portrayal of color contrasts for symbolic purposes, the play of light and dark, the rich color of red against black, the brilliant embroideries[6] on the sable background of the "sad-colored garments."

So far there has been slight mention of the influence of Cotton Mather's writings on *The Scarlet Letter*. These surely require our attention in any study such as this one. Professor Turner believes that certain elements of Mather's *Magnalia Christi Americana*, "and in particular the accounts of God's judgment on adulterers [in II, 397–98], may also have influenced *The Scarlet Letter*. Mather relates [II, 404–05] that a woman who had killed her illegitimate child was exhorted by John Wilson and John Cotton to repent while she was in prison awaiting execution. In like manner, as will be recalled, John Wilson joins with Governor Bellingham and Arthur Dimmesdale in admonishing Hester Prynne to reveal the father of her child."[7] It is possible that an echo of the witch tradition in the *Magnalia Christi Americana* may also be found in *The Scarlet Letter*. "The proposal by Mistress Hibbins that Hester accompany her to a witch meeting is typical of the Mather witch tradition, which included, in accordance with the well known passage in *The Scarlet Letter*, the signing in the devil's book with an iron pen and with blood for ink. . . ."[8] The Black Man mentioned so often by Hawthorne (100, 144, 222–225) [52, 75, 112–14] was familiar to the Puritan settlers of New England. Pearl tells her mother "a story about the Black Man. . . . How he haunts this forest, and carries a book with him,—a big, heavy book, with iron clasps; and how this ugly Black Man offers his book and an iron pen to everybody that meets him here

4. Winsor, I, 484–85. Hawthorne had read the *Acts and Laws . . . of the Massachusetts-Bay in New-England* (Boston, 1726)—see Kesselring, 56.
5. *The American Notebooks*, 109.
6. One of Hawthorne's favorite words—for example, see *The American Notebooks, 97*.
7. Turner, 550; Turner is using the Hartford (1855) edition of the *Magnalia Christi Americana*. See *The Scarlet Letter*, 86–91 [71–74].
8. Turner, 546—see *The Scarlet Letter*, 143–144 [xx], and *Magnalia Christi Americana*, bk. VI, 81: "It was not long before *M. L.* . . . confess'd that *She* rode with her Mother to the said Witch-meeting. . . . At another time *M. L. junior*, the Grand-daughter, aged about 17 Years . . . declares that . . . they . . . rode on a Stick or Pole in the *Air* . . . and that they set their Hands to the Devil's Book. . . ."

among the trees; and they are to write their names with their own blood" (222) [112]. Concerning the Black Man, Cotton Mather had written: "These *Tormentors* tendred unto the afflicted a *Book*, requiring them to *Sign* it, or *Touch* it at least, in token of their consenting to be Listed in the Service of the *Devil*; which they refusing to do, the *Spectres* under the Command of that *Blackman*, as they called him, would apply themselves to Torture them with prodigious Molestations."[9]

Even the portent in the sky, the great red letter A, which was seen on the night of the revered John Winthrop's death (and Dimmesdale's vigil), would not have seemed too strange to Puritan historians. To them it would certainly not have been merely an indication of Hawthorne's gothic interests. Snow had related that when John Cotton had died on Thursday, December 23, 1652, "strange and alarming signs appeared in the heavens, while his body lay, according to the custom of the times, till the Tuesday following" (133).

The idea of the scarlet A had been in Hawthorne's mind for some years before he wrote the novel. In 1844 he had made this comment in his notebooks as a suggestion for a story: "The life of a woman, who, by the old colony law, was condemned always to wear the letter A, sewed on her garment, in token of her having committed adultery."[1] Before that, in "Endicott and the Red Cross," he had told of a "woman with no mean share of beauty" who wore a scarlet A [p. 165 of this Norton Critical Edition]. It has commonly been accepted that the "old colony law" which he had referred to in his notebooks had been found in Felt's *Annals of Salem*, where we read under the date of May 5, 1694: "Among such laws, passed this session, were two against Adultery and Polygamy. Those guilty of the first crime, were to sit an hour on the gallows, with ropes about their necks,—be severely whipt not above 40 stripes; and forever after wear a capital A, two inches long, cut out of cloth coloured differently from their clothes, and sewed on the arms, or back parts of their garments so as always to be seen when they were about."[2]

Exactly when Hawthorne began writing *The Scarlet Letter* is not known, but by September 27, 1849, he was working on it throughout every day. It was finished by February 3, 1850.[3] In the novel there is the same rapid skill at composition which is typical of the notebooks. From the multitude of historical facts he knew he could call forth with severe economy only a few to support the scenes of passion or punishment. Perhaps it does not seem good judgment to claim that Hawthorne wrote *The Scarlet Letter* with a copy of Snow's *History of Boston* on the desk. But it does not appear believable that all these incidental facts from New England histories, the exacting time scheme, the authentic description of Boston in the 1640's, should have remained so extremely clear and perfect in his mind when he was under the extraordinary strain of writing the story. Here the studies of Hawthorne's literary borrowings made by Dawson, Turner, and others must be taken into account. They have shown that in certain of his tales, he "seems to have written with his original open before him."[4] To claim a

9. *Magnalia Christi Americana*, bk. II, 60; see also Massachusetts Historical Society *Collections*, V, 64 (1708); Neal, II, 131, 133–135, 144, 150, 158, 160, 169.
1. *The American Notebooks*, 107 [p. 203 of this Norton Critical Edition].
2. Joseph B. Felt, *The Annals of Salem, from Its First Settlement* (Salem, 1827), 317.
3. Randall Stewart, *Nathaniel Hawthorne* (New Haven, 1948), 93–95.
4. Turner, 547.

firm dependence upon certain New England histories for the background of *The Scarlet Letter* should therefore not seem unreasonable.

The incidents, places, and persons noticed in this article are the principal New England historical references in *The Scarlet Letter.* A study like this of Hawthorne's sources shows something of his thorough method of reading; it reveals especially his certain knowledge of colonial history and his interest in the unusual, obscure fact. But these are side lights of an author's mind. His steady determination was to make the romances of his imagination as real as the prison-house and the grave.

It would be unfair to leave the study of Hawthorne's historical approach here. His final concern in history was the attempt to find the "spiritual significance"[5] of the facts. As his sister Elizabeth had said of the young man: "He was not very fond of history in general."[6] Hawthorne stated concretely his conception of history and the novel in a review (1846) of W. G. Simms's *Views and Reviews in American History*:

> . . . we cannot help feeling that the real treasures of his subject have escaped the author's notice. The themes suggested by him, viewed as he views them, would produce nothing but historical novels, cast in the same worn out mould that has been in use these thirty years, and which it is time to break up and fling away. To be the prophet of Art requires almost as high a gift as to be a fulfiller of the prophecy. Mr. Simms has not this gift; he possesses nothing of the magic touch that should cause new intellectual and moral shapes to spring up in the reader's mind, peopling with varied life what had hitherto been a barren waste.[7]

With the evocation of the spirit of the colonial past, and with a realistic embodiment of scene, Hawthorne repeopled a landscape wherein new intellectual and moral shapes could dwell. The new fiction of Hester Prynne and the old appearances of Mrs. Hibbins could not be separated. Time past and time present became explicable as they were identified in the same profound moral engagement.

FREDERICK NEWBERRY

A Red-Hot *A* and a Lusting Divine: Sources for *The Scarlet Letter*†

While there has been no shortage of studies on Hawthorne's literary borrowings in *The Scarlet Letter*, little has been found concerning historical sources of the letter *A* itself and virtually nothing has been uncovered concerning adulterous figures in Puritan history who might have been the prototypes of Hester Prynne and the Reverend Arthur Dimmesdale. We do

5. Julian Hawthorne, *Hawthorne Reading*, 100.
6. "Recollections of Hawthorne by His Sister Elizabeth," 324.
7. Stewart, "Hawthorne's Contributions to *The Salem Advertiser*," *American Literature*, V (Jan., 1934): 331–332.

† From *New England Quarterly* 60.2 (June 1987): 256–64. Copyright © 1987 *The New England Quarterly*, Inc. Reproduced by permission of the publisher. Page numbers in square brackets refer to this Norton Critical Edition.

know that by 1838, when an early version of Hester appeared in "Endicott and the Red Cross," Hawthorne was aware of the 1694 law enacted in Salem that required a woman convicted of adultery to wear a capital *A* sewn conspicuously on her garments.[1] Although the appearance of this law so late in the century might seem anomalous to the 1634 setting of "Endicott and the Red Cross" or to the 1642–49 setting of *The Scarlet Letter,* we may easily resolve the discrepancy by assuming either that Hawthorne had been influenced instead by the early seventeenth-century case of Goodwife Mendame, sentenced to wear an *AD* on her sleeve, or that, contrary to his usual practice, he felt the need in this instance to take liberties with the historical record.[2] Yet the burning sensation described by the narrator in "The Custom-House" when he places the faded badge on his breast and feels "as if the letter were not of red cloth, but red-hot iron" and Hester's searing torment when observers fix their eyes on the emblem, causing it to be "branded . . . afresh into Hester's soul," may well be based on an actual incident [25, 57].[3] In three separate sources, Hawthorne could have read about a woman who, at a moment very close to the novel's setting, had the letter *A* branded upon her. Perhaps just as curious, this woman was married to a former Puritan minister who had been previously censured for adulterous behavior. Hawthorne was undoubtedly acquainted with the fall of this Puritan divine, the implication being that the adultery of the Reverend Dimmesdale was not entirely the product of Hawthorne's irreverent imagination after all.[4] As the scholarship on Hawthorne's historical works has consistently revealed, the "Actual and the Imaginary" do indeed meet, and "each imbue[s] itself with the nature of the other" (36) [28].

1. Edward Dawson, *Hawthorne's Knowledge and Use of New England History: A Study in Sources* (Nashville: Vanderbilt UP, 1939), 19, first proposed that Hawthorne discovered this source in Joseph Felt, *The Annals of Salem, from Its First Settlement* (Salem, 1827), 317. Hawthorne consulted Felt in 1833, 1834, and 1849. See Marion L. Kesselring, *Hawthorne's Reading, 1828–1850: A Transcription and Identification of Titles Recorded in the Charge-Books of the Salem Athenæum* (1949; reprinted, Folcroft, Pa.: Folcroft Library Editions, 1975), 50.
2. The time frame of *The Scarlet Letter* is well established. See Charles Ryskamp, "The New England Sources of *The Scarlet Letter,*" pp. 279–90 in this Norton Critical Edition. See also H. Bruce Franklin's introduction to *"The Scarlet Letter" and Other Writings* (Philadelphia: Lippincott, 1967), 13–14.

   Austin Warren first mentioned the case of Goodwife Mendame's adultery with an Indian in his introduction to *The Scarlet Letter* (New York: Holt, 1947), vii. Charles Boewe and Murray G. Murphy, "Hester Prynne in History," *American Literature* 32 (1960): 202–4, unconvincingly argue for Hawthorne's potential knowledge of this undated case, based upon the undocumented account in George Willison's *Saints and Strangers* (New York: Regnal and Hitchcock, 1945), 324. Goodwife Mendame was sentenced under a Plymouth law passed in 1636, first noticed and reproduced by Randall Stewart in his edition of Hawthorne's *The American Notebooks* (New Haven: Yale University Press, 1932), 229.

   Neither the Plymouth law nor the case of Goodwife Mendame appears in the histories of Plymouth colony familiar to Hawthorne: Edward Johnson's *The Wonder-Working Providence of Sions Savior in New England,* which appeared in the *Massachusetts Historical Society Collections,* 2d ser., vols. 2–4 (1814–16), and vols. 7–8 (1818–19); William Hubbard's *General History of New England,* which also appeared in *MHSC,* 2d ser., vols. 5–6 (1815); William Prince's *Annals of New England,* which appeared in *MHSC,* 2d ser., vol. 7 (1819) and which concludes its coverage in 1633; and Nathaniel Morton's *New-England's Memorial* (Boston, 1826). For Hawthorne's repeated reading of the *MHSC* and his reading of Morton, see Kesselring, *Hawthorne's Reading,* 56, 57.
3. Cited from *The Centenary Edition of the Works of Nathaniel Hawthorne,* ed. William Charvat, Roy Harvey Pearce, and Claude M. Simpson (Columbus: Ohio State UP, 1962), 32, 86.
4. In his introduction to *The Scarlet Letter,* Charvat says, "No historical equivalents of Dimmesdale and Chillingworth are known, but there are records of scandals in seventeenth-century Massachusetts similar to Dimmesdale's case" (xxvii). As far as I am aware, no one has specified any of these similarities except Michael Colacurcio, who anticipated my independent discovery of the Reverend Stephen Batchellor in Winthrop's *History of New England.* See "'The Woman's Own Choice': Sex, Metaphor, and the Puritan 'Sources' of *The Scarlet Letter,*" in *New Essays on "The Scarlet Letter,"* ed. Michael Colacurcio (New York: Cambridge UP, 1985), 110.

The case of the woman branded for adultery first appeared in the records of York, in what is now Maine. Dated 15 October 1651, the entry reads:

> We do present George Rogers for, & Mary Batcheller the wife of Mr. Steven Batcheller ministr for adultery. It is ordered by ye Court yt George Rogers for his adultery with mis Batcheller shall forthwith have fourty stripes save one upon the bare skine given him: It is ordered yt mis Batcheller for her adultery shall receave 40 stroakes save one at ye First Towne meeting held at Kittery, 6 weeks after her delivery & be branded with the letter A.

Beside that entry, written in the same hand, is the notation, "Execution Done."[5] It appears that Charles Edward Banks, in his *History of York, Maine* (1935), recognized the connection between Hawthorne's novel and this case, for he refers to Mary Batchellor's branding in a section titled "The Scarlet Letter."[6]

Hawthorne did not have to read the original records in order to become acquainted with the punishment of Mary Batchellor. In the first volume of *Collections of the Maine Historical Society* he could have read an account of the sentence passed on George Rogers and Mary Batchellor.[7] We know that Hawthorne had a personal interest in Maine's history. Not only had he attended Bowdoin College during the years immediately following the excitement over Maine's admission to statehood, but his father's family had claims to land there, and his mother's family still lived in Maine.[8] It would not be surprising if, in the course of his research, he came across the reference to Mary Batchellor's sentence.

Still another report of the sentence appears in the second edition of Alonzo Lewis's *History of Lynn* (1844), which also includes a lengthy

5. From a microfilm copy of York County Court Records (1636–1671), roll no. 1, 173–74, located at the County Court House, Alfred, Maine. The original records are deposited in the Maine State Archives. Although I have not located evidence to prove that Hawthorne ever saw these early county records, the possibility does exist. In 1820, soon after Maine was admitted to the Union, the records were moved from York, the crossroads of southern Maine, to the new county seat of Alfred. Using the road map of H. S. Tanner, *Map of the States of Maine, New Hampshire, Vermont, Massachusetts, Connecticut, and Rhode Island* (n.p., 1820), we can see that Hawthorne probably traveled the coastal route from York to Brunswick when he first set out from Salem to attend Bowdoin College in 1821, and thus did not pass through Alfred. The same route would have been the quickest way to reach Augusta in 1837 when he visited his friend Horatio Bridge. It seems unlikely that Hawthorne ventured south to Alfred during college vacations spent with his Manning relations in Raymond. But in 1826, on a trip from Salem to Raymond and back, he may very well have taken the inland route from York to Alfred, and, because it was the largest town in southernmost Maine, he could well have laid over there. For the details of Hawthorne's trips to Maine, see Arlin Turner, *Nathaniel Hawthorne: A Biography* (New York: Oxford UP, 1980), 92–94, 46.
6. Charles Edward Banks, *History of York, Maine*, 3 vols. (Boston: n.p., 1935), 2:241. The Batchellor name is spelled in various ways in historical records: Except in a quotation, I use the spelling as it appears in Winthrop's *History*.
7. *Collections of the Maine Historical Society*, 1st ser., vol. 1 (1831), 276.
8. For the Hawthorne land claims in Maine, see Vernon Loggins; *The Hawthornes* (New York: Columbia UP, 1951), 109, 155, 169–70. For Hawthorne's maternal connection to Maine, see Turner, *Nathaniel Hawthorne*, 13–30. We know, of course, that Hawthorne was familiar with James Sullivan, *The History of the District of Maine*, 2 vols. (Boston: I. Thomas & E. T. Andrews, 1795), because he refers to the work in a footnote to "The Great Carbuncle"—see the *Centenary Edition*, 9:149. And in Thomas Hutchinson, *The History of the Colony and Province of Massachusetts-Bay*, ed. Lawrence Shaw Mayo, 3 vols. (Cambridge: Harvard UP, 1936), 1:150–52, Hawthorne would have read that his first American ancestor, William Hathorne, had been one of the commissioners sent to York in 1651 to resolve a boundary dispute, which eventually brought Kittery and York under the jurisdiction of Massachusetts Bay. For Hawthorne's reading in Hutchinson, see Kesselring, *Hawthorne's Reading*, 53. Hawthorne may have known the fuller details of his ancestor's role in this boundary dispute from William D. Williamson, *The History of the State of Maine*, 2 vols. (Hallowell, Maine: Glazier, Masters, 1832), 1:334–48.

biographical sketch of Mary's husband, the Reverend Stephen Batchellor.[9] Hawthorne, it is true, had read the first edition of Lewis's *History* (1829), which contains most of the sketch on Stephen Batchellor found in the second edition as well as information on his marital troubles with Mary, but the original does not mention Mary's adultery.[1] Nevertheless, Hawthorne may also have consulted the second edition. Having published several volumes of poetry, Alonzo Lewis was both fondly and jokingly known around Boston and Salem as the "Bard of Lynn," and he was a town character frequently subjected to controversy.[2] Hawthorne must have been acquainted with Lewis through local gossip, and he may even have known him by sight, since Lewis habitually walked several miles from his home in Lynn throughout the 1820s and 1830s in order to attend Episcopal services at St. Peter's Church in Salem.[3] These circumstances, in addition to specific reports by word or print, might have elicited Hawthorne's interest in the second edition of the *History*, which was available during the Custom-House period when Hawthorne began rereading historical materials in preparation for writing *The Scarlet Letter.*

One would prefer a more compelling claim than plausibility for Hawthorne's knowledge of Mary Batchellor's case. Indeed, the similarities between Hester Prynne and Mary Batchellor are so outstanding that it is tempting to argue for a direct source. For example, Mary Batchellor's adultery is the only known case involving a child that can be linked to Hester's plight. By postponing execution of the sentence until six weeks after Mrs. Batchellor's delivery, the officials of York obviously considered the health of the unborn child. Hawthorne suggests a similar delay in the novel, for when Hester and Pearl appear in the opening scaffold scene, Pearl is "some three months old" (52) [38]. Although Hester is not physically punished, the account of Mary Batchellor might have provided factual warrant for postponing Hester's sentence to stand exposed to public disgrace and ridicule.

The striking feature of Mary Batchellor's case, however, is the form of punishment. Hawthorne certainly knew that adultery was sometimes a capital offense in Massachusetts Bay. In John Winthrop's *History of New England*, for example, he would have read about James Britton and Mary Latham, who were executed for adultery in 1643. Britton appealed to the General Court for his life, "but they would not grant it, though some of the magistrates spake much for it, and questioned the letter, whether adultery was death by God's law now."[4] Accordingly, in the opening scaffold scene, "the ugliest as well as the most pitiless" of the women spectators says that Hester "has brought shame upon us all, and ought to die. Is there not law for it? Truly there is, both in the Scripture and the statute-book" (51–52) [38].

9. Alonzo Lewis, *The History of Lynn, Including Nahant* (Boston: n.p., 1844), 93–97.
1. See Alonzo Lewis, *The History of Lynn* (Boston: J. H. Eastburn, 1829), 54–57. Hawthorne consulted this edition in 1833. See Kesselring, *Hawthorne's Reading*, 55.
2. See *Dictionary of American Biography*; see also the biographical sketch of Lewis written by James R. Newhall (son-in-law of Lewis) in his updated version of *The History of Lynn* (Boston: John L. Shorey, 1865), 544–48. An anonymous broadside attacking Lewis on several counts can be found among the unnumbered Alonzo Lewis folders at the Lynn Historical Society.
3. See Alonzo Lewis, *Poetical Works*, ed. Ion Lewis (Boston: n.p., 1883), xxi.
4. John Winthrop, *The History of New England from 1630 to 1649*, ed. James Savage, 2 vols. in 1 (1825; reprinted, New York: Arno P, 1972), 2:157–59. For Hawthorne's reading in Winthrop, see Kesselring, *Hawthorne's Reading*, 64.

Another disgruntled woman in this scene would like to see Hester suffer the punishment of Mary Batchellor: "The magistrates are God-fearing gentlemen, but merciful overmuch. . . . At the very least, they should have put the brand of a hot iron on Hester Prynne's forehead" (51) [37]. Although Hawthorne knew that branding was used to punish diverse crimes in early New England, the association of branding with the letter *A* in Mrs. Batchellor's punishment is reflected not only in Hester's sense of the scarlet letter as an "ignominious brand" (86) [57] that is "flaming" (79) [53], which of course also suggests the figurative heat of shame or passion, but also in the narrator's description of the letter as a brand in "The Custom-House."

If Hawthorne was aware of Mary Batchellor's marriage to Stephen Batchellor, it could well have inspired the creation not only of Arthur Dimmesdale but also of Roger Chillingworth. Batchellor himself was no stranger to Hawthorne. In the edition of Winthrop's *History* familiar to him, editor James Savage calls special attention to the "unfortunate" Stephen Batchellor, who arrived in Massachusetts Bay on 5 June 1632 at the age of seventy-one.[5] Batchellor was the subject of two controversies in the 1630s concerning his unsanctioned methods of establishing separate churches at Lynn, but these squabbles were insignificant compared to the one at Hampton in 1641, which Winthrop describes at some length:

> Mr. Stephen Batchellor, the pastor of the church at Hampton, who had suffered much at the hands of the bishops in England, being about 80 years of age, and having a lusty comely woman to his wife, did solicit the chastity of his neighbour's wife, who acquainted her husband therewith; whereupon he was dealt with, but denied it, as he had told the woman he would do, and complained to the magistrates against the woman and her husband for slandering him. The church likewise dealing with him, he stiffly denied it, but soon after, when the Lord's supper was to be administered, he did voluntarily confess the attempt, and that he did intend to have defiled her, if she would have consented. The church, being moved with his free confession and tears, silently forgave him, and communicated with him: but after, finding how scandalous it was, they took advice of other elders, and after long debate and much pleading and standing upon the church's forgiving and being reconciled to him in communicating with him after he had confessed it, they proceeded to cast him out. After this he went on in a variable course, sometimes seeming very penitent, soon after again excusing himself, and casting blame upon others. . . . He was off and on for a long time, and when he had seemed most penitent, so as the church were ready to have received him in again, he would fall back again, and as it were repent of his repentance.[6]

Hawthorne could have found all but the last sentence and clause of this case quoted from Winthrop in the second edition of Lewis's *History*.[7] In the first edition, however, Hawthorne would have learned only that Batchellor had been "excommunicated" in 1641 for "irregular conduct," although this edition does mention that Batchellor was ninety at the time of his remarriage in 1650 to Mary (the "lusty comely" wife of 1641 having died).[8]

5. Winthrop, *History of New England*, 1:78n.
6. Winthrop, *History of New England*, 2:44–45.
7. Lewis, *History of Lynn* (1844), 94.
8. Lewis, *History of Lynn* (1829), 55.

Their union drew the attention of Bay authorities when Batchellor was "fined ten pounds, for not publishing his intention of marriage, according to law," and again, later in 1650, when the General Court ordered the couple to "lyve together as man and wife," thereby denying both of their petitions for divorce.[9] Sometime in 1651, Batchellor returned to England, where he remarried and lived another ten years, his polygamy apparently undetected.[1] One cannot determine from either Lewis's first or second edition whether Batchellor left America before or after Mary's trial for adultery. Within the narrow time margins involved, however, he probably knew that Mary was pregnant from an extra-marital union. The would-be adulterer had himself become a cuckold, and his response was to flee.

Few details in Batchellor's life invite comparison with Hawthorne's Dimmesdale. Indeed, Batchellor's advanced age, his young and wayward wife, and his incorrigibility attracting public censure are more reminiscent of Chillingworth. But Batchellor's attempt to seduce another man's wife links his American experience to Dimmesdale's. Moreover, Batchellor's attempted adultery, followed by his repeated confessions and denials, suggests the major dilemma tormenting Dimmesdale throughout *The Scarlet Letter.* Knowing at the outset that he should confess, yet perhaps fearing that he will be excommunicated (as Batchellor had been for a seemingly lesser offense), Dimmesdale cannot bring himself to reveal his role in Hester's sin until seven years later in the climactic scaffold scene.[2] It is also worth considering that Batchellor's return to England might have given Hawthorne the idea of having Hester propose to Dimmesdale that they escape to the Old World. Alternately, knowing that Mary Batchellor was left with the difficulty of providing for family after her husband's flight, Hawthorne might have seen the need to discover the moral necessity and the future independence of America lying behind Hester's ultimate decision to remain in New England.

Finally, the year of Batchellor's attempted seduction probably influenced Hawthorne's manipulation of the historical time frame of *The Scarlet Letter.* When that attempt took place in 1641, Richard Bellingham was governor. Bellingham is clearly the governor, "the chief ruler," in the opening scaffold scene of *The Scarlet Letter* (64) [45]. And yet, because that scene takes place in June 1642, Hawthorne should have designated Winthrop, who had become governor in May.[3] Hawthorne, who had a high opinion of Winthrop, may have created this anachronism among the otherwise accurate details of the novel's historical setting in order to dissociate Winthrop from the Puritan "sages of rigid aspect" who rule in Hester's case but who are not "capable of sitting in judgment on an erring woman's heart" (64) [45]. Hawthorne surely knew from his reading in Winthrop, however, that the historical Bellingham would have been as unqualified to judge Hester as he was to rule in Batchellor's case. While

9. Lewis, *History of Lynn* (1829), 55–56.
1. Lewis, *History of Lynn* (1829), 56. In 1656, Mary petitioned the Court to be free of her marriage to Batchellor in order to remarry for the sake of her children (56–57). Earlier, however, she had demonstrated that the branding and stripes did not change her behavior, for in March 1652 (1651 old style), she was again sentenced to be whipped for adultery. See York County Court Records, 188. Both her first and second punishments are recorded in *Province and Court Records of Maine* (Portland: Maine Historical Society, 1928), 1:164, 176.
2. For the necessity of public confession in Puritan Massachusetts Bay, see Ernest W. Baughman, "Public Confession and *The Scarlet Letter,*" *New England Quarterly* 40 (1967): 532–50.
3. Winthrop replaced Bellingham on 18 May—see Winthrop, *History of New England*, 2:63.

governor in 1641, not long before he presided over the General Court's arraignment of Batchellor, Bellingham had won the hand of a woman who had previously pledged herself to his friend. The governor not only circumvented the law by failing to publish the banns but also performed his own marriage ceremony.[4] As reported by Winthrop, Bellingham refused to disqualify himself when the General Court convened to take up charges brought against him by the "great inquest." The Court was "unwilling to command him publicly to go off the bench, and yet not thinking it fit he should sit as a judge, when he was by law to answer as an offender."[5] That he subsequently sat on the bench when the Reverend Batchellor's case came before the Court would no doubt have pleased Hawthorne's sense of irony and may further have prompted him to allow the unworthy Bellingham to preside over Hester's public humiliation.

One of the more unique aspects of Hawthorne's fiction is how it sends us back to the record books in search of individuals and events that, through the force of his art, he has made us experience as historically real. We do know that Hawthorne did not entirely invent the circumstances and dilemmas of his characters, but we cannot always be sure that he knew what we have discovered in the historical record available to him. While he almost certainly drew upon the life of the Reverend Batchellor, the case of his ill-fated wife is more problematic. Even if, however, we were to dismiss the possibility that Hawthorne knew about Mary Batchellor—which I do not think we can or should do—the historical analogy remains tantalizing. Had *The Scarlet Letter* never been written, many of us would never have been aware that in mid-seventeenth-century New England even Puritan divines were implicated in cases of adultery and that wayward women faced the threat of being physically as well as socially stigmatized by a burning A. One of Hawthorne's particular gifts is that he not only brings such facts to light but also that from them he spins stories of such psychological and moral power that they have fascinated readers for generations and promise to do so for generations to come.

KRISTIN BOUDREAU

Hawthorne's Model of Christian Charity†

Jefferson's belief that fiction could be used to bind the nation in ideological unity was shared by those who followed him into the nineteenth century. In 1833 the criminal lawyer Rufus Choate addressed a crowd in Salem on the topic of historical novels and their usefulness in "illustrating New-England History." In doing so, he used terms even more charged

4. Without mentioning his source, Franklin, *"The Scarlet Letter" and Other Writings*, 13, mentions these details.

5. The full case and the quotations come from Winthrop, *History of New England*, 2:43. In connection to Hawthorne's likely appreciation of the irony involved in Bellingham's case, it seems worth recording Winthrop's preface to the account: "Query, whether the following be fit to be published."

† From *Sympathy in American Literature: American Sentiments from Jefferson to the Jameses* (Gainesville: UP of Florida, 2002), pp. 49–82, 212–14. Reprinted by permission of the University Press of Florida. Page references in square brackets refer to this Norton Critical Edition.

with overt political motives, even less distinguished by references to moral sentiments than those used by earlier leaders of the republic. The significant historical American romance, he argued, had not yet been written, even though such a work could surpass conventional histories in "speak[ing] directly to the heart and affections and imagination of the whole people" (3). The notion that the affections have cultural and political work to do should by now be familiar—as it was to Choate's audience—though the faith in such a creature as a "whole people" may have been new to Choate's generation. Like Thomas Jefferson and Benjamin Rush, Choate wanted desperately to consider his nation a single entity bound by common sentiments. Indeed, in his address the stakes seem even higher than they were when the country was still very young. A new generation of authors was coming of age, and while the great literary renaissance was still a few years away, writers like Washington Irving had already begun to establish national pride in the native cultural resources of the United States. If the political experiment known as the United States were to fail, it would be a great loss to local literature and the arts, which had only recently begun to develop independently of their European roots. Choate thought these cultural productions might help to unify the nation.

Toward the close of his address, Choate proposed that a series of historical romances like those of Walter Scott "might do *something* to perpetuate the Union itself" (36). The tone of his recommendation seemed to indicate that the union was already in a state of peril. A successful attempt to illustrate our common history in fictional form

> would turn back our thoughts from these recent and overrated diversities of interest,—these controversies about negro-cloth, coarse-woolled sheep, and cotton bagging,—to the day when our fathers walked hand in hand together through the valley of the Shadow of Death in the War of Independence. Reminded of our fathers, we should remember that we are brethren. The exclusiveness of State pride, the narrow selfishness of a mere local policy, and the small jealousies of vulgar minds, would be merged in an expanded, comprehensive, constitutional sentiment of old, family, fraternal regard. (37)

Choate understood that regional interests, some of the same issues that would explode in civil war twenty-seven years later, were "overrated" in comparison with the longer-standing, more natural family sentiment. Like Jefferson appealing to "consanguinity," Choate worried that "diversities of interest" would break through the affections that an earlier generation of cultural leaders had worked so hard to develop. Though Choate didn't indicate whether these affections were natural or produced, he did believe that a step backward in time, taking the people closer to the source of their common experiences, would help to heal the wounds caused by diversity, would help to recreate one people.

Choate called emphatically upon writers to produce literature that could properly be called national, given that it would become the "common property of all the States" (36). He imagined ideal reading communities dotting the American landscape, uniting distant regions. "Poems and romances which shall be read in every parlor, by every fireside, in every school-house, behind every counter, in every printing-office, in every lawyer's office, at every weekly evening club, in all the States of this Confederacy, must do something, along with more palpable if not more powerful agents, toward moulding and fixing that final, grand, complex result,—the national character" (36).

Although the previously dominant term "sensibility" had been replaced by "character," Choate was still referring to the ties of affection that came from common exposure to a carefully constructed and well-regulated model of morality. Faced with regional disputes and the threat of disunion, Choate echoed the sentiments of his eighteenth-century predecessors and called upon novelists to bind an otherwise dispersed nation. The only difference is that Choate considered that the common topic of these romances would be history rather than the American family. Otherwise, his project resembled that of the early novels we have already considered, which exhort their readers to share in the sensibilities of an ideal American.

Was it only coincidence that Choate, speaking to a Salem audience that may have included Nathaniel Hawthorne, suggested "the *old Puritan character*" as a particularly rich topic (21–22)?[1] Implicitly, Choate was raising the possibility that the "particular duty" of the Puritan character might be called for once again on behalf of a nation that, though secularized almost beyond recognition, still needed a disciplined conscience to hold it together. If the Puritan character was once "dissolved . . . into its elements" (22), perhaps it might be summoned forth in a time of national crisis. The novelist was the ideal soldier for a such a task.

Nathaniel Hawthorne's most successful novel, of course, answers Choate's call for a historical romance while developing the genre of the seduction novel. *The Scarlet Letter*, published in 1850, in many ways restores a sense of the past to Hawthorne's nineteenth-century audience. A number of excellent studies have demonstrated the complicated historical issues submerged beneath the romantic plot of the fallen woman Hester Prynne.[2] The novel also speaks to its own historical context, as Sacvan Bercovitch has splendidly shown.[3] Students of Hawthorne are now familiar with the historical details, both accurate and inaccurate, that he wrote into our seven-year encounter with Hester, Dimmesdale, Pearl, and Chillingworth, the only purely fictional characters in the novel. The others, Governors Bellingham and Winthrop, the Reverend Mr. Wilson, Mistress Hibbins, and even the jailer, Master Brackett, are historical figures that Hawthorne drew from his intense reading in colonial history: Increase Mather's *Illustrious Providences*, Cotton Mather's *Magnalia Christi Americana*, George Bancroft's *History of the United States*, Thomas Hutchinson's *History of the Colony and Province of Massachusetts Bay*, Caleb H. Snow's *History of Boston*, John Winthrop's *Journal*.[4]

Hawthorne was also familiar with European and American seduction novels. Although the records of his adolescent reading are scant, he does tell us that as a boy he read any "light books within [his] reach" (J. Hawthorne

1. Although there is no direct evidence that Hawthorne was present at Choate's lecture (his extant journals begin in 1835), there is strong circumstantial evidence that he was at least acquainted with Choate's argument, perhaps having heard an account of the address. In October and December of 1833, and again in February of 1834, Hawthorne's aunt borrowed the works of Walter Scott from the Salem Athenaeum. It is possible that Hawthorne's interest in the historical novel began with Choate's public lecture. For Hawthorne's sources, see Kesselring.
2. Colacurcio has done impressive work here. See also Baughman, Ryskamp, Bell, and Hugh J. Dawson.
3. Others who have explored the antebellum context include Leverenz, Douglas Anderson, Michael T. Gilmore, Larry Reynolds, Railton, and Van Leer.
4. Edward Dawson, 5–6, explains Hawthorne's sources for *The Scarlet Letter*.

1:96). Nina Baym notes that "he particularly loved the books that were popular successes in the early decades of the nineteenth century" (*Shape* 16). Baym cites English novels, but there is every reason to suppose that Hawthorne had also read the most popular American novels. These included a large number of seduction novels, which took part in a "national debate over the uses of a woman's body" (Williams, "Victims" 60). Unfortunately, the detailed records of the Hawthorne and Manning activity at the Salem Athenaeum do not begin until 1828, when Hawthorne had put the lightest reading behind him. By that time, he was probably familiar with the conventional plot of the seduction novel. To be sure, the novelist made many changes in shaping this transhistorical genre to a precise historical context. *The Scarlet Letter* begins after the heroine's fall, so we do not see how Hester Prynne was tempted and how she might have resisted the Reverend Mr. Dimmesdale. The conventional moral, to avoid sins of passion, appears most emphatically only in the early chapters, and is voiced not by the narrator but by an array of characters (magistrates, ministers, and vindictive townspeople) who should not be taken to represent the author's position. Still, *The Scarlet Letter* is indebted to the plot of the seduction novel.

Like those stories, Hawthorne's romance is fairly pessimistic about the American family: marriage ties are dissolved, and the one human relation with "a consecration of its own" does not promise to be renewed after death, whatever Hester's wishes (1:195) [118]. As the dying Dimmesdale tells her, "when we violated our reverence each for the other's soul,—it was thenceforth vain to hope that we could meet hereafter" (256) [151]. Their American offspring, Pearl, does not remain in the New World but returns to the Old, the place of her mother's origins. While Hester returns to counsel other unhappy women, to see the office of the scarlet letter fulfilled, Hawthorne gives us no model family to carry on the New England Way.

But Hawthorne introduced important innovations to the conventional seduction novel. The moral of seduction novels, like that of public punishments, was clear: Spectators must avoid the errors that brought this spectacle to her unhappy end. In Hawthorne's romance, though public scrutiny is certainly fastened upon the erring Hester, we are not meant to read her story merely in order to avoid her trespasses. Hawthorne has other uses in mind for his character and his readers. Hester's story, that is, concerns not her fall but her punishment, her place in the community that condemns her, and the roles of mercy and justice in seeing to the reformation of erring colonists. In this regard Hawthorne wrote the historical romance that Choate had demanded.

Of all the historical figures who enter *The Scarlet Letter*, it is appropriate that John Winthrop presides most influentially over Hawthorne's text, for he was the Puritan most concerned with binding his fellow colonists together in ties of mutual affection.[5] Writing as civil war loomed, but before it seemed inevitable, Hawthorne too was wary of deep political differences: an expanding market economy, violent labor and political disputes,

5. For an excellent analysis of the uses of sympathy in "mediat[ing] between, rather than isolat[ing], individuals who themselves have successfully internalized regulatory functions," see Alkana, 56–81. Throughout this book Alkana traces the nineteenth-century "social self" to Common Sense philosophers who sought to use moral sentiments to unite diverse populations. His chapter on *The Scarlet Letter* considers the novel mainly in the context of Hawthorne's own time, while I am attempting to look more closely at his historical sources. For Hawthorne's debt to Adam Smith's conception of sympathy, see Hunt.

and a culture whose emerging ethos was based on individualism. He turned to Winthrop as a model peacemaker. In recalling the colonial figure most concerned with binding his fellow colonists together in bonds of mutual affection, Hawthorne also participated in the project that Choate and earlier writers had called for. We might say, to use Choate's words, that Hawthorne "mould[ed] and fix[ed] that final, grand, complex result,—the national character."

A Sweet Moral Blossom

Ostensibly, of course, Hawthorne's moral involves a condemnation of Dimmesdale and his deceptive life. "Be true! Be true! Be true!" urges the narrator. "Show freely to the world, if not your worst, yet some trait whereby the worst may be inferred!" (260) [153]. In this exhortation, the narrator resembles the magistrates and ministers, most notably the Reverend Mr. Wilson, who call upon Hester to "lay open her heart's secrets in such broad daylight, and in presence of so great a multitude" (65) [46]. But the directness of this particular moral conceals the ambiguity with which Hawthorne, conveys it, giving us his lesson only obliquely and calling it only one "among many morals" (260) [153]. The author leaves us to guess that other lessons are not laid out so clearly, that we must search for them in the shadows of his text. But one moral is equally clear, though it is put into a metaphor rather than a sentence. The "sweet moral blossom" that the narrator hopes for in the opening chapter is figured by the rosebush outside the prison door: "we could hardly do otherwise than pluck one of its flowers and present it to the reader. It may serve, let us hope, to symbolize some sweet moral blossom, that may be found along the track, or relieve the darkening close of a tale of human frailty and sorrow" (48) [36]. A moral blossom meant to "relieve" the close of the romance: Hawthorne's language is very much like Adam Smith's discussion of sympathy, the only sentiment, Smith claims, that can relieve human suffering (15). Perhaps Hawthorne, anticipating Henry James's criticism, means only aesthetic, rather than moral, relief. As James later wrote, this work is "densely dark, with a single spot of vivid colour in it; and it will probably long remain the most consistently gloomy of English novels of the first order" (*Hawthorne* 87). Although Hawthorne clearly had aesthetic issues on his mind, he was also considering moral, social, and historical ones, and what the metaphorical flower does aesthetically for the reader, the actual flower does sympathetically for the suffering prisoner. The "delicate gems" of the rose-bush, the narrator of *The Scarlet Letter* notes, "might be imagined to offer their fragrance and fragile beauty to the prisoner as he went in, and to the condemned criminal as he came forth to his doom, in token that the deep heart of Nature could pity and be kind to him" (48) [35]. Pitted against other flowers—"the black flower of civilized society, a prison" (48) [35] and the black flower of fate that prevents Chillingworth from forgiving his enemy (174) [106]—this blossom evokes a decades-long debate in the colonies about the treatment of transgressions. That Hawthorne draws on this pitying flower as the image of his moral suggests that whatever lesson we find will tell us something about the justice of mercy. Whatever the attitudes of Hester's fellow townspeople, the narrator draws our attention to the heart of Nature, thereby giving us a model, as John Winthrop called it, of "Christian charity."

A Story and a By-Word

Winthrop's most famous document informs Hawthorne's most famous novel. "A Model of Christian Charity" was delivered in 1630 to the passengers aboard the *Arbella*, as they contemplated their roles in the settlement of the Bay Colony. The Reverend Mr. John Wilson, who appears prominently at Hester's public punishment, was one of those passengers. So was Richard Bellingham, whom Hawthorne designates as governor during the period that includes the opening scene of *The Scarlet Letter*. Hester Prynne, on the other hand, was not present among the charter members of Massachusetts Bay. Hawthorne puts her in New England no earlier than 1640, fully ten years after Winthrop delivered his speech. Whether the Prynnes were Puritans, as is likely from their association with Amsterdam, or whether they had more secular interests in the colony, the principles of these Puritan families applied to all colonists alike. Chillingworth learns that theirs is "a land where iniquity is searched out, and punished in the sight of rulers and people" (62) [43].

As if prophesying the fate of the errant Hester, poised high on a scaffold in a crowded marketplace, Winthrop had told his fellow passengers that "we shall be as a city upon a Hill, the eyes of all people are upon us; so that if we shall deal falsely with our god in this work we have undertaken and so cause him to withdraw his present help from us, we shall be made a story and a by-word through the world" (295). Hawthorne literalizes this image in Hester, who must serve, as Chillingworth observes, as "a living sermon against sin" (63) [44]. The novel opens with a crowd of spectators, "all with their eyes intently fastened on the iron-clamped oaken door" of the prison (49) [36]. Upon the scaffold, Hester feels "the heavy weight of a thousand unrelenting eyes, all fastened upon her, and concentred at her bosom" (57) [41], as she enacts the most memorable lines in Winthrop's address, the lines that insist upon public scrutiny.

As Q. D. Leavis, Larry Reynolds, and others have noted, the novel seems to be structured according to the three scenes located at the scaffold. Before Hester's entrance, moreover, the narrator suggests a range of public punishments and the offenses that might bring a malefactor to the scaffold: heterodoxy, sluggishness, disobedience, idleness, drunkenness, and witchcraft all provoked "the same solemnity" among the spectators, for whom "the mildest and the severest acts of public discipline were alike made venerable and awful" (49–50) [36]. As Winthrop's speech closes with strong words of warning, so Hawthorne's novel opens with the horror of strict justice meted out in a public place. Hyperconscious of the public scrutiny under which they lived, the people of Massachusetts were uneasily aware of the many transgressions that might bring one of them into the marketplace. And while there might seem to be a world of difference between shiftlessness and witchcraft, the shame that followed from the revelation of these sins was equally painful. While Hester considers her position, she fears "she must needs shriek out with the full power of her lungs, and cast herself from the scaffold down upon the ground, or else go mad at once" (57) [41]. The unsympathetic eyes of her spectators drive home to her the almost corporal violence of exhibitionist justice. Using a metaphor drawn from the tradition of punitive mutilations, Hester claims that the letter "is too deeply branded" to be removed (68) [47].

The gathering of Bostonians enacts a central component of Winthrop's famous speech. Even Hester, "knowing well her part," performs according to apparently familiar conventions (55) [40]. And the success of those conventions depends upon the distance between the criminal and her audience. Though a limited identification is necessary for the spectators to fear their own passions and avoid Hester's missteps, they must not feel sympathy for the criminal. The eyes of the crowd must be fastened in judgment alone, lest the spectators feel too deeply for Hester and lay the blame for her suffering upon the prominent men who are responsible for her public exposure. The narrator comes close to such a conclusion: "They were, doubtless, good men, just, and sage," he admits, yet "out of the whole human family, it would not have been easy to select the same number of wise and virtuous persons, who should be less capable of sitting in judgment on an erring woman's heart" (64) [45]. But Hawthorne's narrator is not representative of these practiced Puritans, determined to bestow justice rather than mercy. Typically, he tells us, "the sympathy that a transgressor might look for, from such bystanders at the scaffold," was "meagre, indeed, and cold" (50) [36].

Hawthorne's Puritans understood well the roles that justice called them to perform. But Hawthorne's readers, promised a "sweet moral blossom," might be dismayed to find that mercy is represented by such "meager" and "cold" sympathy. The dominant theme of Winthrop's sermon, Christian charity, seems mostly to have dried up; among most of the citizens, sweet mercy and charity are replaced with adamant justice.

The point of public punishments, of course, was to bind the colonists in a mutual sense of proper behavior. The culprit exhibited upon the scaffold symbolized a breach of law, but in so doing she served to rally her townspeople in an affirmation of the specific law that was broken and the idea of law itself. Public spectacles of discipline helped to remind the people of their consent to the laws that governed them; by publicly greeting the offender with opprobrium, they acted as agents of the law and strengthened the regulations that held the community together (Erikson 6).

What are the implications of linking the affections and the law? A Marxist or feminist reading—or indeed any reading that seeks to understand how literature participates in hegemonic values—might condemn both Hawthorne and Winthrop for masking ideological apparatuses in moral sentiments. A sympathetic reading of Hawthorne and Winthrop, however, can overcome the need to see the forces lurking at the heart of sentimental culture as either subversive or repressive. In reading *The Scarlet Letter* against the backdrop of seventeenth-century political and social crises, we shall see that the association between sentiments and the law is not necessarily oppressive; neither is the deployment of sympathy simply subversive. Hawthorne's novel demonstrates that sympathy, though perhaps a "natural" sentiment, can lie dormant in the face of deep suffering or long-standing self-absorption. Likewise, it can be coaxed back to life in order to revitalize both the individual and the social institutions that serve communities. Though we may understand the relationship between individuals and communities to be antagonistic, even Thoreau never claimed that such antagonism was inevitable.

The Rule of Mercy

As John Winthrop well knew, justice is not the only implement that fortifies a community. For him, the affections that prompted sympathy were at times stronger and more appropriate mechanisms for binding a community that might otherwise be tempted to disperse. His "Model of Christian Charity" distinguishes between the "two rules whereby we are to walk one towards another: JUSTICE and MERCY" (283). Winthrop contended that the second law was inspired by a natural human sentiment, meant "to strengthen defend preserve and comfort" fellow citizens (289). Such affection and sympathy, furthermore, would operate as an invisible bond between the people: "though we were absent from each other many miles, and had our employments as far distant, yet we ought to account our selves knit together by this bond of love, and live in the exercise of it" (292). The rule of justice applied in ordinary times, but a "community of peril" required the rule of mercy, which Winthrop described as "more enlargement towards others and less respect towards ourselves" (287).

Until the end of his life in 1649, Winthrop considered Massachusetts Bay Colony to be one such imperiled community, in need of more mercy than justice. Though his "Model" was well received by his fellow Puritans, his policy on mercy caused him subsequent political problems. A dispute in 1635 between Winthrop and Thomas Dudley, the longtime deputy governor, concerned a difference of opinion regarding the severity of punishments called for by the magistrates. While Dudley advocated severity, Winthrop called for "more lenyte" (*Journal* 165). Although Winthrop, learning that public opinion was against him, eventually modified his position, his words do not indicate a true change of heart. He answered his fellow magistrates "that his speeches & carriage had been in part mistaken, but withall professed, that it was his judgment . . . that in the infancy of plantations justice should be administered with more lenyte than in a settled state, because people were then more apt to transgress partly of ignorance of new laws & Orders, partly through oppression of business & other straits" (167).

The cause of this dispute, according to some historians, was the treatment of Roger Williams, the religious dissident who was forced out of the Bay Colony in 1635. Although the magistrates issued a warrant requiring Williams, a Puritan much more radical than the non-Separatists of Boston, to be returned on the next ship to England, he received advance warning, which allowed him to escape to Rhode Island. James G. Moseley speculates that "the warmth of Williams's life-long affection and respect for Winthrop suggests that he may have been the one who helped Williams elude the Puritans' grasp" (72).

That lifelong affection, the fruit of Winthrop's mercy, was at no time regrettable, either for Winthrop or the colony itself. Although the wording of a parliamentary letter of safe conduct of 1644 may be slightly exaggerated, Winthrop's and Williams's sentiments for each other remained warm: Williams and the Massachusetts magistrates, the letter indicates, "mutually give good testimony each of other, as we observe you do of him, and he abundantly of you" (*Journal* 541).

For all of their doctrinal disputes—and they were significant—Winthrop's affection for Williams, like his public policy, kept pace with the theory of sentiments he had sketched out in his "Model." That theory included two kinds of arguments, those of a fervently religious man and those of a pragmatic political leader. Winthrop was both.[6] The exercise of charity, he argued, helped every person to "have need of other, and from hence they might be all knit more nearly together in the Bond of brotherly affection" ("Model" 283). The alliance between Williams and Winthrop proved the truth of this claim. A second reason, though strictly religious on the face of it, was also pragmatic: "That [God] might have the more occasion to manifest the work of his Spirit . . . in the regenerate in exercising his graces in them, as in the great ones, their love, mercy, gentleness, temperance etc." (283). Here Winthrop's argument resembles Jefferson's a century later. In exercising our charity, we learn to be more charitable. A society of charitable citizens, whose private interests "stand aside" until the good of the whole is served, cannot help but survive (286).

In Hawthorne's novel we see a model of sympathy in the narrator, who, by acknowledging the ministers' and magistrates' inability to judge Hester, implicitly refuses to judge her as they do. We see a clearer model in the young woman who tries to moderate the self-righteous utterances of her companions. Those companions, women who "appeared to take a peculiar interest" in Hester's punishment (1:50) [36], speak with derision of the malefactor and invent even more shameful punishments for her: branding her forehead and replacing her ornate letter with "a rag of mine own rheumatic flannel" (54) [39].

The compassionate woman, in contrast, wishes to alleviate Hester's shame. As she tries to silence her companions, there is gentle scolding in her words: "O, peace, neighbours, peace! . . . Do not let her hear you! Not a stitch in that embroidered letter, but she has felt it in her heart!" (54) [39]. She alone among these women has entered into the feelings of the suffering Hester.

The argument among the ordinary townspeople in this early scene echoes an ongoing debate between the magistrates and deputies of John Winthrop's time. As a number of scholars have pointed out, the period covered by the novel, 1642 to 1649, coincides almost precisely with struggles for political power waged between the petty bourgeoisie, represented by the deputies, and the governing aristocracy, represented by the magistrates.[7] One of those debates, directly reflected in the story of Hester's adultery and public punishment, concerned the degree to which the magistrates should be free to assign penalties for crimes. The issue came before the General Court in March 1638, when three people had been convicted of adultery, a capital offense in Massachusetts.[8] Because the magistrates were uncertain whether the law had been clearly publicized, the three criminals had been sitting in jail since June 1637. As the prisoners awaited sentencing, the magistrates concluded that "if the law had been sufficiently

6. Cotton Mather mentions Winthrop's double sense of vocation: "But though he would rather have Devoted himself unto the Study of Mr. *John Calvin*, than of Sir *Edward Cooke*; nevertheless, the Accomplishments of a *Lawyer*, were those wherewith Heaven made his chief Opportunities to be Serviceable" (2:213–14).

7. See, for instance, Berlant's reading of the novel in the context of these political crises. Wall provides a detailed account of these years. Colacurcio notes that a major theme throughout Winthrop's journal is "democratic excess" ("'The Woman's Own Choice'" 106).

8. In 1631 the Court of Assistants had ruled that "if any man shall have carnall copulacon with another mans wife <be she English or Indian> they both shalbe punished by death" (*Records of the Court* 2:19, 66, 70).

published, they ought to be put to death." But after discussion, "it was thought safest, that these three persons should be whipped and banished; and the law was confirmed and published" (Winthrop, *Journal* 249).

When, in March of 1644, Mary Latham and James Britton were convicted of adultery and executed according to this law, they became the only two persons ever executed for adultery in Massachusetts. Even this one case was controversial, as Winthrop notes: "some of the magistrates thought the evidence not sufficient against her, because there were not two direct witnesses; but the jury [convicted] her" (*Journal* 501). As Hugh J. Dawson explains, "the members of that unhappy couple's court searched the Bible and the Colony's stipulated penalty in an anguished effort to find some escape from imposing the prescribed sentence" (228–29).[9] Winthrop wrote that "some of the magistrates spake much for [the sparing of Britton's life], and questioned the letter, whether adultery was death by God's law now" (*Journal* 502).

Representing the will of the people, the jury condemned this couple to death, even while some of the magistrates, again, representing the will of God, objected.[1] The question again concerned the choice between justice and mercy, particularly, in this case, whether a sentence of death could be just in the absence of irrefutable evidence. Being closer to God than the people were, the magistrates felt that they were better qualified to determine penalties in specific cases.[2]

In other words, while the people may have been more rigid than their elected governors, their legalistic ardor may have limited their capacity to be merciful. One of Hawthorne's characters makes this point explicitly, when she observes that the punishment settled upon by the magistrates was much lighter than what Hester would have received from her peers. "It would be greatly for the public behoof," this "hard-featured dame of fifty" says, "if we women, being of mature age and church-members in good repute, should have the handling of such malefactresses as this Hester Prynne. What think ye, gossips? If the hussy stood up for judgment before us five, that are now here in a knot together, would she come off with such a sentence as the worshipful magistrates have awarded? Marry, I trow not!" (1:51) [37].

Popular opinion, which Winthrop often dismissed as "mere democracy" (*Journal* 456), was just as distasteful to Hawthorne, who in "The Custom House" is deeply critical of democratic excess. Aesthetically and ethically, Hawthorne preferred the "sweet moral blossom" to the shrill cries for justice that greet Hester's entrance into the marketplace. While the woman who clamors to bestow her "rheumatic flannel" upon Hester Prynne may be an extreme example of sanctimonious passion, however, Hawthorne has added other, more complicated voices to the assembly, reenacting among his fictional community the contentions that had engaged Massachusetts governors for an entire decade.

9. Dawson's brief essay contains a remarkably thorough and insightful discussion of the treatment of adulterers and the role that William Hathorne played in insisting on the letter of the law.
1. While the deputies represented the freemen, magistrates held a higher power, one that they sought to protect against the deputies' mounting demand for more influence. In a 1636 letter stating the magistratical position, John Cotton asked, "If the people be governors, who shall be governed?" (Hutchinson 1:415). For divine sanction of the magistrates, see Wall, 78–80.
2. For this reason, perhaps, Massachusetts magistrates generally opposed the use of juries for criminal cases. For an evaluation of seventeenth-century attitudes toward the jury system, see Murrin.

Just as the narrator concedes the godliness of the magistrates but hints that they may be too severe, too little capable of mercy when the crime involves the heart, so the townspeople recognize the righteousness of their leaders. But their conclusion is far different: these men have been too lenient. One woman observes, "The magistrates are God-fearing gentlemen, but merciful overmuch" (51) [37]. Another, "the ugliest as well as the most pitiless of these self-constituted judges," calls for the ultimate penalty: "This woman has brought shame upon us all, and ought to die" (51) [38]. Although Hawthorne may seem to have made it easy for his readers to condemn this merciless woman, attaching ugliness to her hard justice, she speaks with political sense. "Is there not law for it? Truly there is, both in the Scripture and the statute-book. Then let the magistrates, who have made it of no effect, thank themselves if their own wives and daughters go astray!" (51–52) [38].

In his careful reading, Hawthorne encountered this argument repeatedly. The pages of Winthrop's journal are full of references to this question of whether specific penalties should be fixed with certainty. If a law is published but never enforced, how effective a deterrent is the fear of punishment? If punishments are always commuted, why specify them at all? Many New Englanders insisted that specific penalties should be the inevitable result of particular crimes. As early as 1641, according to Winthrop, the General Court "was full of uncomfortable agitations and contentions," as people had been calling for the passage and publication of precise punishments for a number of offenses (*Journal* 376). During a three-week session in 1641, one hundred laws, called the Body of Liberties, were drawn up against the will of John Winthrop with his belief that magistrates should have discretionary power to assign penalties.

Winthrop believed that severity was not necessarily the best instrument of government. Though he had little patience for sin, especially when the stability of the colony was threatened, his compassion toward sinners was genuine. In the words of one biographer, "Winthrop portrayed life as a purposeful process of enduring struggle, rather than as a one-time contest between absolute good and evil" (Moseley 93). During the 1640s a series of bestiality cases occurred both in Boston and in Plymouth, cases that indicate Winthrop's sympathy for the human struggle against sin in contrast to the response of his neighbor, Governor William Bradford of Plymouth. Moseley argues that whereas Winthrop's narrative of the Boston sex offender, Hackett, "expresses the characteristic Puritan concern with the struggle between sin and grace in each individual's life, . . . Bradford's account of Granger's fate articulates a typical Separatist desire to weed out wicked people who may have crept into the community of the saints" (132). While Moseley's comparison may be unfair to Bradford, it may also be too generous to the non-Separatists whom Winthrop represented, especially when we note how often the Massachusetts governor found himself pleading for leniency. The death penalty might indeed assure the colony that a sinner had been eradicated, but Winthrop believed that penitence and conversion were the greater victories. Furthermore, he thought that most virtuous behavior was not motivated by fear. He classified people into two types: those "who are godly & virtuous" and observed the law "for Conscience, & Virtue's sake" and those who "must be held in by fear of punishment" ("Arbitrary Government" 453). Although the social and political community consisted of both types, the covenant with God involved only the first kind of person. Winthrop felt a stronger commitment to the spiritual welfare of this first group.

When the citizens in Hawthorne's romance debate the purpose and severity of punitive treatment, they are repeating the terms of a familiar argument. The ugliest and most merciless woman in the crowd calls for the preservation of social order when she advocates capital punishment. If Hester Prynne's sentence is commuted, she warns, even the wives and daughters of the magistrates themselves will "go astray." She is answered in words very much like those Winthrop might have used: "'Mercy on us, goodwife,' exclaimed a man in the crowd, 'is there no virtue in woman, save what springs from a wholesome fear of the gallows? That is the hardest word yet!' " (52) [38]. The two compassionate characters in the marketplace, this man and the young woman, use rhetorical terms that are loaded with political significance: "mercy" and "peace." Hawthorne has overshadowed the immediate occasion for the novel, the revelation of Hester's adultery, with a broader moral and political debate that engrossed the Massachusetts governors and their subjects. What role does mercy play in the governance of a Christian community? Must peace depend upon justice alone? Not surprisingly, the words "mercy" and "justice" and "sympathy" are far more conspicuous in Hawthorne's novel than the merely nominal problem of adultery.[3]

Although Hawthorne displaces the magistratical debate onto the villagers in the square, we do see signs that a discussion has transpired among the governors and magistrates, who have commuted Hester's sentence from death and are willing to be even more lenient if she speaks the name of her partner. The removal of the letter would indicate visually what Hester's utterance would enact symbolically. As Ernest W. Baughman has explained, public confession was not merely a part of the penalty for sin; it was the entrance into forgiveness, "the means by which an individual can remain a part of society; lacking confession, the sinner ceases to be a part of that society, or he is so much at odds with it that his functioning is seriously impaired" (540). The connection between mercy and confession—urged vehemently upon Hester by the Reverend Mr. Wilson, weakly and ambivalently by Dimmesdale—is made explicit in the New England *Platform of Church Discipline*, published in 1649: "If the Lord sanctify the censure to the offender, so as by the grace of Christ, he doth testify his repentance, with humble confession of his sin, and judging of himself, giving glory unto God; the Church is then to *forgive* him, and to *comfort* him, and to *restore* him to the wonted brotherly communion, which formerly he enjoyed with them" (228).

Although the *Platform* was written by New England church elders and representatives of their congregations, it was done at the request of the General Court, and received unanimous approval from the magistrates,

3. The point that I am arguing—that Hester's adultery is only a specific occasion for the introduction of much broader political and social issues—has an actual historical precedent. As Moseley observes about the many sexual misconduct cases of the 1640s, "However grievous such personal matters were, though, the Bay Colony's problems in these years went well beyond instances of individual immorality. Indeed, upon the heels of their victory in the Pequod war and their resolution of the Antinomian crisis, the Puritans faced a broader, more far-reaching challenge, involving their relationship with England and raising unavoidable questions about the stability, success, and meaning of their entire errand" (87).

including John Winthrop. While Winthrop was not personally responsible for the language of this passage, he most likely endorsed its sentiments and agreed that one important goal of church censure is "the reclaiming and gaining of offending brethren" (227). Sacvan Bercovitch has argued convincingly that the strength of Hawthorne's Puritan community "lies not in coercion but, on the contrary, in their susceptibility to reassessment and change." The Massachusetts leaders have chosen "not to apply the letter of the law . . . but instead to define it through the ambiguities of mercy and justice" (47). In offering to remove Hester's letter, Wilson indicates his eagerness to accept her once again into the community, to suspend strict justice in favor of mercy.

The Politics of Charity

The question of how to treat a single sinner, whether to exile or rehabilitate her, had particular relevance to the larger political issues confronting antebellum America. David S. Reynolds has pointed out that Hawthorne's contemporary culture was dominated by "intensifying moral wars" (*Beneath the American Renaissance* 113). Political and social enthusiasms, evangelical fervor, and moral absolutes abounded in the political and social contests of the day—discussions in which the word "compromise" usually provoked virulent condemnation. Hawthorne's own ruminations on the Civil War sound distinctly familiar when we consider the pedagogical, reformist uses of punishment. In 1861 he wrote: "though I approve the war . . . I don't quite understand what we are fighting for, or what definite result can be expected. If we pummel the South ever so hard, they will love us none the better for it; and even if we subjugate them, our next step should be to cut them adrift."[4]

Bercovitch explains the influence of nineteenth-century domestic political conflicts on *The Scarlet Letter*. Hawthorne clearly recognized these episodes of confrontation—slavery, westward expansion, industrialization, regional disputes, and conflicts over the treatment of women, blacks, laborers, the poor—as the modern versions of political differences that had challenged the colony some two hundred years earlier. When Bercovitch points to the cultural work of *The Scarlet Letter*, he dwells on a political gesture, a ritual of consensus. The word "consensus," like "compromise," is a term based on political expedience—which, Thoreau argued, has no place in the realm of moral issues. In returning to Winthrop, an astute political thinker who drew on the bonds of love and charity to secure political stability, Hawthorne inflects that expedient term "compromise" with the ethically sublime concept of charity. *The Scarlet Letter* is saturated with Winthrop's influence.

Apart from his death, which provides the occasion for a midnight reunion between Hester and Dimmesdale, John Winthrop is absent from the novel. Lauren Berlant is correct in claiming that the "absent" Winthrop is nevertheless "woven into the novel" (74). Berlant is referring to the "struggles over legal representation that fractured the state apparatus of the Massachusetts Bay Colony in the 1630s and 1640s" (73–74). I would add that Hawthorne's interest in Winthrop had less to do with the precise

4. Letter to Horatio Bridge, 26 May 1861, cited in Masur, "*The Real War*," 164.

political arguments of the seventeenth century than with the role of sympathy in binding together an otherwise dispersing population in Hawthorne's own day.

The novel's treatment of the Reverend John Wilson furnishes one such example. Edward Johnson, a contemporary chronicler of the colony, notes that Wilson, pastor of the Second Church of Christ, "made a powerful instrument in [Christ's] hands for the cutting down of Error, and Schism" (67). Johnson's description does not seem to indicate tolerance; rather, it resembles the language used by Hawthorne in "The Maypole of Merry-Mount," where John Endicott ruthlessly hacks away at the sign of error and revelry. Johnson was writing generally about the Antinomian crisis of 1637, but we also have evidence of particular occasions when Wilson was unyielding. In a case of heresy in 1643, John Wilson was the most vehement of the elders in calling for Samuel Gorton's execution, even swaying the more temperate John Cotton. As Robert Emmet Wall Jr. writes, "Cotton may have called for reasonableness initially, but his final preachings before the sentencing had called for the most serious of punishments. In this, he was in full accord with his colleague in the Boston pulpit, John Wilson" (140).[5] Wilson, who constantly clamored in his sermons for Gorton's death, is here used as a model of rigidity. In Hawthorne's romance, however, he appears to be much more lenient. Though the novelist makes Wilson significantly older than he was in 1642, he describes him as "a man of kind and genial spirit" (65) [45]. Of course, Hawthorne tells us that this spirit "was, in truth, rather a matter of shame than self-congratulation with him" (65) [45], as if to note that Wilson's charity is not his most prized quality among the stern leaders of Massachusetts. Nevertheless, the nineteenth-century narrator clearly approves of these sentiments. Like Winthrop, Wilson does not assume that severity is his only course, but considers "what arguments to use, whether of tenderness or terror," to draw a confession from Hester (65) [46]. And it is Wilson himself, finally, who tries to negotiate with Hester by offering to remove her letter.

I have been arguing that John Winthrop, though not actually present as a character in *The Scarlet Letter*, in fact permeates the novel. His character is suggested by the pitying rosebush, which "by a strange chance, has been kept alive in history" (1:48) [35]. If we continue to read this passage metaphorically, we shall see that Hawthorne has an interest in keeping this tradition alive into the nineteenth century, a tradition that comes to us here in the guise of the shadowy John Winthrop. His influence is clear in the softening of John Wilson, the senior church leader at Hester's punishment. The civil representative, Governor Bellingham, also speaks briefly, and though he does not speak of mercy, he hints at it, urging Dimmesdale to extract a confession "as a proof and consequence" of Hester's repentance (66) [46].

It is appropriate, given Hawthorne's indebtedness to Winthrop, that these high-ranking officials recommend mercy, while the people call for justice. During the many years in which he battled the people (through the deputies, their political representatives), Winthrop maintained that the magistrates had a more direct link with God, and thus were better qualified to assign the penalties for crime and sin. Arguing in 1641 for the

5. A compromise was reached, and Gorton was first imprisoned and then exiled. See Wall, 121–56, for an extended discussion of the case.

continuance of discretionary power, he contended that "all punishments, except such as are made certain in the law of God, or are not subject to variation by merit of circumstances, ought to be left arbitrary to the wisdom of the judges" (*Journal* 381). In 1644 he recommended that magistrates "be guided by the word of God" rather than the prescribed penalties written by mere men (*Journal* 554). Winthrop even went so far as to claim that judges "are Gods upon earth: therefore, in their Administrations, they are to hold forth the wisdom & mercy of God, . . . as well as his Justice" ("Arbitrary Government" 448). He made his case most emphatically in 1645, after having survived a petition for his impeachment: "The great Questions that have troubled the Country, are about the Authority of the magistrates & the Liberty of the people: It is your selves, who have called us to this office, & being called by you, we have our Authority from God, in way of an Ordinance, such as hath the image of God eminently stamped upon it" ("Speech on Authority and Liberty" in *Journal* 586).

Winthrop's defense of discretionary power for the magistrates, an argument he was forced to repeat frequently during his many years as governor,[6] was fairly simple. The argument is best articulated in his address "Arbitrary Government," delivered in 1644 after an unpleasant confrontation between the deputies and magistrates. Because there was no clear rule about whether to exercise mercy or justice upon a sinner, Winthrop contended, these decisions should be left open to those who were best able to see as God sees. Juries could be trusted to use discretion in civil cases, but they did not have the requisite wisdom to decide criminal cases, where a person's soul was at stake. Winthrop held that the law of the land, by which "the Officers of this Body politick" must walk, was "the Word of God" ("Arbitrary Government" 445). But the Word of God required interpretation, and the governor believed that interpretation often involved the individual circumstances of a case. A fixed penalty for a specific crime, decided in the abstract, could not take account of mitigating circumstances. "If all penalties were prescribed," he asked, "then what need were there of any special wisdom, learning, Courage, zeal, or faithfulness in a Judge?" (446). Winthrop was particularly troubled by the prospect of setting penalties for crimes whose penalties were unspecified in Scripture: "I would know by what Rule we may take upon us, to prescribe penalties, where God prescribes none" (449).

The deputies were clearly right in understanding that the higher-ranking magistrates wielded considerable power. Winthrop had a definite sense of social and political hierarchy, and he supported the clear differences in power between the several groups. "The determination of Law belongs properly to God," he conceded, but God "hath given power & gifts to men to interpret his Laws: & this belongs principally to the highest Authority in a Commonwealth & subordinately to other magistrates & Judges according to their several places" (453). Winthrop had no interest in seeing these hierarchies leveled. Such distinctions were the self-evident truths of his generation of magistrates, and the deputies could not ask for greater powers without hearing the sneering term "mere democracy."

Still, Winthrop's obstinacy on this issue did not make him a tyrant. Because he believed so firmly that he was one of the "Gods upon earth,"

6. As early as 3 August 1632, Winthrop's *Journal* records such a defense.

he was careful not to abuse his power but to represent God's law as truthfully as he could. For this reason, he was much more apt to call for mercy than some of the other political leaders, who saw their roles largely in secular and social terms, and thus were willing to destroy or exile anyone who threatened the status of the community. Halfway through his discourse on arbitrary government, Winthrop turned to the issue of adultery, which, as we have seen, occupied much of the colony's attention during the 1640s, as sex crimes were becoming more prevalent. The example surely resonated with his audience, many of whom had been present at the double execution for adultery only seven months earlier.

In a significant passage on adultery, Winthrop cites a number of Scriptural cases where the law calling for the death of adulterers was overlooked:

> Adultery & incest deserved death, by the Law, in Jacob's time . . . yet Ruben was punished only with loss of his Birthright, because he was a Patriarch. David his life was not taken away for his Adultery & murder, (but he was otherwise punished) in respect of public interest & advantage, he was valued at 10000 common men. Bathsheba was not put to death for her Adultery, because the Kings desire had with her the force of a Law. . . . But if Judges be tied to a prescript punishment, & no liberty left for dispensation or mitigation in any case, here is no place left for wisdom or mercy. (449)[7]

The point, says Winthrop, is "to show how God hath sometimes (in his wisdom & mercy) dispensed with the rigor of his own Law: & that Princes have sometimes done the like, upon public or other prevalent considerations" (457).

Winthrop does not explain what these considerations might be. As a political leader with deeply religious sentiments, his reasons were probably twofold. First, although he was a civil leader, his authority over the realm of crime and punishment meant that he had a stake in the spiritual condition of all convicted criminals.[8] In Hester Prynne, Hawthorne gives us a fine example of a criminal whose life was spared and soul rehabilitated. Though it takes many years for the scarlet letter to do its office, by the end of the novel she has become an exemplary citizen and Christian, sharing with her townspeople tokens of the same mercy that was exercised upon her. Winthrop was reluctant to practice strict justice while compelling spiritual reasons might be found for substituting mercy and thus for saving both a life and a soul. He felt, moreover, genuine affection for his fellow colonists who, like himself, had forsaken more comfortable lives in Europe in order to participate in a godly mission in a new world. This affection caused him to look mercifully upon their transgressions, to plead for their lives in capital cases, and to look for extenuating circumstances.

A second, more pragmatic reason for his leaning toward mercy had to do with the many battles over political authority waged during what Wall calls the colony's "crucial decade." As a political leader, Winthrop

7. The story of Bathsheba and David is represented on a tapestry that hangs in Dimmesdale's apartment. [See p. 79 of this Norton Critical Edition—*editor's note.*]
8. Early on, the Massachusetts Bay Colony decided upon a separation of religious and civil leadership. Although church elders and court magistrates frequently consulted with each other, Winthrop records in July 1632 that the Boston congregation decided that no civil magistrate could be nominated as a ruling elder (*Journal* 71).

understood the important role of public consent to the authority of a political system and its heads of state. He begins his discourse on arbitrary government by recognizing what John Locke and other theorists would later describe more fully, that governments derive their just authority from the consent of the governed. "Arbitrary government," he writes, "is, where a people have men set over them, without their choice, or allowance: who have power to govern them, & Judge their Causes without a Rule. God only hath this prerogative" (440). Because governors were elected each year, Winthrop was acutely aware of the role of public choice in the development of a healthy political system. He also understood that the colonists of Massachusetts would not long tolerate harsh and judgmental political leaders. As his "Model of Christian Charity" suggests, the welfare of the whole colony was Winthrop's foremost concern, and the colony would never be unified if the people distrusted their leaders.

Resistance might be futile, but desertion was a distinct possibility, and one that Winthrop feared from the beginning. In his "Model" he urged his listeners to consider themselves a body knit together: "though we were absent from each other many miles, and had our employments as far distant, yet we ought to account ourselves knit together by this bond of love" (292). Winthrop had hoped that the absences he alluded to were only temporary. He lived, however, to see the strands of his community come apart, as economic interests and love of liberty prevailed over the sense of charity that he had hoped would hold the community together. One moving passage from his journal is worth citing at length. Following an economic crisis caused by declining immigration, as the people debated their right to leave Massachusetts for "outward advantages," he wrote in September 1642:

> For such as come together into a wilderness, where are nothing but wild beasts and beastlike men, and there confederate together in civil and church estate, whereby they do, implicitly at least, bind themselves to support each other, . . . how they can break from this without free consent, is hard to find, so as may satisfy a tender or good conscience in time of trial. Ask thy conscience, if thou wouldst have plucked up thy stakes, and brought thy family 3000 miles, if thou hadst expected that all, or most, would have forsaken thee there. Ask again, what liberty thou hast towards others, which thou likest not to allow others towards thyself; for if one may go, another may, and so the greater part, and so church and commonwealth may be left destitute in a wilderness, exposed to misery and reproach, and all for thy ease and pleasure. (416)

In this remarkable passage, Winthrop ceases to be a historian of the colony and assumes the role of preacher, appealing to the conscience of his imagined reader, and reminding future generations that a covenant had been broken and a people betrayed. Charity might have held the community together, but selfishness and love of liberty threatened the ruin of the colony. For Winthrop, even the ultimate economic success of Massachusetts would have signified little, so long as that success resulted from the pursuit of private interests. He could take little joy in overall prosperity when it was merely the sum total of private gains.

Winthrop wrote this lament in 1642, the year that Hawthorne chose for the opening scene of his romance. He lived another seven years, long enough to preside over some of the most unsettling political events in the colony's history: the struggle between the deputies and magistrates, prompted by questions about the ownership of a sow (an episode that Hawthorne alludes to in his romance); news of the English civil war, where victory by their Puritan brethren left the New Englanders abandoned as the tides of emigration dried up; and an impeachment trial, which (though the charges were ludicrous and the victory easy) surely caused Winthrop great pain.[9] Through all these episodes, the governor remained true to his policy on mercy, and sincerely regretted having listened to those who had advised greater strictness. Until his dying hour, Winthrop's advocacy of mercy waned only once, in his uncharacteristically severe management of the Antinomian controversy surrounding Anne Hutchinson. Thomas Hutchinson reports that "upon [Winthrop's] death-bed, when Mr. Dudley pressed him to sign an order of banishment of an heterodox person, he refused, saying, 'he had done too much of that work already'" (1:129).

Winthrop's deathbed intimation resonates in Hawthorne's romance, where Hester Prynne is compared to the "sainted" Anne Hutchinson (48) [36]. While it may be tempting to understand this early characterization of the Antinomian leader as Hawthorne's defense of her conduct, such a reading is unconvincing when we consider that Hester is nearly identified with "the image of Divine Maternity" (1:56) [40]. In neither case does the narrator exonerate the woman for her lawlessness. Rather, the association between Hester and Hutchinson reminds us, if only by allusion, that merciful conduct does not come easily. Even the angelic John Winthrop shared some of Chillingworth's least attractive impulses.

By all accounts, Winthrop's death in 1649 signaled the end of charity as a quasi-official policy. According to George Bancroft (whom Hawthorne had read), Massachusetts became much more harsh after losing Winthrop's tolerant leadership, ruthlessly prosecuting Quakers, witches, and other heterodox people. Hutchinson writes of one petty example of Puritan intolerance. "Soon after Mr. Winthrop's death, Mr. Endicott, the most rigid of any of the magistrates, being governor, he joined with the other assistants in an association against long hair" (130). Wall calls the orthodoxy following Winthrop's death "uncompromising and harsh" (233).

A sign of that rigidity might be found in one response to the *Book of General Laws*, finally published in 1648 after years of battle between the magistrates and the deputies. In the end, the magistrates acquiesced in the project. But Winthrop was clearly reluctant, as we have seen. Edward Johnson's *Wonder-Working Providence*, published four years after Winthrop's death, describes the work of the government in compiling the code of laws. His unmistakable enthusiasm for law and order clearly indicates the changes in the colony during the four short years since Winthrop's death:

> but let not any ill-affected persons find fault with [the laws], because they suit not with their own humour, or because they meddle with matters of Religion, for it is no wrong to any man, that a people who have spent their estates, many of them, and ventured their lives for to

9. Wall writes: "Apparently several of the magistrates thought there were grounds for refusing to entertain the petition. Winthrop was accused for doing what he had every right to do, and the charges against him were ridiculous. Winthrop disagreed. He was anxious to clear his name and by so doing to uphold the authority of his fellow magistrates" (106).

> keep faith and a pure conscience, to use all means that the Word of God allows for maintenance and continuance of the same, especially they have taken up a desolate Wilderness to be their habitation, and not deluded any by keeping their profession in huggermug, but print and proclaim to all the way and course they intend, God willing, to walk in. (244–45)

Johnson's ideas, of course, are similar to the ones we find in Winthrop's lament about the breakup of his community: the sacrifices of those who have come to the new world, the dangers posed by individual whims, and the sanctity of conscience. But the tone could not be more different. Johnson defends the use of "all means that the Word of God allows" to protect the community, whereas Winthrop sadly accedes to the dispersal of his people because he cannot bring himself to use any means necessary. Hoping to preserve the community by appealing to charity and affection alone, he found that these instruments were insufficient. And yet, even to the end of his life, he refused to use harsher methods.

The Golden Bonds of Love

In Hawthorne's novel, the death of John Winthrop is notable beyond its use in bringing Hester, Pearl, and Arthur together at the scaffold. While the scaffold remains in its usual place, the emblems of mercy have disappeared. On the day of election, when the people will choose Winthrop's successor, his fictional counterpart, the sympathetic young woman of the opening scene, has also passed on to a more merciful world. These absences are consequential for the long-suffering Hester, whose apparently hopeful words to Pearl are surely ironic. Explaining the election-day events, she tells Pearl that "the children have come from their schools, and the grown people from their workshops and their fields, on purpose to be happy. For to-day, a new man is beginning to rule over them; and so—as has been the custom of mankind ever since a nation was first gathered—they make merry and rejoice; as if a good and golden year were at length to pass over the poor old world!" (229) [136]. We must doubt Hester's faith in the reality of this congregated mirth, since her last encounter with the public holiday, when the children were released from their schoolwork, was the scene of her ignominious display seven years earlier. Then, as now, the children did not understand the political significance of the event, only that they were given a "half-holiday" (54) [40]. The two events, indicating shame and happiness, flow together in Hester's experience. Even now, after years of being accustomed to her scarlet letter, Hester's neighbors rediscover their pleasure in her discomfort. As newcomers stare at the letter for the first time, "the inhabitants of the town (their own interest in this worn-out subject languidly reviving itself, by sympathy with what they saw others feel) lounged idly to the same quarter, and tormented Hester Prynne, perhaps more than all the rest, with their cool, well-acquainted gaze at her familiar shame" (246) [146]. The original scene of Hester's humiliation is nearly replicated, except that she encounters no expressions of mercy, but anticipates that "a new man is beginning to rule over them." Furthermore, as Michael Davitt Bell has pointed out, the "new man" is none other than the despotic John Endicott, who "went on to dominate the government of Massachusetts until his own death in 1665." Though Hawthorne does not mention the name of the new governor, "the suppressed 'truth' about Endicott's succession is intended ironically to undercut the hopefulness of Hester and Dimmesdale" (141). Hester, we must conclude, fully intends all of the ironies of her political lesson to Pearl.

With the death of Winthrop, charity seems to have evaporated among the political and religious leaders of New England; consequently, so have the real bonds of community. We see this point most clearly in Arthur Dimmesdale's election-day sermon, delivered as he is planning his covert departure from the colony. On the face of it, the sermon describes the very preoccupation that had consumed John Winthrop: the future of this godly experiment called New England, and the reward that God meant to bestow upon those who remained faithful to the covenant. "His subject, it appeared, had been the relation between the Deity and the communities of mankind, with a special reference to the New England which they were here planting in the wilderness. And, as he drew towards the close, a spirit as of prophecy had come upon him, constraining him to its purpose as mightily as the old prophets of Israel were constrained; only with this difference, that, whereas the Jewish seers had denounced judgments and ruin on their country, it was his mission to foretell a high and glorious destiny for the newly gathered people of the Lord" (249) [147]. The conclusion of the sermon is one for which Winthrop himself, writing in 1630, could only hope. Dimmesdale's apparent certainty, "constraining" him to this prophecy of congregational victory, is especially incriminating when we consider that the preacher is plotting his escape from the very community that has selected him as its prophet and saint. He is constrained by nothing but his desire to escape the grasp of Chillingworth, accompanied by Hester and Pearl. He composes his inspired sermon in the afterglow of his forest meeting with Hester, in anticipation of his escape. Returning to his study after the encounter, he flings "the already written pages of the Election Sermon into the fire," and "forthwith beg[ins] another." Writing "with earnest haste and ecstasy," he consumes the entire night at his desk and discovers himself, at dawn, with "a vast, immeasurable tract of written space behind him!" (225) [134]. The metaphor clearly indicates that Dimmesdale's mind is fastened not on spiritual matters but on the open space that he plans to put between himself and his tormentor, and also, unavoidably, between himself and the people who share in his covenant with God. We might almost hear Winthrop's lament in the background of Dimmesdale's sermon; here, indeed, is a man ready to take the "liberty of removing for outward advantages." Perhaps Dimmesdale has heard these echoes from the dead Winthrop. After all, his newfound attraction to "ease and pleasure" is just as alienating as his habitual penance and solitude have been. In his case, neither punishment nor its escape brings him any closer to his community. In this gifted minister, Hawthorne has given us a clear sign that Winthrop's New England has forgotten its covenant of charity.

Another sign that true charity is a thing of the past is that those who pretend to it inspire nothing but horror in the observant reader. For all of his malice, Roger Chillingworth is a keen student of Winthrop and understands better than Dimmesdale that community ties cannot be so easily dissolved. Echoing Winthrop and poisoning his meaning, Chillingworth explains to Hester why he cannot leave New England: "elsewhere a wanderer, and isolated from human interests, I find here a woman, a man, a child, amongst whom and myself there exist the closest ligaments. No matter whether of love or hate; no matter whether of right or wrong!" (76) [51–52]. The ligaments that for Winthrop were made of love become for Chillingworth ties of "interests" that have nothing to do with affection. What matters to him is the compulsion to recognize and pursue those claims to human community. As he tells the dying Dimmesdale, "Hadst thou sought the whole earth over, . . . there was no one place so secret,—no high place nor lowly place, where thou couldst have escaped me,—save on this very scaffold!" (253) [149]. However vast the geographic distance the minister might put between himself and his tormentor, only the moral space afforded by Dimmesdale's confession could have severed the ligaments of guilt and retribution connecting the two men.

Chillingworth understands these things because he has sought to master what Winthrop described as the "sensibleness and sympathy of [an]other's conditions" ("Model" 289). To gain this sympathy, Winthrop advised, we must practice "more enlargement towards others and less respect towards ourselves" (287). Chillingworth satisfies this condition in the extreme, exchanging his own subjectivity for whatever elements of another consciousness might flow into his own. He is able to "go deep into his patient's bosom" precisely because he can put his own personality aside (124) [83]. Seeing the success of Chillingworth's practice, the narrator issues this warning:

> A man burdened with a secret should especially avoid the intimacy of his physician. If the latter possess native sagacity, and a nameless something more,—let us call it intuition; if he show no intrusive egotism, nor disagreeably prominent characteristics of his own; if he have the power, which must be born with him, to bring his mind into such affinity with his patient's, that this last shall unawares have spoken what he imagines himself only to have thought; . . . then, at some inevitable moment, will the soul of the sufferer be dissolved, and flow forth in a dark, but transparent stream, bringing all its mysteries into the daylight. (124) [78–79]

The language here is very like that used by Hawthorne in his famous description of mesmerism. To enter into the feelings of another, that is, one must put one's own feelings aside and imagine what the object of the gaze is undergoing; the result is a flowing of the soul of the observed, often unawares, into the grasp of the observer. Only the observer's motives can protect the object of sympathy. If the practitioner of sympathy has pernicious designs on his companion, designs that Hawthorne feared in a mesmerist and dramatized in Chillingworth, the transaction is formidable. Amazed to see sympathy used for malicious ends, Hester articulates some of the awe with which her author regarded complete sympathy. "'Thy acts are like mercy,' said Hester, bewildered and appalled. 'But thy words interpret thee as a terror!'" (76) [51].

Dimmesdale, in contrast, becomes a victim precisely because he is unable to escape his own ego and thereby enter into the feelings and motives of others. On the night of Winthrop's death, the minister suffers from "a highly disordered mental state" because he misreads the token in the sky as a reference to his own private sin. "Rendered morbidly self-contemplative by long, intense, and secret pain," Dimmesdale has "extended his egotism over the whole expanse of nature, until the firmament itself should appear no more than a fitting page for his soul's history and fate" (155) [96]. The sign and the cause of this morbidity are the same: Dimmesdale's secret sin has become the one significant fact in his life, and it colors his perceptions of the world. Preoccupied with his own sin, he assumes that all humans are likewise fallen, likewise suspect. "Trusting no man as his friend, he could not recognize his enemy when the latter actually appeared" (130) [82].

Hester suffers from the same solipsism, but unlike Dimmesdale she recognizes it as a symptom of her own subjectivism rather than as a truth of the universe. When "the red infamy upon her breast would give a sympathetic throb" (87) [57] indicating "the hidden sin in other hearts" (86) [57], Hester resists the conclusion that the world is as fallen as she. "Be it accepted as a proof that all was not corrupt in this poor victim of her own frailty, and man's hard law, that Hester Prynne yet struggled to believe that no fellow-mortal was guilty like herself" (87) [57–58].

Chillingworth, suffering from no such intrusive egotism, knows that his conclusions are accurate. As he explains to Hester, the science of sympathy, learned well, is just as reliable as other sciences. "I shall seek this man, as I have sought truth in books; as I have sought gold in alchemy. There is a sympathy that will make me conscious of him. I shall see him tremble. I shall feel myself shudder, suddenly and unawares. Sooner or later, he must needs be mine!" (75) [51]. Chillingworth is concerned with science, not art, but his language curiously echoes the passage in "The Custom-House" where Hawthorne describes how the letter "on [his] breast" provoked "a sensation not altogether physical, yet almost so, as of burning heat; and as if the letter were not of red cloth, but red-hot iron. I shuddered, and involuntarily let it fall upon the floor" (32) [25]. "Suddenly and unawares," Hawthorne experiences the sensations of a perfect sympathetic encounter, producing an involuntary bodily reaction. Like Chillingworth, the author identifies fully with Hester herself, who experiences the letter as an almost literal wound, "the tenderest spot" of her sore consciousness (84) [56], a spot that seems "to grow more sensitive with daily torture" (86) [57]. Unlike Chillingworth, who rejoices in the shudder that constitutes the proof he has sought, Hawthorne would like to escape the sympathetic encounter embodied in the scarlet letter. As the narrator admits at the close of the novel, he "would gladly, now that it has done its office, erase its deep print out of [his] own brain; where long meditation has fixed it in very undesirable distinctness" (259) [152]. He does not tell us whether such a wish is realistic or whether, like Hester's, his mark is "too deeply branded" to be removed. Like Dimmesdale, the narrator has developed a sympathetic replica of Hester's mark of shame; unlike Dimmesdale, he sees it as a link between himself and the human community, a token of what Winthrop called the "sensibleness and sympathy" of another's condition.

Hester, who comes to be known through the town as a "Sister of Mercy" (161) [99], likewise performs deeds that look like mercy but are really quite different. If, as Winthrop believed, charity arises out of ligaments of affection, then Hester's acts of mercy are not true charity, though they eventually restore her to her community and make true charity possible. While she "came to have a part to perform in the world," Hester does not feel as though she belongs; all of her encounters "implied, and often expressed, that she was banished, and as much alone as if she inhabited another sphere" (84) [56]. Though this banishment is imposed on Hester by the token that she must wear, during her years of exile she comes to internalize the judgment of the people and extricate herself from the grasp of human sympathies. Her face typically shows "the frozen calmness of a dead woman's features; owing this dreary resemblance to the fact that Hester was actually dead, in respect to any claim of sympathy, and had departed out of the world with which she still seemed to mingle" (226) [135]. While she comes to be known by a "blameless purity" that is taken to indicate her "genuine regard for virtue" (160) [99], we learn that she does not share in her community's sense of virtue: "The world's law was no law for her mind" (164) [101].

Whatever motivates her kind deeds, then, they are not acts of charity, since she seems actively to eschew the "sensibleness and sympathy" of another's condition that prompt charitable work. Instead, the narrator dismisses Hester's apparently charitable acts as mere "morbid meddling of conscience with an immaterial matter" (84) [55], like the impulses that drive the minister to fast and scourge himself "as an act of penance" (144) [90]. Hawthorne wishes us to recognize the difference between the appearance of charity and its deeper reality, even if the community does not. Indeed the one person of whose condition she is most apt to be sensible does not gain Hester's immediate sympathy. She feels the solace of Arthur Dimmesdale's "human eye," giving her "momentary relief, as if half of her agony were shared" (86) [57], but the glance does not also offer her entrance into his suffering. Wandering "without a clew in the dark labyrinth of mind" (166) [102], Hester forgets that she is not the only one undergoing torment.

Only when she sees him upon the scaffold, occupying the same place where she had stood in her own ignominy and torment, does Hester begin to understand the depth of her lover's suffering. Hawthorne has nearly literally enacted the scene of sympathy as Adam Smith describes it, where we can understand another person's pain only by imagining how it would feel for ourselves. Only in imaginatively exchanging places with the object of suffering can we enter into that suffering. Dimmesdale attempts to take on his due share of punishment by occupying the same ground where Hester had stood; Hester, having stood there once before, is able to feel the torment that Dimmesdale has endured at the hands of his physician. The moment proves pivotal for the socially estranged woman, who, "in the misanthropy of her own trouble, [had] left the minister to bear what she might picture to herself as a more tolerable doom. But of late, since the night of his vigil, all her sympathies towards him had been both softened and invigorated. She now read his heart more accurately" (193) [116–17]. Hester's egotism and sense of her own estrangement have made her nearly impervious to the sufferings of the one person whose pain she should have intuited. Hardened by years of loneliness, she overlooks the obligations demanded by the "iron link of mutual crime" (160) [98].

It is notable that one of the most arresting statements in the novel—"The scarlet letter had not done its office"—is immediately followed by an indication that Hester is about to acknowledge her debt to Dimmesdale (166) [102]. The moment is significant because, rather than reintegrating her into the community, as punishments should do, the scarlet letter has instead alienated Hester from the sentiments that must seem natural in order to provide a sense of communal identity. Looking at Dimmesdale now with sympathy rather than solipsism, Hester begins to feel the tug of communal loyalties for the first time in seven years:

> The scarlet letter had not done its office.
>
> Now, however, her interview with the Reverend Mr. Dimmesdale, on the night of his vigil, had given her a new theme of reflection, and held up to her an object that appeared worthy of any exertion and sacrifice for its attainment. She had witnessed the intense misery beneath which the minister struggled, or, to speak more accurately, had ceased to struggle. (166) [102]

Hester's newfound sense of responsibility for another person, unlike her other charitable acts, can be considered real charity in Winthrop's sense: she decides to lay aside her own concerns, including her promise to Chillingworth and the consequences of breaking it, and "redeem her error" (167) [102]. One such consequence might well be the deaths of Hester and Arthur. Since the statute calls for the death of adulterers, the revelation that Hester's husband is still alive could result in a double execution. Of course, given the length of time since the crime, it is possible that the magistrates would not insist on the extreme penalty. But Chillingworth, "the man whom [they have] most vilely wronged," has every right to demand the lives of these sinners, and his voice would be a powerful argument in favor of execution (172) [105]. As he tells Hester, "My finger, pointed at this man, would have hurled him from his pulpit into a dungeon,—thence, peradventure, to the gallows!" (171) [104]. Since the death of Winthrop, furthermore, mercy could not be expected at the hands of the law.

As one of the first signs that she has rejoined her Christian community, Hester begins to speak of mercy and charity. Since, for many years, "there seemed to be no longer any thing in Hester's face for Love to dwell upon" (163) [100], Hester's association with these words, rather than with judgment and revenge, signifies a dramatic change, and hints that the letter is finally doing its office. Calling upon Chillingworth to give up his fiendish revenge, Hester asks him to "purge it out of thee, and be once more human. . . . Forgive, and leave [Arthur's] further retribution to the Power that claims it!" (173–74) [106].

Though Chillingworth refuses Hester's appeal, she has gained something from contemplating the benefits of charity and the effects of unmerciful thoughts on her former husband. No longer seeing the world through the single lens of retribution, she considers that mercy might also be part of her lot. Hester has long understood that it is Pearl's role to "connect her parent for ever with the race and descent of mortals" (89) [58], but until she opens her heart to the redemptive power of mercy, the mother recognizes that relation as primarily punitive. Though she admits that Pearl "is my happiness," she dwells on the child's function as a representative of the law. "Pearl punishes me too! See ye not, she is the scarlet letter . . . endowed with a million-fold the power of retribution for my sin" (113) [72].

But after beginning to understand and sympathize with the sufferings of others, Hester recognizes that perhaps Pearl might offer more than "justice and retribution": "never, until now, had she bethought herself to ask, whether, linked with that design, there might not likewise be a purpose of mercy and beneficence. If little Pearl were entertained with faith and trust, as a spirit-messenger no less than an earthly child, might it not be her errand to soothe away the sorrow that lay cold in her mother's heart, and converted it into a tomb?" (180) [109–10]. Hester rejects Pearl's silent offer of sympathy, but the child's work has nevertheless begun. It will be finished at the scene of Dimmesdale's death, where "Pearl's errand as a messenger of anguish was all fulfilled" (256) [151]. Hester's tomb-like heart becomes a scene of resurrection when she allows her sorrows to stir again and become a conduit to other sorrows than her own.

In accepting her link with the human community, moreover, Hester embodies Winthrop's connection between the affections and the social order. Covenants, that is, can only be upheld when the bonds of affection unite all members. As Hawthorne explains, a "general sentiment . . . gives law its vitality" (231) [137]. Before her seven years of alienation, Hester had understood that law was deeply rooted in the affections, not wicked and repressive but gentle and firm. Standing upon the scaffold, she recalls the deepest influence over her past life, the "look of heedful and anxious love" in her mother's face, "which, even since her death, had so often laid the impediment of a gentle remonstrance in her daughter's pathway" (58) [42]. Likewise, Hester seems to intuit the good motives of the magistrates; even during these early days of her humiliation, she regards the law with mixed feelings, understanding that "the very law that condemned her—a giant of stern features, but with vigor to support, as well as to annihilate, in his iron arm—had held her up, through the terrible ordeal of her ignominy" (78) [52]. This law seems almost parental, offering Hester mercy and support when it might have called for her death.

But after years of enduring "the ever relentless vigor with which society frowned upon her sin" (83) [55], Hester comes to confuse official law with the way it takes root in the popular imagination. On the first day of her punishment, these two sources of discipline are not the same. "Amongst any other population," the narrator explains, "the grim rigidity that petrified the bearded physiognomies of these good people . . . could have betokened nothing short of the anticipated execution of some noted culprit, on whom the sentence of a legal tribunal had but confirmed the verdict of public sentiment" (49) [36]. The point is easy to miss: the spectacle will not be an execution, but, just as significantly, the legal tribunal has refrained from confirming the verdict of the public. Rather, as we have seen, the magistrates have been much more gentle, almost parental in their censure.

By conflating these two sources of penance and dismissing abstract law as mere vindictiveness, Hester has fully alienated herself from her community and rejected law altogether. Hester's treatment of Pearl indicates how unlike her own mother she has become. Though "she early sought to impose a tender, but strict, control over the infant immortality that was committed to her charge," she eventually learns that the lawless Pearl will not be governed (91–92) [59]. Pearl's disorderly nature is only one reason why Hester relinquishes her motherly control of the child; for, in her long alienation from human society, Hester has set herself free from the forms of human restraint. "For years past she had looked from this estranged point of view at human institutions, and whatever priests or legislators had established; criticizing all with hardly more reverence than the Indian would feel for the clerical band, the judicial robe, the pillory, the gallows, the fireside, or the church" (199) [120].

The narrator, as we have seen, shares some of Hester's ambivalence about the men who judge her and the peculiar penalty they have assigned. Almost literally calling upon her to bare her soul, requiring her to write her sin upon her bosom, they have, as Dimmesdale says of the physician, "violated, in cold blood, the sanctity of a human heart" (195) [118]. Describing the effect of the pillory, the narrator steps forth with vehement condemnation of an unnatural punishment: "There can be no outrage, methinks, against our common nature,—whatever be the delinquencies of the individual,—no outrage more flagrant than to forbid the culprit to hide his face for shame; as it was the essence of this punishment to do" (55) [40]. Though Hester is spared this particular torment, her punishment is very similar: she cannot hide the mark of her sin, but must become "the general symbol at which the preacher and moralist might point" (79) [53].

Both Hester and the narrator shift between condemnation of the Puritan leaders and a grudging respect for their piety. These shifting positions are also more objectively apparent in the relative tolerance of the magistrates and elders in the opening chapters, as opposed to their "iron framework of reasoning" that prevents them from forgiving Hester as fully as her townspeople do (162) [100]. In the actions of his Puritan characters Hawthorne demonstrates his own complicated attitudes toward the Puritan elders. As Joseph Alkana and Lawrence Buell have persuasively demonstrated, these shifting attitudes toward official Puritan culture coincide with the conventional and ongoing tension among historians of Hawthorne's generation between filiopiety and criticism of the Puritan founding fathers. The unstable attitudes of Hester's neighbors also reflect an antebellum ambivalence toward an earlier generation, whose moral firmness must have seemed to Hawthorne at times admirable, at times disappointing.

Hawthorne's narrator does not oppose the magistrates or impugn their character. His criticism has only to do with the sanctity of a soul and his horror at seeing it exposed. Even as the chronicler of this tragic tale, he often discreetly turns aside from private details. While Dimmesdale contemplates his escape from New England, the narrator admits to an act of narrative omission: "The struggle, if there were one, need not be described. Let it suffice, that the clergyman resolved to flee, and not alone" (201) [121]. At the climax of the novel, while he relays the many accounts of Dimmesdale's dying revelation, he refrains from giving us the true one, explaining that "it were irreverent to describe that revelation" (255) [151].

As if to correct Hester's mistakes, the narrator reiterates the forgotten message of charity. Even toward Roger Chillingworth, he claims he "would fain be merciful" (260) [153], and so he consigns the leech to a benevolent afterlife, where he and the minister "may, unawares, have found their earthly stock of hatred and antipathy transmuted into golden love" (261) [153].

In this image we again hear the words of John Winthrop resonating through Hawthorne's romance, long after we have forgotten the man himself. The narrator helps Hester to transmute her own anger into golden love, tempering her irreverent judgment of the New England authorities with his own description. The magistrates "had fortitude and self-reliance, and, in time of difficulty or peril, stood up for the welfare of the state like a line of cliffs against a tempestuous tide" (238) [141]. Like Hester's anxious mother, the statesmen wish mainly to protect their charges. Although they have not arranged the election and invited the people "on purpose to be happy," neither do the stern rulers wish to impose sorrow on the world. As a sign that Hester's antipathy has also been transmuted into golden love, she once more joins her community and abides by its laws. For Winthrop as for this narrator whom he so deeply influenced, love and law are one and the same.

With the death of John Winthrop, as we have seen, charity all but disappeared from the official policy of New England. But Hawthorne restored Winthrop's most prized sentiments to the New England of his romance and of his antebellum reader. Though mercy and sympathy are not fully developed until the end of the novel, when Hester Prynne returns to New England and assumes the scarlet letter "of her own free will" (263) [154], the seeds of these virtues are first sown at the scaffold on the night of Winthrop's death, when Dimmesdale enacts Hester's penance and Hester first recognizes the extent of his.

Just as the citizens in the opening scenes articulate the public debates over mercy and judgment that have consumed the political leaders, in the closing scene Hawthorne grants to Hester, a woman estranged from the court, the church, and the meeting-house, the power to realize Winthrop's ideal of charity. When she returns to New England, it is to a "real life" (262) [154], not to the "labyrinth of mind" that had been her home for seven years. Her charity, like her life, also becomes real; it springs from common experiences and true sympathy for "the continually recurring trials of wounded, wasted, wronged, misplaced, or erring and sinful passion" (263) [154–55]. No longer ridiculed by those she tries to help, she is instead sought out and finally understood by those who value her sympathetic counsel as coming from "one who had herself gone through a mighty trouble" (263) [154].

In this "Sister of Mercy," Hawthorne offers his nineteenth-century readers a modern model of the politically specific charity that had gone out of circulation with the death of John Winthrop in 1649. By the end of the romance, all of the major characters have learned the lesson of charity. We are given hope that Dimmesdale and Chillingworth, each "dependent for the food of his affections and spiritual life upon [the] other" (260) [153], are bound by the ligaments of "golden love" (261) [153]. As for Pearl, the scene of her father's death has "developed all her sympathies" (256) [151]. The unnamed women of the village, once represented by the most pitiless of all the Puritans, now look to Hester's letter with "awe, yet with reverence too" (263) [154]. And Hester's affections (sorrow and penitence) draw her back into the community she once tried to escape.

When Hawthorne railed against the "mob of scribbling women," he was attacking, I suspect, neither literary women nor what we have come to call "sentimental novels," if by that we mean novels that share a preoccupation with the moral sentiments, particularly sympathy. Clearly *The Scarlet Letter* is one such sentimental novel. Instead, Hawthorne's impatience was directed not only at the literary market but also at the peculiar nineteenth-century American manifestation of sympathy, a vague, free-floating sentimentalism that encouraged inconsequential expressions of feeling. Hawthorne considered moral sentiments deeply significant. If sympathy helped to bring human beings together, then communities governed by sympathy, he believed, were not oppressive, and laws motivated by love should be gratefully upheld.

Both Rufus Choate and John Winthrop, arguably the only early Americans who inspired Hawthorne with unqualified filiopiety, would have approved of Hester's return to New England. In restoring this most alienated Puritan to the land that had made her a story and a byword, and in giving her the task of counseling other disaffected members and thereby helping them to accept their burdens, Hawthorne expressed his hope that, to recall Choate, "overrated diversities of interest" might be overlooked in favor of a larger social good, the "constitutional sentiment of old, family, fraternal regard." In so doing he countered both versions of sentiment prevalent in his own age: the generalized and effete sentimental response to all scenes of pathos, and the politically contentious sympathies often stridently aligned with moral and political absolutes. Rewriting the history of New England, Hawthorne corrected what he saw as the errors of the past, granting the lowliest of our ancestors the task of internalizing and dispersing, rather than losing, the greatest of Winthrop's political contributions: the understanding that law derives its fortitude from general sentiments.

WORKS CITED

Alkana, Joseph. *The Social Self: Hawthorne, Howells, William James, and Nineteenth-Century Psychology.* Lexington: UP of Kentucky, 1977.

Anderson, Douglas: "Jefferson, Hawthorne, and 'The Custom-House.'" *Nineteenth-Century Literature* 46 (1991): 309–26.

Baughman, Ernest W. "Public Confession and *The Scarlet Letter.*" *New England Quarterly* 40 (1967): 532–50.

Bell, Michael Davitt. *Hawthorne and the Historical Romance of New England.* Princeton: Princeton UP, 1971.

Bercovitch, Sacvan. *The Office of* The Scarlet Letter. Baltimore: Johns Hopkins UP, 1991.

Berlant, Lauren. *The Anatomy of National Fantasy: Hawthorne, Utopia, and Everyday Life.* Chicago: U of Chicago P, 1991.

Buell, Lawrence. *New England Literary Culture: From Revolution Through Renaissance.* New York: Cambridge UP, 1986.

Choate, Rufus. "The Importance of Illustrating New-England History by a Series of Romances like the Waverly Novels." In *Addresses and Orations of Rufus Choate.* Boston: Little, Brown, 1878. 1–39.

Dawson, Edward. *Hawthorne's Knowledge and Use of New England History: A Study of Sources.* Nashville: Vanderbilt UP, 1939.

Dawson, Hugh J. "Hester Prynne, William Hathorne, and the Bay Colony Adultery Laws of 1641–42." *ESQ* 32 (1986): 225–31.

Erickson, Kai T. *Wayward Puritans: A Study in the Sociology of Deviance.* New York: Wiley, 1966.

Gilmore, Michael T. "Hawthorne and the Making of the Middle Class." *Discovering Difference: Contemporary Essays in American Culture.* Ed. Christoph Lohmann. Bloomington: Indiana UP, 1993. 88–104.

Hunt, Lester. "*The Scarlet Letter*: Hawthorne's Theory of Moral Sentiments." *Philosophy and Literature* 8 (1984): 78–88.

Hutchinson, Thomas. *The History of the Colony and Province of Massachusetts Bay.* 1764. Ed. Lawrence Shaw Mayo. 3 vols. Cambridge: Harvard UP, 1936.

Johnson, Edward. *Johnson's Wonder-Working Providence 1628–1651.* 1653. Ed J. Franklin Jameson. New York: Scribner's, 1910.
Kesselring, Marion L. *Hawthorne's Reading, 1828–1850.* New York: New York Public Library, 1949.
Leavis, Q. D. "Hawthorne as Poet." *Sewanee Review* 59 (1951): 426–58.
Leverenz, David. "Mrs. Hawthorne's Headache: Reading *The Scarlet Letter.*" *Nineteenth-Century Fiction* 37 (1983): 552–573.
Masur, Louis P., ed. *"The Real War Will Never Get in the Books": Selections from Writers during the Civil War.* New York: Oxford UP, 1993.
Mather, Cotton. *Magnalia Christi Americana, Books I and II.* Ed. Kenneth B. Murdock. Cambridge: Harvard UP, 1977.
Moseley, James G. *John Winthrop's World: History as a Story, the Story as History.* Madison: U of Wisconsin P, 1992.
Murrin, John M. "Magistrates, Sinners, and a Precarious Liberty: Trial by Jury in Seventeenth-Century New England." In *Saints and Revolutionaries: Essays on Early American History.* Ed. David D. Hall, John M. Murrin, and Thad Tate. New York: W. W. Norton and Company, 1984. 152–206.
Railton, Stephen. "The Address of *The Scarlet Letter.*" *Readers in History: Nineteenth-Century American Literature and the Contexts of Response.* Ed. James L. Machor. Baltimore: Johns Hopkins UP, 1993. 138–63.
Reynolds, David S. *Beneath the American Renaissance: The Subversive Imagination in the Age of Emerson and Whitman.* Cambridge: Harvard UP, 1988.
Reynolds, Larry J. "*The Scarlet Letter* and Revolutions Abroad." *American Literature* 57.1 (March 1985): 44–67. Pp. 484–99 in this Norton Critical Edition.
Ryskamp, Charles. "The New England Sources of *The Scarlet Letter.*" *American Literature* 31 (1959): 257–72. Pp. 279–90 in this Norton Critical Edition.
Van Leer, David. "Hester's Labyrinth: Transcendental Rhetoric in Puritan Boston." *New Essays on* The Scarlet Letter, Ed. Michael J. Colacurcio. New York: Cambridge UP, 1985. 57–100.
Wall, Robert Emmet, Jr. *Massachusetts Bay: The Crucial Decade, 1640–1650.* New Haven: Yale UP, 1972.
Williams, Daniel E. "Victims of Narrative Seduction: The Literary Translations of Elizabeth (and 'Miss Harriot') Wilson." *Early American Literature* 28 (1993): 148–70.
Winthrop, John. "Arbitrary Government, Described." 1645. In *Life and Letters of John Winthrop.* Boston: Ticknor and Fields, 1867. 440–59.
———. *The Journal of John Winthrop, 1630–1649.* Ed. Richard S. Dunn, James Savage, and Laetitia Yeandle. Cambridge: Harvard UP, 1996.
———. "A Model of Christian Charity." In *Winthrop Papers.* New York: Russell and Russell, 1968–. 2: 282–95.

LAURA HANFT KOROBKIN

The Scarlet Letter of the Law: Hawthorne and Criminal Justice†

I

In the scene of public witnessing that begins *The Scarlet Letter*, the Governor, magistrates, and elders look down from their balcony to the platform where Hester Prynne stands in proud shame, raised in turn above the grim faces of the milling crowd. The ground-level voices we hear express a resentment quite appropriate to a townspeople both beneath and outside the nexus of unassailable power represented by that balcony, "the place whence proclamations were wont to be made" (64) [45]. Though they hold varying opinions of the letter-wearing sentence, in one thing the townspeople's comments are consistent: not they but the magistrates have had the sole power and authority to deal with Hester Prynne. "This woman has brought shame upon us all and ought to die," rants one woman. "Is there not law for it? Truly there is, both in the Scripture and the statute-book. Then let the magistrates, who have made it of no effect, thank themselves if their own wives and daughters go astray!" (51–52) [38].

In Hawthorne's Puritan world, the only decision-makers standing between Hester and the gallows are the all-powerful magistrates. Their word is law, their discretion untrammeled. If the colony has a fully developed criminal justice system—grand juries returning indictments, juries assessing trial testimony and returning verdicts, pre-determined criminal penalties governing the sentencing of offenders—we don't hear about it. Instead, the entire apparatus of the Puritan Rule of Law in *The Scarlet Letter* is signified by this small group of powerful men, accountable apparently to none but themselves and their God. The first three chapters create the clear impression that the townspeople are wholly excluded from any decision-making role in dealing with crime.[1] The conflation of religious, political, legislative, and judicial power in Hawthorne's early New England is total: the monolith rules and sentences. The people may mutter, but they must also unhesitatingly obey.

As Hawthorne well knew, however, the legal and judicial authority of the magistrates, particularly during the tumultuous decade in which he situates his tale, was anything but unassailable.[2] The colony's criminal justice system had been intentionally structured to counterpoint magistratical power

† From *NOVEL: A Forum on Fiction* 30.2 (Winter 1997): 193–217. Republished by permission of the copyright holder, and the present publisher, Duke University Press. www.dukeupress.edu. Page numbers in square brackets refer to this Norton Critical Edition.

1. Virtually all of the townspeople who comment on Hester's punishment note, in one form or another, "the worshipful magistrates" who have "awarded" Hester's sentence. One says the magistrates are "God-fearing" but "merciful overmuch," while another wishes that they, not the magistrates, had been in charge of determining the penalty (51) [37]. Every comment testifies to the magistrates' power; none mentions a trial.
2. Hawthorne's scholarly familiarity is attested to not only by the historical details used throughout his Puritan fiction, but by existing library records of the Salem Athenaeum, listing the source books he borrowed. See Ryskamp 257–58, Kesselring *passim*, and Colacurcio, "Footsteps" 462 n6. On the persistent efforts of the colonists to circumscribe magistratical power throughout this period, see Haskins 29–42.

with the peer-group power of grand and petit juries and elected legislators who were not magistrates. And, wherever magistrates nevertheless attempted to exert hegemonic power, their authority was under persistent siege. Juries brought in verdicts which frustrated and undermined magistrates' judicial authority. Agitators for legislative reform worked to set prescribed penalties for crimes with the explicit goal of circumscribing the magistrates' discretionary sentencing authority. In short, the historical magistrates were neither as powerful nor as awe-inspiring as their fictional counterparts.

The wealth of excellent recent scholarship focusing on Hawthorne's Puritan sources has explicated a host of issues underlying the Puritan world of Hawthorne's fiction; the religion, sexuality, and politics of prominent Puritans Anne Hutchinson, Richard Bellingham, and John Winthrop have all been persuasively detailed (Colacurcio, *Essays;* Arac; Bercovitch; Berlant). But what has not been adequately explored in this most legal of novels is Hawthorne's ahistorical imaging of the machinery of Puritan criminal law. At the same time that the novel seems obsessed with crime and punishment, it avoids—indeed erases—the institutions and procedures that constitute public criminal process. This essay applies a process of double historicization to the novel's exploration of the relation of crime and law to the private individual. The first level of inquiry compares case histories, statutes, and legal disputes of the 1640s to Hawthorne's fictional Boston, to locate points of significant variance. Hawthorne's occasional inaccuracies have been the subject of critical discussion for more than thirty years; as recently as 1988 Michael Colacurcio declared that the novel's "major historical fabrication" was casting Bellingham rather than Winthrop as governor in June of 1642 (110). I argue that a great deal more was changed or eliminated, and that the text consequently bears the traces of a series of historical figures it struggles to suppress. These include, most significantly, the shadows of three: the townsman juror, the whipped woman, and the political malcontent.

Focusing historical analysis on what Hawthorne's text omits will require a reevaluation of the novel's traditional villains, the "grim-visag'd" magistrates whose sentence binds Hester to the red letter of the law. Hawthorne's narrator characterizes them as rigid, severe, and frostbitten, incapable of judging a woman's heart; surprisingly, critics who specialize in exposing this narrator's ironic doublespeak in other areas have accepted such epithets.[3] Measured against their historical counterparts, however, Hawthorne's magistrates emerge as distinctly progressive; if they are more autocratic, they exercise their power with compassion and restraint.

If Puritan history makes Hawthorne's historical manipulations visible, the politics of 1850 make them comprehensible. My focus here is on two law-related issues *The Scarlet Letter* profoundly meditates: First, what obligations do individuals have to obey laws regulating private behavior, laws that directly conflict with individuals' deeply held principles, and which they have had no hand in making? And second, must submission to

3. See, e.g., Shulman, who argues that Hawthorne placed Hester in a "joyless, punitive society" run by men who are "capable but insensitive, heartless" (86–88), or Berlant, who characterizes Bellingham in the child custody scene as showing "symptomatic Puritan inability to adjudicate properly, due to his oversaturation with juridical consciousness" (78).

such a law be viewed as integrity-destroying cowardice, or can it be understood as courageous and even beneficial to the individual? These questions sounded with particular resonance in 1850, when the legitimacy of public law and criminal process was a matter of intense national debate. In the same month that *The Scarlet Letter* appeared, Senator Daniel Webster irrevocably destroyed his almost mythic reputation by arguing that the Compromise Act of 1850 was a reasonable means to preserve the nation and should be adopted. Abolitionists thundered against him from newspapers, pulpits, and platforms. Citizens were urged to defy the Compromise's Fugitive Slave Law provisions, with violence if necessary.

As Jonathan Arac and Sacvan Bercovitch have shown, *The Scarlet Letter* was written in a time of intense concern about the Fugitive Slave Laws. Like Arac and Bercovitch, I read *The Scarlet Letter* as a response to the political anxieties of 1850. My focus, however, is not (or not exclusively) on Hawthorne's attitude toward slavery or the necessity for broad political compromise, but on the specific crisis faced by Northerners who had to decide whether or not to cooperate with the provisions of the Fugitive Slave Law. That law, passed by federal legislators in Washington, placed all citizens under affirmative legal obligations not only to permit but to assist in the capture and return of fugitive slaves. Significantly, the federal Compromise Act of 1850 conflicted with, and overturned, state-level protections for runaway slaves that had been provided for by such laws as the Massachusetts Personal Liberty Acts of 1843 (Schwartz 191). Throughout most of the 1840s these laws made it virtually impossible to recover a runaway slave in many Northeastern states; after 1850, federal law would force New Englanders to become passive witnesses to the capture, imprisonment, and rendition of the runaways in their midst. Abolitionist fervor intensified in response. The battle cry of violent resistance in the name of a Higher Law was preached to enthusiastic crowds by the Boston Vigilance Committee, which also undertook to rescue imprisoned fugitives. From the pulpit, abolitionist minister Theodore Parker "adjured" his parishioners to "reverence a government that is right, statutes that are right . . . but to disobey every thing that is wrong" in the name of the "higher law," the "law of God" (40).

Recent articles by Jean Fagan Yellin and Jennifer Fleischner have productively explored Hawthorne's refusal to represent slavery in his works and documented his belief that the enslavement of African-Americans was in general neither outrageous nor inhumane. As Yellin has shown, the sufferings and dilemmas that interested Hawthorne were those of whites, not blacks. In this context, I will argue that in designing Hester's punishment, Hawthorne eliminated whatever would have suggested a resemblance between her situation and that of the slave. As a figure for white northerners, Hester's behavior becomes all the more controversial. To a member of the Vigilance Committee, Hester's outward submission to the strictures of Puritan law might well appear a shameful knuckling under, the kind of failure of the will that buys safety at the price of personal integrity. Yet Hawthorne convinces us that Hester's behavioral acquiescence is both truly heroic and intellectually liberating. With Hester as a model, the novel suggests, readers may find the courage not to rebel but to forego rebellion. In the 1838 speech that set the pro-law terms of the debate over obedience to law, Lincoln had warned that if "the laws be continually despised and disregarded" by "mobs" who take it on themselves to judge the acceptability of each law, order, reason and the Government itself must eventually fall ("Address" 22). Yet it was precisely the choice to submit to law rather than to follow the dictates of private conscience that Parker, Thoreau and others found absolutely unacceptable.

After exploring this historical context, the final section of this essay will argue that the debate over obedience to law is crucial to reading *The Scarlet Letter*. By entering such debates through the medium of fiction, a novel can significantly influence readers' responses to the challenges presented by contemporary events without requiring them to read the novel in explicitly political terms. Where contemporary speeches and editorials impelled readers to respond by adopting or refuting the specific position argued in the text, Hester Prynne's hard-won freedom and serenity could elicit admiration and a desire for emulation even from readers who would ordinarily have resisted the novel's underlying attitude toward law and obedience. This is not to reduce the novel to the status of an essay or a narrow political allegory; instead, it is precisely the literary qualities of evocativeness, complexity, and imaginative richness that make its impact on these law-related issues possible, just as those same qualities have produced an extraordinary range of "purely literary" interpretations. But if fiction's fundamental openness to interpretation distinguishes it from most forms of discursive advocacy, I will also argue that it makes it *like* rather than unlike most forms of law, which also depend on a process of continuing interpretation for their power. Though law is often imaged as fixed and rigid in contrast to literature's interpretive fluidity, I will argue that *The Scarlet Letter*'s contribution to the American cultural dialogue about law and obedience owes as much to its narrative specificity, boundedness, and closure as to its interpretive ambiguities.

II

Explaining Hester's appearance on the scaffold to a just-arrived Roger Chillingworth, a townsman says that

> our Massachusetts magistracy, bethinking themselves that this woman is youthful and fair, and doubtless was strongly tempted to her fall;—and that, moreover, as is most likely, her husband may be at the bottom of the sea;—they have not been bold to put in force the extremity of our righteous law against her. The penalty thereof is death. But, in their great mercy and tenderness of heart, they have doomed Mistress Prynne to stand only a space of three hours on the platform of the pillory, and then and thereafter, for the remainder of her natural life, to wear a mark of shame upon her bosom. (62–63) [44]

While Hawthorne's fictional magistrates *are* the criminal justice system, their historical counterparts' authority was significantly limited by the colonial jury's verdict-making powers, on one hand, and the colony's prescribed criminal procedures, on the other. A person accused of crime was not a passive or disempowered object to whom law was "done." Rather, under the 1641 Body of Liberties, a hard-won definition of the civil rights of colonists, the defendant enjoyed a series of important rights, including the right to choose a jury or bench trial, to challenge jurors for cause, to have another person speak on her behalf, to a speedy trial at which written records were made, and to appeal the verdict to a higher court (see paragraphs 29, 30, 36, 41, 64 of the Body of Liberties in Powers's appendix [533–48]). The range of available punishments, and the circumstances in which they could be imposed were also regulated by the Body of Liberties, which guaranteed in its first paragraph that "no mans life shall be taken away, no mans honour or good name shall be stayned, no mans person shall be arrested, restrayned, banished, dismembred, nor any wayes punished . . . unlesse it be by vertue or equitie of some expresse law of the Country waranting the same," and in its forty-sixth outlawed punishments "that are inhumane, Barbarous or cruell." A person accused of adultery would have been tried by a jury of twelve freemen in a trial court where magistrates sat as judges. The jury would have decided, in private, not only the general question whether to convict or acquit but also, because the conviction must be for a specific crime, what crime had been provably committed. Only the return of the verdict triggered the judge's sentencing powers (Konig 33, Haskins 32–36).

Death was the mandatory penalty for adultery in the Massachusetts Bay Colony in 1642, though it was not then a capital crime either in England or in the colony at Plymouth (Powers 78, 261–62, 300). Once the jury returned a verdict for adultery, the magistrate-judges had no discretionary authority to demonstrate "tenderness and mercy" by imposing a lesser penalty. Conversely, if the jury chose to return a verdict for a lesser sexual offense such as "lewd and lascivious behavior," the death penalty could not be imposed.

The power of juries to control sentencing by determining verdicts was frequently commented on at the time. In John Winthrop's Journal, recently argued to be not only Hawthorne's "prime and obvious source" for *The Scarlet Letter*, but to have furnished "the novel's most essential themes" as well (Colacurcio, "Woman's" 103), Winthrop describes a "sad business" that "fell out" in the spring of 1645, in which a churchgoing young man went to England, leaving his wife in the care of another "pious" and "sincere" member,

> who in time grew over familiar with his master's wife, (a young woman no member of the church) so as she would be with him oft in his chamber, etc. and one night two of the servants, being up, perceived him to go up into their dame's chamber, which coming to the magistrate's knowledge, they were both sent for and examined . . . and confessed not only that he was in her chamber with her in such a suspicious manner, but also that he was in bed with her, but both denied any carnal knowledge, and being tried by a jury upon their lives by our law, which makes adultery death, the jury acquitted them of adultery, but found them guilty of adulterous behavior. This was much against the minds of many, both of the magistrates and elders, who judged them worthy of death . . . [A]ll that the evidence could evince was but suspicion of adultery, but neither God's law nor ours doth make suspicion of adultery (though never so strong) to be death; whereupon the case seeming doubtful to the jury, they judged it safest in case of life to find as they did. So the court adjudged them to stand upon the ladder at the place of execution with halters about their necks one hour, and then to be whipped or each of them to pay 20 pounds. (2: 305–06)

Clearly, the locus of power in this case is the jury, not the magistrates; unless the jury convicts, the magistrates cannot execute, however much they may deem an offender "worthy of death." Where the jury in this 1645 case used the "expresse law" requirement to avoid the death penalty many magistrates would have preferred, a jury three years earlier had used its verdict to require an execution the magistrates would gladly have avoided. In 1642, the same year Hester received her red A, Mary Latham and James Britton became the only persons executed for adultery in the history of the Massachusetts Bay Colony. Describing the case in his journal, Winthrop noted that "some of the magistrates thought the evidence not sufficient against her, because there were not two direct witnesses, but the jury cast her, and then she confessed the fact, and accused twelve others, whereof two were married men" (2: 190–91). Britton petitioned the general court for his life, "but they would not grant it, though some of the magistrates spake much for it, and questioned the letter, whether adultery was death by God's law now" (2: 191) No discretion was available once a jury verdict for adultery had been returned. Latham and Britton were hanged.

Did Hester have a trial? If she chose a jury rather than a bench trial, of what specific crime did the jurors convict her? Did the grim crowd that watched her mount the scaffold include jurors whose votes helped put her there? And, perhaps most significantly, why did Hawthorne invent a criminal case in which the determination of guilt or innocence would have been made by a jury and then carefully construct the impression that the magistrates acted as a law unto themselves?

For one thing, we should note that the execution of Britton and Latham was by no means characteristic of Puritan justice. Although indictments on capital offenses were fairly common, convictions were rare.[4] The Body of Liberties specified twelve capital crimes; yet, as Edwin Powers puts it, "time after time, juries refused to bring in verdicts that might have led to the scaffold" (279). Instead, they used their verdict-controlling power as modern juries do, to bring in verdicts for lesser-included offenses when death seemed too extreme a punishment. The Records of the Court of Assistants, the court of general jurisdiction of the Bay Colony, reflect numerous cases in which juries effectively circumscribed judges' sentencing powers, eliminating possible infliction of the death penalty. In cases charging adultery, verdicts were returned reading "not legally guilty but guilty of very filthy carriage, etc.," "not guilty according to Indictment but found him Guilty of vile, filthy and abominably libidinous Actions," and "not legally Guilty according to Indictment but doe find hir Guilty of Prostituting hir body to him to Committ Adultery" (Powers 103, 279).

The real power to determine sentencing in adultery cases thus often lay with the jury, who rarely used it to its harshest capacity. Yet Hawthorne's

4. This clemency did not, unfortunately, extend to cases for witchcraft, as the fate of Mistress Hibbins, the sister of Governor Bellingham who invites Hester to a witches' meeting in the forest, demonstrates. Her case is described in Caleb Snow's *History of Boston*, which Ryskamp has shown was Hawthorne's primary source for historical details. As Snow tells it, "the jury brought her in guilty (of witchcraft), but the magistrates refused to accept the verdict; so the cause came to the general court, where the popular clamour prevailed against her, and the miserable old lady was condemned and executed in June 1656" (qtd. in Ryskamp 267). As this account shows, where conflicts occurred between juries and magistrates the "popular clamour" of those at the juror level could win out, even if it meant executing the governor's sister. Such power in the populace contrasts sharply with the impotence of those who "clamour" against Hester in the novel's opening scene.

townspeople suggest that whipping, branding, or death would have been more appropriate punishment for Hester's crime. Clearly, she is better off in the magistrates' hands than left to the townspeople's mercy. While Hawthorne's narrator suggests that "out of the whole human family" it would be difficult to find persons "less capable of sitting in judgment on an erring woman's heart" than the magistrates (64) [45], he situates Hester, with characteristic irony, above a throng of neighbors whose judgment would have been far harsher and more vengeful. Erasing the Puritan jury not only makes the magistrates the sole source of judgment, it also increases the ambit of their sentencing discretion. This helps explain why the "expresse law" for which Hester is convicted is never mentioned, and why the word "adultery" never appears in the text. Hester is apparently convicted of adultery, a capital crime, but she receives a sentence which could only have been imposed for a much less serious infraction. By implying that the magistrates' range of options included these much harsher punishments, Hawthorne both inflates the magistrates' powers and highlights their compassionate consideration for Hester's circumstances.

This reallocation of power has three important effects. As already discussed, it enlarges and concentrates the magistrates' power over the townspeople while putting in their hands enormous discretionary authority. This erases any semblance of participatory or democratic government and replaces it with an image of authoritarian oligarchy. Second, it exacerbates the distance, in terms of class and power, between the lone woman on the scaffold and those who decide her fate. Finally, it disconnects the townspeople from any involvement in or responsibility for judging and sentencing Hester. Because I will explore the implications of the first two later in this essay, I want to look now at the third point, the non-participatory status of the witnessing bystanders in the opening scene.

To have Hester's guilt determined by a jury of her peers would suggest that the power to judge crime and assign penalties is not limited to those on high but shared or mediated among the community as a whole. And, of course, sharing the power to judge is precisely what juries are all about. As Alexis de Tocqueville noted in his brief paean to the American jury, the jury system

> places the real direction of society in the hands of the governed, or of a portion of the governed, and not in that of the government. . . . The true sanction of political laws is to be found in penal legislation; and if that sanction is wanting, the law will sooner or later lose its cogency. He who punishes the criminal is therefore the real master of society. Now the institution of the jury raises the people itself, or at least a class of citizens, to the bench of judges. The institution of the jury consequently invests the people, or that class of citizens, with the direction of society. (1: 293)

Hawthorne's deliberate elimination of the jury is, in de Tocqueville's terms, an inversion designed to make the magistrates rather than the townspeople "the real master of society." Hawthorne's irony in dubbing one opinionated townswoman a "self-constituted judge" is thus double: her presumption in taking on the magistratical job of judging a neighbor's guilt is an example of comic overreaching as much because she is an ordinary citizen as because she is a woman.

Hawthorne's townspeople judge—but their judgments trigger no consequences. No matter how extreme a punishment they call for, they are not responsible for any sufferings that result. Because their "verdicts" float free from liability for the burdens they impose, they can judge and, more importantly, rejudge Hester, until the harshness of those early judgments melts into later admiration for her modesty and usefulness. By the novel's end, the once reviled adulteress has become a woman respected for her wisdom as well as her sufferings, as those who once reviled her (or their daughters) now seek out her advice.

What I am suggesting is that the townspeople's freedom to interpret Hester is enabled by their position outside the concentration of power up on that balcony. Indeed, the novel insists in a variety of ways that mental development is always fullest and freest when liberated from the crushing weight of behavior's consequences. Readers raised on *The Plague* and *Man's Fate*—or for that matter, on *Middlemarch*—may find paradoxical at best the suggestion that personal development is fully possible only when the thinker is not required to act in consistency with his beliefs. But for Hester, as well as for the townspeople, it is precisely the unlinking of mental freedom from any obligation toward consonant action that makes her intellectual growth possible. In "Another View of Hester" we learn that in the isolation of her lonely cottage Hester's extraordinary "freedom of speculation" has led her to believe that "the whole system of society is to be torn down, and built up anew" (164–65) [101, 102]. Yet Hester's conclusion that "the world's law was no law for her mind" is antinomian only in the theoretical sense. An Anne Hutchinson of the mind alone, Hester does not behave as if she believes that the world's law is no law for her body. She submits. She counsels young women to submit. Whatever she may be feeling, she walks through the streets of Boston with her head modestly down, and her outward obedience buys her the unassaulted privacy in which to continue thinking. She hopes that some bright future time will resolve the world's inequities, but in the here and now her freedom exists in the realm of thought, not action.[5]

There is an important connection between Hester's restricted behavior and her unrestricted thoughts. Her position outside the web of normal social obligation and activity permits her to see and judge society whole. Also, and perhaps more significantly, Hawthorne implies that it is precisely because she need not think about practical action that her mental explorations can be so far-reaching. The energies dammed up by her life of enforced outward conformity can flow freely only in this interior realm, where their penetration of unexplored terrain will not be cut off by the

5. In this context, it is important to note that Hester's one moment of rebellion—the plan to leave the colony with Arthur Dimmesdale—is presented within the novel as not simply doomed but wrongful. Just as Dimmesdale's presence has been a primary reason for Hester's remaining in Boston, so his failing health and spirits under Chillingworth's predatory attentions convince her to leave. Motivated by his personal weakness, characterized by secrecy, fear, and deception, the proposed flight is not a courageous challenge to authority but an attempt to run away from its adverse effects. While we admire Hester's strength and compassion, and yearn, on her behalf, for some means of escape from the pressures of community, law, and responsibility, everything in the novel suggests that such flight is neither viable nor admirable. Because even private sexual misconduct is a violation of community norms and laws, expiation and reintegration can only be achieved through a public process that involves both wrongdoer and the community in which the wrong was committed.

confrontations and compromises that would inevitably ensue were they to be translated into radical action. While Hawthorne suggests that in different circumstances Hester might have been a prophetess or a revolutionary, it is quite clear that in Hester's world, the satisfactions of philosophical exploration are made available through her life of privation and are one of its few compensations. Like the widely various juridical opinions of the milling crowd beneath the scaffold, Hester's philosophical and historical conclusions can be radical, even revolutionary, precisely because they are not muddied by the messy and corrupting process of attempting change in the real world.[6] It is significant too that outward obedience is all that the Puritan rule of law requires; its justice system punishes only acts, not thoughts. Because community stability depends on each member's self-restraint, Hester's conformity to behavioral expectations helps hold the community together even when her thoughts may be at their bitterest. If protecting society from disruptive assaults is a paramount goal, then such a system helps preserve order, while leaving each individual the sanctum of his or her own mind. Whether that very limited freedom is enough is of course quite another question.

III

If Hawthorne's elimination of the Puritan jury suggests that he wanted his magistrates to be more powerful than their historical counterparts, his handling of Hester's sentence suggests that he also wanted them to be more responsive. Let us return briefly to Puritan legal history. In 1641, the general court of the Colony of New Plymouth sentenced Thomas Bray and Anne, the wife of Francis Linceford, for the "crime of adultery and uncleanesse," to which both had confessed publicly. Much of the court's sentence will be familiar:

> [T]he Court doth censure them as followeth: that they be severely whipt immediately at the publik post, & that they shall weare (whilst they remayne in the government) two letters, viz. an AD, for Adulterers, daily, upon the outeside of their uppermost garment, in a most emenent place thereof; and if they shalbe found at any tyme in any towne or place within the government without them so worne upon their uppermost garment as aforesaid, that the constable of the towne or place shall take them, or either of them omitting so to weare the said two letters, and shall forthwith whip them for their negligence and shall cause them to be immediately put on againe, and so worne by them and either of them; and also that they shalbe both whipt at Yarmouth, publikely, where the offence was committed, in such fitt season as shalbe thought meete by Mr. Edmond Freeman & such others as are authorized for the keepeing of the Courts in these partes. (*Records of Plymouth* 2: 28)

6. Like Bercovitch, I place major emphasis on the moment of Hester's return, though my focus does not read it as the defining moment in the creation of a liberal ideology of compromise. While I agree with Arac's argument that the novel "is propaganda—*not* to change your life," I argue that it offers the rule of law rather than simple "character" as the central enabling and positive value. Like Fleischner, I argue that Hawthorne sees dissociation from public political action as privately beneficial; I want to suggest, however, that Hester represents the situation of the ordinary reader rather than, as Fleischner suggests, the artist.

The similarities between Anne Linceford's sentence and Hester Prynne's are obvious. Yet the discovery that Hester's lifetime letter-wearing sentence has a precise historical analogue should not obscure the fundamental difference between the two: like virtually all Puritan sexual offenders, Anne Linceford was publicly stripped and severely whipped, most probably receiving the Biblically allowable maximum of forty lashes. Hester receives no physical punishment. If *The Scarlet Letter* is the painstakingly accurate representation of Puritan life it is often assumed to be, the example of Anne Linceford is troubling. Why did Hawthorne, whose historical research was nit-picking enough to permit him to provide correct street addresses for his historically prominent characters, eliminate what was for the Puritans a fundamental component of all serious punishment? In short, why isn't Hester whipped?

In 1844, Hawthorne noted as a story idea in his journal "the life of a woman, who, by the old colony law, was condemned always to wear the letter A, sewed on her garment, in token of having comitted adultery" (*American* 254) [203]. Two colonial statutes have generally been accepted as Hawthorne's possible source; significantly, each mandates a severe public whipping before the offender begins a life of sartorial humiliation. The penalty not only shames the sinful spirit, but harshly mortifies its fleshly container. The first, a 1658 act of the colony of New Plymouth, provided: "whosoever shall committ Adultery shalbee severely punished by whipping two severall times; . . . and likewise to weare two Capital letters viz. AD cut out in cloth and sewed on theire uppermost garments on their arme or backe; and iff att any time they shalbee taken without the said letter whiles they are in the Government soe worn to bee forthwith taken and publickly whipt" (*Compact* 42–43). The second, a 1694 act described in Joseph Felt's 1827 *Annals of Salem*, provides that adulterers be made "to sit an hour on the gallows, with ropes about their necks, be severely whip't not above 40 stripes and forever after to wear a capital A 2 inches long, cut out of cloth colored differently from their clothes and sewed on the arms or back parts of their garments so as always to be seen when they were about" (317). Whichever statute Hawthorne may have seen, his fascination with the psychological possibilities of the sentence did not extend to its corporal component. Hester is humiliated, first by being made to stand three hours on the scaffold, and then by her lifetime of letter wearing; she is also apparently imprisoned, since Pearl is born in prison, and Hester returns to the prison after her morning exposure. But she is not whipped. Hester's red A, frequently characterized as a happily-vanished instance of Puritan severity, is most noteworthy for its extraordinary *leniency*, its complete avoidance of the physical chastisement so essential to Puritan programs for spiritual correction.

Whipping was the standard colonial punishment for bastardy and other sex crimes. As Powers notes: "a review of the Colonial court records gives one the impression that one who held the sex mores of the times in light esteem would sooner or later have an engagement with the constable at the whipping post" (172). Unless the defendant was rich enough to substitute payment of a fine, a whipping of up to forty lashes might be expected. The *Records of the Court of Assistants* in Boston in 1642, the year Hester mounted her scaffold, contain numerous examples of such punishment. Robert Wyar and John Garland, for instance, for "ravishing" two young girls, "the fact confessed by the girles, & the girles both upon search found to have bin defloured, & filthy dalliance confessed by the boyes," were sentenced to be severely and publicly whipped, the girls to be whipped also, but privately (121). The denouement to the 1645 case from Winthrop's Journal, described earlier, provides a further example. The court sentenced the pair to an hour on the gallows with halters around their necks, and then either to pay 20 pounds, or be whipped. "The husband (although he condemned his wife's immodest behavior, yet) was so confident of her innocency in point of adultery, as he would have paid 20 pounds rather than she should have been whipped; but their estate being but mean, she chose rather to submit to the rest of her punishment than that her husband should suffer so much for her folly. So he received her again, and they lived lovingly together" (2: 258). Even where letter-wearing penalties were inflicted for long periods of time, they began with a whipping.[7] Hawthorne's avoidance of physical punishment is thus stunningly ahistorical, whether considered in light of the typical punishment for lesser sex crimes or as an example of an "old colony law" on adultery.

Whipping was not only the most historically common punishment for adultery and other sex crimes in Boston in 1642, it was also one of the hottest social and literary subjects of Hawthorne's day. As Richard Brodhead has demonstrated in his examination of antebellum America's obsession with the subject, Hawthorne's audience was embroiled in controversy over corporal versus psychological punishment. Psychological punishment, touted as "discipline through love," was associated with progress and reform (70). Originating in family based disciplinary instruction, such methods substituted the infliction of guilt for the infliction of physical pain, thereby internalizing parental authority. Controversies over methods of punishment were played out in every medium, both fictional and journalistic. Throughout the 1840s and 1850s, Brodhead declares, "the picturing of scenes of physical correction emerges as a major form of imaginative activity in America, and arguing the merits of such discipline becomes a major item on the American public agenda" (67). Furors over inflammatory scenes of whipping in novels like *Two Years Before the Mast* (1840) and *Uncle Tom's Cabin* (1852) stimulated book sales enormously.

Why then isn't Hester whipped? Paradoxically, the very topicality of corporal punishment may have prevented Hawthorne from consigning Hester to the whipping post. Brodhead begins his general analysis by establishing that in the ante-bellum period, "whipping *means* slavery . . . and considerable evidence suggests that the more general imaging of such punishment at this time has slavery as its ultimate referent" (68). If we connect this perception about the 1840s to the anomalous absence of a whipping from Hester's 1642 sentence—a juxtaposition Brodhead does not make—we see that to have Hester receive a historically accurate forty lashes would have irrevocably positioned her as a slave surrogate, an object of pity and a spur

7. In 1657, for example, the Pymouth Court again imposed its "old colony law," this time on Katherine Aines, who, "for unclean and lascivious behavior," was sentenced to be twice publicly whipped and then forever after to weare "a Roman B" on her clothes (*Records of Plymouth* 3: 111, Powers 198). In 1639, in Boston, John Davies was sentenced "for grosse offences in attempting lewdness with divers women" to be severely whipped, both in Boston and at Ipswich, and then "to weare the letter V upon his breast upon his uppermost garment untill the Court do discharge him," which it did, in recognition of his good behavior, some six months later (*Records of Court* 2: 81, 87).

to activism. To put it simply, had Hester's punishment begun with a whipping, *The Scarlet Letter* would have been a book about slavery. As a white victim of the type of cruelties associated with slavery, Hester would have made those horrors immediate and accessible. "Hester Prynne" would have been a sign pointing toward an absent and unrepresented character—the American slave—and the novel might have joined the decade's greatest bestseller, *Uncle Tom's Cabin*, as a passionate abolitionist manifesto.

But that was precisely what the novel is structured to prevent. Hawthorne's discomfort with abolitionism is well known; although not a defender of slavery, he rejected the notion of kinship with slaves or identification with their plight. In an 1851 letter to Longfellow, he declared that he never "did, nor ever shall, feel any pre-eminent ardor for the cause" of abolitionism. Two months later, he assured Zachariah Burchmore that he had "not . . . the slightest sympathy for the slaves; or, at least, not half so much as for the laboring whites, who, I believe, as a general thing, are ten times worse off than the Southern negroes" (Mellow 409–10). Where Hawthorne feared to tread, he discouraged his readers from rushing in: his heroine is neither stripped nor whipped. The phantom figure of the oppressed slave is kept out of his text, and his heroine does not become, as Stowe's light-skinned Eliza would two years later, a slave surrogate with whom readers could identify.

If Hawthorne must reinvent history in order to make Hester a stand-in for the reader herself rather than a figure for enslaved African-Americans, he cannot entirely eradicate what Toni Morrison has called the "Africanist presence" in the shadowy margins of canonical American texts (44–46). Two recent critics, Mara Dukats and Caroline Woidat, have identified Hester as a slave figure, and have suggested that contemporary novels such as Maryse Condé's *I, Tituba* and Morrison's *Beloved* rewrite her story with all the violence and racial confrontation the earlier novel suppresses (and, as Sethe's chokecherry-tree back testifies, with all the horrors of whipping made manifest as well). Viewed through the lens of slavery, Hester is indeed the "allegorical figure of patient submission to tyranny" Woidat constructs, "a white, female" Uncle Tom whose refusal to rebel ensures her continuing victimization (537). But if it is important to recognize that the racialized context of *The Scarlet Letter* is unavoidable, it is equally important to explore the consequences for the novel of Hawthorne's insistence that Hester's choices and burdens are those of whites, not blacks.

Morrison's evocative essay articulates the racialized ground in which American literary narratives construct affirmative "American" values such as independence and freedom by contrasting the position of white Americans with those of enslaved blacks. In *The Scarlet Letter*, Hawthorne's efforts to maintain Hester's racial separation are twofold. Not only does he attempt to eliminate those characteristics of Hester's situation—like whipping—that would link her to enslaved African-Americans, he also tries to overcome the assumption, created by the omnipresence of slavery in antebellum American thought, that a person whose behavior is severely restricted by the force of law must be mentally enslaved as well. Castigating fellow Massachusetts citizens as virtual slaves because they would not resist the Fugitive Slave Law, Henry David Thoreau fumed that "there are some, who, if they were tied to a whipping-post, and could but get one hand free, would use it to ring the bells and fire the cannons to celebrate *their* liberty" (30). That Hester, though harshly constrained, both avoids the whipping post and enjoys a substantial measure of intellectual liberty suggests that Hawthorne's point is the inverse of Thoreau's: for Hawthorne, no amount of legal repression can eradicate the essential capacity for inner freedom he believes to be inherent in whites. Nor can it merge their situation with that of slaves.

Through Hester then, white readers may well have encountered their own legal/political situation, de-familiarized, perhaps, but recognizable nevertheless. Like New England citizens, but not like the whipped slave, Hester must decide whether or not to obey a harsh law imposed on her by distant, all-powerful lawmakers who are more concerned with the fate of the entire community than with any erring individual. Like them, but not like the slave, she has the freedom ultimately to *consent*, rather than merely submit, to the weight of Law she neither approves of nor has had any hand in making. She is thus a monitory model for the reader, and her progress toward inner freedom lights the way for our own. That the repugnant laws Hawthorne believed in the necessity of obeying were primarily fugitive slave laws argues even more strongly the importance of excluding the shadow slave from a position of possible reader-identification. In his 1852 campaign biography of Franklin Pierce, Hawthorne would declare that "merely human wisdom and human efforts" could not overcome the evils of slavery "except by tearing to pieces the Constitution, breaking the pledges which it sanctions, and severing into distracted fragments" the whole country (415). From his anxious viewpoint, excessive sympathy for slave-victims might lead to attempted rescues, confrontations, and violence, which in turn would undermine the cohesion and security of the national union. As his rhetoric clearly reveals, abolitionism is firmly equated in Hawthorne's mind with destructiveness; it is a force which "tears," "breaks," and "severs" the already vulnerable legal structures binding the states together.[8]

If readers are to be persuaded that the crushing weight of law should be borne uncomplainingly, those who administer that law must not be monsters. Harsh whippings inflicted on lovely heroines prevent readers from respecting those who call for the lash. Hester is spared her turn at the whipping post because, beneath their "grim-visaged" exteriors, Hawthorne's quasi-federal magistrates must be humane. They eschew violence and retribution in favor of inducements toward self-reformation. Because such psychological punishments were widely perceived as progressive, the magistrates are subtly aligned with forces of reform familiar to readers through debates over educational and military discipline.[9]

8. Arac also discusses aspects of Hawthorne's treatment of slavery in the *Life of Pierce* (253–54), as does Bercovitch (86–88). While the biography was, unlike the novel, an explicitly political act written for a particular campaign, and, on Hawthorne's part, written too with hopes of reaping a political appointment as a reward, its attitude toward the possible consequences of violent abolitionism is consistent with that expressed elsewhere in his writings.

9. Because Brodhead's analysis of 19th century discipline uses antebellum but not Puritan materials, his discussion of the various forms of punishment in *The Scarlet Letter* misses this crucial aspect of Hester's punishment. Because his methodology prevents his recognizing the significance of Hester's unwhipped back—the affirmative marker of meaning here constituted by an unmarked surface—Brodhead mis-characterizes Hester's letter as a harsh instance of "corporal correction," "correction performed through the external, visible marking of the body" (77–78), rather than as a splendid example of the very shift from corporal to psychological punishment which his article so persuasively documents as having occurred in the 1840s and 1850s. Berlant too refers to Hester's "marked body," as if the letter were a brand or tattoo (65).

While Hawthorne's narrator primly recoils from these "sages of rigid aspect," Hawthorne is thus hard at work rewriting history to improve their authority and compassion. They may appear cold, distant, cruel, and unfair, and in the short run, they may be so, but the novel as a whole exhibits a powerful underlying faith in the notions of order, authority, and non-participatory law-making which the magistrates represent. Somehow the whole point is that their immediate unfairness is *not* the point: we must suffer and obey what appears today to be harsh and oppressive law because in the long run maintaining the community is more important than protesting injustice. As several commentators have noted, Hawthorne's defense of the Fugitive Slave Law in *The Life of Pierce* parallels Hester's counsel to the unhappy women who seek her out in the novel's final chapter: systemic injustice cannot be "remedied by human contrivance" but must be left to "Heaven's own time" when it will "vanish like a dream." To rebel against present unfairness is not only arrogant and unnecessary, but dangerous.

IV

Just as the "real world" of Hawthorne's fiction is constructed at least partially by excluding figures like the townsman juror and the whipped woman whose presence would undermine the magistrates' justness and authority, it also suppresses all evidence of popular challenges to the magistrates' powers. 1642, the year Hester mounted her scaffold, saw intense attacks on the magistrates' undemocratic authority, attacks fully characteristic of the decade in which they occurred. At least two such disputes—the infamous squabble over Goody Sherman's sow and the debate over the magistrates' unlimited sentencing discretion—hover influentially just below the surface of Hawthorne's narrative. Nowhere is his historical sleight of hand more visible than in these efforts to strip volatile conflicts of their politically problematic content while recasting them so as to enhance and entrench the power of his magistrates.

When Hester visits the governor's mansion to protest Pearl's rumored removal from her custody, Hawthorne's narrator explains the "ludicrous" involvement of such eminent figures in a small-scale dispute of this kind:

> At that epoch of pristine simplicity, however, matters of even slighter public interest, and of far less intrinsic weight than the welfare of Hester and her child, were strangely mixed up with the deliberations of legislators and acts of state. The period was hardly, if at all, earlier than that of our story, when a dispute concerning the right of property in a pig, not only caused a fierce and bitter contest in the legislative body of the colony, but resulted in an important modification of the framework itself of the legislature. (101) [65]

Hawthorne's narrator smiles a bit condescendingly at this trivial dispute in a charmingly simpler era, memorable as much for the magistrates' magnanimous concern for petty squabbles as for the unspecified legislative reform it occasioned. In fact, it was not only one of the most complex and contorted lawsuits in Massachusetts history, but also, as Winthrop's lengthy account (dated June, 1642, the month Hawthorne's narrative begins) makes clear, "gave occasion to many to speak unreverently of the court, especially of the magistrates."[1] Goody Sherman, a poor widow, apparently egged on by her unscrupulous lodger, one Story, sued Captain Keayne, a sharp and much disliked merchant, for the value of a sow she claimed he had mistakenly slaughtered. Suits (for damages) and countersuits (for libel) wound through an astonishing variety of trial and appeals courts for more than two years. More than seven days of testimony were eventually heard by the full Court of Assistants, comprising 9 Magistrates and 30 Deputies. Sympathy ran high for the widow against the rich merchant, although everyone seems to have known that much of the evidence and testimony had been manufactured after the fact. Although a majority of the deputies sided with Mrs. Sherman, the case was determined in Captain Keayne's favor when the magistrates exercised their "Negative Voice," a device permitting them to nullify the votes of non-magistrate assistants by returning a verdict in favor of the party receiving a majority of the magistrates' votes. This exercise of magistratical supremacy only made matters worse with the public, for, as Winthrop tells it, "the report went, that their negative voice had hindered the course of justice, and that these magistrates must be put out, that the power of the negative voice might be taken away" (qtd. in R. Winthrop 282). Because "there was much laboring under a false supposition," Winthrop published a "declaration of the true state of the cause" and a defense of the Negative Voice (qtd. in R. Winthrop 283). The following year he managed to defeat efforts to abolish it, but the bad taste lingered in many mouths; the consequent separation of the deputies and magistrates into two legislative bodies speaks as much of persistent resentment and conflict as it does of triumphant resolution.

Like widow Sherman, Hester Prynne receives personal attention from powerful magistrates. Unlike the widow whose litigation exploited the availability of public process and judicial review almost beyond human capacity, Hester's custody case is resolved immediately, through an ad-hoc informal conversation in the ex-governor's garden. Hawthorne's magistrates hold neither criminal trials for adultery nor legal hearings on child custody. Applying their discretionary authority to the circumstances of the case at the moment it is presented, they appear rigid but behave with compassion, leaving Hester her child as they left her unmarked skin. Hawthorne's ambivalent handling of the sow case—gratuitously intruding it into his narrative while suppressing its subversive resonance—suggests the ultimate goal of his re-inventions of Puritan history. On the one hand, he wants the aura of historical accuracy and authorial erudition that such references to long-forgotten cases can produce. On the other hand, he wants to present the Puritan magistracy as unassailably secure in their powers, a type of idealized federal bureaucracy. The result is the same as in his handling of the absent jury trial; all challenges to magistratical authority disappear. Instead of providing evidence that the magistrates were under attack, the case is refigured as a comic example of the magistrates' willingness to stoop responsively to aid a poor widow, just as they deign to listen to Hester's plea for continued custody of Pearl.

1. In Robert C. Winthrop's *Life*, "The Stray Sow and the Negative Voice" warrants an entire chapter, which traces in detail the course of the dispute, the extended litigation it triggered, and its consequences within the community (2: 280–95).

The challenge to the magistrates' Negative Voice was not the only, or even the most serious, attack on the powers of the magistrates during the period of Hawthorne's fiction. As Mark D. Cahn has shown, the unlimited discretionary powers of the magistrates—precisely the aspect of their authority Hawthorne is at greatest pains to present as entrenched—were the focus of the intense debates over the codification of non-capital criminal sentences. Colonists outside the inner circle of power fought to circumscribe the magistrate-judges sentencing authority by establishing pre-set penalties for specific crimes. While there were several reasons for the movement to codify, "none . . . was more significant than the freemen's desire to curb the discretionary powers of the magistrates. The freemen feared that unless penalties were established by statute and rules regarding punishment made public, the magistrates could not be trusted to impose penalties fairly and with regularity" (108). Without pre-set penalties, challengers argued, judges imposed widely varying sentences for the same crime, making the outcome of cases dependent on judicial bias or whim, and undermining the system's consistency and predictability. The magistrates (led by the indomitable Winthrop) fought vigorously to retain their discretionary authority, insisting that as God's representatives on earth they were specially empowered, and that through their discretion God's law was harmonized with the diversity of human experience. As Cahn notes, "the leitmotif of Winthrop's political writings was that God intended certain men to be magistrates—and for these individuals, once elected, to govern unimpeded by dictates and restrictions imposed by those less fit to rule" (121). In 1644, Winthrop lost the battle when the General Court voted to accept the validity of prescribed punishments and restrict the penalties magistrates could impose, though the magistrates managed to delay the adoption of an extensive code until 1648.

The period between 1641 and 1648—roughly the period covered by the events of *The Scarlet Letter*—was thus permeated by this bitter challenge to the arbitrariness of magistratical power. Though there is no explicit reference in Hawthorne's text to the battle over codification, I believe that the extraordinary emphasis in the opening scene—and indeed, throughout the novel—on the sole and unlimited discretion of the magistrates in dealing with Hester is more than coincidental. Hawthorne's narrative both illustrates and argues Winthrop's case for the legitimacy of discretionary authority. In Hawthorne's Boston you can trust the powers that be, however far-away, grim-visaged, and sealed off from popular input, because they are ultimately compassionate, reasonable, and devoted to the community's welfare.

The influence of the codification dispute on Hawthorne's text can best be understood by seeing judicial discretion in sentencing as a form of interpretation. Codification supporters were suspicious of authority, which they saw as arbitrary and subjective. So long as magistrates could decide for themselves what penalty should be imposed in a non-capital case, they could "interpret" the circumstances of the case to mete out punishments based on political or personal bias. Inflexible statutory penalties were proposed as means to curb judicial inconsistency by eliminating the judge's power to interpret. Interpretation was thus imaged negatively; consistent justice could be grounded only in the predictability afforded by black-letter pre-set penalties. The same infraction should produce the same penalty, regardless of who was the judge, who the defendant. In direct opposition, the magistrates championed case-by-case interpretation of the law. They presented their authority as objective, skillful, and sensitive. Arguing that it was the defendants, not the magistrates, who were likely to be variable, the magistrates wanted the power to mold sentences to the circumstances of each case. Though the sentence once imposed would not change, it would be the product of an interpretive response to particular circumstances.

In this context, Hester's scarlet A is Winthrop's best case. The letter, a seemingly rigid and unalterable sign of objective authority, in practice represents flexibility and openness to interpretation, both past and future. While the letter A stands for the crime she has committed, it is also "the letter of the law," the penalty for that crime. Created before the novel's action begins through a process of compassionate interpretation, the letter has been carefully shaped to fit the circumstances of Hester's case, a case which calls, not for whipping or execution, but for this sentence only. The letter also opens to a future of reinterpretation, as the seemingly unchangeable A comes over time to signify Hester's admirable qualities rather than her past behavior. The letter itself cannot change, of course, but its meaning can. As Hester's personal sign, the way the letter is read tracks her changing relations with the Puritan community. Ultimately, she claims the letter as the marker of her identity, beyond the power of the magistrates to remove. In a typical Hawthorne paradox, what is most rigid signifies what is most fluid; what is most fungible, abstract, impersonal—a letter of the alphabet—becomes what is most personal, unique, and identity-bound. The A represents the potential for compassionate interpretation which lies within seeming inflexibility, and which can best be achieved by granting broad discretionary authority to magistratical decision-makers.

V

If the changing course of Hester's relationship with the Puritan legal establishment demonstrates that accepting the law's strictures can strengthen both the individual and the community, the terrifying counter-example set by Roger Chillingworth completes the lesson. Upon discovering his wife being publicly punished for adultery, Roger Prynne could have taken action against Hester by divorcing her, an option he is never described as considering.[2] With respect to her lover, however, he was legally obligated to leave the prosecution and punishment of the crime to the colony's courts. Instead of doing so, he devotes his life to the secret discovery and punishment of her partner. In what seems to be an act of kindness, he assures Hester that though he will make it his business to find out the identity of her lover, he will never "betray him to the gripe of human law" (80) [51]. In actuality, his preservation of secrecy is anything but kind. Unlike Hester, who protects Dimmesdale's identity in order to spare him pain, Chillingworth's aim is to monopolize the power to investigate, condemn, and punish the wrongdoer. Policeman, magistrate, judge, jury, and executioner in

2. While divorce was virtually unavailable, and socially unacceptable, in England in the 17th century, colonial American courts occasionally issued divorce decrees from their earliest years, and partners were free to remarry. See Haskins 63, 81, 194–95.

one, Chillingworth usurps every governmental role in the criminal justice system. Though his methods do not include the actual use of whips or stocks, Hawthorne describes his interaction with Dimmesdale as the ongoing infliction of torture: "Would he arouse [his victim] with a throb of agony? The victim was for ever on the rack; it needed only to know the spring that controlled the engine;—and the physician knew it well!" (140) [87]. If we assess Chillingworth's behavior in terms of its attitude toward law, it is clear that his rejection of public magistratical process in favor of fanatical service to the private law of vengeance marks him as the novel's vigilante figure. If Hester models the ultimate benefits to be derived from accepting the workings of legal process, her husband presents the necessary disaster—both to the community and to the vigilante himself—that results when individuals reject public process in favor of private action.

To one who believes in the unifying powers of the rule of law, the vigilante is a destabilizing threat: an anarchic evader of legal structures who is willing to employ violence. Lincoln had warned in the "Address to Young Men" in 1838 that when individuals arrogate to themselves the right to punish even genuine wrongdoers, mob violence is often the result, and, in turn, the destruction of government through a gradual weakening of the people's respect for it. The "lawless in spirit" become "lawless in practice," he charged, until those who have "ever regarded the Government as their deadliest bane . . . make a jubilee of the suspension of its operations; and pray for nothing so much as its total annihilation," while even good men lose their attachment to law and thus "the strongest bulwark of any government" is destroyed (30–31). In Lincoln's apocalyptic vision, dis-interested judgment, due process, indeed, all the forces of cohesion in society disappear when such figures are ascendant. To prevent such harm, Lincoln called for "every American, every lover of liberty" to swear "never to violate in the least particular, the laws of the country," and to teach reverence for the laws in every home, school, and college until it should become "the political religion of the nation" (32).

Significantly, Hawthorne's imaging of the vigilante figure's destructiveness focuses as much on the vigilante himself as on his victim. As a one-man criminal justice system, Chillingworth's rage to punish is unrestricted. None of the colony's institutional safeguards, such as statutorily limited penalties, judges uninvolved in the case, public records and public scrutiny, operate to keep his drive for vengeance within civilized bounds. When he makes what he thinks is a principled decision to retain control of Dimmesdale's prosecution rather than turn it over to the magistrates, he loses all the community-directed restraints that are designed to make the rule of law both rational and humane. The fact that Chillingworth feels so deeply the insult of Hester's adultery, far from conferring an obligation to judge or punish, should disqualify him from both roles. Yet he takes on the roles of judge and executioner, roles which, because his relationship to them is so inappropriate, cause the corruption of his own soul, "striking evidence of man's faculty of transforming himself into a devil, if he will only, for a reasonable space of time, undertake a devil's office" (170) [104]. Instead of a heightened integrity and a strengthened individualism, his actions lead to a loss of perspective, self-control, and qualitative humanness.

It is highly significant that the "devil's office" the good doctor assumes is precisely that role which the rule of law most triumphantly eliminates: that of torturer. The rack and other such "engines" were the kind of "inhumane, barbarous and cruel" punishments outlawed by the 1641 Body of Liberties. Hester, dealt with by the public authorities, receives a humane sentence most striking in its failure to include physical punishment.

In Hawthorne's grim utopia, the justice system is administered by men who rise above any personal drive for revenge, eschew violence, and illustrate the superiority of disinterested lawmaking over the "barbarity" of private justice. The magistrates' cold distance protects Hester from both her husband's and the townspeople's outrage. If the magistrates' sentence inflicts years of lonely suffering on Hester, her submission to the discipline of law also enables her intellectual growth and, the final chapter suggests, something like serenity. Those who evade "the gripe of law" are destroyed, either by falling victim to vengeance and private self-punishment, like Dimmesdale, or by suffering the self-destructiveness of unregulated and inappropriately assumed punitive power, like Chillingworth.[3]

That Chillingworth is positioned as a vigilante suggests that the spectre of violent disdain for law was one of Hawthorne's motivating fears in writing the novel. The public discourse of New England abolitionism in the late 1840s and 1850s preached just such vigilanteism, and it is useful to read Hawthorne's fiction within this larger cultural conversation about decisions to obey or resist the strictures of law. In an 1852 speech, for instance, Frederick Douglass thundered to a largely white political audience that "the only way to make the Fugitive Slave Law a dead letter is to make half a dozen or more dead kidnappers" (*Life* 207). His words were received with laughter and applause. Two years later, when a white federal guard was killed in a botched attempt to rescue Anthony Burns, a fugitive, Douglass defended the act in an article titled "Is it Right and Wise to Kill a Kidnapper?" by arguing that the guard's "slaughter" "was as innocent in the sight of God, as would be the slaughter of a ravenous wolf in the act of throttling an infant. We hold that he had forfeited his right to live, and that his death was necessary, as a warning to others liable to pursue a like course" (*Life* 287). In a sermon preached two days after the incident, Theodore Parker justified the killing, while laying the blame for violence not on those who sought to free Burns but on the state and federal officials who had imprisoned him.[4] While the use of violence was certainly not universally accepted within Abolitionist circles, it was condoned, justified, and encouraged by many as the

3. Chillingworth's attempt to achieve a wholly private revenge on Dimmesdale is ultimately unsuccessful, for the minister not only resists the doctor's continual efforts to elicit for himself alone what in Puritan terms must be a public confession, but "escapes" into the realm of the public to confess and die in the midst of his congregation. In this rule of law analysis, it is fitting that Dimmesdale's ultimate sanctuary is not a private but a public place, where relief from the burden of past deeds is found through permitting the community, however ambiguously, to know and assess his sin. Dimmesdale can defeat—though not survive—Chillingworth's lawless surveillance by mounting the scaffold in the presence of magistrates and townspeople. Similarly, his attempts to take on the governmental role of judge and punisher through self-flagellation must fail, because they do not proceed from a public assessment by uninvolved decisionmakers.

4. It was Parker himself who, on the night of the attempt to free Burns, had roused the crowd at Boston's Faneuil Hail and made the motion to "adjourn to Court Square" where Burns was being held. His sermon, along with journalistic reports of the events and coverage of the legal proceedings, were published in pamphlet form as *Boston Slave Riot and Trial of Anthony Burns*. The sermon is on 30–33.

only way to conscientiously resist the rendition of fugitives. Douglass and others preached that each individual has the obligation to judge man-made law against the Higher Law of right and wrong set by God. Where that law is judged and condemned as profoundly wrong, the individual must not only refuse to obey it, he should be willing to take on himself the infliction of punishment for that wrongdoing, even if such punishment leads to fatal violence. So Douglass calls on fellow abolitionists to execute slaveowning "kidnappers" whose capture of fugitives is lawful under the Compromise Act of 1850, and Parker refuses to condemn those who caused the death of a guard while attempting to free a fugitive awaiting rendition.

Read in the context of such calls for independent judgment of laws and violent resistance to their enforcement, *The Scarlet Letter* becomes a creative brief in support of Lincoln's doctrine of obedience to law. This is not to say that the story of Hester Prynne's willed submission to the harsh but compassionate law of the Puritan magistrates is no more than a response to contemporary political anxieties. If colonial and antebellum politics make this reading visible, they do not restrict its relevance for readers from Hawthorne's day to our own. Along with the many other issues the novel raises, *The Scarlet Letter* is a powerful and imaginative meditation on the larger issues implicated in every individual's continuing relation to governmental law, authority, and enforcement. The power to judge and punish wrongdoing, the novel suggests, must remain with the public authorities because, whereas vigilanteism undermines the peaceful order of society, the rule of law is the best hope of preserving it. Only the magistrates can translate judgment into action without becoming contaminated by passion. The novel suggests that we, the reader-peers figuratively participating in the crowd around Hester's scaffold, should accept the ultimate usefulness to society even of some deeply repugnant laws instead of challenging their immediate harshness.

Positioned within the novel as a sympathetic heroine, Hester certainly does not reap the traditional heroine's rewards of amatory and economic success. She is loved but not permitted to live with or marry her lover, and the wealth that Chillingworth possesses is left to Pearl, not to her. While Hester's heroism in resisting the temptation to rebel does bring such benefits as privacy, intellectual independence, and an apparently useful wisdom, each is achieved as a direct consequence of her suffering, isolation, and shame. If the violent resister proves his commitment by his willingness to suffer imprisonment and public condemnation, perhaps the person who chooses to obey in a time of general resistance must also be prepared to pay a severe price. The novel resists providing any simplistic vision of happiness for those who might follow in Hester's footsteps, insisting instead that this form of heroic compromise is anything but an easy way out.

The novel makes no explicit connections to the Compromise Act of 1850, and there is no evidence that contemporary readers made such connections, or even, for that matter, that Hawthorne himself did.[5] Yet

5. The contemporary reviews of *The Scarlet Letter* collected by Scharnhost suggest that readers of Hawthorne's day singled out the novel's poetry and passion, its study of the human heart and its handling of the consequences of adultery. Hester's heroic obedience and the magistrates' compassion were noted, as when the Portland Transcript applauded Hester's "almost proud

the national controversy that swirled around the Fugitive Slave provisions of the Act, the intense anxieties about possible disruptive violence, sectional fragmentation, and the obligations of individuals to obey unpalatable laws were very much part of the atmosphere in which the text was both written and read, and I believe it is reasonable to suggest that the novel both expressed and shaped the direction of those concerns. To make such a claim raises two important and connected questions with which I want to end this essay. First, how can a novel—especially one that does not make its politics a matter of explicit concern—play an important role in such an ongoing cultural dialogue? And second, if it can, does it do so only by ignoring or sidestepping the novel's essential literariness, reducing it to a species of political discourse? While a full answer to these questions would require another essay, a few points can be noted.

In my view, it is precisely *The Scarlet Letter*'s "literariness" that empowered it to play a critical role in shaping its readers' views, even if the dimensions of that role may be impossible to verify or recover. Obviously, a novel is neither a political essay nor a proposed statute. Perhaps less obviously, its unlikeness from each of these is different. A political essay that argues strongly for a specific position requires its audience to take an equally specific position in response—not just to agree or disagree with its assertions, but to restrict the ambit of that agreement to the particular issue and questions framed for debate. The story of Hester Prynne remains open in precisely the ways that such an essay is circumscribed. Like the letter she wears, the novel itself opens out to a range of interpretive responses among readers and invites a change in such responses over time. We can read the novel as being about passion, repression, and hypocrisy, without being required to accept or reject its implicit claims about civic obligation. Further, because the novel does not force readers to defend previously articulated political stands—does not, in fact, appear to be about contemporary politics at all—it can make new ways of thinking about those issues more easily acceptable. Through Hester readers can come to see freely chosen acceptance of the law's strictures as heroic. The transformation of such submission from cowardice into courage can elicit admiration and evoke a desire for emulation without forcing readers into a debate about the provisions of any statute. Among the many things it accomplishes, *The Scarlet Letter* permits its readers to explore the courage that can sometimes be required to obey the law without conducting that exploration in the already-tainted context of discussions about controversial fugitive slave laws. Once affected by these underlying issues in the novel's powerful story, it seems to me quite reasonable to suggest that, without necessarily attributing it to the novel, a reader might find herself responding differently to subsequent calls for violent resistance to divisive and unpalatable laws.

If evocative power and openness to interpretation distinguish the novel from position papers of any kind, boundedness, narrative closure, and moral inflection mark fiction's difference from most discursive forms of

submission to the indignities inflicted upon her" (24), and Orestes Brownson critiqued Hawthorne's failure to condemn the adultery, declaring that the magistrates' "treatment of the adulteress was far more Christian than [Hawthorne's] ridicule of it" (34–39). I have, however, found no reviews connecting the novel to agitation about the Compromise Act.

law. In juxtaposing law and literature, writers have sometimes imaged law as the kind of fixed text that determines and closes down the process of interpretation, while literature offers an openness to interpretation derived from its imaginative complexity.[6] Such dichotomies, though useful, erase important continuities connecting the two discourses.[7] What is significant is that legal texts, whether statute, constitutional provision or judicial decision, are necessarily part of an ongoing process of interpretation as the language of impersonal rules and previous interpretations is applied in specific, often unforeseeable cases. When the fugitive slave provisions of the Compromise Act of 1850 were drafted, legislators could sketch out a statutory scheme for recovering runaway slaves, but until the law was applied in a variety of states and contexts, its ambiguities, enforceability and consequences could not be known. It required, and received, judicial interpretation in a variety of adversarial contexts, interpretation that varied over time and with the identity of the judicial interpreter. Just as recent Supreme Court decisions have significantly changed "the law" in many areas by reinterpreting such textually fixed Constitutional provisions as the Fourth Amendment's protection against unreasonable search and seizure, so state and federal courts are continually called on to interpret and reinterpret in new contexts the meaning of statutes, codes, and earlier decisions. This openness to a necessary and evolving process of interpretation suggests a profound continuity that links law to literature through a shared dependence on language and rhetoric, with all the richness, complexity and ambiguity inherent in any discursive enterprise. Hester's red "letter of the law" is an apt symbol of that continuity: like the text of the novel in which it appears, it is both the product of and the textual stimulus for continuing interpretation.

But if law and literature both require interpretive readings, literary narratives can provide a degree of specificity and closure unavailable to law. Faced with Hester's adultery, Hawthorne's Puritan magistrates, like Winthrop's contemporaries who fought hard to retain the privilege, had the power to fashion a discretionary sentence responsive to her unique circumstances; they could not, however, predict with certainty its effect on Hester's future life. Similarly, the Compromise Act of 1850 could provide the occasion for debate and dire prediction, but it could not determine the outcome of a particular fugitive's quest for freedom or trace the particular series of events that would implicate that Act in the eventual outbreak of the Civil War. The connection of legal cause to ultimate practical effect, the vivid, particular way that a harsh law will change the lived experience of those it affects, can only be imagined as among the possibilities that may ensue if this or that legal text becomes law. Literary narratives in contrast, provide not just beginnings but endings, sequences

6. See, for instance Tanner, who positions marriage law as non-narrative and rigid, in contrast to the emotional responsiveness of fiction: "In the bourgeois novel we can find a strictness that works to maintain the law, and a sympathy and understanding with the adulterous violator that works to undermine it," he claims, adding later that "you can't have a society without laws, and you can't have a novel without sympathy (or empathy, or understanding, etc.)" (14, 24). Another good example is Dimock, who distinguishes the criminal law's "impulse toward precision" and narrowness, "whose operative terms were to become specific and explicit, without nuance or ambiguity," from the novel's "signifying latitude," which produces a symbolic amplitude and a "continuing (or perhaps even expanding) capacity for symbolization" (215–17).

7. The equal openness of both legal and literary texts to interpretation has been forcefully argued by Fish and explored in Brooks and Gewirtz's recent collection of essays.

of events in which the consequences of actions can be traced through time. If we are used to recognizing that a novel like *The Scarlet Letter* presents a complex and richly imagined world characterized by ambiguity and multiple possibilities for interpretation, it is also true that the story of Hester Prynne, Arthur Dimmesdale, and Roger Chillingworth describes not an infinity of paths but a course of connected events, producing *these* results and conflicts, and, finally, *this* particular form of closure. In 1850 it was the novel, not the statute, that could present a definite story of how the law affects the individual.

WORKS CITED

Arac, Jonathan. "The Politics of *The Scarlet Letter*." *Ideology and Classic American Literature*. Ed. Sacvan Bercovitch and Myra Jehlen. Cambridge: Cambridge UP, 1986. 247–66.

Bercovitch, Sacvan. *The Office of the Scarlet Letter*. Baltimore: Johns Hopkins UP, 1991.

Berlant, Lauren. *The Anatomy of National Fantasy: Hawthorne, Utopia, and Everyday Life*. Chicago: U of Chicago P, 1991.

Boston Slave Riot and Trial of Anthony Burns. Boston: Fetridge, 1854.

Brodhead, Richard H. "Sparing the Rod: Discipline and Fiction in Antebellum America." *Representations* 21 (1988): 67–93.

Brooks, Peter, and Paul Gewirtz, eds. *Law's Stories: Narrative and Rhetoric in the Law*. New Haven: Yale UP, 1996.

Cahn, Mark D. "Punishment, Discretion, and the Codification of Prescribed Penalties in Colonial Massachusetts." *The American Journal of Legal History* 33.2 (1989): 107–36.

Colacurcio, Michael. "Footsteps of Ann Hutchinson: The Context of *The Scarlet Letter*." *ELH* 39 (1972): 459–94.

———. "'The Woman's Own Choice': Sex, Metaphor, and the Puritan 'Sources' of *The Scarlet Letter*." Colacurcio, *Essays* 101–35.

Colacurcio, Michael, ed. *New Essays on* The Scarlet Letter. Cambridge: Cambridge UP, 1985.

The Compact with the Charter and Laws of the Colony of New Plymouth. Boston: Dutton, 1836.

Dimock, Wai Chee. "Criminal Law, Female Virtue, and the Rise of Liberalism." *The Yale Journal of Law & the Humanities* 4 (1992): 209–47.

Dukats, Mara L. "The Hybrid Terrain of Literary Imagination: Maryse Condé's Black Witch of Salem, Nathaniel Hawthorne's Hester Prynne, and Aime Cesaire's Heroic Poetic Voice." *College Literature* 22 (1995): 51–61.

Douglass, Frederick. *Life and Writings of Frederick Douglass*. Ed. Philip S. Foner. New York: International, 1950.

Felt, Joseph. *Annals of Salem*. Salem, 1827.

Fish, Stanley, "Working on the Chain Gang: Interpretation in Law and Literature." *Doing What Comes Naturally: Change, Rhetoric, and the Practice of Theory in Literary and Legal Studies*. Durham: Duke UP, 1989. 87–102.

Fleischner, Jennifer. "Hawthorne and the Politics of Slavery." *Studies in the Novel* 23.1 (1991): 96–106.

Haskins, George Lee. *Law and Authority in Early Massachusetts*. New York: Macmillan, 1960.

Hawthorne, Nathaniel. *The American Notebooks*. Ed. Claude M. Simpson. Columbus: Ohio State UP, 1962.

———. *Life of Franklin Pierce. Works*. Vol. 12. Ed. G.P. Lathrop. Boston: Houghton, 1883.

———. *The Scarlet Letter*. Columbus: Ohio State UP, 1962.

Kesselring, Marion. *Hawthorne's Reading 1828–1850*. New York: NY Public Library, 1949.

Konig, David T. *Law and Society in Puritan Massachusetts: Essex County 1629–1692*. Chapel Hill: U of North Carolina P, 1979.

Lincoln, Abraham. "Address to the Young Men's Lyceum of Springfield, Illinois." *Abraham Lincoln, Speeches and Writings 1832–1858*. New York: Library of America, 1989.

Mellow, James R. *Nathaniel Hawthorne in his Times*. Boston: Houghton, 1980.

Morrison, Toni. *Playing in the Dark: Whiteness and the Literary Imagination*. Cambridge: Harvard UP, 1990.

Parker, Theodore. "The Three Chief Safeguards of Society, Considered in a Sermon." Boston: Crosby, 1851.

Powers, Edwin. *Crime and Punishment in Early Massachusetts*. Boston: Beacon, 1966.

Records of the Court of Assistants of the Colony of the Massachusetts Bay 1630–1692. Vol. 12. Boston: County of Suffolk, 1904.

Records of the Plymouth Colony: Court Orders. Ed. Nathaniel Shurtleff. Boston: White, 1855.

Ryskamp, Charles. "The New England Sources of *The Scarlet Letter*." *American Literature* 31.3 (1959): 257–72.

Scharnhorst, Gary, ed. *The Critical Response to Nathaniel Hawthorne's* The Scarlet Letter. New York: Greenwood, 1992.

Schwartz, Harold. "Fugitive Slave Days in Boston." *New England Quarterly* 27 (1954): 191–212.

Shulman, Robert. "The Artist in the Slammer: Hawthorne, Melville, Poe and the Prison of their Times." *Modern Literary Studies* 14.1 (1984): 79–88.

Tanner, Tony. *Adultery in the Novel: Contract and Transgression*. Baltimore: Johns Hopkins UP, 1979.

Thoreau, Henry David. "Slavery in Massachusetts." *Anti-Slavery and Reform Papers*. Montreal: Harvest, 1963. 26–41.

Tocqueville, Alexis de. *Democracy in America*. 1835. Ed. Phillips Bradley. Trans. Henry Reeve, Francis Bowen. 2 vols. New York: Random, 1945.

Winthrop, John. *The History of New England from 1630 to 1649*. 4 vols. Ed. James Savage. Boston: Little, 1853.

Winthrop, Robert C. *Life and Letters of John Winthrop*. New York: Da Capo, 1971.

Woidat, Caroline M. "Talking Back to Schoolteacher: Morrison's Confrontation with Hawthorne's *Beloved*." *Modern Fiction Studies* 39.3–4 (1993): 527–46.

Yellin, Jean Fagan. "Hawthorne and the American National Sin." *The Green American Tradition: Essays and Poems for Sherman Paul*. Ed. Daniel Peck. Baton Rouge: Louisiana State UP, 1989, 75–97.

LAURA DOYLE

"A" for Atlantic: The Colonizing Force of Hawthorne's *The Scarlet Letter*†

In *The Scarlet Letter,* colonization just happens or, more accurately, has just happened. We might recall, by contrast, how Catharine Maria Sedgwick's novel *Hope Leslie* elaborately narrates the sociopolitical process of making an Indian village into a native English spot. Hawthorne eclipses this drama of settlement. Although Hawthorne, like Sedgwick, sets his plot of sexual crisis in the early colonial period of Stuart political crisis and English Civil War, he places these events in the distant backdrop, as remote from his seventeenth-century characters as his nineteenth-century readers. Meanwhile, he recasts Sedgwick's whimsical heroine, Hope Leslie, as a sober, already arrived, and already fallen woman.

In beginning from this already fallen moment, Hawthorne keeps offstage both the "fall" of colonization and its sexual accompaniment. He thereby obscures his relationship to a long Atlantic literary and political history. But if we attend to the colonizing processes submerged in *The Scarlet Letter,* we discover the novel's place in transatlantic history—a history catalyzed by the English Civil War and imbued with that conflict's rhetoric of native liberty. We see that Hawthorne's text partakes of an implictly racialized, Atlantic ur-narrative, in which a people's quest for freedom entails an ocean crossing and a crisis of bodily ruin. That is, *The Scarlet Letter* fits a formation reaching from *Oroonoko, Moll Flanders, Charlotte Temple,* and Olaudah Equiano's *Interesting Narrative* to *The Monk* and *Wieland* and continuing through such divergent yet fundamentally Atlantic texts as *Billy Budd, Of One Blood, The Voyage Out,* and *Quicksand.*[1]

Critics have long noted the offstage locale of Hester Prynne and Arthur Dimmesdale's act of passion and Hawthorne's choice to keep us at one remove from its catalyzing force. But no one has noted the novel's elision of the original condition for that passionate act: the transatlantic migration of Hester Prynne *alone*. It is this fact that prepares Hester's "fall." And if Hester's journey alone, and into a deeply solitary interiority, emblematizes the exilic effects of Atlantic modernity, the aborted journey of Roger Prynne (aka Chillingworth) into "grievous mishaps by land and sea" and "bond[age] among heathen-folk" emblematizes its violent encounters.[2] But these conditions are placed in the past and only alluded to, so that, as Leslie Fiedler notes, the characters' "whole prehistory remains shadowy and vague."[3] Instead, within the novel, the punishment for adultery becomes the point of origin. This way of placing key events at one remove, gestured toward yet

† From *American Literature* 79.2 (June 2007): 243–73. Republished by permission of the copyright holder, and the present publisher, Duke University Press. www.dukeupress.edu. Page numbers in square brackets refer to this Norton Critical Edition.

1. I trace this extended narrative in my book *Freedom's Empire: Race, Rape, and the Rise of the Novel in Atlantic Modernity, 1640–1940* (Durham, N.C.: Duke Univ. Press, 2007).
2. Nathaniel Hawthorne, *The Scarlet Letter* (New York: Norton Critical Edition, 1988), 44 [43]. Further references are to this edition [edited by Sculley Bradley, Richmond Croom Beatty, E. Hudson Long, and Seymour Gross] and will be cited parenthetically in the text.
3. Leslie A. Fiedler, *Love and Death in the American Novel* (New York: Criterion, 1960), 131.

submerged, characterizes the novel's historical method and its repressed relation to Atlantic history.

At the same time, Hawthorne does implicitly make matters of removal and habitation important to Hester's fall. He does so first, albeit indirectly, in "Introductory: The Custom-House," where he prefaces his story of Hester's "sin" with an account of his own troubled relation to his "native spot," what he calls his "unjoyous attachment to my native town" (11) [12]. He more directly sets up a correlation between Hester's departure from home and her loss of innocence (and thus conforms to an Atlantic narrative tradition that merges sexual and colonial ruin) when, as she stands on the scaffold in Boston, Hester looks back to her "village in rural England" where "stainless maidenhood seemed yet to be in her mother's keeping" but which village is now "foreign to her, by comparison" (56) [53].

Yet Hawthorne most directly points to the Atlantic coloniality that issues in Hester's fall when his narrator announces that "[Hester's] sin, her ignominy, were the roots which she had struck into the soil" (56) [53]. I suggest we take him literally. His words echo those in the "Introductory," when he confesses guilt about "the deep and aged roots which my family has struck into the soil" (8) [10]. Perhaps, after all, the "sin" with which Hawthorne is most preoccupied is neither adultery nor his ancestors' whipping of adulterous women but, rather, colonization itself. Hester's "A" is a layered code. Under "adultress" lie the merged meanings of Anglo-Saxon and Atlantic. And under Hawthorne's *The Scarlet Letter* lie many English-language narratives in which sexual plots of undoing carry, like silenced cargo, transatlantic stories of violent colonization that give rise to an Anglo-Atlantic freedom.

To appreciate the palimpsest that is Hawthorne's narrative, we must first turn back to early-seventeenth-century England, when the story of a potentially ruinous liberty became racialized under revolutionary, transatlantic conditions.

Replotting Race on the Atlantic

During the English Civil War, race unfurled as a freedom myth. Witness Englishman John Hare's Civil War pamphlet in 1647, *St. Edward's Ghost or Anti-Normanism*:

> There is no man that understands rightly what an Englishman is, but knows withal, that we are a member of the Teutonick nation, and descended out of Germany: a descent so honourable and happy, if duly considered, as that the like could not have been fetched from any other part of Europe. . . . In England the whole commonalty, are German, and of the German blood; and scarcely was there any worth or manhood left in these occidental nations, after their long servitude under the Roman yoke, until these new supplies of free-born men from Germany reinfused the same. . . . Did our ancestors, therefore, shake off the Roman yoke . . . that the honour and freedom of their blood might be reserved for an untainted prey to a future conqueror?[4]

4. John Hare, *St. Edward's Ghost or Anti-Normanism* (1647), quoted in Samuel Kliger, *The Goths in England: A Study in Seventeenth and Eighteenth Century Thought* (Cambridge: Harvard Univ. Press, 1952), 136–37.

Hare joins many others who yoke freedom and race in an Anglo-Saxon discourse of resistance to conquerors and tyranny. Over the next century, this discourse yielded the notion that some races are born to seek freedom—and therefore deserve it—and others are not. By the later eighteenth century and until today in Iraq, peoples or races must, from a Western point of view, demonstrate their "capacity" for freedom, or be ruined. In the Western idea of freedom, race and modernity join hands, for the will to freedom is the very essence, according to Hegel and others, of "world-historical," modern races.[5] In modernity, it is above all the capacity for freedom that measures a race.

Yet it's important that we recognize this seventeenth-century rhetoric not just as the seedbed for slavery, Nazism, and U.S. imperialism but also as the postcolonial revolutionary resistance it was intended to be. The early fashioners of the discourse of race and freedom understood themselves to be reclaiming their trammeled native rights from foreign usurpers—Norman, French, and popish. Only when we recognize this old and dissenting genealogy of race and freedom do we understand fully the seductive power and social dynamics of race in the modern West. Herein lies race's promise to offer affiliative bonds, exercised in the quest for freedom. As we attend to this genealogy, we begin to glimpse the depth at which English-language narratives are racial narratives and, in their Anglo-Atlantic forms, hegemonically so—exactly because they are structured by a freedom plot.

It was the Reformation that first gave rise to the Saxonist refashioning of English identity. The search for links to the "primitive" German church predating Christianity's dependence on bishops and popes initiated the turn toward an Anglo-Saxon lineage that would eventually become insistently racialized.[6] Henry VIII authorized Matthew Parker to gather from England and abroad all documents revealing the Germanic and Anglo-Saxon origins of the "true and primitive" Church that predated popery (*OES*, 11). In his preface to *A Testimonie of Antiquities* (1566–67), Parker draws on Saxon materials to offer, he says, "testimonye of verye auncient tyme, wherein is plainly showed what was the judgement of the learned men in thys matter, in the days of the Saxons before the Conquest."[7] This notion of a return to the Saxon ancestors' pre-Catholic simplicity laid the foundation for the later, secular notion of Anglo-Saxonism.

We can trace the turn from more strictly religious Anglo-Saxonism to legal, cultural, and racial Anglo-Saxonism by way of the Society of Antiquaries, originally founded by Tudor kings for religious purposes. Under the Stuart king James I, however, the Society of Antiquaries turned its attention increasingly to old Saxon legal documents. As the Stuarts spoke more and more insistently of their divine right to absolute rule, Parliament members made increasing use of the legal documents being unearthed and

5. W. G. F. Hegel, *The Philosophy of History*, trans. John Sibree (1899; reprint, New York: Dover, 1956).
6. For discussion of how the Reformation in England led to the antiquarian interest in Anglo-Saxons, see Kliger, *The Goths in England*; Roberta Brinkley, *Arthurian Legend in the Seventeenth Century* (Baltimore: Johns Hopkins Univ. Press, 1932); and Eleanor Adams, *Old English Scholarship in England from 1566–1800* (New Haven: Yale Univ. Press, 1917), 24–25. Further references to *Old English Scholarship* will be cited parenthetically as *OES*.
7. Matthew Parker, preface to *A Testimonie of Antiquities* (1566–67), quoted in *OES*, 24–25.

translated by the Society's scholars. Invoking the pre-Conquest Magna Carta and common law traditions, and gathering evidence of Anglo-Saxon law-making councils, which fueled the arguments of Parliamentary lawyers, scholars such as John Selden found themselves censored and imprisoned along with Sir Edward Coke and other Parliamentary lawyers.[8] The Society of Antiquaries was finally disbanded by royal decree.

Matters reached a critical turning point—and the rhetoric of ancient Saxon rights found its legs—when in 1620 the King issued a proclamation restricting Parliament's right to discuss high matters of state. Parliament responded directly, coining a language that would not only become the basis of its 1628 Petition of Right but would also create the heart of Whig politics and Saxon myth that lasted well into the twentieth century:

> The privileges and rights of Parliament are an ancient and indubitable birthright and inheritance of the English, and all important and urgent affairs in Church and State as well as the drawing up of laws and the remedying of abuses, are the proper subjects of the deliberation and resolutions of the Parliament. The members are free to speak upon them in such order as they please, and cannot be called to account for them.[9]

In further exchanges with the King, the Parliament reasserted its "Ancient and Undoubted Right, and an Inheritance received from our Ancestors," until the King "publicly tore these protests from the Journal of the House of Commons and dissolved Parliament."[1] Throughout the 1620s and 1630s Parliament and the Stuart kings reached several such moments of impasse. Finally, in 1629, Charles I dissolved Parliament—and it did not reconvene until 1640.

Meanwhile, however, other forces were gathering. Across the Atlantic, a group of men was building a new commercial network that would eventually help to break the impasse. Ultimately, this development would make the racialized rhetoric of liberty a transatlantic phenomenon, embedding it deep in the structures of English-language narrative. In a sense, the English Civil War and its aftermath, from Cromwell's Commonwealth to Queen Victoria's empire, find their necessary cause in the 1610s and 1620s, in the form of this group of "new men," middling-class and eventually Puritan-affiliated, who initiated the activities and alliances that would reshape the economic balance of power.[2] For with the Parliamentary crisis from 1628 to 1629, culminating in Charles I's eleven-year dissolution of Parliament and renewed persecution of Puritans, a small group of Atlantic merchants who had been accruing land, power, and wealth in the west Atlantic throughout the 1620s joined hands with those interested in building colonies as safe havens for religious refugees. Together, in effect, these men overthrew the King.

8. See Kliger, *The Goths in England*, 126.
9. Quoted in Brinkley, *Arthurian Legend*, 38.
1. Ibid.
2. More accurately, these developments consolidated a shift that had begun with sixteenth-century land redistribution, privatization, and enclosure. My discussion draws from Robert Brenner, *Merchants and Revolution: Commercial Change, Political Conflict, and London's Overseas Traders, 1550–1653* (Princeton, N.J: Princeton Univ. Press, 1993). Further references will be cited parenthetically in the text as *MR*.

This colonial development formed a crucial condition for the Civil War in England; its tobacco and sugar profits, in fact, eventually fueled Parliamentary warships.[3] As Robert Brenner has documented, there evolved "growing ties between the American merchant leadership and the great Puritan aristocrats who ran the Bermuda and Providence Island companies, as well as the lesser gentry who governed the New England colonies" (*MR*, 149). Men such as Maurice Thomson and his brother-in-law William Tucker, who had begun as ship captains, entered the breach left by the retreat of the King's trading companies in Virginia. The absence of Royal Company rules allowed these men to run both exports and imports and to set up shop on both sides of the Atlantic (a practice prohibited in the royal companies). As a result, they quickly monopolized the import of supplies for settlers as well as the export of tobacco, and they accrued huge profits. Working together with a handful of others, they extended their reach south to the West Indies (where they headed interloping invasions against the colonies of other European powers) and north to the Massachusetts Bay Colony, financially backing the Puritan settlement of Massachusetts and helping to organize provisions for colonies both north and south.

These ties eventually laid the foundation for the "transatlantic network of Puritan religio-political opposition to the crown" that included Massachusetts, Connecticut, Rhode Island, and, in the West Indies, Bermuda Island and Providence Island, all of which drew investors for religious and political reasons as well as for profits and all of which served as both "ports of exile and staging posts for revolt" (*MR*, 113, 110). Under these conditions, the pursuit of religious freedom, so touted in American history books, was utterly involved with the pursuit of mercantile freedom, for even when religious motives were paramount, economic "freedoms" were requisite to make the colonial settlements viable, as Karen Kupperman has shown.[4] Furthermore, it was from this base, and for this base, that Thomson and his circle became interlopers in the slave trade and the East Indies trade and then, in turn, began to build the enormously profitable West Indies sugar plantations during the 1640s (*MR*, 161–65).

When Parliament finally reconvened in 1640, a new coalition of members, including Puritans backed by these merchants, succeeded in abolishing the Star Chamber (which had handled licensing and censorship since 1586); purging those members they considered popish or unlawful; exerting powerful resistance to the King's demands; and eventually declaring war. They spoke a liberty rhetoric that loosely blended religious and economic meanings, casting both forms of restriction as "infringements of our Native Liberties."[5] Via the notion of native liberty, the Atlantic economy joined the nation, and, in turn, native liberty extended across the Atlantic.

At the same time, throughout the 1640s, contemporaneous with Hester Prynne's ordeal, liberty rhetoric spread "downward" because of the

3. On this point, see Robert M. Bliss, *Revolution and Empire: English Politics and the American Colonies in the Seventeenth Century* (Manchester, Eng.: Manchester Univ. Press, 1990), 48. Bliss gives a useful overview of these transatlantic conditions, with a more cultural emphasis than Brenner's *Merchants and Revolution*.
4. See Karen Ordahl Kupperman, *Providence Island, 1630–1641: The Other Puritan Colony* (Cambridge, Eng.: Cambridge Univ. Press, 1993), 142.
5. See William Haller, ed., *Tracts on Liberty in the Puritan Revolution, 1638–1647*, 3 vols. (New York: Columbia Univ. Press, 1933), 3:358.

uncensored press and the unregulated preaching of ministers; and its nativist overtones became louder. The Long Parliament had not immediately replaced the Star Chamber with any equivalent censorship organ, and so there circulated increasing numbers of polemical newspapers, pamphlets, and petitions that eventually made it impossible for the entrepreneurs and the Puritans to maintain control of the liberty discourse. Indeed, this is the moment when the Habermasian public sphere becomes a reality in England—briefly yet influentially.[6] Especially as the Puritan-slanted Parliament gained the upper hand in the war, numerous petitions were presented to the House of Commons, expressing the desire of soldiers, soldiers' wives, tradespeople, religious sects, and laborers for relief from painful economic conditions and for fuller representation of their voices. But relief and representation were not forthcoming, and so "the public" printed, agitated, and formed new coalitions. By 1647, the failure to hold new Parliamentary elections with an expanded electorate, to pay soldiers their arrears, to finance support for widows and orphans or for citizens who quartered the soldiers, to break up monopolies of trade in an already debilitated postwar economy, to allow for full religious toleration instead of new preferential treatment of the Puritans, and to repeal the tithes and taxes that weighed heavily on the poorest—all of these failures fed widespread disenchantment among a people who had sustained years of war for the sake of better living conditions.

Increasingly politicized middle-rank women as well as men wrote petitions, held meetings, and joined or led public protests to address these injustices. That is, the Civil War was an event in the history of gender politics as well of class, religious, and racial politics. Early in 1641, 400 women gathered at Parliament to demand a response to a petition on the loss of trade. When they received no satisfactory attention, they penned the "Humble Petition of many hundreds of distressed women, Tradesmens wives, and widdowes" in which they claimed that "we have an interest in the common Privileges with them [who have petitioned for the] Liberty of our Husbands, persons, and estates."[7] Such demonstrations continued to occur, as when in August 1643, some 5,000 to 6,000 women (as numbered by their critics) marched on the Commons for peace. By 1647, petitioning women appeared frequently on the steps of Parliament until the House of Commons enacted an ordinance to clear away "those clamourous women, which were wont to hang in clusters on the staires."[8]

All of these groups, women as well as men, spoke continually of "native rights," "the people's just rights and liberties," the "Nation's freedoms," "the free-born people of England," and the "free-born People's freedoms or rights."[9] Like John Hare, Nathaniel Bacon elaborately laid out the Saxonist

6. For discussion of how such publications shaped the pivotal role that the media would play in modernity, see John B. Thompson, *The Media and Modernity* (Stanford, Calif.: Stanford Univ. Press, 1995). For literature's development of this public sphere in the seventeenth century, see David Norbrook, *Writing the English Republic: Poetry, Rhetoric, and Politics, 1627–1660* (New York: Cambridge Univ. Press, 1999).

7. "Humble Petition. . . . ," quoted in Ann Marie McEntee, "The (Un)Civil Sisterhood of Oranges and Lemons: Female Petitioners and Demonstrators, 1642–53," in *Pamphlet Wars: Prose in the English Revolution*, ed. James Holstun (Buffalo: State University of New York Press, 1992), 93–94.

8. The ordinance is dated 26 August 1647, quoted in McEntee, "The (Un)Civil Sisterhood," 96. McEntee's source is Patricia Higgins, "The Reactions of Women. . . ." in *Politics, Religion, and the English Civil War*, ed. Brian Manning (London: Edward Arnold, 1973), 179–97.

9. See the documents quoted in Stuart E. Prall, ed., *The Puritan Revolution: A Documentary History* (Gloucester, Mass: Peter Smith, 1973), 134, 129, 134, 127, 128, respectively.

historical narrative underlying this nativist rhetoric that would become Whig orthodoxy by the early eighteenth century—including reference to Tacitus. In his *Historical and Political Discourse of the Laws and Government of England*, which addresses the "Debate concerning the Right of an English King to Arbitrary Rule over English Subjects, as Successor to the Norman Conqueror" (1647), Bacon remarks that it is

> both needless and fruitless to enter into the Lists, concerning the original of the Saxons. . . . They were a free people, governed by Laws, and those made not after the manner of the Gauls (as Caesar noteth) by the great men, but by the people; and therefore called a free people, because they are a law unto themselves; and this was a privilege belonging to all the germans, as tacitus observeth. . . . The Saxons fealty to their King, was subservient to the publick safety; and the publick safety is necessarily dependant [*sic*] upon the liberty of the Laws.[1]

Such pronouncements opened the way to more radical thinkers such as the Diggers, who nonetheless invoked the same nativist rhetoric. The Digger Gerard Winstanley echoed it in pronouncing that "the last enslaving conquest which the enemy got over Israel was the Norman over England."[2] The many migrations, rebellions, ironies, crimes—and texts—of English-language Atlantic history (including *The Scarlet Letter*) follow from this inextricable intertwining of the colonial, revolutionary, and nativist roots of the modern notion of freedom.

Equally important to Atlantic history and to Hawthorne's novel, the liberty rhetoric also took what we might call an interior turn. Leveller pamphleteer John Warr signaled the shift when he claimed that "[j]ustice was in men, before it came to be in Laws."[3] It is beyond the scope of this essay to consider the long path by which such claims led to the interiorization of both racial identity and modern narrative, and to the forms of power Michel Foucault analyzes, but suffice it to note that the work of Hegel gives a glimpse of the way the revolutionary nativist vision became an interiorized, racist one. As it did for Gerard Winstanley, for Hegel, too, "Reason" drives the "Universal History" of the world toward "Freedom," but Hegel more hubristically declares German culture to be the ultimate incarnation of this process. "The German spirit," writes Hegel, "is the Spirit of the new World. Its aim is the realization of absolute Truth as the unlimited self-determination of Freedom. . . . The destiny of the German peoples is to be the bearers of the Christian principle . . . of Spiritual Freedom."[4] The movement from the Reformation to the Civil War to Hegel neatly encapsulates how a discourse of race merged, including through this inward turn, with a discourse of freedom and, via the prosperous Atlantic economy, gave rise to an imperial chauvinism.

But in the 1640s no such grand visions were yet conceivable. With the monarch under arrest, women protesting in the streets, families fleeing to colonies across the Atlantic that were themselves in struggle with the

1. Nathaniel Bacon, *Historical and Political Discourse of the Laws and Government of England*, quoted in Kliger, *The Goths in England*, 139.
2. William E. Gerard Winstanley, "The True Levellers' Standard" (1649), quoted in *The Puritan Revolution*, ed. Prall, 179.
3. John Warr, *The Corruption and Deficiency of the Lawes of England* (London, 1649), quoted in Kliger, *The Goths in England*, 269.
4. Hegel, *The Philosophy of History*, 10, 19, 341.

Indian peoples whose land they seized, all while at home the problems of poverty, homelessness, and hunger were finding unbridled expression in a new world of print—under these conditions, as contemporaries reported, "There is a great expectation of sudden destruction" for "the greatest powers in the kingdom have been shaken."[5] It is this crisis—in which English society seems teetering on a cliff—that racialism works to contain and that, in his own period of political and racial crises, Hawthorne kept off of his page.

Hawthorne's Puritan Palimpsest

Criticism on *The Scarlet Letter* makes clear that the novel is a historical palimpsest—with a surface as illegible and in need of translation as the archaic, "gules" A. Not just one but two histories are submerged here, one contemporary with Hester and one with Hawthorne. Or rather, as I will argue, what is ultimately submerged is the deep connection between these two histories—that is, the uninterrupted project of colonization.

Many earlier critics of the novel consider it both a critique and an expression of American Puritanism, and most of these critics share Hawthorne's sense of that legacy as *the* cultural origin of U.S. national history. In his 1879 book, *Hawthorne*, Henry James helped to establish the identification between Hawthorne and the Puritan tradition, invoking the notion of a racial inheritance when he concludes that *The Scarlet Letter* is utterly "impregnated with that after-sense of the old Puritan consciousness of life" and that indeed the "qualities of his ancestors filtered down through generations into his composition," so that "*The Scarlet Letter* was, as it were, the vessel that gathered up the last of the precious drops."[6] This sense of the book as a racial expression hereafter found an echo in critics from William Dean Howells, who suggests that "Hawthorne was writing to and from a sensitive nerve in the English race that it had never known in its English home," to Carl Van Doren, who sees in Hawthorne "the old Puritan tradition that, much as he might disagree with it on occasion, he had none the less in his blood," to Elizabeth Deering Hanscom, who in her Macmillan introduction to the novel concludes that "in his attitude toward life, in his inner thought, [Hawthorne] was bone of the bone, blood of the blood of Puritan New England."[7] By the time of Lloyd Morris's 1928 biography of Hawthorne, *The Rebellious Puritan*, this lineage for Hawthorne had become a critical orthodoxy in the form of the idea that Hawthorne "had sought to liberate himself from his origins and environment, but they and not he had determined the character of that effort for emancipation."[8] Building on the

5. See the transcripts of the Putney debates, as presented in A. S. P. Woodhouse, ed., *Puritanism and Liberty: Being the Army Debates (1647–49), from the Clarke Manuscript* (London: J. M. Dent, 1938), 42, 20.
6. Henry James, *Hawthorne* (New York: Harper and Brothers, 1880), quoted in *The Critical Response to Nathaniel Hawthorne's "The Scarlet Letter,"* ed. Gary Scharnhorst (New York: Greenwood, 1992), 79. See also Scharnhorst's discussion of James's biography in his introduction (xvii–xix).
7. William Dean Howells, "Hawthorne's Hester Prynne," in *Heroines of Fiction* (New York: Harpers, 1901), 1; Carl Van Doren, "The Flower of Puritanism," *Nation*, 8 December 1920, 649–50; and Elizabeth Deering Hanscom, introduction to *The Scarlet Letter* (New York: Macmillan, 1927); all reprinted in *The Critical Response*, ed. Scharnhorst, 102, 140, 146, respectively. As Scharnhorst discusses, other critics who made similar suggestions include Herbert Schneider, *The Puritan Mind* (1930), Yvor Winters, *Maule's Curse* (1938), and, of course, F. O. Matthiessen, *American Renaissance* (1941).
8. Lloyd Morris, *The Rebellious Puritan* (London, 1928), quoted in Jane Lundblad, *Nathaniel Hawthorne and the Tradition of Gothic Romance* (New York: Haskell House, 1964), 10.

notion that Hawthorne's very dissent made him the child of Puritan America, early-twentieth-century scholars tracked Hawthorne's knowledge of Puritan sources and studied his main characters as they suffer under and, perhaps, redeem that legacy.

More recently, however, an increasing number of scholars place the novel explicitly within the political concerns of the volatile 1840s. These critics call attention to the fact that in the decade leading up to Hawthorne's writing of *The Scarlet Letter*, the nation was embroiled in conflict over a range of issues—the Indian Removal Acts, the annexation of western territories and war with Mexico, the Fugitive Slave Law, the 1848 Women's Convention in Seneca Falls, and the spectre (as many felt it) of the European revolutions of 1848. Accordingly, they have considered the novel's drama of law, punishment, dissent, and consent as a coded exploration of a citizen's proper response to these matters. In many of these readings, Hawthorne's vanishing allusions to Indians, his absence of allusions to slavery, and his conservative closure with Hester's final return appear as evidence of his investment in what Sacvan Bercovitch deems a liberal process of compromise and consensus, which ultimately advises that obedience to the law, however flawed the law may be (even if it meant sending escaped African Americans back into slavery), ultimately sets the nation free.[9] Others, however, have highlighted the same ambiguity earlier critics celebrated, finding in the narrator's sinuous movements and undecidable equivocations an invitation to readers to become active interpreters and, by extension, sympathetic, questioning citizens, including of the law.[1]

9. For Sacvan Bercovitch's argument, see his *The Office of the Scarlet Letter* (Baltimore: Johns Hopkins Univ. Press, 1991). Most readings acknowledge the ambiguity of voice and position in Hawthorne's work, but for critics who, like Bercovitch, align him most fully with a traditionalist orientation, see, for instance, David Leverenz, "Mrs. Hawthorne's Headache," *Nineteenth-Century Fiction* 37 (March 1983): 552–75; Myra Jehlen, "The Novel and the Middle Class in America," in *Ideology and Classic American Literature*, ed. Sacvan Bercovitch and Myra Jehlen (Cambridge, Eng.: Cambridge Univ. Press, 1986), 125–144; Jennifer Fleischner, "Hawthorne and the Politics of Slavery," *Studies in the Novel* 23 (spring 1990): 514–33; Larry J. Reynolds, *European Revolutions and the American Literary Renaissance* (New Haven: Yale Univ. Press, 1988), 79–96; Deborah Madsen, "'A' for Abolition: Hawthorne's Bond-Servant and the Shadow of Slavery," *Journal of American Studies* 25 (August 1991): 255–59; Gillian Brown, "Hawthorne, Inheritance, and Women's Property," *Studies in the Novel* 23 (spring 1991): 107–18; Deborah Gussman, "Inalienable Rights: Fictions of Political Identity in *Hobomok* and *The Scarlet Letter*," *College Literature* 22 (June 1995): 58–80; Lucy Maddox, *Removals: Nineteenth-Century American Literature and the Politics of Indian Affairs* (New York: Oxford Univ. Press, 1991); Laura Hanft Korobkin, "The Scarlet Letter of the Law: Hawthorne and Criminal Justice," *NOVEL: A Forum on Fiction* 30 (winter 1997): 193–217; and Renée Bergland, *The National Uncanny: Indian Ghosts and American Subjects* (Hanover, N.H.: University Press of New England, 2000); further references to *The National Uncanny* will be cited parenthetically as *NU*. See also Jamie Barlowe's study of the ways Hawthorne criticism has perpetuated this conservatism in overlooking the work of women scholars on Hawthorne (*The Scarlet Mob of Scribblers* [Carbondale: Southern Illinois Univ. Press, 2000]).

1. Critics who acknowledge Hawthorne's conservative gestures of containment but nonetheless consider his ambiguous narrative voice or his romance form an expression of subversive impulses include Michael Bell, *Hawthorne and the Historical Romance of New England* (Princeton, N.J.: Princeton Univ. Press, 1971); Evan Carton, *The Rhetoric of American Romance: Dialectic and Identity in Emerson, Dickinson, Poe, and Hawthorne* (Baltimore: Johns Hopkins Univ. Press, 1985); Gordon Hutner, *Secrets and Sympathy: Forms of Disclosure in Hawthorne's Novels* (Athens: Univ. of Georgia Press, 1988); Robert S. Levine, *Conspiracy and Romance: Studies in Brockden Brown, Cooper, Hawthorne, and Melville* (New York: Cambridge Univ. Press, 1989); Lauren Berlant, *The Anatomy of National Fantasy: Hawthorne, Utopia, and Everyday Life* (Chicago: Univ. of Chicago Press, 1991), although Berlant seems to give equal emphasis to Hawthorne's double impulses to subvert and conserve; Richard Millington, *Practicing Romance: Narrative Form and Cultural Engagement in Hawthorne's Fiction* (Princeton, N.J.: Princeton Univ. Press, 1992); Emily Budick, *Engendering Romance: Women Writers and the Hawthorne Tradition, 1850–1990* (New Haven: Yale Univ. Press, 1994); Brook Thomas, "Citizen Hester: *The Scarlet Letter* as Civic Myth," *American Literary History* 13 (summer 2001): 181–211; and Peter J. Bellis, *Writing Revolution: Aesthetics and Politics in Hawthorne, Whitman, and Thoreau* (Athens: Univ. of Georgia Press, 2003).

Rich as these many readings are, in substituting Hawthorne's historical surround for Hester's, they risk overlooking the most deeply historical dimension of Hawthorne's novel—his brooding on the relation *between* the 1640s and the 1840s. The research of Michael Colacurcio and Laura Korobkin (extending the suggestions of Amy Schrager Lang's scholarship on Anne Hutchinson and Hawthorne) helps to right this imbalance.[2] Colacurcio and Korobkin bring into sharp relief the work of Hawthorne's text in its own historical present by meticulously probing the (non)correspondence between the facts of seventeenth-century Puritan history and the picture of it that Hawthorne creates. While Colacurcio sees Hawthorne quietly indicting the Puritan elders more than we might at first think—and he crucially unveils the troubled coupling of sexuality and governance in the Puritan period—Korobkin argues that Hawthorne softens the portraits and punitive practices of the Puritan rulers in a way that makes more palatable his closing turn—Hester's resubmission to the law. It is worth briefly considering their arguments, for taken together with scholarship focused on the 1840s, they allow us to place Hawthorne's novel within the history of Atlantic modernity reaching from the seventeenth to the nineteenth centuries.

Most crucial in Colacurcio's and Korobkin's work is their identification of the "constitutional crisis" troubling the colony in the 1640s, the period of the novel's action.[3] Although neither gives any attention to the transatlantic nature of this crisis, their emphasis on Hawthorne's handling of the colonial side lays the foundation for a transatlantic view of Hawthorne's historical work. Colacurcio concludes that Hester is "caught up in the midst of a constitutional crisis," in which sexual misconduct by John Winthrop and others in the colony had thrown the authority of the governors into turmoil, so that, as he puts it, "the whole crisis seems to take Hester's 'adultery' as its fitting symbol."[4] He convincingly suggests that these political conflicts of the 1640s provide the context for "the sex-freedom link in *The Scarlet Letter*" ("F," 188). Yet what Colacurcio never sufficiently acknowledges, but Laura Korobkin explores, is the way that these connections are buried in Hawthorne's novel—so much so that no critic before Colacurcio had unearthed them.

By contrast to Colacurcio, Korobkin argues that Hawthorne suppresses rather than signals the political turmoil of the Puritan community. According to the laws of the day, the nature of Hester's punishment would not have been at the discretion of the magistrates—who, in Hawthorne's rendering, appear as mercifully lenient. There would have been a jury, and the jury would have insured that the magistrates followed the punishment preset for any particular crime—a procedure that had been arranged, after political wrangling, exactly so as to limit the discretion of the magistrates. In the case of adultery, Hester would, at minimum, have been publicly

2. See Michael J. Colacurcio, "Footsteps of Anne Hutchinson: A Puritan Context for *The Scarlet Letter*," in *Doctrine and Difference: Essays in the Literature of New England* (New York: Routledge, 1997), 177–204; and Korobkin, "The Scarlet Letter of the Law," 206–7. Further references to "Footsteps of Ann Hutchinson" will be cited parenthetically as "F." See also Amy Schraeger Lang, *Prophetic Woman: Anne Hutchinson and the Problem of Dissent in the Literature of New England* (Berkeley and Los Angeles: Univ. of California Press, 1987).

3. Michael J. Colacurcio, "'The Woman's Own Choice': Sex, Metaphor, and the Puritan 'Sources' of *The Scarlet Letter*," in *Doctrine and Difference*, 211.

4. Ibid., 210–11.

stripped and whipped. As Korobkin sees it, Hawthorne is "hard at work rewriting history to improve [the magistrates'] authority and compassion,"[5] In short, while for Colacurcio, the details of Puritan history establish that Hawthorne was a closet rebel and woman-sympathizer, for Korobkin they reveal him as an ameliorating apologist for authoritarian law.

It seems clear to me that in *The Scarlet Letter* at least, Hawthorne stills the volatility and veils the violence of the Massachusetts Puritan community for his readers, even as he may coyly signal their suppressed presence. Indeed, he suppresses history even more thoroughly than Korobkin suggests. For operating hand in hand with his muffling of political instability in Massachusetts are his suppressions of this colony's involvement not only in Indian wars but also in a transatlantic political crisis that would culminate with a king's beheading in 1649—the very year that Hester and Dimmesdale's relationship comes to its final crisis and Hawthorne's story-proper ends.[6] In short, Hawthorne's story, as he well knows, takes place in a colony flanked on one side by the peopled and troubled nation of England and on the other side by the peopled and troubled nations of Indian America, but as I will show presently, Hawthorne largely de-peoples these adjacent, interlocking communities. His softening of the violence (toward a woman such as Hester) within the colony extends to making absent the foundational violence of colonization.

That is, just as Hawthorne lifts the magistrates up onto a balcony and lifts Hester up onto a scaffold—neither of which is historically accurate—so he raises his history up out of the mess of Atlantic maneuvering in 1642—and, by extension, also keeps it at one remove from what Bercovitch characterizes as the "deep cultural anxiety" circulating in the 1840s.[7] This process of the "removal" of transatlantic history under the cover of an apparent immersion in history begins in his "Introductory," where he creates a virtual allegory of romance writing as sublimated colonial violence.

Garrison Republic, Native Spot

Hawthorne's "Introductory" tells the story of his own story, the Alpha-origin of his writing and of Hester's "A," as critics have noted. But it does so at one remove, through a logic of substitution and a rhetoric of exposure and confession that veils as much as it reveals. Hawthorne's "Introductory" marks Salem as a native spot that is no longer native and a scene of violence that is no longer violent, productively so for Hawthorne's authorship. His once-removed relation to this violent natality prefigures Hester's removal from her native spot in England while it also narrates such removals as journeys into a native freedom—and, in Hawthorne's case, native writing.

Readers have long recognized that Hawthorne both judges and praises his Puritan ancestors, but he is not simply being judicious. He is carefully managing the "ancestors." When he speaks of his "grave, bearded, sable-cloaked, and steeple-crowned progenitor" as a man "of war and peace . . . soldier, legislator, judge" with "all the Puritanic traits, both good and evil,"

5. Korobkin, "The Scarlet Letter of the Law," 206.
6. The novel opens in June of 1642, and it is "seven long years" later, in 1649 (as Hawthorne mentions more than once), that Dimmesdale gives his Election Day sermon [e.g., 98].
7. Bercovitch, *The Office of the Scarlet Letter*, 152.

Hawthorne at once registers and smoothes over the inherent tension between the qualities of legislator and soldier, and between their conflicting principles of freedom and colonization (9) [11]. Likewise in the novel, after mentioning that Governor Bellingham had led a regiment in the Pequod War, the narrator remarks: "For, though bred a lawyer, and accustomed to speak of Bacon, Coke, Noye, and Finch, as his professional associates, the exigencies of this new country had transformed Governor Bellingham into a soldier, as well as statesmen and ruler" (73) [68]. Much is compacted in the word "exigencies."

Even though Hawthorne makes these passing references to soldiering, neither his story nor his "Introductory" gives any attention to wars between Puritans and Indians; rather, he directs our gaze strictly to intra-community Puritan violence toward religious and moral transgressors like Hester Prynne. Hawthorne decoys any interest in the warring colonial surround exactly by emphasizing the Puritans' "persecuting spirit" (9) [11]. Thus one ancestor, he admits with seeming openness, "made himself so conspicuous in the martyrdom of the witches, that their blood may fairly be said to have left a stain upon him"—"So deep a stain, indeed, that his old dry bones, in the Charter Street burial-ground, must still retain it" (9) [11]. Hawthorne avows that "I, the present writer, as their representative, hereby take upon myself shame for their sakes, and pray that any curse incurred by . . . the race . . . may now be henceforth removed" (9) [11]. He seems unflinchingly to expose ancestral and Puritan violence among a tribe of "Britons" set down in a lonely wilderness.

But of course the "wilderness" was inhabited and the blood soaking the soil was more frequently that of Indian Americans. It is after all because of *this* blood-soaked soil that the Anglo-Saxons' primary "sin, [their] ignominy, were the roots which [they] had struck into the soil" (56) [53]. His rendering performs a double displacement of violence against Indians, in both the seventeenth and the nineteenth centuries. The "removal" he achieves is nicely expressed in his quaint use of the word "race" and in his wish to "henceforth remove" the curse on his own "race." He conjures the word's more archaic, kinship connotations and looks past its contemporaneous saturation by ethnographic, racist meanings. Via such substitutions, Hawthorne does indeed undertake the work of "removing the curse" from his race, but in a different sense than he implies.

At the end of his "Introductory," Hawthorne completes this equivocal turn by which he simultaneously condemns, cleanses, and lays claim to membership in the Anglo-Atlantic community. In the same bantering tone he has used all along to affiliate with while distancing himself from his ancestors and their contemporary incarnations in Salem, Hawthorne describes his relation to the republic that has employed him. In particular, he stresses the bureaucracy's demasculating effects, comically fashioning himself as its victim, now "decapitated" (33) [32]. In these descriptions, and in his allusion to the "political guillotine" (33) [32], Hawthorne implies his awareness of a long Atlantic history that begins with Charles I's beheading in 1649 and reaches through 1789 and 1848. He takes the role of the decapitated king whose spectacular death unleashes liberty, launches colonial "surveyor" projects, and generates history. Although he seems to affiliate himself with the king, by the end of his introduction he will have positioned himself as the republic's renewed native man.

To arrive there, Hawthorne again works through a number of submerged removals, as indicated by his private letters about his loss of position at the Custom House. While in his "Introductory" he uses a revolutionary republican vocabulary, in his private writing, as Renée Bergland has pointed out, he adopts quite a different metaphor for his response to his dismissal: that of an avenging Indian (*NU*, 157). Much of his anger was of course directed at Charles Upham, a one-time friend who had become leader of the Whig party in Essex County and had actively lobbied against Hawthorne's reinstatement.[8] Writing to Horace Mann, Hawthorne reports that he planned to "do my best to kill and scalp him," a plan he carries out in the "Introductory" by exposing the corruption in this key institution in Upham's district.[9] In a letter to Longfellow, Hawthorne similarly shares his plans to "immolate one or two of them,"[1] and in the letter to Mann he again invokes Indian-associated imagery in suggesting that the public responded, he says, as if he had "burned down the Custom-house and quenched its last smoking ember in the blood of a certain venerable personage" (*NU*, 157). In these fantasies, the author himself becomes the "removed" victim (as he similarly identifies with a fugitive slave when he reports that "it stirs up a little of the devil within me, to find myself hunted by these political bloodhounds").[2] This is the complex layering of Hawthorne's colonial work: identifying with the "removed" and violated outsider, taking up the very weapons of that wronged figure, he then occupies the place of that "vanishing" figure. In this way, the founding national violence against Indians is submerged into the story of abused *Anglo* native energy, with Hawthorne as the mock-hero who overcomes this injustice.

The benefits of this substitution are displayed by the fact that Hawthorne's ejection from the Custom House ultimately recovers the native man in himself and in turn enables his creation of the novel *The Scarlet Letter.* He explains that whatever the custom officer's former bravery on the battlefield or at sea, because the officer ensconced at the Custom House "leans on the mighty arm of the Republic, his own proper strength departs from him" (30) [29]. In this state, a man may lose his soul's "sturdy force, its courage and constancy, its truth, its self-reliance, and all that gives emphasis to manly character" (30) [30], for he becomes a servant with "the hang-dog look of a Republican official" (26) [26]. And yet "[i]f he possesses an unusual share of native energy, or the magic of place do not operate too long upon him, his forfeited powers may be redeemable. The ejected officer . . . may return to himself, and become all that he has ever been" (30) [29]. Implicitly, of course, Hawthorne is such a man who gets "ejected" only to discover enough "native energy" and "manly character" to "return to" his Anglo-American self and become again "all that he has ever been." While seeming to understand his ejection as a casting out *from* the republic, Hawthorne at the same time reaffirms a republican individualism in which "self-reliance" makes the native man.

8. See Thomas Woodson, "Hawthorne, Upham, and *The Scarlet Letter,*" in *Critical Essays on Hawthorne's "The Scarlet Letter,"* ed. David B. Kesterson (Boston: G. K. Hall, 1988), 186–87.
9. Hawthorne to Horace Mann, n.d., in *The Letters, 1843–1853*, vol. 16 of *The Centenary Edition of the Works of Nathaniel Hawthorne* (Columbus: Ohio State Univ. Press, 1997), 293; quoted in *NU*, 157.
1. Hawthorne to Henry Wadsworth Longfellow, 5 June 1849, quoted in Woodson, "Hawthorne, Upham, and *The Scarlet Letter,*" in *Critical Essays*, ed. Kesterson, 183.
2. Ibid.

Furthermore, it is within the republic's Custom House that Hawthorne discovers the native past that regenerates his writing, a past that once more sublimates an Indian presence within his own. Before he loses his post, he spends his time looking through old records kept in the second story of the Custom House, regretting the absence of records from the days of Cromwell, which would have "affected me with the same pleasure as when I used to pick up Indian arrow-heads" (23) [23]. The parallel signals the American colonist's double origin in a republic turned military protectorate and a deracinated native culture (23) [23], both of which have now become identity-forging pasts for the republican citizen. Although Hawthorne finds no old records, he is nonetheless pleased to find a substitute (and the logic of substitution, particularly substitution as the work of history, is everywhere): a packet of papers belonging to an eighteenth-century man, Mr. Pue, who is, tellingly, both a surveyor and "local antiquarian" (24) [24].[3] By way of Mr. Pue's papers, Hawthorne gains access to the drama of Hester Prynne and her embroidered scarlet letter, which takes place exactly in the Cromwellian period to which Hawthorne longs to return.

The beautiful red letter at first strikes Hawthorne as "one of those decorations which the white men used to contrive, in order to take the eyes of Indians—I happened to place it on my breast" (25) [25]—a juxtaposition that, together with his earlier allusion to Indian arrowheads, prefigures the full import of his story of the "wild" white colonist, Hester (whose free spirit will be repeatedly compared to that of American Indians). On his chest, the letter burns and it seems to him that the "ancient surveyor" Pue with his "ghostly voice" (who actually is of the eighteenth century but whom Hawthorne now antiquates) exhorts Hawthorne to tell Hester's story: "[D]o this, and the profit shall be all your own! You will shortly need it; for it is not in your days as it was in mine, when a man's office was a life-lease" (26) [26]. Indeed. This republic, which first makes Hawthorne dependent and drains all of his manly strength, will after all present him with a native history and an alternative income. It passes on to him the red "A," this "sign" of a fallen woman, whose story will not only replace and "take the eyes" of the Indian but will also thereby accrue a profit "all your own" to the white male author. The allegory Hawthorne writes is undoubtedly deeper than he realizes.

Interiority as Native History

Hawthorne begins his tale at "The Prison-Door." At this door, we join "[a] throng of bearded men . . . intermixed with women," awaiting the appearance of a fallen woman before the public eye (35) [35]. Our gaze is drawn to the legendary rose bush said to have "sprung up under the footsteps of the sainted Ann Hutchinson, as she entered the prison-door" (36) [36]. The mention of the antinomian rebel Hutchinson as the ghost who presides at the prison door—a threshold between interior and exterior as well as captivity and freedom—calls to mind Colacurcio's comment that "at one primal level, the whole antinomian controversy is about the inner and the outer, the private and the public person," for Hutchinson raised the question of what "our outward works, positive or negative, really reveal

3. For a full contextualization of the work of surveying and its influence on American literature, see Myra Jehlen, *American Incarnation: The Individual, the Nation, and the Continent* (Cambridge: Harvard Univ. Press, 1987).

about our salvation status" ("F," 193). If so, Hawthorne's novel does not simply allude to the antinomian controversy; it enacts it in its ambiguous play at this threshold.

The female prisoner who emerges, although led by the beadle, shows herself akin to the native author, for at "the threshold of the prison-door, she repelled him, by an action marked with natural dignity and force of character, and stepped into the open air, as if by her own free-will" (39) [38]. Like many an Atlantic protagonist before her, including Ann Hutchinson, Hester exudes a natural dignity that is the mark of her free self. Indeed her dignity seems to lift her "out of ordinary relations with humanity, and inclos[e] her in a sphere by herself" (40) [39]. We are given this image of a free, female self absolutely apart, and then, after several pages (during which the narrator hovers outside her, fixed like the spectators by the scarlet letter, and simultaneously piquing his readers' desire to enter, to make that boundary between inner and outer as transparent as ever "our fathers" could wish), the narrator finally takes us—not the townspeople—across the threshold into Hester's consciousness, her interior prison and freedom. Impossibly, we witness her aloneness—an image of our own—and so too receive a reassuring image of our interiority as something witnessed, communal, and free.

And native. For this interior is not only "marked with natural dignity and force of character" but it also contains a history, a familiar Anglo-Atlantic history that has become Hester's and the Anglo reader's psychological history. As Hester stands on the scaffold, the narrator makes us privy to her memories of the modest cottage of her childhood "retaining a half-obliterated shield of arms" in her "native village in Old England" (43) [42]. With her natural dignity and force of character, she is the effaced, perhaps pre-Norman nobility of English history.[4] It is because "the tendency of [Hester's] fate and fortunes had been to set her free" that this Anglo-Atlantic woman crosses the Atlantic, has an affair with the minister, gives birth to an illegitimate child, and nonetheless "step[s] out into the open air" of the New World and lives a long life in the colony (136, 40) [120, 38]. Hester's native self, like that of many an Anglo-Atlantic traveler, manifests a freedom-hunger that appears as essentially interior, individual, and ahistorical.

Only by such a fashioning can Hester stand in her raised position as a paragon of the modern Anglo-Atlantic and national self, implicitly carrying forward the colonizing project with impunity. The novel casts her as a figure of both release and lonely subjectivity living on the colonial Atlantic seashore:

> Standing alone in the world,—alone, as to any dependence on society, and with little Pearl to be guided and protected,—alone, and hopeless of retrieving her position, even had she not scorned to consider it desirable,—she cast away the fragments of a broken chain. The world's law was no law for her mind. It was an age in which the human

4. Frederick Newberry stresses Hawthorne's evocations of "old world" England, noting that "Hawthorne regularly surrounds Hester, Pearl, and Dimmesdale with Old World motifs . . . to emphasize positive historical and cultural continuities" (*Hawthorne's Divided Loyalties: England and America in Hawthorne's Works* [Toronto: Associated University Presses, 1987], 173). Newberry's study reinforces my transatlantic emphasis, although our interpretations differ.

> intellect, newly emancipated, had taken a more active and a wider range than for many centuries before. Men of the sword had overthrown nobles and kings. Men bolder than these had overthrown and rearranged—not actually, but within the sphere of theory, which was their most real abode—the whole system of ancient prejudice, wherewith was linked much of ancient principle. Hester imbibed this spirit. She assumed a freedom of speculation, then common enough on the other side of the Atlantic, but which our forefathers, had they known of it, would have held to be a deadlier crime than that stigmatized by the scarlet letter. In her lonesome cottage by the sea-shore, thoughts visited her. . . . (112–13) [101]

By way of this characterization of Hester's isolation and *imaginative* freedom (of a piece with that of the free-thinking men whose revolutions are mainly in "the sphere of theory"), Hawthorne makes the colony a more innocent place than it was and makes freedom a less materially "levelling" force, In isolating both Hester and the colony, he occludes the active world of transatlantic trade, travel, interloping, and political maneuvering. As we have seen, the Massachusetts colony was fully involved with events and people in England and on the continent; and the rebellious events in England rocked the Puritan colony at every turn. In fact, indirectly and sometimes in body, the Puritan colonists were the very men who, by sword and print, were at this moment overthrowing nobles and kings. While Hawthorne might not have had full knowledge of these networks, his text erases them altogether. The narrator not only places all such rebels "on the other side of the Atlantic," he characterizes "our forefathers" as relatively ignorant of their free thinking. But Puritanism itself entailed "freedom of speculation" in religious as well as legal practice, which is why it was so difficult to draw the line against antinomian innovations—because they were actually extensions of Puritan innovations. And the Puritan freedom of "speculation" was economic and geographical as well as spiritual, legal, and intellectual.

In other words, Hester's new-world adultery—far from representing something the "forefathers" could not in their pristine innocence grasp—is of a piece with this speculative venture that searches out and claims possession of new-world sources of political, financial, and sexual liberty. The fact that Hester comes to live at the center of the community (sewing the official garments of the Governor, presiding at births and deaths, drawing the gaze and taunts of children) taken together with the text's hints about colonial politics and corruption that Colacurcio and Korobkin trace, indicates that Hawthorne at some level understood such women's pivotal role in the colony as embodiments of a "sex-freedom link" requiring carefully contained manipulation.

That is, by making Hester a singular and radically interior self, by quietly dehistoricizing her and casting her as one who doesn't actually want social membership in this blood-tainted and hypocritical community ("In all her intercourse with society, however, there was nothing that made her feel as if she belonged to it" [59] [56]), Hawthorne allows us to embrace her as the rebel-progenitor of "our" community. Thus the seductive if paradoxical racial dream takes hold: as isolated soul, she expresses the essence of the race and becomes the avatar of a free community of readers. And in a further irony, her race essence finds expression exactly insofar as she absorbs and sublimates the "freedom" of American Indians.

Indian-Saxons

It is the ability to live in isolation, to survive in a cottage alone on the shore of a strange continent, with all of her freedom interiorized, that makes Hester the most successful colonist and the queenly ancestor of an Anglo-American reading community. After her public humiliation, as the narrator emphasizes, Hester remains in her community even though she is "free to return to her birthplace, or to any other European land, and there hide her character and identity under a new exterior, as completely as if emerging into another state of being" (56) [53]. But Hester has already emerged into another state of being, and it gives her power in the colony—not least because this new state of being entails her internalization and sublimation, in Hawthorne's rendering, of Indian powers.

With varying degrees of critical distance, readers have remarked on the novel's affiliation of Hester with Indians, beginning at least with Leslie Fiedler, who calls Hester "the wildest Indian."[5] Parallel to the operations of what Toni Morrison calls Africanism in other Anglo-American fiction, *The Scarlet Letter* is one founding text for the practice of Indianism: the Indian's freedom or "wildness" gets absorbed into the stories of white characters, in a racial sleight of hand that enhances, ironically, the nativeness of the whites' free interiors.[6] Hawthorne first of all conjures the possibility that Hester could escape her shame by traveling west, for "the wildness of her nature" is such that it "might assimilate itself with a people whose customs and life were alien from the law that had condemned her" (56) [53]. Hawthorne makes the American Indian a model for Hester's freedom, remarking that "[h]er intellect and heart had their home, as it were, in desert places, where she roamed as freely as the wild Indian in his woods" (136) [120]. And so she adopts the Indian perspective on her culture: "For years past she had looked from this estranged point of view at human institutions, and whatever priests or legislators had established; criticizing all with hardly more reverence than the Indian would feel for the clerical band, the judicial robe, the pillory, the gallows, the fireside, or the church" (136) [120].[7]

And paradoxically, exactly because she adopts the Indian point of view, she is at home as a colonist, so that "[i]t was as if a new birth, with stronger assimilations than the first, had converted the forest-land, still so uncongenial to every other pilgrim and wanderer, into Hester Prynne's wild and dreary, but life-long home." In fact, "even that village of rural England . . . like garments put off long ago—[was] foreign to her, by comparison." Hester is an Anglo-Atlantic creature of modernity who can travel if she chooses, but

5. See the excerpt from Fiedler's *Love and Death in the American Novel* in *Hester Prynne*, ed. Harold Bloom (New York: Chelsea House, 1990), 69.

6. For Toni Morrison's discussion of "American Africanism," see *Playing in the Dark: Whiteness and the Literary Imagination* (New York: Vintage, 1993).

7. Hawthorne may also gesture toward the slave economy "holding up" Hester's interior freedom in his reference to the "iron" arm of the law and in the later comment that "[t]he chain that bound her here was of iron links, and galling her to the inmost soul, but never could be broken" (56) [53]. On Hawthorne, Hester, and slavery, see Fleischner, "Hawthorne and the Politics of Slavery"; Madsen, "'A' is for Abolition"; and Korobkin, "The Scarlet Letter of the Law." For discussions linking Hawthorne's text to slave narratives, see Mara Dukats, "The Hybrid Terrain of Literary Imagination: Maryse Condé's Black Witch of Salem, Nathaniel Hawthorne's Hester Prynne, and Aimé Césaire's Heroic Poetic Voice," in *Order and Partialities: Theory, Pedagogy, and the "Postcolonial,"* ed. Kostas Myrsiades and Jerry McGuire (Albany: State University of New York Press, 1995), 325–40; and Jane Cocalis "The 'Dark and Abiding Presence' in Hawthorne's *The Scarlet Letter* and Toni Morrison's *Beloved*," in *The Calvinist Roots of the Modern Era*, ed. Aliki Barnstone, Michael Tomask, and Carol J. Singley (Hanover, N.H.: University Press of New England, 1997), 250–62.

she does not do so because "[h]er sin, her ignominy" and her emergence "into another state of being" were "the roots which she had struck into the soil" (56) [53]. Much as she seems an outsider to the community, this is not strictly so: "The very law that condemned her—a giant of stern features, but with vigor to support, as well as to annihilate, in his iron arm—had held her up, through the terrible ordeal of her ignominy" (55) [52]. As such, she is the paradigmatic figure for an Anglo nation's future on this land.

While some readers continue to consider the affiliation of Hester with Indians as a mark of her position as a "non-citizen" who "threatens the hegemony" of the Puritan ideology,[8] other recent critics, Bergland most astutely, understand Hester as one of Hawthorne's vehicles for instilling Indian presence into his own writing in a way that authorizes his role as national author. Bergland persuasively argues that "the internalization of Native American qualities was central to [Hawthorne's] process of writing" (*NU*, 156). She tracks the process from Hawthorne's observation (laced with resentment) that "no writer can be more secure of a permanent place in our literature than the biographer of Indian chiefs" through his writing of the next two decades in which Indian characters appear as catalyzing spectres—exactly during the period in which the policy of Indian removal was put into law and, in Illinois, Florida, and Oklahoma, violently enforced, amid loud voices of dissent in Massachusetts.[9] Bergland finds a combination of attraction and repulsion toward Indian "wildness" in these stories, tamed by closing tropes of vanishing Indian presences. In "The Old Manse," where the surrounding land is scattered with Indian relics and haunted with Indian presences, Hawthorne and his companions emulate what they imagine as "freeing" Indian ways: "Strange and happy times were those, when we cast aside all irksome forms and straight-laced habitudes, and delivered ourselves up to the free air, to live like Indians."[1]

Bergland further suggests that in *The Scarlet Letter* "each of the main characters is transformed into an Indian, or at the very least, described as internalizing Indian consciousness" (*NU*, 157). I would add that all three main protagonists follow an Atlantic trajectory that brings them from Europe to America and into association with an Indian presence that at least temporarily enhances their quests for freedom—religious, scientific, or sexual. Reverend Dimmesdale, whom Hawthorne casts as a quintessentially pure Protestant Anglo-Saxon,[2] had come to what the

8. See, for instance, Gussman, "Inalienable Rights," 9.
9. Nathaniel Hawthorne, "Our Evening Party among the Mountains" (1835), in *Tales and Sketches/Nathaniel Hawthorne*, ed. Roy Harvey Pearce (New York: Library of America, 1982), 342–43; quoted in *NU*, 145.
1. Nathaniel Hawthorne "The Old Manse," in *Mosses from an Old Manse* (Boston: Houghton Mifflin, 1882); quoted in *NU*, 155.
2. The narrator also associates Dimmesdale with a native Anglo-Saxon legacy of pristine, pre-Catholic innocence. Like the poetic, brooding, yet honest Anglo-Saxons, he prefers "seclusion," which "kept [him] simple and childlike" and gave his sermons "a freshness"; meanwhile his "high native gifts" and his "dewy purity of thought" are expressed in a voice that sounds "like the speech of an angel"—or perhaps, a sublime Angle, for there was a widespread misconception that the Angles of England first got this name because their blond hair made them look like angels (48) [46]. Furthermore, Dimmesdale's interior, like Hester's, is betokened by ancient Protestant nativeness, as suggested in the scene in which Chillingworth enters Dimmesdale's study to discover his innermost secret. Dimmesdale has been reading a book in the vernacular: a "large blackletter volume [lay] open before him on the table" (95) [86]. In the seventeenth century, "black letter" print was used for the vernacular and was closely associated with the Reformation (starting with the vernacular Bible through which it came into use), and thereafter with the "recovery" of "ancient" native identity through native-language literacy. Dimmesdale's "deep,

narrator calls "*our* wild forest-land" from one of "the great English universities"; once arrived, he of course makes his regular visits into the forest to redeem Indians—along the way meeting with Hester Prynne (48, my emphasis) [46]. Roger Chillingworth comes from England via Germany to America, where, like Hester and Dimmesdale, he mingles his old-world knowledge with his "potent" new-world discoveries. Some colonists imagine that Chillingworth has been "transported . . . bodily through the air" by heaven from "a German university" to work his "cure" upon Reverend Dimmesdale (which Claudia Johnson argues is a potion to induce impotency).[3] He strengthens his scientific powers by combining "knowledge of the properties of native roots and herbs" gained during his "Indian captivity" with the "antique physic" of "European Pharmocopoeia" (82) [76] and his "old studies in alchemy" (51) [49]. As Chillingworth and Dimmesdale take "long walks on the seashore" and Chillingworth gathers native "plants with healing balm in them," Dimmesdale finds himself attracted by the "range and freedom of ideas" that Chillingworth exhibits as "a man of science" (85) [78]. These two men share the impulse to invigorate their free-thinking knowledge through encounters with the "savage."

Yet in the end, Dimmesdale's English "native" interior cannot sustain the encounter with America—at least not insofar as it also entails a homosocial struggle with Chillingworth, especially as the latter draws on the powers of his hybrid "Pharmocopeia." Indeed, in this struggle between Chillingworth and Dimmesdale over Hester—the woman who most enduringly strikes roots into the soil and most successfully makes the colony a place where she can cultivate her free interior—Hawthorne begins to sketch the sexual predicament of the colonial Anglo-American man, including himself.

Sex in the Colony

The Scarlet Letter may after all be most fundamentally concerned with the crisis that colonization provokes in Anglo-Atlantic heterosexuality. From one angle, the novel narrates the aftermath of an Atlantic rupture in which each of the main characters becomes an isolated individual with an interiority that exceeds community membership and so, particularly in the woman, threatens proper patriarchal coupledom. Certainly, among its other effects, colonization created a margin of possibility for being single and for other sexual choices among Anglo women as well as Anglo men, epitomized in the so-called "New England marriages" of single women who chose to live together as well as in those scribbling women Hawthorne lamented. As usual, Hawthorne turns his gaze to what most worries him, for these new-world conditions do seem to bring what, in the essay "Mrs. Hutchinson," he feared would be the end of a "race" of "domestic" Anglo-American women, which occurs by their adoption of an "Indian" freedom, an outcome figured, in that essay, as the independent woman's ultimate "ruin" by Indians and, in the novel, by Hester's

deep slumber" over the book implies his seduction by the Anglo-Protestant dream of a fully manifested interior, which in turn confirms his native character (95) [86]. On black-letter print, see Charles Mish, "Black Letter as a Social Discriminant in the Seventeenth Century," *PMLA* 68 (June 1953): 627–30; and Gerald Newton, "*Deutsche Schrift*: The Demise and Rise of German Black Letter," *German Life and Letters* 56 (April 2003): 183–204.

3. See Claudia Durst Johnson, "Impotence and Omnipotence in *The Scarlet Letter*," *New England Quarterly* 66 (December 1993): 594–612.

lonely but dignified life.[4] The increased independence of women seems required for colonization, however, and colonization is a project that Hawthorne embraces by instinct if not by love. Hawthorne wrestles with this trade-off required by the continuing project of colonization, including in Hawthorne's own day the implicit exchange wherein Anglo-Atlantic men's hold on Indian lands entailed some loosening of their hold on ("Indianized") Anglo-Atlantic women.

In key scenes throughout the novel, we glimpse Hawthorne's narrator grappling with the transformation of women and the reach of his heroine's freedom, especially insofar as, like the writing women of his day, Hester's freedom of thought rivals his.[5] At one point, Hester fully explores in her mind how the ideal of freedom has implications for "the whole race of womanhood" and sees the need for "the whole system of society . . . to be torn down, and built up anew" (113) [101, 102]. In particular, she touches exactly on the distinction between nature and culture on which the system rests, for, she thinks, "the very nature of the opposite sex, or its long hereditary habit, which has become *like* nature, is to be essentially modified, before woman can be allowed to assume what seems a fair and suitable position" (113, my emphasis) [102]. Yet as the narrator also tells us, Hester fears the danger that along the way "she herself shall have undergone a still mightier change; in which perhaps, the ethereal essence, wherein she has her truest life, will be found to have evaporated" (113) [102].

Here is the abyss that opens up under liberty in this colonized land, native only by force and interior fabrication. In this world that seems to offer complete freedom, perhaps eventually no one will have an essential and permanent self. Perhaps we will discover that we have no ultimate bond to, or steady identity within, a community that remains intact over time. Perhaps we will find that what seemed like "nature" is only "hereditary habit." So if we "tear down" the hereditary habit, including the habit of "the opposite sex"—which is exactly what Hawthorne has done—we may lose our moorings.

Tellingly, however, it is at this moment that our narrator steps abruptly out of Hester's consciousness to announce that "[a] woman never overcomes these problems by any exercise of thought," having suggested a bit earlier that Hester had already undergone a "sad transformation," in which "some attribute had departed from her, the permanence of which had been essential to keep her a woman" (113, 112) [102, 100]. At this point we may wonder if it is our narrator, more than Hester, who is "wander[ing] without a clew in the dark labyrinth of mind" (114) [102]—exactly the labyrinth of a mind that on the one hand seeks an attribute "the permanence of which had been essential" and yet on the other unsettles the possibility of that permanence precisely in this restless "wilderness" seeking.

In response, Hawthorne, to steady his hold, imitates Hester's most successful colonial strategy: keeping interiority contained, maintaining a

4. For these phrases in Hawthorne's essay, see "Mrs. Hutchinson," in vol. 23 of *The Centenary Edition of the Works of Nathaniel Hawthorne*, ed. Thomas Woodson et al. (Columbus: Ohio State Univ. Press, 1994), 66–67 [159]. For a full consideration of "Mrs. Hutchinson," especially the closing scene in which Hawthorne imagines Hutchinson's death in an Indian attack, see my *Freedom's Empire*.

5. In "Mrs. Hutchinson," Hawthorne deplores the increasing number of "ink-stained Amazons" because these "prolific" writers threaten to "expel their rivals"—that is, male writers (*Centenary Edition*, 23:67) [159].

threshold between private and public, and only very selectively opening the door to cross it. All of Hawthorne's tales ultimately rest on a narrative opacity, keeping the veil over an obscure interiority. On one hand, his narrators pursue enigmatic characters with the tenacity of Chillingworth, who "strove to go deep into his patient's bosom." Hawthorne similarly seems to seek what this novel's narrator calls Chillingworth's "power . . . to bring his mind into such affinity with his patient's, that this last shall unawares have spoken what he imagines himself only to have thought" (86) [78, 79]. On the other hand, Hawthorne also practices a more measured artistry that succeeds where Chillingworth's fails. Dimmesdale collapses under Chillingworth's too-close scrutiny. Hawthorne, by contrast, never takes us over the threshold into Dimmesdale's interior. In the crucial scene in which Dimmesdale falls asleep over his ancient "black-letter" book, Chillingworth thrusts back the vestment and sees, apparently, a sign, but we readers never do. Hawthorne shows us only the disturbingly gleeful face of Chillingworth as it comprehends what he sees. Hawthorne shows us, that is, the *desire* to penetrate to the deepest interior, simultaneously heightening and constraining his (and our) desire to see, by keeping that interior cloaked. Similarly with the meteoric sign in the sky, we learn only the townspeople's fully cathected speculations about it. Hawthorne keeps uncertain the reality of the meteor, and in the process keeps open the question of God's endorsement or condemnation of Dimmesdale, and, by extension, keeps in shadow the degree of Dimmesdale's colonial guilt. Hawthorne *keeps* the secret of these characters' "sin"—the colonial one, that is, of striking roots into stolen soil—and he accordingly does the same for his audience.

In short, by emulating not Chillingworth but the diffident Hester, Hawthorne finds a more effective way to write his way into Atlantic modernity as a man. Like Hester, he successfully re-nativizes this New England colonial spot and keeps its protective threshold intact. He understands that in founding a native community on stolen soil, one may penetrate a woman's interior in order to establish and protect a man's. Fictional women may be cast as having a free interiority, and that free interiority may justify colonization, all of which—if distilled into allegory—can accrue to the male author, just as Surveyor Pue promised.

Thus I suggest that we might read the final image of Hester as she returns—an image in which her interior remains inaccessible—as an image of our author, Hawthorne. For after all, as he confesses in his "Introductory," it is he who must return and plant himself in the New England village if he is to write a myth of Anglo-American origins. Appropriately, in the novel's final scene, Hawthorne positions us once again at the threshold, on the Atlantic shore, with Hester at her cottage door. We meet her as the figure turning between two worlds, a woman choosing the abode of colonization but whose free interior after all remains mostly in shadow—apolitical, latent. Hawthorne makes a "native" woman's interior freedom the veiled vessel of Anglo-Atlantic colonization.

The Custom-House

JOHN FRANZOSA

"The Custom-House," *The Scarlet Letter*, and Hawthorne's Separation from Salem†

> I desire to set before my fellows the likeness of a man in all the truth of nature, and that man myself. Myself alone! I know the feelings of my heart, and I know men. I am not made like any of those I have seen. I venture to believe that I am not made like any of those who are in existence. If I am not better, at least I am different.
>
> —*Confessions*, Rousseau

On March 21, 1850, the Salem *Register* printed a long review of Nathaniel Hawthorne's *The Scarlet Letter*, published just five days previous. In the opening paragraph, the reviewer praised the "exquisiteness of Hawthorne's genius" and the "affluence of imagination and bold and striking thought" which sustained his narrative.[1] This much took up less than a quarter of the review; for the rest, the reviewer felt that "justice compels us to notice some other things": the "small sneers at Salem," and the "calumnious caricatures of inoffensive men," one of whom was a "venerable gentleman, whose chief crime seems to be that he loves a good dinner, has preserved a youthful flow of cheerfulness, and can tell a graphic story." The reviewer likened such "refinement of cruelty" to the vengeance of Roger Chillingworth but asserted that Hawthorne had no "visible motive for so much malice."

There was, of course, sufficient motive in the matter of Hawthorne's removal from the Custom House amid charges of untoward partisanship and corruption.[2] Hawthorne had threatened his revenge in a letter to Longfellow, and in later correspondence he wrote somewhat approvingly of the notoriety this sketch had won for him.[3] His response to the *Register*'s

† From *ESQ* 24 (1978): 57–71. Reprinted by permission. Page numbers in square brackets refer to this Norton Critical Edition.

1. This review, probably written by John Chapman, the paper's editor and a virulent Whig, is reprinted in Benjamin Lease, "Salem vs. Hawthorne: an Early Review of *The Scarlet Letter*," *New England Quarterly* 44 (1971): 110–117.
2. For the particulars of this case, including a reprint of the "Memorial" signed by Charles W. Upham for the Salem Whig committee which brought these charges against Hawthorne, see Winfield S. Nevins, "Nathaniel Hawthorne's Removal from the Salem Custom House," *Essex Institute Historical Collections* 53 (1917): 97–132. Ephraim Miller, the old general's son, was the main target of the Whigs and not Hawthorne, who was felt to be "the abused instrument of others" (118). See also B. Bernard Cohen, "Edward Everett and Hawthorne's Removal from the Salem Custom House," *American Literature* 27 (1955): 245–49; and Kenneth W. Cameron, "New Light on Hawthorne's Removal from the Salem Custom House," *Emerson Society Quarterly* 23 (2nd quarter 1961), 2–5.
3. See Frank MacShane, "The House of the Dead: Hawthorne's 'The Custom-House' and *The Scarlet Letter*," *New England Quarterly* 35 (1962): 98; Horatio Bridge, *Personal Recollections of Nathaniel Hawthorne* (New York: Harper & Bros., 1893), 114; and Lease, 112. See also William Charvat, "Introduction," *The Scarlet Letter*, Centenary Edition (Columbus: Ohio State UP, 1962), xxiii. All references to Hawthorne's writing will note volume and page of this edition.

criticism in his preface to the second edition was "surprise" and "amusement" at the "respectable community" for having been aroused to such a pitch of "unprecedented excitement" by the "atrocities" committed in his introduction. He defended the "general accuracy" of his characterizations, affirmed the "genuine good humor" of the sketch and—perhaps mocking the *Register*'s compelling need to attack him—felt "constrained" to reprint the entire sketch "without the change of a word" (I, 1–2) [5].

In this preface, Hawthorne demonstrates a precious obstinacy which underscores the *Register*'s complaints through his diffident rejection of them. Perhaps the "outrageous personalities" portrayed had served him as objects on which to "vent his spite"; perhaps ineffectual old men had been vilified for their dependency and particularly their love of food; and perhaps a "strange antipathy toward the aged" had manifested itself and had apparently infected the romance as well. As evidence of that antipathy, a passage had been quoted at length which appeared to satirize the piety of a widow with a "heart full of reminiscences about her dead husband and children" who had "fed herself" with "religious consolations" (I, 218) [130]. More accurately than many modern readings of the romance, I think, this first criticism not only demonstrated the unity of this sketch and the narrative it introduces, but unwittingly identified the important psychological themes operating in *The Scarlet Letter.* Where psychoanalytic and less doctrinaire "Romantic" readers have discussed the individual's rebellion against authority, the self's struggle to mediate desire and guilt, the isolated individual's effort to realize a viable self against an alienating community, this reviewer responded to themes of dependence—nourishment, loss, and rejection—themes which because of their developmental priority shape more mature relationships between self and other.[4]

4. In its psychological implications, most Hawthorne criticism—and criticism of "The Custom-House" in particular—proceeds from an essentially Freudian view of conflict. MacShane; and David Stouck, "The Surveyor of the Custom-House: A Narrator for *The Scarlet Letter,*" *The Centennial Review* 15 (1971): 309–329, are most explicit in this regard. But other critics, such as Nina Baym, "The Romantic *Malgré Lui:* Hawthorne in the Custom House," *ESQ* 19 (1973): 14–25; Dan McCall, "The Design of Hawthorne's Custom-House," *Nineteenth-Century Fiction* 21 (1967): 349–58; Charles R. O'Donnell, "Hawthorne and Dimmesdale: The Search for the Realm of Quiet," *Nineteenth-Century Fiction* 14 (1960): 317–32; and Marshall Van Deusen, "Narrative Tone in 'The Custom-House' and *The Scarlet Letter,*" *Nineteenth-Century Fiction* 21 (1967): 61–71, also frame their arguments in terms of a struggle between authority and the beleaguered individual. Paul John Eakin, "Hawthorne's Imagination and the Structure of 'The Custom-House,'" *American Literature* 43 (1971): 346–58, on the other hand, views this authority as dead: there is a necessary journey to the source of creative power, a necessary "communion with the dead," that precedes Hawthorne's "salvation." See also Jean Normand, *Nathaniel Hawthorne: An Approach to an Analysis of Artistic Creation*, trans. Derek Coltman (Cleveland: Case Western Reserve UP, 1970). James M. Cox, "*The Scarlet Letter*: Through the Old Manse and the Custom House," *Virginia Quarterly Review* 51 (1975): 432–47; Thomas H. Pauly, "Hawthorne's Houses of Fiction," *American Literature* 48 (1976): 271–91; and Harry C. West, "Hawthorne's Editorial Pose," *American Literature*, 44 (1972): 208–221, though explicitly non-psychoanalytic—in fact, Cox carefully distinguishes his position from that of Frederick Crews, *The Sins of the Fathers* (New York: Oxford UP, 1966)—have offered the most useful insights for a psychoanalytic reading of "The Custom-House." All three see Hawthorne in a sense as a man "whose art caused his life"; all three comment on the way Hawthorne focused (and still focuses) discussion on the processes of writing and not on self-revelation. I think implicit in their views is a psychological argument that as a person, Hawthorne was first and foremost a performer. A glance at his early letters to his mother, sisters, uncles, and wife should confirm the primacy of literary performance in even his most intimate social and personal relationships.

I

According to contemporary psychoanalytic theory, the key relationship in the first year of life is the "symbiotic dyad." Within weeks after its biological separation from the mother, the infant is "seduced" into a psychological symbiosis with a maternal figure. Words cannot adequately describe this relationship, since it predates the infant's acquisition of language and use of symbols; and the term "relationship" is misleading because in this experience there is no differentiation between self and other.[5] Using a biological metaphor, Heinz Lichtenstein describes this dyadic relationship as a "schema" or a maternal *Umwelt*—not an environment, but a range of experience in which the infant figures as an organ within a larger organism. The mother "imprints" the infant with a specific mode of being for her, a specific way of relating. This experience will have a determining effect on the infant's separation from the maternal *Umwelt*, its sense of itself as different from its mother and its ability to individuate itself as a separate entity, a self. At a later time, the child enters into triadic relationships—typically the oedipal configuration—when sexual differences are realized, gender identity is formed, language is acquired, and the experiences of conflict and ambivalence arise. For a boy, the father is not merely the prohibitive Victorian censor but a mediator between boy and mother—an object which the boy's emergent identity requires to complement his identification with his mother and to complete the process of separating from her and individuating as a male person in his own right. Throughout life, as Lichtenstein argues, identity remains most fundamentally an "experience of a potential instrumentality for another."[6]

As *The Scarlet Letter* opens, a "throng of bearded men . . . intermixed with women" is seen gathered outside the prison door from which Hester Prynne and her infant will emerge. It is the figure of maternity which immediately captures interest: the metaphoric, the literal, and, I suggest, the latent maternal women who, through the figure of Hester, provide the matrix from which characters, motives, and meanings will separate, clarify, conflict, and resolve. The roses near the threshold of the prison seem outward signs of the "deep heart" of "Nature," just as the prison itself is described as the "black flower of civilization." That these roses might have "sprung up under the footsteps of the sainted Ann Hutchinson" suggests perhaps a "civilized" matrix born of a more natural one, or perhaps the possibility that history mediates between repressive civilizing forces and natural, creative ones. At any rate, the figure of maternity seems to provide the ground on which the drama is played out, be it the "native soil" (or the "infertile soil" on which Hester lives in exile), the legacy of the ancestral martyr, the larger comprehensiveness of Nature, the influence of the spectral Mistress Hibbins, the piety of Dimmesdale's housekeeper, or the community of vindictive Puritan women.

5. Heinz Lichtenstein, "Identity and Sexuality," *Journal of the American Psychoanalytic Association*, 9 (1961): 200. See also Margaret Mahler, *On Human Symbiosis and the Vicissitudes of Individuation* (New York: International UP, 1966); Michael Balint, *The Basic Fault: Therapeutic Aspects of Regression* (London: Tavistock Publications, 1968); and Phyllis Greenacre, *Emotional Growth* (New York: International UP, 1971).
6. Lichtenstein, 203.

From the beginning, *The Scarlet Letter* offers a particular type of maternal figure which embodies both masculine and feminine characteristics. As the reader awaits Hester's entrance, Hawthorne provides examples of criminals likely to be punished, all of them implying transgressions of dependency relationships. Those guilty of such private crimes would ordinarily be punished in public, according to Hawthorne, because religious and legal authority were "almost identical" and "interfused" (I, 50) [36] in the Puritan theocracy. Symbolically, the law is associated with masculine characters and images such as "iron" and "oak"; religion is associated more with women: Anne Hutchinson, the Quaker woman whipped by Major Hathorne, and the "whore of Babylon," the Catholic Church. The social structure suggests a composite parental figure, and so does history: Anne Hutchinson—whose footsteps germinate roses—and Queen Elizabeth.[7] Fed on the "beef and ale of their native land," the "not unsubstantial" Puritan women of Boston "stood within less than half a century of the period when the man-like Elizabeth had been the not altogether unsuitable representative of the sex" (I, 50) [37].[8]

These conflated parental figures suggest a fantasy of a "man-like" or phallic mother, a fantasy which dominates this romance. Such a fantasy ordinarily serves to deny the fact of sexual difference by fusing both parents in one image. In cases where the father is absent (as Hawthorne's was), it can serve to deny this loss and allow a boy to identify with his lost father by identifying with a fantasy of his mother somehow embodying his father.[9] Such a strategy, however, can obviously lead to role and gender confusions. "Plying her needle," Hester Prynne "possessed an art" which "might have been a mode of expressing and therefore soothing, the passion of her life," except that she "rejected it as sin" (I, 84) [55]. As it is, her "rich, voluptuous and Oriental" character found nothing else to "exercise itself upon" but the "exquisite productions of her needle," commissioned by the more affluent women of Boston. Such "delicate toil" is a pleasure "incomprehensible to the other sex" (I, 83) [55]. In contrast to her phallic, active qualities, the dominant features of her partner in sin, Arthur Dimmesdale, are his "tremulous lips" and his hand compulsively clutching his heart. It is Hester's needle which affords her—in her privacy—a "part to perform in the world" even if, in her "intercourse with society," there was "nothing that made her feel as if she belonged to it."

With the absence of Pearl's father, Hester as mother has been forced to become active and self-reliant—in nineteenth-century terms, "manly." And the effect of such a maternal figure on succeeding generations is painfully obvious to Hawthorne. The particular importance of the mother's influence on the child is demonstrated as Hester recalls her mother's face with the "look of heedful and anxious love which it always wore," a look

7. Footsteps, particularly when combined with thresholds, usually imply a phallic significance for Hawthorne: in one of his earliest stories, "My Kinsman, Major Molineux," a lady with scarlet petticoats takes the young Robin and draws "his half-willing footsteps nearly to the threshold" before the "keeper of midnight order" scares her away; and one of the unfinished romances concerns a blood-stained footprint on a threshold, the sign of a lost ancestor. Characters do not merely walk through doorways—Jaffrey Pyncheon in *The House of the Seven Gables*, for instance, intrudes, penetrates, or thrusts himself into others' private spaces.
8. The "man-like Elizabeth" suggests an implied pun, since, as he wrote, Hawthorne "stood within less than half a century" of the time when his mother had been Elizabeth Manning.
9. See Greenacre, I, 81–82; also Melanie Klein, *The Psychoanalysis of Children* (London: Hogarth Press, 1932), 188–190. Hawthorne's father, a sea captain, died when the boy was four years old.

which since her death had proved an "impediment" (I, 58) [42]. And if Hester's own "impassioned state" was the "medium through which were transmitted to the unborn infant the rays of its moral life" (I, 91) [59], we can see the problem of identity when the maternal *Umwelt* includes not only a maternal presence but an absent father. Being in the world, for Hester, means activity, particularly the alienated activity of her needlework, and this affects Dimmesdale as well as her daughter. "Exchange this false life of thine for a true one. . . . Preach! Write! Act!" she charges him as he lies prostrate on a "luxuriant heap of moss" (I, 198) [119–20]. But of course, Dimmesdale cannot act on his own, condemned as he is to "eat his unsavory morsel at another's board" and "to warm himself by another's fire." "Think for me, Hester," he pleads: "Be thou strong for me" (I, 196) [118, 119]. There is no authentic core to Dimmesdale's character; thus to "be . . . for" Hester is to fuse with her will and to identify with a fantasy of her "masculine" qualities, and to "be . . . for" himself is to be isolated and empty.

Central to *The Scarlet Letter* is the figure of the maternal woman, not only the matrix from which characters separate and act, but also a type of phallic mother. To this extent, the maternal figure is a denial of loss—an important consideration for Hawthorne, who had lost his position at the Salem Custom House and shortly thereafter had lost his mother, another Elizabeth, with whom his relationship had been admittedly ambivalent.[1] The romance is framed by allusions to "the man-like Elizabeth" (I, 50) [37] and to "the sunny richness of the Elizabethan epoch" (I, 230) [137], and, like the figures of maternal women, it contains a paternal absence: the central chapter, "The Minister's Vigil," takes place on the night of John Winthrop's death which occurs beyond the periphery of the setting. One recent critic, noting that this date (1649) fixes the chronology of *The Scarlet Letter*, has argued that the romance, "above all else, is about history," specifically about the Puritans' definitive turning away from their English (although Hawthorne emphasizes Elizabethan) past and its "gentle" qualities.[2] More importantly, *The Scarlet Letter* is also about the resolution of a personal conflict in 1849.[3] As he looks back on what has been lost—whether mother, Salem, Custom House, or Elizabethan gaiety—Hawthorne finds those "gentle" qualities intermixed with a more disturbing absence, all figured in the "man-like" woman. *The Scarlet Letter* and "The Custom-House" are representations of Hawthorne's attempt to come to terms with a significant loss, to articulate the "basic fault" in a particular maternal relationship, and to frame a profound self-discovery.[4]

1. In *The American Notebooks*, Hawthorne writes: "I love my mother; but there has been, ever since my boyhood, a sort of coldness of intercourse between us, such as is apt to come between persons of strong feelings, if they are not managed rightly" (VIII, 429).
2. Frederick Newberry, "Tradition and Disinheritance in *The Scarlet Letter*," *ESQ* 23 (1977): 1–26.
3. Hawthorne's penchant for anniversaries—using historical material from fifty, one hundred or two hundred years previous in order to illuminate contemporary conflicts—may be seen in such tales as "My Kinsman, Major Molineux," "Roger Malvin's Burial," and "Endicott and the Red Cross." See, for example, Robert Daly, "History and Chivalric Myth in 'Roger Malvin's Burial,'" *Essex Institute Historical Collections* 109 (1973): 99–115; and Roy Harvey Pearce, "Hawthorne and the Sense of the Past or, the Immortality of Major Molineux" (1954), rpt. in *Historicism Once More: Problems & Occasions for the American Scholar* (Princeton, N.J.: Princeton UP, 1969).
4. Balint, 21, describes this "fault" in the self—which derives from the experience in the dyadic relationship—as a "sudden irregularity in the overall structure, an irregularity which in normal circumstances might be hidden, but, if strains and stresses occur, may lead to a break, profoundly disrupting the overall structure." Except for the aspect of "shame" which Balint discusses elsewhere, this might be Hawthorne writing of a "secret guilt."

II

When Hawthorne apologizes for his "autobiographical impulse" in his introductory sketch, it is more likely that he had failed in *The Scarlet Letter*—and not in "The Custom-House"—to keep the "inmost Me behind a veil," and that this sketch is intended to dispel the "stern and somber aspect" of self-discovery. His claim that he was happy while "straying through the gloom of these sunless fantasies" (I, 43) [32] is sheer fabrication. According to his wife, he apparently "came near a brain fever" when his mother died; a month later he was writing "*immensely*" so that she was "almost frightened about it."[5] The moth-eaten scarlet letter, whose discovery among the dusty bins of the Custom House places the sketch "within the law of literary propriety," is a more obvious fiction, despite Hawthorne's curiously adamant contention: the letter and manuscript "are still in my possession, and shall be freely exhibited" (I, 33) [26]. Such fictions—even such a time-worn convention as the recovered lost manuscript—are significant in the context of a sketch whose most insistent concern is "authenticity." They deflect attention from Hawthorne the man and artist toward Hawthorne the perceiver and editor; thus, psychological and moral issues seem to evaporate in mists of aesthetics and epistemology.[6] In ironic self-deprecation, Hawthorne equates the inevitability of his returns to Salem with that of a "bad half-penny," and I think "The Custom-House" must be read in that spirit: its truths must be taken seriously but not at face value. Hawthorne's fictions do not conceal truths, but, like the "scrubby trees" that hide Hester Prynne's cottage, they "denote that here was some object which would fain have been, or at least ought to be, concealed" (I, 81) [54].

In contrast to Hawthorne's reticence concerning self-exposure, his initial description of the Custom House confronts the reader with an interpretation of its essential qualities. The dominant symbol of the Custom House and the republic for which it is an emblem is "an enormous specimen of the American eagle" which "hovers" over the entrance (I, 5) [8]. Despite her expansive wings to which many flock for "shelter," Hawthorne finds "the fierceness of her beak and eye" posing a threat of "mischief to

5. Julian Hawthorne, *Nathaniel Hawthorne and His Wife* (Boston: J. R. Osgood & Co., 1884), I, 352–54.

6. The ethical questions raised by "The Custom-House" are treated by Larzer Ziff, "The Ethical Dimension of 'The Custom-House,'" *Modern Language Notes* 73 (1958): 338–44, who finds Hawthorne's idea of the "good life" involves "an enrichment of the material by the inner self, an appreciation of the present through a consciousness of the past" (341). Van Deusen reads "The Custom-House" as an epistemological problem. Despite the fine readings that their approaches yield, there is a danger of being led away from the heart of the matter by following Hawthorne's intellectual speculation. West and Pauly, for instance, see a complex process of transforming facts into significance, but their arguments rest on unexamined assertions of "spiritual significance" in Hawthorne's material, or Hawthorne's "intentions." In the famous aesthetic passage on the "neutral territory" in "The Custom-House," it is interesting that the "material" is enriched by the "inner self" only after an elaborate abstraction. Hawthorne begins with "my imagination," then "the imaginative faculty"; but finally it is "the Imaginary" that intermixes with "the Actual." Hawthorne is distanced from the fusion that takes place and participates only visually: among the "domestic scenery of the well-known apartment" transfigured by moonlight, we may "look about us and discover a form, beloved, but gone hence, now sitting quietly" (I, 34–36) [27–28]. Contrast this with his more immediate experience a little earlier on finding the scarlet letter. Imagining the decoration to be—of all things—a contrivance to "take eyes of Indians," "I happened to place it on my breast"; he imagines the reader smiling as he reports a "sensation not altogether physical, yet almost so, as of burning heat" (I, 32) [25]. Both passages imply a recovery of something lost, one by conjuring a "ghost," the other by identification with a lost maternal figure: but though framed in a tentative fiction, the latter experience communicates a fear of fusion that the former defends against through a process of intellectualizing.

the inoffensive community." While others may imagine that "her bosom has all the softness and snugness of an eiderdown pillow," Hawthorne notes a "shield before her breast" and a "bunch of intermingled thunderbolts and arrows in each claw." Left alone to rule her roost, this "unhappy fowl" has incorporated the emblems of patriarchal authority, but the "dignity" of "his Majesty" has been transformed into a "bunch" of destructive weapons. So awful are these nether regions, apparently, that Hawthorne equivocates in his description, prefacing his remarks with "if I recollect aright." Even so, Americans are drawn to her in search of shelter, as if they must return to her to actualize their identities as her instruments. The "patriarchal" civil servants are reduced to "dependents"; for their "half-torpid systems," the Custom House has become "the sole principle of life."

By his own admission, Hawthorne's relation to Salem is similar to that dependency he patronizes in his fellow civil servants. Salem is his "native place" and he returns as if under a "spell." His desire to return is "not love, but instinct," as if Salem were "the inevitable centre of the universe" for him (I, 12) [12, 13]. Like his paternal ancestors who "followed the sea," Hawthorne has obtained a post related to maritime commerce; but unlike them, he has remained home, near his "native soil."[7] His post places him in relation to both paternal and maternal qualities, yet the paternal elements—like those of the Custom House eagle—are merged, engulfed, or incorporated into a larger matrix. For example, the once bustling wharf is now dilapidated: the "tide often overflows" it at one end and grass has grown up around it at the other (I, 5) [8].[8] Hawthorne's "affection" for Salem is "assignable to the deep and aged roots which my family has struck into the soil" (I, 8) [10]. The dust of his immigrant ancestor is mingled in that soil and "the attachment of which I speak is the mere sensuous sympathy of dust for dust." That "deep and aged roots" had been "struck into" the "native soil" suggests that, though the fathers are absent, their phallic qualities remain present—if only as "dust" within an essentially maternal configuration, Hawthorne's "natal spot." The return to Salem thus implies not only a regression, but a search through a maternal configuration for a lost paternal ancestor.[9] It is not quite accurate to speak of a search, however, for Hawthorne is not really distinct from Salem (he must distinguish himself first as "surveyor"), nor are the objects of his search separate or individualized (hence, there is no "love" but "instinct"). Rather, there is a range of experience from which Hawthorne must differentiate himself and other objects.

Symptomatic of the relationships possible with the republican eagle is the "frozen thought and benumbed language" which pervades the Custom House. The "dependent mode of life" (I, 16) [15] Hawthorne finds among

7. According to Randall Stewart, *Nathaniel Hawthorne: A Biography* (New Haven: Yale UP, 1948), 76, Hawthorne had turned down several appointments—clerk in the Charlestown Navy Yard and naval storekeeper in Portsmouth were two—because he apparently wanted a post in Salem.

8. The "wheel-track" at the Old Manse was also "overgrown with grass" (X, 3), but there the image was of nourishment, not decay. That grass afforded "dainty mouthfuls to two or three vagrant cows" (X, 3); and the orchard planted by the late minister suggested the "idea of an infinite generosity and exhaustless bounty, on the part of our Mother Nature" (X, 13). The garret of the Old Manse contained "nooks, or rather caverns of deep obscurity, the secrets of which I never learned" (X, 16); and consequently, Hawthorne produced "no profound treatise of ethics—no philosophic history—no novel" (X, 34).

9. Hawthorne's ancestral roots had been "planted deep, in the town's earliest infancy and childhood" (I, 10) [12]. This suggests the process of "imprinting" an identity and ties that identity to a fantasy or a fused parental image.

the "patriarchs" there has apparently obviated any need for speech or action. They would view his nervous pacing as useful only insofar as it served him "to get an appetite for dinner" (I, 34) [27]; their discourse has been reduced to "pass words and countersigns," to "frozen witticisms," its content more likely to center on "their morning's breakfast, or yesterday's, to-day's or to-morrow's dinner" than "the shipwreck of forty or fifty years ago" (I, 15–16) [14–15].[1] Their debased language implies that their relationships have regressed to a stage of development characterized by the symbiotic fusion of self and object, and for Hawthorne the problem is compounded by the general ineffectiveness of language even in describing such regressed states: "A better book than I shall ever write was there" (I, 37) [29].[2] Yet having returned to his "native place," Hawthorne realizes, I think, that this area, describable only by indirection, is central to his experience. It remains for him to separate himself and to establish a relationship between himself as surveyor and the world as a cluster of distinct objects. As with any separation, however, there is a danger that the psychological separation of self and object may not conform to the "real" separation of two physical bodies: some aspects of the world may appear as elements of himself, while other aspects of himself may be projected onto the world. In his native town of Salem, for instance, his "name is seldom heard" and his "face hardly known." Yet, as surveyor, his name is stenciled on dutiable merchandise to be shipped across the globe. In Salem, he has a sense of identity based on his "attachment" to his ancestors, an identity which is not confirmed by his social experience; outside Salem, he has a social role as surveyor of customs which does not feel real—a "false self." In Salem, he is "in possession" of a secret identity which he defends against the intrusion of the social world; outside Salem his "autobiographical impulse" might itself be considered an "intrusion."

As surveyor, Hawthorne finds himself balanced precariously between the themes of possession and intrusion, engaged now in a dependency relationship with the republican eagle. For Hawthorne, possession and intrusion are really two aspects of the same theme. Since the separation of self and other appears problematic, the self cannot feel complete unless it

1. Hawthorne's father had died at sea forty-one years before.
2. It is possible to view this quotation as Hawthorne's choice—despite himself—to write romance rather than realism, to express the "inner light" rather than adapt to authoritarian conventions (see Baym, 16). But if *The Scarlet Letter* speaks "a reality far more powerful than the quotidian" (16), why should a novel of the Custom House have been a "better book?" Is that merely the guilty Hawthorne heaping conventional judgments on himself? In part, yes, but there is more involved: Hawthorne cannot write of the Custom House unless he can satisfactorily separate himself from it. His art (and most psychoanalysts would include language generally) requires distance, a separation of self from other objects which language mediates by providing a space where self and objects can merge while retaining their distinctness. Inside the Custom House, this distance is insufficient—Hawthorne is overwhelmed by an otherness and his vestigial literary capacities atrophy (I, 34 [27], 38 [29]); outside the distance is too great to be bridged by language—"it ceases to be a reality of my life" (I, 44) [33]. To write a "better book" Hawthorne would have to find a language (and by implication, a mode of relating) which could present his world and himself in it—in a sense, he would have to be able to stand alternately inside and outside himself and other objects. More than "the individual in conflict with authority" (Baym, 18), the issue of "The Custom-House" concerns the establishing of a "true relation"; and from the moment of Hawthorne's instinctive return to his "natal spot," the ontological basis of the "individual" is called into question. If he cannot accept otherness (not merely "conventional modes," but the existence of objects outside himself), if he cannot establish an authentic relation with "individuals unlike himself" (I, 24) [20], Hawthorne is forced to people his world with parts of himself: his ancestors, Surveyor Pue, even Hester Prynne. What Baym celebrates as Hawthorne's "turning into the self" (19) is really a symptom of his inability to "open an intercourse with the world" (IX, 6).

can intrude on the other and possess those contents which it fantasizes as part of itself. At the same time, the self may harbor contents which it fantasizes as parts of the other, which it fears the other may reclaim by some act of intrusion. Such an experience, then, cannot be analyzed (much less moralized) unless the objects differentiate from each other and one becomes the intruder and the other the object of intrusion. Hawthorne attempts to separate and individuate these themes in the persons of the Inspector and Collector (as well as in the romance at times between the Reverend Mr. Wilson and Hester Prynne, for instance, or Chillingworth and Dimmesdale). The Inspector, as we might expect, is a hyper-intrusive surveyor, whereas the Collector is an object of intrusion: "The closer you penetrated to the substance of his mind, the sounder it appeared" (I, 20) [18]. The themes of possession and intrusion, within the dependency mode, have a two-fold significance: some characters use their own or others' dependence as a way of intruding (at the risk of being intruded upon), while other characters strive to remain independent as a way of retaining their possessions (at the risk of cutting themselves off from others).

The Inspector is emblematic of the first type and significantly is identified only by his role.[3] He "possessed no power of thought, no depth of feeling, no troublesome sensibilities; nothing, in short, but a few commonplace instincts" (I, 17) [16]. A man of roughly eighty years, he appears "as ready for sport as any unbreeched infant." He seems "earthy and sensuous," the quality of his life little better than a beast's, except in one "advantage over his four-footed brethren"—"his ability to recollect the good dinners" which he had eaten. This "contrivance of Mother Nature," too shallow to mourn the loss of three wives and most of his twenty children, too devoted to "the delight and profit of his maw" to have corrupted any "spiritual endowment," is human only in his ability to recollect and tell stories. "I have heard him smack his lips over dinners, every guest at which, except himself, had long been food for worms." With life reduced to the proposition that one must eat or be eaten, Hawthorne apparently finds it necessary to place certain limits on such unbridled sensuality. "The chief tragic event of the old man's life . . . was his mishap with a certain goose" which was "so inveterately tough that the carving-knife would make no impression on its carcass" (I, 18–19) [17]. Hawthorne seems bound to assert an impenetrable object with contents safe from even an inspector's intrusion.

One such impenetrable object is the Collector, identified by name as General James F. Miller, hero of the Battle of Lundy's Lane in the War of 1812. Hawthorne is able to "sketch only the merest outline" of the general, only "the main points of his portrait": his "step was palsied" and only with aid could he "slowly and painfully ascend the Custom-House steps."

3. Hawthorne must have known the response this characterization would engender. When Evert Duyckinck requested an excerpt of Hawthorne's new novel for his *Literary World*, Hawthorne's publisher advised him to use this one, the "Old Gourmand." Hawthorne preferred the passage on General Miller, but the former was used over his objections (Charvat, xvii). The Collector's portrait is obviously more reverential and its stance differs little from Upham's in "The Memorial": "His powers of locomotion are wholly destroyed, and his articulation rendered quite difficult, but his mental faculties have not shared the decrepitude of his physical frame" (in Nevins, 115). The Whigs approved of Miller, appointed as he was "before the parties that now divide the country had been formed" (110), and advocated his re-appointment as Collector at the expense of his son's (and Hawthorne's) removal from office. Hawthorne's reverence is part of a complex relation to others, and Duyckinck as well as Chapman at the *Register* were astute enough to read through Hawthorne's pieties to find the barbs.

Like Fort Ticonderoga, some areas of his aspect were "almost complete" but others could be aptly described as "a shapeless mound, cumbrous with its very strength." Nonetheless, Hawthorne finds a "light within him," a wholeness of character discoverable when his "notice was sought" and a discourse established. This wholeness includes not only a "stubborn and ponderous endurance," but "benevolence," and a foundation of "innate kindliness," "humor" and "native elegance, seldom seen in the masculine character after childhood and early youth." He seemed, for instance, to have a "young girl's appreciation of the floral tribe." For the surveyor, it is much more satisfying "standing at a distance and watching" the Collector than "engaging him in conversation" which is a "difficult task." Thus distanced, he is safe to surmise that the old warrior "lived a more real life within his own thoughts" and recollections. Indeed, it is the recollection of his famous "I'll try, Sir" (I, 20–23) [17–20] which not only provides Hawthorne with the contour of the general's character, but a spur to his own activity: when he imagines Surveyor Pue's ghost charging him to dispose properly of his tale, he responds, "I will" (I, 34) [26].

The quality of recollection, then, would seem to be that quality which humanizes the beast by distancing him from his objects. Even the Inspector rises above other animals because of this quality; and because of his delight in memories, he stands far above the chorus of torpid "patriarchs." This quality is also the "bright gleam" along the blade of the "rusty . . . old sword" which is General Miller. Similarly, the documents of the part-time antiquarian, Surveyor Pue, provide "petty activity" for Hawthorne's mind and keep it from being "eaten up with rust" (I, 30) [24]. Dredging up and collecting bits of the past afford an escape from being metaphorically consumed in the present. Hawthorne, speaking of the positive aspects of the Custom House, had noted his desire to "nourish myself with food for which I had hitherto had little appetite" (I, 25) [21]. But such a "change of diet" has placed him in the relation of a dependent, and while being nourished he is also being consumed. Recollection, it seems, is his escape: "there would have been something sad, unutterably dreary in all this, had I not been conscious that it lay at my own option to recall whatever was valuable in the past" (I, 26) [21].

III

Dependence on an external object weakens a person's "own proper strength" and condemns him to a search for "support external to himself." Only a "slight . . . taste of office" infects one with a "singular disease" which torments the victim even after life has ceased. Public office robs the soul of its "sturdy force, its courage and constancy, its truth, its self-reliance, and all that gives the emphasis to manly character" (I, 39) [30]. Yet it is difficult to take this prating on self-reliance at face value, considering Hawthorne's own dependence on either "our common Uncle" or his maternal uncle, Robert Manning, for almost every position he ever held. An impenetrable goose may frustrate the Inspector but does not mediate the processes of consumption; recollection may provide temporary escape but cannot finally ameliorate the condition of decay. The surveyor, trapped in these processes which threaten to disintegrate him—he fears his intellect is "dwindling away; or exhaling . . . like ether out of a phial" (I, 38) [29], begins to drift toward paranoia.

Recollection is not only limited in its positive effect, but reveals a darker side as well. Hawthorne discovers the tale of the scarlet letter through "poking and burrowing," "unfolding" and "unbending" and "prying." He finds "traces of Mr. Pue's mental part, and the internal operations of his head" (I, 29–30) [24]. As if to absolve himself of a wrongful "intrusiveness" more characteristic of Chillingworth's "delving," "prying" and "probing" (I, 124) [78], he fantasizes the old surveyor delivering the manuscript to his hand. His intrusion into Surveyor Pue's "private" documents is sanctioned by his "filial duty and reverence" towards the surveyor's ghost. He becomes a collector of documents, not an inspector; and as if to discourage further intrusion, he vouches only for the "authenticity" of the "outline" of the tale, claiming nothing for its contents. Obsession with secrets implies not only the keeping of one's own but the disclosing of others', just as an obsession with eating implies not only consuming but being consumed. The two-person relationship, which allows for fusion and separation between self and other, cannot provide for a mediation between the two.

Hawthorne's attempt to mediate the processes of possession and intrusion forces him to become the "inevitable centre of the universe"; he is the one chosen by Surveyor Pue to bring Hester Prynne's tale to light and the one to assume his family's shame and to absolve the Hathornes of their "curse." Refusing to become "thoroughly adapted" to the Custom House, like the nameless and titleless "man of talent," Hawthorne recalls his Puritan ancestors, the witchcraft judge and the persecutor of the Quakers. When he takes their "shame" upon himself (I, 10) [11], he asserts for himself an objective identity, a self that distances him from his contemporaries even as it makes him visible to them. Through his family name, the spelling of which he altered slightly when he became a writer, he claims continuity with the Hathornes whose "persecuting spirit" had made them "conspicuous." Their shame protects the civil servant from anonymity, from being consumed by the present: even though an identification with his ancestors affords him a rather negative distinction, it makes him, if not noteworthy, then perhaps notorious.[4] It provides him with a kind of social reality, though perhaps incongruous or inappropriate, which suggests the possibility of a "real" Hawthorne. Though he takes upon himself their two-hundred-year-old shame, he only hints at the "strong traits of their natures"—perhaps a "hard severity" or a "persecuting spirit"—which remain hidden in his otherwise "inoffensive" character. The identification with his "great-grandsires" is further complicated, qualified by his professed guilt and inadequacy, his confessing himself an "idler" and his works "worthless, if not positively disgraceful."[5] He imagines their "compliments": "Why, the degenerate fellow might as well have been a fiddler!" (I, 10) [11–12]. Inward guilt in the face of his ancestors, like outward shame in the face of his community, seems to

4. Heinz Lichtenstein, "The Dilemma of Human Identity: Notes on Self-Transformation, Self-Objectivation, and Metamorphosis," *Journal of the American Psychoanalytic Association* 11 (1963): 215: "Shame occurs in such situations where the burden of identity and separateness, the loneliness of autonomy has become unbearable: the temptation to abandon it, to give up one's will, to become a slave, a physical thing, triumphs over our defenses."

5. Although the context renders these judgments ironic, Hawthorne did refer to *The Scarlet Letter* in a more serious vein as "positively a hell-fired story." Bridge, 112.

provide Hawthorne with a coherent sense of self, if coherent only insofar as the self focuses attacks from the community outside or from the guilt inside. As Frederick Crews has remarked, "Hawthorne's evocations of Puritan times gave him *a guilty identity, which was better than none.*"[6] Hawthorne himself explains his need for social animosity: since, as a civil servant, "his interests are within the control of individuals who neither love nor understand him, . . . he would rather be injured than obliged" (I, 40) [30].

Throughout the sketch, he seems to invite notoriety, identifying himself with radical Jacksonians (as a "Loco-Foco Surveyor") and needling the Whigs whenever possible. Alternately, he identifies his interests with the loyal monarchist, Surveyor Pue, or his own shameful ancestors. These are, of course, so many more fictions, masks manipulated by Hawthorne. It is as if each pose, each objective identity, were calculated to entice the reader to infer the essential subjective Hawthorne, but each pose reveals a mass of contradiction on closer inspection. The Salem Whigs, led by the Reverend Charles W. Upham, accused Hawthorne of partisanship, of belonging to certain radical Democratic clubs, and of marching in torchlight parades. Hawthorne's defense was that he not only never belonged to such organizations, but he rarely even voted in elections—an admission which lost much of what little support he had among local Democrats. This is not to say that Hawthorne was a lukewarm Democrat or that he was politically naive; rather, he would prefer to present and manipulate an objective identity of his own. Careful readers may have found in the words of Hawthorne's "great-grandsires" not his ancestors' view of storytellers, but a close paraphrase of the "Loco-Foco" attack on anti-Democratic literature. When Hawthorne's friend, John L. O'Sullivan, assumed full control of *The Democratic Review*, he began by publishing literary manifestoes, one of which denounced aristocratic poets as "fiddlers." Petrarch was a prime example: a "passion suddenly conceived . . .. by a priest . . . for the wife of another"—such a topic was "too nonsensical for a Yankee imagination" and could "never, under any circumstances, become popular" in America. The poetry, though "exquisite," was thus as "worthless . . . as a penny-whistle."[7] Hawthorne seems to have effectively criticized his own efforts through the words of his political allies and distanced himself—at least in his own mind, perhaps—from his most ardent political and literary supporters. They had cited Hawthorne as one of the "writers of the first class" who "have . . . expressed themselves plainly in the terms of the democratic creed," the foundation of which was the belief that "there is no hereditary line" of genius, intellect, or manly character.[8] Thus, it is doubly ironic that Hawthorne's ancestors, from whom he had inherited "strong traits," should condemn him on Democratic principles.

There is a further irony in the "shame" Hawthorne assumes on behalf of his ancestors. Upham had accused Hawthorne specifically of fraud and corruption and had compounded the insult by charging Hawthorne

6. Crews, 38.
7. "Petrarch," *The Democratic Review*, 11 (1842): 279–80.
8. "Democracy and Literature," *The Democratic Review*, 11 (1842), 199.

to have been the "abused instrument of others."[9] In his *Lectures on Witchcraft*, however, Upham had virtually ignored Hawthorne's ancestors, only twice even mentioning Judge John Hathorne. Years later, he would praise the Hawthorne family: "No one in our annals fills a larger space" than William Hathorne, and John "succeeded him in all his public honors." "The name is indelibly stamped on the hills and meadows of the region, as it was in the civil history of that age, and has been in the elegant literature of the present."[1] Hawthorne would probably have known Upham's earlier work, although the ironic nature of his family's shame would certainly have been lost on his readers. It is evident, however, that the shame he takes upon himself is largely of his own invention. Since he is dealing with those who "neither love nor understand him" and must be either injured or obliged (I, 40) [30], he at least chooses the shameful identity he will assume, the particular injury he will suffer. Of primary importance is self-preservation. He will not seem to be the "abused instrument" of petty grafters or political hacks, but an estranged Salemite, brooding guiltily over a shameful past. In this way, Hawthorne presents an invented self which seems to denote an almost inaccessible secret self that would rather be, or ought to be, hidden. The shameful, objective identity he adopts, then, allows him to balance between the threats of the intrusion of a hostile community and the isolation of absolute self-possession.

This compromise, which allows him to exist—albeit inauthentically—within the political community, is short-lived, however. The dependence which consumes the "manly" strength of the individual and the process of recollection, which is ultimately the self's intrusion into the contents of the past, are the positive aspects of life with the republican eagle. Negatively, the barbed arrows and thunderbolts in her claws threaten the young from their nest and this "inoffensive" public official with decapitation. Hawthorne seems to prefer this alternative, however, for half a man seems better than none and persecution does imply a certain measure of notoriety. Life in the Custom House was "an unnatural state," and, like a person contemplating suicide who meets with "the good hap to be murdered," Hawthorne is freed from a false relation blamelessly when he is turned out of office. His "autobiographical impulse," too immodest for a living person, can "be excused in a gentleman who writes from beyond the grave" (I, 44) [32]. The "true relation" to the audience as a negation of the self, however, seems to defeat its purpose.

9. Nevins, 118. This further insult must have been the cruelest blow for Hawthorne. Not only was he blamed for wrongdoing; he was not even given credit for having willfully done wrong. A similar situation had developed when Hawthorne was caught gambling while a college student, and his response is indicative of his need to control his own self-presentation. The Reverend William Allen, President of Bowdoin College, wrote to Mrs. Hawthorne, informing her of her son's transgressions but blaming them on "the influence of a student whom we have dismissed." Hawthorne wrote a rather perfunctory letter to his mother, admitting the charge, asking her not to show the official letter of reprimand to anyone, and adding a more serious charge of his own—the "Card Players" had gambled for a quart of wine. If to his mother he was unrepentant, to his sister he was defiant: "I was fully as willing to play as the person he suspects of having enticed me, and would have been influenced by no one. I have a great mind to commence playing again, merely to show him that I scorn to be seduced by another into anything wrong." Letters quoted in Manning Hawthorne, "Nathaniel Hawthorne at Bowdoin," *New England Quarterly* 13 (1940): 260–62.

1. Charles W. Upham, *Lectures on Witchcraft* (Boston: Carter, Hendee and Babcock, 1831), 24, 71–80; *Salem Witchcraft* (Boston: Wiggin and Lunt, 1867), I, 99–100.

But Hawthorne is not content to leave himself dead and outcast, if true and free at last. The "real human being," he says, resides elsewhere. The Custom House and its circumstances are disconnected from his "recollections": "it ceases to be a reality of my life. I am a citizen of somewhere else" (I, 44) [33]. He resurrects himself to reject the Salem that rejected him and dreams of a future time when the locality of "The Town-Pump" will have been immortalized by one of his tales. The reversal is complete: through a fortuitous decapitation, Hawthorne has become central to Salem, eminently famous and totally untouchable.

IV

When Dimmesdale returns from the forest, possessed by an uncharacteristic energy, he views the settlement from a new perspective, as if all before had been a dream, or as if this new view were a dream. He realizes "wicked impulses" toward an old man and a devout widow. The voice of the "arch-fiend" tempts him, on seeing a young maiden, to "drop into her bosom a germ of evil that would be sure to blossom darkly soon, and bear black fruit betimes" (I, 220) [131].[2] He feels a secret camaraderie with "heaven-defying" seamen and with Mistress Hibbins, the "bitter-tempered widow of the magistrate." He returns home to his unfinished sermon, separates from Chillingworth his leech ("no more of your drugs"), and instead eats "with ravenous appetite" and gives vent to an "impulsive flow of thought and emotion," feeling himself all the while a "foul" medium for this energy (I, 224–25) [134]. All Hester can hear of this sermon on Election Day is an "indistinct, but varied, murmur,"[3] yet her sympathy is aroused in fairly sensual terms without benefit of the "grosser medium" of words. The "tone and cadence" of his "vocal organ" (a "rich endowment") communicates immediately "the complaint of a human heart, sorrow laden, perchance guilty, telling its secret" (I, 243) [144]. It is a sermon whose significance cannot adequately be expressed in words—only in a "tongue native to the human heart"—whose undercurrent of emotion cannot be mediated by the social and cultural institutions which frame it.

The background to this revelation is the resurgence of Elizabethan custom, not merely the conflated religious and legal rituals of the first chapters, but the melding of law and license in the carnival, the fusion of town and country, of sober Indians and drunken mariners, of law-abiding Puritans and men of the sea "unregulated by human law." At the center of this tumult, Hester is saved from "indistinctness" by the "talisman" of her letter, her public shame.[4] Pearl, too, as the "gem on her mother's unquiet

2. In his letter to Longfellow, Hawthorne threatened: "I may perhaps select a victim and let fall one little drop of venom on his heart, that shall make him writhe before the grin of the multitude for a considerable time to come" (MacShane, 98).

3. Compare *The American Notebooks:* Hawthorne's dying mother "could only murmur a few indistinct words. . . . I found the tears slowly gathering in my eyes. I tried to keep them down; but it would not be—I kept filling up, till, for a few moments, I shook with sobs. For a long time, I knelt there, holding her hand; and surely it is the darkest hour I have ever lived" (VIII, 429). Here, as in the romance, there is a communication immediately heard and felt, not distanced through sight or words.

4. In his fine historical reading, Newberry argues that *The Scarlet Letter* concerns the conflict between a dominant Puritan militancy and recessive—primarily English—forces of "sympathy, charity, gaiety, respect for tradition, and appreciation of art" (1). "Hawthorne unmistakably presents English forms and traditions as positive historical origins from which, by and large, the Puritans have dissociated themselves" (19). Unfortunately, this view takes Hawthorne's nostalgia at

bosom," stands in a "charmed circle," a space created by the simultaneous "attraction and repugnance of the mystic symbol" (I, 246) [136]. Dimmesdale, like an "oracle," speaks and enrapts his listeners in a "high spell"; yet, as they "gush forth from the doors of the church" all in a "babble" with "applause for the minister," he appears spent, "deathlike." An uncharacteristic energy (which "might be spiritual") had propelled Dimmesdale's body in the procession to the meeting house: "There was no feebleness of step . . . his frame was not bent" (I, 238) [141]. But leaving the church, he walks toward Hester with "the wavering effort of an infant, with its mother's arms in view, outstretched to tempt him forward" (I, 251) [148].

Dimmesdale will not be "tempted forward" in the way that Hester had wished, however, because he has annihilated himself and all objects. He proclaims himself "the one sinner of the world," and through this act not only does he escape from Chillingworth, but from Hester too, to whom his feeble body finally "crept." There is little hope for reunion hereafter, he tells her rather unfeelingly (I, 256) [150, 151], and Hawthorne surmises that even their dust has "no right to mingle" in the cemetery (I, 264) [155]. Dimmesdale's identification with a heavenly father provides the hope which is the necessary motive for resolving his torment in self-annihilation.[5] Within the religious schema, he has grasped God's will and thus has become a type of His justice and mercy; at the same time, he has delved within himself and exposed his secret to the world. As Hester supports Dimmesdale's body ("twine thy strength around me," he had asked), she becomes a type of medium through which these complementary revelations are accomplished. But when inner contents (His mercy, Dimmesdale's guilt) are revealed, Dimmesdale dies and with him all the relationships which those contents had "twined" together so hopelessly. Chillingworth's "life seemed to have departed" already on the scaffold and soon Hester and Pearl "disappeared," Hester to return later to her exile near Boston and Pearl to live forever "elsewhere." Dimmesdale's revelation itself becomes garbled in a host of conflicting reports and contradictory interpretations, and Hawthorne urges us to "Be true" by showing the world some "trait by which our worst may be inferred" (I, 260) [153]. At least, of all the morals which Dimmesdale's end brings to mind, this injunction only is "put . . . into a sentence." Relationships wind up split or fragmented with the disclosure of the secret.

---

face value. The recessive (actually, *regressive*) forces do not always yield "fond" memories of an English past (6); they provide Hester with an escape from the here-and-now of the scaffold in Boston, but they present her with—among other images—a "decayed house," her father's "bald brow," her mother's "look of heedful and anxious love" (I, 58) [42]. Origins are really the not-now for Hawthorne, just as England is the not-here. Threatened with an "intolerable burden of identity and separateness," as Lichtenstein would put it, Hester denies her reality: "Standing on that miserable eminence, she saw again her native village" (I, 58) [42]. Election Day presents the reverse situation: the Elizabethan festival threatens to dissolve separate identities, transforming individuals into a motley throng: "rough figures" in "deer-skins" (I, 226) [134], groups of "painted barbarians," sailors with their "animal ferocity" (I, 232) [138], and the brown- and gray-clad Puritans. Here, Hester's gray cloth "had the effect of making her fade" while the scarlet letter "brought her back from this twilight indistinctness" (I, 226) [135]. These festivities—saved from their traditional excesses by the grim keepers of Puritan law—seem more threatening than gay; and it is only Hester's shameful though exquisite production (whether her child or her art) that allows her an identity. For the community at large—as for Hawthorne personally—the choices are extreme: engulfment or isolation.

5. David Holbrook, "R. D. Laing & the Death Circuit," *Encounter* 31 (1968): 35–45, describes what he calls "schizoid suicide," a delusion that the suicide harbors (Sylvia Plath and Dylan Thomas are his prime examples): that he (or someone else) will rescue his true self from the torments of inauthenticity through his suicide.

Dimmesdale's experience is the tragic complement to Surveyor Hawthorne's and the decapitation and resurrection of the latter should alert us to the truths expressed in *The Scarlet Letter* as well as "The Custom-House." Unlike Dimmesdale, whose loss of self is irrevocable, Hawthorne seems content to give up pieces of himself—not his heart "delicately fried, with brain-sauce" (X, 33) to be sure—but enough of himself to entice his audience to infer his character from the traits he has offered. This is not Romantic self-expression by any means but an attempt to achieve fame without exposing the "inmost Me," to invent a self which appears real but in which he has a minimal investment. Hawthorne's insight into this problem of inauthenticity and his ambivalence concerning fusion and separation are particularly noteworthy because he is able to faithfully render such a dilemma while exerting a remarkable control over his material and his audience. His presentation is clearly manipulative—there is little feeling for a person suffering intolerably because he does not feel real, a person driven to suicide in the delusional hope that the "real human being" can be released from the sway of a false self. But this real human being did read Dimmesdale's rejection of life and Hester Prynne to Sophia Hawthorne shortly after writing it—"tried to read it, rather, for my voice swelled and heaved, as if I were tossed up and down on an ocean, as it subsides after a storm."[6] The next day he wrote his friend Horatio Bridge: the conclusion of the romance "broke her heart, and sent her to bed with a grievous headache, which I look upon as a triumphant success."[7]

I think Hawthorne realized at some level that, given his psychology, the alternative to self-annihilation was a denial of the claims of the real world—a rejection on the order of Dimmesdale's though perhaps on a smaller scale. With his mother gone, there was no tie to Salem or to his paternal ancestors; and in the letter to Bridge quoted above, he expresses his desire to leave Salem, to invigorate himself elsewhere. Significantly, he asks Bridge not to mention his health in his own letters since "my wife already sermonizes me quite sufficiently on my habits; and I never own up to not feeling perfectly well." He never returned to Salem, and although he bought and christened "The Wayside" in Concord after his year in the Berkshires, he never felt as if he had a home again.[8]

With "The Custom-House," Hawthorne claims to give up that part of himself which has identified with Salem, with the ancestral dust in the native soil. But in the next year, he wrote *The House of the Seven Gables* from his rented house in Lenox, taking his revenge on Upham thinly disguised as Jaffrey Pyncheon; he tried to escape the hold of this fantasy of a dead patriarch within a maternal configuration, this time through an almost ritualistic repetition and undoing. His characters all leave Salem, but Hawthorne could never escape from that fantasy; it haunts his fiction, obsessively so in *The Marble Faun* and the unfinished romances. Salem might be rejected, as feared maternal figures could be rejected, but these

6. *The English Notebooks*, ed. Randall Stewart (New York: MLA, 1941), 225. The passage continues, "But I was in a very nervous state, then, having gone through a great diversity and severity of emotion, for many months past. I think I have never overcome my own adamant in any other instance."
7. Hawthorne to Bridge, February 4, 1850; quoted in Bridge, 111.
8. According to Julian Hawthorne, I, 429, "after freeing himself from Salem, Hawthorne never found any permanent rest anywhere. . . . A novelist would say that he inherited the roving disposition of his seafaring ancestors."

are types of a deep fault which provided the design for his best fiction as it bound him to its psychological demands. Serving up "traits" to be analyzed, claiming nothing for their truth and denying any single interpretation, Hawthorne was able to maintain his "intercourse with the world"; but like Dimmesdale's more drastic loss of self, such a strategy could not resolve his dilemma. For Hawthorne, the problem seems only aggravated: to live is to exist inauthentically; to give up the falsehood, to expose the secret, to "Be true," is to perish.

ERIC SAVOY

Nathaniel Hawthorne and the Anxieties of the Archive†

Twenty-five years ago, Robert K. Martin published an influential essay on the gendered anxieties that permeated Hawthorne's career. Under the auspices of the Men in Feminism project, he shifted critical focus from Hawthorne's literary treatment of women—a line of inquiry that had generated divergent opinion—toward the problematic construction of male authorial subjectivity.[1] Martin's context is at once biographical, historical, and social: if Hawthorne's early career was "genteel and feminine" (123), as evidenced by the publication of tales and sketches for a female readership in such venues as *Godey's Lady's Book*, his move to novel writing entailed an entry into a professionalized, competitive, "masculine" sphere that, ironically, was dominated by women. Like Hepzibah Pyncheon in the cent shop [in *The House of the Seven Gables*], his hand would forthwith bear the stain of commerce. Martin succinctly delineates Hawthorne's predicament: "Even though [he] triumphed in this new arena, his triumph came at considerable psychic cost, since it meant betraying the fathers by abandoning the gentility of anonymity and domestic seclusion and by becoming—as publishing author and emblazoned artist—the scarlet woman" (123). As gender roles solidified in Jacksonian America and after, it would seem that *neither* of the identificatory positions available to Hawthorne as fiction writer was immune to the charge of effeminacy.[2] To write from within the increasingly feminized sphere of domestic retirement, and thus to be

† From *Canadian Review of American Studies* 45.1 (2015): 39–66. Reprinted with permission from University of Toronto Press (www.utpjournals.com). Page numbers in square brackets refer to this Norton Critical Edition.

1. See, for example, Louise DeSalvo's *Nathaniel Hawthorne*: DeSalvo explores the female body as the site of violence. Leland S. Person—whose work Robert K. Martin follows—offers a considerably more nuanced reading of Hawthorne's gender politics. Building on Nina Baym's work, Person argues that the canonical masculine poetics of the American mid-century were "feminized"—that authors such as Hawthorne "use female characters to test and challenge the limits of verbal, imagistic, and literary form—and thereby the limits of male authority" (7). In taking up the complex issue of authorial subjectivity, Person advances the idea that "male writers tried to accommodate the feminine . . . by identifying the creative vitality of the works with the women embedded in them" (8). Person's reading of Hawthorne is an important precursor of my own analysis of the complexity and contradictory impulses in Hawthorne's archival thought experiment: for Hawthorne's creativity entailed having a "haunted mind" (12) or, in words "an identification of the self with the otherness of the work and a surrender of control over the creative or writing process" (10). My project explores the figurative poetics of such a "surrender" in the gothic matrix of "The Custom-House."
2. Suspension between identificatory positions is, indeed, the deconstructive matrix of Hawthorne's mapping of the fictional/autobiographical character, "Nathaniel Hawthorne," in "The Custom-House," which I approach as a psychodrama generated by archival imperatives.

faithful to a class privilege that was in any case unsustainable, would not have been manly. At the same time, to join the fray of commercial publishing was not only vulgar but meant becoming one of the company that Hawthorne dismissed, notoriously, as the "damned mob of scribbling women," a professional class that, as Martin points out, he both "scorned and feared" (122).

What is most compelling about what Martin terms the "psychic cost" of Hawthorne's coming-out as a public man of letters is not primarily the sacrifice of secluded, gentlemanly anonymity for public exposure; nor is it, essentially, the "betrayal" of the male ancestors by choosing a feminized profession. These are, at best, convenient images for historians of American class structure. It does, however, have to do with gender formation. At issue is the *psychic viability* of an ethics of male authorship—an ethics that would be accountable to the Puritan tradition of public service and would impel a literary intervention of some practical use to American society at mid-century. In order to map the terms of Hawthorne's negotiation of such an ethical vocation, I shall return (once again) to the psychic emplotment of "The Custom-House." In my earlier work on this endlessly fascinating text, I focused on Hawthorne's gothic poetics—that is, the semiotics of ghostly revenants, the literal and metaphorical images of corpses, and the consequent lamination of graveyards to dusty archives—to illuminate the figurative economy within which the author accepts the exhortation of Surveyor Pue to respect his "filial duty" by delivering to the public the historical romance of Hester Prynne. This time around, I am concerned with what Jacques Derrida conceptualizes as the *event* of the archive—that is, the quotidian, ritualistic, and performative ways in which subjects are constituted precisely *as* subjects, in conformity with the regulatory ideals of nation, religious tradition, and gender. If there is no subject without a subtending and largely mythical archive of cultural practice, then "The Custom-House" is a spectacular *mise-en-scène*—rhetorically rich and charged with affect—of the ritualistic protocols of *assujettissement*. In this psychodrama, the identities of masculinity and authorship converge under the signs of "authenticity" and "usefulness"—and are consequently beset by homosocial anxiety. Martin notes that symptoms of such anxiety in "The Custom-House" are "a primitive fear of the empowered woman" (124) and, more specifically, a bitter "scorn for his colleagues and superiors [,] his recognition of their uselessness [which] barely conceals his recurring anxiety about filial relationships and authority" (125). I would argue, however, that subtending these localized manifestations of gendered anxiety is a more exacting literary problem: why did Hawthorne choose to situate himself as a character in a gothic tale of the ghostly revenant and to fictionalize both the germ of his novel and his subject-position as editor/rewriter of Pue's nonexistent manuscript, rather than to write a straightforward preface about the postcolonial imperatives of the American writer? Why did he say anything *at all* about the genesis of his text or about the intertwined genealogies of the archive and the family? I take, as my point of departure, Martin's observations that "'The Custom-House' is above all an essay in sexual politics" and that its "dilemma . . . is one that concerns not only men's relationship to women, but men's relationship to other men" (124).

Relier de Nouveau *(to Bind Anew)*: *On Repetition as Restitution*

The Scarlet Letter intertwines the trials of Hester Prynne with the futile attempts of the Puritan community to stabilize the import of the letter A. Its narrative career, inevitably, is one of repetition and return—and of failure. As the letter circulates, eliciting conjectural interpretations that slide from "adultery" to "angel" to "able" and to murkier territory beyond, the novel accrues interest as a master lesson in the hermeneutic circuitry of *différance*.[3] Hawthorne's conclusion that "the scarlet letter had not done its office" (261) [109] refers simultaneously to the inconclusive allegory of reading that the symbol had generated and to its ineffectuality as a technique of discipline and punishment: for Hester, far from being recalled to Puritan notions of sin and repentance, "assumed a freedom of speculation . . . which our forefathers, had they known of it, would have held to be a deadlier crime than that stigmatized by the scarlet letter" (259) [107].

There is another—and for my purposes, vitally important—way in which the *writing* of *The Scarlet Letter* had gone awry and had strayed precipitously from the authorial mission that Hawthorne had accepted from his ghostly, patriarchal interlocutor. Instead of serving as "a living sermon against sin" (171) [46], the novel and its protagonist bear witness to the resilience and recalcitrance of the oppositional human spirit embodied by a woman and thus *betray* the archival imperative to awaken and to renew the Puritan model of subjectivity. In terms of the construction—that is to say, the contract—of authorship, "Nathaniel Hawthorne" had not fulfilled *his* office, for he (presumably) did not repeat either the spirit or the letter of his archival and authenticating source; rather, he rewrote it. It is *here* that I would locate the volatile field of Martin's "psychic cost" of authorship; for it is precisely the straying from archival origin—or the erosion of initial "intention" in the spirals of *différance*—that tests severely what I have termed the psychic viability of gendered authorial identity. By "Nathaniel Hawthorne," I mean the ficto-biographical figure of the author who is engendered, who binds himself to patriarchal tradition, at the hand of Surveyor Pue in the ficto-literal archive of the custom house. In the overarching economy of the book, the encounter with Pue must be considered as arising from the novel's absolute primal scene, which conflates a hypothetical exhumation of the forefather with the protocols of archival reading. I shall have more to say about "Hawthorne" in the custom house and about the temporality of primal scenes, but first, I need to clarify the function of the archival *mise-en-scène* in a more conceptual manner.

3. I take it as axiomatic that *The Scarlet Letter* constitutes a master class, or a workshop, in Jacques Derrida's theory of *différance* and that it does so more exhaustively than any other canonical American text. *Différance* condenses the meanings of the French verbs "to differ from" and "to defer." Particularly in the scenes of reading of the scarlet letter itself, but also in the general economy of the narrative, Hawthorne's writing both conceptualizes and demonstrates that the meaning of the sign is never present in the present, for the *instance* of the sign is but a repetition of its wider circulation in the text and beyond, and these incidental meanings shift in the *déroulement* of the text. Consequently, the recurrence of the sign is caught up in a temporal matrix: it points to a future time, at which we never arrive, when all will be clarified. The best reading of *différance* in Hawthorne's work is J. Hillis Miller's *Hawthorne and History*, which explores the temporality of repetition and deferral in "The Minister's Black Veil."

For Derrida, the material archive—the institute, the collection, the repository—is the substrate upon which national or cultural belonging is both anchored and predicated. Given that its function is to conserve, its political agency is conservative: all archives are houses of custom, and their spatial configuration is essentially narrative—here, it says, are your origins and your history; this is who you are; this is where you come to *know* who you are. The archive, then, is at once a domicile and an ideological apparatus; at once the rule of law and the symbolic order. If the documentary archive is localized, its regulatory effects are everywhere because it grounds and authenticates the interpellation of the subject. Derrida's concept of the archive is essentially a theory of the *temporality* of the subject: the subject does not exist in any coherent way until he or she is called to order by an *event* of the archive. This temporal model has certain things in common with Judith Butler's model of the performative dimensions of gender: the gendered subject materializes through continually reiterated "citational practices" that stabilize the subject's gendered identity. Gender is, for Butler, a cultural archive of performatives that, through the repetition of entirely familiar and traditional codes, enacts the seeming naturalness of gender as custom. Crucially, for Butler, there is no preexisting, volitional subject who "acts"; there is merely a reiterated acting that consolidates both the gendered subject and the ideology of gender itself. If, for Butler, there is no "origin" for gender apart from its continual materialization in the present, then this is where she parts company with Derrida, for whom the archive is a literal text that precedes the subject and that guarantees the *affiliation* of the subject with *being-as-dutiful-imperative*, or being as doing. To put this another way, Butler's archive is predicated upon the unconscious and unintentional repetition of an archive that does not exist apart from custom, whereas Derrida's archive is predicated upon a determined protocol of repetition: it remembers and revivifies a commandment that issues from the *circulation* of a founding text, that is, a mythical point of commencement.

In *Mal d'archive* (translated in English as *Archive Fever*), Derrida offers an example of an archival event that, in its *recall* of the subject to cultural belonging, is strikingly congruent with "Nathaniel Hawthorne's" encounter with Surveyor Pue in "The Custom-House." On 6 May 1891, the occasion of Sigmund Freud's thirty-fifth birthday, his father presented him with the Philippsohn Bible that Sigmund had studied in his youth.[4] The gift bears an inscription in the father's hand: "In the seventh in the days of the years of your life [the day of Freud's circumcision], the Spirit of the Lord began to move you and spoke within you . . . Since then the book has been stored like the fragments of the tablets in an ark with me. For the day on which your years were filled to five and thirty I have put upon it *a cover of new skin* and have called it: 'Spring up, O well, sing ye unto it!' And I have presented it to you as *a memorial and a reminder*" (qtd. in

4. For an explanation of the cultural and archival status of the Philippsohn Bible, see Alan T. Levenson's *The Making of the Modern Jewish Bible*. Levenson's erudite and wonderfully readable text explores the following question: "With the Bible having been bequeathed to Christianity and Islam and Western culture in general, can the Bible still be 'Jewish'?" (1).

Derrida 23; emphasis added). As commemoration and "reminder" of an anniversary—that of Sigmund Freud's circum-inscription into Jewish belonging in accordance with the commandment—the event of the gift is patri-archival. It is, in a manner, at once performative, coercive, literal, and figurative, entirely and coherently *restorative*. It literally restores the biblical substrate to Freud's adult hand, and this "book of books" carries the father's exhortation.

Most striking, however, is the event's figurative dimension, which laminates together or "rebinds" all of the others: as Derrida writes, "the father restores it to him, after having made a present of it to him; he restitutes it as a gift, *with a new leather binding* (21; emphasis added). In order to grasp the figurative import of the new binding, it is helpful to turn to Derrida's French as he addresses the father's inscription: "*Celui-ci parle à ce sujet d'une peau neuve: 'new skin' dit la traduction anglaise de l'hébrou, 'une nouvelle housse en peau,' selon les traducteurs francais de la traduction anglaise*" (41; author's translation). Working among the languages of translation and transmission, from Jakob Freud's Hebrew to the English translation to the French, Derrida puts into tacit relation the following terms: on the one hand, the Bible's literal "new leather binding" or "*une nouvelle housse en peau*" (literally, a new cover of skin), and on the other hand, in the shift from the noun to the verb—the interpellative act of "rebinding" a subject to a tradition arising from the literal act of rebinding an archival book. In French, this act involves the relation of the noun "*reliure*" (the binding of a book, noun-as-gerund, or indeed, the bound book) and "*relier*" (the infinitive form of "to bind"). He writes, "*[I]l la lui restitue en cadeau, avec une nouvelle reliure de cuir. Relier de nouveau, c'est an acte d'amour. D'amour paternel*" (41; author's translation, emphasis added)—or, in English, "To bind anew: this is an act of love. Of paternal love" (21).

Derrida notes the multifaceted complexity of this archival moment: "The foliaceous stratification, the pellicular superimposition of these cutaneous marks seems to defy analysis. It accumulates so many *sedimented archives*, some of which are written right on the epidermis of a body proper, others on the substrate of an 'exterior' body" (20; emphasis added). Although Derrida does not tease out the significance of the "sedimentation" of the literal and figurative axes of the rebound Bible and the father's exhortation, my argument requires me to do so. I would argue that, subtending all of the "re-"prefixed nouns—recall, renewal, rebinding—is the hypothetical archival temporality of *return* figured as *restitution*. It is *as if* the symbolically charged "*nouvelle housse en peau*" returns the adult Sigmund Freud to the time after his birth and before his circumcision, to that historical moment immediately prior to his circumscription in Jewish belonging—*as if* by figuratively restituting his foreskin, he would not only be returned to the gravity of that primal scene, that constitution of the filial subject as such, that moment of origin, but also and by corollary implication re-member his "member" as a member, and thus return to membership in Jewish tradition and reconsecrate his life. *Relier de nouveau*, then, in addition to its status as an archival event that rebinds the subject to the archive of commencement and commandment—the origin and ground of subjectivity—solicits a response akin to that of "Nathaniel Hawthorne" to Surveyor Pue: "I will" (147) [28].[5]

The event of the patriarchal gift—invariably enacted within the temporality of repetition-as-restitution—confers authenticity as it consigns the subject to the archive, and vice versa. As event, it is the *charnière*—the hinge—between opposed temporal currents: the retrospective, or the recall of mythical origin; and the prospective, or the promise of the tradition's fulfilment in the time to come. It is not only history, but also the *dispositif* of historical understanding. History, then, and the subject's place in its unfolding, are circumscribed in the economy of the archive, which Derrida conceptualizes as "the archontic dimension." As cultural practice and as the field of the human sciences, it follows

> the logic and the semantics of the archive, of memory and of the memorial, of conservation and of inscription which put into reserve ("store"), accumulate, capitalize, stock a quasi-infinity of layers, of archival strata that are at once superimposed, overprinted, and enveloped in each other. *To read*, in this case, requires working at geological or archaeological excavations, on substrates or under surfaces, old or new skins, the hypermnesic and hypnomesic epidermises of books or penises. (22; emphasis added)

Relier de Nouveau: *On the Revenant as Recall*

Derrida's linkage of "books [and] penises" within the interlined "strata" of the archive condenses metonymically an aspect (the performative ritual) of what Butler would call the "materialization" of the gendered body as a citational practice. Clearly, Derrida owes a great deal to Foucault, who, in *The Archaeology of Knowledge*, was the first to detach the archive from its primary definition of an institutional collection of documents and to relocate it in the event—in this case, the enunciation. Foucault proposes to call "the archive" "the density of discursive practices, systems that establish statements as events" (128). As *dispositif*, or "the law of what can be said," the archive inheres in "the system of discursivity, in the enunciative possibilities that it lays down" (129). The archival event, then, is a speaking-forth authenticated by the law; and if the archive acts as guarantor, defining "at the outset *the system of its enunciability*," then it is also, and concomitantly, "that which defines the mode of occurrence of the statement-thing; it is *the system of its functioning*" (129; original emphasis). Foucault's repeated insistence that the archive is fundamentally a system of coherent and binding knowledge production turns, in due course, to the role of the subject within it—that is to say, the subject as the object of the archive's regulatory reach. At this point in his argument, Foucault's discourse achieves a remarkable expository clarity, within a rhetoric that is, somewhat paradoxically,

5. If Freud, Derrida, and Hawthorne had been Catholics, or indeed members of any church that practices infant baptism, this performative would be a sacrament of the Church, like that of Confirmation—which "renews" the binding effect of baptism, which in turn marks the subject's entry into the community of the Church. The point of this comparison is to demonstrate that "rebinding" the subject occurs in a temporal "moment" that (*pace* Derrida) is more retrospective than prospective. In archival terms, a Jewish bris or a Catholic baptism points beyond the subject to a "historicity" whose origin is at once lost in the mists of time and prescribed by archival protocol.

strikingly figurative—as though he had arrived at the limit of what could be communicated literally and directly. Here, the *temporal* predication of enunciative events—the enabling constraints imposed by tradition—is represented metaphorically as the subject's *spatial* positioning. The archive

> is at once close to us, and different from our present existence; it is the border of time that surrounds our presence, which overhangs it, and which indicates it in its otherness; it is that which, outside ourselves, delimits us. The description of the archive deploys its possibilities (and the mastery of its possibilities) on the basis of the very discourses that have *just ceased to be ours*; its threshold of existence is established by the discontinuity that separates us from what we can no longer say, and from that which falls outside our discursive practice; it begins with the outside of our own language; its locus is the gap between our own discursive practices. (130–1; emphasis added)

Foucault's spatial positioning of the archive might seem to be both self-contradictory and resistant to the argument about the historical weight of the archival event that I have been developing: it is "the border of time," the "threshold of existence," "the gap," and "the outside" of our discursive practice; yet, at the same time, it "surrounds our presence," "overhangs," and "delimits us." Temporally distant and spatially "outside," it nonetheless frames the possibilities of our being, the conditions of our utterances and, consequently, the meaning of event. The dialectic of the archive's simultaneous status of outside and inside may point to Lacan's utterance that this "it" is "in you more than you."[6] At issue, it seems to me, is the affective charge that takes place at this very "border of time," at this "threshold of existence." What is this nebulous affect? To clarify, I return to Jakob Freud's gift of the rebound Philippsohn Bible to his son, which was simultaneously a transmission of the archival imperative and what Foucault calls "a description of the archive": it was evidently motivated by precisely an *anxiety* about "discontinuity" and about the rupture in his son's life with the discourse of Jewish identity that "had just ceased to be [his]."

The father's goal in restituting the archival substrate and, figuratively, the new skin, cannot have been other than to suture anew that discontinuity and to close "the gap" between Sigmund Freud's present and his immediate genealogical past. The spatiotemporal positioning of the subject by the archival event is of crucial, and fascinating, importance in the system of the subject's restoration. As I argued above, in my analysis of Derrida, the timing of Jakob Freud's present (in the present) collapses the interval between 1891 and 1856, returning Sigmund Freud to the moment when he was (is) just about to undergo the rite of the briss, that inscription of the covenant when, as Jakob writes, "a vision of the Almighty did see you; you heard and strove to do, and you soared on the wings of the Spirit" (Derrida

6. See chapter twenty of Lacan's *The Four Fundamental Concepts of Psychoanalysis*. "Ask the faithful, or even the priests, what differentiates confirmation from baptism—for indeed, if it is a sacrament, if it operates, it operates on something. Where it washes away sins, where it renews a certain pact—I would put a question mark here—is it a pact? Is it something else? What passes through this dimension? . . . In all the answers we get, we will always find this mark, by which is invoked the beyond of religion, operational and magical. We cannot invoke this operational dimension without realizing that within religion . . . it is this that is marked with oblivion" (265). In taking up the repetitive temporality of the sacramental ritual in relation to a lost origin, Lacan anticipates Derrida's argument about archival logics.

23). For Derrida, the Freudian presenting is exemplary of the archival system because it repositions the subject both retrospectively and, more importantly, prospectively: to return to the scene of circumcision is to renew what Jakob Freud terms the imperative of "[striving] *to do*," or the resumption of filial duty that reorients the subject toward an observant time to come. Derrida repeatedly insists that

> the question of the archive is not . . . a question of the past. It is not the question of a concept dealing with the past that might *already* be at our disposal or not at our disposal, *an archivable concept of the archive*. It is a question of the future, the question of the future itself, the question of a response, of a promise and of a responsibility for tomorrow. The archive: if we want to know what that will have meant [*aura voulu dire*: literally, "will have wanted to say"], we will only know it in times to come. (36; original emphasis)

The relation of archival event to the past is the domain of Foucault, whose interest concerns the system itself and its disciplinary uses for the present: hence the father's anxiety to establish his son's connection to a discourse "that had just ceased to be ours." But the relation of the archival event to an open futurity under the seal of filial duty determines the anxiety, the tyrannical superego, of the son: it is precisely this linkage of (re)commencement and commandment that reveals, for Derrida, the burden (*fardeau*) of the archive. If the archive is, according to the father, the very ground of the subject's being, then its psychic viability is contingent upon the trials of anxiety, which are suggested, not only by the escalating *gravitas* of Derrida's nouns—"the question of a response, of a promise, and of a responsibility"—but also by the complex, queer temporality of the *future anterior* of the son's hermeneutic mission to figure out what the archive "will have wanted to say." It is this latter duty that subtends, and haunts the margins of, Derrida's figurative explanation (quoted in the passage just prior to this section of the essay) of the verb "to read" as an "archaeological excavation." To read, in this sense, is not unlike the French reflexive verb, which needs a reflexive pronoun to indicate that the subject is performing the action of the verb upon itself: *se lire*. If such a grammar of the subject is the field of Freudian psychoanalysis, then it is also, and inevitably, the fundamental narrative trajectory of the archive, conceptualized simultaneously as the document, the tradition, the law, and the historical unconscious. It is here, in this place, that the psychodrama unfolds and where the subject orients himself toward the past, assumes the burden of future responsibility—and worries. For if, as Foucault asserts, the archive is that which "overhangs" us, can it be other than a bad hangover?

My *portage* through the territory of Derrida and Foucault has served to construct a conceptual frame in which I can situate Hawthorne's meditation on the uses of the past. From the get-go, he represents the archival overhang as a melancholy haunting that, although consistently figurative, accrues, through sheer repetition, the illusion—and the anxious affect—of the literal. While this illusion is basic to Hawthorne's narrative mode of the spectral gothic, his persona remains fixed in place and firmly anchored to the present life of Salem. For instance, he acknowledges that his American origin—"the figure of that first ancestor"—"still haunts me, and induces a sort of home-feeling with the past" (126) [11]. Yet, this comfortable sense of the *heimlich* is thoroughly interlined with the *unheimlich*, the uncanny moment of recognition of the gruesome and the unsettling. Hawthorne wastes no time in launching the figurative register, which entails an entirely speculative *mise-en-scène* of a *hypothetical* archaeological excavation. Chronicling the "persecuting spirit" of one of his Puritan forebears, particularly his cruelty toward the witches, he conjectures that "their blood may *fairly be said* to have left a stain upon him. So deep a stain, indeed, that his dry old bones, in the Charter Street burial-ground, must still retain it, if they have not crumbled utterly to dust!" (126 [11]; emphasis added). The expression "fairly be said" is a prime index of Hawthorne's speculative, spectral poetics: it denotes the (poetic) justice of the utterance, but it also connotes its merely relative accuracy: it is the discourse of a literary thought experiment that takes place in a sort of neutral territory between the literal and the figurative, as the one slides toward the other. As a hypothetical archaeological event, this crucial passage in "The Custom-House" demonstrates the astuteness of Foucault's claim, cited earlier, that the archive is "the border of time that surrounds our presence, . . . and which indicates it in its otherness." What takes place at this archaeological site is, of course, writing, but it is writing that is *figured as reading*, for the text of the "dry bones" bears the inscription, the "stain," of the martyrs' blood.

This is, I say, and for several reasons, a vitally important passage. First of all, it constitutes Foucault's archival "border of time" as such, as that which "surrounds our presence" in the present; moreover, it does so by figuring (and hypothetically undoing) the spatial border between the reader above ground and the text below. Second, it invites us to look again at Foucault's sentence: what is the antecedent of "it" that is "indicate[d] in its otherness?" Is "it" "time"? or is it "our presence" that is rendered other at the border? It's clearer in the French original: *"[C]'est la bordure du temps qui entoure notre présent, qui le surplombe et que l'indique dans son altérité"* (179). Note the important difference in focus that arises between the original and the received translation: Foucault writes that the archive "is the border that surrounds our *present*, which overhangs it and which indicates it in its otherness." "Our present" has a temporal acuity that is lost in the nebulous and multifaceted "presence," although it surely orients the meaning of our presence toward the past. In Hawthorne's scene of imaginary excavation, it is the present that is made other as it is overhung, haunted, indeed compelled to re-member that disintegrated patriarchal corpse and to read the lesson of the stain as the mark of Puritan ideology, of public service and civic duty. Third, and most importantly, this passage is essentially proleptic: it is, I suggest, the primal scene of writing, for, in its speculative figuration, it shadows forth and anticipates both the discovery of Hester's scarlet letter and "Nathaniel Hawthorne's" assumption of the filial duty to recount that history.[7] For what is the status of the blood stain, if not as the original mark of Puritan misogyny and thus the scarlet letter *avant la lettre*?

7. The classic model of the primal scene is, of course, Freud's. In the "Wolf Man" case history, the primal scene is the founding, traumatic impression upon the subject that can never be consciously remembered; rather, its repression is generative of the presenting symptom. Reconstructed in analysis, the primal scene is of enormous *narrative* importance: "it is indispensable to a comprehensive solution, . . . [because] all the consequences radiate out from it, just as all the

The foregoing analysis of *The Scarlet Letter*'s primal scene of composition at the archaeological or material border of the archive—the grave as the repository of consignment—raises the critical question of Hawthorne's figurative method, which revivifies the dusty residue of the grave, the "consigned," as sign. It also returns my project to the question I posed at the outset of this essay: why did Nathaniel Hawthorne cast "Nathaniel Hawthorne" as a character in this ficto-critical preface? What motivated this attempt to fictionalize the genesis of his novel and to theorize poetically about the provenance of his historical romance? The answer, as I have anticipated by linking Robert K. Martin's historical survey of masculine identity to Derrida's theory of archival event, has to do with what I termed, at the outset, the psychic viability of an ethics of American authorship. "The Custom-House" is less a fiction or an autobiography than it is a psychodrama, a dream-like thought experiment that frames and tests the viability of that identity. Its method is to raise the dead and to animate them in the echo chamber of the superego; it is to attend to the revenant. And in the logics of archival figuration, it is to *extend one's hand* across time to clasp the hand of the progenitor who wielded, in his day, "his Bible and his sword" (126) [11], as surely as Sigmund Freud was recalled to his proper place in the world by the circumscription of his father's hand, by the uncanny *pressure* of that hand. As Derrida writes, "a spectral messianicity is at work in the concept of the archive and ties it, like religion, like history, like science itself, to a very singular experience of the promise" (36). The "experience of the promise" is, of course, never present in the present, but is, indeed, the responsibility to realize it in the future. As such, then, "spectral messianicity" is *the* thing that is caught up in *différance*, and its representation can only be figurative; for "the structure of the archive . . . is spectral *a priori*: neither present nor absent 'in the flesh,' neither visible nor invisible, a trace always referring to another *whose eyes can never be met*, no more than those of Hamlet's father" (84; emphasis added).

Having disinterred the sternest of the forefathers, Hawthorne proceeds to personify them and to situate "Nathaniel Hawthorne" as the object of their scorn. This personification falls short of a fully realized *prosopopoeia*, for it generates no face to be faced; as Derrida suggests, the spectral father is, at this point in the narrative, "neither visible nor invisible," for there are no eyes to be met in this nascent economy of the gaze. That will come later. For the moment, in the proximity of the hypothetical archaeological excavation, we attend aurally to the voices: "'What is he?' murmurs one grey shadow of my forefathers to the other. 'A writer of storybooks! What kind of a business of life,—what mode of glorifying God, or being serviceable to mankind in his day and generation,—may that be?'" (127) [11]. While the intervening generations followed the

threads of the analysis have led up to it" (289). The narrative function of the primal scene has been explored richly by Peter Brooks: taking up Freud's uncertainty about the historical reality of the primal scene, and writing under the influence of Derridean *différance*, he argues that "so spectacular a moment of origin doubles back on itself to question that origin and indeed to displace the whole question of origins, to suggest another kind of referentiality, in that all tales may lead back, not so much to events as to other tales, to man as *a structure of the fictions he tells about himself*" (277; emphasis added). Brooks's explanation is resonant with my argument about Hawthorne's archival fiction as a thought experiment.

manly occupation of sea-faring, "Nathaniel Hawthorne" stands at the end of the line as the exception, as a "degenerate fellow" (127) [12]. This scene is essentially an animation of the paternal superego that, in archival terms, asserts the law of gender conformity essential to generational continuity; its office is to regulate the subject through the tactics of shaming. To attach oneself to the label degenerate and to have lapsed in the cross-generational imperative is to bear a psychic scarlet letter; it is to occupy the status of the *abject*. For shame, as Eve Kosofsky Sedgwick has explained, is a radical emptying-out of subjectivity in the present; yet, at the same time, it "generates and legitimates the place of identity—*the question of identity*—at the origin of the impulse to the performative, but does so without giving that identity-space the standing of an essence. It constitutes itself as to-be-constituted" (14; original emphasis).

The archaeological moment—the speculative *descent* into the ancestral grave, the reading of the import of the scarlet stain, and the attentiveness to the shaming voices that arise therein—marks the lowest and most depressive point of Hawthorne's vocational meditation. This atmosphere broadens, above ground, as Hawthorne—the Surveyor of Customs Revenue—surveys the stifling ennui of the civil servant's life, the banality of official custom, and the lethargy of his companions. It is as if Hawthorne's turn toward a remunerative but useless occupation sustains and prolongs the abjection he suffered; it is a kind of living death, for "a gift, a faculty, if had not departed, was suspended and inanimate within me" (141) [21]. "Nathaniel Hawthorne's" task is not only to reanimate his gift, but also and more importantly to *justify* the vital importance of his craft in the eyes and the ears of the spectral fathers as a means of "being serviceable to mankind in his day and generation." It is to recover, and to assert, the ethical subjectivity that is his vocation. It is, by implication, to promise to take up Derrida's "experience of the promise" and to carry it toward its ultimate fulfilment. The accomplishment of this task is framed by Hawthorne's spatial poetics when they arrive at equilibrium: the *archaeological descent* into the primal scene of the ancestral grave is now counter-balanced by an *architectural ascent* to the archive proper. Here, on the second story of the Custom House, at a remove from soporific business, Hawthorne will unfold his second story in successive scenes of reading—and conjuring. The rites of ascent are continuous with Hawthorne's archaeological figuration, for here, too, the archive is littered with textual corpses. Much of the archive is of little interest because everything pertaining to the colonial era was carried off to Halifax prior to the Revolution. It isn't a real archive, then, because it has lost the connective thread of its commencement; it's at best the detritus of history, a pile of "heaped-up rubbish" (143) [23]. As the "degenerate" abject—a rubbish son, a rubbish writer—"Hawthorne" would seem to have met his appropriate object, for nothing comes to hand to summon up the inspiration for a useful, workable project. Instead, as he unfolds dreary documents, he glances at them "with the saddened, weary, half-reluctant interest which we bestow on the corpse of dead activity" (144) [23–24].

One day, however, "Nathaniel Hawthorne" "chanced to lay [his] hand on a small package" (144) [24]. Stylistically, this is an important image for, in keeping with what I have defined as the logics of archival figuration, he performatively *extends his hand* toward the past to grasp its import (and will, in due course, extend his hand further in the exercise of writing). The package, left behind during the clearing-out of the Salem archive long ago, turns out to contain a document within a document within a document. The first, enfolding the others, is "a commission, in favor of one Jonathan Pue," appointing him to the government office that Hawthorne himself now holds. Although Jonathan Pue died and was buried eighty years prior to this archival scene, Hawthorne aligns him with his own forefathers, through the repetition of the gruesome archaeological excavation of the grave. This time around, the scene is literal rather than hypothetical, for he "remembered to have read . . . in a newspaper of recent times, an account of the digging up of his remains in the little grave-yard of St. Peter's Church, during the renewal of that edifice" (144) [24]. It is an uncanny repetition in two ways: first, it requires us to read retrospectively back to the image's original occurrence; second, it renders *real* what was a speculative figure. Although Hawthorne lingers, in a fetishistic manner, upon the little that remained of Pue's corpse—some bones, a few scraps of apparel, and a majestic wig—the second document reveals "*traces* of Mr. Pue's mental part, and the internal operations of his head" (144 [24]; emphasis added). Let me recall, at this point, Derrida's argument that the structure of the archive "is spectral *a priori*: neither present nor absent 'in the flesh,' neither visible nor invisible, a *trace always referring to another*." Leaving the matrix of spectrality aside for the moment (while noting that Pue has already been drawn into the sequential figurative economy of disinterment, reading, personification, and haunting), I would underscore the primordial function of the trace as *différance* in "Nathaniel Hawthorne's" archival adventure. He clasps, in his hand, the material trace of Pue's hand: the writing "on several foolscap sheets, containing many particulars respecting the life and conversation of one Hester Prynne, who . . . had flourished during a period between the early days of Massachusetts and the close of the seventeenth century" (146) [25]. Pue's tracing of the traces left by Hester Prynne's life relied on no archival document, but rather upon the "oral testimony" of "aged persons, alive in the time of Mr. Surveyor Pue [who] remembered her, in their youth, as a very old, but not decrepit woman" (146) [25]. His writing, then, is an impressionistic trace that weaved together a series of impressionistic traces: as biography, it was generated in the temporal gaps between writing and listening to "testimony," and between remembering and witnessing. However, the historical reality of Hester Prynne is not at issue: what matters (in what is, after all, a fictive scene of historiography) is the archival transmission of a particular legend—which gives rise, in turn, to the renewed custom of *surveillance*, by the Surveyor, of a mythical and unruly woman. Pue's work was to weave this legend into the national symbolic by assuming, consciously or unconsciously, the duty to repeat an ethos of masculine comportment that was present at the beginning.

Although I have called Pue's text the second document (for reasons that have to do with the continuity of my argument), it is actually the third—for, at the centre of the package, it supplements the illegible "affair of red cloth, much worn and faded" (145) [25] that is wrapped around it. Once again, we are presented with an archival object that is the trace of a trace. It forms the capital letter A, barely legible: "there were traces about it of gold embroidery, which, however, was greatly frayed and defaced." Clearly an ornamental article of dress, the ragged text remains stubbornly elusive: "what rank, honor, and dignity, in by-past times, were signified by it, was a riddle which . . . [he] saw little hope of solving" (145) [25]. Fascinated by the promise of "some deep meaning" in it, "Nathaniel Hawthorne" approaches the scarlet letter, not as a signifier, but as a "mystic symbol" (146) [25], replete with sensibility, but somehow beyond analysis—for even its material composition is but the trace of a "now forgotten art, not to be recovered even by the process of picking out the threads" (145) [25]. Within the framework of Derridian archival structure, the scarlet letter—at once historical artefact, text, signifier, and symbol of some mysterious import—is a trace that refers to another. However, Pue's textual elaboration is caught up in the circuits of supplementarity: it clarifies the origin and provenance of the artefact—that it has something to do with the patriarchy's relation to a transgressive woman—but Pue can only situate the letter in historical time; he cannot explain it. That mission now falls to "Nathaniel Hawthorne," which is to say, as Lacan puts it, a letter always arrives at its destination. The scene of its discovery-as-arrival is, of course, the very point—the "Go" on the board game of "The Custom House." Hawthorne will turn, immediately, to reframe, scenically, his archival discovery as a *delivery* (and a *deliverance* from the dead-letter office of the custom house proper); he will do so in accordance with Derrida's model of "spectral messianicity" and by repeating his figurative economy; that is, the turn to the personification of the dead as a form of thought experiment. For he will extend his hand to receive the archival substrate from the ghostly hand of the literary father who wrote it.

The recessive and recalcitrant nature of the scarlet letter, like that of Hester Prynne herself, will persist beyond the Custom House archive to the very end of the novel. Were it transparently lucid, it would offer no occasion for the profound meditation that is Hawthorne's narrative writing. At the same time, however, it offers "Nathaniel Hawthorne" an opportunity to inscribe himself in the patriarchal tradition of public service, precisely, as a surveyor of custom and thus to transform his passive abjection into an active, vocational subjectivity. This is psychically viable within the structure of the archive, for it will suture a venerable past to a lesson for the present and a gift for the future realization of the Puritan promise. It supremely matters, then, that the letter arrives with no clear import and that it is delivered by the patriarchal revenant who exhorts him to do his duty. At various moments in Jacques Lacan's career, he conceptualized the signifier as a persistent, meaningless letter that "marks the destiny of the subject and which he must decipher" (Evans 100). The letter, then, is essentially "that which returns and repeats itself; it constantly insists on inscribing itself in the subject's life" (100). If Lacan's return to Freud emphasizes the subject's need to interpret the letter's concealed message, he relies, at other moments, on linguistic models of circulation—most notably, in his seminar on Poe's story "The Purloined Letter"—to argue that the import of the letter is less important than the fact that it assigns a position to whoever possesses it; or rather, whoever is possessed by it. It is interesting and instructive to consider what I have termed a patriarchal anxiety (whether that of Jakob Freud or of Hawthorne's ghostly ancestors) to rebind the son, under the auspices of the archival event, as concentrated in the *scenic* delivery of a letter. As Lacan explains, in his essay "Seminar on 'The Purloined Letter,'"

> we are quite simply dealing with a *letter* which has been *detoured*, one whose trajectory has been *prolonged*, . . . or, to resort to the language of the post office, a letter *en souffrance* (awaiting delivery or unclaimed) . . . While the letter may be *en souffrance*, [the subjects] are the ones who shall suffer from it. By passing beneath its shadow, they become its reflection. By coming into the letter's possession—an admirably ambiguous bit of language—its meaning possesses them. (21; original emphasis)

Lacan's model of the circulation of the letter and its belated arrival at its destination—its "detour"—is remarkably congruent with the archival economy as such, for one might say that Hester Prynne's scarlet letter and Pue's text, like Freud's Philippsohn Bible, remained *en souffrance*—in abeyance—awaiting delivery. And the consequence of its arrival, for the addressee, is that he must suffer the burden—that is, be possessed by the imperative to decipher the letter that marks his destiny.

My lamination of Derrida's archival event to Lacan's theory of the trajectory arises from the fact that they share an interest in the letter, both as material artefact and as a positioning—an *assujettissment* or a call to subjectivity—that constitutes the subject's identity in the symbolic order. This, indeed, is the point at which they intersect, for Derrida's understanding that the gift of the letter is an instance of *relier de nouveau*, the binding anew of the subject into the generations, is but another term for Lacan's grasp of the ramifications of "*souffrance*." Essentially, the work of both theorists arose from their reading in the Freudian archive, particularly, the aspects of Freud's metapsychology having to do with the repetition compulsion, or what Lacan terms the "repetition automatism" ("Seminar")—that is, the symbolic system that lies beyond the subject's volition but that determines the subject's status and role. But there is an important difference between them: Lacan explores the function of the letter's circulation within a purely psychic economy, whereas for Derrida—and for Hawthorne—the letter's arrival is a ritualistic event that is framed by the protocols of gender politics.

In both political and psychoanalytic terms, it is worth exploring the climactic scene of "The Custom-House" narrative—the personification of Surveyor Pue as revenant—as simultaneously an instance of Lacan's "repetition automatism" and a repetition compulsion within Hawthorne's overarching poetics of the archive. Hawthorne's discourse is always already *both*, for the scene, from its outset, coalesces under the sign of *recall*. The event of the letter's arrival at its destination made such an impression, he writes, that it "recalled my mind to its old track" (147) [26]. "Track"—in French, *trace*—condenses several interpretive possibilities. The metaphor refers, in the first instance, to Hawthorne's literary vocation: the event has put him back on track. By implication, this track will not be what it was before because "Nathaniel Hawthorne" needs to justify that vocation to his ghostly, dismissive, abjecting fathers—this is the depressive track that he has traced, autobiographically, from the graveyard to the custom house. Finally, and most importantly, it recalls him to his poetic practice—rehearsed in the archaeological primal scene of this narrative—of figuring archival reading as the personification of the dead. And while the delineation of the primal scene traced, as I have argued, the vertical axis of descent into depression and abjection, the ensuing scene (the corollary to the primal scene—or, one might argue, its therapeutic correction) crowns the axis of ascent into the archive with the conferral of an authentic subjectivity, albeit contingent upon the arrival of a "meaningless" letter.

Once again, Hawthorne embarks upon a speculative thought experiment: "It impressed me *as if* the ancient Surveyor, in his garb of a hundred years gone by, and wearing his immortal wig,—which was buried with him, but did not perish in the grave,—had met me in the deserted chamber of the Custom-House" (147) [26]. The details of Surveyor Pue's attire fulfil the requirements of the hypothetical "as if," for they exert a pull toward the literal, the actual, and the credible—*as if* this spectral figuration were as tangible as the archival documents themselves. Such, I would argue, is the appeal of Hawthorne's gothic articulation: it conjures even as it conjectures. It is irresistible. It endows the surveyor with an entirely *portentous* and authoritative subjectivity, for "in his *port* [i.e., comportment] was the dignity of one who had borne his Majesty's commission, and who was therefore illuminated by a ray of the splendor that shone so dazzlingly about the throne," which stands in sharp contrast to the abject non-entity who comports himself with "the hang-dog look of a republican official, who, as the servant of the people, feels himself less than the least, and below the lowest" (147) [26]. As the spectral embodiment of a "ray of splendor," which is his firm connection to the old patriarchal order, Surveyor Pue's office allowed him to confer office and, by implication, a subject-position within that symbolic order.[8] He accomplishes this by extending his hand to the errant apostate and by presenting the work of his hand (as did Jakob Freud to his son) as "*a memorial and a reminder*." The scene takes the form of a ritual handing-over, a call to duty; as a kind of circum-inscription of the subject, it conjoins the expert hand with the rather cutting voice:

> With his own ghostly hand, the obscurely seen, but majestic, figure had imparted to me the scarlet symbol, and the little roll of explanatory manuscript. With his own ghostly voice, he had exhorted me, on the sacred consideration of my filial duty and reverence toward him,—who might reasonably regard himself as my official ancestor,—to bring his mouldy and moth-eaten lucubrations before the public. "Do this," said the ghost of Mr. Surveyor Pue, emphatically nodding the head that looked so imposing within its memorable wig, "do this, and the profit shall be all your own! . . . But I charge you, in this matter of

8. Robert K. Martin explains the ironies of Pue's conferring office within a superannuated symbolic order: "American history, then, is founded upon a violation of authority, a refusal of the king-father's rule; and yet descent from those original parricides, ironically, becomes a mark of legitimating authority" (125).

old Mistress Prynne, give to your predecessor's memory the credit which will be rightfully its due!" And I said to the ghost of Mr. Surveyor Pue,—"I will!" (147) [26]

Relier de Nouveau? *On Origins and Departures*

In other words, *do this in memory of me*. Given the ritualistic context of this scene, "Nathaniel Hawthorne's" acceptance of his mission not only seals the archival event but constitutes a binding performative, a speech act that is transformative of his subjectivity. No longer the "hang-dog" public servant, he is henceforth authorized as an official historian of the Puritans. But something goes awry in the retelling of Hester Prynne's story. We do not have access to Pue's text; for, despite "Nathaniel Hawthorne's" fictional promise (his lie) that "the original papers . . . shall be freely exhibited to whomsoever . . . may desire a sight of them" (147) [26], the missing archival origin takes its place in the narrative's general economy of displacement and deferral. However, it may be fairly conjectured that *The Scarlet Letter* diverges from its putative source: if Pue, writing in temporal proximity to the Puritan era, may be surmised to have composed a straightforward account of crime and punishment—in a narrative that did not call into question the justice of the magistrates' sentence—then Hawthorne's elaboration clearly does not follow any of the Puritan lines. For the historical romance does not toe the line of the law; rather, it uncannily repeats the experience of transcendence by which a "degenerate" abject arises to the status of an ethical and fully integrated subjectivity, which is the essential story of "Nathaniel Hawthorne's" ascent in the custom house. Although Hawthorne is not uncritical of Hester Prynne's spiritual independence, which is acquired at the expense of nineteenth-century conventions of femininity, the narrative a*lign*ment with Hester's inward trials is both sympathetic and identificatory. Let's say, for the purposes of argument, that Hawthorne's relationship with his protagonist is ambivalent, for his sympathetic attachment is fissured by his anxiety about strong and purposeful women. Such ambivalence would go far in explaining the ambiguity with which Hester's scarlet letter is continually imbued. If, for Pue, the letter was merely a transparently legible historical artefact that authenticated his historiography, Hawthorne's letter departs from such a certain *line*. Shifting in import as it circulates in the Puritan community, its interpretive detour ultimately arrives in the form of a letter that has been overwritten: as a palimpsest, it never loses its traces of "a stigma which attracted the world's scorn and bitterness"; yet it "became a *type of something* to be sorrowed over, and looked upon with awe, [and] with reverence too" (344) [154]. What, then, went awry in Hawthorne's archival mission? What became of the forefathers' "spectral messianicity" and the "experience of the promise"? I shall set out various possibilities.

The narrative's inability to reduce the scarlet letter to a coherent, symbolic meaning that would rebind historical imperative to the filial duty of *vigilant* surveillance points to the psychic inviability of Hawthorne's archival burden. It does so in two ways, the first of which is evident: if Hawthorne remains suspended between Puritan belonging and identification with the abject, then the letter remains suspended in its multiple and contradictory significations and ultimately slides toward catachresis—the figure that voids the referential contract of figure. Continually reread and reinterpreted by the Puritan community in the unfolding of the narrative, it is finally held out to us as the material "type" of a recessive "something," which, in turn, is located in a sorrowful affect—a human universal worthy of our deepest respect. And if the writerly burden is not viable, caught up as it is in *différance*, then it has certain implications for the archival event itself. Consider: the nebulous hermeneutic circuitry of the scarlet letter merely *repeats* and perseverates the affect induced by "Nathaniel Hawthorne's" initial encounter with it. Attracted by the "riddle" of its "deep meaning" that "subtly communicat[ed] itself to [his] sensibilities but evad[ed] the analysis of [his] mind," "Nathaniel Hawthorne" automatically placed it on his breast, under the uncanny pressure of the historical unconscious, and experienced an uncanny effect: "[I]t seemed to me, then, that I experienced a sensation not altogether physical, yet almost so, as of burning heat; and as if the letter were not of red cloth, but red-hot iron" (145–6) [25]. Hawthorne's recourse to the hypothetical "as if" partakes of his resourceful gothic figuration and seems to literalize the *psychic impression* as a painful branding. This spectacular moment should not blind us, however, to the curiously confused amalgam of reception that registers the letter's mark simultaneously upon the affectual sensibilities, the analytic faculties, and the physical body. Moreover, the fact that "Nathaniel Hawthorne" "involuntarily shuddered, and let it fall upon the floor" (146) [25] should alert us to the letter's simultaneous charges of attraction and repulsion. Such a force field of the signifier cannot fail to be deterministic and thus would account for the inability of the subject to, as Dylan Evans puts it above, "decipher the letter" that has "insisted on inscribing itself into the subject's life." *Différance*, then, would be less a literary symptom than a psychic one, as the letter—despite its arrival—remains perpetually *en souffrance*.

One might say that the psychic viability of "Nathaniel Hawthorne's" archival *assujettissement* is always already compromised by his physical and "sensible" attachment to the archival object, or by its very proximity. For this is not a transparently serviceable object, or a regular call to the renewal of identity, that was conveyed from father to son in Derrida's account of Freud's Philippsohn Bible. Rather, it has excited the narrator's passions within its complex force field; he cannot achieve the proper, objective distance from the object necessary to resituate it in either its historical veracity or semiotic transparency. It is, perhaps, too tightly bound into the subject's identificatory matrix—and such binding comes at the cost of filial duty, for interlined with "Nathaniel Hawthorne's" obligations to the forefathers is *already* an attachment to the transgressive mother.[9] Has he not donned her garment? If, for Derrida, the archival event is a suturing of the "epidermises of books and penises" (22; qtd. above), then for Hawthorne, it entails the uncanny relocation of the mystical talisman upon the

9. Robert K. Martin notes the split or what I have called the "suspension" of "Hawthorne's" identificatory impulses: "Telling this tale . . . constitutes a curious act of fidelity to the fathers; for it transforms the assignment of the letter by the ancestral fathers into a punishment with ample recompense, an *A* that can be made over into a proclamation of worth rather than a badge of shame. But telling the tale is also the metaphorical equivalent of putting on the letter. . . . Hawthorne's assumption of the letter aligns him . . . more significantly with Hester, for it is she above all who both wears and 'writes' the letter" (127).

breast—a gestural, if not a fully performative, identification of the son with the mother. The narrative and imagistic reach of this original moment of discovery is both transformative and proleptic because it not only overwrites the archaeological primal scene and mitigates the iron grasp of the forefathers but also anticipates the novel's primary engagement, *not* with the paternal phallus, but with the sorrowful mysteries of the catachrestic human heart.

Despite its scenic plausibility as a thought experiment, the archival event of reception, transmission, and oath-swearing at the hand of Surveyor Pue may turn out to have been what Lacan calls a "missed encounter" (*Four* 55) rather than its purported consolidation of identity and vocation. Lacan returns to the implications of the letter *en souffrance* when he takes up the phenomenon of melancholy repetition in *The Four Fundamental Concepts of Psychoanalysis.* His text is Freud's *The Interpretation of Dreams,* specifically the "dream of the burning child." In Freud's account, a bereaved man keeps vigil in a room next to the one in which his dead son lies; he falls asleep and dreams that he is awakened by his son, who "takes him by the arm and whispers to him reproachfully, 'Father, can't you see that I am burning?'" (Lacan, *Four* 58). The father then literally awakens to find that the vigil candle has fallen and has burned the corpse of his son. For Lacan, Freud's story reveals the function of what he conceptualizes as the *tuché*, or the encounter with the traumatic real, "in so far as it may be missed, in so far as it is essentially the missed encounter" (55). The missed reality—the incomprehensible fact of the child's death—is expressed in the son's reproachful utterance. Lacan writes, "Is not the dream essentially, one might say, an act of homage to the missed reality—the *reality that can no longer produce itself except by repeating itself endlessly, in some never attained awakening?"* (58; emphasis added). And again, if "the encounter, forever missed, has occurred between dream and awakening," then "only *a rite, an endlessly repeated act*, can commemorate [it]" (59; emphasis added).

An entire essay could be written on the resonances and the overlap between Lacan's theory of the *tuché* and Hawthorne's archival *mise-en-scène.* I shall confine myself to the main points of correspondence. As a speculative thought experiment haunted by spectres, "The Custom-House" and its gothic excesses unfold in an essentially dreamlike structure of the image. In response to the personified forefathers, "Nathaniel Hawthorne's" oath of fealty would seem to house a concealed, burning resentment, figuratively impressed by the scorching heat of the scarlet letter upon his breast, the traces of which he retains in his literary heart. Suspended between paternal and maternal allegiances, between bearing witness to the justice of the law and identifying with the victim of its cruelty, his archival mission turns out to be a sufferance and a suffering. Having failed to grasp the full import of the archive, its traumatic and violent reality, and seduced by the promise of vocational belonging, he can but repeat *from the outside* the encounter with the scarlet letter that he has forever missed. The spiralling *différance* of the letter's narrative return constitutes "a rite, an endlessly repeated act" that both "commemorates" the missed encounter and enables it to "produce itself by repeating itself endlessly" (Lacan, *Four* 58–9).

Alternatively, the tensions and competing allegiances that subtend "Hawthorne's" psychic viability may arise from his choosing not to choose and thereby to arrive at a more complex ethical position. Retrospectively, then, "The Custom-House" can be read as a dream allegory of such anxieties. Derrida's political critique of the archival pressure upon the subject to conform to the law is bound up with his recognition that any matrix of repetition is expressive of the death drive. Consequently, and paradoxically, then, this "death drive, sometimes aggression drive, sometimes destruction drive" works always and insidiously against the archive itself: "It works *to destroy the archive: on the condition of effacing* . . . its own proper traces." Silent, untraceable, and ubiquitous, the death drive "seems not only to be anarchic" but "above all, *anarchivic*, one could say, or *archiviolithic*. It will always have been archive-destroying" (11; original emphasis). At issue for Derrida is the archival injunction to rebind the subject by remembering and by repetition, for such a gathering of the subject into the One is always exclusionary of, and violent toward, the Other. He focuses his critique on Yosef Hayim Yerushalmi's *Zakhor: Jewish History and Jewish Memory*, writing, "I would have liked to spend hours, in truth an eternity, meditating while trembling before this sentence: 'Only in Israel and nowhere else is the injunction to remember felt as a religious imperative to an entire people'" (76). Such a deployment of the archive, he continues, "situates the place of all violences," for "the totalizing assemblage ('to an entire people')," or "the gathering into itself of the One is never without violence, nor is the self-affirmation of the Unique, the law of the archontic, the law of *consignation* which orders the archive" (77–8).

If the archive is, indeed, both the site and the agency of the death drive, then Hawthorne would seem to stand both inside and outside its logics. Too close to the archive's gothic sensationalism to read it with objective distance, his stubborn attachment to Hester's scarlet letter may impel a countervailing *detachment* from the wasting rigours of Puritan ideology. Yet, the letter's detour through the indeterminate field of *différance* conforms to the archival logic of return as circular repetition: for the scarlet stain originates in the forefather's grave, and the scarlet letter finds its ultimate inscription on Hester's tombstone, where it remains as elusive as ever. And if, finally, the imagistic interlinearity of corpses and texts in the multiple archives of "The Custom-House" remains indissoluble—as if such images are too tightly bound in their gothic expression—then this might explain the oddly unsatisfying ghosts that people *The Scarlet Letter*. For despite "Nathaniel Hawthorne's" burning attachment,

> the characters of the narrative would not be warmed and rendered malleable, by any heat that I could kindle at my intellectual forge. They would take neither the glow of passion nor the tenderness of sentiment, but retained all the rigidity of dead corpses, and stared at me in the face with a fixed and ghastly grin of contemptuous defiance. "What have you to do with us?" (148) [27]

The figurative economy of personification—now arrived at a fully expressive prosopopoeia—turns out, ironically, to be ineffectual. The archive is burning.

What, then, became of Hawthorne's authorial vocation, and the archival promise? "The reader may choose among these theories" (Hawthorne 340) [152].

WORKS CITED

Brooks, Peter. *Reading for the Plot: Design and Intention in Narrative.* Cambridge, MA: Harvard UP, 1984.

Derrida, Jacques. *Mal d'archive : une impression freudienne.* Paris: Éditions Galilée, 1995.

Derrida, Jacques. *Archive Fever: A Freudian Impression.* Trans. Eric Prenowitz. Chicago: U of Chicago P, 1996.

DeSalvo, Louise. *Nathaniel Hawthorne.* Atlantic Highlands, NJ: Humanities, 1987.

Evans, Dylan. *An Introductory Dictionary of Lacanian Psychoanalysis.* London: Brunner-Routledge, 1996.

Foucault, Michel. *L'archéologie du savoir.* 1969. Paris: Gallimard, 2011.

Foucault, Michel. *The Archaeology of Knowledge.* 1972. Trans. A.M. Sheridan Smith. New York: Random, 2010.

Freud, Sigmund. "From the History of an Infantile Neurosis." [The 'Wolf Man'] *Sigmund Freud: Case Histories II.* Trans. James Strachey. Ed. Angela Richards. London: Penguin, 1991. 227–366.

Hawthorne, Nathaniel. *Novels.* Library of America ed. Ed. Millicent Bell. New York: Literary Classics, 1983.

Lacan, Jacques. *The Four Fundamental Concepts of Psychoanalysis.* Trans. Alan Sheridan. New York: Norton, 1978.

Lacan, Jacques. "Seminar on 'The Purloined Letter.'" *Écrits: the Complete Edition in English.* Trans. Bruce Fink et al. New York: Norton, 2006. 6–48.

Levenson, Alan T. *The Making of the Modern Jewish Bible: How Scholars in Germany, Israel, and America Transformed an Ancient Text.* Lanham, MD: Rowman, 2011.

Martin, Robert K. "Hester Prynne, *C'est Moi:* Nathaniel Hawthorne and the Anxieties of Gender." *Engendering Men: The Question of Male Feminist Criticism* Ed. Joseph Boone and Michael Cadden. New York: Routledge, 1990. 122–39.

Miller, J. Hillis. *Hawthorne and History.* Cambridge, MA: Blackwell, 1991.

Person, Leland S., Jr. *Aesthetic Headaches: Women and a Masculine Poetics in Poe, Melville, and Hawthorne.* Athens, GA: U of Georgia P, 1988.

Savoy, Eric. "'Filial Duty': Reading the Patriarchal Body in *The Scarlet Letter.*" *Studies in the Novel* 25 (1993): 397–417.

Savoy, Eric. "Necro-filia, or Hawthorne's Melancholia." *English Studies in Canada* 27 (2001): 459–80.

Sedgwick, Eve Kosofsky "Queer Performativity: Henry James's *The Art of the Novel.*" *GLQ* 1 (1993): 1–16.

Nineteenth-Century Backgrounds

MICHAEL WINSHIP

Hawthorne and the "Scribbling Women": Publishing *The Scarlet Letter* in the Nineteenth-Century United States†

> Besides, America is now wholly given over to a d——d mob of scribbling women, and I should have no chance of success while the public taste is occupied with their trash—and should be ashamed of myself if I did succeed. What is the mystery of these innumerable editions of the Lamplighter, and other books neither better nor worse?—worse they could not be, and better they need not be, when they sell by 100,000.

It may well be that no single passage written by Nathaniel Hawthorne is better known than this or, at least over the past few decades, more widely quoted.[1] An extraordinary possibility, especially as the passage comes from the middle of a rather long private letter, written to his publisher and friend William D. Ticknor on January 19, 1855, and first published only in 1910.[2] Nevertheless, this passage has resonated through recent discussions of American literary history, for it raises questions that are key to our understandings of that tradition: What is the relationship between popular success and literary quality? What role do gender politics play in our assessment of a work? In what ways have the economic factors facing authors and publishers fostered or discouraged authorship in the United States? And how is it that during the 1850s, a decade that came to be dubbed the "American Renaissance," sentimental novels could have enjoyed such popular success, while the "classics" by Hawthorne, Melville, Thoreau, and Whitman did not?

Although he could hardly have thought in such terms, clearly these issues bothered Hawthorne as he pondered in what direction to continue his literary career. He returned to the subject in his very next letter to Ticknor, written two weeks later, but here at least he selects one of that

† From *Studies in American Fiction* 29.1 (Spring 2001): 3–11.

1. Letter, Nathaniel Hawthorne to William D. Ticknor, January 19, 1855, in Nathaniel Hawthorne, *The Letters, 1853–1856*, ed. Thomas Woodson et al., *The Centenary Edition of the Works of Nathaniel Hawthorne*, Vol. 16 (Columbus; Ohio State Univ. Press, 1987), 304. Hereafter cited parenthetically by volume and page number.
2. In Nathaniel Hawthorne, *Letters of Hawthorne to William D. Ticknor, 1851–1864* (Newark; Cateret Book Club, 1910), 73–76, and again in Caroline Ticknor, *Hawthorne and His Publisher* (Boston: Houghton Mifflin, 1913), 141.

"scribbling mob," Fanny Fern, for praise.[3] His original outburst had been directed specifically at the work of another, Maria Susannah Cummins, whose best-selling novel *The Lamplighter* was making a tremendous success. Published in early March 1854, this work is reported to have sold 20,000 copies in twenty days, and 40,000 copies in eight weeks, By year's end, nearly 75,000 copies had been produced; by the end of the decade, total sales in the United States were somewhere around 90,000.[4] Nevertheless, Hawthorne clearly exaggerated when in his exasperation he claimed that books written by women were selling by the hundred thousand. Although sales of *The Lamplighter* approached that figure, its success was exceptional, and its sales were not matched by other novels of the decade. The exception, of course, was Harriet Beecher Stowe's *Uncle Tom's Cabin*, which indeed did sell in the hundred thousands—around 310,000 copies during the 1850s.[5]

Hawthorne's frustration is understandable. Consider his most popular work, *The Scarlet Letter*, which was published in March 1850: only 11,800 copies had been produced by 1860. For the short term at least, sales of his works had to be reckoned in the thousands instead of tens of thousands, much less hundreds of thousands. But as we pass the sesquicentennial of the original publication of *The Scarlet Letter*, it pays to look at the longer term. What was the publication history of the work for the remainder of the nineteenth century? And how does this history compare to that of *Uncle Tom's Cabin*? What can the comparison tell us about these works' subsequent histories and reputations?

The story of the composition and original publication of *The Scarlet Letter* is well known.[6] Hawthorne, who was an established writer of short stories and sketches, began work on the manuscript sometime—probably late summer—during 1849, the same year that he was dismissed from his job at the Salem Custom House. Before year's end, the Boston publisher James T. Fields called on Hawthorne in Salem and came away with a draft of "The Scarlet Letter," which Hawthorne imagined as one of several stories in a collection to be called *Old-Time Legends* (or possibly *The Custom-House*). Fields encouraged Hawthorne to consider expanding the work for separate publication, and Hawthorne eventually agreed. On January 15, 1850 Hawthorne sent the revised manuscript to Fields, including the introductory "Custom-House" sketch but missing three chapters, which were sent on to Boston on February 3. In the meanwhile, Fields had gone ahead with production, putting typesetters to work, and by February 18 he was able to

3. Letter, Nathaniel Hawthorne to William D. Ticknor, February 2, 1855, *CE* 16:307–8.
4. See advertisements and notices in *Norton's Literary Gazette* April 1, 1854, May 1, 1854, and Dec, 15, 1854. The total sales are difficult to know precisely: the work's publisher, John P. Jewett, lists the eighty-ninth thousand in his catalog dated April 1, 1858; *American Publishers' Circular* 4 (Aug. 14, 1858), 391, gives the total as 90,000. These figures do not account for the many thousands that must have been produced and sold in Great Britain.
5. *American Publishers' Circular* 4 (Aug, 14, 1858), 391, gives the following totals for American fiction: Stowe's *Uncle Tom's Cabin* (1852), 310,000; Cummins's *The Lamplighter* (1854), 90,000; Fanny Fern's *Fern Leaves* (1853), 70,000, and *Ruth Hall* (1855), 55,000; Martha Stone Hubbell, *The Shady Side* (1853), 42,000; Marion Harland's *Alone* (1854), *The Hidden Path* (1855), and *Moss Side* (1857), 25,000 each.
6. The following account is based on William Charvat, "Introduction," in *The Scarlet Letter*, ed. William Charvat et al., *The Centenary Edition of the Works of Nathaniel Hawthorne*, Vol, 1 (Columbus: Ohio State Univ. Press, 1962), xv–xxviii, supplemented with information taken from *The Cost Books of Ticknor and Fields and Their Predecessors, 1832–1858*, ed. Warren S. Tryon and William Charvat (New York: Bibliographical Society of America, 1949) and C. E. Frazer Clark, *Nathaniel Hawthorne: A Descriptive Bibliography* (Pittsburgh: Univ. of Pittsburgh Press, 1978).

include the sheets "as far as printed" in a parcel sent to the London publisher Richard Bentley. On March 16 the first edition of 2,500 copies appeared at a retail price of 75 cents and bound in the characteristic Ticknor and Fields binding of brown ribbed T cloth.

As Fields had hoped but to Hawthorne's apparent surprise, the work was both well received and a moderate success.[7] A second edition of 2,500 copies was issued on April 22, containing a new preface by Hawthorne dated "March 30, 1850," and a third edition of 1,000 copies, for the first time printed from stereotype plates, followed on September 9. By year's end, Hawthorne had earned $663.75 in royalties—his royalty was reckoned at fifteen percent of the retail price—while the publisher's profits came to roughly $900, even after paying for the cost of the stereotype plates. These results were in part due to Fields's talents as a publisher, for he was skilled at managing the publicity of announcements, advertisements, and a network of sympathetic reviewers to push his firm's publications.[8]

In one regard, though, Fields fell short, for in rushing the work into publication, he failed to allow time for arrangements for an authorized English edition. British copyright law required that such an edition appear before or simultaneously with the American edition, but by the time that Bentley, the London publisher that Fields had approached, received the entire text, it had already been published in Boston. Bentley reported that two other firms were preparing unauthorized editions and declined to publish the work. In any event the work was not reprinted in England until May 1851, though imported copies of the American sheets had been available earlier.

The wish to rush the work into publication may also explain in part why the firm printed the first two editions from type instead of from stereotype plates, although this is not as surprising as it may at first appear. Only a few years later, the firm's standard practice would become to print most of its new publications from plates; in 1850, however, only four of its eighteen new works were stereotyped for the first printing. The reason may have been financial, as the firm was in the process of expanding its list and may have wished to avoid the extra investment that plates entailed. The cost of producing stereotype plates nearly doubled the cost of composition: in the case of *The Scarlet Letter* composition for the first two editions came to $130.11 and $121.57, respectively, whereas the cost of composition and stereotyping for the third was $233.39. Clearly it would have been more economical to produce the plates immediately, though the firm may not have expected the work to have such success.[9]

Despite these oversights, Hawthorne was surely pleased with Fields's handling of the work's publication, for over the next several years Fields's firm, Ticknor and Fields, was to reissue many of Hawthorne's earlier

7. The reception history of *The Scarlet Letter* can be traced in *The Critical Response to Nathaniel Hawthorne's* The Scarlet Letter, ed. Gary Scharnhorst (New York: Greenwood Press, 1992). For a more general discussion of Hawthorne's reputation over time, see Richard H. Brodhead, *The School of Hawthorne* (Oxford: Oxford Univ. Press, 1986).
8. *Cost Books of Ticknor and Fields*, entries A173a, A79a, Al89a; see also William Charvat, "James T. Fields and the Beginnings of Book Promotion," in *The Profession of Authorship in America, 1800–1870: The Papers of William Charvat*, ed. Matthew J. Broccoli (Columbus: Ohio State Univ. Press, 1968), [168]–189.
9. *Cost Books of Ticknor and Fields*, entries A173a, A79a, A189a. For a general discussion of the firm's practice in regard to printing from plates, see Michael Winship, *American Literary Publishing in the Mid-Nineteenth Century. The Business of Ticknor and Fields* (Cambridge: Cambridge Univ. Press, 1995), 103–10, 142–47. It should be noted that the relatively large press runs for the first two editions reflect the fact that they were printed from type instead of plates.

works and to publish his new works as they were finished. Hawthorne himself was to form close personal ties with both partners, and Ticknor and Fields and its successor firms remained Hawthorne's primary publisher for the rest of the century. Thus, Hawthorne's works formed a key part of the core list of canonical American literary works—including those of Emerson and Thoreau—that modern scholars have come to associate with Houghton, Mifflin & Co., the firm into which Ticknor and Fields evolved.

Harriet Beecher Stowe was less fortunate in her original choice of publisher for *Uncle Tom's Cabin*, Boston's John P. Jewett. Despite the work's tremendous initial success, and despite the skillful promotional efforts of its publisher, demand fell off markedly after little over a year. Shortly thereafter Stowe fell out with Jewett over contract terms, and for future works she turned for a second time to another Boston publisher, Phillips, Sampson & Co., a firm that had originally declined to publish her anti-slavery masterpiece. After the break with Stowe, Jewett remained the publisher of *Uncle Tom's Cabin,* but it cannot have brought him much profit: he nearly failed during the Panic of 1857 and finally dissolved his publishing business in 1860. In 1859 Stowe's chief publishers, Phillips, Sampson & Co., also went out of business, and in consequence she approached Fields to act as her publisher. When Stowe's works joined those of Hawthorne on the list of Ticknor and Fields in 1860, *Uncle Tom's Cabin* had been, for all intents and purposes, out of print for many years.[1]

In the meanwhile Hawthorne's *Scarlet Letter* had remained happily in print with steady sales, which declined only slightly as time passed. The investment in stereotype plates for the third edition of September 1850 allowed the firm to produce small impressions over time as demand required. A second printing from these plates—the fourth printing of the work over all—of 800 copies was produced in June 1851, and between then and Hawthorne's death on May 19, 1864, the plates were used for thirteen impressions of 500 copies each, a total of 6,500 copies at an average of one impression per year.[2]

Hawthorne's death occurred a little over a month after that of William D. Ticknor, the senior partner of Ticknor and Fields, which had occurred on April 10, 1864, while Hawthorne and Ticknor were on a vacation trip to the South that, it was hoped, would revive Hawthorne's failing health. Inevitably, these deaths had an effect on the publication of Hawthorne's works, including *The Scarlet Letter.* Hawthorne's business relations with Ticknor and Fields had been complicated, based on a series of verbal agreements that stipulated that the royalties on his works were set at varying terms, at ten percent of the retail price for some and fifteen percent for others. Once the firm was reorganized and Fields firmly in charge as senior partner, he arranged to regularize matters, and it was agreed that the firm would in the future pay a flat sum of 12 cents for each copy sold of any of Hawthorne's works.[3]

1. Fields was in no hurry to rush it back into print: his firm's first printing, a mere 270 copies, was not completed until November 1862. For the history of the publication of *Uncle Tom's Cabin*, see Michael Winship, "'The Greatest Book of Its Kind': A Publishing History of *Uncle Tom's Cabin*," *Proceedings of the American Antiquarian Society*, 109 (1999), 309–32.
2. Ticknor and Fields, Cost Books, fair, A-D, in the Houghton Library, Harvard University (MS Am 2030–2, items 14–17).
3. For this and the next paragraph, see Ellen B. Ballou, *The Building of the House: Houghton Mifflin's Formative Years* (Boston: Houghton Mifflin, 1970), 142–56; Tryon, 333–49.

The fairness of this new arrangement is difficult to assess. At the time, it certainly seemed generous, for 12 cents a copy represented an increase in royalty on *The Scarlet Letter*, from 11-1/4 cents to 12 cents. Similarly, it meant an increase in royalty for all of Hawthorne's other works save two—*The House of the Seven Gables* and *Our Old Home*—that were earning 15 cents per copy under the old arrangement. The problem, however, arises from the shift in the method of determining royalties from a percentage basis to a flat fee. Retail prices for books had remained remarkably stable throughout the 1840s and 1850s, but the Civil War had brought about a period of inflation and a consequent increase in book production costs, which in turn inevitably led to an increase in retail prices, a result that Fields must have foreseen. By the late 1860s, retail prices on all of Hawthorne's works had risen to $1.50 or $2.00, which meant that the royalty on *The Scarlet Letter*, for example, would have risen under the old agreement to 30 cents a copy. From this perspective, Fields had struck a very hard bargain indeed.

Conflict was inevitable. Hawthorne's widow, Sophia, raising three children alone, found herself strapped financially and sensed that perhaps Hawthorne's royalties were less than they should be. Her suspicions seemed confirmed in 1868, when Gail Hamilton, another Ticknor and Fields author whose royalty terms had been changed in 1864 in a manner similar to Hawthorne's, began to raise questions. Upset, Sophia Hawthorne even went so far as to threaten to transfer future rights in her husband's works to another firm. Fields reacted quickly: he prepared an explanation, backed by figures, and offered to submit the matter to arbitration. Eventually Sophia Hawthorne's sister Elizabeth Peabody intervened. After examining the firm's accounts, she came to the conclusion that, despite several clerical errors and other carelessness, the firm's records were consistent with each other, but she also noted that demand had been such that neither author nor firm had received as much as $1000 per year on average in income from Hawthorne's books. Ticknor and Fields was technically vindicated, but relations were soured. In an attempt to placate Sophia Hawthorne, Fields offered to pay in future a royalty of ten percent on all of Hawthorne's works. These terms were agreed to and remained in force until 1875, when Hawthorne's heirs accepted the firm's offer of a regular annuity of $2,000 in lieu of royalties.[4]

Hawthorne's death in 1864 was to have another important impact on the publication of *The Scarlet Letter*, for in the fall of that year the firm issued the first collected edition of Hawthorne's works. *The Scarlet Letter*, printed from the 1850 plates, appeared as volume six of fourteen in this so-called "Tinted Edition" (A16.3.q; B1). A second collected edition was issued in 1871, and the same plates were used for *The Scarlet Letter*, which appeared bound with *The Blithedale Romance* as volume four in a twelve-volume "Illustrated Library Edition" (A16.3.w; B2). This trend continued, for when new plates for *The Scarlet Letter* were finally cast in 1875, they were for use in the twenty-three-volume "Little Classic Edition" of Hawthorne's collected works (A16.8.a; B5). A third set of plates, cast in 1883, was prepared for the "Riverside Edition" (A16.13.a; B9),

4. Ballou, 242, states that the offer was not accepted due to wrangling among Hawthorne's children but the firm's records indicate otherwise; see "Agreement No. 2," May 1, 1875, in the uncataloged Houghton Mifflin contract files in the Houghton Library, Harvard University (MS Storage 288, box 34).

where *The Scarlet Letter* appeared, again bound with *The Blithedale Romance*, as volume five of twelve. These sets of Hawthorne's collected works were expanded as posthumous works appeared and, over the years, were repackaged and reissued in a variety of formats and bindings at a range of prices, but for the rest of the century *The Scarlet Letter* was generally speaking marketed as part of his collected works, and not singly as Hawthorne's greatest masterpiece.[5]

Hawthorne and Stowe shared the same publisher from 1860, but the pattern of publication of their works was quite different: *Uncle Tom's Cabin* clearly stood out among Stowe's works, not just in terms of importance but also income: by the end of the 1880s her earnings from *Uncle Tom's Cabin* equaled nearly two and a half times the combined royalty on all her other books. Unlike *The Scarlet Letter*, which had been chiefly available as part of a set of Hawthorne's collected works since his death in 1864, *Uncle Tom's Cabin* was chiefly sold by itself. Stowe, who survived Hawthorne by over thirty years, continued to produce new works through the 1870s, and there was no collected edition of Stowe's works until after her death in 1896.

And what of the sales of the two works? *The Scarlet Letter* started out at considerable disadvantage, but as time passed and sales increased, the difference grew less striking. During the 1860s, roughly 6,500 copies of *The Scarlet Letter* were produced, compared to 8,000 copies of *Uncle Tom's Cabin*; during the 1870s, roughly 20,000 copies, compared to 26,000 copies of *Uncle Tom's Cabin*.[6] In 1878, with the formation of a new business partnership, the stereotype plates of the two works were inventoried and valued, a figure that served as a guide to estimating the worth of the rights to their publication: the plates of *The Scarlet Letter* were valued at $4,792.38; those for *Uncle Tom's Cabin* at $4,524.60.[7]

Although *The Scarlet Letter* was chiefly marketed as part of Hawthorne's collected works, it was also issued from time to time in separate editions. In late 1877, as the end of the original copyright term of twenty-eight years approached, James R. Osgood & Co., a successor firm to Ticknor and Fields, issued the first new separate edition of *The Scarlet Letter* (A16.10). With rather lavish illustrations by Mary Hallock Foote, this was an expensive volume at $4 in cloth, $9 in leather. The illustrations may have been intended to support the firm's claim in the work, though the copyright in the text could be and was renewed and protected for a further fourteen years. In 1879 Houghton, Osgood & Co., another successor firm, issued for $10 F. O. C. Darley's *Compositions in Outline from Hawthorne's* Scarlet Letter, a series of 12 illustrated prints, each accompanied by a page of text extracted from Hawthorne's work.[8]

During the 1880s, both Hawthorne's works and *Uncle Tom's Cabin* continued as steady sellers, although developments in the book trade were troubling. Throughout the decade the market for books was bedeviled by

5. See Frazer Clark, *Nathaniel Hawthorne*, for details; references to entries to this bibliography are given in parentheses in text.
6. These figures are based on Ticknor and Fields & Houghton, Osgood & Co., Sheet Stock Books, in the Houghton Library, Harvard University (fMS Am 2030.2, item 22–23, & fMS AM 2030, item 16).
7. Ticknor and Fields, "Plate Inventory (1873–77)" in the Houghton Library, Harvard University (fMS Am 2030.2, item 25).
8. For a full discussion of the illustrated editions of *The Scarlet Letter*, see Rita K. Gollin, "The Scarlet Letter" in "From Cover to Cover: The Presentation of Hawthorne's Major Romances," *Essex Institute Historical Collections*, 127 (1991), 12–30.

pirates and undersellers, an emerging group of publishers and booksellers who took advantage of the lack of international copyright on British works and the increasing use of trade sales as a means of dumping surplus or out-of-date stock to flood the market with cheap books. As an established trade publisher, Houghton, Mifflin & Co., the final inheritor of the rights to the works of both authors, was well aware of the losses caused by these practices. The threat was exacerbated by the fact that early in the 1890s the copyrights on both *The Scarlet Letter* and *Uncle Tom's Cabin* were due to expire, and these works were destined to enter the public domain.[9]

During fall 1891, the editors at Houghton, Mifflin discussed strategies for the continued publication of both works while they planned new and cheap editions that they hoped would maintain their control of the market. The main issue under discussion was the timing for issuing these new editions, but things came to a head in spring 1892, when the firm learned that new plates of both works were already being prepared for sale to the publishers of cheap publications. Legal advice was sought, and although the copyright technicalities were obscure, the firm put up a bold front and succeeded in driving off the competition, but only for the time being.[1]

By early 1892, Houghton, Mifflin had prepared and issued new and cheap separate editions of both *The Scarlet Letter* and *Uncle Tom's Cabin*, for the most part printed from plates that were already in use. In its spring announcement of March, the firm listed two new separate editions of *The Scarlet Letter*: the "Universal Edition" (printed from plates of the "Riverside Edition"),[2] which cost 50 cents in cloth, 25 cents in paper, and an even cheaper "Salem Edition" (A16.8.d; printed from the plates of the "Little Classic Edition"), at 30 cents in cloth, 15 cents in paper. These were followed in May by an expensive edition, illustrated with photogravures based on Darley's outline drawings: the "trade edition" cost $2.50 (A16.13.g; printed from plates of the "Riverside Edition"), and also issued in a special "Large Paper Edition," limited to 200 numbered copies and bound in vellum, $7.50 (A16.13.h). These joined the "Popular Edition" of *The Scarlet Letter* (A16.13.d; printed from the plates of the "Riverside Edition"), the only other separate edition that had been issued, which had been in print since 1885 and was priced at $1 in cloth, 50 cents in paper. When the work entered the public domain, its authorized publisher Houghton, Mifflin & Co. made sure it was available separately in a range of formats and prices that appealed to as broad a market as possible.

The copyright on *Uncle Tom's Cabin* expired in May 1893. As with *The Scarlet Letter*, the firm had already issued a range of new editions, both cheap and expensive, in an attempt to maintain their hold over the

9. Curiously, Houghton, Mifflin & Co. seems to have been confused about the actual date that both works were destined to become public property. At this period legal copyright in a work was established when it was registered and a pre-publication title page was filed in the District Court Clerk's office: the original term of twenty-eight years thus established could then be renewed for a further fourteen years. The copyright on *Uncle Tom's Cabin* was thus destined to expire on May 12, 1893, though in late 1891 Henry Oscar Houghton was under the mistaken impression that the date was 1892. Similarly, the firm maintained that the copyright on *The Scarlet Letter* expired on November 15, 1892, though it seems very probable that the title page had been deposited in late 1849, meaning that it would enter the public domain in 1891 rather than 1892.

1. See letters, Francis Jackson Garrison to Henry Oscar Houghton, July 14, 1891 to December 4, 1891, in the Houghton Library, Harvard University (Ms Am 1648 [330]), and letters regarding the Liberty Book Company in the Houghton Mifflin contract files.

2. This "edition" is not listed in Frazer Clark, *Nathaniel Hawthorne*, but most likely it is that listed as A16.13.f.

market. After both works entered the public domain, however, they were quickly reprinted in unauthorized editions. By century's end, separate editions of *The Scarlet Letter* were available from many of the firms that specialized in cheap publishing—Altemus, Bay View, Burt, Caldwell, Coates, Crowell, Donohue, Hill, Hurst, Lupton, McKay, Mershon, Ogilvie, Page, Rand, Stokes, Truslove, Warne, Ziegler—a list that closely matches that for *Uncle Tom's Cabin.*[3] Houghton, Mifflin & Co. continued as an important publisher of both works, but was no longer able to control the ways in which they were packaged and marketed.

For much of the twentieth century, critical opinion of the two works followed different paths. *Uncle Tom's Cabin* came to be viewed as flawed, overly sentimental, and frankly racist, in fact an embarrassment—an assessment that has only recently been revised. In the meanwhile, *The Scarlet Letter* emerged, along with Melville's *Moby-Dick*, as one of the "two most nearly undisputed classics of American fiction."[4] Clearly, reception history cannot be explained only by a work's publication and marketing, but it is interesting to speculate on the extent to which they have influenced the critical understanding of the importance of *The Scarlet Letter* in Hawthorne's oeuvre, for as the twentieth century dawned it was for the first time readily and widely available to readers not as one volume from his collected works, but as a distinct and separate work.

BROOK THOMAS

Citizen Hester: *The Scarlet Letter* as Civic Myth†

Early in *The Scarlet Letter* (1850), as Hester Prynne faces public discipline, the narrator halts to comment, "In fact, this scaffold constituted a portion of a penal machine, which now, for two or three generations past, has been merely historical and traditionary among us, but was held, in the old time, to be as effectual an agent in the promotion of good citizenship, as ever was the guillotine among the terrorists of France" (55) [40]. In a subtle reading of this passage Larry Reynolds notes the anachronistic use of "scaffold"—the normal instruments of punishment in the Massachusetts Bay Colony were the whipping post, the stocks, and the pillory—to argue that Nathaniel Hawthorne self-consciously alludes to public beheadings, especially the regicidal revolutions in seventeenth-century England and eighteenth-century France. But none of Hawthorne's many critics has noted the anachronistic use of *good citizenship*, a phrase that suggests the rich historical layering of Hawthorne's nineteenth-century romance about seventeenth-century New England Puritans.

3. See *The United States Catalog: Books in Print, 1899 (Part 1: Author Index)*, ed. George Flavel Danforth and Marion E. Potter (Minneapolis: H. W. Wilson, 1900), 290, 631.
4. Charvat, "Introduction" to *The Scarlet Letter*, xv. This introduction is printed in the very first volume completed and published of what—as the Center for Editions of American Authors—became an ambitious attempt to produce critical texts of the American classics by applying the same methodologies developed for editing the works of the English renaissance, especially those of Shakespeare.

† From *American Literary History* 13.2 (2001): 181–211. Reprinted by permission of Oxford University Press. Page numbers in square brackets refer to this Norton Critical Edition.

Of course, *citizen* existed in English in the seventeenth century, but it was used primarily to designate an inhabitant of a city, as Hawthorne does when he mentions "an aged handicraftsman . . . who had been a citizen of London at the period of Sir Thomas Overbury's murder, now some thirty years agone" (127) [80]. The official political status of residents of Boston in June 1642 was not that of citizens, but subjects of the King, a status suggested when Hester leaves the prison and the Beadle cries, "Make way, good people, make way, in the King's name" (54) [40]. Historically resonant itself, this cry reminds us that it was precisely in June 1642 that civil war broke out in England (Ryskamp, Newberry). In fact, the book's action unfolds over the seven years in which the relation between the people and their sovereign was in doubt, the years generally acknowledged as the time when "the Englishman could develop a civic consciousness, an awareness of himself as a political actor in a public realm" (Pocock 335); that is, as a citizen as those in the nineteenth century would have understood the term.[1] Even so, it was not until after the French and American Revolutions that *good citizenship* came into common use.

When Hawthorne inserts the nineteenth-century term *good citizenship* into a seventeenth-century setting he subtly participates in a persistent national myth that sees US citizenship as an outgrowth of citizenship developed in colonial New England. Hawthorne's participation in this myth is important to note because much of his labor is devoted to challenging its standard version. According to the standard version, conditions for democratic citizenship flourished the moment colonists made the journey to the "New World." If the people in the 13 colonies were officially subjects of the king, the seeds of good citizenship were carried across the Atlantic, especially by freedom-loving Pilgrims, who found a more fertile soil for civic participation than in England. A recent example of this version of the story comes in the work of the noted historian Edmund S. Morgan. Describing "the first constitution of Massachusetts" in 1630 when the assistants of the Massachusetts Bay Company were "transformed from an executive council to a legislative body," Morgan writes, "the term 'freeman' was transformed from a designation for members of a commercial company, exercising legislative and judicial control over that company and its property, into a designation for the citizens of a state, with the right to vote and hold office. . . . This change presaged the admission to freemanship of a large proportion of settlers, men who could contribute to the joint stock nothing but godliness and good citizenship" (*Puritan Dilemma* 91).[2] When

1. Thomas Hobbes in *De Cive*, published in Latin in 1642 and translated into English in 1651, did use *citizen* to designate membership in a commonwealth. But he did not use it as Aristotle did to designate a member of a republic who has the capacity to both rule and be ruled. Instead, like the French absolutist Jean Bodin, he distinguished citizens, who had specific benefits, from other subjects, like denizens, who did not have all or any of them. In *Leviathan* (1651) Hobbes uses *citizen* more in the sense of a city dweller. For instance, he writes of a man: "Let him therefore consider with himself, when taking a journey, he armes himselfe, and seeks to go well accompanied; when going to sleep, he locks his dores; when even in his house he locks his chests; and this when he knows there bee Lawes, and publike Officers, armed to revenge all injuries shall be done him; what opinion he has of his fellow subjects, when he rides armed; of his fellow Citizens, when he locks his dores; and his children, and servants, when he locks his chests" (186–87). "Fellow Citizens" are clearly those "fellow subjects" who dwell in close proximity to the man.
2. Morgan also uses the term *good citizen* when he acknowledges that the Puritans' phrase would have been a "civil man" (*Puritan Family* 1).

Morgan designates freemen *citizens*, he projects onto Puritan New England his awareness of political changes still to come just as most studies of colonial American literature project the country's present political boundaries backward and treat only the 13 colonies that eventually became the US.

This tendency to read the Puritan past teleologically is a product of the antebellum period. For instance, in his multivolume *History of the United States*, which found its way into nearly a third of New England homes (Nye, *George* 102), George Bancroft attributed the "political education" of people in Connecticut "to the happy organization of towns, which here, as indeed throughout all New England, constituted each separate settlement as a little democracy in itself. It was the natural reproduction of the system, which the instinct of humanity had imperfectly revealed to our Anglo-Saxon ancestors. In the ancient republics, citizenship had been an hereditary privilege. In Connecticut, citizenship was acquired by inhabitancy, was lost by removal. Each town-meeting was a little legislature, and all inhabitants, the affluent and the more needy, the wise and the foolish, were members with equal franchises." Quoting this passage, an anonymous reviewer for the *American Jurist and Law Magazine* enthusiastically adds that in colonial New England's "institutions lies the germ of all that distinguishes our government from others, which are more or less founded in individual freedom" (230).

Clearly, the "mild" and "humane" laws of Bancroft's Puritans are not those of Hawthorne's (229). Indeed, much of Hawthorne's notorious irony is directed against the idealization of New England ancestors by Bancroft and others.[3] For instance, if Bancroft celebrates New England as the breeding ground of democratic citizenship because of the people's civic participation in town hall meetings and the like, Hawthorne's image of the scaffold reminds us that good citizenship requires obedience. If Bancroft stresses the freedom entailed in good citizenship, Hawthorne reminds us of the repressions required to produce good citizens. Hawthorne's irony reaches a peak late in the book when he calls Chillingworth, the book's villain, a "reputable" "citizen" (233) [139]. Truly good citizens, it seems, cannot be distinguished from those who simply appear respectable.

In different ways some of Hawthorne's best historically minded critics have noted his challenge to the standard version of the Puritan origins of US citizenship. But for all of their brilliance, none have noted Hawthorne's anachronistic use of the term *citizen*. On the contrary, like Hawthorne, some of these same critics refer to Puritans in seventeenth-century Boston as citizens in the political sense of the term (Berlant; Colacurcio, "Woman's Own Choice"; and Pease), just as does the allegedly ahistorical Frederick Crews (149). In doing so they unconsciously participate in the very myth they think they are demystifying, a participation that makes it impossible for them to recognize Hawthorne's important contribution to it.[4]

3. For an excellent summary of speeches by people like Daniel Webster, Joseph Story, and Edward Everett that share Bancroft's view of the Puritans' republican institutions, see John P. McWilliams, 25–36. On Bancroft, see Levin.

4. In noting that many of Hawthorne's critics remain as much within the myth of the Puritan origins of US. citizenship as he does, I am not implying that I somehow can stand outside of and above myth to expose it as an ideological distortion. Whereas I fully recognize that *The Scarlet Letter*, as a work of fiction, does not give us a historically accurate account of seventeenth-century Puritan society and political thought, to dismiss it as mere ideology does not get us very far. On the contrary, since according to today's critical commonplace we are always within ideology, it is not enough to expose persistent national myths as ideological, which is how the

We can start to identify that contribution by noting that Hawthorne employs little or no irony at the end of his romance when Hester returns to Boston and devotes herself to serving the unfortunate. Having "no selfish ends," not living "in any measure for her own profit and enjoyment," counseling those bringing to her "their sorrows and perplexities" (263) [154], Hester in her unselfish commitment to her community has by most measures earned the label *good citizen*. By most, but not by all. For instance, Judith Shklar identifies the two most important attributes of US citizenship as the right to vote and the right to earn a living. Although Hester earlier earned her keep with her needlework, economic self-sufficiency is not a defining aspect of her citizenship. Nor is the right to vote. Indeed, as a woman, Hester in the seventeenth (even in the early nineteenth) century could not fit definitions of *good citizenship* in either the economic or the political spheres.

Even so, rather than abandon the concept of *good citizenship*, Hawthorne through Hester expands our notion of what it can entail by stressing the importance of actions within what political scientists call *civil society*, "a sphere of social interaction between economy and the state, composed of the intimate sphere (especially the family), the sphere of associations (especially voluntary associations), social movements, and forms of public communication" (Cohen and Arato ix).[5] Acutely aware that the stress on civic participation could obscure important interior matters of the heart and spirit, Hawthorne does not, as many critics argue he does, retreat from public to private concerns, but instead tells the tale of how a "fallen woman" finds redemption by helping to generate within a repressive Puritan community the beginnings of an independent civil society. In telling that tale Hawthorne provides more than a civics lesson. He participates in and helps to shape the contours of a powerful civic myth.[6]

1. *Working on/with Myth*

But what is a civic myth? The term comes from *Civic Ideals: Conflicting Visions of Citizenship in US History* (1997), Rogers Smith's exhaustive study of how the law both reflects and helps to produce attitudes toward citizenship in the US. In Smith's complex account, US citizenship has been determined not only by liberal civic ideals, but also by civic myths, which he defines as "compelling stories" that explain "why persons form a people, usually indicating how a political community originated, who

present generation of critics of American literature has generally distinguished itself from the myth and symbol school. What we need to do as well is to evaluate the effect of various myths in terms of what Kenneth Burke called "equipment for living." Such work on/with myth might help to generate a revitalized political criticism that once again, like Aristotle, sees politics as the art of the possible.

5. Informed by events in the former Soviet bloc in 1989, where the economic sphere was controlled by the state, this and other current definitions do not include the economic in civil society, as did Adam Smith, Adam Ferguson, and Hegel.
6. My point is not that Hawthorne set out to write a story arguing for the importance of an independent civil society in the way that a political economist might. His goal was to write the most compelling story that he could. Nonetheless, in its reception, especially the role it has played in education in the US, *The Scarlet Letter*, with its representation of people's desires and how those desires can best be fulfilled, imparts certain attitudes, values, and structures of feeling that coincide with the attitudes, values, and structures of feeling associated with civil society arguments. Furthermore, even if Hawthorne did not self-consciously set out to make an argument for an independent civil society, he would have known about such arguments through the Scottish Enlightenment figures of Adam Smith and Ferguson.

is eligible for membership, who is not and why, and what the community's values and aims are" (33).

Literature's potential to generate civic myths was the topic of an 1834 speech called "The Importance of Illustrating New England History by a Series of Romances Like the Waverley Novels," which was given in Hawthorne's home town of Salem by the Whig lawyer Rufus Choate. Alluding to the Scottish nationalist Andrew Fletcher's often-quoted statement that "I know a very wise man . . . [who] believed if a man were permitted to make all the ballads, he need not care who should make the laws of the nation" (108), Choate argues that a proper literary treatment of the past would mold and fix "that final, grand, complex result—the national character."[7] In doing so it would make the country forget its "recent and overrated diversities of interest" and "reassemble, as it were, the people of America in one vast congregation." "Reminded of our fathers," he argues, "we should remember that we are brethren" (1: 344).

Choate understands how works of literature can serve as civic myths, but he also reveals why Smith worries about the effects of civic myths and their "fictional embroidery" (33). The stories Choate advocates would not, he admits, be a full disclosure of the past. A literary artist should remember that "it is an heroic age to whose contemplation he would turn us back; and as no man is a hero to his servant, so no age is heroic of which the whole truth is recorded. He tells the truth, to be sure, but he does not tell the whole truth, for that would be sometimes misplaced and discordant" (1: 340).[8] Aware that "much of what history relates . . . chills, shames and disgusts us," producing "discordant and contradictory emotions," Choate, therefore, counsels writers to leave out accounts of the "persecution of the Quakers, the controversies with Roger Williams and Mrs. Hutchinson" (1: 339). Literature as civic myth would seem to allow authors to avoid altogether those embarrassing national events that historians should not ignore, even if there is, as Herman Melville puts it, "a considerate way of historically treating them" (55).

But the Hawthorne that Melville so admired presents a more complicated case.[9] He had, for instance, already written about precisely the topics

7. The quotation comes from *An Account of a Conversation Concerning a Right Regulation of Governments for the Common Good of Mankind in a Letter to the Marquis of Montrose, the Earls of Rothes, Roxburg and Haddington* (1703). It might seem ironic that in making a plea for the US to unite and to forget regional differences Choate quotes a Scottish nationalist. At the same time, Fletcher advocated a federal union of Scotland and England, so he could be said to have anticipated the federal system of the US.

8. Choate is echoed by Will Kymlicka, the contemporary theorist of multicultural citizenship, who argues that finding a shared national identity in history "often requires a very selective, even manipulative retelling of that history" (189).

9. On Hawthorne and a national literature, see Doubleday. See also Arac, "Narrative Forms," who argues that *The Scarlet Letter* is an aesthetic narrative, not a national narrative, and that it became representative of the nation only through a retrospective process of canonization that devalued national narratives. Arac's provocative argument reminds us that literature can do many more things than give us compelling stories about national membership and values. Indeed, many works do not even have the potential to become civic myths. Nonetheless, *The Scarlet Letter*'s engagement with the myth of Puritan origins does give it that potential. More important, Arac focuses on a work's form and content, but form and content alone do not make a narrative national; its reception plays a role as well. Whereas a study of *The Scarlet Letter*'s reception is beyond the scope of this essay (see Brodhead), the important question for me is why Arac's "aesthetic narratives," like *The Scarlet Letter* and *Adventures of Huckleberry Finn* (1884), become civic myths while more explicitly national or patriotic narratives have lost favor over time. Simply to raise that question is to suggest that literature's relation to nationalist ideologies is a complicated one. Answering it might also demonstrate, as I try to do in this essay, that those ideologies are themselves more complicated than the ones that many recent literary critics are so intent on demystifying.

that Choate says should be avoided, evidence of a critical attitude toward the past that has caused so many critics to focus on his ironic demystifications. But, as we have seen, Hawthorne does more than demystify prevailing myths. As George Dekker shrewdly puts it, Hawthorne's "best hope for both short-and long-term success was to make the great American myths his own" (148). Hawthorne is neither solely a mythmaker nor a critical demystifier. Instead, to use Hans Blumenberg's phrase, he "works on/with myth." Effectively working on/with the myth of the nation's relation to its Puritan past, *The Scarlet Letter* as civic myth does not advocate obedience to the state or even primary loyalty to the nation.[1] Instead, it illustrates how important it is for liberal democracies to maintain the space of an independent civil society in which alternative obediences and loyalties are allowed a chance to flourish. It should come as no surprise then that the novel's power comes more through its love story than through its politics, or perhaps better put, its politics reminds us of the importance love stories have for most citizens' lives.

Of course, most readers of *The Scarlet Letter* do not need to be reminded that its mythopoetic power lies in its love story. All the more noteworthy, therefore, that recent political readings of the novel have tended to divert our attention from the love story or downplay its significance.[2] What those readings fail to acknowledge is that the love plot is a vital part of Hawthorne's civic vision because it is in the love plot that he explores the possibilities of life in civil society. He does so by working on/with the great exceptionalist myth that America offers the hope for a radical break with the past and the promise of a new start.

2. *Begin All Anew*

Hawthorne's romance is an extended account of various efforts to begin anew. It starts with reflections on the Puritans' attempt to establish a fresh

1. Some of the most provocative—if conflicting—accounts of Hawthorne's relation to the Puritans are Baym; Bell; Bercovitch; Colacurcio, "Footsteps" and "Woman's Own Choice"; and Pease.
2. Almost three decades ago Colacurcio began turning critics' attention away from the novel's love story to examine the chapter "Another View of Hester" and Hester's final return. Nonetheless, he pointedly remarks on the danger of turning "away from the richness and particularity of Hester's own love story" ("Footsteps" 461). In contrast, Berlant sees the love story as a retreat from the book's more important political concerns: "Now the tale of Hester and Dimmesdale, a political scandal, is reduced to a mere love plot" (154). According to Bercovitch the dramatic reunion of Hester and Dimmesdale in the forest "is a lovers' reunion, a pledge of mutual dependence, and no doubt readers have sometimes responded in these terms, if only by association with other texts. But in *this* text the focus of our response is the individual, not the couple (or the family)" (122). In contrast, see Millington's claim that Bercovitch's "erasure of the book's emotional investments" is "characteristic of the present moment in the history of Americanist criticism" (6, 2).

   Millington's observation helps me to address what might seem to be a contradiction in my argument. If, as I claim in note 9 [on p. 418], a work's status as civic myth depends in large part on its reception, do not all readings contribute to that status? And if they do, how can I claim that some readings are misreadings? There is no easy answer to those questions, but I can at least suggest the direction that an answer might take. First, it is partially true that the entire reception of a work—including its misreading—helps to give a work the status of civic myth, since without a widespread reception the book could not serve as myth. Nonetheless, the fact should not keep us from recognizing that very often popular readings tend to perpetuate commonplace myths and miss how a novel or story also works on those myths. Take, for instance, the Demi Moore film of *The Scarlet Letter*. By completely sympathizing with the lovers against a harsh Puritan society it misreads the novel as much as many undergraduates do. If the book were indeed that simple-minded, it would not have had a very long reception history. Even so, by responding to this emotional aspect of the book, such misreadings do give us a sense of the book's popular power that critical dismissals of the love plot miss. A novel or story that simply works on myth without working with it will have little chance of having a popular reception. My reading of *The Scarlet Letter* as civic myth tries to account for both its long and its popular reception.

start in the New World and the narrator's whimsical comment that "[t]he founders of a new colony, whatever Utopia of human virtue and happiness they might originally project, have invariably recognized it among their earliest practical necessities to allot a portion of the virgin soil as a cemetery, and another portion as the site of a prison" (47) [35]. It then opens the second half with "Another View of Hester" and Hester's realization that the radical reforms she imagines would require "the whole system of society to be torn down, and built up anew" (165) [102]. Hester's radical speculations are in turn linked to the book's emotional climax in the forest scene when Hester pleads to Dimmesdale, "Leave this wreck and ruin here where it hath happened! Meddle no more with it! Begin all anew!" (198) [119].

Even though each of these efforts is frustrated, much of the story's emotional tension has to do with readers' hopes—secret or not—that one or the other—or all—will succeed. Of all the attempts, however, that of Hester and Dimmesdale has awakened most readers' hopes. Confronted with a book of memorable scenes, readers past and present have found the forest meeting between Dimmesdale and Hester the most memorable.[3] It is so powerful that, as anyone who has taught the book knows, students have to be carefully guided to those passages in which the narrator in fact condemns the lovers' sentiments. To understand *The Scarlet Letter* as civic myth, we need to understand why, after marshaling all of his rhetorical force to make us sympathize with his lovers, Hawthorne does not allow them a new beginning.

Puritan authorities might have answered that question by relying on John Winthrop's distinction between natural and civil liberty. "The first is common to man with beasts and other creatures. By this, man, as he stands in relation to man simply, hath liberty to do what he lists; it is a liberty to evil as well as to good." In contrast, civil liberty has to do with the "covenant between God and man, in the moral law, and the politic covenants and constitutions, amongst men themselves. This liberty is the proper end and object of authority, and cannot subsist without it; and it is a liberty to that only which is good, just, and honest" (83–84). Significant for a novel about adultery, Winthrop's analogy for political covenants is marriage. Assuming the common law doctrine of coverture in which husband and wife become one corporate body with the husband granted sole legal authority, Winthrop compares a woman's willing subjection in marriage to an individual's subjection to the magistrates who govern the political covenant to which he consents. "The woman's own choice makes such a man her husband; yet being so chosen, he is her lord, and she is to be subject to him, yet in a way of liberty, not of bondage; and a true wife accounts her subjection her honor and her freedom. . . . Even so brethren, it will be between you and your magistrates" (238–39). In turn, both marriage and political covenants are analogous to "the covenant between God and man, in the moral law" in which a Christian can achieve true liberty only through total submission to Christ. For the Puritans, the political institutions of civil society and the civil ceremony of marriage

3. Of this moment, when the two "recognize that, in spite of all their open and secret misery, they are still lovers, and capable of claiming for the very body of their sin a species of justification," William Dean Howells writes, "There is greatness in this scene unmatched, I think, in the book, and I was almost ready to say, out of it" (105, 108).

are governed by the moral law because they have God's sanction. A political covenant is not simply a contract among men; like the marriage contract between a man and woman, it needs God's witness.

To apply this doctrine of covenant theology to *The Scarlet Letter* is to see that for the Puritans Hester's greatest sin would not have been her adultery, whose visible evidence they see in the birth of Pearl, but a remark that Dimmesdale alone hears her make: her defiant cry that what the two lovers did "had a consecration of its own" (195) [118]. Resonating with so many readers, this proclamation is in fact sinful because it implies that Hester's and Dimmesdale's love is a self-contained act, not one in need of God's sanction. As such their love exists in the realm of natural, not civil, liberty and must be contained.

The nineteenth-century version of Winthrop's distinction between natural and civil liberty is the distinction often made in political oratory between license and liberty. *The Scarlet Letter* is a civic myth about the importance of civil society, not about the glories of natural man or woman, because Hawthorne, despite the sympathy that he creates for his lovers, recognizes with Winthrop the dangers of natural liberty. But if Hawthorne shares Winthrop's distrust of natural liberty, he does not share the Puritans' belief that the only way for political subjects to achieve civil liberty is through absolute submission to civil authority.

Because Winthrop speaks of a political subject's participation in a covenant rather than of his relation to a monarch and because of Hawthorne's own reference to *good citizenship*, critics who evoke Winthrop while writing on *The Scarlet Letter* have assumed that he is describing the situation of citizens, not subjects.[4] But he is not. Winthrop's subjects are still subjects, and citizens for him remain residents of a city, as is the case for John Cotton, who in 1645 declared that the best way to unite or combine people together into "one visible body" was a "mutual covenant" between "husband and wife in the family, Magistrates and subjects in the Commonwealth, fellow Citizens in the same citie" (qtd. in Norton 13). This distinction between subjects and citizens is not just a quibble over terms. As a political category, not simply a resident of a city, *citizen* implies the capacity to rule as well as be ruled. The relative independence of citizens would, therefore, undercut Winthrop's analogy between the wife in a marriage under coverture and the subjects of a commonwealth. For instance, as Linda Kerber has shown, covenant theology's strict analogy between marriage and political covenants broke down in the Revolutionary era. On the one hand, independence generated an ideological disjunction. Founded on the principle that the terms of political obligation of British subjects could be renegotiated to create US citizens, the nation was ruled, nonetheless, by men who for the most part wanted to retain a family structure in which a wife owed her husband eternal obedience (13). On the other hand, the rhetoric of citizenship generated a new republican model of marriage that challenged the doctrine of coverture. As Merril Smith puts it, "Tyranny was not to be considered in public or private life, and marriage was now to be considered a republican contract between wives and husbands, a contract based on mutual affection" (51).

4. See Colacurcio, "Woman's Own Choice," and his student Berlant, who asserts that for Winthrop the citizen is a woman. Citizens, we need to remember, are subjects, but not all subjects are citizens.

Hawthorne's challenge to Winthrop's belief in the absolute authority of magistrates is thus as important a part of *The Scarlet Letter*'s function as civic myth as is their shared distrust of the potential dangers of natural liberty. Winthrop claims absolute authority because he lives in a theocracy in which, as Hawthorne puts it, "forms of authority were felt to possess the sacredness of divine institutions" (64) [45]. Distrustful of granting civil authority divine sanction, Hawthorne questions the capacity of the Puritan magistrates to judge Hester. Their problem is not that they are evil men. "They were, doubtless, good men, just and sage" (64) [45]. Their problem is that "out of the whole human family, it would not have been easy to select the same number of wise and virtuous persons, who should be less capable of sitting in judgment on an erring woman's heart, and disentangling its mesh of good and evil" (64) [45]. Assuming the moral position of God, the magistrates lack what Hester develops over the course of the book: the "power to sympathize" (161) [99].[5] That power causes a political dilemma. If on the one hand Hawthorne appeals to sympathy to temper the rigid and authoritarian rule of a system in which "religion and law were almost identical" (50) [36], on the other he warns of the dangers of having that sympathy lapse into a sentimental embrace of natural liberty with all of its potential dangers.

That dilemma is, of course, precisely the dilemma Hawthorne's readers confront when they sympathize with his two lovers in the forest. Hawthorne's answer to it is not, as critics too often assume, to advocate absolute submission to the existing civil authority. It is instead to imagine alternative possibilities for human relations within the civil order by drawing on the power to sympathize. Both that capacity of the imagination and the power to sympathize flourish best in Hawthorne's world in the space of civil society not directly under state supervision, a space prohibited in the Puritan theocracy at the beginning of Hawthorne's novel.

Hester will help generate that space, but she first has to acknowledge the importance of civil society by recognizing her sin. For Hawthorne that sin is not so much—as it would have been for Winthrop—a sin against God's law as it is a sin against the intersubjective agreements that human beings make with one another. Indeed, her adultery is another example of a premature effort to begin anew. After all, Hester's adultery with Dimmesdale takes place with her assuming, before the fact, that her husband is dead. When Chillingworth appears, therefore, he appears not only as a vengeful, cuckolded husband but also as a figure from a not-yet-buried past prepared to block Hester and Dimmesdale from achieving her dream of starting anew. To expand our understanding of why Hawthorne does not allow that new beginning, we need to look again at that "reputable" "citizen" (233) [139], Hester's husband.

3. *Another View of Mr. Prynne*

Hawthorne may elicit our sympathy for Hester and Dimmesdale while condemning their adultery, but he generates little sympathy for Hester's husband. From Chaucer's January to various figures in Shakespeare to Charles Bovary to Leopold Bloom, the cuckolded husband has been

5. The best discussion of sympathy in the novel is Hutner.

treated with varying amounts of humor, pathos, sympathy, and contempt. Few, however, are as villainous as Roger Chillingworth. Hawthorne's treatment of him starkly contrasts with the sympathetic treatment some courts gave to cuckolded husbands in the 1840s, when various states began applying the so-called unwritten law by which a husband who killed his wife's lover in the act of adultery was acquitted. Arguments for those acquittals portrayed avenging husbands as "involuntary agents of God." In contrast, lovers were condemned as "children of Satan," "serpents," and "noxious reptiles" with supernatural power allowing them to invade the "paradise of blissful marriages" (Ireland, "Libertine" 32).[6]

In *The Scarlet Letter* this imagery is reversed. It is the avenging husband who stalks his wife's lover with "other senses than [those ministers and magistrates] possess" and who is associated with "Satan himself, or Satan's emissary" (75, 128) [51, 81]. In the meantime, we imagine Arthur, Hester, and Pearl as a possible family (Herbert 201). The narrator so writes off Chillingworth as Hester's legal husband that he refers to him as her "former husband" (167) [102], causing Michael T. Gilmore to follow suit (93) and D. H. Lawrence to designate Mr. Prynne Hester's "first" husband. A legal scholar writing on adultery goes so far as to call Hester an "unwed mother" (Weinstein 225).

By reversing the sympathy that courts gave to cuckolded husbands taking revenge into their own hands, Hawthorne draws attention to the importance of seeking justice within the confines of the written law. Feminist historians have, for good reasons, stressed the ideological function of laws condemning adultery as a way to guarantee the legitimacy of patriarchal lineage. As accurate as this account is, in the nineteenth century an alternative account, stressing the law's positive function, was available.[7] Oliver Wendell Holmes, Jr., summarized much nineteenth-century writing on law's anthropological function when he wrote: "The early forms of legal procedure were grounded in vengeance" (2). Adultery is a case in point. Prior to the sixth century, revenge for adultery in England was carried out by the wronged husband and his kinship group. This reliance on *vendetta* resulted in long-standing blood feuds. To stop the social disruption caused by cycles of revenge, Aethelberht created his Code of Dooms that gave responsibility for punishing adultery and other crimes to the state. In his famous *Ancient Law: Its Connection with the Early History of Society and Its Relation to Modern Ideas* (1861), published a decade after *The Scarlet Letter*, Sir Henry Maine drew on similar evidence from Roman law to argue that criminal law served the social order by taking from individuals responsibility for punishing wrongdoers. The hoped-for result was a state monopoly on violence, for if the state alone could resort to violence to mete out justice, socially disruptive cycles of violent revenge could be avoided (Weinstein).

Dramatizing the dangers of achieving justice outside the law, Chillingworth illustrates natural liberty's potential for evil as well as for good. On the one hand, it prompts Hester to question the law in the name of a

6. For more on cases involving the "unwritten law," see Ireland, "Insanity"; Hartog; and Ganz.
7. Adultery has a complicated history in Anglo-American law. It was not a criminal act in common law, but was dealt with in ecclesiastical courts. As a result, the legal fiction of criminal conversation developed to allow common-law courts to rule on adultery, even if under an assumed name (Korobkin). Then in 1650, Puritans in England criminalized adultery. Even before that event a number of colonies, including Massachusetts, had criminalized the act. For the actual laws of adultery in seventeenth-century New England, see Dayton; Hull; Koehler; Norton; and Ramsey.

more equitable social order. On the other, it can allow Chillingworth to take the law into his own hands for personal revenge. If Hester's desire to create the world "anew" suggests utopian possibilities, Chillingworth's revenge, driven by "new interests" and "a new purpose" (119) [75], suggests the potential for a reign of terror. Hawthorne links these two seeming opposites through the secret pact that Hester and her husband forge on his return. Hester's dreams of a new social order result from her having "imbibed . . . a freedom of speculation" growing out of a new way of thinking that challenges "the whole system of ancient prejudice" (164) [101], but during her prison interview with her husband she imbibes a draught he has concocted out of the "many new secrets" he has learned in the wilderness from Indians (72) [49]. Chillingworth's "new secrets" might be associated with a "primitive" realm that Hester's vision of an enlightened future hopes to overcome, but the "promise of secrecy" that once again binds husband and wife suggests a possible connection between the two (170) [104]. Their secret bond in turn parallels the secret bond of natural lovers that Hester and Dimmesdale contemplate in their meeting in the forest. The two bonds even have structural similarities. For instance, just as Hester's new bond with her husband can be maintained only because he has taken on a new name, so Hester counsels her lover, "Give up this name of Arthur Dimmesdale, and make thyself another" (198) [120]. More importantly, the secrecy in which both bonds are made isolates everyone involved from the human community. As such, both are in stark contrast to the bond created by the civil ceremony of marriage whose public witness links husband and wife to the community.

Much has been made of Hester's adulterous violation of her marriage vows. Not much attention, however, has been paid to her husband's violation of his vows, even though the narrator comments on it. For instance, in prison Hester asks her husband why he will "not announce thyself openly, and cast me off at once?" His reply: "It may be . . . because I will not encounter the dishonor that besmirches the husband of a faithless woman. It may be for other reasons. Enough, it is to my purpose to live and die unknown" (76) [52]. In legal terms, Chillingworth's fear of dishonor makes no sense inasmuch as he has committed no crime. But if some antebellum courts displayed great sympathy to cuckolded husbands through the unwritten law, there was a long tradition—still powerful in the seventeenth century—of popular and bawdy rituals mocking cuckolded husbands (Ramsey 202–07). No matter what other motives Chillingworth might have, the narrator makes clear that the man "whose connection with the fallen woman had been the most intimate and sacred of them all" resolves "not to be pilloried beside her on her pedestal of shame" (118) [75]. That resolve explains "why—since the choice was with himself—" he does not "come forward to vindicate his claim to an inheritance so little desirable" (118) [75].

According to coverture, that undesirable inheritance was not only Hester, but also her child. Fully aware of his husbandly rights, Chillingworth tells his wife, "Thou and thine, Hester Prynne, belong to me" (76) [52]. Nonetheless, he refuses to acknowledge his inheritance, telling Hester in the same scene, "The child is yours,—she is none of mine,—neither will she recognize my voice or aspect as a father" (72) [49]. The doctrine of coverture was clearly a patriarchal institution; nonetheless, it was not solely to the advantage of the husband. It was also a means to hold him responsible for the well-being of his wife and children. Chillingworth might not be Pearl's biological father, but he was her father in the eyes of the law. That legal status adds another dimension to the recognition scene that occurs when Chillingworth walks out of the forest and finds his wife on public display for having committed adultery. "Speak, woman!" he "coldly and sternly" cries from the crowd. "Speak; and give your child a father!" (68) [47]. Commanding his wife to reveal the name of her lover, the wronged husband also inadvertently reminds us that at any moment Hester could have given Pearl a legal father by identifying him. Even more important, Chillingworth could have identified himself. But the same man who knows his legal rights of possession as a husband refuses to take on his legal responsibilities as a father.

Pearl, in other words, has not one but two fathers who refuse to accept their responsibilities. Having lost his own father as a young boy and doubting his ability financially to support his children on losing his job at the Custom House, Hawthorne was acutely aware of the need for fathers to live up to their name. In fact, by the end of the novel he ensures Pearl's future by having her two fathers finally accept their responsibilities. At his death Dimmesdale publicly acknowledges his paternity, eliciting from Pearl a "pledge that she would grow up amid human joy and sorrow, nor for ever do battle with the world" (256) [151]. At his death Chillingworth bequeaths to his once-rejected inheritance "a considerable amount of property, both here and in England" (261) [153]. Even so, the book's emphasis on failed fathers raises the possibility that Hester will earn her claim to good citizenship through her role as a mother.

4. *A Mother's Rights*

The Scarlet Letter, according to Tony Tanner, is a major exception to the "curiously little interest" the novel of adultery pays to the child of an illicit liaison, "even on the part of the mother (or especially on part of the mother)" (98). Indeed, Hester's relation to Pearl is a major part of Hawthorne's story. Accompanying her mother in almost every scene in which Hester appears, Pearl embodies a major paradox: although there is perhaps no better symbol of the hope for a new beginning than the birth of a child, Hester's daughter continually reminds her mother of her sinful past. Like the scarlet letter to which she is frequently compared, Pearl serves therefore as an agent of her mother's socialization. Part of Hester's socialization is in turn to socialize her daughter. Worried that Pearl is of demon origin or that her mother is not doing a proper job of raising her, some of the "leading inhabitants" are rumored to be campaigning to transfer Pearl "to wiser and better guardianship than Hester Prynne's" (100–01) [64]. In response Hester concocts an excuse to go to the governor's hall, only to find Governor Bellingham and Reverend Wilson convinced of their plan when Pearl impiously responds to their interrogations. Desperately turning to Dimmesdale, Hester implores: "I will not lose the child! Speak for me! Thou knowest,—for thou hast sympathies which these men lack!—Thou knowest what is in my heart, and what are a mother's rights" (113) [72].

As much an anachronism as Hawthorne's evocation of the concept of good citizenship, Hester's appeal to a mother's rights helps to locate

Hawthorne's attitude toward motherhood. In the seventeenth century no mother threatened with losing custody of her child could have successfully evoked the idea of a mother's rights. On the contrary, as we have seen, under the doctrine of coverture the child belonged legally to the father. In fact, in custody disputes between husband and wife a common law court did not grant custody to the mother until 1774. Even in this landmark case Chief Justice Lord Mansfield acknowledged the "father's natural right" while ruling that "the public right to superintend the education of its citizens" had more weight (qtd. in Grossberg 52). Mansfield's seemingly revolutionary ruling, in other words, would have confirmed the Puritan elders' sense that for her own good and that of the commonwealth Pearl, who had no father willing to claim her, could be taken from her mother. It was not until the courts were convinced that the education of children as citizens was best accomplished by their mothers that the idea of a mother's right to her child could gain force.

That process began in a few highly publicized cases in the US just before Hawthorne began writing *The Scarlet Letter.* These cases in which a mother won custody from a father coincided with a challenge to coverture posed by the rise of republican rhetoric that opposed coverture's image of marriage as a corporate body presided over by the husband with the image of marriage as a contractual relation, with husband and wife bringing to the union complementary, if not identical, duties and obligations. Not yet willing to grant women an active role in the political sphere of the new republic, this rhetoric still gave them an important role to play, that of raising children as citizens in service of the nation. Emphasizing the nurturing role of the mother, this cult of republican motherhood bolstered a wife's claim to gain custody of her child, especially one of "tender" years. Indeed, in the D'Hauteville case, one of the most publicized custody battles, the wife's lawyers contrasted the increasingly progressive republican nature of marriage in the US to the outmoded feudal concept of coverture maintained by her Swiss husband (Grossberg).

In her plea for a mother's rights Hester echoes the antebellum rhetoric of republican motherhood, which, like Hester's appeal to Dimmesdale, emphasized the capacity for sympathy. A product of "paternal" and "maternal" qualities, a proper republican citizen was not simply the obedient subject produced under the paternal regime of both coverture and seventeenth-century Puritanism. Instead, a good citizen should also have the moral quality of sympathy nurtured through a mother's love. Hawthorne dramatizes the marriage of these two qualities in the final scaffold scene when Dimmesdale, the biological father, elicits Pearl's pledge of obedience, a pledge that comes in the form of tears produced because the scene has "developed all her sympathies" (256) [151].

It would, nonetheless, be a mistake to assume that Hester becomes a model citizen by the end of *The Scarlet Letter* through her role as a mother. If republican mothers were supposed to raise citizens for the nation, Pearl does not become a "citizen" of Boston. Whereas, in typical Hawthornian fashion, we are not completely certain where Pearl ends up, circumstantial evidence indicates that she has successfully married and lives somewhere in Europe, most likely on the continent, not even in England. Measured by the most important standard of success for a republican mother, therefore, Hester fails. Rather than raise a child inculcated in proper values to serve the nation/commonwealth, Hester raises a child who finds "a home and comfort" in an "unknown region" (166, 262) [102, 154], just as Hawthorne ends "The Custom-House" imagining himself a "citizen of somewhere else" (44) [33].

To the patriotically minded, Hester's failure to produce a representative of the new generation bound by loyalty to the nation would seem to disqualify her as a model citizen. In fact that failure helps to suggest how Hawthorne expands our sense of good citizenship. As the cult of republican motherhood demonstrates, the republican challenge to authoritarian forms of government involved more than exchanging the political status of subjects for that of citizens. It also tempered hierarchical rule with sympathy, which because of its capacity for identification across barriers of status is a decidedly unhierarchical emotion. Even so, much republican rhetoric continued to channel sympathy into service of the state by implying that sympathies cultivated in the family would lead to local and regional ones before culminating in identification with all members of the nation. Within this developmental narrative, the function of the state is to enforce the civil order in the name of "the people" sympathetically bound together as a nation. In contrast, Hawthorne suggests an interactive, not developmental, model for the relation between sympathy and the state. Also stressing the need to temper harsh, hierarchical rule with a capacity for sympathy, Hawthorne does not see sympathetic identification with members of the nation as necessarily an expansion of the moral capacity of individual citizens. On the contrary, his continual stress on the importance of local attachments suggests that the state should guarantee a civil order in which such attachments can be cultivated because they are valuable in themselves, not because they will eventually lead to an attachment to the nation. National sympathies for him are not inevitably of a higher order than more local ones.[8]

Hawthorne's interactive model is compatible with a belief shared by many, if not all Americans, indeed, by many, if not all citizens of liberal democracies. The primary goal for them is not necessarily to produce citizens who display loyalty to the state as representative of "the people" bound together as a nation. The goal instead is to produce independent citizens capable of choosing where they can best develop their capacities. To be sure, this goal is in part conditioned by the ideology of liberal democracies, like the US, which values freedom of choice. In the US of Hawthorne's day that freedom was officially endorsed through the government's support of a citizen's right to expatriation, whereas British subjects owed perpetual allegiance to their sovereign (Tsiang; James).[9]

Of course, it is one thing to emphasize freedom of choice and quite another to provide the conditions making it possible. If much recent

8. See Berlant's excellent discussion of how Hawthorne adjudicates "the different claims for federal, state, local, and private identity that circulate through the American system" (203). See also Carey McWilliams and note 9 [on p. 418], below.
9. Hawthorne's celebration in "The Custom-House" of the renewing powers of "frequent transplantation" indirectly lends support to arguments for the right to expatriation (8–9) [10–11]. Although he describes what, "in lack of a better phrase," he must call "affection" for his "native place" of Salem (8) [10], he also insists, "Human nature will not flourish, any more than a potato, if it be planted and replanted, for too long a series of generations, in the same worn-out soil. My children have had other birthplaces, and so far as their fortunes may be within my control, shall strike their roots into unaccustomed earth" (11–12) [12].

criticism of US literature has used this disparity to question how "free" freedom of choice "really is," we should also not forget that it is one thing to acknowledge that a preference for freedom of choice is in part a product of ideology and quite another to claim that such a preference makes no difference. The power of *The Scarlet Letter* as civic myth has to do with its dramatization of the difference that a preference for freedom of choice can make and how important the existence of an independent civil society is for its cultivation. That difference is most poignantly dramatized in Hester's decision to return to Boston at the end of the book.

That decision is freely chosen in the sense that no one forces Hester to make it, but it is certainly not a decision made without pressure from many complicated historical and psychological factors, just as one's decision as to where to maintain or seek citizenship is not simply a rational choice about possibilities for political or economic freedom but one conditioned by numerous factors that one cannot control, such as where one was born and where one's intimate ties are located. In this regard Hester's return is especially important because she returns no longer primarily defined by relations of status that so governed the women of her time; that is, the status of lover, mother, or wife. On the contrary, with her lover and husband dead and her child apparently married and in another country, she returns as a woman, a woman devoted, nonetheless, not to individual fulfillment but to the interpersonal relations of civil society. It is in this space, which incorporates "many of the associations and identities that we value outside of, prior to, or in the shadow of state and citizenship" (Walzer, Introduction 1), that Hester provides us paradoxically with a model of good citizenship that no liberal democracy can afford to do without.

5. *Hester's Unexceptional Return*

At the start of the novel the scarlet letter has "the effect of a spell" on Hester, "taking her out of the ordinary relations with humanity, and inclosing her in a sphere by herself" (54) [39]. As the novel goes on, however, it assumes the scaffold's role of promoting good citizenship. With Hester's return and her willing resumption of the letter—"for not the sternest magistrate of the period would have imposed it" (263) [154]—the scarlet letter has, as Sacvan Bercovitch has argued, finally done its office. But just as Hester's actions change the meaning that people give to the scarlet A, so too they alter the sense of good citizenship with which the book begins.

The book begins with an image of good citizenship as the sort of absolute obedience that Winthrop wanted his subjects to give to their magistrates. The distance Hawthorne moves away from that image can be measured by a comparison between Dimmesdale and Hester. Tempted in the forest to break completely with the dictates of civil authority, Dimmesdale goes back on his resolve and instead seeks salvation by submitting totally to the existing civil order through participation in the civic activities of the election-day ceremonies. His submission culminates in his sermon that teleologically projects a utopian vision of a cohesive—and, it is important to emphasize, closed—Puritan community into the future. Dimmesdale, in other words, becomes the obedient subject that Winthrop desires. He is joined during these public ceremonies by almost the entire Puritan crowd, which submits "with childlike loyalty" to its rulers (250) [148].

Hester, however, is not among that crowd. Her good citizenship comes because of, rather than despite of, her failure to submit so loyally.

Through her return Hester acknowledges the civil law in a way that she did not in her rebellious earlier days. Nonetheless, she does not, as Dimmesdale does, submit totally to the state. On the contrary, she receives the Puritan magistrates' toleration of—and even admiration of—actions that are not directly under their supervision. Concerned with counseling and comforting those who feel marginalized by official Puritan society, especially women whose attempts at intimacy had failed, those activities extend the parameters of good citizenship to an interpersonal realm concerned with affairs of the heart that no affairs of state seem capable of remedying. If Dimmesdale simply channels his capacity for sympathy into total service to the state, Hester dramatizes how important it is for the state to promote spaces in which the capacity for sympathy can be cultivated while simultaneously guarding against the dangers of natural liberty. Thus, even though Hester has no place within the civic sphere, she, unlike Dimmesdale, helps to bring about a possible structural realignment of Puritan society by having it include what we can call the nascent formation of an independent civil society.

Stressing the importance of the civil order, the Puritans, as represented by Hawthorne, had no place for an independent civil society because they felt the need to control all aspects of life. As the narrator notes regarding the concern over Pearl's upbringing, "Matters of even slighter public interest, and of far less intrinsic weight than the welfare of Hester and her child, were strangely mixed up with the deliberations of legislatures and acts of the state" (101) [65]. Indeed, the relative independence granted to Hester at the end of the book markedly contrasts with an earlier description of her cottage, which she could possess only "by the license of the magistrates, who still kept an inquisitorial watch over her" (81) [54].[1]

If the Puritan theocracy, like all absolutist forms of government, has no room for an independent civil society, such a society is an essential feature of liberal democracies. In Michael Walzer's words, "It is very risky for a democratic government when the state takes up all the available room and there are no alternative associations, no protected social space, where people can seek relief from politics, nurse wounds, find comfort, build strength for future encounters" (Introduction 1). It is so risky that one of the functions of the state in liberal democracies is to ensure that alternative associations and protected spaces exist. In dramatizing the importance of their existence, Hester's activities on her return to Boston indicate the kinds of nonpolitical transformations that for Hawthorne were necessary for democratic rule to emerge from the Puritans' authoritarian rule.

By emphasizing the Puritans' authoritarianism rather than their democracy, Hawthorne works on/with the antebellum myth of the Puritan origins of American democracy. That myth has been perpetuated by both

1. Pease makes a number of interesting points about *The Scarlet Letter* in relation to civic duty, but he neglects the crucial role of civil society and confines himself to a discussion of the "reciprocity between the public and private worlds" (82). His sense of community also depends upon Rousseau's "general will," which Pease equates with an American notion of a "public will" (24). But an independent civil society is important because it allows for associations that resist potentially tyrannical conformity enforced in the name of an abstract "general will," the most obvious example being the "reign of terror." It is no accident that Pease champions Hawthorne's Puritans, finding in them a positive "unrealized vision of community" (53).

supporters of the country's claim to foster democratic rule and critics of it, such as Bercovitch and Lauren Berlant, the two best recent readers of Hawthorne's politics. Like most recent critics, including myself, both Bercovitch and Berlant read *The Scarlet Letter*'s seventeenth-century moment of representation as a comment on its antebellum moment of production.[2] But, unlike me, they do so by turning a nineteenth-century liberal democracy into a secular version of the Puritans' seventeenth-century theocracy.[3] Fitting *The Scarlet Letter* into the project he has conducted throughout his distinguished career, Bercovitch plots a complicated narrative of secularization in which the New England Way becomes the American Way, while Berlant without elaboration simply posits the continuities of the "Puritan/American project" (158). Supplementing her narrative of secularization with Louis Althusser's account of the ideological interpellation of subjects, Berlant reads Hawthorne's portrayal of seventeenth-century Boston allegorically to solve the "problem of understanding national citizenship in early national America" (6), assuming that Winthrop's subjects have the same relation to the state as citizens in a nineteenth-century democracy, a relation that "in theory" allows "neither a private part to which the state is not privy, nor a thought outside of the state's affairs" (98). But if Althusser's model, in which all aspects of civil society are simply part of the state's ideological apparatus, might conceivably work for a seventeenth-century theocracy and its demands for absolute obedience, it does not work for liberal democracies. Convinced that it does, Berlant feels compelled to look for resistance to the total control that Althusser attributes to the state and finds it, as do I, in Hawthorne's portrayal of "the scene of everyday life relations and consciousness" (95). What she fails to realize is that, far from challenging the ideology of liberal democracies, locating the potential for resistance in such everyday associations is a vital part of that ideology.

Bercovitch is acutely aware of how resistance to the state can serve the ideology of liberal democracies. Nonetheless, his need to see the US's nineteenth-century democracy as a secularized version of the Puritans' seventeenth-century theocracy betrays his otherwise magnificent reading of Hawthorne's novel in two important ways. First, since the great crisis for Puritanism was the antinomian controversy, Bercovitch needs to assert that "the only plausible modes of American dissent are those that center on the self" and then to read *The Scarlet Letter* as a book about Hester's individualism (31). But Hester, I hope I have established, is defined much more by her commitment to interpersonal relations than by her individualism, which is not to say that Hawthorne does not value the independence that she displays in contrast to the "childlike loyalty" of other Puritan subjects. But that independence for Hawthorne is not a product of a naturally self-

2. According to Colacurcio, "If the plot leaves Hester Prynne suspended between the repressive but obsolescent world of Ann Hutchinson and the dangerous new freedoms of the world of Margaret Fuller, the theme of the romance takes us very surely from the high noon of the Puritan theocracy to the dawn of the Romantic Protest in the nineteenth century" (*Province* 32). Both Gilmore and Herbert argue that *The Scarlet Letter*'s world may be Puritan New England but that its major characters have a nineteenth-century moral outlook. Both follow Baym, who claims that Hawthorne "has created an authoritarian [Puritan] state with a Victorian moral outlook" (215). Baym's comment is extremely important since it reminds us of the extent to which Hawthorne's representation of the Puritans is work on/with myth, not an accurate representation.

3. Bercovitch's narrative of secularization necessarily minimizes important developments in the eighteenth century, such as the structural transformation of the public sphere and its relation to the rise of a relatively independent civil society (Habermas).

sufficient self; it is instead bred and cultivated in the associational activities of an independent civil society.[4]

Bercovitch's second misreading has to do with Hawthorne's attitude toward the nation. Certainly many Americans see the US as fulfilling a divine mission, just as the Puritans saw themselves as the chosen people. But Hawthorne's work on/with that exceptionalist myth is too powerful to be confined by Bercovitch's narrative of secularization, however subtle and complicated that narrative is. On the contrary, Hawthorne's well-documented skepticism about revolutionary reformers questions the sacred mission they grant to themselves. For instance, both the Puritan Revolution in England and the French Revolution toppled sovereigns claiming divine authority, and yet Cromwell's mission in England was to establish a New Jerusalem while the revolutionaries of France transferred the king's claim to absolute authority to the nation and, with religious zeal, condemned to death anyone opposed to its new principles. From a Hawthornian perspective, the danger is not that America will stray from its divine mission, but that it will follow the path of other revolutions and believe too fervently that it has such an exceptional destiny.

That ever-present danger means that, although there is a structural difference between an antebellum democracy and a seventeenth-century theocracy, perpetual work is required to guard against a patriotism that "loses all sense of the distinction between State, nation, and government" (Bourne 357). Hawthorne accomplishes that work in his introductory sketch of "The Custom-House" as well as in his novel. Hawthorne, Stephen Nissenbaum has documented, was heavily involved in local partisan politics and fought extremely hard to retain his civil service post in the Custom House. Nonetheless, his fictional version of his dismissal tells a different story. If, as Gordon Hutner puts it, Hawthorne "introduces his novel about the public history of private lives with his private history of public lives" (20), in both the novel and the sketch he ends by locating his protagonists in the space of civil society between the public and private. And just as the novel looks ironically at various ideals of good citizenship, so does the sketch. For instance, Hawthorne's portrayal of the ex-military men working at the Custom House undercuts the ideal of the citizen-soldier, an ideal that contributed to the election of military hero Zachary Taylor as president and thus indirectly led to Hawthorne's dismissal. Taylor's election is a perfect example of the failure of a second ideal: people displaying and cultivating their virtue through participation in the political process. Far from a realm in which citizens sacrifice their own interests for the good of the nation, politics in "The Custom-House" has degenerated into a battle of self-interest. Its debilitating effects are most prominently displayed in the spoils system, which, especially in Hawthorne's hands, puts a lie to a third ideal: the good citizen as devoted civil servant.

4. "Rather than revealing that . . . 'the only plausible modes of American dissent are those that center on the self,' *The Scarlet Letter* seems to demonstrate that the only form of selfhood worth having is generated by reciprocal connection to others—and that one may choose constraints because there are no meanings without them. Hester's deepest yearning, this is to say, is not for freedom but for a reimagined social life (the very thing that in Bercovitch's account, consensus ideology removes from view) for a lover and for a community able to accommodate the forms of connection she envisions for them" (Millington 6). To which I add: the relative freedom from state supervision provided by an independent civil society enhances the possibility of imagining and working toward that different social life.

Presided over by a flag that marks it as "a civil, and not a military post of Uncle Sam's government" (5), the Custom House is occupied by people who fail to heed the fierce look of the American eagle over its entrance that warns "all citizens, careful of their safety, against intruding on the premises which she overshadows with her wings" (5) [8]. Instead, they seek "to shelter themselves under the wing of the federal eagle" (5) [8], not so much to serve the country as to be guaranteed a comfortable livelihood. The expectation that the federal eagle's "bosom has all the softness and snugness of an eider-down pillow" (5) [8] is the mirror-image of the childlike loyalty that causes the Puritan crowd uncritically to submit to its magistrates' rule. Choosing neither the nation's maternal protection nor its paternal authority, Hawthorne weaves a fiction in which he best serves the country not as a civil servant paid by the state but as a nonpartisan writer located in an independent civil society. Thus he portrays himself as happily leaving the Custom House so that he could once again take up his pen. The novel that he subsequently wrote, which more than any other work has become part of the "general incorporation of literature into education" and thus part of the "channel through which the [national] ethos is disseminated and . . . the means by which outsiders are brought inside it," gives substance to thc cliché that democracy is a way of life as well as a political system (Brodhead 61, 60).

6. *Conclusion*

If *The Scarlet Letter* suggests that political institutions alone cannot make a democracy, its emphasis on good citizenship in the civil as well as in the civic sphere is by no means a solution to all of the country's problems. The issue of race, for instance, marks an important limit to that emphasis. Conflicted between loyalties to an individual state and to the federal union, Hawthorne searched for a reason to fight the Civil War.[5] Writing to his friend Horatio Bridge, he identified the issue of slavery. "If we are fighting for the annihilation of slavery . . . it might be a wise object, and offers a tangible result, and the only one which is consistent with a future Union between North and South. A continuance of the war would soon make this plain to us; and we should see the expediency of preparing our black brethren for future citizenship by allowing them to fight for their own liberties, and educating them through heroic influences" (*Letters* 381). Whereas the annihilation of slavery was indeed the basis for restoring the Union, a truly equitable citizenship for blacks was, as we know, derailed by the reconciliation of white North and South.

Even though he died in 1864, Hawthorne unintentionally anticipates a reason for those derailed efforts in his metaphoric descriptions of the scarlet letter. The letter is called variously a mark, a brand, a badge of shame, and a stigma. What Hawthorne could not have known was that a

5. In an essay that caused some to question his patriotism Hawthorne wrote: "The anomaly of two allegiances (of which that of the State comes nearest home to a man's feelings, and includes the altar and the hearth, while the General Government claims his devotion only to an airy mode of law, and has no symbol but a flag)" means that "[t]here never existed any other government, against which treason was so easy, and could defend itself by such plausible arguments" ("Chiefly" 416). He added: "In the vast extent of our country—too vast, by far, to be taken into one small human heart—we inevitably limit to our own State, or, at farthest, to our own Section, that sentiment of physical love for the soil which renders an Englishman, for example, so intensely sensitive to the dignity and well-being of his little island" ("Chiefly" 416–17).

few years after *The Scarlet Letter* appeared, Justice Taney in the Dred Scott case would use similar metaphors to deny citizenship to anyone of African descent—free or slave. Since in a republic there is only one class of citizens, Taney argued, "the deep and enduring marks of inferiority and degradation" implanted on blacks had so "stigmatized" them that they were excluded from the sovereign body constituting the nation (416). In an effort to undo the damage done by Dred Scott, the Supreme Court after the Civil War ruled that the Thirteenth Amendment forbade not only slavery but also all "badges and incidents" of slavery. The difference between a badge and a stigma is significant. A badge can be removed; a stigma, coming from the Greek word for a brand, is implanted for a lifetime—and for Taney could be passed from generation to generation (Thomas).

The Scarlet Letter ends by giving Hester a choice of whether to wear her "badge of shame" (161) [99]. She willingly chooses to wear it, in part because through her own agency the letter has "ceased to be a stigma" (263) [154]. In contrast, the possibility of achieving the status of model citizen through individual effort was denied African Americans because their race meant that, as a group, they inherited a badge of slavery, whose stigma persisted. The civil society argument about "uncoerced human associations" by itself is not adequate to deal with that problem (Walzer, "Concept" 7). Instead a much more traditional argument about active citizen participation in the political sphere would seem to be called for.

Clearly, a danger of an exclusive emphasis on good citizenship in civil society is political quietism of the sort that Hawthorne succumbed to in the 1850s when he argued that slavery would wither and die of its own accord. As regrettable as Hawthorne's quietism was, however, that biographical fact should not be, as some critics make it, the final word on "the politics" of his most famous novel (Cheyfitz; Arac, "Politics"). *The Scarlet Letter* does not so much reject civic notions of good citizenship as question empty platitudes about them while expanding our sense of what they can entail. That expanded sense of good citizenship is by no means sufficient to solve issues of racial inequality—as if any one course of action is—but it may be an important component of any solution. Indeed, it is not simply an accident that the movement agitating for first-class citizenship for African Americans was called the *civil* rights movement. To be sure, civil rights by definition are guaranteed by the state, and to be effective they have to be enforced by the state. Nonetheless, agitation for civil rights reminds us that one of the most important goals of political activism is the creation of a space where the voluntary associations located in civil society exist according to principles of equity and fairness. As Walzer puts it, "Only a democratic state can create a democratic civil society; only a democratic civil society can sustain a democratic state. The civility that makes democratic politics possible can only be learned in the associational networks; the roughly equal and widely dispersed capabilities that sustain the networks have to be fostered by the democratic state" ("Concept" 24).[6]

6. Walzer continues, "The state can never be what it appears to be in liberal theory, a mere framework for civil society. It is also the instrument of the struggle, used to give particular shape to the common life." Nonetheless, he adds that it is not necessary to find "in politics, as Rousseau urged, the greater part of our happiness. Most of us will be happier elsewhere, involved only sometimes in affairs of state. But we must leave the state open to our sometime involvement" ("Concept" 24).

What Walzer does not do, however, is give us a concrete sense of what a democratic civil society looks like. Thus his helpful, but too balanced, formulation needs to be supplemented by the observation that a major debate within democratic politics is how to define a democratic civil society. *The Scarlet Letter* does not provide that definition, but it does contribute to democratic politics by implying an answer to another question raised by Walzer's formulation: which comes first, democratic state or democratic civil society? Challenging the standard account that locates the seeds of a later democracy in the political institutions of seventeenth-century New England, *The Scarlet Letter* implies that the nascent formation of an independent civil society precedes and helps to generate a democratic state. If that implied narrative has limits—and like all narratives it does—it has also served as a powerful and enabling civic myth for many, like those whom Hester counsels, whose failed efforts at sympathy make them feel marginalized by the existing—not so—civil order.[7]

WORKS CITED

Althusser, Louis. "Ideology and Ideological State Apparatuses (Notes Toward an Investigation)." *Lenin and Philosophy, and Other Essays.* Trans. Ben Brewster. New York: Monthly Review Press, 1971. 127 86.

Anon. "Bancroft's History of the United States." *American Jurist and Law Magazine* 2 (1838): 229–31.

Arac, Jonathan. "Narrative Forms." *The Cambridge History of American Literature.* Ed. Sacvan Bercovitch. vol. 2. New York: Cambridge UP, 1995. 605–777.

———. "The Politics of *The Scarlet Letter.*" *Ideology and Classic American Literature.* Ed. Sacvan Bercovitch and Myra Jehlen. New York: Cambridge UP, 1986. 247–66.

Baym, Nina. "Passion and Authority in *The Scarlet Letter.*" *New England Quarterly* 43 (1970): 209–30.

Bell, Michael Davitt. *Hawthorne and the Historical Romance of New England.* Princeton: Princeton UP, 1971.

Bercovitch, Sacvan. *The Office of "The Scarlet Letter."* Baltimore: Johns Hopkins UP, 1991.

Berlant, Lauren. *The Anatomy of National Fantasy: Hawthorne, Utopia, and Everyday Life.* Chicago: U of Chicago P, 1991.

Blumenberg, Hans. *Work on Myth.* Cambridge: MIT P, 1985.

Bourne, Randolph. *The Radical Will: Selected Writings, 1911–1918.* Ed. Olaf Hansen. New York: Urizen, 1977.

Brodhead, Richard. *The School of Hawthorne.* New York: Oxford UP, 1986.

Burke, Kenneth. "Literature as Equipment for Living." *The Philosophy of Literary Form.* Baton Rouge: Louisiana State UP, 1941. 293–304.

Cheyfitz, Eric. "The Irresistibleness of Great Literature: Reconstructing Hawthorne's Politics." *American Literary History* 6 (1994): 539–58.

Choate, Rufus. "The Importance of Illustrating New England History by a Series of Romances Like the Waverley Novels." *The Works of Rufus*

7. I am grateful for the comments provided by Jayne Lewis, Robert Milder, Frederick Newberry, Steven Mailloux, and audiences at the University of Oregon, the University of Washington, and the Kennedy Institute for North American Studies in Berlin.

Choate with a Memoir of His Life. Ed. Samuel Gilman Brown. 2 vols. Boston: Brown, 1862. 1: 319–46.

Cohen, Jean, and Andrew Arato. *Civil Society and Political Theory*. Cambridge: MIT P, 1992.

Colacurcio, Michael J. "Footsteps of Ann Hutchinson: The Context of *The Scarlet Letter*." *ELH* 39 (1972): 459–94.

———. *The Province of Piety: Moral History in Hawthorne's Early Tales*. Cambridge: Harvard UP, 1984.

———. "'The Woman's Own Choice': Sex, Metaphor, and the Puritan 'Sources' of *The Scarlet Letter*." *Doctrine and Difference*. New York: Routledge, 1997. 205–28.

Crews, Frederick C. *The Sins of the Fathers: Hawthorne's Psychological Themes*. New York: Oxford UP, 1966.

Dayton, Cornelia. *Women Before the Bar: Gender, Law, and Society in Connecticut, 1639–1789*. Chapel Hill: U of North Carolina P, 1995.

Dekker, George. *The American Historical Romance*. New York: Cambridge UP, 1987.

Doubleday, Neal Frank. "Hawthorne and Literary Nationalism," *American Literature* 12 (1942): 447–53.

Dred Scott v. Sandford. 19 Howard 393 (1857).

Fletcher, Andrew. *Andrew Fletcher of Saltoun: Selected Political Writings*. Ed. David Daiches. Edinburgh: Scottish Academic Press, 1979.

Ganz, Melissa J. "Wicked Women and Veiled Ladies: Gendered Narratives of the McFarland-Richardson Tragedy." *Yale Journal of Law and Feminism* 9 (1997): 255–303.

Gilmore, Michael T. "Hawthorne and the Making of the Middle Class." *Discovering Difference*. Ed. Christoph K. Lohman. Bloomington: Indiana UP, 1993. 88–104.

Grossberg, Michael. *A Judgment for Solomon: The D'Hauteville Case and Legal Experience in Antebellum America*. New York: Cambridge UP, 1996.

Habermas, Jürgen. *The Structural Transformation of the Public Sphere*. Cambridge: MIT P, 1989.

Hartog, Hendrik. "Lawyering Husbands' Rights, and 'the Unwritten Law' in Nineteenth-Century America." *Journal of American History* 84 (1997): 67–96.

Hawthorne, Nathaniel. *The Scarlet Letter*. Centenary Edition. Vol. 1. Columbus: Ohio State UP, 1962.

———. "Chiefly about War Matters. By a Peaceable Man." *Miscellaneous Prose*. Centenary Edition. Vol. 23. Columbus: Ohio State UP, 1994. 403–42.

———. *The Letters, 1857–1864*. Centenary Edition. Vol. 18. Columbus: Ohio State UP, 1987.

Herbert, T. Walter. *Dearest Beloved: The Hawthornes and the Making of the Middle–Class Family*. Berkeley: U of California P, 1993.

Hobbes, Thomas. *On the Citizen*. Trans. Richard Tuck and Michael Silverthorne. Cambridge: Cambridge UP, 1998.

———. *Leviathan*. 1651. New York: Penguin, 1986.

Holmes, Oliver Wendell, Jr. *The Common Law*. Cambridge: Belknap P of Harvard UP, 1963.

Howells, William Dean. "Hawthorne's Hester Prynne." *The Critical Response to Nathaniel Hawthorne's* The Scarlet Letter. Ed. Gary Scharnhorst. Westport, CT: Greenwood Press, 1992. 101–09.

Hull, N. E. H. *Female Felons: Women and Serious Crime in Colonial Massachusetts.* Urbana: U of Illinois P, 1989.

Hutner, Gordon. *Secrets and Sympathy: Forms of Disclosure in Hawthorne's Novels.* Athens: U of Georgia P, 1988.

Ireland, Robert M. "The Libertine Must Die: Sexual Dishonour and the Unwritten Law in the Nineteenth-Century United States." *Journal of Social History* 23 (1989): 27–44.

———. "Insanity and the Unwritten Law." *American Journal of Legal History* 32 (1988): 157–72.

James, Alan G. "Expatriation in the United States: Precept and Practice Today and Yesterday." *San Diego Law Review* 27 (1990): 853–905.

Kerber, Linda K. *No Constitutional Right 'to Be Ladies': Women and the Obligations of Citizenship.* New York: Hill and Wang, 1998.

Koehler, Lyle. *A Search for Power: The "Weaker Sex" in Seventeenth-Century New England.* Urbana: U of Illinois P, 1980.

Korobkin, Laura Hanft. *Criminal Conversations: Sentimentality and Nineteenth-Century Legal Stories of Adultery.* New York: Columbia UP, 1998.

Kymlicka, Will. *Multicultural Citizenship.* Oxford: Clarendon, 1995.

Lawrence, D. H. *Studies in Classic American Literature.* Garden City, NY: Doubleday, 1951.

Levin, David. *History as Romantic Art.* Stanford: Stanford UP, 1959.

Maine, Sir Henry Sumner. *Ancient Law: Its Connection with the Early History of Society and Its Relation to Modern Ideas.* 1861. New York: Dorset, 1986.

McWilliams, John P., Jr. *Hawthorne, Melville, and the American Character: A Looking-Glass Business.* New York: Cambridge UP, 1984.

McWilliams, W. Carey. *The Idea of Fraternity in America.* Berkeley: U of California P, 1973.

Melville, Herman. *Billy Budd, Sailor (An Inside Narrative).* Ed. Harrison Hayford and Merton M. Sealts, Jr. Chicago: U of Chicago P, 1962.

Millington, Richard. "*The Office of 'The Scarlet Letter'*: An 'Inside Narrative'?" *Nathaniel Hawthorne Review* 22 (1996): 1–8.

Morgan, Edmund S. *The Puritan Dilemma: The Story of John Winthrop.* Boston: Little, Brown, 1958.

———. *Puritan Family.* New York: Harper, 1966.

Newberry, Frederick. *Hawthorne's Divided Loyalties: England and America in His Work.* Rutherford: Fairleigh Dickinson UP, 1987.

Nissenbaum, Stephen, ed. Introduction. *The Scarlet Letter and Selected Writings of Nathaniel Hawthorne.* New York: The Modern Library, 1984. vii–xlv.

Norton, Mary Beth. *Founding Mothers and Fathers: General Power and the Forming of American Society.* New York: Knopf, 1996.

Nye, R. B. *George Bancroft.* New York: Knopf, 1945.

———. Introduction. *History of the United States* (abridged). By George Bancroft. Chicago: U of Chicago P, 1966.

Pateman, Carole. *The Sexual Contract.* Stanford: Stanford UP, 1988.

Pease, Donald E. *Visionary Compacts: American Renaissance Writings in Cultural Context.* Madison: U of Wisconsin P, 1987.
Pocock, J. G. A. *The Machiavellian Moment: Florentine Political Thought and the Atlantic Republican Tradition.* Princeton: Princeton UP, 1975.
Ramsey, Carolyn B. "Sex and Social Order: The Selective Enforcement of Colonial American Adultery Laws in the English Context." *Yale Journal of Law & the Humanities* 10 (1998): 191–228.
Reynolds, Larry J. *European Revolutions and the American Literary Renaissance.* New Haven: Yale UP, 1988.
Ryskamp, Charles. "The New England Sources of *The Scarlet Letter.*" *American Literature* 31 (1960): 237–72.
Shklar, Judith. *American Citizenship: The Quest for Inclusion.* Cambridge: Harvard UP, 1991.
Smith, Merril D. *Breaking the Bonds: Marital Discord in Pennsylvania, 1730–1830.* New York: New York UP, 1991.
Smith, Rogers. *Civic Ideals: Conflicting Visions of Citizenship in U.S. History.* New Haven: Yale UP, 1997.
Tanner, Tony. *Adultery in the Novel: Contract and Transgression.* Baltimore: Johns Hopkins UP, 1979.
Thomas, Brook. "Stigmas, Badges, and Brands: Discriminating Marks in Legal History." *History, Memory, and the Law.* Ed. Austin Sarat and Thomas R. Kearns. Ann Arbor: U of Michigan P, 1999. 249–82.
Tsiang, I-Mien. *The Question of Expatriation in America Prior to 1907.* Baltimore: Johns Hopkins UP, 1942.
Walzer, Michael. "The Concept of Civil Society." *Toward a Global Civil Society.* Ed. Michael Walzer. Providence: Berghahn, 1995. 7–27.
———. Introduction. *Toward a Global Civil Society.* Ed. Michael Walzer. Providence: Berghahn, 1995. 1–4.
Weinstein, Jeremy D. "Adultery, Law, and the State: A History." *Hastings Law Journal* 38 (1986): 195–238.
Winthrop, John. *Winthrop's Journal: "History of New England."* Ed. J. K. Hosmer. Vol. 2. New York: Scribners, 1908.

MICHAEL RYAN

"The Puritans of Today": The Anti-Whig Argument of *The Scarlet Letter*†

> To fetter the freedom of man is not only to act the part of tyranny, but to inflict a gross wrong . . . [against] the essential equality of men. . . . Endued, as they are, with the same appetites and desires, with conscience, reason, and free will . . . sharers of the same beautiful existence, handiwork of the same God, children of a common destiny, hastening on to an eternal world, who shall . . . affix the mark which shall debar either this one or that one from the full fruition of every blessing of existence—every gift of God?
>
> —"Democracy" (attr. John L. O'Sullivan)

† From *Canadian Review of American Studies* 38.2 (2008): 202–225. Reprinted by permission from University of Toronto Press (www.utpjournals.com). Page numbers in square brackets refer to this Norton Critical Edition.

> We all know then, what moral government is, and that men cannot exist in society without it. In that form of it called civil government, the lowest culprit in his prison knows its general nature, its principles, its end, and its absolute necessity to this end, as well as the judge who condemns him. . . . Now the object of a perfect moral governor is not merely to secure right moral action, but to secure it . . . by a peculiar influence—the influence of his authority . . . It is to bring his subject to . . . an act of obedience . . . involving . . . recognition of [the governor's] right to rule.
>
> —Nathaniel Taylor, *Lectures on the Moral Government of God* (1859)

Scholars of Nathaniel Hawthorne's *The Scarlet Letter* (1850) have paid surprisingly little attention to the ambient religious debates from which the novel draws much of the substance for its internal polemic.[1] A conflict between those who would use legislation to control personal morality and those who opposed such intrusiveness in the name of the doctrine of natural revelation is central both to the novel and to the culture wars of the era in which Hawthorne wrote. Historian Daniel Walker Howe argues that "without an understanding of the religion of the middle period, there can be no understanding of its politics" ("Religion" 121), and the same might be said of a work such as *The Scarlet Letter* that is concerned with moral issues. To understand it fully, we must take into consideration the religious debates of the era.[2]

The leading public intellectuals of the early nineteenth century were theologians such as Lyman Beecher, Francis Wayland, Asa Burton, Nathaniel Taylor and Horace Bushnell. Those intellectuals who were not professional theologians—Ralph Waldo Emerson, Henry David Thoreau, Margaret Fuller, George Bancroft and John L. O'Sullivan—were nevertheless steeped in religious ideas, and their writings resonate with a sense of the unquestioned truth of the doctrines of natural revelation. For the orthodox Protestant theologians such as Beecher and Taylor, God was a distant being who oversaw a sin-prone humanity that needed guidance from churchmen in order to attain salvation. In *Their Brothers' Keepers: Moral Stewardship in the United States, 1800–1865,* Clifford Griffin writes, "These latter-day Calvinists still argued that only the Almighty could save souls—that men by their own acts alone could not merit redemption" (47–8). For intellectuals such as Bancroft and O'Sullivan, in contrast, as well as for more liberal theologians such as Burton and Bushnell, God was a presence in nature which guaranteed that human passion was intrinsically moral and that human striving could lead to salvation. In the chapter "The Rise of Religious Liberalism," in his book *Churchmen and Philosophers,* Bruce Kuklick notes that, at this time,

> [t]he issues of sovereignty, responsibility, grace, and depravity all found their critical substantive locus in the question of the will's freedom—the most important recurring theme in the literature. . . . [N]atural theology was increasingly stressed. Newtonianism implied

1. Several scholars have commented on religion in Hawthorne's work, although always in regard to religious themes rather than to the contemporary debates; see, most recently, Denis Donohue; Agnes McNeill Donohue; also, Fick; Warren; Simonson. Harvey Gable discusses overlaps between the work of Horace Bushnell and Hawthorne, but he is concerned primarily with language and typology, not politics or theology.
2. See Bodo; Cole; Griffin; Bushman; Marty; Handy; Welch; Hammond; Kuklick; Stavely; Hatch; Noll.

> that knowledge of God was contained in nature. . . . For the liberals, at bottom, religion rested on the revelations to which the biblical miracles testified. Other bases for Christianity—experience, tradition, the authority of the church—were set aside. (44, 82–3)

These theological differences were not merely intellectual. The two major political parties of the era, the Whigs and the Democrats, appealed to religious ideas to justify their contending positions regarding everything from moral legislation to economic policy. A major difference between anticlerical Democrats and Whig evangelical conservatives was in their views of the proper relation of church and state.[3] The Whigs, whom Daniel Howe characterizes as "the evangelical united front in the polling place" (Howe, *Political Culture* 18), laid claim to the Puritan theocratic legacy that authorized public supervision of private morality. Wishing to create a "Righteous Empire" in the United States, in which what they called "moral government" would merge law and religion, Whigs sought to use legislation to bring about "moral purity and reform" (Kelley, *Cultural Pattern* 163). Modernizers, they believed in using government to promote business-friendly economic development through monopolies, charters, tariffs and publicly funded canals and roads. Having witnessed the disestablishment of the Christian churches in all the states, Whig religious conservatives turned, in mid-century, to education, especially in the Sunday School and public school movements, to ensure that Christian values and norms prevailed in America.[4] Historian Ronald Formisano writes,

> Henry Cabot Lodge once called the Federalists "the Puritan party," and he might have said the same of the Whigs. The Federal and Whig parties both expressed the ancient Puritan concern for society as a corporate whole; both attempted to use the government to provide for society's moral and material development. (*Transformation* 289)

Whigs found justification for their program of moral government in the reformed Calvinist theology that emerged in the early nineteenth century. While taking into account the emphasis on free will in post-Enlightenment America, theologians such as Beecher and Taylor nevertheless held that humans were innately prone to sinfulness. Beecher argued "that man is desperately wicked, and cannot be qualified for good membership in society without the influence of moral restraint" (qtd. in Bodo 153). In his famous 1828 sermon, *Concio ad Clerum: A Sermon On Human Nature, Sin, and Freedom*, Nathaniel Taylor preached,

> What then are we to understand, when it is said that mankind are depraved *by nature?* I answer—*that such is their nature, that they will sin and only sin in all the appropriate circumstances of their being*. . . .

3. See Kelley, *Transatlantic*; McCormick; Benson; Formisano, *Birth*; Kelley, *Cultural Pattern*; Howe, *Political Culture*; Formisano, *Transformation*; Holt.
4. Historian Louise Stevenson characterizes the Whigs as economic and social modernizers:

> Whiggery stood for the triumph of the cosmopolitan and national over the provincial and local, or rational order over irrational spontaneity, of school-based learning over traditional folkways and customs, and of self-control over self-expression. Whigs believed that every person had the potential to become moral or good if family, school, and community nurtured the seed of goodness in his moral nature. (6)

> By the moral depravity of mankind I intend generally, the entire sinfulness of their moral character. (qtd. in Ahlstrom 220–2)

This vision of the world justified the Whig claim that an elite of morally superior people should exercise moral government over others through the political state. Lee Benson notes that "[Whig] Horace Greeley's *New York Tribune* argued that the state was a proper instrument for the regulation of every 'evil' in society" (207). Michael Holt, in *The Rise and Fall of the American Whig Party*, points out that "to a far greater degree than Democrats, Whigs backed state intervention to regulate social behavior: temperance legislation, Sunday blue laws, and the creation of state-run public schools" (68).

Democrats opposed the Whig desire to merge church and state and to use law to regulate personal behaviour. "The Puritans of today," a Democratic newspaper editor wrote in 1852, "like the Puritans of 1700, conceive themselves to be better and holier than others, and entitled—by divine right as it were—to govern and control the actions and dictate the opinions of others" (Kelley, "Portrait" 79). If Whigs thought of themselves as "the party of decency and respectability, the guardians of piety, sober living, proper manners, thrift, steady habits, and book learning" (Kelley, *Cultural Pattern* 166), Democrats portrayed them as religious zealots who constituted a threat to civil liberties. William Cullen Bryant held that "our civil and religious liberty exists, not in consequence, but in spite of the spirit and genius of Puritanism" (qtd. in Howe, *Political Culture* 89). In one fictional debate between a Democrat and a Whig in a pamphlet from 1854, the Democrat contends that "*all legislation* having any other object but the protection of rights is not only injurious to morality, but is in itself immoral and wicked" (qtd. in Benson, 206–7). Robert Kelley notes, regarding these debates,

> Central to the ideology of Anglo-American, freethinking intellectuals was a conviction that clerics must be kept out of politics, that their moral preachments were arrogance and their attempts to control the lives of others a continuing danger to freedom of thought and belief. This led many Democratic intellectuals to reject abolitionism, for it emerged out of the camp of the enemy: zealous, moralistic, church-and-state Yankeeism. (*Cultural Pattern* 171)

If Whig political theory drew on reformed neo-Calvinist theology to justify its ideal of moral government, Democratic political theory turned instead to the theological doctrine of natural revelation. Democratic intellectuals such as O'Sullivan and Bancroft believed that nature, because infused with divinity, should be allowed to follow its own course without theocratic interference. Morality arose spontaneously from natural processes, and no legislation and no moral discipline were needed to control passions that, because natural and therefore divinely inspired, could not be sinful. Natural revelation located divinity in everyone, without distinction of "rank," an appealing idea in an era when egalitarian ideals fuelled immigrant aspirations. Bancroft writes, "The barbarian who roams our Western prairies has like passions and like endowments with ourselves. He bears within him the instinct of Deity, the consciousness of a spiritual nature, the love of beauty, the rule of morality" (414). By placing the "rule of morality" in nature rather than institutions or laws, Democrats sought to undermine the Whig justification for moral government by a holy elite.

Democrats used similar arguments to oppose Whig pro-business economic policy. They believed that a benevolent nature, if left on its own, would make all right with the economic world.[5] Whig "improvements," especially banks and chartered monopolies that favoured the economic elite, did more harm than good by making people dependent on government instead of on their own natural talents. Democrats argued for a model of economic self-government that would approximate the self-regulating laws of nature that are God's creation: "[T]he voluntary principle, the principle of Freedom, suggested to us by the analogy of the divine government of the Creator; the natural laws which will establish themselves and find their own level are the best laws" ("Introduction" 7). America should place trust in "the same fundamental principle of spontaneous action and self-regulation which produces the beautiful order of [nature]" ("Introduction" 7). At the core of the theory was an ideal of self-dependent manhood: "[By] throwing men upon their own energies for success, [free trade] would accustom them to the practice of self-dependence and train them to habits of perseverance and economy" ("History" 305).

Sexual morality was a crucial site of conflict between the contending parties. In 1843 in Michigan, when Democrats "amended and loosened the laws relating to adultery and fornication which had made those sins criminal offenses," Whigs characterized the change as an outrage that struck "at the foundations of our social system" (Formisano, *Birth* 124). If Whigs devoted their energies to institutions such as the Magdalen Society that spread moral discipline amongst prostitutes and domestic workers, Democrats were more inclined to be critical of moral institutions that discriminated along gender lines in the allocation of moral punishment. In an 1846 essay—"The Legal Wrongs of Women"—in *The Democratic Review*, the anonymous writer argues, "The injustice is sufficiently flagrant which permits a man, whether single or married, to lead a licentious life without losing caste, while a poor girl, betrayed through her affections into guilt, almost inevitably becomes castaway through public scorn" (482). That this argument is central to *The Scarlet Letter* should alert us to how steeped the novel is in these contemporary debates.

As an inhabitant of a Whig stronghold—Salem, Massachusetts—Hawthorne would have been keenly aware of the Whigs' use of Puritanism as a justification for a strong government role in promoting business-friendly economic development and the moral "improvement" of the population. In 1849, while Hawthorne was composing *The Scarlet Letter*, regular advertisements for a history of the Puritans in the town's Whig newspaper, the *Salem Gazette*, made explicit links between Puritanism and Whig policies. According to the advertisements, the book, a republication of William Hubbard's *General History of New England*, praised "the hardy Puritan pioneers who for God, for Conscience, and for Humanity braved the perils of the ocean . . . [and] planted on this rocky shore the principles that have

5. See also, Welter 92, 88. Welter notes that Democratic economic theory "owed more to 'nature' understood as morality" than to laissez-faire theory. "Unlike the Whigs, they were committed to an economy of principle rather than an economy of consumption. . . . [They were] convinced that moral laws governed the economy."

made us what we are." Those "principles" sustained "the best" of New England, "its Manufactures, its Commerce, its Public Buildings . . . its contributions to the public and private nature of the great stack of New England industrial and productive accumulations" (*Salem Gazette*).[6] As the Whigs styled themselves neo-Puritans, the Puritans, according to the *Advertiser*, were proto-Whigs. The Sunday School Union, a typical Whig evangelical organization, which emphasized rote learning, Puritan-style, of religious doctrine, also held its annual meeting in Salem in the summer of 1849. And the Whig governor of Massachusetts, George Briggs, made clear, with calls for public days of fasting in November of the same year, that he felt religion and government served a common purpose: "The Holy Scriptures declare, that if men will 'acknowledge God in all their ways, he will direct their paths'" (*Salem Gazette*). It must have been a trying place to live for a Democrat who was "beyond the church," as his sister-in-law Louisa Hawthorne put it in a letter to Sophia Peabody (qtd. in Fick 155).

What this polemical context suggests is that, in 1850, a depiction of Puritans as less moral than they claimed to be or as being as prone to the very passions they condemned in others would have been an especially appealing polemical gesture for a Democrat to make. Kelley notes that "Most Whigs believed . . . [t]he nation was rightfully to be thought of as a single moral community to be welded together in holy living, to be fashioned in the image of New England's 'citty on a hill'" (*Cultural Pattern* 163). *The Scarlet Letter*—I will argue—by asking how Puritans can pretend to govern others' morality if they themselves are subject to the same natural passions as everyone else, implies a similar question regarding the Whigs who sought to establish moral government in America in the mid-nineteenth century.

The story of *The Scarlet Letter* is framed by a rejection of moral government, "whatever priests or legislators ha[ve] established" (199) [120]. The anachronistic "legislators" (there were none in America in the 1640s) should have alerted readers that, when Hester points Dimmesdale the way to freedom, away from "the clerical band . . . or the church" (199) [120], her gesture has a contemporary resonance. The major themes and concerns of the novel acquire an added significance when considered in this polemical context. The Custom House sketch comes to seem a meditation on the negative effect of Whig economic policies on the people's moral health. Dimmesdale's election-day sermon becomes more pointedly Democratic in tenor and substance, and his death, rather than seeming a gesture of abdication, instead takes on the air of a martyrological indictment of Whig religious institutions that, in the eyes of Democrats, fostered hypocrisy and promoted an unhealthy sense of guilt and self-punishment regarding natural inclinations. Hester's fate, in the polemical context, seems less a model for a vague sense of compromise with slavery than a more affirmative and positive example of non-institutional religion and of a natural piety that owes nothing to the moral guidance of orthodox churchmen.

6. Hawthorne would have had ample contact with Whig culture and with neo-Puritanism growing up in Salem, which, along with Newburyport, was a centre of Federalist and Whig power in the state of Massachussets; see Hartford 1–64.

In reconsidering the novel in light of the culture war between Whigs and Democrats, I will concentrate on the Custom House chapter, the conflict between Hester, Pearl and the Puritan authorities, and Dimmesdale's fate.

The Scarlet Letter opens with an odd meditation on the negative moral effects of people's economic dependence on Uncle Sam. Whig economic policy called for the government to subsidize economic development that, according to Democrats, served the interests of the Whig economic elite. In Whig hands, government had been used to create, as Hawthorne puts it in *The Life of Franklin Pierce*, "commerce where it did not exist" (29). Government assistance for business was also seen by Democrats as an obstacle to the free development of the talents of the industrious classes—the labourers, artisans, farmers and small businessmen who were natural Jacksonian constituencies.

Commentators on the Custom House have focused on Hawthorne's account of his Puritan ancestors and on the description of the romance aesthetic. Yet these topics occupy only a few pages of the text. The rest of the sketch is concerned with the negative moral consequences of dependence on government for one's economic well-being. According to Rush Welter, Democrats were especially concerned with the moral effects of politics on society, and in their eyes, Whigs, by asking for government subsidies for business, were undermining the ideal of self-dependence that was the foundation of moral citizenship. Democrat Gideon Welles, in 1847, articulated this position: "One man looks to government for assistance, his neighbor relying on his own energies asks only for protection. . . . [T]heir fundamental opinions . . . distinguish between the democrat and the anti-democrat" (qtd. in Brock 23).

In his account of life in the Custom House, Hawthorne portrays those who accept government support as having lost contact with the spirit in nature. He compares them to quadrupeds, whose "torpid," "sluggish," and "dependent" behaviour contains "a very trifling admixture of moral and spiritual ingredients" (14) [15, 16]. Hawthorne notes that "the greater part of my officers were Whigs" (13) [14], and he is especially severe in his evaluation of the Inspector: "He had no soul, no heart, no mind." His father "had created an office for him," and like the others, he suffered "moral detriment from this peculiar mode of life" (18) [16, 17]. The negative consequence of having a powerful government make jobs for people by fostering economic development is that it makes people lose contact with their own "native energy," "original nature," and capacity for "self-support." To Democrats such as Bancroft and O'Sullivan, such loss was the same as losing contact with revealed divinity in nature. For this reason, perhaps, Hawthorne compares such a dependent state of being to "a quality of enchantment like that of the Devil's wages" that deprives one of one's "soul" (39) [30]. In a recognizably Democratic gesture, Hawthorne characterizes Whig policies that interfere with nature as immoral.

Hawthorne offers himself as an example of why government support of any kind in economic life is detrimental to the "moral and intellectual health" of people. Having become dependent for support on the government by taking the surveyor's job at the Custom House, Hawthorne suffers the same moral harm as his Whig co-workers. Like the Whigs, he loses contact with divinity revealed in nature: "Nature,—except it were human nature,—the nature that is developed in earth and sky, was, in one sense, hidden from me; and all the imaginative delight, wherewith it had been spiritualized, passed away out of my mind" (26) [21]. In this light, the account of romance aesthetics acquires a political resonance. Hawthorne twice uses the word "spiritualize"—the same word he uses to characterize the alienation from nature brought about by dependence on government—to describe how romance transforms reality. When "spiritualized," life's details "seem to lose their actual substance, and become things of intellect" (39) [27]. That Hawthorne regains his ability to "spiritualize" nature once he leaves the Custom House is proof that freedom from dependence on government is morally salutary. When he puts aside "the strong arm of Uncle Sam" that raises and supports him, he is able to regain his "original nature, the capability of self-support," and this takes the form of his exercising his natural talent as a writer. He regains "his soul . . . its sturdy force, its courage and constancy, its truth, its self-reliance, and all that gives the emphasis to manly character" (39) [30].

What is striking about the Custom House sketch is the identification of theology, politics and aesthetics. Whig economic policies are criticized on Democratic theological grounds. They distance the individual from nature and from natural revelation. The romance aesthetic, by abstracting imaginatively from the facts of sense perception—the basis, according to orthodox neo-Puritan epistemology, of moral truth—allows access to spirit in nature.[7] Exercising one's natural talents in an economic sense, similarly "spiritualizes" the world. Spontaneous natural action, for Democrats, mimes divinity in nature. The Democratic objection to Whig economic policy is that it denies the individual the ability to achieve the moral good through self-dependence by drawing on her or his own natural (and therefore divinely inspired) talents. The novel continues this argument by portraying the Whig policy of merging church and state as having a similarly immoral effect on people whose natural moral powers are extinguished by moral supervision. The "consecration" that is their natural impulse is reviled rather than revered.

The novel opens by evoking what, in Democratic eyes, was the Whig ideal of moral government, a world in which, as Hawthorne puts it, "religion and law were almost identical" (50) [36]. From the outset, Hawthorne, by using terms such as "natural dignity" and "free will," characterizes Hester in terms that suggest Democratic resistance to such government. Hester fulfils the Democratic ideal of a nature at odds with restrictive moral rules and given over to the free expression of self-regulating laws. She grudgingly accepts her punishment, but her extravagant creativity in stitching the scarlet letter is suggestive of imaginative resistance. That Hester exercises her imagination to transform a stigmatic emblem into an object that embodies "spirit" also aligns her character with Hawthorne's theological ideal of revealed divinity in nature accessible through the imagination—"artistically done, and with so much fertility and gorgeous luxuriance of fancy [that it] seemed to express the attitude of her spirit" (53) [39].

7. For an account of "common sense" realism, see Charvat 36–9; Martin 91, 151–65. Hawthorne mocks such "common sense" in "The Snow Image."

Perhaps the most interesting word, for my argument, that Hawthorne associates with Hester is "christianize": "It was the exhilarating effect—upon a prisoner just escaped from the dungeon of his own heart—of breathing the wild, free atmosphere of an unredeemed, unchristianized, lawless region" (201) [121]. Ronald Formisano describes the "evangelical impulse to Christianize America, that ubiquitous energy radiated by New England Protestantism to create the moral, homogeneous, commonwealth. . . . Thus, the Whig party . . . became the evangelicals' best hope to Christianize America" (*Transformation* 61, 104). Hester is often read as embodying a general antinomianism, but a word like "unchristianized" lends her resistance a more specific historical inflection as a fictional expression of the Democratic argument that the Whig ideal of moral government was an offence to civil liberties.

Given how important the Puritan legacy was to Whigs, perhaps the most interesting moment in Pearl's characterization is when she dances on the graves of the Puritans. She is described as embodied spirit, and that for Hawthorne meant that she is aligned with the idea of revealed divinity in nature. Like nature, her spontaneous actions are inherently spiritual. Toward Dimmesdale, for example, she gives "marks of childish preference, accorded spontaneously by a spiritual instinct" (124) [74]. Pearl embodies the idea that natural impulses, however at odds with moral laws, cannot be crimes if divinity resides in nature: "It was as if she had been made afresh, out of new elements, and must perforce be permitted to live her own life, and be a law unto herself, without her eccentricities being reckoned to her for a crime" (135) [85]. The antithesis of the Whig theological concept of innate sinfulness, she represents the idea that passionate natural behaviour should not be subject to restrictive moral laws or branded with harmful moral judgements. In her repeated calls to Dimmesdale to "be true," she articulates the Democratic conviction that, as long as one is true to nature, either in the form of revealed divinity or as one's own "original nature," one cannot but be moral.

It would be a mistake to suggest that the novel is an allegory in which all terms match up with the contemporary debate between Democrats and Whigs. Nevertheless, the fact that the Whigs had a positive conception of Puritan moral government, while Democrats found it alien to their own ideals, suggests that a Democratic writer's portrait of Puritan governors, especially, might evoke Democratic arguments against the Whigs. The governors are, in fact, described in terms that are evocative of Whig ideas and culture. Democrats accused Whigs of aspiring to create an aristocracy in America and of being overly loyal to English cultural models. In the Province House tales that Hawthorne published in the early issues of the *Democratic Review,* he uses the term "rank" to characterize English governors who trample on the rights of Americans. The term reappears in the novel in association with the fictional governors. He describes them, in terms that echo Democratic characterizations of Whigs, as aristocratic "men of rank" who are charged with "the guardianship of the public morals" (162; see Ashworth 34–47). The governor's mansion is characterized in terms that recall English models, and a significantly placed *Chronicles of England* sits on a table in the hallway. Moreover, the governor behaves toward them in a way that embodies the high-church Yankeeism that historians ascribe to Whigs. He asks Pearl to prove that she has submitted to church discipline by reciting the Catechism, and he threatens to deprive an unwed mother of her child.[8]

Roger Chillingworth is, in some respects, the double of the governors. He demonstrates that their apparent distance from passion in fact conceals passions far worse than those they condemn in others. Chillingworth takes pleasure in disciplining others for their pleasures. Hawthorne is quite clear in his evaluation of the difference. Hester and Dimmesdale's passion is counted a "consecration" in natural religious terms, while Chillingworth, in equally religious terms, is condemned for having abused the "sanctity" of the human heart. This way of characterizing Chillingworth has a contemporary resonance with Democratic arguments against Whig moral government. Democrats characterized Whig intrusiveness in others' moral lives as a species of abuse. One Democrat writing in the Boston *Globe* on 22 April 1841, described Whigs as "men who can fathom all the mysteries of the human heart; who have studied all the direct and indirect ways of approaching the citadel of integrity, and all the means of undermining or sapping its integrity" (qtd. in Kohl 32).[9]

In the character of Chillingworth, Whig moral government is represented as an assault on the theological foundations of Democratic political theory. To pursue others' sins, as Whigs would have government do, is to pervert nature and harm natural spirituality. Given the prominent role the world "spiritualize" plays in the articulation of Hawthorne's argument, it is noteworthy that Chillingworth is characterized as having "lost the spiritual view of existence" (119) [75]. As elsewhere in the novel, such a loss of access to spirit in nature is associated with a departure from the naturalist principles of democracy. Hawthorne uses a vocabulary that draws on Chillingworth's name to describe the negative effect of Puritanism on Hester and Dimmesdale's aspirations for freedom from moral supervision. The word "chill" occurs in the forest scene as the antithesis of natural passion, natural freedom, and by implication, natural divinity: "Arthur Dimmesdale put forth his hand, chill as death, and touched the chill hand of Hester Prynne" (190) [115]. The word "chill" suggests the denial of divinely inspired natural impulses of the kind that are evident especially in Pearl and the refusal to recognize their true "worth," a phrasing that recalls Hawthorne's earlier description of spirit in nature as "the true and indestructible value."

"Chill," in the novel's political typology, also stands opposed to terms such as "heart" that Hawthorne associates with the democratic multitude. He uses it on the first pages of the book in regard to divinity: "the one heart and mind of perfect sympathy" (3) [7]. And throughout the novel, it is used to characterize democracy. In the election sermon, Dimmesdale's voice, for example, is described as having "blended into one great voice by the universal impulse which makes likewise one vast heart of the many" (250)

8. Kelley writes,

> Parishioners were drilled in the specifics of doctrine, not let to apply them to the ills of the world. "Instruction in the Catechism," Charles A. Briggs of Union Theological Seminary in New York City later observed, "was almost universal. . . ." The reigning mode of Biblical interpretation was sternly fundamentalist. (*Transatlantic Persuasion* 313)

9. This was a dimension of Whig moral government that even some Whigs criticized: "In the solemn affairs of religion, moreover, instead of looking into our own sins, we are striving to look into the hearts of others." (qtd. In Kohl 32). In contrast, democracy, O'Sullivan suggests, "respects the human soul" ("Introduction" 11).

[148]. In *The Life of Franklin Pierce,* he uses the same metaphor to characterize Pierce's democratic effect on his audiences: "It was the influence of a great heart pervading the general heart, and throbbing with it in the same pulsation" (50). "Chill," in contrast, is used to characterize harm to democracy: "This frankness, this democracy of good feeling, had not been chilled by the society of politicians" (*Life* 18). Chillingworth's abuse of the sanctity of the human heart is thus an offence to the natural divinity that underwrites the Democratic ideal of an egalitarian community. To exercise moral supervision over private morality is to harm democratic equality by presuming moral superiority over others.

Initially a servant and a captive of moral government, Dimmesdale comes eventually to be associated with Democratic values. He moves from an acceptance of restraints on his natural, passionate self to the freely chosen revelation of his spiritual nature. As with Hawthorne himself in the Custom House, this rediscovery of "original nature" occurs as a recovery of a lost natural talent, a theme that evokes the Democratic economic argument against Whig governmentalism—that it suppresses the natural talents of working men in favour of what Hawthorne calls "monopolized labor" (7) [9]. As Hawthorne was able, finally, to write his novel, Dimmesdale is able to write his election-day sermon "with such an impulsive flow of thought and emotion, that he fancied himself inspired" (225) [134]. That it is "inspired" suggests that it draws on the same well of natural revelation that sustains Democratic values. His walk to the church is characterized in similar terms: "[H]is strength seemed not of the body. It might be spiritual, and imparted to him by angelic ministrations" (238) [142].

The election-day sermon is a work of Democratic Party oratory, and the election day itself is, of course, a democratic event, "the day on which the new Governor was to receive his office at the hands of the people" (226) [134]. Hawthorne characterizes the market place as a site of contention between popular passions and moral government in a way that echoes the contemporary cultural battles between Democrats and Whigs that often came down to conflicts between immigrant holiday customs, such as public beer drinking, and Protestant Sabbatarianism. On election day, most entertainments are outlawed, and when an illegal entertainment breaks out, "much to the disappointment of the crowd, . . . the town beadle, who had no idea of permitting the majesty of the law to be violated by such an abuse of one of its consecrated places" interposes (232) [138].

Dimmesdale's sermon both evokes Democratic ideals and produces democratic effects. He speaks democratically "to the great heart of mankind," and he speaks to everyone equally: "Like all other music, it breathed passion and pathos, and emotions high or tender, in a tongue native to the human heart, wherever educated" (243) [144]. This egalitarian ideal, while crucial to Democrats' vision of America's unique destiny among nations, contradicts Whig assumptions regarding the necessary segregation of the respectable from the reprobate. Hawthorne implicitly denies such distinctions when he notes how the sermon provokes a democratic reaction:

> This [shout] . . . was felt to be an irrepressible outburst of the enthusiasm kindled in the auditors by that high strain of eloquence which was yet reverberating in their ears. Each felt the impulse in himself, and, in the same breath, caught it from his neighbor. . . . [E]ven that

> mighty well of many voices, blended into one great voice by the universal impulse which makes likewise one vast heart out of the many. (250) [148]

The ideal of equality is the principle that both Bancroft and O'Sullivan foreground as the source of America's uniqueness. In "The Great Nation of Futurity," O'Sullivan uses the word "destiny" to describe the country's identification with equality, and Hawthorne assigns to Dimmesdale O'Sullivan's word, as well as the idea with which it is associated: "[A]s he drew towards the close, a spirit as of prophecy had come upon him . . . [I]t was his mission to foretell a high and glorious destiny for the newly gathered people of the Lord" (249) [147]. Dimmesdale's sermon is characterized as being almost a direct expression of divinity ("spirit of prophesy"). And it gives rise to a pre-civil sound that is akin to unmediated nature speaking. For Democrats, such expressions of nature always have an element of divine sanction to them. According to Andrew Jackson, in his first Inaugural Address (1828), when "the people" spoke, God spoke (qtd. in McLoughlin 139).

But if Dimmesdale is aligned with Democratic values and especially with the idea of revealed divinity in nature, how should we read his final act of confession? It has been interpreted as an act of abdication and of compromise with the Puritan moral authorities. But given the ethos of the election sermon, it would be inconsistent for Dimmesdale, who has just inspired so strong a sense of democratic egalitarianism in his audience, to embrace the inegalitarian ideology of Whig moral government. It would be more in keeping with Hawthorne's argument up to this point for his confession to represent a rejection of the distinction between the holy and the reprobate.

I have suggested that the contest between theocratic Whigs and anticlerical Democrats hinged on the question of whether an elite of moral guardians had the right to supervise the morality of others. Yankees, one Democrat argued, referring to New England Whigs, felt "they had a presumptive right to impose their politics, their habits, manners, and dogmas on the sister states" (Kelley, *Portrait* 79). They did so because they believed divinity was not present in nature and not accessible, as liberal theology claimed, to human striving. Mark Noll, in *America's God,* notes that liberal theology "amounted to an open invitation for others to postulate an autonomy of action for the faculty of the will. . . . Against this rising tide of rights talk, the Calvinists were trying to stand firm" (284). Because the neo-Calvinists held that divinity was distant from human life, they felt moral government by churchmen was necessary to assist people towards "regeneration."[1] Regarding this neo-Puritan theological position, historian Bruce Kuklick writes, "[T]he New England Theologians assumed a great divide between God and man, and between the realms of nature and grace. Both distinctions were reflected in the persistent exploration of human responsibility for sin and God's sovereignty over grace" (44).

Much of Hawthorne's argumentative labour in the latter part of the novel is occupied with dispelling this notion. His most significant move in this argument is his depiction of Dimmesdale as being able to achieve atonement on his own, without assistance or direction from his church.

1. On regeneration, see Beecher.

God, rather than be a distant being, appears instead, in the forest scene, as a spirit in nature who is available to human striving. And Hester, after rejecting the Whig ideal of moral government and of restrictive moral laws that would seek to imprint morality on her from without, finds in her own nature and in her own natural labours a more substantial morality.[2] The most striking move in this part of Hawthorne's fictional argument is Dimmesdale's revelation of his own "sinfulness." It suggests that, if the holy are no higher than the reprobates, the justification for moral government both in the seventeenth and in the nineteenth centuries loses force. The holy cannot govern the reprobates, Hawthorne argues, because they are all equally natural, passionate and "sinful." Hawthorne has prepared such a reading in the course of the novel by noting of Dimmesdale, for example, that he feels "sympathies so intimate with the sinful brotherhood of mankind; so that his heart vibrated in unison with theirs" (142) [89]. The point is made more polemically in Hester's characterization:

> Sometimes, the red infamy upon her breast would give a sympathetic throb, as she passed near a venerable minister or magistrate, the model of piety and justice, to whom that age of antique reverence looked up, as to a moral man in fellowship with angels. 'What evil thing is at hand?' would Hester say to herself. (87) [57]

Dimmesdale's gesture in the market place suggests the possibility of spontaneous, self-achieved, and unguided atonement outside the jurisdiction of moral government. Hawthorne's reference to "Christian nurture" in the novel links his thinking to that of the revisionist theologian Horace Bushnell, whose book of that title (*Views of Christian Nurture*) appeared in 1847.[3] Bushnell's major work, *God in Christ,* appeared in the spring of 1849, and it seems clear that Hawthorne read it and was aware of the controversy it generated.[4] *The Scarlet Letter* is the only one of his works in which he makes repeated references to typology ("hieroglyph,"

2. Rather than be seen in the end as an emblem of compromise, Hester should be read as resisting moral government by withdrawing from its control, pursuing a life of "Christian-style simplicity," and creating a more authentic religious community of her own. Marty describes contemporary left-wing Christianity in this way: "These idol-smashers found a target in the alliance between evangelicals and other defenders of the political or economic existing order . . . They criticized the ties between clergy and men of wealth or of middle-class aspirations, in the interest of primitive Christian-style simplicity or authentic human community" (113).
3. The following passage seems to prefigure Hawthorne's description of how Pearl inherits Hester's "unquiet elements":

> [I]f we examine the relation of parent and child, we shall not fail to discover something like a law of organic connection, as regards character, subsisting between them. Such a connection as makes it easy to believe, and natural to expect that the faith of the one will be propagated in the other. Perhaps I should rather say, such a connection as induces the conviction that the character of one is actually included in that of the other, as a seed is formed in the capsule . . . And the parental life will be flowing into him all that time. (Bushnell, *Views* 109)

For an account of Bushnell's work, see Barnes.

4. Published in April of 1849, by mid-May, *God in Christ* was already generating discussion in the *Boston Daily Advertiser.* Moreover, stories in the orthodox press suggest that Bushnell had become an object of much lively discussion in the spring and summer of 1849. In the 30 March issue of the *Boston Recorder,* an anonymous reviewer writes that "Dr. Bushnell has gone mad with panic and is for killing all the harmless and useful dogmas which venture to show themselves in the street without their muzzles on." A writer in the *Puritan Recorder* of 8 November 1849 writes that Bushnell "has become a kind of 'chartered libertine'" who

> has acquired the right to do as he likes with impunity. . . . Still, it is a most solemn consideration that *somebody* must be responsible for all the disturbance of the peace of the churches, this distraction of the minds and Christians from the quiet work of saving men, and the agitation and unsettlement of the faith of many in the great concerns of redemption.

"figure," "symbol"), and Bushnell's book is famous for its first chapter, a "Preliminary Dissertation on Language," that contains a lengthy discussion of Christian typology.[5] Moreover, while orthodox theologians believed that a sense of one's own sinfulness was necessary for salvation, Bushnell contended that the sense of having sinned impedes rather than assists redemption. He was critical of the "self-accusing spirit of sin," which he felt inspired unhealthy self-punishment in the form of "vigils . . . tortures . . . to ease the guilt of the mind" (*God in Christ* 212–3). It is a conception that Hawthorne seems to evoke in his portrait of Dimmesdale as someone who is "conscious that the poison of one morbid spot was infecting his heart's entire substance" (140) [88]. He characterizes the scarlet letter as having a similar effect on Hester: "This morbid meddling of conscience with an immaterial matter betokened . . . something doubtful, something that might be deeply wrong, beneath" (84) [55]. In typically Democratic fashion, Hawthorne assigns "wrong" to what moral government does to people ("morbid meddling") rather than to the natural acts that moral government brands as sinful.

Dimmesdale finally sheds his self-punitive sense of sinfulness and begins moving toward a public act of atonement when Hester pledges her love for him in the forest. When Bushnell describes atonement in *God in Christ,* he does so in terms that resonate with the forest scene and with crucial words such as "joy" and "sacred love" in the novel's conclusion. Bushnell argues that a "sinner" cannot attain atonement alone:

> [T]o remove this disability, God needs to be manifested as Love. The Divine Object rejected by sin and practically annihilated as a spiritual conception, needs to be imported into sense. Then when God appears in His beauty, loving and lovely, the good, the glory, the sunlight of soul, the affections, previously dead, wake into life and joyful play. (Bushnell 212)

Hester's avowal of love for Dimmesdale allows him to imagine freeing himself from the power of moral government. And the ensuing burst of sunshine would seem, in Bushnell's understanding of the divine character of sunlight, to add divine sanction to the event—"All at once, as with a sudden smile of heaven, forth burst the sunshine. . . ." (203) [122].

In the final scaffold scene, Dimmesdale brings the natural principle of the forest to the very place that is most associated with moral government's disciplinary practice of shame. He embraces the natural family that his sense of sinfulness had obliged him to abandon. That family has been made to suffer at the hands of a moral government that would put formal rules before natural principles. The distinction is registered in the difference between the natural, if doctrinally "sinful," family of Dimmesdale, Hester and Pearl and the "unnatural," if doctrinally correct, one of Hester and Chillingworth. In the scaffold scene, Dimmesdale affirms the

5. Bushnell advocates a figural conception of language and of the world: "There is a logos in the form of things, by which they are prepared to serve as types or images of what is inmost in our souls" (*God in Christ* 23). Bushnell's influence on Hawthorne would explain the expression of spirit in physical characteristics, such as Chillingworth's deformity or Dimmesdale's letter. Those events seem to exemplify Bushnell's ideas concerning the relation between physical form and spirit: "For the body is living logos, added to the soul, to be its form, and play it forth into social understanding. . . . [S]ubjective truths often find objective representations" (*God in Christ* 23, 250); see Gura 156; Roger.

pre-eminence of the natural to the doctrinal, of the spontaneous principles of natural spirituality to the theological doctrines of the church. Natural principles are also given primacy over formal doctrine in the revelation of Dimmesdale's spontaneously generated "A." The gesture of revealing the letter constitutes a rejection of the emotional economy of sin that would stigmatize natural passion and make it an occasion for shame. By publicly assuming responsibility for his actions, Dimmesdale rejects the imposition by moral government of shame, falseness and self-concealment on people who are made to feel guilt regarding natural impulses. He is finally "true," to use Pearl's term. Having achieved atonement through his own striving and through contact with divinity in nature, he bears proof, appropriately generated spontaneously on his natural body, that the ideology of moral government is flawed. By depicting atonement as attainable through free will and human striving, Hawthorne sides with the liberal position in the religious debates of the period.

Hawthorne proposes, then rejects, a Whig interpretation of Dimmesdale's death: "It was to teach [the people], that the holiest among us has but attained so far above his fellows as to discern more clearly the Mercy which looks down, and repudiate more utterly the phantom of human merit, which would look aspiringly upward" (259) [152]. In few other places in the novel does Hawthorne so clearly describe the hierarchical theological assumption of Whig moral government. God is a distant being, and people cannot atone and achieve salvation on their own. They should instead submit to their governors. Hawthorne immediately questions this interpretation and calls it "stubborn." It is contradicted, he writes, by "proofs, clear as the mid-day sunshine on the scarlet letter, [that] establish [Dimmesdale] a false and sin-stained creature of the dust" (259) [152]. This characterization of Dimmesdale as being no holier than anyone else suggests an equality comparable to that inspired by his sermon and seems more consistent with Hawthorne's argument so far than does the suggestion that Dimmesdale compromises in the end with the Puritan authorities and with Calvinist theological assumptions. Hawthorne's point would seem to be, rather, that even moral governors are creatures of passion, and they, as a result, cannot claim a right to supervise the moral lives of others. Indeed, the point of the final chapter would seem to be that true moral worth exists apart entirely from the institution of the church, whose instrument of moral government—the scarlet letter—eventually becomes an ironic sign of moral virtue rather than of shame.

A similar irony governs the novel's entire argument. The truly virtuous are shamed and punished, while the emblems of Whig intrusiveness in others' moral lives are honoured. The emblem of moral government is compared to the devil, while the innocent child, whose criminal and lawless nature he would dissect, is compared to divinity itself. The irony pointed to by Democratic arguments at the time was that moral government was itself immoral, while the sins such government sought to criminalize were natural passions that, because they were natural, were imbued with divinity. In the context of this Democratic framework, the moral wrong in the novel is thus not that someone committed adultery. Indeed, at the time, Democrats were active, as in Michigan in 1843, in efforts to decriminalize the practice. Rather, the moral wrong is the stigmatizing of natural passion by both seventeenth and nineteenth century

Puritanism that betrays natural divinity and obliges people to be false about feelings that, because they are natural, are spiritual. In this light of this argument, it is noticeable that the narrator's final injunction—"Be true!"—is not a call to avoid such "sins" as adultery. It is, rather, a call not to be false about one's passionate nature. For Democrats of the time, like Hawthorne, the true immorality in the novel would not have been the "sin" of adultery but rather the imposition of moral laws that treat passion as a crime rather than a manifestation of "spiritual nature."

WORKS CITED

Ahlstrom, Sydney, ed. *Theology in America: The Major Protestant Voices from Puritanism to Neo-Orthodoxy.* Indianapolis: Bobbs-Merill, 1967.

Ashworth, John. *"Agrararians" and "Aristocrats": Party Political Ideology in the United States, 1837–1846.* Cambridge: Harvard UP, 1987.

Bancroft, George. *Literary and Historical Miscellanies.* New York, 1855. 408–35.

Barnes, Howard. *Horace Bushnell and the Virtuous Republic.* Metuchen: Scarecrow, 1991.

Beecher, Lyman. *Views in Theology.* New York, 1836.

Benson, Lee. *The Concept of Jacksonian Democracy.* Princeton: Princeton UP, 1961.

Bodo, John. *The Protestant Clergy and Public Issues, 1812–1848.* Princeton: Princeton UP, 1954.

Brock, William R. *Parties and Political Conscience: American Dilemmas 1840–1850.* New York: KTO Press, 1979.

Bushman, Richard. *From Puritan to Yankee: Character and the Social Order in Connecticut, 1690–1765.* Cambridge: Harvard UP, 1967.

Bushnell, Horace. *God in Christ: Three Discourses, delivered at New Haven, Cambridge, and Andover, with a Preliminary Dissertation on Language.* Hartford, 1849.

———. *Views of Christian Nurture.* 1847. *Horace Bushnell: Selected Writings on Language, Religion, and American Culture.* Ed. David L. Smith. Chico: Scholars' Press, 1984.

Charvat, William. *The Origins of American Critical Thought 1810–1835.* Philadelphia: U of Pennsylvania P, 1936.

Cole, Charles. *The Social Ideas of the Northern Evangelists, 1826–1860.* New York: Octagon, 1966.

"Democracy." Attr. John L. O'Sullivan. *United States and Democratic Review* 27 (1840): 215–29.

Donohue, Agnes McNeill. *Hawthorne: Calvin's Ironic Stepchild.* Kent: Kent State UP, 1985.

Donohue, Denis. "Hawthorne and Sin." *Christianity and Literature* 2 (2003): 215–32.

Fick, Leonard. *The Light Beyond: A Study of Hawthorne's Theology.* Westminster: Newman Press, 1955.

Formisano, Ronald. *The Birth of Mass Political Parties, Michigan, 1827–1861.* Princeton: Princeton UP, 1971.

———. *The Transformation of Political Culture.* New York: Oxford UP, 1983.

Gable, Harvey. *Liquid Fire: Transcendental Mysticism in the Romances of Nathaniel Hawthorne.* New York: Lang, 1998.

"The Great Nation of Futurity." Attr. John L. O'Sullivan. *United States and Democratic Review* 23 (1839): 426–30.

Griffin, Clifford. *Their Brothers' Keepers: Moral Stewardship in the United States, 1800–1865.* New Brunswick: Rutgers UP, 1960.

Gura, Philip F. *The Wisdom of Words: Language, Theology, and Literature in the New England Renaissance.* Middletown: Wesleyan UP, 1981.

Hammond, John L. *The Politics of Benevolence: Revival Religion and American Voting Behavior.* Norwood: Ablex, 1979.

Handy, Robert A. *Christian America: Protestant Hopes and Historical Realities.* New York: Oxford UP, 1971.

Hartford, William F. *Money, Morals, and Politics: Massachusetts in the Age of the Boston Associates.* Boston: Northeastern UP, 2001.

Hatch, Nathan O. *The Democratization of American Christianity.* New Haven: Yale UP, 1989.

Hawthorne, Nathaniel. *The Scarlet Letter. Works of Nathaniel Hawthorne.* Centenary ed. vol. 1. Columbus: Ohio State UP, 1962.

———. *Life of Franklin Pierce.* Boston, 1852.

"The History and Moral Relations of Political Economy." Attr. John L. O'Sullivan. *United States and Democratic Review* 34 (1840): 291–311.

Holt, Michael. *The Rise and Fall of the American Whig Party.* New York: Oxford UP, 1999.

Howe, Daniel Walker. *The Political Culture of the American Whigs.* Chicago: U of Chicago P, 1979.

———. "Religion and Politics in the Antebellum North." *Religion and American Politics: From the Colonial Period to the 1980s.* Ed. Mark Noll. New York: Oxford UP, 1990.

"Introduction." Attr. John L. O'Sullivan. *United States and Democratic Review* 1 (1837): 1–15.

Kelley, Robert. *The Cultural Pattern in American Politics.* New York: Knopf, 1979.

———. "A Portrait of Democratic America in the Mid-Nineteenth Century." *Democrats and the American Idea.* Ed. Peter B. Kovler. Washington: Center for National Policy Press, 1991.

———. *The Transatlantic Persuasion: The Liberal-Democratic Mind in the Age of Gladstone.* New York: Knopf, 1969.

Kohl, Lawrence. *The Politics of Individualism: Parties and the American Character in the Jacksonian Era.* New York: Oxford UP, 1989.

Kuklick, Bruce. *Churchmen and Philosophers from Jonathan Edwards to John Dewey.* New Haven: Yale UP, 1985.

"The Legal Wrongs of Women." *United States and Democratic Review* 14 (May 1844): 477–84.

Martin, Terence. *The Instructed Vision: Scottish Common Sense Philosophy and the Origins of American Fiction.* Bloomington: Indiana UP, 1961.

Marty, Martin E. *Righteous Empire: The Protestant Experience in America.* New York: Dial Press, 1970.

McCormick, Richard. *The Party Period and Public Policy.* New York: Oxford UP, 1986.

McLoughlin, William G. *Revivals, Awakenings, and Reform: An Essay on Religion and Social Change in America, 1607–1977.* Chicago: U of Chicago P, 1978.
Noll, Mark. *America's God from Jonathan Edwards to Abraham Lincoln.* New York: Oxford UP, 2002.
Puritan Recorder 8 Nov. 1849.
Roger, Patricia. "Taking a Perspective: Hawthorne's Concept of Language and Nineteenth Century Language Theory." *Nineteenth Century Literature* 4 (1997): 433–54.
Salem Gazette. 19 Nov. 1849: 1.
Simonson, Harold. *Radical Discontinuities: American Romanticism and Christian Consciousness.* Rutherford: Fairleigh Dickinson UP, 1983.
Stavely, Keith. *Puritan Legacies: Paradise Lost and the New England Tradition, 1630–1890.* Ithaca: Cornell UP, 1987.
Stevenson, Louise. *Scholarly Means to Evangelical Ends: The New Haven Scholars and the Transformation of Higher Learning in America 1830–1890.* Baltimore: Johns Hopkins UP, 1986.
Taylor, Nathaniel. *Lectures on the Moral Government of God.* New York, 1859.
Warren, Austin. "*The Scarlet Letter:* A Literary Exercise in Moral Theology." *The Southern Review* 1 (1965): 22–45.
Welch, Claude. *Protestant Thought in the Nineteenth Century.* New Haven: Yale UP, 1972.
Welter, Rush. *The Mind of America 1820–1860.* New York: Columbia UP, 1975.

SACVAN BERCOVITCH

The A-Politics of Ambiguity†

The drama of Hester Prynne's return has gone unappreciated, no doubt because it is absent from the novel. At a certain missing point in the narrative, through an unrecorded process of introspection, Hester abandons the high, sustained self-reliance by which we have come to identify her, from her opening gesture of defiance, when she repels the beadle and walks proudly "into the open air" (162) [38], to the forest scene seven years later, when she casts off her A and urges Dimmesdale to a new life—choosing for no clear reason to abandon her heroic independence and acquiescing to the A after all. Voluntarily she returns to the colony that had tried to make her (she once believed) a "life-long bond-slave," although Hawthorne pointedly records the rumors that Pearl "would most joyfully have entertained [her] . . . mother at her fireside" (313–14, 344) [135, 154]. And voluntarily Hester resumes the letter as a "woman stained with sin, bowed down with shame," although, he adds, "not the sternest magistrate of that iron period would have imposed it" (344) [154]. As in a camera obscura, isolation and schism are inverted into vehicles of moral, political, and historical continuity:

† From *The Office of* The Scarlet Letter (Baltimore: Johns Hopkins UP, 1991), pp. 1–31. Reprinted by permission of the Bercovitch Estate. Page numbers in square brackets refer to this Norton Critical Edition.

> Women, more especially . . . came to Hester's cottage, demanding why they were so wretched, and what the remedy! Hester comforted and counseled them, as best she might. She assured them, too, of her firm belief, that, at some brighter period, when the world should have grown ripe for it, in Heaven's own time, a new truth would be revealed, in order to establish the whole relation between man and woman on a surer ground of mutual happiness. Earlier in life, Hester had vainly imagined that she herself might be the destined prophetess, but had long since recognized the impossibility that any mission of divine and mysterious truth should be confided to a woman stained with sin, bowed down with shame, or even burdened with a life-long sorrow. The angel and apostle of the coming revelation must be a woman, indeed, but lofty, pure, and beautiful; and wise, moreover, not through dusky grief, but the ethereal medium of joy; and showing how sacred love should make us happy, by the truest test of a life successful to such an end! (344–45) [154–55]

The entire novel tends toward this moment of reconciliation, but the basis for reconciliation, the source of Hester's revision, remains entirely unexplained. The issue is not that Hester returns, which Hawthorne does account for, in his way: "There was a more real life for Hester Prynne, here, in New England" (344) [154]. Nor is it that she resumes the A: we might anticipate that return to beginnings, by the principles of narrative closure. What remains problematic, what Hawthorne compels us to explain for ourselves (as well as on Hester's behalf), is her dramatic change of purpose and belief. Throughout her "seven years of outlaw and ignominy," Hester had considered her A a "scorching stigma" and herself "the people's victim" (291, 331, 313–14) [121, 146, 135]. Only some "galling" combination of fatalism and love, Hawthorne tells us early in the novel, had kept her from leaving the colony at once, after her condemnation (188) [53]. She had been "free to return" to England; she had also had

> the passes of the dark, inscrutable forest open to her, where the wildness of her nature might assimilate itself with a people whose customs and life were alien from the law that had condemned her. . . . But [Hester was possessed by] . . . a fatality, a feeling so irresistible and inevitable that it ha[d] the force of doom. . . . Her sin, her ignominy, were the roots which she had struck into the soil. It was as if a new birth, with stronger assimilations than the first, had converted the forest-land . . . into Hester Prynne's wild and dreary, but life-long home. . . . The chain that bound her here was of iron links, and galling to her inmost soul. . . . What she compelled herself to believe,—what, finally, she reasoned upon, as her motive for continuing a resident of New England,—was half a truth, and half a self-delusion. Here, she said to herself, had been the scene of her guilt, and here should be the scene of her daily punishment; and so, perchance, the torture of her daily shame would at length purge her soul, and work out another purity than that which she had lost; more saint-like, because the result of martyrdom. (186–87) [53–54]

Something of that force of necessity attends Hester's return, together with that earlier self-denying, self-aggrandizing quest for martyrdom. But it now conveys a far less "wild and dreary" prospect. Hester chooses to make herself not only an object of the law, "saint-like" by her resignation to "daily punishment," but more largely an agent of the law, the sainted guide toward "another purity," "some brighter period" of "sacred love" foreshadowed by her agon (344–45) [155]. What had been half-truth, half-delusion is rendered whole as a vision of progress through due process. And the bond she thus forges anew with the community lends another moral interpretation to her "new birth" as American. It recasts her adopted "forest-land" into the site of prophecy, home-to-be of the "angel and apostle of the coming revelation"; it reconstitutes Hester herself, *as a marginal dissenter,* into an exemplum of historical continuity (344–45) [155].

We accept all this as inevitable, as readers did from the start, because Hawthorne has prepared us for it. His strategies of ambiguity and irony *require* Hester's conversion to the letter. And since the magistrates themselves do not impose the A; since the community has long since come to regard Hester as an "angel or apostle" in her own right; since, moreover, we never learn the process of her conversion to the A (while her development through the novel tends in exactly the opposite direction); since, in short, neither author nor characters help us—we must meet the requirement ourselves.

"The scarlet letter had not done its office," and, when it has, its office depends on our interpretation—or, more precisely, on our capacities to respond to Hawthorne's directives for interpretation. The burden this imposes can be specified by contrast with Dimmesdale's metamorphosis, earlier in the story, from secret rebel into prophet of New Israel. Hawthorne details the state of despair in which the minister agrees to leave, elaborates the disordered fantasies that follow, and yet leaves it to us to explain Dimmesdale's recantation. In this case, however, the explanation emerges directly from character and plot. "The minister," Hawthorne writes, "had never gone through an experience calculated to lead him beyond the scope of generally received laws; although, in a single instance, he had so fearfully transgressed one of the most sacred of them. But this had been a sin of passion, not of principle, nor even purpose" (290) [120]. When, accordingly, Dimmesdale decides to leave with Hester, he does so only because he believes he is "irrevocably doomed" (291) [121], and we infer upon his return that he has regained his faith after all—that he has made peace at last with the Puritan ambiguities of mercy and justice, good and evil, head and heart, which he had abandoned momentarily in the forest.

The reasons for Hester's reversal are far more complex. It takes the whole story to work them through. To begin with, there is the problem of form, since in her case (unlike Dimmesdale's) the reversal so conspicuously defies tradition. I refer to the genre of tragic love to which *The Scarlet Letter* belongs. Had Hester returned for love alone (the A for Arthur) or under the cloud of disaster abroad (the A for adversity), we could follow her reasoning readily enough. But Hawthorne asks us to consider the disparity between these familiar tragic endings and Hester's choice. The familiar endings, from *Antigone* and *Medea* through *Antony and Cleopatra* and *Tristan and Isolde,* are variations on the theme of love against the world. Hester's return merges love *and* the world. In this aspect (as in others) it offers a dramatic contrast with European novels of adultery, which narrative theorists have classified in terms either of subversion or of containment, as implying "a fatal break in the rigid system of bourgeois realism," or as "work[ing] to subvert what [the novel] aims to celebrate," or else (because of the "nearly universal *failure* of the adulterous affair") as serving "closurally to reinstate social norms."[1] *Madame Bovary* and *Anna Karenina* can be said to fit any of these descriptions. *The Scarlet Letter* fits none. Hester neither reaffirms her adulterous affair nor disavows it; her actions neither undermine the social order nor celebrate it; and at the end she neither reinstates the old norms nor breaks with them. Instead, she projects her dream of love onto some "surer ground" in the future, when "the whole relation between man and woman" can be reestablished. In other words, her return deliberately breaks with tradition by its emphasis on the political implications of process as closure.

The political emphasis is appropriate for the same reason that it is problematic: Hawthorne's portrait of Hester is a study of the lover as social rebel. Not as antinomian or witch, as he explicitly tells us, and certainly not as adulteress—if anything Hester errs at the opposite extreme, by her utter repression of eros. This emphasis on the non- or even antierotic is also to highlight sexual transgression, of course, but wholly by contrast; and it is to reinforce the contrast that Hawthorne insinuates by his often-remarked parallels between Hester and "unnatural" Anne Hutchinson, mother of "monstrous misconceptions",[2] as well as imperious, "bitter-tempered" Mistress Hibbins (217) [74]. Hawthorne remarks, with a note of disgust, that Hester had lost her "womanly" qualities, had become almost manlike in her harshness of manner and feature:

> Even the attractiveness of her person had undergone a . . . sad transformation . . . [so that] there seemed to be no longer anything in Hester's face for Love to dwell upon; nothing in Hester's face for Love to dwell upon; nothing in Hester's form, though majestic and statue-like, that Passion could ever dream of clasping in its embrace; nothing in Hester's bosom, to make it ever again the pillow of Affection. Some attribute had departed from her, the permanence of which had been essential to keep her a woman. (258–59) [100]

Hester errs, then, not in her sexual transgression but in her "stern development" as an individualist of increasingly revolutionary commitment (259) [100]. At the novel's center is a subtle and devastating critique of radicalism that might be titled "The 'Martyrdom' of Hester Prynne." It leads from her bitter sense of herself as victim to her self-conscious manipulation of the townspeople, and it reveals an ego nourished by antagonism; self-protected from guilt by a refusal to look inward; using penance as a refuge from penitence; feeding on shame, self-pity, and hatred; and motivated by the conviction that society is the enemy of the self.

Let me recall the scene I began with, in the chapter midway through the novel. Seven years have passed and the townspeople have come to regard Hester with affection, admiration, even reverence. On her part, Hester

1. Joseph Allen Boone, *Tradition Counter Tradition: Love and the Form of Fiction* (Chicago: U of Chicago P, 1987), 48; Tony Tanner, *Adultery in the Novel: Contract and Transgression* (Baltimore: Johns Hopkins UP, 1979), 13.
2. Amy S. Lang, *Prophetic Woman: Anne Hutchinson and the Problem of Dissent in the Literature of New England* (Berkeley: U of California P, 1987), 58, 67.

has masked her pride as humility, has repeatedly reminded them, by gesture and look, of her "saint-*like*" suffering, and in general has played upon their guilt and generosity until "society was inclined to show its *former* victim a more benign countenance than she cared to be favored with, or, perchance, than she deserved" (187, 257; my emphasis) [54, 100]. And like other hypocrites in Hawthorne's work, Hester pays a heavy price for success. "All the light and graceful foliage of her character had been withered up," he tells us, "leaving a bare and harsh outline, which might have been repulsive, had she possessed friends or companions to be repelled by it" (258) [100]. She has none because she wants none. The "links that united her to the rest of human kind had all been broken," save for "the iron link of mutual crime" (255) [98]. She considers Pearl, whom she loves, an instrument of "retribution" (273) [109]. Concerning those to whom she ministers—not only "her enemies" but also those for whom "the scarlet letter had the effect of a cross on a nun's bosom"—Hawthorne points out that Hester "forebore to pray for [them], lest, in spite of her forgiving aspirations, the words of the blessing should stubbornly twist themselves into a curse" (258, 191) [100, 56].

It is worth stressing the severity of Hawthorne's critique. After seven years Hester has become an avenging angel, a figure of penance unrepentant, a so-called Sister of Mercy who not only scorns those who call her so but who has developed contempt for all "human institutions," "whatever priests or legislators had established" (257, 290) [99, 120]. Despairing, therefore, of any improvement short of tearing down "the whole system of society" and doubtful even of that "remedy," she turns her energies first against "the world's law" and then against her daughter and herself (260, 259) [102, 101]. Her heart, Hawthorne tells us,

> had lost its regular and healthy throb, [and she] wandered without a clew in the dark labyrinth of mind; now turned aside by an insurmountable precipice; now starting back from a deep chasm. There was wild and ghastly scenery all around her, and a home and comfort nowhere. At times, a fearful doubt strove to possess her soul, whether it were not better to send Pearl at once to heaven, and go herself to such futurity as Eternal Justice should provide. (261) [102]

Here is the allegorical landscape of misguided rebellion: a wild, self-vaunting independence leading by a ghastly logic of its own to the brink of murder and suicide. No wonder Hawthorne remarks at this point that "the scarlet letter had not done its office."

I do not mean by this to deny the obvious. Hester is a romantic heroine. She is endowed with all the attributes this term implies of natural dignity, generosity of instinct, and what Hawthorne calls "a woman's strength" (257) [99]. Although she persistently abuses or represses these qualities, nonetheless they remain potential in her—dormant but felt in her every thought and action—and Hawthorne clearly means them to move us all the more forcefully for the contrast. As he remarks after detailing her "sad transformation," "She who has once been a woman, and ceased to be so, might at any moment become a woman again, if there were only the magic touch to effect the transfiguration. We shall see whether Hester Prynne were ever afterwards so touched, and so transfigured" (259) [101]. While we wait to see, Hester persistently invites our pity and praise, and by and large she succeeds, as she did with the Puritans. But to take her point of view is to prevent the scarlet letter from doing its office. It leads us, as it did Hester, into conflict—compels us to choose between the reasons of the heart and the claims of institutions—and conflict is precisely what the letter is designed to eliminate.

Again, a distinction is called for. Conflict is also a form of process, of course, but one that assumes inherent antagonism; it derives from a partiality that inspires partisanship. Conflict forces us to take positions and thus issues in active oppositions: one certainty against another, one generation against the next, one class or gender against another. Process (for Hawthorne) is a form of partiality that accepts limitation, acknowledges its own incompleteness, and so tends toward tolerance, accommodation, pluralism, acquiescence, inaction.

The contrary tendency toward conflict is the dark side of Hawthorne's chiaroscuro portrait of Hester. Her black eyes and hair—always a danger signal in Hawthorne's (culture's) symbolic system—are complemented, so to speak, by his relentless critical commentary on her every misstep into independence. We feel it the moment she crosses the prison threshold to his gently mocking "*as if* by her own *free-will*" (162; my emphasis) [38]. We see it detailed in her radical speculations, when her mind wanders

> without rule or guidance, in a moral wilderness as vast, as intricate and shadowy, as the untamed forest. . . . Her intellect and heart had their home, as it were, in desert places, where she roamed as *freely* as the wild Indian in his woods. . . . Shame, Despair, Solitude! These had been her teachers,—stern and wild ones,—and they had made her strong, but taught her much amiss. (290; my emphasis) [120]

This running gloss on the ways that the letter has not done its office reaches its nadir in her forest meeting with Dimmesdale. Amidst the fallen autumn leaves, Hester discards the A in a gesture of defiance for which (Hawthorne reminds us) her entire seven years had been the preparation. "The past is gone!" she exclaims. "With this symbol, I undo it all, and make it as it had never been!" (292) [122]. And the narrator adds, with characteristic irony (characteristic, among other things, in that the irony borders on moralism):

> O exquisite relief! She had not known the weight, until she felt the *freedom*! . . . All at once, *as with* a sudden smile of heaven, burst forth the sunshine.
>
> Such was the sympathy of Nature—that wild, heathen Nature of the forest never subjugated by human law, nor illumined by higher truth—with the bliss of these two spirits! (292–93; my emphasis) [122]

The narrator's ironies are not Hawthorne's precisely, and the difference, as we shall see, allows for a significant leeway in interpretation. But even within this larger perspective the contrast in forms of process is unmistakable. The radicalization of Hester Prynne builds on the politics of either/or. Hawthorne's symbolic method requires the politics of both/and. To that end, in the forest scene, Pearl keeps Hester from disavowing the office of the A, as earlier she had kept her from becoming another antinomian Anne or Witch Hibbins. Indeed, it is worth digressing for a moment to point out how closely for these purposes Pearl is bound to the A—with what painstaking care this *almost* anarchic figure is molded into a force for integration. Hawthorne presents in Pearl a profound challenge to the boundaries of socialization,[3] but he also details her restraining role with a consistency that verges on the didactic. He sustains this technique through virtually all her dialogues, with their conspicuously emblematic messages. And he reinforces it with his every definition of Pearl: as "imp" of the "perverse" *and* "pearl of great price," as "demon offspring," "Red Rose," "elf-child," and "mother's child" (Hester's "blessing" and "retribution" all in one); as the image simultaneously of "untamed nature" and the "angel of judgment," and, at the climactic election-day ritual, as (successively) "sin-born child," "witch-baby," the quintessential outsider who engages with and so weaves together all sections of the diverse holiday crowd—"even as a bird of bright plumage illuminates a whole tree of dusky foliage"—and, finally, as the fully "human" daughter who breaks once and for all the "spell" of mutual isolation (208, 211, 210, 215, 202–203, 205, 329, 330, 336, 339) [68, 64, 61, 64, 70, 73, 74, 73, 149, 145, 145, 151]. Throughout this *development*—in effect, our developing sense of Pearl as "the scarlet letter endowed with life"—Pearl serves increasingly to underscore what is wrong with Hester's radicalism, what remains "womanly" about Hester despite her manlike "freedom of speculation," and what sort of politics Hester must adopt if she is to help effect the changes that history calls for (204, 210, 259) [65, 100, 101].

No other character in the novel, not even the shadowy Roger Chillingworth, is more carefully orchestrated into the narrative design or more single-mindedly rendered a means of orchestration. Midway through the story, at the midnight scaffold, Hawthorne pointedly presents us with a *figura* of things to come: "There stood the minister, with his hand over his heart; and Hester Prynne, with the embroidered letter glimmering on her bosom; and little Pearl, herself a symbol, and the connecting link between those two" (251) [95]. At the last scaffold scene Pearl kisses the minister, now openly her father at last, and Hawthorne remarks: "Towards her mother, too, Pearl's errand as a messenger of anguish was all fulfilled" (339) [151]. And with that office accomplished—by the one character, it

3. * * * It may be noted here, as intrinsic to Hawthorne's mode of ambiguity, that Pearl, who forces Hester to restore the A, is among other things an incarnation of Emersonian whim (not to say Poesque perversity)—a figure of "infinite variety," "mutability," and "caprice," with "wild, desperate, defiant" proclivities; the very spirit of negation, toward her inwardly rebellious mother no less than toward the apparent consensus of Puritan Boston (269–74) [108–10]—and that all these traits, including the most "freakish" ("fiend-like," "demon off-spring"), give symbolic substance to the "imperious gesture" with which Pearl asserts her "authority" in the forest scene (298–99) [125]:

> Pearl still pointed with her forefinger; and a frown gathered on her brow; the more impressive from the childish . . . aspect of the features that conveyed it. . . .
>
> "Hasten Pearl; or I shall be angry with thee!" cried Hester Prynne. . . .
>
> But Pearl, not a whit startled at her mother's threats, any more than mollified by her entreaties, now suddenly burst into a fit of passion, gesticulating violently, and throwing her small figure into the most extravagant contortions. . . . Seen in the brook, once more, was the shadowy wrath of Pearl's image, crowned and girdled with flowers, but stamping its foot, wildly gesticulating, and, in the midst of it all, still pointing its small forefinger at Hester's bosom! . . .
>
> "Come thou and take it up!" (298–300) [125–26]

Pearl's reflection in the brook is a memorable representation of the reciprocities of process and telos. It stands for nature (and the natural) as an office of repression. Equally and *simultaneously*, it stands for the demands of social conformity, indifferent to threat and entreaty, and conveyed through an impassioned willfulness. It is the letter of the law conceived in the spirit of resistant individuality, and vice versa.

will bear repeating, who might be imagined to offer an alternative vision in the novel—Hester can choose in due time to become the agent of her own domestication.

There is a certain irony here, to be sure, but it functions to support Hester's choice by reminding us of the burden of free will, when freedom is properly willed, for, although the burden is a tragic one, it alone carries the prospect of progressive (because incremental, nonconflictual) change. * * * Let me say * * * that [this irony] pertains above all to historical process and that it is perhaps especially prominent in Hawthorne's tales of the Puritans. The obvious contrast to Hester's return in this respect is the fate of Young Goodman Brown. Unlike Hester, Brown insists on alternatives when he rejoins the settlement—innocence or guilt, the truths of the town or those of the forest—and so finds himself in a hermeneutical impasse, a paralysis of thought and action whose issue is unambiguous "gloom." The no less obvious parallel (in view of Hester's propensities for the unillumined "sympathy of Nature") is the lovers' choice that ends "The May-Pole of Merry Mount." Strictly speaking, it is John Endicott, "the Puritan of Puritans," who forces the former "Lord and Lady of the May" into history. But in fact, Hawthorne emphasizes, they had started on that harsh, necessary road to progress long before, of their own free will: "From the moment they truly loved, they had subjected themselves to earth's doom of care, and sorrow, and troubled joy, and had no more a home at Merry Mount."[4]

Much the same might be said of Hester and Dimmesdale, although they must learn the lesson for themselves, separately, and offstage as it were—Dimmesdale, in the privacy of his study (following a "maze" of Goodman Brown–like temptations); Hester across the ocean, in the "merry old England" that the Puritans had rejected together with the Maypole (303, 211) [128, 70]. Like Dimmesdale, she comes back home as a mixed figure of "pathos" and promise (333) [147]—"angel of judgment" and mercy, "messenger of anguish" and hope. Hawthorne writes of the fully humanized Pearl, the former "wild child" who has at last "developed all her sympathies," that she would no longer need to "do battle with the world, but [could] be a woman in it" (339) [151]. It might be said of Hester upon her return that she can leave Pearl behind because she has taught herself to play Pearl to her own former Hester. She no longer needs restrictions because, after her long battle with the world, she has learned how to restrict herself—how to obviate the conflict between self and society, between the certainty of love and certain prospects of social change, between prophetic hope and politics as usual. As a woman in the world, she has learned to deflect, defuse, or at least defer that inherently explosive conflict and at best to transmute it, freely, into a faith that identifies continuity with progress.

This political level of meaning is closely connected to the moral. What I just called Hawthorne's politics of both/and is directly based upon his concept of truth. Critics often remark on the moral he draws from Dimmesdale's experience: "Be true! Be true! Be true!" (341) [153]. But, as

4. Hawthorne, "The May-Pole of Merry Mount," in *Tales and Sketches*, ed. Roy Harvey Pearce (New York: Library of America, 1982), 370, 367, 363.

usual with Hawthorne, it is hinged to the narrative by ambiguities. He tells us that he has culled the moral ("among many" others) from "a manuscript of old date," which he has "*chiefly* followed" (341; my italics) [153]. And he prefaces the moral with a dazzling variety of reports about the scarlet letter (or the absence of it) on the minister's breast. For Hawthorne, partiality is to process what multiplicity is to truth—a series of limited perspectives whose effectiveness depends on their being partial without becoming exclusive and partisan in such a way as gradually, by complementarity rather than conflict, to represent the whole. His political meaning here points us toward the premises of liberal society. His moral meaning is grounded in the premises of Puritan thought. The connection between the two is that between the Hobbist and the Calvinist meanings of the Fall. *The Scarlet Letter* is a story of concealment and revelation, where the point of revelation is not to know the truth but to embrace many truths and where concealment is not a crime, but a sin.

Not crime, but sin: Hawthorne adopts this fine theological distinction for his own liberal purposes. A crime pertains to externals, and, as a rule, it involves others, as in the case of murder or adultery. A sin pertains to the spiritual and internal, to an act of will. It may or may not involve crime, just as a crime (murder, for example, or adultery) *may* not involve sin. It depends on the inner cause, the motive. The issue, that is, is guilt, not shame: not the deceiving of others, but the skewing of one's own point of view. The political office of the A is to make partisanship an agent of reciprocity. Its moral office is to lead from the willful self-binding of a truth—paradigmatically, a truth of one's own—to the redemptive vision of many possible truths.

In the next-to-last chapter, that office is rendered (as the chapter title tells us) through "The Revelation of the Scarlet Letter" (332) [146]. The action centers on the scaffold, as it does twice previously. The first time is at Hester's midday "public exposure" (172) [44], where the A denotes various kinds of division (within the community, within Dimmesdale, and, most dramatically, between Hester and the community). The second scaffold scene comes midway through the novel, in the midnight meeting that draws the main characters together, and by implication, the townspeople as well, for the A that flashes across the night sky lights up the entire town "with the distinctness of mid-day . . . [lending] another moral interpretation to the things of this world than they had ever borne before . . . as if [this] . . . were the light that is to reveal all secrets, and the day-break that shall unite all who belong to one another" (251) [95].

In short, the novel tends increasingly toward reconciliation through a series of ambiguous unveilings, each of which might be titled "The Revelation of the Scarlet Letter." In that penultimate chapter Dimmesdale reconciles himself with his guilt, with Pearl, with Hester, with Chillingworth, and, in "words of flame," with the destiny of New Israel (332) [147]. Now it only remains for Hester to join the telos in process. When she does so, in the conclusion, her moral interpretation of things past and future may be seen to reverse her first misstep across the prison threshold. Indeed, the scene deliberately echoes that initiation into concealment so as emphatically to invert it. When Hester returns, she pauses "on the threshold" of her old home—as many years before she had paused "on the threshold of the prison door"—long enough to display to the onlookers a scarlet letter on her breast (343, 162) [154, 38]. It is a nice instance of liminality serving its proper conservative function at last. Then, at the start of her trials, Hester had repelled the beadle, representative of "the Puritanic code of law" (162) [38], in order to assert "her own free-will." Now she returns as representative of the need for law and the limits of free will. Having abandoned the hope of erasing the past, Hester internalizes the past in all its shame and sorrow. Franz Kafka's penal colony requires a fatal mechanism of authority in order to make the prisoner accept his guilt; Hester preempts the mechanism by authorizing her own punishment and inscribing her guilt upon herself. In a gesture that both declares her independence *and* honors her superiors, she reforms herself, voluntarily, as the vehicle of social order.

This moral design parallels the political process I outlined, but with an important difference. Hester's radicalism sets her apart and sustains her marginality to the end. The sin she commits (her double act of concealment, first of her lover, then of her husband) links her to everyone else. She is unique as a rebel but typical as a liar. Indeed, telling lies is the donnée of the novel as for the Puritans the prison is the donnée of their venture in utopia. It establishes the terms of human possibility in an adulterated world. Directly or indirectly—as deception, concealment, or hypocrisy, through silence (in Hester's case), cunning (in Chillingworth's), eloquence (in Dimmesdale's), or perversity (in Pearl's)—lies constitute the very texture of community in *The Scarlet Letter.* But the texture itself is not *simply* evil. All of Hawthorne's main characters are good people trapped by circumstance, all are helping others in spite of themselves, and all are doing harm for what might justifiably be considered the best of reasons: Hester for love, Dimmesdale for duty, and the Puritan magistrates for moral order. Even Chillingworth, that least ambiguous of villains, is essentially a good man who has been wronged, who lies in order to find the truth, who prods his victim to confess (partly, perhaps, through love), and who, in leaving his wealth to Pearl (gratuitously), provides the basis for whatever there is of a happy ending to the story.

Hawthorne owes this *complex* view of evil—good and evil entwined, the visible "power of blackness" symbiotically augmented by the pervasive if sometimes oblique power of light—to Puritan theology. As the New England primer put it, Adam's fall did much more than fell us all. It also brought the promise of grace through Christ, the Second Adam. Justice *and* mercy, law *and* love: from these twin perspectives, the Puritans built the scaffold and imposed the A. Restrictions were necessary because the Fall had sundered the affections from the intellect; it had set the truths of the heart at odds with the truths of the mind. Now only faith could reconcile the two kinds of truth. They who bound themselves to a single view, *either* justice *or* mercy, were entering into a Devil's pact. They were committing themselves to a lie by concealing a part of reality from themselves, including the reality of the self in all its ambiguity, both human and divine—hence, the degeneration of Chillingworth, "demon of the intellect" (321), and Dimmesdale, until he manages to harmonize the minister's gospel of love with the lover's self-punishment. Hence, too, Pearl's fragmented identity: she is a shifting collage of retribution and love, seeking integration; and, hence, the reciprocal movement of Hester and the community, from opposition to mutuality. As she acts the Sister of Mercy toward those who merely judged her, and so judged too harshly, Hester increasingly touches the people's "great and warm heart" (226) [80]. At the end, after she has passed judgment on herself, Hester gains a fuller, more generous vision of reality than she dreamed possible in the forest. Then it was love with a consecration of its own. Now her love has the consecration of justice, morality, and community.

I rehearse this familiar pattern in order to point out that nothing in it is random or arbitrary. Not a single aspect of this apparent multiplicity (reversals, revisions, and diverse points of view) permits free choice. Hawthorne's celebrated evasiveness comes with a stern imperative. Penitence, he would urge us, has more substance than the absolutism of either/or. Drab though it seems, the morality of both/and heightens personal vision by grounding it in the facts of experience. It takes more courage to compromise. It is a greater act of self-assertion to recognize our limits—to "be true" to what we most deeply are while admitting the fragmentary quality of our truth—to keep faith in our boldest convictions while acknowledging the incompleteness of those convictions, and so to discipline ourselves, of our "own free-will," to the pluralist forms of progress.

It amounts to a code of liberal heroics. Hawthorne's focus is first and last upon the individual; his emphasis on perspective assumes faith in ambiguity; and his ambiguities compel resolution through the higher laws of both/and. Through those higher laws we learn how to sustain certain ideals *and* deny the immediate claims of their certainty upon us; how to possess the self by being self-possessed (which is to say, to hold the self intact by holding it in check); and, from both of these perspectives, how voluntarily to embrace gradualism and consensus in the expectation that, gradually, "when the world should have grown ripe for it," consensus will yield proximate justice for the community and, for the individual, the prospect of unadulterated love.

The prospect leads from the moral to the aesthetic level of the novel, Again, Hawthorne himself provides the link—in this case through the parallel he assumes between moral bivalence and symbolic ambiguity. Consider the title he gives to that chapter midway through the novel. "Another View of Hester" means an inside view of her secret radicalism; it also means a public view of Hester through her acts of charity, which in turn involves a distinction between the view of the many, who consider her "angelic," and the view of the few, "the wise and learned men" who were reluctant to give up earlier "prejudices" (257) [100]. "Another view" means a true sight of Hester, as she really is (rather than as she appears), *and* it means a glimpse of Hester in medias res, in the process of development. Above all, it means another view in the sense of differences of interpretation: interpretation in the form of rumor and legend (the A that magically protects Hester "amid all peril"); interpretation as a mode of sacralization (the A as a nun's cross); interpretation as agent of social change; and interpretation as vehicle of manipulation (Hester "never raised her head to receive their greeting. If they were resolute to accost her, she laid her finger on the scarlet letter, and passed on. This might be pride, but was so like humility that it produced all the softening effects of the latter quality on the public mind"); and, of course, interpretation as the avenue to multiple meanings—the A as sign of infamy, pillow for the sick, shield against Indian arrows, "glittering" and "fantastic" work of art (257–58, 255) [99, 100].

All this and more. No critical term is more firmly associated with *The Scarlet Letter* than ambiguity. What has not been adequately remarked, and questioned, is the persistent, almost pedantic pointedness of Hawthorne's technique. F. O. Matthiessen defined Hawthorne's ambiguity as "the device of multiple-choice"[5]—and so it is, if we recognize it as a device for enclosure and control. That strategy can be traced on every page of the novel, from start to finish, in Hawthorne's innumerable directives for interpretation: from the wild rose he presents to his readers in chapter I—in a virtuoso performance of multiple choice that is meant to preclude choice (for it instructs us *not* to choose between the local flower, the figural passion flower, and the legacy of the ambiguously "sainted Anne Hutchinson")—to the heraldic device with which the novel ends: the "engraved escutcheon" whose endlessly interpretable design (one "ever-glowing point of light gloomier than the shadow" but a source of relief nonetheless) "*might* serve for a motto *or* brief description of our now concluded tale" (345; my emphasis) [155]. Concluded *then*, but, by authorial direction, it is *now* in process, a prod to our continuing speculations. The "curious investigator may still discern [it]," Hawthorne remarks, "and perplex himself with the purport" (345) [155], and the interplay between our perplexity and its purport, like that between process and telos in the description of the rose ("It *may* serve, let us hope, to symbolize *some* sweet moral blossom, that *may* be found along the track, or relieve the darkening close tale of human frailty and suffering"), tells us that meaning, while indefinite, is neither random nor arbitrary; rather, it is gradual, cumulative, and increasingly comprehensive (159; my emphasis) [36].

The Scarlet Letter is an interpreter's guide into perplexity. As critics have long pointed out, virtually every scene in the novel is symbolic, virtually every symbol demands interpretation, and virtually every interpretation takes the form of a question that opens out into a variety of possible answers, none of them entirely wrong, and none in itself satisfactory. But the result (to repeat) is neither random nor arbitrary. It is a strategy of pluralism—issuing, on the reader's part, in a mystifying sense of multiplicity—through which each set of questions and answers is turned toward the same solution: all meanings are partly true, hence, interpreters must choose as many parts as possible of the truth and/or as many truths as they can possibly find in the symbol.

Let me illustrate my point through the single most straightforward instance of choice in the novel. Describing Hester's "sad transformation" (midway through the story), Hawthorne remarks that her "rich and luxuriant hair had either been cut off, or was . . . completely hidden by the cap" (259) [100]. For once, it seems, we have a plain truth to discover. Something has been hidden, a question about it has been raised, and we await the moment of disclosure; that moment reveals, of course, in "a flood of sunshine" (292) [122], that Hester had *not* cut off her hair. But of course, too, Hawthorne means for us to recognize that in some sense she *had*—had cut off her "essential womanhood, had cut herself off from community, and

5. F. O. Matthiessen, *American Renaissance: Art and Expression in the Age of Emerson and Whitman* (New York: Oxford UP, 1941), 276.

had cut away her natural luxuriance of character by willfully hiding it beneath an Odysseus' cloak of conformity. These are metaphors, not facts. But in Hawthorne's ambiguous world a main function of choice is to blur the commonsense lines between metaphor and fact, and nowhere is that blurring process better demonstrated than at the moment of revelation, during her forest meeting with Dimmesdale, when Hester discards the A:

> By another impulse, she took off the formal cap that confined her hair; and down it fell upon her shoulders, dark and rich, with at once a shadow and a light in its abundance, and imparting the charm of softness to her features. There played around her mouth, and beamed out of her eyes, a radiant and tender smile, that *seemed* gushing from the very heart of womanhood. . . . Her sex, her youth, and the whole richness of her beauty, came back from *what men call* the irrevocable past. (292–93; my emphasis) [122]

Shadow and light, seemed and was, irrevocable and renewed, womanhood cut off/hidden/lost/restored: *The Scarlet Letter* is a novel of endless points of view that together conspire to deprive us of choice. We are enticed by questions so that we can be allowed to see the polarity between seeking *the* answer, any answer, and undertaking an interpretation. The option is never one thing or another; it is all or nothing. We are offered an alternative, not between different meanings, but between meaning or meaninglessness, and it is meaning in that processual, pluralistic, and therefore (we are asked to believe) progressivist sense that Hester opts for when she returns to New England.

In that option lies the moral-aesthetic significance of Hawthorne's representation of crime as sin. Crime involves social transgression, as in the tradition of the detective story, which centers on the discovery of the criminal. Or, more equivocally, it might involve a conflict of rights that must be decided one way or another, as in the tradition of the novel of adultery, which opposes the claims of the heart to those of civic order. Hawthorne makes use of both kinds of plot, only to absorb them—climactically, through Hester's return—into a story about the trials and triumphs of ambiguity. Through the office of the scarlet letter, all particulars of the criminal act, together with the conflicts they entail, dissolve into a widening series of reciprocities. We come to see that the issue is not a breach of commandment, but (as Hawthorne signals by the conspicuous absence from the novel of the word "adultery") an incremental process of interpretation by which we discern the purport of the broken law for ourselves, and we do so by turning speculation against the tendency either to take sides or to view conflicting sides as irreconcilable.

To represent crime as sin is first of all to universalize the legal problem. It forces us to read a particular transgression in terms of innate human defects and the recurrent conflict of good and evil. But more comprehensively it makes the universal itself a curious object of interpretation—not in order to demystify it, not even to analyze it (in any cognitive sense), but, on the contrary, to invest it with richer significance and a more compelling universality. The ambiguities of *The Scarlet Letter* lead us systematically forward, from the political to the moral or religious to the aesthetic levels, toward what we are meant to understand is a series of broader and ampler meanings. *Always* ampler, and therefore at any given point indefinite: a spiral of ambiguities whose tendency to expand in scope and depth is all the more decisive for the fact that the process occurs in unexpected ways. The result is a liberal hierarchy of meaning, a series of unfoldings from simple to complex, "superficial" to "profound," which is as schematic, comprehensive, and coercive as the medieval fourfold system. Hawthorne's representation of crime as sin requires us to remain vague about all issues of good versus evil (except the evils of partiality and partisanship) in order to teach us that the Puritans' final, relatively nonconflictual view of Hester is deeper than the single-minded judgment reflected in the governor's iron breastplate, just as her final, relatively nonconflictual position toward their bigotry opens the way for both personal and historical development.

I have been using the term "option" in connection with Hester's return in order to stress the overriding distinction in Hawthorne's "device of multiple-choice" between making choices and having choice. His point is not that Hester finally makes a choice against adultery. It is that she has no choice but to resume the A. To make choices involves alternatives; it requires us to reject or exclude on the ground that certain meanings are wrong or incompatible or mutually contradictory. To have choice (in Hawthorne's fiction) is to keep open the prospects for interpretation on the grounds that reality never means either one thing or another but, rather, is Meaning fragmented by plural points of view, for, although the fragmentation is a source of many a "tale of frailty and sorrow," such as *The Scarlet Letter*, it is also, as *The Scarlet Letter* demonstrates, the source of an enriched sense of unity, provided we attend to the principles of liberal exegesis. And by these principles, to opt for meaning in all its multifariousness—to have your adulterous love and do the work of society too—is to obviate not only the conflicts embodied in opposing views but also the contradictions implicit in the very act of personal interpretation between the fact of multiple meaning and the imperative of self-assertion.

In other words, to interpret is willfully, in the interests of some larger truth, *not to choose*. Ambiguity is a function of prescriptiveness. To entertain plural possibilities is to eliminate possible divisions. We are forced to find meaning in the letter, but we cannot choose one meaning out of many: Chillingworth's fate cautions us against that self-destructive act of exclusion. Nor can we choose to interpret any of the novel's uncertainties as contradictions: the antagonism between Hester and the townspeople (or between Chillingworth and Dimmesdale, or between the minister and his conscience) cautions us repeatedly against that abuse of free will. What remains, then, is the alternative that symbols are lies, multiple choice is a mask for absence of meaning, and the letter is an arbitrary sign of transient social structures. And Hester's incipient nihilism cautions us at every turn against that flight from responsibility: in the first scene, by her instinctive attempt to conceal the letter; then, three years later, by concealing its meaning from Pearl (to Hawthorne's suggestion that "some new evil had crept into [her heart], or some old one had never been expelled"); later, in the forest scene, by flinging the letter "into infinite space" and drawing an (infinitely illusory) "hour's free breath"; and, finally, at the election day ritual, by gloating secretly at the prospect of its annihilation, a prospect that

Hawthorne opens to her imagination so that, by absorbing it into what the entire novel makes us think *must* be some larger, truer interpretation, we can effectually exclude it as an alternative from our own (162, 274, 300) [38, 110, 126].

If we refuse to exclude it—if we are tempted like Hester in the forest to reject meaning, if we make Chillingworth's choice at the scaffold against mercy or Dimmesdale's in his "secret closet" for contradiction (242) [90]—then interpretation has not done its office. And lest, like these characters, we find ourselves wandering in a maze, Hawthorne points us toward the true path, midway in our journey through the novel. In "Another View of Hester" he impresses upon us: the need for personal interpretation; the inevitably partial nature of such interpretation; the richly varied experiential bases of interpretation; the tendency of these partial and shifting interpretations to polarize into symbolic oppositions, such as rumor and event, metaphor and fact, natural and supernatural, good and evil, head and heart, concealment and revelation, fusion and fragmentation; the need to recognize that these polarities, because symbolic, are never an inherent source of conflict, but instead they are always entwined in symbiotic antagonism and therefore mutually sustaining; and, as the key to it all; the *clavis symbolistica*, the need for faith both in the value of experience (shifting, private, and partial though it is) and in some ultimate hermeneutical complementarity, as in an ideal prospect that impels us toward an ever-larger truth.

That faith involves a *certain* activity on the reader's part. We need to make sense of the entire process for ourselves so that the process can in turn make sense of our partial contributions. The text elicits personal response in order to allow each of us to contribute to the expanding continuum of liberal reciprocity. It is a hermeneutics designed to make subjectivity the primary agency of change while keeping the subject under control, and it accomplishes this double function by representing interpretation as multiplicity flowing naturally into consensus. For, as oppositions interchange and fuse in the text, they yield a synthesis that is itself a symbol in process, an office not yet done. It is a richer symbol now than it was before, a higher office, but still veiled in the winding *perhapses, ors,* and *mights* that simultaneously open new vistas of meaning and dictate the terms of closure.

It may be helpful to distinguish this strategy from others to which it has been compared. Hawthorne does not deconstruct the A; he does not anticipate the principle of indeterminacy; and he offers neither an aesthetics of relativism nor a dialectics of conflict. We might say that in some sense he is doing all of these things, but only in the sense that to do so is to dissipate the integral force of each. His purpose is to rechannel indeterminacy into pluralism, conflict into correspondence, and relativism into consensus. Insofar as terms such as "instability" and "self-reflexiveness" apply to *The Scarlet Letter,* they are agencies of a certain kind of interpretation, much as private enterprise and self-interest were said to be agencies of the general good in antebellum America. Frank Kermode's claim for Hawthorne's modernity—"his texts . . . are meant as invitations to co-production on the part of the reader"—is accurate in a sense quite different from that which he intended. Kermode speaks of "a virtually infinite set of questions."

The Scarlet Letter holds out that mystifying prospect, much as Jacksonian liberals held out the prospect of infinite possibility, in order to implicate us as coproducers of meaning in a single, coherent moral-political-aesthetic design.[6]

This contrast pertains even more pointedly to Mikhail Bakhtin's concept of the dialogic imagination, which it has recently become fashionable to apply to American novels, and *The Scarlet Letter* in particular. Dialogics is the process by which a singular authorial vision unfolds as a "polyphony" of distinct voices. It entails a sustained open-ended tension between fundamentally conflicting outlooks. They are said to be conflicting insofar as they are *not* partial reflections (such as good or evil) of a more complex truth but each of them, rather, the expression of a separate and distinct way of understanding, a substantially different conception or configuration of good and evil. And they are said to be open-ended because the tension this involves is sustained *not* through the incremental layers of meaning but through the dynamics of diversity itself, which is, by definition, subversive of any culturally prescribed set of designs, including those of group pluralism. Bakhtin's dialogics denies telos through a "modernist" recognition of difference. Hawthorne's ambiguities imply telos through the evasion of conflict. They are modernist in the sense of modern middle-class culture—which is to say, in their *use* of difference (including marginality, complexity, and displacement) for purposes of social cohesion. Recent theorists such as Paul Ricoeur and Hans Blumenberg tell us that the novel (the genre par excellence of the dialogic) "legitimates the aesthetic qualities of *novitas*, . . . removes the dubiousness from what is new, and so *terra incognita*, or the *munda novus*, becomes possible."[7] Hawthorne seeks precisely to rein in what becomes possible. Aesthetically, it is the letter's office, as *novitas*, to enclose "the new world," whether as alternative order or as Bakhtinian carnival, within culture, *as* culture.

We might term this strategy the "monologics of liberal ambiguity." It serves to mystify hierarchy as multiplicity and diversity as harmony in process. Dialogics unsettles the link between process and closure. Hawthorne details the manifold discrepancies between process and closure in order to make discrepancy itself—incompleteness, concealment, the distance between penance and penitence—a vehicle of acculturation. To that end he guides his readers (as he does his errant heroine) to a *certain* belief in the unity of the symbol. He shows us that, precisely by insisting on difference, we can fuse an apparently (but not really) fragmented reality. Augustine's answer to Manichaean dualism was to redefine evil as the absence of good. Hawthorne's answer to the threat of multiplicity is to redefine conflict as the absence of ambiguity, and ambiguity, therefore, as the absence of conflict.

Ambiguity is the absence of conflict: Hawthorne's logic is as simple in theory as it is complicated in application. Historical facts tend toward fragmentation, but ambiguity brings this tendency under control, gives

6. Frank Kermode, *The Classic: Literary Images of Permanence and Change* (New York: Knopf, 1975), 43.

7. Mikhail Bakhtin, *Problems in Dostoevsky's Poetics*, ed. and trans. Caryl Emerson (Minneapolis: U of Minnesota P, 1984), passim; Hans Blumenberg, "The Concept of Reality and the Possibility of the Novel," in *New Perspectives in German Literary Criticism*, ed. Richard Amacher and Victor Lange (Princeton: Princeton UP, 1979), 32.

it purpose and direction, by ordering the facts into general polarities. Fragmentation itself thus becomes a function of consensus. For once the fragments have been ordered into polarities, the polarities can be multiplied ad infinitum, since each polarity entails or engenders other parallel, contrasting, or subsidiary sets of polarities. The process is one of endless variation upon a theme. And vice versa: it is a process of variation endlessly restricted to a single theme, because (in Hawthorne's fiction) all polarity is by definition ambiguous, all ambiguity is symbolic, and all symbols tend toward reconciliation—hence, the distinctly narrowing effect of Hawthorne's technique, in spite of his persistent allusions and deliberate elusiveness. He himself wrote of *The Scarlet Letter* to his publisher, James T. Fields, on November 3, 1850, that since the novel was "all in one tone" it could have gone on "interminably."[8] We might reverse this to say that what makes the novel hermeneutically interminable also makes it formally and thematically hermetic. In that sustained counterpoint between endlessness and monotone lies the dynamic behind Hawthorne's model of pluralist containment. Process for him is a means of converting the *threat* of multiplicity (fragmentation, irreconcilability, discontinuity) into the pleasures of multiple choice, where the implied answer, "all of the above," guarantees consensus.

The process of conversion follows the symbolic logic of the scarlet letter. It is the office of the A to demonstrate that naturally, organically, pluralism tends to absorb differences into polar opposites, and that bipolarity, properly interpreted, tends of its own accord toward integration. So conceived, the monologics of ambiguity in *The Scarlet Letter* extend to structures of gender, religion, history, psychology, aesthetics, morality, and epistemology. One instance, a minor one but suggestive of Hawthorne's range, is the imaginary "Papist" at the first scaffold scene (166) [40], who sees Hester as the Virgin Mother and who seems to offer an option—an oppositional view, in Raymond Williams's sense, or, more accurately (in light of Hawthorne's emphasis on the relative newness in 1642 of the Reformation), a residual view—that goes deeper than personal and partial differences of perspective.[9] But here, as elsewhere, Hawthorne's point is to intrigue us with notions of conflict in order to disperse them. He can be said to have invented the Papist (that Puritan symbol of irreconcilable antagonism) on our behalf as an early step in our education in ambiguity, and the education proceeds through our recognition, in due time, that the putative contrast is really just one pole in the reciprocity between justice and love. Thus, Catholic and Protestant outlooks merge, midway through the novel, in the townspeople who interpret the A sympathetically as "a cross on a nun's bosom" and, more powerfully, at the final scaffold scene, in the apparent pietà, where Hester (in an image that prepares us for her final role as prophet) plays Sacred Mother to Dimmesdale's Christ (258, 339) [100, 151].

It makes for a rich fusion of polarities, with multiple implications for Hawthorne's symbolic method. For example, the papist perspective (if I may call it so) clearly parallels the compassionate view of Hester expressed by the young mother at the prison door, and clearly, Hawthorne presents

8. Hawthorne, *Letters, 1843–1853*, ed. Thomas Woodson, L. Neal Smith, and Norman Holmes Pearson, in *Works*, Centenary Edition, 23 vols. (Columbus: Ohio State UP, 1985), 16:371.
9. Raymond Williams, *Marxism and Literature* (Oxford: Oxford UP, 1977), 121–27.

her view mainly for purposes of contrast, as he does the Papist's, to highlight the harshness of Puritan judgment, whether from magistrates or from "matrons" (161) [37]. In each case the contrast turns out to be a form of symbolic doubling. The "young wife, holding a child by the hand," in some sense mirrors Hester; the embittered matrons in some sense preview Hester's later "injustice"—her impenitent, sometimes brutal judgments (variously motivated by "scorn," "hatred," and "asperity") of her perceived "enemies," her husband, and even her daughter (162, 269, 274) [38, 107, 110]. The result is a spiral of symbolic reciprocities, reinforced by principles of psychology (head-heart) and morality (good-evil), which grow increasingly comprehensive in their image of womanhood—increasingly comprehensive and, proportionately, increasingly positive. They find their high point in Hester the prophet: the rehumanized (because refeminized) heroine whose fall, though it warrants "strict" and "severe" censure (274) [110], augments the promise she represents of future good things. The revelation is still to come, but Hester at last has reached the proper *womanly* vantage point for perceiving something of its import; she has earned the privilege of paying homage, if not directly to Hawthorne's "little Dove, Sophia," as several critics have argued, then to the dream of "sacred love," which Sophia shared with Nathaniel and which largely derived from the mid-nineteenth-century cult of domesticity (344) [155].[1]

A similar strategy of incorporation applies to the parallel between the Papists and the other non-Puritan culture represented at the first scaffold scene. I refer to the local Indians, who judge from Hester's "brilliantly embroidered badge" that she "must needs be a personage of high dignity among her people" (330) [146]. Hawthorne invests the story's Indians with much the same processual-symbolic effect as he does the Catholic. He juxtaposes the outsider's perspective, in both cases, to that of the Puritans in order to absorb historical difference into what we are meant to think of as broader, universal categories. To that end he deploys the keywords of savagism: "stone-headed" implements, "snake-like" features, "savage finery," "painted barbarians," and, most frequently, "wild"—"wild Indian," "wild men . . . of the land," the "wildness" of their "native garb" (318, 329, 315, 330, 169) [146, 138, 145, 42].

It makes for an all-too-familiar Romantic-Jacksonian configuration: the primitive as an early stage of social growth, which the civilized state not only supersedes but (in the process) ingests, so that its best society combines the "natural" state with the "higher" advantages of culture. Hence, the Indian aspect of Pearl, the "wild child" (329) [145], and, above all, the Indian wildness of Hester's radicalization:

> Her intellect and heart had their home, as it were, in desert places, where she roamed as freely as the wild Indian in his woods . . . criticizing all with hardly more reverence than the Indian would feel for the clerical band, the judicial robe, the pillory, the gallows, the fireside, or the church. (290) [120]

An entire culture is represented in these cunningly compressed polarities. Hawthorne appreciates the natural freedom of the "red men" (287)

1. T. Walter Herbert, Jr., "Nathaniel Hawthorne, Una Hawthorne, and *The Scarlet Letter*: Interactive Selfhoods and the Cultural Construction of Gender," *PMLA* 103 (1988): 285–97.

[119], just as he deplores the civilized excesses of the Puritan pillory—and vice versa; he recognizes the dangers of "desert places" just as he acknowledges the need for fireside and church. It is an ambiguity that effectually deprives the Indians of both nature and civilization, a high literary variation on an imperial rhetoric that ranges from Francis Parkman's elegies for a "noble," "primitive," "dying" race to what Herman Melville satirized as "the metaphysics of Indian-hating."[2] Here it serves to empty the "savages" of their own history so as to universalize them as metaphors for Hester's development.

As all of these examples suggest, the basic symbolic opposition in *The Scarlet Letter* is that between self and society. I said earlier that Hawthorne portrays Hester as an individualist of increasingly radical commitment. I might as well have said a radical of increasingly individualist commitment, for Hawthorne's aim is to counter the dangerously diverse social possibilities to which she has access, in fantasy or fact—Indian society, witch covens, Elizabethan hierarchy, Leveler and Ranter utopia (289, 313–30) [119, 134–46]—to bring all such unruly alternatives under control, rhetorically and hence morally and politically, by implicating them all under the symbol of the unrestrained self.

No symbol was better calculated to rechannel dissent into the gradualism of process. And no symbol was more deeply rooted in the culture. As *The Scarlet Letter* reminds us, it served as a major Puritan strategy of socialization, through a process of inversion that typifies all such strategies. Society in this polar opposition became the symbol of unity, and the unsocialized self was designated the symbol of chaos unleashed—"sin, in all its branches," as the Reverend John Wilson details them in the first scaffold scene for "the poor culprit's" sake (176) [48]. Or, mutatis mutandis, the unsocialized self was a morass of "monstrous misconceptions," as John Winthrop labeled Anne Hutchinson, and society stood not just for legal order (as against antinomianism), but for Order at large—"the laws of nature and the laws of grace" (to quote Winthrop again) through which "we are bound together as one man."[3]

In either case, the polarity of self and society remained central through the successive discourses of libertarianism, federalism, republicanism, and Jacksonian individualism. Its negative pressures, implicit in Hawthorne's reference to "the sainted Anne Hutchinson" (and explicit in his essay on "Mrs. Hutchinson")—as well as in recurrent charges of antinomianism against those who were said to have "sprung up under her footsteps" [36], from Edwards through Emerson—are memorably conveyed in Alexis de Tocqueville's contrast between "traditional" and "modern" modes of control: "The ruler no longer says: 'You must think as I do or die.' He says: 'You are free to think differently, and to retain your life, your property, and all that you possess; but from this day on you are a stranger among us.'" Its positive form can be inferred from Edwin Chapin's Massachusetts election day

2. Francis Parkman, *The Jesuits in North America in the Seventeenth Century* (1867), in *France and England in North America,* ed. David Levin (New York: Library of America, 1983), 1:343, 461, 466; Herman Melville, *The Confidence-Man: His Masquerade,* in *Pierre, Israel Potter, The Confidence-Man, Tales, and Billy Budd,* ed. Harrison Hayford (New York: Library of America, 1984), 994.

3. John Winthrop, *The History of New England,* ed. James Savage (Boston: Phelps and Farnham, 1825), 1:166; and "A Model of Christian Charity" (1630), in *Winthrop Papers,* ed. Stewart Mitchell (Boston: Massachusetts Historical Society, 1931), 2:124.

sermon of 1844, *The Relation of the Individual to the Republic*. The self, Chapin argues, denotes "matters of *principle*" and society entails "matters of *compromise*," but in the American way of "self-government" (as nowhere else) it is "compromise not *of* principle but *for* principle."[4]

The Scarlet Letter is the story of a stranger who rejoins the community by compromising for principle, and her resolution has farreaching implications about the symbolic structures of the American ideology. First, the only plausible modes of American dissent are those that center on self: as stranger or prophet, rebel or revolutionary, lawbreaker or Truth seeker, or any other adversarial or oppositional form of individualism. Second, whatever good we imagine must emerge—and, properly understood, *has* emerged and is continuing to emerge—from things as they are, insofar as these are conducive to independence, progress, and other norms of group pluralism. And third, radicalism has a place in society, after all, as the example of Hester demonstrates—radicalism, that is, in the American grain, defined through the ambiguities of both/and, consecrated by the tropes of theology ("heaven's time," "justice and mercy," "divine providence"), and interpreted through the polar unities at the heart of American liberalism: fusion and fragmentation, diversity as consensus, process through closure.

MILLICENT BELL

The Obliquity of Signs: *The Scarlet Letter*†

It is not wrong to identify in this famous short novel the subjects that lie so clearly upon its surface—the effect of concealed and admitted sin, or the opposed conditions of isolation and community, or the antithetic viewpoints of romantic individualism and puritan moral pessimism or the dictates of nature and law. But—and perhaps it is the current self-consciousness of literature that makes this so—it may now be possible to find in this work a primary preoccupation with the rendering of reality into a system of signs. Hawthorne may have had similar reasons to our own for questioning—while performing—the interpretation of experience as a species of message. It is a general human impulse to seek coherence—a syntax—in life, but it is the artist above all who does so most heroically, who is the champion of our general endeavor. When that endeavor becomes dubious, art itself becomes questionable. Like ourselves, Hawthorne may have come to feel that the universe at large speaks an incomprehensible babble in which it merely amuses us to suppose we hear communicating voices, explanation—even consolation.

The very title of the book is a sign, the smallest of literary units, the character "standing for" no more than a speech sound. The letter "A" is the first letter, moreover, of the alphabet, which Pearl recognizes as having seen in her horn book, and represents the beginning, therefore, of literacy. Reading will be given the broadest meaning in this novel. It will

4. Alexis de Tocqueville, *Democracy in America*, ed. J. T. Mayer, trans. George Lawrence (Garden City, N.Y.: Doubleday, 1969), 72; Edwin Chapin, *The Relation of the Individual to the Republic* (Boston: Dutton and Wentworth, 1844), 27, 31.

† From *The Massachusetts Review* 23.1 (Spring 1982): 9–26. Reprinted by permission of the publisher. Page numbers in square brackets refer to this Norton Critical Edition.

become a trope for the decipherment of the world as a text. *The Scarlet Letter*, then, is, as much as any work of fiction can be, an essay in semiology. Its theme is the obliquity or indeterminacy of signs. From this source comes an energy present in every part of the book; from it derives the peculiar life of those other themes which might otherwise seem lacking in modern interest.

That the status of signs is especially important to Hawthorne is evident in a peculiar stylistic feature of *The Scarlet Letter.* Though the reader has the impression of a constant encouragement to symbolic interpretation, it turns out, upon examination, that Hawthorne's prose contains only occasional metaphor or simile and no true allegorical cohesion. What in fact happens is something else: we are frequently *asked to consider* things as symbolic; objects, persons and events are *called* signs rather than being silently presented as such. Hawthorne undertakes a narrative putatively historical, to begin with, introducing it in the Custom House Preface as a redaction from a documentary record, to reinforce the sense of a reconstructed literal past. But again and again he deliberately declares that the actualities of his tale are or may be taken as signs, and he uses repeatedly such words as "type," "emblem," "token," or "hieroglyph." All these words are used in a sense roughly synonymous.

"Type" is almost invariably employed to mean "that by which something is symbolized or figured; anything having a symbolic signification; a symbol, emblem" (*OED*), a sense which had been current already in English during the Renaissance and can be found in one of Hawthorne's favorite older writers, Spenser. The word was still used in this way in the mid-nineteenth century when the meaning more common with us, of a general form or of a kind or class, arose, and Hawthorne, who is conservative in language, almost always seems to be employing the older rather than the newer of these two senses. He even occasionally hints the special theological usage which identifies in the Old Testament events in the New of which they are "types"—or rather he employs a reversed adaptation of this which labels something in his story a "type" of a Bible element—as when Hester Prynne is called "a scarlet woman and a worthy type of her of Babylon" [70]. But it should be observed that this particular description comes not from the narrator but from one of the tale's seventeenth century Puritans who might be expected to typologize in this way, just as it is the Puritan authority that has affixed upon Hester the signifying letter which is invariably described as being not red but scarlet. She is called a "type" in a non-scriptural sense by the Hawthorne-narrator. At such times she can be associated with traditional figures of moral personification when he comments, "It may seem marvellous that this woman should call that place her home, where, and where only, she must needs be the type of shame" [53]—which still implicitly refers to the viewpoint of the Boston community or, again, when Chillingworth is said to have come home to behold "the woman, in whom he hoped to find embodied the warmth and cheerfulness of home, set up as a type of sin before the people" [75]. Other occurrences of the term, however, are closer to the simpler meaning of a symbol. Such is the early designation of the infant Pearl as "a forcible type, in its little frame, of the moral agony which Hester Prynne had borne throughout the day" [48].

"Token," i.e., "something that serves to indicate a fact, event, object, feeling, etc.; a sign, a symbol" (*OED*) also serves to indicate a sign, with the added implication that the sign is an evidence, even a consequence of the signified. Dimmesdale's distaste for Chillingworth's appearance is "a token, implicitly to be relied on, of a deeper antipathy in the breast of the latter" [88]. "Emblem" is another name for a symbolic signifier, more exclusively visual, deriving from the seventeenth century taste for expressing abstractions by means of objects or pictured objects, but since used as another synonym for symbol as well as for an armorial device or even for a badge that might be worn on clothing. Hester's "A" is all these—a badge she wears, a device for the escutcheon on her tombstone—"On a field sable, the letter A, gules" [155] and the "emblem of her guilt" [65].

Finally, there is "hieroglyph," which more than any of the terms just glanced at suggests the art of writing at the same time that it suggests the pictorial figure, in the reference to the picture-writing of the Egyptians. By extension, too, a hieroglyph is "a figure, device, or sign, having some hidden meaning; a secret or enigmatical symbol" (*OED*), and so more than any of the others expresses Hawthorne's feelings about the signifiers he has marked out in his tale. Such a mystery is the child Pearl, as we shall shortly consider. As Hester and Dimmesdale watch her in the forest, it is observed, "She had been offered to the world, these seven years past, as the living hieroglyphic, in which was revealed the secret they so darkly sought to hide,—all written in this symbol,—all plainly manifest,—had there been a prophet or magician skilled to read the character of flame" [124]. Pearl, the animate letter or character, is truly the hieroglyphic figure which hides an elusive meaning.

Like a rhetorician, Hawthorne has, in the examples I have given, labelled his subjects as though they were figures of speech in a spoken or written text. But, of course, these types, emblems, tokens, and hieroglyphs are not really supposed to be products of the human imagination. They belong to the category of privileged signs deriving from a transcendent presence. They are "written" by a spiritual force which expresses itself in the secret language of appearances. To read such texts one must be gifted with a prophet's or a magician's power to see beyond actuality. As Chillingworth says of the "riddle" ("a question or statement intentionally worded in a dark or puzzling manner" [*OED*]) of the identity of Hester's lover, "the Daniel who shall expound it is yet a-wanting" [44]. The surface of life which he beholds is thus compared to the most famous of dark texts, the writing on the wall at Belshazzar's feast.

For us, there may no longer be a center, as Jacques Derrida would call it, to assure to such—or any—appearances the status of signs. With our loss of confidence in the sacred grounding of signs we have lost confidence in their objectivity, and see them only as games of the mind. Hawthorne may have been at the threshold of our condition, though he was still formally committed to older views. The Puritan ontology as well as the Puritan morality haunted the American mind in Hawthorne's day, and haunted his in particular. We think more usually of the moral imperatives of Puritanism as a lingering presence in Hawthorne's writing—and where more than in this tale of transgression and penance? But it is the Puritan understanding of the relation of natural to divine reality that was more important to him. The Puritans regarded reality textually; a long tradition of Christian thought which spoke through them analogized the world as a book which might be compared to scripture as an act of divine writing. What God had written in the creation was a cryptic language, yet one could be confident nonetheless that no phenomenon but had its sacred sense. Such a viewpoint was older than Christianity, having its roots in platonism. It was, too, enjoying a new life in the secularized religion of romantic transcendentalism of which Hawthorne was aware at close hand.

Hawthorne knew perfectly well his difference and distance from the Puritans though "strong traits of their nature," he said, had "intertwined themselves" [12] with his. He was skeptical as well about the convictions of his Concord neighbors, Emerson and Thoreau. His temperamental nominalism, which is so visible in the determined abstention from all interpretation practiced in his Notebooks with their tireless recording of trivial realities, made him a man for whom the world is exactly what it is and no more. Yet, as for so many mid-nineteenth century minds the loss of the visionary sense, the draining of significance from the mundane, was felt with a certain anguish, at best a wry humor, and the viewpoint of science seemed to him pitiably meager and even morally dangerous. In *The Scarlet Letter* he gives play to all of his mingled feelings—his tenderness for the poetry of a lost faith in essences, his ironic detachment and disbelief, and his fear of such disbelief in himself or others.

The agency of these complex feelings is, in the novel, a persona about whom too little has been said. His divided attitudes are made clear in the Custom House Preface—making the Preface a necessary part of the fictional whole, giving a character to the narrating voice. This narrator appears to us in the Preface as a man undecided in his view of reality between the Puritan-transcendental conviction that the invisible speaks ceaselessly behind the visible and the materialism that finds the explanation of things merely in accident and physical laws. He admits his deviation from the beliefs of his "grave, bearded, sable-cloaked and steeple-crowned" [11] progenitors, yet declares a legitimate descent from them. He values his experiences at the Custom House as an antidote to transcendental associations and inclinations. Even the old inspector, a personality of unillumined materiality, was, he tells us, "desirable as a change of diet, to a man who had known Alcott" [21]. Yet his final and most moving words are a tribute to the art that discerns the spirit essence in the quotidian.

In this well-known section of the Preface he discusses his aesthetic problems while striving, in the Custom House, to overcome his creative torpor. But it should be noted that his problem is as much ontological as aesthetic: it involves his unsuccessful struggle to attain the transcendental sense. In the nighttime vigil in the parlor of his Salem house, the moonlight "making every object so minutely visible, yet so unlike a morning or noontide visibility," the homely details of the room were completely seen, he recalls, "yet spiritualized by the unusual light" [27]. The room became a neutral territory where "the Actual and the Imaginary may meet" [28]. It was the sort of meeting he would have liked to bring about in his writing yet could not, though the "wiser effort would have been, to diffuse thought and imagination through the opaque substance of today . . . to seek, resolutely, the true and indestructible value that lay hidden in the petty and wearisome incidents, and ordinary characters" [28–29]. His failure was no matter merely of skill or of artistic imagination, as it has seemed to most readers. The requisite imagination that he lacked was the prophet or magician's—or if the poet's, then the romantic poet's seer-like power to discern higher truth. Unable to find essence in his surroundings he could only retreat to the unsubstantiality of the past or the fanciful, in which one might play with the idea of significance in the mode of romance.

Nevertheless, nothing is more serious than *The Scarlet Letter*, despite the charge of Henry James that its faults are "a want of reality and an abuse of the fanciful element—of a certain superficial symbolism" which "grazes triviality." James did not see that Hawthorne's method in the book was to express his own profoundest problem. In a way that is seldom understood and seems sometimes merely coy, he offers and withdraws, denies and provides the sense of the spirituality of life—and so suggests the opacity or unreliability of its signs. Many a reader has been irritated by the narrator's reluctance to decide what, if anything, Chillingworth saw on Dimmesdale's bosom, or what, if anything, was seen in the sky during the night-scaffold scene in Chapter XII or what, if anything, was seen on Dimmesdale's bosom, again, by the assembled multitude in the final scaffold scene. These are only the most memorable instances of Hawthorne's reluctance to settle a simple question of appearances. More important, however, is his refusal to help us to assign final significance to these phenomena, even if granted. Repeatedly, he seems only willing to say, as at the conclusion of the final scene after summarizing the conflicting reports of witnesses, "The reader may choose among these theories" [152].

Nowhere is this insistent ambiguity more conspicuous than in the central scaffold scene—which James, it may be noted, particularly disliked. Here are duplicated the conditions of the moonlit chamber of the Preface; the scene is bathed in a supernal light which makes each detail both completely visible and radiant with meaning. In the light cast from the sky during the minister's night-vigil, he sees for the first time that Roger Chillingworth is no friend, he pierces the veil. Yet this is also the occasion for the narrator's most skeptical discussion of the delusiveness of signs. He comments upon the "messages" read into nature by man and the egotism of the assumption that they are addressed to our particular selves. "We impute it, therefore, solely to the disease in his own eye and heart that the minister, looking upward toward the zenith, beheld there the appearance of an immense letter,—the letter A" [96], he seems to conclude. But, immediately after, we hear that the sexton reported the next day that "a great red letter in the sky,—the letter A" [98], was seen by others also, and by them taken to stand for Angel, to signify the governor's passing. So, what are we to make of the reading of signs? The sexton, who has found Dimmesdale's glove on the scaffold, says that Satan must have dropped it there, intending—falsely—to impute that Dimmesdale belongs where evil-doers are set up to public shame. Signs may be only the mischief-making of Satan, then, and no true tokens? Except, of course, that this token *is* well placed!

Hester's letter is the central example of the almost infinite potentialities of semantic variety. A material object, a piece of embroidered cloth held in the finder's hand, it is the one irreducible reality which connects the intangible historic past with the narrator's present sensation; it authenticates, is an evidence of its vanished substantiality. As an abstract sign on Hester's bosom, it purports to speak both for the nature of her past and for the present condition of the wearer. It is a letter of the alphabet, but also, presumably, an initial, a sign of a sign, since it represents a word, the next larger linguistic unit after the letter. But "adultery" is never "spelled out." The word, like the act it designates, is invisible in the text—the act held inaccessibly out of the reader's sight while the word only hovers in his mind. The merely implied word becomes somehow less explicit, and when we are told that the letter is a "talisman" (a magic object generally engraved with figures or characters) of the Fiend, we suspect a more generalized significance. It is said to throb in sympathy with all sin of whatever kind beheld by Hester. It seems to represent an absolute and undenotable evil.

The letter may indicate the presence in Hester of Original Sin, and refer to a common corruption which requires no outward demonstration, which does not manifest itself in true signs, which even the most virtuous in deed must share. The old Calvinist mystery is really the mystery of signs—there is an inner reality that cannot be signified by deed, while deeds, good works or the reverse, are without inner meaning. Trapped in this disjunction the Puritans themselves forget the original significance of Hester's letter and take it to stand for "able"—which is, unlike "adultery," enunciated in the text—because of her good works. But Hester, when the magistrates consider removing the stigma, says, "were I worthy to be quit of it, it would fall away of its own nature, or be transformed into something that should speak a different purport" [103]. She seems, still, to insist upon its relation to her inner self. Yet she will try to comfort Dimmesdale by pointing out *his* good works—"Is there no reality in the penitence thus sealed and witnessed by good works?" [116]—until he tells her that *his* scarlet letter still "burns in secret" [116].

On Dimmesdale, where the letter may be guessed to have appeared for a similar signifying function as on Hester, it is, however, as invisible as the act or condition it refers to. Society has placed no token upon him and when Chillingworth opens the sleeping minister's vestment he sees "something" which is not pictured or named for the reader. Even in the final scene when he tears his own garment from his chest we are told only, "It was revealed! But it were irreverent to describe that revelation" [151], and the reader is cheated again of the confirming spectacle. Although some spectators testified to having seen that the minister did bear a letter like Hester's, others saw nothing. And the sign, if it had really been there, might, anyhow, our narrator remarks, have been only the medical symptom of Dimmesdale's psychic distress, "the effect of the ever active tooth of remorse gnawing from the inmost heart outwardly" [152], an instance of psychosomatic symptomology (another sign theory which greatly interested Hawthorne, as we shall see in a moment).

Pearl, the asker of so many preternaturally pertinent questions, asks her mother to explain the meaning of the sign she wears and is not answered—plausibly because the answer would be beyond her grasp but also so that the reader may still not hear the signified, the unutterable. When she asks, "What does the letter mean, mother," Hester says evasively, "I wear it for the sake of its gold thread" [110]. Pearl says that she has been told that the scarlet letter is the Black Man's mark, an expression she repeats when she asks if the minister holds his hand upon his bosom because the Black Man has "set his mark" [112] there. "Mark," for which the root sense is, once again, token or sign, implies here, as a special meaning, a signature, the personal sign of a signer set in stead of his name. As such it signifies not the wearer but the writer, the author of all sin. Pearl connects this guessed-at sign with Hester's A when she asks, "Is this his mark?" [112] and extracts from her mother the acknowledgement, "Once in my life I met the Black Man! This scarlet letter is his mark!" [113]. And Pearl then guesses, when she sees the minister's hand over his heart, "Is it because, when the minister wrote his name in the book, the Black Man set his mark in that place?" [114].

Both symbol and consequence of Hester's and Dimmesdale's sin, Pearl is herself an instance of the ambiguity of signs. She is the animate letter, the child dressed in gold-embroidered scarlet, "the scarlet letter in another form, the scarlet letter endowed with life" [65]. Yet when she dances about in the final scene in the city square, "her dress, so proper was it to little Pearl, seemed an effluence or inevitable development and outward manifestation of her character, no more to be separated from her than the many-hued brilliancy from a butterfly's wing, or the painted glory from the leaf of a bright flower" [135]. Her appearance, once the sign, the "effluence" of her parents' sin, is now an exterior organically developed from her own airy nature, as the rest of nature's signs emanate from transcendent being. Earlier, she reverses or nullifies Hester's sign when she places it, made of eel grass, upon herself. It is the color of nature, green, the eidetic image of her mother's token, and Pearl waggishly reflects as though to mock the meaning-searcher, "I wonder if mother will ask me what it means!" [108]. But Hester refuses to see it as a sign, and says, "The green letter on thy childish bosom has no purport" [108].

The mystery of meaning is expressed in the obliquity of Pearl's own answers to the question of what she is. Hester wonders, "Child, what are thou?" and is answered, "O, I am your little Pearl" [63], which is no answer for her name is her sign not her significance. Hester asks, "Tell me, then what thou art and who sent thee hither?" [63] and then answers this question herself, but Pearl demurs, "I have no Heavenly Father" [64], the animate sign denying its source in the divine. A little later, the Reverend Wilson asks again, "Who made thee?" [71] and Pearl's answer is that she has not been made but plucked from the prison rose-bush, an answer at once improbably arch and informed with a pantheistic view of nature, dispensing with the myth of express creation. Hawthorne's ironic dubeity can be felt in his presentation of the Governor's shocked, "a child of three years old, and she cannot tell who made her!" [72]—for how many among his readers would have had perfect confidence in the catechism reply? Of course, all the while that we have had this play of alternative semiologies, of Puritan and transcendental explanations of origin, it is obvious that Pearl's pert remarks are naturalistically explicable; she has just seen the roses in the governor's garden, has already been called "red rose" [70] by Wilson himself, who also calls her "little bird of scarlet plumage" [70], the natural creature she will be likened to in the last scene.

Our first view of Hester and her child occurs, nevertheless, when they emerge from the prison door passing the rose-bush and the weeds "which evidently found something congenial in the soil that had so early borne the black flower of civilized society, a prison" [35]. Weeds and prison are linked by a resemblance that is not merely metaphor but attributable to the generative force of which they are both products. Nearby, the rose-bush holds up "delicate gems" which "might be imagined to offer" sweetness to the condemned "in token that the deep heart of Nature could pity and be kind to him" [35]. Nature, the symbolizer, proffers a token from the realm of spirit in the same way as the Christian godhead has sent Pearl as "emblem and product of sin" [61]. But Hawthorne does not assert either source of signification uncontrovertibly: his weak copula, "might be imagined" [35], is a reminder that such symbolizing may be only the result of the human imagination.

It is quite "significant," therefore, that the artistic imagination appears centrally in Hester herself who is an artist of needlework, the only medium available to a woman in her day. Her works are distinguished by their power of symbolic exhibition, her first oeuvre of note having been her letter. She is afterwards called upon to show the meaning of other human situations, the pomp of public ceremonies, the sorrow of funerals which she would "typify by manifold emblematic devices" [55]. Her art is also *self*-expressive, "a mode of expressing and therefore soothing the passion of her life" [55]. Pearl has something of her mother's instinct: her creativity operates upon "a stick, a bunch of rags, a flower" [61], adapting them to her inner drama. Her art is harmless play. But Hester collaborates with Puritan society in converting Pearl herself into a symbol by clothing her in her symbolizing, signifying costume. She thus does violence to the irreducible being of the child who is shown repeatedly to be a natural phenomenon, a whimsical child and nothing else. All art, all symbolizing, is reductive.

In the mirror of art the truth is distorted from its natural proportions, as Hester's own image is when she sees herself in the polished armor in the Governor's house. The monstrously enlarged "A" upon her breast, her face reduced to insignificance by the convex surface of the breastplate, represent her reduction as the woman behind the scarlet letter. And in time, "all the light and graceful foliage of her character had been withered by this red hot brand, and had long fallen away, leaving a bare and harsh outline" [100], the person *becoming* the symbol. Yet the narrative shows at the same time that Hester resists this simplification, remaining a complex, developing personality. An opposite process takes place in the case of Chillingworth who, as his history advances, becomes more and more an abstract symbol of infernal malice until at the end he simply shrivels to nothing, all his humanity gone.

But Hawthorne does not dismiss or disparage the reading of signs altogether. He continues throughout the narrative to find ways of exploring the relation of phenomenon and meaning, of outerness and inwardness; his narrative discovers and tests other pairs of terms that represent signifier and signified. One example is the theory of disease by which he anticipates psychosomatic medicine. Chillingworth, it will be recalled, ascribes his patient's malady to a spiritual cause. "He to whom only the outward and physical evil is laid open knoweth, oftentimes, but half the evil which he is called upon to cure. A bodily disease, which we look upon as whole and entire within itself, may, after all, be but a symptom of some ailment in the spiritual part . . . a sickness, a sore place, if we may so call it, in your spirit, hath immediately its appropriate manifestation in your bodily frame" [85–86]. Bodily disease, then, is a "manifestation" of spirit, another instance of the sign language of all phenomena. Hawthorne's interest in the general science of signs, extends, logically, to the branch of medicine having to do with symptoms, which is also called semiology. Older medical concepts and even modern ones, of course, imply a dualism in the patient whose disease is defined as a manifestation of some hidden meaning—and a meaning that was truly inaccessible, for the most part, before the germ theory and modern knowledge of physiology. And so the source of disease, though presumed to be spiritual in Dimmesdale's case, will not, after all, be accessible to the probing of his physician-enemy.

Dimmesdale himself subscribes to his physicians' theories when he attributes his own distrust of Chillingworth to his inner spiritual disorder—"the poison of one morbid spot was infecting his heart's entire substance" [88]. In fact, his perceptions are accurate. But his inner condition does produce hallucinations, delusive signs. These seem to demonstrate, again, Hawthorne's view of the effects of the Puritan-transcendental view of a superior spiritual reality, his preference for the matter-of-fact: "It is the unspeakable misery of a life so false as his, that it steals the pith and substance out of whatever realities there are around us . . . To the untrue man, the whole universe is false,—and it is impalpable,—it shrinks to nothing within his grasp" [91].

All things hidden and all things exposed become antonyms in the novel to reflect the opposition of outer and inner. The forest, where the lovers meet alone save for little Pearl who does not understand what she sees except by occult instinct, is the place of a seclusive truth, difficult to read; the forest path, like a hard text, is "obscure" [118]. The public scenes in which Hester and Dimmesdale are together are the locus of communal truth, that which is perceived by all. "We must not talk in the market-place of what happens to us in the forest" [142], Hester warns Pearl, distinguishing between the unutterable inner world and the world of speech. Hester's "A", Mistress Hibbins says, is a "token" [143] that Hester has been to the forest many times, but the minister's visit is ultimately incommunicable. His election sermon is best understood when, in fact, its *words* are indistinguishable and only the mournful tone of his voice conveys his state to Hester as she stands outside. Language, by implication, misleads us, tells us nothing of the heart, which has no language. Dimmesdale's unintelligible murmur is like Pearl's babble or the gibberish she speaks in his ear in the night scaffold scene—perhaps a sacred speaking in tongues, perhaps the non-sense of a message-less world.

Nature, too, only babbles. The forest keeps its secrets though the babbling brook would seem to want to "speak" them: "All these giant trees and boulders of granite seemed intent on making a mystery of the course of this small brook: fearing, perhaps, that, within its never-ceasing loquacity, it should whisper tales out of the heart of the old forest whence it flowed, or mirror its revelations on the smooth surface of a pool" [113]. Pearl asks what the brook says, but Hester replies that if she had a sorrow of her own the brook would tell her of it, "even as it is telling me of mine!" [113], implying that she has understood the brook as the brook has understood her, but also that one hears in Nature's babble what one's own experience suggests. And as Pearl continues to play by the side of the brook her own cheerful babble mingles with its melancholy one, we are told, and "the little stream would not be comforted and still kept telling its unintelligible secret of some mournful mystery" [114].

Hawthorne's antithesis between the solitary soul and society is a variation upon the theme of an inexpressible inner reality. Hester is one of the great American isolatoes, who cannot speak the language of community. At his extremest, this loner is Melville's Bartleby, who withdraws from language altogether. By embracing silence he acknowledges the lapse of a common truth which unites not only men with one another but which, by a language of signs, unites the universe to mankind. Hester's sin is not only unutterable but involves a name, that of her partner, which she refuses to utter. Her sexual history is so private that it cannot be imagined when we gaze at her in the chaste aftermath of Hawthorne's novel. And yet that privacy has its public manifestation, the child Pearl. And Hester's sin is outrageously publicized by her exposure in the most public of places, the town pillory. The opening scene of the novel draws thrilling intensity from this paradox. From the hidden interior of her prison cell, from the secrecy of her own heart, Hester emerges with the child upon her arm, isolated and silent, to stand upon the most public site in Boston. A special piece of cruel machinery, a vise to hold the head upright, is available on the scaffold so that the condemned may be forced to face those who look upon him, and Hawthorne comments, "There is no outrage more flagrant than to forbid the culprit to hide his face for shame" [40]. But Hester voluntarily faces her viewers. Nevertheless, her exposure reveals nothing. Indeed, the spectator is prompted to find her an "image of Divine Maternity" [40], to read the scarlet woman as her opposite.

"Secret" is a key word in *The Scarlet Letter.* All the principal personages have secrets—Hester, the identity of her lover, Dimmesdale his sin, Chillingworth his own identity and motive. Chillingworth's name is like Hester's sin in never being enunciated in the text—though we may guess that it is Prynne. Perhaps the most important of these secrets, in terms of the progressive tension of the plot, is Dimmesdale's. Chillingworth's struggle to bring to the surface what lies hidden in the minister's heart is the primary conflict of the story (James was right in saying that the essential drama is there, between the two men, and not in Hester). This is also because more is involved in their struggle than the story tells: theirs is the contest between two views of the communicability of meaning. Chillingworth had asked Hester to name her lover and she had refused, eliciting from him the comment, "there are few things,—whether in the outward world, or the invisible sphere of thought,—few things hidden from the man, who devotes himself earnestly and unreservedly to the solution of a mystery" [51]. By profession a scientist, an investigator of nature and mankind, he is confident that he can compel all mysteries to yield to him. From the "prying multitude," from even the magistrates and ministers, Hester's secret may be hidden, but, Chillingworth declares, "I come to the inquest with other senses than they possess. I shall seek this man as I have sought truth in books, as I have sought gold in alchemy" [51].

Chillingworth is defined as a materialist, one of a species of men who have lost the sense of spiritual meanings. "In their researches into the human frame, it may be that the higher and more subtle faculties of such men were materialized, and that they lost the spiritual view of existence amid the intricacies of that wondrous mechanism which seemed to involve art enough to comprise all of life within itself" [75–76]. The newcomer becomes the community physician, replacing the aged deacon and apothecary, "whose piety and deportment were stronger testimonials in his favor" than a medical diploma, for he is learned in both "antique physic" [76] and the Indians' homeopathic medicine.

He believes, consequently, that the inner condition of his patient, the meaning of his disease can be understood. The narrator seems to agree: "Few secrets can escape an investigator who has opportunity and license to undertake such a quest" [78]. Like a researcher into a difficult scientific problem he is described "prying into his patient's bosom, delving among his principles, prying into his recollections, proving everything with a cautious touch, like a treasure-seeker in a dark cavern" [78]. Hawthorne even goes on to say, "A man burdened with a secret should especially avoid the intimacy of his physician" for "at some inevitable moment, will the soul of the sufferer be dissolved, and flow forth in a dark and transparent stream, bringing all its mysteries into the daylight" [78–79]. The doctor—more investigator than therapist, is said to be "desirous only of truth, even as if the question involved no more than the air-drawn lines and figures of a geometrical problem, instead of human passions, and wrongs inflicted on himself" [81].

His contest with Dimmesdale, who steadfastly protects his secret, is dramatically illustrated in their conversation in the graveyard in Chapter Ten. Upon a grave without identifying tombstone the physician finds weeds that "grew out of the [buried man's] heart and typify, it may be, some hideous secret that was buried with him" [82]. They have sprung up there, Chillingworth declares, "to make manifest an unspoken crime" [83]. Dimmesdale, however, insists upon the inaccessibility and sacredness of the dead man's secrets. "There can be . . . no power, short of the Divine mercy, to disclose, whether by uttered words, or by type or emblem, the secrets that may be buried with a human heart. The heart . . . must perforce hold them, until the day when all hidden things shall be revealed. . . . These revelations . . . are meant merely to promote the intellectual satisfaction of all intelligent beings, who will stand waiting, on that day, to see the dark problem of this life made plain" [83]. Not merely, then, does he not choose to tell his secret; it cannot ever be revealed to men until Judgment Day. It is a mystery too profound for us before that. And, Hawthorne's language seems to suggest, it is a mystery which is only part of the general mystery of "hidden things" for which "type or emblem" [83]—the language of appearances—provide no clue. As the methodical indeterminacy of *The Scarlet Letter* suggests, there is no present disclosure of "the dark problem of this life" [83].

To presume otherwise by trying to penetrate the mystery of another soul is Chillingworth's sin, as Dimmesdale tells Hester. He "violated in cold blood the sanctity of a human heart" [118]. This statement is not usually understood, though invariably quoted in discussions of the novel. We tend to think that Chillingworth has sinned because he has criminally used the knowledge he has gained in order to manipulate and destroy the minister. But this is not what the words say. The insistence upon illicit discovery, the assault by Chillingworth upon sacred knowledge, is itself illegitimate.

Here perhaps is the pious man's reply to the problem of the obliquity of signs. Hawthorne may have felt that it was his only stay against skepticism to believe in an ultimate revelation, an ultimate deciphering of what is beyond our comprehension in this life. But he may also have entertained the suspicion that no ultimate meanings exist. Perhaps he sometimes felt bold enough to share the thought expressed by Melville in a letter he got from him only a year after *The Scarlet Letter* was published: "If any of those other Powers choose to withhold certain secrets, let them; that does not impair my sovereignty in myself; that does not make me tributary. And perhaps after all, there is *no* secret."

LARRY J. REYNOLDS

The Scarlet Letter and Revolutions Abroad†

When Hawthorne wrote *The Scarlet Letter* in the fall of 1849, the fact and idea of revolution were much on his mind. In "The Custom-House" sketch, while forewarning the reader of the darkness in the story to follow, he explains that "this uncaptivating effect is perhaps due to the period of hardly accomplished revolution and still seething turmoil, in which the story shaped itself."[1] His explicit reference is to his recent ouster from the Salem Custom House, his "beheading" as he calls it, but we know that the death of his mother and anxiety about where and how he would support his family added to his sense of upheaval. Lying behind all these referents, however, are additional ones that have gone unnoticed: actual revolutions, past and present, which Hawthorne had been reading about and pondering for almost twenty consecutive months. These provided the political context for *The Scarlet Letter* and shaped the structure, characterizations, and themes of the work.

I

ROME YET UNCONQUERED! FRANCE TRANQUIL LEDRUROLLIN NOT TAKEN. THE HUNGARIANS TRIUMPH! GREAT BATTLE NEAR RAAB! THE AUSTRIANS AND RUSSIANS BEATEN. CONFLICTS AT PETERWARDEIN AND JORDANOW. SOUTHERN GERMANY REPUBLICAN. BATTLE WITH THE PRUSSIANS AT MANHEIM. RESULT UNDECIDED. These are the headlines of the *New York Tribune* for 5 July 1849; and because they are typical, they suggest the excitement and interest generated in America by the wave upon wave of revolution that swept across Europe during the years 1848 and 1849. In Naples, Sicily, Paris, Berlin, Vienna, Milan, Venice, Munich, Rome, and nearly all the other cities and states of continental Europe, rulers and their unpopular ministers were overthrown, most notably Louis Philippe and Guizot in France, Ferdinand I and Metternich in Austria, and Pope Pius IX and Rossi in the Papal States.[2] Meanwhile revolutionary leaders such as

† From *American Literature* 57.1 (March 1985): 44–67. www.dukeupress.edu. Page numbers in square brackets refer to this Norton Critical Edition.

1. *The Scarlet Letter*, ed. William Charvat, Roy Harvey Pearce, and Claude Simpson (Columbus: Ohio State UP, 1962), 43 [32]. Hereafter cited parenthetically in the text.
2. Useful overviews of the revolutions of 1848–49 are provided by *The Opening of An Era: 1848 An Historical Symposium*, ed. François Fejto (1948; rpt. New York: Howard Fertig, 1966), and *The Revolutions of 1848–49*, ed. Frank Eyck (New York: Barnes & Noble, 1972).

Lamartine, Kossuth, and Mazzini became heroes in American eyes as they tried to institute representative governments and alleviate the poverty and oppression that precipitated the revolutions.

By the fall of 1849, all of the fledgling republics had been crushed by conservative and reactionary forces, and this fact explains in part why the influence of the revolutions upon *The Scarlet Letter* in particular and the American literary renaissance in general has been overlooked. Unlike the American Revolution (whose influence has received thorough study), the revolutions of 1848–49 came to naught, making them appear inconsequential in retrospect. In addition, the excitement generated in America, while intense, was short lived and soon forgotten; national attention soon turned to the turmoil generated by the slavery issue, which obscured Europe's role as the previous focus of this attention. A third explanation for the neglect is that studies of the literature of this period have tended to focus on native themes and materials. Concomitantly, reference works such as James D. Hart's *Oxford Companion to American Literature* and John C. Gerber's *Twentieth Century Interpretations of "The Scarlet Letter"* have provided chronological indexes that correlate only American history with the lives and works of American authors, despite the fact that the major newspapers of the day devoted three-fourths of their front-page coverage to European events.

Although the European revolutions all failed, from the spring of 1848 to the fall of 1849, the American public displayed its interest and sympathy by mass gatherings, parades, fireworks, proclamations, speeches, and constant newspaper coverage, which swelled with the arrival of each steamer.[3] Members of the American *literati*, Hawthorne's friends among them, also responded with ardor. To celebrate the French Revolution, Lowell wrote two poems, "Ode to France, 1848," in which he linked American Freedom with the fires burning in the streets of Paris, and "To Lamartine, 1848," in which he sang the praises of the poet-statesman who headed the new provisional government. Evert Duyckinck, Hawthorne's editor at Wiley & Putnam's, declared himself *"en rapport"* with the French Revolution;[4] and S. G. Goodrich, Hawthorne's former publisher, who witnessed events in Paris, wrote an enthusiastic account for the *Boston Courier*. Emerson, who visited Paris in May 1848, expressed reservations about the posturings of the mobs in the streets but was impressed by Lamartine and sympathized with the social activists. "The deep sincerity of the speakers," he wrote, "who are agitating social not political questions, and who are studying how to secure a fair share of bread to everyman, and to get the God's justice done through the land, is very good to hear."[5]

3. A comprehensive study of the American response to the revolutions has yet to be published; however, the specialized studies of Elizabeth B. White, *American Opinion of France, from Lafayette to Poincaré* (New York: Knopf, 1927), Arthur James May, *Contemporary American Opinion of the Mid-Century Revolutions in Central Europe* (Philadelphia: U of Pennsylvania P, 1927), and Howard R. Marraro, *American Opinion of the Unification of Italy, 1846–1861* (New York: Columbia UP, 1932), when placed side by side, cover most of the salient features of this response, as it revealed itself publicly.
4. Letter to George Duyckinck, 18 March 1848. All of the letters from the brothers Duyckinck are quoted with the kind permission of the Duyckinck Family Papers, Rare Books and Manuscript Division, The New York Public Library, Astor, Lenox and Tilden Foundations.
5. *The Letters of Ralph Waldo Emerson*, ed. Ralph L. Rusk, IV (New York: Columbia UP, 1939), 73–74.

Margaret Fuller, who served as one model for Hester, became, as is well known, more intently engaged in the European revolutions than any of her countrymen. As a witness to the rise and fall of the Roman Republic, she wrote impassioned letters to the *New York Tribune* praising the efforts of her friend Mazzini, describing the defense of Rome, and pleading for American support. "The struggle is now fairly, thoroughly commenced between the principle of democracy and the old powers, no longer legitimate," she wrote in the spring of 1849. "Every struggle made by the old tyrannies, all their Jesuitical deceptions, their rapacity, their imprisonments and executions of the most generous men, only sow more dragon's teeth; the crop shoots up daily more and more plenteous." When the battle of Rome was fought, Fuller served tirelessly as a nurse and watched the warfare surrounding her. "Men are daily slain," she wrote on June 21, "and this state of suspense is agonizing. In the evening 't is pretty, though terrible, to see the bombs, fiery meteors, springing from the horizon line upon their bright path, to do their wicked message." After the French had invaded the city, she wrote, "I see you have meetings, where you speak of the Italians, the Hungarians. I pray you *do something*. . . . Send money, send cheer,—acknowledge as the legitimate leaders and rulers those men who represent the people. . . ."[6]

As Hawthorne was defending himself from the attacks of the Salem Whigs and battling to be reinstated as surveyor, the developments in Italy received a predominant amount of American attention. On June 20, the *Boston Daily Advertiser* reported that the French, in order to restore the power of the Pope, were marching on Rome with 80,000 men, and it quoted Mazzini's declaration that "We shall fight to the last against all projects of a restoration." The following day, alongside of Hawthorne's public letter to Hillard, this same newspaper reported Garibaldi's arrival upon Neapolitan territory and printed Louis Napoleon's lengthy speech explaining his government's support of the Pope. During the next two months, as Hawthorne ceased careering through the public prints in his decapitated state, accounts of the defeat of the Roman revolutionaries made their way to the United States, where they were greeted by most with sadness or outrage.

Although Margaret Fuller's former devotee Sophia Hawthorne (in her dutifully childlike manner) expressed approval of the republican successes in Europe as they were occurring in 1848,[7] her husband most likely shared neither her optimism nor the enthusiasm of their literary friends, particularly Fuller. In fact, the book that he wrote in the wake of the revolutions in 1849 indicates that they reaffirmed his scepticism about revolution and reform and inspired a strong reactionary spirit which underlies the work.

Revolution had been a fearful thing in Hawthorne's mind for some time, even though he found the ends it wrought at times admirable.[8] Violent

6. *At Home and Abroad, or Things and Thoughts in America and Europe*, ed. Arthur B. Fuller (1856; rpt. Port Washington, N.Y.: Kennikat, 1971), 380, 381, 409, 421.
7. In a December 1848 letter to her mother, Sophia declared, "What good news from France! . . . There seems to be a fine fresh air in France just now. . . . it is very pretty when the people do not hurt the kings, but merely make them run. Since Prince Metternich has resigned, I conceive that monarchy is in its decline," quoted in Julian Hawthorne, *Nathaniel Hawthorne and His Wife,* I (1884; rpt. n.p.: Archon Books, 1968), 331.
8. See Celeste Loughman, "Hawthorne's Patriarchs and the American Revolution," *American Transcendental Quarterly,* 40 (1979), 340–41, and John P. McWilliams, Jr., "'Thorough-Going Democrat' and 'Modern Tory': Hawthorne and the Puritan Revolution of 1776," *Studies in Romanticism,* 15 (1976), 551.

reform and the behavior of mobs particularly disturbed him,[9] as the final scene of "My Kinsman, Major Molineux" makes clear. This story may celebrate the beginnings of a new democratic era, as some have suggested, but it cannot be denied that Molineux is presented as a noble victim of a hellish mob. "On they went," Hawthorne wrote, "like fiends that throng in mockery round some dead potentate, mighty no more, but majestic still in his agony."[1] Similarly, in "The Custom-House" sketch, Hawthorne presents himself as the victim of another "bloodthirsty" mob, the Whigs, who, acting out of a "fierce and bitter spirit of malice and revenge," have struck off his head with the political guillotine and ignominiously kicked it about. This presentation, humorous in tone but serious in intent, gives *The Scarlet Letter* its alternate title of "THE POSTHUMOUS PAPERS OF A DECAPITATED SURVEYOR" and foreshadows the use and treatment of revolutionary imagery in the novel proper [32].

This imagery, of course, is drawn from the French Revolution of 1789, which was at the forefront of Hawthorne's mind for several reasons. First of all, the spectacular excesses of that revolution provided the language and metaphors used by conservatives to describe events in 1848–49. In a letter to the *New York Courier and Enquirer*, Bishop Hughes, a supporter of Pope Pius IX, denounced the revolutionaries of Rome and claimed, "They have established, according to what I regard as the truest accounts, a reign of terror over the Roman people, which they call a government." Alluding to Margaret Fuller, Hughes added that "no ambassador from foreign countries has recognized such a republic, except it be the female plenipotentiary who furnishes the *Tribune* with diplomatic correspondence."[2] Evert Duyckinck, keeping his brother George (who was in Paris) abreast of American attitudes in the spring of 1848, reported that "People look at this Revolution with recollections of the Era of Robespierre and suspect every revival of the old political phraseology of that period. An article attributed to Alison is going the rounds from Blackwoods in which he sets Satan grinning over the shoulders of Lamartine."[3] The *Blackwood's* article referred to had echoed the theme of "Earth's Holocaust" as it declared, "Experience will prove whether, by discarding all former institutions, we have cast off at the same time the slough of corruption which has descended to all from our first parents. We shall see whether the effects of the fall can be shaken off by changing the institutions of society; whether the devil cannot find as many agents among the Socialists as the Jacobins; whether he cannot mount on the shoulders of Lamartine and Arago as well as he did on those of Robespierre and Marat."[4]

9. Hawthorne's reservations about the behavior of revolutionary mobs can also be seen in his sketches "The Old Tory" (1835) and "Liberty Tree" (1840). His manuscript "Septimius Felton" contains some of his final thoughts on the subject. "In times of Revolution and public disturbance," he writes, "all absurdities are more unrestrained; the measure of calm sense, the habits, the orderly decency, are in a measure lost. More people become insane, I should suppose; offenses against public morality, female license, are more numerous; suicides, murders, all ungovernable outbreaks of men's thoughts, embodying themselves in wild acts, take place more frequently, and with less horror to the lookers-on." See *The Elixir of Life Manuscripts*, ed. Edward H. Davidson, Claude M. Simpson, and L. Neal Smith (Columbus: Ohio State UP, 1977), 67.
1. "My Kinsman, Major Molineux," in *The Snow-Image and Uncollected Tales*, ed. J. Donald Crowley (Columbus: Ohio State UP, 1974), 230.
2. Rpt. *Boston Post*, 29 June 1849, p. 1, col. 6.
3. Letter to George Duyckinck, 18 April 1848; Duyckinck Family Papers, New York Public Library.
4. "Fall of the Throne of the Barricades," *Blackwood's*, 63 (1848): 399.

The bloody June Days of 1848 seemed to confirm such scepticism, and even George Duyckinck, an ardent supporter of the French people, was reminded of the Reign of Terror and the role women played in it as he reflected upon recent events. "Human nature," he wrote his brother, "seems to be the same it was sixty years ago. Heads were stuck on pikes or swords and women danced about them as they did then and who can doubt but that if the insurgents had succeeded the guillotine would have been as busily at work today as it was then."[5] Although use of the guillotine had been discontinued (General Cavaignac used the firing squad during the June Days, when an estimated 10,000 died), the shadow of that instrument loomed over all, and after Louis Napoleon came to power, it became unwise even to mention this symbol of revolution. Under Napoleon's administration in 1849, the owner of a Paris newspaper called *Le Peuple*, for an article entitled "The Restoration of the Guillotine," was fined and sentenced to five years imprisonment, while the proprietor of *La Revolution Democratique et Sociale*, for an article entitled "The Political Scaffold," was fined and sentenced to three years imprisonment.[6] In America, the guillotine and the scaffold carried not quite so much import, except, of course, in the mind of one decapitated surveyor.

Predictably, the American press drew careless comparisons between the European revolutions and the American political scene. When Zachary Taylor began his series of political appointments in the spring and summer of 1849, they were reported in the Democratic papers as revolutionary acts, as symbolic beheadings of Democratic party members. Some seven times in May and June, for example, the *Boston Post* printed, in conjunction with the announcement of a political appointment and removal, a small drawing presumably of General Taylor standing beside a guillotine, puffing a cigar, surrounded by heads (presumably of Democrats) at his feet. One of these drawings appeared on 11 June and on the following day, a letter to the editor appeared objecting to Hawthorne's removal from the Salem Custom House. "This is one of the most heartless acts of this heartless administration," the anonymous writer declared. "The head of the poet and the scholar is stricken off to gratify and reward some greedy partizan! . . . There stands, at the guillotine, beside the headless trunk of a pure minded, faithful and well deserving officer, sacrificed to the worth of party proscription, Gen. Zachary Taylor, now President." As Arlin Turner has pointed out, this letter was probably a source of Hawthorne's "beheading" metaphor;[7] however, behind the reference were two years of revolutionary events in Europe, two years of revolutionary rhetoric and imagery.

II

Such rhetoric and imagery appeared not only in the newspapers, of course, but also in contemporary books, some of which dealt with revolution in a serious historical manner. Although *The Scarlet Letter* has often been praised for its fidelity to New England history, the central setting of the

5. Letter to Evert Duyckinck, 30 June 1848; Duyckinck Family Papers, New York Public Library.
6. See the review "Lamartine's Histoire des Girondins," *Southern Quarterly Review*, 16 (1849): 58.
7. *Nathaniel Hawthorne: A Biography* (New York: Oxford UP, 1980), 181.

novel, the scaffold, is, I believe, an historical inaccuracy intentionally used by Hawthorne to develop the theme of revolution. The Puritans occasionally sentenced a malefactor to stand upon a shoulder-high block or upon the ladder of the gallows (at times with a halter about the neck),[8] but in none of the New England histories Hawthorne used as sources (viz., Felt, Snow, Mather, Hutchinson, and Winthrop) are these structures called scaffolds. In fact, I have been unable to find the word "scaffold" in them. The common instruments of punishment in the Massachusetts Bay Colony were, as Hawthorne shows in "Endicott and the Red Cross," the whipping post, the stocks, and the pillory. (The gallows, located in Boston at the end of town,[9] was used for hangings and serious public humiliations.) Although Hawthorne in his romance identifies the scaffold as part of the pillory, his narrator and his characters refer to it by the former term alone some twenty-six times, calling it the scaffold of the pillory only four times and the pillory only once.[1]

As early as 1557 and then later with increasing frequency during the first French Revolution, the word "scaffold" served as a synecdoche for a public beheading—by the executioner's axe or the guillotine. And, because of its role in the regicides of overthrown kings, the word acquired powerful political associations, which it still retains.[2] When King Charles I was beheaded with an axe following the successful rebellion led by Cromwell, Andrew Marvell in his "An Horation Ode" used the word in the following tribute to his king:

> . . . thence the royal actor born
> The tragic scaffold might adorn:
> While round the armed bands
> Did clap their bloody hands.
> *He* nothing common did or mean
> Upon that memorable scene:
> But bowed his comely head
> Down, as upon a bed.[3]

One hundred and forty-four years later, when Louis XVI became a liability to the new French republic, he too, of course, mounted what was termed the "scaffold" and there became one of the victims of the new device being advocated by Dr. Guillotin. The association of a scaffold with revolution and beheading, particularly the beheading of Charles I and Louis XVI, explains, I think, why Hawthorne uses it as his central and dominant

8. See Joseph B. Felt, *The Annals of Salem, from Its First Settlement* (Salem: W. & S. B. Ives, 1827), 176, 317.
9. See Caleb H. Snow, *A History of Boston, the Metropolis of Massachusetts* (Boston: A. Bowen, 1825), 169.
1. See John R. Byers, Jr., and James J. Owen, *A Concordance to the Five Novels of Nathaniel Hawthorne*, II (New York: Garland, 1979), 667, 579.
2. *Oxford English Dictionary*, IX (Oxford: Clarendon Press, 1933), 159. Beheading was not a common form of punishment in the Massachusetts Bay Colony. The only mention I have found of it involved the punishment of an Indian found guilty of theft and of striking a settler's wife in the head with a hammer, causing her to lose her senses. Neither a block nor a scaffold was used in his execution, however. "The executioner would strike off his head with a falchion," John Winthrop reported, "but he had eight blows at it before he could effect it, and the Indian sat upright and stirred not all the time," *The History of New England from 1630 to 1649*, ed. James Savage, 2 vols. (1825–1826; rpt. New York: Arno Press, 1972), II, 189.
3. *The Complete English Poems*, ed. Elizabeth Story Donno (New York: St. Martin's Press, 1972), 56.

setting. It links the narrator of "The Custom-House" sketch with the two main characters in the romance proper, and it raises their common predicaments above the plane of the personal into the helix of history.

Hawthorne's desire to connect his narrative with historic revolutions abroad is further shown by the time frame he uses. The opening scenes of the novel take place in May 1642 and the closing ones in May 1649.[4] These dates coincide almost exactly with those of the English Civil War fought between King Charles I and his Puritan Parliament. Hawthorne was familiar with histories of this subject and had recently (June 1848) checked out of the Salem Atheneum François Guizot's *History of the English Revolution of 1640, Commonly Called the Great Rebellion*.[5] Guizot, Professor of Modern History of the Sorbonne when he wrote this work, became, of course, Louis Philippe's Prime Minister whose policies provoked the French Revolution of 1848. During the spring of 1848 Guizot's name became familiar to Americans, and probably the man's recent notoriety led Hawthorne to a reading of his work in the summer of 1848.

Examination of the simultaneity between fictional events in *The Scarlet Letter* and historical events in America and England verifies that the 1642–1649 time frame for events in the romance was carefully chosen to enhance the treatment of revolutionary themes. When Hester Prynne is led from the prison by the beadle who cries, "Make way, good people, make way, in the King's name" [40], less than a month has passed since Charles's Puritan Parliament had sent him what amounted to a declaration of war. Five months later, in October, 1642, the first battle between Roundheads and Cavaliers was fought at Edgehill, and word of the open hostilities reached America in December.[6] Then and in the years that followed, the Bay Colony fasted and prayed for victory by Parliament, but these became times of political anxiety and stress in America as well as England. According to one of Hawthorne's sources, Felt's *Annals of Salem*, in November 1646 the General Court (presided over by Messrs. Bartholomew and Hathorne) ordered "a fast on Dec. 24th, for the hazardous state of England . . . and difficulties of Church and State among themselves, both of which, say they, some strive to undermine."[7] By the final scenes of the novel, when Arthur is deciding to die as a martyr, Charles I has just been beheaded (on 30 January 1649); thus, when Chillingworth sarcastically thanks Arthur for his prayers, calling them "golden recompense" and "the current gold coin of the New Jerusalem, with the King's own mint-mark on them" (224) [134], Hawthorne adds to Chillingworth's irony with his own. Furthermore, given the novel's time frame, the tableau of Arthur bowing "his head forward on the cushions of the pulpit, at the close of his Election Sermon" (250) [147], while Hester stands waiting beside the scaffold, radiates with ominous import, particularly when one recalls that Arthur is not a graduate of Cambridge, as most of the Puritan ministers of New England were,[8] but rather of Oxford, the center

4. See Edward Dawson, *Hawthorne's Knowledge and Use of New England History: A Study of Sources* (Nashville: Joint Univ. Libraries, 1939), 17.
5. Marion L. Kesselring, *Hawthorne's Reading, 1828–1850* (1949; rpt. New York: Norwood, 1976), 52.
6. See Winthrop, II, 85.
7. Felt, 175.
8. See Frederick Newberry, "Tradition and Disinheritance in *The Scarlet Letter*," *ESQ*, 23 (1977): 13.

of Laudian and Royalist sympathies and the place of refuge for King Charles during the Revolution.

By thus setting events in an age when "men of the sword had overthrown nobles and kings" (164) [101], Hawthorne provides a potent historical backdrop for the revolutionary and counter-revolutionary battles fought, with shifting allegiances, among the four main characters and the Puritan leadership. Furthermore, his battle imagery, such as Governor Bellingham's armor and Pearl's simulated slaying of the Puritan children, draws upon and reflects the actual warfare abroad and thus illuminates the struggles being fought on social, moral, and metaphysical grounds in Boston.

Bearing upon the novel perhaps even more than its connections with the English "Rebellion" and its attendant regicide are its connections with the first French Revolution and the execution of Louis XVI. In the romance itself, Hawthorne first alludes to one tie when he describes the scaffold in the opening scenes; "it constituted," he writes, "a portion of a penal machine, which . . . was held, in the old time, to be as effectual an agent in the promotion of good citizenship, as ever was the guillotine among the terrorists of France" (55) [40]. This allusion may be derived from the imagery appearing, as discussed above, in the contemporary press; but it is also shaped, in a more profound way, by an overlooked source of *The Scarlet Letter*, Alphonse de Lamartine's *History of the Girondists*, a history of the first French Revolution published in France in 1847, translated into English by H. T. Ryde and published in the United States in three volumes in 1847–48.[9]

Lamartine, the poet-statesman who had risen to the head of the Provisional Government in Paris following the February 1848 Revolution, became a well-known figure in America during 1848–49 and was widely admired for his idealism, courage, and eloquence. Numerous Americans expressed high regard for him;[1] the *New York Herald*, deviating from its usual format, ran an engraving of him on its front page,[2] and one New York City speculator, trying to dignify a venture, even named a street of sixpenny shanties "Lamartine's Row."[3] After he had fallen from power due to his unwillingness to align himself with the radical republicans or the right-wing Bonapartists, Lamartine was treated as a noble martyr in the American press. The *New York Evening Post* on 23 June 1849, the day after Bryant's editorial on Hawthorne's behalf appeared there, devoted two and a half front-page columns to a glowing summary of Lamartine's literary and

9. The relationship between the French and English revolutions is one point emphasized in Lamartine's work; he points out that "Louis XVI had read much history, especially the history of England. . . . The portrait of Charles I., by Van Dyck, was constantly before his eyes in the closet in the Tuileries; his history continually open on his table. He had been struck by two circumstances; that James II. had lost his throne because he had left his kingdom, and that Charles I. had been beheaded for having made war against his parliament and his people," (New York: Harper, 1847–48) I, 52. (This edition of Lamartine will hereafter be cited parenthetically in the text.)
1. Evert Duyckinck in a letter to his brother George, reported that from the point of view of Americans "Lamartine stands out nobly" and speculated that if the new republic progressed well then Lamartine "will be the Washington of France" (3 April 1848, Duyckinck Family Papers, New York Public Library). Emerson, while in Paris, attended a session of the National Assembly and heard Lamartine's speech on Poland. "He did not speak . . . with much energy," Emerson wrote his wife, "but is a manly handsome greyhaired gentleman with nothing of the rust of the man of letters, and delivers himself with great ease & superiority" (*Letters*, IV, 77).
2. On 29 March 1848.
3. See *The Diary of George Templeton Strong*, ed. Allan Nevins and Milton Halsey Thomas, I (New York: Macmillan, 1952), 344.

political career. "[His] brief administration," the article concluded, "born of the barricades of February, expired amidst the roar of the cannon of June," proved "that Lamartine is too righteous a man to be a politician." "He was no demagogue; he appreciated the crisis, approved the revolution, but dreaded its excess. To save his country from terrorism and communism, he cheerfully laid down his popularity, as he would have laid down his life."

Before this political martyrdom, which would have engaged Hawthorne's sympathy, Lamartine's career had been advanced by his writings; his *Histoire des Girondins* established his credentials as a republican leader, helped inspire the Revolution of 1848 that he struggled to lead and moderate, and acquired much international renown. "We doubt whether this is not already the most popular *book*, as its author is the most popular *man* of the day," a reviewer for the *New York Courier and Enquirer* proclaimed,[4] when the English translation appeared. Unlike Guizot, Lamartine was not a scholarly historian, and his account of the first French Revolution is an imaginative and dramatic construct that gains much of its power from its sympathetic treatment of Louis XVI and its suspenseful narrative structure, which includes a tableau at the scaffold as its climactic scene. Throughout the first volume and a half of his history, Lamartine, while detailing the political infighting of the National Assembly and their struggle with the king for power, generates sympathy for Louis. He and his family are seldom free from danger, and the two high points of Volume I are their unsuccessful attempt to flee the country and their confrontation with a mob of thousands at the Chateau of the Tuileries. In Volume II, Lamartine shows the situation of the royal family becoming more desperate and the king acquiring strength and character as his fate unfolds. In terms a decapitated surveyor could appreciate, Lamartine observes that "all the faults of preceding administrations, all the vices of kings, all the shame of courts, all the griefs of the people, were accumulated on his head and marked his innocent brow for the expiation of many ages" (I, 27). "He was the scape-goat of olden time, that bore the sins of all" (II, 323).

Lamartine shifts from third-person omniscient narration to third-person limited after the National Assembly renders its verdict of guilty and its judgment of death. Thus, unlike Carlyle's clipped, brusque, and almost sarcastic account of the regicide, Lamartine's treatment generates sympathy; the reader is beside the king for some thirty intensely moving pages—as he parts with his family, as he rides in the carriage with his priest, who hears his confession, and as he sees and enters the Place de la Révolution to be beheaded. "There," Lamartine writes, "a ray of the winter's sun . . . showed the place filled by 100,000 heads, the regiments of the garrison of Paris drawn up round all sides of the scaffold, the executioners, awaiting the victim, and the instrument of death prominent above the mob, with its beams and posts painted blood-color. It was the guillotine!" (II, 370). Stationed around the scaffold are "unscrupulous and pitiless ruffians," who desire "the punishment should be consummated and applauded" (II, 371). In contrast, the king steps forward composed and aloof. Humiliated by being bound, he regains his composure, mounts the scaffold, faces the multitude, casts a farewell glance on his priest, and meets his death. "The plank sunk, the blade glided, the head fell" (II, 373), Lamartine writes, as

4. 17 May 1848, p. 2, c. 3.

this chilling and memorable scene comes to an end. Appearing in almost the exact center of the narrative, on the 867th page of 1578, the scene dominates the history; all that goes before anticipates it; all that follows refers back to it. The rest of Volume II and all of III detail the excesses of the Revolution: the assassination of Marat, the Reign of Terror, the wave upon wave of bloodletting, and so on, all of which become horrifyingly repetitive.

Lamartine's stirring treatment of revolutionary events and political martyrdom and especially his unprecedented use of the scaffold as both a dramatic setting and a unifying structural device lead one to speculate that Hawthorne may have read this work before he wrote *The Scarlet Letter*; however, speculation is unnecessary. He did. The records of the Salem Atheneum reveal that on 13 September 1849, he checked out the first two volumes of Lamartine's *History*.[5] Moreover, Sophia Hawthorne's letters to her sister and mother, combined with Hawthorne's notebook entries, reveal, as no biographer has yet pointed out, that it was about ten days later, most likely between 21 September and 25 September, that Hawthorne began work in earnest on *The Scarlet Letter*.[6] On 27 September he checked out the third volume of Lamartine's *History*, and on that date Sophia, in an often-quoted letter, informed her mother, "Mr. Hawthorne is writing morning & afternoon. . . . He writes immensely—I am almost frightened about it—But he is well now & looks very shining."[7] (He returned the first two volumes of the *History* 6 November and the third volume 12 November.) This correlation in dates plus Hawthorne's allusions to the terrorists of France suggests that what has become one of the most celebrated settings in American literature, the scaffold of *The Scarlet Letter*, was taken from the Place de la Révolution of eighteenth-century Paris, as described by Lamartine, and transported to the Marketplace of seventeenth-century Boston, where it became the focal point of Hawthorne's narrative. Along with it came, most likely, a reinforced scepticism about violent reform.

III

Recognition that revolutionary struggle stirred at the front of Hawthorne's consciousness as he wrote *The Scarlet Letter* not only accounts for many structural and thematic details in the novel but also explains some of the apparent inconsistencies in his treatment of his characters, especially Hester and Arthur. The issue of the degree and nature of Hawthorne's sympathies in the novel has been debated for years, at times heatedly, and I have no hope of resolving the debate here; however, I think the revolutionary context of events provides a key for sorting out Hawthorne's sympathies, or

5. Kesselring, 42.
6. Hawthorne and his wife spent much of the last half of August and the first part of September househunting, first on the Atlantic shore near Kittery Point, and then in the Berkshires near Lenox. Hawthorne may have worked on *The Scarlet Letter* during the second week in September after Sophia returned from Lenox, but if he did, it was not with the commitment he later displayed, for on 17 September he set out with his friend Ephraim Miller on a leisurely three-day journey to Temple, New Hampshire. Assuming he rested on the 20th, the day after his return, and knowing it was the 27th when Sophia first said he was writing "immensely" mornings and afternoons, it seems likely that between 21 September and 25 September he became absorbed in the writing of his romance. All of the letters from Sophia Hawthorne to her mother Elizabeth P. Peabody and her sister Mary Mann are quoted with the kind permission of the Henry W. and Albert A. Berg Collection, The New York Public Library, Astor, Lenox and Tilden Foundation.
7. Letter to Elizabeth P. Peabody (mother); Berg Collection, New York Public Library.

more accurately those of his narrator (whose biases closely resemble Hawthorne's). The narrator, as a member of a toppled established order, an *ancien régime* so to speak, possesses instincts that are conservative and antirevolutionary, consistently so, but the individuals he regards undergo considerable change, thus evoking inconsistent attitudes on his part. Specifically, when Hester or Arthur battle to maintain or regain their rightful place in the social or spiritual order, the narrator sympathizes with them; when they become revolutionary instead and attempt to overthrow an established order, he becomes unsympathetic.[8] The scaffold serves to clarify the political and spiritual issues raised by events in the novel, and the decapitated surveyor of the Custom House, not surprisingly, identifies with whoever becomes a martyr upon it.

Hawthorne's use of the scaffold as a structural device has long been recognized; in 1944 Leland Schubert pointed out that the novel "is built around the scaffold. At the beginning, in the middle and at the end of the story the scaffold is the dominating point.[9] The way in which the scaffold serves as a touchstone for the narrator's sympathies, however, has not been fully explored, particularly with reference to the matter of revolution.

As every reader notices, at the beginning of the story, Hester is accorded much sympathy. Her beauty, her courage, her pride, all receive emphasis; and the scaffold, meant to degrade her, elevates her, figuratively as well as literally. The narrator presents her as an image of Divine Maternity, and more importantly, as a member of the old order of nobility suffering at the hands of a vulgar mob. Her recollection of her paternal home, "poverty-stricken," but "retaining a half-obliterated shield of arms over the portal" (58) [42] establishes her link to aristocracy. Furthermore, although she has been sentenced by the Puritan magistrates, her worst enemies are the coarse, beefy, pitiless "gossips" who surround the scaffold and argue that she should be hanged or at least branded on the forehead. The magistrates, whom Hawthorne characterizes as "good men, just, and sage" have shown clemency in their sentence, and that clemency is unpopular with the chorus of matrons who apparently speak for the people.

Through the first twelve chapters, half of the book, the narrator's sympathies remain with Hester, for she continues to represent, like Charles I, Louis XVI, and Surveyor Hawthorne, a fallen aristocratic order struggling in defense of her rights against an antagonistic populace. The poor, the well-to-do, adults, children, laymen, clergy, all torment her in various ways; but she, the narrator tells us, "was patient,—a martyr, indeed" (85) [56]. It is Pearl, of course, who anticipates what Hester will become—a revolutionary—and reveals the combative streak her mother possesses. "The warfare of Hester's spirit," Hawthorne writes, "was perpetuated in Pearl" (91) [59], and this is shown by Pearl's throwing stones at the Puritan children ("the most intolerant brood that ever lived" [94] [61]), her smiting and uprooting the weeds that represent these children, and her splashing the Governor himself with water. "She never created a friend, but seemed

8. Nina Baym in her discussion of "The Custom-House" sketch posits that "like Hester, [Hawthorne] becomes a rebel because he is thrown out of society, by society. . . . The direct attack of 'The Custom-House' on some of the citizens of Salem adds a fillip of personal revenge to the theoretical rebellion that it dramatizes," *The Shape of Hawthorne's Career* (Ithaca: Cornell UP, 1976), 148–49. Hawthorne's attack, I think, can be more accurately termed a counterattack and seen as dramatizing not a rebellion but his reaction to a rebellion.
9. *Hawthorne the Artist* (Chapel Hill: U of North Carolina P, 1944), 137–38.

always to be sowing broadcast the dragon's teeth, whence sprung a harvest of armed enemies, against whom she rushed to battle" (95) [62]. (The echo here of Margaret Fuller's dispatch from Rome is probably not coincidental.)

Hester's own martial spirit comes to the fore in the confrontation with Bellingham, but here she fights only to maintain the *status quo* and thus keeps the narrator's sympathies. She visits the Governor not to attack him in any way but to defend her right to raise Pearl. Undaunted by Bellingham's shining armor, which "was not meant for mere idle show," Hester triumphs, because she has the natural order upon her side and because Arthur comes to her aid. Drawing Pearl forcibly into her arms, she confronts "the old Puritan magistrate with almost a fierce expression"; and Arthur, prompted into action by Hester's veiled threats, responds like a valiant Cavalier. His voice, as he speaks on her behalf, is "sweet, tremulous, but powerful, insomuch that the hall reechoed, and the hollow armour rang with it" (114) [73].

In the central chapters of the novel, when the narrator turns his attention toward Arthur and evidences antipathy toward him, it is not only because of the minister's obvious hypocrisy but also because of the intellectual change that he has undergone at Chillingworth's hands. Subtly, Arthur becomes radicalized and anticipates Hester's ventures into the realm of speculative and revolutionary thought. "There was a fascination for the minister," Hawthorne writes, "in the company of the man of science, in whom he recognized an intellectual cultivation of no moderate depth or scope; together with a range and freedom of ideas, that he would have vainly looked for among the members of his own profession" (123) [78]. And if Arthur is the victim of the leech's herbs and poisons, he is also a victim of more deadly intellectual brews as well. The central scene of the novel, Arthur's "vigil" on the scaffold, is inspired, apparently, by the "liberal views" he has begun to entertain. "On one of those ugly nights," we are told, "the minister started from his chair. A new thought had struck him" (146) [91]. This thought is to stand on the scaffold in the middle of the night, but by so doing he joins the ranks of Satan's rebellious legions. As he indulges in "the mockery of penitence" upon the scaffold, his guilt becomes "heaven-defying" and reprehensible, in the narrator's eyes. Rather than seeking to reestablish his moral force, which has been "abased into more than childish weakness," Arthur, in his imagination, mocks the Reverend Wilson, the people of Boston, and God himself. Furthermore, as Henry Nash Smith has pointed out, the "lurid playfulness" Arthur indulges in upon the scaffold, calls into question "the very idea of a solid, orderly universe existing independently of consciousness.[1] The questioning remains Arthur's, however, not the narrator's, and the scene itself, with the scaffold as its setting, serves to reveal the cowardice and licentiousness Arthur has been reduced to. The blazing A in the sky, which Arthur sees "addressed to himself alone," marks Governor Winthrop's death, according to the townspeople, and thus further emphasizes (by its reference to Winthrop's famous leadership and integrity) the nadir Arthur has reached by his indulgence in defiant thought and behavior.

1. *Democracy and the Novel: Popular Resistance to Classic American Writers* (New York: Oxford UP, 1978), 25.

The transformation Hester undergoes in the middle of the novel (which only appears to be from sinner to saint) is a stronger version of that which Arthur has undergone at her husband's hands; she too becomes, like the French revolutionaries of 1789 and the Italian revolutionaries of 1849, a radical thinker engaged in a revolutionary struggle against an established political-religious order. And as such, she loses the narrator's sympathies (while gaining those of most readers). The transformation begins with her regaining, over the course of seven years, the goodwill of the public, which "was inclined to show its former victim a more benign countenance than she cared to be favored with, or, perchance, than she deserved" (162) [100]. The rulers of the community, who "were longer in acknowledging the influence of Hester's good qualities than the people," become, as time passes, not her antagonists but rather the objects of her antagonism. We first see her impulse to challenge their authority when Chillingworth tells her that the magistrates have discussed allowing her to remove the scarlet letter from her bosom. "It lies not in the pleasure of the magistrates to take off this badge" (169) [103], she tells him. Similarly, when she meets Arthur in the forest several days later, she subversively asks, "What hast thou to do with all these iron men and their opinions? They have kept thy better part in bondage too long already!" (197) [119].

The new direction Hester's combativeness has taken is political in nature and flows from her isolation and indulgence in speculation. In a passage often quoted, but seldom viewed as consistent with the rest of the novel, because of its unsympathetic tone, the narrator explains that Hester Prynne "had wandered, without rule or guidance, in a moral wilderness. . . . Shame, Despair, Solitude! These had been her teachers,—stern and wild ones,—and they had made her strong, but taught her much amiss" (199–200) [120]. Hester's ventures into new areas of thought link her, significantly, with the overthrow of governments and the overthrow of "ancient prejudice, wherewith was linked much of ancient principle." "She assumed," the narrator points out, "a freedom of speculation, then common enough on the other side of the Atlantic, but which our forefathers, had they known of it, would have held to be a deadlier crime than that stigmatized by the scarlet letter" (164) [101]. Referring for the second time to the antinomian Anne Hutchinson, whom Hawthorne in another work had treated with little sympathy, the narrator speculates that if Pearl had not become the object of her mother's devotion, Hester "might, and not improbably would, have suffered death from the stern tribunals of the period, for attempting to undermine the foundations of the Puritan establishment" (165) [101].[2]

Although Hester does not lead a political-religious revolt against the Puritan leadership, these speculations are quite relevant to the action which follows, for Hawthorne shows her radicalism finding an outlet in her renewed relationship with Arthur, which assumes revolutionary form. When they hold their colloquy in the forest, during which she reenacts her role as Eve the subversive temptress, we learn that "the whole seven years of outlaw and ignominy had been little other than a preparation for

2. For excellent discussions of Hawthorne's attitudes toward women activists, see Neal F. Doubleday, "Hawthorne's Hester and Feminism," *PMLA*, 54 (1939): 825–28; Morton Cronin, "Hawthorne on Romantic Love and the Status of Women," *PMLA*, 69 (1954): 89–98; and Darrel Abel, "Hawthorne on the Strong Dividing Lines of Nature," *American Transcendental Quarterly* 14 (1972): 23–31.

this very hour" (200) [121]. What Hester accomplishes during this hour (other than raising the reader's hopes) is once again to overthrow Arthur's system and undermine his loyalty to the Puritan community and the Puritan God. She establishes a temporary provisional government within him, so to speak, which fails to sustain itself. Although Hester obviously loves Arthur and seeks only their happiness together, her plan, which most readers heartily endorse, challenges, in the narrator's eyes, the social order of the community and the spiritual order of the universe, and thus earns his explicit disapproval.

When Hester tells Arthur that the magistrates have kept his better part in bondage, the narrator makes it clear that it is Arthur's better part that has actually kept his worse and lawless self imprisoned. For some time the prison has proved sound, but "the breach which guilt has once made into the human soul is never, in this mortal state, repaired," the narrator declares. "It may be watched and guarded; so that the enemy shall not force his way again into the citadel. . . . But there is still the ruined wall" (200–201) [121]. Thus, as Hawthorne draws upon the popular revolutionary imagery of 1848–49 to present Hester as a goddess of Liberty leading a military assault, she prevails; however, her victory, like that of the first Bastille day, sets loose forces of anarchy and wickedness. Arthur experiences "a glow of strange enjoyment" after he agrees to flee with her, but to clarify the moral dimensions of this freedom, Hawthorne adds, "It was the exhilarating effect—upon a prisoner just escaped from the dungeon of his own heart—of breathing the wild, free atmosphere of an unredeemed, unchristianized, lawless region" (201) [121].

Unlike the earlier struggle that Hester and Arthur had fought together to maintain the *status quo*—the traditional relationship between mother and child—this struggle accomplishes something far more pernicious: "a revolution in the sphere of thought and feeling." And because it does, it receives unsympathetic treatment. "In truth," Hawthorne writes, "nothing short of a total change of dynasty and moral code, in that interior kingdom, was adequate to account for the impulses now communicated to the unfortunate and startled minister. At every step he was incited to do some strange, wild, wicked thing or other, with a sense that it would be at once involuntary and intentional" (217) [130].

Donald A. Ringe among others has suggested that this abrupt change in Arthur's system is beneficent, a fortunate fall, in other words, that gives him insight and powers of expression;[3] however, the narrative emphasizes that it is unfortunate and unholy. Arthur's impulses to blaspheme, curse, and lead innocence astray are a stronger version of those seen during his vigil, and they confirm the narrator's assertion that the minister has acquired "sympathy and fellowship with wicked mortals and the world of perverted spirits" (222) [133]. It is important to notice also that the success of Arthur's sermon, which is so eloquent, so filled with compassion and wisdom, depends ultimately not upon his new revolutionary impulses but upon older counter-revolutionary sources that are spiritually conservative. He draws upon the "energy—or say, rather, the inspiration which had held him up, until he should have delivered the sacred message that brought its own strength along with it from heaven" (251) [148].

3. "Hawthorne's Psychology of the Head and Heart," *PMLA* 65 (1950): 129.

The final change of heart and spirit that Arthur undergoes and that leads him to his death on the scaffold is foreshadowed by events in the marketplace prior to his sermon. There the exhibition of broadswords upon the scaffold plus Pearl's sense of "impending revolution" suggests that while the minister's better self has been overthrown, it will reassert itself shortly. The procession in which Arthur appears dramatizes the alternative to the lawless freedom Hester has offered. Here, as Michael Davitt Bell has observed, we have "the greatest tribute in all of Hawthorne's writing to the nobility of the founders."[4] The people, we are told, had bestowed their reverence "on the white hair and venerable brow of age; on long-tried integrity; on solid wisdom and sad-colored experience; on endowments of that grave and weighty order, which gives the idea of permanence, and comes under the general definition of respectability" (237–38) [141]. These are the qualities that distinguish Bradstreet, Endicott, Dudley, Bellingham, and their compeers. And, although we are not told who the new governor is (it was Endicott), we know that his election represents orderly change, in contrast to the rebellion and regicide that has recently occurred in England. "Today," Hester tells Pearl, "a new man is beginning to rule over them" (229) [136], and, in harmony with this event, Arthur acts to reestablish his place within the order of the community and within the order of the kingdom of God.

During the sermon Arthur seems to regain some of his spiritual stature and is described as an angel, who, "in his passage to the skies, had shaken his bright wings over the people for an instant,—at once a shadow and a splendor" (249) [147]. Because Arthur is still a hypocrite, considerable irony exists within this description; however, when the minister walks to and mounts the scaffold, the narrator's irony turns to sincerity. Arthur attempts, before he dies, to regain God's favor, and as he nears the scaffold, where Hester and Chillingworth will both oppose his effort to confess, we are told that "it was hardly a man with life in him, that tottered on his path so nervelessly, yet tottered, and did not fall!" (251) [148]. The exclamation mark indicates the double sense of "fall" Hawthorne wishes to suggest, and at the end Arthur seems to escape from the provisional control over him that both Chillingworth and Hester have had.

"Is not this better than what we dreamed of in the forest?" he asks Hester, and although she replies "I know not! I know not!" (254) [150] the revolutionary context of the novel, the bias toward restoration and order, indicate we are supposed to agree that it is.[5] Arthur's final scene upon the scaffold mirrors Hester's first scene there, even though he proceeds from the church whereas she had proceeded from the prison. But, unlike Hester, Arthur through humility and faith seems to achieve peace, whereas she, through "the combative energy of her character," had achieved only "a kind of lurid triumph" (78) [52]. In the final scaffold scene, Pearl acts as an ethical agent once again and emphasizes Hawthorne's themes about peace and battle, order and revolt. At the moment of his death, Arthur kisses Pearl, and the tears she then sheds are "the pledge that she would grow up amid human joy

4. *Hawthorne and the Historical Romance of New England* (Princeton: Princeton UP, 1971), 140.
5. A number of critics have read this scene as ironic and seen Dimmesdale as deluded or damned; however, the *Pietà* tableau, Arthur's Christlike foregiveness of Chillingworth, and Hawthorne's own emotional response to the scene (when he read it to Sophia) make it difficult to agree with such a reading.

and sorrow, nor for ever do battle with the world, but be a woman in it" (256) [151]. In what seems to be a reward for her docility, she marries into European nobility (thereby accomplishing a restoration of the ties with aristocracy her maternal relatives once enjoyed); similarly, Hester at last, we are told in a summary, forsakes her radicalism and recognizes that the woman who would lead the reform movements of the future and establish women's rights must be less "stained with sin," less "bowed down with shame" than she. This woman must be "lofty, pure, and beautiful, and wise, moreover, not through dusky grief, but the ethereal medium of joy" (263) [155].

More than one reader has correctly surmised that this ending to the novel constitutes a veiled compliment to Hawthorne's little Dove, Sophia, and a veiled criticism of Margaret Fuller, America's foremost advocate of women's rights and, at the time, one suffering from a sullied reputation due to gossip about her child and questionable marriage. Hawthorne's long and ambivalent relationship with Fuller and his response to her activities as a radical and revolutionary in 1849 had a decided effect upon the novel. There are several parallels which indicate Fuller served as a model for Hester: both had the problem of facing a Puritan society encumbered by a child of questionable legitimacy; both were concerned with social reform and the role of woman in society; both functioned as counsellor and comforter to women; and both had children entitled to use the armorial seals of a non-English noble family. All of these Francis E. Kearns has pointed out;[6] however, a more important parallel Kearns fails to mention is that for Hawthorne both women were associated with the ideas of temptation and revolution, with the figures of Eve and Liberty. Fuller was not only the most intelligent, articulate, and passionate woman Hawthorne had ever spent so many hours alone with,[7] she was also, as he began *The Scarlet Letter*, an ardent revolutionary supporting the overthrow of the most prominent political-religious leader in the world.

Certainly Hawthorne's knowledge of and interest in the New England past were considerable; however, as Thomas Woodson has pointed out, his interest in his contemporary world was far greater than the critical emphases of recent decades would indicate.[8] In his writing of *The Scarlet Letter* he drew upon the issues and rhetoric he was encountering in the present, especially those relating to himself as a public figure. Moreover, he responded strongly and creatively to accounts of foreign revolutions and revolutionaries that he found in the newspapers, the periodicals, and books new to the libraries. Although to most of his countrymen the overthrow of kings and the triumph of republicanism were exhilarating events, to a man of Hawthorne's temperament, the violence, the bloodshed, the extended chaos that accompanied the revolutions of 1848–49 were deeply disturbing. Associated in his own mind with his personal plight, they, along with his reading in Guizot and Lamartine, shaped *The Scarlet Letter* in Burkean ways the reader of today finds difficult to accept. We value too highly Thomas Paine and the rights of woman.

6. "Margaret Fuller as a Model for Hester Prynne," *Jahrbuch für Amerikastudien*, 10 (1965): 191–97.
7. The most revealing information about the Hawthorne-Fuller relationship is contained in Margaret Fuller's 1844 Commonplace Book, which deserves to be edited and published. The manuscript is on deposit by Mrs. Lewis F. Perry at the Massachusetts Historical Society.
8. "Hawthorne's Interest in the Contemporary," *Nathaniel Hawthorne Society Newsletter*, 7 (1981): 1.

EMILY MILLER BUDICK

Hawthorne, Pearl, and the Primal Sin of Culture†

In his long critical essay entitled simply "Hawthorne" (published in 1879), Henry James narrates the story of his own coming to know Hawthorne's most famous work of fiction, *The Scarlet Letter.* Speaking in an impersonal third person, James, "who was a child at the time," explains that he

> remembers dimly the sensation that book produced, and the little shudder with which people alluded to it, as if a peculiar horror were mixed in its attractions. He was too young to read it himself, but its title, upon which he fixed his eyes as the book lay upon the table, had a mysterious charm. . . . Of course it was difficult to explain to a child the significance of poor Hester Prynne's blood-coloured *A.* But the mystery was at last partly dispelled by his being taken to see a collection of pictures (the annual exhibition of the National Academy), where he encountered a representation of a pale, handsome woman, in a quaint black dress and white coif, holding between her knees an elfish-looking little girl, fantastically dressed and crowned with flowers. Embroidered on the woman's breast was a great crimson *A,* over which the child's fingers, as she glanced strangely out of the picture, were maliciously playing. I was told that this was Hester Prynne and little Pearl, and that when I grew older I might read their interesting history. But the picture remained vividly imprinted on my mind . . . and when, years afterward, I first read the novel, I seemed to myself to have read it before, and to be familiar with its two strange heroines. (402)

James's experience of *The Scarlet Letter* can stand for the experience of many American readers (and non-American readers as well) from James's time to our own, in relation to what is probably the most important mid-nineteenth-century work of American fiction, the founding text for many Americanists of at least one major American literary tradition. By the time most of us have come to read the book (often in high school or college), we have already encountered its representations throughout the culture. Therefore we feel ourselves to "have read it before." We have already experienced the "shudder," "horror," and "charm" that the book produces. And we have looked forward to that day, which has now suddenly arrived, when we will actually come to read the novel and decipher its "mystery."

I invoke this moment in American cultural history for two reasons. The first is specifically relevant to my own status as an expatriate American. My own "first reading" of *The Scarlet Letter,* which was, of necessity, already a rereading, occurred for me in the United States, before I emigrated to Israel. Nonetheless, my writing about *The Scarlet Letter*—all those subsequent rereadings, including this one—has always taken place from foreign shores, with at least the pretense of some objective distance between

† From *Journal of American Studies* 39.2 (2005): 167–85. Reprinted by permission of the publisher. Page numbers in square brackets refer to this Norton Critical Edition.

my subject and me (see especially my *Engendering Romance,* 13–29). One question I want to probe in this essay is what it means to separate out from one's culture, to see it from afar and abroad—as did James in most of his fiction. Indeed, in *The Portrait of a Lady,* James's most clear rewriting of Hawthorne's *Scarlet Letter,* he specifically alludes to the very figure that in Hawthorne's own novel takes up the position of expatriated American: Pearl. I will be putting Pearl at the center of my reading as a way of examining the author himself, who, like her, and unlike her mother, is also a child of the Puritans, but who, like her mother and unlike Pearl, will *not* (as he tells us quite emphatically in "The Custom-House") depart his native ground. We have in recent years developed an extensive vocabulary for thinking about otherness—the other in society, ourselves as other (as women, African Americans, Jews). Hawthorne's subject is the subject of sameness, where by "subject" I mean an idea of selfhood, subjectivity—the processes by which an individual defines or constructs himself or herself as subject. A subject is always other, and always sees others as other as well.

And this leads to my other reason for invoking James. I want to make a point concerning the contexts of culture: the way in which we are always inside the culture we would decode, subject to its horrors and charms and mystery long before we can render these features of our cultural experience conscious or interpretable and long after we imagine we have successfully resolved them. This is true whether we literally remove ourselves from our culture or remain within it, to take up a critical or oppositional stance. I take this to be the point not only of James's memoir (and therefore of his own art, especially in novels like *The Portrait of a Lady*), but of Hawthorne's novel as well. It is one of the triumphs of Nathaniel Hawthorne's first major romance (published in 1850) that such clarification as James desires never occurs, either for James or for the generations of readers after James who have tried to decode *The Scarlet Letter.* Who is to say what the *A* actually, definitively, means—for the Puritans who condemn Hester to wear it in punishment for her sin of adultery (although the word "adultery" never once appears in the novel); for Hester herself, who embroiders it into a fantastic work of Art (perhaps also in devotion to her lover Arthur, his name so purposively chosen to replicate not only the letter *A* but the word "Art" as well); for Pearl, who is, we are told, the letter incarnate ("it," the texts says, impersonalizing the child the same way that James, in his memoir, impersonalizes himself, as if people too were signifiers rather than subjects within their culture), "was the scarlet letter in another form; the scarlet letter endowed with life"—102 [65]), for Hawthorne himself, the literal Artist/*Art*ist/Arthur whose novel is *The Scarlet Letter* as nothing less than text itself; and finally, and perhaps even more importantly, for us as readers, especially those of us who are American readers, for whom the text is the text of our culture—the original primal sin or scene of our birth into intellectual consciousness. That for Americans America is *our* culture is simply, tautologically true, whether or not we feel ourselves to be the sinners of that culture (or their descendants)—the Puritan patriarchs as it were, and which of us, even white Protestant males, feel themselves the direct descendants of those Puritans?—or, (more likely) whether we imagine ourselves as the sinned against: the marginalized Puritan minority, like Hester or Pearl, or their descendants, such as Hawthorne; or whether we belong to another category of the marginalized and sinned against: African and Native Americans, immigrant Americans, and other Americans who were not even around at the founding moment and have not subsequently been represented in the American story. For all of the above, America is our America. What, then, is our scarlet letter? And what does it mean to refuse to wear that letter, to imagine, not only that the letter is not our scarlet letter after all, but that it is itself the stigma and mark of original sin from which we most have to distance ourselves? Are the sinned against ever not themselves sinners against something? And might not the sinning against be configured in exactly the refusal to take responsibility for the letter, which is to say the culture and its texts?

For *The Scarlet Letter* is no more today the governing text in the American tradition than the Puritans were the ruling force in Hawthorne's nineteenth-century America, when he chose to glance back to New England's cultural origins and (in at least some readings of the novel) brand his Puritan ancestors as the perpetrators of a new original sin in a new American Eden. This is the way Frederick Crews (in a landmark essay of 1988) has described the transformation in American literary criticism that took place in the 1980s and continues today. "The questioning of absolutes," he says by way of introduction, "is now being conducted in all branches of literary study. It reflects an irresistible trend in the academy toward spurning unified schemes and hierarchies of every kind." But, he goes on,

> What gives the New Americanist critique a special emotional force . . . is its connection both to our historical national shames—slavery, "Indian removal," aggressive expansion, imperialism, and so forth—and to current struggles for equal social opportunity. When a New Americanist shows, for example, that a canonical work such as *Huckleberry Finn* indulges in the stereotypical "objectifying" of blacks, Native Americans, women, or others, a double effect results. First, the canon begins to look less sacrosanct and is thus readied for expansion to include works by long-dead representatives of those same groups. Second, their contemporary descendants are offered a reason for entering into an academic dialogue that had previously slighted them. In short, the New Americanist program aims at altering the literary department's social makeup as well as their dominant style of criticism. (68–69)

Any glance at the contents of literary journals and conferences over the last twenty years will attest to the validity of Crews's assertion that not only has the subject of American literary criticism changed but that the change represents an attack, sometimes explicit, sometimes more modulated, against the literary and scholarly tradition that until the 1970s reigned supreme in the American canon. The founding texts seemed to many recent critics no less than an original scene of primal sin than the Puritans had, perhaps, seemed to Hawthorne, or than Hester had seemed to the Puritans.

Of course that new critical tradition itself was a fairly recent development of the 1940s, 1950s, and 1960s. The idea governing canon formation, especially in the years following World War II, was a celebration of those American values—democracy, pluralism, and freedom—that European fascism and communism seemed to preclude and that seemed to account for America's success as a nation and as a people. The nineteenth-century tradition of Hawthorne and Melville, which gradually expanded to include Poe, James, and Faulkner as well, came to seem a paragon of an anti-ideological or ideologically resistant literature and, therefore, a product to be exported for the betterment of those cultures that had become the victims of totalitarian regimes. That democracy, pluralism, and freedom are themselves ideologies no one seemed quite to notice, or care. Nor were the proponents of what became known as the romance tradition of American fiction (the Hawthorne, Melville, James, Faulkner tradition) particularly attentive to what American democracy, pluralism, and freedom had themselves excluded: notably African Americans and women, not to mention a host of other others (Native Americans, for example), who, in recent years, have also made their presence felt on the American cultural scene. By inserting into the canon such other texts, authors, and subjects as had been excluded, the critical establishment would transform literary studies. It would finally permit the realization of just those values for which Americans had, politically, always stood, and that the romance tradition had actually once seemed, to some, to promote. (I discuss many of these ideas in greater detail in my book on *Nineteenth-Century American Romance*.)

The question that I raise here is whether the romance tradition as exemplified by a text like *The Scarlet Letter*, even if it did not embody the pluralism it seemed to encourage, is not, when all is said and done, nonetheless still a vehicle of such pluralism in American culture (as the old vanguard of romance critics had maintained). And I want to add to that claim the further suggestion that the romance tradition, both as a canon of primary texts and as a reading strategy, was a better vehicle than the identity politics and multiculturalism that have in recent decades displaced it. For, when all is said and done, what is the subject of Hawthorne's *Scarlet Letter* if not, as I have already begun to suggest, "national shame"—just that subject that has emerged as central to the New Americanism; and what is Hawthorne's romance questioning if not such "absolutes" (another good *A* word) as characterized the Puritans themselves, and as might, the novel recognizes, characterize the novel as well?

And this is where the fictional text differentiates itself from the kind of identity politics that characterized the Puritans and has reemerged in contemporary literary studies—almost as a kind of return of the repressed, as we shall see. "National shame" is the Puritans' target. And their response is to brand Hester as "the type of shame," the word "type" and the definite article "the" in the phrase producing that quality of the absolute which is realized quite literally, materially, in the literal letter itself: "*the* scarlet letter." But *The Scarlet Letter* is also the title of Hawthorne's book. National shame, as I have suggested, is its subject: "Wherefore is she here set up to public shame?" asks Chillingworth when he first sees Hester on the scaffold (61) [43], and the text continues to dance around the word "shame," as if this were its major subject. Though we might like to think that the title of the novel simply alludes to or registers an accusation against that other letter, the Puritans' letter, which the novel is undoubtedly critiquing, the case is not quite that simple. To some significant degree the book itself is also a scarlet letter (albeit a voluminous one), making the same accusation of shame against the Puritans that they levy against Hester and replicating the same structure of absolutes—in the sense of beliefs, commitments, a moral code—which the book is condemning in them.

What we seem to have here is an uncomfortable repetition of some childish form of name-calling. Shame on you, say the Puritans to Hester. Shame on you, says Hawthorne to the Puritans. Shame on you, say the New Americanists to Hawthorne. And Shame on you, say I to the New Americanists. Since all of us in this chain of accusation are members of, and, after the Puritans, inheritors of, the same culture, there is more than a small amount of repression and projection being exhibited here. So where does the name-calling stop, or at least yield to something more noble, or at very least more useful? It is at this moment that one wants to invoke what we might think of as the aesthetic defense: that what distinguishes literary texts from other forms of writing is that they cannot, by definition, deal in absolutes, that Hawthorne's *Scarlet Letter,* despite the "the" in the title, does not, like "*the* scarlet letter" of the Puritans (and perhaps of the New Americanists as well) stand for one thing and one thing only; it does not stand simply for "shame," but rather, as numerous critics of the novel have noted, for everything from "artist" to "angel" to "able" to "America" itself. At the very least, literary texts complicate our notion of what and how things mean and what does and does not constitute sin, crime, and shame.

And yet this is a bit too easy and facile, since however much Hawthorne, like his protagonist, embroiders the letter and tries to free it from the literalism of his Puritan ancestors, nonetheless it does deliver its moral message just as emphatically as the Puritans deliver theirs. Hawthorne's *A* may broaden the range of its meanings, but it does *mean* in a very literal way—where by "literal" I mean also to point to the letter as letter, as signifying in an alphabet of meanings. At the same time, it also misses lots of things, as the New Americanism has charged.

Hawthorne's *A*, in other words, is as much a symbol as the Puritans' *A* or, for that matter, Hester's herself. Indeed, Hawthorne embodies his own problem as an artist, trying to wrestle free of the meanings imposed upon him by custom and tradition (as an employee of "The Custom House") through the struggles of his major protagonist and counterpart in the novel: Hester. Hester is this book's artist, not to mention its surveyor of customs, par excellence, where by "customs" is meant not simply traditions (and what does Hester become if not *our* tradition?), but duties and obligations as well. And what we are shown throughout the book is how Hester's *A*, especially as it comes to be replicated in the lavishly decorated child Pearl, who is, we are told, nothing less than the *A* incarnate, incorporates much of the Puritans' own ideology concerning the legibility of the world's visible signs and of our moral responsibilities as a consequence of that legibility. Hester, as the victim of the Puritans' reading practices, to be sure resists the content of the Puritans' ideology. She does not, however, despite the apparent difference between their straitlaced flannel letter and her extravagantly embroidered one, resist the reading practice by which the community represents that content. And her attenuating the letter's meaning, making it more literary than literal, does not get her out of the bind of repetition, for what is the literary but the literal amplified and multiplied and, in a word, embroidered, but always, nonetheless, tied to the literal, indeed controlled by it, and finally in its service. The Puritans might well have stopped staring at Hester's bosom if she had not made the sign of her sexuality so incredibly appealing an artwork. And she, not to mention the Puritans, might well have faded from public consciousness were it not for the greatness of Hawthorne's novel.

The only way out of this bind of the literal would seem to be to get rid of the letter altogether. This is what we might understand the New Americanism to be doing, when it gets rid of such texts as *The Scarlet Letter* as founding texts and simultaneously imports into the canon of American literature texts heretofore excluded and therefore outside the self-reflecting mirror of culture. Such revisionism, I hasten to add, is part of the very healthy process by which a culture expands and enlarges—though whether, even through such a process, it ever leaves behind the founding letter is exactly my question. At one moment in the text Hester does try to throw away the letter. But, ironically perhaps, at no moment in the novel is Hester more in conformity with Puritan practice than when, during her conversation with Dimmesdale in the forest, she discards the *A*, insisting that "The past is gone!"—introducing a refrain that will echo all along the corridors of American cultural history. And she continues in words that say it all: "With this symbol, I undo it all, and make it as it had never been!"—the word "this" referring equally to the letter and to her action of throwing it away (202) [122]. Hester would reject the symbolism by which she is labeled, but she would not reject symbolism as such. Hester is a perfect embodiment of her culture. She may discard their interpretations but she totally accepts their reading practices.

Furthermore, although no Puritan theologian of the time would have put it this way, getting rid of the past was in some sense what the Puritan migration to New England was all about. On the intellectual level the Puritans would have insisted emphatically on the continuity of history. "In Adam's fall we sinn'd all" is the first line for the first letter—the letter *A*—in *The New England Primer.* Nonetheless, the very point of producing a community of saints in the New World was to escape one history in order to enter another. And the reason for putting the *A* visibly on Hester's bosom is to suggest that, by contrast, it is *not* on every bosom. Some people, those blessed by divine grace, have been relieved of their letter—at least in the hereafter. Hester's casting off the *A* is her final, consummate gesture of rebellion against the Puritans. It is also, however, her ultimate act of repetition whereby, in resisting Puritan theology, she replicates it. It reveals the deep degree to which, in resistance to them, she is nonetheless one of them—she denies her *A* and declares herself blessed. *We* never violated the sanctity of a human heart, she declares to Dimmesdale. She is as wrong about her sanctity as the Puritans were about theirs.

Hawthorne's point is that in New England everyone wears the letter *A* even, or rather especially, when they do not. For this reason Hawthorne reverses Hester's, and by extension the Puritans', gesture of casting the *A* aside, when he himself, in "The Custom-House," takes up the letter and puts it to his own breast. This action is repeated in the novel proper (even as, historically, it is preceded or anticipated by it) when Hester herself resumes the letter, twice. The first time is in the forest scene above, when, immediately after casting it off, she puts it back on again, in response to Pearl's unwillingness to recognize her mother without the letter. In a very real way Pearl's existence in the world is inseparable from the letter. Throw away the letter signifying the adulterous union by which Pearl was conceived and you throw away Pearl as well, which might not be a bad way for all of us to think about our relationship with culture, especially when we are about to discard its *Scarlet Letter.* The second time Hester takes up the *A* is at the end of the book, when Hester "return[s]" to New England "and resume[s],—of her own free will, for not the sternest magistrate of that iron period would have imposed it,—resume[s] the symbol of which we have related so dark a tale" (263) [154]. As I have argued at greater length elsewhere (in *Engendering Romance*), resuming the letter—the word "resume" is repeated in the passage—is an act of taking responsibility, of making a conscious choice to accept the law of the land, which is replicated even as it is anticipated by Hawthorne's action in the Custom-House, where it is represented in exactly the right theological terms: "I, the present writer," says Hawthorne, "as their representative [i.e. representative of his sinning Puritan ancestors], hereby take shame upon myself for their sakes, and pray that any curse incurred by them . . . may be now and henceforth removed" (10) [11]. ("I am a dying man," Dimmesdale says in his final words to Hester; "So let me make haste to take my shame upon me" (254) [150].)

One can disown one's relationship with shame, or one can acknowledge it, whether the shame is one's own or someone else's—even the shame of culture's treatment of you rather than your treatment of them. In "The Custom-House" Hawthorne is taking upon himself the Adamic role of representative man, not as the initiator of humankind's fallen condition, nor (as is the case with that second Adam, Christ) as humankind's redeemer, but, rather, and more simply, in acknowledgement of the truth of the assertion that, however we would prefer to think otherwise, in Adam's fall we sinn'd All. Culture is a collectivity, in which we are all, whether we like it or not, implicated, even when we are ostensible victims. This is what Hester acknowledges as well at the end of the novel when she resumes the *A*. It is also what is implicit in her earlier resumption of the letter in the forest, when her daughter, for simpler biological, familial, and human reasons, makes her mother take up the letter once again. Victim, she is victimizer as well, and of no less precious a human being than her own daughter—purchased at the price of everything she held dear.

Pearl's position in all this provides an important insight into how we might understand Hawthorne's taking up of the scarlet letter in "The Custom-House" and his writing the novel he comes to call by that title. To make Hawthorne's putting the letter on his breast the equivalent of Hester's is to make Hawthorne a bit too noble, a bit too self-sacrificing in a specifically Christ-like sort of way. After all, unlike Hester, or, for that matter, Adam, Hawthorne is not himself guilty of any moral wrongdoing, at least not that we as readers of his book know of. His gesture seems then a bit presumptuous, histrionic, and also hollow—what does he risk, taking up the letter, but to become America's first and foremost creator of a literary tradition? While Hester, whatever the limits of her artistry, displays profound moral and social courage at the end of the book, Hawthorne's taking up the letter in "The Custom-House" seems a cheap parlor trick at best. (The "torpor" of the Custom-House, writes Hawthorne, toward the end of the sketch, which prevents him from being a writer, does not "quit" him even "when, late at night, I sat in the deserted parlour, lighted only by the glimmering coal-fire and the moon, striving to picture forth imaginary scenes" (35) [27].)

Here it is useful to observe an interesting difference between Hester and Hawthorne, which turns out to be an equally interesting similarity between Hawthorne and the other major female character in the book, Pearl. Although we the readers never see Hester without the letter, except for that brief moment in the forest when she tosses it away, Hester is, in fact, not born into the world of the scarlet letter. She comes from outside the world in which she sins and is punished. This is not the case for Pearl and Hawthorne. As Hawthorne makes very clear in the "Custom-House" sketch, *The Scarlet Letter* is very much about his own position as child of the Puritans, the inheritor and descendant of their customs and traditions: "disgraceful" and "degenerate" (10) [11, 12], like Pearl, and not, in his words, the "legitimate son of the revenue system, dyed in the wool, or rather, born in the purple" (16) [15, 16]. Hawthorne's taking up the letter in "The Custom-House" less signifies his identification with his ostensible heroine Hester (though it does signify that as well; Hawthorne experiences a "red-hot" "burning" sensation when he takes up the letter (32) [25], which recalls Hester's own relation to the letter—"not a stitch in that embroidered letter," we are told by one of the Puritan matrons, "but she has felt it in her heart" (54) [39]) than it marks him as her son, her other illegitimate, degenerate, and disgraceful child. It is Pearl who, in the course of the novel, is seen, like Hawthorne in the "Custom-House" sketch, to take up the letter and try to fathom its meaning. Like the book itself and her portrait in the James essay, Hester appears to us in the novel and to her daughter in the world of the novel with the letter already stitched on her breast. It is inseparably a part of our and her experience of Hester, which is in part of course why Pearl insists that her mother put the letter back on after she has thrown it away. We, like Pearl, and like her author, are born into the world of the letter. Pearl's story is therefore the story of our birth into a culture already given—the story James tells in his essay on Hawthorne. When Hawthorne takes up the letter in "The Custom-House" he less recalls Hester than he does Pearl, the inheritor and product of her mother's sin, who nonetheless must take up the letter, fashion it, and interpret it for herself.

At the end of the novel Pearl will decide that living under the sign of the letter is not for her and off she will go to be an heiress in Europe, the inheritor not of the letter *A* (her mother's letter, the letter branded on her chest of her biological father Dimmesdale, the letter that stands for *A*merica), but of her legal father Chillingworth's money. Hawthorne is the child who stays. And he invites the reader to share with him this condition, of being inside their own culture, subject to its laws and conditions, free to leave and yet required, if they stay, to take up the burden of the letter and do something with it.

Hawthorne could not make his intentions clearer when he titles the first chapter of the book "The Prison-Door" (47) [35]. This chapter immediately follows the autobiographical sketch entitled "The Custom-House." We exit what is clearly presented as a nineteenth-century prison-house of customs—the idea of *duties* itself, as I have suggested, carrying the same double sense of the word *customs*—into the even more dramatically imprisoning world of the seventeenth-century Puritans—represented not only by the literal prison that occupies our attention at the beginning of the novel but also by the restrictive, imprisoning social realm of Puritan culture itself. All these prisons are really quite comparable to one another, and all have their counterparts in the book itself. The words "the prison-door" are, after all, our point of entry into the book itself. The book, its author is telling us, is also a prison, perhaps even a grave—those two features of the human society, the cemetery and the prison, that are, Hawthorne tells us explicitly at the beginning of Chapter 1, the essential institutions without which no society can exist. The material text is no less a prison/grave (books, after all, do in their three-dimensional rectangularity evoke prison-houses and tombstones). And this imprisoning, death-dealing quality of literature is an integral part of what constitutes it a cultural or literary artifact.

Lest we miss the congruity of text with prison-house and grave, Hawthorne returns at the very end of the chapter to the idea of a **portal** or **threshold**, which is represented as belonging as much to the book as to the world that the book describes. Thus, as we exit from the chapter entitled "The Prison-Door," moments before Hester herself, within the story, exits that prison-door, the narrator plucks a rosebud from a bush located within the world of the fiction and presents it to the reader *through* the story, transgressing ontological space and time (itself a sin, we might feel). "On one side of the portal, and rooted almost at the threshold, was a wild rose-bush, covered, in this month of June, with its delicate gems. Where this rose-bush originates," the narrator says rather coyly, he will "not take upon us to determine." But,

> finding it so directly on the threshold of our narrative, which is now about to issue from that inauspicious portal [of the story, as much as of the jail], we could hardly do otherwise than pluck one of its flowers and present it to reader. It may serve, let us hope, to symbolize some sweet moral blossom, that may be found along the track, or relieve the darkening close of a tale of human frailty and sorrow. (48) [36]

And to complete this process whereby the text and prison-house merge and to associate both as well with the second term in Hawthorne's first chapter—the graveyard—the rose does indeed reappear at the end of the narrative, associated with that very tombstone that bears the A that is also the title of this novel:

> On this simple slab of slate—as the curious investigator may still discern, and perplex himself with the purpose—there appeared the semblance of an engraved escutcheon. It bore a device, a herald's wording of which might serve for a motto and brief description of our now concluded legend; so sombre is it, and relieved only by one ever-glowing point of light gloomier than the shadow:-
>
> ON A FIELD, SABLE, THE LETTER A, GULES (264) [155]

Though the rose seems initially to align the narrative with everything the world of the Puritans opposes (nature, sweetness, light, beauty, freedom, life), its reemergence at the end of the novel in the form of the word *gules*—which means "rose," among other things—does not finally "relieve the darkening close of a tale of human frailty and sorrow," nor does it derail it from its "track." Indeed, that ever-glowing point of light is even gloomier than the shadow it seems to relieve, as if our fantasy of escape from the letter is more damaging, more damning, than our acceptance of it. And this is as true of the narrative as it is of the world that the narrative purports to describe. As Evan Carton has put it, "bondage and compulsive repetition are, throughout Hawthorne's career, intimately and uncomfortably associated with the act of representation"—a bondage Hawthorne himself breaks through by his own dialectical engagement with his materials (155). Throughout Hawthorne's book we are in danger of becoming its prisoners. We could, of course, choose, like Pearl, to leave *The Scarlet Letter* behind altogether, but then we might simply become prisoners of some other world, some other imprisoning reality—as James makes vividly clear in *The Portrait.* How far can we run? Where is there an outside of culture to which we can escape? If we travel the book to its conclusion, we arrive at its end, which constitutes our own boundaries and limits, as readers of this text. There is something about admitting this that is, for Hawthorne, an essential requirement of citizenship.

The question the text is raising is what it means to take up one's culture and consent to it—whether we are its villains or its victims, Hawthorne occupying both of these positions, at least in his own interpretation of things. We can, of course, try to discard our culture altogether or escape it. But what does it mean if we remain? Does it mean simply accepting everything that our culture tells us to do and think—what Emerson and Thoreau call conformity? Does it mean the opposite: rejecting everything, starting as it were from scratch, as if that were possible? For Hawthorne even the rosebush has origins outside the purely natural world: it is rumored to have sprung up in the footsteps of Ann Hutchinson. This is not to say that purely natural rosebushes do not exist for Hawthorne, but that they do not exist in any meaningful way within human experience, which is always interpreting the objects of its perception through a vocabulary of meanings that precede our experience of the objects themselves.

It is to make this point that the novel gives us two long scenes concerning Pearl, both of which are crucial to understanding what it means to be born into a culture and yet to consent to it nonetheless, not without reservations and criticism, but acknowledging that one is a part of this culture, implicit in its sins—even its sins against us—and caught within the mirror of repetition we would most especially like to break free of. These scenes are important indices of what Hawthorne imagines culture to be and how he imagines our relationship with it to come about. More critically perhaps they set the ground not for the novel's thematic interest concerning our induction into culture, but, rather, for the cultural work in which the text is directly engaging. As James's essay verifies, the American reader's exposure to the letter is a re-exposure. We are already trained in the vocabulary of culture before we set about learning that vocabulary. "Words," writes Stanley Cavell,

> come to us from a distance. They were there before we were; we are born into them. Meaning them is accepting that fact of their condition. To discover what is being said to us, as to discover what we are saying, is to discover the precise location from which it is said; to understand why it is said from just there, and at that time. The art of fiction is to teach us distance—that the sources of what is said, the character of whomever says it, is for us to discover. (63)

In the first, and perhaps simpler, of the two scenes,

> Hester [has] bade little Pearl run down to the margin of the water, and play with the shells and tangled seaweed, until she should have talked awhile with yonder gatherer of herbs. So the child flew away like a bird, and, making bare her small white feet, went pattering along the moist margin of the sea. Here and there, she came to a full stop, and peeped curiously into a pool, left by the retiring tide as a mirror for Pearl to see her face in. Forth peeped at her, out of the pool, with dark, glistening curls around her head, and an elf-smile in her eyes, the image of a little maid, whom Pearl, having no other playmate, invited to take her hand and run a race with her. But the visionary little maid, on her part, beckoned likewise, as if to say—"This is a better place! Come thou into the pool!" And Pearl, stepping in, mid-leg deep, beheld her own white feet at the bottom; while, out of a still lower depth, came the gleam of a kind of fragmentary smile, floating to and fro in the agitated water. (168) [103]

The second scene, which is simply a longer replay of the first, goes as follows:

> Pearl, whose activity of spirit never flagged, had been at no loss for amusement while her mother talked with the old gatherer of herbs. At first, as already told, she had flirted fancifully with her own image in a pool of water, beckoning the phantom forth, and—as it declined to venture—seeking a passage for herself into its sphere of impalpable earth and unattainable sky. Soon finding, however, that either she or the image was unreal, she turned elsewhere for better pastime. She made little boats out of birch-bark, and freighted them with snail-shells, and sent out more ventures on the mighty deep than any merchant in New England; but the larger of them foundered near the shore. She seized a live horseshoe by the tail, and made prize of several five-fingers, and laid out a jelly-fish to melt in the warm sun. Then she took up the white foam, that streaked the line of the advancing tide, and threw it upon the breeze, scampering after it with winged footsteps, to catch the great snow-flakes ere they fell. Perceiving a flock of beach-birds that fed and fluttered along the shore, the naughty child picked up her apron full of pebbles, and, creeping from rock to rock after these small sea-fowl, displayed remarkable dexterity in pelting them. One little gray bird, with a white breast, Pearl was almost sure, had been hit by a pebble, and fluttered away with a broken wing. But then the elf-child sighed, and gave up her sport; because it grieved her to have done harm to a little being that was as wild as the sea-breeze, as or wild as Pearl herself.
>
> Her final employment was to gather sea-weed, of various kinds, and make herself a scarf, or mantle, and a head-dress, and thus assume the aspect of a little mermaid. She inherited her mother's gift for devising drapery and costume. As the last touch to her mermaid's garb, Pearl took some eel-grass and imitated, as best she could, on her own bosom, the decoration with which she was so familiar on her mother's. A letter,—the letter A,—but freshly green, instead of scarlet! The child bent her chin upon her breast, and contemplated this device with strange interest; even as if the one only thing for which she had been brought into the world was to make out its hidden import. (177–78) [108]

Generally, we think of play as that activity whereby children learn to manipulate reality, including its symbols, so as to bring it under private authority and make it an expression of their private will. Yet in the first instance what children do, when they imitate adult behavior, is less control the adult world than make themselves subject to its authority. Since Hawthorne himself used the example of his own daughter Una in constructing Pearl, let me make my point about what Hawthorne is doing here through the example of my grandson. When he was a mere twelve months old, he was playing with some scarves and hats I had laid out in a box for him. Suddenly, he pulled out one very long, rectangular-shaped scarf, with fringes at the end, and put it on his shoulders. No sooner had he done this, than he popped up, raced on his still wobbly toddler legs, into the other room, and removed a book from the shelf. It happened to be a prayer book, another item familiar to him from everyday experience, and with the scarf around his shoulders and his prayer book in his hands, we began to shuckle back and forth, as he had seen his father do every morning, when he recited the daily Jewish prayers. The scarf—no more than an ordinary scarf, which I had worn many times on my head and around my neck—reminded him, not as it did Pearl or Hawthorne, of a mermaid's garment, but of the Jewish prayer shawl that his father donned every morning. Therefore he put it on his shoulders, not (as I would have expected) on his head, and the moment he did that, he needed a prayer book to complete the outfit, and he needed to pray in order to complete the activity prompted by these props. My grandson was not playing at praying. He was following out the logic of a certain set of cultural artifacts that were no less determinative of his behavior than hunger or exhaustion.

I choose this particular scene not only because it so aptly enacts what is going on in Hawthorne's text, but because it once more verifies that Hawthorne's truths are homely truths we have all experienced. I also choose it for the fact of its Jewish, and therefore for some of my readers at least, unfamiliar cultural context. Just like learning the culture happens from within, so reading its evidences also comes from within. For me, within this particular Jewish religious culture, it was obvious what my grandson was enacting. For someone outside the culture, the enactment might well have been less obvious or downright unreadable.

What Hawthorne shows us in his text is that, like all children, Pearl imitates the adult world around her, and in this way tries to bring it within her comprehension and control. She builds little ships and sets them to sail upon the water and she tosses pebbles at the birds. What Pearl's play teaches her, however, is not boundless authority and freedom, but, rather, failure and a moral conscience; her boats founder, her jellyfish die, and one of her birds is wounded by the pebbles she has thrown.

The message Hawthorne only seems to be endorsing is that, left alone to his or her own devices, the child will, indeed, independently come upon those moral and psychological truths by which the human relation to the world, including to other people, ought to be governed. This would harmonize with the romantic leanings of *The Scarlet Letter*, the way in which it sides with the natural world (the rosebush) against society, at least Puritan society and, by extension, Hawthorne's own nineteenth-century America. But Hawthorne is no romantic, even if he is a romancer.

Like the rose the narrator proffers to the reader, so the scene is double-edged. Pearl is no pagan child of the forests, and even if her sense of sin and shame are innate, the forms of its expression will be determined by symbols that precede her own experience of it.

Therefore the scene's action, as we are told, takes place on the margins of culture: the "margin of the water," the "margin of the sea." It is also very deliberately positioned against the background of Hester's and Chillingworth's conversation with each other. Hence Hawthorne's splitting the scene in two. At the core of this scene, splitting it apart, is the relationship between the husband and wife, between the individual and the social contracts into which he or she enters. In other words, surrounding this scene, structuring it, is the relationship between nature, on the one hand, and society and culture, on the other. Pearl does not invent the forms of her play. Like my grandson, she borrows them from the culture, which controls them. Most especially she imitates, in order to try to learn, its language, which not only she but we the readers discover is also not within her control.

Pearl is neither an infant nor a toddler when she fashions her letter A. But she is an infant in an earlier scene, which acts as a prelude to this later moment in the text. "The very first thing," we are told, which baby Pearl "noticed, in her life, was—what?—not the mother's smile. . . . By no means! . . . [T]hat first object of which Pearl seemed to become aware was—shall we say it?—the scarlet letter on Hester's bosom!" (96) [62]. The child's behavior seems so unutterably perverse that the narrator can hardly get out his words. And we might be tempted to agree with him, until we consider, for a moment, where the scarlet letter is placed: "on Hester's bosom." From the moment Hester walks out on the scaffold, the community's eyes are fixed on her bosom. There is something decidedly lascivious about these Puritanical Puritans. Her bosom is not where their eyes should be, especially given their accusations against her. But Pearl's gaze is something else again. For any infant, what precedes even the mother's smile is, of course, her breast, Hester's bosom is exactly the right place for the child's eyes to light. And yet, as our narrator so well expresses it, this seems rather perverse, an indication, perhaps, of something like our human fallen-ness after all.

Not accidentally does Hawthorne in this way link the female breast and language. Human beings are from the start, for Hawthorne, those creatures who learn language (which is to say, culture as well) along with their mother's milk. But—and this is the relevant point—this somehow seems retrospectively like a violation, a sin, like sex itself. For the narrator, as for Hester herself, there is something inappropriately sexual about the child's relation to her mother. Perhaps she is the devil's child after all. But that is because for them the breast has already assumed a sexual significance. It has already entered into the realm of the not-purely-natural, the realm of the symbolic and the cultural. They are already ashamed of their adult sexuality, their sexualized relationship with the breast, as well might the Puritans themselves be when they are staring at Hester on the scaffold. And so they project their guilt instead onto the child (Hester herself does the same thing when she dresses the child in such a way as to make her a version of the letter, a fallen woman of scarlet in miniature), as Hawthorne—or his narrator, at least—projects it onto the

Puritans, as the New Americanists, let us say, project it onto Hawthorne and other nineteenth-century writers, and so on and so forth. What we have here is an infinite return of the repressed. The sin is always someone else's for us to condemn and, in condemning, absolve ourselves of guilt.

But Hawthorne does something else as well. Like Pearl, he also takes up the letter, to discover what it means, to discover its relevance to him, how he himself is implicated in what is not so much a primal scene or an original sin as his very birth into his very own culture—his birth into what I am calling the primal sin of culture. Without the act of adultery that brings her into the world, there is literally no Pearl, which is why Pearl insists so heartily that her mother take up the letter she has so boldly, passionately discarded in the forest. To throw away the *A* is to de-create the child, make her as if she had never been. And this, Hester realizes, is (to allude to Hester's own reasons in naming Pearl) too great a price to pay. Without his Puritan ancestors there literally can be no Hawthorne. And without Hawthorne, and without this text, there cannot be American culture as we know it, for us, if we are Americans, to interpret, critique, reject, and finally, if we choose (like Hester and Hawthorne) to remain within this culture, assume responsibility for it. *The Scarlet Letter* literally founds a culture, however much we might later wish to toss it away. The *A* that Hawthorne takes up in the Custom-House is that *A* that Hester resumes at the end of the novel. And it makes of Hawthorne neither a rebel nor a type of shame, but the fully willing inheritor of all his ancestors' sins (Hester's and the Puritans' both), and their representative, who will take upon himself both the crimes of culture and their endless and only partial working through.

This returns to us Henry James's discussion of the most famous work of his most famous literary ancestor. Before we are able to understand what culture means, before we have any picture of it in our mind, or any words with which to analyze that picture, we are within that culture, subject to its laws and language. Hawthorne's *Scarlet Letter* does not stand outside or apart from the other letters it repeats—the Puritans', Hester's, and even little Pearl's. Rather it is implicated in the tyranny of language, which is to say culture. Yet it is also a part of the "mystery" of language and culture, which defies our ever knowing precisely what language and culture mean.

At the very end of *The Scarlet Letter,* the narrator tells us, and generously imputes the words and thoughts to Hester herself, that, "earlier in life, Hester had vainly imagined that she herself might be the destined prophetess" of the "new truth" whereby "the relation between man and woman on a surer ground of mutual happiness . . . might be revealed." But, the text continues, she "had long since recognized the impossibility that any mission of divine and mysterious truth should be confided to a woman stained with sin, bowed down with shame, or even burdened with a life-long sorrow." The "new truth" would be "revealed" only "when the world should have grown ripe for it, in Heaven's own time" (263) [155]. Thus is the modesty of Hester Prynne, and of Hawthorne. Since for any reader reading this text, as for the writer writing it, or Hester living it, Heaven's own time must always be, by definition, outside human time, it must always be in the future of the moment in which we are now reading or writing or living these words.

And this is what Henry James, viewing his culture, not only retrospectively but from abroad, from England, also understands in his memoir, in which he places his reading of *The Scarlet Letter* both in the future and, obviously, already in the past as well. "Vividly imprinted" on the mind, to quote James's words once again, our culture is what is already "familiar" to us. It is what we have already read before we read it for the "first time," which is to say that there never is a first time. Nonetheless we do, if we choose, at some point sit down and intentionally, consciously, read this text again, which is to say, there is a moment when we take it up, of our own free will, and consent to reading it, consent to being (like Pearl in the forest) the child of that culture whose text is our own. James reads Hawthorne's *Scarlet Letter*, Hawthorne reads Hester's, we read James's and Hawthorne's and Hester's. And Hester peering down at her own bosom at the end of the book (as Pearl had peered down at hers) reads her own letter as well, and, like her daughter, her son (Hawthorne), and her grandson (James), tries to fathom its meaning. That is perhaps what reading means: taking up the letter and putting it on, to see how it fits, to see what it means to us.

Or not.

The question Hawthorne raises is whether choosing "not" to take up the letter and read it is, somehow, like choosing not to enter consciousness at all. Hence the importance of those scenes of Pearl entering into cultural, which is to say symbolic, consciousness. I will not explore the Lacanian dimensions of these scenes here (see Mellard 69–106). Suffice it to say that Pearl becomes conscious through and by the letter. She can no more discard it, except at great price, than can her mother. Nonetheless those moments by the seaside are no more demonic than that moment even earlier when she nursed at her mother's breast. Only later do we read back onto the moments of encroaching consciousness the evil that is consequence of the denial of those moments, not its incorporation into our being. As we all know, and as Hawthorne himself dramatizes in the novel itself, through the figure of Dimmesdale, consciousness once acquired is gotten rid of only at unbearable cost—which is distributed onto the community as well. To discard the letter, which is to say, to deny consciousness itself, opens us up—individually and collectively—to a return of the repressed far more imprisoning than the prison-house of culture itself, and far more damaging and destructive.

Whoever we would brand with the scarlet letter, it is on every breast, including our own. Better to begin reading—as does Hawthorne in "The Custom-House" and Hester when she returns to New England—by reading our own letter before we turn to the equally adulterated and "lettered intercourse" (27) [22], of the other cultures in which we reside and take our consciousness as well as our moral being.

REFERENCES

Budick, Emily Miller, *Engendering Romance: Women Writers and the Hawthorne Tradition, 1850–1990* (New Haven: Yale University Press, 1994).

Budick, Emily Miller, *Nineteenth-Century American Romance: Genre and the Democratic Construction of Culture* (New York: Twayne/Macmillan, 1996).

Carton, Evan, *The Rhetoric of American Romance: Dialectic and Identity in Emerson, Dickinson, Poe, and Hawthorne* (Baltimore: Johns Hopkins University Press, 1985).

Cavell, Stanley, *The Senses of Walden* (New York: Viking Press, 1972).

Crews, Frederick, "Whose American Renaissance?", *New York Review of Books,* 27 October 1988, 68–81.

Hawthorne, Nathaniel, *The Scarlet Letter,* the Centenary Edition of the Works of Nathaniel Hawthorne, ed. William Charvat et al., Vol. 1, Columbus, Ohio, Ohio State University Press, 1962.

James, Henry, "Hawthorne," in *Literary Criticism: Essays on Literature, American Writers, English Writers,* ed. Leon Edel (Cambridge: Library of America, 1984), 315–457.

Mellard, James M, *Using Lacan, Reading Fiction* (Urbana: University of Illinois Press, 1991).

Gender and Sexuality

LOUISE DESALVO

Nathaniel Hawthorne and the Feminists: *The Scarlet Letter*†

On Saturday, 27 July 1844, Nathaniel Hawthorne made an entry in his notebook that was the germ for his most famous and most critically acclaimed novel, *The Scarlet Letter*: 'The life of a woman, who by old colony law, was condemned always to wear the letter A, sewed on her garment, in token of her having committed adultery.'[1]

Between the publication of *Fanshawe* in 1828, until the time he began working on *The Scarlet Letter*, in 1849, when he was in his mid-forties, Hawthorne had confined his writing to short works,[2] many of which were published in periodicals such as *The Token* and the *Salem Gazette*, and later collected in *Twice-Told Tales* and *Mosses from an Old Manse*.[3] Although Hawthorne was prolific, he found it impossible to support his

† From *Nathaniel Hawthorne* (Atlantic Highlands, NJ: Humanities Press International, 1987), pp. 57–76. Copyright © 1987 by Louise DeSalvo. Reprinted by permission of the author. Page numbers in square brackets refer to this Norton Critical Edition.

1. N. Hawthorne, *The American Notebooks*, 254 [203].
2. N. Arvin, Introduction to *Hawthorne's Short Stories*, 254.
3. For a history of the publication of these volumes, see F. Bowers's 'Textual commentary' in N. Hawthorne, *Twice-Told Tales, The Centenary Edition of the Works of Nathaniel Hawthorne*, vol. IX, and J. D. Crowley's 'Historical commentary' in *Mosses from an Old Manse, The Centenary Edition of the Works of Nathaniel Hawthorne*, vol. X. The final collection to appear in Hawthorne's lifetime was *The Snow-Image* (1851). *See* J. D. Crowley, 'Historical commentary' in *The Snow-Image and Uncollected Tales, The Centenary Edition of the Works of Nathaniel Hawthorne*, vol. XI.

Virtually all of Hawthorne's short fiction is of interest to the feminist reader, and a complete study of his short works from a feminist perspective is sorely needed. * * * A considerable number of the stories deal with the issue of love and marriage, such as 'The Wedding Knell', 'The Minister's Black Veil', 'The Birth-mark', 'The New Adam and Eve', to name but a few, and they offer superb insights, not only into Hawthorne's attitudes, but also to the prevailing attitudes of the time. Others, such as 'Rappaccini's Daughter' and 'The Snow-Image,' deal with the issue of the condition of daughterhood within the patriarchal family as Hawthorne conceived it, which illuminates his treatment of Una. Still others, like 'The Gentle Boy' and 'My Kinsman, Major Molineux,' express Hawthorne's often ambivalent attitudes to his patriarchal forebears. Certain tales, such as 'Alice Doane's Appeal,' treat the issue of incest. A sketch which castigates a woman for modesty, and which illustrates Hawthorne's vacillating attitude to women's sexuality, in addition to demonstrating his voyeurism, is 'The Canal-Boat' which, according to J. R. Mellow, presents a 'devastating sketch' (54) of the American woman but which also presents as devastating a sketch of Hawthorne's persecutory attitude towards women. In addition, a number of the tales discuss the relationship of women to learning and education: among them, 'The Birth-mark' and 'Rappaccini's Daughter'. Several tales deal with the witchcraft era in American history: 'Young Goodman Brown', 'Alice Doane's Appeal', 'The Hollow of the Three Hills', 'The Prophetic Pictures', 'Edward Randolph's Portrait', 'Drowne's Wooden Image', for example. According to the feminist critic, N. Baym, a number of the tales uncover the psychopathology of the rejection of women, or of the feminine—'Rappaccini's Daughter', 'The Birth-mark', 'Wakefield', 'The Man of Adamant', 'The family on the paltry amounts he earned from his writing—he was paid only 'an average of $1.00 per page'[4] for his contributions to *The Token*—so that by the time he began *The Scarlet Letter*, he and Sophia had experienced periods of financial destitution, which had forced them to move in with Hawthorne's mother, then living on Mall Street in Salem, and had led Hawthorne to accept, on 3 April 1846, the position of Surveyor of the Custom-House in Salem.[5]

Hawthorne worked at the Custom-House until his dismissal, on 8 June 1849, as a result of the Whigs electoral victory. His wife, Sophia, 'greeted the first news of his dismissal with the remark, "Oh, then, you can write your book!"'

It was an extremely difficult time for him, for, on 30 July, his mother died. On the day before her death, in a journal entry, Hawthorne described the time that he spent at her bedside as 'the darkest hour I ever lived'. After her death he became seriously ill, but in September, after he recovered, he began writing *The Scarlet Letter* in 'response to his mother's death',[6] and as a kind of elegy to her. Like his heroine Hester Prynne, Elizabeth Hawthorne had become pregnant with her first child outside of marriage and had been socially ostracized; like Hester, his mother was husbandless, a single parent, raising her offspring alone, on the fringes of society. But it is also likely that his mother's death provided a kind of creative release, for he worked 'with an intensity that almost frightened his wife, and with a speed that brought the book to completion before the year ended'.[7] His wife Sophia certainly made the writing of the novel possible because without the money she had saved, 'a hundred and fifty dollars in bills, in silver, even in coppers',[8] her earnings from her decorative work, Hawthorne would have had to find another form of employment. The novel was composed quickly, and was completed on 3 February 1850.[9]

'The Custom-House' which serves as a introduction to and provides a frame for *The Scarlet Letter* was intended by Hawthorne to be a deliberate act of aggression and revenge against the Whigs who were responsible for his dismissal from his post. As author of the piece, he saw himself as a 'hunter', and he described the act of writing as the loading of a 'gun with a bullet and several buckshot'; reading his work, moreover, would 'kill the public . . . with my biggest and heaviest lump of lead'.[1] 'The Custom-House' is a savage, even vicious, satire against government service, and the kind of

---

Prophetic Pictures', 'Roger Malvin's Burial', 'Young Goodman Brown', 'The Minister's Black Veil', 'The Shaker Bridal'.

For illuminating discussion of the tales, see N. Baym, 'Hawthorne's women'; M. D. Bell, *Hawthorne and the Historical Romance of New England*; C. M. Bensick, *La Nouvelle Béatrice*; F. Crews, *The Sins of the Fathers*; J. Fetterley, *The Resisting Reader* (on 'The Birthmark'); J. R. Mellow, *Nathaniel Hawthorne*; N. F. Doubleday, *Hawthorne's Early Tales*.

4. J. D. Crowley, 'Historical commentary' to *Twice-told Tales, The Centenary Edition of the Works of Nathaniel Hawthorne*, 497.
5. R. Stewart, *Nathaniel Hawthorne*, 79. *See* G. C. Erlich, *Family Themes and Hawthorne's Fiction*.
6. H. H. Hoeltje, 'The writing of *The Scarlet Letter*', 342; N. Baym, 'Nathaniel Hawthorne and his mother', 20, 21. For a discussion of the effect of his mother's death on the creation of the novel, see, especially, N. Baym and G. C. Erlich. * * *
7. N. Baym, 1. See G. C. Erlich for a discussion of how his mother's death freed him to write his masterpiece.
8. L. H. Tharp, *The Peabody Sisters of Salem*, 190.
9. H. H. Hoeltje, 344.
1. W. Charvat, 'Introduction to *The Scarlet Letter*', *The Scarlet Letter. The Centenary Edition of the Works of Nathaniel Hawthorne*, xxii [214]. Citations from this edition will be placed within parentheses throughout this essay.

men who enter it. The Custom-House officers talk 'with that lack of energy that distinguishes the occupants of alms-houses, and all other human beings who depend for subsistence on charity . . . or any thing else but their own independent exertions' (7) [9]; the veterans under Hawthorne's orders, largely Whigs, whom he has not dismissed, although he should have, 'go lazily about what they termed duty, and, at their own leisure and convenience, betake themselves to bed again' (13) [13]. Hawthorne clearly feels that if he has tolerated them, and allowed them to continue in government service, because he 'could never quite find in my heart to act upon the knowledge' (14) [14] that they should be dismissed, the Whigs should have allowed him to retain his position.

Hawthorne reserves his most acid attack for the Permanent Inspector, animal-like, with 'no soul, no heart, no mind; nothing, as I have already said, but instincts' (18) [16] whose main attribute was his ability to recollect 'the ghosts of bygone meals', 'while all the subsequent experience of our race . . . had gone over him with as little permanent effect as the passing breeze' (19) [17].

Although Hawthorne makes it quite clear that he believes that he has been betrayed by *male* politicians, and that he intends to take his revenge upon them, none the less, the image which introduces 'The Custom-House' and dominates it, is the image of the American eagle, which Hawthorne depicts as a negative *female* image. It is worth quoting in its entirety, because the tale that Hawthorne tells of betrayal by the political system is enacted against the backdrop of this image of female betrayal. Over the entrance to the Custom-House,

> hovers an enormous specimen of the American eagle, with out-spread wings, a shield before her breast, and, . . . a bunch of intermingled thunderbolts and barbed arrows in each claw. With the customary infirmity of temper that characterizes this unhappy fowl, she appears, by the fierceness of her beak and eye and the general truculency of her attitude, to threaten mischief to the inoffensive community; and especially to warn all citizens, careful of their safety, against intruding on the premises which she overlooks with her wings. Nevertheless, vixenly as she looks, many people are seeking, at this very moment, to shelter themselves under the wing of the federal eagle; imagining, I presume, that her bosom has all the softness and snugness of an eider-down pillow. But she has no great tenderness, even in her best of moods, and, sooner or later, . . . is apt to fling off her nestlings with a scratch of her claw, a dab of her beak, or a rankling wound from her barbed arrows. (5) [8]

This image is extraordinary not only because it transforms the American eagle (the symbol of American might and strength) into a *female* image, but also because the effect of this transformation is subtly yet bitterly misogynist. It states that this female eagle is potentially and unpredictably vicious 'to the inoffensive community' precisely because she is female, and it is in the nature of females to be unpredictable; moreover, this female eagle is doubly vicious and unpredictable because she is an *unhappy* female who might lash out, with her 'thunderbolts' and her 'barbed arrows' at any one, at any time, whether or not they deserve it.

What Hawthorne accomplishes by rendering the national American symbol as female, is, in effect, a shifting of the responsibility and blame for his dismissal from the Custom-House away from the men who were responsible (and, by extension, away from the male-dominated patriarchal political system). Instead, despite his acerbic and rancorous remarks about men in government service, it is clear from this image of the female federal eagle, that, at some deep level, Hawthorne experienced his dismissal as *maternal* rejection, rather than as the result of a male political wrangle. Thus, the female federal eagle becomes the bad mother. She does not offer 'great tenderness, even in her best of moods', but rather flings her nestlings (one of whom was Hawthorne himself) from the nest 'with a scratch of her claw', or, even worse, 'a rankling wound from her barbed arrows' (5) [8]. This image subtly, but effectively, indicates that, at some deep level, Hawthorne irrationally blamed his mother for having abandoned him through her death, and that he also blamed her for losing his position at the Custom-House. What the feminist reader of 'The Custom-House' must note is how this image misrepresents political power as female, and how it blames a maternal figure for what is, in reality, the action of a male-dominated political machine.

Near the beginning of 'The Custom-House', Hawthorne announces that one of his reasons for writing the sketch is to 'put myself in my true position as editor' (4) [8] of Hester Prynne's tale, and to explain 'how a large portion of the following pages came into my possession, and as offering proofs of authenticity of a narrative contained therein' (4) [7–8]. Hawthorne describes finding barrels in the second story of the Custom House, containing bundles of documents of Jonathan Pue, a surveyor and local antiquarian, which could be used for 'a regular history of Salem' (31) [24]. In the bundle, Hawthorne discovers 'a certain affair of fine red cloth, much worn and faded . . . [which] on careful examination assumed the shape of a letter. It was the capital letter A' (31) [25] together with 'a small roll of dingy paper' containing 'a reasonably complete explanation of the whole affair' (32) [25].

This fictive posture of Hawthorne as editor of an historical document is an attempt to persuade the reader that the story that he tells, though embellished, is in effect an authentic, if not true account of Hester Prynne's fate in Puritan New England, and that the novel is the 'representation of a mode of life not heretofore described' (4) [8]. 'What I contend for is the authenticity of the outline' (33) [26]. Thus Hester's story comes to represent a model for describing the ways in which women who deviated from Puritan law were treated in Puritan New England and a model for understanding the character of the male élite Puritan oligarchy.

Although, ostensibly, *The Scarlet Letter* is a novel about a woman, what is insisted upon in the purported autobiography at the beginning of the novel, is that writing the history of a culture, that writing the history of a woman's fate within a culture, is, and ought to be, a strictly male enterprise. Attributing Hester's tale to the researches of Jonathan Pue effectively serves to write out his wife Sophia's, his mother Elizabeth's, and his daughter Una's relationship to the creation of the novel. If Hawthorne's prologue is, in fact, what it purports to be—an autobiographical account of how *The Scarlet Letter* came into being—it seems curious that Hawthorne would create Surveyor Pue and then credit him for the novel, rather than the women to whom he was indebted. Why bother to invent the fiction of the autobiographical frame at all if the autobiographical frame is, in fact, a fiction? In fact, what Hawthorne is doing, although he may not be aware of it, is creating his own autobiography by disavowing the facts of his life: he is creating what he would like the reader to know about himself: he is creating himself even as he creates Hester Prynne.

So convincing is Hawthorne's ploy that the reader rarely notices that the introduction provides a fascinating account of how the writing of romance or history *must* be articulated as if it proceeds from male to male even when (or perhaps especially when) women are the subject of that history, even when women have, in fact, made the writing of that work possible. In one very important scene, Hawthorne has a fantasy about Pue handing him Hester's story, and he describes Pue here as a ghostly father:

> With his own ghostly hand, the obscurely seen, but majestic figure had imparted to me the scarlet symbol, and the little roll of explanatory manuscript. With his own ghostly voice, he had exhorted me, on the sacred consideration of my filial duty and reverence towards him,—who might reasonably regard himself as my official ancestor to bring his mouldy and moth-eaten lucubrations before the public. (33) [26]

Thus, Hawthorne does not perceive the function of telling his tale as serving the causes of women's history; rather, Hawthorne is using one woman's story to serve the purposes of *male* history, both his own and men in general—a fact that has been overlooked by feminist literary critics. As Pue's ghost instructs Hawthorne: '. . . I charge you, in this matter of old Mistress Prynne, give to your predecessor's memory the credit which will be rightfully its due!' (33–34) [26].

The fundamental assumption about the nature of history that is embedded here is that woman's history is, and ought to be, the *property* of the male historian. Indeed, in the first moments of the novel, as Hester is about to emerge from the prison, Hawthorne uses the word 'narrative' to describe her: just as she is about to emerge from the jail, so is 'our narrative, . . . about to issue from that inauspicious portal' (48) [36]. She is not a character, she is a narrative, and in the language that Hawthorne insists upon, Hester and the narrative are, in fact, the same.

Moreover it is absolutely necessary that this history be presented as if it were authentic *especially if* that account grossly misrepresents woman's history, as the life story of Hester Prynne in *The Scarlet Letter* grossly distorts the fate of women who committed adultery in Puritan New England. In one very important sense, depicting Hester's strength and her resilience in the face of her punishment serves to nullify the effects of such persecution. If Hester could endure, and triumph (as women who were persecuted for adultery surely did not), then the negative consequences of the persecution itself are blunted, and the persecuting fathers rendered less virulent than they in fact were.

Critics, such as Charles Ryskamp and William Charvat have documented the fact that Hawthorne read widely in the history of Puritan New England of the 1640s, and that his notebooks indicate his concern with getting the details of that time precisely correct. Sources such as Caleb H. Snow's *History of Boston* and Joseph Felt's *Annals of Salem* were read and used 'to create an authentic picture of the seventeenth century'.[2] But despite Hawthorne's concern for accurate details, what has been glossed is the fundamental inaccuracy of the substance of Hawthorne's tale. Although Hawthorne would have known that Plymouth law decreed two whippings and the wearing of the letters 'AD' on the arm or back of the adulteress,[3] he has omitted the whippings from his romance and has Hester embellish the 'A' into an object of great beauty; although 'Governor John Winthrop's journal of 1644 records the execution, in the Bay Colony, of Mary Latham, eighteen, who having married an "ancient man . . . whom she had no affection unto," committed adultery with "divers young men"'[4] as the fate of the adulteress, Hawthorne writes, instead, how Hester's punishment is to endure the stares of the townspeople as she stands on the scaffold; although a real Hester Craford, in '1668 was found guilty of fornication', a 'lesser sin than adultery', and was 'whipped, and had her child taken from her', Hester is neither whipped, nor deprived of her child (xxvii); although Felt's *Annals* records that in 1694, 'adultery was punishable by an hour on the gallows, forty stripes' (xxvii), and the wearing of a capital 'A', Hawthorne has omitted the forty stripes from his tale. At the end of the novel, Hawthorne has Hester returning to the place of her punishment and *voluntarily* resuming her persecution, for 'not the sternest magistrate of that iron period would have imposed it,—[she] resumed the symbol of which we have related so dark a tale' (262) [154]. This is certainly a bizarre touch—to state that even the most iron-handed Puritan magistrate would not persecute her is falsifying history; to state that she willingly puts on the letter 'A' makes her *self*-persecuting, rather than persecuted. If Hawthorne, as he himself stated, was striving for *authenticity* in writing Hester's tale, *The Scarlet Letter* must be adjudged a woefully inaccurate failure.

In fact, even rendering the character of Dimmesdale as so pathetic, so ineffectual, so self-destructive effectively serves to dim the ferocity of his historical counterparts: it is impossible to read Hawthorne's Dimmesdale and conceptualize the Puritan oligarchy as avenging avatars. Instead of a persecuting angel, inspired by the wrath of the righteous, we are given the portrait of a bumbling lover, a portrait of a man who beats himself with 'a bloody scourge' (144) [90], who punishes *himself* for hiding his sin, rather than a man who persecutes others.

Hawthorne substitutes Dimmesdale's refusal to acknowledge the fact of his paternity (which is surely interesting in light of Hawthorne's biography) and Chillingworth's probing into the secrets of Dimmesdale's heart as

2. C. Ryskamp, 'The New England sources of *The Scarlet Letter*,' 269 [288].
3. W. Charvat, xxvi.
4. W. Charvat, xxvii. An entry in Hawthorne's *The American Notebooks* for 28 August 1837 details a visit to Eben Hathorne, who might have been the model for Surveyor Pue:

> The pride of ancestry seems to be his great hobby; he had a good many old papers in his desk at the custom-house, which he produced and dissertated upon, and afterwards went with me to his sister's in Howard place; and showed me an old book, with a record of the first emigrants (who came over 200 years ago) children in his own handwriting. . . . As we walked, he kept telling stories of the family. . . . (74)

An entry for 15 June 1838 describes going to the burial-ground in Charter Street, and seeing a gravestone 'to the memory of "Colonel John Hathorne, Esq.," who died in 1717. This was the witch-judge. . . . Other Hathornes lie buried in a range with him on either side' (172). Hawthorne concludes: 'It gives strange ideas, to think how convenient to Dr. Peabody's family this burial-ground is,—the monuments standing almost within arm's reach of the side windows of the parlor' (172). In my judgement, Nathaniel's marriage to Sophia Peabody was, in part, an attempt to forge a connection with his own past, and I use, as evidence, this quotation from his diaries.

greater evils than the evils suffered by those persecuted by Puritan justice! And so, in the context of *The Scarlet Letter*, in a fascinating reversal of the facts of history, Dimmesdale, the representative of the Puritan state and Puritan power in the novel, becomes more sinned against than sinning—he is described as the 'victim . . . for ever on the rack' (140) [87]—and his victimization at the hands of Chillingworth becomes of greater consequence and has more dire results than Hester's punishment! Hawthorne, in his revisionist history, thus substitutes a portrait of a male victim for an accurate portrait of a female victim of the Puritan oligarchy.

One of his primary arguments that blunts the effect of the persecution of women is that grief and suffering, rather than being destructive to a woman, is, in fact, ennobling, so long as she does not become a social reformer as a result of it: Hester wears 'a halo of misfortune' (53) [39]; when Hester sees Pearl's gaiety, what she wants for her, is 'a grief that should deeply touch her, and thus humanize and make her capable of sympathy' (184) [112].

As the novel progresses, Hawthorne subtly shifts the blame for what happens to Chillingworth and to Dimmesdale onto the shoulders of Hester. The effect of this is to render Hester completely responsible for the physical, emotional and spiritual well-being of the men in her life. Chillingworth tells her 'Woman, woman, thou art accountable for this' (194) [117] and Dimmesdale repeatedly insists that his salvation is her responsibility: 'Think for me Hester! Thou art strong' (196) [118]. In one very important scene, Chillingworth blames Hester, and not the rigid system of Puritan justice, or his own actions, for Dimmesdale's slow demise: he tells her 'you [Hester] cause him to die daily a living death' (171) [104]; and she accepts the blame for Chillingworth's obsession with revenge—when he asks her who is responsible, she says 'It was myself' (173) [105] just as she accounts herself responsible for Dimmesdale: she becomes 'sensible of the deep injury for which she was responsible to this unhappy man, in permitting him to lie for so many years' (192) [116]. Thus, Hester, the person with the least amount of real power in the novel is made, symbolically, the person with the most power, and the most responsibility for the outcome of the tale. This move deflects attention away from the reality of Hester's utter powerlessness in the Puritan scheme. Just as Hawthorne fixes our attention on Dimmesdale's dying, and on Hester's heroism and her responsibility for the well-being of the men in her life, to blunt the facts of Puritan history, so he employs a similar strategy when, in 'The Custom-House', his symbolic language presents *his* persecution at the hands of the Whigs as having a greater consequence than Hester's suffering: he aligns himself with the benighted Dimmesdale when he describes writing Hester's story 'from beyond the grave' (44) [32], and he states, in bold, attention-getting letters, that her tale 'may be considered as the POSTHUMOUS PAPERS OF A DECAPITATED SURVEYOR' (43) [32]; the 'A' that he finds in the pack of old papers sears *his* breast, not hers.

In 'The Custom-House', Hawthorne describes his very real connection with the historical figures in the novel by relating his family history, 'the deep and aged roots which my family has struck into the soil' (8) [10]. It is important to note that he conceptualizes his family history in completely male terms; in the family tree that he outlines here for himself, not one woman is mentioned. He describes the 'figure of that first ancestor, invested by family tradition with a dim and dusky grandeur, was present to my boyish imagination, as far back as I can remember' (9) [11] and how the figure of this 'grave, bearded, sable-cloaked, and steeple-crowned progenitor,—who came so early with his Bible and his sword' haunted his imagination. Hawthorne describes this forebear as a 'bitter persecutor', and he relates how his son 'inherited the persecuting spirit' by making 'himself so conspicuous in the martyrdom of the witches' (9) [11].

Although Hawthorne admits that his male ancestors were 'bitter persecutors', and although he admits that they indulged in 'cruelties', he blunts the effect of this admission by various narrative strategies. One strategy is to describe actual historical events as hypothetical events. For example, he does not describe the persecution of heterodox believers and so-called witches as *having* taken place: he describes those persecutions as if they *might* have taken place: 'It might be, that an Antinomian, a Quaker, or other heterodox religionist, was to be scourged out of town . . . It might be, too, that a witch like old Mistress Hibbins . . . was to die upon the gallows' (48) [36]. Or he blunts the effect of persecution, when he describes it, by employing neutral language: offenders are not beaten, they are 'corrected at the whipping-post' (49) [36]; the pillory is 'an agent in the promotion of good citizenship' (55) [40]; Hester's 'A' is not a punishment, but a 'fitting decoration to the apparel which she wore' (53) [39].

Thus, the effect of Hawthorne's description of the Puritan oligarchs is to render them as having been simply intolerant men, rather than as sadists and misogynists, the only conclusion that can be reached after reading a history of the Hawthorne family, such as that by Vernon Loggins. In Loggins, the bearded, sable-cloaked Hathorne's typical punishments are described: he 'ordered a constable to cut off a convicted burglar's ear and brand the letter B on his forehead'; he sentenced a woman accused of fornication to a whipping, but simply fined a man who routinely beat his wife; he ordered into slavery the children of Quakers; he ordered 'hangings, the cutting off of men's ears, the boring of holes through women's tongues with red hot irons' and the starving to death of imprisoned Quakers (63). Nor can Hathorne's punishments be explained away by stating that they were typical of the time: indeed, so excessive was his zeal, that in 1661, because of his behaviour, Charles II commanded that any case involving a Quaker should be transferred to English courts. Hathorne's son was equally guilty of excesses; by 1692, as the magistrate responsible for the preliminary hearings in the Salem witchcraft trials, he had crowded the prisons with 'supposed witches and wizards'; on 19 July 1692 five women were hanged.[5]

Hawthorne's purpose in writing *The Scarlet Letter* is overtly announced in 'The Custom-House'. He states:

> I know not whether these ancestors of mine bethought themselves to repent, and ask pardon of Heaven for their cruelties; or whether they are not groaning under the heavy consequences of them, in another state of being. At all events, I, the present writer, as their representative, hereby take shame upon myself for their sakes, and pray that any curse incurred by them . . . may now and henceforth be removed. (9–10) [11]

5. V. Loggins, *The Hawthornes*, 130–303.

The act of writing the novel, will, therefore, exculpate Hawthorne's Puritan forebears 'now and henceforth'. It is no wonder, therefore, given his self-admitted reason for writing the novel, why he revises what Hester's fate would have been into the tale that he tells. In order to remove the burden of guilt and responsibility from his male forebears, he must rewrite woman's history. Instead of presenting a vicious tale of brandings and beatings, what Hawthorne presents instead is the pargeted[6] tale of Hester Prynne: 'her beauty shone out, and made a halo of the misfortune and ignominy in which she was enveloped' (53) [40].

Hawthorne's principal strategy, at the beginning of the novel itself, is to deflect attention away from the Puritan patriarchs who have voted Hester's punishment, the patriarchs who have made the laws and who enforce them. They scarcely exist as far as the novel is concerned. Rather, as Hester Prynne emerges from the jail, Hawthorne focuses all of his narrative attention, and for several pages, upon the vengeful response of the Puritan *women*. They take 'a peculiar interest in whatever penal infliction might be expected to ensue' (50) [36]. In the novel, it is not the oligarchs, Hawthorne's forebears, who punish, it is the goodwives who demand justice:

> 'Goodwives', said a hard-featured dame of fifty, 'I'll tell ye a piece of my mind. It would be greatly for the public behoof, if we women, being of mature age and church-members in good repute, should have the handling of such malefactresses as this Hester Prynne. What think ye, gossips? If the hussy stood up for judgment before us five, that are now here in a knot together, would she come off with such a sentence as the worshipful magistrates have awarded? Marry, I trow not!' (51) [37]

Although Hawthorne appears to be arguing for gender-determined punishment, he is also stating that the male judges who did exist were fairer than any woman would have been. A man in the crowd who overhears the women says 'is there no virtue in woman, save what springs from a wholesome fear of the gallows?' (52) [38]. And Hawthorne states that the uglier a woman is, the more vengeance she would exact from criminals: 'the ugliest as well as the most pitiless of these self-constituted judges' says 'This woman has brought shame upon us all and ought to die. Is there not law for it?' (51) [38]. All the vengeance at the beginning of the novel has been female; when a beadle finally appears, he seems far less severe than the women; in contrast to their graphic desire to 'put the brand of a hot iron on Hester Prynne's forehead' (51) [37], he simply 'prefigured and represented in his aspect the whole dismal severity of the Puritanic code of law' (52) [38].

This is a fascinating strategy. Hawthorne denies history by misrepresenting Hester's punishment as if it would have been essentially fair and judicious. Then he argues that if women had had political power, they would have been harsher to adulteresses than his fictional Puritan leaders had been to Hester. Hawthorne therefore creates a romance about judicious Puritan rule which denies the reality of the abuse of power by Puritan rulers, and then he uses the fiction he has created to argue that men

6. J. Marcus has used the term in reference to Virginia Woolf's process of composition in *The Years*. *See* J. Marcus, '*The Years* as Greek drama, domestic novel, and Götterdämmerung'.

are essentially more fair-minded than women would be! This literary strategy, though highly persuasive, is extraordinarily illogical and misleading for Hawthorne draws ethical conclusions about justice being fair-minded if it is male, and vengeful if it is female from a universe which he himself has created, and which is a misrepresentation of historical reality.

In the context of the Puritan cosmology developed in *The Scarlet Letter*, Hester Prynne is enormously concerned about what will happen to her child Pearl, as well she should be because the children of miscreants were not treated well in Puritan New England. Throughout the novel, Pearl is repeatedly associated with the devil, with evil, with sin, and with witchcraft: her looks are 'perverse' and 'malicious' (92) [60]; she is an 'imp of evil' (93) [61], 'fiend-like', an 'evil spirit' (97) [63], 'a shadowy reflection of evil' (94) [61], a 'demon offspring' (99) [64], a 'demon-child' (100) [64]; there is a 'fire in her' (101) [65]; her cries are 'a witch's anathemas in some unknown tongue' (94) [61], she is a 'little baggage [who] hath witchcraft in her' (116) [74], there is 'witchcraft in little Pearl's eyes' (154) [96]; she is 'a shadowy reflection of evil' (94) [61], the 'effluence of her mother's lawless passion' (165) [101]; her imaginary playmates are 'the puppets of Pearl's witchcraft' (95) [61]. And Hawthorne makes it clear that she has inherited these tendencies from her mother: although, *in utero*, Pearl's character was at first unblemished, her 'mother's impassioned state had been the medium through which were transmitted to the unborn infant the rays of its mortal life; and, however white and clear originally, they had taken the deep stains of crimson and gold, the fiery lustre, the black shadow' (91) [59].

The character who is repeatedly associated with Pearl is Mistress Hibbins, who is based upon an actual woman, who, the narrative records, will be accused of witchcraft and who will 'die upon the gallows' (49) [36]. In the forest, Dimmesdale likens Pearl's cries to 'the cankered wrath of an old witch, like Mistress Hibbins' (210) [126]. Nor can Hester civilize Pearl. Hawthorne suggests that, without a man in the house, Hester is incapable of controlling Pearl's bad behaviour—raising Pearl is described as a process which is identical to an exorcism which Hester is not capable of performing, an attitude towards childrearing that Hawthorne certainly manifested in his journal descriptions in reference to his daughter Una who is described in terms very like those used to describe Pearl: 'The child could not be made amenable to rules', her 'elements were perhaps beautiful and brilliant, but all in disorder' (91) [59].[7]

The process of caring for Pearl, however, leads to Hester's salvation, because through caring for Pearl she avoids becoming a latter-day Anne Hutchinson, and accepts her womanly role, which, according to Hawthorne, is essential if a woman is to be saved. But it is the absent father Dimmesdale who is responsible for Pearl's salvation: as Pearl kisses him at the end of the

7. T. Walter Herbert, 'Nathaniel Hawthorne, Una Hawthorne, and *The Scarlet Letter*: Interactive Selfhoods and the Cultural Construction of Gender,' *PMLA* 103 (1988): 285–97. Hawthorne's entries in *The American Notebooks* from 19 March 1848 to 30 July 1849 are illuminating, as Herbert argues brilliantly in his article. *See, especially*: 'The children have been playing ball together; and Una, heated by the violence with which she plays, sits down on the floor, and complains grievously of warmth—opens her breast. This is the physical manifestation of the evil spirit that struggles for the mastery of her; he is not a spirit at all, but an earthy monster, who lays his grasp on her spinal marrow, her brain, and other parts of her body that lie in closest contiguity to her soul; so that the soul has the discredit of evil deeds' (420–1) [206]. It seems bizarre to see the spectacle of the devil inhabiting one's daughter as she undoes her clothes because she has been playing with gusto.

novel, she feels grief for the first time, and it is this grief, and not her mother's care, which humanizes her:

> Pearl kissed his lips. A spell was broken. The great scene of grief, in which the wild infant bore a part, had developed all her sympathies; and as her tears fell upon her father's cheek, they were the pledge that she would grow up amid human joy and sorrow, nor for ever do battle with the world, but be a woman in it. (256) [151]

This scene effectively obliterates all the years of Hester's mothering. Just as Hawthorne has written his own mother and his wife out of 'The Custom-House', so he writes Hester out of the cause for Pearl's salvation in *The Scarlet Letter.* It is not all the years of Hester's toil which saves Pearl from a life of evil in Puritan New England, or from being persecuted as a witch, like Mistress Hibbins. Rather, Pearl becomes a happy woman because of this single moment that she shares with her father Dimmesdale which unlocks her ability to feel grief. Salvation comes, not as a result of Pearl and Hester working together through the years to make a reasonably good life for themselves despite persecution. No, salvation comes, in Hawthorne's world, from being humanized as a result of feeling sorry for the suffering your *father* has experienced! And even the reprehensible Chillingworth, in leaving Pearl 'a very considerable amount of property, both here and in England' (261) [153] is made even more responsible for Pearl's good fortune, than all the years of Hester's toil as a single parent, raising her child alone. Hawthorne, therefore, privileges the effect of the absent father upon the good fortune of the child over the labour of the present mother.

It is no wonder, then, that heaven is described in the novel as an all-male paradise, with no room for Hester, a paradise where no women are permitted. Although Hester is buried near Dimmesdale, in paradise, however, in 'the spiritual world, the old physician and the minister—mutual victims as they have been—may . . . have found their earthly stock of hatred and antipathy transmuted into golden love' (261) [153].

* * *

* * * In 'The Custom-House', when Hawthorne picks up Hester's scarlet A and places it on his breast, he 'experienced a sensation not altogether physical, yet almost so, as of burning heat; and as if the letter were not of red cloth, but red-hot iron', he 'shuddered, and involuntarily let it fall upon the floor' (32) [25]. Although branding was, in fact, a rather typical Puritan punishment for transgressors, it is important to note that Hawthorne records the branding as having occurred (symbolically) to himself, and not to Hester, which serves to focus attention upon himself, and to deflect attention away from her. As she is an adulteress, he becomes an adulterator, as he misleads his audience into thinking that his Puritan forebears were simply intolerant men, rather than sadistic persecutors who repeatedly sentenced adulteresses either to banishment (which meant certain death) or to 'thirty stripes from a knotted whip',[8] which usually led to festering wounds, serious infection, and a long, lingering, painful illness, often resulting in death. Hawthorne has therefore used the novel to control his own past, to rewrite his past into a version that would provide him with less virulent

8. V. Loggins, p. 69.

male ancestors and that would present them to the world as less sadistic than they in fact were. And Hawthorne's desire in writing *The Scarlet Letter*, to remove the curse from his Puritan forebears 'now and henceforth' (10) [11] was so immensely successful that his rewritten, highly inaccurate version of Puritan history, which blunts the reality of the persecutions of the time, is the version that most Americans believe, because most Americans learn their Puritan history, not through a history which graphically describes the savagery of the Hathornes, but instead, through reading *The Scarlet Letter*. *The Scarlet Letter* has, indeed, absolved Hawthorne's forebears from guilt, 'now and henceforth'.

THOMAS R. MITCHELL

"Speak Thou for Me!": The "Strange Earnestness" of *The Scarlet Letter*†

> You speak, my friend, with a strange earnestness.
>
> Roger to Arthur in *The Scarlet Letter*

Many fine arguments have been made that would locate the creative origins of *The Scarlet Letter* in Hawthorne's personal crises in the summer of 1849—the scandal over his "decapitation" at the Custom House and the greater personal crisis of his mother's death.[1] As influential as these events may have been to the impetus to write again and as much as they may have influenced the narrative, they do not allow or answer a crucial question about the origins of the narrative. Why write about adultery? Specifically on the consequences of an adulterous moment seven years in the past? And why meditate on the "motives and modes of passion that influence" the characters in the narrative to such a degree that Sophia would be astonished by, even resentful of, the fury with which Hawthorne worked (1:33) [26]?[2]

† From *Hawthorne's Fuller Mystery* (Amherst: University of Massachusetts Press, 1998), pp. 125–58. Unless otherwise specified, page numbers in square brackets refer to this Norton Critical Edition.

1. Stephen Nissenbaum, introduction to Nathaniel Hawthorne, *The Scarlet Letter and Selected Writings*, Modern Library Edition (New York: Random House, 1984), vii–xlii, has made the strongest argument for the influence of Hawthorne's "firing" on the "fiction" of "The Custom-House" and on the "*real* autobiography" of *The Scarlet Letter* (xix); see also his "Firing of Nathaniel Hawthorne," *Essex Institute Historical Collections* 114 (1978): 57–86. Nina Baym, "Nathaniel Hawthorne and His Mother: A Biographical Speculation," *American Literature* 54 (1982): 1–27, has made the best case for the influence of Hawthorne's mother and her death on the writing of *The Scarlet Letter*; along similar lines, see also, more recently, Miller, *Salem Is My Dwelling Place*, 278–98. Erlich, *Family Themes*, on the other hand, argues that Louisa and Elizabeth, along with the mother, inform Hawthorne's conception of Hester (99). See also, James M. Cox, "*The Scarlet Letter*: Through the Old Manse and the Custom House," *Virginia Quarterly Review* 51 (1975): 432–47.
2. Sophia was perhaps more resentful than worried. The only passages from Sophia's 27 September 1849 letter to her mother (Berg) that are frequently quoted are her statements that Hawthorne "is writing morning & afternoon" and that he "writes immensely—I am almost frightened about it." Sophia makes these statements, however, to explain why *she* has had no time of her own to devote to her drawing and why she needs her parents to send writing paper and ink. Having asked her mother to send two more yards of material, she writes:

> I have been wholly absorbed by making this dressing-gown myself—cutting it out and all—Tell Livy that I have to get ready for winter before I can draw. Also tell her that if I did *not* have to sew now, I still could not do anything yet—because Mr. Hawthorne is writing morning & afternoon, & I have no time yet. I must take care of the children now all day long—& sew at the

The source studies that find the romance's narrative origins in Hawthorne's meditation on New England history, particularly on Anne Hutchinson, fail also to account for the passion with which Hawthorne wrote.[3] Can we really read the romance as the profound but detached artistic product of a purely historical imagination? Hawthorne could not. When he read, or "tried to read," the just-completed manuscript to Sophia, as he confessed to himself years later in his journal, he found himself so moved by his own words during the final scenes that, like the voice of Arthur in its metaphorical effect on Hester, his own "voice swelled and heaved as if I were tossed up and down on an ocean, as it subsided after a storm." "I was in a very nervous state, then," he reminds himself, "having gone through a great diversity and severity of emotion, for many months past. I think I have never overcome my own adamant in any other instance."[4] Hawthorne would fail to mention his own reaction in a letter written the next day to his friend Horatio Bridge, but he does report proudly that Sophia, after the reading, had gone to bed with "a grievous headache," the effect, he implies, of the work's power (16:311) [p. 311 of the Ohio State UP Centenary Edition vol. 16: *The Letters, 1843–1853*].[5] Given his later, private account of the unbearable emotion with which he read, we may well suspect that Sophia's headache was brought on not only by what Hawthorne wrote of Arthur and Hester but also by what Sophia inferred from the passion with which he wrote it and read it. As Herbert has so thoroughly demonstrated, however, Sophia consistently and resolutely avoided facing disturbing truths, suppressing any suggestion that Hawthorne was not her altogether happy Apollo.[6] When her less-worshipful sister Elizabeth passed along approvingly the observation that Hawthorne in writing *The Scarlet Letter* had "purified himself by casting out a legion of devils into imaginary beings," Sophia insisted vehemently that "it was a work of the imagination wholly & no personal experience, as you well know."[7] Was Hawthorne's romance the creation of the sympathetic artist's imaginative identification

---

same time. . . . Will you ask father to buy half a ream of good letter paper for as cheap as possible. I have no paper. And I want some yellow envelopes. I have no ink down stairs & cannot disturb Mr. Hawthorne. He writes immensely—I am almost frightened about it. But he is well now & looks very shining.

Herbert, *Dearest Beloved*, does not quote this letter, but he might well have, for it seems to support much of what he says about the tensions within the Hawthorne marriage.

3. The most noted of these studies are Charles Ryskamp, "The New England Sources of *The Scarlet Letter*,"*American Literature* 31 (1959): 257–72; and Michael J. Colacurcio, "Footsteps of Anne Hutchinson: The Context of *The Scarlet Letter*," *ELH* 39 (1972): 459–94. See also Colacurcio's essay "'The Woman's Own Choice': Sex, Metaphor, and the Puritan 'Sources' of *The Scarlet Letter*," in Michael J. Colacurcio, ed., *New Essays on "The Scarlet Letter"* (Cambridge: Cambridge University Press, 1985), 101–35; and Frederick Newberry, "A Red-Hot *A* and a Lusting Divine: Sources for *The Scarlet Letter*," *New England Quarterly* 60 (1987): 256–64.
4. Nathaniel Hawthorne, *The English Notebooks*, ed. Randall Stewart (New York: Russell and Russell, 1941), 225. Hawthorne recalled this moment on 14 September 1855.
5. Person, *Aesthetic Headaches*, finds in this episode a metaphor for the creative origins and purposes of Hawthorne's, Poe's, and Melville's art.
6. Herbert, *Dearest Beloved*, 151–52. Herbert attributes Sophia's headache to the "thunderbolt" of Hawthorne's "depiction of the burdens" imposed on women by the domestic ideal praised in the final sentence of the penultimate paragraph (209).
7. Sophia Hawthorne to Elizabeth Peabody in a 21 June 1850 letter (Berg). Elizabeth had apparently written to Sophia the comments of a Mr. Bellows and endorsed them. Sophia's letter responds not only to that comment but to another similar one by Ellery Channing that Elizabeth reported to her. Sophia's explanation is that Hawthorne "sees men & he sees passions & crimes & sorrows by the intuition of genius, & all the better for the calm, cool, serene height from which he looks." In the next sentence, however, she states: "Doubtless all the tendencies of powerful, great natures lie deep in his soul; but they have not been waked, & sleep fixedly, because the noblest only have been called into action." I contend, of course, that these "tendencies" were

with sin and guilt, as Sophia would have it, or was it, like Arthur's passionate sermons, another author's concealed confession? Elizabeth, as Sophia reminded her, did not know; however, she could infer what Sophia would never admit. Perhaps the moral of *The Scarlet Letter* is, in fact, the moral the narrator says it is: "Be true! Be true! Be true! Show freely to the world, if not your worst, yet some trait whereby the worst may be inferred!" (1:260) [153].

What is often overlooked in Hawthorne's famous image of himself as romancer in a moonlit, coal-fired familiar room "glancing at the looking-glass" and recording the meeting of the "Actual and Imaginary" (1:35–36) [28] is what is not directly revealed but may be seen—what may be, that is, inferred. Hawthorne directs our attention to the "neutral territory" that lies "deep within" the "haunted verge" of the mirror, but if we shift our focus from the margins of the glass to its anything-but-neutral center we may see a face staring into the mirror's margins at the "ghosts" that, once transformed by the light of his imagination, no longer "affrighten" him, a face staring also into the mirror's center, at his own face, transformed also, ghostly too in the light (1:36) [28]. It is this face that will not be seen in the mirror nor recognized behind the "veil" by "most of his schoolmates and lifemates," those who think they are closest to him but who understand him less than the "one heart and mind of perfect sympathy," the "kind and apprehensive, though not the closest friend" who will listen to him and in recognizing his voice as the voice speaking "of the circumstances that lie around" them and "even of himself" will be the "genial consciousness" that will enable him to "complete his circle of existence by bringing him into communion with" the "divided segment" of his "own nature" (1:3–4) [7].

The self-riven Arthur in the romance speaks as does this author of the romance.[8] The "strange earnestness" of Arthur's concealed confessions finds also only "one heart and mind of perfect sympathy" capable of recognizing that what sounds like the voice of an imaginary self, a persona conceived in the desire for rhetorical effect, is, in fact, his actual voice being true, as openly as he finds possible, to his own nature. Perhaps we join the

---

awakened more than she was willing to admit and took "action" in art. Sophia's other sister, Mary Peabody Mann, agreed with Elizabeth, writing her son Horace that incidents in Hawthorne's life inevitably found themselves "bye and bye in books, for he always put himself into his books; he cannot help it" (Antioch College Library; qtd. in Miller, *Salem is My Dwelling Place*, 9). Elizabeth's approval of the assessment of Hawthorne's art as an exorcism of private demons accords with Hawthorne's own frequent "complaint" that the demonic seemed to overtake him in the act of writing, a fear first expressed jokingly to his mother in the 13 March 1821 letter in which he announced that he would become a writer (15:139) [p. 139 of the Centenary Edition, vol. 15: *The Letters, 1813–1843*].

8. In his rhetorical, not biographical, study, Gordon Hutner, *Secrets and Sympathy: Forms of Disclosure in Hawthorne's Novels* (Athens: University of Georgia Press, 1988) identifies Dimmesdale's Election Day sermon as a paradigm for the purposes and methods of *The Scarlet Letter* and, indeed, of art in general (25–26). Hawthorne's identification with Arthur, of course, has often been made by critics who read the romance as veiled autobiography. Nissenbaum, introduction to *The Scarlet Letter and Other Writings*, for instance, argues that Arthur Dimmesdale, as "priest-artist" (Arthur/author and Arthur) embodies Hawthorne's guilt for having compromised through ambition his artistic integrity ("celibacy") in the politics of the Custom House (xxviii–xxxvi). Evan Carton, "'A Daughter of the Puritans,'" reads Dimmesdale as epitomizing the "contradictions" in the novel between the "diverse sexual and familial roles" plaguing Hawthorne and informing his characterizations not only of Dimmesdale but also of Chillingworth, Hester, and Pearl (222). Miller, *Salem is My Dwelling Place*, argues for Hawthorne's "being the sum total" of all four of his characters (296–97). See also William C. Spengemann, *The Forms of Autobiography: Episodes in the History of a Literary Genre* (New Haven: Yale University Press, 1980), who discusses the work within the generic context of American autobiography (132–65).

interpretative community of Arthur's congregation—of Hawthorne's "schoolmates and lifemates"—when we read as mere rhetorical pose Hawthorne's claim that "thoughts are frozen and utterance benumbed" unless he has a "true relation" with that audience of "one heart and mind of perfect sympathy" (1:4) [7]. Perhaps the pose is not, after all, a pose. With his first chapters at press and his final three chapters yet to be written, Hawthorne concluded the romance by imagining in Arthur's triumphant eloquence the irony of his own anticipated success with the impending publication of his "scarlet letter" to the world, its title page garishly blazoned like Hester's breast with Arthur's concealed sign of his own guilt, guilt self-inscribed beneath the cover of his clothing but exposed and finally read on the body of Arthur within the body of the book. A member of the "priesthood" of literature—as he had named himself recently in a letter to Longfellow (16:270–71) [pp. 270–71 of the Centenary Edition, vol. 16]—Hawthorne makes Arthur's Election Day sermon a metaphor for the "passion and pathos," the power and purpose, of his own art (1:243) [144].[9]

Arthur speaks two messages to two audiences. Those near him in the congregation hear the "grosser medium" of Arthur's words, the meaning that "clogs" the "spiritual sense" (1:243) [144]. At a distance, outside the walls of Arthur's church, Hester, the second audience, the audience of "one mind and heart of perfect sympathy," listens "with such intentness" and "so intimately" that she hears the personal message within the sermon, the sermon's "spiritual sense," the "profound and continual undertone" of "the complaint of a human heart, sorrow-laden, perchance guilty, telling its secret, whether of guilt or sorrow, to the great heart of mankind; beseeching its sympathy or forgiveness,—at every moment—in each accent" (1:243–44) [144]. "Guilt or sorrow," "sympathy or forgiveness"—even if we hear the confessional undertone sustaining the sermon's or the romance's power, we may, like Arthur's auditors, like witnesses to his final evasive confession on the scaffold, interpret the brand of the "red-hot iron," the "burning heat" of the "scarlet letter" on Hawthorne's own breast, when he takes it up, as the mark of the author's passionate sorrow and sympathy (1:32) [25]. Or we may, like "the one mind and heart of perfect sympathy," like Hester of Arthur, interpret Hawthorne's "shudder," his "involuntary" failure to hold the burning letter long to his breast, as the sign of his own identification with the scarlet letter, of his own guilt, which he, like Arthur, cannot show freely to the world. We may note as well that just as Arthur did not achieve his greatest artistic power until he burned the first draft of his Election Day sermon and transformed his anguish into a passionate, public oration that did and did not reveal its creative origins in the self, so Hawthorne could not "warm" exclusively at the "intellectual forge" of his "imagination" the "figures" of his tale until he entered the coal-fired, moonlit "familiar

9. Nissenbaum, introduction to *The Scarlet Letter*, explores in some depth Hawthorne's identification of himself with Dimmesdale as both artist and "priest." As more of a "priest" than a "minister," Dimmesdale in committing "adultery" violates his vows of "chastity," and his subsequent hypocrisy is rooted in his professional ambition. Nissenbaum argues that in Dimmesdale, Hawthorne expresses his own guilt over violating his professional integrity as an artist by soiling himself politically and then hypocritically proclaiming his innocence (xxix–xxxvi). I argue that Hawthorne's identification of Dimmesdale as a "priest" arises more from personal rather than professional guilt and is also associated with an "adultery" in Rome, not New England, by a Catholic, not a Protestant.

room" of the actual and confronted in its heat and light the "neutral territory" of the real transformed by the imaginary, until he confronted, that is, the "ghosts" in the mirror (1:34–36) [26–28].

1

Speak thou for me!
Hester to Arthur in *The Scarlet Letter*

In Hawthorne's authorial fiction, the "ghosts" of the real speak through the imaginative voice of Hawthorne as "editor" of a briefer tale based on an actual event already once-told by a former surveyor with antiquarian and literary interests—by, in other words, something of the "ghost" of Hawthorne himself. As editor, Hawthorne acknowledges that he has expanded imaginatively on "the modes and motives of the passions" of the brief original but declares that "the authenticity of the outline" of the event remains true to its unedited origins (1:33) [26]. In "Rappaccini's Daughter," Hawthorne had adopted the persona of "translator" as a veil for his transformation of private experience into public art. As it anticipates the characters and themes of *The Scarlet Letter*, "Rappaccini's Daughter" attempts to work out "the riddle" that Beatrice holds for Giovanni's "existence." In *The Scarlet Letter* Hawthorne adopts the persona of editor for identical purposes—to examine a "riddle," as he claims, that he "saw little hope of solving," except perhaps through art (1:31) [25]. It is the riddle, of course, of Hester's character and Arthur's obsession. As editor, he may transform the original text of the tale through extensive revision while disavowing personal responsibility for its now-edited origins. He may claim to be its writer but not its author. Because he presents himself to us behind the veil of editor, we dismiss, as Arthur's congregation did his confessions, Hawthorne's pose as a transparently rhetorical fiction. Of course he is its author. But he is its author in the deepest, most personal sense, and that is why he must also be its editor, why he must revise the original narrative, retaining while concealing its origins, saying but not saying. As Hawthorne brought the romance to a close, he summoned Arthur to the pulpit and the scaffold to make his confession. He seems to summon himself as well. Providing one highly edited autobiographical account of the origins of *The Scarlet Letter* in "The Custom-House," in the confessional subtext of Arthur's sermon, Hawthorne closes the romance by providing a revelatory metaphor for the deeply personal origins and power of his art, the letter to be heard in the "undertone" of the literal letter of his words, the letter to be seen on the scaffold, beneath the cover of and inscribed on the very body of its progenitor. But the revelation continues. At the most crucial—and currently the most critically contested point—Hawthorne must account for Hester's future without Arthur, and he does so by becoming, in fact, the editor rather than the author of the tale. In so doing, he provides us with an essential revelation of part of the tale's origins and of his role as both its author and its editor.

Quite simply, Hawthorne did not author Hester's fate. Margaret Fuller did. In 1843, almost seven years before Hawthorne wrote the ending to *The Scarlet Letter*, Fuller had boldly praised and yet condemned Georges Sand in "The Great Lawsuit" in the terms that Hawthorne would have his narrator employ to judge Hester both in the chapter "Another View of Hester" and in the penultimate paragraph of the romance.[1] Sand, Fuller wrote, was "rich in genius, of most tender sympathies, and capable of high virtue and a chastened harmony," but she suffered the fate of many such women, women who "ought not to find themselves by birth in a place so narrow, that in breaking bonds they become outlaws," who because they cannot find "much room in the world" for themselves "run their heads wildly against its laws."[2] Fuller then concludes as Hawthorne would conclude:

> Women like Sand will speak now, and cannot be silenced; their characters and their eloquence alike foretell an era when such as they shall easier learn to lead true lives. But though such forebode, not such shall be the parents of it. Those who would reform the world must show that they do not speak in the heat of wild impulse; their lives must be unstained by passionate error; they must be severe lawgivers to themselves. As to their transgressions and opinions, it may be observed, that the resolve of Eloisa to be only the mistress of Abelard, was that of one who saw the contract of marriage as a seal of degradation. Wherever abuses of this sort are seen, the timid will suffer, the bold protest. But society is in the right to outlaw them till she has revised her law, and she must be taught to do so, by one who speaks with authority, not in anger and haste.[3]

So said Margaret Fuller—when, as Hawthorne commented on Hester, she "had vainly imagined that she herself might be the destined prophetess" (1:263) [155]. As Hawthorne wrote the last paragraphs in early February 1850, he anticipated Margaret Fuller's imminent return of "her own free will" (1:263) [154] from the failures of a "hardly accomplished revolution" (1:43) [32], in Italy to a still puritanical New England, where, in the end, as Hester had at the beginning, she and her presumably illegitimate baby, Angelo, would have to confront public censure and humiliation.[4] She would also have to face what Hester did not. As Hawthorne envisions

1. Fuller's praise for George Sand was truly bold, but as Chevigny, *Woman and the Myth*, points out it was always hedged by qualifiers while Fuller remained in America and remained a virgin and praised chastity. In Europe her attitudes changed toward both Sand and virginity, and she praised Sand, in fact, for having "bravely acted out her nature" (300–301).
2. Fuller, "The Great Lawsuit," 29–30. Fuller revised and expanded "The Great Lawsuit" during the fall and winter of 1844 into *Woman in the Nineteenth Century* (1845). Except where noted, all the citations from "The Great Lawsuit" may be found extant in *Woman in the Nineteenth Century*, the most recent and accessible edition being Steele's in *The Essential Margaret Fuller*, 243–378.
3. Fuller, "The Great Lawsuit," 30. In *Woman in the Nineteenth Century,* Fuller attempts to explain what she had meant by "severe lawgivers to themselves" by revising the paragraph following that sentence to read:

> They must be religious students of the divine purpose with regard to man, if they would not confound the fancies of a day with the requisitions of eternal good. Their liberty must be the liberty of law and knowledge. But, as to the transgressions against custom which have caused such outcry against those of noble intention, it may be observed, that the resolve of Eloisa to be only the mistress of Abelard, was that of one who saw in practice around her, the contract of marriage made the seal of degradation. Shelley feared not to be fettered, unless so to be was to be false. Wherever abuses are seen. . . . (286)

4. Fuller, Ossoli, and their son, Angelo, were not to set sail for America, however, until 17 May 1850. At the time Hawthorne completed *The Scarlet Letter*, Fuller and Ossoli were living in Florence after having fled Rome in July and Rieti in September. With Ossoli cut off from his inheritance and with Fuller struggling to complete her manuscript on the recent revolutions, they were entirely dependent on loans and gifts from family and friends. Their politics also kept them at some risk, for they were kept under surveillance during their stay in Florence (see Joseph Jay Deiss, *The Roman Years of Margaret Fuller* [New York: Thomas Y. Crowell, 1969], 278–307; and

Fuller's ordeal on the scaffold of public opinion in nineteenth-century New England, he reminds us repeatedly and pointedly during his description of Hester's ordeal two centuries before of the greater cruelty that Hester would have faced in a New England of "our days" (1:50) [36]—the "heartlessness" of becoming "only a theme for jest" (1:56) [41]—of suffering only "mocking infamy and ridicule" (1:50) [36]. Almost nine years later, eight after Fuller's death, Hawthorne would claim that it would be from such "ridicule" that "Providence" had been "kind" in saving Fuller (14:156–57).

In February 1850, however, Hawthorne would anticipate Fuller's return by imagining the possibility, and perhaps even advocating, a different reception for Fuller, one in which, as he says of Hester, "the scarlet letter" had "ceased to be a stigma which attracted the world's scorn and bitterness" and had become, through the sympathetic agency of the romance itself, "a type of something to be sorrowed over, and looked upon with awe, yet with reverence too" (1:263) [154]. He imagined her returning, as Hester did, to resume her work as counselor to wronged women, living long enough to read her words on Sand restated as the ironic prophecy of her own inability, by her own standards, to become the "destined prophetess."[5] And yet confronting women's questions as to "why they were so wretched," she would continue to teach them to identify the sources of

Blanchard, *From Transcendentalism to Revolution,* 314–30). Though Fuller did not begin making actual arrangements for a return to America until the spring of 1850, she wrote of her general plans to return during the fall of 1849 (see, for example, *Letters,* 5:300–301). Her increasingly precarious financial and political situation in Italy since midsummer made a return not only likely but virtually inevitable. Her friends—among them Emelyn Story, William Channing, and Caroline Sturgis Tappan—had anticipated as much and, as Blanchard says, had "all discreetly warned her of what she might have to face at home" (318). Her friends in New England had long been confronting the gossip on her behalf. As a Swedish visitor to New England in early 1850, Frederika Bremer wrote of the attacks against Fuller's character and the vehement defenses by her friends caused by the gossip of "a Fourierest or Socialist marriage, without the external ceremony" (qtd. in Chevigny, *The Woman and the Myth,* 393). Her friends were hard-pressed in their defense, however, for as Sarah Clarke noted in a blunt letter to Fuller, without any evidence of a marriage, they found themselves "in a most unpleasant position" in responding to "the world," which "said such injurious things of you which we were not authorized to deny." Clarke herself had decided that "it seemed that you were more afraid of being thought to have submitted to the ceremony of marriage than to have omitted it" (qtd. in Chevigny, *The Woman and the Myth,* 393–94). "What you say of the meddling curiosity of people repels me," Fuller wrote to Caroline Sturgis in December 1849 (*Letters,* 5:303).

5. While Fuller's dispatches from Europe eloquently condemn broader economic, social, and political injustices, clearly her passion for rectifying the wrongs committed against women had also intensified. Though the following passage from a *Tribune* dispatch written on 2 December 1848 reflects Fuller's exhaustion and despondency over having to leave her three-month-old baby in Rieti in order to return to Ossoli and Rome, it clearly reveals that Fuller planned to keep working to transform "the whole relation between men and women." It may also suggest in her references to the need for a woman "younger and stronger" and "more worthy" to take up the "battle" on behalf of women that she anticipated that the scandal of her new status as an unwed mother would compromise her effectiveness as an advocate for women:

> Another century, and I might ask to be made Ambassador myself . . . , but woman's day has not come yet. They hold their clubs in Paris, but even George Sand will not act with women as they are. They say she pleads they are too mean, too treacherous. She should not abandon them for that, which is not nature but misfortune. How much I shall have to say on that subject if I live, which I hope I shall not, for I am very tired of the battle with giant wrongs, and would like to have some one younger and stronger arise to say what ought to be said, still more to do what ought to be done. Enough! If I felt these things in privileged America, the cries of mothers and wives beaten at night by sons and husbands for their diversion after drinking, as I have repeatedly heard them these past months, the excuse for falsehood, "I *dare not* tell my husband, he would be ready to kill me," have sharpened my perception as to the ills of Women's condition and remedies that must be applied. Had I but genius, had I but energy, to tell what I know as it ought to be told! God grant them me, or some other more worthy woman, I pray. ("*These Sad but Glorious Days,*" 245–46)

their sorrow not in their "sin" nor their "shame" but in the very unjust, very "unsure" nature of "the whole relation between men and women" (1:263) [155]. "Destined prophetess" or not, she would continue, as Richard Millington has recently argued, to prophesy, and her prophecy would continue to subvert rather than reconfirm the patriarchal culture that had condemned her and made wretched other women.[6]

If the "office" of the scarlet letter, as Sacvan Bercovitch has claimed, is to subdue Hester to a gradualist liberal consensus, to have her accept that subjugation freely, and to have her counsel other women to do the same, it fails.[7] For she comes back to New England unrepentantly not only to counsel others to reject the very cultural values that condemn them but also, defiantly, to be near the site of her memories and of the body of the very person she had at one time, despite her suffering, refused to leave. She will join him finally in the same cemetery, and though their society will not allow even their "dust" to "mingle," they, finally, will be united through inscription, through the scarlet letter, their now common legend on the tombstone, in the romance (1:264) [155]. It should be noted that the "old and sunken" grave (1:264) [155] of the man who had once gained "the very proudest eminence of superiority" for prophesying "a high and glorious destiny for the newly gathered people of the Lord" (1:249) [147] is unmarked—apparently forgotten, that is—until it is conjoined by the grave of the woman who prophesied that these "newly gathered people of the Lord" needed to learn a "new truth" about the "whole relation between men and women" (1:263) [155] before they could, in effect, merit their "high and glorious destiny."

6. Richard Millington, *Practicing Romance: Narrative Form and Cultural Engagement in Hawthorne's Fiction* (Princeton: Princeton University Press, 1992), 100–103. Millington's argument is that for Hawthorne "freedom of mind" required both understanding "the sense in which the meaning of one's own life—even to oneself—belongs to the community" but refusing "nevertheless to accede to the coercive patterns of mind that the community attempts to enforce." Thus Hester "remains faithful to her acts of rebellion by choosing again the context that gave those acts their meaning" (100).

7. Sacvan Bercovitch, *The Office of "The Scarlet Letter"* (Baltimore: Johns Hopkins University Press, 1991). Bercovitch's argument for a chastened Hester finally integrated into the community and liberal ideology is weakened by his general failure to acknowledge that in practicing the Christian ethic that Dimmesdale only professed she became more a living part of the community than Dimmesdale, its hermetic ideological hero. A more common interpretation of the ending is that Hawthorne attempts to constrain Hester and the sympathies that he has unleashed on her behalf by inserting her squarely within the ideology of domesticity and condemning her, by contrast, with her foil, the "domestic angel." Reynolds, *European Revolutions,* terms it "a veiled compliment to Hawthorne's little Dove, Sophia" (79). For Milton R. Stern, *Contexts for Hawthorne: "The Marble Faun" and the Politics of Openness and Closure in American Literature* (Urbana: University of Illinois Press, 1991), Hawthorne's sudden evocation of "the unfallen spotless heroine of the marketplace ideologies" is a "failure of nerve," the "voice of the one who would belong, unmaking in political rhetoric what he has painstakingly created in image, characterization, and event" (157–58). Millington's specific argument against the view that Hawthorne turns on Hester and his novel or that he engages Hester and the reader in the compromises of patience counseled by liberal consensus is that such views ignore Dimmesdale's torture at the hands of his "unexamined conformity to a dominant ideology," assume that Hester's advice to wronged women is "palliative" when in fact Hester herself has never repented of her own sin with Dimmesdale, and disregard the fact that even talking about the need for a "social transformation" would have been extraordinarily unwelcome to the patriarchy of seventeenth-century Puritan New England (*Practicing Romance,* 101–3). Millington's argument follows essentially Nina Baym's earlier contention, in *The Shape of Hawthorne's Career,* that in returning, Hester "does not acknowledge her guilt" but "admits that the shape of her life has been determined by the interaction between that letter, the social definition of her identity, and her private attempt to withstand that definition," an attempt that is successful in that she eventually brings "the community to accept that letter on her terms rather than its own" and thus brings "about a modest social change" (129–30).

Because no one has fully recognized that Hawthorne speaks for and about Fuller at the close of the romance, the irony of the ending has gone largely unappreciated.[8] He closes the romance as he began it, by writing of Hester's fate two centuries before from the double perspective of her time and his. Hester's words in the seventeenth century are Fuller's words in the nineteenth century. Nothing has changed. "Heaven's own time" has clearly not come, not for women. But then, "heaven's own time" had not come for the "newly gathered people of the Lord" in New England and America. Their "high and glorious destiny" had not arrived, largely because, as Fuller herself had written in "The Great Lawsuit," the chosen people of America, like the Jews "when Moses was leading them to the promised land," had done everything that "inherited depravity could, to hinder the promise of heaven from its fulfillment," the "cross" having been planted in America "only to be blasphemed by cruelty and fraud"—to "the red man, the black man," and, as she later makes abundantly clear, to all women.[9]

Two centuries after Hawthorne would have a fraudulent Arthur elevate himself temporarily to his culture's highest eminence by envisioning a "high and glorious destiny" for America, Fuller would deplore the proliferation in her own age of such "'word heroes' . . . word-Christs" as Hawthorne's Arthur, protesting that because "never were lungs so puffed with the wind of declamation, on moral and religious subjects, as now," she feels "tempted to implore" them "to remember that hypocrisy is the most hopeless as well as the meanest of crimes, and that those must surely be polluted by it, who do not keep a little of all this morality and religion for private use." She would look back on the "ages of failure" in American history to achieve "freedom and equality" for women as well as for "the red man, the black man" and yet still be able to maintain that though it might be "given to eternity to fulfill . . . this country is as surely destined to elucidate a great moral law, as Europe was to promote the mental culture of man."[1] Arthur's prophecy and Hester's intersect in Fuller's.

The ending of *The Scarlet Letter* originates where it had begun—with Fuller—and with Hawthorne's renewed confrontation with the riddle of her character and of their relationship. Fuller has for some time, of course, been linked loosely with Hawthorne's Hester. First to argue persuasively

8. Reynolds does claim that the closing reference to the feminist prophetess and "angel" of "sacred love" is both a "veiled compliment" to his "little Dove, Sophia" and "a veiled criticism of Margaret Fuller" (79), but he does not note Hawthorne's editing of Fuller's text. Charles Swann, "Hester and the Second Coming: A Note on the Conclusion to *The Scarlet Letter*," *Journal of American Studies* 21 (1987): 264–68, comes closer to this recognition. In countering Colacurcio's seventeenth-century contextual reading of the ending, Swann mentions that Fuller's "Great Lawsuit" "equally clearly bears on Hester's case" and quotes one sentence ("Those who would reform the world. . . ."), but he immediately moves on to consider Mother Ann's relevance without making any further claims for Fuller's personal or authorial influence on Hawthorne (265). His interpretation of the ending as Hester's vision of a literal Second Coming of Christ as a woman is clearly far removed from what Fuller or Hawthorne had in mind. Donna Dickenson, introduction to Margaret Fuller, *Woman in the Nineteenth Century and Other Writings,* World's Classics (New York: Oxford University Press, 1994), cites the penultimate paragraph of *The Scarlet Letter* and Fuller's comments on George Sand to assert that "Fuller anticipates Hawthorne's belief that the female Messiah must herself be pure" (vii–xxix; xiii). More recently, Robert Milder, "*The Scarlet Letter* and Its Discontents," *Nathaniel Hawthorne Review* 22 (spring 1996): 9–25, cites the passage as evidence that Fuller had prophesied her own (and Hester's) fate as a feminist who commits a sexual transgression (12). When I first made the observation in a 1989 seminar paper for Professor Larry J. Reynolds that Hawthorne's penultimate paragraph paraphrased Fuller, it was original. As Fuller's work becomes better known, however, that is no longer quite the case.
9. Fuller, "The Great Lawsuit," 7–8, and *Woman in the Nineteenth Century,* 253.
1. Fuller, "The Great Lawsuit," 8, and *Woman in the Nineteenth Century,* 253–55.

for Fuller as a model for Hester, Francis E. Kearns noted the parallels between Fuller's and Hester's lives as mothers of illegitimate children who are or become linked with the non-English aristocracy, as social reformers and feminists, as counselors to women, and as nurses to the dying. Reynolds explored the link in more depth, arguing that, more than that of any of the other suggested models for Hester, Fuller's life "served" Hawthorne "most provokingly" for both personal and ideological reasons, Fuller and Hester representing in Hawthorne's mind "the figures of Liberty and Eve," the "ideas of revolution and temptation, which lie at the heart of the novel." As "Eve," Reynolds suggests, Fuller during her intimate friendship with Hawthorne at Concord had unwittingly become the object of Hawthorne's sexual interest. Hawthorne's "guilt and anger" over his own "attraction to her" provide the best explanation, argues Reynolds, for the motivation behind Hawthorne's sudden, inexplicable denunciation of Fuller in 1858 when he heard gossip about her relationship with Ossoli. As "Liberty," Fuller, "a female revolutionary trying to overthrow the world's most prominent political-religious leader," merged with "Eve" in Hawthorne's mind to represent "a freethinking temptress who had almost subverted his right-minded thoughts and feelings." More recently, Sacvan Bercovitch has built on Reynolds's original exploration of Hawthorne's conservative reaction to European revolutions and has expanded that context to include Hawthorne's anxieties about potential revolutions within the home and within the nation prompted by radical advocates of women's rights and abolitionism. Though he does not explore the subject in any depth, Bercovitch follows Reynolds by endorsing the view that Fuller provided the model for Hester as the embodiment of many of Hawthorne's concerns.[2]

I contend that Fuller figured much more deeply in Hawthorne's imagination before and during the writing of *The Scarlet Letter* than anyone has suspected. In all the aforementioned studies, Fuller is cited as the model for the socially and sexually threatening Hester that the narrator of *The Scarlet Letter* condemns. I argue, however, that Fuller informs Hawthorne's total conception of Hester, the Hester who inspires Hawthorne's sympathetic admiration and respect as well as his fears and guilt. Hawthorne did not simply decide suddenly in September 1849 to retrieve a character from "Endicott and the Red Cross," write a romance about a seventeenth-century Puritan mother of an illegitimate baby, and then draw upon his friend's life to flesh out his characterization of Hester's radical potential. Fuller was at the heart of Hawthorne's very conception of Hester. Through Hester, Hawthorne, on one level at least, continues his now-distant dialogue with Fuller and attempts to represent, if not actually to solve, the riddle of Fuller and their relationship.

2

Hawthorne began *The Scarlet Letter* seven years and one month after that fateful afternoon in the woods of Sleepy Hollow with Fuller, the

2. Francis E. Kearns, "Margaret Fuller as a Model for Hester Panne," *Jahrbuch für Amerikastudien* 10 (1965): 191–97; Reynolds, *European Revolutions,* 79–80. Reynolds's explanation for the underlying causes of Hawthorne's sudden denunciation of Fuller in the 1858 notebook passage was made earlier, but less explicitly, by Blanchard, *From Transcendentalism to Revolution* (195), whom Reynolds acknowledges. Bercovitch, *The Office of "The Scarlet Letter,"* 85.

moment when their friendship intensified into an ambivalent intimacy that would haunt Hawthorne for years, the very moment he had puzzled over in his representation of Giovanni's first encounter with Beatrice in Rappaccini's garden. In the seven years that had passed since that moment, in the five years since confronting it in "Rappaccini's Daughter," much had changed for both Hawthorne and Fuller. Their world then had been Emerson's Edenic garden in Concord; seven years later they each found themselves, though a continent apart, in a troubling world of personal crises and political strife, in a world where the garden, it seemed, for all of its own shadows, had now become a fully dark forest.

While Hawthorne fought a very public and humiliating battle during the late spring and summer to retain his position in the Salem Custom House, a position he had gained by following fellow Democrats in their brief return to power and had lost to the resurgence of a Whig party led by Zachary Taylor, Fuller fought a grander and more dangerous political battle on behalf of the revolutionary republican government of Rome, besieged during June and early July by French troops fighting to restore an overthrown Papacy and foreign hegemony.[3] She had publicly and privately committed herself entirely to the battle. Having written as a correspondent for the *New-York Daily Tribune* firsthand accounts celebrating the inception of the revolution, Fuller risked her life to remain in the city during the nightly French artillery bombardment and the daily fighting to describe to America those "sad but glorious days" when the republican forces fought a desperate battle to save a doomed revolution that had become, in Fuller's proud words, "now radical" in its determination to bring republican government to all of Italy, to make "the idea," "the destiny of our own great nation," the destiny of all Europe.[4] As a participant, she aided the wounded at the hospital on Tiber Island and described the terrible mutilations of the young. Though sickened by the suffering and the destruction, she nevertheless took on the persona of prophetess of Liberty defiantly chronicling the tragic victory of "tyranny" over "democracy," presenting herself as being more than willing to be a martyr to the good cause if it would "transport" her soul "to some sphere where Virtue and Love are not tyrannized over by egotism and brute force."[5] Watching the young die, describing a pair of skeletal legs that "protruded from a bank of one barricade," imagining her own death as republican martyr, Fuller in Rome, like Hawthorne in Salem beside his mother's sickbed, confronted the terrors of death during July 1849 as neither of them had before.[6]

By the end of the summer both of them, in suffering devastating political and personal losses, would confront extraordinarily uncertain futures. In mid-July, Fuller by military edict would be ordered to leave her adopted home in Rome. She went first to Rieti, the mountain town where she had left her baby in the care of a wet nurse, finding him near death from

3. Fuller, *"These Sad but Glorious Days,"* 238–47. Dates of publication in the *New-York Daily Tribune* are given parenthetically in the notes occasionally when the appearance of those columns seems to me important in terms of Hawthorne's writing of *The Scarlet Letter.*
4. Ibid., 285 (24 July 1849), 278 (23 June 1849), 154 (25 December 1847). For an account of the great excitement with which Americans read Fuller's dispatches for news of the revolution, see Reynolds's *European Revolutions,* 1–24, 54–78, 137–39, and his and Smith's introduction to *"These Sad but Glorious Days,"* 1–2.
5. Ibid., Fuller, *"These Sad but Glorious Days,"* 303 (11 August 1849).
6. Ibid., 310 (11 August 1849).

malnutrition. By October she found temporary sanctuary in Florence, uncertain for a time whether to return to America. In early September, Hawthorne, recovering from his "brain fever" after his mother's death and his firing from the Custom House, would also begin the search for another home, an exile from Salem looking to be a "citizen of somewhere else" (1:44) [33].[7] Both faced poverty with no certain prospects for any immediate relief. Fuller would place her hopes on completing and publishing the history of the Italian revolution that she had announced as early as December 1848.[8] She would finish it in Florence, claim it as her masterpiece, and apparently, lose it at sea during the shipwreck that cost her life. Placing his own hopes on writing, Hawthorne would follow Fuller's way but not her course. Both would write histories, but histories of different kinds. If she could be said to be following in autobiographically based history what Hawthorne called the "wiser effort" of diffusing "thought and imagination through the opaque substance of today" in order to find "the true and indestructible value" within a troubled world (1:37) [28, 29], Hawthorne chose the ghostly light of a historical romance of the seventeenth century to illumine the "opaque substance" of his and Fuller's past, present, and anticipated future.

As Reynolds has noted, Hawthorne began his historical romance less than two weeks after learning in early September through Caroline Sturgis Tappan that Fuller had become the mother of an apparently illegitimate baby.[9] Hawthorne reacted to this final shock of an unsettling summer by writing his "scarlet letter" to and about Fuller and himself. Fuller inspired not only the subject and the character but also the private audience for Hawthorne's "confidential depths of revelation" (1:3) [7]. Hawthorne acknowledged that Fuller possessed the sympathetic power to understand him, an admission that as far as we can tell he made to no one else, including Sophia. Following the lead of his earlier response to Fuller in his first published preface, "Writings of Aubépine"—in which he had represented his audience as an "individual or possibly isolated clique"—Hawthorne again presents himself as addressing "only and exclusively" an audience of a single friend, "the one heart and mind of perfect sympathy" who will listen to his "talk" of the "circumstances that lie around us" as he searched in dialogue, by the act of writing, for "the divided segment" of his own being in the natures of Arthur, Roger, and Hester, hoping to "complete his circle of existence by bringing" himself "into communion with it" (1:3–4) [7]. He seems to attempt, in other words, to solve once again what Giovanni had called "the riddle of his own existence," the riddle that he had located in

7. Sophia referred to Hawthorne's "brain fever" in a 1 August 1849 letter to her mother (Berg; qtd. in Miller, *Salem is My Dwelling Place,* 273).
8. Fuller, *"These Sad but Glorious Days,"* 237 (19 January 1849).
9. Hawthorne may have heard gossip about Fuller's baby before early September, but as Reynolds points out, Caroline Sturgis Tappan would almost certainly have informed Sophia during Sophia's visit with her in the Berkshires during 3–8 September 1849, if not earlier in their exchange of letters during the summer of 1849. Fuller had informed Caroline of her baby and of his father, Giovanni Angelo Ossoli, an Italian marquis, in the early spring of 1849, months before informing anyone else in America (*European Revolutions,* 187, n. 2). The original letter in which Fuller informed Caroline of her baby was lost or destroyed. The earliest extant letter describing the baby (not announcing his existence) is Fuller's letter to Caroline on 16 March 1849 (*Letters,* 5:207–11). As a revelation of the "gossip circuit" between New England and Rome, Fuller acknowledges in the same letter to Caroline that she had heard of Caroline's recent marriage in December long before Caroline announced it to her in her last letter. Reynolds argues persuasively that Hawthorne began writing *The Scarlet Letter* between 21 and 25 September (*European Revolutions,* 189, n. 30).

"the mystery" of Beatrice—"the riddle" that Fuller had become and as late as 1858 still remained.

3

> Under the appellation of Roger Chillingworth, the reader will remember, was hidden another name, which its former wearer had resolved should never more be spoken.
>
> *The Scarlet Letter*

The figure of the branded woman condemned and scorned by Puritan society, punished by the humiliation of wearing the scarlet A, was first introduced, of course, in the 1837 tale "Endicott and the Red Cross." When Fuller wrote her first review of Hawthorne in 1841 to praise *Grandfather's Chair* yet to urge him to continue to draw from his "deep well" for "the older and sadder," of all the tales Hawthorne had written, "Endicott and the Red Cross" was the one tale Fuller singled out as representing the "power so peculiar" to his "genius."[1] Hawthorne, as I have argued, gave Fuller's reviews his most serious attention. As Hawthorne began "Rappaccini's Daughter" in October 1844, meditating deeply on his relationship with Fuller, translating in part her review of *Twice-Told Tales* into his translator's preface, he thought of her earlier review and the tale she had praised and he considered another narrative in which he could confront and yet conceal his relationship with Fuller. On 13 October, seven years after he had first created her, he suddenly recalls in his notebook, and without further comment, the woman, the letter, and the sin that he later would not describe or name: "The life of a woman, who, by the old colony law, was condemned always to wear the letter A, sewed on her garment, in token of her having committed adultery" (8:254) [p. 254 of the Centenary Edition, vol. 8: *American Notebooks*] [203]. Written earlier on the same day, another entry reveals him meditating with a Giovanni-like self-contempt on the nature of his enterprise in "Rappaccini's Daughter." As he considers the origins of an author's works in his life, he expresses his disgust at those writers, like Byron, who too transparently, too artlessly reveal to the public their innermost lives, who "serve up their own hearts, duly spiced, and with brain-sauce out of their own heads, as a repast for the public" (8:253) [p. 253 of the Centenary Edition, vol. 8].[2] Rather than deter him from an autobiographical impulse, his disgust worked to strengthen his determination to conceal the fundamental confessional nature of his work from all but the most sympathetic.

Months before beginning *The Scarlet Letter*, before he hears gossip of scandal, we find Hawthorne again thinking of Endicott and his own family's role in the cruel persecution of a woman, not an adulteress but the outspoken Quaker radical Ann Coleman. Published in May 1849 in Elizabeth Peabody's *Aesthetic Papers*, positioned as the lead piece in a trilogy of politically critical articles—S. H. Perkins's "Abuse of Representative Government" and Henry David Thoreau's "Resistance to Civil

1. Fuller, review of *Grandfather's Chair*, 58.
2. Hawthorne admired Byron enough to have his portrait look down upon him from the walls of the Manse. In a 30 August–4 September 1842 letter to her mother (Berg), Sophia describes her progress in decorating the Manse, mentioning a portrait of Byron, a statue of Napoleon, and a statue of Apollo (this given as a wedding gift by Caroline Sturgis).

Government"—Hawthorne's "Main-street," in itself and in the setting that Peabody gave it within *Aesthetic Papers*, suggests that Hawthorne's sympathetic portrait of Ann Coleman's "bold" denunciation of "established authority, . . . the priest and his steeple-house" may have been informed by his sympathetic reading of Fuller's increasingly outspoken defense of the Roman revolutionary republicans.[3] Coleman's "wild, shrill voice" denouncing established, intolerant authority "appalls" those in authority and provokes them to brutal suppression precisely because of the revolutionary effect of her words on the people, the "living truth" that she told, which seemed, "for the first time," to have "forced its way through the crust of habit" and "reached their hearts and awakened them to life."[4]

If in the spring Fuller's outspoken support of the revolution in Rome and her withering criticism of an America that had betrayed its "nobler spirit" informed Hawthorne's depiction of Ann Coleman, by the fall Fuller's status as an apparently unwed mother led Hawthorne once again at a crucial moment to associate Fuller with the woman condemned by the letter A. Drawing, I believe, upon Thoreau's doctrine of the radical power of passive individual resistance, Hawthorne would combine both figures in Hester. She would greet humiliating persecution not with a "wild, shrill" cry of condemnation but with a defiant silence. Her silence; however, is not assent, for in her bold speculation and counsel to women she is not the meek, submissive figure of pity Hawthorne had envisioned in 1837.

What Hawthorne had heard in 1849 about Fuller's infant son, Angelo, or his father, the Marquis Giovanni Angelo Ossoli, we do not know. We do know, of course, that he chose to model Hester's child after his own.[5] But in choosing the name "Pearl" for Una's fictional counterpart, Hawthorne provides yet another suggestion of Fuller's intimate involvement in *The Scarlet Letter*. Hawthorne first mentions the possibility of naming a character "Pearl" in a long, undated passage in his notebook between entries dated

3. Nathaniel Hawthorne, "Main-street," in *Aesthetic Papers*, ed. Elizabeth P. Peabody (1849; rpt. New York: AMS Press, 1967), 145–74; 163. Peabody's positioning of "Main-street" establishes an ideological foundation within a historical context for the arguments of the two essays that immediately follow it—S. H. Perkins's "Abuse of Representative Government" and Thoreau's "Resistance to Civil Government [Civil Disobedience]." Read as a unit, Hawthorne's historical indictment of Puritan New England's "hard, cold, and confined . . . system," the "iron cage" of "that which they called Liberty" (153), leads into Perkins's condemnation of the intolerances and brutalities of contemporary partisan politics, where individuals and minorities, where principle itself, are sacrificed for power, and Perkins's indictment, of course, provides a powerful introduction for Thoreau's radical solution to the problem. Thoreau's essay influenced Hawthorne to some extent in his portrayal of Hester's "silence" on the scaffold and of her silence (for a time, at least) about her increasingly radical intellectual resistance to the "untrue" ground on which the relations between men and women have been established and institutionalized. Her resistance is more, not less, active in the closing view of her counseling other women.
4. Hawthorne, "Main-street," 163. Baym, *The Shape of Hawthorne's Career*, argues that in "Main-street," Hawthorne envisions the golden ages of New England history in the pre-European matriarchy of an Indian culture in harmony with nature and in the first phase of independent Puritan families, in which "personal freedom and human relation combine in a natural world free from social institution" (120–21). In subsequent generations, as the Puritans establish communities and oppressive institutions, "the matriarchy and the life of the yeoman family" are destroyed as, in Hawthorne's words, "the pavements of Main-street" are "laid over the red man's grave" (*The Shape of Hawthorne's Career*, 120; "Main-street," 150). It is "to the influence of these children and grandchildren" of the original Puritans, claims Baym, that "Hawthorne attributes much of the worst in American life and character even in the nineteenth century" (121). "Let us thank God," the narrator of "Main-street" urges, "for having given us such ancestors; and let each successive generation thank him, not less fervently, for being one step further from them in the march of ages" (162).
5. For excellent discussions of the implications of Hawthorne's decision of modeling "Pearl" on Una, see Carton, "'A Daughter of the Puritans,'" and Herbert, *Dearest Beloved*, 202–8.

1 June 1842 and 27 July 1844: "Pearl—the English of Margaret—a pretty name for a girl in a story" (8:242) [p. 242 of the Centenary Edition, vol. 8]. Written during the period in which Hawthorne's friendship with Fuller was at its most intense, the notation on the origins of the name "Margaret" was almost certainly inspired by his conversations with Fuller, for she habitually informed others of her name's meaning and was fond of meditating on the implications its symbolism held for her life.[6] The association between Fuller, Una, and the character Pearl, however, was deeper in Hawthorne's mind than merely comparable names. The extraordinarily close relationship that Fuller and Una established with each other between July and late September 1844—between Una's fifth and seventh months—created an association between the two in Hawthorne's mind that would not only inform his characterization of Hester and Pearl but also subtly influence his characterization of the Fuller-like Miriam in 1858–59, when the first collapse of Una's physical and mental health struck terror in his heart at the very moment that he was writing of Fuller's own "collapse."

The intimate relationship between Una and Fuller established during this brief period in Una's life could not be widely known until the 1991 publication of Fuller's 1844 journal. Previously, the only published description of Fuller with Una had been Sophia's comment in the Hawthornes' joint notebook quoted in Arlin Turner's 1980 biography. In that description, Sophia emphasized Una's uncanny ability to recognize and admire genius in others. She describes Una at first staring at Fuller "with earnest and even frowning brow" because she recognized that she was in the presence of "a complex being, rich and magnificent, but difficult to comprehend and of a peculiar kind, perhaps unique." Once Fuller took her in her arms, however, the frown quickly disappeared. Una "smiled approvingly" once she had comprehended "her greatness and real sweetness and love" and then "trusted in her wholly," remaining "with full content by the hour" in Fuller's arms. Without access to Fuller's 1844 journal, Turner quotes that passage as evidence of Sophia's "continuing, half-playful argument" about Fuller with Hawthorne and, in Turner's animus toward Fuller, even goes so far as to speculate that Sophia "perhaps concocted" the story "to prove that a child could discern Margaret's virtue."[7]

Fuller's 1844 journal leaves no doubt that the relationship between Fuller and Una was indeed extraordinary. Fuller's visit to Concord in July 1844 was occasioned in large part by the births of two children and the expected birth of a third. Fulfilling the expectation that Hawthorne had

6. In a passage that Hawthorne might well have recalled in his writing of *The Scarlet Letter*, Fuller praised William Godwin for writing "like a brother" in defense of his wife, Mary Wollstonecraft, one of those, like Sand, whom Fuller had described, in the present state of society, as becoming the world's "outlaws" for "breaking bonds." Of Sand, Wollstonecraft, and Godwin, Fuller wrote in *Woman in the Nineteenth Century*: "They find their way, at last, to light and air, but the world will not take off the brand it has set upon them. The champion of the Rights of Woman found, in Godwin, one who would plead that cause like a brother. He who delineated with such purity of traits the form of woman in the Marguerite, . . . a pearl indeed . . . was not false in life to the faith by which he had hallowed his romance. He acted as he wrote, like a brother" (284). In a poem in her 1844 journal, for instance, Fuller defines the meaning of "Marguerite" as the fusion of "love, grief, hope and fear / In that one century-hallowed tear," which she then identifies as "a pearl beyond all price so round and clear / For which must seek a Diver, too, without reproach or fear" ("'The Impulses of Human Nature,'" 112).

7. Manuscript joint notebook, 1843–44, 8 (qtd. in Turner, *Nathaniel Hawthorne*, 148); Turner, *Nathaniel Hawthorne*, 148.

expressed when he and Sophia had selected Concord for their first home, Fuller arrived in Concord to be the houseguest of the Hawthornes and visit with them and their firstborn, Una. She also came to visit with her sister, Ellen, her brother-in-law, Ellery, and their recent firstborn, her niece and namesake, Margaret "Greta" Channing. When Fuller arrived at the Hawthornes', Lidian Emerson was also just days away from delivering her second son, Edward Waldo Emerson. Between their many walks through the night woods and their boat rides on the Concord, Hawthorne was to see Fuller with babies in her arms frequently during her monthlong visit, particularly Una. In fact, during her stay with the Hawthornes, she and Hawthorne would spend much of the day together baby-sitting Una while Sophia was away at the Channings serving as Greta's wet nurse, Ellen having proven incapable of nursing the baby herself.[8] Whatever preternaturally perceptive and trusting relationship Una may have felt toward Fuller in Sophia's eyes, clearly, in Fuller's eyes at least, there was a powerful and reciprocal bond with Una, whom she described on the day she first met her as "a most beautiful child," her beauty being both "noble and harmonious," both "strong" and "sweet."[9] Nine days later she recounts an early evening walk with Hawthorne by the river and comments that "I love him much, & love to be with him in this sweet tender homely scene" though "I should like too, to be with him on the bold ocean shore"; she then describes the "homely scene" that took place on their return and of a bond between herself and Una that Fuller narrates as being stronger—at least on this night—than that of parent and child:

> When we came back Una was lying on the sofa all undrest. She acted like a little wild thing towards me, leaning towards me, stretching out her arms whenever I turned. Her mother tried to attract her attention, in vain, her father took my place, she looked on him and smiled, but discontinued this gesture, the moment I came she resumed it. She has daily become more attached to me; she often kisses me in her way, or nestles her head in my bosom. But her prettiest and most marked way with me is to lean her forehead upon mine. As she does this she looks into my eyes, & I into hers. This act gives me singular pleasure: it is described in no initiation. I never saw any body prompted to do it as a caress. It indicates I think great purity of relation.[1]

Fuller then considers the "treasury of sweet pictures of this child" that she has stored in her "mind" and concludes: "Never was lovelier or nobler little creature! Next to little Waldo I love her better than any child I ever saw." In this often troubled summer, her relationship with Una, as well as with Una's father, seemed to bring Fuller what peace she was to find. That night, after spending the early evening with Hawthorne and then with Una, Fuller describes going out into the night and lying "in the avenue

8. On 13 July, for instance, Fuller records "playing with the beautiful Una, reading." The next day she "staid with Una while H. & Sophia took a walk & then S. went to Ellen." In the following day's entry, she refers again to baby-sitting Una with Hawthorne after Sophia had left and records: "We had most pleasant communion. He is mild, deep and large" ("'The Impulses of Human Nature,'" 84–85; see also, 93).
9. Ibid., 81–82.
1. Ibid., 89.

for hours, looking up at the stars." As "the trees whispered," she records, "How happy, even pure I felt—"[2]

The bond Fuller felt with Una exceeded that with her niece, of whom she writes, "this child interests but does not attach me yet." Though she would describe Emerson's three-year-old daughter Edith as "like a seraph" with a "poetic and tender" smile, she worried that she was "too frail a beauty for this world," inferior to Una, whose "noble and harmonious beauty seems as strong as sweet, as if she might stay here always."[3]

Spending her last full day in Concord with the Hawthornes before returning to Cambridge, Fuller records that her regret in having to depart is centered on Una and Hawthorne: "O it is sad that I shall see Una no more in this stage of her beauty. When I *do* see her again she will be quite another child." The emphasis that Fuller placed on "do" suggests, of course, her determination to reestablish the bond with Una, even if she is "quite another child." In the same journal entry Fuller expressed a similar determination to reestablish and deepen an already intimate relationship with Hawthorne. Describing getting lost in the forest's "long paths, dark and mystical" with Hawthorne and concluding that she felt with Hawthorne that he "might be a brother" to her more than she had ever felt "with any man before," she writes, "Yet with him it is though sweet, not deep kindred, at least, not deep yet."[4]

As she began revising "The Great Lawsuit" that fall, Fuller paid tribute to Una and to Hawthorne by expanding her conception of the development of the individual and of the possibilities of intellectual and spiritual union in marriage to include the enormous influence of parenthood in marriages based on equality. In her journal, Fuller had written on 18 July that Una was "the child of a [blank space in manuscript] holy and equal marriage" and that she would "have a good chance for freedom and happiness in the quiet wisdom of her father, the obedient goodness of her mother."[5] In her revision of "The Great Lawsuit," she adds the following paragraph to *Woman in the Nineteenth Century* immediately after her discussion of friendship between men and women and before her discussion of four types of marriages; it would be Fuller's single greatest revision in her conception of marriage:

> What deep communion, what real intercourse is implied by the sharing [of] the joys and cares of parentage, when any degree of equality is admitted between the parties! It is true that, in a majority of instances, the man looks upon his wife as an adopted child, and places her to the other children in the relation of nurse or governess, rather than of parent. Her influence with them is sure, but she misses the education which should enlighten that influence, by being thus treated. It is the order of nature that children should complete the education, moral and mental, of parents, by making them think what is needed for the best culture of human beings, and conquer all faults and impulses that interfere with their giving this to these dear objects, who represent the world to them. Father and

2. Ibid. The "Waldo" to whom Fuller referred was Emerson's first child, whom she had adored.
3. Ibid., 90, 82.
4. Ibid., 108.
5. Ibid., 89.

mother should assist one another to learn what is required for this sublime priesthood of nature. But, for this, a religious recognition of equality is required.[6]

When Hawthorne heard that Fuller had become a mother herself and he looked into the "tarnished mirror" of memory to recreate her story and his story in that of the woman bearing the letter A and of her unconfessed lover, the "ghosts" that he would see in that "familiar room" would appear as they did five years before. Hawthorne saw his child in Fuller's arms, and as Sophia recognized and as Fuller claimed, he witnessed the extraordinary bond Fuller established with Una during the very summer in which his own relationship with Fuller reached its greatest intimacy. In reconceiving his own child as Hester's child and Margaret's namesake, Hawthorne unites Margaret and Una in "Pearl" and suggests the depth of his sympathetic identification with Hester as Pearl's "other," "actual" parent. By so doing, he also strengthens the confessional implications of his decision to name Pearl's father the "imaginary" Arthur and then to edit the presence of that hidden presence of himself as the child's "actual" father, yet leave a trace of his method of concealment by selecting a last name that comments on the "dim" figure of the author in the first name—Arthur Dimmesdale.

The Scarlet Letter would have the "Actual" and the "Imaginary" meet on many such levels. It is a historical romance of New England set in the remote past of another revolution across the Atlantic, 1642 through 1649, the seven-year span of Hester and Arthur's union, separation, and then reunion, and it is also, it seems, Hawthorne's meditation on his own recent past, on the seven years between 1842 and 1849, between that moment in the woods of Sleepy Hollow in August 1842 when he discovered, through conversation, that his friendship with Fuller was capable of, and indeed had already developed into, a deeper intimacy, and that moment when, as he anticipated, Fuller would return from the European revolutions, with baby in arms, to confront public scorn and ridicule but also, of course, to confront him after a five-year absence from his life. He, like all her friends, would have to discover the grounds on which he could respond to the new challenge that Fuller posed to their friendship and their values.

His response, to himself and to Fuller, is to write the "scarlet letter" for which he at once and on several levels both claims and disclaims authorship through the "edited" narrative of Hester, Arthur, Pearl, and Roger. He fathered the imaginary child "Pearl" in Una, named her after the "Margaret" whose life and character authored Hester, and at once both exposed and concealed her paternity by naming her father "Arthur." If such covert creative strategies reveal Hawthorne meditating on the implications of the "motives and modes of passion" that drew him into an intimate friendship with Fuller seven years in the past, his decision to name Hester's injured and injuring husband and interrogator Roger Chillingworth reveals him employing a similar strategy in his meditation on his present and future relationship, as both a man and an artist, with Fuller.

"Chillingworth" appropriately names, as many have noted, Hawthorne's own disgust with the cold interrogation of the heart, the penetration and

6. *Woman in the Nineteenth Century*, 282. Compare to "The Great Lawsuit," 28.

mastery of self and other that is the center of Hawthorne's own creative obsessions. "Prynne" is the name that Roger would conceal. But Hawthorne also conceals Roger's association with "Prynne." In its obscure way, "Prynne" may be as appropriately descriptive as are the names "Dimmesdale" and "Chillingworth." As has been noted, but rarely, Hawthorne named Roger after his historical contemporary William Prynne (1600–1669).[7] The historical Prynne was a Presbyterian lawyer and writer whose criticism of the king in 1634 and Bishop Laud in 1637 led him to be imprisoned, stripped of his Oxford degree, disbarred, disfigured by the cutting off first of his ear lobes and then their stumps, and most notably for Hawthorne's purposes, branded on both his cheeks with S. L. for "seditious libeler," which he in turn transformed into a badge of honor, as Hester was to do with her letter, by reinterpreting it, the brand becoming for him *Stigmata Laudis*, the brutal signature of his enemy Laud. Prynne's claim to martyrdom at the hands of state and church tyranny, however, was dissipated by his later betrayal of fellow Presbyterians. Once he was elected to the House of Commons, he accused the Commonwealth government of moral laxity, joined the king's side, and later became in fact the champion of the state and church that he had once opposed, writing *Vindication of Ecclesiastical Jurisdiction of the English Kings* (1666–70). Of particular thematic importance for his namesake's function in *The Scarlet Letter* is the fact that he originally earned the displeasure of Charles I, whose wife was something of an amateur actress, by writing *Histrio-Mastix* (1633), an attack on makeup, long hair, and primarily lewd entertainment, particularly plays, in which he indexed the names of actresses under the heading "notorious whores." Prynne was also renowned for his vindictiveness, particularly in the persecution of his old enemy, Laud. He tampered with witnesses, personally searched Laud's rooms, rifled through his pockets, published Laud's private diary in mutilated form, and prior to his trial, wrote an account of Laud's "crimes" entitled "Hidden Works of Darkness Brought to Public Light." Most importantly, for our concerns at least, he is also noted for attacking John Milton's ideas on divorce and for provoking Milton to answer him in *Colasterion* and to allude to him contemptuously in *Means to Remove Hirelings* as a "hot querist for tithes . . . a fierce reformer once, now rankled by a contrary heat."[8]

Hawthorne's identification of Roger "Chillingworth" Prynne with William Prynne and of both with that part of himself he felt compelled to condemn within the concealed scaffold of his art is, to say the least, complex.

7. See Alfred S. Reid, *The Yellow Ruff and "The Scarlet Letter"* (Gainesville: University of Florida Press, 1955); 96–97, and especially Mukhtar Ali Isani's "Hawthorne and the Branding of William Prynne," *New England Quarterly* 45 (1972): 182–95. Isani explores some parallels between Hester and Prynne and Roger and Prynne but generally confines himself to the implications of Prynne's conflict with Laud and does not explore Prynne's opposition to Milton's views of marriage and divorce.

8. My account of Prynne's life is based on Sir Leslie Stephen and Sir Sidney Lee, eds., *Dictionary of National Biography*, 22 vols. (London: Oxford University Press, 1922), 16:432–37; *Encyclopaedia Britannica*, 15th ed., s. v. "Prynne, William" and "History of England and Great Britain: Charles I"; and Clarence L. Barnhart, *The New Century Cyclopedia of Names* (New York: Appleton-Century-Crofts, 1954), 3264–65.

John Milton, *Means to Remove Hirelings*, in Frank Allen Patterson, ed., *The Student's Milton*, rev. ed. (New York: Appleton-Century-Crofts, 1933), 878–98; 886. Patterson notes of this allusion that Milton "never condescends to call him by name" ("Glossary," 38). James Holly Hanford and James G. Taaffe, *A Milton Handbook*, 5th ed. (New York: Appleton-Century-Crofts, 1970), identify Prynne as Milton's most explicitly identified target in *Colasterion*. Prynne "had stigmatized" Milton's argument for divorce as a "monstrous heresy of 'divorce at pleasure'" (75–76).

Like his historical counterpart branded by the scarlet scar tissue of the letters S. L., the fictional Prynne sees himself—through both his false marriage to Hester and his current relationship with her as cuckold—as being equally branded by her S. L., her scarlet letter A. In a powerfully ironic dramatization of the historical Prynne's equation of stage actresses with "notorious whores," the long-absent fictional Prynne's first glimpse of Hester on his return is the one he obtains by joining the audience to watch Hester's defiant performance of her shame on the public stage that is the scaffold. Though he was unable to expose himself—to hold metaphorically the letter to his own chest, as Hawthorne also claims to have been unable to do—he, like Hawthorne, is compelled to interrogate Hester and Arthur, to penetrate to the "motives and modes" of their "passion" and to violate, in order to expose, the sanctity of the self. Within Hawthorne's fictional world, Prynne thus "authors" the action in the same way and for the same motives that Hawthorne authors the tale. To do so, both must become, like William Prynne, "seditious libelers" who betray themselves as they betray others. When Prynne takes on the false identity of "Chillingworth" so that he may "seditiously" expose Arthur's tormented self while claiming in the role of detached anatomist and physician of the heart that he would cure Arthur by provoking him to a damning confession, he but practices the same "arts of deception" as Hawthorne.[9] He dramatizes within the tale Hawthorne's act of writing the tale and becomes the living embodiment of Hawthorne's contempt for the very origins of his art in the brutal dissection and assiduously concealed exposure of the self, the self betrayed. The scarlet scar of the historical Prynne and the brand appropriate for that part of Hawthorne masquerading in the fictional Prynne's false identity as "Chillingworth" is indeed S.L.—the "seditious libeler" who, like that other part of Hawthorne invested in Arthur, would be known for the esteem accorded to the triumph born of his greatest confession and deception—*The Scarlet Letter*.

Hawthorne, I suspect, selected the name "Prynne" for at least two other reasons—Prynne's betrayal of his political principles and former allies and, more importantly, Prynne's opposition to Milton's views of marriage. In both cases, Hawthorne's identification of Chillingworth with Prynne and in turn Hawthorne's identification of a part of himself with Chillingworth reveal Hawthorne, once again, expressing contempt for his own deceptions and betrayals. In both cases, also, we find Hawthorne meditating on Fuller and the meaning of their past and present relationship.

Hawthorne, of course, presented himself in "The Custom-House" as a somewhat sanguine political martyr to a brutal government, a "decapitated surveyor," a contemporary of the other victims of injustice and intolerance portrayed in "Main-street," which, we must remember, he still planned at the time that he wrote "The Custom-House" to include among the tales to be published with *The Scarlet Letter*. But while Fuller metaphorically exposed herself in the summer of 1849 to the "guillotine" of a once-revolutionary French government now crushing its fellow republican revolutionaries in

9. The phrase "arts of deception" is Michael Davitt Bell's. His essay "Arts of Deception: Hawthorne, 'Romance,' and *The Scarlet Letter*," in *New Essays on "The Scarlet Letter,"* 29–56, is a fine analysis of Hawthorne's duplicitous strategies for making acceptable his engagement with the imaginative fictions of "romance," whose "delusions" were "clearly dangerous" to a culture that valued "reason or judgment," for it served "to undermine the basis of psychological and social order, to alienate oneself from 'the real businesses of life'" (37).

Italy, Hawthorne fought with all his might to retain his position within the government.[1] Influenced by his other radical friend, Thoreau, Hawthorne would almost paraphrase "Resistance to Civil Government" to condemn his own loss of self, of "proper strength," in leaning "on the mighty arm of the Republic," and he would have us believe that, while he in fact clung desperately to the office, he had already begun—before the ax fell—to consider leaving the Custom House in order to preserve what remained of his self-reliant manhood (1:38) [29–30]. A part of Hawthorne did, in fact, see himself as betrayed and publicly humiliated, another martyr to the brute force of state power, a power that in "The Custom-House" he insists he did not use when he controlled the "guillotine." He did sympathetically identify with the political losses and personal scandal of Fuller, and he expressed that sympathy in the narrator's admiration for the proud defiance of Hester on the scaffold. But a part of him also acknowledged his hypocrisy and confronted his own betrayal of himself and his friends.[2] If he positions his narrator and his reader alongside Hester on the scaffold, he also positions Arthur Dimmesdale above and Roger Prynne below. One carries out his duties to the state in judging her, and one stands with the multitude in condemning her. Both should stand with her and speak on her behalf, but both betray her with silence, a silence that one asks her to break in order to expose him and that one signals her to keep in order to conceal him.

In the more transparently veiled autobiography of "The Custom-House," Hawthorne would present his betrayal as a sign "of a system naturally well balanced" (1:25) [21]. Temporarily at least, he could abandon his

1. Reynolds, "*The Scarlet Letter* and Revolutions Abroad," *American Literature* 57 (1985): 44–67 (also, *European Revolutions*, 79–96), is the first to demonstrate the importance of "revolutionary" imagery and themes to Hawthorne's imagination as he wrote "The Custom-House" and *The Scarlet Letter.* Of particular importance are Hawthorne's references to the guillotine in "The Custom-House" and his association of it with the scaffold in *The Scarlet Letter.* For Hawthorne, according to Reynolds, the Jacobin mobs of the original French Revolution came to be associated with the revolutionary mobs of Paris during the "Bloody June Days" of 1848 and, in turn, with the Whig "mobs" out for his own head. As a representative of the spirit of Liberty as well as Eve, Hester's influence on Arthur is "revolutionary" and, as Reynolds argues (based on Arthur's unleashed passions after their meeting in the forest), destructive. While I agree that Hawthorne feared the anarchy of mobs, I contend that though Hawthorne indeed associated the guillotine with French revolutionaries, he would have specifically associated it in the fall of 1849 with those revolutionaries, the Jacobins, who on obtaining civil authority used that authority to betray their principles and their fellow republicans, destroying one tyranny in order to establish an even greater one. When the French, after their revolutions of 1848, marched on the fledgling Republic of Rome to reestablish a reactionary Papal government and foreign hegemony in Italy, they betrayed their republican principles and fellow revolutionaries, as Fuller so vehemently condemned them. As Hawthorne wrote that fall, French armies occupied Rome under martial law. While the Whigs were anything but revolutionaries, the "mob" of Whigs after Hawthorne's head, from his point of view, at least, had betrayed their promises to reform in the name of justice and tolerance what they had defined as the Democrats' practice of automatically replacing political appointees, promising instead to replace only those who had been maleficent in office (see Nissenbaum, "The Firing of Nathaniel Hawthorne," 65). As Hawthorne would portray them, once the Whigs gained authority, they too abandoned principle for the privileges of power, as he makes clear: "There are few uglier traits of human nature than this tendency—which I now witnessed in men no worse than their neighbors—to grow cruel, merely because they possessed the power of inflicting harm" (1:40–41) [31]. Similarly, the Puritans, fleeing oppression in England, had established a government every bit as oppressive and intolerant as the one they had fled, establishing as one of their first institutions the "black rose" of the prison and the scaffold to extirpate the "red rose" of America. As Hawthorne was with the Whigs and Fuller with the apostate republicans of the French army, so Hester is with the Puritan authorities. She is a victim not of the anarchy of revolution but of the oppressive power of institutionalized authority. That such authority wields that power hypocritically is reenforced not only by Arthur's public role in her persecution and humiliation but also by Hawthorne's deliberate historical anachronism in making Bellingham the chief civil authority as governor presiding over her punishment.
2. See Nissenbaum, introduction to *The Scarlet Letter*, for the relationship between Arthur's guilt, hypocrisy, and need for confession and Hawthorne's political and artistic guilt (xxviii–xxxvi).

stimulating friends in Concord—as well as nature, books, literature, and "a gift, a faculty," all "imaginative delight"—and join without a "murmur" the "living dead" old men and the soulless inspector and be the better for what he admits is a "corrupt" and "corrupting" service to the state (1:25–26) [21]. As long as he did not live too long as someone "other than . . . [he] had been," he could "recall" and thus redeem his truer self (1:26) [21]. Informed by Thoreau's classification of those who serve the state, Hawthorne could indict in the barely living old men and the soulless inspector those who serve the state only with their bodies and are, as Thoreau wrote, on a "level with wood and earth and stones . . . [and] horses and dogs." He could include himself, at least temporarily, with those who serve chiefly with their minds but not their consciences and are, in Thoreau's words, thus "as likely to serve the devil, without *intending* it, as God." And he could seek redemption by realigning himself with and writing in defense of Thoreau's "heroes, patriots, martyrs, reformers" who are "commonly treated as enemies" by the state because they serve it by resisting it "with their consciences."[3] He would write first "Main-street" and then, recalling his old self and his old friend, who was now truly something of an "enemy" to the old order of state and domestic politics, he would write *The Scarlet Letter*.

The intersection between state and domestic politics, of course, is marriage, and Hawthorne's selection of "Prynne" is especially appropriate in that the historical Prynne's attack on Milton's views of marriage and divorce are parallel to the two conceptions of marriage that, as T. Walter Herbert has so persuasively argued, are at issue in *The Scarlet Letter*.[4] Milton had argued that marriage consisted of a sacred bond of love between a man and a woman in the eyes of God, that it was instituted by God as a union of spirits meant to prevent or remedy the solitude of the self, and that once this bond had ceased to exist the marriage had ended. The state simply recognizes—in marriage or in divorce—what the couple and God have already recognized. For Milton, the single civil cause then recognized for divorce—adultery—was the "last and meanest" cause, in fact "a perverse injury" to God's intent for marriage, for adultery destroyed only those unions that were based not on an intellectual and spiritual bond but on "a sublunary and bestial burning, which frugal diet, without marriage, would easily satisfy." Milton condemns Protestants for having rejected Catholicism's elevation of marriage to a sacrament only to make it an "idol" with which they "invest . . . such an awful sanctity and give it such adamantine chains to bind with, as if it were to be worshipped like some Indian deity, when it can confer no blessing upon us, but works more and more to our misery."[5] The historical Prynne's objection to Milton may be heard in Roger "Chillingworth" Prynne's insistence on his legal—rather than emotional or sacred—claim to Hester: "Thou and thine, Hester Prynne, belong to me" (1:76) [52]. He makes this claim, of course, immediately after admitting

3. Henry David Thoreau, "Resistance to Civil Government [Civil Disobedience]," in Carl Bode, ed., *The Portable Thoreau*, rev. ed. (New York: Penguin, 1964), 109–37; 112–13.
4. Herbert, *Dearest Beloved*, 184–211. Though Herbert does not establish a connection between Milton, Fuller, and Hawthorne, he identifies the essential conflict between the civil and sacred conceptions of marriage at work in middle-class nineteenth-century culture and at issue in *The Scarlet Letter*. I find his cultural and biographical analysis persuasive, as my own views will amply demonstrate in their debt to his, but I find the biographical and literary context to be broader than Herbert presents.
5. John Milton, *Doctrine and Discipline of Divorce*, in *The Student's Milton*, 573–626, especially 582; quotations on 591, 594.

that her "wrong" to him had been the consequence of his original "betrayal" of her into a loveless marriage.[6] Hester's claim to Arthur that their relationship "had a consecration of its own" (1:195) [118] suggests that she seeks to redefine marriage, as Milton did, as a sacred bond established by love, not civil contract. She may sever the bonds of a false marriage by breaking her civil obligation to "belong" to Prynne, but she remains faithful to the higher "consecration" of her "marriage" to Arthur, except, significantly, when she allows Prynne's claim to a civil right over her to persuade her to keep his identity secret. Hester's crime against the state and against official morality, of course, is that she broke the vows of her civil marriage.

When Hawthorne edited the passage from Margaret Fuller's "Great Lawsuit" and *Woman in the Nineteenth Century* to write Hester's fate, he prophesied that the "angel and the apostle of the coming revelation" of the new order between men and women will not be stained by "sin" and "shame" but will show "how sacred love should make us happy" (1:263) [155]. Because "sacred love" is usually taken to mean "married love," critics have read that phrase to be Hawthorne's resolution of his ambiguity toward Hester. Under the terms of the mid-nineteenth-century's discourse on the religious sanctity of marriage, he decides, finally, to condemn her, just as Fuller, the source of and for whom that passage was written, had argued that, despite "the contract of marriage" being often "a seal of degradation" that "the timid will suffer" and "the bold protest," "society is in the right to outlaw them till she has revised her law," which she "must be taught to do . . . by one who speaks with authority, not in anger and haste."[7] Hawthorne's "resolution" of his ambiguity, however, may just as well be read as his final, ironic gesture toward Hester, and through her to Fuller, if we complicate, as both he and Fuller did, the often disjunctive relationship between "marriage" and "sacred love."

Hawthorne was drawn to the acrimonious marriage debate between Prynne and Milton in the seventeenth century because it provided an appropriate historical parallel for the dialogue he and Fuller, and indeed the entire Concord circle, had once had over the nature of marriage, a dialogue that informed Fuller's views of marriage and celibacy in "The Great Lawsuit" and *Woman in the Nineteenth Century*, and a dialogue that Hawthorne reinitiates in *The Scarlet Letter*.[8] In the conversations that took

6. From Milton's point of view, "Chillingworth's" desire to assuage the pangs of loneliness and solitude through marriage would be appropriate but impossible since there was no "real" union between himself and Hester, and only an authentic union can vanquish solitude. Hester's physical "adultery" was thus inevitable, in fact, was faithful in its way to the absence of union that was the nature of that "marriage."
7. Fuller, "The Great Lawsuit," 30, and *Woman in the Nineteenth Century*, 286.
8. Reynolds, "From *Dial* Essay to New York Book," demonstrates the importance of Fuller's dialogue with her friends in 1842 and again in 1844 in Concord (including Hawthorne but especially Emerson) as the impetus for her articulation of her views of marriage and male-female friendships. Milton's vision of the Garden of Eden and the marriage of Adam and Eve, of course, was very much on Hawthorne's mind during his Old Manse days, as was Fuller's conception of his and Sophia's marriage, specifically her prediction in her letter of July 1842 that he and Sophia would develop the highest form of marriage. In Hawthorne's 1 February 1843 letter to Fuller, he mentions that he and Sophia had been reading "through Milton's Paradise Lost, and other famous books." He then states, significantly, that "it sometimes startles me to think how we, in some cases, annul the verdict of applauding centuries, and compel poets and prosers to stand another trial, and receive condemnatory sentence at our bar" (15:671) [p. 671 of the Centenary Edition, vol. 15: *The Letters, 1813–1843*]. Though Hawthorne may have been thinking of Milton's literary reputation, I contend that, within the context of his lengthy description of his own marriage, he was thinking of Milton's conception of marriage and divorce and the "condemnatory sentence" he would have received at the "bar" of Hawthorne's own age.

place in Concord, Hawthorne and Fuller both held essentially Miltonic views of marriage as a sacred union. Emerson, however, represented the views of Prynne. In this, and in other ways, Hawthorne recalls for reexamination the triangular tensions in Concord between himself, Fuller, and Emerson that he had earlier "translated" into Rappaccini's garden.

Hester's justification of her union with Arthur as having a "consecration of its own" that supersedes civil recognition follows to its inevitable end the argument that Hawthorne and Fuller had both made for marriage. A full three years before Hawthorne and Sophia signed the civil contract in a public ceremony, Hawthorne had "consecrated" his relationship with Sophia as a "marriage" that, as he explained to her, "God himself has joined," for they had established "a bond between our Souls, infinitely stronger than any external rite" (15:329) [p. 329 of the Centenary Edition, vol. 15: *The Letters, 1813–1843*]. Throughout his courtship letters to Sophia over the next three years, he refers to himself as her "husband" and to her, his Dove, as his "wife." Indeed, much of the tension in Hawthorne's premarital "marriage" talk in those letters arises from his anxiety that a resistant, "naughty Sophie" threatened to disrupt the idealized union he had created between them when he cast her in the role of his redemptive "Dove." We also know that long before he and Sophia actually married and moved to Concord he had attempted to describe to Fuller, if not his own family, the intimacy of his relationship with Sophia (15:612) [p. 612 of the Centenary Edition, vol. 15]. Though Hawthorne had complained to Sophia that he could not describe satisfactorily their bond to others, "not even Margaret," Fuller seems to have understood him rather well.

In her reply to Sophia's announcement that she and Hawthorne were finally to be married formally, Fuller expresses her faith in Sophia's ability to maintain precisely the kind of love Hawthorne had long insisted his Dove had given him in their "marriage," a love that Fuller describes as "wise and pure and religious." Fuller had also become convinced through their many conversations that Hawthorne possessed a unique balance of the masculine and feminine that made him capable of responding to and sustaining such a love: "I think there will be great happiness," she predicted to Sophia, "for if ever I saw a man who combined delicate tenderness to understand the heart of a woman, with quiet depth and manliness enough to satisfy her, it is Mr Hawthorne." Though Sophia's love for Hawthorne was "wise, pure, and religious," she imagines them capable of an even higher form of union—an "intellectual friendship" of two artists that surpasses "love merely in the heart" or even "the common destiny of two souls."[9] When Hawthorne and Sophia finally moved to Concord, Hawthorne found, at least initially, that they had come to represent, in his mind as well as Fuller's, an alternative Eden to the one proposed by Emerson's vision of self-reliant individualism.[1]

For several years Fuller and Emerson had skirmished over the nature and possibilities of friendship and marriage. Voiced often in an emotionally charged undertone, their debate centered on whether the self-sovereign

9. *Letters*, 3:66.
1. For an analysis of Hawthorne's response to the personal, marital, and creative conflicts that followed upon his entry into "Emerson's" Concord, see Reynolds, "Hawthorne and Emerson in 'The Old Manse.'" See also Herbert, *Dearest Beloved*, 109–60.

individual could ever really unite intimately with another soul. Fuller had insisted that such unions were possible, and Emerson had been equally insistent that they were not. Fuller records in her journal on 1 September 1842, for instance, an afternoon walk with Emerson in which the subject of marriage was once again discussed. First observing that Emerson "has little sympathy with mere life," Fuller illustrates her point by summarizing Emerson's views on marriage:

> We got to talking, as we almost always do, on Man and Woman, and Marriage.—W. took his usual ground. Love is only phenomenal, a contrivance of nature, in her circular motion. Man, in proportion as he is completely unfolded is man and woman by turns. The soul knows nothing of marriage, in the sense of a permanent union between two personal existences. The soul is married to each new thought as it enters into it. If this thought puts on the form of man or woman[,] if it last you seventy years, what then? There is but one love, that for the Soul of all Souls, let it put on what cunning disguises it will, still at last you find yourself lonely,—*the Soul*. There seems to be no end to these conversations.[2]

And indeed there was no end. Eight days later, Emerson entered Fuller's bedroom to read what he had written in his journal about marriage, and their debate began again, but Emerson was, as Fuller phrased it, "nowise convinced." But Fuller would not drop the subject either. Reading through his journals later, she quotes two of Emerson's statements about marriage and then vows that she "shall write to him about it." One of the statements illustrates clearly just how far apart his views of marriage were from Fuller's and Hawthorne's and just how close they are to "Chillingworth" Prynne's: "Is it not enough that souls should meet in a law, in a thought, obey the same love, demonstrate the same idea. These alone are the nuptials of minds[.] I marry you for better, not for worse, I marry impersonally."[3] As Fuller challenged Emerson's attempt to ground his unhappiness in philosophy, she had to contend with the consequences in Emerson's "mere life." Lidian's great unhappiness and her jealousy of Fuller's intimacy with Emerson erupted one night at the dinner table in an embarrassing scene. Lidian's problem, Fuller wrote, was that she still hoped Emerson's "character" would one day "alter" and he would "be capable of an intimate union." By now, however, Fuller had come to know better. Her "expectations" of a more intimate friendship with Emerson were "moderate now," she had written soon after arriving for her stay at the Emersons' that summer.[4]

In Hawthorne, however, her expectations at this time were clearly on the ascendent. Fuller began to see in Hawthorne the possibility of establishing the type of intimate friendship she had sought with Emerson. In contrast to Emerson, Hawthorne seemed capable of responding to nature, marriage, and friendship with an intelligence warmed by a depth and quiet passion impossible to Emerson. On Saturday night, 20 August 1842, the

2. Fuller, "Margaret Fuller's 1842 Journal," 330.
3. Ibid., 335. See Emerson, *The Journals and Miscellaneous Notebooks of Ralph Waldo Emerson*, 7:336, 8:144, 7:532–33.
4. Fuller, "Margaret Fuller's 1842 Journal," 331–32, 326.

day before their afternoon-long conversation in the woods of Sleepy Hollow, Fuller describes Hawthorne during their walk taking in the beauty of the moon and responding to the moment by speaking to her of his marriage, telling her that he "should be much more willing to die than two months ago, for he had had some real possession in life, but still he never wished to leave this earth: it was beautiful enough." Hawthorne, Fuller then writes, "expressed, as he always does, many fine perceptions. I like to hear the lightest thing he says!"[5]

During the winter after that summer, as Reynolds has observed, Fuller continued her debate with Emerson on marriage by writing "The Great Lawsuit," attempting, as she had noted in a letter to Emerson in the fall, to prove "that permanent marriage cannot interfere with the soul's destiny."[6] Fuller's conception of marriage is informed by her contrasting conversations and experience with Emerson and Hawthorne and is framed within the terms of the Milton and Prynne debate, namely, "whether earthly marriage is to be a union of souls, or merely a contract of convenience and utility."[7] Of the four types of marriage she identifies in "The Great Lawsuit," her opinion of Emerson's second marriage shapes her description of the lowest type, a practical, civil marriage between a provider and a housekeeper who feel for each other merely a "mutual esteem" and "mutual dependence."[8] Sophia's "wise, pure, and religious" love and Hawthorne's capacity to develop an "intellectual friendship" with a woman provide Fuller with an example of a marriage that seems capable of reaching its highest and fullest potential as the fourth type of marriage—a "religious" marriage of a man and a woman on a "pilgrimage towards a common shrine," a marriage that incorporates all other types, including the marriage of "intellectual companionship" just below it on Fuller's scale."[9]

As the friendship between Fuller and Hawthorne deepened during the summer of 1844, they resumed their dialogue over the relationship between men and women in friendship and in marriage just as, only weeks later, Fuller was to resume her dialogue with the public over these same issues in revising "The Great Lawsuit" into *Woman in the Nineteenth Century*. After one of her many walks alone with Hawthorne through Sleepy Hollow, Fuller considers her "place" in life and finds that though she is able to take a "superior" view of it, she knows that "the deep yearnings of the heart & the baffling of time will again be felt, & then I shall long for some dear hand to hold." She quells the impulse to self-pity, however, by recalling a recent conversation with Hawthorne: "But I shall never forget that my

5. Ibid., 325.
6. *Letters*, 3:96. See Reynolds, "From *Dial* Essay to New York Book," for an account of the influence of the 1842 and 1844 conversations between Fuller and Emerson on Fuller's views of marriage in "The Great Lawsuit" and her revisions in *Woman in the Nineteenth Century*.
7. Fuller, "The Great Lawsuit," 27. In *Woman in the Nineteenth Century*, Fuller revises the passage to make her position clearer that a "union of one with one is believed to be the only pure form of marriage" (281).
8. Fuller, "The Great Lawsuit," 28, and *Woman in the Nineteenth Century*, 282. Reynolds makes this same observation in "From *Dial* Essay to New York Book."
9. *Letters*, 3:66; "The Great Lawsuit," 28–32, and *Woman in the Nineteenth Century*, 282–87, 289. The debate between Fuller and Emerson continued, in a sense, after her death. To her good friend William Henry Channing's belief that Fuller's "view of a noble life" would have prevented her from compromising and submitting to "a legal tie" with Ossoli, Emerson responded in his journal that he believed Fuller would have sacrificed her principles once faced with the "practical question" and "a vast public opinion, too vast to brave" (*The Journals and Miscellaneous Notebooks*, 11:463). Without evidence, Emerson, of course, presented her in the *Memoirs* as "married."

curse is nothing compared with that of those who have entered into those relations but not made them real: who only *seem* husbands, wives, & friends. H. was saying as much the other evening."[1] Fuller does not identify Emerson as the subject of her and Hawthorne's condemnation of the false spouse and friend; but when Emerson later read that passage as he was preparing Fuller's *Memoirs*, he clearly considered himself the topic of their talk and did what he could to obscure that fact.[2]

In a sense, Hawthorne continues that very conversation in *The Scarlet Letter*, and he, like Emerson, does what he can to obscure that fact. As Fuller's paraphrase and endorsement of Hawthorne's statement seems to imply, both she and Hawthorne considered not only Emerson's marriage with Lidian as a false union but also his friendship with them as less "real" than the friendship they had "entered into."

In "Chillingworth" Prynne, Hawthorne would embody a segment of his being that he despised, but he would also shape Prynne's character, to a great extent, in the image of the Emerson he had come to see not only as his rival in friendship with Fuller but also as the embodiment of that part of himself that, in the name of his own masculine self-sufficiency, resisted making his "relations" with the feminine "real." In Prynne's alias, "Chillingworth" is the husband, friend, and philosopher that Emerson had come to seem to both Fuller and Hawthorne, the man whose vision of life (in an essay called, at the time, in fact, "Life") Fuller had condemned in 1844 as being "beautiful, and full and grand" yet "oh, how cold." "Nothing but Truth in the Universe, no love, and no various realities." In a statement that Hester could have easily made of Chillingworth, Fuller then rebukes herself for having been "foolish . . . to be grieved at him for showing towards" her "what exists toward all," and reminds herself that she must never again trust him, as Lidian still did, to be capable of making a relationship "real": "But lure me not again too near thee, fair Greek, I must keep steadily in mind what you are." In his scholarly solitude and justification for the claims of the unchecked self, in his unorthodox study of nature, in his "seeming" friendship, and of course, in his failure to seek, much less form, a "sacred" union in his marriage, Emerson's "ghost" in Hawthorne's past inhabits his imaginative vision of Chillingworth.[3]

1. Fuller, "'The Impulses of Human Nature,'" 92. Sophia also criticized Emerson's marriage in much the same terms as Hawthorne did to Fuller. As Reynolds has demonstrated ("Hawthorne and Emerson in 'The Old Manse,'" 73–77), Hawthorne had managed to convert Sophia, who had once idolized Emerson, to his and Fuller's opinion of Emerson's emotional deficiencies. Writing to her mother (6 June 1843 [Berg]) about a letter in which her sister Elizabeth had employed Emerson's conception of the "self-sufficiency" of the individual in marriage to praise the marriage of Sam and Ana Barker Ward, Sophia defends her own marriage and challenges her sister's praise of the Wards' marriage. In a true marriage, she insists, neither partner "is wholly independent of the other, except intellectually" because "heart & spirit are forever, undissolubly one." Emerson cannot understand this, she asserts, because he "knows not much of love" and "has never yet said any thing to show that he does." "He is an isolation," Sophia concludes. "He has never yet known what union meant with any soul." Citing this letter, Herbert, *Dearest Beloved*, concludes, rightly, I think, that Sophia's definition of the "oneness" of the "true husband and wife" and her critique of Emerson is, in effect, Sophia's condemnation of the Emersons' marriage as "an adulterous legal marriage," parallel to the marriage of Roger and Hester (188–89).
2. In his journal, Emerson obscured the target of Fuller's and Hawthorne's criticism by attributing Hawthorne's remark to himself. He changed the "H." in Fuller's journal to a "W." in his. In the *Memoirs* he continued the deception, but more "honestly," by simply leaving out the entire last sentence of the passage so it appears that Fuller is writing generally about friendship and marriage rather than paraphrasing a conversation about a specific person. See Emerson, *The Journals and Miscellaneous Notebooks*, 11:463; and *Memoirs*, 2:292.
3. Fuller, "'The Impulses of Human Nature,'" 83.

4

> He knew not whether it were a woman or a shadow. It may be, that his pathway through life was haunted thus, by a spectre that had stolen out from among his thoughts.
>
> Arthur on seeing Hester in the forest, again, in *The Scarlet Letter*

As he did in October 1844, when he searched for the narrative vehicle that became "Rappaccini's Daughter," in September 1849, with Fuller and baby in "pleasant" Italy (to use Hester's adjective), Hawthorne followed a symmetrical logic of narrative deception in recalling the woman branded by the stigma of the letter A. The woods of Concord had become the garden of Italy in "Rappaccini's Daughter," and now the garden of a decidedly "unpleasant," revolution-torn Italy would become the forests of the New World during the time (1642–49) of another upheaval across the seas, a time parallel two hundred years ago to the upheaval in the lives of both Fuller and Hawthorne. Traces of Hawthorne's method explain, for instance, why his narrator would think of Hester and her infant on the scaffold in terms of an Italianate, Catholic Madonna confronting the severity of a paternalistic Puritan morality—a reversal of Protestant Fuller's predicament at the time in Rome but appropriate to the likely perception of Fuller on her return from Italy with her child and her Italian–Catholic husband—why also he would consistently allude to the man who was her secret lover as a "priest" associated with the imagery of Catholicism rather than as a "minister."[4] If and when Fuller and her baby, Angelo, stepped off the ship from Italy, as Hester and Pearl stepped into the light from the seclusion of prison, she too would have to confront the ancestors of Hester's judges. She would have to explain what marriage had served to "consecrate" the conception of that child, and she would be expected to name the father.

She would have to confront as well the two men who, in Hawthorne's mind, had once been closest to her and rivals for her intimacy, the two men who would no longer be imagined as the mentor-father Dr. Rappaccini and the faithless friend Giovanni, but—given the revelation of Fuller's bold sexuality—imagined as the manipulative and loveless husband Chillingworth and the guilt-obsessed, self-absorbed lover Dimmesdale. The garden becomes the forest where Beatrice—still pure if not virginal—becomes Hester meeting her Arthur.

4. Despite Fuller's virulent anti-Catholic attacks on papal politics and particularly against Jesuits, Hawthorne was almost certain to have associated Fuller's Italian "husband" with Catholicism regardless of how little he had heard about him. He may well have heard that the family of the Marquis Ossoli was directly associated with the Pope and the Papal Guard, though Ossoli went against his own family in opposing the Pope during the revolution. Hawthorne was more likely to have heard that Fuller had named her son Angelo, which suggests that Hawthorne's choice of "Angel" as the people's epithet for the charitable Hester was inspired at least in part for its value as a covert allusion. Though Nissenbaum, in the introduction to *The Scarlet Letter*, pursues an entirely different interpretation, he does identify many of the key images of Catholicism associated with Dimmesdale (xxix–xxx). It should be noted, as well, that James Lowell reported in a 12 June 1860 letter to Jane Norton that Hawthorne had considered having Dimmesdale confess to a priest: "I have seen Hawthorne twice. . . . He is writing another story. He said that it had been part of his plan in 'The Scarlet Letter' to make Dimmesdale confess himself to a Catholic priest. I, for one, am sorry he didn't. It would have [been] psychologically admirable" (qtd. in Henry G. Fairbanks, "Hawthorne and Confession," *Catholic Historical Review* 43 (1957): 38–45; 40).

Hawthorne, in effect, rewrites "Rappaccini's Daughter" as *The Scarlet Letter.* Inspector Pue's brief tale, as Hawthorne claimed, is indeed edited and expanded. As he did in "Rappaccini's Daughter," Hawthorne revisits that afternoon in the woods of Sleepy Hollow when his and Fuller's relationship had deepened into intimate friendship.[5] In a tale of adultery in which the adulterous moment does not happen, in which in fact the very act is not named, Hawthorne imagines having realized a relationship with Fuller that did not happen, a relationship that his guilt-haunted imagination will not permit him to confront except in its moralized aftermath, in a retrospective art. Seven years had passed, and Hawthorne and Fuller had long been separated. But just as Hawthorne imagines that he will soon have to confront Fuller again and meet her this time as a sexually experienced, scandal-tainted mother with child, so Hawthorne describes in the "tarnished mirror" of his imagination Arthur and Hester's moment in the forest as a reunion of "ghosts" in which both confront in the "mirror" of each other and themselves the forces within their natures that had led them to this moment in their lives: "Each a ghost, and awe-stricken at the other ghost! They were awe-stricken likewise at themselves; because the crisis flung back to them their consciousness, and revealed to each heart its history and experience, as life never does, except at such breathless epochs. The soul beheld its features in the mirror of the passing moment" (1:190) [115].

What they, and we, see are two very different natures, but they are, essentially, the same two natures we saw in Beatrice and Giovanni. Once again, Hawthorne stages a confrontation between man's desire to submit to the force of a liberating, feminine nature and his perversion of that desire through fear and suppression—in himself and in "human law." As Giovanni with Beatrice, Arthur discovers the poison of Hester's love not within her, but within his own "unsacred" desire for, and fear of, her forbidden body and the "wild, heathen Nature of the forest" and "Love, whether newly born or aroused," which have come to be represented by her body (1:203) [122]. If it is within Hester's nature to "consecrate" herself to a kind of "sacred love" that would endure seven years of silence and ignominy and yet draw her back to his side even after his death, it is not within Arthur's nature. His love is compounded by the kind of "lurid intermixture" of emotions that erupt in the solitude of his study even as he attempts to flay them into suppression, the same emotions he finds wickedly liberated during his walk back to town after his second encounter with Hester in the forest. Nature, his nature, does not blossom into beauty nor express itself as a transforming, life-giving love of the self for the Other, the very "sacred love" that God, according to Milton, had made possible in the marriage of one soul to another as a means of liberating us from the pangs of solitude. Arthur's love, on the contrary, more nearly resembles what Milton

5. A few of the physical details of the setting in which Arthur encounters Hester in the forest even parallel the setting in which Hawthorne encountered Fuller in Sleepy Hollow on that Sunday afternoon in August 1842. Hawthorne came upon Fuller unexpectedly in a small clearing just off the pathway. Hawthorne emphasizes the fact that the clearing was obscured from the path by a surrounding ridge and the forest and that she reclined in the grass while he sat beside her (8:342–43) [pp. 342–43 of the Centenary Edition, vol. 8: *American Notebooks*] [204]. As Arthur walks along the forest path, of course, he encounters a waiting Hester, and they retire to a "little dell" that is obscured from the pathway by "a leaf-strewn bank rising gently on either side" and by the forest. They sit on the ground as they talk (1:186, 190).

terms "a sublunary and bestial burning," which Arthur cannot "chasten," as Milton claims he should be able to do.[6] As with Giovanni, Arthur's passion, once it is no longer directed at suppressing the despised self, is directed at others, expressing itself in the desire to use its cruel power to infect others with a share of its misery, and infect them, revealingly, by a brutal candor rather than by covert confession. Significantly, Arthur resists this temptation and remains silent. For Arthur, such passion must indeed be contained within the walls of his heart and "subjugated by human law," though ironically it is from such containments that his passion was originally perverted. Only in the act of writing and of speaking as an artist of the pulpit does he find a culturally sanctioned means to contain and yet to liberate his desire—both for abject confession and for power.

In Arthur's "shattered and subdued" spirit and in his dependency on Hester's bold courage to speak and act upon his own desires, Hawthorne complicates the confrontation between the masculine and feminine here and elsewhere in the romance by reversing the roles—Arthur in a conventionally "feminine" role and Hester in a conventionally "masculine" one. The implications of such a reversal gain greater significance if read within the context of another topic of dialogue among the Concord group—the fluidity of gender. In their ongoing conversation about the nature of marriage and friendship and about the development of the self-reliant individual, Emerson and Fuller had agreed that the fully realized individual crossed, at some level, the boundaries of gender. For Fuller, the ideal to be strived for was a harmony of both masculine and feminine qualities, a balance, as noted in her letter to Sophia, that she thought Hawthorne approached. One of the most famous passages of "The Great Lawsuit" originates from these conversations. Fuller writes that "male and female represent the two sides of the great radical dualism," each "perpetually passing into one another" so that "there is no wholly masculine man, no purely feminine woman." In *Woman in the Nineteenth Century,* she identifies the characteristics of this "radical dualism" as "Energy and Harmony, Power and Beauty, Intellect and Love," the first of each pair being traditionally associated with the masculine, the second of each with the feminine.[7]

In the seven years prior to her reunion with Arthur in the forest, in order to endure, Hester has had to suppress the feminine in her life, if not in her art. Once in the forest again with Arthur and giving herself up to the liberating influence of "Nature" and "Love," Hester undergoes a transformation that seems to bring each of the three pairs of masculine and feminine traits into equilibrium, finding, for a moment, what Fuller described as most rare, and fleeting, "perfect harmony in human nature."[8] Transformed by love, "her sex, her youth, and whole richness of her beauty, came back" (1:202) [122]. She is once again beautiful in a conventionally feminine way, but in confronting the ruin of Arthur and the possibility of transforming present misery into future happiness, she loses none of the masculine qualities that have sustained her—her energy, power, and intellect. At that moment of harmony within Hester's nature, when she reclaims the feminine that she has suppressed, "all at once, as with a sudden smile of

6. Milton, *Doctrine and Discipline of Divorce,* 582, 591.
7. Fuller, "The Great Lawsuit," 43, and *Woman in the Nineteenth Century,* 343.
8. Fuller, *Woman in the Nineteenth Century,* 343.

heaven," a sympathetic Nature seems resurrected. Sunlight "floods" shadows, green leaves "gladden," and dying leaves "transmute" into gold (1:202–3) [122]. The generative, feminine principle of passion and beauty in Nature that Fuller had celebrated in her flower sketches and that Beatrice had embodied for Giovanni transforms both Hester and the forest, but not Arthur. For Arthur, for the male narrator, and for the patriarchal order they represent, the feminine power of the natural world of the forest and of its human expression in love inspires both desire and terror.[9] If Hester is resurrected by such a power, Arthur experiences it as a "wild, free atmosphere of an unredeemed, unchristianized, lawless region" that must be "subjugated by human law," the "human" law, of course, of the men who make it (1:201, 203) [121, 122]. Before the interview, Arthur had deployed his increasingly depleting "masculinity," not to endure, much less overcome, his own shame, but to exacerbate it. Possessing little if any of Fuller's six traits of gender during much of the interview with Hester, Arthur, in his walk back home and in his study that night, is also transformed, but instead of a harmony of the masculine and feminine, he finds himself once again "a man," but a man in whom the masculine traits of energy, power, and intellect express themselves in a passion untempered by feminine "love, beauty, harmony." It is a violent, potentially destructive passion that is ignited not by his desire for nor submission to the feminine power that transformed Hester but by his fear of it.[1] Ironically, though Arthur and the narrator himself may fear the "wild, heathen Nature" associated with Hester's love, as Giovanni feared Beatrice's "poison," the destructive power at work here, as in "Rappaccini's Daughter," is not feminine, but masculine.

And that power is encoded in a "human law" that mystifies its gendered origins. As Michael J. Colacurcio has demonstrated, Hawthorne rewrites New England history to make this very point, deliberately creating a historical anachronism by making Bellingham, as governor, the chief legal authority enforcing Hester's punishment. Hawthorne has Governor Bellingham, as the highest representative of civil authority and "human law," punish Hester for a sexual offense similar to one he had committed—taking to himself a wife without the benefit of an official marriage ceremony and doing so with impunity, if not without some scandal.[2]

Thus, in order to return to town, where Nature has indeed been subjugated by law, such law at least as represented by Bellingham, Hester must destroy the equilibrium attained in the forest. She must once again suppress the feminine and take on the decidedly masculine. And she must take it on through an act of physical suppression—pinning back her luxuriant hair beneath her cap, pinning back the badge of masculine judgment on her breast.

If Hawthorne would have Hester find in her masculine nature the strength and courage to endure her estrangement from society and from Arthur, he would also associate her bold and increasingly radical speculation with the imbalance she must maintain between the masculine and

9. For a general analysis of the importance of masculine obsession and terror to Hawthorne's work, see Nina Baym's important revisionist essay "Thwarted Nature."

1. Leverenz, *Manhood and the American Renaissance*, makes the point that both Chillingworth and Dimmesdale maintain "their intellectual or spiritual self-control by rejecting intimacy" (269). Leverenz sees the narrator as obsessed by a fear that both "Hester's passionate loving, like Chillingworth's passionate hating, leaves the self wide open to demonic possession" (264).

2. Colacurcio, "'The Woman's Own Choice,'" 101–35, especially 109–11.

feminine. She who had "once been woman, and ceased to be so" in order to survive had suppressed her "tenderness," her "passion," and her "feeling" and turned to "thought," the "world's law" becoming "no law for her mind" (1:164) [101]. The "shame, despair, solitude" of her position "had made her strong, but taught her much amiss" (1:199) [120].

While it may certainly seem that Hawthorne is clearly condemning Hester's, and indirectly Fuller's, bold feminism, he is actually in accord here with Fuller. Almost a decade before he had met Fuller or formed his first truly intimate relationship with a woman, with Sophia, Hawthorne, then twenty-six and struggling in obscurity as a writer, had used the introduction to his unsigned biographical sketch "Mrs. Hutchinson" for the *Salem Gazette* to express his resentment against critics who were encouraging "a girlish feebleness to the tottering infancy of our literature" by praising too uncritically the work of a growing number of women writers. Such critics, he charged, labor under a misplaced "courtesy" and "a false liberality, which mistakes the strong division-lines of Nature for arbitrary distinctions."[3] The "strong division-lines of Nature" so uncomplicatedly distinct to the sexually naive twenty-six-year-old were considerably less distinct or uncomplicated to the sexually experienced forty-five-year-old who in the two decades since writing that subordinate clause had married, fathered a daughter, become intimate friends with the leading feminist of the day, and spent much of the 1840s participating—as a man and as a writer—in the debate reexamining the relations between the sexes.[4] By 1849 Hawthorne could join Fuller in condemning cultural constructions of gender that provide women, according to Fuller, "a place so narrow, that, in breaking bonds, they become outlaws." Confined to the claustrophobic sphere of the "strong-division lines" of a "Nature" defined by human law and society, such gifted, intelligent women as the Fuller-like feminist in Hawthorne's "Christmas Banquet" (1844), find that "in the world," as opposed to the home, there is "nothing to achieve, nothing to enjoy, and nothing even to suffer." She, like all such women, finds her "unemployed energy" thrown back on itself, driving her "to the verge of madness by dark broodings over the wrongs of her sex, and its exclusion from a proper field of action" (10:303) [p. 303 of the Centenary Edition, vol. 10: *Mosses from an Old Manse*]. Were society to provide such women with sufficient "room in the world" to develop fully their masculine and feminine natures, as Fuller says, with George Sand and Mary Wollstonecraft as her examples, "they would not," according to Fuller, "run their heads so wildly against the walls, but prize their shelter rather."

3. Nathaniel Hawthorne, "Mrs. Hutchinson," *Tales, Sketches, and Other Papers*, vol. 12 of *The Works of Nathaniel Hawthorne* (Boston: Osgood, 1883), 218. Hawthorne's sketch of Anne Hutchinson originally appeared in the *Salem Gazette*, 7 December 1830, 4 [159].

4. Yes, you might say, but what about those letters to his publishers in the mid-and late 1850s denigrating that "d——d mod of scribbling women" (17:304)? Those letters, of course, cannot be ignored, but they cannot be assumed to present a clear notion of Hawthorne's attitudes toward women or women writers. As James D. Wallace, "Hawthorne and the Scribbling Women Reconsidered," *American Literature* 62 (1990): 201–22, has demonstrated so persuasively, Hawthorne praised women writers as profusely as he sometimes condemned them. Both the praise and the condemnation centered on just those qualities in their writings that characterized Hawthorne's own works and that caused him profound ambivalence. I would also add to Wallace's argument that any reading of those letters to his publishers in the 1850s should be placed within the context of Hawthorne's long period of creative inactivity (perhaps, creative sterility) and of their audience, written as they were to publishers whose list was made up overwhelmingly of male writers.

"George Sand smokes, wears male attire, wishes to be addressed as Mon frère," Fuller had written in "The Great Lawsuit," but "perhaps, if she found those who were as brothers indeed, she would not care whether she were brother or sister." As Hester was to do in transforming the sign of the scarlet letter to read "Able" and "Angel," such women as Sand and Wollstonecraft, even without such "room," eventually "find their way, at last, to light and air" though "the world will not take off the brand it has set upon them."[5]

Fuller's plea in "The Great Lawsuit" and *Woman in the Nineteenth Century* is for the "fair and suitable position" for women that Hester sought, and it was for a "position" that allowed women the "room" to develop the full potential of their natures, masculine and feminine, without suppressing either. For Fuller, as well as Hawthorne, this meant that what had come to be seen as "the very nature of the opposite sex, or its long hereditary habit, which has become like nature" had "to be essentially modified" and that women must undergo "a still mightier change" in developing the masculine half of their dual nature. In doing so, however, they risked, as in George Sand's case, the creation of another imbalance, losing "the ethereal essence wherein"—according to contemporary constructions of the feminine that both Hawthorne and to a lesser extent Fuller assented—"she has her truest life" (1:165–66) [102]. The "whole relation between men and women" needed to be established "on a surer ground of mutual happiness" (1:263) [155], as Hawthorne has Hester say, but until that time, as Fuller wrote and Hawthorne edited for Hester, "the timid will suffer, the bold protest . . . but society is in the right to outlaw them till she has revised her law, and she must be taught to do so, by one who speaks with authority, not in anger and haste." To speak with such "authority," Fuller wrote, "those who would reform the world must show that they do not speak in the heat of wild impulse; their lives must be unstained by passionate error; they must be severe lawgivers to themselves."[6]

In 1843 Fuller had condemned both the "seal of degradation" branded on women by "the contract of marriage" and the "passionate error" of those who broke "bonds" and spoke "in the heat of wild impulse." Though Fuller would grant in "The Great Lawsuit" that "any elevation, in the view of union, is to be hailed with joy," rather than accepting the imperfections of unions in the present, she would be a "severe lawgiver" to herself in proclaiming "celibacy as the great fact of the time . . . from which no vow, no arrangement, can at present save a thinking mind." Fuller concludes "The Great Lawsuit," in fact, by speculating that given the present state of marriage and woman's subjugation to her husband, the "prophetess" who would "vindicate" the "birthright for all women" might have to speak as a virgin.[7] Though Hawthorne's penultimate paragraph in *The Scarlet Letter* edits and redeploys Fuller's earlier passage on George Sand and Mary Wollstonecraft, this final passage in Fuller's essay informs much of Hawthorne's response to Fuller in his depiction of Hester:

5. Fuller, "The Great Lawsuit," 29–30, and *Woman in the Nineteenth Century*, 284.
6. Fuller, "The Great Lawsuit," 30, and *Woman in the Nineteenth Century*, 286.
7. Fuller, "The Great Lawsuit," 30, 44, 47, and *Woman in the Nineteenth Century*, 286, 312, 347. Steele, *Representations of the Self*, demonstrates thoroughly how Fuller's concept of the power of virginity, symbolized by the goddess Minerva, is an attempt to relocate "the idea of woman" and her "independent spiritual authority" within "women's souls" by "advocating female self-reliance outside of male-female relations," a frontal assault "at nineteenth-century faith in motherhood as the ideal of female being" (127).

> A profound thinker has said "no married woman can represent the female world, for she belongs to her husband. The idea of woman must be represented by a virgin."
>
> But that is the very fault of marriage, and of the present relation between the sexes, that the woman does belong to the man, instead of forming a whole with him. Were it otherwise there would be no such limitation to the thought.
>
> Woman, self-centred, would never be absorbed by any relation; it would be only an experience to her as to man. It is a vulgar error that love, *a* love to woman is her whole existence; she also is born for Truth and Love in their universal energy. Would she but assume her inheritance, Mary would not be the only Virgin Mother. Not Manzoni alone would celebrate in his wife the virgin mind with the maternal wisdom and conjugal affections. The soul is ever young, ever virgin.
>
> And will not she soon appear? The woman who shall vindicate their birth-right for all women; who shall teach what to claim, and how to use what they obtain? Shall not her name be for her era Victoria, for her country and her life Virginia? Yet predictions are rash; she herself must teach us to give her the fitting name.[8]

The fitting name that Hawthorne would have his Madonna-like heroine give to her vision of the "destined prophetess" of "Love and Truth," and a new truth about love itself, would be the prophetess not of a "chaste" but of a "sacred love." Fuller was not to be the "virginal" prophetess that she had once "vainly imagined herself to be," for she had turned to the chaste life, as Hester had chosen it after 1642, not as an ideal, but as a temporary alternative to the present "seal of degradation" imposed on women by "the contract of marriage." Like Hester, like Hawthorne, Fuller all along had acknowledged that not all marriages were based on "sacred love" and that "sacred love" itself, as Hawthorne had explained to Sophia and as Hester would explain to Arthur, had a "consecration all its own" that superseded the "external rites" of civil marriage. The "sacred love that *should* make us happy" is, as Fuller defined it, a marriage of the masculine and feminine within the self and between a man and a woman, a marriage in which the woman does not "belong to the man" but forms "a whole with him."

Chastity is indeed after 1642 the "fact of the time" for Roger and Arthur as well as Hester, but in each case, the chaste life must be maintained by an isolation that splits the self by prohibiting a union—within or without—between masculine and feminine, man and woman. In each, the masculine subjugates the feminine, and in each that willful suppression destroys what could have been redeemed. For Hawthorne, men without women and women without men reinstate within the individual the very masculine subjugation of the feminine that chastity seeks to avoid in marriage.

If in Hester's time, and in Fuller and Hawthorne's, such a love found itself in conflict with the "long hereditary habit" of gender constructions that had come to seem "like nature" and in conflict all too often as well with the sacramentalization of the civil contract of marriage, such a love could be safely expressed only in an edited language that translated the actual into the imaginary, creating an art in which Hawthorne could confront the ghosts of the past in a room where they no longer "affrighten,"

8. Fuller, "The Great Lawsuit," 47, and *Woman in the Nineteenth Century*, 347.

give them speech again, say again differently what had already been said, and say also what had not and could not be said in any other language.

NINA BAYM

Revisiting Hawthorne's Feminism†

In this essay I swim against the tide to argue—again—for Hawthorne as a feminist writer from *The Scarlet Letter* onward. I argued this in essays published throughout the 1970s and 1980s, as well as in *The Shape of Hawthorne's Career* (1976). My readings have been contested, debated, revised, extended; the idea of Hawthorne as a feminist has been overwhelmingly rejected. In what follows I'll describe my original positions, summarize the important critical challenges to them, and finally—in view of these challenges—propose the idea of a feminist Hawthorne anew.

* * *

The Shape of Hawthorne's Career pointed to a significant change in Hawthorne's thematic emphasis around the time of *Mosses from an Old Manse* (1846). He began publishing as a Neoclassicist strongly critical of non-conforming individualism, became increasingly Romantic during the 1840s, and emerged with *The Scarlet Letter* into full daylight as a partisan of his outcast heroine, Hester Prynne. The character type faulted by the Neoclassicist, however, did not now earn the approval of the Romanticist. That Hester was a woman made all the difference. With one exception (Catharine in "The Gentle Boy"), the denigrated Romantic characters had been men. The later, admired Romantic characters were women.

Guided by my perception of this important distinction, I made several claims about Hawthorne and women characters, about Hawthorne and the feminine (defining the feminine as an assemblage of traits associated with persons socially identified as women), and about Hawthorne and feminism. I proposed that as the fiction increasingly centered on sympathetic women characters struggling against a murderous male authority, Hawthorne frequently contrasted such characters to an alternative female type. This structure deployed the traditional literary contrast between "dark" and "fair" ladies. I called the dark lady a "real" woman and the fair lady a "social myth" invented by patriarchal culture to discipline "real" women; I proposed that real women embodied creative force while fair women, if favored by men, destroyed not only the women they had bested but also the men who chose them. I also suggested that Hawthorne linked artistic creativity, including his own, with traits that might be fairly called "feminine" because they were represented by these sympathetic, embattled "real" women as natural expressions of their womanhood. I further proposed in two essays—"Hawthorne's Women" (1971) and "Thwarted Nature" (1982)—that, because Hawthorne represented women simultaneously as embodiments of misogynistic male fantasies about women and

† From *Nathaniel Hawthorne Review* 30 (Spring & Fall 2004): 32–55. Reprinted by permission of the author. Page numbers in square brackets refer to this Norton Critical Edition.

as "real" women struggling against these fantasies, it was appropriate to call him a feminist.

I excepted Hawthorne from feminist arguments that the canonized authors were patriarchal sexists, even though it was he who infamously referred to literary women as a "d——d mob of scribbling women." No true patriarch, I thought, could have invented Hester. In proposing Hester as the novel's protagonist—initially in "Passion and Authority" (1970)—I was not offering a new reading but reviving an approach that had languished under Perry Miller's powerful influence. The "Miller" line understood Hawthorne to be writing about Puritan theology from a perspective close to the Puritans themselves, and criticizing his own culture for foolish optimism about human nature. This approach assumed that Arthur Dimmesdale was *The Scarlet Letter's* protagonist. If *The Scarlet Letter* retold Genesis, Hester was the temptress Eve; if it retold Puritan history, she was Anne Hutchinson. Yes, Hawthorne began *The Scarlet Letter* with reference to "the sainted Ann Hutchinson"; but that was irony, as recourse to his early, uncollected "Mrs. Hutchinson" (1830) made clear. No matter that twenty years had elapsed between the works; Hester was no more than a sexual and doctrinal temptress whose scheming led poor Arthur Dimmesdale, the novel's beset hero, astray.

Having interpreted the novel as a story about Hester (an interpretation which, no matter how my feminist angle has been rejected, is now standard in discussions of the novel), I then reread "The Custom-House" as an autobiographical allegory about a blocked artist breaking out by identifying himself with an imaginary, stigmatized woman ("The Romantic Malgré Lui," 1973). When "Hawthorne" wanders into the upper floor of the Custom-House, finds the letter, puts it on his chest, and feels it burn, he is saying in several senses, "this is my character." He associates the creative flood that sweeps him out of the Custom-House backwater with the life force inhering in this admirable woman. For, as the novel unfolds, the letter, intended by the Authorities to signify harsh but just condemnation, is made by Hester to signify something entirely different—able, admirable. From this perspective, I thought at the time, the novel's feminism was self-evident.

I interpreted Dimmesdale as Hester's foil—weak, orthodox, conventional—and as her temptation rather than she his. As an actor in his own story, Dimmesdale, like so many other Hawthorne male characters, rejects the woman at his own cost as well as hers. In *The Scarlet Letter*, oppression or rejection of women, rather than surrender to them, led to male downfall. This motif, incipient in earlier stories like "Young Goodman Brown" (1835) and "The Minister's Black Veil" (1836), became dominant in the 1840s; this change gave Hawthorne's career its "shape."

How to explain this development? Between 1841 and 1846 he encountered the Transcendentalists via his eight-month sojourn at Brook Farm and several years in Concord. He came into close contact with Romantic ideas as they were vividly articulated by living people. The iconoclastic idealists who became his friends—Thoreau, Fuller, Emerson—had an intellectual vigor and courage he could not help but admire. These provocations coincided with his marriage. They also coincided with his recognition that New England literary culture was itself moving in a more liberal direction.

Then, in 1849, having returned to Salem three years earlier to work in a dull job secured through political patronage, he suffered two bereavements. He lost this boring yet remunerative occupation when his party lost the elections, and his mother died. According to Sophia, Hawthorne's wife, and also according to Hawthorne's courtship letters, his mother was reclusive and remote. Julian Hawthorne's biography of his parents affirmed the characterization; but Hawthorne's journals, along with his letters to his mother and sisters, testify to her strong and much-loved presence. Her loss was shattering. In "Hawthorne and His Mother" (1982) I proposed that this death, in complex ways, might have motivated *The Scarlet Letter.* I imagined Hester as a version of the mother, and the novel as allying with her—she who had always supported his literary ambitions—rather than with the stultifying, untrustworthy authority of Uncle Sam's Custom-House.

* * *

Challenges to these readings have countered that Hawthorne's use of the feminine was always in the service of masculinity issues, that he viewed assertive or aggressive or rebellious women as threats to masculinity; that his inevitable punishing or containing of truant women demonstrated deep hostility to them and a profoundly conservative view of their proper place; that this penalizing reflected discomfort over the ongoing feminization of the literary profession; that women characters are tools of the state, upholding or instilling conservative, bourgeois values, "naturalizing" the capitalist agenda by a politics of domesticity. These points recur and overlap throughout six critical approaches, of which I now offer representative examples: Puritan/traditionalist, masculinity, gay/queer, feminist, political, and biographical studies.

Puritan/traditionalist criticism, masculinity studies, and gay/queer studies converge in Hawthorne studies by agreeing that men are central and women ancillary in his writings. Traditionalist studies of Hawthorne took the male center for granted; Baym (as I will now call myself) questioned this. Traditional critics now started insisting that, yes, Hawthorne's main characters were, indeed, male. Michael Colacurcio's "Woman's Own Choice" (1985) reads *The Scarlet Letter* as Dimmesdale's overcoming Hester's literalist temptation. Saying that "the fiercely logical structure of *The Scarlet Letter* points to Dimmesdale as the indubitable center of literary organization" (119), that "structural analysis . . . must always privilege Dimmesdale" (133), Colacurcio thinks Hester's choice of "a human lover" rather than "the figure of salvation in covenant" (Dimmesdale's choice) demonstrates her female incapacity for abstract thought. When her moment to choose comes, "she passionately chooses the literal. To which she almost converts even Dimmesdale" (124). Dimmesdale, however, demonstrates his manhood and admirable orthodoxy by rejecting her temptation.

Masculinity studies and gay/queer studies also focus on men. Masculinity critics identify "feminine" tendencies in the self or the culture, and consider how masculinity is defined as their opposite. Some critics in this mode—e.g., Leland S. Person, who writes that Coverdale, Kenyon, and Dimmesdale briefly "enjoy creative relationships with women before repressing the self-creative impulse within themselves which the woman evokes" (115)—take the Baym-like position that men need the feminine;

others, like T. Walter Herbert (*Dearest Beloved*) and David Leverenz, argue that rejecting women is (unfortunately) necessary for men dedicated to Jacksonian masculinity. Leverenz writes: "story after story presents conventional manliness as aggressive, insensitive, and murderously dominant" (231); "as male rivalry takes center stage, the narrators disengage from their heroines. . . . They put up a fog of ambiguities, ironies, self-consciousness, and multiple points of view to screen their covert participation in the men's struggle for narcissistic self-empowering" (246–47). Some critics propose that the biographical Hawthorne is uncomfortable about operating in an increasingly feminine profession (the opposite of Baym's point that a good male artist has to accept the inner feminine). Millicent Bell observes that "his authorial mode was that of the feminized male author who knows he has entered a female world in becoming a writer rather than a businessman or politician" (15). Kenneth Egan proposes that Hawthorne made Hester an adulteress specifically to express the author's sense that to be a male writer in his culture "was necessarily to be an adulteress or feminized adulterer of the truth" (41). James D. Wallace says that Hawthorne's hostile reaction to women writers originates in his uneasy identification with them. Herbert brings the lessons of *The Scarlet Letter* home to men: "Men whose manhood compels them to hate what they conceive as the womanhood in them lead psychic lives that are characterized by a continuous internal assault that is gendered male-on-female. . . . [Men] may find inspiration in Hester, as we seek to rework the manhood that dooms us to pornographic enchantments and cripples our capacity for intimacy with real women" ("Pornographic," 116, 119).

Gay/queer criticism is male-centered by definition. Homosocial and homoerotic moments are excavated and attributed either to Hawthorne's own suppressed sexual inclinations or to the sublimated, affect-laden idealizations of male-male relationships in antebellum culture. Robert K. Martin points to "the dynamics of male relationship" in *The Scarlet Letter* and *The Blithedale Romance*—noting for example the "extraordinarily erotic moment" When Chillingworth half-undresses the sleeping Dimmesdale, and claiming that Coverdale's disquisitions on polymorphous sexuality at Blithedale register his attraction to Hollingsworth (132). Lauren Berlant also finds Coverdale's attraction to Hollingsworth central in *The Blithedale Romance* ("Fantasies"). Scott S. Derrick identifies the erotic connection between Dimmesdale and Chillingworth as the affective center of *The Scarlet Letter*; Leverenz writes, "the cuckold and the lover rise together to an all-male paradise, while Hester mutely returns to Boston" (275). Karen L. Kilcup argues that the relationship of author to reader in Hawthorne's work is erotically charged and—supposing that Hawthorne imagines his reader as male—specifically homoerotic.

Although academic feminism is rife with internal dissention, most self-identified Hawthorne feminists agree in rejecting Baym's readings. Stressing the way Hawthorne inevitably punishes and/or silences unconventional women, they find his plots antifeminist, reinforcing a culturally conservative agenda, and testimony to authorial misogyny. A preponderance of feminist criticism thinks Hawthorne felt threatened by emergent feminism, female activism, female sexuality, and women writers—in fact, by any sign of women's desire to improve their lot or determine their own lives. For this group, Hawthorne's narratorial hand-wringing over the intractability of social institutions and the misery of women is a sentimental sham.

The first scholar to make this argument was probably Judith Fetterley, who sees "The Birth-mark" as illustration of "the great American dream of eliminating women" (24) [709]. Louise DeSalvo concedes that Hawthorne "portrayed, with superb accuracy, the condition of women in the nineteenth century and the psychological processes of men who could not tolerate the notion of female equality" but decides that because "he shared the misogynistic view of his age, he could not condemn what he saw and drew back from defining the implications, for society, of what he so astutely observed about the reality of women's lives" (121–22). Barbara Bardes and Suzanne Gossett use *The Blithedale Romance* and *The Scarlet Letter* to observe that while Hawthorne's novels reveal "distrust of almost anyone with the power of speech," he was especially wary of "women with 'tongues' because he connected women's public speech with sexual exposure and expression" (59). Joyce W. Warren finds him especially severe on strong women, who "are never allowed to pursue what might seem to be the implications of their characters; they do not become heroic leaders or independent public figures. . . . This hesitation is owing in part to Hawthorne's belief in a conventional image of feminine behavior" (189). Jean Yellin thinks *The Scarlet Letter* "seriously considers the new feminist definitions of womanhood and, rejecting them, replicates traditional imagery and endorses patriarchal notions" (126). Amy Schrager Lang complains about Hester's unfeminist subsidence into a "vague politics of patience and inaction" (191).

More recently Jamie Barlowe's "Rereading Women" (1997) and follow-up *Scarlet Mob* (2002) coins the term "Hester Prynnism" to denote anti-feminist silencing and ignoring of women's voices, of which the first and most egregious practitioner was Hawthorne himself. Barlowe idealizes the Hester of Roland Joffé's 1995 film; she sees Joffé's plot—in which Hester spends a lot of time naked, Dimmesdale professes undying love, and the couple (with Pearl recently born rather than seven years old) successfully run off together—as a feminist way to tell Hester's story. Although the idea of an entirely different Hester in an entirely different narrative (along with an entirely different Dimmesdale) may seem bizarre, and although not every feminist will share Barlowe's idea of feminism, most feminist criticism does fantasize a different ending to Hester's story. Sandra Tomc, invoking Hawthorne's "notoriously conservative ideas about women [and] feminism," agrees that Hawthorne "repudiates both his heroine and the genre [of seduction novel] that inspired her" (469, 472). Todd Onderdonk imputes an obsession with sexual purity to Hawthorne deriving from the author's conviction that women have only one "necessary function—motherhood" (97).

Political, new-historical, and ideological critics perceive Hawthorne's women as servants of bourgeois capitalism. Gendered spheres, here, are key to the consolidation of middle-class, capitalist culture in Hawthorne's era; women were pressed into bourgeois service by persuading them that their natural abilities suit them for the domestic sphere, where they are cherished and protected. And Hawthorne went along with this project, which relies heavily on what such critics consider sentimental hocus-pocus. Michael Davitt Bell calls sentimentalism "a medium of psychological repression and social control" that, in his view, the conservative

Hawthorne deployed against women's rights and other movements for social change (191). Gillian Brown argues that a female domestic subculture "based on self-denial and collectivity—the ethos of sympathy customarily and disparagingly called sentimentalism" (6)—has not produced "an antithetical model of selfhood" but only further domesticated "an already domesticated selfhood. . . . Far from an account of the female subject, domesticity signifies a feminization of selfhood in service to an individualism most available to (white) men" (7). The novel most suited to this approach is *The House of the Seven Gables*; but other novels also figure. Lori Merish, for example, reads *The Blithedale Romance* as an exposure of "the extent to which the bourgeois ideology of romantic love and consensual familial ties produces new species of patriarchal power, reconfiguring, rather than undermining, women's construction as property" (172). Ellen Weinauer says that "Despite his effort to situate himself in an alternative and indeed antipatriarchal history, *The Scarlet Letter* taken as a whole suggests that . . . Hawthorne needs the possessed female body" (107).

Psychologizing this political class-and-gender analysis, Joel Pfister thinks Hawthorne's plots about "the feminization of women" (8) reflect and contribute to a middle-class discourse of sexuality. Still, Pfister says that Hawthorne recognizes the artificiality of this discourse: he is "product, agent, and critic of an emerging middle-class interiority, and is aware that his participation in reproducing the forms of subjectivity of his class is political" (183). Herbert (*Dearest Beloved*) interweaves Hawthorne's fiction with family letters to show how internalized social ideas of true woman and manly man are experienced as authentic. He develops a poignant portrait of a Sophia who has violently disciplined herself to become the woman Nathaniel required as the condition of his adoration. And he shows a Nathaniel imprisoned in turn by the surveillance of the monstrous woman he created. According to Herbert, this paradigm implicates the entire culture; the Hawthornes "vividly exemplified the domestic ideal of family relations that became dominant in the early nineteenth century" (xvi). For Herbert, Hawthorne is both participant and critic of the represented scene, seeking to "contain his material within the rhetoric of the domestic ideal, even as he lays open the dilemmas intrinsic to that ideal" (199).

Three important ideological readings make Hester into a problem or opportunity for the state. Sacvan Bercovitch reads Hester's return to Boston and resumption of the letter as the novel's most important event. In a "woman's own choice" quite unlike Colacurcio's, Hester elects "to become the agent of her own domestication" (11); "in the most carefully prepared-for reversal in classic American literature, Hester herself imposes the symbol" (92). Her choice, he says, illustrates the workings of liberal capitalism, a political form "designed to make subjectivity the primary agency of change while keeping the subject under control," a double function accomplished "by representing interpretation as multiplicity flowing naturally into consensus" (23–24). Where Bercovitch finds Hester a type of the citizen in her free acceptance of social control, Berlant describes her citizen type less forgivingly as a "mutual articulation of woman, privacy, and political submission" (*Anatomy*, 106) and proposes that "woman" is a "manifest public problem of law and order" that is solved by containment (108). Brook Thomas likens Hester's final community of women to Habermas's public sphere, a zone of independent activity that guarantees the state's democratic functioning. Curiously, of these three readings only Berlant makes a point of Hester's gender, and even she perceives "the male body" (Dimmesdale's male body) as "the major stress point in the social order" (115).

All lines of criticism fret over Hawthorne's self-awareness. Even Bercovitch, in a troubled and troubling footnote, remarks that Hawthorne, not liberal culture, ultimately "compels Hester to resume the A"; this "sense of compulsion . . . adds a discordant note to Hawthorne's orchestration of pluralist points of view" (157). For if Hawthorne knew about the woman-centered reform initiatives in antebellum culture, if he knew well some of the women actively involved in such efforts, if he knew that the literary market he sought to enter and dominate was filled with women authors, then it becomes likely that his strategies of plot containment represent his own intentions rather unthinkingly reproducing social platitudes. Hawthorne was well acquainted with, even close to, many women who lived undomestic lives but neither killed themselves nor kept silent. He knew women scholars, writers, artists, and activists: among others, Elizabeth Peabody, Mary Peabody Mann, Sarah Josepha Hale, Grace Greenwood, and, perhaps above all, Margaret Fuller. Thomas R. Mitchell has written compellingly about traces of Fuller in Hawthorne's writing; he sees her in Hester, Zenobia, and Miriam. He finds sentences in *The Scarlet Letter* appropriating "The Great Lawsuit" almost verbatim. It would seem to follow that if he created powerful women only to silence them, he did so as a way of rejecting the signs of the times that he saw all around him.

* * *

To suggest a feminist Hawthorne in the face of all this criticism is foolhardy. Nevertheless, I will give it a try. I grant at the outset—a point that criticism usually ignores, perhaps because it's so obvious—that he stopped with white American women resident in New England, presumably heterosexual, of English descent. Hawthorne knew his own limits in this regard; when he wrote "women this" and "men that," he asked to be understood as an interpreter of New England life and character, not the universe. I grant too that if a feminist position must be manifested by plots wherein women live happily ever after on their own terms, Hawthorne fails. Happy endings on such terms are impossible in his novels—men are not what women imagine. One might argue that, given the kind of man Dimmesdale is, *The Scarlet Letter* is far more feminist for refusing the bogus finale of a happy escape than had it allowed the two (three, actually) to make their getaway. The mismatch between women's illusions and the realities of male character is a leading motif in the later romances.

Hawthorne's observation that smart women often make foolish choices is not erroneous; the problem is still with us. There are, of course, other imaginable endings than a happy marriage; much feminist writing today exists to show how women might get along without men (or, more strategically, to show what social reforms are necessary so that women can actually get along without men). In which case, since Hawthorne's work exactly fits this description, his feminism should be obvious. Smart, independent women like Hester, Zenobia, and Miriam are his heroines in *The Scarlet Letter, The Blithedale Romance*, and *The Marble Faun*. But even the more conventional women—Phoebe, Priscilla, and Hilda, not to mention the amazing

Hepzibah, who deserves much more careful appraisal than I have space for here—are women used to living alone and fending for themselves.

They can and do get along without men, but they would prefer not to. No question: this author is preoccupied with the heterosexual couple and the obstacles that make of a supposedly fulfilling social form something so fraught with misery. Why this is so, and what to do about it, are questions his plots repeatedly ask. But if his plots do not—cannot—lead to happy outcomes in a conventional romantic sense, neither do they end in utter social futility. They offer limited, incremental change, although often at a great cost to the agent of such change.

By the time he wrote *The Scarlet Letter* Hawthorne had become aware of and deeply involved in ideas about women's grievances as they were expressed by articulate women he knew, especially Margaret Fuller. He understood that, since he had chosen to write fiction, many if not most of his readers would be women; he anticipated their thoughtful familiarity with these ideas. Virtually alone among male authors of his generation, he made space for feminist analysis in his fiction. Hester is made to introduce feminist ideas at least twice in *The Scarlet Letter*: once in chapter 13, where, brooding over the situation of woman, she wonders whether existence was "worth accepting, even to the happiest among them" (165) [101], and again in chapter 24, where she tells sorrowful women about her hope for a new truth that would "establish the whole relation between man and woman on a surer ground of mutual happiness" (263) [155].

The subject runs constantly through *The Blithedale Romance*, especially in chapter 14, where Zenobia insists that "when my sex shall achieve its rights, there will be ten eloquent women, where there is now one eloquent man" and goes on to say that "thus far, no woman in the world has ever once spoken out her whole heart and her whole mind! The mistrust and disapproval of the vast bulk of society throttles us, as with two gigantic hands at our throats!" (120). True, Coverdale attributes "the animosity with which she now took up the general quarrel of women against men" to "Zenobia's inward trouble"—but this is at least to concede that such a quarrel exists. When Coverdale adds that he personally would be happy to give woman "all she asks, and add a great deal more, which she will not be the party to demand, but men, if they were generous and wise, would grant of their own free motion" (121), he is saying men are neither generous nor wise. Hollingsworth's behavior (and his own) bear this out.

Feminist ideas typical of Hawthorne's era also circulate through *The Marble Faun*, as in chapter 13, when Miriam tells Kenyon that men wrongly believe "Nature has made women especially prone to throw their whole being into what is technically called Love. We have, to say the least, no more necessity for it than yourselves—only, we have nothing else to do with our hearts. . . . I can think of many women, distinguished in art, literature, and science—and multitudes whose hearts and minds find good employment in less ostentatious ways—who lead high, lonely lives, and are conscious of no sacrifice, so far as your sex is concerned" (121). Miriam's reference to distinguished professional women, which Kenyon does not contest, alludes to a situation that few male authors of Hawthorne's day recognized in print.

Certainly, Hawthorne's novels point to errors that reform-minded women typically make—mainly through lack of self-understanding and unwitting complicity in the conditions they deplore. But unless a feminist perspective requires asserting that women are perfect, his observations might be seen as attempts to intervene on the feminist side. That feminist ideas are uttered by flawed women, that activist women generalize from their own situations, does not invalidate the ideas. The imperfect women who utter feminist sentiments are treated sympathetically and admiringly. They have enormous courage and considerable intellect. Even when defeated, they make things happen. Bercovitch presumes that the liberal polity absorbs Hester's dissent; but I think her resumption of the letter permanently changes her public world. Hawthorne writes: "The scarlet letter ceased to be a stigma which attracted the world's scorn and bitterness, and became a type of something to be sorrowed over, and looked upon with awe, yet with reverence too" (263) [154].

Had Hester never returned, she would have been forgotten. Had she returned and not worn the letter, she would have escaped the letter's meaning without changing it. Returning and wearing the letter when "not the sternest magistrate of that iron period would have imposed it" alters the way the letter is perceived, changes its definition. People look on the letter "with reverence." Hester preaches a model of change involving sudden revelation from a sinless apostle of joy, but she practices a model of slow change brought about by persons visibly scarred by battles fought and ignominy endured. This is not entirely a sentimental outcome, resulting from pity ("sorrowed over"); it is also the outcome of reverence, of awe. Re-entering civil life adorned with the letter by her own choice, Hester moves Puritan Boston from the Dark Ages towards enlightened modernity. Now, a century and a half after *The Scarlet Letter*, popular references to a scarlet letter invariably imply unjust stigmatization. In the wider world, this novel has performed that cultural work.

One concedes that Hester and Hawthorne understand women's plight in reference to intimacy with men: This was the overwhelming focus of feminist reformers of his day, the overwhelming subject of the novel (his chosen genre), and (after all) the overwhelming concern of feminist reformers now. The underlying sources of the difficulty in thinking of a feminist Hawthorne, as I see it now, are his beliefs in such a thing as human nature and, worse still, in essential differences between human nature in the two sexes. The key point is that, for Hawthorne, the age-old, relentless history of male exploitation of women could only be explained as the outgrowth of essential characteristics of human selves.

If the two sexes are essentially different, as I now think Hawthorne thinks they are, then the social change sought by feminism would seem impossible to achieve. But, oddly, Hawthorne's work does not offer these essentialist beliefs as conservative excuses to keep women down. On the contrary, it struggles to plot a way in which, given these "facts" about the human, women may be raised up. His novels are replete with objections against social arrangements that tell so heavily and unjustly against women. I used to think that Hawthorne presented romantic love as a social myth invented to ensnare women, because woman after woman in his texts suffers from what might be called "a broken heart." Passages in the novels seem to support this idea of love as a social myth; for example, Coverdale, at the end of *Blithedale*, laments that "It is nonsense, and a miserable wrong—the result, like so many others, of masculine egotism—that the success or failure of woman's existence should be made to depend wholly on the affections, and on one species of affection; while man has such a multitude of other chances, that this seems but an incident. For its own sake, if it will do no more, the world should throw open all its avenues to the passport of a woman's bleeding heart" (241). But now I notice that this social asymmetry stems from "masculine egotism," not a constructed condition but a fact about men, and see that these sentences argue explicitly for social change.

Hawthorne's point, one might say, is not that essentialism makes social change impossible but that it makes social change impossible unless differences between women and men are taken into account. The slow, uneven pace of social change occupies Hawthorne much in these later works wherein—unlike so much antifeminist rhetoric of his day, of which Hollingsworth's tirade in chapter 14 of *The Blithedale Romance* is a splendid example—inequality of physical strength is unimportant. What matters is the internal difference to which Hollingsworth refers when, after threatening violence against female reformers ("petticoated monstrosities," as he calls them), he says that force is unnecessary because "the heart of true womanhood knows where its own sphere is, and never seeks to stray beyond it" (123). That the heart of womanhood knows its sphere, as Hollingsworth would define it, is repeatedly given the lie in Hawthorne's writings; but that women have more heart than men is something he counts on. In women's excess of heart, compared to men—or in men's deficiency of heart, compared to women—lies the explanation for an asymmetrical culture.

Male egotism and its associated cold-heartedness had been a Hawthorne obsession from far back in his career: "Wakefield" (1835), implicitly in "Young Goodman Brown" (1835) and "The Minister's Black Veil" (1837), "The Man of Adamant" (1837), and "The Shaker Bridal" (1838); on through the 1840s: "Egotism; or, the Bosom Serpent" (1843), "The Birth-mark" (1843), and "Rappaccini's Daughter" (1844). Still prominent in three of the four later romances (the contrast between Hepzibah and Clifford introduces it into *Seven Gables* as well)—these traits are now considered entirely in terms of their meaning for women's lives. In *The Blithedale Romance*, Coverdale's speculations about Hollingsworth in chapter 7—"Sad, indeed, but by no means unusual" that he has mistaken his "terrible egotism" for "an angel of God" (55)—and in chapter 9: "Men who have surrendered themselves to an over-ruling purpose" are almost never able to "recognize the process, by which godlike benevolence has been debased into all-devouring egotism" (70–71)—acquire their significance because of the way this man behaves towards women. In chapter 28, as noted above, he deplores the masculine egotism that makes women's success wholly dependent on love (241). Most memorably, in chapter 14, he interprets Hollingsworth's antifeminist tirade as an

> outrageous affirmation of what struck me as the intensity of masculine egotism. It centred everything in itself, and deprived woman of her very soul, her inexpressible and unfathomable all, to make it a mere incident in the great sum of man. Hollingsworth had boldly uttered what he, and millions of despots like him, really felt. Without intending it, he had disclosed the well-spring of all these troubled waters. (123)

Masculine self-absorption is the source, but it prevails because women lack a corresponding self-absorption. When, rather than rising to

Hollingsworth's challenge, Zenobia mildly counters that, if man were "but manly and godlike," woman would be "only too ready to become to him what you say" (124), she demonstrates the other side of the intractable equation. Coverdale, appalled, wonders whether her words express women's "nature" or "the result of ages of compelled degradation" and whether, in either case, it could "be possible ever to redeem them." He does not, however, notice that this meditation has been sparked by his peevish awareness that the women are more attracted to Hollingsworth than to him: "how little did these two women care for me, who had freely conceded all their claims, and a great deal more, out of the fullness of my heart" (124). Hawthorne lets readers see that Coverdale's self-labeled feminism is also exploitative of women; he is not exempt from the situation that he analyzes so cannily.

Nor is Zenobia exempt; her capitulation here puts a feminist conundrum squarely before us. If millions of men behave like Hollingsworth and millions of women behave like Zenobia (and their gendered behavior is reproduced more palely in Coverdale and Priscilla), what can be hoped for? The intransigency of the problem, rather than their weakness of intellect, baffles his women characters. The more acute their minds, the more they recognize the conundrum. When the narrator comments on Hester's meditations, he says: "A woman never overcomes these problems by any exercise of thought. They are not to be solved, or only in one way. If her heart chance to come uppermost, they vanish" (166) [108]. This passage has been read as a misogynistic dismissal of female intellectuality. But it can be read to mean, not that women lack intellect, but that nobody can solve these problems by sheer intellectual force, and that a woman happily in love will likely stop worrying about them. Yes, but how many women are happily in love in Hawthorne's world? Too, if unhappiness makes women into reformers, it does the same with men. As Holgrave says, "The world owes all its onward impulse to men ill at ease. The happy man inevitably confines himself within ancient limits" (306–07).

Thus, to say that Hawthorne's women have more heart than his men does not imply that they have less brain. Mind does not unsex a woman; Zenobia is smarter than Hollingsworth, Miriam than Kenyon. Hester leaves Dimmesdale in the dust. True, all that thinking in Hester's case is associated with an apparent loss of beauty, insofar as female beauty means sexual attractiveness to heterosexual men. That Hawthorne might have reconsidered this facile association is inferable from the fact that neither Zenobia nor Miriam is less sexually appealing because of her intellect. With these two later heroines Hawthorne seems to propose that beauty and braininess are different expressions of the same force, a force lacking in the counter-heroines Priscilla and Hilda whose appeal lies in both their physical plainness and intellectual timidity.

But if Hester's thinking makes her less conventionally sexy, it does not make her less loving, and if motherhood is Hawthorne's test for true womanhood, then Hester is a paragon. I read *The Scarlet Letter* as an "antiseduction" novel. The protagonist does not die; she succeeds as a single mother supporting herself and her child and, when her child is grown, becomes a valued and valuable member of the community.

In the plot's unfolding, the significant result of Hester's independent thinking is a profoundly womanly decision (in Hawthorne's terms) to "rescue" Dimmesdale by telling him who Chillingworth is. Thus the forest meeting where Dimmesdale's cold heart triumphs over her warm one, his appeal to the great heart of womanhood effectively seducing Hester for a second time.

Herbert, like other critics, says "Hester has already contrived the plan that she now persuades Arthur to adopt. She wants them to leave the colony" (*Dearest Beloved*, 201). No; she wants *him* to leave the colony. Horrified by Dimmesdale's condition as she sees it in the second scaffold scene, inferring that Chillingworth is somehow to blame, Hester determines first to "meet her former husband"—not her former lover—and "do what might be in her power for the rescue of the victim on whom he had so evidently set his gripe" (167) [102]. "I must reveal the secret," she tells Chillingworth; "what may be the result, I know not" (173) [106]. Chillingworth releases her from her oath of secrecy, freeing her to follow her conscience: "Go thy ways, and deal as thou wilt with yonder man" (174) [106].

As chapter 16 tells us, Hester's one intention is "to make known to Mr. Dimmesdale, at whatever risk of present pain or ulterior consequences, the true character of the man who had crept into his intimacy" (182) [110]. Beyond this she has no plan. Their forest meeting makes her, at last, "fully sensible of the deep injury for which she was responsible to this unhappy man" (192) [116]. No plan already decided on, even if she had one, could possibly be useful in view of what she sees Dimmesdale to be: "She now read his heart more accurately" (193) [117].

Dimmesdale then commandeers the conversation in typical passive-aggressive fashion by pleading for her help: "Be thou strong for me! . . . Advise me what to do" (196) [119]. She answers by counseling him to leave Boston—whether heading west into the forest or east over the ocean—in ringing rhetoric that implies no intention to go with him: "Leave this wreck and ruin here where it hath happened! Meddle no more with it! Begin all anew. . . . The future is yet full of trial and success. There is happiness to be enjoyed! There is good to be done! . . . Up, and away!" (198) [119–20]. But then Dimmesdale replies, "There is not the strength or courage left me to venture into the wide, strange, difficult world, alone!" and, as the narrator stresses, repeats the word *alone*. Hester hears his unvoiced plea and responds to it: "Thou shalt not go alone!" (198) [120]. In this exchange Dimmesdale manipulates Hester—no doubt sincerely—through a sentimental appeal to (as he put it earlier) the "wondrous strength and generosity of a woman's heart" (68) [47]. Only a woman's heart can be so worked upon.

Dimmesdale's thoughts in this scene are all for himself, Hester's all for him. He is, after all, a true man as she is a true woman. One might suppose that it was always thus for these two—the woman in love with the man, the man in love with himself. Few seduced women get the chance to express the point with more intensity than Zenobia in her great attack on Hollingsworth: "It is all self! . . . Nothing else but self, self, self! . . . I see it now! I am awake, disenchanted, disenthralled! Self, self, self!" (218). Certainly, Hester never approaches this level of feminist rage, even though Dimmesdale's death amounts to yet another disavowal of human love. What Hester achieves through Dimmesdale's death, however, is unique in that it is a loving—that is, womanly—life without a man. It prefigures and registers the eruption of women into the public sphere through various forms of culturally acceptable womanly activity, what historians have come to call "domestic feminism."

Herbert writes that *The Scarlet Letter* "does not work out a solution to the male-on-female sexual abuse that it so pervasively depicts, but in Hester Prynne [Hawthorne] dramatizes the struggle of women to disentangle themselves from this enslavement, and to find lives of independent self-respect, and to define an autonomous sexual selfhood" ("Pornographic," 119). What Herbert calls male-on-female sexual abuse, Hawthorne would call the inequality of egotism versus love; I see no male character in the whole Hawthorne repertory who "loves" a woman the way a woman loves him. However, that Hester is struggling to find a life of independent self-respect is not so clear, still less that she is trying to define an "autonomous sexual selfhood"; on the contrary, Hawthorne proposes, exactly her disregard for self (including, yes, her celibacy) changes the public mind. Because Hester "had no selfish ends, nor lived in any measure for her own profit and enjoyment," people look up to her (263) [154].

This change is not a reversal, but a continuation of what had been happening over the seven years of her earlier sojourn in Boston:

> She was self-ordained a Sister of Mercy; or, we may rather say, the world's heavy hand had so ordained her, when neither the world nor she looked forward to this result. The letter was the symbol of her calling. Such helpfulness was found in her—so much power to do, and power to sympathize—that many people refused to interpret the scarlet A by its original signification. They said that it meant Able; so strong was Hester Prynne, with a woman's strength. (161) [99]

Pearl, too, is crucial to understanding Hester as a loving woman in the less heterosexualized sense that Hawthorne represents in *The Scarlet Letter.* Much has been written about Pearl as her mother's disciplinarian, the externalization of her mother's sin, and so on (e.g., Emily Budick, Franny Nudelman). But Pearl is also a human person, somebody for Hester to love, her child. The quiet and outwardly compliant Hester explodes furiously when the Governor threatens to take Pearl from her. "Hester caught hold of Pearl, and drew her forcibly into her arms, confronting the old Puritan magistrate with almost a fierce expression. Alone in the world, cast off by it, and with this sole treasure to keep her heart alive, she felt that she possessed indefeasible rights against the world, and was ready to defend them to the death. . . . 'Ye shall not take her! I will die first!'" To Dimmesdale she cries out, 'thou knowest what is in my heart, and what are a mother's rights, and how much the stronger they are, when that mother has but her child and the scarlet letter! Look thou to it! I will not lose the child! Look to it!" (113) [72].

Leaving the governor's hall, her mother's rights confirmed, Hester declines Mistress Hibbens's invitation to a witches' gathering. "I must tarry at home, and keep watch over my little Pearl. Had they taken her from me, I would willingly have gone with thee into the forest, and signed my name in the Black Man's book too, and that with mine own blood!" (117) [74–75]. Had Pearl been taken from her, Hester would have embraced her social identity as a bad woman; Pearl keeps Hester's loving heart—her identity as a good woman—alive. She thinks, the narrator reports: "How strange, indeed! Man had marked this woman's sin by a scarlet letter. . . .

God, as a direct consequence of the sin which man thus punished, had given her a lovely child, whose place was on that same dishonored bosom, to connect her parent for ever with the race and descent of mortals, and to be finally a blessed soul in heaven!" (89) [58]. Not incidentally, that same dishonored bosom is where people in extreme situations come over time to rest: "Hester's nature showed itself warm and rich; a well-spring of human tenderness, unfailing to every real demand, and inexhaustible by the largest. Her breast, with its badge of shame, was but the softer pillow for the head that needed one" (162) [99].

Hester's character cannot be explained through allusions to ideology or social expectations. If anything, the course of her life shows that ideology and social expectations should have made her a very different person from the one she is. As a seduced ingénue, she should have died; as an evil temptress, she should have recanted, gone mad, or been murdered by the mob. Instead, she survives as love's incarnation. Yet, or because of this, she does not find happiness with one man. A new model of love between woman and man is needed, a new kind of couple—"sacred love" Hawthorne calls it at the novel's end (263) [155]. If only a love of this sort can guarantee mutual happiness between men and women, only a love like this could underpin a society of equals that might supersede present-day exploitation. As long as there are two sexes in the world, a just and humane polity must perceive each as equal to the other. But the very difference between the sexes that demands better forms of human intimacy also impedes their realization.

Awareness of the recalcitrant reality of human nature occupies Hester's meditations on reconstructing the social system, which include the tasks—in order—of tearing down society and rebuilding it, altering what in men is either their "very nature or its long hereditary habit, which has become like nature"; and finally producing, somehow, a "still mightier change" in woman herself—a change "in which, perhaps, the ethereal essence, wherein she has her truest life, will be found to have evaporated" (165–66) [102]. This "ethereal essence" is her capacity of loving. Should that be destroyed, then the social formation of which Hester dreams is an impossibility, because it is grounded in exactly that love, the only counter to male self-love. The "whole new truth," the "coming revelation" of which Hester later speaks (263) [155], would then be beyond human reach.

What, then, is to be done? The only possible changes Hawthorne can imagine—and he does imagine them—are palliative and far from romantic. Society can begin to compensate for the trials and consequences of woman's greater heart. Women's lot can be eased; they can be helped rather than punished; their humanity can be affirmed by recognizing their equality with men, their intellects respected, possibilities opened for them other than the domesticity that has failed them. Above all, perhaps, the institution of marriage needs to be freely elected; neither men nor women should be required to marry. *The Scarlet Letter*, *The Blithedale Romance*, and *The Marble Faun* all make space for emancipated female rhetoric, place transgressing women at the center, insist on women's equality with men, and deny the universal applicability of domestic ideals. But none imagines different men and different women from the ones Hawthorne believes people foundationally are, or opportunistically describes social outcomes that the author thinks cannot possibly come to pass.

WORKS CITED

Bardes, Barbara, and Suzanne Gossett. *Declarations of Independence: Women and Political Power in Nineteenth-Century American Fiction.* New Brunswick: Rutgers UP, 1990.

Barlowe, Jamie. "Rereading Women: Hester Prynne-ism and the Scarlet Mob of Scribblers." *American Literary History* 9 (1997): 197–225.

———. *The Scarlet Mob of Scribblers: Rereading Hester Prynne.* Carbondale: Southern Illinois UP, 2000.

Baym, Nina. *The Shape of Hawthorne's Career.* Ithaca and London: Cornell UP, 1976.

———. "Passion and Authority in The Scarlet Letter." *The New England Quarterly* 43 (1970): 209–30.

———. "Hawthorne's Women: The Tyranny of Social Myths." *Centennial Review* 15 (1971): 250–72.

———. "The Romantic Malgré Lui: Hawthorne in the Custom-House." *ESQ* 19 (1973): 14–25.

———. "Thwarted Nature: Nathaniel Hawthorne as Feminist." In *American Novelists Revisited: Essays in Feminist Criticism.* Ed. Fritz Fleischmann. Boston: G. K. Hall, 1982, 58–77.

———. "Nathaniel Hawthorne and His Mother: A Biographical Speculation." *American Literature* 54 (1982): 1–27.

Bell, Michael Davitt. *The Development of American Romance: The Sacrifice of Relation.* Chicago: U of Chicago P, 1980.

Bell, Millicent. "Introduction." *New Essays on Hawthorne's Major Tales.* New York: Cambridge UP, 1993.

Bercovitch, Sacvan. *The Office of the Scarlet Letter.* Baltimore: The Johns Hopkins UP, 1991.

Berlant, Lauren. "Fantasies of Utopia in *The Blithedale Romance.*" *American Literary History* 1 (1989): 30–62.

———. *The Anatomy of National Fantasy.* Chicago: U of Chicago P, 1991.

Brown, Gillian. *Domestic Individualism: Imagining Self in Nineteenth-Century America.* Berkeley: U of California P, 1990.

Budick, Emily Miller. *Engendering Romance: Women Writers and the Hawthorne Tradition, 1850–1990.* New Haven: Yale UP, 1994.

Colacurcio, Michael J. "'The Woman's Own Choice': Sex, Metaphor, and the Puritan 'Sources' of *The Scarlet Letter.*" In *New Essays on* The Scarlet Letter. Ed. Colacurcio. New York: Cambridge UP, 1985, 101–36.

Derrick, Scott S. "'A Curious Subject of Observation and Inquiry': Homoeroticism, the Body, and Authorship in Hawthorne's *The Scarlet Letter.*" *Novel* 28 (1995): 308–26.

DeSalvo, Louise. *Nathaniel Hawthorne.* Atlantic Highlands, NJ: Humanities Press International, 1987.

Egan, Ken. "The Adulteress in the Market-Place: Hawthorne and *The Scarlet Letter.*" *Studies in the Novel* 27 (1995): 26–41.

Fetterley, Judith. *The Resisting Reader: A Feminist Approach to American Fiction.* Bloomington: Indiana UP, 1978.

Hawthorne, Nathaniel. *The Scarlet Letter.* Vol. 1 of the *Centenary Edition of the Works of Nathaniel Hawthorne.* Ohio State UP, 1962.

———. *The House of the Seven Gables.* Vol. 2 of the *Centenary Edition of the Works of Nathaniel Hawthorne.* Ohio State UP, 1965.

———. *The Blithedale Romance*. Vol. 3 of the *Centenary Edition of the Works of Nathaniel Hawthorne*. Ohio State UP, 1964.

———. *The Marble Faun*. Vol. 4 of the *Centenary Edition of the Works of Nathaniel Hawthorne*. Ohio State UP, 1968.

Herbert, T. Walter. *Dearest Beloved: The Hawthornes and the Making of the Middle-Class Family*. Berkeley: U of California P, 1993.

———. "Pornographic Manhood and *The Scarlet Letter*." *Studies in American Fiction* 29 (2001): 113–20.

Kilcup, Karen L. "'Ourself behind Ourself, Concealed': The Homoerotics of Reading in *The Scarlet Letter*." *ESQ* 42 (1996): 1–28.

Lang, Amy Schrager. *Prophetic Woman: Anne Hutchinson and the Problem of Dissent in the Literature of New England*. Berkeley: U of California P, 1987.

Leverenz, David. *Manhood and the American Renaissance*. Ithaca: Cornell UP, 1989.

Martin, Robert K. "Hester Prynne, *C'est Moi*: Nathaniel Hawthorne and the Anxieties of Gender." In *Engendering Men: The Question of Male Feminist Criticism*. Ed. Joseph A. Boone and Michael Cadden. New York: Routledge, 1990, 122–39, 304–306.

Merish, Lori. *Sentimental Materialism: Gender, Commodity Culture, and Nineteenth-Century American Literature*. Durham: Duke UP, 2000.

Mitchell, Thomas R. *Hawthorne's Fuller Mystery*. Amherst: U of Massachusetts P, 1998.

Nudelman, Franny. "'Emblem and Product of Sin': The Poisoned Child in *The Scarlet Letter* and Domestic Advice Literature." *Yale Journal of Criticism* 10 (1997): 193–213.

Onderdonk, Todd. "The Marble Mothers: Hawthorne's Iconographies of the Feminine." *Studies in American Fiction* 31 (2003): 73–100.

Person, Leland S. *Aesthetic Headaches: Women and a Masculine Poetics in Poe, Melville, and Hawthorne*. Athens: U of Georgia P, 1988.

Pfister, Joel. *The Production of Personal Life: Class, Gender, and the Psychological in Hawthorne's Fiction*. Stanford: Stanford UP, 1991.

Thomas, Brook. "Citizen Hester: *The Scarlet Letter* as Civic Myth." *American Literary History* 13 (2001): 181–211.

Tomc, Sandra. "A Change of Art: Hester, Hawthorne, and the Service of Love." *Nineteenth-Century Literature* 56 (2002): 466–94.

Wallace, James D. "Hawthorne and the Scribbling Women Reconsidered." *American Literature* 62 (1990): 201–22.

Warren, Joyce W. *The American Narcissus: Individualism and Women in Nineteenth-Century American Fiction*. New Brunswick: Rutgers UP, 1984.

Weinauer, Ellen. "Considering Possession in *The Scarlet Letter*." *Studies in American Fiction* 29 (2001): 93–112.

Yellin, Jean Fagan. *Women and Sisters: The Antislavery Feminists in American Culture*. New Haven: Yale UP, 1989.

Race and Slavery

JEAN FAGAN YELLIN

The Scarlet Letter and the Antislavery Feminists†

> Picture . . . Hawthorne's grand woman, in all her native dignity, standing calm and self-poised.
>
> —*Elizabeth Cady Stanton*[1]

Perhaps the most complex and influential literary work that uses the antislavery women's iconography to reject their ideology is *The Scarlet Letter.*[2] Tracing the abolitionists' discursive codes in literary works is, obviously, different from tracing them in sculptural icons. In Hawthorne's romance, the emblem of the antislavery women functions as a subtext to Hester Prynne's emblem.

† From *Women and Sisters: The Antislavery Feminists in American Culture* (New Haven: Yale UP, 1989), pp. 125–150. Reprinted by permission of the author. Page numbers in square brackets refer to this Norton Critical Edition.

1. Elizabeth Cady Stanton, "Hester Vaughn," *The Revolution* (10 December 1868): 360–61.
2. Hawthorne scholarship has been transformed by the appearance of *The Centenary Edition of the Works of Nathaniel Hawthorne*, ed. William Charvat, Roy Harvey Pearce, and Claude M. Simpson, 23 vols. (Columbus: Ohio State UP, 1962–). References to *The Scarlet Letter* and other works by Hawthorne are to this edition. References to the *American Notebooks* are to volume 8 of this edition. See also *Hawthorne's Lost Notebook, 1835–1841*, transcribed by Barbara S. Mouffe (University Park: Pennsylvania State UP, 1978).

   Hawthorne's letters are designated by dates and addresses in the text or in the notes; quoted passages follow the texts established for *The Centenary Edition*. I am grateful to L. Neal Smith, associate textual editor of *The Centenary Edition*, who made typescripts and notes relating to this correspondence available to me before these texts appeared in print.

   Building on Robert Cantwell's *Nathaniel Hawthorne, the American Years* (New York: Rinehart, 1948), relevant biographical studies include Arlin Turner's "Hawthorne and Reform," *New England Quarterly* 15 (1942): 700–14; his "Needs in Hawthorne Biography," *Nathaniel Hawthorne Journal* 2 (1972): 43–45; and his *Nathaniel Hawthorne: A Biography* (New York: Oxford UP, 1980); as well as James R. Mellow, *Nathaniel Hawthorne in His Times* (Boston: Houghton Mifflin, 1980). In addition to works mentioned in the notes that follow, I have found the following critical studies particularly useful: Nina Baym's "Hawthorne's Women: The Tyranny of Social Myths," *Centennial Review* 15 (1971): 250–72; her "The Significance of Plot in Hawthorne's Romances," in G. R. Thompson et al., eds., *The Ruined Eden of the Present: Hawthorne, Melville, Poe* (Lafayette, Ind.: Purdue UP, 1981), 44–70; her "Passion and Authority in *The Scarlet Letter,*" *New England Quarterly* 43 (1970): 209–30; and her full-length study, *The Shape of Hawthorne's Career* (Ithaca, N.Y.: Cornell UP, 1976); Michael D. Bell's *Hawthorne and the Historical Romance of New England* (Princeton: Princeton UP, 1971); Millicent Bell, "The Obliquity of Signs: *The Scarlet Letter,*" *Massachusetts Review* 23 (1982): 9–26; Sacvan Bercovitch's *The American Jeremiad* (Wisconsin: U of Wisconsin P, 1978); Richard Brodhead's *Hawthorne, Melville and the Novel* (Chicago: U of Chicago P, 1976); Michael Colacurcio's "Footsteps of Anne Hutchinson: The Context of *The Scarlet Letter,*" *ELH* 39 (1972): 459–94; and his *The Province of Piety* (Cambridge, Mass.: Harvard UP, 1984); Ronald J. Gervais, "A Papist Among the Puritans: Icon and Logos in *The Scarlet Letter,*" *Journal of the American Renaissance* 25

The narrator of *The Scarlet Letter* addresses, and invites his audience to address, the issue of signification by overtly playing with the meanings of Hester's "A." Intended to brand her as an adultress, after seven years it means *able* to a portion of Boston's population. Hester herself, the narrator tells us, has been encoded to signify "woman's frailty and sinful passion," and at times this is how she understands herself (79) [53]. But behind these shifting interpretations of Hester and her emblem lie the shifting interpretations of the emblem of the female slave. Dramatizing the notion that womanhood is not a natural but a conventional construct—as Sojourner Truth, Harriet Jacobs, and the other female slaves excluded from this category testified—Hawthorne's book begins by presenting a woman publicly exposed, a figure made familiar by the abolitionists. Hester Prynne's ordeal on the scaffold, and the events that follow, signal her exclusion from the category of true womanhood. Although its focus later shifts, *The Scarlet Letter* recurrently presents the structures of the antislavery women's discourse. Repeatedly addressing the concerns of the antislavery women, the book explores Hester's ideas, and those of the narrator, in relation to her identity (her womanhood) and her membership in the community (her sisterhood).

In *The Scarlet Letter*, as in the speeches, writings, and images of the antislavery feminists, enchaining and exposure signify woman's oppression. Hawthorne, however, does not use them as a signal that woman should mount a public effort for self-liberation. Although his book ends with a kind of restoration of Hester as a Woman and a Sister, she does not achieve this identity by asserting the antislavery feminists' ideology or by struggling for her rights in the public sphere. Instead, learning that she should accept her lot, confident that someday, somehow, things will change for the better, Hester conforms at last to patriarchal definitions of womanhood. *The Scarlet Letter* is, on one level, a critique of patriarchal ideologies and structures, but it is a critique that seriously considers the new feminist definitions of womanhood and, rejecting them, replicates traditional imagery and endorses patriarchal notions. Hawthorne's narrator openly discusses the reencoding of both the embroidered emblem and of Hester herself. But the presence of the emblem of the enchained female slave, which functions as a subtext of Hawthorne's emblem, is never mentioned; its reencoding is covert.

Recently, studies of Hawthorne's life and of the selfhood that he expressed in his writings have explored beyond the walls of his legendary "dismal and squalid chamber" at Salem. They have, for example, established that, like other white Americans of his place and time, he had everyday contact with African-Americans and that, in addition to sharing the common awareness of the slavery question, he was more than routinely familiar with Cuban slavery and something of an expert concerning African colonization and the slave trade.[3]

(1979): 11–16; Lawrence S. Hall, *Hawthorne, Critic of Society* (New Haven: Yale UP, 1944); Harry B. Henderson III, *Versions of the Past: The Historical Imagination in American Fiction* (New York: Oxford UP, 1974); and Tony Tanner, *Adultery in the Novel: Contract and Transgression* (Baltimore: Johns Hopkins UP, 1979).

3. See my "Hawthorne and the American National Sin," in *The Green Tradition in American Culture*, ed. H. Daniel Peck (Baton Rouge: Louisiana State UP, 1989).

As a writer of fiction, Hawthorne chose to center his work on moral issues; repeatedly, on one level, his subject was psychological slavery. Although the antislavery activists were denouncing slavery as a tyranny of both flesh and spirit—and his brother-in-law Horace Mann was condemning slavery on the floor of Congress as a system in which "man claims authority over the body, mind, and soul of his fellowmen"—Hawthorne never did connect slavery with the bondage that was his chosen subject. He never did identify the "unpardonable sin" that obsessed him with "the American national sin" that the Garrisonians were denouncing.[4] Where Hiram Powers had distanced an enchained white woman in space and called her a *Greek Slave*, Nathaniel Hawthorne distanced an enchained white woman in time and called her Hester Prynne.

Hawthorne was, of course, familiar with the abolitionists. Even conservative Salem, although not a Garrisonian center like Boston, had its share of antislavery controversy. His black townswomen had been first in the nation to organize a female antislavery society, and they later joined an interracial organization that sent delegates to all three Conventions of American Women Against Slavery. At the 1839 convention, Clarissa C. Lawrence, a black Salem delegate, took the floor to point out that she, and other blacks, "meet the monster prejudice *every where*." She urged the white delegates, "Place yourselves, dear friends, in our stead." Four years earlier, racist, antiabolitionist white Salemites had mobbed an antislavery meeting planned for Howard Street Church. When the Grimkés had come to town in 1837, however, local antislavery women successfully scheduled four full days of activities. Angelina Grimké wrote that she and her sister spoke at the Friends' Meeting House, met with "colored members of the Seaman's and Moral Reform Society," addressed an audience of more than a thousand at the Howard Street Meeting House, and talked with children and adults at the "colored Sabbath School."[5]

Although Sophia Peabody, the townswoman Hawthorne would later marry, represented just the sort of person that the abolitionist activists were trying to influence, she did not involve herself in the female antislavery movement in Salem or elsewhere. Long before her marriage, the invalided Sophia had lived for a year and a half on a Cuban sugar plantation. In the "Cuba Journals" (letters she and her sister Mary, her companion, had written home to Salem), the Peabody sisters recorded their experiences and impressions of slavery. They commented on the overwhelming black presence and on the picturesqueness of the Africans at play, on the oppressiveness of slave labor, on the brutality of the system, on the sexual exploitation of slave women, and even on the practice of infanticide. This Cuban experience motivated Mary Peabody to become an active abolitionist, but it prompted Sophia to decide not even to think about slavery. Dwelling on this subject, she wrote, "would certainly counteract the beneficent influences" of her Cuban visit; and her faith in God

4. Horace Mann, speech of 23 February 1849, *Slavery: Writings and Speeches* (New York: Burt Franklin, 1969), 153–54.
5. Although well documented in the holdings of the Essex Institute, the history of the Salem, Massachusetts, Female Anti-Slavery Society has not been written. See *The Liberator*, 7 January 1832 and 17 November 1832; *Proceedings, Third Anti-Slavery Convention of American Women 1839* (Philadelphia: Merrihew and Thompson, 1839), 8–9; and Angelina Grimké to Jane Smith, 16 July 1837, Grimké-Weld papers, Clements Memorial Library, University of Michigan.

reassured her "that he makes up to every being the measure of happiness which he loses thro' the instrumentality of others. I try to realize how much shorter time is, than eternity and then endeavour to lose myself in other subjects of thought."[6]

Back in Boston, before her marriage, Sophia Peabody was part of a grouping that included Maria White, who converted her fiancé James Russell Lowell to abolitionism, and the Sturgis sisters, members, as was White, of the Female Anti-Slavery Society. At the bookstore of her sister Elizabeth, she met with other "advanced" women to attend Margaret Fuller's Conversations and discuss topics such as art, ethics, great men—and Woman. After marrying Hawthorne and moving to Concord, her circle included the Alcotts and Lydian Emerson, members of the local society. But Sophia Peabody Hawthorne distanced herself from abolitionism. She expressed her hostility to women's antislavery activities when, during her pregnancy, she wrote to her mother that she planned to engage "the ladies of the antislavery society" to sew a layette because "I have no manner of scruple about making them take as little as possible; while I could not think of not giving full and ample price to a poor person, or a seamstress by profession."[7]

She rejected not only the work of the female abolitionists, but also the feminist ideas Margaret Fuller voiced in "The Great Lawsuit," the most important statement on the "woman question" to follow Sarah Grimké's *Letters on the Equality of the Sexes*. Commenting to her mother about Fuller's polemic, Sophia Peabody Hawthorne writes that Fuller fails to understand that "Had there never been false and profane marriages, there would not only be no commotion about woman's rights, but it would be Heaven here at once. Even before I was married, however, I could never feel the slightest interest in this movement [women's rights]."[8] Sophia Peabody Hawthorne then voiced her disapproval of women who, following in the Grimkés' footsteps, brave the public sphere and speak from the platform: "it was always a shock to me to have women mount the rostrum. Home, I think, is the greatest arena for women."[9]

Years later, she indignantly responded when her sister Elizabeth Peabody, who had written an antislavery pamphlet, mailed it to the Hawthornes' young daughter, Una. Attacking radicals like the abolitionists and the feminists who advocated breaking oppressive laws in the name of a higher law, Sophia Peabody Hawthorne wrote, "I consider it a very dangerous and demoralizing doctrine and have always called it 'transcendental slang.'" Announcing that she had not shown the pamphlet to Una and that she would not, she stated that she was familiar with reports of slave

6. "Cuba Journal," 1; Sophia Peabody letters, 20 December [1833]–2 July [1834], Mary Peabody letters, 8 January–31 May [1834], Sophia Peabody to Mrs. Elizabeth Palmer Peabody, 21 March [1834], Henry W. and Albert A. Berg Collection, The New York Public Library, Astor, Lenox and Tilden Foundations. Used by permission.
7. Sophia Peabody Hawthorne to her mother, Mrs. Elizabeth Peabody, 15 November 1843. Unless otherwise stated, quoted passages from Sophia Peabody Hawthorne's correspondence are from The Henry W. and Albert A. Berg Collection, The New York Public Library, Astor, Lenox and Tilden Foundations, and are used by permission.
8. "The Great Lawsuit: Man versus Men; Woman versus Women," *The Dial* (July 1843); revised and republished as *Woman in the Nineteenth Century*, ed. Arthur B. Fuller (New York: Tribune P, 1845). Sophia Peabody Hawthorne to her mother, Elizabeth P. Peabody, quoted in Julian Hawthorne, *Nathaniel Hawthorne and His Wife: A Biography* (Boston: Houghton Mifflin, 1884), 1:257.
9. Sophia Peabody Hawthorne to her mother, in Julian Hawthorne, *Nathaniel Hawthorne and His Wife*, 1:257.

sales and did not believe her sister's accusations that slave women were routinely subjected to sexual abuse: "And you would display before . . . [my daughter's] great, innocent eyes a naked slave girl on a block at auction (which I am sure is an exaggeration for I have read of those auctions often and even the worst facts are never so bad as absolute nudity)."[1] This correspondence reveals that Sophia Peabody Hawthorne followed neither Angelina Grimké's advice to act against slavery nor Catharine Beecher's advice to influence men to take action against it.

Some of the Hawthornes' neighbors, however, were deeply involved in antislavery feminist activities. In Salem in the spring of 1848—shortly after Hawthorne's persistence had won him an appointment as surveyor of the Custom House and the family had moved from Concord back to their home town—the first women's rights petitions in Massachusetts were circulated. For six years, with the help of Phebe King of Danvers, Mary Upton Ferrin drafted, circulated, and submitted petitions. Literally standing in Angelina Grimké's footsteps, in 1850 she addressed a committee of the Massachusetts legislature. Using what was by now a staple of feminist rhetoric, Ferrin likened the condition of woman to that of the slave. Although widely ridiculed, Ferrin was encouraged by a local minister turned politician, Rev. Charles W. Upham—the same man who led the local Whig attack against Hawthorne's federal appointment in 1849. Aiding him was Phebe King's son Daniel, a U.S. congressman. Salem politics dictated that feminism's strongest supporters were Hawthorne's bitterest political enemies.[2]

The reformers made consistent gains in their efforts to bring the slavery question home to New England throughout the years of Hawthorne's literary apprenticeship. They succeeded in alerting everyone to the 1835 attack on the antislavery women, and many agreed with Angelina Grimké that the mobbed women were martyrs. This was the term that the British writer Harriet Martineau, who had met the members of the Boston Female Anti-Slavery Society, used to describe them in the *London and Westminster Review*. Attacked in the newspapers for endorsing the women's antislavery principles, Martineau cut short her American excursion shortly after the riot. Four years later, when the women were again mobbed in Philadelphia, reform Boston gossiped about Maria Weston Chapman's collapse after her failed attempt to speak at Pennsylvania Hall—her voice drowned out by the mob surrounding the building and her red shawl flaming against the Quaker gray worn by the besieged abolitionists trapped within.[3]

1. Sophia Peabody Hawthorne to her sister Elizabeth P. Peabody [Spring 1860], quoted in Louise Hall Tharp, *The Peabody Sisters of Salem* (Boston: Little, Brown, 1950), 288; and in Rose Hawthorne Lathrop, *Memories of Hawthorne* (Boston: Houghton Mifflin, 1923), 358.
2. Independently of Ferrin's efforts, the Know-Nothings, who dominated the legislature, passed a Married Woman's Property Act in 1853; a liberalized divorce law followed in 1855. See *History of Woman Suffrage*, ed. Elizabeth Cady Stanton et al., 6 vols. (New York, 1881–1922), 1:208–15. For Hawthorne and Upham, see Hawthorne to Horace Mann, Salem, 26 June 1849 and 8 August 1849; and to Burchmore, 17 September 1850, and 7 April 1851, in *Letters*, 16:284–86, 291–95; 364–66; 415–16. Also see Stephen Nissenbaum, "The Firing of Nathaniel Hawthorne," *Essex Institute Historical Collections* 114 (April 1978): 57–86.
3. For commentary that blamed the abolitionists for inciting the mob, see "Reported Riot in Boston" in the *American Monthly Magazine*, which Hawthorne would edit the following year, vol. 2 (1835): 164. Harriet Martineau, "The Martyr Age of the United States," *London and Westminster Review* (December 1838; republ. Newcastle upon Tyne: Finlay and Charlton, 1840): 43; *Autobiography of Harriet Martineau*, ed. Maria W. Chapman, 3 vols. (Boston: J. R. Osgood, 1877), 1:347–57. For Garrison's description of Chapman at Pennsylvania Hall, and of her subsequent illness, see his letter to his mother dated 19 May 1838; and to George Benson, 25 May 1838, in the *Letters of William Lloyd Garrison*, ed. Walter M. Merrill

Other well-publicized incidents involved abolitionist efforts to rescue fugitive slaves. In 1845, Jonathan Walker, a white Massachusetts sea captain, was arrested and jailed in Florida for aiding fugitives. Convicted, Walker was punished by being enchained, displayed on the pillory, and branded on the hand with the letters SS (slave stealer). Back home, he went on the antislavery lecture circuit, recounting his harrowing experience and displaying his "branded hand," which was daguerrotyped by Southworth and Hawes. Walker's ordeal inspired John Greenleaf Whittier to write a poem that elevates the brand by likening it to armorial hatchments and reversing its signification from negative to positive, from "slave stealer" to "salvation to the slave." In 1848, the slave emblem was transformed into a living tableau by the capture of the schooner *Pearl*. Mary and Emily Edmondson, two young black sisters, had been seized when the schooner was caught transporting a group of fugitive slaves to freedom. To raise money for their manumission, Henry Ward Beecher invited his congregation to attend a mock slave auction where the audience could simultaneously experience the delights of beneficence by helping emancipate slaves while experiencing the delights of sin by bidding for females standing on the block. The widely publicized Plymouth Church "auction" of the Edmondsons—and later, of other fugitive slave women and girls—caused a sensation. The Edmondsons' would-be rescuers, however, were not so fortunate as Beecher's "slaves." After a celebrated trial, Captain Edward Sayer and Mate Daniel Dreyton were convicted. Sentenced to heavy fines they could not pay, the men spent four years in jail before being pardoned.[4]

Two of these sensational events personally involved members of Sophia Peabody Hawthorne's family. In 1835, Hawthorne's future sister-in-law Elizabeth Peabody had been a guest in the house where Martineau was staying and had burned the Boston newspapers to spare the Englishwoman the embarrassment of reading their attacks. Martineau later wrote that Peabody had begged her to modify her antislavery sentiments in order to retrieve at least a degree of respectability. When Martineau refused, her American tour collapsed. In the case of the *Pearl*, Hawthorne's brother-in-law Horace Mann defended Sayers and Drayton in court. As a literate New Englander, Hawthorne would doubtless have been aware of

and Louis Ruchames, 6 vols. (Cambridge, Mass.: Harvard UP, 1971–1981), 2:363, 366. Also see Lydia Maria Child to Caroline Weston, 28 July 1838, in *Selected Letters, 1817–1880*, ed. Milton Meltzer and Patricia G. Holland (Amherst: U of Massachusetts P, 1982), 79–82; Child to Louisa Loring, 3 June 1838; and to Lydia B. Child, 7 August 1838, in *The Collected Correspondence of Lydia Maria Child, 1817–1880*, ed. Patricia G. Holland and Milton Meltzer (Millwood, N.Y.: Kraus Microform, 1980).

4. For Whittier's poem, "The Branded Hand," see *The Complete Poetical Works of John Greenleaf Whittier*, Cambridge Edition (Boston and New York: Houghton Mifflin, Riverside P, 1894), 296. Walker later lectured with Harriet Jacobs's brother, John S. Jacobs. Daniel Drayton, too, later spoke for the abolitionists; he published an account of his ordeal in *Personal Memoirs* (Boston: Bella Marsh, 1855). For Mann's defense of Drayton, see Louisa Hall Tharp, *Until Victory: Horace Mann and Mary Peabody Mann* (Boston: Little, Brown, 1953), 224–34. For the *Pearl* refugees, see Harriet Beecher Stowe, *Key to Uncle Tom's Cabin* (1853; reprint, New York: Arno, 1968), 306–30. For Beecher's "slave sales," see William C. Beecher and Rev. Samuel Scoville, *A Biography of Henry Ward Beecher* (New York: C. L. Webster, 1888), 292–300; and William G. McLaughlen, *The Meaning of Henry Ward Beecher* (New York: Knopf, 1970), 200–201. Eastman Johnson, who had painted Hawthorne's portrait in 1846, made the Plymouth Church "slave sale" the subject of his oil painting *The Freedom Ring* after the magazinist Rose Terry Cooke, who was in the church, heard Beecher's plea and dropped her ring into the collection plate.

these highly publicized incidents; as Sophia Peabody's husband, he had direct knowledge of two of them.[5]

Hawthorne knew the reformers well. He did not like them much. An 1835 journal entry testifies to his lack of sympathy for them and for their causes:

> A sketch to be given of a modern reformer—a type of the extreme doctrines on the subject of slaves, cold-water, and all that. He goes about the street haranging most eloquently, and is on the point of making many converts, when his labors are suddenly interrupted by the appearance of a keeper of a madhouse, where he has escaped. Much may be made of this idea.[6]

After his brief stay at Brook Farm, Hawthorne sketched the reformers in more than a half-dozen short pieces. "Earth's Holocaust," which condemns them most fully, presents a vision of ultimate destruction not uncommon in his time, and commonplace in our own. Here the world is destroyed not by accident or by oppressors desperate to maintain their power, but by reformers whose attacks on social corruption ultimately consume mankind's most valuable achievements.[7]

Women's rights reformers appeared repeatedly in Hawthorne's writings years before he created in Zenobia of *The Blithedale Romance* a fictional feminist who betrays both her sister and herself. As early as 1830, he published a biographical piece so unsympathetic to Anne Hutchinson that it mandates a careful examination of his apparent approval of her in *The Scarlet Letter*. In this early work, Hawthorne's narrator states that he is repelled by women who make themselves intellectually visible—even by writing for publication: "There is a delicacy . . . that perceives, or fancies, a sort of impropriety in the display of a woman's natal mind to the gaze of the world, with indications by which its inmost secrets may be searched out" [159].[8]

In "The Gentle Boy" (1832), Hawthorne cast as a woman the most prominent Quaker exemplar of "unbridled fanaticism." Ilbrahim's mother, Catharine, "neglectful of the holiest trust which can be committed to a woman," abandons her son in response to her driving need to testify against persecution (and to experience it).[9] In "A New Adam and Eve" (1843),

5. *The Autobiography of Harriet Martineau*, vol. 2, 33–35. For the Manns, see Louise Tharp, *Until Victory*, especially 224–34; and Jonathan Messerli, *Horace Mann* (New York: Knopf, 1972).
6. 1835 entry, *American Notebooks,* 10.
7. Hawthorne lived at Brook Farm from April to November 1841. For his critical private views of reformers, see, for example, *American Notebooks,* 10, 136. For his critical public views, see, for example, "The Hall of Fantasy," *Pioneer* (February 1843), and a series of pieces in *The Democratic Review,* including "The New Adam and Eve" (February 1843); "The Procession of Life" (April 1843); "The Celestial Rail-road" (May 1843); "The Christmas Banquet" (January 1844); "The Intelligence Office" (March 1844); and "A Select Party" (July 1844). "Earth's Holocaust" appeared in *Graham's Magazine* in May 1844. All of these were collected in the 1846 edition of *Mosses from an Old Manse*.
8. *The Blithedale Romance* (1852); "Mrs. Hutchinson," Salem *Gazette*, 7 December 1830, republished in *Biographical Sketches*, Riverside Edition, vol. 17, 217–26. Evidently Hawthorne did not think that female artists exposed themselves as female writers did. Yet while courting Sophia Peabody, long before *The Scarlet Letter,* he alternately fantasized about a life in which both of them would create art and expressed concern that efforts to do serious work as an artist would jeopardize her health. See Hawthorne to Sophia Peabody, 21 and 23 August 1839; 3 January 1840; 15 April 1840; 12 August 1841, in *The Letters,* vol. 15. 339, 397, 440, 557. For Sophia Peabody Hawthorne as an artist, see Josephine Withers, "Artistic Women and Women Artists," *Art Journal,* 35 (1976): 330–36.
9. "The Gentle Boy" first appeared in *The Token* for 1832; it was collected in *Twice-Told Tales*. For an informed discussion, see Michael J. Colacurcio, *The Province of Piety*; 63, 66–68 suggest connections among this text, "Mrs. Hutchinson," and *The Scarlet Letter.*

Hawthorne follows a pair of newly created innocents through nineteenth-century Boston. Entering the legislature, Adam seats Eve in the speaker's chair, and the narrator comments that "he thus exemplifies Man's intellect, moderated by Woman's tenderness and moral sense! Were such the legislation of the world, there would be no need of State Houses, Capitols, Halls of Parliament." These words, although appearing to endorse political feminism, also enunciate a gendered ideology that assigns intellect to males and morality and emotion to females.[1]

The opposition to feminism is more direct in "The Hall of Fantasy," also published in 1843. Despite a disclaimer, by identifying only Abigail Folsom among the reformers, Hawthorne's narrator suggests that women who defy tradition are insane. (Famous for speechifying at antislavery meetings, Folsom was generally considered quite mad; Emerson called her "the flea of the conventions.") Female reformers concerned with women's rights are openly criticized in Hawthorne's "A Christmas Banquet" (1844). Here the narrator includes among his misfits an ideological feminist who has "driven herself to the verge of madness by dark broodings on the wrongs of her sex, and its exclusion from a proper field of action."[2]

In light of this hostility to feminist reformers, it is surprising that in *The Scarlet Letter* the first view of Hester standing on the pillory recalls the antislavery feminists and their emblem of the Woman and Sister. Of course, Hawthorne's terms are different from theirs: he identifies a different cast of characters and locates them within a different temporal framework.

Yet the women we have been examining had repeatedly related their struggle for self-definition to similar struggles in the seventeenth century, the period Hawthorne chose for his romance. Responding to the clerical attack against women's participation in public life, in *Letters on the Equality of the Sexes* Sarah Grimké had likened the Puritan persecution of "witches" to the abuse heaped on her and the other antislavery feminists. Grimké predicted that in a freer future, "the sentiments contained in the Pastoral Letter will be referred to with as much astonishment as . . . that judges should have sat on the trials of witches, and solemnly condemned nineteen persons and one dog to death for witchcraft."[3] Similarly, Abby Kelley Foster, appearing before hostile audiences, had routinely discussed the martyrdom that Mary Dyer and others had suffered at the hands of the New England Puritans in relation to the hostility directed against her for deviating from established norms: "A century ago my Quaker ancestors were acquainted with the Deacons of New England. Their backs were stripped and whipped until the skin was torn off, and their ears were cut off, and they were sometimes even put to death. . . . I have no doubt you would commit similar barbarities upon my person if you thought public sentiment would allow it."[4]

1. "The New Adam and Eve" first appeared in the *United States Magazine and Democratic Review* (12 February 1843); it was collected into *Mosses* in 1846 and 1854; see *The Centenary Edition,* vol. 10.
2. "The Hall of Fantasy," which first appeared in *The Pioneer* (1843), and "A Christmas Banquet," which first appeared in *Democratic Review* in 1844, were collected into *Mosses* in 1846 and 1854; for all, see *The Centenary Edition,* vol. 10. For Abby Folsom, see *The National Cyclopaedia of American Biography* (New York: James T. White, 1921), 2:394. Hawthorne's reference to Folsom—like his other references to contemporaries—was dropped when this piece was anthologized.
3. Sarah M. Grimké, *Letters on the Equality of the Sexes* (Boston: Isaac Knapp, 1838), letter 3, 14.
4. Margaret Hope Bacon, *I Speak for my Slave Sister: The Life of Abby Kelley Foster* (New York: Thomas Crowell, 1974), 131.

The Scarlet Letter encodes not only connections between repression in seventeenth-century Boston and in nineteenth-century Salem—between the public punishment of Hester Prynne and the exposure and "beheading" of the "Custom House" narrator—it also encodes the public attacks on the antislavery women, Hawthorne's contemporaries.[5] One of their central icons is displayed in the opening scene: the figure of a woman forcibly exposed in public. Although Hester is not marked by an iron chain but by a piece of needlework, recurrent references to the scarlet letter as a brand force the connections between the embroidered symbol and the instruments of slavery. Later, presenting abolitionist iconography in its fullness, the narrator irrevocably links Hester, his seventeenth-century adultress, to the antislavery feminists, his contemporaries, by using their image of an enchained woman to describe Hester's condition in Boston: "The chain that bound her . . . was of iron links and galling to her inmost soul" (80) [53].

Perhaps the opening view of Hester on the scaffold of the pillory seemed familiar to the antislavery feminists among Hawthorne's earliest readers. Some might have reacted to the color of Hester's embroidery by recalling the red shawl of Maria Weston Chapman at Pennsylvania Hall; others perhaps associated Hester's symbolic branding to Jonathan Walker's barbarous punishment. Certainly Elizabeth Cady Stanton was responding to the feminist subtext of *The Scarlet Letter* when she evoked Hester as "Hawthorne's . . . grand woman, in all her native dignity, standing calm and self-poised through long years of dreary isolation from all her kind."[6] This regal conquered female figure is best depicted, in nineteenth-century American sculpture, by Harriet Hosmer's *Zenobia*. Shown exhibited as a trophy by the Roman emperor Aurelian and wearing the chains of her conquerers, Hosmer's Queen of Palmyra submits to public display.[7] In Hawthorne's writings, however (despite her name), it is not the disappointed suicide—the Zenobia of *Blithedale*—but the defiant adultress, Hester of *The Scarlet Letter*, who embodies the energy of the captive queen.

When the sculptor Howard Roberts created his full-length figure of Hester, he, too, incorporated the abolitionists' iconography of link chains. Like Hawthorne's creation, Roberts's Hester stands at once fully disclosed and completely unrevealed. Although she is forcibly displayed before us, her gaze does not meet ours. She is not nude but fully clothed. Her left arm cradles her sleeping baby, whose head rests against a large badge ornamented with a capital *A* on Hester's breast. Roberts's Hester does not try to hide either the letter or the babe. Like Hawthorne's creation, she is apparently concentrating on some painful reality, and her

5. For echoes in *The Scarlet Letter* of the French Revolution, and of the European revolutions of 1848, see Larry J. Reynolds, "*The Scarlet Letter* and Revolutions Abroad," *American Literature* 57 (March 1985): 44–67 [pp. 484–49 in this Norton Critical Edition—*editor*]. For the "decapitated surveyor" Hawthorne and Hester, see Stephen Nissenbaum, "The Firing of Nathaniel Hawthorne," *Essex Institute Historical Collections* 114 (April 1978): 57–86.
6. "Hester Vaughn," *The Revolution* (10 December 1868): 360–61.
7. For Harriet Hosmer's *Zenobia* (1858), see *Harriet Hosmer: Letters and Memories,* ed. Cornelia Carr (London: J. Lane, The Bodley Head, 1913), 191–93, 199–204, 363–68; Lorado Taft, *History of American Sculpture* (New York: Macmillan, 1930); and Margaret F. Thorp, *The Literary Sculptors* (Durham: Duke UP, 1965), 87–88. A version of this work is at the Wadsworth Atheneum, Hartford. The subject of Zenobia, Queen of Palmyra, was of interest to the American feminists; Child discusses her in *History of the Condition of Woman*, 2 vols. (Boston: J. Allen, 1835). William Ware's 1837 novel is best known by its second title, *Zenobia: or, the Queen of Palmyra. The Blithedale Romance*, Hawthorne's comment on Brook Farm, was published in 1852.

posture and expression suggest strain. Standing on rough wooden boards, she rests her right hand on a wooden post, and below that hand, Roberts carved the two hanging links of chain. These links connect his pilloried Puritan with Powers's *Greek Slave* and Hosmer's *Zenobia*—and all of them with antislavery iconography.[8]

From the moment Hawthorne's Hester emerges from prison, her regal impression underscored both by her person, "tall, with a figure of perfect elegance, on a large scale," and by her manner, "characterized by a certain state and dignity," she is presented in terms of the conquered queen (53) [39]. The description of the letter branding her emphasizes this identification. "Glittering like a . . . jewel" (202) [122], it is characterized as an ornament appropriate for "dames of a court" (81) [54], as "fitting decoration" (53) [39] for royalty. In response to the magnificence of her badge and her regal presence, those unfamiliar with Puritan customs judge Hester "a great lady in the land" (104) [67], "a personage of high dignity among her people" (246) [146]. Like Hosmer's *Zenobia*, who wears her manacles as if they were bracelets, and named for Esther, the captive biblical queen whose defiant courage the antislavery women admired, Hawthorne's Hester wears her embroidered brand like conquered royalty.

The Scarlet Letter is linked to the discourse of the antislavery women not only by these iconographic motifs, but also by its central concerns. In *Letters on the Equality of the Sexes*, Sarah Grimké had argued that after the fall in Eden woman was the first victim. Although the action of Hawthorne's book takes place in America, almost with his first words the narrator characterizes this new world as fallen. Describing Hester's appearance on the pillory, he notes that while "a Papist" "might have seen in this beautiful . . . woman with the infant at her bosom, an object to remind him of the image of Divine Maternity," she actually resembles Mary only "by contrast" [56]. Hester, we are told, is not like that second sinless Eve, but the first, and it is as a type of fallen woman, he tells us, that she is condemned to be identified by the people. Hawthorne's opening pages present a repressive new society practicing institutionalized violence against a woman.

The first scaffold scene showing Hester, her infant in her arms, displayed in public as punishment for the crime of adultery, of course focuses a number of issues. The problem Hawthorne's Puritan theocracy addresses (a problem central in nineteenth-century American fiction) is that the family on the scaffold is incomplete. This woman lacks a husband; this child lacks a father. The solution to this problem, the emergence of the absent male figure, like the solution to the problem of Arthur Dimmesdale, the polluted priest, is figured in the three scaffold scenes as the story unfolds.

But this initial view of Hester on the scaffold also dramatizes the issues that the antislavery women were addressing. *The Scarlet Letter*, no less than *Uncle Tom's Cabin*, published three years later, portrays an American society where publicly, officially, and institutionally, one of God's reasoning creatures is transformed into a thing; and where, privately, one individual tyrannizes over another in this world and threatens his salvation in the next. Here a human being is branded, displayed before the community, and

8. Roberts's 1869 sculpture is in the collection of the Library Company of Philadelphia. See *Philadelphia: Three Centuries of American Art. Bicentennial Exhibition,* catalog (Philadelphia: Philadelphia Museum of Art, 1976), 34.

dehumanized. It is perhaps not surprising that, written as the abolitionists labored to convince Americans that slavery was the national sin and as the feminists intensified their characterization of woman's condition as slavery, *The Scarlet Letter* dramatizes the problem of the institutional violation of an individual, the problem central to the notion of slavery, in the person of a woman. Hawthorne's intermittent use of the discourse of the antislavery women to present the twin issues of Hester's dehumanization and her isolation—which relate to the issues of womanhood and sisterhood the antislavery women were raising—results in a series of images that periodically push against the static symmetry of the scaffold scenes.

At the beginning of the book, the drama in Hawthorne's Market Place is heightened because the community condemning Hester is shown as monolithic. While Native Americans are present, here their society is not seen as an alternative to Boston (although Hester will later suggest this to Arthur). Further, as Hawthorne knew, in addition to the Native Americans and transplanted Europeans his narrator shows in the Market Place, the population of seventeenth-century New England had included another group: the Africans. By obliterating this historic black presence, Hawthorne's narrator helps guarantee Hester's absolute isolation.[9]

He does, of course, suggest a black community of a kind. *The Scarlet Letter* presents a classic displacement: color is the sign not of race, but of grace—and of its absence. Black skin is seen as blackened soul. Instead of presenting an African-American alternative to white Boston, it hints at a diabolical reversal. This satanic conspiracy, like establishment Boston, is male-dominated; it is ruled by the Black Man whose purpose seems to be to enslave others—particularly women. (That he wants their bodies as well as their souls is suggested by Hester's statement to Pearl that the Black Man is the child's father. It is tempting to play with connections between dark unpredictable Pearl and the untamed exotic dark females of American nineteenth-century letters. Subtexts concerning women and the sinfulness of female sexuality on the one hand, and blacks and the sinfulness of black sexuality on the other, lend added significance to the nineteenth-century phrase coupling "women and Negroes").

Members of this subversive group live in the town, but their activities in the forest suggest the international slave trade with the colors of the participants reversed: the names of whites are signed in a book belonging to the Black Man, and whites are branded with the Black Man's mark. They suggest, too, white reports of black Africans—of wild dancing in the woods.[1]

9. For blacks in seventeenth-century Massachusetts, see, for example, Robert C. Twombly and Robert H. Moore, "Black Puritan: The Negro in Seventeenth-Century Massachusetts," *William and Mary Quarterly*, 3d series (April 1967): 224–42; and George H. Moore, *Notes on the History of Slavery in Massachusetts* (1866; reprint, New York: Negro Universities P, 1968). For Hawthorne's early awareness of this historic black presence, see, for example, *American Notebooks* 21, 150; his continued interest in the history of African-Americans is expressed (550). Also see "Old News," *The New-England Magazine* 1–3 (February–May 1835) in *The Snow Image* (1852); and *Grandfather's Chair* 2 (1841) in *The Stories: Writings for Children, Centenary Edition*, vol. 6; for more, see my essay "Hawthorne and the American National Sin," *The Green American Tradition*, ed. H. Daniel Peck.

1. The convergence of western structures signifying diabolism and spiritual enslavement, and western structures signifying Africans and the African slave trade, is discussed in *The Image of the Black in Western Art*, vol. 2, *From the Early Church to the "Age of Discovery,"* pt. 1, Jean Devisse, *From the Demonic Threat to the Incarnation of Sainthood* (New York: William Morrow, 1979); and in *The Image of the Black in Western Art*, vol. 2, *From the Early Christian Era to the "Age of Discovery,"* pt. 2, Jean Devisse and Michel Mollat, *Africans in the Christian Ordinance of the World*, Fourteenth to the Sixteenth Century (New York: William Morrow, 1979).

Although this diabolical society is never taken seriously, when "black" is read as describing skin color and not moral status, the text of *The Scarlet Letter* reveals the obsessive concern with blacks and blackness, with the presence of a dangerous dark group within society's midst, that is characteristic of American political discourse in the last decades before Emancipation.

By choosing to obliterate the historic black presence and by choosing not to show Native American culture as an alternative to the society of white Boston, Hawthorne's narrator helps guarantee Hester's absolute isolation. With the negation of these potential alternative communities, her ostracism will be complete if the women—the one group within Boston society who might possibly sympathize with her—endorse Hester's official condemnation.

The opening pages of *The Scarlet Letter* dramatize these women rejecting their common sisterhood with Hester. Demonstrating that they have thoroughly internalized the patriarchal values of the community, the women in the Market Place judge Hester's punishment not as too harsh but as too lenient. Even while condemning the criminal more severely than do the male officials of the colony, however, they implicitly acknowledge that although they do not recognize Hester as one of themselves, they are nevertheless her sisters in the eyes of the patriarchy: "This woman has brought shame upon us all, and ought to die" (51) [38]. Further, they acknowledge that as a representative female, Hester must be punished if society is to continue to control the women. If she is reprieved, "Then let the magistrates, who have made . . . [the law] of no effect, thank themselves if their own wives and daughters go astray!" (52) [38]. Their fury is an index of their oppression. One of the clearest measures of the lack of true community in the Boston of *The Scarlet Letter* is the women's determination to deny Hester's sisterhood.[2]

As the narrative progresses, the issue of sisterhood takes several forms. After Hester's release from jail, isolated and reviled, the narrator writes that she intermittently "felt or fancied . . . [that] a mystic sisterhood would contumaciously assert itself." This sisterhood would apparently link her to other sinners, particularly to women who, although never accused, seemed to her to have engaged in forbidden sexual activity (86–87) [57]. But while noting that she felt herself a member of a sisterhood of sinners, the narrator comments that she fought against this sense of community, "struggling to believe that no fellow-mortal was guilty like herself" (87) [58].

He later discusses another kind of sisterhood. The people of Boston, he writes,

> perceived . . . that . . . Hester . . . was quick to acknowledge her sisterhood with the race of man, whenever benefits were to be conferred. . . . She came, not as a guest, but as a rightful inmate, into the household that was darkened by trouble . . . as if its gloomy twilight were a medium in which she was entitled to hold intercourse with her fellow-creatures. . . . She was a self-ordained Sister of Mercy. (160–61) [99]

2. These women are so punitive that an anonymous man chides them. One, however—a young mother—sympathizes with Hester. Her comment about the badge—"Not a stitch in that embroidered letter, but she has felt it in her heart" (54) [39]—recalls a motto of the sewing circles organized by the female antislavery societies: "May the points of our needles prick the slaveholders' consciences."

Hester willingly shares a sense not of mutual sin, but of sickness; she shares the sorrow common to fallen humans in a fallen world.

Although Hester denies any knowledge of the community led by the Black Man, at least one Boston woman welcomes her membership in this grouping. This diabolical secret society is an assembly to which many thought Hester belonged. The reputed witch Mistress Hibbins, believing Hester's brand a sign of her membership and affirming their common sisterhood, interprets Hester's rejection as a clumsy effort to preserve organizational secrecy.

Despite these actual or imagined communities, however, near the end of *The Scarlet Letter*, when Hester stands at the foot of the scaffold as Arthur preaches his Election Sermon, she is again the center of a "magic circle of ignominy," and the women who had earlier condemned her are again among those surrounding her (246) [146]. Their "cool, well-acquainted gaze at her familiar shame," along with that of the other townspeople, "tormented Hester Prynne, perhaps more than all the rest" (246) [146]. Seven years—and twenty chapters—after her branding on the pillory, Hester remains rejected as a sister by the women of Boston.[3]

In addition to dramatizing the women's informal denial of Hester's sisterhood, the first scaffold scene also dramatizes the institutional denial of her identity. Hester's official transformation from woman to thing enacts the negation central to the institution of slavery—the denial of the humanity of one of God's reasoning creatures, the denial the abolitionists saw as sin. "Giving up her individuality," the narrator says, "she would become the general symbol at which the preacher and the moralist might point, and in which they might vivify and embody their images of woman's frailty and sinful passion" (79) [53].

The discourse of the antislavery feminists also functions as a subtext of *The Scarlet Letter* in connection with the question of Hester's womanhood. This is most apparent in the chapter called "Another View of Hester." At this point in the narrative, Hester, shocked by Arthur's distraught appearance in the second scaffold scene, has decided to try to rescue him from Roger's torture. Here the narrator pauses to explore Hester's identity seven years after her sentence and punishment. He places his examination within the context of a general discussion of the nature and condition of woman—a topic which by 1849 the antislavery feminists had been considering for more than a dozen years and which some of them had recently addressed in new, revolutionary ways at Seneca Falls.

Using a series of buried images and convoluted formulations, Hawthorne's narrator characterizes womanhood as conditional and woman as passive, fragile, and endangered. Imaging Hester first in terms of a group of static free-standing vertical objects—a blasted tree, a neoclassical sculpture—and generalizing on her nature, he comments that as a result of undergoing "an experience of peculiar severity" (figured as being crushed), a woman frequently loses "some attribute . . . the permanence of which has been essential to keep her a woman" (163) [100]. He assures us that

3. If anything, Hester's isolation is now more complete. Although some have come to believe her A signifies "Able," we are not told that the women have stopped behaving like witches and distilling their poison for her ears; the tender-hearted young mother who initially sympathized with her has died in the seven-year interval.

this lost womanhood can, however, be restored by another physical contact, "the magic touch to effect the transfiguration" (164) [101].

This description of Hester, which notes her "marble coldness" (seen as a result of her life having "turned . . . from passion and feeling, to thought") and observes that she is "standing alone," suggests a sculpture (164) [101]. Activating this freestanding marble-like figure, the narrator evokes Hiram Powers's enchained *Greek Slave* as he describes Hester engaging in action: "She cast away the fragments of a broken chain. The world's law was not law for her mind" [101].[4] This apparently revolutionary act, however, does not free her. Hester's chains, the narrator explains, were already broken.

> It was an age in which the human intellect, newly emancipated, had taken a more active and a wider range than for many centuries before. Men of the sword had overthrown nobles and kings. Men bolder than these had overthrown and rearranged—not actually, but within the sphere of theory, which was their most real abode—the whole system of ancient prejudice, wherewith was linked much of ancient principle. (164) [101]

Hester is presented as the beneficiary of these male revolutionaries, not as a revolutionist herself. Likened to an intellectual whose radicalism stops short of "the flesh and blood of action," she might have been a revolutionist, it is suggested, but for the birth of her child (164) [101]. The characterization of this event as "providential" implies that it may be a blessing that woman's reproductive role prevents her from becoming an activist. Had Pearl not been born, the narrator suggests, Hester might have become a religious leader or prophet, might have "come down to us in history, hand in hand with Ann Hutchinson, as the foundress of a religious sect. She might, in one of her phases, have become a prophetess. She might, and not improbably would, have suffered death from the stern tribunals of the period, for attempting to undermine the foundations of the Puritan establishment" (165) [101]. But Hester does not become a link in this extraordinary chain of female revolutionists, the foremothers Abby Kelley Foster had eulogized, whom Hawthorne had condemned as fanatics in "The Gentle Boy." Instead, learning the lesson Ilbrahim's mother had failed to learn, Hester focuses her energies on motherhood.

Ironically, however, her efforts to fulfill this traditional female role inevitably raise precisely the line of revolutionary inquiry that the narrator suggests characterized the thought of women who assumed untraditional public roles. Concerning Pearl, "in bitterness of heart" Hester asks "whether it were for ill or good that the poor little creature had been born at all" (165) [101]. It might seem peculiar that she had not become desperate much earlier. Hester had given birth to an illegitimate child before *The Scarlet Letter* begins. However, unlike the drowned Martha Hunt (whom Hawthorne and the others had pulled out of the dark pond in 1845), and unlike the countless deceived maidens of American nineteenth-century

4. Miner Kellogg's promotional pamphlet *Powers' Statue of The Greek Slave* (New York: R. Craigshead, 1847) underscores Hester's connections with Powers's work; see especially p. 125.

fiction, she evidently had not attempted abortion, infanticide, or suicide. Only after seven years of dehumanization and ostracism does Hawthorne's female sexual rebel apparently consider adopting the destructive and self-destructive patterns assigned to "fallen women" in nineteenth-century American life and letters.

The questions she poses make clear that, despite her continued ostracism, she does not see her condition as fundamentally different from that of other women. "Indeed, the same dark question often rose into her mind, with reference to the whole race of womanhood. Was existence worth accepting, even to the happiest among them? As concerned her own individual existence, she had long ago decided in the negative, and dismissed the point as settled" (165) [101].

In the discussion that follows, the narrator initially intimates that the origins of the "woman question" may be social; this implies that public activity, like that of the antislavery feminists, could change woman's situation. But while suggesting that intense activity might result in reforms, this passage does not present woman as an active agent. Note the use of the passive voice:

> As a first step, the whole system of society is to be torn down, and built up anew. Then, the very nature of the opposite sex, or its long hereditary habit, which has become like nature, is to be essentially modified, before woman can be allowed to assume what seems a fair and suitable position. Finally, all other difficulties being obviated, woman cannot take advantage of these preliminary reforms, until she herself shall have undergone a still mightier change, in which, perhaps, the ethereal essence, wherein she has her truest life, will be found to have evaporated. (166) [102][5]

This analysis of woman's condition—that she is not permitted "a fair and suitable position"—appears similar to the analysis of the antislavery feminists. But here the cause of woman's oppression is identified differently, and the solution proposed is different from theirs.

Feminists such as Sarah Grimké, arguing that male domination is the source of woman's problems, had proposed that women actively oppose patriarchal restrictions and reassert their God-given role as corulers of the earth. Hawthorne's narrator suggests, however, that woman's condition is somehow a consequence of her female essence, of her essential nature. Accordingly, she can rectify her oppressed social status neither by acting (like those "men of the sword" who had toppled aristocratic political structures), nor by thinking (like the "bold" men who had transformed ancient ideological structures). Instead, using an absolute negation and an equally absolute assertion, the narrator suggests that women remain passive and trust their physiology and their luck. Although women are ephemeral and in danger of "evaporation," he writes, their problems are equally

5. The danger of women's evaporation was again voiced following the October 1851 Woman's Rights Convention at Worcester. When an article in the *Christian Inquirer* queried, "Place woman unbonneted and unshawled before the public gaze and what becomes of her modesty and her virtue?" feminist Ernestine Rose responded: "In [the writer's] benighted mind, the modesty and virtue of woman is of so fragile a nature, that when it is in contact with the atmosphere, it evaporates like chloroform. Such a sentiment," she continued, "carries its own deep condemnation" (*History of Woman Suffrage,* 1:245–46).

ephemeral: "A woman never overcomes these problems by any exercise of thought. They are not to be solved, or only in one way. If her heart chance to come uppermost, they vanish" (166) [102].

Hester—"whose heart," we are told, "has lost its regular and healthy throb," now is seen not as a defeated queen or an enchained sculpture, but as a lost and fallen Eve (166) [102]. The consequence of her single act toward self-liberation—throwing away her broken ideological chains—is not freedom, as it had been for Angelina Grimké. Although no longer bound by a repressive ideology, she is enslaved by her own nature. Tyrannized by her own ideas, tortured by thoughts of infanticide and suicide, she now wanders alone, a fallen Eve in a fallen world: "There was wild and ghastly scenery all around her, and a home and comfort nowhere" (166) [102]. This entire passage, which recalls Angelina Grimké's early sense of bewilderment at finding herself on an "untrodden path," demonstrates the complex use, in *The Scarlet Letter*, of the antislavery feminists' structures of ideology and discourse. Unlike Grimké, who pursued her innovative course and redefined herself, Hester cannot successfully achieve womanhood and sisterhood unless Hawthorne's narrator changes his definition of woman or presents a narrative that contradicts these constructions.[6] Despite this, however, in the balance of the text we are intermittently presented with fragmentary views of Hester that figure her efforts to achieve these dual goals.

Hester's attempts to act out traditional female roles involve her with a number of issues Hawthorne's feminist contemporaries were raising. Her situation as Roger's wife, for example, dramatizes their demand for more adequate divorce laws, and her plight as a mother accused of unfitness addresses the feminist issue of a woman's right to her child.[7]

Hester's relationship to Arthur, it would appear, is by definition unconventional. Instead, however, it follows a standard pattern. Her functional role as Arthur's true wife becomes clear in the forest scene when she tells him that Roger is her legal husband. Here the question of a wife's duty, a question of deep interest to the feminists, is raised. Earlier, the narrator has revealed that Hester felt connected to Arthur in ways that the patriarchy mandated that true women be connected to their husbands—as primarily defined by this relationship, as primarily wives, and not (as the antislavery women urged) as primarily God's creatures or as citizens. In this scene, the narrator articulates Hester's lack of a sense of autonomy, her crucial dependence on her perception of Arthur's estimate of her: "All the world had frowned on her. . . . Heaven, likewise, had frowned upon her. . . . But the frown of this pale, weak, sinful, and sorrow-stricken man was what Hester could not bear, and live!" (194–95) [117–18]. When Hester deliberately risks Arthur's anger in an attempt to rescue him, she risks what—as every traditional wife knows—amounts to annihilation. This explains the wildness of her plea, following his fury: "Let God punish! Thou shalt forgive!" (194) [117]. Arthur's response, "I freely forgive you now. May God forgive us both!" articulates his acceptance of the

6. Angelina Grimké to Jane Smith, 9 May–5 June–7 June 1837. For Nathaniel Hawthorne's ambivalence concerning the construction of gender, see Walter Herbert, Jr., "Nathaniel Hawthorne, Una Hawthorne, and *The Scarlet Letter*," *PMLA* 103 (May 1988): 285–97.

7. "Declaration of Sentiments," *History of Woman Suffrage*, 1:70–73.

traditional husband's role as mediator between his erring wife and the Creator (195) [118].[8]

Their reconciliation, however, results in a reversal of the traditional marital relationship. Fearing Arthur is endangered both physically and spiritually as a result of his prolonged torture, Hester initially acts as a traditional wife by urging him to save himself. But when Arthur claims he is unable to choose, to act independently, when he urges her to assume a dominant role—"Think for me . . . thou art strong. Resolve for me!" (196) [118]; she does so—"Thou shalt not go alone!" (198) [120].

Following this exchange, the narrator pauses to present another detailed view of Hester. She is no longer seen as a fallen, wandering Eve, but as the quintessential enemy of Puritan civilization, an Indian at home in the vast wild. "Her intellect and heart had their home, as it were, in desert places, where she roamed as freely as the wild Indian in his woods" (199) [120]. Hester's new perspective recalls her shift in vision on the scaffold seven years earlier. Then, she had reviewed her life in an effort to comprehend her current situation and to validate her immediate perceptions. Now, however, Hester looks outward. Her new vision of society is perhaps analogous to the prospect viewed by Hawthorne's narrator after he was forced from the Custom House, or to the prospect from Henry David Thoreau's jail cell. "It was," Thoreau had written, "like travelling into a far country. . . . It was to see my native village in light of the middle ages. . . . It was a closer view of my native town. . . . I had not seen its institutions before."[9]

From this new perspective, Hester identifies the institutional structures of Boston, domestic as well as religious and political, with enemy eyes. She now examines, we are told, "whatever priests or legislators had established; criticizing all with hardly more reverence than the Indian would feel for the clerical band, the judicial robe, the pillory, the gallows, the fireside, or the church" (199) [120]. Hester is no longer like other women. The narrator explains: "The tendency of her fate and fortunes had been to set her free. The scarlet letter was her passport into regions where other women dared not tread" (199) [120].[1]

8. For a contemporary feminist discussion of the duties of a wife and of the traditional role of a Christian husband as mediator between his wife and her Creator, see Sarah Grimké's *Letters on the Equality of the Sexes*. For a very different reading of Hester's relationship with Arthur and of the book's conclusion, see Nina Baym, "Thwarted Nature: Nathaniel Hawthorne as Feminist," *American Novelists Revisited: Essays in Feminist Criticism*, ed. Fritz Fleischmann (Boston: G. K. Hall, 1982), 58–77.

Hester's self-identification as Arthur's true wife is demonstrated by her refusal to name him in the first scaffold scene and in her prison interview with Roger; by the narrator's insinuation that her decision to remain in Boston after being released from prison is at least partially based on her feeling "connected in a union, that, unrecognized on earth, would bring them together before the bar of final judgment, and make that their marriage-altar" (80) [53]; by her tenacious appeal to Arthur to intervene on her behalf with the Governor when her custody of Pearl is jeopardized; and by her sense of a responsibility and commitment to Arthur, her only "significant link" to anyone in the community. This commitment leads her to tell Roger that she has decided to break her promise and reveal his identity to Arthur.

Hester's revelation of Roger's identity comments not only on her earlier promise to keep that identity secret and her refusal to name her lover to him, it also underscores her sense of herself as Arthur's wife by reversing the stock literary situation in which an adulterous wife finally identifies a secret lover to her husband.

9. In 1848, Thoreau had included these ideas in his talk on "Civil Disobedience" at the Concord Lyceum; Elizabeth Peabody had published his essay—along with Hawthorne's sketch "Main Street"—in *Aesthetic Papers* (May 1849); reprint, Gainesville, Fla.: Scholar's Facsimiles and Reprints, 1957.

1. The narrator does not entirely approve. He comments, "Shame, Despair, Solitude! These had been her teachers,—and they had made her strong, but taught her much amiss" (199–200) [120].

If, at this stage, Hester's development were to replicate the pattern described by the antislavery feminists, her new-found ability to identify the patriarchal society as her enemy would spark a new assertion of her identity, her womanhood. But Hester does not remain a wild Indian for long. Although after her agreement with Arthur she casts aside the dehumanizing scarlet letter and in the forest with him again becomes a woman, this renewal is presented as a consequence of her restored relationship with a man, as a result of his "magic touch." It is not shown as the result of her reawakened sense of her own identity—a consciousness that, in the case of the deeply religious antislavery feminists, grew from a sense of a renewed relationship with the Creator. Nevertheless, in the forest Hester appears an unfallen Eve in tune with Nature. When she again takes up the scarlet letter to suffer the "dreary change" into a thing, it is, she thinks, merely to "bear its torture . . . only a few days longer," until she can drown the dehumanizing brand in the deep sea (211) [127, 126].

Her return is quickly followed by the third scaffold scene, which presents the carefully prefigured resolution of many of the problems the first scaffold scene had figured. Arthur confesses his crime in public; the spell binding Pearl is broken; the family is made complete; and Hester's relationship with Arthur is reordered to conform to a conventional marital pattern. When he calls, "Hester . . . come hither!" (252) [149], the woman who seven years earlier had defied the demands of husband, church, and state to name her partner in adultery—the woman who, we have just been told, sees all of Boston through the eyes of a "wild Indian"—obeys wordlessly "as if impelled by inevitable fate, and against her strongest will" (252) [149]. When Hester momentarily hesitates before reaching his side, Arthur reformulates and repeats his command: "Hester Prynne . . . in the name of Him, so terrible and so merciful, who gives me grace . . . come hither now, and twine thy strength about me!" (253) [149]. Arthur's language gains added significance in the context of this display of female obedience and of Hester's unanswered question, "Shall we not spend our immortal life together?" (256) [151]. His choice of words underscores the impression that this display of the completed nuclear family figures the standard wedding scene that climaxes much of our nineteenth-century fiction.

As Hawthorne wrote, use of the trope of the marriage of vine and elm became common in discussions of the woman question. The authors of the 1837 Pastoral Letter had used it to attack the Grimké sisters:

> If the vine, whose strength and beauty is to lean upon the trellis-work, and half conceal its clusters, thinks to assume the independence and the overshadowing nature of the elm, it will not only cease to bear fruit, but fall in shame and dishonor into the dust. We can not, therefore, but regret the mistaken conduct of those who encourage females to bear an obtrusive and ostentatious part in measures of reform, and countenance any of that sex who so far forget themselves as to itinerate in the character of public lecturers and teachers.[2]

2. The 1837 Pastoral Letter of "The General Association of Massachusetts (Orthodox) to the Churches under Their Care" is extracted in Stanton, *History of Woman Suffrage*, 1:81–82. The figure of the marriage of weak female vine and strong male elm appeared in classical writings by Virgil, in medieval emblem books, and in English masterworks by Spenser, Shakespeare, and Milton. I am indebted to A. Bartlett Giamatti for bringing to my attention Peter Demetz's "Elm and Vine: Notes Toward the History of a Marriage Topos," *PMLA* 73 (1958): 521–32.

In response, Sarah Grimké had presented a startling feminist variant. In her version, phallic objects that suggest both pastoral and military life apparently guard woman against some unnamed threat; but actually, they are woman's oppressors, not her defenders. Grimké's prose suggests that women who look to men for protection discover that males are at best weak and impotent, at worst dangerously rapacious. "Ah! How many of my sex feel in the dominion, thus unrighteously exercised over them, under the gentle appelation of *protection*, that what they have leaned upon has proved a broken reed at best, and oft a spear."[3]

The version of the trope presented in *The Scarlet Letter* is peculiar because in the third scaffold scene traditional gender roles are apparently reversed. The female vine is physically strong and the male elm, weak. Nonetheless, the physically stronger female submits against her will to the weaker male, who acts out the traditional husband's role of intermediary between his wife and the Creator. Thus Arthur: "Thy strength, Hester; but let it be guided by the will which God hath granted me. . . . Come, Hester, come! Support me up yonder scaffold!" (253) [149].[4]

Critics have pointed out that the final scaffold scene resolves many of the issues implicit in the first. But this scene fails to figure the resolution of the problems encoded in the earlier image of Hester exposed and enchained. Nor does it resolve the issues we have been examining: the official denial of Hester's womanhood and the informal denial of her sisterhood. As if acknowledging this, in the last chapter the narrator again turns to the twin issues of Hester's womanhood and sisterhood. Long after Pearl's humanization and Arthur's death on the scaffold, and long after Hester had taken Pearl away from Boston, he says, "Hester Prynne had returned, and taken up her long-forgotten shame. . . . [T]here was more real life for Hester Prynne, here, in New England, than in that unknown region where Pearl had found a home. Here had been her sin; here, her sorrow; and here was yet to be her penitence" (262–63) [154]. Hawthorne's feminist readers might find this promising. Despite the narrator's earlier statements about women, perhaps here, in contrast to most nineteenth-century fiction, a woman's life will not be seen as defined by a single event; perhaps here a female character will finally be treated as autonomous. While Hester's womanhood has been seen as contingent on her relationship with a man, and while her sisterhood is still denied, despite stirrings in the community, surely now that the narrator has returned to these issues, he will resolve them—although the comment that she must still become penitent perhaps presents a problem.

But the book ends on the next page. Her "real life" (262) [154]—inevitably involving her womanhood and sisterhood, inevitably involving her

3. Sarah Grimké *Letters on the Equality of the Sexes* (Boston: Isaac Knapp, 1838), 21.

4. The comments of Hawthorne's narrator about the events following Arthur's spectacular death reinforce the impression that Hester, like a true wife, has grounded her sense of self in her relationship with her man. Reporting that after the minister's death Roger "positively withered up" because revenge was "the very principle of his life" and with Arthur's death there was "no further material to support it" (260) [153], he comments on the similarities between love and hate and defines both as parasitic: "Each renders one individual dependent for the food of his affections and spiritual life upon another; each leaves the passionate lover, or the no less passionate hater, forlorn and desolate by the withdrawal of his object" (260) [153]. References to connections between marriage and mosses and references to wives as "gentle parasites" recur throughout Hawthorne's work. Here his account of Hester's disappearance after Arthur's death underscores the notion that she is a true wife, a gentle parasite deprived of her host.

"sin . . . sorrow and . . . penitence" (263) [154], inevitably involving the sequence of views of Hester that we have been charting—is disposed of in eleven sentences in the next to last paragraph of *The Scarlet Letter.*

This passage begins with an announcement that the signification of her brand has been transformed. The "scarlet letter ceased to be a stigma which attracted the world's scorn and bitterness, and became a type of something to be sorrowed over, and looked upon with awe, yet with reverence too" (263) [154]. Its changed signification, however, alters neither the form nor the force of Hester's dehumanizing brand. Although she is no longer identified as a threat but as a source of support, as "one who had herself gone through a mighty trouble," Hester evidently has not again regained the womanhood stripped from her on the scaffold and so briefly restored during her forest meeting with Arthur (263) [154]. Apparently in the world of *The Scarlet Letter,* only a man's potent touch can restore lost womanhood; here a woman is defined neither as God's moral creature nor as a female member of society, but as the object of a man's love.

This passage reveals a triple vision. At its center, in a scene that finally does function as a pendant to the opening scene of Hester exhibited in the Market Place, the narrator shows her inside her hut, surrounded by the women of Boston. This glimpse of Hester among the women appears to resolve many of the tensions figured by the denial of her sisterhood in that first scaffold scene, and it perhaps even suggests the kind of women's discussion group that Margaret Fuller had created, which Sophia Peabody Hawthorne had attended. Actually, however, it is very different. What troubles the "wretched" "demanding" females surrounding Hester are not the complexities of the woman question that we are told had once tortured her, complexities that, by 1849, feminist intellectuals and activists had been addressing for years: the nature of woman as God's creature and the character of woman's oppression by the patriarchy. Instead, the women around Hester focus on private problems resulting from their sexual experiences with men: on the consequences of breaking the patriarchal rules restraining female "passion," or on the sense of worthlessness they feel because they are "unvalued and unsought" by men (263) [155]. Like the women surrounding her, Hester now focuses on these private and domestic aspects of women's lives. And like them, she does so in private.

Leaving implicit an assertion with which the feminists certainly agreed—that "the whole relation between man and woman" requires basic change—Hester makes explicit a series of ideas about how this change is to come about, ideas with which they certainly disagreed (263) [155]. What she says counters feminist assertions that women are competent to analyze their situation and to conceptualize the changes needed, that the time for these changes is now, and that women acting together can successfully achieve the necessary reforms. Instead, Hawthorne's narrator writes that Hester assures the women around her that the inevitable change must be based on a new revelation, that it will not occur now, and that it will not result from any actions of hers or of their own: "at some brighter period, when the world should have grown ripe for it, in Heaven's own time, a new truth will be revealed" (263) [155]. Hawthorne would repeat both the meter and the matter of these phrases. Four years later, rejecting antislavery activism in his campaign biography of Franklin Pierce, he would echo his rejection of feminist activism in *The Scarlet Letter.* A wise man, he would assert, "looks upon slavery as one of those evils which divine Providence does not leave to be remedied by human contrivances, but which, in its own good time, by some means impossible to be anticipated, but of the simplest and easiest operation, when all its uses shall have been fulfilled, it causes to vanish like a dream."[5]

At the end of *The Scarlet Letter*, we are told that Hester sketches, for the women around her, alternative versions of the female figure who, she asserts, will herald the "new truth." She dismisses the first: It is a vision of Hester herself as a "destined prophetess" (263) [155] publicly proclaiming a revolution in "the whole relation between man and woman." She rejects this figure although it would appear to climax and to culminate the fragmentary series of views of Hester that we have been tracing—as queen, as woman, and as sculpture, all in chains; as fallen, lost Eve; as wild Indian; and as prelapsarian Eve. And she rejects it although the narrator's comment that she herself had once "vainly imagined" (263) [155] she might fulfill this role underscores the intellectual and aesthetic inevitability of Hester's transformation into the "destined prophetess."

In refusing this vision of herself, Hester evidently repudiates the feminist subtext—the language and iconography of the antislavery women, the images of women in chains, of female figures erect in space—that has fueled *The Scarlet Letter.* Clearly, she now repudiates tactics like those of the antislavery feminists who were defying social taboos in an effort to move other women to action, polemicizing, lecturing, and preaching in public, and prophesying a change in "this whole relation between man and woman."

Just as clearly, she repudiates their ideology. Although finally surrounded by women and acknowledging her connections with these everyday sinners (and thus by extension with women like those who, in Hawthorne's time, were held in slavery and with those who, although legally free, identified so deeply with their sisters in chains that they figured themselves as slaves), she now denies the central assertion of the antislavery feminists: "that any mission of divine and mysterious truth should be confided to a woman stained with sin, bowed down with shame, or even burdened with a lifelong sorrow" (263) [155]. With these words, Hester denies that any of them (she herself, the women around her, or by extension nineteenth-century women literally and figuratively in chains) can act to end patriarchal oppression, can break her own chains and the chains of her sisters.

Instead, we are told that Hester has endorsed a different ideology. She now asserts that a woman unlike herself and her audience will function as "the angel and apostle of the coming revelation" (263) [155]. In accordance with this new idea, she projects a new iconography. The figure she envisions, the antithesis of the woman in chains, is a divine female rescuer "lofty, pure, and beautiful; and wise . . . through . . . the ethereal medium of joy" (263) [155]. Instead of proselytizing in public like the "destined prophetess"—and like the antislavery feminists who figured themselves as self-liberated liberators—this rescuer will deliver her message in private. Instead of engaging in debate and agitation like the antislavery feminists, the rescuer will present her message by example, simply showing "how

5. *Life of Franklin Pierce, Centenary Edition,* 5:416–17 [220].

sacred love can make us happy, by the truest test of a life successful to such an end!" (263) [155]. Five years after publication of *The Scarlet Letter*, Coventry Patmore's poem would provide this female culture figure with a name. The image Hester here envisions and endorses, a superhuman, privatized, and domesticated version of the Liberator of the double antislavery emblem, is the patriarchy's paradigm of true womanhood, the Angel in the House.[6]

Ironically, Hester's new vision does not resolve the division between women that the first scaffold scene had dramatized; it simply reverses it. Now on one side is a happy lone figure; on the other, among women "stained . . . bowed down . . . and burdened," crouches Hester. She counsels those around her to comfort each other in private, to be patient, and to have faith. Formulated in the terms of the antislavery emblem, she advises them to function as sisters, as members of their human community, but not as women, God's autonomous moral creatures.

The final lines of *The Scarlet Letter* powerfully reinforce its opening scene. The free-standing vertical "slab of slate"—placed not over Hester's grave but between it and Arthur's, marked not with her name or his or even the word naming their relationship, but with the brand that signaled, for her, the denial of both womanhood and sisterhood—this gravestone recalls our first view of Hester Prynne (264) [155]. Then, on the scaffold of the pillory, a young woman dressed in gray and branded with red was forced to expose herself as a punishment for breaking patriarchal laws restricting sexuality and as a device for controlling the behavior of other women in the colony. As then in the Market Place, so now in the cemetery, both her womanhood and her sisterhood are denied. Marked with a symbol whose meanings "perplex" (264) [155] their nineteenth-century viewers (and become available to us only through the intervention of Surveyor Pue and his decapitated successor, male officials of governments succeeding the Puritan theocracy), this "slab of slate" represents the ultimate denial of Hester's humanity and her membership in the community.[7] What we remember is the reassertion of the iconography of the antislavery feminists, now used to counter their definitions of true womanhood; we remember Hester's final exclusion from sisterhood with the dead, her final reduction from woman to nameless thing.

With the *Greek Slave* and *The Scarlet Letter*, the emblem of a woman exposed and enchained, of a woman pleading, again signifies a female victim, as it had before Angelina Grimké's excited realization that to appeal is to assert power. The antislavery feminists had recorded the supplicant image to express woman's struggle against oppression and to announce

6. Coventry Patmore, *The Angel in the House*, 2 vols. (London: Macmillan, 1863); Patmore's poem had been published in sections in 1854, 1856, 1860, and 1863. Hawthorne, who was familiar with Patmore's poetry, wrote that "The Angel in the House" was "a favorite" of his and of Sophia; he judged it "a poem for married people to read together." See Mellow, *Nathaniel Hawthorne in His Times*, 40, 439; and *English Notebooks*, 620. I am deeply grateful to Milton R. Stern, whose comments, made so long ago, moved me to attempt an adequate reading of this passage, and who explores some of the complexities of Hawthorne's thought and production in "Nathaniel Hawthorne: Conservative After Heaven's Own Fashion," in Joseph Waldmeier, ed., *Essays in Honor of Russel B. Nye* (East Lansing, Mich.: Michigan State UP, 1978), 195–225.

7. For an awareness of the role of the government officials, I am indebted to Nina Baym's "George Sand in American Periodicals" (Paper delivered at the meetings of the Nathaniel Hawthorne Society, New York, 1983).

their own self-liberation. But the emblem of the female in chains was reappropriated and again recorded. Over time, as patriarchal discourse became utterly dominant, the speeches and writings of the antislavery feminists were marginalized. Then they disappeared from the page.

JAY GROSSMAN

"A" Is for Abolition?: Race, Authorship, *The Scarlet Letter*†

I

> The pre-eminent reputation of *The Scarlet Letter* has obscured the fact that Hawthorne's three other novels all deal with contemporary material. It is important to correct the prevailing conception of him as the re-creator of a dim past, primarily because such a view usually carries with it the belief that he thus failed to fulfil the major obligation of the artist, the obligation to confront actual life . . .
>
> (F. O. Matthiessen)[1]

It has become standard operating procedure to begin a historical revaluation of some element of classic American literature by smirking while quoting from the Oedipal father F. O. Matthiessen and the Oedipal text, *American Renaissance*, and I have followed at least half of that procedure here. But I want to take Matthiessen's situation as a cautionary tale about our interpretations of literature and our historicist praxis. As Eric Cheyfitz has recently reminded us, the problem in *American Renaissance* is not so much that Matthiessen does not engage history, as what happens when he does. '[T]he unconscious rhetorical strategy of *American Renaissance*', Cheyfitz writes, no sooner 'approaches a subject like slavery or class conflict [than it] sublimates the political issue in a "larger" or more "complex" aesthetic or metaphysical issue.'[2] That is why Matthiessen, in my epigraph, links artistic achievement to an engagement with 'actual life', but later, in a chapter interestingly entitled 'A dark necessity', puts the equation in these terms:

> The importance of that sense [of the past] for an artist is that by it alone can he *escape* from *mere* contemporaneity, from the superficial and journalistic *aberrations* of the moment, and come into possession of the primary attributes of man, . . . what is essentially human.
>
> (p. 320; my emphases)

In this essay, I am most interested in what we can learn from the uneasy tension these two passages register—the artist's and his or her art's oscillation between engagement and escape—especially as the critic and the critical text may register a similar oscillation. I suppose I do not smirk, then,

† From *Textual Practice* 7 (1993): 13–30. Reprinted by permission of the publisher. Page numbers in square brackets refer to this Norton Critical Edition.

1. F. O. Matthiessen, *American Renaissance: Art and Expression in the Age of Emerson and Whitman* (New York: Oxford University Press, 1941 (1968)), p. 192. Hereafter cited in text.
2. Eric Cheyfitz, 'Matthiessen's *American Renaissance:* Circumscribing the revolution', *American Quarterly*, 41 (1989), p. 357.

because I wish to unsettle the notion that we at the present moment have successfully thrown off our Oedipal father.[3]

Instead, I want in this paper to reconsider the Author as the privileged category for organizing our interpretations of American Literature by attempting to account for a specific recurring image in *The Scarlet Letter*: that of the black man. Despite Matthiessen's view that Hawthorne's 'three *other* novels all deal with contemporary material', I contend that *The Scarlet Letter* is itself profoundly implicated in 'contemporary material'—specifically, antebellum discourses of miscegenation. The representation (or lack of representation) of adulterous sexual relations at the novel's centre draws specifically upon antebellum fears about miscegenational sexual union by figuring sexual misconduct in distinctly racial terms. But the black man as a marker of the novel's participation in these discourses has largely remained invisible, and I want in this essay to speculate upon the reasons why that invisibility has been the case. After demonstrating the black man's presence in Hawthorne's novel, then, I offer an extended review of some recent criticism of the novel that has overlooked the black man, sometimes in essays that seem to set out precisely in search of such a figure. I conclude by exploring the ramifications of this figure of the black man for a theory of literary production and the role of the author in the American Renaissance.[4]

II

> [I]t is . . . plain that a very different-looking class of people are springing up at the south, and are now held in slavery, from those originally brought to this country from Africa; and if their increase will do no other good, it will do away the force of the argument, that God cursed Ham, and therefore American slavery is right. If the lineal descendants of Ham are alone to be scripturally enslaved, it is certain that slavery at the south must soon become unscriptural . . .
>
> (Frederick Douglass)[5]

The Scarlet Letter is a novel obsessed with origins, and not only because it opens with 'The Custom-House', a fictionalized account of the 'birth' of

3. Although the Oedipal metaphors are my own, Michael J. Colacurcio shares my wonder while providing one of the most inclusive accounts of the pervasive influence of Matthiesssen's model of the American Renaissance on subsequent literary criticism; see 'The American-Renaissance Renaissance', *New England Quarterly*, 64 (1991), pp. 445–93.
4. The word 'miscegenation' that I utilize in this paper to denote interracial sexual mixing has an unusual history that it may be useful briefly to detail. The word was not available at the time of the publication of *The Scarlet Letter* in 1850; according to the *OED*, 'miscegenation' was coined in 1864 in an anonymous pamphlet published in New York City entitled *Miscegenation: The Theory of the Blending of the Races, applied to the American White Man and Negro*. In the strongest possible terms, and by drawing upon a wide variety of evidence, this pamphlet advocated 'the intermarriage of diverse races [as] indispensable to a progressive humanity', although it now seems likely that the pamphlet was actually intended as a biting political attack perpetrated by partisan Democratic journalists as a parody of what they perceived to be the Republican party's (and President Lincoln's) racial agenda. Still, in the context of this paper, it's fascinating to note that these parodists needed to invent a new word for their mock endorsement of interracial mixing, so heavily laden were the old terms (particularly 'amalgamation') with the widest range of antebellum racist antipathies. On party politics surrounding miscegenation, see Eric Foner, *Free Soil, Free Labor, Free Men: The Ideology of the Republican Party before the Civil War* (New York: Oxford University Press, 1970), especially pp. 237–41: on *Miscegenation* (the pamphlet) and the various responses it engendered, see J. M. Bloch, *Miscegenation, Melaleukation, and Mr. Lincoln's Dog* (New York: Schaum Publishing Co., 1958); for a history of race and sexuality in the United States, see John D'Emilio and Estelle Freedman, *Intimate Matters: A History of Sexuality in America* (New York: Harper & Row, 1988), ch. 5.
5. Frederick Douglass, *Narrative of the Life of Frederick Douglass An American Slave* (1845; reprinted New York: Signet, 1968), p. 24; ch. 1.

the novel out of its author's fortuitous rummaging in the Custom-House attic. More specifically, the novel's catalytic question—the one with which the Puritan fathers are obsessed—is, of course, a question of paternity: 'I charge thee to speak out the name of thy fellow-sinner and fellow-sufferer!' [47], urges Dimmesdale from his perch high above Hester in the market-place.[6] In the reading I am about to flesh out, Hester is a victimized woman and Pearl the illegitimate child of a father-master whose identity we do not know when the novel opens. This is surely a common enough 'real-life' scenario in the mid-nineteenth-century South, as the quotation from Douglass's *Narrative* suggests, and as he knew at first hand. The fact of miscegenation before the Civil War has been voluminously documented, most recently by John D'Emilio and Estelle Freedman, who show that interracial sex was a great (and in some ways hypocritical) rallying cry of Northern abolitionists. But if the abolitionists who denounced miscegenation in the South 'attributed to slavery a form of sexual exploitation that occurred in the free-labor society of the North, where prostitution grew visibly by mid-century and working women had to contend with the sexual advances of their employers',[7] the novel's depiction of miscegenation does not merely reproduce the terms of the Southern confrontation between a white master and a female slave. Rather, the novel shifts the genders of that equation, with the effect ultimately of revealing the white fears that linked North and South: a shared belief in the unbridled sexuality of African men and the vulnerability of white women, a shared panic when confronted with the possibilities of racial mixing.

While Hester, the abused female 'slave', is on numerous occasions described by the narrator as 'chained',[8] at the pivot point of this reading is Pearl, whose presence leads us irrevocably into the heart of this novel, if only because hers is a heart so difficultly discerned. If the central question out of which the novel grows is that of paternity, not even Hester, who presumably understands Pearl's origins, can explain where Pearl comes from or what precisely she is. No fewer than six times in the chapter which bears the child's name, Hester asks a variant of the question that has plagued critics of the novel as well. 'Child, what art thou?' Hester repeatedly asks; the problem has led more than one critic to settle things by locating Pearl's origins in a supposedly determining biographical fact: Hawthorne's own troubled daughter, Una.[9]

But Pearl is a mixture, or more acutely, a *mixed-breed*, about whom the novel never seems to tire of attempting to describe. She possesses at once 'the wild-flower prettiness of a peasant-baby' and 'the pomp . . . of an infant princess' (p. 114; ch. 6) [59], but her mother can't help wondering whether

6. Nathaniel Hawthorne, *The Scarlet Letter* (New York: Penguin, 1983), p. 93; ch. 3. Hereafter cited in text by page number and chapter.
7. D'Emilio and Freedman, *Intimate Matters*, p. 101.
8. Among numerous examples, one must select a very few: the narrator tells us Hester is linked to Dimmesdale by 'the iron link of mutual crime' (p. 178; ch. 13) [98], and this image recurs in one of the novel's climaxes, when Arthur, Hester and Pearl stand together on the scaffold and form 'an electric chain' (p. 172; ch. 12) [95]. Images of bondage are hardly less ubiquitous; Hester is 'the people's victim and life-long bond-slave' (p. 242; ch. 21) [135] and as such, she seems able to see how Arthur is similarly hounded by the Puritan orthodoxy: 'what hast thou to do with all these iron men, and their opinions?' she asks him in the forest. 'They have kept thy better part in bondage too long already!' (p. 215; ch. 17) [119].
9. T. Walter Herbert, Jr has recently taken this approach in 'Nathaniel Hawthorne, Una Hawthorne, and *The Scarlet Letter*: Interactive selfhoods and the cultural construction of gender', *PMLA*, 103 (1988), pp. 285–97.

'Pearl [is] a human child' (p. 116; ch. 6) [60] and not some 'little elf' or worse yet, 'fiend-like' and possessed by an 'evil spirit' (p. 120; ch. 6) [63]. What's more, as the narrator tells us,

> The child could not be made amenable to rules. In giving her existence, a great law had been broken; and the result was a being, whose elements were perhaps beautiful and brilliant, but all in disorder.
>
> (p. 114; ch. 6) [59]

What is the Great Law that has been broken in this book in which, as we know, the words 'adultery' and 'adulterer' never appear? And whence does Pearl's devilish behaviour derive? The book offers at least two answers, one of which I suggest has not been thoroughly analysed until now. So let us take the more familiar solution first.

To do so, we need to plunge for a moment into the editorial notes in the recent Penguin edition of the novel. Accompanying Hester's question to her husband, Roger—'Art thou like the Black Man that haunts the forest round about us?' (p. 102; ch. 4) [52]—and specifically attached to that phrase 'the Black Man', Thomas E. Connolly gives this circumscribing reading in an endnote:

> [W]itchcraft sprang from primitive religions that expressed belief in the incarnation of a god in a human or an animal. This god was always called a devil by the Christians and it appeared disguised as an animal or dressed *inconspicuously* in black; hence the Devil is called the black man.
>
> (p. 281; my emphasis)

This note stands in a direct ancestral lineage from Matthiessen's belief that Hawthorne had engaged contemporaneous events in his other novels, but not in *The Scarlet Letter*. When like Connolly we read 'black man' selectively, in a quasi-allegorical mode, contemporaneity collapses under the overbearing claims of timelessness and the novel is taken to reflect just one more version of, in Nina Baym's words, 'the conflict between repressive societies and defiant individuals'.[1] Pearl's curiously mixed behaviour then only reiterates the Puritan belief in deviant behaviour as devil-inspired, and Hester is the heroine of another classically American confrontation between self and society. This closed interpretive circle is complete when 'history' (so-called) enters and *The Scarlet Letter* can be made to *stand in* as another version of *The Puritan Origins of the American Self*, to quote the title of the definitive study on the topic.

But I would argue that in 1850, in hyper-racialized America, North or South, blackness never appears *inconspicuously*—nor is there anything inconspicuous about the presence of blackness in *The Scarlet Letter*. In fact, once we begin to read the word other than allegorically—once we notice that the novel compulsively figures the Other and Other-ness as black, and that the *OED* reports use of the word 'black' in reference to those of African heritage as early as the year 890—we begin to see how a standard and seemingly harmless footnote in the standard textual apparatus cuts off a *racial* reading of *The Scarlet Letter*.

1. See her introduction to the Penguin edition of the novel, p. 19.

To return now to a further delineation of the black man's presence in this mid-century 'Puritan' tale, and the second (and now nearly apparent) explanation of Pearl's devilish behaviour, we might remember that Pearl, too, is obsessed with her origins, and teases Hester constantly—in one instance, particularly poignantly—about this mystery. On their way into the forest, Pearl asks to be told a story.

> A story, child!' said Hester. 'And about what?'
>
> 'O, a story about the Black Man!' answered Pearl, taking hold of her mother's gown, and looking up, half earnestly, half mischievously, into her face. 'How he haunts this forest, and . . . offers his book and an iron pen to every body that meets him . . . and they are to write their names with their own blood. And then he sets his mark on their bosoms! Didst thou ever meet the Black Man, mother?'
>
> 'And who told you this story, Pearl?' asked her mother, recognizing a common superstition of the period.
>
> (p. 202; ch. 16) [112]

As Pearl presses the point further, Hester at last gives in:

> 'Wilt thou let me be at peace, If I once tell thee?' . . .
>
> 'Yes, if thou tellest me all', answered Pearl.
>
> 'Once in my life I met the Black Man!' said her mother. This scarlet letter is his mark!'
>
> (p. 203; ch. 16) [112–13]

In Hester's admission is the strongest evidence for miscegenation, especially when placed beside the text's obsessive figuring of Dimmesdale (and now the name begins to resonate) as black. Dressed in black and wearing black gloves, Dimmesdale speaks of sinful men as 'black and filthy' (p. 153; ch. 10) [83], describes his own fallen state as 'the black reality' (p. 209; ch. 17) [116], is burdened by a 'black secret' (p. 164; ch. 11) [90], and thanks Hester for bringing hope to his 'sick, sin-stained and sorrow-blackened' self (p. 219; ch. 18) [121]. And in the most extraordinary example, a renewed Dimmesdale returning to town is tempted by 'the arch-fiend' to do it all again, that is, to 'drop into [a young maiden's] tender bosom a germ of evil that would be sure to blossom darkly soon, and bear black fruit betimes' (p. 235; ch. 20) [131].[2]

Pearl shows concern not merely about her origins, however; she also wants to know when she will be recognized, when the minister will confess to his role in her life. 'Will he go back with us, hand in hand, we three together, into the town?' (p. 228; ch. 19) [127], she asks, in one of the more pointed instances, as the three leave the forest. From within the paradigm of miscegenation, Pearl's questions take on an added significance, especially when considered in relation to her refusal in the forest to approach the minister and Hester, who has let down her hair and taken the 'A' off and thrown it 'to a distance among the withered leaves' (p. 219; ch. 18) [122].

2. The image of the black fruit first appears just before Hester emerges into the market-place from within 'the black flower of civilized society, a prison' (p. 76; ch. 1) [35]. Pearl brings the reader back to this passage when later she tells Mr Wilson 'that she had not been made at all, but had been plucked by her mother off the bush of wild roses, that grew by the prison-door' (p. 134; ch. 8) [72].

> But Pearl . . . now suddenly burst into a fit of passion, gesticulating violently, and throwing her small figure into the most extravagant contortions. She accompanied this wild outbreak with piercing shrieks, which the woods reverberated on all sides; so that, alone as she was in her childish and unreasonable wrath, it seemed as if a hidden multitude were lending her their sympathy and encouragement.
> (p. 226; ch. 19) [126]

It is not the first time, we should note, that Pearl has seemed to embody a multiplicity within her sole self. Earlier, in a line that echoes Enobarbus's description of another figure of racial otherness, Cleopatra, and that seems to point toward the composite, varied nature of Pearl's appearance, the narrator tells us that 'Pearl's aspect was imbued with a spell of infinite variety; in this one child there were many children . . .' (p. 114; ch. 6) [59].[3] Pearl has here stumbled upon a kind of primal scene in the forest, that unbridled space where the black man as representative of everything uncivilized and unstructured has his free rein. But why does she react so violently? In this reading, she speaks as a mulatto child whose existence can be dismissed ostensibly as easily as Hester has cast off the Scarlet Letter—the marker that is always compared and equated by the narrator to Pearl herself.[4] Thus, Pearl reads the discarded 'A' as a discarded Pearl. To bring Pearl back, Hester replaces the letter and hides her hair:

> 'Wilt thou come across the brook, and own thy mother, now that she has her shame upon her—now that she is sad?'
> 'Yes; now I will!' answered the child . . .
> In a mood of tenderness that was not usual with her, she drew down her mother's head, and kissed her brow and both her cheeks. But then—by a kind of necessity that always impelled this child to alloy whatever comfort she might chance to give with a throb of anguish—Pearl put up her mouth, and kissed the scarlet letter too!
> (p. 228; ch. 19) [127]

Not even this momentary reconciliation is immune from Pearl's insistent doubleness: once more she 'alloys' rare comfort with a mischievous, and very nearly simultaneous, gesture that elicits pain.

The chapter 'The revelation of the Scarlet Letter' has as its most significant consequence Arthur's acknowledgment of his role in Pearl's life, signalled first by his public announcement—'Hester . . . come hither! Come, my little Pearl!' (p. 265; ch. 23) [149]—and then by hers, a reversal of Dimmesdale's earlier kiss, which Pearl had washed off her forehead:

> Pearl kissed his lips. A spell was broken. The great scene of grief, in which the wild infant bore a part, had developed all her sympathies; and . . . her tears fell upon her father's cheek. . . .
> (p. 268; ch. 23) [151]

3. *Antony and Cleopatra*, ed. M. R. Ridley (London and New York: Methuen, 1986), II.ii.236.
4. Cf., for example, pp. 124–5; ch. 7 [65]: 'But it was a remarkable attribute of [Pearl's] garb, and, indeed, of the child's whole appearance, that it irresistibly and inevitably reminded the beholder of the token which Hester Prynne was doomed to wear upon her bosom. It was the scarlet letter in another form; the scarlet letter endowed with life!'

In the novel's final pages, the rivalry between the two men—each of whose sins has seemed 'blacker' to the other[5]—collapses into equivalence. The novel ultimately equates Dimmesdale and Chillingworth when each publicly accepts Pearl as his own: Dimmesdale on the scaffold, and Chillingworth in his will (pp. 272–3; ch. 24) [153].[6] Pearl receives an inheritance from her mother's cuckolded husband and becomes 'the richest heiress of her day' (p. 273; ch. 24) [153], but it is a bequest the mixed-breed child can gain and enjoy only in England, the Old World newly free of slavery. In 1850, such a fairy-tale ending is impossible in the New—although it is interesting to note that the novel's ending may share with the abolition of American slavery a certain unpredictability, as Hawthorne himself famously explained:

> [Slavery is] one of those evils which divine Providence does not leave to be remedied by human contrivances, but which, in its own good time, by some means impossible to be anticipated, but of the simplest and easiest operation, when all its uses have been fulfilled, it causes to vanish like a dream. [234][7]

The profound passivity Hawthorne insists upon in this oft-quoted passage—the sense one gets in reading it of the apparent inconsequence of individuated, human agency—serves our purposes here as an ironic counterpoint to the myth of the transcendent author that Hawthorne is often said to embody and that I examine in this essay's remaining sections.

III

> It is singular, however, how long a time often passes before words embody things; and with what security two persons, who choose to avoid a certain subject, may approach its very verge, and retire without disturbing it.
>
> (Nathaniel Hawthorne)[8]

In what follows, I consider two recent articles about *The Scarlet Letter* that investigate Hawthorne's engagement with the issue of slavery, but that nevertheless provide no account of the black man in the text. I have in mind recent essays by Jonathan Arac and Jean Fagan Yellin, each of which I take up in turn after a closer look at Matthiessen's *American Renaissance*. In general, I want to speculate upon the continuing predominance of the quasi-allegorical reading of the black man in *The Scarlet Letter* by considering some of the conceptions that frame our readings of this highly valued, canonical work.

To begin such an analysis, I would return us to the assumptions that underlie Matthiessen's critical method in *American Renaissance*:

5. As Dimmesdale tells Hester in the forest: 'That old man's revenge has been blacker than my sin. He has violated, in cold blood, the sanctity of a human heart' (p. 212; ch. 17) [118].
6. The novel foreshadows this equivalence with descriptions of the two men that are virtually interchangeable—a fact about which Pearl is once more the most scrupulous reader. Looking up at the window through which Chillingworth looks down, she exclaims: 'Come away, mother! Come away, or yonder old Black Man will catch you!' (p. 155; ch. 10) [84]. Pearl speaks as if she knows the rumours spreading that the doctor's 'visage was getting sooty with the smoke' (p. 149; ch. 9) [81] from the fires in his laboratory where he develops his 'black devices' (p. 160; ch. 11) [87].
7. Quoted in Matthiessen, *American Renaissance*, p. 317.
8. *The Scarlet Letter*, p. 239; ch. 20 [134].

> [M]y main subject has become the conceptions held by five of our major writers concerning the function and nature of literature, and the degree to which their practice bore out their theories.
>
> (p. vii)

Here Matthiessen announces his governing interpretative tautology: the terms of his analysis derive in full from the minds of the authors who become at once creators and critics of their own texts. We would want to note this formulation's overriding valorization—even fetishization—of the minds of these five authors; to do so is to recognize the degree to which one of the constitutive documents of American literary criticism is embedded fully in paradigms of the author as a unique, generative agent, creator and controller of meaning.

Indeed, precisely this obsession with the author links together the two passages from *American Renaissance* that I quoted in the first section and that seemed from another perspective to offer contradictory positions about the relation between an author and history. Whether Matthiessen depicts the relationship as 'the major obligation of the artist . . . to confront actual life', or, alternatively, as the author's choice to 'escape from mere contemporaneity', in either case the author 'confronts' or 'escapes' history; in both cases, he is an agent somehow separated from it (pp. 192 and 320). Within this paradigm, I want to argue, it has proven difficult to acknowledge and account for the presence of the black man in *The Scarlet Letter*.

In 'The politics of *The Scarlet Letter*', Jonathan Arac seeks deliberately to revise our common contextual understanding of Hawthorne's novel. Citing Walter Benjamin's challenge to the cultural historian to 'brush history against the grain', Arac attempts '[t]o raise up to prominence what is usually smoothed over' by acknowledging at the outset that '[s]lavery was the issue that agitated American politics most deeply in Hawthorne's time, and abolitionism made the young Henry Adams feel that Boston in 1850 was once again revolutionary.'[9] Arac proposes 'to define a relation between *The Scarlet Letter* and the political response to masters' barbarism and slaves' anonymous toil' (p. 248). He also brushes against the grain Hawthorne's claims to an artistic space immune from politics, in part by bringing to the surface the constructedness of that myth:

> Poe may have preceded Hawthorne in the attempt to establish such an artistic space, but Hawthorne was the first to do it effectively, to make it stick, in a way recognized by his contemporaries and for the future.
>
> (p. 248)

But in spite of these preliminary unmaskings, Arac's reading of the politics of *The Scarlet Letter* reinscribes one of the sustaining myths of the American ideology he self-consciously sets out to interrogate: namely, the myth of the empowered and self-possessed individual, whose infinite potential as a maker of history and author of meaning is his inalienable American

9. Jonathan Arac, 'The politics of *The Scarlet Letter*', in *Ideology and Classic American Literature*, ed. Sacvan Bercovitch and Myra Jehlen (Cambridge: Cambridge University Press, 1986), p. 248. Hereafter cited in text.

birthright. Thus, while Arac may be brushing history against the grain, he never challenges the primary agent of that history.[1]

I want to say at this point that I find Arac's reading of the replacement of action by character and stasis in *The Scarlet Letter*, 'The Custom-House', and *The Life of Franklin Pierce* engaging and convincing. I am nevertheless taking issue with the degree to which Arac's analyses depend upon the figure of the author as the mechanism that enables his manoeuvres around and across these texts. Arac's inquiry, in fact, derives from a related assumption that raises important questions he of necessity must disregard. Arac tells us how his '[s]tudents marvel that the author of "The Custom-House" was in less than three years to write *The Life of Franklin Pierce*' (p. 251). As reasonable as this observation appears, it nevertheless warrants examination. We are implicitly asked to presume that these students' concerns are somehow natural or innocent—worthy of our contemplation and even our wholesale appropriation for originating from such untutored and guileless readers. But the issue of common ground between *The Scarlet Letter* and the political biography actually resists a number of questions with regard to the function of authorship: for example, what precisely is the nature of the presumed contiguities between texts produced by the 'same author', and how can that 'sameness' best be understood? Rather than demonstrating the relative naïveté of the students, their observation actually demonstrates how embedded they are in the ideology of the individually distinct and empowered agent author. And while it might be argued that such ideological embeddedness rising to the surface (or presented) as naïveté is the very mark of ideological interpellation, an informed poststructuralist critique would want nevertheless to interrogate such 'natural' assumptions.

Foucault's well-known essay on the author provides a point from which to launch this critique. To do so, it is important to see that the students' implicit demand for continuity (or integrity) in the author-function—their question, after all, grows out of their disbelief that the author who abhors politics in 'The Custom-House' could turn around and benefit from it a short time later by writing a campaign biography—demonstrates their full subjection (I choose the word carefully) within a system that makes of the author 'not an indefinite source of significations which fill a work' but rather 'a certain functional principle by which, in our culture, one limits, excludes, and chooses; in short, by which one impedes the free circulation, . . . composition, decomposition, and recomposition of fiction.'[2] The contradiction these students believe they have uncovered in the works of the 'same' author forces them (and Arac) to re-investigate the 'received understanding' of the author's biography.[3] As Foucault says, 'we are accustomed to presenting the author as a genius . . . we make him

1. Cf. Arac's opening sentence: 'If the study of American literature is not merely to reproduce the American ideology, it must engage directly with the debates of literary theory, which allow us to raise basic questions about the values and practices at stake in reading, studying, and teaching American literature and culture' (p. 247). C. B. MacPherson's *Political Theory of Possessive Individualism* (Oxford: Clarendon Press, 1962) is, of course, the classic statement on the ideology and origins of possessive individualism.
2. Michel Foucault, 'What is an author?', *The Foucault Reader*, ed. Paul Rabinow (New York: Pantheon Books, 1984), pp. 118–19. Hereafter cited in text.
3. This is, in fact, Arac's starting place: 'In arguing for a specific interpretation of *The Scarlet Letter* that is neither authorial in the "interpretationist" sense nor mystifying, as I find indeterminism, I begin from several concrete problems in our received understanding of Hawthorne' (p. 250).

function in exactly the opposite fashion' (p. 119), and the author as the privileged term of agency performs in the example from Arac's students its highest organizing function by deflecting critique from the global concept to the local example. Adjusting Hawthorne's biography permits these readers to account for the contradictions without seeing through to what the system of authorship obfuscates: namely, the embeddedness of artistic creation within social systems, or (in Raymond Williams's words), the fact that '[c]onsciousness . . . is social being'.[4] Arac's students merely point out an incoherence that becomes *the exception that proves the rule* for authorship as the necessary governing frame for the interpretation of texts, rather than an exception that forces other questions about the 'naturalness' or 'logic' of the system in the first place.

Arac's own understanding of politics and *The Scarlet Letter* grows out of a similar concern for the continuity of the author: 'The problem is to determine a relation, perhaps even a common ground, between the writing of *The Scarlet Letter* and that of *The Life of Pierce*' (p. 251). The geography of Hawthorne's career is the landscape upon which Arac will measure out that 'common ground.' But we might notice of this goal that it shares certain characteristics with another of Arac's students' observations: that '*The Scarlet Letter* [is] an intransitive "work of art", unlike, say *Uncle Tom's Cabin*, which is "propaganda" rather than "art", for it aims to change your life' (p. 251). Both Arac's and his students' observations beg the question of the canon. That is, Arac's assumption in seeking out 'common ground' via the figure of the author is reinforced by the assumption that generically, *inherently*, *The Scarlet Letter* and *Uncle Tom's Cabin* represent different modes of writing that can only be associated at the level of their authors, or their authors' intentions. Arac's 'resolution' of his students' second observation reinscribes the same dichotomy: 'If recent revaluation has shown that *Uncle Tom's Cabin* is also art, may it not be equally important to show that *The Scarlet Letter* is also propaganda' (p. 251).

But more important for our purposes, the distinction Arac's students make between *The Scarlet Letter* and *Uncle Tom's Cabin* recurs in the implicit distinction he draws between Hawthorne's novel and the political biography. The task as Arac has it is to link the artfulness of one with the politics of the other via Hawthorne, rather than to work from the assumption that the relationship might be determined in terms of some other criteria—for example, as documents existing at the intersection of a variety of historical or socio-political discourses and produced in relation to them. Arac's essay does not allow for the possibility that factors other than an author's biography might be the coordinates upon which to map relations between texts, whether within an individual corpus, or across and between different authors. Instead, Arac's methodology valorizes his students' conception of the canon and of 'artistic' production to precisely the extent that at the fundamental level he utilizes the political biography to access 'The politics of *The Scarlet Letter*'.

Arac continues to focus upon the author when he turns to the issue of the relationship between 'The Custom-House' and the novel it introduces. 'There is always some doubt what we mean when we say, "The

4. Raymond Williams, *Marxism and Literature* (New York: Oxford University Press, 1986), p. 41.

Scarlet Letter"' Arac reminds us: do we mean only the twenty-four chapters of the novel 'proper', the novel including 'The Custom-House', or the actual embroidered article sewn to Hester's dress?' (p. 251). Arac resolves this ambiguity by linking biographical facts in Hawthorne's life to the 'situations' of characters in the novel[5]—a process he describes in terms of allegory,[6] and that begins, significantly, with a claim of ownership: 'By taking possession through "The Custom-House" of the (physical) scarlet letter as his property, the author of "The Custom-House" personalizes the narrative' (p. 252). Arac's conclusion about the effect of possessing the letter is by no means the only conclusion one might reach. From the same act, it is in fact possible to conclude just the opposite: that the author by taking possession of the 'A' *de-personalizes* the narrative because both the letter and the written history he finds in the attic are (he insists) objects of his embellishment in *The Scarlet Letter*, rather than his own *original* productions.[7]

Indeed, one can say even more: in the calculus of identity politics as Arac deploys it, taking possession of an object represents the enabling activity for the construction of individuality; Hawthorne 'personalizes' the narrative and takes possession of it with the same gesture, thus constituting himself in the terms of the possessive individualist ideologies that underwrite his status as an author in the first place. But from the point of view of Arac's announced desire to interrogate 'the American ideology' and to reveal slavery as the issue at the core of political praxis in the antebellum period, this reliance upon the tenets of possessive individualism is, to say the least, ironic. Most importantly, as we will see, Arac's depiction of Hawthorne's ownership of his text puts him in no better position to see slavery (or the metonymic figures within his own text that represent it) than the critics he is presumably revising.

For while he argues in his opening pages that he wishes to reintegrate slavery into the fuller context of antebellum America, Arac's primary definition of slavery as 'masters' barbarism and slaves' anonymous toil' rewrites race as social class and economics, and in so doing restricts his conceptualization of slavery to the realm of political economy in which the ideologies of possessive individualism gain their most profound validations. This rewriting proves consequential at precisely the instant when the black man makes an appearance in Arac's text: 'Hester is described as not true to the letter when she analyzes it contractually, as the mark of her meeting with *the black man* in the woods' (p. 261; my emphasis). The black man here remains 'anonymous' because Arac has validated not so much Hawthorne's 'barbarism', as its culturally-inscribed opposite: Hawthorne's valuable labour as an individual and an intentioned author. In this way Arac's

5. 'The many correspondences between the authorial figure of "The Custom-House" and the characters of *The Scarlet Letter*—for example, the disapproval shown to both Hester and Hawthorne by an imagined crowd of Puritan authorities, the dual status Dimmesdale and Hawthorne share of a passionate inner life wholly at odds with their "official" public position, the work both author and Chillingworth do as analysts of character—allow us to naturalize the presence of "The Custom-House" and justify its excess' (p. 252).
6. The correspondences between Hawthorne and his characters 'undermine the self-sufficiency of "The Scarlet Letter"—making it an allegory of the writer's situation in 1850' (p. 252).
7. Indeed, the narrator of 'The Custom-House' makes no claims of ownership in regard to what he delineates as the historically verifiable, and therefore "original" (in the sense of temporally prior) parameters of the story. Quite to the contrary, he insists upon 'the authenticity of the outline' (p. 63; 'The Custom-House') [26].

essay participates in precisely the system it set out to uncover—the workings of 'the American ideology'. The black man must remain invisible when viewed from within the discourses of possessive individualism that underwrite the essay's explication of indeterminacy—what Arac tellingly calls 'Hawthorne's own authorial meaning'.[8]

A brief glance at Jean Fagan Yellin's 'Hawthorne and the American national sin' reveals that many of the paradigms we have been considering recur, although the essay puts them to a slightly different purpose, and exhibits, as a consequence, a peculiar kind of success. By 'success' I am referring to the language of morality that first appears in Yellin's title, and that permits her ultimately to denounce Hawthorne for failing to 'respond imaginatively to the centrality of race and slavery in America' until 'long after he had produced his great romances, in which any recognition of these issues is conspicuously lacking'.[9] Besides some entries in the Notebooks of the 1840s and the essay 'Chiefly about war matters' (1862), Yellin demonstrates that Hawthorne seems to have troubled himself relatively little about slavery; indeed, we have already seen his well-known insistence that slavery would 'vanish like a dream'. This Yellin rightly (if implicitly) deprecates as a paltry substitute for the vocal denunciation of, say, a Whittier, and it is clear from her article that she wishes to have found an equally emphatic moral stand from one of the century's great novelists.

But I have called Yellin's success 'peculiar'—a better word would perhaps be 'partial'—because she relies upon assumptions about artistry and authorship that actually undercut her attempts to see the black man as he *does* in fact make his appearance in Hawthorne's best-known novel. That she adopts a traditional understanding about the methods of artistic production emerges when Yellin names as her focus 'the essential facts of chattel slavery', but then distances herself (and by extension, Hawthorne) from an engagement with that 'essential fact' by suggesting that slavery 'might naturally be expected to illustrate [the romances'] major theme of psychological bondage' (p. 76). This re-naming, however 'natural', actually does a disservice to Yellin's interpretative efforts, and in her conclusion she can find no links between 'metaphorical slavery and the literal enslavement of blacks' (p. 88). Yellin searches for Hawthorne's 'recognition', as well as for some signs that he 'finally did respond imaginatively', but the terms 'recognition' and 'response', no less than the adjective 'great', delimit Yellin's capacity to account for the presence of the black man because these words carry a wide range of commonsensical assumptions about literary agency and an author's turning inward to compose solely out of the stuff of his own isolated psychology and individual experience. Such terminology, along with the argument that the 'studied ambiguity' of these works represents 'deliberate artistic decisions' bespeak Yellin's assumption that in the romances she will find only *conscious* translations of the 'essential facts of slavery', and that the absence of such evidence in *The Scarlet Letter* marks, then, 'a strategy of avoidance and denial' (p. 97) on Hawthorne's part. Blinded by canonicity and by a genius-centred model of

8. 'My argument has tried to show that Hawthorne's own authorial meaning establishes an "indeterminacy" that is not merely a modern critical aberration' (p. 261).
9. Jean Fagan Yellin, 'Hawthorne and the American national sin', in *The Green American Tradition: Essays and Poems for Sherman Paul*, ed. H. Daniel Peck (Baton Rouge: Louisiana State University Press, 1989), p. 96. Hereafter cited in text.

literary production, Yellin, too, cannot see the black man, and once again, Matthiessen's binarism recurs: there is either conscious engagement or conscious escape. Yellin blames Hawthorne where Arac credits him—but both read his novel guided by an assumption that what they find between its covers represents the deliberate choices of its individuated author.

In a sense, Yellin's essay registers a pattern Barbara Herrnstein Smith identifies as the cycle of the canonical within the academy:

> [B]y providing . . . 'necessary backgrounds', teaching . . . 'appropriate skills', 'cultivating . . . interests', and, generally, 'developing . . . tastes', the academy produces generation after generation of subjects for whom the objects and texts thus labeled do indeed perform the functions thus privileged, thereby insuring the continuity of mutually defining canonical works, canonical functions, and canonical audiences.[1]

Canonical texts reproduce us and we reproduce them, and canonized along with the privileged text is a certain range of permissible interpretations restricted by that privileged status. Yellin is certain most of all of the canonical status of the works she studies, and it is that very notion of canonicity that makes it impossible for her to see the simple, untranslated presence of a black man in Hawthorne's most famous novel.

IV

Of course, my reading of the presence of the black man in *The Scarlet Letter* also makes claims upon notions of artistry and intention, but primarily by negation. In place of the canonical model of the empowered genius, I want now to take the unnoticed and seemingly unbidden presence of the black man in *The Scarlet Letter* as a point of departure for describing another mode of literary production in the American Renaissance. To do this, I will look briefly at one other text—an 1851 review of *The Scarlet Letter*—the language of which points up the need for reconceiving literary creation in terms of the discursive conditions that structure the production of writing within particular historical frames.

The review, entitled 'The Writings of Hawthorne', appeared in the January 1851 issue of *The Church Review*, a periodical associated with the Protestant Episcopal Church and published in New Haven, Connecticut. Apparently written by the conservative theologian Arthur Cleveland Coxe, who later became an Episcopal bishop, the review decribes all kinds of immoral literature, and, with reference to Hawthorne, specifically attacks 'any toleration [of] . . . a popular and gifted writer, when he perpetrates bad morals', especially because 'stories should always be of moral benefit'.[2] But a single passage in this otherwise predictable review serves our purposes, for it highlights precisely what is at stake in shifting

1. Barbara Herrnstein Smith, 'Contingencies of value', in *Canons*, ed. Robert von Hallberg (Chicago: University of Chicago Press, 1984), p. 27.
2. My information on Coxe derives from a biographical note appended to a partial reproduction of the review in B. Bernard Cohen (ed.), *The Recognition of Nathaniel Hawthorne* (Ann Arbor: University of Michigan Press, 1969). I do not know how Cohen verified Coxe's authorship; the article appears to be anonymous in *The Church Review*—which, in a sense, is precisely my point. Page numbers refer to the original periodical; these two quotations appear on pp. 502 and 501 respectively. Hereafter cited in text.

the focus away from an author-centred interpretative paradigm and toward a neo-Marxist assumption about the embeddedness of artistic production within socially constituted networks of meaning. Coxe writes that

> the language of [Hawthorne], like patent blacking, 'would not soil the whitest linen', and yet the composition itself, would suffice, if well laid on, to Ethiopize the snowiest conscience that ever sat like a swan upon that mirror of heaven, a Christian maiden's imagination. (p. 507) [259]

Two points about this excerpt must be made. The first merely acknowledges this review's position in the line of works we have been considering in which meaning belongs to, and is determined by, the author. There is no sense that anyone or anything other than Hawthorne can be responsible for the effects his book produces.

But the second point is by far the more significant, and focuses our attention upon the striking metaphor the review employs to demonstrate the effect of Hawthorne's novel on the morally pure 'maiden'. The passage relies at its core upon an image of the (female) reader as a 'mirror'—as defenceless as she is pure—and able only to reflect passively whatever is placed before her. This sense of helplessness is extended by one of the operative, contemporaneous meanings the *OED* gives for the action of 'laying on': related to the practices of the publishing house, the action refers specifically to the passive imprinting of a (white) sheet of paper as it is placed upon inked type in a press. Confronting this defenceless maiden as mirror and as blank page is 'the composition', Hawthorne's novel, spectacularly anthropomorphized with the verb 'Ethiopize', for which the *OED* offers no entry, and which we could best define as 'to make black'. Thus, *The Scarlet Letter* and the review project versions of what is often taken to be the nightmare image of American race relations: in the review, 'the composition' figured as black threatens to overpower and 'Ethiopize' a 'white' woman in a thinly veiled scene of reading as rape; in the novel, the black man unequivocally succeeds, as Hester herself admits.[3] And once the spectre of miscegenation enters the review, 'heaven' itself is soon threatened, so dire are the consequences of the insidiously corrupt novel Hawthorne has written.

I place this review alongside the black man in the novel in order to demonstrate an organizing principle that lies beyond the reach of both Coxe and Hawthorne—but more generally, beyond any individual, especially insofar as 'individual' marks always the presumed opposite term to 'society'. For the fundamental element in both the novel and this excerpt from the review is the fact that their racialized metaphors occur in essentially non-racial contexts that do not reveal to us a 'local' or 'artistic' reason why such metaphors should be appropriate. The pervasiveness of these metaphors across divergent genres (novel, 'propaganda', review) and seemingly disparate subject positions (novelist, theologian, abolitionist) reveals this culture's propensity for speaking about sexual immorality in racial terms that derive their force from the (im)possibility of miscegenation.

3. I borrow the diction of 'nightmare' from Myra Jehlen. The noun 'Ethiop', and the adjective 'Ethiopian', both, of course, have long histories.

Thus, the black man in *The Scarlet Letter* forces us to re-evaluate the place of the author's presiding genius within our understanding of literary production.[4] The image's recurrence in the review and the novel compels us to re-examine the discursive power of the fact of slavery, its powerful position as a metaphor operative in texts ostensibly separate from the concerns of slavery *per se*. Here, in 'literary' and 'theological' (rather than 'political') territory, slavery rises to the surface, linguistic proof of shared consciousness and of a culture drenched in racialized symbologies. To account for the conjunction, we must rethink the model of transcendence that scholars of Hawthorne (and of American literature more generally) have insisted upon virtually since the emergence of an American tradition in letters. The co-incidence of the racial metaphors has the effect of disintegrating the autonomy of the agents who (presumably) 'produced' them, by revealing the social dimensions of literary production.

At another point in his essay Coxe derides the fact that *The Scarlet Letter* is 'a book made for the market, and that the market has made it merchantable' (p. 507) [259]. Understanding 'market' here in its broadest sense—as a metaphor for a discursive space of shared meanings and exchanged discourses, as a figure for the interpellation of subjects (including authors and theologians) within particular temporal and cultural arenas—and so remembering that the review no less than the novel is 'merchantable' in precisely this sense, we could do worse than to take Coxe at his word. To do so, I have argued, is to take Hawthorne's word, and the words of countless others, as well.

4. My argument here shares certain aspects with Toni Morrison's 'Unspeakable things unspoken: The Afro-American presence in American literature'—which I came upon after I had already presented an early version of this paper—particularly her call to re-examine 'founding nineteenth-century works . . . for the ways in which the presence of Afro-Americans has shaped the choices, the language, the structure—the meaning of so much American literature' (p. 11). However, Morrison appears less willing to abdicate the notion of an empowered author, as this quotation suggests: 'The spectacularly interesting question is "What intellectual feats had to be performed by the author or his critic to erase [the Afro-American presence] from a society seething with [it], and what effect has that performance had on the work?"' (pp. 11–12). In general, I have been less concerned with 'intellectual feats' and 'performance' than I have been with what Foucault calls (in *The Order of Things* (New York: Vintage, 1973)) epistemic systems of shared meaning and language. Morrison's article appears in *Michigan Quarterly Review*, 28 (1989), pp. 1–34.

The Scarlet Letter Films

JAMIE BARLOWE

Demi's Hester and Hester's Demi(se): *The* (New) *Scarlet Letter* and Its Spectators†

The 1995 cinematic adaptation of *The Scarlet Letter*, produced and directed by Roland Joffé (*Killing Fields, The Mission, City of Joy*), is the eighth in a line beginning with silent versions in 1911, 1913, and 1917.[1]

† From *The Scarlet Mob of Scribblers: Rereading Hester Prynne* (Carbondale: Southern Illinois University Press, 2000), pp. 80–120, 139–144. Copyright © Southern Illinois University Press, 2000. Reprinted by permission of the publisher. Page numbers in square brackets refer to this Norton Critical Edition.

1. The 1911 silent version starred King Baggot as Arthur Dimmesdale and Gene Gauntier as Hester Prynne. This film was produced by Kalem Studios (the name a combination of the initials of George Kleine, Samuel Long, and Frank J. Marion) and directed by Sidney Olcott, whom Gauntier "credited . . . with psychic powers and a virtually hypnotic control over his actors" (Wagenknecht *Movies* 53). During Gauntier's reign as "leading lady" of Kalem, she made over five hundred films, some finished in a day. In addition to acting in these films, she

> herself had written all except half a dozen. When you are sometimes required to turn out three scenarios during a single day, it is a great help to have standard literature to draw on. Miss Gauntier began by drawing on *Tom Sawyer.* She adapted *As You Like It, Evangeline, Hiawatha,* and *The Scarlet Letter.* (Wagenknecht, *Movies* 53–54)

Gauntier, like Demi Moore, had her own production company (Seger 9). The silent 1913 version of *The Scarlet Letter* starred D. W. Griffith's first wife, Linda Arvidson, as Hester Prynne. In addition to *The Scarlet Letter,* Arvidson acted in 118 films between 1908 and 1916. The 1917 film, produced by Fox Film Corporation and directed by Carl Harbaugh, starred Mary Martin as Hester Prynne and Stuart Holmes as Arthur Dimmesdale. Mary Martin's other films include *The Tiger Woman* (also 1917), *Eternal Sappho* (1916), *The Vixen* (1916), *Devil's Daughter* (1915), and *Great Love Hath No Man* (1915) (Internet Movie Database). The next film adaptation of *The Scarlet Letter* was another silent version in 1926, produced by MGM and directed by Victor Sjöström, followed by a 1934 sound film, produced by Majestic Film Studios and directed by Robert G. Vignola (both are discussed in the text of the chapter). A German version in 1972 known as *Der Scharlachrote Buchstabe* (also released in Spanish and English) was directed by Wim Wenders and starred Senta Berger as Hester Prynne, Lou Castel as Arthur Dimmesdale, and Hans Christian Blech as Roger Chillingworth (see Gollin, "Wim Wenders's *Scarlet Letter*"). A PBS four-hour mini-series appeared in 1979—produced and directed by Rick Hauser, filmed in Salem, Massachusetts, and starring Meg Foster as Hester Prynne, John Heard as Arthur Dimmesdale, and Kevin Conway as Roger Chillingworth. The version was adapted for television by Allan Knee and Alvin Sapinsley. Hawthorne's *The House of Seven Gables* was also translated to film in 1940 by Universal Pictures and directed by Joe May. The cast included George Sanders as Jaffrey Pyncheon, Margaret Lindsay as Hepzibah, Vincent Price as Clifford, Nan Gray as Phoebe, and Dick Foran as Holgrave. Lester Cole and Harold Greene collaborated on the screenplay. Several of Hawthorne's stories/tales have also been made as films, including "Feathertop" (1912), "Rappaccini's Daughter" (1980), and "Young Goodman Brown" (1993); the latter two have been shown on television. In 1963 Admiral Pictures produced (under the genre of "Horror") one of Hawthorne's short-story collections, *Twice-Told Tales,* which was directed by Sidney Salkow. Vincent Price played Alex Medbourne, Rappaccini, and Gerald Pyncheon, Brett Halsey played Giovanni Guasconti, and Joyce Taylor played Beatrice Rappaccini. The cast also included Sebastian Cabot, Beverly Garland, Richard Denning, Mari Blanchard, and Jacqueline DeWit. The screenplay was written by Robert E. Kent (Internet Movie Database).

Before it was released, I assumed it would participate in the Hollywood tradition of translating Hawthorne's novel to the screen—a tradition that partakes primarily of the melodrama, sometimes with added comic scenes and sometimes deviating completely from Hawthorne's "script." For example, the 1911 film, starring Gene Gauntier as Hester Prynne—also adapted for the screen by Gauntier—ends, as "the censors required," with the marriage of Hester and Arthur; Edward Wagenknecht says in response, "one wonders what the story can possibly have been about" (*Movies* 54). And the official description of the 1917 film, starring Mary Martin as Hester Prynne, says that it "tells the story of a noble but poor woman who arrives at Boston in the seventeenth century. There she marries an old but quite rich doctor but does not become happy" (Boehm).

Having read many of the prerelease reviews, I also assumed the new film, because it stars Demi Moore, would function, like the literary text, in terms of Hester Prynne as spectacle * * *—her threat contained at the level of narrative by her punishment and alienation from the community, allowing for voyeuristic containment, and at the level of cinematic apparatus by fetishizing her, allowing for scopophilic containment; in other words, the pleasure in looking *at* Demi Moore:

> [V]oyeurism . . . has associations with sadism: pleasure lies in ascertaining guilt . . . asserting control, and subjecting the guilty person through punishment or forgiveness. This sadistic side fits well with narrative. Sadism demands a story, depends on making something happen, forcing a change in another person, a battle of will and strength, victory/defeat, all occurring in linear time . . . Fetishistic scopophilia, on the other hand, can exist outside linear time as the erotic instinct is focused on the look alone. (Mulvey 64)

Early feminist film theory, like that of Laura Mulvey, long ago argued that both kinds of containment are necessary because the "female figure as spectacle can . . . provoke the very anxiety [in the male viewer] it was intended to contain" (Penley on Mulvey, *Future* 42). As illustrative of this anxiety, one prerelease reviewer from *Newsweek* wrote in an essay entitled "Hester Prynncesse" that "[t]he producers, presumably, had chosen [Demi] Moore knowing that the audience would spend 90 percent of the movie staring right at the big red 'A' on her chest" (Adler 58).

Further, I assumed, following some of Mary Ann Doane's work, that the film would probably not attempt to construct a female spectator beyond the one who would become, rather than resist or reject, the classical image of Hester-Prynne-ism; in other words, "[f]or the female spectator . . . to posses the image through the gaze is to become it" ("*Caught*" 199). The image of Hester-Prynne-ism is that of the bad/good woman—bad, because she violates society's codes of confession and paternity, and good, because of her

> inherent virtue and her indomitable will . . . transcend[ing] society's view of her; she achieves an heroic stature because of her good works. . . . She does not become self-pitying, spiteful, or hostile. She endures and, by so doing, using Faulkner's phrase, she prevails. (Sochen 12 * * *)

And, finally, I assumed that the new film would compound the containment already at work in Hawthorne's text—a containment of the objectification and sexual violence that exist at its core and are replicated in the mainstream academic critical tradition I have discussed so far in this book * * *. However, once I saw it, I realized that something much more complicated and interesting occurs in this film I now call *The* (New) *Scarlet Letter*.

Thus, in this chapter I will reread *The* (New) *Scarlet Letter*—in part, a prequel to Hawthorne's text; in part, an historical contextualizing of the novel; and in part, a revisioning of its ending—as an attempt to rewrite Hester Prynne by exposing the literary text's implicit violence and voyeurism. That is, the film represents the historical and sexual excesses that a Hawthornian semiotics tries to contain. Containment, as I am using it, is an attempt to hide the failures of cultural repression—the excesses. As Penley says, "repression is never complete. For, in fact, we only know of repression through its failures; if repression were total, nothing would remain to make us aware of what had been repressed or the act of repression itself" (52). The practice of containment was also attempted, I will show, in earlier film versions of *The Scarlet Letter*, not only at the level of the content and technique, but also in the choice of actors who have portrayed Hester Prynne. I will also argue that *The* (New) *Scarlet Letter* functions to destabilize and disrupt—to trouble the cinematic frame—rather than to reinscribe male desire. Jacqueline Rose has argued that "cinema appears as an apparatus which tries to close itself off as a system of representation, but that there is always a certain refusal of difference, of any troubling of the system, an attempt to run away from the moment of difference, and to bind it back into the logic or perfection of the film system itself" (qtd. in Penley 44). But, using Constance Penley's use of Rose's phrase, then, I am saying that the film attempts to refuse the "refusal of difference," or following Tania Modleski, I would say that this film disrupts the "fetishistic disavowal in the male . . . the means by which the psyche avoids facing the fact of woman's difference, the fact of her *being* a woman" (22). I am arguing further that all of the reviews and critiques, in an effort to re-contain what the film exposes, do reinscribe male desire by fetishizing Demi Moore, and by extension, Hester Prynne, and thus participate in the established tradition of Hester-Prynneism * * *. These reviews, along with my discussion of the old films and actors, as well as the feminist film theory, function as part of the context in which I reread *The* (New) *Scarlet Letter*.

The critics' hysteria, nostalgia, and fetishization of Demi Moore. The examples of the critical hysteria and efforts at containment that introduced this film to the viewing public are numerous. In addition to the *Newsweek* article I just quoted, other critics discuss Demi Moore's chest and body, derisively dismissing actor, producer, director, and film, perhaps oblivious to its attempts (though Disneyized at times, as Joyce Carol Oates suggests) to expose rather than participate in the ongoing American pastime of fetishizing Hester Prynne. This time the critics had two targets, Prynne and Demi Moore, but they missed the revelation of their own nostalgia and desire as they cried out for the "real" Hester Prynne of Hawthorne's text, the beautiful but silent one who is fully clothed, her body forever imaginary. For example, one review calls Demi Moore's Hester a "pious hoochie-koochie girl" and the film a "superficial story [with] cardboard characters who are about as complex as the imbeciles on 'Melrose Place.'" Unable, seemingly, to stop himself, this reviewer goes on to say that Moore's Hester Prynne "turns the New World into the Nude World" and further, that "from the looks of Hester's chest, her doctor husband must be a plastic surgeon." Marjorie Rosen, in the *New York Times*, reveals even more (about herself) when she says that Demi "tripped up in 'The Scarlet Letter' and 'Striptease'" and thus, "might give her thoroughbred thighs and perky implants a rest" by doing a female film version of *The Nutty Professor* in the style of Eddie Murphy (37).

Some critics decry the loss of Hawthorne's classic story; for example, Jim Welsh in *Film and Lit Quarterly* calls the film "Roland Joffé's foolishly updated . . . version of . . . *The Scarlet Letter.*" Welsh goes so far as to call the film "an insult to literature of the highest order." He says, "Only an ignoramus would advise stuffy purists to 'lighten up' when confronting this film. Anyone who really cares about literature will be upset when an important novel is corrupted beyond endurance and almost beyond recognition" (299). And Rita Kempley of the *Washington Post* called it a "dumbed down version of Nathaniel Hawthorne's tale of sexual misadventure among the Puritans." Anthony Lane, in the *New Yorker,* also regrets the loss of Hawthorne's text when he says that "Roland Joffé's film is, in the words of the opening credits, 'freely adapted from the novel by Nathaniel Hawthorne,' in the same way that methane is freely adapted from cows." He goes on, after saying that he doesn't "object to films that take liberties," to say that there "is more suspense, more dramatic torque, in one page of Hawthorne's heart-racked ruminations on the Christian conscience than in all Demi Moore's woodland gallops and horizontal barn dancing" (114). Joyce Carol Oates in the *New York Times* calls the film "a backlash against every great American prose classic in which happy endings are denied"; she adds that the film "represents American filmmaking at its most spectacularly superficial" and Hester Prynne as changed "into a patronizing, predictable figure whose independence and single-mother feistiness would have been absurd in Hawthorne's theocratic, thoroughly patriarchal Puritan community: anyone who behaved as she does would have been broken, driven away, her baby taken from her."

Interestingly, these reviews echo the textual nostalgia at work in the reviews of the 1934 *Scarlet Letter*, a B-movie produced by one of the most reputable and respected B-movie studios, Majestic (which, a year later, merged with five others into Republic Studios; see Balio). This version of Hawthorne's tale, produced by Larry Darmour and directed by Robert Vignola, starred Colleen Moore as Hester Prynne and featured Hardie Albright as Dimmesdale. A review in *Variety* argued that, although the screenplay was well written (including its scenes of comic relief, which features added characters such as Bartholomew Hockins [played by Alan Hale], Abigail Crakstone [played by Virginia Howell], and Samson Goodfellow [played by William Kent]), and argued further that although the originary novel is "dismally dark . . . it is at least one of the most outstanding examples of early American writing and should not be tampered with to make it conform to the Hollywood tradition" (Scharnhorst 252). This reviewer, however, concludes by praising the acting in the film, saying that "[i]t would be difficult to imagine a more happy choice for Hester than Colleen Moore. Her work is informed by gentle humility which gives the part dignity and appeal" (Scharnhost 252–53). The *Hollywood Reporter* reviewer, however, cries out:

> Pity Nathaniel Hawthorne! The production that Majestic Pictures have given his . . . 'Scarlet Letter' almost succeeds in making the old classic ridiculous. . . . The fault lies primarily with the direction . . . and with the screen play by Leonard Fields and David Silverstein, which treats the old tale with no respect at all and won't even hand it some crutches when it falls down. (Scharnhost 251)[2]

The Women Who Have Played Hester Prynne in Films

Colleen Moore. Colleen Moore, the star of the 1934 film adaptation of *The Scarlet Letter*—like Lillian Gish and Demi Moore, who also play Hester Prynne and whom I will discuss later—can be seen in the context of her successful status as a woman actor. Like Gish, Colleen Moore's career first began in the "extras ranks" in Chicago (C. Moore 247–48), and then with D. W. Griffith's studio in 1917, when she was still a teenager. Although she never appeared in a Griffith film, she was given a six-month trial contract as a favor Griffith owed her uncle, "Walter Howey, the Chicago *Examiner* editor who helped him clear 'Birth of a Nation' and 'Intolerance' through the censors" (Stephan; see also MacCann 247). Her fame came after years "of playing supporting roles with Tom Mix and others" (MacCann 247), but "only after she had her hair cut and played a 'flapper' in *Flaming Youth*" in 1923 (MacCann 196), the same year "she married the first of her four husbands" (Stephan). Some film historians call her the "original screen flapper . . . who was making $12,500 a week . . . compared with Clara Bow's $2500" (MacCann 204). By 1926 and 1927, according to the Exhibitors Herald Box Office poll of 2500 theater owners, Colleen Moore was judged to be the female actor who brought in the most money (MacCann 9). She invested much of her salary in the stock market, later writing a book on investing, as well as marrying two stockbrokers (Stephan). Moreover, she made fifty-eight films in all—six sound films, the last of which was *The Scarlet Letter*, then "separated herself from Hollywood . . . [beginning] a super-elaborate hobby, known as Colleen Moore's Doll House, which . . . became a touring event to benefit crippled children" (MacCann 197).[3]

Moore's flapper status, which might have allowed her to be sexually objectified, is mitigated and contained by her comedic status; as Molly Haskell explains:

> [I]n the twenties, perhaps in defense of the assault on the "woman's domain" by silent comedy, there were a great many female comics of varying types . . . mimics, pranksters, and buffoons [such] as Colleen Moore, Gloria Swanson, Bea Lillie, Marie Dressler, Mabel Normand, Bebe Daniels, Clara Bow, and Marion Davies. . . . There was a

2. The 1926 film also received some negative reviews, particularly from *Photoplay*. "Their reviews of *La Boheme* (May, 1926) and *The Scarlet Letter* (Oct., 1926) were disgraceful" (Wagenknecht, *Movies* 237 n10). According to actor Louise Brooks, James Quirk of *Photoplay* waged a "war against Lillian Gish" (Wagenknecht, *Movies* 237 n10).
3. Mary Martin also played Hester Prynne as her last role—in the 1917 film version of *The Scarlet Letter*.

> tradition of the cutup or personality girl, who was more down to earth and (theoretically) less beautiful than the romantic heroine. They, too, were divided by sexual stereotyping into "good girls" and ["bad girls"] but the categories were less hierarchical . . . the "good girl" comedienne . . . actresses like Colleen Moore [particularly in the film *Ella Cinders*] . . . represent[ed] the true, back-home values that were being threatened. (62–65)

Richard Dyer MacCann adds to this description:

> Clara Bow, Joan Crawford, Colleen Moore, and Louise Brooks were the leading bobbed-hair "flappers" of the screen . . . acting out their freedoms in part by smoking, drinking and dancing the Charleston. They were modified vamps, perhaps, not menacing or mysterious like Theda Bara and her ilk, but open and, as they said, "fast." A good title writer could provide them with "wisecracks," and sex was obviously on their minds. (MacCann 137)

Moore was also protected from societal sexualization by her position as an oppressed, long-suffering wife of a manipulative, alcoholic, suicidal husband. In fact, the famous 1937 film *A Star Is Born*, a remake of *What Price Hollywood* (screenplay by Adela Rogers St. John), is "drawn in part from the life of Colleen Moore," as the tragic hero[ine]/top star of First National studio, married to its handsome, but alcoholic publicity man—later to be head of production, John McCormick (MacCann 186, 196).

In such a cultural and cinematic space, the choice of Colleen Moore as Hester Prynne is neither as surprising nor puzzling as some film historians have claimed. Like most films in the 1920s and 1930s, this film version of *The Scarlet Letter* confirms American values; during these decades, as Sochen explains, "[t]he movies did not become a daring medium in which new social ideas were explored or old ideas were challenged. Rather, in a continued effort to be popular and profitable, they confirmed the culture's values" (5). Even the opening credits include this statement: "This is more than the story of a woman—it is the portrait of the Puritan period in American life. Though to us, the customs seem grim and the punishments hard, they were a necessity of the times and [the beginnings] of the doctrine of a nation."

Colleen Moore's Hester Prynne assumes not just guilt and shame about Pearl, but all of the fault and responsibility. In a scene that does not appear in Hawthorne's novel, Dimmesdale visits Prynne shortly after the first scaffold scene, begging her to marry him. Hester, who is not distraught after her day's ordeal, but merely and melodramatically sad, refuses Dimmesdale, saying that "[m]y salvation and yours can come only from heaven." She then pats his hand, asking him to leave her alone. As another example, late in the film, the character of Roger Chillingworth, played by Henry B. Walthall (who also played Roger in the 1926 film *The Scarlet Letter*), tells Hester that the Puritan council has debated whether or not to allow her to take off the scarlet A; she replies, "Were I worthy to be rid of it, it would fall off of its own nature." In the last scene of the film—not of the novel—Dimmesdale's confession occurs on the scaffold on Election Day. As he begins to confess, Colleen's Hester rushes to quiet him, her admonitions echoed by Chillingworth, who of course desires Dimmesdale's silence. Dimmesdale confesses, nevertheless, exposing the A carved into his chest, collapses, kisses and blesses Pearl, and dies. A bell tolls. The community in unison remove their hats and bow their heads—fade to black as the film ends.

Lillian Gish. There is even less of a problem or puzzle in the choice of Lillian Gish as the star of the 1926 silent version of *The Scarlet Letter*. Gish, professionally molded by director D. W. Griffith, played Hester Prynne (not for Griffith, but for MGM) in the wake of her successes with Griffith in *Birth of a Nation* (1915), *True Heart Susie* (1919), and *Way Down East* (1920). It has been argued that Griffith's films often presented "typically Victorian fantasies of delicate, idealized girls tormented by brutish males" (MacCann 78). In Gish, Griffith "constructed a symbol of fragility and he delight[ed] in putting it in the direst jeopardy, fairly tantalizing the audience, capitalizing on the lurid fascination with peril" (Affron 52). In *True Heart Susie,* Gish played "a true-hearted country girl" (Naremore 77), and in *Way Down East* she portrayed a self-sacrificing, long-suffering character named Anna Moore, who is seduced by a scoundrel and abandoned during her pregnancy (Sochen 7; see also Lucas).[4] The audiences, especially the female ones, were said to have loved this silent, melodramatic film that focused on pregnancy and motherhood, represented as "woman's unique experience, burden, and pain . . . her agony, as well as the sign to the world of immorality, if the mother is without a husband" (Sochen 9). Such a character, who "appeared fragile, [but] was a very tough woman" (Sochen 6), has been categorized by film historians as a version of the Hollywood type, the "Girl-Woman . . . Lillian Gish's portrayal of Anna Moore in *Way Down East* and Hester Prynne in *The Scarlet Letter*, displayed th[is] . . . type: a victim pure of heart who is also strong and enduring, the melodramatic heroine" (Sochen 11–12)—somewhat like Colleen Moore's portrayal of Prynne, but without the comedic or latent flapper attitudes and their suggestion of freedom. As Lucas explains "Gish had the gift for externalizing [an] interior drama . . . [She had] the physical attributes and skill to convey softness, delicacy, and vulnerability readily; but she also suggests an implicit strength" (40)—or, as Haskell puts it, Gish was "delicate as a figurine but durable as an ox" (qtd. in MacCann 3).

By Hollywood standards, Gish was considered far more beautiful than Colleen Moore, which situates her Hester Prynne differently as well.

4. Gish left Griffith's production company after the completion of *Orphans of the Storm*; as Affron tells us:

> One of Gish's major accomplishments is the amount of personality and individuality she brings to the caricature [of the woman in peril], along with frightening dedication. *Orphans of the Storm* is the final step, where the pattern is pushed to its limit. Gish is introduced as the familiar wide-eyes ingenue . . . separated from her blind sister (Dorothy Gish), and Griffith fabricates for her the recognition/reunion scene par excellence. . . . Griffith spares no detail; photographing her in all possible positions, he fairly chortles over the scope of [the] final predicament. The logical conclusion of Gish's career with Griffith is the image of her head jutting out of that [guillotine] brace . . . she is offered up, as it were, to camera and posterity. If the blade doesn't drop in *Orphans of the Storm,* it does in the Griffith-Gish collaboration. There is little more to ask of an actress, but one shudders with apprehension over the outcome if it had. "Yes, Mr. Griffith, whatever you say." In their first film, *An Unseen Enemy* (1912), Lillian and Dorothy Gish were locked in a room and menaced by a gun that appeared through a hole in the wall. The dangers became more sophisticated and elaborate in the intervening eleven years, and they were punctuated with scenes of tenderness and reflection. Yet Lillian Gish had to leave Griffith to get out of that locked room, to get her neck out of the guillotine. She had to grow up. Her subsequent films in the twenties prove that a fascinating and challenging woman was lurking all the time in those pinafores and beneath those funny hats. The dedication, force, sense of the camera, and above all, the belief in form developed with Griffith were to be channeled by Gish into some of the most remarkable films of the late silent era. No other actress ever passed through a more grueling initiation to earn her stardom. (52–58)

Richard Dyer argues that even the cinematic techniques of *Way Down East, True Heart Susie,* and *The Scarlet Letter* function to make Gish the "supreme instance of the confluence of the aesthetic-moral equation of light, virtue and femininity with Hollywood's development of glamour and spectacle. She may also be its turning point," he goes on to say, because very "soon, the radiance of femininity came to be seen as a trap for men, not a source of redemption" (4). Gish's version of Hester Prynne was directed by Victor Sjöström (known in the United States as Seastrom) and costarred Lars Hanson as Arthur Dimmesdale. He was a great director, according to Edward Wagenknecht, but

> made the best showing . . . in his two films with Lillian Gish, *The Scarlet Letter,* and *The Wind.* Like Miss Gish herself, Sjöström was a great human being as well as a great artist; each instantly perceived kinship in the other and drew upon their own and each other's deepest resources, and they remained close friends to the end of Sjöström's life. (*Movies* 205–6)

Gish came to MGM, after "her two independently produced films, *The White Sister* and *Romola* . . . Despite the stamp of big studio lavishness and care in production, there is a non-MGM quality about *The Scarlet Letter,* as well as the last Gish-Hanson-Seastrom collaboration, *The Wind* (1928). Both films are deadly serious; neither is designed to appeal to the Saturday-night movie audience out for a thrill or a laugh or a cry" (Affron 66, 78)—unlike Griffith's films. Seastrom's *Scarlet Letter* focuses often on the public spaces of the Puritan community, alternately filling and clearing them of people and using them as the site of conflict and punishment. There is even a scene when Hester Prynne is pilloried that perhaps unintentionally invokes Gish on the brace of the guillotine in her last film with Griffith.

> There is a calm about Gish in these [public] moments that bespeaks her understanding of dramatic potential of situations in which the dynamics of mass pivots around the actress. The hysteria of Griffith's crowd scenes took on contour in the reach of [Gish's] body, which she stretched beyond possibility. The hush of *The Scarlet Letter's* Puritanism is drawn inside the actress's eyes and enfolding arms. (Affron 80)

As in many of her earlier films, Gish portrays a character who is "the angel of light," redeeming the male character's "carnal yearning" (Dyer 4).[5] Lars Hanson's Dimmesdale in *The Scarlet Letter* is the dark version of

5. Affron argues that in "projecting the [Hester Prynne] role's sexuality, [Gish] is abetted by Seastrom's invention and fervor. After the avowal of her previous marriage, Seastrom sends [Gish's Hester] out into the snow hysterical, and then caps the sequence with a strategically projected shadow upon her apron," evoking the film's "symbolic sexuality" (80). Another film critic, Raymond Durgnat, views the sexuality in this film similarly, as he discusses Gish's Prynne:

> Every evening she sits, very much alone, in her cottage, spinning, the starched white collar of her Puritan dress covers her breasts and the fire flings the whirling shadows of her spinning wheel across her apron. The contrast between her rich femininity and her loneliness; between the stiff dress and the tremulous delicacy of Lillian Gish; her placidity and industriousness; has . . . superb erotic solidity. (qtd. in Affron, 80 nl)

Like the critics of the other films of *The Scarlet Letter,* these critics may be saying more about themselves as spectators of Gish, caught in their own desire, than about the representation of Prynne by Gish.

carnal desire, saved by the light of Lillian Gish as Hester Prynne. In the scenes between them, the clothing of Gish, and the lighting behind her, envelope her to provide the associations with the strength of redemptive purity and femininity. In some of these films, Dyer claims, the male character, dressed in black, "rears up out of the darkness," but Gish's characters are "already in the light. That light comes from behind his head, magically catching the top of his hair but falling full on her face, itself an unblemished surface of white make-up which sends the light back on to his face" (4). Affron claims that, in fact, the success of *The Scarlet Letter* is

> guaranteed by Seastrom, by the integrity of his vision and the full use of decor, lighting, camera placement, and his sensitivity to landscape and texture. This 'Scarlet Letter' may not be pure Hawthorne, but it emerges with a stylistic unity that sustains the conflict between desire and society. (82–83)

Despite Colleen Moore's also being trained by Griffith for silent films, significant in assessing differences between her 1934 Hester Prynne and Gish's 1926 portrayal is that Gish's film is silent:

> In silent pictures the conception of acting was somewhat different . . . the players were artists by definition, "poets" who suggested through pantomime more than they actually stated. . . . "We are forced to develop a new technique of acting before the camera," Griffith wrote. "People who come to me from the theatre use the quick broad gestures and movements which they have employed upon the stage. I am trying to develop realism in pictures by teaching the value of deliberation and repose" . . . [an] artless conception of acting. (Naremore 77)

Gish's own version of Griffith's techniques show up most clearly when Hester Prynne is distraught, for example, when Pearl is ill in the prison and Chillingworth comes to visit:

> Gish transforms her face into a succession of symbols for despair. Such gestures as hands freely mingling with and blocking out areas of face are exemplary of Gish's ability and willingness to translate expression into disharmonious physical states. This is not overacting, but acting that is more figurative than purely life-like. Gish balances these two extremes of representation. . . . (Affron 82)

In these films, as Naremore argues, "Gish specialized in child women with a strong maternal streak—a description which already suggests some of the[se cultural and cinematic] oppositions [embodied in women that] she was able to contain" (79)—both in terms of the roles she played and her position as star, which in that system of filmmaking linked her inevitably to her roles (77). Blake Lucas explains,

> [T]he public would be lured by something they believed to be real in the star's personality; and, that aspect of the star's personality was usually seized upon and became the basis of the star's enduring popularity. . . . Whether by design or personal limitation, stars generally play roles within a relatively narrow range. (38)

Even as a never-married woman, Gish was able to maintain her incredible stardom in a culture that demanded that women marry. As Sochen says:

> The ideal of the New Woman, first announced during the 1910s as a harbinger of a radically different type of woman, did not fulfill its promise (or fear) in the 1920s . . . Indeed, the vital feminist movement of the first decade seemed dormant during the 1920s . . . The majority of both sexes still believed that the woman's primary adult role was as wife and mother; the majority still believed in chastity for females before marriage . . . [even when some] social realities [i.e., education, credit buying, urban living] were challenging those beliefs. (3–4)

Perhaps Gish's nonmarried status further emphasized the public's image of her as ethereally beautiful, pure and feminine, but strong and enduring.

Like Colleen Moore, Gish began her film career as a teenager, at first helped by Mary Pickford, also a child actor, to get jobs as an extra in films produced by Biograph, then, after she was introduced to D. W. Griffith by Pickford, as a featured actor (Fontana; MacCann 3; see also Sochen 5). Her film career, however, followed her earlier stint as a working child actor, along with her sister Dorothy, in their mother's act, which she began as a way to "support the family . . . where [the] restless father was frequently absent" (Fontana). Gish began working at age five because her father,

> unsatisfied and unsuccessful as the owner of a candy store, had left the family in Baltimore and gone to New York. Her mother tried to join him there, but found herself alone, making a poor living as demonstrator in a department store. She then turned to the theatre, where she made $15 a week. (MacCann 2–3)

Later, Gish toured with another acting company that "needed a child performer"; she was "entrusted to an actress friend" by her mother (3).

But unlike Colleen Moore and Mary Martin, who played Hester Prynne in the 1917 film, Gish did not leave Hollywood after the completion of *The Scarlet Letter*, except to take "the five-day train trip across the country to go to London to be with her mother" (MacCann 7). Gish had a long screen history and life, living until 1993, and she made the last of her more than eighty films in 1987, *Whales of August,* sharing the lead with Bette Davis. In fact, Gish stayed in the film industry most of her life, producing pictures at Metro that made money (quote from Loretta Young in *When Women* 41), and even directing one film—*Remodeling Her Husband* (1920)—although in

> the late 20's, [her] star began to wane [as] sound pictures became the rage with the viewing public. Lillian would resist the new sound pictures as she believed that silent pictures had a greater power and impact on audiences . . . [and she] was released by MGM . . . [making only two films in the 1930s]. In the forties she again appeared in a handful of "talkies" and received a Best Supporting Actress Oscar nomination for her role . . . in "Duel in the Sun" (1946). (Fontana)

Gish made three films in the 1950s and five in the 1960s. "In 1970 she received a special Academy Award 'for superlative artistry and distinguished contributions to the progress of motion pictures'" (Fontana). Gish's status in Hollywood allowed her always to claim "the right to make her roles over to suit" herself, as Edward Wagenknecht wrote in 1927 (qtd. in Naremore 77).

Demi Moore. Demi Moore's status in Hollywood is both similar to and different from that of these earlier cinematic "daughters" of Hester Prynne, though no less connected to her particular cultural moment. She is like them in that she enjoys immense box-office success: from 1981 to 1997, Demi Moore made thirty films and produced five. However, she is not a construction of the purified, rarefied, even comedic good girl; instead, she is an object of desire, often of derision, as well as an object of controversy in her subject-position as a strong-willed, capable woman. As one newspaper reporter put it, "It's no accident that Demi Moore is the highest-paid actress in Hollywood. You can tell by the way she walks . . . that she's a woman in control of her own destiny . . . She [now] runs her own production company and served as producer on *Now and Then*" (Uricchio 1). Moore thus follows in the footsteps of other, earlier female actors, such as Mary Pickford, who also set records for her salary and established her own studio, The Mary Pickford Company, an affiliate of Adolph Zukor's Famous Players Company. Moore's marriage to Bruce Willis and their status as doting parents has only minimally complicated the culturally constructed sexualization of Demi Moore, especially after *Vanity Fair*'s "attempt to boost sales with [its] . . . controversial cover photograph of the nude and heavily pregnant Demi Moore" (Francke 149).

Interestingly, with Colleen Moore, Demi Moore shares an interest in dolls and dollhouses; she has a collection of two thousand dolls (Cerio et al.). With Gish, she shares the experience of an unstable life as a child:

> Her father left her mother . . . before Demi was born. Her stepfather Danny Guynes didn't add much stability . . . either. He frequently changed jobs, made the family move a total of 40 times. The parents kept on drinking, arguing, and beating, until Guynes finally committed suicide. (Zoerner)

Moore has called herself a

> trailer park kid . . . [but p]overty wasn't Moore's only problem. At 12, she developed a crossed right eye, which required two operations to correct. At 14, she discovered that Guynes was not her biological father. . . . That revelation, and the constant drinking and fighting of her parents, set her adrift. "I got lost," she [said]. . . . "I had an essence in my life that I was nothing." (Cerio et al.)

Like both Colleen Moore and Gish, Demi Moore began working early; she "quit school at the age of 16 to work as a pin-up girl" (Zoerner). Another biography of Moore says that she left high school at age sixteen, "got a job at a debt collection agency, took her own apartment and did some modeling" (Cerio et al.).

By eighteen she was married to rock musician Freddy Moore, but her marriage lasted only four years.

> At 19 she became a regular on the TV show "General Hospital" (1963). From the first salaries she started celebrating parties and sniffing cocaine. That lasted more than 3 years, until director Joel Schumacher fired her from the set of *St. Elmo's Fire* (1985) when she turned up high. She got a withdrawal treatment and returned clean after a week . . . and stayed clean. (Zoerner)

Cerio et al. claim that after Moore was clean, she

> never looked back. Her mother, who has a long record of arrests for crimes, including drunken driving and arson, has not been so lucky. Virginia [Guynes] posed nude in 1993 for *High Society* magazine, imitating her daughter's *Vanity Fair* pregnancy cover. Through it all, Moore tried to get help for her mother, but when Virginia walked from a rehab stay Moore had paid for . . . Moore broke off contact with her.

Moore's consistent refusal to deny or apologize for her past, ambition, outspokenness, power, and sexuality also contextualizes her Hester Prynne as completely as those of Lillian Gish and Colleen Moore. Part of Demi Moore's context is, of course, the locating of male desire in and on her body rather than in men who objectify her through their desire and in women whose gaze at other women is also constructed by male desire. Unlike Gish or Colleen Moore, but like Hester Prynne, Demi Moore has been blamed for her own objectification, choices, and resistances; she has also, like Hester Prynne, been publicly demeaned. So, despite academic and media claims to the contrary, Moore seems quite the perfect choice for this role in the late 1990s.

Rereading The *(New)* Scarlet Letter

Significantly for this study, Demi Moore's status and sexuality are not contained in *The* (New) *Scarlet Letter.* In fact, the film's content and cinematic apparatus expose the attempts at containment in the earlier films, as well as the containment of a radical woman by Hawthorne in his novel and by "malestream" academic critics who have written about Hester-Prynne-as-woman while excluding women's scholarship. Director Joffé's insistence on Demi Moore as Hester Prynne recontextualizes the earlier cinematic versions and Hawthorne's text, and thus allows for a rereading of the actors Colleen Moore and Lillian Gish in relation to Demi Moore. Such a rereading destabilizes the picture of these actors as fully contained by directors, studio codes and rules, and social constructions, and recognizes them as women who were personally and professionally trapped inside the cultural economy of Hester-Prynne-ism, but also outside it because of their life choices, rebellions and resistances, public positions and status. These three women actors—Moore, Gish, and Moore—play Hester Prynne differently. Colleen Moore's and Lillian Gish's versions of Prynne participate in the cultural conditioning of women to accept all responsibility and blame and then to redeem men, yet did not reflect their own lives. Demi Moore plays Hester Prynne in ways similar to her own life, as she pursues her career goals, remains single-minded, and resists all attempts to contain her, her out-spokenness, or her self-styled sexuality. Joffé's casting of Demi Moore, then, inevitably rewrites Prynne, and thus, Hester-Prynne-ism.

In fact, Joffé's film rewrites most of Hawthorne's novel, and this rewriting has been the basis for much of the negative response to the film. Therefore, in order to reread the film against the critical responses and in terms of the cultural, psychological, literary, and cinematic gender issues

I have so far set up—particularly Hester-Prynne-ism—I must also examine *The* (New) *Scarlet Letter's* rewriting of Hawthorne's text through its representation of the Wampanoags (the Algonquin nation most closely associated with the Plymouth colonies), as well as its representations of Mituba (a mute slave woman), Harriet Hibbins, Mary Rollings, and a Quaker woman. The film's twentieth-century representations, which are suggestive of the historical contexts that informed American Puritan communities of the seventeenth century, have been attacked and demeaned by the critics of Joffé's film, in part because of their significant differences from Hawthorne's nineteenth-century relationship to the historical and social issues, incidents, and persons that shaped his representations. In examining and discussing these differences I am not claiming that Joffé's twentieth-century film is an accurate version of seventeenth-century historical and social conditions in the Plymouth colonies and that Hawthorne's nineteenth-century novel is not. Neither am I claiming that Joffé transcends his own time, and Hawthorne could not—or that Joffé cannot and Hawthorne did. Instead, my examination of the film allows me to destabilize and recontextualize Hawthorne's novel and his character Hester Prynne * * * particularly as his text assiduously avoids the contexts that Joffé's film articulates. Their differences cannot be simply explained by their differing personal, aesthetic, and political sensibilities, nor merely by twentieth-century artistic freedoms and nineteenth-century restrictions, although of course these issues inform the representations by Joffé and Hawthorne. Significantly for my project, though, I am not making claims about the intentions of either of these male artists—or deciding who is better or worse, but rather examining particular kinds of consequences for marginalized individuals and groups that are exposed by avoidances, absences, and uncontained excesses in artistic renderings. These discussions will allow me to make claims about the extent to which Joffé's film functions as a revisioning of Hester-Prynne-ism—whatever his intention, about the critical and cultural efforts to recontain what his film exposes, about the fear and desire that a new representation of Hester Prynne elicits, and about the ways readers/viewers of various versions of *The Scarlet Letter* have been constructed.

Rereading the film in its context of Native American history. By beginning his screenplay at the funeral of Massasoit—his death dated in various historical accounts as 1660, 1661, and 1662—followed by Hester Prynne's arrival in the Massachusetts colony in 1666 (rather than in the 1640s as in Hawthorne's novel), Douglas Day Stewart (known best for *An Officer and a Gentleman*) can represent the unstable environment within and without the Massachusetts Bay Colony. He can also imagine Prynne's and Dimmesdale's early connections, as well as Hester Prynne's strength and outspokenness (or as Dimmesdale puts it during their second meeting: "Your tongue knows no rules, Mistress Prynne"). Moreover, Stewart can set up an eventual avenue of escape for them during Metacomet's raid on the colony to free the imprisoned "praying Indians." His screenplay also depicts how the community reacts to Hester prior to her emergence from the prison onto the scaffold (the point at which the novel begins), how the Puritan communities are inevitably contextualized by their dependence on and relationship with the Algonquin nations, particularly the Wampanoags (a point the novel ignores), and how Hester Prynne suffers and endures (a point the novel eventually trivializes, as do earlier films).[6]

For example, in the opening scene of the film Dimmesdale and Governor Bellingham express to Metacomet their appreciation for earlier assistance by the Wampanoags and their condolences on his father's, grand sachem Massasoit's, death. Metacomet responds, through an interpreter—Johnny Sassamon, one of Dimmesdale's converted "praying Indians"—"My father should have let you die." He turns then to Dimmesdale and says, "You are the only one who comes to us with an open heart, but your people have murdered my father with their lies." One history of Native Americans tells us that the English who landed at Plymouth in 1620 would not have survived if the Indians had not saved them:

> A Pemaquid named Samoset and three Wampanoags named Massasoit, Squanto, and Hobomah became self-appointed missionaries to the Pilgrims. All spoke some English, learned from explorers who had touched ashore in previous years. Squanto . . . and the other Indians regarded the . . . colonists as helpless children. . . . By the time Massasoit . . . died in 1662 his people were being pushed back into the wilderness. (Brown 3)[7]

Another historical account of the Wampanoags adds information about the death of Metacomet's older brother, Wamsutta, renamed Alexander by the General Court at Plymouth. Wamsutta became grand sachem after his father's death, but the English, who "were not pleased with his independent attitude . . . invited him to Plymouth for 'talks.'" After one

6. Perhaps more than any of the other film translations of *The Scarlet Letter*, Rick Hauser's 1979 PBS production is guided by Hawthorne's text. As Hauser has said,

> Hawthorne's original keeps my feet on the ground. I repeatedly go back to his masterwork to check impressions, to gauge our collective success at creating his people and their world. His story, remains—a wry marvel, continually fascinating. What television viewers will see, of course, is one of innumerable *possible* "Scarlet Letters," this one filtered through the imaginative perception of the people who caused the thing to take the form it will have on the television screen. . . . Authentic we were not and could not be; faithful I dare to believe we were. ("Viewer's Guide")

7. Rick Hauser's 1979 PBS version of *The Scarlet Letter* imagined the Puritans' relationship to the wilderness differently—holding, one could say, the same historically romantic view as Hawthorne. As the "Viewer's Guide" to the film states:

> Nothing in [the Puritans'] experience had prepared them for the sight of a forest so thick and dark it threatened to extend endlessly and to harbor endless menace. Colonial diarists from Massachusetts to Carolina insisted on a single word—"howling"—to describe this wilderness at their backs. They seemed to mean by it something more than the cries of wolves and bears or the "outrageous roaring and whooping" of angry Indians: some profound subjective fear howled inwardly long after the range of objective dangers had been identified. And yet, despite such fears, New England's Puritans were uniquely prepared for this "vast and empty chaos."

This description, although mirroring canonical American literature's romanticized representation of life in the wilderness, is hardly more than another attempt at containment, failing to account, again, for the Native Americans' generous saving of the colonists from death and complete destruction. Also, the Puritans' inner turmoil and fears—which were immense, I agree—were considered more significant to discuss than their tendency to project these fears onto the external landscape, rather than, as this account does, assume their ability to separate themselves from and identify "the range of objective dangers." Moreover, this "range of objective dangers"—primarily caused by the Puritans' insensitivity to the land and to the human beings who inhabited it, consistently refusing to see the Native populations as anything more than howling, whooping, roaring, angry savages—is ignored in Hauser's film version, as it is in the novel.

of the meals offered during the "talks," Wamsutta died. "The Wampanoag were told he died of a fever, but the records from the Plymouth Council at the time make a note of an expense for poison 'to rid ourselves of a pest.'" The English were not much happier with Metacomet, "King Philip," when he became grand sachem, although they did not poison him.

> Philip does not appear to have been a man of hate, but under his leadership, the Wampanoag attitude towards the colonists underwent a drastic change. Realizing that the English would not stop until they had taken everything, Philip was determined to prevent further expansion of English settlement, but this was impossible for the Wampanoag by themselves since they were down to only 1,000 people by this time. Travelling from his village at Mount Hope, Philip began to slowly enlist other tribes for this purpose. Even then it was a daunting task, since the colonists in New England by this time outnumbered the natives better than two to one (35,000 versus 15,000). Philip made little attempt to disguise his purpose, and through a network of spies (Praying Indians), the English knew what he was doing. (Sultzman)

Running Moose, renamed Johnny Sassamon, Dimmesdale's friend and primary convert in the film, also has an historical counterpart in John Sassamon, "a Christian Indian informer" whose murdered body was discovered in January 1675:

> Three Wampanoag warriors were arrested, tried for the murder, and hanged. After this provocation, Philip could no longer restrain his warriors, and amid rumors the English intended to arrest him, Philip held a council of war at Mount Hope. He could count on the support of most of the Wampanoag except for those on the off-shore islands. For similar reasons, the Nauset on Cape Cod would also remain neutral, but most Nipmuc and Pocumtuc were ready for war along with some of the Pennacook and Abenaki. The Narragansett, however, had not completed preparations and had been forced to sign a treaty with the English. (Sultzman; see also Gussman's parable of "A Good Indian's Dilemma")

By mid-1675 the tensions between the colonists and the Native Americans increased. Metacomet's men began to kill livestock near Swansea, "convincing many that an assault was imminent. After a Swansea boy killed a warrior, the Indians attacked the town" (Mandell 375). "King Philip's War" escalated and included an alliance of Native nations; even the Narragansetts "joined the uprising after English forces attacked their village" (375). When, overpowered by the increasingly effective colonial forces and ravaged by disease and hunger, the alliance began to break apart by the spring of 1676, Metacomet "headed for home after his allies threatened to send his head to the English as a peace offering" (375). In August, he was killed by colonial forces and beheaded; his head was "exhibited in the fort at Plymouth, Massachusetts, for twenty-five years" (374).

These accounts are pertinent to the film's narrative because they discuss the general historical context that was used to shape *The* (New) *Scarlet Letter's* specific narrative action. For example, although Johnny Sassamon is not killed in the film, Roger Prynne (Chillingworth) commits a murder and attempts to make it look as though it was committed by Indians. Chillingworth mistakenly assumes he is killing Dimmesdale, as well as mistakenly assuming he is functioning like a Native warrior. Instead, Duvall's Prynne/Chillingworth has "gone native" in the worst sense of that phrase; that is, Prynne releases his own repressed savagery, naming it (and blaming it) on his association with the Tarantines and the Wampanoags. In one particularly memorable scene, Prynne so completely loses his inhibition and dances so crazily with a dead deer on his head during a Native ceremony that it frightens the Wampanoag. One of the Native women says, "He has a ghost in him," and another answers, "He'll bring us bad luck. Send him home." But, as a consequence of the murder (and scalping of the victim), the "praying Indians" who live inside the confines of the Puritan community are accused of savagery and murder and imprisoned. Dimmesdale sends Johnny to urge Metacomet to free them. In the ensuing skirmish between the Puritan soldiers and the Wampanoag warriors, Dimmesdale, Hester Prynne, Harriet Hibbins, and the other women are saved from hanging.

Hawthorne and the Native Americans: Containment and Excesses

Hawthorne lived in Salem, Boston, and Concord, and spent time as a child at Nahant, a nearby seaside resort, as well as in Raymond, Maine, with his uncle Richard Manning. Although he seems to have developed a deep respect for nature, he, like most nineteenth-century white, privileged Americans, does not appear to have had extensive knowledge of the histories and struggles of the native populations who had been defeated in order for him to roam freely in nature and for his family to own property in these locations. He seems locked in a romanticized notion of Native American freedom from all restraints, in contrast to the restrictiveness of Puritan life. Hawthorne did not acknowledge Native Americans to any great extent in his narratives (see Barnett, *Ignoble*),[8] although he often included Puritan historical figures and incidents that shaped his stories and informed his characters' lives. In other words, his historical context was primarily white and colonialist. When Native Americans appear in Hawthorne's narratives, they are most often background figures steeped in the sensationalized fear of the "savage," "barbarous," "heathen" Other, even if and when Hawthorne may have been problematizing such racist labels. For example, in Hawthorne's "Mrs. Hutchinson"—first published in 1830 and the tale most often assumed as precursor to *The Scarlet Letter*—his narrator sensationalizes Hutchinson's death, along with a group of like-minded colonists, at the hand of Indians:

> Her final movement was to lead her family within the limits of the Dutch Jurisdiction, where, having felled the trees of a virgin soil, she became herself the virtual head, civil and ecclesiastical, of a little colony. . . . Her last scene is as difficult to be portrayed as a shipwreck, where the shrieks of the victims die unheard along a desolate sea, and a shapeless mass of agony is all that can be brought home to the imagination. The *savage* foe was on the watch for blood. Sixteen persons assembled at the evening prayer; in the deep midnight, their cry rang through the forest; and daylight dawned upon the lifeless

8. See Louise Barnett's *Ignoble Savage* for her careful analysis of the narratives in which Hawthorne includes Native American characters or references to Indian life.

> clay of all but one. It was a circumstance not to be unnoticed by our stern ancestors, in considering the fate of her who had so troubled their religion, that an infant daughter, the sole survivor amid the terrible destruction of her mother's household, was bred in a *barbarous* faith, and never learned the way to the Christian's Heaven. (24, emphasis mine) [163]

In Hawthorne's *Scarlet Letter,* the Native Americans are a shadowy presence—staring with "their snake-like black eyes" (*SL* 246) [146], more absent than present, appearing only occasionally on the fringes of the narrative action—as also, for example, in "Roger Malvin's Burial," "The May-Pole of Merry Mount," "Old Ticonderoga," and "Young Goodman Brown," where the focus is on the whites in relation to Indian battles or incidents. As Louise Barnett argues, although

> Hawthorne usually sees the Indian as one of the many victims of Puritanism. . . . Like other Americans, [he] could not repudiate white conquest. In his history of Salem, he writes: "Even so shall it be. The payments of the Main Street must be laid over the red man's grave." The same sketch encapsulates in the figure of a drunken Indian "the vast growth and prosperity of one race, and the fated decay of another." (*Ignoble* 155)

Lucy Maddox says that at the end of *The Scarlet Letter* "Hester chose submission and assimilation, and the 'race' of American womanhood has flourished; the Indians refused assimilation, and their race is, in Hawthorne's view, well on its way to predictable extinction" (120). Barnett reminds us, too, of Hawthorne's words in "The Old Manse":

> Finding an arrowhead in the vicinity of the Old Manse establishes a rapport between Hawthorne and the red hunter whose hand touched it centuries ago: "Such an incident builds up again the Indian village and its encircling forest, and recalls to life the painted chiefs and warriors, the squaws at their household toil, and the children sporting among the wigwams, while the little wind-rocked papoose swings from the branch of the tree. It can hardly be told whether it is a joy or a pain, after such a momentary vision, to gaze around in the broad daylight of reality and see stone fences, white houses, potato fields, and men doggedly hoeing in their shirt-sleeves and homespun pantaloons. But this is nonsense. The Old Manse is better than a thousand wigwams." (*Ignoble* 156)

That Native Americans appear at all in Hawthorne's texts—that he found it necessary to romanticize and appropriate them—is evidence of the kind of excess I am discussing, an excess represented by the Other that most American historical and fictional narratives have attempted to contain through omission, disregard, stereotyping, and marginalization. As Julia Kristeva and others have argued, various texts contain "the *excess* of meaning that constantly threatens to disrupt the boundaries of . . . defined identities and expose the fiction of any imposed 'truth'" (Morris 138, emphasis mine). Such containments have functioned to normalize the narrative of white, male Americanness * * *, making it seem progressive or radical when an author like Hawthorne bothers to include references to Native Americans—or as Barnett explains, even "conjures up the Indian

Wappacowet" in "Main Street" (*Ignoble* 157)—or when he focuses on a woman condemned by the dominant Puritan community, such as Anne Hutchinson or Hester Prynne.

The omissions and containments are repeated in much of the critical mainstream's historical contextualizings of *The Scarlet Letter*, which consists primarily of additional Puritan history or that of other white settlers in Massachusetts. The history of the Other, like women's history and their scholarship discussed throughout this book, has been generally omitted, although sometimes the Antinomians and witch trials are mentioned—not in terms of the women who were victims of that particular cultural moment, but, instead, as the cultural hysteria implicating Hawthorne's ancestors, such as William Hathorne, a judge who persecuted Quakers, and John Hathorne, a judge involved in the witch trials in Salem in 1692. Even in textbooks used in classrooms—for example, St. Martin's "Case Studies" of *The Scarlet Letter*—the section entitled "Biographical and Historical Background" discusses Puritan history as the novel's context, but fails even to mention the Puritan attempts to convert to Christianity the indigenous populations of America, whom these Puritans called "savages." The "Viewer's Guide" to the 1979 PBS film of *The Scarlet Letter*, directed by Rick Hauser and primarily shown in classrooms, gives extensive Puritan history as "background" to the textual narrative and the film, explaining religious, social, and family life in New England with almost no references to the Other.[9]

9. Rick Hauser's 1979 film maintained its focus on what was interpreted as Hawthorne's ability to recognize the "connections between the enforced uniformity of his ancestors' lives and the bland conformism of his contemporaries." The "Viewer's Guide" for the film thus suggests about the Puritans' relationship to nonconformists the following:

> not everybody in the colony qualified as a brother [or sister?]. Those who did not conform—Quakers or atheists or deviant Puritans like Anne Hutchinson—were considered subversives. Having barely escaped the persecuting English Church, the authorities of the Massachusetts Bay proceeded to persecute every dissenter stubborn enough to speak out. They set Catholic sympathizers in the stocks, lashed Quakers through the streets before banishing them from the territory and executed heretics in formal public hangings. From our vantage point the men and women who established Massachusetts in the name of "holy liberty" can seem arrogant fanatics, pure and simple. It takes a great deal of effort of the imagination to see them as they saw themselves: engaged in a sacred experiment to live and worship perfectly and to make their Commonwealth the saving model for a Christian universe.

It is this kind of demand for continuous focus on the intentions of the persecutors and oppressors that I have been discussing * * * because, in so doing, the focus on those who have been oppressed has been so minimal as to be distorted when presented. While I am not arguing that we forget or lose this dominant historical heritage, I propose that we *also* listen to and tell the stories and histories of those who were oppressed, erased, dominated, even destroyed, by patriarchal groups like the Puritans of New England. Instead, the "Viewer's Guide," like almost every class I had in which *The Scarlet Letter* was taught—from high school through graduate school—encouraged me to "forget" my own perspective in relationship to this novel and its oppressed, explicit in Hester Prynne and implicit in the absence of other Others, and try my best to imagine how the Puritan patriarchs felt. I was encouraged to "understand" them and their intentions, not in addition to my own or those of Others, but in place of them. At best, my various fellow students and I were asked to think about how "we" were still influenced by the Puritans—although in no way connected with the oppressed. The "Viewer's Guide" for the 1979 film concurs with this *view* and *guides* its viewers as follows: "And, however different from us the Puritans at first appear, we might inquire in our time, as Hawthorne did in his, how the Puritan legacy still colors America's idea of itself." While a worthwhile goal, it is *how* we inquire, *how* we define "America's idea of itself," and *who* we focus on in such an inquiry which I have been trying to question * * *. Put more simply, it is what assumptions we find when we conduct such inquiries that I have been encouraging. If we continue to come to the same answer, generation after generation, yet claim progress and change, then there is no real inquiry, other than the kind always already guided by a GUIDE, always already steeped in the assumptions which have guided and shaped us. Roland Joffé's film of *The Scarlet Letter* functions in terms of the consequences of inquiries made by the production staff, writers, and actors—guided by expert Others, including Native American consultants to the film—into the histories and intentions of those oppressed by Puritan hegemony and focuses our gaze on the *consequences*, rather than the intentions, of their legacy.

Moreover, this film's depictions of Native Americans further calls into question Hawthorne's romanticizings and omissions in relation to nineteenth-century white America's responses to the "Indian question"—a reductive phrase that supposedly considered the lives and futures of Native Americans. One of the answers to that "question" was calculated genocide and another was the policy of removal. For example, in the 1830s, just a decade before Hawthorne wrote *The Scarlet Letter,* the "Five Civilized Tribes"—the Cherokees, Choctaws, Chickasaws, Creeks, and Seminoles—were forcibly removed "from their ancient homeland in the East to present-day Oklahoma" (Hoxie 639). The removal, known as "The Trail of Tears" because thousands died of exposure and smallpox, began in 1831 and continued throughout the decade, although the removal of the Seminoles, who "fought when federal authorities insisted they honor [a] fraudulent treaty," was not completed until 1859, when the "last band of Seminoles were forced westward in chains" (Hoxie 640). By 1850, the year of publication of *The Scarlet Letter,* all land east of the Mississippi, as well as in the states that formed the western borders of that river, had been transferred to the U.S. government (Edmunds 291). These few examples barely touch the extent of the savage and brutal treatment of Native Americans during Hawthorne's lifetime, but which was ignored or considered necessary by most of white America, including Hawthorne. By denying the savagery perpetrated on millions of Indians—and instead projecting savagery onto them—white America could avoid accountability and responsibility.

In the film *The* (New) *Scarlet Letter* the issue of savagery is explored through various namings and manifestations; for example, Thomas Cheever, newly arrived from England, says to Dimmesdale: "So like home. And yet beyond those trees, I suspect a savage land with savage passions, dark and untamed." More significantly, the issue of savagery is focused, as I earlier mentioned, on the character of Roger Prynne/Chillingworth—chillingly portrayed in terms of his most evil potential and worth by Robert Duvall. The film's version of Chillingworth represents the excesses that are implicit in the novel's version of him. Lucy Maddox argues that in the novel, in the Election Day scene, Hawthorne equates Native Americans with the "sun-blackened sailors." According to Maddox, quoting Hawthorne, as

> "wild as were these painted barbarians, [they] were [not] the wildest feature of the scene. This distinction could more justly be claimed by some mariners,—a part of the crew of the vessel from the Spanish Main." . . . These sailors are "the swarthy-cheeked wild men of the ocean, as the Indians were of the land." . . . The very color of the . . . sailors in this scene links them with the darkskinned Indians in what is apparently for Hawthorne a sinister opposition to the whiteness of Puritan Boston, since the dark physiognomies are an index of the moral nature of both the sailors and the Indians. (122) [138]

Maddox argues further that Hawthorne also links the Indians and sailors with Chillingworth, "whose own complexion has darkened from dusky to black as he has pursued his campaign of revenge against Dimmesdale. . . . Chillingworth has apparently learned two things during his Indian captivity . . . herbal medicine, and the pleasures of revenge" (122–23). Thus, Chillingworth's cold, calculating obsessiveness, his complete lack of compassion, his desire for revenge, and his constant surveillance of

Dimmesdale in the novel are characteristics and actions that attempt to contain, even hide, his underlying anger, resentment, bitterness, loss, fear, and desire, and his potential for brutality. All are exposed in the film as physical, rather than only mental, actions, delinking him from Native Americans. It is as though the film literalizes Hawthorne's words: "a terrible fascination, a kind of fierce, . . . necessity seized the old man within its gripe, and never set him free again, until he had done all its bidding" (*SL* 129) [81].

Hawthorne's own fears of becoming isolated—of functioning only in terms of a morbid curiosity and without sympathy, of committing the "unpardonable sin"—are well known. His series of male characters (besides Chillingworth, for example, Hollingsworth, Ethan Brand, Rappaccini, Judge Pyncheon, Westerfelt, Wakefield, Aylmer, the painter in "The Prophetic Pictures," Peter Hovenden, and to a lesser extent Coverdale, Clifford Pyncheon, Holgrave, Kenyon, Reverend Hooper, Young Goodman Brown, Owen Warland, Drowne, Giovanni Guasconti, even Dimmesdale) are, then, exposed as representations of the excesses that threatened Hawthorne, ones that he attempted to theorize and contain, particularly by creating these male characters with sensibilities varying along the spectrum from sympathetic and involved to cold and calculating (see also Amy Louise Reed). The excessiveness of his worries and fears about failures of sympathy and human engagement were perhaps also recognition at some level—certainly not consciously—of the immensity of his, and the rest of white America's, denial of its failure of common humanity, of its ongoing commission of the unpardonable sin in the name of democratic ideals.

Rereading the Film in the Context of Slavery and Witchcraft

The absent presence of slavery in American literature. Containment in relationship to the slavery of Africans was also most severely practiced in canonical American literature, including that written by Hawthorne in the nineteenth century. In seventeenth-century America, the time of Hawthorne's *The Scarlet Letter*—as well as in Hawthorne's own time—slavery was the legal and social norm. In fact, the first Africans—three of them women—were brought to Jamestown, Virginia, in 1619 and sold as slaves (Davidson and Wagner-Martin 950). In the year of publication of *The Scarlet Letter*, 1850, the Fugitive Slave Act was passed. This act, which confirmed the legal condition of slaves as chattel, allowed slave owners or their representatives to enter free states and seize freed and escaped slaves. Only two years later, Hawthorne's friend Franklin Pierce was elected as president.

> Hawthorne did his share by writing the official campaign biography, in which he extols Pierce as "the statesman of practical sagacity—who loves his country *as it is*, and evolves good from things *as they exist*"—and he defends Pierce's support of the Fugitive Slave Act. (Bercovitch, "Hawthorne's A-Morality" 7)

In the campaign biography, called *Life of Pierce*, Hawthorne explains that slavery is

> one of those evils which divine Providence does not leave to be remedied by human contrivances, but which, in its own good time, by some means impossible to be anticipated, but of the simplest and easiest operation, when all its uses shall have been fulfilled, it causes to vanish like a dream. (qtd. in Bercovitch, "Hawthorne's A-Morality" 8). [220]

As Bercovitch argues, "Only the security of commonplace could allow for this daring inversion in logic, whereby slavery is represented, symbolically, as part of the 'continued miracle' of America's progress. Like the scarlet letter, Hawthorne's argument has the power of a long-preserved cultural artifact" (8).

Yet slavery is never mentioned in Hawthorne's novel, and no slaves or black characters appear, even as shadowy background figures like the Native American characters. As Toni Morrison argues,

> For some time now I have been thinking about the validity or vulnerability of a certain set of assumptions conventionally accepted among literary historians and critics and circulated as "knowledge." This knowledge holds that traditional, canonical American literature is free of, uninformed, and unshaped by the four-hundred-year-old presence of, first, Africans and then African-Americans in the United States. It assumes that this presence—which shaped the body politic, the Constitution, and the entire history of the culture—has had no significant place or consequence in the origin and development of that culture's literature. Moreover, such knowledge assumes that the characteristics of our national literature emanate from a particular "Americanness" that is separate from and unaccountable to this presence. There seems to be a more or less tacit agreement among literary scholars that, because American literature has been clearly the preserve of white male views, genius, and power, those views, genius, and power are without relationship to and removed from the overwhelming presence of black people in the United States. This agreement is made about a population that preceded every American writer of renown and was, I have come to believe, one of the most furtively radical impinging forces on the country's literature. The contemplation of this black presence is central to any understanding of our national literature and should not be permitted to hover at the margins of the literary imagination. (*Playing* 4–5)

In Joffé's *The* (New) *Scarlet Letter* slavery is made evident in the character of Mituba, a mute slave woman. From the slave narratives by women, as well as the historical, critical, and theoretical work done on them, we have learned that

> the institution of slavery subverted and deformed all aspects of black female life in the attempt to colonize the African woman as worker and as producer of workers in a grand experiment of transportation of free labor . . . they were obliged to take on the role of "surrogate men" and to become "breeders." (Scott 814)

Early in the film, after Hester Prynne has *chosen* her land and house on the cliffs overlooking the sea (rather than, as in the novel, forced to live away from the community), she buys the time of two indentured men who can help her clear her land and plant crops. The trader, after insulting Hester

Prynne about her gender, offers to "throw in" Mituba as part of the deal. When Prynne objects, saying that Mituba is a slave, the trader explains how she "don't speak, if that be a problem—born that way." Demi's Prynne decides to take Mituba, who is not only mute, but cowering and fearful. Although there is no further direct dialogue about Mituba's condition, the next time we see her, she is no longer cowering or dirty, and, later, we see that she lives in the house with Hester Prynne, both of them sharing the domestic duties, as well as sharing the farming duties with the two indentured male servants. The film, then, seems to be working against another recurrent problem of slavery—that slave women "had no protection and little pity from their mistresses. The slave woman became the scapegoat in the domestic politics of the master" (Scott 816). Whatever the film's partial successes in exposing earlier attempts at containment, as Bercovitch argues, in Hawthorne's *The Scarlet Letter*, "[n]o doubt, the overall tendency is toward evasion" ("Hawthorne's A-Morality" 8) of the issue of slavery.

Female sexuality and desire in relation to Mituba and Hester. In addition to repressing slavery and racial and gender opposition as the shaping force of canonical American literature and its national narratives and characters, that literature—and for my purposes here, specifically Hawthorne's *The Scarlet Letter*—has also repressed female sexuality as feared and desired (and therefore rendered taboo), * * * . In the film female sexuality is represented and played out in relation to both Hester and Mituba—once, in a scene I will discuss more fully later, Hester's bathing and mirror scene, and, again, during the love scene between Hester and Arthur. In the first scene, Mituba stands with a container of hot water, prepared, yet afraid, to fill Hester's bathing tub (already part of the community's myth about Hester; one of the women sarcastically says when she hears of the bathing tub: "What is she? French?"). In answer to Mituba's fear, Hester explains that it is only a bathing tub, not a "toy of Satan."

While Hester bathes, Mituba watches through the keyhole, although not voyeuristically. She sees Hester's desire ignited as she thinks of Arthur and examines her own reflection in the mirror. Later, when Dimmesdale and Prynne make love for the first time—after receiving word that Roger Prynne has died during a Tarantine attack on his ship (although his body is not recovered)—Mituba is aware of them in the barn. She uses the time to bathe in the water that she has drawn for Hester. The scene moves between the lovemaking of Hester and Arthur and Mituba's eroticized bath. By using extremely close-in camera shots, Joffé avoids objectification of their bodies and refuses to offer the spectator voyeuristic opportunities and satisfaction, yet he also manages to represent something of female desire and sexuality, both denied by the Puritan society and condemned, misunderstood, or reinscribed as male desire in nineteenth-century America, and ever since.

By showing that Hester's and Mituba's desire and sexuality are similar and strong, Joffe's film also exposes the received belief of slaveholders and citizens throughout the history of slavery in the United States who claimed slave women to be, like Native American women, "savage" and "oversexed"—different from white women. This received belief of course justified slave owners' systematic rapes of slave women, placing the blame for the rapes on the women's sexuality. Joyce Hope Scott adds to this abusive picture:

> Generally, slave plantations were structured in such a way as to reflect a type of extended family. The master served as master-father to his own family as well as to his slaves, with his wife assuming the role of mistress-mother. Given their precarious positions within such a "dysfunctional" family, slave women experienced incestuous rape, child abuse, and neglect. (816)

Mituba and Tituba. Mituba's presence in the film calls to mind a particular slave woman, Tituba—

> sold in Barbados to the Reverend Samuel Parris, and brought to the American colonies to serve his family. Alleged to have bewitched the minister's daughter and his niece, causing the children severe fits of hysteria, and initiating the accusations, interrogations, and executions of the Salem witches, Tituba has become inseparably linked with the beginning of this notoriously odious chapter of American history. (Dukats 325–26)

Reconstructing Tituba's silenced story, Maryse Condé's first-person narrative *I, Tituba*

> has inserted extracts from Tituba's official deposition—the one safely deposited for posterity in the Essex County Archives in Salem, Massachusetts. Ironically, this recontextualization of Tituba's interrogation and of the words she is recorded as having spoken, serves to underscore her voicelessness—her lack of access to a voice at the very moment when she is called upon to speak. (Dukats 326)

As Dukats says of Condé—that she "allows Tituba a new hearing" (327)—one could say of this new film version of *The Scarlet Letter*, as it allows for the visual representation of the muted history of a slave woman, particularly of one later implicated in witchcraft, as Mituba will also be. Representing Mituba as mute in the film could, of course, be read (as some critics have) as evidence of Joffé's further silencing. In my rereading I see it instead as exposing and depicting the attempts at historical silencing of slave women. It exposes as well attempts by canonical authors like Hawthorne to contain the significance of slavery and racial opposition—though repressed and misrecognized by whites—as allowing for the creation of a self-narrative of the (so-called) white race in the United States. As I argue in my introduction, the space for the creation of self-narrative of whites in the United States has been in the opposition between their freedom and the enslavement of Africans, and the space for the creation of the self-narrative of white men has been in the opposition between their public power and women's domesticity and reproductive capacity. That is, without the opposition of the Other, always tautologically defined as the absence of the privileged defined, the privileged defined cannot be defined—nor its narratives told and retold * * *. As Dukats argues, quoting from Morrison's *Playing in the Dark*, "the Africanist presence has provided the 'arena for the elaboration of the quintessential American identity" (329)—whatever century has been depicted.

Dukats goes on to say that "*I, Tituba* presents itself as a corrective for historical oblivion" (329). In her novel, Condé incorporates Hester Prynne into the narrative, saying that Hester "enabled [her] to think about the way that her heroine, Tituba, had enabled Hawthorne to produce the 'true beginning of American prose fiction.' Condé thus returns to this canonized 'point of origin . . . in order to reclaim Tituba's part in this origin, and to reaffirm Tituba's presence in what inevitably becomes a 'wider landscape,'" following Morrison (Dukats 331). In *I, Tituba*,

> Tituba strives to follow Hester's recommendations and give the magistrates "their money's worth" . . . describing to them exactly what they believe, even if it has nothing to do with the truth. Hester thus advises Tituba to legitimate the beliefs of her persecutors, and not without good reason, because, by law, the life of a witch was spared if she confessed. By confessing then, Tituba escapes death, yet legitimates the authority of her oppressors. (331)

The opposite occurs in the film for Mituba. After delivering a note to Dimmesdale from Hester and while returning with a note for Hester, Mituba is stopped in the woods by Chillingworth. He slaps at her, verbally tortures her, and then takes her to the Puritan ministers who interrogate her about Hester Prynne's activities and about the presence of the devil—that is, women's sexuality and desire. The interrogation is brutal, motivated by their desire for titillating information, although their self-narrative concerns the discovery and elimination of witchcraft in their community. In fact, they have interpreted all of the community's problems, at the suggestion of Chillingworth—harsh weather, bad crops, Hester Prynne's adultery, threats of Native uprisings—as evidence of witchcraft. Joffé's depictions, however, leave no doubt of the desire and the sexual excitement felt by these ministers as they overpower Mituba until she confesses. Of course, since Mituba is mute, they must interpret her signs, putting words into her mouth, as it were, as they hover over her. One says, "[the devil] made you strip naked before him," and the other cries out, in obvious excitement, "Totally naked." Finally, Chillingworth stops them as they sweat, pant, and yell (reminiscent of the Senate committee at the Thomas/Hill hearing).

Mituba later consents to meet Chillingworth when he summons her, because she wants to undo any damage she might have done to Hester Prynne and because she is called as a witness against Mistress Hibbins, known in the film as Harriet Hibbins (played by Joan Plowright), who is seized as a witch. Chillingworth, however, kills Mituba; her "confession" to Chillingworth and the Puritan ministers ensures her death. The ministers and magistrates, as well as the community—hysterical by this point in the witch trial—interpret Mituba's death as further evidence of witchcraft. The fate of Hibbins is sealed, though, when at that moment Chillingworth, who knows of Pearl's birthmark on her stomach, exposes it as a witchmark—the consequence, he says, of Hibbins's midwifing at Pearl's birth. As Marjorie Pryse says of the town's interpretation of Hester in the novel: they "disregard Hester as a symbol of human imperfection . . . and view her instead as the actual flaw. She possesses their 'birthmark,' which must be revealed, in an attempt to erase it in themselves" (23). Pearl is also marked, becoming the flaw that must be erased.

Pearl's birth and her mother; Una Hawthorne and her father. In the film, there is a rather remarkable and lengthy depiction of Pearl's birth in the prison, where Hester has been held without examination or trial for five months; this scene functions as another exposure of the excesses almost contained in the novel. The painful birth in filthy conditions is attended only by Hibbins, who is both solicitous and efficient; she tells Hester that she must have a "will of iron." During labor Hester cries out to ask Harriet if God is punishing her. How simple it has always been for readers not to consider what Hester must have faced during her imprisonment and childbirth, or any real woman who faced similar circumstances in the Puritan colonies. As Delese Wear explains, "Women, especially midwives . . . attended deliveries. The patriarchal society, church, and government, which had assigned strict domestic roles to women, regarded pregnancy as a private women's matter" (111). Wear continues,

> [T]he presence of pain during labor represented to men and women alike an infliction by God for women's assumed perdition and moral frailty. As Cotton Mather's solemn and frightening words to expectant mothers suggest, childbirth was a serious travail with high infant mortality rates: mother might "need no other linnen . . . but a *Winding Sheet*, and have no other chamber but a *grave*, no neighbors but *worms*." Private writings show an acceptance of pain as punishment and the fear and terror it inspired. (111)

Anne Bradstreet, though, in her poetry shows that the pain of birth "can't be told by tongue." For centuries, even women writers like Anne Bradstreet accepted the supposed sinfulness of sex and the guilt imposed on them by unrelenting patriarchs (Wear 111).

Although Hawthorne was confined by social values and rules, it is unlikely, given his temperament, fears, sexual guilt, "addict[ion] to passive enjoyment" (Herbert, *Dearest* 123) and intense focus on sensitive, creative, intelligent, fearful, guilty (in other words, super-ego-driven) men like Dimmesdale (like himself) that he would have, even without the social constraints, included the scene of the birth or the sex between Dimmesdale and Prynne that necessarily preceded the pregnancy. In fact, Herbert argues in *Dearest Beloved* that the "Hawthornes' marital relation bears a striking analogy to the adultery portrayed in *The Scarlet Letter*" (115). Herbert also explains that the

> erotics of middle-class purity were inherently precarious, and for the Hawthornes the characteristic tensions were unusually strong, so that Sophia envisions sexuality as an all-but-forbidden fruit of wonderful deliciousness that is available only when the purity of the sexual partners is absolute. . . . Sophia believed Nathaniel as pure as she herself, having a communion like her own with absolute right, which canceled merely private will. . . . Nathaniel's insistence on Sophia's absolute pond-lily purity aided him in overcoming his dread of sexual pollution through her; and her insistence on his 'heavenly health' allayed her own sharp misgivings at the threat of subordination to him, including sexual subordination. (146–48)

With regard to childbirth and children, Herbert notes that for the Hawthornes,

> to contemplate having a baby brought unacknowledged conflicts to a new pitch of intensity. . . . The Hawthornes met the anxieties of parenthood through a massive deployment of their characteristic strategy, sublimating discord into an exaltation of their union. . . . The baby administered a powerful shock to the Hawthorne's relationship, as

> revealed by Sophia's abrupt announcement that Nathaniel is sending her home to her mother. (150–51)

Although Sophia invented justifications for Nathaniel's desire for her to go home for a fortnight,

> [b]oth Sophia and Nathaniel were temporarily confounded, and each ascribed the collapse of their oneness to the other. . . . His letters to Sophia during her absence indicate how acutely he desired a restoration of the exclusive relation he had enjoyed before the child was born. . . . The fusion of contraries that was present in the union of Sophia and Nathaniel was thus focused on their newborn child. Una is made to represent the central tension of her parent's relationship. (152)

Hester Prynne's Pearl is of course based on Una, just as the relationship between Dimmesdale and Prynne reflects Sophia and Nathaniel Hawthorne's.

The time of the writing of *The Scarlet Letter* was further troubled for Hawthorne because of his firing from the Salem Custom-House and the death of his mother: "Nathaniel's mental life was shaken to its foundations by his mother's death. His literary identity, with its fragile aura of sacred privilege, was grounded in a prior selfhood that still bound him to her" (Herbert, *Dearest* 167). When she was dying,

> after his sobbing ended, Hawthorne stood at the window . . . "[and] saw . . . Una . . . then I looked at my poor dying mother; and seemed to see the whole of human existence at once . . . Oh what a mockery!" . . . Looking to Una for divine consolation, Nathaniel is startled by glimmers of the fiendlike. Not only does the child speak bluntly of death, but she is also fascinated by the slow failure of his mother's body. Nathaniel is appalled. . . . Returning Nathaniel to the darkest hour he ever lived, Una appeared to him an unaccountable compound of the divine and the demonic . . . carried forward into the writing of *The Scarlet Letter*. . . . In modeling the character of Pearl on Una, Hawthorne sought to make sense of the enigma he saw in her . . . [and] [t]he "hell-fire" in which the book was written had cast its glow on the hearthside of his Salem household. (Herbert, *Dearest* 169–70).

Hawthorne's attempts to control and contain his grief about his mother's death and the loss of his job and family income—as well as his attempts to repress his revulsion about women (including his wife and daughter[s]), marriage, and sex—are also exposed in *The Scarlet Letter* in the excessive exaltedness of the relationship between Hester and Arthur, in the presence of Pearl as devil and daughter, and in the absent presence of sex and childbirth, never mentioned, but always there shaping the narrative.

Rereading the film in the context of women's history. The (New) *Scarlet Letter* also provides us with a site for contestation and debate about other gender issues that remain merely implicit in Hawthorne's text, in the older films and their female stars, in the critics' responses, and in the cultural constructions and views of women. As Roland Joffé, the director, stated, "the book is set in a time when the seeds were sown for the bigotry, sexism, and lack of tolerance we still battle today . . . yet it is often looked at merely as a tale of nineteenth-century moralizing, a treatise against adultery" (qtd. in Ebert)—or, I would add, as a tale of nineteenth-century moralizing, which was far more about social propriety and middle-class manners than moral issues like genocide, removal policies, slavery, and the subjugation of women of all races. In fact, Hawthorne's kind of nineteenth-century moralizing depended on denial of these moral issues. Further, by displacing his moralizing onto his mythical Puritan community, he could escape accountability yet remain complicit, as was discussed earlier.

The film opens up the site of contestation and debate about gender issues and female sexuality and subjectivity through its representation of the male Puritan's community's fear of (but desire for) Hester, and its depictions of rape and sexual abuse, including the legally sanctioned processes by which women were accused of witchcraft, brutally interrogated, and indicted—even killed. Thus, this film allows us to examine how typical social and cinematic constructions of the male and female generate and perpetuate notions of desire and desiring and of the female imago. At the level of narrative content, not only does this new film represent the community's sexual violence directed at Hester Prynne, but also at Mistress Hibbins, at the prostitute Sally Short, at Mary Rollings who was a captive of Native Americans, and at Mituba. By focusing on the persecution of more women than Hester Prynne—and by making fully explicit the hatred and desire directed at her in the punishments she endures—Hawthorne's efforts in his novel to contain Hester's agony and the community's violence are revealed, as well as Hawthorne's primary focus on Dimmesdale's angst-driven, self-tortured, over-determined, over-valued phallus, and on his own nineteenth-century privileged position as a member of various dominant groups, including that of white, educated men. In this film a small community of heretical and ostracized women, which includes Prynne and Hibbins, provides a marked contrast to the male-Puritan-identified women, who (as Hawthorne does show) accuse and persecute out of fear and in order to justify and maintain their own social and religious subject-positions. In the film, it is such a woman, the wife of a minister, who thinks up the punishment of the scarlet A, and (Nancy Reagan–like) whispers her plan into the ear of her husband; she gets his attention first, though, by telling him to release Hester from prison after Pearl is born. She says, "You don't put her in the prison; you put the prison in her, so that each time someone sets eyes on her, her sin will be marked into her soul afresh." Later, Hester confirms the power of this punishment when she says to Hibbins (who is perhaps based on Ann Hibbins, executed as a witch in Boston in 1656): "I never imagined how cruel and cunning their punishment could be . . . [men] preaching at me in the streets, the people pointing and shouting, even the children, and that horrible drummer boy following me everywhere." In despair, she continues: "I wonder if existence as a woman is worthwhile at all, even for the happiest of women." She pauses, then says, "What if everything I believed so strongly was a lie?"

The film's Hester Prynne, like the male characters and Mituba, can be reread in light of historical incidents, not only in terms of birth statistics and witch trials, persecution, and hangings, but also in terms of education and intellectual work—for example, when Demi's Hester brings books to Arthur, a scene set against their earlier, brief encounter in the town's library. This library is primarily full of tracts, pamphlets, religious treatises, and books on animal husbandry, all of which both Hester and Arthur have read. Hester also displays her knowledge of the Bible, quoting occasionally from it—first, in a scene just after she arrives in the colony, upsetting the ministers who do not allow women publicly to display their knowledge (or even to have it)—and, later, she theorizes its teachings, particularly in the film's scene of the women's gathering at Hester's home.

Anne Hutchinson. In the film, the gatherings in Hester Prynne's home may be based on those held after church by Anne Hutchinson, beginning in 1635 until she was expelled from the Massachusetts Bay Colony in 1637 (see Davis, DeSalvo), unlike Hawthorne's novel in which, as Amy Schrager Lang argues, "Hester Prynne is not the antithesis but the fictional embodiment of a 'fictional' Anne Hutchinson . . . a figure bearing her name, created by the Puritan chroniclers and kept alive by their heirs" (165). Not twice-removed historically from Anne Hutchinson, Demi's Hester further displays her knowledge of the scriptures during her various trials and hearings before the Puritan ministers and governor, just as Anne Hutchinson did when she was tried for heresy and banished from the colony and the church. Hutchinson

> aroused the suspicion of John Winthrop and other political and clerical leaders when she began holding "prophesyings" to discuss the sermons of her minister, John Cotton. Through these informal meetings of lay members of the church, Hutchinson became the head of a religious faction that believed in direct revelations from God or, in their own terms, "assurance of grace." (Kibbey 410)

Like Demi's Hester, "Hutchinson was vilified as a monstrous threat to public order, the epitome of evil female sexuality unwilling to submit to male authority. . . . The language of Hutchinson's enemies linked female sexuality and antinomianism," as does the film (Kibbey 410). The language of the enemies of women's rights in the nineteenth century linked their fears of women to their laws, social codes, and representations—for example, in their vilifications of the suffrage movement, especially when Elizabeth Cady Stanton and Lucretia Mott rewrote the *Declaration of Independence* as the *Declaration of Sentiments*, calling for social and legal reforms for women, and presented it at the 1848 Women's Rights Convention in Seneca Falls, New York (see earlier discussions of Hawthorne's relationship to Margaret Fuller and to the women's movement).

Mary Rowlandson's story and other captivity narratives by women. In the film, at the gathering at Hester's house—later used as evidence against Hester at her hearing for heresy and adultery—various women speak, including Mary Rollings, a Native captive, who says that the Indians treated her "real fair and square. If truth be told, what's cruel is how you folks have treated me since I come home." Perhaps this character is loosely based on Mary Rowlandson, although it was not until February of 1676 that Rowlandson was taken captive by Metacomet when the Wampanoag burned Lancaster, Massachusetts. Her eleven-week captivity was chronicled in her "principal work, *The Soveraignty and Goodness of God, Together, with the Faithfulness of His Promises Displayed (1682)*" (Amore 770–71). Unlike the film's Mary Rollings's verbal account, captivity narratives, especially early ones, "emphasized God's role in spiritual conversion and deliverance; later ones focused more on woman captives' attempts to cope or to flee through their own wits" (Amore 148). As Amore explains about Mary Rowlandson, she saw

> herself as part of the New Israel's Elect. Thus, when she and her three children were captured . . . she turned to God and the Scriptures for mercy and deliverance. She records events in the same way biblical scholars did, and presents the wilderness as hell and the Indians as kin to Satan . . . her religious sense dominates all she experiences. Throughout her trial, Rowlandson prays for deliverance, and when it is granted through ransom [her husband pays twenty pounds ransom], she sees her freedom as an act of God. (148)

Accounts of captivity were also written and published by other women, for example, Hannah Duston, Susannah Johnson, Elizabeth Hanson, and Jane Adeline Wilson, although not all followed Rowlandson's tone or shape. And as Amore reminds us, "History fails to provide a clear record of Indian women who were killed, held captive, or suffered at the hands of white men" (148–49). Laura Tanner complicates this when she argues that a

> revelation of culture's double standards emerges in Sarah Winnemucca Hopkins' *Life Among the Piutes* (1883). This Native American perspective on frontier life reveals the Indian woman's fears of rape by white frontiersmen, and thus provides an important counter to a genre of Indian captivity narratives in which the virtue of white women is repeatedly threatened by the savagery of their captors. (741)

For example, in Hawthorne's essay on Hannah Duston, Lucy Maddox argues that

> by the end of the essay Mrs. Duston and the Indians have changed places: she is the one who is "raging" and "bloody" while the Indians have become a "copper-colored" version of what the Dustons once were, a peacefully sleeping family. Although Hawthorne does not explain *how* the good woman becomes a savage (hence the reader's surprise), the essay does at least suggest that the transformation begins at the moment Mrs. Duston is forced to "follow the Indians into the dark gloom of the forest. . . ." When Mrs. Duston is removed from her domestic place and separated from her husband and children, she is simultaneously removed from all the constraints that made her the good woman. Forced to exchange the cottage for the dark forest, she rapidly ceases to be "woman" (since "woman" does not, *by definition*, kill children, even Indian ones) and becomes "hag" or witch, capable of out-Indianing the Indians: they killed one of her children, but she kills six of theirs. Their savagery pales before hers. (118)

In the film, representing such fears of white Puritan women, Goody Hunter says, almost wistfully—and certainly not convincingly, "The thought of being taken by a savage. It makes me sick to my stomach." In fact, in 1639 Mary Mandame was "convicted of a 'dallyance' with a Native American in Plymouth, Massachusetts, and after being whipped, [was] sentenced to wear a badge of shame on her sleeve" (Davidson and Wagner-Martin 950; see also Lydia Maria Child's novel, *Hobomok;* Gussman).

Quaker women and heresy. As history has recorded, the Quakers were persecuted, even killed, with state sanction by the Puritans. In 1656 Quakers first arrived in Massachusetts and were quickly banished. By 1658 there was a death penalty for Quakers that stayed in place until 1661, when Charles II stopped such executions in New England. Mary

Dyer, a Quaker woman who was Anne Hutchinson's friend, was imprisoned for preaching in Boston. "She wrote *A Call from Death to Life* (1660), in which she reports her last-minute reprieve from hanging. She continued her preaching, faced a second trial, and was hanged" (Amore 754). Many other Quakers, like Ann Curwen and her husband, were "beaten and imprisoned . . . [for] preaching equality. Curwen defended the right of slaves as well, and demanded they be allowed to attend Quaker meetings" (Amore 754). Hawthorne's relationship to the Quakers can be seen in his story, "The Gentle Boy," in which

> the Quaker woman Catherine allows her religious "fanaticism" to draw her away from her child; the abandoned boy is taken in by a Puritan family, but they cannot protect him from the abuse of other children, who despise him because he is a Quaker. While her child suffers and declines, Catherine continues to "wander on a mistaken errand, neglectful of the holiest trust which can be committed to woman." She leaves the domestic place to wander, physically and intellectually, and the ultimate result is the death of her child. (Maddox 118)

In the film, also present at Hester Prynne's gathering is an unnamed Quaker woman who says, during the women's discussion of the Bible: "We Quakers believe that the scriptures be not religion in themselves, but only the ceremony and history of it." Hester agrees with this description of the scriptures, saying, "For are not the laws of men but the imagination of mortals, and our inner spirit the true voice from Heaven." Her comment is considered blasphemous because it implies that Hester speaks "directly to the Diety," as one woman then queries Hester. In response, Hester admits that she has talked to God since she was a small child, and Hibbins warns her to be careful or "they will be talking of thee the way they talk of me."

During the gathering, the women also use humor, even sexual humor, and offend Goody Hunter so completely that she leaves. Later, it is she who tells the Puritans about Hester's gatherings and her probable pregnancy, and testifies against Hester. At Hester's hearing for heresy and adultery, when one of the ministers says that the kind of talk that has gone on at these gatherings is "what comes when there is no qualified man present to guide these women in their untutored chattering," Hester answers that if the "discourse of women is untutored chattering, why, then, does the Bible tell us that women shall be the teachers of women?" The ministers ask Hester to cease the gatherings, but she refuses, functioning not like Hawthorne's version of Anne Hutchinson, but like the historical Hutchinson.

Mrs. Hopkins, Anne Bradstreet, Mary Oliver, Mary Hammon, and Goodwife Norman. The film's fictional scenes are played out in relationship to many historical moments and figures, including writings such as Governor John Winthrop's 1645 account of "the sad fate of a Mrs. Hopkins, who was driven mad by 'giving herself wholly to reading and writing'" (Perkins, Warhol, and Perkins 8). As a fictional intellectual, Hester Prynne, then, like a real Anne Bradstreet, had "reason to be wary of 'each carping tongue who says my hand a needle better fits'" (qtd. in Perkins, Warhol, and Perkins 8). But like Bradstreet who was "writing poetry within a year of her arrival in the New World and [who] continued to write verse and prose throughout her life" (8), Demi's Hester Prynne refuses to give up her intellectual life or its consequences.

The film's representation of Prynne also evokes Mary Oliver, "Salem's own heretic," eventually banished by William Hathorne (Davis 193). Unlike Hawthorne's novel, in which Hester's intellectualism is turned into a problem and a marker of her loss of femininity, in the film her abilities and knowledge serve her well, not only because they are depicted as partial reasons for the attraction between Hester and Arthur, but also because they allow her to argue her case and to stand up for the other women who are accused of witchcraft. Of course, the more these women logically and wisely present their cases and positions, the more persecuted they become. Because their abilities belie the social constructions of women as lesser than men, the women threaten the attempts at containment at work in those constructions. And like all fictions about women imposed on them by dominant paradigms and social codes, these fictions must be continually protected and shored up, like a dam with holes that must be plugged again and again. It can only be imagined how frightened the Puritans must have been of lesbian sexuality; there is a recorded case of Goodwife Norman and Mary Hammon who were tried for lesbianism in the Massachusetts Bay Colony in 1649. Norman was "forced to make a public confession" and found guilty. Her confession allowed charges to be dropped against sixteen-year-old Mary Hammon. "Norman is believed to be the first woman in America convicted of lesbianism" (Swade's Tribal Voice 2; see also Davidson and Wagner-Martin 950).

Arthur suffers; Hester endures. As in the book, Dimmesdale suffers in the film, though not to the same extent. During the film's scene where Mistress Hibbins is captured while hiding in Hester's house and accused of witchcraft, Arthur arrives. Other than the scene in which he stands on the scaffold, rubbing his hands raw against its post, the one in Hester's house is the only other scene in the film in which Arthur is presented as weak and self-flagellating as he is in the novel. As Louise DeSalvo says of the novelistic Dimmesdale:

> [E]ven rendering the character of Dimmesdale as so pathetic, so ineffectual, so self-destructive effectively serves to dim the ferocity of his historical counterparts: it is impossible to read Hawthorne's Dimmesdale and conceptualize the Puritan oligarchy as avenging avatars. . . . And so, in the context of *The Scarlet Letter,* in a fascinating reversal of the facts of history, Dimmesdale, the representative of the Puritan state and Puritan power in the novel, becomes more sinned against than sinning—he is described as the "victim. . . ." Hawthorne, in his revisionist history, thus substitutes a portrait of a male victim for an accurate portrait of a female victim of the Puritan oligarchy. (64 * * *) [522–23]

At Hester's house, just as Hibbins is captured by the Puritan militia, Dimmesdale shows his power and his complicity in the Puritan patriarchy. He asks Hester Prynne why she continues to risk "further anger from the elders." She replies, "Because Mistress Hibbins is no witch, and she has committed no crime beyond speaking her mind." Dimmesdale argues, then, that if "she is innocent, I assure you no harm will befall her." Astounded, Hester responds, "Arthur, after all that has happened, how can you still trust these iron men? Do you not see what is happening. Last night, they took Sally Short in for questioning." She continues her list. Growing defensively angry, Dimmesdale cries out that the elders are "just questioning them. What is the crime in that?" Hester answers that the crime is that "they had done nothing. Do you not see that this is all part of some malevolence. What has become of you?" Arthur begins then a tale of self-pitying woe, shouting at Hester about the hell he is in, ending with "I am a pollution, Woman. I am a lie." Hester says, "*They* are the pollution. *They* are the lie. But you are allowing them to destroy everything that is good in you. What has happened to the man I loved? Does he not still live inside thee?" Arthur then repudiates Hester and their love, saying quietly but with angry passion, "Our love, Woman, was a folly, and that voice we heard . . . we have been justly punished for listening to it." Not content with what he has thus far said to hurt Hester, he looks out the window to see that Hibbins has been captured and turns to Hester, insinuating her complicity with witchcraft by saying, "What have *I* become?"

Yet, amidst threats, her first imprisonment, punishment, isolation, brutal words and actions from Roger Prynne, and, finally, Dimmesdale's repudiation, Hester endures. Even after the murder of Mituba and after Hester and Arthur reconcile in a later scene in the forest, where Arthur states his undying love for her and claims Hester and Pearl as his family, Hester endures an attempted rape by a soldier, a witch trial, her second imprisonment, and a verdict of hanging, and grows stronger—and more outspoken. In direct opposition to Hawthorne's Hester, Demi's Hester does not remain silent, except about Arthur Dimmesdale's identity. Throughout this film, he implores her to be silent to avoid incurring the wrath of the Puritan fathers. Each time she answers him with such comments as: "I had to speak out; I couldn't stop myself." And when he asks her how far she will go, she says, "As far as my strength will take me." This new Hester never assumes, guilt, shame, or blame, consistently resisting the Puritans' constructions of women and of her. During the first scaffold scene, she claims, when asked if she believes she has sinned, "I believe I have sinned in your eyes."

The only time Dimmesdale speaks for her or about his relationship to her, she is again on the scaffold, gagged and ready to be hanged, convicted finally of witchcraft, along with Sally Short, Mary Rollings, and Harriet Hibbins. When Arthur confesses to the crowd at that moment, taking the noose from her neck, the crowd calls out for his hanging, and Hester's rope is placed around his neck. At the end of the film, after she, Dimmesdale, and the other women escape their hangings because of the Wampanoag's freeing of the imprisoned "praying Indians," and after Governor Bellingham decides to pardon Hester—allowing her to take off the A, she says, "This letter has served a purpose, but not the one they had intended. So, why would I stay here, to be accepted by them, to be tamed by them?" She leaves, and Dimmesdale finally hops aboard the little cart holding Pearl and their possessions. As they drive away, Pearl drops the scarlet letter on the ground, and the wheels of the cart roll over it. The relationship to the scarlet letter for Demi's Hester is far different from that of Colleen Moore and Lillian Gish, and certainly from that of other actors who have played Prynne, including Gene Gauntier (1911), Mary Martin (1917) and Meg Foster (1979). Thus, even though Joffé's film and Stewart's screenplay could be, and have been, accused of also romanticizing historical and cultural problems—and particularly of idealizing the love between Prynne and Dimmesdale—in this film narrative they rewrite Hawthorne's rewriting of history, retold in this film by Pearl in a voice-over (see also Welch).[1]

Some Cinematic Techniques of The *(New)* Scarlet Letter

The bathing and mirror scene. Although many of the film's scenes function to destabilize and reread Hawthorne's original narrative, as I have so far described, many also destabilize the cinematic frame, especially calling attention to those techniques typically used to direct the gaze of the spectator at women on the screen. One instance of this revisioning occurs in the scene where Demi/Hester gazes at herself in a mirror while bathing. Typically, the spectator would look at the mirror image of the female character—that is, look at what she looks at—the mirror-image of the film's image of her. This gaze is made possible by positioning the spectator psychically between the camera's gaze and the mirror image. In this scene, however, the spectator's gaze is directed at Hester looking, but never into the mirror, never seeing the image she sees; in fact, the camera is positioned behind the mirror, which mostly blocks the spectator's view of her body. The spectator watches Hester's desire, excited by her own body and by her memories of Arthur. For most of that scene—itself interrupted by cuts to Arthur, as she remembers him in certain instances, and to Mituba, who peers through the keyhole—the spectator's view of Demi/Hester is from her shoulders up. Finally, though, the camera swings to a long upward side shot of her, allowing a quick but gratuitous glance at one side of her naked body.

The scaffold scene. Another instance of cinematic revisioning occurs in one of the only scenes that attempts to follow Hawthorne's text: the first scaffold scene. As such, this scene disrupts nostalgic memories of Hawthorne's novel and his constructed reader. In his text, Hawthorne describes the platform of the pillory on which Hester must stand before the gathered crowd; he says,

> above it rose the framework of that instrument of discipline, so fashioned as to confine the human head in its tight grasp, and thus hold it up to public gaze . . . the Governor, and several of his counsellors, a judge, a general, and the ministers of the town . . . sat or stood in a balcony of the meeting-house, looking down on the platform . . . the crowd was somber and grave. The unhappy culprit sustained herself as best a woman might, under the heavy weight of a thousand unrelenting eyes, all fastened upon her, and concentered at her bosom . . . she was the most conspicuous object. (55–57) [40–41]

When Dimmesdale speaks to her in the novel, Hawthorne says, "the eyes of the whole crowd [were] upon the Reverend Mr. Dimmesdale" (66), as he leans over the balcony to speak down to her. The perspective shifts from the crowd to ministers to governor to Hester and finally to Chillingworth

1. In June 1996, at the conference of the American Literature Association in San Diego, California, I was on a panel with Roland Joffé at a session sponsored by the Hawthorne society, and chaired by T. Walter Herbert, which focused on Joffé's film. During Joffé's extemporaneous talk, he mentioned that part of his deal with the studio was that he had final say over each day's script and rewrites, as well as over the dailies of the film. He did not, however, have control over the film's final cuts. During his presentation he mentioned that the director's cut of the film includes many scenes, helpful to the progression of his narrative, which did not make it through the studio's final cutting sessions.

who arrives at the edge of the crowd. Consistently, though, Hester is the spectacle on whom all gazes rest. She is positioned between the crowd and the Puritan patriarchs, and all, including the constructed reader, look at her * * *.

In the film, however, the differences are noticeable: Hester Prynne stands on a platform above the Puritan fathers, who are seated in front of, but below her, yet on the scaffold—as is Dimmesdale. Behind her are other people in the Puritan community, including wives. Most significantly, the camera follows Hester Prynne's gaze. Even when Dimmesdale speaks, he is slightly below her, as though the gaze remains hers. And his back is to the crowd and to the other ministers and governor. Even during one shot when she is captured in his beseeching gaze, the perspective is hers, and the shots of the crowd remain in the context of her looking at them rather than their looking at her. We can even see Pearl's head and the edge of her blanket intruding onto the images of Dimmesdale and the crowd—demonstrating that the camera's gaze is from the Hester's perspective; that is, we sees what she sees. The whole sequence is one of the film's most disruptive uses of a version of the shot-reverse-shot technique, which has generally functioned to objectify the female image. Although Hester Prynne is objectified by the crowd and the Puritan patriarchs, in this scene the camera functions to displace the spectators' voyeuristic and fetishizing gaze.

The *(New)* Scarlet Letter *and Feminist Cinema*

Thus Demi's Hester, as Beverle Houston argues about women characters in other films, "asserts a will to subjecthood" (273). The very different plot, dialogue, and representation of Hester Prynne in this new version of *The Scarlet Letter* function to interrupt the cultural and discursive hegemony of Hawthorne's novel, without even having to stray from representing the excesses that his novel, his avoidance of the history of the Other, his other texts, and his life attempt to contain. Although I do not want to suggest that *The* (New) *Scarlet Letter* functions as what Sandy Flitterman-Lewis calls feminist cinema, I think, as I have said, that it disrupts traditional classical constructions of the male and female spectator (as well as the constructions of the reader of Hawthorne's novel that are already implied in the cinematic constructions). To discover such disruptions, it is necessary to examine the historical and social context of the films and their producers, directors, actors, and viewers, as well as examine, as Flitterman-Lewis urges, the entire cinematic apparatus, what she describes as "its functioning, and its specific determinations for figuring female sexuality. This apparatus," she argues, "is designed to produce and maintain a fascinating hold on the spectator by mobilizing pleasure—the unconscious desire of the subject—through interlocking systems of narrativity, continuity, point of view, and identification. But it is not simply an individualized and self-contained process . . . [because of] the social inscription of the cinema," what Christian Metz has called its "'dual kinship' with the psychic life of the spectator" (Flitterman-Lewis 3). *The* (New) *Scarlet Letter*, then, I am claiming, demobilizes the pleasure of the look and destabilizes the point of view of the novel and of the classic film versions of *The Scarlet Letter*. Thus, what E. Ann Kaplan has called "historical spectators" come to this new film with layers of expectations generated by the novel and by earlier films, and find themselves unseated, so to speak, by the deconstructing of the "hypothetical" spectator-position, and without the means to control and consume this film-product. Further following Kaplan, the "contemporary female spectator whose reading of the film might be inflected by a female consciousness" may not be unseated, but instead resituated by this disruption of classic cinematic techniques (Flitterman-Lewis on Kaplan 9).

Not surprisingly, the film has been soundly rejected by unseated academics, film critics, and in their wake, the general viewing public. Although the film manages to expose the excesses of barbarism, savagery (not in the Native Americans but in the Puritans), slavery, violence, witchcraft trials, and a profound fear and hatred of, but desire for, women (Hester-Prynne-ism), the critical commentary and academic response has attempted to recontain them. The success of these efforts can be seen in the loss of acclaim and revenue (only $10.3 million). Roland Joffé's (New) *Scarlet Letter* has been named a failure. Demi Moore has been refetishized, carrying on her body the scarlet mark of the woman. Hester-Prynne-ism prevails, and Hester Prynne is in her place once again, as the spectacle, the image, not the bearer or maker of the gaze (Mulvey 62–63).

Other Writings

AMY SCHRAGER LANG

Anne Hutchinson†

In 1830 the Salem *Gazette* printed a biographical sketch of Anne Hutchinson written by Nathaniel Hawthorne, native son and aspiring young author. The sketch, entitled "Mrs. Hutchinson," is a curious one. Not only does Hawthorne assume that his readers will recognize his Mrs. Hutchinson and remember her history but he assumes as well the immediacy of her story. Nowhere in the sketch does he detail the story of Hutchinson's excommunication from the First Church of Boston or her banishment from Massachusetts Bay in 1638. Instead, "Mrs. Hutchinson," ostensibly a portrait of the seventeenth-century heretic, begins with an attack on the "public women" of the nineteenth century, in whose ranks one might, Hawthorne suggests, find Hutchinson's "living resemblance." Hawthorne acknowledges the peculiarity of this opening—excusing "any want of present applicability" by the "general soundness of the moral"—but insists nonetheless on the aptness of his comparison. Hutchinson, that public woman of the past "whereof one was a burthen too grievous for our fathers,"[1] has been succeeded by a new breed of woman no less burdensome and far more numerous. Paradoxically, the religious heretic of the 1630s is reincarnated in the female sentimental novelist of the 1830s.

Twenty-five years before writing his famous letter to William Ticknor bemoaning the domination of the literary marketplace by the "d—d mob of scribbling women," Hawthorne introduces his reader to Anne Hutchinson by deploring the "irregularity" of the "ink-stained Amazons" who present themselves at the bar of literary criticism. In Hawthorne's estimation the "false liberality and . . . courtesy" accorded these women by their reviewers combine to add a girlish feebleness to the tottering infancy of [American] literature" (*TS*, 18) [159]. All too clearly, the Amazons will "expel their rivals . . . and petticoats [will] wave triumphantly over all the field" (*TS*, 18) [159].

† From *Prophetic Woman: Anne Hutchinson and the Problem of Dissent in the Literature of New England* (Berkeley: U of California P, 1987), pp. 1–14. Reprinted by permission of the publisher and author. Page numbers in square brackets refer to this Norton Critical Edition.

1. Nathaniel Hawthorne, *Tales and Sketches* (New York: Library of America, 1982), 18. Additional references to this edition will be cited in the text and identified by the abbreviation *TS* followed by the page number.

Hawthorne himself realized that the juxtaposition of seventeenth-century antinomian and nineteenth-century author demanded explanation, particularly since the latter, rather than promulgating "strange and dangerous opinions," consistently urged the virtues of home and hearth. What links these disparate figures is "feminine ambition." Both Hutchinson and her literary counterpart of the nineteenth century have abandoned their embroidery for careers as public speakers. As prophetess in one case and novelist in the other, each has stepped out of her appointed place and indelicately displayed her "naked mind to the gaze of the world, with indications by which its inmost secrets may be searched out" (*TS*, 19) [159]. This unseemly exposure of a private self to the public gaze—indecorous at best and at worst positively lewd—is urged on antinomian and sentimental author alike by an apparently irresistible "inward voice." Like the errant Anne Hutchinson, who, according to Hawthorne, confused "carnal pride" with the gift of prophecy, so the misguided author yields to the "impulse of genius like a command of Heaven within her" (*TS*, 19) [159].

The similarity between antinomian and sentimentalist, both of whom appeal to an inner voice to rationalize their intrusion into the public arena, casts the literary critic in the role of the Puritan magistrate. The court of literary opinion must, Hawthorne insists, "examine with a stricter . . . eye the merits of females at its bar, because they are to justify themselves for an irregularity which men do not commit in appearing there" (*TS*, 19) [159]. Likewise, Hawthorne refuses to dismiss the Puritans' judgment against Anne Hutchinson as the action of an "illiberal age." On the contrary, "worldly policy and enlightened wisdom" would also dictate her banishment (*TS*, 21) [161]. In the interest of the rising culture, the literary critics of the nineteenth century would do well to banish from the American literary scene those whose "slender fingers" enfeeble it.

The energy of Hawthorne's sketch comes from his intuitive recognition of a resemblance between Anne Hutchinson and the scribbling women. The story of "the Woman," as Hawthorne invariably calls Hutchinson, is offered as the quintessential story of female empowerment—of its origin in an erroneous "inward voice," its unseemly public expression, and its disastrous social effect. Inevitably, that story is fraught with tension. In fact, insofar as one term points toward the broad masculine realm of autonomous action in the world while the other calls us home again, the phrase "public women" captures a crucial contradiction. This contradiction is similarly reflected in the competing images of unnatural strength and equally unnatural weakness in Hawthorne's sketch. The female author, with an Amazon-like disregard for feminine decorum, expels her male rivals from the literary field, but the flag her "slender fingers" hoist over that field is only, after all, a "petticoat": the woman warrior writes domestic fiction.

On the one hand, the act of female authorship constitutes an assertion of autonomy and, thus, a challenge to authority as dramatic as Hutchinson's antinomianism. By choosing "the path of feverish hope, of tremulous success, of bitter and ignominious disappointment" (*TS*, 19) [159], the woman writer defies her place. On the other hand, it is not strength but rather the "girlish feebleness" of the sentimental domestic novel penned by the Amazon that threatens to undermine the national literature. In other words, the trouble with the Amazon, ink-stained or otherwise, is that she remains a woman and, as Woman, implies an order that, as Amazon, she violates. Insofar as Woman contains in herself the possibility of Amazonian defiance, she suggests the further—and more frightening—possibility that men too might step out of their places.

In this sense the gender-specific problem of the public woman figures the larger dilemma of maintaining the law in a culture that simultaneously celebrates and fears the authority of the individual. That dilemma has long been identified with antinomianism, but Hawthorne's sketch calls our attention to the fact that the problem of individual autonomy is especially problematic when the individual is female. The fact that Anne Hutchinson, the classic American representative of a radical and socially destructive self-trust, is a woman compounds and complicates her heresy. In "Mrs. Hutchinson" the problem of antinomianism is propounded as the problem of Anne Hutchinson, which is, in turn, the problem of the public woman.

It is precisely this sequence in which I am interested. What little information exists about the historical Anne Hutchinson has long since been unearthed; likewise, the term *antinomian* has been appropriated by scholars in a wide variety of disciplines to mark the outer limits of American individualism. But in all the work that has been done on antinomianism, both as Puritan and as American heresy, the fact that the antinomian is embodied as a woman has received scant attention. Another history of antinomianism needs to be written, one focusing on the special relevance of "the Woman" to the nature of antinomian heresy and depending less on the historical record than on the "literary" one. In the context of this history, Hawthorne's sketch seems less odd than inevitable.

The contours of Hutchinson's story are familiar. In 1634 Anne Hutchinson, a woman in middle age, left England with her large family to follow the much-admired Reverend John Cotton to New England. Admitted to the church, her husband elected to the high office of deputy to the Massachusetts General Court, Hutchinson set about establishing herself in her new home. Her first prominence was as a nurse-midwife and spiritual adviser to women, but sometime during her first two years in Boston she began to hold weekly gatherings for the purpose of reviewing and commenting on the sermon of the previous Sunday. These meetings, first attended exclusively by women, quickly grew to include men and soon drew a regular attendance of sixty or more of the town's inhabitants, including the young governor, Henry Vane, and a number of other men of power and prominence. As her following changed in both size and prestige, so too, apparently, did Hutchinson's message. Rather than simply recapitulating the weekly sermon, she undertook to reproach the Massachusetts clergy. The leaders of the church, she claimed, had fallen into a covenant of works. "Legalists" all, they mistakenly took sanctification—the successful struggle of the saint against sin—as evidence of election, failing to understand that works and redemption bear no necessary connection. In essence Hutchinson spoke for a doctrine of free grace, characterized by the inefficacy of works and the absolute assurance of the saint. Until the arrival in 1636 of her brother-in-law and supporter, the Reverend John Wheelwright, John Cotton alone was exempt from her criticism.

Hutchinson's followers, convinced that the Massachusetts Bay ministers were guilty of preaching a covenant of works, were moved to action. Efforts were made to replace the Reverend John Wilson, then pastor of the Boston church, with John Wheelwright. The animosity between Hutchinson's supporters and her opponents grew until, in January 1637, a Fast Day was set aside in an effort to restore the peace. In a conciliatory move John Wheelwright was asked to deliver the Fast Day sermon. His sermon provoked a charge of sedition, and this charge, in turn, brought petitions on his behalf to the General Court. Accusations of antinomianism from one side were met with thinly veiled charges of papism from the other. Between the two the errand of the New Israel seemed doomed. As disruption and contention spread, affecting participation in colonial elections and the conduct of the Pequot War, not even the prominence of Hutchinson's followers could protect her. The authorities of the Bay moved into action, first meeting privately with Hutchinson, Cotton, and Wheelwright to inquire into their difference with the orthodox ministers. Discontented with the answers they received, the ministers convened a synod—the first in the colonies—for the purpose of formulating and responding to the errors of the antinomians. The General Court followed suit with sterner measures: the leaders among the antinomians were variously disenfranchised and banished, their male supporters disarmed. Considered by most the ringleader, Hutchinson was herself brought to trial by the Court in the fall of 1637 and by the church the following spring. Exiled and excommunicated, she fled to Rhode Island in 1638, moving five years later to New York, where, apparently in providential vindication of her judges, she and all but one member of her family were killed in an Indian raid.

The rapidity with which the Massachusetts Bay community fell into dissension has captured the interest of a wide range of modern scholars. In a period in which scarcity and inflation were intensified by a steady increase in population, the disproportionate affiliation of merchants with the "Hutchinsonians" has been explained as a response to the "insupportable pressures" suffered by those who would be both pious and successful in commerce. The merchants, it is claimed, used antinomianism as their way to rebel against an authoritarian Puritan regime, which tended to constrain their economic behavior.[2] The special appeal of Hutchinson's "primitive feminism" to women has also been explored. In her "new theology," one scholar argues, "*both* men and women were relegated, vis-à-vis God, to the status that women occupied in Puritan society vis-à-vis men."[3] Social and intellectual historians have mined the antinomian controversy for information about the limits of Puritan orthodoxy, social "boundary-marking" in the colonies, Puritan attitudes toward dissent, and the status of women in early New England. Literary critics have adopted the term *antinomian* to describe the oppositional quality they find in the classic literature of the American Renaissance and have taken

2. The most thorough exploration of the role of merchants in the antinomian controversy can be found in Emory Battis, *Saints and Sectaries: Anne Hutchinson and the Antinomian Controversy in the Massachusetts Bay Colony* (Chapel Hill: University of North Carolina Press, 1962). Other versions of this argument appear in Bernard Bailyn, *The New England Merchant in the Seventeenth Century* (New York: Harper and Row, 1955), and Larzer Ziff, *Puritanism in America: New Culture in a New World* (New York: Viking, 1973).
3. Lyle Koehler, *A Search for Power: The "Weaker Sex" in Seventeenth-Century New England* (Chicago: University of Illinois Press, 1980), 221.

the antinomian impulse to enable literary production in a Puritan culture. Important as it is, however, this wealth of historical data and literary conjecture does relatively little to advance our understanding of the symbolic value of the story of Anne Hutchinson beyond reinforcing what the early narratives tell us more forcefully—that is, that antinomianism represented a rejected alternative to the New England Way.

While the orthodox struggled to bring together citizen and saint, to establish a connection between private and public realms of experience, the antinomian, building on the ambiguous status of the individual in Reformed theology, proposed a new relationship between the two central facts of Christianity—the unmerited redemption of mankind by Christ and the continued existence of sin and misery in the world. Justification, the antinomian claimed, is a gift freely given to fallen man without which all his pious endeavor is to no purpose. At the moment of conversion, the saint, like an empty vessel, is cleansed and filled with Christ's love, and by this motion of the divine, the chosen are at once freed from accountability under the Law and assured of election. Individual identity is subsumed in divinity; the works of the saint are one with Christ's. As one English antinomian explained it, "in all your acts Christ acts, and in all Christ acts within you, you act . . . and in your lowest acts Christ acts as well as in your highest."[4] Man's sinful nature remains unaltered by the influx of grace, but the antinomian's perception of his sinfulness is radically changed. Secure in his election, the antinomian ascribes neither real nor symbolic value to deeds: sin exists only "that there may be a place for faith." Instead of the orthodox notion of "visible sainthood," which proposed that grace would emanate in good works and which thus nicely accommodated the exigencies of community, antinomianism offered a perfectionist theology wherein election is witnessed and sealed by the spirit and cannot be tested by outward means.

Even in this brief description of the heresy, it is possible to discern both the paradoxical nature of the antinomian as he presents himself to the world and the larger problem of authority that antinomianism engages. The antinomian regards the saint as indissolubly joined with Christ. Individuality is merely the medium through which God exerts himself. Because this is equally true of all the saints and because the nature of divinity is necessarily constant, the experiences of the saints answer one another "as face answers face in a glass."[5] Thus, those who "belong to God" are able to distinguish between God's people and the reprobate. On these grounds the American antinomians rejected the colonists' view of themselves as a chosen people, bound by covenant to fulfill God's work in the New World, and offered in its place the notion of a mystical community of the elect. The system of rewards and punishments adduced from the Law and embodied in temporal authority is, for the antinomian, irrelevant. Sin and sanctity alike are transient, inconsequential events, their significance lost in the very moment of the Spirit's invisible witness to election.

From this moment forward, the antinomian lives in a world free of the sometimes productive, sometimes disabling anxiety characteristic of the Puritan saint. Election is, for him, a condition of self-abnegation, his

4. Quoted in Gertrude Huehns, *Antinomianism in English History* (London: Cresset Press, 1951), 44.
5. Quoted in Huehns, *Antinomianism*, 93.

individuality no longer individual, his labors at an end, his destiny secure. In this sense, as the Puritans were wont to point out, antinomianism "quenches all endeavor." Yet rather than appearing passive, the antinomian seemed, even to the Puritans, to indulge the furthest extremes of self-assertion. For those who did not share their belief in the immediate union of the elect with Christ, the antinomian's claim to invisible witness, absolute assurance, and exemption from the Law could only seem like sheer arrogance. Abandoning the social for the teleological, then, the antinomian elevates the self to a new status precisely by insisting on the dissolution of the self in Christ.

The theological idea that the covenant of grace releases the elect from accountability under the Law is not indigenous to the United States. The idea is generally traced back to Paul's epistles, while the term *antinomian* is said to have been coined by Luther, who used it in reference to the German Johannes Agricola.[6] The charge of antinomianism was leveled against John of Leyden and the Anabaptist leaders of the Münster Rebellion in 1533, and concern with the heresy burgeoned during the English Civil War as writers like John Eaton, the "father of English antinomianism," produced tract after tract with titles like *The Honey-combe of free Justification by Christ Alone*. The persistence of the term into the nineteenth century likewise is not peculiar to America, nor is its appearance in literary guise. James Hogg's 1824 *Confessions of a Justified Sinner* relates the fall into antinomianism of the unrepentant Robert Colwan; Mike Hartley, the "Antinomian weaver," makes his appearance twenty-five years later in Charlotte Brontë's *Shirley*. Yet there is reason to speak of American antinomianism as a separate phenomenon, for here the heresy is encoded in the story of Anne Hutchinson's conflict with the authorities of the Massachusetts Bay as this was repeated and elaborated from one generation to the next. Tied to the figure-legend of Anne Hutchinson, the local history of antinomianism is distinct from its universal one. Moreover, it is gendered female from the outset.

New World antinomianism was suppressed in 1638 and the authority of the ministers and magistrates restored. Yet the recurrent and pejorative use of both the figure of Anne Hutchinson and the term *antinomian* into the nineteenth century suggests that the tensions ostensibly resolved at the time lingered on, transformed. Celebratory or even apologetic accounts of the antinomian controversy are rare even though precedents exist for the celebration of if not heresy then at least certain heretics. The Romantic historian George Bancroft reconstructed Roger Williams, for example, as a "lark" giving voice to the "clear carols" of intellectual liberty, thus suggesting that some forms of Puritan heresy could be absorbed into a future consensus. Hutchinson resists just such renovation. Arrogant, rebellious, enthusiastic, the American Jezebel remains "tinged with fanaticism" throughout her history.[7] She is striking largely for her ability to rouse anxiety. If only because it remained so disturbing to Americans for so long, Hutchinson's story tells us a great deal about the unspoken concerns of her

6. The biblical text most generally cited is Romans 6. For a discussion of the history of antinomianism as a term, see Ronald A. Knox, *Enthusiasm* (New York: Oxford University Press, 1950), and Huehns, *Antinomianism*.
7. George Bancroft, *History of the United States of America* (Boston: Little, Brown, 1879), 6 vols., 1:298, 296.

countrymen. But to account for its longevity and its force, we will need terms other than those commonly used. Most crucially, we must distinguish between the antinomian controversy itself—a historical event important largely because it sheds light on the social reality of the early colonies—and the narrative accounts of that controversy—the stories told about it (or more loosely, drawn from it), sometimes in self-defense, sometimes in indignation, but always to caution against present or future danger. While representing themselves for the most part as "histories," these narratives contribute less to our knowledge of the "real" controversy than they do to our understanding of how Americans have formulated and lent meaning to a series of events, local in nature and removed in time.

John Winthrop, for instance, goes to some lengths to insist that the Hutchinsonians were summoned before the Court not as heretics but because they "disturb[ed] the Churches . . . interrupt[ed] the civill Peace . . . and began to raise sedition amongst us, to the indangering of the Commonwealth." "Hereupon for these grounds named, (and not for their opinion . . .)," he goes on, "for these reasons (I say) being civill disturbances, the Magistrate convents . . . and censures them."[8] Winthrop's vehemence on this point has quite reasonably been taken as a response to English accusations that the colonists prosecuted cases of conscience as civil crimes, accusations he feared would be politically damaging to the Bay colony. Like his ministerial colleagues, the official historian of the controversy knew that in antinomianism he was faced with an interpretation of doctrine that proposed not only an alternative conception of the self but also, by implication at least, a radically different order of society.

It would seem that gender is of no concern here, but the immediate political motive and the larger ideological one do not entirely explain Winthrop's narrative. They fail to account for the zeal with which Winthrop details such matters as the physical deformities of Hutchinson's friend Mary Dyer's stillborn child or Hutchinson's own "misconceptions." In order to talk about these elements of Winthrop's narrative, we must think not of history but of story, not of the heretic but of the woman who promises a stable, familial community yet remains capable of producing monsters.

Even in Winthrop's narrative, the historical Anne Hutchinson is swallowed up in the monitory figure of the American Jezebel as narrative history is overwhelmed by cautionary tale. Much as the accounts of the antinomian controversy respond to changes in political climate, shift in their emphasis, and vary in their form, they nonetheless document an urgent and continuing need to choose against antinomianism that can best be understood if we consider the evolution and the place of Hutchinson's story in an American pattern of meaning. Like that story, this study begins with history and ends with literature. It does not treat every account of the antinomian controversy, nor does it trace all representations of the antinomian impulse in the literature of the United States. It focuses instead on the cautionary tale of Anne Hutchinson, a tale that does not simply recast the past for the purposes of the present but continually reenlivens the female heretic only to banish her once again. As

8. David D. Hall, ed., *The Antinomian Controversy, 1636–1638: A Documentary History* (Middletown, Conn.: Wesleyan University Press, 1968), 213–14.

an exploration of the strategies Americans used to contain antinomianism, this study is designed not to move away from the larger issues of American individualism engaged by the antinomian controversy but to ground these in a specific story. Insofar as antinomianism presumes, as one scholar has put it, "that social benefit [can] arise from an a-social orientation of the component parts of . . . society,"[9] the story of Anne Hutchinson sets in relief a continuing tension in American culture over the relationship between private and public realms of experience expressed concurrently in the figure of the antinomian and that of "the Woman."

The concurrence of these figures is particularly apparent in Hawthorne's "Mrs. Hutchinson." Hawthorne organizes his sketch into two vignettes, one centering on Hutchinson's crime, the other on her punishment. The first vignette is set at dusk in the Hutchinsons' crude house, newly built at the "extremity of the village," on a street "yet roughened by the roots of the trees, as if the forest . . . had left its reluctant footprints behind" (*TS*, 20) [160]. As we enter the "thronged doorway," we sense that everything about this scene is wrong. Positioned at the furthest margin of the settlement where the wilderness still encroaches, "the Woman" faces her distinguished and predominantly male audience: Governor Henry Vane, in whose eyes Hutchinson's "dark enthusiasm" is mirrored; John Cotton, "no young and hot enthusiast" but nonetheless "deceived by the strange fire now laid upon the altar"; the Reverend Hugh Peters, "full of holy wrath"; the "shuddering and weeping" women, and the young men, "fiery and impatient, fit instruments for whatever rash deed may be suggested" (*TS*, 20–21 [160–61]). No one, Hawthorne emphasizes, is indifferent to Hutchinson's words, yet the meeting proceeds in utter silence as far as the reader is concerned. This silencing of Hutchinson measures the tension in Hawthorne's sketch.

The threat of an Anne Hutchinson is captured not in overt statement but in the small disjunctive elements that fill the scene. Hutchinson's proper attire, for example, is belied by the inappropriateness of her preaching. Likewise, her eloquence stands in sharp contrast to the "quiet voice of prayer" we overhear elsewhere in the settlement. Her house is thronged with disciples but devoid of domestic comforts. The household hearth has been supplanted by a "strange fire" that inflames rather than warms; the family board is replaced by an altar. Instead of offering maternal reassurance to the "infant colony," Hutchinson reduces her audience to frightened "children who . . . enticed far from home . . . see the features of their guides . . . assuming a fiendish shape" (*TS*, 21) [161]. The woman of this house is no Daughter of Zion but instead a "disturber of Israel" unfolding "seditious doctrines" designed to persuade her neighbors that "they have put their trust in unregenerate and uncommissioned men, and have followed them into the wilderness for naught" (*TS*, 21) [161–62].

Like the first vignette, the second one begins not with Hutchinson but with another audience, this one sitting in judgment, not in awe. Once again the mood is ominous: a "sleety shower beats fitfully against the windows, driven by the November blast, which comes howling onward from the northern desert" (*TS*, 22) [162]. And once again the events of the

9. Huehns, *Antinomianism*, 169.

controversy are associated with that threatening wilderness against which the Puritan community did constant battle. Before us are ranged those "blessed fathers of the land" whom we Americans, Hawthorne reminds us with characteristic irony, rank next only to the "Evangelists of Holy Writ" (*TS*, 22) [162]. Hutchinson faces this venerable company with "a flash of carnal pride half hidden in her eye, as she surveys the many learned and famous men whom her doctrines have put in fear" (*TS*, 23) [162]. Hutchinson's eloquence is, in some sense, the subject of the first vignette; her arrogance is the theme of the second. We watch first as Hutchinson exults in her contest with the "deepest controversialists" of the day and then as exultation leads inevitably to self-incrimination. With victory in sight, she arrogates to herself the role of judge, claiming "the peculiar power of distinguishing between the chosen of man and the Sealed of Heaven. . . . She declares herself commissioned to separate the true shepherds from the false, and denounces present and future judgments on the land, if she be disturbed in her celestial errand. Thus," says Hawthorne, "the accusations are proved from her own mouth" (*TS*, 23) [163]. He means, of course, that Hutchinson's words reveal her antinomian tendencies, but the impact of the vignette depends heavily on the Woman's usurpation of man's place as judge and on her arrogant assertion of a superior knowledge of God's will.

Like the words that inspire her audience to turn against minister and magistrate, the theological demonstration that stands behind Hutchinson's condemnation of the Bay leaders—that is, antinomianism as the Puritans understood it—is absent from Hawthorne's sketch. Both omissions work to define Hutchinson's crime as social rather than theological and, in this way, to broaden its relevance. She is guilty of resistance to lawful authority, on the one hand, and of refusing to play the part of woman, on the other. The failure of domesticity that Hawthorne associates with the eloquence of "the Woman," because he regards it as a violation of the law that governs her nature, points inevitably to heresy and sedition. In Hawthorne's sketch, then, antinomianism names a pattern of opposition shared by the "public woman" of the nineteenth century and the heretic of the seventeenth.

The antinomian lays claim to an unassailable inner knowledge, and the moment when Hutchinson alleges her peculiar powers is both the moment of her condemnation and the dramatic climax of Hawthorne's sketch. As Hawthorne understood, knowledge of the kind Hutchinson claimed respects no authority outside itself and is susceptible to no proof. Witnessed by an invisible spirit taken to be divine, it supersedes all other forms of knowledge. The antinomian regards this knowledge as a product of the influx of divinity and herself merely as the organ of its expression. Potentially, at least, the inner certainty this knowledge lends empowers the individual to act without reference to external authority or even against it. But the antinomian's knowledge transcends her condition. She must continue to live in the world and act the part of the sinner. Having relinquished the old natural self to God, however, she no longer attributes the usual significance to her words and deeds, for these cannot adequately represent the new self that belongs to God. Her deeds are hers only in the most provisional sense.

The uncomfortable resemblance between the language of antinomianism and a dominant American rhetoric committed to the values of individualism and the unfettered expression of the self in the political and economic arenas as well as the private one has encouraged scholars to generalize the problem of antinomianism into one simply of authority, of how the claims of the individual are to be balanced against those of the community. And, in fact, translated into the secular language of self and community, the problem of antinomianism might be said to stand at the heart of liberal ideology. But this shift in terms suppresses the role of gender in our formulation of the heresy and, in this way, replicates earlier versions of Hutchinson's story in which gender is likewise the suppressed issue. In other words, by generalizing the problem of antinomianism, we set aside the very story that has defined antinomianism for Americans. By placing "the Woman" in the "centre of all eyes" (*TS*, 23) [162], Hawthorne's sketch recalls us to that story. It reminds us that the problem of antinomianism and the problem of female empowerment are entwined from the very beginning.

JOHN NICKEL

Hawthorne's Demystification of History in "Endicott and the Red Cross"†

Since the "return to history," one of Hawthorne's tales of Puritan history, "Endicott and the Red Cross" (1837), has received increased critical attention. Michael Colacurcio, in his study of Hawthorne's role as a moral historian in the early tales, argues that "'Endicott and the Red Cross' dramatizes, from the seventeenth-century point of view . . . that two very different sorts of specifically religious separations have been unleashed on the colony of the Massachusetts in the 1630's": the figure of John Endicott, according to Colacurcio, represents the "Puritanic" idea of separation from popery, while the character of Roger Williams primarily stands for the more radical separation of spiritual concerns from social and political powers.[1] John Franzosa, employing an Eriksonian model of psychohistory, suggests that Hawthorne sought fame by writing historical tales such as "Endicott and the Red Cross" and that the figure of Williams serves as a "vehicle for Hawthorne's self-presentation."[2] Joining these recent interpreters of the tale, Harold Bush states that, through the negative portrayal of Endicott, Hawthorne "presented historical challenges and corrections to the prominent and pervasive romantic constructions of Puritanism."[3] In their studies all three critics focus on Hawthorne's depiction of the Puritans, on his participation in the historical debate over the import of New England's forebears.

"Endicott and the Red Cross," however, instructs its readers not solely about Puritan history but also about the writing of history itself.

† From *Texas Studies in Literature and Language* 42.4 (2000): 347–62. Page numbers in square brackets refer to this Norton Critical Edition.

1. Michael Colacurcio, *The Province of Piety: Moral History in Hawthorne's Early Tales* (Cambridge: Harvard UP, 1984), 230.
2. John Franzosa, "Young Man Hawthorne: Scrutinizing the Discourse of History," in *Self, Sign, and Symbol*, ed. Mark Neuman and Michael Payne (Lewisburg, Penn.: Bucknell UP, 1987), 86.
3. Harold Bush, "Re-Inventing the Puritan Fathers: George Bancroft, Nathaniel Hawthorne, and the Birth of Endicott's Ghost," *ATQ* 9 (June 1995): 132.

Composing the tale during the first boom period of U.S. historiography, Hawthorne is wary of contemporary histories' tendency to produce myths, to represent what Hawthorne considers to be subjective interpretations of the past as revelations of natural, eternal laws.[4] This tendency characterizes romantic historiography, the dominant, coeval mode of narrating the past, exemplified in the writings of George Bancroft and William Prescott. As we will see, by employing and subverting the conventions of romantic historiography in the tale, Hawthorne exposes their biased character and process of producing myth. Hawthorne's own narration of the past, I believe, both converges with and diverges from romantic historiography.

My reading of Hawthorne departs, in turn, from the few studies that examine his fiction in relation to the historiography of the time in that I see Hawthorne directly engaging with and challenging contemporary modes of constructing history.[5] A pointed satire of romantic historiography, "Endicott and the Red Cross" presents us with an opportunity to explore how Hawthorne intervenes in the social arena through his romance art, which was once thought to be hardly concerned with "*history as history*."[6] Only by situating "Endicott and the Red Cross" in the context of contemporary historiography can we understand the curious formal features of the tale: deliberate self-contradiction in the conclusion; narration of a historical event that appears more like an imaginative reverie than a historical account; and abundance of symbols including the red cross, which is defaced because of its symbolism.

Imagination, Objectivity, and the Writing of History

The most influential romantic history in the U.S., of course, is Bancroft's *History of the United States*, the first volume of which was published in

4. Hawthorne's view of romantic historiography resembles Roland Barthes's description of myth as a system of communications that hides its ideological message. In contrast to a straightforward political statement, a myth relies on images or formulas whose associations communicate the myth's critique of romantic meaning. As we will see, moreover, Hawthorne's critique of romantic histories bears similarities to what Barthes describes in *Mythologies* as a way to attack myths: "the best weapon against myth is perhaps to mythify it in its turn, and to produce an *artificial myth*: and this reconstituted myth will in fact be a mythology." The "artificial myth," according to Barthes, differs from other myths in that it reveals itself to be a myth, a mode of communication. For this effect to occur, another point of view is offered that enables the readers to see the myth as not innocent but motivated; thus, Barthes writes, the artificial-myth maker "has strewn his reconstitution with supplementary ornaments which demystify it" (Roland Barthes, *Mythologies*, trans. Annette Lavers [New York: Hill and Wang, 1972], 136).
5. See Harry B. Henderson, *Versions of the Past: The Historical Imagination in American Fiction* (New York: Oxford UP, 1974), 91–126; and Susan Mizruchi, *The Power of Historical Knowledge: Narrating the Past in Hawthorne, James, and Dreiser* (Princeton: Princeton UP, 1988), 83–134.
6. Seymour Gross's often-quoted formulation—"*history as history* had but very little meaning for Hawthorne artistically"—has been challenged by many Hawthorne critics (Seymour Gross, "Hawthorne's 'My Kinsman, Major Molineux': History as Moral Adventure," *Nineteenth-Century Fiction* 12 [September 1957]: 99). Some discussions of the treatment of history in Hawthorne's fiction include Emily Miller Budick, *Fiction and Historical Consciousness: The American Romance Tradition* (New Haven: Yale UP, 1989), 36–54, 79–142; Alide Cagidemetrio, *Fictions of the Past: Hawthorne & Melville* (Amherst: Institute for Advanced Study in the Humanities, 1992), 3–58; Robert Daly, "'We Have Really No Country at All': Hawthorne's Reoccupations of History," *Arachne* 3 (1996): 67; George Dekker, *The American Historical Romance* (Cambridge: Cambridge UP, 1987), 129–85; John P. McWilliams, *Hawthorne, Melville, and the American Character: A Looking-Glass Business* (Cambridge: Cambridge UP, 1984), 25–48; and Alan O. Weltzien, "The Picture of History in 'The May-Pole of Merry Mount,'" *Arizona Quarterly* 45 (Spring 1989): 29–48.

1834 and quickly became a success. A biographer of Bancroft, Russel Nye, notes that "within a year [of the first volume's publication] Bancroft's book had found its way into nearly a third of the homes of New England, and the author's name was well on the way to becoming a household word."[7] One of the first U.S. historians to study in Germany, where he was taught by Heeren, Hegel, Savigny, and Böckh, Bancroft narrates in his *History* the struggle for freedom in the colonization and revolution. For the patriotic and melioristic Bancroft, the role of the historian is to reveal his country's divinely inspired liberation, as he insists in the famous introduction to the 1834 volume of *History of the United States*:

> It is the object of the present work to explain how the change in the condition of our land has been accomplished; and, as the fortunes of a nation are not under the control of blind destiny, to follow the steps by which a favouring Providence, calling our institutions into being, has conducted the country to its present happiness and glory.[8]

Bancroft's *History,* with its treatment of national progress, serves to defend the present state of affairs of Jacksonian America and presents an idealized vision of the future.

Like other romantic historians, especially Prescott, Bancroft views past events through a transcendental, systematic framework, locating what he believes to be moral laws operating in the past and then tracing their evolution. Discussing U.S. histories at this time, de Tocqueville notes their tendency to present a "single great historical system"; he writes that "historians who live in democratic ages are not only prone to assign a great cause to every incident, but they are also given to connect incidents together to deduce a system from them."[9] Similarly, David Levin, in his study of U.S. romantic historians, states that they tried to abstract a principle behind every major action, and this principle or moral is what gave unity to the past and what the historian kept constantly before the reader.[1]

To recreate the past, romantic historians use their imagination as well as literary conventions found in novels such as symbolism, character types and sketches, and picturesque scenery. As part of their effort to appeal to audiences, romantic historians also employ dramatic techniques, including dialogue and stirring incidents: Bancroft even received critical praise for his dramatic ability.[2] While often emulating the writing techniques of historical fiction, romantic historians make sure to distinguish their history from historical romance, noting, for example, their avoidance of "imaginary" conversations and their fidelity to documents.[3] The rigorous, systematic examination of historical documents, for romantic historians, gives their endeavors scientific authenticity. As Richard Vitzthum writes, Bancroft, for instance, posits that "the historian approaches a historical subject the way an astronomer views an

7. Russel B. Nye, *George Bancroft: Brahmin Rebel* (New York: Knopf, 1945), 102.
8. George Bancroft, *History of the United States from the Discovery of the American Continent*, vol. 1 (1834; rpt., Boston: Little, Brown, and Co., 1855), 4.
9. Alexis de Tocqueville, *Democracy in America*, ed. Richard Heffner (New York: Mentor, 1956), 186.
1. David Levin, *History as Romantic Art: Bancroft, Prescott, Motley, and Parkman* (New York: AMS, 1967), 27.
2. On romantic historians' use of literary conventions, see Levin, *History*, 7–22.
3. Levin, *History*, 11.

unknown constellation or a biologist studies a new species of plant": the historian observes the subject as accurately as possible, exhaustively analyzes and compares all available documents, and then extracts the cause of historical events and, most importantly, the moral law present in the past.[4] A romantic historian like Bancroft believes that the historical narrative is a natural extension of the scientific analysis of documents, an objective, truthful representation of the discovered natural laws of the past. Conceived as such, romantic histories—especially in the narrative proper—hide rather than call attention to their invented or constructed character.

The romantic histories of New England and the U.S. were read extensively by Hawthorne, as Marion Kesselring's well-known account of Hawthorne's reading reveals. In fact, according to Kesselring, a few months before he wrote "Endicott and the Red Cross," Hawthorne borrowed the first volume of Bancroft's *History* from the Salem Athenaeum (from April 13 to May 6, 1837).[5] Based on the assumption, shared by romantic historians, that a historical re-creation will most engage and involve the reader in the past, Hawthorne's own mode of historical writing also uses imagination to recreate history. In the opening of an early sketch, "Sir William Phips" (1830), Hawthorne defends his fictional writing of history. Historical figures, he states, "seldom stand up in our imaginations like men." "So," he asserts,

> a license must be assumed in brightening the materials which time has rusted and in tracing out the half-obliterated inscriptions on the columns of antiquity: fancy must throw her reviving light on the faded incidents that indicate character, whence a ray will be reflected, more or less vividly, on the person to be described.[6]

In this work, Hawthorne distinguishes his historical writing from the chronicle or record of past events, as he does in "My Kinsman, Major Molineux" (1831), where he states that his tale differs from the "long and dry detail of colonial affairs" (11:209).

Whereas romantic historians like Bancroft view their histories as objective representations of the past, Hawthorne acknowledges and embraces his historical writing as a subjective creation. Since original events in the past as they actually occurred are inaccessible and history is received only through textual traces, Hawthorne believes the imagination to be crucial, not just in the process of writing but in the very act of constructing meaning for the past. Hawthorne, writes Johannes Kjørven, "regards

4. Richard C. Vitzthum, *The American Compromise: Theme and Method in the Histories of Bancroft, Parkman, and Adams* (Norman: U of Oklahoma P, 1974), 34. Vitzthum discusses Bancroft's claim that history in general and his *History* in particular are scientific, drawing on many of the ten volumes of Bancroft's *History* (32–36).
5. Kesselring's study shows the dates that Hawthorne checked out Bancroft's *History* (Marion L. Kesselring, *Hawthorne's Reading, 1828–1850: A Transcription and Identification of Titles Recorded in the Charge-Books of the Salem Athenaeum* [1949; rpt., Folcraft, Penn.: Folcraft, 1969], 39). Following Neal Frank Doubleday, Lea Bertani Vozar Newman notes that the tale was probably composed in the early fall of 1837. See Neal Frank Doubleday, *Hawthorne's Early Tales: A Critical Study* (Durham: Duke UP, 1972), 101; and Lea Bertani Vozar Newman, *A Reader's Guide to the Short Stories of Nathaniel Hawthorne* (Boston: G. K. Hall, 1979), 89.
6. Nathaniel Hawthorne, "Sir William Phips," in *Miscellaneous Prose and Verse,* ed. Thomas Woodson, vol. 23 of *The Centenary Edition of the Works of Nathaniel Hawthorne* (Columbus, Ohio: Ohio State UP, 1994), 59. All the following quotations from Hawthorne's works are from this edition and will be cited parenthetically (Nathaniel Hawthorne, *The Centenary Edition of the Works of Nathaniel Hawthorne*, ed. William Charvat, 23 vols. [Columbus: Ohio State UP, 1963–1994]).

any experience of a historical fact in terms of a reading imposed upon the fact."[7] Imaginative historical interpretations, though, for Hawthorne, should not completely divorce themselves from past reality. Attempting to connect the past and imaginative associations of it, Hawthorne's historical romance is written, as he states in the "Custom-House" (1850), in "a neutral territory, somewhere between the real world and fairy-land, where the Actual and the Imaginary may meet, and each imbue itself with the nature of the other" (1:36) [28]. Although Hawthorne's stories are not literally true and do not objectively represent the "Actual," Hawthorne would claim that his mode of telling history still contains truth about the past—an imaginative truth.

A Critique of Romantic Historiography

Hawthorne would agree with Hayden White that historical narratives are "verbal fictions, the contents of which are as much invented as found and the forms of which have more in common with their counterparts in literature than they have with those in science."[8] Romantic histories in particular, Hawthorne perceives, are constructs that often present distorted views of the past, masked as natural and divine laws, and socialize their readers to these views. In "Endicott and the Red Cross," first published in *The Salem Gazette* on November 4, 1837,[9] Hawthorne employs conventions of romantic historiography and reveals them for what they are—literary conventions manipulated to suit the author's subjective perspective. Hawthorne, in telling the tale, provides his readers with interpretive tools to recognize the literary and rhetorical techniques of romantic historiography and immunize themselves against the power of these techniques to spread ideology. In this way, Hawthorne uses his historical romance to teach his audience how to read romantic history.[1]

One literary device that romantic histories often rely on—and one that Hawthorne employs subversively in his tale—is the representative hero, a figure who embodies the pervading principle of the time. Levin describes the representative hero in romantic historiography:

> The ideally representative man . . . was the incarnation of the People. He represented national ideals. He acted in the name of the People, and they acted through him. The relationship was emotional, often mystical. However lofty the leader was, he loved the People. When he had to, he reprimanded them, and he often rejuvenated

7. Johannes Kjørven, "Hawthorne and the Significance of *History*," in *Americana Norvegica: Norwegian Contributions to American Studies*, vol. 1, ed. Sigmund Skard and Henry Wasser (Philadelphia: U of Pennsylvania P, 1966), 112.
8. Hayden White, "The Historical Text as Literary Artifact," in *The Writing of History: Literary Form and Historical Understanding*, ed. Robert Canary and Henry Kozicki (Madison: U of Wisconsin P, 1978), 42. One of the foremost critics on historical writing, White discusses the literary nature of historical narratives in this essay. For an extensive analysis of the rhetorical tropes employed by the leading European historians of the nineteenth century, see his *Metahistory: The Historical Imagination in Nineteenth-Century Europe* (Baltimore: John Hopkins UP, 1973).
9. "Endicott and the Red Cross" was also printed in the 1838 edition of *The Token*, which appeared in 1837, and in the second series of *Twice Told Tales*, which came out in 1842.
1. By Hawthorne's "readers" and "audience," I am referring to his ideal audience. The satirical critique, we will see, of romantic histories in "Endicott and the Red Cross" is subtle, and, as a reviewer of this essay noted, it is uncertain whether Hawthorne's intention would have been generally recognizable to an actual contemporary audience.

> them in a moment of peril. Every one of the historians iterated a cliché that dramatizes this relationship: in battle after battle the leader "infused his spirit" into his men, or "animated them with his own spirit," or "inspired them with his own energy," or "breathed his own spirit into them."[2]

Bancroft, following the lead of the many biographies of George Washington published in the early part of the century, treats Washington as such a hero, particularly in the ninth volume when Bancroft describes Washington rallying his troops after their loss in the Battle of Long Island. To the same effect, Prescott in *History of the Reign of Ferdinand and Isabella, the Catholic*, which appeared in three volumes from 1837 to 1841, describes the ability of the Spanish commander, Gonsalvo de Cordova, to "infuse heart into his followers" and display "that inflexible constancy which enables the strong mind in the hour of darkness and peril to buoy up the sinking spirits around it."[3]

Hawthorne parodies this convention in his treatment of Endicott as a representative hero, with Endicott resembling more an ill-tempered tyrant bent on securing his own authority than a true leader of the people. As Endicott, who is in charge of the Salem community and its militia, reads Governor Winthrop's letter that reports the plans of Charles I to send a governor-general to the colonies, a "wrathful change came over his manly countenance," and his "blood glowed through it, till it seemed to be kindling with an internal heat; nor was it unnatural to suppose that his breastplate would likewise become red hot, with the angry fire of the bosom which it covered" (9:437) [166]. As opposed to the heroes depicted in romantic histories who keep their composure, remaining calm in drastic situations, Endicott allows anger to overwhelm him—anger that, along with his desire to assert power, prompts his supposedly courageous deed.

After summoning the villagers and militia, he delivers an emotion-filled speech, in which he tells them about the letter and calls on them to resist the king's tyranny. Endicott's speech is interrupted by a male villager being punished for offering his own interpretation of the New Testament, different from that of the church leaders, and forced to wear the label, "a wanton gospeller." Endicott "in the excitement of the moment, shook his sword wrathfully at the culprit,—an ominous gesture from a man like him" and threatens, "Break not in upon my speech; or I will lay thee neck and heels this time till tomorrow" (439) [167]. At the continuation of Endicott's agitative speech, according to the narrator, a "deep groan from the auditors,—responded to the intelligence" (440) [168]. Declaring the colony's independence just prior to his triumphant act—the rending of the red cross from the militia's royal banner—Endicott "gazed round at the excited countenances of the people, now full of his own spirit" (440) [168]. The crowd has become imbued with his "spirit," though not because of his heroism but because of his fanatical and menacing character. Through this depiction of Endicott, Hawthorne, in effect, warns his readers against uncritically accepting romantic histories' portrayals of heroic figures.

2. Levin, *History*, 50–51.
3. William H. Prescott, *History of the Reign of Ferdinand and Isabella, the Catholic*, vol. 3 (1841; rpt., Philadelphia: Lippincott, 1872), 131.

Symbolizing the principle of liberty, the representative hero's actions in romantic histories usually foreshadow future events in which liberty triumphs over tyranny. The romantic historians' treatment of the hero's deeds functions as part of the historians' application of typology—the Christian theological doctrine that events of the Old Testament predict events in the New Testament—to historical writings. Romantic historians view New England colonial history in light of the revolution, searching for types of the victory of liberty. In Michael Davitt Bell's words, "Each instance of the struggle between liberty and tyranny, each emergence of embryonic democracy, could be regarded as a type of the great culminating example of the victory of liberty over tyranny—the American Revolution."[4] While reviewing a volume of Bancroft's *History*, Prescott, for instance, claims that the "principle" of American colonial history was the "tendency to independence." "It is this struggle with the mother-country," Prescott states, "this constant assertion of the right of self-government, this tendency—feeble in the beginning, increasing with increasing age—towards republican institutions, which connects the Colonial history with that of the Union and forms the true point of view from which it is to be regarded."[5] To Prescott's comments Bancroft would add that Jacksonian democracy, in particular, is a further extension of revolutionary teleology.

Rendering Endicott as a precursor of the American revolution, Hawthorne unveils how romantic historians impose their revolutionary interpretations on the past. At times, in fact, the tale's narrator, antiquarian and usually uncritical, resembles a romantic historian. After dramatically describing the increasing tyranny of Charles I at the outset of the tale, the narrator maintains: "There is evidence on record, that our forefathers perceived their danger, but were resolved that their infant country should not fall without a struggle, even beneath the giant strength of the King's right arm" (9:433) [164]. The narrator reveals himself to be a patriotic member of the New England community, one who might not be objective in narrating a historical event of the Salem colony. The bias of the narrator's history, however, becomes clear in Endicott's incendiary speech. The rhetoric of the speech is anachronistic to New England in the 1630s. Endicott's invocations of "liberty" and "civil rights" and his rhetorical questions—"Who shall enslave us here?" and "What have we to do with England?"—belong rather to the 1760s and 1770s,[6] the implication being that the narrator has embellished Endicott's speech by inserting this revolutionary language (9: 439–40) [168].

The prejudiced, constructed nature of the tale's typological history is also apparent in its conclusion, where, after Endicott has excised the flag's red cross because it represents royal tyranny and popery, the narrator proclaims:

> We look back through the mist of ages, and recognise, in the rending of the Red Cross from New England's banner, the first omen of that deliverance which our fathers consummated, after the bones of the stern Puritan had lain more than a century in the dust. (9:441) [168]

4. Michael Davitt Bell, *Hawthorne and the Historical Romance of New England* (Princeton: Princeton UP, 1971), 8.
5. Quoted in Michael Davitt Bell, *Hawthorne*, 8.
6. Colacurcio, *Province of Piety*, notes the historically incorrect rhetoric of the speech (232).

The tale's earlier depiction of the leader's own tyranny points to the unreliability of the narrator's final claim, that Endicott's action prefigures the revolution. Moreover, the reference to Endicott's "bones" in the "dust" undercuts the loftiness of the theme, revealing a discrepancy between the Salem leader's actual legacy, a decayed corpse, and its idealized glorification by the narrator.

Hawthorne further subverts romantic historians' use of typology in his treatment of Endicott's breastplate, which Sacvan Bercovitch identifies as the "dominant symbol" of the tale. The breastplate, Bercovitch notes, appears in the Old Testament as an emblem of Israelite theocracy. Moses gave his brother, Aaron, a breastplate in fulfillment of God's commandment to establish a sanctuary for God among God's people.[7] On one hand, then, the breastplate in Hawthorne's tale signals that Endicott is a descendent, a typological fulfillment, of the Israelites who is carrying out God's mission in the New Canaan of America. But in the tale, the breastplate does not just function symbolically; it reflects the scene in the village—"This piece of armour was so highly polished, that the whole scene had its image in the glittering steel" (9:434) [164]. When looking at Endicott, one sees the breastplate's reflection of the fierce punishments in the village: on the steps of the church are the "wanton gospeller" and a woman wearing a "cleft stick on her tongue" because she criticized the church elders; a suspected Catholic and a royalist are confined to a pillory and the stocks, respectively; and in the crowd in the square are villagers with cropped ears, branded cheeks, and slit and seared nostrils, and a woman who, resembling Hester Prynne, is forced to wear the letter "A" on her dress (9:435) [165]. Showing Endicott to be a cruel fanatic who is not divinely inspired or fulfilling a providential mission, the breastplate, as a self-contradictory symbol, casts suspicion on the mythic symbols and typology employed in historical accounts.

In the portrayal of Endicott, Hawthorne deliberately constructs a false myth (with enough traces of Endicott's cruelty to enable his readers to deconstruct it). Hawthorne's historical sources for the tale clearly show that Endicott's reason for rending the red cross was not political but religious. For New England Puritans, it was idolatrous to use any sacred symbol, and the cross, furthermore, was associated with the Roman Catholic Church.[8] Hawthorne's knowledge of Endicott's true motivation appears in the account of Endicott's action in *Grandfather's Chair*, Hawthorne's book of stories of New England history for children, which was published in 1840. In this work, when the child asks her grandfather whether Endicott's action was "meant to imply that Massachusetts was independent of England," the elderly storyteller replies: "I doubt whether he had given the matter much consideration except in its religious bearing" (6:24).

In Hawthorne's tale the subjectively constructed nature of mythic histories is illustrated not only in the narrator's account of Endicott's action but also in Endicott's telling of the colony's history in his speech. Like

7. Sacvan Bercovitch, "Endicott's Breastplate: Symbolism and Typology in 'Endicott and the Red Cross,'" *Studies in Short Fiction* 4 (Summer 1967): 289–91.

8. Doubleday, *Hawthorne's Early Tales*, analyzes Hawthorne's relation to his historical sources in the early tales and provides an informative discussion of the subject in "Endicott and the Red Cross" (101–108). For an examination of the religious contexts of Endicott's action, though one that only addresses Hawthorne's tale by way of introduction, see Francis J. Bremer, "Endicott and the Red Cross: Puritan Iconoclasm in the New World," *Journal of American Studies* 24 (April 1990): 5–22.

the narrator and romantic historians, Endicott narrates a highly rhetorical and evocative history, first asking the colonists,

> Wherefore, I say, have we left the green and fertile fields, the cottages, or, perchance, the old gray halls, where we were born and bred, the church-yards where our forefathers lie buried? Wherefore have we come hither to set up our tombstones in a wilderness?

"A howling wilderness it is!" he continues, invoking threatening images of the colony's surroundings. "The wolf and the bear meet us within halloo of our dwellings. The savage lieth in wait for us in the dismal shadow of the woods" (9:438) [167]. Though Endicott treats his account as objective, the reader of "Endicott and the Red Cross" cannot help but notice its misrepresentation of the colony's situation. The head of a "slain" wolf, for instance, has just been nailed to the steps of the church—"The blood was still plashing on the door-step" (434) [164]—evincing that the colony endangers wild animals and not the converse. In addition, American Indians are not hiding in the forest but openly watching Endicott lead the militia in its drills. Earlier in the tale, the narrator stated that the Indians' "flint-headed arrows were but childish weapons, compared with the matchlocks of the Puritans," and Endicott even boasted of the militia's clear military superiority to the Indians (436) [165].

Endicott's interpretation of the colony's history is soon contested openly. Endicott, similar to romantic historians, ascribes ideal motives to the founding of the colony: "Wherefore, I say again, have we sought this country of a rugged soil and wintry sky? Was it not for the enjoyment of our civil rights?" He then adds, "Was it not for the liberty to worship God according to our conscience?" To which, the "wanton gospeller" responds, "Call you this liberty of conscience?" (439) [167]. Roger Williams's reaction to this interruption is a "sad and quiet smile," an indication that he also believes that Endicott's actions belie his historical interpretation. After threatening the young man, as we have seen, Endicott, resuming his history, symbolically calls the continent a "new world" where they can "seek a path from hence to Heaven" (439) [167].

By incorporating a figure of a mythmaker in Endicott, Hawthorne takes his early nineteenth-century readers back to the 1630s to show them that the grand, ideal mission of the New England colony's founding was not inherent in the spirit of the colony but was invented. This skillful maneuver demystifies history by historicizing the myth of American colonization, revealing—albeit fictionally—when and how it originated.[9] The specific, historically determined context of Endicott's mythic history of the colonization is his fear of losing his authority in the Salem community, and the import of his history is that he should remain in power in order to lead the colony's resistance. Hawthorne, then, implicitly criticizes not only Endicott's role as a leader but also his manner of history telling: the hegemonic Endicott constructs a deceitful analogy between the history's images and message,

9. I echo Richard Slotkin's statement on the relation of history and myth, influenced by Barthes's *Mythologies*: "We can only demystify our history by historicizing our myths—that is, by treating them as human creations, produced in a specific historical time and place, in response to the contingencies of social and historical life" (Slotkin, "Myth and the Production of History," in *Ideology and Classic American Literature*, ed. Sacvan Bercovitch and Myra Jehlen [Cambridge: Cambridge UP, 1986], 80).

between the revolutionary associations of the colony and his self-serving agenda. By invoking the colonists' collective experience of their journey to the "new world" and promoting values and images that seemingly transcend the differences between himself and the other colonists, Endicott conceals his own social and political dominance. "Endicott and the Red Cross" evinces that mythic histories—such as those of Endicott and romantic historians—have social implications and an underlying ideological meaning despite appearing to express common sense or universal values. Consenting to Endicott's reading of history for his Salem audience is, in effect, to accept his position of power, just as agreeing with romantic historians' interpretations of the past is also to assent to their praising view of present affairs.

Endicott's mythic history resembles romantic histories in another way, the relation between himself as a historian and the audience. Both the romantic historian and Endicott narrate monological histories, imposing their interpretation of the past on the reader by suppressing differing viewpoints and providing the reader only with their perspective. Vitzthum describes the relation between the speaker and the reader in Bancroft's *History of the United States*: "As in all the volumes of the *History*, the speaker in Volume I stands between the reader and the past, telling him in no uncertain terms what it means, how it should be judged, how much it is worth."[1] Endicott, after reading Winthrop's letter, tells Williams that "if John Endicott's voice be loud enough, man, woman, and child shall hear" the contents of the letter (9:438) [167]. We soon realize that Endicott's voice has to be loud enough to drown out competing voices, such as the "voice" of the "wanton gospeller" (439) [167]. Already punished for giving his own interpretation of the Bible, the young man will be punished further if he offers his own history of the colony's past. When Williams rebukes Endicott for crudely referring to the beheading of Queen Mary Stuart, Endicott "imperiously" responds: "Hold thy peace, Roger Williams!" (439) [167]. A symbolic tyrant, Endicott suppresses dialogue in order to protect his mythic interpretation of the colony's history and to construct for the colonists their connection with the past.

Historical Romance

What Hawthorne earnestly seeks to avoid in "Endicott and the Red Cross" is forcing his subjective view of the past on the reader. In order to prevent the mythicizing of his history and the reader's uncritical acceptance of the narrative as an accurate reflection of the past, Hawthorne highlights the fictional quality of his historical writing. In "Endicott and the Red Cross," the narrator first describes the village as it appears through the "mirrored picture" of Endicott's breastplate, and it remains unclear whether the whole tale is told from this perspective (9:434) [164]. Endicott's breastplate, like the breastplate hanging in the governor's mansion in *The Scarlet Letter*, is convex, presenting a distorted reflection of the surrounding scene, signaling that the tale is not a transparent medium for the expression of the Salem colony's past. Furthermore, while romantic histories do not refer to the historian in the narrative, providing no signs that the author is organizing the history's materials,

1. Vitzthum, *American Compromise*, 44.

Hawthorne's tale calls attention to the antiquarian narrator as a producer of meanings: "There happened to be visible, at the same noontide hour, so many other characteristics of the times and manners of the Puritans, that we must endeavor to represent them in a sketch" (434) [164]. This tale, in contrast to romantic histories, does not give its readers the sense of a direct relation between themselves and the past. The constituted nature of the tale is heightened in the conclusion when the narrator states that "[w]e look back through the mist of ages" at Endicott's action, suggesting that the tale may have been a dream or an imaginative meditation of the past, which has been perceived through a cloudy "mist" (441) [168].

Another strategy Hawthorne utilizes to keep his readers from passively accepting his representations of history is to refuse to provide them with a reliable moral or interpretation of past events. In the preface to *The House of Seven Gables* (1851), Hawthorne distinguishes himself from writers who "relentlessly . . . impale the story with its moral" (2:2). "When romances do really teach anything," he insists, "or produce any effective operation, it is usually through a far more subtile process than the ostensible one" (2:2). The "subtile process" in "Endicott and the Red Cross" is to encourage readers to create their own morals from history, to participate in the production of meaning. By leaving the readers with a conclusion that contradicts the rest of the tale—a conclusion that ironically celebrates a fanatic tyrant—Hawthorne invites the readers to scrutinize the judgments of the antiquarian narrator and become historians, in effect, who sort through the evidence in the narrative. Complicating the reader's task, Hawthorne presents a dialogic or polyphonic history, one that offers several different ideological voices or perspective—those of Endicott, the narrator, the "wanton gospeller," and Williams, among others—for the reader to compare and evaluate.[2] In doing so, Hawthorne rejects romantic historians' tendency to reduce American colonial history to a set of essences, and, unlike these historians, he involves his readers in a hermeneutic problem, in which they must create history rather than simply reading about it.[3]

Hawthorne's history-making reader also becomes a symbol producer. The symbol, for Hawthorne, expresses the imaginary relation to the past—it is a literary construct used to endow a past event with meaning.[4]

2. Hawthorne's tale is similar to what Mikhail Bakhtin describes as the chief characteristic of Dostoevsky's novels: the "plurality of independent and unmerged voices and consciousnesses, a genuine polyphony of fully valid voices" (M. M. Bakhtin, *Problems of Dostoevsky's Poetics*, ed. and trans. Caryl Emerson [Minneapolis: U of Minnesota P, 1984], 6). For Bakhtin, multivocal or polyphonic works, like Dostoevsky's novels, allow the voices of the main characters as much authority as the narrator's voice, which engages in dialogue with the characters' voices. For an extended discussion of his theory of polyphony and dialogue, see M. M. Bakhtin, "Discourse in the Novel," in *The Dialogic Imagination: Four Essays by M. M. Bakhtin*, trans. Caryl Emerson and Michael Holquist (Austin: U of Texas P, 1981), 259–422.
3. My thinking on the relation between reading and the performance of history has been influenced by J. Hillis Miller's interpretation of "The Minister's Black Veil." Miller argues that the act of reading the tale involves the process of producing history (*Hawthorne* and *History: Defacing It* [Cambridge, Mass.: Basil Blackwell, 1991], 45–132).
4. I do not intend to enter the much discussed debate over whether Hawthorne was more of a symbolist than an allegorist, or vice versa. Millicent Bell has recently summarized the different critical views on this subject and states: "[i]t would appear that for Melville and Hawthorne, as for Emerson, allegory was not distinguished from symbolism with any precision" (introduction to *New Essays on Hawthorne's Major Tales*, ed. Millicent Bell [Cambridge: Cambridge UP, 1993], 23). For a discussion of Hawthorne's role as a symbolist or allegorist, see, for example, F. O. Matthiessen, *American Renaissance: Art and Expression in the Age of Emerson and Whitman* (London: Oxford UP, 1941), 242–315; Charles Feidelson, *Symbolism and American Literature* (Chicago: U of Chicago P, 1953), 6–16; and Ursula Brumm, *American Thought and Religious Typology* (New Brunswick: Rutgers UP, 1970), 11–128.

Hawthorne's use of symbolism is evident in his favorable depiction of William's entrance into the Salem village: with "apostolic dignity" Williams appears as a pious shepherd and Christ-like figure and drinks from a natural spring fountain, a traditional emblem of wisdom and divinity (9:436) [166]. In contrast to romantic historians' employment of mythic symbolism, the always self-reflexive Hawthorne calls attention to his symbols—as well as other literary conventions—as creations. In "Endicott and the Red Cross," the reader is asked to analyze the tale's symbols and select those appropriate to represent the past. In this way the reader partakes in the process of historical symbol invention, as Roy Harvey Pearce argues. Symbolism in Hawthorne's fiction, Pearce suggests, is "expressed as process rather than as source or product."[5]

Through the many symbols in the tale,[6] Hawthorne conveys that human experience relies on sign and symbol production to make sense of the world. The punishments in the colony, for instance, all involve physical manifestations of the purported crime; in the case of the young man accused of providing his own interpretation of the New Testament, the punishment is literally a sign, which reads "a wanton gospeller." The church at first appears to lack a signifier: it has "neither steeple nor bell to proclaim it,—what nevertheless it was,—the house of prayer" (9:434) [164]. The narrator then points out the wolf's head nailed to its porch, which, recalling the animals sacrificed to consecrate altars in the Old Testament, reveals that the church has a symbol after all. When Endicott defaces the royal banner because of the symbolism of its red cross, he deliberately performs an act that stands for the colony's resistance to royal tyranny: the act of destroying a symbol thus becomes a symbol. The readers of the tale, Hawthorne hopes, will recognize that symbols are inevitable and constructed and will then consciously create their own symbols.

To reinforce the importance of an active, creative mode of interpretation, Hawthorne, we can say finally, presents two characters as his ideal readers. The "wanton gospeller," as we have seen, is a resistant reader, who views Endicott's history as a misrepresentation, not a seamless whole, and gives his own interpretation of the Bible, "unsanctioned by the infallible judgment of the civil and religious leaders" (9:435) [164–65]. The second ideal reader is the woman with the symbolic letter A imposed on her. Though "even her own children knew what the initial signified," she refuses to accept passively the meaning attached to it (435) [165]. As the narrator states,

> Sporting with her infamy, the lost and desperate creature had embroidered the fatal token in scarlet cloth, with golden thread and the nicest art of needle-work; so that the capital A might have been thought to mean Admirable, or any thing rather than Adulteress. (435) [165]

5. Roy Harvey Pearce, "Romance and the Study of *History*," in *Hawthorne Centenary Essays,* ed. Pearce (Columbus: Ohio State UP, 1964), 238.
6. Stephen Orton, in "De-centered Symbols in 'Endicott and the Red Cross,'" *Studies in Short Fiction* 30 (Fall 1993): 565–73, provides a Derridean deconstructive reading of some of the symbols in "Endicott and the Red Cross." While this interpretation helps explain on a linguistic level how Hawthorne subverts some of the tale's negative symbols, such as the breastplate and royal flag, it cannot explain more positive symbols, such as those in the description of Williams.

The reference to the "art" of the needleworker identifies her as an artist, an interpreter who recreates symbols. Depicted in situations of reading, these characters come closest to expressing Hawthorne's pedagogic message about the proper way to interpret romantic histories.[7]

An astute critic of contemporary writings and reading assumptions and expectations, Hawthorne sought to disrupt the interpretive conventions of antebellum audiences and give them new reading strategies. While modern critics have often contrasted Hawthorne with contemporary writers who are more overtly socially and politically oriented such as Harriet Beecher Stowe, Hawthorne shares with these writers a strong didactic motivation, an intention to effect cultural change.[8] Though Hawthorne does not provide "Endicott and the Red Cross" with a reliable, central message, we, as the tale's interpreters, can: instead of reading romantic histories in complicity with their authors, we should follow the examples of the "wanton gospeller" and needleworker, challenge these histories' symbolic representations and masked ideologies, and construct our own interpretation of the past.

JOHN RONAN

"Young Goodman Brown" and the Mathers[†]

Nathaniel Hawthorne's tale of a young Puritan of Salem Village who loses his Faith after attending—or dreaming that he has attended—a witch meeting in or around 1692 manifestly draws upon Cotton Mather's *The Wonders of the Invisible World* (1692).[1] The narrator of "Young Goodman Brown" (1835) quotes Mather's infamous "*Memorandum*" from *Wonders* in which he describes Martha Carrier as a "Rampant Hag";[2] the story features material found in Mather's trial histories, most notably a witches' sabbath;[3] and the initials of Hawthorne's protagonist were

7. We can begin to reexamine the ubiquitous figures of reading in Hawthorne's fiction as responses to specific coeval discourses. Given the similarities between the needleworker and Hester Prynne, *The Scarlet Letter* seems a likely place to start, and an interpretation would assume that Hester is a resistant reader. In an influential interpretation of the romance, Bercovitch argues to the contrary and suggests that Hester, as she reveals in her return to the Puritan village, abandons her earlier self-reliance and defiance, internalizes society's constricting values, and "has no choice but to resume the A" (*The Office of The Scarlet Letter* [Baltimore: Johns Hopkins UP, 1991], 21). Bercovitch's reading, which is more complex than I have presented it, does not consider, though, that Pearl's departure to an "unknown region" (1: 262) [154] represents an alternative to the consensus-dominated, Puritan village. Recognizing this alternative before her return, Hester, I would argue, continues to perceive the limits of hegemonic views of her and attach her own meanings to the scarlet letter.
8. In her well-known study, Jane Tompkins discusses the rise of Hawthorne's literary reputation and argues that to Hawthorne's contemporaries "his fiction did not distinguish itself at all clearly from that of the sentimental novelists—whose work we now see as occupying an entirely separate category" (*Sensational Designs: The Cultural Work of American Fiction, 1790–1860* [New York: Oxford UP, 1985], 17). For reasons that Tompkins explains, modern critics have tended to oppose his writing to the popular fiction of his time, praising Hawthorne for his psychological depth and density of composition. See Tompkins, 3–39.

† From *The New England Quarterly* 85.2 (June 2012): 253–80. Page numbers in square brackets refer to this Norton Critical Edition.

1. See Cotton Mather, *The Wonders of the Invisible World* (1692; repr., Boston: Benjamin Harris, 1693), p. 138. See also "Young Goodman Brown," in *The Centenary Edition of the Works of Nathaniel Hawthorne*, ed. William Charvat et. al., 23 vols. (Columbus: Ohio State University Press, 1962–97), 10:86 [176]. All quotations of Hawthorne's work will be from this edition and will be indicated by volume and page number in the text.
2. G. H. Orians, "New England Witchcraft in Fiction," *American Literature* 2 (March 1930): 65.
3. Arlin Turner, "Hawthorne's Literary Borrowings," *PMLA* 51 (June 1936): 545–46.

probably inspired by Mather's epithet for George Burroughs, G. B.[4] Furthermore, "Young Goodman Brown" alludes to the witchcraft narratives Mather wrote shortly after the Salem trials ended in 1693—"A Brand Pluck'd out of the Burning," and "Another Brand Pluck'd out of the Burning"—it likens the devil to Cotton's father, Increase, and it refers to his grandfather John Cotton's catechism, *Milk for Babes*.[5] Evidently, Cotton Mather and his family were very much on Hawthorne's mind when he wrote his famous story about the Salem witchcraft crisis.

Yet, to my knowledge, Michael J. Colacurcio is the only critic to have offered a full-bodied explanation of Mather's robust presence in "Young Goodman Brown." When the devil first appears in the story, Colacurcio notes, he is said to have "an indescribable air of one who knew the world, and would not have felt abashed at the governor's dinner-table, or in King William's court, were it possible that his affairs should call him thither" (10:76) [170]; Satan, he infers compellingly, must in some important way represent Increase Mather, who in 1692 procured the appointment of Sir William Phips as governor of the Massachusetts Bay province from William III in England. Since the Evil One is also described as a fifty-year-old version of Goodman Brown in "expression," dress, and "manner," so that the two might be "taken for father and son" (10:76) [170], Colacurcio concludes that Brown is meant to be a "fictional metonym" for Increase Mather's son, the last great champion of the New England Way and its requirement of "visible sanctity" for church membership. According to this reading, Brown, like Cotton Mather and the magistrates of the Salem Court of Oyer and Terminer, accepts spectral evidence—the testimony of allegedly bewitched persons that they have been harmed physically and psychologically by the supernatural apparitions of other individuals—as sufficient grounds for suspecting even reputable Christians of witchcraft because Congregationalism has taught him to trust appearances; and his consequent fall from faith symbolizes Mather's fall from grace in New England after supporting the trials.[6]

Although Colacurcio offers a plausible account of much that is in the story, Goodman Brown is not at all like Cotton Mather, who was a renowned clergyman, a powerful politician, the author of more than four hundred published works, and arguably the most learned man in New

4. Michael J. Colacurcio, "Visible Sanctity and Specter Evidence: The Moral World of Hawthorne's 'Young Goodman Brown,'" *Essex Institute Historical Collections* 110 (1974): 297.
5. During Goodman Brown's journey to the witches' sabbath, the devil and either Deacon Gookin or the Salem Village minister stop to "pluck" branches, evoking Mather's two "Brand" narratives (10:80, 81) [172, 173]. "A Brand" was not published until 1914, but Hawthorne would have known of its existence through references to it in "Another Brand" in Robert Calef's *More Wonders of the Invisible World* (1700; repr., London: William Carlton, 1796), pp. 19–42. Colacurcio points out Hawthorne's allusion to Increase Mather in his description of the devil (*The Province of Piety: Moral History in Hawthorne's Early Tales* [Cambridge: Harvard University Press, 1984], pp. 311–12). Robert C. Grayson outlines Hawthorne's use of *Milk for Babes* ("Curdled Milk for Babes: The Role of the Catechism in 'Young Goodman Brown,"' *Nathaniel Hawthorne Review* 16 [Spring 1990]: 1–5).
6. Colacurcio, *Province of Piety*, pp. 283–313. Colacurcio's chapter on "Young Goodman Brown" in the *Province of Piety* is the second of three versions of his essay on Hawthorne's tale—the first is cited above in n. 4—and, strikingly, he does not identify Goodman Brown with Cotton Mather in the third and final draft ("'Certain Circumstances': Hawthorne and the Interest of History," in *New Essays on Hawthorne's Major Tales*, ed. Millicent Bell [New York: Cambridge University Press, 1993], pp. 37–66). For an extensive discussion of spectral evidence, the Salem witch trials, and the origins of democracy in America, see Nancy Ruttenburg, *Democratic Personality: Popular Voice and the Trial of American Authorship* (Stanford: Stanford University Press, 1998), pp. 31–82 and passim.

England at the end of the seventeenth century.[7] By contrast, Brown is a "simple husbandman" (10:77) [171] and a false convert who seems not to understand some of the most basic tenets of Calvinist Christianity.[8] What is more, the few biographical details the narrative actually provides about Brown's family indicate that he is a Hathorne, not a Mather. Satan informs Brown that his grandfather "lashed [a] Quaker woman . . . through the streets of Salem" (10:77) [170]; Hawthorne's great-great-great grandfather, Major William Hathorne, ordered the whipping of the Quaker Ann Coleman through the streets of Salem, Boston, and Dedham. In addition, the Evil One asserts that Brown's deceased father fought in King Philip's War, as did Major Hathorne's son Captain William Hathorne, who died more than a decade before the witch trials. The captain's widow, Sarah Ruck, married George Burroughs, who would be hanged by William's brother (and Hawthorne's great-great grandfather) John Hathorne's court in 1692. The fit is by no means perfect—William Hathorne, for instance, did not whip Coleman himself—and the devil's testimony is always suspect. Nevertheless, it seems likely that Hawthorne thought of Goodman Brown as a fictitious son of Captain William Hathorne and thus as the stepson of George Burroughs.[9]

Goodman Brown is, I will argue, first and foremost an "Everyman figure"[1] for late seventeenth-century New England Puritans who, in Hawthorne's view, were led to the Salem disaster by Increase and Cotton Mather. Specifically, as I will contend, in "Young Goodman Brown" Hawthorne proposes that Increase Mather's book *An Essay for the Recording of Illustrious Providences* (1684) laid the cultural groundwork for 1692 when it introduced Restoration demonology into America. Then, after the Salem outbreak had developed into a full-fledged crisis, Hawthorne asserts, Cotton aggravated it when, in a sermon of 4 August 1692 published in *The*

7. In Hawthorne's "Famous Old People" (1841), Grandfather says that in 1707 Mather was "the most learned man . . . that had ever been born in America" (6:92).
8. Claudia G. Johnson argues persuasively for Goodman Brown's being a false convert ("'Young Goodman Brown' and Puritan Justification," *Studies in Short Fiction* 11 [Spring 1974]: 200–203). Benjamin Franklin V demonstrates that Brown has not learned "the fundamental beliefs of his faith," as articulated in Cotton's catechism ("Goodman Brown and the Puritan Catechism," *ESQ: A Journal of the American Renaissance* 40 [1994]: 82). And Jane Donahue Eberwein, noting Brown's ignorance of "authentic Puritan theology," contends that he "inhabit[s] a fictively Puritan world which seems, at base, not really Christian" ("My Faith Is Gone! 'Young Goodman Brown' and Puritan Conversion," *Christianity and Literature* 32 [October 1982]: 30).
9. For the Burroughs-Ruck-Hathorne connection, see Enders Robinson, *The Devil Discovered: Salem Witchcraft 1692* (New York: Hippocrene Books, 1991), pp. 87–91. Captain William Hathorne (1645–78) died childless (Vernon Loggins, *The Hawthornes: The Story of Seven Generations of an American Family* [New York: Columbia University Press, 1951], p. 84; my thanks to Sam Coale for locating this reference). The narrator of "Young Goodman Brown" says that the devil looks "about fifty years old" (10:76), the age of Judge Hathorne (1641–1717) when the Salem trials began; but Hawthorne makes a point of noting that Brown's father is dead in the tale, and Captain Hathorne would have been forty-seven in 1692 had he lived, which is certainly "about fifty." Hawthorne undoubtedly saw Mather's reference in *Wonders* to Salem court testimony that Burroughs "had been the Death of" his first two wives, including Ruck, who died sometime before 1690; he must have speculated, therefore, about John Hathorne's emotions when he examined the man in 1692 (see Mather, *Wonders of the Invisible World*, pp. 98, 101–2). However, given his extremely sympathetic portrayal of Burroughs in "Alice Doane's Appeal" and "Main-street" (1849), it is unlikely that Hawthorne believed the charges himself when he wrote "Young Goodman Brown." See Hawthorne, *Works*, 11:76–77, 279. For more on Burroughs, Ruck, and Hathorne, see Mary Beth Norton, *In the Devil's Snare: The Salem Witchcraft Crisis of 1692* (2002; repr., New York: Vintage Books, 2003), pp. 125, 130–31.
1. Eberwein suggests that Brown is such a figure but argues that his function in the story is to illustrate the connection between New England's decline in faith over the course of the seventeenth century, its "increasingly exacting standards of piety," and the spectral evidence theory of guilt that fueled the Salem trials ("My Faith is Gone!" pp. 30–31).

Wonders of the Invisible World, he attempted to use the panic to launch a religious revival in Massachusetts.

A Salem native and the great-great grandson of a judge, Hawthorne (1804–64) undoubtedly imbibed family and town lore about the witchcraft crisis from a young age. But the author's perspective on the Salem trials in "Young Goodman Brown" is drawn mainly from a number of books and shorter histories that he read in the late 1820s and early 1830s. Major accounts of the crisis that offer both analysis and lengthy selections from seventeenth-century sources—Francis Hutchinson's *An Historical Essay Concerning Witchcraft* (1718), Daniel Neal's *The History of New-England* (1720), Thomas Hutchinson's *The History of the Province of Massachusetts-Bay* (1767), and Charles W. Upham's *Lectures on Witchcraft* (1831)—could alone have supplied Hawthorne with almost all of the information he includes in his fictional treatments of the Salem tragedy, from "Alice Doane's Appeal" (1835) to *The House of the Seven Gables* (1851).[2] Before writing "Young Goodman Brown," Hawthorne had read Robert Calef's *More Wonders of the Invisible World* (1700).[3] Among the other contemporary sources he had perused were Thomas Brattle's "Letter" (1692) and Increase Mather's *Cases of Conscience Concerning Evil Spirits Personating Men* (1693).[4] He had probably taken a look at Deodat Lawson's *A Brief and True Narrative* (1692).[5] And, of course, he knew Cotton Mather's *The Wonders of the Invisible World* as well as his *Magnalia Christi Americana* (1702).[6]

2. Hawthorne withdrew Francis Hutchinson's and Daniel Neal's histories from the Salem Athenaeum in 1827 and checked out Thomas Hutchinson's book in 1826 and 1829. (He read the 1747 edition of Neal and the 1795 version of Hutchinson, published as the second volume of *The History of New-England*). In addition to these sources, Hawthorne read William Bentley's brief account of the Salem witch hunt in his essay "A Description and History of Salem," which appeared in volume 6 (1800) of the *Collections of the Massachusetts Historical Society,* and he read the overview of the crisis featured in Joseph Felt's *Annals of Salem* (1827). Hawthorne withdrew the volume containing Bentley's history in 1826, 1827, and 1830, while he borrowed Felt in 1833, 1834, and 1849 (see Marion Kesselring, *Hawthorne's Reading: 1828–1850* [New York: New York Public Library, 1949], pp. 50, 53, 56, 58). As for Charles Upham's *Lectures on Witchcraft,* Colacurcio establishes that Hawthorne alludes to them in "Alice Doane's Appeal," written before "Young Goodman Brown" (*Province of Piety,* p. 88).
3. "Alice Doane's Appeal" contains a slightly inaccurate version of Calef's account of the 19 August executions of George Burroughs, John Proctor, John Willard, George Jacobs Sr., and Martha Carrier (see Hawthorne, *Works,* 11:278–79, and Calef, *More Wonders,* pp. 220–23). Hawthorne also alludes to Mather's "Brand" narratives in "Young Goodman Brown," the second of which, as noted above, was published in *More Wonders.*
4. Brattle's "Letter" was published in the *Collections of the Massachusetts Historical Society,* vol. 5 (1798), which Hawthorne withdrew from the Salem Athenaeum in 1830 (see Kesselring, *Hawthorne's Reading,* p. 56). Hawthorne cites "The Return of Several Ministers," found in the "Postscript" of Increase Mather's *Cases of Conscience,* in his 1830 sketch "Sir William Phips" (see Hawthorne, Works, 23:62, and Mather's *Cases of Conscience Concerning Evil Spirits Personating Men* [Boston: Benjamin Harris, 1693], pp. 73–74).
5. Goodman Brown's astonishment at seeing that Goody Cloyse is an old friend of the devil's and is on her way to the witches' sabbath (Hawthorne, *Works,* 10:78–80) [171–73] is reminiscent of Abigail Williams's testimony before the Salem court that she was surprised to see Sarah Towne Cloyce taking communion at a witch meeting, testimony that, as far as I am aware, Hawthorne could have found only in Lawson's text (*A Brief and True Narrative,* in *Narratives of the Witchcraft Cases 1648–1706,* ed. George Lincoln Burr [New York: Barnes and Noble, 1914], p. 161).
6. Hawthorne withdrew Cotton Mather's *Magnalia* twice in 1827 and once in 1828 (see Kesselring, *Hawthorne's Reading,* p. 56). The Salem Athenaeum did not own a copy of Mather's *The Wonders*—or, for that matter, Calef's *More Wonders*—before 1860, so there is no way of knowing exactly when Hawthorne read this text (see the *Catalogue of the Books Belonging to the Salem Athenaeum* [Salem: W. Palfray, 1818, 1826], the *Catalogue of the Books Belonging to the Salem Athenaeum* [Salem: Office of the Gazette, 1842], and *A Catalogue of the Library of the Salem Athenaeum* [Boston: John Wilson and Son, 1858]). Unfortunately, I have not been able to determine whether or not Hawthorne examined surviving documents and evidence from the Salem court, which, as Charles Upham remarks, were held by the Essex County clerk at the time (*Lectures on Witchcraft, Comprising a History of the Delusion in Salem, in 1692* [Boston: Carter, Hendee and Babcock, 1831], p. 45).

Both Hutchinsons and Upham were strongly influenced by Calef, who in his *More Wonders* all but blames Cotton Mather (1663–1728) and, to a lesser degree, Increase Mather (1639–1723) for the Salem witch hunt. Cotton Mather was, to be sure, deeply interested in witchcraft in the 1680s and early 1690s, a topic about which he preached and published while trying to involve himself in as many cases as he could. And though the Mathers played no official role in the Salem trials, one can easily support Calef's charge that Cotton encouraged the judges—four of whom were friends—when, after the first execution in June 1692, he urged them on to a *"speedy and vigorous Prosecution of such as have rendered themselves obnoxious."*[7] More damning still, however, Calef accuses both Increase and Cotton Mather of setting the stage for the events of Salem during the previous decade by promulgating "Signs and Lying Wonders"—a reference to 2 Thessalonians 2:9—by which he likens the Mathers to the biblical Man of Lawlessness.[8] Calef's point, picked up by the historians mentioned above, is that Increase published *Illustrious Providences* in 1684 and Cotton followed with *Memorable Providences, Relating to Witchcrafts and Possessions* in 1689 to convince credulous New Englanders that they were surrounded and besieged by devils and witches—a deed, in Calef's view, more fiendish than any performed by the "witches" of Salem.[9] Moreover, Calef maintains, after the executions had stopped, Cotton Mather tried to instigate a new panic by publishing stories of witchcraft in *The Wonders of the Invisible World* and *The Life of His Excellency Sir William Phips* (1697) and by preparing a narrative, "Another Brand Pluck'd out of the Burning," detailing the diabolic affliction of the seventeen-year-old Margaret Rule. Spicing his claims with an eyewitness account of the Mathers' misconduct in the Rule case, Calef reports that on 13 September 1693 he saw both Cotton and Increase rubbing the teen's stomach and breast before a crowd of thirty or forty and alleges that Cotton tried to coax spectral accusations against her neighbors from her.[1] Finally, and perhaps most unforgettably, Calef writes that when former Salem Village pastor George Burroughs was hanged on 19 August 1692 after having moved the crowd on Gallows Hill to tears with a solemn declaration of his innocence and a final prayer, "Mr. Cotton Mather, being mounted upon a Horse, addressed himself to

7. After the execution of Bridget Bishop on 10 June 1692, Cotton Mather, acting as a spokesman for a group of influential Boston minsters, produced "The Return of Several Ministers" in response to concerns about the proceedings of the Court of Oyer and Terminer. While suggesting caution, he nonetheless recommended that the magistrates continue their prosecutions (Mather, *Cases of Conscience*, p. 74). Calef notes that when Mather reprinted the "Return" in his 1697 biography of Phips, subsequently included in the *Magnalia*, he removed his instigation to a "speedy and vigorous prosecution" (*More Wonders*, pp. 309–12). For Cotton Mather's connection to the judges, see Robinson, *The Devil Discovered*, pp. 32, 193.
8. Calef, *More Wonders*, pp. [1–14], 307–8, 312–18. In a footnote to his abridged edition of *More Wonders*, Burr deduces (correctly, I think) that the "Signs and Lying Wonders" mentioned by Calef were those published by the Mathers (*Narratives of the Witchcraft Cases, 1648–1706*, p. 302). Calef attacks Cotton Mather and his works on witchcraft throughout this section of *More Wonders*, and the sentence attributing the Salem delusion to signs and wonders contains a reference to the title of Increase Mather's *An Essay for the Recording of Illustrious Providences* (*More Wonders*, p. 8).
9. See Francis Hutchinson, *An Historical Essay Concerning Witchcraft, &c.* (London: R. Knaplock, 1718), p. 77. Thomas Hutchinson and Upham largely blame Cotton Mather for drawing Massachusetts into the events of 1692 (see Hutchinson, *The History of the Province of Massachusetts-Bay, From the Charter of King William and Queen Mary, in 1691, Until the Year 1750* [Boston: Thomas and John Fleet, 1767], pp. 18–23, and Upham, *Lectures on Witchcraft*, pp. 104–8, 183, 189).
1. Calef, *More Wonders*, pp. 42–45.

the People, partly to declare, that [Burroughs] was no ordained Minister, and partly to possess the People of his guilt; saying, That the Devil has often been transformed into an Angel of Light; and this did somewhat appease the People, and the Executions went on."[2]

Calef's assertions about the impact of the Mathers' books and his testimony regarding their public behavior are no longer given the credit they received in Hawthorne's sources.[3] Yet, the Mather-centric account of the witch trials presented in *More Wonders* and reinforced by the Hutchinsons and Upham indelibly influenced Hawthorne's writing about the crisis, even though his understanding of Cotton Mather grew more nuanced over time.[4] At the beginning and end of his career as a writer of short fiction, Hawthorne retells Calef's story about Mather's appearance on horseback at the 19 August executions, both times peppering his tales with embellishments and bitter sarcasms no doubt inspired by Upham.[5] In "Alice Doane's Appeal," completed in the early 1830s, Hawthorne describes Mather riding at the rear of the procession to Gallows Hill as "sternly triumphant" and "proud of his well-won dignity, as the representative of all the hateful features of his time; the one blood-thirsty man, in whom were concentrated those vices of spirit and errors of opinion, that sufficed to madden the whole surrounding multitude" (11:279). In "Main-street," written in the late 1840s, the crowd on Gallows Hill pauses to reconsider Burroughs's guilt after listening to his heart-rending final words; then, it quickly concludes, "Ah! no; for listen to wise Cotton Mather, who, as he sits there on his horse, speaks comfortably to the perplexed multitude, and tells them that all has been religiously and justly done, and that Satan's power shall this day receive its death-blow in New England. Heaven grant it be so!—the great scholar must be right! so, lead the poor creatures to their death!" (11:77). Although Hawthorne does not explicitly assign responsibility for the Salem tragedy in "Alice Doane's Appeal" or "Main-street," in "Famous Old People" Grandfather declares that Cotton Mather was "the chief agent of the mischief" (6:94).

The Salem witchcraft crisis was clearly inseparable from the figure of Cotton Mather in Hawthorne's imagination. Not surprisingly, therefore, the author of "Young Goodman Brown" seems to have asked himself the

2. Calef, *More Wonders*, pp. 220–21.
3. A case in point, Norton's *In the Devil's Snare*, the most comprehensive and convincing account of the Salem crisis to date, does not even entertain the question of whether or not the Mathers were to blame for the trials. For insightful discussions of Cotton Mather's role in the tragedy, see Kenneth Silverman, *The Life and Times of Cotton Mather* (New York: Harper and Row, 1984), pp. 87–90, 109–111; David Levin, *Cotton Mather: The Young Life of the Lord's Remembrancer, 1663–1703* (Cambridge: Harvard University Press, 1978), pp. 195–222; and Richard H. Werking, "'Reformation Is Our Only Preservation': Cotton Mather and Salem Witchcraft," in *Witchcraft in Colonial America*, ed. Brian P. Levack (New York: Garland Publishing, 1992), pp. 233–42. Regarding the probable unfairness of Calef's claims about the Mathers and Margaret Rule, see Chadwick Hansen, *Witchcraft at Salem* (New York: George Braziller, 1969), pp. 190–94. For a complete, annotated, and chronologically arranged edition of all surviving legal documents pertaining to the Salem witch trials, see *Records of the Salem Witch-Hunt*, ed. Bernard Rosenthal (New York: Cambridge University Press, 2009).
4. Hawthorne read William Peabody's *Life of Cotton Mather* (1836) sometime before he finished writing "Famous Old People," published in 1841. Under the influence of Peabody, who is cited in the text, Hawthorne tempers his Calefian remarks about Mather's responsibility for the Salem disaster with praise for his stance during the Boston smallpox inoculation controversy of 1721–22 (see Hawthorne, *Works*, 6:99–105, and Peabody, *Life of Cotton Mather*, vol. 2 of *American Biography*, ed. Jared Sparks [New York: Harper and Brothers, 1902], pp. 51–53, 61–62, 65, 81, 88).
5. Upham, *Lectures on Witchcraft*, pp. 102–3.

following question: Why was Mather so keenly, so fatally, interested in witchcraft? Two answers appear to have suggested themselves to the story-writer: first, both Increase and Cotton Mather were driven to seek out and inform the public about cases of demonic activity because they felt it offered irrefutable proof of the reality of Christianity against a rising tide of impiety in Europe and America; second, Cotton Mather thought that the descent of witches upon Salem signaled the millennium's imminence, and thus it was his duty as a special envoy of Christ's to call for reformation.

Increase Mather's *An Essay for the Recording of Illustrious Providences*, which Hawthorne withdrew from the Salem Athenaeum three times between 1827 and 1834, is a collection of purportedly factual narratives displaying extraordinary interventions by God and Satan—with God's permission—in Europe and America.[6] In publishing the book, Mather joined a movement of English Protestant theologians committed to defending Christianity from the spread of atheism, a scourge evident, they believed, in the lax morals of high society and the court of Charles II, the mechanistic materialism of Thomas Hobbes's *Leviathan* (1651), the wit and science of coffeehouse virtuosi, and the skepticism of Thomas Ady, John Wagstaffe, and John Webster.[7] Whether taking on the "practical atheism" of the gentry or the theoretical "Sadducism" of philosophers and scientists, Anglicans such as Henry More and Joseph Glanvill as well as Puritans like Richard Baxter thought that tales of demonic activity offered objective proof of the existence of the spirit world and thus of the truth of Christianity. A sane seventeenth-century Englishman had much to gain by persuading himself or by lying to others that he had undergone a conversion experience, but there was no obvious reward for pretending to have been attacked by the devil and his minions. In addition, while it was impossible to see into the hearts of professors to saving faith, the Evil One's activities were sometimes manifest to witnesses in the world and thus could be presented as empirical evidence to a learned culture that privileged rational inquiry over religious enthusiasm. For this reason, in part, the Royal Society, whose members included Glanvill, was receptive to the work of the demonologists. Finally, witchcraft in particular was thought to provide strong rational proof of the reality of the spirit world because it tied preternatural activity to real human agents who would be executed if found guilty of the crime. What more convincing evidence could there be than the confession of a witch against his or her own life?[8]

Concerned about the situation in England and undoubtedly worried that it was only a matter of time before the Sadducees would begin to

6. Kesselring, *Hawthorne's Reading*, p. 56.
7. See G. E. Aylmer, "Unbelief in Seventeenth-Century England," in *Puritans and Revolutionaries: Essays in Seventeenth-Century History Presented to Christopher Hill*, ed.Donald Pennington and Keith Thomas (Oxford: Oxford University Press, 1978), pp. 22–46; Stuart Clark, *Thinking with Demons: The Idea of Witchcraft in Early Modern Europe* (New York: Oxford University Press, 1997), pp. 238–43, 294–311, 520; Michael Hunter, *Science and Society in Restoration England* (New York: Cambridge University Press, 1981), pp. 162–87; and Michael P. Winship, *Seers of God: Puritan Providentialism in the Restoration and Early Enlightenment* (Baltimore: Johns Hopkins University Press, 1996), pp. 60–70, 118–23.
8. See Thomas Harmon Jobe, "The Devil in Restoration Science: The Glanvill-Webster Witchcraft Debate," *Isis* 72 (1981): 342–56, and Moody E. Prior, "Joseph Glanvill, Witchcraft, and Seventeenth-Century Science," *Modern Philology* 30 (1932): 167–93.

make inroads into New England, where Congregational churches were already grappling with a substantial declension of piety among the second and third generations of English settlers, Mather dedicates significant portions of the preface as well as six of the twelve chapters of *Illustrious Providences* to demonology, where he treats demons, cacodemons, demoniacs, witches, specters, possessions, and Satanic thunderstorms.[9] Regarding witchcraft specifically, Mather tells the story of Ann Cole's bewitchment in Hartford in 1662. He addresses pneumatological questions about witches, such as whether Satan or his spirits can impregnate them with real human offspring or whether they can turn themselves or others into animals. He defends the Puritans' interpretation of the Witch of Endor story in 1 Samuel 28:3–25—that is, it signifies that witches covenant with the devil in exchange for the use of familiar spirits. He discusses the religious legality of the ordeal by water and popular protective measures against witches involving blood, urine, and horseshoes. And should anyone doubt the reality of witchcraft, he menacingly recalls the fate of William de Lure, who argued against the existence of witches only to be accused and convicted of the crime himself.[1] But although witchcraft is an integral part of *Illustrious Providences*, it is raised in only three of the six chapters on demonology and, thus, is not the book's central focus. Mather gives more space to flying objects launched by hidden malefactors than to case histories of bewitchment.

Like his father, Cotton Mather felt the urgency of establishing the reality of the demonic as a bulwark against a growing secularism in Europe and America. In his homily *A Discourse on Witchcraft*, included in *Memorable Providences* (1689), Mather argues that "since there are *Witches* and *Devils*, we may conclude that there are also immortal souls. *Devils* would never contract with *Witches* for their Souls if there were no such things to become a prey unto them." Yet more ominous, he warns, if skepticism about witchcraft were to prevail, "we shall come to have no Christ."[2] In "Another Brand," Mather mocks "the learned witlings of the coffee-house" for their ignorance about the spirit world and asserts that witchcraft narratives, such as his own, offer "glorious evidence for the being of a God."[3] Unlike his father, Cotton devoted most of his attention to witchcraft in his demonological works, and in *Memorable Providences* he followed Glanvill's lead by bolstering Christianity with several accounts of witches appearing spectrally as themselves; a fact that suggests at least the possibility of a link between Mather's text and the Salem trials, where the men and women accused of sorcery were charged based on the court's belief that one could identify a guilty witch from the specters

9. For seventeenth-century New England's decline of faith and the resulting halfway covenant controversy, see Perry Miller, *The New England Mind: From Colony to Province* (Cambridge: Belknap Press of Harvard University Press, 1953), pp. 19–146.
1. Increase Mather, *An Essay for the Recording of Illustrious Providences* (Boston: Samuel Green, 1684), pp. 174–75.
2. Cotton Mather, *Memorable Providences, Relating to Witchcrafts and Possessions* (Boston: R. P., 1689), pp. 16, 14. The Salem Athenaeum did not own a copy of *Memorable Providences* during the 1820s and 1830s, and there is no conclusive evidence to suggest that Hawthorne saw the book. However, he would have known quite a bit about it because Upham summarizes, comments upon, and excerpts quotations from the long section on the Goodwin children's bewitchment (*Lectures on Witchcraft,* pp. 182–90).
3. Mather, "Another Brand," in Calef, *More Wonders*, pp. 35–36.

that appeared to his or her victims. (By contrast, Increase's *Illustrious Providences* features just one such story.)[4] Becoming personally involved in three witchcraft cases, Mather spent untold hours ministering to bewitched children while recording every detail he could recall about their behavior to "confute the Saducism of this debauched age."[5]

Cotton Mather believed that God sanctioned witchcraft in great part to goad the unregenerate to convert, and in 1692 he told his Boston congregation that the Salem outbreak demonstrated that natural men and women were almost out of time. As Sacvan Bercovitch has shown, New England divines were, "technically speaking," pre-millennialists who preached that the chiliad was imminent from the moment John Winthrop's fleet left England in 1630 and consistently thereafter. Like John Cotton, who read *God's Promise to His Plantation* on the occasion of the fleet's departure, they envisioned that New England would be the site of the New Jerusalem as prophesied in the Book of Revelation.[6] The Mathers were vigorous proponents of this view, and by 1689 a series of portentous events had convinced Cotton Mather that he would not only live to see Christ return and bind Satan, thus beginning His and His saints' one-thousand-year reign on earth, but that he himself had been appointed "Herald of the Lord's Kingdome now approaching," a latter-day John the Baptist.[7] He began preaching and publishing more frequently about the coming apocalypse, he called upon his congregation formally to re-own the covenant, and he urged them to accept halfway members so that as many souls as possible might be saved.[8] On 10 April 1692, as the accusations mounted in Salem Village, Mather delivered *A Midnight Cry* at the Second (North) Church exhorting his flock to reform and proclaiming that the millennium would in all likelihood commence in 1697.[9]

When the witch hunt spread from Salem Village to Andover, Massachusetts, in the summer of 1692, Mather concluded that the event was

4. I would add that Increase's story of an unnamed witch appearing in her own shape in *Illustrious Providences* is offset by the more detailed account of Elizabeth Knap's encounter with a specter masquerading as an innocent person (*Illustrious Providences*, pp. 140–42, 197–98). The case histories Joseph Glanvill recounts that feature spectral appearances of witches as themselves are found in part 2 of *Saducismus Triumphatus: or, Full and Plain Evidence Concerning Witches and Apparitions* (London: J. Collins, 1681), pp. 120, 123, 124–25, 128, 135, 139–42, 149–50, 157–58, 162–65, 170–71, 182, 192, 194, 195, 196. Cotton Mather's *Memorable Providences* contains four unequivocal references to witches revealing themselves spectrally, but in the lengthy stretch of the text describing Martha Goodwin's affliction by a group of specters called "*Her Company*," Mather hints more than once that the girl knew her tormenters by their apparitions (pp. 9, 11, 13, 18–40, 56). Cotton Mather's case histories in *The Wonders of the Invisible World* are filled with instances of spectral evidence but these are cited, for the most part, to defend the Salem magistrates rather than Christian faith (pp. 95, 96, 98, 99, 105, 106–9, 111, 112, 113, 116, 118–19, 124, 129, 133–34, 135, 137). On the rare occasions when Cotton wrote about spectral evidence after the trials ended in early 1693, he emphasized that one can be deceived by apparitions (see "A Brand Pluck'd out of the Burning," in Burr, *Narratives*, p. 274; "Another Brand Pluck'd out of the Burning," in Calef, *More Wonders*, pp. 24, 38; and *Magnalia Christi Americana; or The Ecclesiastical History of New-England, From its First Planting, in the Year 1620, Unto the Year of our Lord 1698*, 2 vols. [1702; repr., New York: Russell and Russell, 1967], 1:206–12).
5. Mather, *Memorable Providences*, p. 18. Mather's cases were those of the Goodwin children (1688–89), Mercy Short (1692–93), and Margaret Rule (1693), which he described in *Memorable Providences*, "A Brand," and "Another Brand."
6. Sacvan Bercovitch, *The American Jeremiad* (Madison: University of Wisconsin Press, 1978), pp. 8–9, 86, 94–100.
7. Quoted in Bercovitch, *American Jeremiad*, p. 86; see also Levin, *Cotton Mather*, pp. 175–76, 202.
8. See Levin, *Cotton Mather*, pp. 174–204, and Silverman, *The Life and Times of Cotton Mather*, pp. 118–20.
9. Cotton Mather, *A Midnight Cry* (Boston: John Allen, 1692), pp. 24, 63.

connected to the looming apocalypse. And on 4 August, after hearing a report that a devastating earthquake in Jamaica had killed over two thousand people, he went public with his theory in a sermon on Revelation 12:12: "Wo to the Inhabitants of the Earth, and of the Sea; for the Devil is come down unto you, having great Wrath; because he knoweth, that he hath but a short time."[1] The Bible makes it clear, Mather told his Second Church audience, that "in the last days, perillous times shall come" (2 Timothy 3:1), meaning, he said, that at the approach of the chiliad "the times will grow more full of *Devils,* and therefore more full of *Perils,* than ever they were before." The Evil One will strike humanity with a series of plagues, wars, storms, and earthquakes, and he will cause an unprecedented number of possessions, obsessions, and bewitchments to take place.[2] Since January, eighty-one complaints, examinations, and confessions of witchcraft had been brought before the Salem Court, and now there was the Jamaica earthquake, which could only mean that Satan was actively recruiting followers in New England for the final battle with Christ.[3] Now more than ever, Mather warned, one must secure his or her salvation. Mather published his homily, in redacted form, in *The Wonders of the Invisible World* just days before the Court of Oyer and Terminer was dissolved by Governor Phips on 29 October 1692.

"Young Goodman Brown" suggests that Hawthorne—swayed by the histories of Calef, the Hutchinsons, and Upham—probed the Mathers' work for answers about Salem and was struck by their use of demonic evidence to uphold Christianity. In the first part of the story, Goodman Brown leaves Faith, his wife of three months, to go to a witches' meeting with the devil in the forest outside Salem Village. The newlywed tells himself that one night of diabolic entertainment will be enough and that afterward he will "cling to [Faith's] skirts and follow her to Heaven" (10:75) [169]. But Brown has not ventured far into the woods when his "scruples" (10:76) [170] halt his progress and he announces that he wants to turn back. Satan, who has encountered the young man part way through his journey, chides him; every Puritan who has ever lived, the devil claims, has been one of his associates, including Brown's deceased father and grandfather. Continuing to protest, Brown insists on going home for Faith's sake; again Satan presses his case, assailing Brown with spectral evidence that his catechist, minister, and a deacon are all witches.[4] The Puritan holds his ground, vowing to resist the Evil One "with Heaven above, and Faith below" (10:82) [174]. Just then, however, a black cloud appears overhead, and Brown hears what he thinks are the voices of fellow Salemites urging Faith on as she asks for some illicit favor. He calls to her in vain, he hears a scream, and as the cloud moves away it releases a pink ribbon—apparently from Faith's cap—that floats down and lands on a nearby tree branch. Brown seizes the ribbon and looks at it before exclaiming, "My Faith is gone! . . . There is no good on earth; and sin is

1. Silverman, *The Life and Times of Cotton Mather,* pp. 104–9.
2. Mather, *Wonders of the Invisible World,* pp. 23, 26, 28.
3. See Norton, *In the Devil's Snare,* pp. 121, 253.
4. All readers of "Young Goodman Brown" owe a debt of gratitude to David Levin for his groundbreaking essay on the tale's engagement with spectral evidence. For the examples mentioned above, see his "Shadows of Doubt: Specter Evidence in Hawthorne's 'Young Goodman Brown,'" *American Literature* 34 (November 1963): 348–49.

but a name. Come, devil! for to thee is this world given" (10:83) [174]. Faith, Brown believes, has covenanted with Satan. Abandoning all hope, he takes off for the witches' sabbath "at such a rate, that he seemed to fly along the forest-path, rather than to walk or run" (10:83) [174].

In these scenes, the devil is said physically to resemble Goodman Brown, his father, and his grandfather, a disguise fashioned, in all likelihood, to suggest to the young Puritan that it would be natural for him to join Satan as well. Moreover, as noted earlier, he represents Increase Mather; for, as the narrator observes, he possesses the "indescribable air of one who knew the world, and would not have felt abashed at the governor's dinner-table, or in King William's court." In effect, the first section of "Young Goodman Brown," just described, reads like an allegory of Mather's attempt to awaken his countrymen in *Illustrious Providences,* with the devil playing the role of the Boston clergyman. Satan invites Everyman Brown to boost his "poor little Faith" (10:75) [169] with a demonic spectacle; the Puritan accepts, confident that afterward he will make his way to Heaven; the Wicked One presents Brown with a series of proofs from the spirit world to win his conversion. What is more, the ribbon scene, which brings this portion of "Young Goodman Brown" to a close, is lifted directly from *Illustrious Providences*. In the preface to that text, Mather tells the story of a young French scholar who had signed a contract with the devil when he was in need of money. Realizing his mistake in due course, the young man tries to kill himself but is saved by a group of ministers who hold a day of fasting and prayer for him in the field where the fatal pact was signed. The clergymen, Mather says, prayed "earnestly to the Lord to make known his power over Satan, in constraining him to give up that contract," and "after some hours continuance in Prayer, a Cloud was seen to spread it self over them, and out of it the very contract signed with the poor creatures Blood was dropped down amongst them; which being taken up and viewed, the party concerned took it, and tore it in pieces."[5]

In his *Lectures on Witchcraft*, Charles Upham observed that the Mathers and their English colleagues "seem to have persuaded themselves into the belief, that the doctrines of demonology were essential to the gospel, and that the rejection of them was equivalent to infidelity."[6] Hawthorne takes the point a step further in the first part of "Young Goodman Brown" when he implies that Increase Mather, in propping up Christianity with demonology, was essentially telling the impious halfway generations to place their faith in Satan and his works *before* God. That Mather succeeded in attracting converts, in Hawthorne's view, is evident in the way the ribbon scene rewrites its original: in Mather's story of the French scholar, evidence (the contract) from the dark world, when examined, establishes God's and the ministers' power; in Hawthorne's version of Mather's story, Goodman Brown's reliance upon such evidence (the pink ribbon) reveals his lack of faith in Christ and prompts him to embrace the devil.

Although Hawthorne unmistakably engages Increase Mather's demonology in the first section of "Young Goodman Brown," he also hints there that, with his *Memorable Providences,* Cotton Mather contributed to his

5. Mather, *Illustrious Providences,* [pp. 4–5].
6. Upham, *Lectures on Witchcraft*, p. 216.

father's misguided campaign. Before Brown's discovery of the pink ribbon, he identifies three potential witches—Goody Cloyse, Deacon Gookin, and the minister—by their supernatural appearance. (As was mentioned earlier, specter evidence of guilt is exceedingly rare in *Illustrious Providences* but common in *Memorable Providences.*) It is likely, too, that Hawthorne's allusions to Cotton's "Brand" narratives in the same portion of the story are intended to support this insinuation: for though he did not read "A Brand," Hawthorne could not have missed in "Another Brand" Mather's vehement denial of accusations that he had helped to bring about the Salem tragedy by encouraging "excessive credit of spectral accusations."[7]

The second part of "Young Goodman Brown," the witch meeting, invokes Cotton Mather's depiction of the Salem crisis as the fruit of the devil's apocalyptic fury in the redacted 4 August sermon on Revelation 12:12, *A Discourse on the Wonders of the Invisible World.* In *A Discourse,* which comprises the largest division of *The Wonders of the Invisible World*, Mather asserts that when Christ is ready to "set up His Kingdom" on earth, "it is a Thousand to One, but *the Devil* will in sundry *parts of the world*, assay *the like* for Himself, with a most Apish Imitation." Indeed, as Mather notes, confessing Salem witches have revealed in court that they "have met in Hellish *Randezvouzes*," where they engage in "Diabolical Sacraments, imitating the *Baptism* and the *Supper* of our Lord." In fact, as he remarks in a later section of *Wonders of the Invisible World*, penitent witches have admitted to blasphemously imitating a wide range of "Divine Things," from organizing themselves "much after the manner of *Congregational Churches*" to copying the miracles of Christ and "His Prophets."[8] Satan, Mather suggests, has begun his final offensive—luring the weak in faith to his side with spurious rites and ceremonies that are perilously enticing because they resemble genuine Christian ones.[9]

In Hawthorne's tale, Goodman Brown is led to a "Hellish *Randezvouz*" in a clearing deep in the woods, organized for the sole purpose of committing the young Puritan and his Faith to the devil. Satan, who has changed his disguise, presides over the ceremony "in the garb and manner . . . of some grave divine of the New England churches" (10:86) [176]. The scene is lit by four pines burning "like candles at an evening meeting" (10:84), [175] and the assembled faithful sing a "dreadful anthem" (10:85–86) [176] of sin to the tune of a popular hymn. Underscoring the allusion to Mather, Hawthorne stresses that the Evil One apishly imitates "the *Baptism* and the *Supper*" of Christ at the witches' sabbath. The devil welcomes the two "converts" to "the communion of [their] race" (10:86) [176] and then, standing before "a rock, bearing some rude, natural resemblance either to an altar or a pulpit" (10:84) [175], prepares to "lay the mark of baptism" upon them with a crimsoned liquid that appears to be, or is, blood (10:88) [177]. Presenting the ceremony of admission as a fusion of the two Puritan

7. See "Another Brand" in Calef, *More Wonders*, pp. 37–42. Mather's rebuttal of his critics is also excerpted in Upham, *Lectures on Witchcraft*, pp. 108–9.
8. Mather, *Wonders of the Invisible World*, pp. 28, 49, 139–41.
9. Not coincidentally, therefore, in the case histories Mather expresses particular disdain for Burroughs, referring to him as G. B. because he cannot bring himself to utter the man's name. For to Mather, not only was Burroughs "an Head Actor" (*Wonders of the Invisible World*, p. 94) at the witches' gatherings but, as Norton has pointed out, as the former Salem pastor, he embodied diabolic hypocrisy and temptation (personal communication, 21 May 2011).

sacraments, held at a crude "altar or pulpit," Hawthorne underscores Mather's view that the devil's rites are insubstantial approximations of Christian originals designed to ensnare the unregenerate.

At the end of *A Discourse on the Wonders of the Invisible World,* Mather offers "An Hortatory and Necessary Address" in which he defends the Salem court's controversial admission of spectral testimony in its proceedings, declaring that such evidence has been used by the judges only to initiate "further Enquiries into the *Lives* of the persons accused" and never to convict them. He admits that a large number of individuals, some of previously unblemished character, have been charged with witchcraft chiefly on the basis of spectral evidence, but he avers that the court has made no decisions in these cases. Moreover, he intimates, the "*Multitude* and *Quality*" of those accused at Salem on the strength of spectral witness is not the judges' fault; they have never encouraged such complaints. Instead, it is proof that the devil, spurred on by the approaching chiliad, is assaulting New England with a craft and vigor never seen before: either he is cruelly tricking accusers by tormenting them with spirits that have assumed the form of the innocent, or he is enabling an alarming number of New Englanders (several of good repute) to harm bewitched persons spectrally in their own shapes, or he is doing both. Whatever the truth of the matter, Mather says, the Evil One has succeeded brilliantly in turning the community against itself, and he has put the court in a terrible position: it cannot ignore such a large number of accusations and leave the province unprotected; yet, spectral evidence by itself is not to be trusted since it is "founded in the *Dark World*." "The whole business is become hereupon so *Snarled*," Mather concludes, "that our Honourable Judges have a Room for *Jehoshaphat's* Exclamation; *We know not what to do!*"[1]

When Goodman Brown arrives at the witches' sabbath, the narrator emphasizes the *multitude* and *quality* of the people he can make out in the light of Satan's ritual fires. Brown is able to identify respected government officials, august clergymen, members of high society, and countless others of good character from across the Province of Massachusetts Bay. At the same time, he sees "men of dissolute lives and women of spotted fame, wretches given over to all mean and filthy vice, and suspected even of horrid crimes" mingling shamelessly with the saints, who for their part "[shrink] not from the wicked" (10:85) [175]. What Brown sees, however, is manipulated by Satan: "As the red light arose and fell, a numerous congregation alternately shone forth, then disappeared in shadow, and again grew, as it were, out of the darkness, peopling the heart of the solitary woods at once" (10:84) [175]. Since the Father of Lies is the author of Brown's vision, there is no way for the Puritan to be certain that any of the individuals he recognizes are actually witches; indeed, he never does find out. Brown's situation is, in other words, that of the Salem magistrates, the afflicted accusers, and other innocent New Englanders in August 1692, according to Cotton Mather: the devil, in his millenarian rage, has assumed such complete control over the world of appearances that no one is exempt from suspicion of witchcraft, but no earthly judge can separate the guilty from the innocent.

1. Mather, *Wonders of the Invisible World*, pp. 53, 52, 53, 52–53.

For Mather, the gravity of New England's plight in 1692 was, paradoxically, cause for great hope because it presaged the Second Coming. Consequently, he exhorted his auditors to "learn some *good,* even from the *Wicked One* himself":

> When the *Divel* perceives his *Time* is but *short,* it puts him upon *Great Wrath.* But how should it be with *us,* when we perceive that our *Time* is but *short?* why, it should put us upon *Great Work.* The motive which makes the *Divel* to be more full of *wrath,* should make us more full of *warmth,* more full of *watch,* and more full of *All Diligence to make our Vocation, and Election sure.*[2]

Taking his cue from Satan, therefore, Mather concludes his sermon with an application urging the people of Massachusetts to repent, to be charitable to one another, and to make their peace with God because "REFORMATION! REFORMATION! Has been the Repeated *Cry,* of all the Judgments, that have hitherto been upon us."[3]

When the "grave divine of the New-England churches"—the devil's new avatar, Cotton Mather—addresses Goodman Brown, he offers the young man a very different interpretation of the diabolic activity described in *A Discourse,* namely, that the number and quality of the individuals represented spectrally in Salem means that "virtue" is "all a dream" and thus that the young Puritan should formally renounce Christianity and pledge his allegiance to Satan (10:87–88) [177]. As in his exposé of Increase Mather's *Illustrious Providences* in the first part of "Young Goodman Brown," then, Hawthorne shows that in using demonic evidence to inspire Christian faith, Cotton Mather unwittingly serves as an advocate for the devil.

Robert Calef had made the point almost a century and a half earlier when comparing the Mathers' demonological publications to the apocalyptic work of the Man of Lawlessness. In the second chapter of 2 Thessalonians, Paul cautions the Christian community at Thessalonica that the "day of Christ" will not arrive until there is a widespread "falling away" from God and a "man of sin" is revealed (2:3). As the chiliad approaches, he writes, "the mystery of iniquity" (*anomia*—also lawlessness, disobedience, or sin) that is already active in the world but held in check by a restraining force will be unleashed (2:7). God will send the unrighteous "a strong delusion" and Satan will empower the Lawless One, who will establish himself in God's temple and proclaim himself God, to seduce them to Hell with "signs and lying wonders" (2:11, 9). At the end of the appointed period of tribulation, Paul prophesies, Christ will consume and annihilate this man with "the spirit of his mouth" and "the brightness of his coming" (2:8).[4] Calef, who wished to divorce religion from wonders, did not mean to be taken literally when he referred to the Mathers as the Man of Sin.[5] And, of course, he knew that Increase's *Illustrious Providences,* unlike Cotton's *Wonders of the Invisible World,* is not expressly eschatological. The point of the allusion is, I believe, to insinuate that if

2. Mather, *Wonders of the Invisible World,* p. 32.
3. Mather, *Wonders of the Invisible World,* p. 62.
4. Here and elsewhere in this essay, I cite the King James translation of the Bible, though the Geneva Bible's versions of these texts are similar.
5. For Calef's view of special providences, see Winship, *Seers of God,* pp. 128–29.

1692 really had signaled the beginning of the apocalypse, the Mathers, who unintentionally advanced the devil's end with their moral deceptions and spurious miracles before and during the trials, would have been its Antichrists.[6]

At the witches' meeting in "Young Goodman Brown," the "Shape of Evil" (10:88) [177]—Cotton Mather—presents himself as a minister of God's temple in New England. He tells the protagonist and his wife, Faith, that his is the only real power in the world, and he tempts them by demonstrating his control over the province through a supernatural spectacle. He promises the newlyweds that once they have consecrated their bond with him, they will possess the ability "to penetrate, in every bosom, the deep mystery of sin, the fountain of all wicked arts" (10:87) [177]. And as the "grave divine" dips his hand into the red baptismal liquid, the narrator affirms that the rite will make the two "partakers of the mystery of sin, more conscious of the secret guilt of others, both in deed and thought, than they could now be of their own" (10:88) [177]. Although the references to 2 Thessalonians 2 are unmistakable, Hawthorne diverges from traditional interpretations of Paul's chapter when he proposes that the "mystery of *anomia*" is the knowledge of evil in the hearts of others, disclosed by the Man of Sin to those whose souls he has won for Satan.[7]

In the first section of "Young Goodman Brown," the devil is said to carry a staff "formerly lent to the Egyptian Magi" (10:79) [172], which suggests that he represents not the Man of Lawlessness but the false teachers of the tribulation period described in the third chapter of 2 Timothy—a text, as noted earlier, that Cotton Mather uses in *A Discourse on the Wonders of the Invisible World* to frame the witchcraft crisis as evidence of Satan's chiliastic wrath. The author of this letter warns that "in the last days" wise men like "Jannes and Jambres" will arise to oppose God's truth and lead the faithless to their destruction, much as the Lawless One is expected to do (3:1–9). But the teachers of 2 Timothy, like their predecessors, the Egyptian magicians of Exodus 7, are indisputably lesser enemies of God's than the Man of Sin, the Evil One's chief agent on earth.[8] Although Increase Mather was partially responsible for the events of 1692, the text intimates, Cotton Mather was the principal catalyst of the trials.

Capitalizing on a clue from Calef, Hawthorne portrays the Mathers as types of God's end-time enemies in "Young Goodman Brown." Increase Mather was New England's first purveyor of counterfeit wonders; then, with much of Massachusetts already firmly in the grip of a "strong delusion," his son sealed the colony's fate when he preached "the mystery of

6. The New Testament contains a number of references to figures like the Man of Sin, including the Antichrists of 1 and 2 John, the "beast" of Revelation 13:11–18, and "the false Christs and false prophets" of Mark 13:22 and Matthew 24:24. While the Man of Lawlessness and the Antichrist(s) are often equated with one another, there is considerable debate among New Testament scholars about whether the two are actually one in the same. Abraham J. Malherbe, for instance, contends that the Man of Lawlessness is an Anti-God rather than an Anti-Christ figure (*The Letters to the Thessalonians,* ed. Abraham J. Malherbe, in the Anchor Bible [New York: Doubleday, 2000], pp. 430–34), while F. F. Bruce believes he is the Antichrist (*1 & 2 Thessalonians,* ed. F. F. Bruce [Waco, Tex.: World Books, 1982], pp. 179–88).
7. For more conventional views of the passage, see Malherbe, *The Letters to the Thessalonians,* pp. 432–33.
8. For the identity of these figures, see *The First and Second Letters to Timothy,* ed. Luke Timothy Johnson, in the Anchor Bible (New York: Doubleday, 2001), pp. 407–8, 410–11, and *The Writings of St. Paul, A Norton Critical Edition,* 2nd ed., ed. Wayne A. Meeks and John T. Fitzgerald (New York: Norton, 2007), p. 134.

sin" in *A Discourse*. Mather delivered his sermon on 4 August, with eight of the twenty victims of the Salem court newly executed and a day before George Burroughs, John and Elizabeth Proctor and possibly John Willard were tried and sentenced to hang.[9] Since the homily carries a date, Hawthorne would have known how much was at stake when it was first uttered—in fact, he may have believed that all of the executions at Salem took place in August and September 1692.[1] Moreover, he would have seen that Mather reiterated his chiliastic call for a revival, based on the ongoing events at Salem, in *Wonders of the Invisible World,* published in mid-October when the court had executed twenty and expected to try many more in the coming months. Instead of using his considerable influence to stop the trials, Mather, Hawthorne implies, spurred the magistrates and the community on by teaching them to forget Christian charity and become "more conscious of the secret guilt of others," to suspect, now that the apocalypse was at hand, that anyone could be a witch.[2] A writer of "agnostic" short fiction that occasionally examines Christian faith in an "unorthodox light,"[3] Hawthorne's biblical depiction of the Mathers is figurative. But if, according to Hawthorne, ever there were false teachers and divines who led their contemporaries to perdition by publishing the devil's power, Increase Mather and, particularly, his son Cotton fit that bill.

The final section of "Young Goodman Brown" offers a brief synopsis of the protagonist's unhappy life after 1692, omitting commentary on the executions that followed Mather's presentation of *A Discourse* on 4 August. Goodman Brown cries out to his wife to resist the devil an instant before being baptized at the witches' sabbath, and he awakes to find himself alone in the forest. Perhaps, the narrator suggests, he "only dreamed a wild dream of a witch-meeting" (10:89) [178]. Regardless, the experience transforms Brown into "a stern, a sad, a darkly meditative, a distrustful, if not a desperate man," who suspects that everyone he saw at the sabbath is in fact a member of Satan's fold. On Sundays, when the congregation sings "a holy psalm," he can't hear it, "because an anthem of sin rush[es] loudly upon his ear" (10:89) [178]. "When the minister [speaks] from the pulpit" about "the sacred truths" of Christianity, he blanches, fearing that the roof will fall on "the gray blasphemer and his hearers" (10:89) [178]. Brown "[shrinks] from the bosom of Faith" when he wakes in the night, and he "scowl[s] and mutter[s] to himself" (10:89) [178] when he sees her at prayer. When, at last, many years later, Goodman Brown is "borne to his grave, a hoary corpse," the narrator reports that "no hopeful verse" would be inscribed "upon his tomb-stone; for his dying hour was gloom" (10:90) [178].

9. While she is certain of the date of Burroughs's and the Proctors' trials, Norton suggests that Willard may also have been arraigned on 5 August (*In the Devil's Snare,* pp. 315–16).
1. See Hawthorne, *Works,* 6:78. Mather mentions in *A Discourse* that "some of the *Witch Gang* have been fairly Executed" (p. 53); Hawthorne may have simply forgotten this when he came to write "Famous Old People" a few years later.
2. On one occasion in *A Discourse,* Mather, citing 1 Corinthians 13:5, asks his audience to exercise Christian charity when considering the cases of witches accused only on the basis of spectral evidence. However, his plea is tacked on to a more lengthy and impassioned request for sympathy for the magistrates, and it is all but drowned out by the rest of his discussion, which is far less charitable to the accused (pp. 69–72).
3. For Hawthorne's "agnostic" treatment of Christianity in his early fiction, see Bill Christophersen, "Agnostic Tensions in Hawthorne's Short Stories," *American Literature* 72 (September 2000): 596, 597, and passim.

The finale of "Young Goodman Brown" is devoted entirely to its protagonist's religious experience after the crisis. However, the pervasive ambiguity of this portion of the text makes it exceedingly difficult to settle on its meaning. Readers are invited to speculate about Goodman Brown's eternal destiny but are given no conclusive evidence to confirm their hypotheses. Expressing concern for Faith's welfare instead of his own at the sabbath, Brown repels the devil in dramatic fashion. But one good deed early in life is not proof of an individual's salvation in Calvinism—or, for that matter, in any branch of Christianity—and it is hard to find another act of charity on Brown's part in subsequent scenes.[4] Aware of his spiritual distress, which appears to have accompanied him to his "dying hour," Brown's family, friends, and fellow church members presume that there is no hope for him in Heaven. Yet, unlike "The Man of Adamant" (1837) and "Ethan Brand" (1850), "Young Goodman Brown" features no postmortem evidence of its central character's doom. Brown is permanently scarred by his encounter with the Mathers, as his hardened doubt of his wife and neighbors attests; but, given the theology of sin presented in the tale, the fact that he suspects rather than knows that evil lurks in their hearts implies that however much he may be tortured on earth, he is not necessarily damned in the hereafter.

Hawthorne challenges his readers further by presenting Goodman Brown as both an Everyman and an outsider in the narrative's conclusion. A "simple husbandman," who is neither a minister, magistrate, accused witch, or afflicted accuser, Brown appears to be a figure for New Englanders who fell under the Mathers' spell—an experience depicted as a "wild dream" in the story—but who were not involved in the trials.[5] (Conceivably for this reason, neither the Salem court proceedings nor the executions are mentioned explicitly in the tale.) But Brown's mistrust of his wife, religious leaders, and congregation, together with his community's doleful verdict on his soul, strongly mark him as an isolato. The inhabitants of Salem Village seem to have moved on with their lives and left Goodman Brown behind to grapple with the significance of 1692. Or, have they? The narrator, after all, does not reproduce any of their words and provides scant information about their thoughts and feelings. Furthermore, when Brown dies, his body is followed to its grave by "a goodly procession" (10:90) [178] of children and grandchildren, together with many of his neighbors, which suggests that, at least outwardly, he too had led a conventional life after the crisis. Although he was believed to be faithless, he was still allowed to remain a member of the Salem congregation. In short, Goodman Brown might not be the only one in Hawthorne's post-1692 Salem Village to have been changed inwardly by the tragedy.

Goodman Brown's problematic status as an Everyman may be linked, in part, to the devil's hint that he is the son of Captain Hathorne and the stepson of George Burroughs, G. B. Given the amount of research

4. Perhaps Brown's "snatch[ing] away" (10:89) [178] of a catechumen from Goody Cloyse upon his return to Salem Village from the forest can be construed as charitable. Eberwein comes to a similar conclusion about Brown's rejection of the devil at the witch meeting ("My Faith is Gone!" p. 28).

5. This suggestion is reinforced by Hawthorne's echoes of Upham's description of the bewilderment of "the people" of Massachusetts at the end of the crisis in his account of Brown's first waking moments in the forest and his reentry into Salem Village. See Hawthorne, *Works*, 10:88, and Upham, *Lectures on Witchcraft*, p. 33.

Hawthorne invested in "Young Goodman Brown," there can be little doubt of his concern for historical accuracy, a concern that would have led him to own, as he does in "The Custom-House" sketch in *The Scarlet Letter* (1850), the personal nature of his interest in Puritan New England in general and the Salem witchcraft crisis in particular. Hawthorne perused the 1692 archive not as an objective historian but as the scion of Judge Hathorne and the heir of his guilt, an experience he allegorizes in "Young Goodman Brown" by making his Everyman protagonist both a Hathorne and a son of Burroughs—"the holy man" (11:76) whose hanging is portrayed in "Main-street" as the ultimate symbol of the witch trials' injustice. Brown's anger at his wife and neighbors for pretending, he believes, to be Christians after 1692 was probably Hawthorne's anger at his ancestor and his community for bringing about the tragedy and for returning to their Puritan lives afterwards, as if nothing essential had changed in Massachusetts.

If so, one could argue that in "Young Goodman Brown," Hawthorne attempts to represent the Mathers' seduction of New England as faithfully as he can by making Brown a seventeenth-century Everyman *and* a figure for himself, a biased nineteenth-century reader of source texts about the crisis. Additionally, one might add, in order to remain true to his biographically and historically circumscribed outlook on 1692, Hawthorne says as little as possible about Brown's contemporaries' responses to the Salem tragedy in the closing section of the narrative, signaling, thereby, his awareness of the distance separating his view of the event from the event itself. Finally, Hawthorne's refusal to pronounce a sentence upon Goodman Brown's soul, while allowing his family and congregation to do so, can be attributed to his cognizance of the subjectivity of both his own and his characters' perspectives. As Colacurcio has claimed, 1692 marked the end of the Puritan age for the author of "Young Goodman Brown."[6] Too late for Puritanism but too early for the Enlightenment, Goodman Brown is trapped in a historical moment, a moment in which Calvinist salvation and damnation may already be cultural realities of a bygone era, a moment that nonetheless condemns him to a nether state in this world—and perhaps beyond.

J. HILLIS MILLER

The Problem of History in "The Minister's Black Veil"[†]

* * *

Just why does the simple act of wearing the black veil cause all this devastation in the little community of Milford? The catastrophic effect seems outrageously incommensurate with its trivial cause. The wearing of the veil, I answer, suspends two of the basic assumptions that make society

6. See Colacurcio, *The Province of Piety*, p. 312.

† From *Hawthorne and History: Defacing It* (Oxford: Basil Blackwell, 1991), pp. 92–106. Copyright © J. Hillis Miller. Reprinted by permission of the author. Page numbers in square brackets refer to this Norton Critical Edition.

possible: the assumption that a person's face is the sign of his selfhood and the accompanying presumption that this sign can in one way or another be read. A whole series of presuppositions accompany those assumptions: the presupposition that there are natural as opposed to arbitrary signs, in this case the face; the presupposition that the face as exterior and visible natural sign refers to an interior, nonlinguistic entity, the consciousness, subjectivity, soul, or selfhood of the person who presents that face to the world; the presupposition that the procedure whereby we read a person's selfhood by his or her face is paradigmatic for sign-reading in general. The reading of person by face can then be universally extended to the reading of all natural and supernatural entities, all entities not persons—the absent, the inanimate, the dead. This reading would be expressed by those most basic of tropes, prosopopoeia and apostrophe, as in Wordsworth's opening address in "The Boy of Winander": "There was a boy: ye knew him well, ye cliffs / And islands of Winander!" It is all very well to say that of course we know that reading a personality by a face is a precarious dependence on an unreliable trope, but we go on knowing, choosing, and deciding in daily life as if this were not the case. Hawthorne's story shows that if the originary figure of reading self by face is put in question, then the whole set of assumptions making individual and social life possible are suspended.

When he puts on the black veil the Reverend Hooper is as if he were already dead. Or, rather, he seems already to have withdrawn to that realm where signs cannot reach, for which "death" is one name. Or, rather, it is as if the simple act of putting on the black veil had revealed the unverifiable possibility that each of us already dwells in that realm, both as we are for other people and even as we are for ourselves. The black veil reveals in these effects the possibility that unveiling, apocalyptic or otherwise, is impossible.

The most literal and direct effect of the veil is to suspend for Hooper's parishioners access to his subjectivity. His "figure" becomes ambiguous, disquietingly attractive, fascinating, just because his face has become invisible. This is expressed by a regular distinction between "face" and "figure" that Hawthorne borrows from common parlance: "Strangers came long distances to attend services at his church, with the mere idle purpose of gazing at his figure, because it was forbidden them to behold his face" (381) [185].[1]

Hooper's last words, "I look around me, and, lo! on every visage a Black Veil" (384) [187], assert that the face itself is a veil. Hooper's corpse mouldering in the earth, still veiled, is a veiled veil, a veil on top of a veil. There is no reason to assume that even the most extravagant series of unveilings would ever reach anything but another veil. Death, as Paul de Man says, is "a displaced name for a linguistic predicament."[2] This is the predicament of never being able to name the realities we most want to name—the self, nature, God, the realm beyond the borders of life—except in that unverifiable trope called a catachresis. Catachresis often

1. Quotations from Hawthorne's short fiction are from *Tales and Sketches*, ed. Roy Harvey Pearce (New York: The Library of America, 1982), and will be cited by page number in the text [*editor's note*].
2. Paul de Man, "Autobiography as De-Facement," *The Rhetoric of Romanticism* (New York: Columbia University Press, 1984), p. 81.

takes the form of a prosopopoeia, as in "face of a mountain" or "eye of a storm." Such a trope defaces or disfigures in the very act whereby it ascribes a face to what has none.

Hawthorne gives striking typological expression to this predicament in his image of the veiled face as a veil behind a veil. The black veil is literally a de-facement or dis-figurement. It deprives Hooper of the face whereby his neighbours assume they know him. However one wishes to describe it generically, as allegorical personification, or as parabolic realism, or as apocalyptic prophecy, the veil as type or symbol de-faces that for which it stands. At the same time the veil disfigures its referent in another way. The veil between us and that for which it is a type and a symbol is an enigmatic sign that appears to give access to what it stands for while forbidding the one who confronts it to move behind it by any effort of hermeneutic interpretation. If Hooper's face behind the veil, like that of all his neighbours, is yet another veil, then it can be said that the real face too is not a valid sign but another de-facement. The face de-faces . . . it.

Systematic narrative and figurative notations in the story of the things that are covered by a black veil extend the meaning of veiling to cover the whole repertoire of those entities that are the outside grounding of social life: nature, God, death, or the realm we shall enter after death. It is as if the inaccessibility of what the black veil covers makes it spread out to include not just Hooper's face as the sign of his selfhood but the whole array of things that are the threatening exterior of social life, while at the same time presumed to be its secure foundation.

When Hooper's subjectivity becomes inaccessible by way of his face, his veil covers a kind of floating location of the unlocatable. The spectator's speculations about what may be behind the veil drifts from consciousness to the place of death, to God, to nature. Hawthorne's story implicitly recognizes that prosopopoeia is the primary means by which mankind names and tames all that is outside the human. We give nature, God, or death a human face in order to give ourselves the illusion that we can have access to them, understand them, appropriate them as the grounds of our social intercourse. But Hooper's wearing the black veil, by suspending that primary "literal" prosopopoeia whereby we interpret a person's facial features as the signs of his or her selfhood, suspends also those extensions of prosopopoeia that are ordinarily so taken for granted as not even to be recognizable as tropes, for example when we call nature "she."

When Hooper is affrighted by his own face in the mirror during the wedding service, he rushes forth into the darkness: "For the Earth, too, had on her Black Veil." "Dying sinners" shudder when Hooper puts his veiled face near their own, "such were the terrors of the black veil, even when Death had bared his visage" (381) [185]. In his deathbed speech Hooper speaks of the way man now does "vainly shrink from the eye of his Creator" (384) [187]. All these prosopopoeias—Earth as a woman, Death as man, God as possessing an all-seeing eye—discreetly signalled by capitals, by pronouns, or by the projection on what is not human of parts of the human body, are so inextricably woven into everyday speech as to be almost invisible. They are almost effaced or "dead" metaphors. Of course we speak of the earth as "she." Of course we speak of being face to face with death, or of being under the eye of God. How could we speak at all of these things otherwise? Such universal tropes become visible only when they are suspended. Prosopopoeia is essential to allegory, as in the capitalizations of Earth, Death, and Creator here. How could there be allegory without abstractions personified and capitalized, "Orgoglio" in Spenser, or "Caritas" and the rest in Giotto's Allegory of the Virtues and Vices at Padua? Prosopopoeia is the catachrestic trope that covers our ignorance of nature, death, and God. Prosopopoeia makes everything we say of these, like what we say of the human heart, an allegory. They are allegorical in the sense of being simultaneously an unveiling (speaking of Mother Earth opens up the possibility of incorporating nature into our discourse), and a veiling (speaking of Mother Earth covers over the otherness of nature by ascribing to it a spurious similarity to ourselves).

Much is at stake in being able to go on seeing these effaced prosopopoeias as valid. At stake is our ability to go on living with a modest sense of security as mortals in an alien and threatening universe. At stake also is even our sense of ourselves *as* selves, since to question those ubiquitous personifications of nature, God, and death is, by a reciprocal putting in question, to suspend that "literal" prosopopoeia whereby a human face, our own or that of another, is an index to a self behind the face. It is no wonder the good citizens of Milford are appalled.

Hooper performs all this putting-in-question not by a disarticulating process using language against language. Such an effort always fails by smuggling back into the effort of disarticulation the very thing that is being disarticulated, as in my almost effaced prosopopoeia in "smuggling." Hooper's act works because it is done in perhaps the only way such an act can be effectively performed: in a silent "gesture" that is not really a gesture, since it is not part of a usual system of bodily movements, and by the proffering of a sign that is not really a sign, since its referent and its signification remain forever unverifiable. He appears wearing a black veil.

The performative effect of this silent act can be compared to the equally devastating effect of Bartleby's "I would prefer not to," in Melville's "Bartleby the Scrivener."[3] In both cases, once by an act of language, once by an act outside of language, language is brought to a stop, rendered powerless. This inhibition includes all kinds of language: narrative language, language conceptual, dialectical, critical, historical, biographical, and so on. In the "The Minister's Black Veil" it can be said that the efforts of Hooper's parishioners, of Hooper himself, of the narrator of his story, of Hawthorne, and of all readings of the story, including this one, are unavailing attempts to find language adequate to reincorporate into our everyday world the mute sign Hooper displays as an affront to his community.

It is now possible to answer the question,"Of what, then, is 'The Minister's Black Veil' a parable?" The story is not simply a parable of the working of parable, as opposed to being the parabolic expression of a "spiritual" meaning, a meaning capable of being expressed in no other way. Biblical scholars and critics of secular parables, such as those of Kafka, have observed that all parables tend to be about their own working. Jesus's parable of the sower is the paradigmatic example. This would not be enough to

3. I have discussed Melville's story in *Versions of Pygmalion* (Cambridge: Harvard University Press, 1990), pp. 141–78.

say about "The Minister's Black Veil." Of this story it would be better to say that it is the indirect, veiled expression of the impossibility of expressing anything verifiable at all in parable except the impossibility of expressing anything verifiable.

The veil is the type and symbol of the fact that all signs are potentially unreadable, or that the reading of them is potentially unverifiable. If the reader has no access to what lies behind a sign but another sign, then all reading of signs cannot be sure whether or not it is in error. Reading would then be a perpetual wandering or displacement that can never be checked against anything except another sign. If the artwork should be, in Kant's formulation, the indispensable bridge between epistemology and ethics, from knowledge to justified action, Hawthorne's story, it can be said, puts all its readers together on that bridge, stuck there without entrance or egress, able to go neither forward nor backward, neither back to certain knowledge of what the story means nor forward to conscientious ethical or political action in the real world.

This situation is intolerable. To live is to act, to need to act, and to need to act with a sense that we are justified in what we do. We would do anything to escape from this situation or to persuade ourselves that we are not in it. "The Minister's Black Veil" presents the reader with a full repertoire of the ways this attempt can be made. All critical essays on the story are so many more attempts to put something verifiable behind the veil, to make the veil the type and symbol of something definite one can confront directly, face to face, *through* the veil, by means of the veil. Each of these attempts proffers an hypothesis about the meaning of the veil. Each proposes or posits some entity there. This is followed, in each case, by unsuccessful attempts to verify this hypothesis, or by an implicit recognition in the act of positing the hypothesis that it is intrinsically unverifiable.

I have already cited passages in which Hooper's parishioners imagine that some stranger may have changed places with their pastor. I have also cited the discussion with Elizabeth in which Hooper answers the direct request to "take away the veil" if not from his face then from the enigmatic or "mysterious" words he uses to explain the veil. He answers only in terms of riddling "ifs": "Know, then, this veil is a type and a symbol . . . *If* it be a sign of mourning . . . *If* I hide my face for sorrow . . . and *if* I cover it for secret sin" (378–79, my italics) [184]. But Elizabeth, Hooper's parishioners, the narrator, Hooper himself, and the reader do not want "ifs" and "perhapses". We want certainty. In response to that hermeneutic need the good citizens of Milford suppose there must be some specific cause or explanation for Hooper's taking the veil. They conclude, for example, from the fact that he avoids looking at his own veiled face that, like the Reverend Moody of York, though by intent rather than by accident, he must have performed some deeply guilty act: "This was what gave plausibility to the whispers, that Mr Hooper's conscience tortured him for some great crime, too horrible to be entirely concealed, or otherwise than so obscurely intimated" (380) [185]. On the other hand, according to another hypothesis, they suppose that Hooper must be possessed and may be consorting with the devil behind the veil: "It was said, that ghost or fiend consorted with him there" (380) [185]. Another possibility, proposed to herself earlier by Elizabeth, is that the wearing of the veil "was perhaps a symptom of mental disease" (379) [184]. The operative words here are *perhaps*, *obscurely*, and *or*. It may be this or it may be that. There is no way to tell. Whatever hypothesis anyone makes about what is behind the veil, whatever proposal, proposition, or positing anyone makes, remains just that, an unverifiable hypothesis. There is no way, in this life, once you have accepted the complex ideology of the veil, to get behind the veil to find out what is really going on back there, though this is what that ideology leads us to want to do.

If, after death, good Mr Hooper's face "mouldered beneath the Black Veil," this suggests the inextricable involvement of the inaccessibility of death in the ideological system of veiling. It confirms that death is indeed a displaced name for a linguistic predicament, the predicament of being able to posit or project names freely, in primal personifying apostrophes, but unable to validate those names by any direct experience of what is named. The positing itself erects a barrier or veil. Such naming is premimetic or pre-representational, that is, it does not point toward anything that can be directly experienced. At the same time such naming forbids ever entering a representational or mimetic domain where words can be matched with things known directly, prior to language. The face is a defacing, as the *pro* ("in front of, before") in *prosopon* or *prosopopoeia* suggests. *Prosopon* means face *or* mask, the face as mask put in front of an unfathomable enigma. The figure of the face as that which is "in front" of something behind is present still in all our English words in "front": "confront," "affront," "frontal," and "front" itself. These come from Latin *frons*: "forehead," "brow." The title of Hardy's "In Front of the Landscape," for example, is already a covert prosopopoeia, as is a colloquial phrase like "the front of the house." The most disquieting effect of Hooper's veil, as the story makes clear in Hooper's last speech, is to show that the face itself is already an impenetrable veil. A veiled face is a veil over a veil, a veiling of what is already veiled.

Even for Hooper, who lives behind the veil and should therefore know what it typifies, the sight of his veiled face is terrifying. This is the case not because the veil signifies a secret guilt of which he is aware, nor because he knows that he consorts with the devil behind it—no textual support is given to these hypotheses—but because for him too its meaning cannot be specified and then verified. Though he is behind the veil, when he catches a glimpse of his veiled face in the mirror he is as much outside the veil as anyone else. For Hooper too the meaning of the veil is a matter of "if," of "perhaps," and of the "or" of ambiguity.

Nor is Hooper himself exempt from the irresistible temptation to make the veil typify something definite in order to escape from the unbearable suspension of not knowing for sure. His proposal involves speculations not about what is within his own hidden subjectivity, except insofar as it is hidden even from himself, but rather speculations about what is beyond the grave, beyond even that apocalyptic unveiling when all shall be revealed: "What, but the mystery which it obscurely typifies, has made this piece of crape so awful? When the friend shows his inmost heart to his friend; the lover to his best-beloved; when man does not vainly shrink from the eye of his Creator, loathsomely treasuring up the secret of his sin; then deem me a monster, for the symbol beneath which I have lived, and die!" (384) [187].

The black veil is the presentation not so much of a secular symbol as of a spiritual symbol that has only an individual authority, just as, within

Protestantism generally, or New England Puritanism in particular, every man may be his own priest, his own validation for a testimony that goes beyond biblical precedent and institutional authority. The minister's veil extrapolates beyond the biblical texts about veils, just as Hawthorne's story is a parable added in supplement to the canonical parables of Jesus in the New Testament. * * * Hooper nowhere claims that his authority for wearing the veil is some special mission, election, or calling that has commanded him to do so as witness to some peculiar insight mediated to his congregation by means of the veil. He is conspicuously silent where he might speak out, by saying "God commanded me to do it," or "A still small voice told me to do it," or even, "The devil made me do it." He just does it. Though Hooper becomes an awesome power in the New England church, a famous preacher who strikes religious terror into the hearts of all who hear him, that church dispatches a representative to his deathbed to try (unsuccessfully) to persuade Hooper to remove the veil before he dies. His stubborn refusal is seen as a scandal by his church.

Moreover, the traditional theological terminology of Hooper's refusal (in the words "mystery" and "obscurely typifies") is displaced to name the linguistic predicament I have identified. The emphasis is on that particular form of this predicament so fascinating to Hawthorne: the incommensurability of solitary consciousness and any language whatsoever that "may be understood and felt by anybody, who will give himself the trouble to read it" (Preface to *Twice-told Tales*, 1152). The logic of Hooper's formulation turns on "when" and "then." *When* each of us does not hide his inmost heart from God, from those closest to him, even from himself, or as Hooper has put it in his initial sermon after he dons the Black Veil, when we no longer cover "those sad mysteries which we hide from our nearest and dearest, and would fain conceal from our own consciousness, even forgetting that the Omniscient can detect them" (373) [180], *then* "deem me a monster, for the symbol beneath which I have lived, and die!" The now of that "then", however, has not yet come. In this life it remains the imminence of a perpetual "not quite yet" within which "every visage" is a Black Veil, as impenetrable as Hooper's literal veil of crape. Within the time of waiting for that perpetually deferred uncovering, Hooper is not a monster, or not yet a monster, unless all others are monsters too, though it *would* be monstrous to wear the black veil still, after the universal unveiling at the apocalypse.

Monster: the word means "showing forth," the demonstration of something hideously unlawful or unique, for example, a monstrous birth. Now Hooper is not a monster because all men and women are monsters. All manifest, in spite of themselves, the sign of the nameless and unattainable secrets all hide in their hearts, secrets monstrously different in each case. The singularity of selfhood, its uniqueness, the impossibility of fitting selfhood into any categories of genre or species, and the impossibility of saying anything definite about it are, as I have said, perhaps that "Unpardonable Sin" Ethan Brand seeks everywhere in the world and then finds in his own heart. The unpardonable sin is that sin beyond the reach of language, beyond even that particular form of performative language called a pardon, beyond even God's speech of pardon, if not beyond God's all-seeing eye. How awful to be visible to God but beyond the reach of God's pardoning word!

Hooper dies not only still veiled, but still with "a faint smile lingering on the lips" (384) [187]. This dimly glimmering smile is the sign of his characteristic irony, meaning by irony a perpetual suspension of definite meaning. Hooper's smile accompanies the unresolvable ambiguity of the veil itself and of everything that is said about it, by the narrator, by the people of Milford, and by Hooper himself, however desperate all of these are to put an end to that ambiguity by saying something definite and verifiable about the meaning of the veil. Hooper's neighbours, the narrator, and the readers of the story are driven to extravagant unverifiable hypotheses by the juxtaposition of that faint smile and the surmounting blank black veil, marked only by its fold. I suggested earlier that the fold in the veil may perhaps be related to the twice-telling of this tale and to the way a secondary parabolic meaning is superimposed on the primary literal meaning. It would be just as plausible to relate the folding to the double meaning of irony. The two signs, the dim smile without a face and the folded veil above it, would then mean the same thing or would double one another. They would be the type and symbol of the radical undecidability of all ironic expression, even of that form of ironic expression that is not verbal but facial. Irony keeps its own counsel. It responds to our interrogations only with a further ironic smile or with an ominously permissive, "Of course, if you say so."

Insofar as Hooper's sin is the sin of irony, it is appropriate that the story should end with his death, since death and irony have a secret and unsettling alliance. Though Hooper, unlike Socrates, is not put to death for being an ironist, in both cases irony is shown to be lethal. It is deadly both for the ironist and for those on whom the irony is inflicted. Irony puts both the ironist and his victims in proximity to death, but it ironically survives the death of the ironist to go on through perhaps centuries of human history effecting its deadly work of the suspension of that definite meaning for which we all long and which we all think we ought to have. The putting to death of Socrates did not put an end to the effect of Socrates's irony, as the citizens of Athens may have hoped. Quite the contrary, as any good reader of the Platonic dialogues knows. And the citizens of Milford, like the narrator, who in his last sentence places the events he has been telling at a firm historical distance ("The grass of many years has sprung up and withered on that grave" [384] [187]), are still haunted by the memory of Hooper's smile and by the image of his face mouldering beneath the veil.

The attempt by the characters and by the narrator to put an end to painful hermeneutic suspension is continued by all the commentaries on Hawthorne's story that propose some definite explanation of it. One example would be an explanation in terms of history: "The Minister's Black Veil" is a representation by Hawthorne of the historical situation of New England Puritanism surviving into a Franklinian society. Another explanation would appeal to the psychology of the author: "The Minister's Black Veil" expresses Hawthorne's obsession with the theme of secret sin or guilt. Another explanation would be based on intertextual analogies: "The Minister's Black Veil" is to be read in terms of its echoes of similar themes and figures in other works by Hawthorne, for example the motif of the veil in *The Blithedale Romance* and *The Marble Faun*, or the motif of secret sexual transgression in *The Scarlet Letter*, or the theme of unpardonable sin in "Ethan Brand." D. A. Miller's Foucauldian interpretation of the story * * * argues that the story is made definite in meaning when it is placed in the context of nineteenth-century ideas about sexual secrets. My reading differs in principle from all these in being an unveiling and putting in question of the ideology of unveiling that inveigles Hooper, his community, and most readers of the story into believing that there must be something definite behind the veil—both Hooper's veil and the veil of the text as the words on the page—and that our business as readers is to identify it.

"The Minister's Black Veil," both the veil itself, *in* the story, and the text of the story in the sense of the materiality of the letter, the words there on the page, patiently endures all these positings and projections of meaning, but it does not unequivocally endorse any of them. It offers itself to be read. If there is a veil in the text that all those inside the story want desperately to pierce or to lift so they can name once and for all what is behind it, for readers of "The Minister's Black Veil," here and now in 1990, the text itself is a veil we would pierce or lift. This desire to establish a definitive meaning for the black veil by relating it to something behind it for which it stands is an example of the hermeneutic desire as such. This desire would put a stop to the endless drifting of interpretation by saying, once and for all, "The veil means so and so." The reader shares this desire with Hooper's congregation, with his fiancée, with Hooper himself. This might be expressed by saying that the story is an allegory of the reader's own situation in reading it. If this hermeneutic desire could be appeased, then example could be a confirmation of theory, or a means of adjusting it so it could be confirmed. Allegory and realism would then be reconciled, since the realistic story would be the unambiguous carrier of a definite allegorical meaning. Language and history would be brought to touch one another, merge, overlap.

This happy reconciliation, this crossing over by means of parable into the land of parable, behind the veil, does not in this case occur. One remembers Lewis Carroll's poem of "The Walrus and the Carpenter." To the final interrogation of the oysters, "answer came there none, / Which was not surprising since / They'd eaten every one." Of all our interrogations of the veil of "The Minister's Black Veil," as of the interrogations of the veil itself within the story, it can be said "answer came there none." This is not because the text is a self-consuming artifact that eats itself up through some internal contradiction or undecidability. Rather, the attempt to turn the opacity of parabolic symbol into transparent concept is the eating up of the text. The text says what it says, if it says anything, in the way parable does, that is, by way of opaque symbols that resist translation into perspicuous concepts. To say of the veil it is a symbol of sin, it is sorrow, it is madness, it is New England Puritanism in a Franklinian culture, or it is the cover for sexual secrets is to receive no response from the text.

The text remains silent. It gives no answer to our questions, though it endures being translated into the unverifiable concepts which eat it up. Such translations make the story disappear from the page and become those blank pages in the sunlight Hawthorne feared all his works were. To alter the metaphor again: "The Minister's Black Veil," like Bartleby in Melville's story, answers to all our demands: "I should prefer not to." Like Bartleby's phrase, with its conditional "should," its gently indecisive "prefer," both inhibiting the "not" from being the negative of some positive and thereby something we can make part of some dialectical reasoning, "The Minister's Black Veil" is neither positive nor negative. It is patiently neutral. It says neither yes nor no to whatever hypotheses about it the reader proposes. The text offers neither confirmation nor disconfirmation of any speculative formulation about its meaning.

In this the text is like the black veil itself. The performative efficacy of "The Minister's Black Veil" lies in this similarity. It works. Like the veil, the story is a strange kind of efficacious speech act. It is a way of doing things with proffered signs. But it does to undo, to take away foundation or authority from anything any reader can say of it.

JOHN F. BIRK

Hawthorne's Mister Hooper: The Veil of Ham?†

Scholars now generally acknowledge that Nathaniel Hawthorne was no deeply secluded artist, but a man and citizen keenly aware of contemporary social and political issues. Over three decades ago, Arlin Turner maintained that Hawthorne wrote in response not only to his extensive reading in American history but to the burgeoning nationalism of the 1820s and 1830s.[1] We cannot fail to note the growing ascendancy during this period of that one moral issue which would soon come to eclipse all others and draw the nation into its bloodiest conflict—that of slavery. Moreover, the very seedbed of the antislavery movement in these 1830s, the home and headquarters of its fiery central leader, lay but a few miles from Hawthorne's doorstep.

The Abolitionist crusade reached new heights of extremism in Boston in 1831, when William Lloyd Garrison, armed with his newspaper *Liberator*, conducted a holy war on behalf of the more than 130 antislavery societies across the nation. A willful, resolute, unyielding commander, Garrison deemed himself and his disciples "soldiers of God" equipped with a "*gospel of peace*" (his italics).[2] The markedly religious tincture of his campaign cannot be denied. At the same time, Garrison exhibited, as Hilary A. Herbert has put it, "no perspective, no sense of relation or proportion."[3] His fervor simply knew no bounds. In his famous opening salvo in *Liberator*, a declaration that, as George M. Fredrickson has observed, "shocked many by its severity and forthrightness," Garrison boldly averred, "I am in earnest—I will not equivocate—I will not retreat a single inch—AND I WILL BE HEARD."[4] Nor were the good citizens of Massachusetts wholly innocent of this distasteful practice carried on outside

† From *Prospects* 21 (1996): 1–11. Copyright © 1996 Cambridge University Press. Reprinted by permission of the publisher. Page numbers in square brackets refer to this Norton Critical Edition.

1. Arlin Turner, *Nathaniel Hawthorne: An Introduction and Interpretation* (New York: Holt, Rinehart, and Winston, 1961), 27.
2. Estimate in the *Philanthropist* of March 1828, quoted in Hilary A. Herbert's *The Abolitionist Crusade and Its Consequences* (New York: Charles Scribner's Sons, 1912), 56.
3. Herbert, *Abolitionist Crusade*, 57.
4. George M. Fredrickson, ed., *Great Lives Observed: William Lloyd Garrison* (Englewood Cliffs, N.J.: Prentice-Hall, 1968), 22; and Garrison, quoted in Fredrickson, *Great Lives*, 23.

their state. "We are all alike guilty," Garrison asserted in his 1829 Fourth of July Address: "Slavery is strictly a national sin."[5]

Answering this clarion call, Garrison's "New Abolitionists" went forth across state borders to assail slavery and distribute antislavery texts and treatises by the hundreds of thousands.[6] In August 1831, the Nat Turner rebellion took the lives of fifty-five whites.[7] By the middle of the decade, the entire South stood gripped by fear of an imminent general insurrection.[8]

The burden now lay squarely on the shoulders of Northern conservatives. North and South were drifting perilously at odds. To allow such self-righteous frenzy to persist would be to jeopardize the very existence of the Union. Accordingly, in August 1835, those elite Bostonians most cognizant of the crisis assembled at Faneuil Hall to voice their opposition to Garrison's crusade.[9] The group's official censure of the New Abolitionists, as recorded in the *Boston Transcript,* declared in part a policy to "leave to their respective States the jurisdiction pertaining to the relation of master and slave within their boundaries" and to "hold in reprobation all attempts . . . to coerce any of the United States to abolish slavery by *appeals to the terror of the master or the passions of the slave*" (their italics).[1]

By now, such fears were not just Boston's, but the nation's. In a December 1835 message to Congress, President Andrew Jackson found himself compelled to renounce the Abolitionists' "attempts to circulate through the mails *inflammatory appeals addressed to the passions of the slaves*" (his italics).[2]

Now, we might ask: With the burgeoning magnitude of the slavery debate, with its rapid eclipse of virtually all other public affairs and its so deeply embedded moral argument of human freedom and dignity—and so much of it made rudely manifest in Salem's very backyard—could this, the "greatest social and political issue of his lifetime,"[3] escape all but passing mention by Nathaniel Hawthorne, one of America's foremost writers?

The traditional reply has seemingly been in the affirmative. Allen Flint has written that Hawthorne was "little affected by the surge of abolitionist excitement" and that "there is little evidence that he worried about the problems of Negroes and slavery."[4] H. Bruce Franklin has observed that the slavery issue simply repulsed Hawthorne.[5] While Turner has held that on the issue Hawthorne "expressed himself fully,"[6] it is in vain that we page through the bounty of his work for a definitive statement regarding the moral question of human bondage in America. Then, must we conclude that this American voice so celebrated for its championing of the dispossessed in such tales of compassion as "The Gentle Boy" opted to remain mum on the subject? Oddly, the evidence seems to say so.

5. Garrison, quoted in Fredrickson, *Great Lives*, 23.
6. Dwight Lowell Dumond, *Antislavery Origins of the Civil War* (Ann Arbor: University of Michigan Press, 1959), 48.
7. Fredrickson, *Great Lives,* 24.
8. Herbert, *Abolitionist Crusade*.
9. Ibid., 63–64.
1. *Boston Transcript,* quoted in Herbert, *Abolitionist Crusade*, 64.
2. President Andrew Jackson, quoted in Herbert, *Abolitionist Crusade*, 63.
3. Turner, *Nathaniel Hawthorne*, 46.
4. Allen Flint, "Hawthorne and the Slavery Crisis," *New England Quarterly* 41 (1968): 393.
5. H. Bruce Franklin, *Future Perfect: American Science Fiction of the Nineteenth Century* (New York: Oxford University Press, 1978), 4.
6. Turner, *Nathaniel Hawthorne*, 46.

Or does it?

In 1836, amid the white heat of Garrison's campaign and the same year in which Hawthorne quit Salem's relative seclusion to move to Boston and there edit the *American Magazine of Useful and Entertaining Knowledge*,[7] Hawthorne published "The Minister's Black Veil," a tale featuring one of his pet themes—the plight of the zealot who, amid a blinding obsession with a single, overriding program that he assumes will elevate all humankind, only isolates himself more from his fellow beings. Though most scholars have been content to peer back into the gloom of 17th-century colonial history for clues regarding Hawthorne's lingering fascination with models of this particular set of mind, might it not prove fruitful, instead, to consider a figure just down the road, that so-fervent leader of the New Abolitionists, William Lloyd Garrison himself?

"The Minister's Black Veil" centers on one Reverend Hooper, a man of the cloth who dons and refuses ever to remove a hideous veil of crepe. His uncompromising fanaticism shocks and alienates his parishioners. At last, lonely, rejected, and still misunderstood, Hooper carries his veil with him to the grave.

Readers have been as perplexed as Hooper's own congregation. Just what *is* the import of this veil? Does such token secret sin? The mask worn by all humankind? The sorrow innate in this black vale of tears? The ordeal of the serious artist? Recent scholarship has traced the broad range of critical interpretations of this tale through the last century and a half, through its assessment by, as Lea Bertani Vozar Newman has put it, "'expressive' Romantics, the 'objective' New Critics, a variety of psychologically oriented readers, Existentialists, and historicists of every hue, including a Marxist and a feminist."[8] None of these views has come close to nominating the reverend, a man of "about thirty" (184) [179],[9] as a portrait of the crusading Garrison, a man (like Hawthorne himself) thirty years old in 1835[1] and brandishing firmly before every Bostonian, New Englander, and American a perpetual reminder of the guilt that *all* white citizens must acknowledge for allowing "this peculiar institution" to flourish on American soil.

"I determined, at every hazard, to lift up the standards of emancipation in the eyes of the nation," Garrison declared in the most widely cited statement of his career: "That standard is now unfurled; and long may it float. . . . [L]et all enemies of the persecuted blacks tremble."[2] The strategy was not without effect. Fredrickson has observed that the man's "harsh and uncompromising mode of expression that publicized his cause through its shock effect" elevated the slavery debate to a "new form" and prompted "philanthropists and reformers to re-examine their premises," as all the while Garrison "continued up to the time of emancipation to play an indispensable role as a moral gadfly, keeping the ideal ever in sight of those engaged in confronting the actual."[3]

7. Ibid., ix.
8. Lea Bertani Vozar Newman, "One Hundred and Fifty Years of Looking At, Into, Through, Behind, Beyond, and Around "The Minister's Black Veil,'" *Nathaniel Hawthorne Review* 13 (1987): 8.
9. References from this tale come from Nathaniel Hawthorne, *Hawthorne: Selected Tales and Sketches*, introduction by Hyatt H. Waggoner (San Francisco: Rinehart, 1970); page numbers are hereafter cited in the text.
1. Fredrickson, *Great Lives*, 1.
2. Garrison, quoted in Fredrickson, *Great Lives*, 22.
3. Fredrickson, *Great Lives*, 2–3.

The good reverend proves equally obstinate and uncompromising in holding his ideal before his public. "Know, then, this veil is a type and a symbol, and I am bound to wear it ever. . . . No mortal eye will see it withdrawn," he avows (191) [184]. Just as Garrison's fanatical stance "succeeded only in making the antislavery movement ridiculous in the sight of many conservative people," as Dwight Lowell Dumond has put it,[4] and forced such conservatives to convene to take more direct and concerted action to counter his efforts (Fredrickson has seen Garrison's importance to lie precisely in this intense antipathy he aroused),[5] so the parson's "taking the veil" shocks and alienates many of the villagers. "He has changed himself into something awful," declares one on sighting him. "Our Parson has gone mad!" exclaims another (184) [179]. Even Hooper's betrothed looks upon his donning of the crepe as "perhaps a symptom of a mental disease" (192) [184]. As for Garrison's hope to make "all the enemies of the persecuted blacks tremble," Hooper's orations from behind the veil make his "hearers [quake]" (186) [180]; when he proffers a gentle funeral prayer, the "people trembl[e]" (188) [182]; he bends over a dead maiden and soon rumor has it that "the corpse had slightly shuddered" (188) [181]; he beholds his image in a mirror and even his own selfsame "frame shudder[s]" and "his lips [grow] white" (189) [182]. As the worried leaders of Boston assembled at Faneuil Hall in a futile attempt to curb Garrison, so a gathering of concerned congregational leaders confronts, in vain, the zealous parson (190) [183].

To pursue the analogy down another related avenue, we might even for a moment view Hooper as a self-appointed representative of the black race, as a man forced by intention to endure the stigma borne by what Garrison himself pointed out numbered fully one-sixth of the nation's population.[6] Repeatedly Hawthorne emphasizes how it is Hooper's "black face" alone that relegates him to a site beyond the fringes of society. Similarly, Garrison labeled the program of slavery's apologists "EXPULSION OF THE BLACKS."[7] In his *Thoughts on Racial Prejudice* (1832), he cited one who had written "The exclusion of the free blacks from the civil and literary privileges of our country, depends upon another circumstance than that of character. . . . This circumstance is—*he is a black man!*"[8] Garrison elaborates the plight of blacks:

> Let them toil from youth to old age in the honorable pursuit of wisdom—let them store their minds with the most valuable researches of science and literature—and let them add to highly gifted and cultivated intellect, a piety pure, undefiled, and unspotted from the world, *it is all nothing*—they would not be received into the *very lowest walks of society*—admiration of such uncommon beings would mingle with *disgust!* (his italics)[9]

This is precisely the case with the reverend. To underscore the degree to which the parson's exclusion is based solely on appearance with disregard for any underlying character, Hawthorne is careful to present Hooper not as a mere linear-minded fanatic, but as a man of "kind and loving" nature

4. Dumond, *Antislavery Origins*, 86.
5. Fredrickson, *Great Lives*, 3.
6. Garrison, quoted in Fredrickson, *Great Lives*, 34.
7. Ibid., 36.
8. Ibid., 34.
9. Ibid., 32.

and life "irreproachable in outward act" (194) [186]. Despite these virtues, he is "shrouded in dismal suspicion," is "dimly feared," and abides as "a man apart from men, shunned in their health and joy" (194) [186]. "This dismal shade must separate me from the world!" he laments to his betrothed (191) [184] and adds, "It is but a mortal veil—it is not for eternity! O! you know now how lonely I am, and how frightened, to be alone behind my black veil. Do not leave me in this miserable obscurity forever!" (192) [184]. In the case of both skin hue and veil, blackness posed before the public prompts the isolation of the wearer, who loses all authenticity as a human being. The surface veneer alienates beyond repair; the sensitive individual beneath remains submerged, unacknowledged, unappreciated.

As well, we cannot fail to note the fundamentally religious impetus driving the two men. Regarding the cause Garrison took to his public, Dumond has observed, "No other reform movement is quite like the antislavery crusade, because it was based upon an appeal to the consciences of men."[1] True to this plea to conscience *surtout*, Garrison founded his campaign on New Testament love. "Hence, when smitten on one cheek they turn the other also, being defamed they entreat, being reviled they bless," he spoke of his disciples.[2] Such a tactic served to highlight the slavery issue as never before by laying bare the "moral core of the problem as no one else had done," Fredrickson has maintained.[3] Similarly, as another instance of what Norman German has called "Hawthorne's penchant for etymological punning,"[4] the Reverend Hooper's donning of the crepe renders him doubly a "man of the cloth." Garrison and the reverend: Both are social reformers of uncommonly strong religious convictions based on the principle of Christian love.

The cases of the abolitionist leader and the parson share additional features. In the first edition of *Liberator*, Garrison included verses hinting the danger of a general insurrection should white America fail to mend its ways: "Wo if it come with storm, and blood, and fire, / When midnight darkness veils the earth and sky!"[5] Accordingly, in the act of toasting newlyweds, Hooper inadvertently glimpses himself in a mirror and "rushe[s] forth into the darkness," we read, "For the earth, too, had on her black veil" (189) [182]. In each case, an ominous and more pervasive darkness emerges as the public partakes of its most pronounced secular joys. As well, the scenarios of Garrison and the parson illustrate the keener ability of womanhood to acknowledge the truth behind their campaigns. In his 1829 Fourth of July Address, Garrison, a fervent supporter of feminists,[6] called on New England women to "form charitable associations to relieve the degraded of their sex" and added, "As yet, an appeal to their sympathies was never made in vain. They outstrip us in every benevolent race." He concluded, "Females are doing much for their cause at the South; let their example be imitated, and their exertions surpassed, at the North."[7] While Hooper's mask proves a disconcerting sight, one villager alone does

1. Dumond, *Antislavery Origins,* 37.
2. Garrison, quoted in Herbert, *Abolitionist Crusade,* 57.
3. Fredrickson, *Great Lives,* 2.
4. Norman German, "The Veil of Words in 'The Minister's Black Veil,'" *Students in Short Fiction* 25 (1988): 41.
5. Garrison, quoted in Fredrickson, *Great Lives,* 24.
6. Fredrickson, *Great Lives,* 3.
7. Garrison, quoted in Fredrickson, *Great Lives,* 20

remain "unappalled by the awe with which the black veil . . . impresse[s] all besides herself." It is she, the reverend's "plighted wife," who speaks to him far more intimately than does the learned citizens' committee regarding his donning of the crepe. Despite his "gentle, but unconquerable obstinacy," she, "of a firmer character than his own," does if only for a moment pierce through to the veil's profounder import. With her eyes "fixed insensibly on the black veil . . . like a sudden twilight in the air, its terrors [fall] around her" (190–92) [183–84].

Finally, the ordeal of Garrison and other reformers of his time no doubt augmented Hawthorne's skepticism not only toward overzealousness but toward democracy itself, and prompted him to sketch the parson as a genuinely well-meaning man face to face with a stubbornly unperceiving public. In 1835, Garrison was attacked by a mob, a not-rare fate to befall an abolitionist.[8] "On seeing me, three or four of the rioters, uttering a yell, furiously dragged me to the window, with the intention of hurling me from that height to the ground," he told of it later.[9] In this same line, a careful reading of Hooper's tale yields a set of sympathies that do not always lie so unequivocally with the parishioners. We see the villagers portrayed as the name-calling, rumor-spreading, nonmeditative ruminants the human species on occasion can be. Few, if any, attempt to strike through the reverend's mask. Tales such as "The Gentle Boy" and "My Kinsman, Major Molineux," composed during these same years, illustrate quite graphically Hawthorne's abiding awareness and distress about this same tyranny of public opinion. Moreover, some of the strokes that Hawthorne provides do indeed focus less on the reverend's own lack of guidance than on that of his insular-minded flock. Significantly, the tale's advent, from the perspective of the villagers convening for worship, includes children "tripp[ing] merrily," bachelors peeking "sidelong at the pretty maidens," and the elderly "stooping along the street" (183) [178]. Few have their minds set on higher things. What is more, the issue of blackness posed before them comes to dominate their thoughts much as the slavery issue did the public mind in 1830s America: "The next day, the whole village of Milford talked of little else than Parson Hooper's black veil" (189) [182]. The congregation's response is not at all unlike that of Northerners who refused to consider in earnest the slavery debate and too easily place its burden on others: "It was remarkable that all of the busybodies and impertinent people in the parish, not one ventured to put the plain question to Mr. Hooper, wherefore he did this thing" (189) [182]. To a person, they shirk the issue: None "cho[o]se to make the black veil a subject of friendly remonstrance." The veil incites "a feeling of dread, neither plainly confessed nor carefully concealed, which cause[s] each to shift the responsibility upon another" (190) [183]. To any American reader of the 1830s even vaguely attuned to the nation's state of affairs, such reactions to an image of blackness would most conspicuously correspond to the popular mood and response to the crisis of slavery.

Of value here is the testimony of Hawthorne's fellow artist Herman Melville, who wrote how Hawthorne seemed to "feel a touch of a shrink" when hearing of the former's less-fettered vision of a "ruthless democracy

8. Fredrickson, *Great Lives,* 38.
9. Garrison, quoted in Fredrickson, *Great Lives,* 42.

on all sides."[1] In his famous essay *"Hawthorne and His Moses,"* Melville expressly warned against misreading the import of Hawthorne's story titles, for such are often "directly calculated to deceive—egregiously deceive."[2] German, too, has pointed to Hawthorne's use of puns both "artistically and systematically."[3] From this perspective, "The Minister's Black *Vale*," a title hinting more of Hooper's own tribulations in encountering the darkness of his earthbound flock and the refusal of others to confess their own limited powers of perception, might not be so inappropriate as might at first appear.

That slavery finds no outright mention in "The Minister's Black Veil" we can attribute to Hawthorne's well-known reluctance to sketch his intentions amid the full sunshine where "meanings" scintillate indisputably. More recently (1993), Thomas R. Moore, for one, has examined Hawthorne's exposure while a student at Bowdoin College to Hugh Blair's text, *Lectures on Rhetoric and Belles Lettres,* the "most popular rhetoric" of the time in American colleges, and has cited Hawthorne's "skeptical attitude" regarding Blair's advice that a writer strive for simplicity. Hawthorne's "apparent stylistic simplicity is a veil" and his visible compliance with Blair's dictum "masks a socially and culturally variant subtext that undercuts," Moore has maintained, to the degree that Hawthorne's "opaque style is set squarely against these ideals of clarity and concision." While several contemporaries failed to spy such "subversive undertones" (Melville being the prime exception), we have over the years come to perceive such discourse as one "containing a secondary layer of anthracite smoke; ambiguity, and subversion."[4] Carolyn L. Karcher has argued that Melville likewise broached the topic of slavery less obtrusively, communicating in his work the suffering he had endured as a common sailor before the mast and thence "generaliz[ing] about slavery by analogy" by employing whites and other racial types (Polynesians, Native Americans) as "fictional counterparts" for blacks.[5] In that Melville so admired Hawthorne's art—lauding "Nathaniel of Salem" in his only known literary essay and dedicating to him his masterpiece *Moby-Dick*—might we not assume that he recognized the benefits of this same strategy of indirection in the works of the "Man of Mosses"? Perhaps another manifestation of such indirection lies in the parson's disarming character. As apologists for slavery or other whites unwilling to enter the debate found a solace in the stereotypical notion of the "Negro's proverbial cheerfulness,"[6] we find black-faced Hooper customarily sporting a smile—"nodding kindly" (184) [179], with a "sad smile" (187) [181] and a "placid cheerfulness" (188) [182]. Of such unflagging geniality we are told, "There was no quality of his disposition which made him more beloved than this" (188) [182]. Then, the parson's principal

1. Herman Melville, *The Letters of Herman Melville,* ed, Merrill R. Davis and William H. Gilman (New York: Yale University Press, 1960), 126–27.
2. Herman Melville, "Hawthorne and His Mosses," in *Moby-Dick; or, the Whole* (New York: W. W. Norton, 1967), 549.
3. German; "Veil of Words," 47.
4. Thomas R. Moore, "'A Thick and Darksome Veil": The Rhetoric of Hawthorne's Sketches," *Nineteenth-Century Literature* 48 (1998): 310–11, 317, 325
5. Carolyn L. Karcher, *Shadow Over the Promised Land: Slavery, Race and Violence in Melville's America* (Baton Rouge: Louisiana State University Press, 1980), 2.
6. Ibid., 21.

outward trait is that same one of the stereotypical black that most effectively masks the true nature behind it.

We would expect not the geographical or emotive but the *moral* dimension of slavery to elicit Hawthorne's keenest attention, of course, and it was in the 1830s that many churches performed an astounding about-face—donned masks of their own, so to speak—regarding the issue of bondage. As a matter of national expediency, such groups by and large condemned abolition and yet somehow found biblical sanction for the practice of human slavery. Such a particular moral turnabout, as Herbert has described, "has no parallel in all the history of the Christian churches" and bears a significance that "cannot be overstated."[7] Could an artist of such abiding interest in moral questions as Nathaniel Hawthorne let such hypocritical veiling by the nation's clergy pass without comment?

A curious confession and several commentaries of later years may shed added light on Hawthorne's less obtrusive intention in his famous tale of the parson. In *The English Notebooks* for 1888, Hawthorne offered this frank admission: "On the whole, I find myself more of an abolitionist in feeling than in principle."[8] This remark in itself suggests an ambivalence toward slavery, much like that of the villagers toward the veil. In his 1852 biography of former Bowdoin classmate Franklin Pierce, Hawthorne dubbed "wise" the view that "looks upon slavery as one of those evils which divine Providence does not leave to be remedied by human contrivances." Hawthorne elaborated: "There is no instance, in all history, of the human will and intellect having perfected any great moral reform by methods which it adapted to that end" [220].[9] This sentiment, too, is of a piece with the ultimate futility of the parson's gesture. In 1856, after conversing with a black man, Hawthorne wrote that he had found the other "rather shy—reserved, at least, and undemonstrative, yet not harshly so," and added that "I felt, or thought I felt, that his color was continually before his mind, and that he walks cautiously among men, as conscious that every new introduction is a new experiment," that "there may be miles and miles of depth in him, which I saw nothing of."[1] In light of these remarks, would it be strange for Hawthorne to comment obliquely on the slavery issue by employing the equation of a gaping, unperceiving public and a righteous man who, because of his "blackness," remains irrevocably set apart from such? Indeed, Hawthorne's own confessed inability to fathom the nature of a black man standing before him similarly hints his sketch of a black-faced parson and his puzzled parishioners to be in the tradition of a roman à clef, much as Arthur Miller's *The Crucible* a century later tapped the Puritan past to cast light on contemporary political dynamics.

Is it possible that Hawthorne's skepticism mellowed with time? The coming of the conflict linked to slavery brought him near euphoria. Perhaps human endeavor *could* undo a horrible wrong? As he wrote Horatio Bridge in 1861,

> The war, strange to say, has had a beneficial effect upon my spirits, which were flagging woefully before it broke out. . . . If we are fighting

7. Herbert, *Abolitionist Crusade,* 67–69.
8. Nathaniel Hawthorne, *The English Notebooks, ed.* Randall Stewart (New York: Russell and Russell, 1962), 48.
9. Nathaniel Hawthorne, *The Life of Franklin Pierce* (New York: MSS Information, 1970), 113–14.
1. Hawthorne, *English Notebooks,* 849–50.

> for the annihilation of slavery, to be sure, it may be a wise object, and offers a tangible result, and the only one which is consistent with a future Union between North and South. A continuance of the war would soon make this plain to us; and we should see the expediency of preparing our black brethren for future citizenship by allowing them to fight for their own liberties, and educating them through heroic example.
>
> What ever happens next, I must say that I rejoice that the old Union is smashed. We never were one people, and never really had a country since the Constitution was formed.[2]

In "Chiefly About War Matters," appearing in *Atlantic* in 1862, Hawthorne again broached the topic, this time from the perspective of poor Southern whites. Here again, however, ambiguity pervades, to the extent that this essay is one in which the author's own feelings lie "deeply buried beneath layers of irony," writes Deborah L. Madsen. Yet Madsen has managed to glean from Hawthorne's words the belief that human bondage is "something that eludes such simple oppositions as North and South, black and white, rich and poor" and that, in this view, the American Civil War was perhaps no less than "a genuine revolution" meant "to overthrow . . . the psychological shackles that have operated historically to keep Americans, black and white, enslaved."[3]

At the same time, there are still hints through the years of that darker vision all too apparent to friend Melville, who praised not only Hawthorne's willingness to broach those topics so often veiled by appearance but his "blackness, ten times black."[4] In *The Civil War World of Herman Melville* (1993), Stanton Garner has observed Hawthorne's reluctance to accord full rights to blacks after the war due to their lack of education and unfamiliarity with freedom, an attitude prevalent among even the most open-minded of Democrats.[5]

Whatever Hawthorne's private sentiments might have been, an interpretation of the parson's donning of the black veil as a metaphor for the abolitionist cause in a sketch in which a mote of misunderstanding lies not only in the secreted eye of the cleric but in that of the beholder (be it parishioner or even reader) once again brings us face to face with blackness and ambiguity—those twin hallmarks of the Man of Salem.

2. Nathaniel Hawthorne, *The Letters, 1857–64, ed.* Thomas Woodson et al. (Columbus: Ohio State University Press, 1987), 380–81.
3. Deborah L. Madsen, "'A for Abolition': Hawthorne's Bond-Servant and the Shadow of Slavery," *Journal of American Studies* 25 (1991): 255–56, 258.
4. Melville, "Hawthorne and His Mosses," 540.
5. Stanton Garner, *The Civil War World of Herman Melville* (Lawrence: University of Kansas Press, 1993), 154.

JUDITH FETTERLEY

Women Beware Science: "The Birthmark"†

The scientist Aylmer in Nathaniel Hawthorne's "The Birthmark" provides another stage in the psychological history of the American protagonist. Aylmer is [Washington] Irving's Rip and [Sherwood] Anderson's boy [in "I Want to Know Why"] discovered in that middle age which Rip evades and the boy rejects. Aylmer is squarely confronted with the realities of marriage, sex, and women. There are compensations, however, for as an adult he has access to a complex set of mechanisms for accomplishing the great American dream of eliminating women. It is testimony at once to Hawthorne's ambivalence, his seeking to cover with one hand what he uncovers with the other, and to the pervasive sexism of our culture that most readers would describe "The Birthmark" as a story of failure rather than as the success story it really is—the demonstration of how to murder your wife and get away with it. It is, of course, possible to read "The Birthmark" as a story of misguided idealism, a tale of the unhappy consequences of man's nevertheless worthy passion for perfecting and transcending nature; and this is the reading usually given it.[1] This reading, however, ignores the

† From *The Resisting Reader; A Feminist Approach to American Fiction* (Bloomington: Indiana UP, 1978), pp. 22–33. Page numbers in square brackets refer to this Norton Critical Edition.

1. See, for example, Brooks and Warren, *Understanding Fiction* (New York: Appleton-Century-Croft, 1943), pp. 103–106: "We are not, of course, to conceive of Aylmer as a monster, a man who would experiment on his own wife for his own greater glory. Hawthorne does not mean to suggest that Aylmer is depraved and heartless. . . . Aylmer has not realized that perfection is something never achieved on earth and in terms of mortality"; Richard Harter Fogle, *Hawthorne's Fiction: The Light and The Dark*, rev. ed. (Norman, Okla. University of Oklahoma Press, 1964), pp. 117–31; Robert Heilman, "Hawthorne's 'The Birthmark': Science as Religion," *South Atlantic Quarterly* 48 (1949), 574–83: "Aylmer, the overweening secientist, resembles less the villain than the tragic hero: in his catastrophic attempt to improve on human actuality there is not only pride and a deficient sense of reality but also disinterested aspiration"; F. O. Matthiessen, *American Renaissance* (New York: Oxford University Press, 1941), pp. 253–55; Arlin Turner, *Nathaniel Hawthorne* (New York: Holt, Rinehart, and Winston, 1961), pp. 88, 98, 132; "In 'The Birthmark' he applauded Aylmer's noble pursuit of perfection, in contrast to Aminadab's ready acceptance of earthiness, but Aylmer's achievement was tragic failure because he had not realized that perfection is not of this world." The major variation in these readings occurs as a result of the degree to which individual critics see Hawthorne as critical of Aylmer. Still, those who see Hawthorne as critical of Aylmer locate the source of this criticism in Aylmer's idealistic pursuit of perfection—e.g., Millicent Bell, *Hawthorne's View of the Artist* (New York: State University of New York, 1962), pp. 182–85: "Hawthorne, with his powerful Christian sense of the inextricable mixture of evil in the human compound, regards Aylmer as a dangerous perfectibilitarian"; William Bysshe Stein, *Hawthorne's Faust* (Gainesville: University of Florida Press, 1953), pp. 91–92: "Thus the first of Hawthorne's Fausts, in a purely symbolic line of action sacrifices his soul to conquer nature, the universal force of which man is but a tool." Even Simon Lesser, *Fiction and the Unconscious* (1957; rpt. New York: Vintage-Random, 1962), pp. 87–90 and pp. 94–98, who is clearly aware of the sexual implication of the story, subsumes his analysis under the reading of misguided idealism and in so doing provides a fine instance of phallic criticism in action: "The ultimate purpose of Hawthorne's attempt to present Aylmer in balanced perspective is to quiet our fears so that the wishes which motivate his experiment, which are also urgent, can be given their opportunity. Aylmer's sincerity and idealism give us a sense of kinship with him. We see that the plan takes shape gradually in his mind, almost against his conscious intention. We are reassured by the fact that he loves Georgiana and feels confident that his attempt to remove the birthmark will succeed. Thus at the same time that we recoil we can identify with Aylmer and through him act out some of our secret desires. . . . The story not only gives expression to impulses which are ordinarily repressed; it gives them a sympathetic hearing—an opportunity to show whether they can be gratified without causing trouble or pain. There are obvious gains in being able to conduct tests of this kind with no more danger and no greater expenditure of effort than is involved in reading a story." The one significant dissenting view is offered by Frederick Crews, *The Sins of the Fathers* (New York: Oxford University Press, 1966), whose scattered comments on the story focus on the specific form of Aylmer's idealism and its implications for his secret motives.

significance of the form idealism takes in the story. It is not irrelevant that "The Birthmark" is about a man's desire to perfect his wife, nor is it accidental that the consequence of this idealism is the wife's death. In fact, "The Birthmark" provides a brilliant analysis of the sexual politics of idealization and a brilliant exposure of the mechanisms whereby hatred can be disguised as love, neurosis can be disguised as science, murder can be disguised as idealization, and success can be disguised as failure. Thus, Hawthorne's insistence in his story on the metaphor of disguise serves as both warning and clue to a feminist reading.

Even a brief outline is suggestive. A man, dedicated to the pursuit of science, puts aside his passion in order to marry a beautiful woman. Shortly after the marriage he discovers that he is deeply troubled by a tiny birthmark on her left cheek. Of negligible importance to him before marriage, the birthmark now assumes the proportions of an obsession. He reads it as a sign of the inevitable imperfection of all things in nature and sees in it a challenge to man's ability to transcend nature. So nearly perfect as she is, he would have her be completely perfect. In pursuit of this lofty aim, he secludes her in chambers that he has converted for the purpose, subjects her to a series of influences, and finally presents her with a potion which, as she drinks it, removes at last the hated birthmark but kills her in the process. At the end of the story Georgiana is both perfect and dead.

One cannot imagine this story in reverse—that is, a woman's discovering an obsessive need to perfect her husband and deciding to perform experiments on him—nor can one imagine the story being about a man's conceiving such an obsession for another man. It is woman, and specifically woman as wife, who elicits the obsession with imperfection and the compulsion to achieve perfection, just as it is man, and specifically man as husband, who is thus obsessed and compelled. In addition, it is clear from the summary that the imagined perfection is purely physical. Aylmer is not concerned with the quality of Georgiana's character or with the state of her soul, for he considers her "fit for heaven without tasting death" [197]. Rather, he is absorbed in her physical appearance, and perfection for him is equivalent to physical beauty. Georgiana is an exemplum of woman as beautiful object, reduced to and defined by her body. And finally, the conjunction of perfection and nonexistence, while reminding us of Anderson's story in which the good girl is the one you never see, develops what is only implicit in that story: namely, that the only good woman is a dead one and that the motive underlying the desire to perfect is the need to eliminate. "The Birthmark" demonstrates the fact that the idealization of women has its source in a profound hostility toward women and that it is at once a disguise for this hostility and the fullest expression of it.

The emotion that generates the drama of "The Birthmark" is revulsion. Aylmer is moved not by the vision of Georgiana's potential perfection but by his horror at her present condition. His revulsion for the birthmark is insistent: he can't bear to see it or touch it; he has nightmares about it; he has to get it out. Until she is "fixed," he can hardly bear the sight of her and must hide her away in secluded chambers which he visits only intermittently, so great is his fear of contamination. Aylmer's compulsion to perfect Georgiana is a result of his horrified perception of what she actually is, and all his lofty talk about wanting her to be perfect so that just this once the potential of Nature will be fulfilled is but a cover for his central emotion of revulsion. But Aylmer is a creature of disguise and illusion. In order to persuade this beautiful woman to become his wife, he "left his laboratory to the care of an assistant, cleared his fine countenance from the furnace smoke, washed the stains of acid from his fingers" [188]. Best not to let her know who he really is or what he really feels, lest she might say before the marriage instead of after, "You cannot love what shocks you!" [188]. In the chambers where Aylmer secludes Georgiana, "airy figures, absolutely bodiless ideas, and forms of unsubstantial beauty" [193] come disguised as substance in an illusion so nearly perfect as to "warrant the belief that her husband possessed sway over the spiritual world" [193]. While Aylmer does not really possess sway over the spiritual world, he certainly controls Georgiana and he does so in great part because of his mastery of the art of illusion.

If the motive force for Aylmer's action in the story is repulsion, it is the birthmark that is the symbolic location of all that repels him. And it is important that the birthmark is just that: a birth *mark*, that is, something physical; and a *birth* mark, that is, something not acquired but inherent, one of Georgiana's givens, in fact equivalent to her.[2] The close connection between Georgiana and her birthmark is continually emphasized. As her emotions change, so does the birthmark, fading or deepening in response to her feelings and providing a precise clue to her state of mind. Similarly, when her senses are aroused, stroked by the influences that pervade her chamber, the birthmark throbs sympathetically. In his efforts to get rid of the birthmark Aylmer has "administered agents powerful enough to do aught except change your entire physical system" [197], and these have failed. The object of Aylmer's obsessive revulsion, then, is Georgiana's "physical system," and what defines this particular system is the fact that it is female. It is Georgiana's female physiology, which is to say her sexuality, that is the object of Aylmer's relentless attack. The link between Georgiana's birthmark and her sexuality is implicit in the birthmark's role as her emotional barometer, but one specific characteristic of the birthmark makes the connection explicit: the hand which shaped Georgiana's birth has left its mark on her in *blood*. The birthmark is redolent with references to the particular nature of female sexuality; we hardly need Aylmer's insistence on seclusion, with its reminiscences of the treatment of women when they are "unclean," to point us in this direction. What repels Aylmer is Georgiana's sexuality; what is imperfect in her is the fact that she is female; and what perfection means is elimination.

In Hawthorne's analysis the idealization of women stems from a vision of them as hideous and unnatural; it is a form of compensation, an attempt to bring them up to the level of nature. To symbolize female physiology as a blemish, a deformity, a birthmark suggests that women are in need of some such redemption. Indeed, "The Birthmark" is a parable of woman's relation to the cult of female beauty, a cult whose political function is to remind women that they are, in their natural

2. In the conventional reading of the story Georgiana's birthmark is seen as the symbol of original sin—see, for example, Heilman, p. 579; Bell, p. 185. But what this reading ignores are, of course, the implications of the fact that the symbol of original sin is female and that the story only "works" because men have the power to project that definition onto women.

state, unacceptable, imperfect, monstrous. Una Stannard in "The Mask of Beauty" has done a brilliant job of analyzing the implications of this cult:

> Every day, in every way, the billion-dollar beauty business tells women they are monsters in disguise. Every ad for bras tells a woman that her breasts need lifting, every ad for padded bras that what she's got isn't big enough, every ad for girdles that her belly sags and her hips are too wide, every ad for high heels that her legs need propping, every ad for cosmetics that her skin is too dry, too oily, too pale, or too ruddy, or her lips are not bright enough, or her lashes not long enough, every ad for deodorants and perfumes that her natural odors all need disguising, every ad for hair dye, curlers, and permanents that the hair she was born with is the wrong color or too straight or too curly, and lately ads for wigs tell her that she would be better off covering up nature's mistake completely. In this culture women are told they are the fair sex, but at the same time that their "beauty" needs lifting, shaping, dyeing, painting, curling, padding. Women are really being told that "the beauty" is a beast.[3]

The dynamics of idealization are beautifully contained in an analogy which Hawthorne, in typical fashion, remarks on casually: "But it would be as reasonable to say that one of those small blue stains which sometimes occur in the purest statuary marble would convert the Eve of Powers to a monster" [189]. This comparison, despite its apparent protest against just such a conclusion, implies that where women are concerned it doesn't take much to convert purity into monstrosity; Eve herself is a classic example of the ease with which such a transition can occur. And the transition is easy because the presentation of woman's image in marble is essentially an attempt to disguise and cover a monstrous reality. Thus, the slightest flaw will have an immense effect, for it serves as a reminder of the reality that produces the continual need to cast Eve in the form of purest marble and women in the molds of idealization.

In exploring the sources of men's compulsion to idealize women Hawthorne is writing a story about the sickness of men, not a story about the flawed and imperfect nature of women. There is a hint of the nature of Aylmer's ailment in the description of his relation to "mother" Nature, a suggestion that his revulsion for Georgiana has its root in part in a jealousy of the power which her sexuality represents and a frustration in the face of its inpenetrable mystery. Aylmer's scientific aspirations have as their ultimate goal the desire to create human life, but "the latter pursuit, however, Aylmer had long laid aside in unwilling recognition of the truth—against which all seekers sooner or later stumble—that our great creative Mother, while she amuses us with apparently working in the broadest sunshine, is yet severely careful to keep her own secrets, and, in spite of her pretended openness, shows us nothing but results. She permits us, indeed, to mar, but seldom to mend, and, like a jealous patentee, on no account to make" [191]. This passage is striking for its undercurrent of jealousy, hostility, and frustration toward a specifically female force. In the vision of Nature as playing with man, deluding him into thinking he

3. Vivian Gornick and Barbara K. Moran, eds., *Woman in Sexist Society: Studies in Power and Powerlessness* (New York: Basic Books, 1971), p. 192.

can acquire her power, and then at the last minute closing him off and allowing him only the role of one who mars, Hawthorne provides another version of woman as enemy, the force that interposes between man and the accomplishment of his deepest desires. Yet Hawthorne locates the source of this attitude in man's jealousy of woman's having something he does not and his rage at being excluded from participating in it.

Out of Aylmer's jealousy at feeling less than Nature and thus less than woman—for if Nature is woman, woman is also Nature and has, by virtue of her biology, a power he does not—comes his obsessional program for perfecting Georgiana. Believing he is less, he has to convince himself he is more: "and then, most beloved, what will be my triumph when I shall have corrected what Nature left imperfect in her fairest work! Even Pygmalion, when his sculptured woman assumed life, felt not greater ecstasy than mine will be" [191]. What a triumph indeed to upstage and outdo Nature and make himself superior to her. The function of the fantasy that underlies the myth of Pygmalion, as it underlies the myth of Genesis (making Adam, in the words of Mary Daly, "the first among history's unmarried pregnant males"[4]), is obvious from the reality which it seeks to invert. Such myths are powerful image builders, salving man's injured ego by convincing him that he is not only equal to but better than woman, for he creates in spite of, against, and finally better than nature. Yet Aylmer's failure here is as certain as the failure of his other "experiments," for the sickness which he carries within him makes him able only to destroy, not to create.

If Georgiana is envied and hated because she represents what is different from Aylmer and reminds him of what he is not and cannot be, she is feared for her similarity to him and for the fact that she represents aspects of himself that he finds intolerable. Georgiana is as much a reminder to Aylmer of what he is as of what he is not. This apparently contradictory pattern of double-duty is understandable in the light of feminist analyses of female characters in literature, who frequently function this way. Mirrors for men, they serve to indicate the involutions of the male psyche with which literature is primarily concerned, and their characters and identities shift accordingly. They are projections, not people; and thus coherence of characterization is a concept that often makes sense only when applied to the male characters of a particular work. Hawthorne's tale is a classic example of the woman as mirror, for, despite Aylmer's belief that his response to Georgiana is an objective concern for the intellectual and spiritual problem she presents, it is obvious that his reaction to her is intensely subjective. "Shocks you, my husband?" queries Georgiana [188], thus neatly exposing his mask, for one is not shocked by objective perceptions. Indeed, Aylmer views Georgiana's existence as a personal insult and threat to him, which, of course, it is, because what he sees in her is that part of himself he cannot tolerate. By the desire she elicits in him to marry her and possess her birthmark, she forces him to confront his own earthiness and "imperfection."

4. Mary Daly, *Beyond God the Father* (Boston: Beacon P, 1973), p. 195. It is useful to compare Daly's analysis of "Male Mothers" with Ellmann's discussion of the "imagined motherhood of the male" in *Thinking About Women*, pp. 15ff. It is obvious that this myth is prevalent in patriarchal culture, and it would seem reasonable to suggest that the patterns of cooptation noticed in "Rip Van Winkle" and "I Want to Know Why" are minor manifestations of it. *An American Dream* provides a major manifestation, in fact a tour de force, of the myth of male motherhood.

But it is precisely to avoid such a confrontation that Aylmer has fled to the kingdom of science, where he can project himself as a "type of the spiritual element." Unlike Georgiana, in whom the physical and the spiritual are complexly intertwined, Aylmer is hopelessly alienated from himself. Through the figure of Aminadab, the shaggy creature of clay, Hawthorne presents sharply the image of Aylmer's alienation. Aminadab symbolizes that earthly, physical, erotic self that has been split off from Aylmer, that he refuses to recognize as part of himself, and that has become monstrous and grotesque as a result: "With his vast strength, his shaggy hair, his smoky aspect, and the indescribable earthiness that incrusted him, he seemed to represent man's physical nature; while Aylmer's slender figure, and pale, intellectual face, were no less apt a type of the spiritual element" [192]. Aminadab's allegorical function is obvious and so is his connection to Aylmer, for while Aylmer may project himself as objective, intellectual, and scientific and while he may pretend to be totally unrelated to the creature whom he keeps locked up in his dark room to do his dirty work, he cannot function without him. It is Aminadab, after all, who fires the furnace for Alymer's experiments; physicality provides the energy for Aylmer's "science" just as revulsion generates his investment in idealization. Aylmer is, despite his pretenses to the contrary, a highly emotional man: his scientific interests tend suspiciously toward fires and volcanoes; he is given to intense emotional outbursts; and his obsession with his wife's birthmark is a feeling so profound as to disrupt his entire life. Unable to accept himself for what he is, Aylmer constructs a mythology of science and adopts the character of a scientist to disguise his true nature and to hide his real motives, from himself as well as others. As a consequence, he acquires a way of acting out these motives without in fact having to be aware of them. One might describe "The Birthmark" as an exposé of science because it demonstrates the ease with which science can be invoked to conceal highly subjective motives. "The Birthmark" is an exposure of the realities that underlie the scientist's posture of objectivity and rationality and the claims of science to operate in an amoral and value-free world. Pale Aylmer, the intellectual scientist, is a mask for the brutish, earthy, soot-smeared Aminadab, just as the mythology of scientific research and objectivity finally masks murder, disguising Georgiana's death as just one more experiment that failed.

Hawthorne's attitude toward men and their fantasies is more critical than either Irving's or Anderson's. One responds to Aylmer not with pity but with horror. For, unlike Irving and Anderson, Hawthorne has not omitted from his treatment of men an image of the consequences of their ailments for the women who are involved with them. The result of Aylmer's massive self-deception is to live in an unreal world, a world filled with illusions, semblances, and appearances, one which admits of no sunlight and makes no contact with anything outside itself and at whose center is a laboratory, the physical correlative of his utter solipsism. Nevertheless, Hawthorne makes it clear that Aylmer has got someone locked up in that laboratory with him. While "The Birthmark" is by no means explicitly feminist, since Hawthorne seems as eager to be misread and to conceal as he is to be read and to reveal, still it is impossible to read his story without being aware that Georgiana is completely in Aylmer's power. For the subject is finally power. Aylmer is able to project himself onto Georgiana and to work out his obsession through her because as woman and as wife she is his possession and in his power; and because as man he has access to the language and structures of that science which provides the mechanisms for such a process and legitimizes it. In addition, since the power of definition and the authority to make those definitions stick is vested in men, Aylmer can endow his illusions with the weight of spiritual aspiration and universal truth.

The implicit feminism in "The Birthmark" is considerable. On one level the story is a study of sexual politics, of the powerlessness of women and of the psychology which results from that powerlessness. Hawthorne dramatizes the fact that woman's identity is a product of men's responses to her: "It must not be concealed, however, that the impression wrought by this fairy sign manual varied exceedingly, according to the difference of temperament in the beholders" [189]. To those who love Georgiana, her birthmark is evidence of her beauty; to those who envy or hate her, it is an object of disgust. It is Aylmer's repugnance for the birthmark that makes Georgiana blanch, thus causing the mark to emerge as a sharply defined blemish against the whiteness of her cheek. Clearly, the birthmark takes on its character from the eye of the beholder. And just as clearly Georgiana's attitude toward her birthmark varies in response to different observers and definers. Her self-image derives from internalizing the attitudes toward her of the man or men around her. Since what surrounds Georgiana is an obsessional attraction expressed as a total revulsion, the result is not surprising: continual self-consciousness that leads to a pervasive sense of shame and a self-hatred that terminates in an utter readiness to be killed. "The Birthmark" demonstrates the consequences to women of being trapped in the laboratory of man's mind, the object of unrelenting scrutiny, examination, and experimentation.

In addition, "The Birthmark" reveals an implicit understanding of the consequences for women of a linguistic system in which the word "man" refers to both male people and all people. Because of the conventions of this system, Aylmer is able to equate his peculiarly male needs with the needs of all human beings, men and women. And since Aylmer can present his compulsion to idealize and perfect Georgiana as a human aspiration, Georgiana is forced to identify with it. Yet to identify with his aspiration is in fact to identify with his hatred of her and his need to eliminate her. Georgiana's situation is a fictional version of the experience that women undergo when they read a story like "Rip Van Winkle." Under the influence of Aylmer's mind, in the laboratory where she is subjected to his subliminal messages, Georgiana is co-opted into a view of herself as flawed and comes to hate herself as an impediment to Aylmer's aspiration; eventually she wishes to be dead rather than to remain alive as an irritant to him and as a reminder of his failure. And as she identifies with him in her attitude toward herself, so she comes to worship him for his hatred of her and for his refusal to tolerate her existence. The process of projection is neatly reversed: he locates in her everything he cannot accept in himself, and she attributes to him all that is good and then worships in him the image of her own humanity.

Through the system of sexual politics that is Aylmer's compensation for growing up, Hawthorne shows how men gain power over women, the power to create and kill, to "mar," "mend," and "make," without ever having to relinquish their image as "nice guys." Under such a system there need be very few power struggles, because women are programmed to deny the validity of their own perceptions and responses and to accept male illusions as truth. Georgiana does faint when she first enters Aylmer's laboratory and sees it for one second with her own eyes; she is also aware that Aylmer is filling her chamber with appearances, not realities; and she is finally aware that his scientific record is in his own terms one of continual failure. Yet so perfect is the program that she comes to respect him even more for these failures and to aspire to be yet another of them.

Hawthorne's unrelenting emphasis on "seems" and his complex use of the metaphors and structures of disguise imply that women are being deceived and destroyed by man's system. And perhaps the most vicious part of this system is its definition of what constitutes nobility in women: "Drink, then, thou lofty creature," exclaims Aylmer with "fervid admiration" as he hands Georgiana the cup that will kill her [198]. Loftiness in women is directly equivalent to the willingness with which they die at the hands of their husbands, and since such loftiness is the only thing about Georgiana which does elicit admiration from Aylmer, it is no wonder she is willing. Georgiana plays well the one role allowed her, yet one might be justified in suggesting that Hawthorne grants her at the end a slight touch of the satisfaction of revenge: "'My poor Aylmer,' she repeated, with a more than human tenderness, 'you have aimed loftily; you have done nobly. Do not repent that with so high and pure a feeling, you have rejected the best the earth could offer'" [199]. Since dying is the only option, best to make the most of it.

CINDY WEINSTEIN

The Invisible Hand Made Visible: "The Birth-mark"†

"The Custom-House" places us firmly in the world of antebellum America, even as "The Market-Place," chapter 2 of *The Scarlet Letter*, situates us "not less than two centuries ago" (77) [36]. The culture of the nineteenth century unmistakably takes precedence over that of the seventeenth. "The Birth-mark," to the contrary, directs us quite clearly back to "the latter part of the last century" (764) [188]. It is thus not beside the point to consider the ways in which economic analyses of the eighteenth century, particularly those relating to the issue of circulation, might pertain to the economic circulations in "The Birth-mark." Another reading of the birthmark that adds to the debate about the "Crimson Hand" (766) [189], the "spectral Hand" (766) [190], the "odious Hand" (767) [190], and the "Bloody Hand" (765) [189] yet one more hand could be construed as this critic's obsessive reproduction of Aylmer's fetishism. To this end, a

† From *The Literature of Labor and the Labors of Literature: Allegory in Nineteenth-Century American Fiction* (New York: Cambridge UP, 1995), pp. 82–86. Copyright © 1995 Cambridge University Press. Reprinted by permission of the publisher. Page numbers in square brackets refer to this Norton Critical Edition.

glance at the invisible hand of Adam Smith, which makes its appearance in perhaps this most famous passage from *The Wealth of Nations*, permits us to view Aylmer's actions from a different and enlightening perspective: "By preferring the support of domestic to that of foreign industry, he intends only his own security; and by directing that industry in such a manner as its produce may be of the greatest value, he intends only his own gain, and he is in this, as in many other cases, led by an invisible hand to promote an end which was no part of his intention."[1] According to Smith, consequences often have little to do with one's intentions, because one's self-interested intention "frequently promotes [the interest] of the society more effectually than when [one] really intends to promote it" (28). This is clearly not the case in "The Birth-mark," where Aylmer's self-interested intentions bring about self-interested results that do everything to preserve "his own security" and nothing to promote "the public good" (28). Have the goals of 1776, the year which saw the publication of *The Wealth of Nations* and the birth pangs of an American nation, been both forsaken by and made unavailable to America in the 1840s? If so, are we to read "The Birth-mark" as a nineteenth-century corrective to the misguided optimism of political economists like Adam Smith and John Locke, who believed that self-interest and the public good were not mutually exclusive in a society that functioned according to a laissez-faire market economy?[2] Tempting as this reading might be, Hawthorne's story seems less an indictment of a laissez-faire economy than of an economy that isn't laissez-faire enough. The tension at the heart of "The Birth-mark" is this: Aylmer's desired end is the invisible hand of Smith's market economy, but the means he deploys in achieving it fly in the face of Smith's economic directives.

What Hawthorne thought of Smith or whether he even read *The Wealth of Nations* has unfortunately not been documented. Sacvan Bercovitch, however, has recently claimed that the brand of irony at work in Hawthorne's representations of the Puritan past is an "historiographical equivalent of laissez-faire," a "counterpart to Adam Smith's concept of the invisible hand."[3] In following Bercovitch's lead, I want to argue that the free-market ideology at work in Smith's ideal of the invisible hand is, in part, what motivates Aylmer to erase the visible hand that is Georgiana's birthmark. But in living up to Smith's principles, Aylmer uses all the wrong strategies: not only is his task deeply intentional, which is antithetical to the unintentionality that governs the marketplace in *The Wealth of Nations*, but Aylmer's active intervention into Georgiana's body is, to say the least, the furthest thing from a policy of laissez-faire. Aylmer's

1. Adam Smith, *An Inquiry into the Nature and Causes of the Wealth of Nations*, ed. James Rogers (Oxford: Clarendon Press, 1880), vol. 2, 28.
2. Although I emphasize Smith for the obvious reason that the figure of the hand so powerfully conjoins *The Wealth of Nations* and "The Birth-mark," the presence of Locke should also be noted. In *The Second Treatise of Government* (New York: Bobbs-Merrill, 1952), Locke notes that "a state of liberty . . . is not a state of license; though man in that state have an uncontrollable liberty to dispose of his person or possessions, yet he has not liberty to destroy himself, or so much as any creature in his possession" (5). The best discussion of Locke and antebellum configurations of the self can be found in Howard Horwitz's *By the Law of Nature: Form and Value in Nineteenth-Century America* (Oxford: Oxford Univ. Press, 1991).
3. Bercovitch, *The Office of The Scarlet Letter*, 41. For a brilliant discussion of Smith's thought, see Jean-Christophe Agnew, *Worlds Apart: The Market and the Theater in Anglo-American Thought, 1550–1750* (Cambridge: Cambridge Univ. Press, 1986), esp. 149–94. For a brief discussion of Smith's impact on American economic theory and the late-nineteenth-century novel, see Horwitz, *By the Law of Nature*, 126–8.

antimarket means, in other words, will make it impossible for him to attain what he desires, the invisible hand and the subsequent power of the market, or something will go awry in the attempt to fulfill his wishes.

My discussion of the circulations in "The Birth-mark" and Georgiana's body, in particular, has thus far focused on the problematics of gender and signification raised by this state of instability. In moving to a discussion of the economic issues suggested by this thematic of circulation, it will be useful to consider another somewhat lengthier passage from *The Wealth of Nations* in which Smith figures the economic circulations of late-eighteenth-century Great Britain in blatantly physiological terms:

> In her present condition, Great Britain resembles one of those unwholesome bodies in which some of the vital parts are overgrown, and which, upon that account, are liable to many dangerous disorders scarce incident to those in which all the parts are more properly proportioned. A small stop in that great blood-vessel, which has been artificially swelled beyond its natural dimensions, and through which an unnatural proportion of the industry and commerce of the country has been forced to circulate, is very likely to bring on the most dangerous disorders upon the whole body politic. . . . The blood, of which the circulation is stopped in some of the smaller vessels, easily disgorges itself into the greater, without occasioning any dangerous disorder; but, when it is stopped in any of the greater vessels, convulsions, apoplexy, or death are the immediate and unavoidable consequences. (186–87)

According to this description, Georgiana's birthmark could be registering a deeply disordered market. And this is indeed the case; not, however, because Georgiana's fluctuating blood supply manifests any disorder (her paling and blushing would be evidence of a healthy and mobile physiological state) but rather because Aylmer's inability to focus on anything but Georgiana's birthmark brings about a state not unlike the one described by Smith, in which "convulsions, apoplexy," *and* death are the "unavoidable consequences." We are frequently reminded of Aylmer's Ahab-like monomania: "Without intending it—nay, in spite of a purpose to the contrary—[he] reverted to this one disastrous topic" (766) [189] and a page later, "He had not been aware of the tyrannizing influence acquired by one idea over his mind" (767) [190].[4] The problem with the birthmark is not its instability or its uncontrollability or its mobility but the fact that Georgiana has it and Aylmer seems to want it. The birthmark registers Georgiana's ineluctable and successful participation in the market economy. As something Georgiana possesses, the birthmark is also Georgiana, and as such it represents Georgiana's capacity to possess more, and thus it becomes what Aylmer must have. Whereas Georgiana both possesses the "charm" (764) [188] of the birthmark and is possessed by it, Aylmer is clearly possessed by it but receives none of the benefits of possession. He possesses it at the end of the story, when, after receiving assurances from Georgiana that she will drink a potentially fatal elixir, Aylmer has appropriated the power of the market that had been located in the birthmark: "His spirit was ever on the march—ever ascending—and each instant required something that was beyond the scope of the instant before" (777) [197]. Capturing the spirit of [Henry

4. Whereas Aylmer depends upon the fluctuating meanings of the birthmark to exert his power, Ahab attempts to empower himself by hypostatizing the meanings of Moby Dick.

Ward] Beecher and [William Ellery] Channing, Aylmer has recaptured his ever mobile spirit and transcendent identity by immobilizing and appropriating Georgiana's.

He succeeds in doing this by strategically manipulating the competitive principles of the market economy that inform the relations among Georgiana, himself, and her/the birthmark. Aylmer's anxieties about the hermeneutic fluctuations of the birthmark are, as has already been suggested, further exacerbated by the fact of its proprietary indeterminacy; in other words, to whom does the birthmark belong? Is it Georgiana? Is Georgiana the birthmark's? These questions underscore the inextricable relation between matters of economy and the self at the same time as they bring us back to the task of defining Hawthorne's economics of allegory. Aylmer's commitment to erasure is also a commitment to ownership, requiring precisely those hermeneutic and proprietary indeterminacies that had seemed most worrisome. Anxiety producing as they may be, these indeterminacies nevertheless enable him to sustain the belief that Georgiana and her property, that is the birthmark, can be disengaged from one another through a process of disembodiment and thus permit Aylmer's territorial raids upon and into Georgiana's body. Aylmer's relation to Georgiana illustrates what Macpherson has called a "possessive market society," where "a man's energy and skill are his own, yet are regarded not as integral parts of his personality, but as possessions."[5] According to this logic, Aylmer assumes that Georgiana cannot both be the birthmark and possess it; therefore, he has an opportunity to own it. Similarly, in order for Georgiana to own the birthmark, she cannot be the birthmark. It is only by not owning the birthmark that she has a chance of owning it. Because her body has been constituted in the name of private property, the issue arises as to whose property she is now and whose she might become; Georgiana's body therefore functions as the site upon which the competitive spirit of the market economy plays itself out. The birthmark is Georgiana's property, and as property its ownership is transferrable or vulnerable, in this case to scientific experimentation. Yet as the ending of the story makes painfully clear, Georgiana *is* her property, or the birthmark. She both is it and owns it. Property that is not alienable is ultimately self-destructive. Because possession and identity are inextricable in the case of the birthmark, Georgiana commits a grave mistake in hoping that they might be separate. Interestingly enough, Aminadab puts his own finger on this logic when he first sees Georgiana lying unconscious in the laboratory: "If she were my wife, I'd never part with that birth-mark" (770) [192]. Aminadab might simply be communicating his aesthetic preference for Georgiana with a birthmark as opposed to Georgiana without one, but one can also hear in this sentence the inextricable connection between Georgiana's identity and the birthmark. She thinks, however, that in giving up the birthmark she can still be a person: "Either remove this dreadful Hand, or take my wretched life!" (768) [191]. Only when she realizes that her "or" will have to be an "and" will she understand what is at stake in the removal of the birthmark. The problem is not with alienable property but with property that won't be alienable.

5. Macpherson, *The Political Theory of Possessive Individualism: Hobbes to Locke* (Oxford: Oxford UP, 1962), 48.

Whereas in "The Celestial Rail-road" the disjunction between name and character constituted allegorical subjects who either were controlled by the market or transcended it (and thereby most fully exemplified it), "The Birth-mark" produces allegorical subjects as a consequence of applying the principles of the market economy to the relation between persons and (their) bodies. Georgiana's birthmark marks the surplus of meaning generated by the circulations of her body, which are then transformed by Aylmer into a problem both of allegorical interpretation and economic possessiveness. What does her birthmark signify, and who owns it? That these two questions follow from one another, at least for Aylmer, suggests that allegory and the market economy share the mechanism of generating and containing surplus meaning (or value) in order to make that surplus available for possession. Once Georgiana's proprietary relation to her body unravels as a consequence of Aylmer's successful manipulation of the rules of property, an economics of allegory reveals a configuration in which the omnipotence and transcendence of one character, in this case Aylmer, depend upon the geographical and characterological immobility of others, namely Georgiana and to a lesser extent Aminadab. Like Mr. Smooth-it-away [in Hawthorne's "The Celestial Rail-road"] who in seeming to have escaped the exigencies of the market most clearly represented it, Aylmer controls the instabilities of the market, which were most clearly embodied by Georgiana, but the market cannot exist without precisely those instabilities that continually present the occasions for his acts of transcendence.

In making visible the circulations of the physiological and economic systems that define the late-eighteenth-century world of Aylmer and Georgiana, the birthmark locates upon Georgiana's body a version of the market's circulatory system, whose movements are nicely depicted as Georgiana experiences "a stirring up of her system,—a strange indefinite sensation creeping through her veins, and tingling, half painfully, half pleasurably, at her heart" (773–4) [194]. Having destroyed the birthmark as well as his wife, Aylmer has returned the market to its rightful owner—himself. His antipathy to the birthmark was never really the fact of its uncontrollability but rather the fact that Georgiana was both the possessor of and the one possessed by the market's powerful uncontrollability. In the true spirit of Adam Smith, Aylmer has restored the invisibility to the hand that Georgiana had made visible.

Nathaniel Hawthorne: A Chronology

1804 Born on July 4 in Salem, Massachusetts.

1808 Hawthorne's father, a ship captain, dies of yellow fever in Suriname (Dutch Guinea). Hawthorne's mother moves her three children in with her parents, the Mannings.

1813 Receives a mysterious leg injury that lays him up for months.

1816 Hawthornes visit the family property in Raymond, Maine, where Hawthorne enjoys hunting and fishing.

1820 With his sister Louisa, publishes a neighborhood paper, the *Spectator*.

1821 Enters Bowdoin College in Brunswick, Maine, where his classmates include lifelong friends Franklin Pierce, Horatio Bridge, and Henry Wadsworth Longfellow, among others.

1825 An undistinguished student, Hawthorne graduates from Bowdoin in September just barely in the top half of his class; returns to Salem after graduation and lives with his mother and sisters in the Manning household.

1828 Pays $100 to have his first novel, *Fanshawe*, published, but would subsequently repudiate the novel, which is loosely based on his experiences at Bowdoin.

1830 Publishes sketches (including "Mrs. Hutchinson") and tales ("The Hollow of Three Hills" and "An Old Woman's Tale") in the *Salem Gazette*.

1831–36 Publishes many tales and sketches, most of them in *The Token*, an annual gift book by Samuel Goodrich, but is unsuccessful in finding a publisher for three different collections of tales during this period (*Seven Tales of My Native Land*, *Provincial Tales*, and *The Story Teller*).

1836 Edits the short-lived *American Magazine of Useful and Entertaining Knowledge*. Publishes Peter Parley's *Universal History on the Basis of Geography* for Samuel Goodrich.

1837 The American Stationers Company of Boston publishes *Twice-Told Tales*, a collection of eighteen tales and sketches, including "The Gentle Boy," "Wakefield," "The May-Pole of Merry Mount," and "The Minister's Black Veil." Franklin Pierce tries to have Hawthorne appointed historiographer on the South Seas Expedition commanded by Charles Wilkes. Hawthorne meets Sophia Peabody.

1838 Begins publishing tales in the *United States Magazine and Democratic Review*, founded and edited by John Louis O'Sullivan. Hawthorne becomes romantically involved with Mary Silsbee and considers challenging O'Sullivan to a duel on her account. During the summer, Hawthorne spends two months in western Massachusetts. He also begins his courtship of Sophia Peabody.

1839 Appointed Inspector in the Boston Custom House at an annual salary of $1,100. Publishes *The Gentle Boy: A Thrice-Told-Tale*, featuring an engraving by Sophia.

1840 Publishes *Grandfather's Chair*, a collection of historical sketches for children.

1841 Publishes *Famous Old People* and *Liberty Tree*, two more children's books. Leaving his job at the Boston Custom House, Hawthorne buys two $500 shares in Brook Farm, a utopian community founded by George Ripley, with the idea that he and Sophia can live there when they marry. Hawthorne lives at Brook Farm from April to November.

1842 A second, expanded edition of *Twice-Told Tales* appears. Marries Sophia Peabody (July 9) and moves into the Old Manse in Concord, Massachusetts, adjacent to the field where the first battle of the Revolutionary War was fought. Henry David Thoreau had planted a vegetable garden for the newlyweds. During his three years in Concord, Hawthorne gets to know Ralph Waldo Emerson, Margaret Fuller, Ellery Channing, and Thoreau, among others.

1844 The Hawthornes' daughter Una is born (March 3).

1845 Publishes the *Journal of an African Cruiser*, an edited account of his Bowdoin friend Horatio Bridge's journey to Africa. Leaves Concord in November, largely for financial reasons, and moves his family back to Salem.

1846 Appointed Surveyor of the Salem Custom House (April) at an annual salary of $1,200. The Hawthornes' son Julian is born (June 22). Publishes *Mosses from an Old Manse*, including such tales as "Young Goodman Brown," "The Birthmark," "The Artist of the Beautiful," "Rappaccini's Daughter," and "Roger Malvin's Burial."

1847 Moves his family into a larger house on Mall Street in Salem.

1849 Removed from his job in the Custom House by the newly elected Whig administration. Mother dies on July 31. Begins writing *The Scarlet Letter.*

1850 *The Scarlet Letter* is published on March 16 and goes through three editions in nine months. The Hawthornes move to Lenox in western Massachusetts, where Hawthorne meets Herman Melville on August 5. Melville's essay, "Hawthorne and His Mosses," appears in the August 17 and 24 issues of Evert Duyckinck's *Literary World.*

1851 Publishes *The House of the Seven Gables, The Snow Image and Other Twice-Told Tales*, and *The Wonder Book*. The Hawthornes' daughter Rose is born (May 20). The Hawthornes leave Lenox in November for West Newton, Massachusetts, outside of Boston.

1852	Publishes *The Blithedale Romance* and a campaign biography for his old friend, Presidential candidate Franklin Pierce. Buys the only house he will ever own in Concord, Massachusetts, and names it The Wayside.
1853	Publishes *Tanglewood Tales*. Appointed U.S. Consul in Liverpool by President Pierce. The Hawthornes leave for England in July.
1857–59	After resigning the consulship, Hawthorne briefly tours the British Isles and France before settling in Italy (Rome and Florence), prolonging his stay when Una contracts "Roman fever."
1859	Returns to England.
1860	Publishes *The Marble Faun*. Returns to the United States and to The Wayside in Concord after seven years abroad.
1862	Journeys to Washington, D.C., with publisher William Ticknor. Meets President Abraham Lincoln. Publishes "Chiefly About War Matters" in the *Atlantic Monthly*.
1863	Publishes *Our Old Home*, a book of reminiscences about England, and dedicates it to Franklin Pierce.
1864	Dies on May 19 in Plymouth, New Hampshire, where he had traveled with Franklin Pierce. Buried in Concord's Sleepy Hollow cemetery on May 23.

Selected Bibliography

• indicates works included or excerpted in this Norton Critical Edition.

WEBSITES

Hawthorne in Salem. www.hawthorneinsalem.org/
Nathaniel Hawthorne Society. www.tamiu.edu/hawthorne/

BIBLIOGRAPHY AND OTHER SCHOLARLY RESOURCES

Boswell, Jeanetta. *Nathaniel Hawthorne and the Critics: A Checklist of Criticism, 1900–1978.* Metuchen, NJ: Scarecrow, 1982.

Clark, C. E. Frazer, Jr. *Nathaniel Hawthorne; A Descriptive Bibliography.* Pittsburgh, PA. U of Pittsburgh P, 1978.

Cohen, B. Bernard, ed. *The Recognition of Nathaniel Hawthorne.* Ann Arbor: U of Michigan P, 1969.

Crowley, J. Donald, ed. *Hawthorne: The Critical Heritage.* New York: Barnes and Noble, 1970.

Gale, Robert L. *A Nathaniel Hawthorne Encyclopedia.* Westport, CT: Greenwood, 1991.

Idol, John L., Jr., and Buford Jones, eds. *Nathaniel Hawthorne: The Contemporary Reviews.* New York: Cambridge UP, 1994.

Muirhead, Kimberly Free. *Nathaniel Hawthorne's* The Scarlet Letter: *A Critical Resource Guide and Comprehensive Annotated Bibliography of Literary Criticism, 1950–2000.* Lewiston, NY: Edwin Mellen, 2004.

Person, Leland S. "Nathaniel Hawthorne," *Prospects for the Study of American Literature (II).* Ed. Richard Kopley and Barbara Cantalupo. New York: AMS Press, 2009. 26–49.

Ricks, Beatrice, Joseph D. Adams, and Jack O. Hazlerig, eds. *Nathaniel Hawthorne: A Reference Bibliography, 1900–1971.* Boston: G. K. Hall, 1972.

Scharnhorst, Gary, ed. *The Critical Response to Nathaniel Hawthorne's* The Scarlet Letter. Westport, CT: Greenwood, 1992.

———, ed. *Nathaniel Hawthorne: An Annotated Bibliography of Commentary and Criticism before 1900.* Metuchen, NJ: Scarecrow, 1988.

BIOGRAPHY

Bridge, Horatio. *Personal Recollections of Nathaniel Hawthorne.* New York: Harper, 1893.

Cantwell, Robert. *Nathaniel Hawthorne: The American Years.* New York: Rinehart, 1948.

Coale, Samuel Chase. *The Entanglements of Nathaniel Hawthorne: Haunted Minds and Ambiguous Approaches.* Rochester, NY: Camden House, 2011.

Conway, Moncure D. *Life of Nathaniel Hawthorne.* New York: Lovell, 1890.

Fields, James T. *Yesterdays with Authors.* Boston: Houghton Mifflin, 1871.

Hawthorne, Julian. *Nathaniel Hawthorne and His Circle.* New York: Harper & Brothers, 1903.

———. *Nathaniel Hawthorne and His Wife: A Biography.* 2nd ed. 2 vols. Boston: James Osgood, 1885.

Herbert, T. Walter. *Dearest Beloved: The Hawthornes and the Making of the Middle-Class Family.* Berkeley: U of California P, 1993.

Lathrop, Rose Hawthorne. *Memories of Hawthorne* (1897). Boston: Houghton Mifflin, 1923.

Loggins, Vernon. *The Hawthornes: The Story of Seven Generations of an American Family.* New York: Columbia UP, 1951.
Marshall, Megan. *The Peabody Sisters: Three Women Who Ignited American Romanticism.* New York: Houghton Mifflin, 2005.
Mellow, James R. *Nathaniel Hawthorne in His Times.* Boston: Houghton Mifflin, 1980.
Miller, Edwin Haviland. *Salem Is My Dwelling Place: The Life of Nathaniel Hawthorne.* Iowa City: U of Iowa P, 1991.
Moore, Margaret. *The Salem World of Nathaniel Hawthorne.* Columbia: U of Missouri P, 1997.
Scharnhorst, Gary. *Julian Hawthorne: The Life of a Prodigal Son.* Urbana, Chicago, and Springfield: U of Illinois P, 2014.
Stewart, Randall. *Nathaniel Hawthorne: A Biography.* New Haven, CT: Yale UP, 1948.
Ticknor, Caroline. *Hawthorne and His Publisher.* Boston: Houghton Mifflin, 1913.
Turner, Arlin. *Nathaniel Hawthorne: A Biography.* New York: Oxford UP, 1980.
Valenti, Patricia Dunlavy. *Sophia Peabody Hawthorne, A Life, Volume 1, 1809–1847.* Columbia: U of Missouri P, 2004.
———. *Sophia Peabody Hawthorne, A Life, Volume 2, 1848–1871.* Columbia: U of Missouri P, 2015.
———. *To Myself a Stranger: A Biography of Rose Hawthorne Lathrop.* Baton Rouge: Louisiana State UP, 1991.
Van Doren, Mark. *Nathaniel Hawthorne: A Critical Biography.* New York: William Sloane, 1949.
Wineapple, Brenda. *Hawthorne: A Life.* New York: Knopf, 2003.
Woodberry, George E. *Nathaniel Hawthorne* (1902). New York: Chelsea House, 1980.

SELECTED CRITICISM OF *THE SCARLET LETTER*

Abel, Darrel. "Hawthorne's Pearl: Symbol and Character." *ELH* 18 (1951): 50–66.
———. *The Moral Picturesque: Studies in Hawthorne's Fiction.* West Lafayette, IN: Purdue UP, 1988.
Adams, Fred C. "Blood Vengeance in *The Scarlet Letter.*" *Nathaniel Hawthorne Review* 32.2 (2006): 1–12.
Anderson, Douglas. "Jefferson, Hawthorne, and 'The Custom-House.'" *Nineteenth-Century Literature* 46 (1991): 309–26.
Arac, Jonathan. "The Politics of *The Scarlet Letter.*" *Ideology and Classic American Literature.* Ed. Sacvan Bercovitch and Myra Jehlen. New York: Cambridge UP, 1986. 247–66.
Arvin, Newton. *Hawthorne.* Boston: Little, Brown, 1929.
• Barlowe, Jamie. *The Scarlet Mob of Scribblers: Rereading Hester Prynne.* Carbondale: Southern Illinois UP, 2000.
Barnett, Louise K. "Speech and Society in *The Scarlet Letter.*" *ESQ* 29 (1983): 16–24.
Bayer, John. "Narrative Technique and the Oral Tradition in *The Scarlet Letter.*" *American Literature* 52 (1980): 250–63.
Baym, Nina. "Passion and Authority in *The Scarlet Letter.*" *New England Quarterly* 43 (1970): 209–30.
• ———. "Revisiting Hawthorne's Feminism." *Nathaniel Hawthorne Review* 30 (Fall & Spring 2004): 32–55.
———. "The Romantic *Malgré Lui*: Hawthorne in 'The Custom-House.'" *ESQ* 19 (1973): 14–25.
———. The Scarlet Letter: *A Reading.* Boston: Twayne, 1986.
———. *The Shape of Hawthorne's Career.* Ithaca, NY: Cornell UP, 1976.
Becker, John E. *Hawthorne's Historical Allegory: An Examination of the American Conscience.* Port Washington, NY: Kennikat, 1971.
Bell, Michael Davitt. "Arts of Deception: Hawthorne, 'Romance,' and *The Scarlet Letter.*" *New Essays on* The Scarlet Letter. Ed. Michael J. Colacurcio. New York: Cambridge UP, 1985: 29–56.
———. *Hawthorne and the Historical Romance of New England.* Princeton, NJ: Princeton UP, 1971.
Bell, Millicent, ed. *Hawthorne and the Real: Bicentennial Essays.* Columbus: Ohio State UP, 2005.
———. *Hawthorne's View of the Artist.* New York: State U of New York P, 1962.
• ———. "The Obliquity of Signs: *The Scarlet Letter.*" *Massachusetts Review* 23 (1982): 9–26.

Bensick, Carol M. "Dimmesdale and His Bachelorhood: 'Priestly Celibacy' in *The Scarlet Letter*." *Studies in American Fiction* 21 (1993): 103–10.

———. "His Folly, Her Weakness: Demystified Adultery in *The Scarlet Letter*." *New Essays on* The Scarlet Letter. Ed. Michael J. Colacurcio. New York: Cambridge UP, 1985. 137–159.

Bentley, Nancy. *The Ethnography of Manners: Hawthorne, James, Wharton*. Cambridge: Cambridge UP, 1995.

• Bercovitch, Sacvan. *The Office of* The Scarlet Letter. Baltimore, MD: Johns Hopkins UP, 1991.

———. "*The Scarlet Letter*: A Twice-Told Tale." *Nathaniel Hawthorne Review* 22.2 (1996): 1–20.

Bergland, Renee. *The National Uncanny: Indian Ghosts and American Subjects*. Hanover, NH: UP of New England, 2000.

Berlant, Lauren. *The Anatomy of National Fantasy: Hawthorne, Utopia, and Everyday Life*. Chicago: U of Chicago P, 1991.

Bernstein, Cynthia. "Reading *The Scarlet Letter*: Against Hawthorne's Fictional Interpretive Communities." *Language and Literature* 18 (1993): 1–20.

Boewe, Charles, and Murray G. Murphy. "Hester Prynne in History." *American Literature* 32 (1960): 202–204.

Boudreau, Kristin. "*The Scarlet Letter* and the 1833 Murder Trial of the Reverend Ephraim Avery." *ESQ* 47 (2001): 89–112.

• ———. *Sympathy in American Literature: American Sentiments from Jefferson to the Jameses*. Gainesville: UP of Florida, 2002. 49–82.

Branch, Watson. "From Allegory to Romance: Hawthorne's Transformation of *The Scarlet Letter*." *Modern Philology* 80.2 (Nov. 1982): 145–60.

Brodhead, Richard H. *Cultures of Letters: Scenes of Reading and Writing in Nineteenth-Century America*. Chicago: U of Chicago P, 1993.

———. *Hawthorne, Melville, and the Novel*. Chicago: U of Chicago P, 1976.

———. *The School of Hawthorne*. New York: Oxford UP, 1986.

Bronstein, Zelda. "The Parabolic Ploys of *The Scarlet Letter*." *American Quarterly* 39 (1987): 193–210.

Brooke-Rose, Christine. "A for But: 'The Custom-House' in Hawthorne's *The Scarlet Letter*." *Word and Image* 3 (1987): 143–55.

Brown, Gillian. *Domestic Individualism: Imagining Self in Nineteenth-Century America*. Berkeley: U of California P, 1990.

———. "Hawthorne, Inheritance, and Women's Property." *Studies in the Novel* 23 (Spring 1991): 107–18.

Browner, Stephanie P. "Authorizing the Body: Scientific Medicine and *The Scarlet Letter*." *Literature and Medicine* 12 (1993): 139–60.

Budick, Emily Miller. *Engendering Romance: Women Writers and the Hawthorne Tradition, 1850–1990*. New Haven, CT: Yale UP, 1994.

• ———. "Hawthorne, Pearl, and the Primal Sin of Culture." *Journal of American Studies* 39.2 (2005): 167–85.

———. "Hester's Skepticism, Hawthorne's Faith: Or, What Does a Woman Doubt? Instituting the American Romance Tradition." *New Literary History* 22 (1991): 199–211.

Buell, Lawrence. "Hawthorne and the Problem of 'American' Fiction: The Example of *The Scarlet Letter*." *Hawthorne and the Real: Bicentennial Essays*. Ed. Millicent Bell. Columbus: Ohio State UP, 2005. 70–87.

Cameron, Sharon. *The Corporeal Self: Allegories of the Body in Melville and Hawthorne*. Baltimore, MD: Johns Hopkins UP, 1981.

Carpenter, Frederic I. "Scarlet A Minus." *College English* 5 (1944): 173–80.

Carton, Evan. "'A Daughter of the Puritans' and Her Old Master: Hawthorne, Una, and the Sexuality of Romance." *Daughters and Fathers*. Ed. Lynda E. Boose and Betty S. Flowers. Baltimore, MD: Johns Hopkins UP, 1989. 208–32.

———. *The Rhetoric of American Romance*. Baltimore, MD: Johns Hopkins UP, 1985.

Champagne, Lenora. "Outside the Law: Feminist Adaptations of *The Scarlet Letter*." *Feminist Theatrical Revisions of Classic Works: Critical Essays*. Jefferson, NC: McFarland, 2009. 169–88.

Cheyfitz, Eric. "The Irresistibleness of Great Literature: Reconstructing Hawthorne's Politics." *American Literary History* 6 (Fall 1994): 539–58.

Clark, C. E. Fraser. "'Posthumous Papers of a Decapitated Surveyor.'" *Studies in the Novel* 2 (1970): 395–419.

Cline, Irina. "Transgression as a Foundation for Progress: *The Scarlet Letter* and the Politics of Individual Liberty." *The Image of the Outlaw in Literature, Media, and Society*. Pueblo: Colorado State UP, 2011. 234–43.

Coale, Samuel Chase. *The Entanglements of Nathaniel Hawthorne: Haunted Minds and Ambiguous Approaches.* Rochester, NY: Camden House, 2011.

———. *Mesmerism and Hawthorne: Mediums of American Romance.* Tuscaloosa: U of Alabama P, 1998.

———. "*The Scarlet Letter* as Icon." *American Transcendental Quarterly* 6 (1992): 251–62.

Colacurcio, Michael J. "Footsteps of Ann Hutchinson: The Context of *The Scarlet Letter.*" *ELH* 39 (1972): 459–94.

———. "Woman's Heart, Woman's Choice: The 'History' of *The Scarlet Letter.*" *Poe Studies/Dark Romanticism* 39–40 (2006): 104–114.

———. "The Woman's Own Choice: Sex, Metaphor, and the Puritan 'Sources' of *The Scarlet Letter.*" *New Essays on* The Scarlet Letter. Ed. Michael J. Colacurcio. Cambridge: Cambridge UP, 1985: 101–35.

Coleman, Dawn. "Critiquing Perfection: Hawthorne's Revision of Salem's Unitarian Saint." *Nathaniel Hawthorne Review* 37.1 (2011): 1–19.

Cottom, Daniel. "Hawthorne versus Hester: The Ghostly Dialectic of Romance in *The Scarlet Letter.*" *Texas Studies in Literature and Language* 24 (1981): 47–67.

Cox, James M. "*The Scarlet Letter*: Through the Old Manse and the Custom House." *Virginia Quarterly Review* 51 (1975): 432–47.

Crain, Patricia. *The Story of A: The Alphabetization of America from the New England Primer to* The Scarlet Letter. Stanford, CA: Stanford UP, 2000.

Crews, Frederick, *The Sins of the Fathers: Hawthorne's Psychological Themes.* New York: Oxford UP, 1966.

Cronin, Morton. "Hawthorne on Romantic Love and the Status of Women." *PMLA* 69 (1954): 89–98.

Cuddy, Lois A. "Mother-Daughter Identification in *The Scarlet Letter.*" *Mosaic* 19 (1986): 101–15.

Daniels, Cindy Lou. "Hawthorne's Pearl: Woman-Child of the Future." *American Transcendental Quarterly* 19.3 (2005): 221–36.

Dauber, Kenneth. *Rediscovering Hawthorne.* Princeton, NJ: Princeton UP, 1977.

Davis, Clark. *Hawthorne's Shyness: Ethics, Politics, and the Question of Engagement.* Baltimore, MD: Johns Hopkins UP, 2005.

Davis, Sarah I. "Another View of Hester and the Antinomians." *Studies in American Fiction* 12 (1984): 189–98.

Dekker, George. *The American Historical Romance.* Cambridge: Cambridge UP, 1987.

Derrick, Scott S. "'A Curious Subject of Observation and Inquiry': Homoeroticism, the Body, and Authorship in Hawthorne's *The Scarlet Letter.*" *Novel* 28 (1995): 308–26.

• DeSalvo, Louise. *Nathaniel Hawthorne.* Atlantic Highlands, NJ: Humanities P, 1987. 57–76.

Deutsch, Helen. "The Scaffold in the Marketplace: Samuel Johnson, Nathaniel Hawthorne, and the Romance of Authorship." *Nineteenth-Century Literature* 68.3 (2013): 363–95.

Diehl, Joanne Feit. "Re-Reading *The Letter*: Hawthorne, the Fetish, and the (Family) Romance." *New Literary History* 19 (1988): 655–73.

Diffee, Christopher. "Postponing Politics in Hawthorne's *Scarlet Letter.*" *Modern Language Notes* 111 (1996): 835–71.

Dolezal, J. "The Medical Palimpsest of *The Scarlet Letter*: An Interdisciplinary Reading." *Medical Humanities* 31.1 (2005): 17–22.

Dolis, John. *The Style of Hawthorne's Gaze: Regarding Subjectivity.* Tuscaloosa: U of Alabama P, 1993.

Donahue, Agnes McNeil. *Hawthorne: Calvin's Ironic Stepchild.* Kent, OH: Kent State UP, 1985.

• Doyle, Laura. "'A' for Atlantic: The Colonizing Force of Hawthorne's *The Scarlet Letter.*" *American Literature* 79.2 (2007): 243–73.

Dryden, Edgar A. *Nathaniel Hawthorne: The Poetics of Enchantment.* Ithaca, NY: Cornell UP, 1977.

Dunne, Michael. *Hawthorne's Narrative Strategies.* Jackson: UP of Mississippi, 1995.

Eakin, Paul John. "Hawthorne's Imagination and the Structure of 'The Custom-House.'" *American Literature* 43 (1971): 346–58.

Egan, Ken. "The Adultress in the Market-Place: Hawthorne and *The Scarlet Letter.*" *Studies in the Novel* 27 (1995): 26–41.

Elbert, Monica M. *Encoding the Letter "A": Gender and Authority in Hawthorne's Early Fiction.* Frankfurt: Haag & Herchen, 1990.

———. "Hester on the Scaffold, Dimmesdale in the Closet: Hawthorne's Seven-Year Itch." *Essays in Literature* 16 (1989): 234–55.

———. "Hester's Maternity: Stigma or Weapon?" *ESQ* 36 (1990): 175–208.
Elder, Marjorie J. *Nathaniel Hawthorne: Transcendental Symbolist*. Athens: Ohio UP, 1969.
Erlich, Gloria C. *Family Themes and Hawthorne's Fiction: The Tenacious Web*. New Brunswick, NJ: Rutgers UP, 1984.
Faust, Bertha. *Hawthorne's Contemporaneous Reputation: A Study of Literary Opinion in America and England, 1828–1864*. New York: Octagon Books, 1968.
Feidelson, Charles, Jr. "*The Scarlet Letter.*" *Hawthorne Centenary Essays*. Ed. Roy Harvey Pearce. Columbus: Ohio State UP, 1964. 31–77.
Fick, Rev. Leonard J. *The Light Beyond: A Study of Hawthorne's Theology*. Westminster, MD: Newman, 1955.
Fleischner, Jennifer. "Hawthorne and the Politics of Slavery." *Studies in the Novel* 23 (1991): 96–106.
Fogle, Richard H. *Hawthorne's Imagery: The "Proper Light and Shadow" in the Major Romances*. Norman: U of Oklahoma P, 1969.
Foster, Dennis. "The Embroidered Sin: Confessional Evasion in *The Scarlet Letter.*" *Criticism* 25 (1983): 141–63.
• Franzosa, John. "'The Custom-House,' *The Scarlet Letter*, and Hawthorne's Separation from Salem." *ESQ* 24 (1978): 57–71.
Freed, Richard C. "Hawthorne's Reflexive Imagination: *The Scarlet Letter* as Compositional Allegory." *American Transcendental Quarterly* 56 (1985): 31–54.
Fryer, Judith. *The Faces of Eve: Women in the Nineteenth-Century American Novel*. New York: Oxford UP, 1976.
Gable, Harvey L., Jr. *Liquid Fire: Transcendental Mysticism in the Romances of Nathaniel Hawthorne*. New York: Peter Lang, 1998.
Garlitz, Barbara. "Pearl: 1850–1955." *PMLA* 72 (1957): 689–99.
Gartner, Matthew. "*The Scarlet Letter* and the Book of Esther: Scriptural Letter and Narrative Life." *Studies in American Fiction* 23.2 (1995): 131–51.
Gilmore, Michael T. *American Romanticism and the Marketplace*. Chicago: U of Chicago P, 1985.
———. "Hawthorne and the Making of the Middle Class." *Discovering Difference: Contemporary Essays in American Culture*. Ed. Christoph Lohmann. Bloomington: Indiana UP, 1993. 88–104.
———. "Hidden in Plain Sight: *The Scarlet Letter* and American Legibility." *Studies in American Fiction* 29 (Spring 2001): 121–28.
Ginsberg, Lesley. "The ABCs of *The Scarlet Letter.*" *Studies in American Fiction* 29 (Spring 2001): 13–31.
Goddu, Teresa A. "Letters Turned to Gold: Hawthorne, Authorship, and Slavery." *Studies in American Fiction* 29 (Spring 2001): 49–76.
Gollin, Rita K. *Nathaniel Hawthorne and the Truth of Dreams*. Baton Rouge: Louisiana State UP, 1979.
———, and John L. Idol, Jr., eds. *Prophetic Pictures: Nathaniel Hawthorne's Knowledge and Uses of the Visual Arts*. Westport, CT: Greenwood Press, 1991.
Green, Carlanda. "'The Custom-House': Hawthorne's Dark Wood of Error." *New England Quarterly* 53 (1980): 184–95.
Greven, David. *Men Beyond Desire: Manhood, Sex, and Violation in American Literature*. New York: Palgrave Macmillan, 2005. 117–30.
• Grossman, Jay. "'A' is for Abolition?: Race, Authorship, *The Scarlet Letter.*" *Textual Practice* 7 (1993): 13–30.
Hall, Lawrence Sargent. *Hawthorne: Critic of Society*. New Haven, CT: Yale UP, 1944.
Harris, Kenneth Marc. *Hypocrisy and Self-Deception in Hawthorne's Fiction*. Charlottesville: UP of Virginia, 1988.
Harshbarger, Scott. "A 'H-ll-Fired Story': Hawthorne's Rhetoric of Rumor." *College English* 56 (1994): 30–45.
Heddendorf, David. "Anthony Trollope's *Scarlet Letter.*" *Sewanee Review* 121.3 (2013): 368–375.
Hennelly, Mark M., Jr. "*The Scarlet Letter*: 'A Play-Day for the Whole World?'" *New England Quarterly* 61 (1988): 530–54.
Herbert, T. Walter. "Nathaniel Hawthorne, Una Hawthorne, and *The Scarlet Letter*: Interactive Selfhoods and the Cultural Construction of Gender." *PMLA* 103 (1988): 285–97.
———. "Pornographic Manhood and *The Scarlet Letter.*" *Studies in American Fiction* 29 (Spring 2001): 113–20.
Hewitt, Elizabeth. "Scarlet Letters, Dead Letters: Correspondence and the Poetics of Democracy in Melville and Hawthorne." *Yale Journal of Criticism* 12.2 (1999): 295–319.

Hilgers, Thomas L. "The Psychological Conflict Resolution in *The Scarlet Letter.*" *American Transcendental Quarterly* 43 (Summer 1979): 211–24.
Hodges, Elizabeth Perry. "The Letter of the Law: Reading Hawthorne and the Law of Adultery." *Law and Literature Perspectives*. Ed. Bruce L. Rockwood and Roberta Kevelson. New York: Peter Lang, 1996. 133–68.
Hoeltje, H. H. *Inward Sky: The Mind and Heart of Nathaniel Hawthorne*. Durham, NC: Duke UP, 1962.
Hoffman, Daniel G. *Form and Fable in American Fiction*. New York: Oxford UP, 1961.
Hull, Richard. "Sent Meaning vs. Attached Meaning: Two Interpretations of Interpretation in *The Scarlet Letter.*" *American Transcendental Quarterly* 14.2 (2000): 143–58.
Hutner, Gordon. *Secrets and Sympathy: Forms of Disclosure in Hawthorne's Novels*. Athens: U of Georgia P, 1988.
Idol, John L., Jr., and Melinda Ponder, eds. *Hawthorne and Women: Engendering and Expanding the Hawthorne Tradition*. Amherst: U of Massachusetts P, 1999.
Irwin, John T. *American Hieroglyphics: The Symbol of the Egyptian Hieroglyphics in the American Renaissance*. New Haven, CT: Yale UP, 1980.
Isani, Mukhtar Ali. "Hawthorne and the Branding of William Prynne." *New England Quarterly* 45 (1972): 182–95.
Jacobson, Richard J. *Hawthorne's Conception of the Creative Process*. Cambridge, MA: Harvard UP, 1965.
James, Henry. *Hawthorne*. London: Macmillan, 1879.
Johnson, Claudia Durst. "Impotence and Omnipotence in *The Scarlet Letter.*" *New England Quarterly* 66 (1993): 594–612.
———. *The Productive Tension of Hawthorne's Art*. Tuscaloosa: U of Alabama P, 1981.
———, ed. *Understanding* The Scarlet Letter: *A Student Casebook to Issues, Sources, and Historical Documents*. Westport, CT: Greenwood, 1995.
Kamuf, Peggy. "Hawthorne's Genres: The Letter of the Law *Appliquée.*" *After Strange Texts: The Role of Theory in the Study of Literature*. Ed. Gregory S. Jay and David L. Miller. Tuscaloosa: U of Alabama P, 1985. 69–84.
Kesselring, Marion L. *Hawthorne's Reading, 1828–1850*. New York: New York Public Library, 1949.
Kesterson, David B., ed. *Critical Essays on Hawthorne's* The Scarlet Letter. Boston: G. K. Hall, 1988.
Kilcup, Karen L. "'Ourself behind Ourself, Concealed—': The Homoerotics of Reading in *The Scarlet Letter.*" *ESQ* 42 (1996): 1–28.
Kimball, Samuel. "Countersigning Aristotle: The Amimetic Challenge of *The Scarlet Letter.*" *American Transcendental Quarterly* 7 (1993): 141–58.
Kopley, Richard. "Hawthorne's Transplanting and Transforming 'The Tell-Tale Heart.'" *Studies in American Fiction* 23 (1995): 231–41.
———. *The Threads of* The Scarlet Letter: *A Study of Hawthorne's Transformative Art*. Newark: U of Delaware P, 2003.
• Korobkin, Laura Hanft. "The Scarlet Letter of the Law: Hawthorne and Criminal Justice." *Novel* 30 (1997): 193–217.
Kowalski, Philip J. "Catching Little Pearl: Hawthorne's Hybridized Daughters." *Literature in the Early American Republic* 4 (2012): 175–98.
Kramer, Michael P. "Beyond Symbolism: Philosophy of Language in *The Scarlet Letter.*" *Imagining Language in America: From the Revolution to the Civil War*. Princeton, NJ: Princeton UP, 1992. 162–97.
Kreger, Erika M. "'Depravity Dressed Up in a Fascinating Garb': Sentimental Motifs and the Seduced Hero(ine) in *The Scarlet Letter.*" *Nineteenth-Century Literature* 54.3 (1999): 308–335.
Lang, Amy Schrager. *Prophetic Woman: Anne Hutchinson and the Problem of Dissent in the Literature of New England*. Berkeley: U of California P, 1987. 161–92.
Last, Suzan. "Hawthorne's Feminine Voices: Reading *The Scarlet Letter* as a Woman." *Journal of Narrative Technique* 27.3 (1997): 349–76.
Lathrop, George P. *A Study of Hawthorne*. Boston: Osgood, 1876.
Lawrence, D. H. *Studies in Classic American Literature.* 1923. Garden City, NY: Doubleday, 1953.
Lee, A. Robert. *Nathaniel Hawthorne: New Critical Essays*. New York: Barnes and Noble, 1982.
Lefcowitz, Allan. "*Apologia* Pro Roger Prynne: A Psychological Study." *Literature and Psychology* 24 (1974): 34–44.
Leverenz, David. *Manhood and the American Renaissance*. Ithaca, NY: Cornell UP, 1989.
———. "Mrs. Hawthorne's Headache: Reading *The Scarlet Letter.*" *Nineteenth-Century Fiction* 37 (1983): 552–73.

• Levine, Robert S. "Antebellum Feminists on Hawthorne: Reconsidering the Reception of *The Scarlet Letter.*" An essay written for the first edition of this Norton Critical Edition. New York: Norton, 2005.
Loebel, Thomas. "'A' Confession: How to Avoid Speaking the Name of the Father." *Arizona Quarterly* 59.1 (2003): 1–29.
Loving, Jerome. "Hawthorne's Awakening in the Customhouse." *Lost in the Customhouse: Authorship in the American Renaissance*. Iowa City: U of Iowa P, 1993. 19–34.
Luedtke, Luther S. *Nathaniel Hawthorne and the Romance of the Orient*. Bloomington: Indiana UP, 1989.
Lundblad, Jane. *Nathaniel Hawthorne and European Literary Tradition*. New York: Russell and Russell, 1965.
Madsen, Deborah L. "'A for Abolition': Hawthorne's Bond-servant and the Shadow of Slavery." *Journal of American Studies* 25 (1991): 255–59.
Male, Roy R. *Hawthorne's Tragic Vision*. New York: Norton, 1957.
Martin, Robert K. "Hester Prynne, *C'est Moi*: Nathaniel Hawthorne and the Anxieties of Gender." *Engendering Men: The Question of Male Feminist Criticism*. Ed. Joseph A. Boone and Michael Cadden. New York: Routledge, 1990. 122–31.
Martin, Terence. "Dimmesdale's Ultimate Sermon." *Arizona Quarterly* 27 (1971): 230–40.
———. *Nathaniel Hawthorne*. Revised Edition. Boston: Twayne, 1983.
Matthiessen, F. O. *American Renaissance: Art and Expression in the Age of Emerson and Whitman*. New York: Oxford UP, 1941.
McGill, Meredith. "The Problem of Hawthorne's Popularity." *Reciprocal Influences: Literary Production, Distribution, and Consumption in America*. Ed. Steven Fink and Susan S. Williams. Columbus: Ohio State UP, 1999.
McNamara, Anne Marie. "The Character of Flame: The Function of Pearl in *The Scarlet Letter.*" *American Literature* 27 (1956): 537–53.
McPherson, Hugo. *Hawthorne as Myth-Maker: A Study in Imagination*. Toronto: U of Toronto P, 1969.
McWilliams, John P., Jr. *Hawthorne, Melville, and the American Character: A Looking-Glass Business*. Cambridge: Cambridge UP, 1984.
Mellard, James M. "Inscriptions of the Subject: *The Scarlet Letter.*" *Using Lacan, Reading Fiction*. Urbana: U of Illinois P, 1991. 69–106.
Milder, Robert. *Hawthorne's Habitations: A Literary Life*. Oxford and New York: Oxford UP, 2012.
———. "*The Scarlet Letter* and Its Discontents." *Nathaniel Hawthorne Review* 22.1 (1996): 9–25.
Millington, Richard H., ed. *The Cambridge Companion to Nathaniel Hawthorne*. Cambridge: Cambridge UP, 2004.
———. *Practicing Romance: Narrative Form and Cultural Engagement in Hawthorne's Fiction*. Princeton, NJ: Princeton UP, 1992.
• Mitchell, Thomas R. *Hawthorne's Fuller Mystery*. Amherst: U of Massachusetts P, 1998.
Mizruchi, Susan L. *The Power of Historical Knowledge: Narrating the Past in Hawthorne, James, and Dreiser*. Princeton, NJ: Princeton UP, 1988.
Moers, Ellen. "*The Scarlet Letter*: A Political Reading." *Prospects* 9 (1985): 49–70.
Nevins, Winfield S. "Nathaniel Hawthorne's Removal from the Salem Custom House." *Essex Institute Historical Collections* 53 (1917): 97–132.
Newberry, Frederick. *Hawthorne's Divided Loyalties: England and America in His Works*. Rutherford, NJ: Fairleigh Dickinson UP, 1987.
• ———. "A Red-Hot 'A' and a Lusting Divine: Sources for *The Scarlet Letter.*" *New England Quarterly* 60.2 (June 1987): 256–64.
———. "Tradition and Disinheritance in *The Scarlet Letter.*" *ESQ* 23 (1977): 1–26.
Nissenbaum, Stephen. "The Firing of Nathaniel Hawthorne." *Essex Institute Historical Collections* 114 (1978): 57–86.
Norman, Jean. *Nathaniel Hawthorne: An Approach to an Analysis of Artistic Creation*. Trans. Derek Coltman. Cleveland, OH: Press of Case Western University, 1970.
Nudelman, Franny. "'Emblem and Product of Sin': The Poisoned Child in *The Scarlet Letter* and Domestic Advice Literature." *Yale Journal of Criticism* 10.1 (1997): 193–213.
Obenland, Frank. "Intertextuality and History: America's Colonial Past in *The Scarlet Letter.*" *Zeitschrift für Anglistik und Amerikanistik: A Quarterly of Language, Literature and Culture* 53.3 (2005): 211–23.
Otten, Thomas J. "Hawthorne's Twisted Letters." *Modern Language Quarterly* 70.3 (2009): 363–86.
Pearce, Roy Harvey, ed. *Hawthorne Centenary Essays*. Columbus: Ohio State UP, 1964.

Pease, Donald E. "Hawthorne in the Custom-House: The Metapolitics, Postpolitics, and Politics of *The Scarlet Letter.*" *Boundary 2* 32.1 (2005): 53–70.
Person, Leland S. *Aesthetic Headaches: Women and a Masculine Poetics in Poe, Melville, and Hawthorne*. Athens: U of Georgia P, 1988. 122–38.
———. *The Cambridge Introduction to Hawthorne*. New York: Cambridge UP, 2007. 66–81.
———. "The Dark Labyrinth of Mind: Hawthorne, Hester, and the Ironies of Racial Mothering." *Studies in American Fiction* 29 (Spring 2001): 33–48.
———. "Hester's Revenge: The Power of Silence in *The Scarlet Letter.*" *Nineteenth-Century Literature* 43 (1989): 465–83.
———. "*The Scarlet Letter* and the Myth of the Divine Child." *American Transcendental Quarterly* 44 (1979): 295–309.
Pfister, Joel. *The Production of Personal Life: Class, Gender, and the Psychological in Hawthorne's Fiction*. Stanford, CA: Stanford UP, 1991.
Pimple, Kenneth D. "'Subtle, but Remorseful Hypocrite': Dimmesdale's Moral Character." *Studies in the Novel* 25 (1993): 257–71.
Pirnajmuddin, Hossein, and Omid Amani. "The Carnivalesque in Nathaniel Hawthorne's *The Scarlet Letter.*" *Teaching American Literature: A Journal of Theory and Practice* 6.1 (2013): 106–119.
Pisano, Frank. "Dimmesdale's Pious Imperfect Perverseness: Poe's 'The Imp of the Perverse' and *The Scarlet Letter.*" *Poe Writing/Writing Poe*. Ed. Jana L. Argersinger and Richard Kopley. New York: AMS, 2013. 143–57.
Pringle, Michael. "The Scarlet Lever: Hester's Civil Disobedience." *ESQ* 53.1 (2007): 31–55.
Pryse, Marjorie. *The Mark and the Knowledge: Social Stigma in Classic American Fiction*. Columbus: Ohio State UP, 1979. 15–48.
Quirk, Tom. "Hawthorne's Last Tales and 'The Custom-House.'" *ESQ* 30 (1984): 220–31.
Ragussis, Michael. *Acts of Naming: The Family Plot in Fiction*. New York: Oxford UP, 1986. 65–86.
Railton, Stephen. "The Address of *The Scarlet Letter.*" *Readers in History: Nineteenth-Century American Literature and the Contexts of Response*. Ed. James L. Machor. Baltimore, MD: Johns Hopkins UP, 1993. 138–63.
Reed, Jon B. "'A Letter,—the Letter A': A Portrait of the Artist as Hester Prynne." *ESQ* 36.2 (1990): 79–107.
Reid, Alfred S. *The Yellow Ruff and The Scarlet Letter: A Source of Hawthorne's Novel*. Gainesville: U of Florida P, 1955.
Reid, Bethany. "Narrative of the Captivity and Redemption of Roger Prynne: Rereading *The Scarlet Letter.*" *Studies in the Novel* 33 (Fall 2001): 247–67.
Reynolds, David S. "Hawthorne's Cultural Demons: History, Popular Culture, and *The Scarlet Letter.*" *Novel History: Historians and Novelists Confront America's Past (and Each Other)*. Ed. Mark C. Carnes. New York: Simon & Schuster, 2001. 229–34.
Reynolds, Larry J. *Devils and Rebels: The Making of Hawthorne's Damned Politics*. Ann Arbor: U of Michigan P, 2008.
———. *European Revolutions and the American Literary Renaissance*. New Haven, CT: Yale UP, 1988.
———, ed. *A Historical Guide to Nathaniel Hawthorne*. New York: Oxford UP, 2001.
• ———. "*The Scarlet Letter* and Revolutions Abroad." *American Literature* 57.1 (March 1985): 44–67.
Riss, Arthur. *Race, Slavery, and Liberalism in Nineteenth-Century American Literature*. New York: Cambridge UP, 2006. 111–35.
Rozakis, Laurie N. "Another Possible Source of Hawthorne's Hester Prynne." *American Transcendental Quarterly* 59 (1986): 63–71.
Ruetenik, Tadd. "Another View of Arthur Dimmesdale: Scapegoating and Revelation in *The Scarlet Letter.*" *Contagion: Journal of Violence, Mimesis, and Culture* 19 (2012): 69–86.
• Ryan, Michael. "'The Puritans of Today': The Anti-Whig Argument of *The Scarlet Letter.*" *Canadian Review of American Studies* 38.2 (2008): 201–225.
• Ryskamp, Charles. "The New England Sources of *The Scarlet Letter.*" *American Literature* 31.3 (1959): 257–72.
Sandeen, Ernest. "*The Scarlet Letter* as a Love Story." *PMLA* 77 (1962): 425–35.
Sanderlin, Reed. "Hawthorne's *Scarlet Letter*: A Study of the Meaning of Meaning." *Southern Humanities Review* 9 (1975): 145–57.
Savoy, Eric. "'Filial Duty': Reading the Patriarchal Body in 'The Custom-House.'" *Studies in the Novel* 25 (1993): 397–417.

• ———. "Nathaniel Hawthorne and the Anxieties of the Archive." *Canadian Review of American Studies* 45.1 (Spring 2015): 38–66.

Scheiber, Andrew J. "Public Force, Private Sentiment: Hawthorne and the Gender of Politics." *American Transcendental Quarterly* 2.4 (1988): 285–99.

Schiff, James. *Updike's Version: Rewriting* The Scarlet Letter. Columbia: U of Missouri P, 1992.

Schubert, Leland. *Hawthorne the Artist: Fine-Art Devices in Fiction*. Chapel Hill: U of North Carolina P, 1944.

Schwab, Gabriele. "Seduced by Witches: Nathaniel Hawthorne's *The Scarlet Letter* in the Context of New England Witchcraft Fictions." *Seduction and Theory: Readings of Gender, Representation, and Rhetoric*. Ed. Diane Hunter. Urbana: U of Illinois P, 1989. 170–91.

Small, Michel. "Hawthorne's *The Scarlet Letter*: Arthur Dimmesdale's Manipulation of Language." *American Imago* 37 (1980): 113–23.

Smith, Allan Gardner Lloyd. *Eve Tempted: Writing and Sexuality in Hawthorne's Fiction*. Totowa, NJ: Barnes and Noble, 1984.

Söderlind, Sylvia. "Branding the Body American: Violence and Self-Fashioning from *The Scarlet Letter* to *American Psycho*." *Canadian Review of American Studies* 38.1 (2008): 63–81.

Stein, William Bysshe. *Hawthorne's Faust: A Study of the Devil Archetype*. Gainesville: U of Florida P, 1953.

Sterling, Laurie A. "Paternal Gold: Translating Inheritance in *The Scarlet Letter*." *American Transcendental Quarterly* 6 (1992): 17–30.

Stoehr, Taylor. *Hawthorne's Mad Scientists: Pseudoscience and Social Science in Nineteenth-Century Life and Letters*. Hamden, CT: Archon, 1978.

Stubbs, John Caldwell. *The Pursuit of Form: A Study of Hawthorne and the Romance*. Urbana: U of Illinois P, 1970.

Swann, Charles. *Nathaniel Hawthorne: Tradition and Revolution*. New York: Cambridge UP, 1991.

Sweeney, Susan Elizabeth. "The Madonna, the Women's Room, and *The Scarlet Letter*." *College English* 57 (1995): 410–25.

• Thomas, Brook. "Citizen Hester: *The Scarlet Letter* as Civic Myth." *American Literary History* 13 (Summer 2001): 181–211.

———. "Love and Politics, Sympathy and Justice in *The Scarlet Letter*." *The Cambridge Companion to Nathaniel Hawthorne*. Cambridge: Cambridge UP, 2004. 162–85.

Thrailkill, Jane F. "*The Scarlet Letter*'s Romantic Medicine." *Studies in American Fiction* 34.1 (2006): 3–31.

Tomc, Sandra. "A Change of Art: Hester, Hawthorne, and the Service of Love." *Nineteenth-Century Literature* 56.4 (2002): 466–94.

———. "'The Sanctity of the Priesthood': Hawthorne's 'Custom-House.'" *ESQ* 39 (1993): 161–84.

Tompkins, Jane. *Sensational Designs: The Cultural Work of American Fiction, 1790–1860*. New York: Oxford UP, 1985.

Traister, Bryce. "The Bureaucratic Origins of *The Scarlet Letter*." *Studies in American Fiction* 29 (Spring 2001): 77–92.

Valenti, Patricia Dunlavy. "'Then, all was spoken!' What 'The Custom-House' and *The Scarlet Letter* Disclose." *Nathaniel Hawthorne Review* 40.2 (Fall 2014): 19–39.

Van Leer, David. "Hester's Labyrinth: Transcendental Rhetoric in Puritan Boston." *New Essays on* The Scarlet Letter. Ed. Michael J. Colacurcio. New York: Cambridge UP, 1985. 57–100.

von Abele, Rudolph. *The Death of the Artist: A Study of Hawthorne's Disintegration*. The Hague: Martinus Nijhoff, 1955.

Waggoner, Hyatt H. *Hawthorne: A Critical Study*. Rev. ed. Cambridge, MA: Harvard University Press, 1963.

———. *The Presence of Hawthorne*. Baton Rouge: Louisiana State UP, 1979.

Wamser, Garry. "The Scarlet Contract: Puritan Resurgence, the Unwed Mother, and Her Child." *Law and Literature Perspectives*. Ed. Bruce L. Rockwood and Roberta Kevelson. New York: Peter Lang, 1996. 381–406.

Weinauer, Ellen. "Considering Possession in *The Scarlet Letter*." *Studies in American Fiction* 29 (Spring 2001): 93–112.

Weldon, Roberta F. "From 'The Old Manse' to 'The Custom-House': The Growth of the Artist's Mind." *Texas Studies in Literature and Language* 20 (1978): 36–47.

Whelan, Robert Emmet, Jr. "Hester Prynne's Little Pearl: Sacred and Profane Love." *American Literature* 39 (1967): 488–505.

• Winship, Michael. "Hawthorne and the 'Scribbling Women': Publishing *The Scarlet Letter* in the Nineteenth-Century United States." *Studies in American Fiction* 29 (Spring 2001): 3–11.

Wolter, Jürgen C. "Southern Hesters: Hawthorne's Influence on Kate Chopin, Toni Morrison, William Faulkner, and Tennessee Williams." *Southern Quarterly* 50.1 (2012): 24–41.

Yellin, Jean Fagan. "Hawthorne and the American National Sin." *The Green Tradition: Essays and Poems for Sherman Paul*. Ed. H. Daniel Peck. Baton Rouge: Louisiana State UP, 1989. 75–97.

• ———. *Women and Sisters: The Antislavery Feminists in American Culture*. New Haven, CT: Yale UP, 1989. 125–50.

Zwart, Jane. "Initial Misgivings: Hawthorne's *Scarlet Letter* and the Forgery of American Origin." *ESQ* 59.3 (2013): 411–38.

STUDIES OF HAWTHORNE'S TALES AND SKETCHES

Bell, Millicent, ed. *New Essays on Hawthorne's Major Tales*. New York: Cambridge UP, 1993.

Colacurcio, Michael J. *The Province of Piety: Moral History in Hawthorne's Early Tales*. Cambridge, MA: Harvard UP, 1984.

Doubleday, Neil Frank. *Hawthorne's Early Tales: A Critical Study*. Durham, NC: Duke UP, 1972.

Easton, Alison. *The Making of the Hawthorne Subject*. Columbia: U of Missouri P, 1996.

Fogle, Richard H. *Hawthorne's Fiction: The Light and the Dark*. Norman: U of Oklahoma P, 1964.

Newman, Lea Bertani Vozar. *Reader's Guide to the Short Stories of Nathaniel Hawthorne*. Boston: G. K. Hall, 1979.

Thompson, G. R. *The Art of Authorial Presence: Hawthorne's Provincial Tales*. Durham, NC: Duke UP, 1993.

von Frank, Albert J., ed. *Critical Essays on Hawthorne's Short Stories*. Boston: G. K. Hall, 1991.

SELECTED CRITICISM FOR "MRS. HUTCHINSON"

Colacurcio, Michael J. *The Province of Piety: Moral History in Hawthorne's Early Tales*. Cambridge, MA: Harvard UP, 1984. 63–68.

Davis, Sarah I. "Another View of Hester and the Antinomians." *Studies in American Fiction* 12 (1984): 189–98.

• Lang, Amy Schrager. *Prophetic Woman: Anne Hutchinson and the Problem of Dissent in the Literature of New England*. Berkeley: U of California P, 1987.

Maddox, Lucy. *Removals: Nineteenth-Century American Literature and the Politics of Indian Affairs*. New York: Oxford UP, 1991. 110–25.

SELECTED CRITICISM FOR "ENDICOTT AND THE RED CROSS"

Bercovitch, Sacvan. "Endicott's Breastplate: Symbolism and Typology in 'Endicott and the Red Cross.'" *Studies in Short Fiction* 4 (1967): 289–99.

Bremer, Francis J. "'Endicott and the Red Cross': Puritan Iconoclasm in the New World." *Journal of American Studies* 24.1 (1990): 5–22.

Bush, Harold K. "Re-Inventing the Puritan Fathers: George Bancroft, Nathaniel Hawthorne, and the Birth of Endicott's Ghost." *American Transcendental Quarterly* 9.2 (1995): 131–52.

Colacurcio, Michael J. *The Province of Piety: Moral History in Hawthorne's Early Tales*. Cambridge, MA: Harvard UP, 1984.

Franzosa, John. "Young Man Hawthorne: Scrutinizing the Discourse of History." *Bucknell Review* 30.2 (1987): 72–94.

Newberry, Frederick. "The Demonic in 'Endicott and the Red Cross.'" *Papers on Language and Literature* 13 (1977): 251–59.

• Nickel, John. "Hawthorne's Demystification of History in 'Endicott and the Red Cross.'" *Texas Studies in Literature and Language* 42.4 (2000): 347–62.

Orton, Stephen. "De-Centered Symbols in 'Endicott and the Red Cross.'" *Studies in Short Fiction* 30.4 (1993): 565–74.

Royer, Diana. "Puritan Constructs and Nineteenth-Century Politics: Allegory, Rhetoric, and Law in Three Hawthorne Tales." *Worldmaking.* Ed. William Pencak. New York: Peter Lang, 1996. 211–40.

SELECTED CRITICISM FOR "YOUNG GOODMAN BROWN"

Berkove, Lawrence I. "'Reasoning as We Go': The Flawed Logic of 'Young Goodman Brown.'" *Nathaniel Hawthorne Review* 24.1 (1998): 46–52.

Boyer, Paul, and Stephen Nissenbaum, eds. *Salem-Village Witchcraft: A Documentary Record of Local Conflict in Colonial New England.* Boston: Northeastern UP, 1972, 1993.

Calef, Robert. *More Wonders of the Invisible World* 1700; repr., London: William Carlton, 1796.

Clark, James W. "Hawthorne's Use of Evidence in 'Young Goodman Brown.'" *Essex Institute Historical Collections* 111 (1975): 12–34.

Colacurcio, Michael J. "Visible Sanctity and Specter Evidence: The Moral World of Hawthorne's 'Young Goodman Brown.'" *Essex Institute Historical Collections* 110 (1974): 259–99. Revised version in Colacurcio, *The Province of Piety: Moral History in Hawthorne's Early Tales.* Cambridge, MA: Harvard UP, 1984. 283–313.

Cook, Reginald L. "The Forest of Goodman Brown's Night: A Reading of Hawthorne's 'Young Goodman Brown.'" *New England Quarterly* 43 (1970): 473–81.

Crews, Frederick. *The Sins of the Fathers: Hawthorne's Psychological Themes.* New York: Oxford UP, 1966. 98–106.

Eberwein, Jane Donahue. "'My Faith Is Gone!': 'Young Goodman Brown' and Puritan Conversion." *Christianity and Literature* 32 (1982): 23–32.

Franklin, Benjamin V. "Goodman Brown and the Puritan Catechism." *ESQ* 40 (1994): 67–88.

Hostetler, Norman H. "Narrative Structure and Theme in 'Young Goodman Brown.'" *Journal of Narrative Technique* 12 (1982): 221–28.

Jayne, Edward. "Pray Tarry with Me Young Goodman Brown." *Literature and Psychology* 29 (1979): 100–13.

Keil, James C. "Hawthorne's 'Young Goodman Brown': Early Nineteenth-Century and Puritan Constructions of Gender." *New England Quarterly* 69.1 (1996): 33–55.

Kelley, James B. "Leading Students Down Dark Paths: How Teachers Talk about Nathaniel Hawthorne's 'Young Goodman Brown.'" *Teaching American Literature: A Journal of Theory and Practice* 4.3 (2011): 63–85.

Levin, David. "Shadows of Doubt: Specter Evidence in Hawthorne's 'Young Goodman Brown.'" *American Literature* 34 (1962): 344–52.

Levy, Leo B. "The Problem of Faith in 'Young Goodman Brown.'" *Journal of English and Germanic Philology* 74 (1975): 375–87.

Loving, Jerome. "Pretty in Pink: 'Young Goodman Brown' and New-World Dreams." *Critical Essays on Hawthorne's Short Stories.* Ed. J. Albert von Frank. Boston: G. K. Hall, 1991. 219–31.

Morris, Christopher. "Deconstructing 'Young Goodman Brown.'" *American Transcendental Quarterly* 2 (1988): 23–33.

Mosher, Harold F. "The Sources of Ambiguity in Hawthorne's 'Young Goodman Brown': A Structuralist Approach." *ESQ* 26 (1980): 16–25.

Paulits, Walter J. "Ambivalence in 'Young Goodman Brown.'" *American Literature* 41 (1970): 577–84.

Reynolds, Larry J. "Melville's Use of 'Young Goodman Brown.'" *American Transcendental Quarterly* 31 (1976): 12–14.

• Ronan, John. "'Young Goodman Brown' and the Mathers." *New England Quarterly* 85.2 (2012): 253–80.

Shuffleton, Frank. "Nathaniel Hawthorne and the Revival Movement." *American Transcendental Quarterly* 44 (1979): 311–23.

Stoehr, Taylor. "'Young Goodman Brown' and Hawthorne's Theory of Mimesis." *Nineteenth-Century Fiction* 23 (1969): 393–412.

Wright, Elizabeth. "The New Psychoanalysis and Literary Criticism: A Reading of Hawthorne and Melville." *Poetics Today* 3 (1982): 89–105.

Zanger, Jules. "'Young Goodman Brown' and 'A White Heron': Correspondences and Illuminations." *Papers on Language and Literature* 26.3 (Summer 1990): 346–57.

Zapf, Hubert. "The Rewriting of the Faust Myth in Nathaniel Hawthorne's 'Young Goodman Brown.'" *Nathaniel Hawthorne Review* 38.1 (2012): 19–40.

SELECTED CRITICISM FOR "THE MINISTER'S BLACK VEIL"

Barry, Elaine. "Beyond the Veil: A Reading of Hawthorne's 'The Minister's Black Veil.'" *Studies in Short Fiction* 17 (1980): 15–20.

Benoit, Raymond. "Hawthorne's Psychology of Death: 'The Minister's Black Veil.'" *Studies in Short Fiction* 8 (1971): 553–60.

• Birk, John F. "Hawthorne's Mister Hooper: The Veil of Ham." *Prospects* 21 (1996): 1–11.

Boone, N. S. "'The Minister's Black Veil' and Hawthorne's Ethical Refusal of Reciprocity: A Levinasian Parable." *Renascence* 57.3 (2005): 165–76.

Carnochan, W. B. "'The Minister's Black Veil': Symbol, Meaning, and the Context of Hawthorne's Art." *Nineteenth-Century Fiction* 24 (1969): 182–92.

Coale, Samuel. "Hawthorne's Black Veil: From Image to Icon." *CEA Critic* 55 (1993): 79–87.

Colacurcio, Michael J. "Parson Hooper's Power of Blackness: Sin and Self in 'The Minister's Black Veil.'" *Prospects* 5 (1980): 331–411. Also in Colacurcio, *The Province of Piety: Moral History in Hawthorne's Early Tales*. Cambridge, MA: Harvard UP, 1984. 314–85.

Crews, Frederick. *The Sins of the Fathers: Hawthorne's Psychological Themes*. New York: Oxford UP, 1966. 106–11.

Crie, Robert P. "'The Minister's Black Veil': Mr. Hooper's Symbolic Fig Leaf." *Literature and Psychology* 17 (1969): 211–18.

Danow, David K. "The Semiotic Significance of 'The Minister's Black Veil.'" *Semiotica* 113.3–4 (1997): 337–46.

Davis, Clark. "Facing the Veil: Hawthorne, Hooper, and Ethics." *Arizona Quarterly* 55 (Winter 1999): 1–19.

Franklin, Rosemary F. "'The Minister's Black Veil': A Parable." *Studies in Short Fiction* 56 (1985): 55–63.

Freedman, William. "The Artist's Symbol and Hawthorne's Veil: 'The Minister's Black Veil' Resartus." *Studies in Short Fiction* 29 (1992): 353–62.

German, Norman. "The Veil of Words in 'The Minister's Black Veil.'" *Studies in Short Fiction* 25 (1988): 41–47.

McCarthy, Judy. "'The Minister's Black Veil': Concealing Moses and the Holy of Holies." *Studies in Short Fiction* 24 (1987): 131–38.

• Miller, J. Hillis. *Hawthorne and History*. Cambridge: Basil Blackwell, 1991. 73–102.

Morsberger, Robert E. "'The Minister's Black Veil': 'Shrouded in a Blackness, Ten Times Black.'" *New England Quarterly* 46 (1973): 454–63.

Newberry, Frederick. "The Biblical Veil: Sources and Typology in Hawthorne's 'The Minister's Black Veil.'" *Texas Studies in Literature and Language* 31 (1989): 169–95.

Newman, Lea Bertani Vozar. "One Hundred and Fifty Years of Looking at, through, behind, beyond, and around 'The Minister's Black Veil.'" *Nathaniel Hawthorne Review* 13.2 (1987): 5–12.

Ostrowski, Carl. "The Minister's 'Grievous Affliction': Diagnosing Hawthorne's Parson Hooper." *Literature and Medicine* 17.2 (1998): 197–211.

Quinn, James, and Ross Baldessarini. "Literary Technique and Psychological Effect in Hawthorne's 'The Minister's Black Veil.'" *Literature and Psychology* 24 (1974): 115–23.

Randall, Dale B. "Image-Making and Image-Breaking: Seeing 'The Minister's Black Veil' through a Miltonic Glass, Darkly." *Resources for American Literary Study* 23.1 (1997): 19–27.

Reece, James B. "Mr. Hooper's Vow." *ESQ* 21 (1975): 93–102.

Seigel, Catherine F. "Jumping Through Hawthorne's Hoop(er)s." *Short Story* 2.1 (1994): 79–88.

Wallace, James D. "Stowe and Hawthorne." *Hawthorne and Women: Engendering and Expanding the Hawthorne Tradition*. Ed. John L. Idol, Jr., and Melinda M. Ponder. Amherst: U Massachusetts P, 1999. 92–103.

SELECTED CRITICISM FOR "THE BIRTHMARK"

Arner, Robert D. "The Legend of Pygmalion in 'The Birthmark.'" *American Transcendental Quarterly* 12 (1972): 168–71.

Benziman, Galia. "Challenging the Biological: The Fantasy of Male Birth as a Nineteenth-Century Narrative of Ethical Failure." *Women's Studies: An Interdisciplinary Journal* 35.4 (June 2006): 375–95.

Bromell, Nicholas K. "'The Bloody Hand' of Labor: Work, Class, and Gender in Three Stories by Hawthorne." *American Quarterly* 42 (1990): 542–64. Also, as "Women Carved of Oak and Korl," in Bromell, *By the Sweat of the Brow: Literature and Labor in Antebellum America*. Chicago: U of Chicago P, 1993. 99–119.

Burns, Shannon. "Alchemy and 'The Birth-Mark'." *American Transcendental Quarterly* 42 (1979): 147–58.

Eckstein, Barbara. "Hawthorne's 'The Birthmark': Science and Romance as Belief." *Studies in Short Fiction* 26 (1989): 511–19.

Elbert, Monika. "The Surveillance of Woman's Body in Hawthorne's Short Stories." *Women's Studies* 33.1 (2004): 23–46.

• Fetterley, Judith. *The Resisting Reader: A Feminist Approach to American Fiction*. Bloomington: Indiana UP, 1978. 22–33.

Gatta, John. "Aylmer's Alchemy in 'The Birthmark.'" *Philological Quarterly* 57 (1978): 399–413.

Johnson, Barbara. "Is Female to Male as Ground Is to Figure?" *Feminism and Psychoanalysis*. Ed. Richard Feldstein and Judith Roof. Ithaca, NY: Cornell UP, 1989. 255–68.

Marshall, Megan. "Sophia's Crimson Hand." *Nathaniel Hawthorne Review* 37.2 (2011): 36–47.

McKenna, John J. "Lessons about Pygmalion Projects and Temperament in Hawthorne's 'The Birthmark.'" *Eureka Studies in Teaching Short Fiction* 7.1 (2006): 36–43.

McMurray, Price. "'Love Is as Much Its Demand, as Perception': Hawthorne's 'Birthmark' and Emerson's 'Humanity of Science.'" *ESQ* 47.1 (2001): 1–31.

Micklaus, Robert. "Hawthorne's Jekyll and Hyde: The Aminadab in Aylmer." *Literature and Psychology* 29 (1979): 148–59.

Person, Leland S. *Aesthetic Headaches: Women and a Masculine Poetics in Poe, Melville, and Hawthorne*. Athens: U of Georgia P, 1988. 108–12.

Pfister, Joel. *The Production of Personal Life: Class, Gender, and the Psychological in Hawthorne's Fiction*. Stanford, CA: Stanford UP, 1991. 29–48.

Pribek, Thomas. "Hawthorne's Aminadab: Sources and Influence." *Studies in the American Renaissance* (1987): 177–86.

Proudfit, Charles L. "Eroticization of Intellectual Functions as an Oedipal Defence: A Psychoanalytic View of Nathaniel Hawthorne's 'The Birthmark.'" *International Review of Psycho-analysis* 7 (1980): 375–83.

Quinn, James, and Ross Baldessarini. "'The Birth-mark': A Deathmark." *University of Hartford Studies in Literature* 13 (1981): 91–98.

Reid, Alfred S. "Hawthorne's Humanism: 'The Birthmark' and Sir Kenelm Digby." *American Literature* 38 (1966): 337–51.

Rosenberg, Liz. "'The Best That Earth Could Offer': 'The Birthmark,' a Newlywed's Story." *Studies in Short Fiction* 30 (1993): 145–51.

Rucker, Mary F. "Science and Art in Hawthorne's 'The Birthmark.'" *Nineteenth Century Literature* 41 (1987): 445–61.

Shakinovsky, Lynn. "The Return of the Repressed: Illiteracy and the Death of the Narrative in Hawthorne's 'The Birthmark.'" *American Transcendental Quarterly* 9 (1995): 269–81.

Smith, Allan Gardner Lloyd. *Eve Tempted: Writing and Sexuality in Hawthorne's Fiction*. Totowa, NJ: Barnes & Noble, 1983. 95–100.

Smith, Andy. "Sensitive Emulsions: Hawthorne's Proto-Photography." *Nathaniel Hawthorne Review* 33.1 (2009): 46–69.

Van Leer, David. "Aylmer's Library: Transcendental Alchemy in Hawthorne's 'The Birthmark.'" *American Transcendental Quarterly* 25 (1975): 211–20.

Weinstein, Cindy. "The Invisible Hand Made Visible: 'The Birthmark.'" *Nineteenth-Century Literature* 48 (1993): 44–73.

• ———. "The Invisible Hand Made Visible: 'The Birthmark.'" In Weinstein, *The Literature of Labor and the Labors of Literature: Allegory in Nineteenth-Century American Fiction*. New York: Cambridge UP, 1995. 82–86.

West, Andrew Christopher. "'Candles Lighting The Dark': The Birth-Mark's Antinomian Method." *Textual Practice* 27.4 (2013): 617–49.

Westbrook, Ellen E. "'Probable Improbabilities': Verisimilar Romance in Hawthorne's 'The Birth-mark.'" *American Transcendental Quarterly* 3 (1989): 203–17.

Wiegman, Robyn. "The Anatomy of Lynching." *Journal of the History of Sexuality* 3 (1993): 445–67.

Youra, Steven. "'The Fatal Hand': A Sign of Confusion in Hawthorne's 'The Birth-mark.'" *American Transcendental Quarterly* 60 (1986): 43–51.

Zanger, Jules. "Speaking the Unspeakable: Hawthorne's 'The Birthmark.'" *Modern Philology* 80 (1983): 364–71.